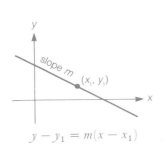

$$y - y_1 = m(x - x_1)$$

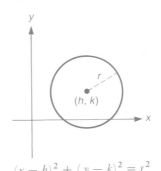

$$(x - h)^2 + (y - k)^2 = r^2$$

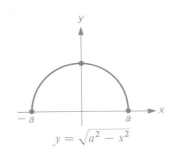

$$y = \sqrt{a^2 - x^2}$$

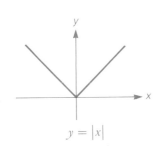

$$y = |x|$$

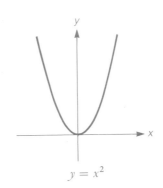

$$y = x^2$$

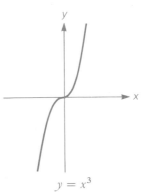

$$y = x^3$$

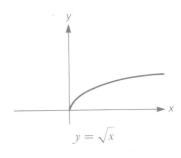

$$y = \sqrt{x}$$

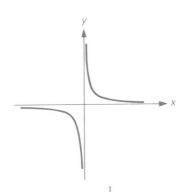

$$y = \frac{1}{x}$$

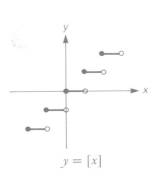

$$y = [x]$$

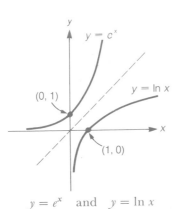

$$y = e^x \quad \text{and} \quad y = \ln x$$

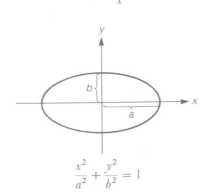

$$\frac{x^2}{a^2} + \frac{y^2}{b^2} = 1$$

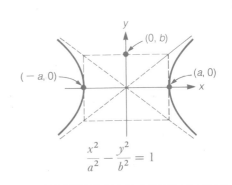

$$\frac{x^2}{a^2} - \frac{y^2}{b^2} = 1$$

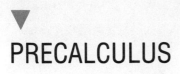

PRECALCULUS

PRECALCULUS

A PROBLEMS-ORIENTED APPROACH

THIRD EDITION

DAVID COHEN

DEPARTMENT OF MATHEMATICS
UNIVERSITY OF CALIFORNIA, LOS ANGELES

WEST PUBLISHING COMPANY

ST. PAUL ▼ NEW YORK ▼ LOS ANGELES ▼ SAN FRANCISCO

TO ▼ MY SON ANDREW

Composition: Syntax International
Copyediting: Susan Gerstein
Technical illustrations: Rolin Graphics and Katherine Townes Books
Text design: Janet Bollow
Cover design: Kristen M. Weber
Cover image: Taken from the spacecraft *Challenger 6*, just before sunrise
over the Pacific Ocean, about 2,100 miles east of Tokyo, Japan. The bands
of color show the various layers of aerosols which surround the Earth.
The brilliant red layer is the atmosphere; the overlap between the red
and blue layers is the stratosphere; the blue layer is the ionosphere. With
increased altitude, the electrons and ions are reduced in number, leaving
nothing but the blackness of space. Courtesy of NASA.

Library of Congress Cataloging-in-Publication Data

Cohen, David, 1942 Sept. 28–
    Precalculus, a problems-oriented approach/David Cohen.—3rd ed.
        p. cm.
    Rev. ed. of: Precalculus, a problems-oriented approach. 2nd ed.
© 1987.
    ISBN 0-314-66790-3
    1. Mathematics. 2. Problem solving. I. Cohen, David, 1942 Sept. 28–
Precalculus, a problems-oriented approach. II. Title.
QA39.2.C64 1990   89-28593
512.1—dc20   CIP

CONTENTS

PREFACE TO THE THIRD EDITION

AUDIENCE

This text is for students who are preparing to take calculus or other courses requiring a background in precalculus mathematics. As in the earlier editions, the presentation is student-oriented in three specific ways. First, I've tried to talk to, rather than lecture at, the student. Second, examples are consistently used to introduce, to explain, and to motivate concepts. And third, all of the initial exercises for each section are carefully correlated with the worked examples in that section.

In writing this book, I have assumed that the students who use it have been exposed to intermediate algebra, but have not necessarily mastered that subject. Also, for many precalculus students, there may be a gap of several years between their previous mathematics courses and the present one. For these reasons, the review material in Chapter 1 and in Appendix A4 is unusually thorough. This allows the instructor two genuine options: cover the review material (or a portion of it) in class; or, assign it to students for study on their own and begin the precalculus course with Chapter 2, *Functions*. (In my own teaching, I tend to mix these two options. I begin the precalculus course with Chapter 2, but I include problems from Chapter 1 and Appendix A4 on nearly every homework assignment.)

CONTENT AND FEATURES

A glance at the precalculus texts published over the past decade shows that the topics, and even their order, are somewhat standard. I think this simply reflects the fact that we (as calculus teachers, for the moment) have a pretty good idea of what we would like our incoming students to know. Where precalculus texts do differ, it seems to me, is more in matters of approach and emphasis. Areas that receive particular emphasis in this text are these:

1. *Word Problems and Applications.* Word problems and strategies for solving them are explained and developed throughout the book. Maximum-minimum problems relating to quadratic functions are discussed in detail in Section 3.4. The preceding section introduces some strategies for approaching these problems. To ensure that precalculus students gain appropriate practice and experience with these important strategies, the initial exercises in Section 3.3 make specific references to the corresponding worked examples in the text. In general, applications are integrated throughout the text. One complete section (Section 4.5) is devoted to applications of the exponential function.

2. *Graphing.* Graphs and techniques for graphing are developed throughout the text. Graphs and symmetry are used to explain and reinforce algebraic concepts. (See, for example, Section 4.3.)

3. *Analytic Geometry.* The basic equations for circles and lines are introduced in Chapters 1 and 2 and used throughout the book. In Chapter 8, conic sections are discussed in greater detail than is found in most other precalculus books. In the text and exercises, properties of the tangents to these curves are also developed using precalculus (as opposed to calculus) techniques. (The exposition and exercises on tangents are arranged so that this material

can be omitted without loss of continuity, if the instructor so desires.) In the past, analytic geometry traditionally served as the capstone in the student's preparation for calculus. It was here that the student had the opportunity to really sharpen his or her algebraic and analytic skills. Chapter 8 is intended to reaffirm, rather than dismiss, that tradition.

4. *Trigonometry.* The development of most topics in this text proceeds from the specific to the general, and the trigonometry follows this pattern too. Thus, the first two sections on trigonometry (Sections 5.1 and 5.2) introduce the trigonometric functions in the context of right triangles. The unit-circle approach follows directly thereafter in Section 5.3. Section 5.4 is an important transitional section providing examples and drill work in the algebraic manipulations needed in trigonometry. This helps ease the way for students when they study the more analytical portions of trigonometry in Chapter 6.

5. *Graded Exercise Sets.* Whether a precalculus text is more than a mere catalog of functions and their properties depends to a large extent on the quality and the quantity of the exercise sets. In this book, most of the exercise sets (except for the chapter review sets) are divided into three catagories: A, B, and C. The group A exercises are based directly on the examples and definitions in that section of the text. Moreover, these problems treat topics in roughly the same order in which they appear in the text. The group B exercises serve several key functions. Some of them require students to use several different techniques or topics in a single solution. Some Group B problems, while not conceptually more difficult than their group A counterparts, require lengthier calculations or algebraic manipulations. Topics that further prepare the student for calculus are often developed in the group B exercises. Finally, some group B exercises are included simply because they are interesting or because they are intended to challenge the students. The group C exercises contain the more challenging problems.

6. *End of Chapter Material.* Each chapter concludes with a detailed chapter summary and an extensive chapter-review exercise set. The first fifteen or so problems in each review exercise set constitute a sample test based on the group A problems. In general, the review exercise sets play several roles. First, they provide a source of routine drill-type problems in a context that is not explicitly linked to a specific section of the chapter. Thus, just as on an exam, the student must decide for himself or herself which tools to bring to bear on a problem. Second, the review exercises provide a rich source of problems at various levels, ranging from the elementary to the challenging. Some of these approach the core material from perspectives that differ slightly from that in the exposition. This is important for students who are planning to apply their precalculus skills in subsequent courses.

7. *Calculator Exercises.* Most students entering a precalculus course own a scientific calculator, and they are eager to use it. The calculator exercises in this text, indicated by the symbol Ⓒ, are designed to take advantage of this situation. There are four broad types of calculator exercises here:
(i)   Exercises that reinforce and supplement core material;
(ii) Exercises with surprising results that motivate subsequent questions;
(iii) Exercises that introduce notions from calculus;
(iv) Exercises demonstrating that the use of a calculator cannot replace thinking or the need for mathematical proofs.

In addition to these specially marked calculator exercises, there are many routine problems, integrated throughout the text, for which either a calculator or tables will suffice. Such problems usually are not preceded by the symbol ⓒ.

CHANGES IN THIS EDITION

Comments and suggestions from students, instructors, and reviewers have helped me to revise this text in a number of ways that I believe will make the book more useful to the instructor and more accessible to the student. One important change throughout much of the text has been to increase the number of drill exercises and to use odd-even exercise pairs where feasible. Other major changes occur in the following areas.

Chapter 1 *Algebra and Coordinate Geometry for Precalculus.* The treatment of inequalities, now in Section 1.7, has been expanded to include polynomial and rational inequalities (which used to appear at the end of Chapter 3.) One advantage to this reorganization is that the techniques for solving such inequalities can be reinforced in Chapter 2 in the course of computing the domains of functions. Section 1.9 on complex numbers has been completely rewritten. The development is now more computational and less axiomatic.

Chapter 2 *Functions and Graphs.* In Section 2.1, the use of symmetry in graphing is given greater emphasis and used more consistently in the examples. The material on slopes and lines, formerly two sections, is now in one section (2.2). (The applications of slope, such as marginal cost and velocity, now appear in Chapter 3 in the context of linear functions). Section 2.4, *The Graph of a Function* has been expanded slightly to include the geometric ideas of increasing and decreasing functions, turning points, and maximum and minimum values. The material on techniques in graphing in Section 2.5 is completely reworked, and the presentation is now, I believe, more unified and cohesive.

Chapter 3 *Polynomial and Rational Functions. Applications to Optimization.* The treatment of linear functions at the beginning of Chapter 3 has been expanded in this edition to a full section. The idea of slope as a rate of change is emphasized through examples involving marginal cost and velocity. The concepts of a scatter diagram and the least-squares line are introduced in the text rather than the exercises, and the treatment is expanded. In Sections 3.2 and 3.4, we now determine the vertex of a parabola by completing the square rather than by relying on the formula $x = -b/2a$. In Section 3.5 on graphs of polynomial functions, the behavior of $f(x)$ when $|x|$ is very large is given greater emphasis both in the text and in the exercises. (See, for Example, Figure 4 and Example 1 on page 182.)

Chapter 4 *Exponential and Logarithmic Functions.* The material introducing the number e and the exponential function $y = e^x$ is now contained in a separate section (Section 4.2), and the number of worked examples and drill exercises has been increased.

Chapter 5 *Trigonometric Functions of Angles.* In the previous edition, the initial introduction to right-triangle trigonometry contained too much material to be covered in one lecture. This material has been reorganized and spread out over two sections. The unit-circle definitions now precede rather than follow the section *Algebra and the Trigonometric Functions.* There were two reasons for this switch: first, the desire to present the unit-circle definitions at an earlier point; second, with the unit-circle definitions completed, the identities in the next section needn't carry the restriction that θ be an acute angle. In the section on the laws of sines and cosines, an additional page of word problems has been added to the exercises. (See page 295.)

Chapter 6 *Trigonometric Functions of Real Numbers.* The material on graphing is introduced much earlier, in Section 6.3. The presentation has been rewritten so that the graphs of $y = A \sin(Bx - C)$ and $y = A \cos(Bx - C)$ are obtained in a more natural way, relying on the techniques of translation developed in Section 2.5. Also, a section has been added to provide a detailed development of the graphs of the tangent and the reciprocal functions.

Chapter 7 *Systems of Equations.* A new section on matrix inverses has been added to this chapter. The idea of a matrix inverse is carefully developed, and we show how inverses are used in solving certain systems of equations.

Chapter 8 *Analytic Geometry.* Translations of the standard conics are now treated in the same sections that introduce these curves, rather than in a separate section. This economy is possible because of the expanded treatment of translation back in Section 2.5.

SUPPLEMENTARY MATERIALS

1. *Student's Solutions Manual,* by Ross Rueger, contains complete solutions for the odd-numbered exercises and for the sample test questions at the beginning of each chapter review set.

2. *Instructor's Solutions Manual,* by Ross Rueger, contains answers or solutions for every exercise in *Precalculus,* 3e.

3. *The West MathTest II* is a computer-generated testing program that is available to schools adopting *Precalculus,* 3e. (This version is a significant upgrade of the computerized testing package that was available with the previous edition of *Precalculus.*)

4. *Test Bank,* by Charles Heuer. For each chapter in *Precalculus,* 3e., this package contains five versions of a chapter test, as well as two multiple choice versions.

5. *Natural Language Mathematics Software,* by Mathens, Inc. (University of Georgia). This algorithm-based software package can be used as an interactive tutorial for students, or as a test generator. The package (which runs on IBM PCs or compatibles) is free to qualified adopters.

6. *GraphToolz,* by Tom Saxton, is a software program for graphing and evaluating functions. This ingenious easy-to-use program, along with *Drill and Enrichment Exercises* by David Cohen, can make many of the topics in *Precalculus,* 3e, really seem to come alive. *GraphToolz* is free to qualified adopters. (Available for the Macintosh family of computers.)

7. *Transparency Masters* for many of the key figures or tables appearing in the text are available to schools adopting *Precalculus*, 3e.

ACKNOWLEDGEMENTS

Many students and colleagues have made useful constructive suggestions about the text and exercises, and I thank them for that. Particular thanks go to Professor Charles Heuer for his careful work in checking the text and the exercise solutions for accuracy. I am also grateful for the help and the valuable suggestions given to me by the following reviewers for this third edition.

J. Curtis Chipman
Oakland University

John Cross
University of Northern Iowa

Barbara Duch
University of Delaware

Donald Estep
Georgia Institute of Technology

Ray Glenn
Tallahassee Community College

Stuart Goldenberg
California Polytechnic State University

Richard Pilgrim
University of California, San Diego

Peter Rice
University of Georgia

David Winslow
Louisiana State University

I would also like to thank two people in particular for their help and advice: Ross Rueger of College of the Sequoias who wrote the supplementary manuals and created the answer section for the text, and Susan Gerstein, who served as copyeditor. Finally, to Peter Marshall, Deanna Quinn, and Maralene Bates at West Publishing Company: thank you for your help and encouragement in bringing this third edition into print.

David Cohen
Palos Verdes, California
December 1989

CHAPTER ONE

ALGEBRA AND COORDINATE GEOMETRY FOR PRECALCULUS

Perhaps Pythagoras was a kind of magician to his followers because he taught them that nature is commanded by numbers. There is a harmony in nature, he said, a unity in her variety, and it has a language: numbers are the language of nature.

Jacob Bronowski

If you want to understand nature, you must be conversant with the language in which nature speaks to us.

Richard Feynman (1918–1988)

INTRODUCTION In this first chapter we review several key topics from algebra and coordinate geometry that form the foundation for our work in precalculus. Although you are probably already familiar with some of this material from previous courses, do not be lulled into a false sense of security. Now, in your second exposure to these topics, you have the opportunity to really master them. Take advantage of this opportunity; it will pay great dividends both in this course and in calculus.

1.1 ▼ THE REAL NUMBERS

Here, as in your previous mathematics courses, most of the numbers we deal with are *real numbers*. These are the numbers used in everyday life, in the sciences, in industry, and in business. Perhaps the simplest way to define a real number is this: A **real number** is any number that can be expressed in decimal form. Some examples of real numbers are

$$7 \ (= 7.000\ldots) \qquad -\frac{2}{3} \ (= -0.\overline{6}) \qquad \sqrt{2} \ (= 1.4142\ldots)$$

(The bar above the 6 in the decimal $-0.\overline{6}$ indicates that the 6 repeats indefinitely.)

Certain sets of real numbers are referred to often enough to be given special names. These are summarized in the box that follows.

PROPERTY SUMMARY | **SETS OF REAL NUMBERS**

NAME	DEFINITION AND COMMENTS	EXAMPLES
Natural numbers	These are the ordinary counting numbers, 1, 2, 3, and so on.	1, 4, 29, 1066
Integers	These are the natural numbers along with their negatives and zero.	$-26, 0, 1, 1989$
Rational numbers	As the name suggests, these are the real numbers that are *ratios* of two integers (with nonzero denominators, of course). It can be proved that a real number is rational if and only if its decimal expansion terminates (e.g., 3.15) or repeats (e.g., $2.\overline{43}$).	$4 \ (= \frac{4}{1}), -\frac{2}{3}$ $1.7 \ (= \frac{17}{10})$ $4.\overline{3}, 4.1\overline{73}$
Irrational numbers	These are the real numbers that are not rational. It can be shown that any number of the form $\sqrt{n}$, where n is a natural number that is not a perfect square, is irrational.* Also, any number that is the sum or the (nonzero) product of a rational number and an irrational number is irrational.	$\sqrt{2}, 3 + \sqrt{2}, 3\sqrt{2}$ $\pi, 4 + \pi, 4\pi$

---

* The fact that $\sqrt{2}$ is irrational (that is, not a ratio of integers) was known to Pythagoras more than two thousand years ago. Section A.3 of the Appendix contains a proof of the fact that the number $\sqrt{2}$ is irrational. The proof that π is irrational is more difficult. The first person to prove that π is irrational was the Swiss mathematician J. H. Lambert (1728–1777).

According to the comments in the box, the repeating decimal number $2.\overline{4}$ is rational. In other words, there is some fraction p/q such that $2.\overline{4} = p/q$. How can we find this fraction? Example 1 shows one way to do this.

EXAMPLE 1 Express the repeating decimal $2.\overline{4}$ in the form p/q, where p and q are integers and $q \neq 0$.

Solution Let $x = 2.444\ldots$. Then $10x = 24.44\ldots$, and we have

$$10x - x = 24.\overline{4} - 2.\overline{4} = 22$$

or

$$9x = 22 \quad \text{and, therefore,} \quad x = \frac{22}{9}$$

We now have $2.\overline{4} = \frac{22}{9}$, as required. ▲

Actually, a rigorous justification of each step used in Example 1 would require certain techniques from calculus. Nevertheless, you can easily verify for yourself that the answer in Example 1 is correct; just divide 22 by 9 and see what you find.

As you've seen in previous courses, the real numbers can be represented as points on a *number line*, as shown in Figure 1. As indicated in Figure 1, the point associated with the number zero is referred to as the **origin**.

FIGURE 1

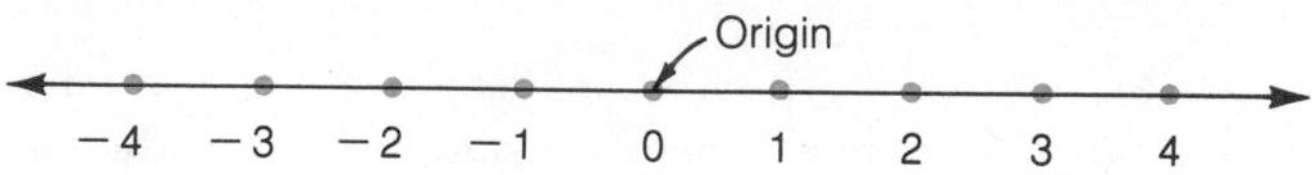

Origin

$$\begin{array}{ccccccccc} -4 & -3 & -2 & -1 & 0 & 1 & 2 & 3 & 4 \end{array}$$

The fundamental fact here is that there is a **one-to-one correspondence** between the set of real numbers and the set of points on the line. This means that each real number is identified with exactly one point on the line; conversely, with each point on the line we identify exactly one real number. The real number associated with a given point is called the **coordinate** of the point. As a practical matter, we're usually more interested in relative locations than precise locations on a number line. For instance, since π is approximately 3.1, we show π slightly to the right of 3 in Figure 2. Similarly, since $\sqrt{2}$ is approximately 1.4, we show $\sqrt{2}$ slightly less than halfway from 1 to 2 in Figure 2.

In some cases it is convenient to use number lines that show reference points other than the integers used in Figure 2. For instance, Figure 3(a) displays a number line with reference points that are multiples of π. In this case, of course, it is the integers that we then locate approximately. For example, in Figure 3(b) we show the approximate location of the number 1 on such a line.

By definition, a real number is *positive* if it lies to the right of the origin and *negative* if it lies to the left of the origin. (We're assuming that the number line is oriented as in Figure 1, 2, or 3.) The number zero is neither positive nor negative. Two of the most basic relations for real numbers are *less than* and *greater than*, symbolized by $<$ and $>$, respectively.* For ease of reference, we define these and the two related symbols $\leq$ (less than or equal) and $\geq$ (greater than or equal) in the box that follows.

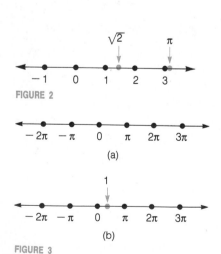

FIGURE 2

(a)

(b)

FIGURE 3

---

* The symbols $<$ and $>$ were invented by the English mathematician and astronomer Thomas Harriot (1560–1621).

PROPERTY SUMMARY — NOTATION FOR LESS THAN AND GREATER THAN

NOTATION	DEFINITION AND COMMENTS	EXAMPLES
$a < b$ (a is less than b.)	The quantity $b - a$ is positive. On a number line oriented as in Figure 1, 2, or 3, this is equivalent to saying that a lies to the left of b.	$2 < 3$; $-4 < 1$
$a \leq b$ (a is less than or equal to b.)	Either $a < b$ or $a = b$.	$2 \leq 3$; $3 \leq 3$
$b > a$ (b is greater than a.)	This is just another way to express the fact $a < b$. Geometrically, it means that b lies to the right of a on a number line.	$3 > 2$; $0 > -1$
$b \geq a$ (b is greater than or equal to a.)	Either $b > a$ or $b = a$.	$3 \geq 2$; $3 \geq 3$

In general, relationships involving real numbers and any of the four symbols $<, \geq, >$, and $\geq$ are called **inequalities**. One of the simplest uses of inequalities is in defining certain sets of real numbers called *intervals*. Roughly speaking, any uninterrupted portion of the number line is referred to as an **interval**. In the definition that follows, you'll see notation such as $a < x < b$. This means that *both* of the inequalities $a < x$ and $x < b$ hold; in other words, the number x is between a and b.

DEFINITION

Open Intervals and Closed Intervals

(a) The open interval (a, b) contains all real numbers from a to b, excluding a and b.

(b) The closed interval $[a, b]$ contains all real numbers from a to b, including a and b.

FIGURE 4

The **open interval** (a, b) consists of all real numbers x such that $a < x < b$. See Figure 4(a).

The **closed interval** $[a, b]$ consists of all real numbers x such that $a \leq x \leq b$. See Figure 4(b).

Notice that the brackets in Figure 4(b) are used to indicate that the numbers a and b are included in the interval $[a, b]$, whereas the parentheses in Figure 4(a) indicate that a and b are excluded from the interval (a, b). At times you'll see notation such as $[a, b)$. This stands for the set of all real numbers x such that $a \leq x < b$. Similarly, $(a, b]$ denotes the set of all real numbers x such that $a < x \leq b$.

EXAMPLE 2   Show each interval on a number line, and specify an inequality describing the numbers x in each interval.

$$[-1, 2] \qquad (-1, 2) \qquad (-1, 2] \qquad [-1, 2)$$

Solution   See Figure 5.

FIGURE 5

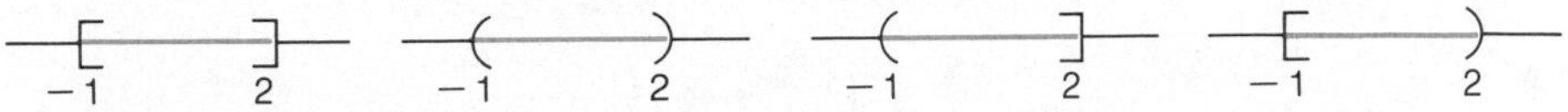

$[-1, 2]$
$-1 \leq x \leq 2$

$(-1, 2)$
$-1 < x < 2$

$(-1, 2]$
$-1 < x \leq 2$

$[-1, 2)$
$-1 \leq x < 2$

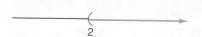

FIGURE 6
The set of all real numbers x such that $x > 2$.

In addition to the four types of intervals shown in Figure 5, we can also consider **unbounded intervals**. These are intervals that extend indefinitely in one direction or the other, as shown, for example, in Figure 6. We also have a convenient notation for unbounded intervals. As an example, we indicate the unbounded interval in Figure 6 with the notation $(2, \infty)$.

Comment and Caution   The symbol ∞ is read *infinity*.* It is not a real number, and its use in the context $(2, \infty)$ is only to indicate that the interval has no right-hand boundary. In the box that follows we define the five types of unbounded intervals. Notice that the last interval, $(-\infty, \infty)$, is actually the entire real number line.

PROPERTY SUMMARY

UNBOUNDED INTERVALS

NOTATION	DEFINING INEQUALITY	EXAMPLE
(a, ∞)	$x > a$	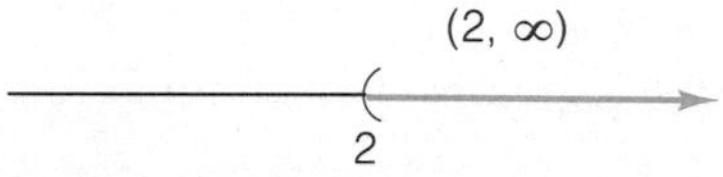 $(2, \infty)$ at 2
$[a, \infty)$	$x \geq a$	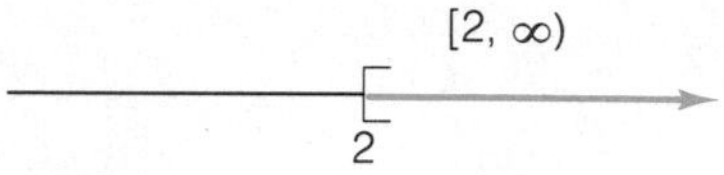 $[2, \infty)$ at 2
$(-\infty, a)$	$x < a$	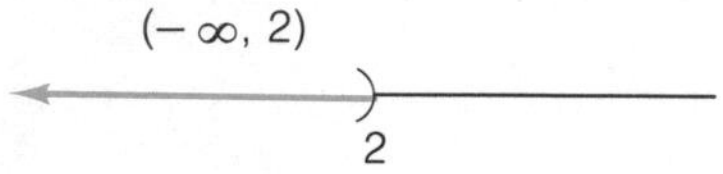 $(-\infty, 2)$ at 2
$(-\infty, a]$	$x \leq a$	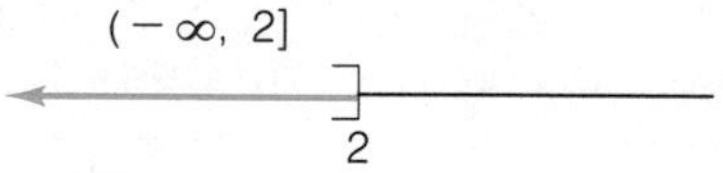 $(-\infty, 2]$ at 2
$(-\infty, \infty)$		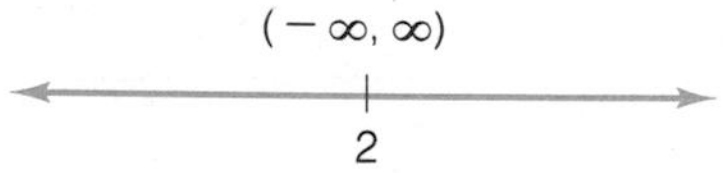 $(-\infty, \infty)$ at 2

EXAMPLE 3   Indicate each set of real numbers on a number line.

(a) $(-\infty, 4]$ **(b)** $(-3, \infty)$

Solution

(a) The interval $(-\infty, 4]$ consists of all real numbers that are less than or equal to 4. See Figure 7.

(b) The interval $(-3, \infty)$ consists of all real numbers that are greater than -3. See Figure 8.   ▲

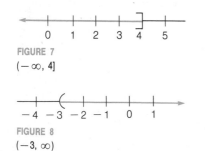

FIGURE 7
$(-\infty, 4]$

FIGURE 8
$(-3, \infty)$

We conclude this section by mentioning that our treatment of the real number system has been rather informal, and we have not derived any of the rules of arithmetic and algebra using the most basic properties of the real numbers. However, we do list those basic properties and derive some of their consequences in Section A.2 of the Appendix.

---

* The mathematician John Wallis (1616–1703) introduced the symbol ∞ in 1655. Some historians, knowing that Wallis was also a classical scholar, have suggested that he derived the infinity symbol from CIↃ, which was one of the Roman notations for 1000.

▼ EXERCISE SET 1.1

A

In Exercises 1–12, specify the category (or categories) to which each number belongs.

EXAMPLES

(a) -27; integer, rational number

(b) $\sqrt{2}$; irrational number

1. **(a)** 7   **(b)** -7
2. **(a)** -203 **(b)** $\frac{204}{2}$
3. **(a)** $\frac{27}{4}$ **(b)** $\sqrt{\frac{27}{4}}$
4. **(a)** $\sqrt{7}$ **(b)** $4\sqrt{7}$
5. **(a)** 10^6 **(b)** $10^6/10^7$
6. **(a)** 8.7 **(b)** $8.\overline{7}$
7. **(a)** 8.74 **(b)** $8.\overline{74}$
8. **(a)** $\sqrt{99}$ **(b)** $\sqrt{99}+1$
9. $3\sqrt{101}+1$ 10. $\dfrac{\sqrt{5}+1}{4}$
11. $(3-\sqrt{2})+(3+\sqrt{2})$ 12. $\dfrac{0.1234}{0.5677}$

In Exercises 13–16, the given number is rational because the decimal expansion either terminates or repeats. In each case, express the number in the form p/q, where p and q are integers and $q \neq 0$.

13. **(a)** 5.4 **(b)** $5.\overline{4}$
14. **(a)** 2.8 **(b)** $2.\overline{8}$
15. **(a)** 0.99 **(b)** $0.\overline{9}$
16. **(a)** 0.66 **(b)** $0.\overline{6}$

In Exercises 17–26, draw a number line similar to the one shown in Figure 1. Then indicate the approximate location of the given number. Where necessary, make use of the approximations $\sqrt{2} \approx 1.4$ and $\sqrt{3} \approx 1.7$. (The symbol $\approx$ means is approximately equal to)

17. $\dfrac{11}{4}$ 18. $-\dfrac{7}{8}$ 19. $1+\sqrt{2}$
20. $1-\sqrt{2}$ 21. $\sqrt{2}-1$ 22. $-\sqrt{2}-1$
23. $\sqrt{2}+\sqrt{3}$ 24. $\sqrt{2}-\sqrt{3}$ 25. $\dfrac{1+\sqrt{2}}{2}$
26. $\dfrac{2\sqrt{3}+1}{2}$

In Exercises 27–36, draw a number line similar to the one shown in Figure 3(a). Then indicate the approximate location of the given number.

27. $\dfrac{\pi}{2}$ 28. $\dfrac{3\pi}{2}$ 29. $\dfrac{\pi}{6}$ 30. $\dfrac{7\pi}{4}$
31. -1 32. 3 33. $\dfrac{\pi}{3}$ 34. $\dfrac{3}{2}$

35. $2\pi+1$ 36. $2\pi-1$

C **37.** Show the approximate location of the irrational number $\dfrac{\sqrt{139}-5}{3}$ on a number line.

C **38.** Show the approximate location of the irrational number $\sqrt{\dfrac{3-\sqrt{5}}{2}}$ on a number line.

In Exercises 39–48, say whether the statement is true or false. (In Exercises 45–48, do not use a calculator or tables; use instead the approximations $\sqrt{2} \approx 1.4$ and $\pi \approx 3.1$.)

39. $-5 < -50$ 40. $0 < -1$ 41. $-2 \leq -2$
42. $\sqrt{7}-2 \geq 0$ 43. $\dfrac{13}{14} > \dfrac{15}{16}$ 44. $0.\overline{7} > 0.7$
45. $2\pi < 6$ 46. $2 \leq \dfrac{\pi+1}{2}$ 47. $2\sqrt{2} \geq 2$
48. $\pi^2 < 12$

In Exercises 49–62, show the given interval on a number line.

49. $(2, 5)$ 50. $(-2, 2)$ 51. $[1, 4]$
52. $[-\frac{3}{2}, \frac{1}{2}]$ 53. $[0, 3)$ 54. $(-4, 0]$
55. $(-3, \infty)$ 56. $(\sqrt{2}, \infty)$ 57. $[-1, \infty)$
58. $[0, \infty)$ 59. $(-\infty, 1)$ 60. $(-\infty, -2)$
61. $(-\infty, \pi]$ 62. $(-\infty, \infty)$

B

In Exercises 63–66, the given number is rational because the decimal expansion repeats. In each case, express the number in the form p/q, where p and q are integers and $q \neq 0$.

63. $0.\overline{19}$ *Hint:* Let $x = 0.\overline{19}$ and consider $100x - x$.

64. $61.\overline{26}$ (See the hint given for Exercise 63.)

65. $0.3121212\ldots$ *Hint:* Let $x = 0.3\overline{12}$ and consider $1000x - 10x$.

66. $0.\overline{142857}$ (Adapt the hint given in Exercise 63.)

C **67.** From grade school on, we all acquire a good deal of experience in working with rational numbers. Certainly we can tell when two rational numbers are equal, although it may require some calculation, as, for example, in the case of $\frac{129}{31} = \frac{2193}{527}$. The point of this exercise is to show you that, in working with irrational numbers, sometimes it's hard to tell (or even to show) that two numbers are equal. In each equation, use a calculator to verify that the quantities on both sides

---

* $\boxed{C}$ denotes a calculator exercise.

agree, to six decimal places. (In each case it can be shown that the two numbers are indeed equal.)

(a) $\sqrt{6} + \sqrt{2} = 2\sqrt{2 + \sqrt{3}}$

(b) $\sqrt{3 + \sqrt{5}} + \sqrt{3 - \sqrt{5}} = \sqrt{10}$

(c) $\sqrt{\sqrt{6 + 4\sqrt{2}}} = \sqrt{2} + \sqrt{2}$

(d) $\sqrt{\dfrac{3}{2} + \dfrac{2}{\sqrt{2}}} + \sqrt{\dfrac{3}{2} - \dfrac{2}{\sqrt{2}}} = 2$

C **68.** The value of the irrational number π, correct to ten decimal places (without rounding off), is 3.1415926535. By using a calculator, determine to how many decimal places each of the following quantities agrees with π.

(a) $\dfrac{22}{7}$ (b) $\dfrac{355}{113}$ (c) $\dfrac{63}{25}\left(\dfrac{17 + 15\sqrt{5}}{7 + 15\sqrt{5}}\right)$

Remark: A simple approximation that agrees with π through the first 14 decimal places is $\dfrac{355}{113}\left(1 - \dfrac{0.0003}{3533}\right)$.

This approximation was discovered by the Indian mathematician Srinivasa Ramanujan (1887–1920). For a fascinating account of the history of π, see the book by Petr Beckmann, *A History of π*, 3rd ed. (New York: St. Martin's Press, 1974).

C

In Exercises 69–71, give an example of irrational numbers a *and* b *such that the indicated expression is* (a) *rational and* (b) *irrational.*

69. $a + b$ **70.** ab **71.** a/b

72. (a) Give an example in which the result of raising a rational number to a rational power is an irrational number.

(b) Give an example in which the result of raising an irrational number to a rational power is a rational number.

73. Can an irrational number raised to an irrational power yield an answer that is rational? This problem shows that the answer here is "yes." (However, if you study the following solution very carefully, you'll see that even though we've answered the question in the affirmative, we've not pinpointed the specific case in which an irrational number raised to an irrational power is rational.)

(a) Let $A = (\sqrt{2})^{\sqrt{2}}$. Now, either A is rational or A is irrational. If A is rational, we are done. Why?

(b) If A is irrational, we are done. Why? *Hint:* Consider $A^{\sqrt{2}}$.

1.2 ▼ ABSOLUTE VALUE

There has been a real need in analysis for a convenient symbolism for "absolute value" . . . and the two vertical bars introduced in 1841 by Weierstrass, as in $|z|$, have met with wide adoption; . . .

Florian Cajori in *A History of Mathematical Notations, Vol. 1* (La Salle, Illinois: The Open Court Publishing Company, 1928)

As an aid in measuring distances and locating positions on the number line, we review the concept of *absolute value*. We will begin with a definition of absolute value that is geometric in nature. After you have developed some familiarity with the concept, we will explain a more algebraic approach that is often useful in analytical work.

DEFINITION

Absolute Value

> The **absolute value** of a real number x, denoted by $|x|$, is the distance from x to the origin.

For instance, because the numbers 5 and -5 are both five units from the origin, we have $|5| = 5$ and $|-5| = 5$. Here are three more examples of absolute values:

$$|17| = 17 \qquad \left|-\dfrac{2}{3}\right| = \dfrac{2}{3} \qquad |0| = 0$$

In dealing with an expression such as $|-5 + 3|$, the convention is to compute the quantity $-5 + 3$ first and then to take the absolute value. We therefore have, in this case,

$$|-5 + 3| = |-2| = 2$$

EXAMPLE 1 Evaluate each expression.

(a) $5 - |6 - 7|$ (b) $\big||-2| - |-3|\big|$

Solution (a) $5 - |6 - 7| = 5 - |-1|$ (b) $\big||-2| - |-3|\big| = |2 - 3|$

$= 5 - 1 = 4$ $= |-1| = 1$ ▲

As we said at the beginning of this section, there is an equivalent, more algebraic way to define absolute value. According to this equivalent definition, the value of $|x|$ is x itself when $x \geq 0$, and the value of $|x|$ is $-x$ when $x < 0$. We can write this symbolically as follows:

DEFINITION

Absolute Value

$$|x| = \begin{cases} x & \text{when } x \geq 0 \\ -x & \text{when } x < 0 \end{cases}$$

By looking at examples with specific numbers, you should be able to convince yourself that both definitions yield the same results. We use the algebraic definition of absolute value in Examples 2 and 3.

EXAMPLE 2 Rewrite each expression in a form that does not contain absolute values.

(a) $|\pi - 4| + 1$ (b) $|x - 5|$ given that $x \geq 5$

(c) $|t - 5|$ given that $t < 5$

Solution (a) The quantity $\pi - 4$ is negative (since $\pi \approx 3.14$), and therefore its absolute value is equal to $-(\pi - 4)$. In view of this, we have

$$|\pi - 4| + 1 = -(\pi - 4) + 1 = -\pi + 5$$

(b) Since $x \geq 5$, the quantity $x - 5$ is nonnegative, and therefore its absolute value is equal to $x - 5$ itself. Thus, we have

$$|x - 5| = x - 5 \qquad \text{when } x \geq 5$$

(c) Since $t < 5$, the quantity $t - 5$ is negative, and therefore its absolute value is equal to $-(t - 5)$, which in turn is equal to $5 - t$. In view of this, we have

$$|t - 5| = 5 - t \qquad \text{when } t < 5$$ ▲

EXAMPLE 3 Simplify the expression $|x - 1| + |x - 2|$, given that x is in the open interval $(1, 2)$.

Solution Since x is greater than 1, the quantity $x - 1$ is positive and, consequently,

$$|x - 1| = x - 1$$

We are also given that x is less than 2. Therefore, the quantity $x - 2$ is negative, and we have

$$|x - 2| = -(x - 2) = -x + 2$$

Putting these results together now, we can write

$$|x - 1| + |x - 2| = (x - 1) + (-x + 2)$$
$$= -1 + 2 = 1$$ ▲

In the box that follows, we list several basic properties of absolute value. Each of these properties can be derived from the geometric and algebraic definitions. (With the exception of the *triangle inequality*, we shall omit the derivations. For a proof of the triangle inequality, see Exercise 63.)

PROPERTY SUMMARY

PROPERTIES OF ABSOLUTE VALUE

1. For all real numbers x, we have
   (a) $|x| \geq 0$
   (b) $x \leq |x|$ and $-x \leq |x|$

2. For all real numbers a and b, we have
   (a) $|ab| = |a||b|$ and $|a/b| = |a|/|b|$ $\quad (b \neq 0)$
   (b) $|a + b| \leq |a| + |b|$ $\quad$ (the triangle inequality)

EXAMPLE 4 Write the expression $|-2 - x^2|$ in an equivalent form that does not contain absolute values.

Solution We have

$$
\begin{aligned}
|-2 - x^2| &= |-1(2 + x^2)| \\
&= |-1||2 + x^2| \quad \text{using Property 2(a)} \\
&= |2 + x^2| \\
&= 2 + x^2
\end{aligned}
$$

The last equality follows from the fact that x^2 is nonnegative for any real number x and, consequently, the quantity $2 + x^2$ is positive. ▲

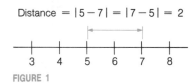

Distance $= |5 - 7| = |7 - 5| = 2$

FIGURE 1

If we think of the real numbers as points on a number line, the distance between two numbers a and b is given by the absolute value of their difference. For instance, as indicated in Figure 1, the distance between 5 and 7, namely 2, is given by either $|5 - 7|$ or $|7 - 5|$. For reference, we summarize this simple but important fact as follows.

PROPERTY SUMMARY

DISTANCE ON A NUMBER LINE

The **distance** between a and b is $|a - b| = |b - a|$.

EXAMPLE 5 Rewrite each statement using absolute values.

(a) The distance between 12 and -5 is 17.
(b) The distance between x and 2 is 4.
(c) The distance between x and 2 is less than 4.
(d) The point t is more than five units from the origin.

Solution (a) $|12 - (-5)| = 17$ or $|-5 - 12| = 17$
(b) $|x - 2| = 4$ or $|2 - x| = 4$
(c) $|x - 2| < 4$ or $|2 - x| < 4$
(d) $|t| > 5$ ▲

EXAMPLE 6 The set of real numbers that satisfies each inequality is one or more intervals on the number line. Show the intervals on a number line.

(a) $|x| < 2$ (b) $|x| > 2$ (c) $|x - 3| < 1$ (d) $|x - 3| \geq 1$

Solution (a) The given inequality tells us that the distance from x to the origin is less than two units. So, as indicated in Figure 2(a), the point x must lie in the open interval $(-2, 2)$.

(b) The condition $|x| > 2$ means that x is more than two units from the origin. Thus, as indicated in Figure 2(b), the point x lies either to the right of 2 or to the left of -2.

(c) The given inequality tells us that x must be less than one unit away from 3 on the number line. Looking one unit to either side of 3, then, we see that x must lie between 2 and 4 and x cannot equal 2 or 4. See Figure 2(c).

(d) The inequality $|x - 3| \geq 1$ says that x is at least one unit away from 3 on the number line. This means that either $x \geq 4$ or $x \leq 2$, as shown in Figure 2(d). [Here's an alternate way of thinking about this. The numbers satisfying the given inequality are precisely those numbers that *do not* satisfy the inequality in part (c). So for part (d), we need to shade that portion of the number line that was not shaded in part (c).]

FIGURE 2

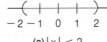

(a) $|x| < 2$

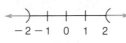

(b) $|x| > 2$

(c) $|x - 3| < 1$

(d) $|x - 3| \geq 1$

▼ EXERCISE SET 1.2

A

In Exercises 1–16, evaluate the expression.

1. $|3|$
2. $3 + |-3|$
3. $|-6|$
4. $-6 - |-6|$
5. $|-1 + 3|$
6. $|-6 + 3|$
7. $\left|-\frac{4}{5}\right| - \frac{4}{5}$
8. $\left|\frac{4}{5}\right| - \frac{4}{5}$
9. $|-6 + 2| - |4|$
10. $|-3 - 4| - |-4|$
11. $\left|-8| + |-9|\right|$
12. $\left||-8| - |-9|\right|$
13. $\left|\dfrac{27 - 5}{5 - 27}\right|$
14. $\dfrac{|27 - 5|}{|5 - 27|}$
15. $|7(-8)| - |7||-8|$
16. $|(-7)^2 + |-7|^2 - (-|-3|)^3|$

In Exercises 17–26, rewrite the expression in a form that does not contain absolute values.

17. $\left|\sqrt{2} - 1\right|$
18. $\left|\sqrt{10} - 3\right|$
19. $|x - 3|$ given that $x \geq 3$
20. $\left|y - \frac{1}{2}\right|$ given that $y > \frac{1}{2}$
21. $|x - 3|$ given that $x < 3$

22. $\left|y - \frac{1}{2}\right|$ given that $y < \frac{1}{2}$
23. $|t^2 + 1|$
24. $|x^4 + 6|$
25. $\left|-\sqrt{3} - 4\right|$
26. $\left|-\sqrt{3} - \sqrt{5} - 1\right|$

In Exercises 27–30, rewrite the expression in a form that does not contain absolute values, assuming that the given restriction on x holds.

27. $|x - 3| + |x - 4|$
    (a) $x < 3$ (b) $x > 4$ (c) $3 < x < 4$
    (d) $x = 4$ (e) $x = 3$
28. $|x - 1| - |x - \pi|$
    (a) $x < 1$ (b) $x > \pi$ (c) $1 < x < \pi$
    (d) $x = 1$ (e) $x = \pi$
29. $|x + 1| + 4|x + 3|$ given that $-\frac{5}{2} < x < -\frac{3}{2}$
30. $|x + 1| + 4|x + 3|$ given that $x < -3$

In Exercises 31–42, rewrite the statement using absolute values, as in Example 5.

31. The distance between x and 4 is 8.
32. The distance between x and 6 is 0.01.
33. The distance between x and -3 is 1.
34. The distance between x and -2 is $\frac{1}{2}$.

35. The distance between x and 2
(a) is $\frac{1}{2}$ (b) is less than $\frac{1}{2}$
(c) is at least $\frac{1}{2}$ (d) exceeds $\frac{1}{2}$

36. The distance between x and -4
(a) is 2 (b) is less than 2
(c) is at least 2 (d) exceeds 2

37. The number y is less than three units from the origin.

38. The number y is less than one unit from the number t.

39. The distance between x^2 and a^2 is less than M.

40. The sum of the distances of a and b from the origin is greater than or equal to the distance of $a + b$ from the origin.

41. The distance between x and a is less than δ.

42. The distance between y and L is less than ε.

In Exercises 43–54, the set of real numbers satisfying the given inequality is one or more intervals on the number line. Show the intervals on a number line.

43. $|x| < 4$ **44.** $|x| < 2$ **45.** $|x| > 1$

46. $|x| > 0$ **47.** $|x - 5| < 3$ **48.** $|x - 4| < 4$

49. $|x - 3| \le 4$ **50.** $|x - 1| \le \frac{1}{2}$ **51.** $|x + \frac{1}{3}| < \frac{3}{2}$

52. $|x + \frac{\pi}{2}| > 1$ **53.** $|x - 5| \ge 2$ **54.** $|x + 5| \ge 2$

B

55. In parts (a) and (b), sketch the interval or intervals corresponding to the given inequality.
(a) $|x - 2| < 1$
(b) $0 < |x - 2| < 1$
(c) In what way do your answers in parts (a) and (b) differ? (The distinction is important in the study of *limits* in calculus.)

56. Given two real numbers a and b, the notation **max(a, b)** denotes the larger of the two numbers. For instance, if $a = 1$ and $b = 2$, then $\max(1, 2) = 2$. In cases where $a = b$, then $\max(a, b)$ denotes the common value of a and b. It can be shown (see Exercise 61) that $\max(a, b)$ can be expressed in terms of absolute value as follows:

$$\max(a, b) = \frac{a + b + |a - b|}{2}$$

Verify this equation in each case.
(a) $a = 1$ and $b = 2$
(b) $a = -1$ and $b = -5$
(c) $a = b = 10$

57. Given two real numbers a and b, the notation **min(a, b)** denotes the smaller of the two numbers. In cases where $a = b$, then $\min(a, b)$ denotes the common value of a and

b. It can be shown (see Exercise 62) that $\min(a, b)$ can be expressed in terms of absolute value as follows:

$$\min(a, b) = \frac{a + b - |a - b|}{2}$$

Verify this equation in each case.
(a) $a = 6$ and $b = 1$
(b) $a = 1$ and $b = -6$
(c) $a = b = -6$

58. Show that for all real numbers a and b, we have

$$|a| - |b| \le |a - b|$$

Hint: Beginning with the identity $a = (a - b) + b$, take the absolute value of each side and then use the triangle inequality.

59. Show that

$$|a + b + c| \le |a| + |b| + |c|$$

for all real numbers a, b, and c. *Hint:* The left-hand side can be written $|a + (b + c)|$. Now use the triangle inequality.

60. Explain why no real number satisfies the equation $|x^2 + 4x| = -12$.

C

61. (As background for this exercise, you'll need to have worked Exercise 56.) Prove that

$$\max(a, b) = \frac{a + b + |a - b|}{2}$$

Hint: Consider three separate cases: $a = b$; $a > b$; $b > a$.

62. (As background for this exercise, you'll need to have worked Exercise 57.) Prove that

$$\min(a, b) = \frac{a + b - |a - b|}{2}$$

63. Complete the following steps to prove the triangle inequality.
(a) Let a and b be real numbers. Which property in the box on page 8 tells us that $a \le |a|$ and $b \le |b|$?
(b) Add the two inequalities in part (a) to obtain $a + b \le |a| + |b|$.
(c) In a similar fashion, add the two inequalities $-a \le |a|$ and $-b \le |b|$ and deduce that $-(a + b) \le |a| + |b|$.
(d) Why do the results in parts (b) and (c) imply that $|a + b| \le |a| + |b|$?

1.3 ▼ INTEGER EXPONENTS AND nTH ROOTS

I write a^{-1}, a^{-2}, a^{-3}, etc., for $\dfrac{1}{a}$, $\dfrac{1}{aa}$, $\dfrac{1}{aaa}$, etc.

Isaac Newton (June 13, 1676)

In algebra, you became familiar with the definition and basic properties of the **exponential notation** a^n, where a is a real number and n is a natural number. Since that definition and those properties are so fundamental to all of our subsequent work, we repeat them here in the two boxes that follow.

DEFINITION 1

Base and Exponent

Given a real number a and a natural number n, we define a^n by

$$a^n = \underbrace{a \cdot a \cdot a \cdot \cdots \cdot a}_{n \text{ factors}}$$

In the expression a^n, a is the **base** and n is the **exponent**.

PROPERTY SUMMARY

PROPERTIES OF EXPONENTS

PROPERTY	EXAMPLES
1. $a^m a^n = a^{m+n}$	$a^5 a^6 = a^{11}$; $(x+1)(x+1)^2 = (x+1)^3$
2. $(a^m)^n = a^{mn}$	$(2^3)^4 = 2^{12}$; $[(x+1)^2]^3 = (x+1)^6$
3. $\dfrac{a^m}{a^n} = \begin{cases} a^{m-n} & \text{if } m > n \\ \dfrac{1}{a^{n-m}} & \text{if } m < n \\ 1 & \text{if } m = n \end{cases}$	$\dfrac{a^6}{a^2} = a^4$; $\dfrac{a^2}{a^6} = \dfrac{1}{a^4}$; $\dfrac{a^5}{a^5} = 1$
4. $(ab)^m = a^m b^m$; $\left(\dfrac{a}{b}\right)^m = \dfrac{a^m}{b^m}$	$(2x^2)^3 = 2^3 \cdot (x^2)^3 = 8x^6$; $\left(\dfrac{x^2}{y^3}\right)^4 = \dfrac{x^8}{y^{12}}$

Each of these properties is a direct consequence of the definition of a^n. For instance, according to the first property we have $a^2 a^3 = a^5$. To verify that this is indeed correct, we note that

$$a^2 a^3 = (aa)(aaa) = a^5$$

Now we want to extend our definition of a^n to allow for exponents that are integers but not necessarily natural numbers. We begin by defining a^0.

DEFINITION 2

Zero Exponent

For any nonzero real number, a,

$$a^0 = 1$$

(0^0 is not defined)

EXAMPLES

(a) $2^0 = 1$

(b) $(-\pi)^0 = 1$

(c) $\left(\dfrac{3}{1 + a^2 + b^2}\right)^0 = 1$

It's easy to see the motivation for defining a^0 to be 1. Assuming that the exponent zero is to have the same properties as do exponents that are natural numbers, we can write

$$a^0 a^n = a^{0+n}$$

that is,

$$a^0 a^n = a^n$$

Now we divide both sides of this last equation by a^n to obtain $a^0 = 1$, which agrees with our definition.

Our next definition assigns a meaning to the expression a^{-n} when n is a natural number.

DEFINITION 3

Negative Exponent

$$a^{-n} = \frac{1}{a^n}$$

where $a \neq 0$ and n is a natural number

EXAMPLES

(a) $2^{-1} = \frac{1}{2^1} = \frac{1}{2}$

(b) $\left(\frac{1}{10}\right)^{-1} = \frac{1}{(\frac{1}{10})^1} = 10$

(c) $x^{-2} = \frac{1}{x^2}$

(d) $(a^2 b)^{-3} = \frac{1}{(a^2 b)^3} = \frac{1}{a^6 b^3}$

(e) $\dfrac{1}{2^{-3}} = \dfrac{1}{\dfrac{1}{2^3}} = 2^3 = 8$

Again, it's easy to see the motivation for this definition. We have

$$a^n a^{-n} = a^{n+(-n)} = a^0 = 1$$

that is,

$$a^n a^{-n} = 1$$

Now divide both sides of this last equation by a^n to obtain $a^{-n} = 1/a^n$, in agreement with Definition 3.

It can be shown that the four properties of exponents that we listed earlier continue to hold for all integer exponents. We make use of this fact in the next two examples.

EXAMPLE 1   Simplify the expression $(a^2 b^3)^2 (a^5 b)^{-1}$. Write the answer in such a way that only positive exponents are used.

Solution   FIRST METHOD

$$(a^2 b^3)^2 (a^5 b)^{-1} = (a^2 b^3)^2 \cdot \frac{1}{a^5 b}$$

$$= \frac{a^4 b^6}{a^5 b} = \frac{b^{6-1}}{a^{5-4}} = \frac{b^5}{a}$$

ALTERNATE METHOD

$$(a^2 b^3)^2 (a^5 b)^{-1} = (a^4 b^6)(a^{-5} b^{-1})$$

$$= a^{4-5} b^{6-1} = a^{-1} b^5$$

$$= \frac{b^5}{a}$$

▲

EXAMPLE 2   Simplify the expression $\left(\dfrac{a^{-5}b^2c^0}{a^3b^{-1}}\right)^3$, writing the answer so that negative exponents are not used.

Solution
$$\left(\frac{a^{-5}b^2c^0}{a^3b^{-1}}\right)^3 = \frac{a^{-15}b^6}{a^9b^{-3}} = \frac{b^{6-(-3)}}{a^{9-(-15)}}$$
$$= \frac{b^9}{a^{24}}$$

▲

Now that we've completed our review of integer exponents, we are ready to consider roots. Here is the basic definition.

DEFINITION 4

*n*th **Roots**

Let *n* be a natural number. If *a* and *b* are real numbers and $$a^n = b$$ then we say that *a* is an ***n*th root** of *b*. When $n = 2$ and when $n = 3$, we refer to the root as a **square root** and **cube root**, respectively.	**EXAMPLES** Both 3 and -3 are square roots of 9 because $3^2 = 9$ and $(-3)^2 = 9$. Both 2 and -2 are fourth roots of 16 because $2^4 = 16$ and $(-2)^4 = 16$. 2 is a cube root of 8 because $2^3 = 8$. -3 is a fifth root of -243 because $(-3)^5 = -243$.

As the examples in the box suggest, square roots, fourth roots, and all *even* roots of positive numbers occur in pairs, one positive and one negative. In these cases, we use the notation $\sqrt[n]{b}$ to denote the positive, or **principal**, *n*th root of *b*. As examples of this notation, we can write

$$\sqrt[4]{81} = 3 \qquad \text{(The principal fourth root of 81 is 3.)}$$
$$\sqrt[4]{81} \neq -3 \qquad \text{(The principal fourth root of 81 is not } -3.)$$
$$-\sqrt[4]{81} = -3 \qquad \text{(The negative of the principal fourth root of 81 is } -3.)$$

The symbol $\sqrt{}$ is called a **radical sign**, and the number within the radical sign is the **radicand**.* The natural number *n* used in the notation $\sqrt[n]{}$ is called the **index** of the radical. For square roots, as you know from basic algebra, we suppress the index and simply write $\sqrt{}$ rather than $\sqrt[2]{}$. So, for example, $\sqrt{25} = 5$.

As we saw in the examples, cube roots, fifth roots, and, in fact, all *odd* roots occur singly, not in pairs. In these cases we again use the notation $\sqrt[n]{b}$ for the *n*th root. The definition and examples in the following box summarize our discussion up to this point.

---

* The radical sign $(\sqrt{})$ was introduced in 1525 by the German mathematician Christoff Rudolff.

DEFINITION 5

Principal *n*th root

EXAMPLES

1. Let n be a natural number. If a and b are nonnegative real numbers, then

$$\sqrt[n]{b} = a \quad \text{if and only if} \quad b = a^n.$$

The number a is the **principal *n*th root** of b.

2. If a and b are negative and n is an odd natural number, then

$$\sqrt[n]{b} = a \quad \text{if and only if} \quad b = a^n.$$

$\sqrt{25} = 5; \quad \sqrt{25} \neq -5$

$-\sqrt{25} = -5$

$\sqrt[4]{\dfrac{1}{16}} = \dfrac{1}{2}; \quad \sqrt[3]{8} = 2$

$\sqrt[5]{-\dfrac{1}{32}} = -\dfrac{1}{2}$

Five properties of nth roots are frequently used in simplifying certain expressions. The first four are similar to the properties of square roots that are developed in elementary algebra. For reference, we list these properties side by side in the following box. Property 5 is listed here only for the sake of completeness; we'll postpone discussing it until Section 1.4.

PROPERTY SUMMARY

PROPERTIES OF *n*TH ROOTS	**CORRESPONDING PROPERTIES FOR SQUARE ROOTS**
Suppose that x and y are real numbers and that m and n are natural numbers. Then each of the following properties holds, provided only that the expressions on both sides of the equation are defined (and so represent real numbers).	
1. $(\sqrt[n]{x})^n = x$	$(\sqrt{x})^2 = x$
2. $\sqrt[n]{xy} = \sqrt[n]{x}\,\sqrt[n]{y}$	$\sqrt{xy} = \sqrt{x}\,\sqrt{y}$
3. $\sqrt[n]{\dfrac{x}{y}} = \dfrac{\sqrt[n]{x}}{\sqrt[n]{y}}$	$\sqrt{\dfrac{x}{y}} = \dfrac{\sqrt{x}}{\sqrt{y}}$
4. n even: $\sqrt[n]{x^n} = \|x\|$ $\quad$ n odd: $\sqrt[n]{x^n} = x$	$\sqrt{x^2} = \|x\|$
5. $\sqrt[m]{\sqrt[n]{x}} = \sqrt[mn]{x}$	

Our immediate use for these properties will be in simplifying expressions involving nth roots. In general, we try to factor the expression under the radical so that one factor is the largest perfect nth power that we can find. Then we apply Property 2 or Property 3. For instance, we can simplify the expression $\sqrt{72}$ as follows:

$$\sqrt{72} = \sqrt{(36)(2)} = \sqrt{36}\sqrt{2} = 6\sqrt{2}$$

In this procedure, we began by factoring 72 as $(36)(2)$. Note that 36 is the largest factor of 72 that is a perfect square. If we had begun instead with a different factorization, say $72 = (9)(8)$, we could still arrive at the same answer, but it would take longer. (Check this for yourself.)

As another example, let us simplify $\sqrt[3]{40}$. First, what (if any) is the largest perfect cube factor of 40? Since the first few perfect cubes are

$$1^3 = 1 \qquad 2^3 = 8 \qquad 3^3 = 27 \qquad 4^3 = 64$$

we see that 8 is a perfect cube factor of 40, and we write

$$\sqrt[3]{40} = \sqrt[3]{(8)(5)} = \sqrt[3]{8}\sqrt[3]{5} = 2\sqrt[3]{5}$$

EXAMPLE 3   Simplify: **(a)** $\sqrt{12} + \sqrt{75}$ **(b)** $\sqrt{\dfrac{162}{49}}$

Solution   **(a)** $\sqrt{12} + \sqrt{75} = \sqrt{(4)(3)} + \sqrt{(25)(3)}$
$$= \sqrt{4}\sqrt{3} + \sqrt{25}\sqrt{3}$$
$$= 2\sqrt{3} + 5\sqrt{3}$$
$$= 7\sqrt{3}$$

(b) $\sqrt{\dfrac{162}{49}} = \dfrac{\sqrt{162}}{\sqrt{49}} = \dfrac{\sqrt{81}\sqrt{2}}{7} = \dfrac{9\sqrt{2}}{7}$

EXAMPLE 4   Simplify: $\sqrt[3]{16} + \sqrt[3]{250} - \sqrt[3]{128}$

Solution $\sqrt[3]{16} + \sqrt[3]{250} - \sqrt[3]{128} = \sqrt[3]{8}\sqrt[3]{2} + \sqrt[3]{125}\sqrt[3]{2} - \sqrt[3]{64}\sqrt[3]{2}$
$$= 2\sqrt[3]{2} + 5\sqrt[3]{2} - 4\sqrt[3]{2}$$
$$= 3\sqrt[3]{2}$$

EXAMPLE 5   Simplify each expression by removing the largest possible perfect square factor from within the radical.

(a) $\sqrt{8x^2}$ **(b)** $\sqrt{8x^2}$ assuming $x \geq 0$
(c) $\sqrt{18a^7}$ assuming $a \geq 0$ **(d)** $\sqrt[3]{16y^5}$

Solution   **(a)** $\sqrt{8x^2} = \sqrt{(4)(2)(x^2)} = \sqrt{4}\sqrt{2}\sqrt{x^2} = 2\sqrt{2}\,|x|$

(b) $\sqrt{8x^2} = \sqrt{4}\sqrt{2}\sqrt{x^2} = 2\sqrt{2}\,x$ $\sqrt{x^2} = x$ because $x \geq 0$

(c) $\sqrt{18a^7} = \sqrt{(9a^6)(2a)} = \sqrt{9a^6}\sqrt{2a} = 3a^3\sqrt{2a}$

(d) $\sqrt[3]{16y^5} = \sqrt[3]{8y^3}\sqrt[3]{2y^2} = 2y\sqrt[3]{2y^2}$

EXAMPLE 6   Simplify: $\dfrac{1}{\sqrt{2}} - 3\sqrt{50}$

Solution   First, we rationalize the denominator in the fraction $\dfrac{1}{\sqrt{2}}$:

$$\frac{1}{\sqrt{2}} = \frac{1}{\sqrt{2}} \cdot 1 = \frac{1}{\sqrt{2}} \cdot \frac{\sqrt{2}}{\sqrt{2}} = \frac{\sqrt{2}}{2}$$

Next, we simplify the expression $3\sqrt{50}$:

$$3\sqrt{50} = 3\sqrt{(25)(2)} = 3\sqrt{25}\sqrt{2} = (3)(5)\sqrt{2} = 15\sqrt{2}$$

Now, putting things together, we have

$$\frac{1}{\sqrt{2}} - 3\sqrt{50} = \frac{\sqrt{2}}{2} - 15\sqrt{2} = \frac{\sqrt{2}}{2} - \frac{30\sqrt{2}}{2}$$

$$= \frac{\sqrt{2} - 30\sqrt{2}}{2} = \frac{(1 - 30)\sqrt{2}}{2}$$

$$= \frac{-29\sqrt{2}}{2}$$

EXAMPLE 7 Rationalize the denominator: $\dfrac{4}{2 + \sqrt{3}}$

Solution We multiply by 1, writing 1 as $\dfrac{2 - \sqrt{3}}{2 - \sqrt{3}}$:

$$\frac{4}{2 + \sqrt{3}} \cdot 1 = \frac{4}{2 + \sqrt{3}} \cdot \frac{2 - \sqrt{3}}{2 - \sqrt{3}}$$

$$= \frac{4(2 - \sqrt{3})}{4 - (\sqrt{3})^2} = \frac{4(2 - \sqrt{3})}{4 - 3} = \frac{8 - 4\sqrt{3}}{1}$$

$$= 8 - 4\sqrt{3}$$

Note Check for yourself that multiplying the original fraction by $\dfrac{2 + \sqrt{3}}{2 + \sqrt{3}}$ does *not* eliminate radicals in the denominator. ▲

In the next example, we are asked to rationalize the numerator rather than the denominator. This is useful in calculus.

EXAMPLE 8 Rationalize the *numerator*: $\dfrac{\sqrt{x} - \sqrt{3}}{x - 3}$ $(x \ge 0, x \ne 3)$

Solution $\dfrac{\sqrt{x} - \sqrt{3}}{x - 3} \cdot 1 = \dfrac{\sqrt{x} - \sqrt{3}}{x - 3} \cdot \dfrac{\sqrt{x} + \sqrt{3}}{\sqrt{x} + \sqrt{3}}$

$$= \frac{(\sqrt{x})^2 - (\sqrt{3})^2}{(x - 3)(\sqrt{x} + \sqrt{3})} = \frac{x - 3}{(x - 3)(\sqrt{x} + \sqrt{3})}$$

$$= \frac{1}{\sqrt{x} + \sqrt{3}}$$ ▲

The strategy for rationalizing numerators or denominators involving nth roots is similar to that used for square roots. To rationalize a numerator or a denominator involving an nth root, we multiply that numerator or denominator by a factor that yields a product that itself is a perfect nth power. The next example displays two instances of this.

EXAMPLE 9 **(a)** Rationalize the denominator: $\dfrac{6}{\sqrt[3]{7}}$

(b) Rationalize the denominator: $\dfrac{ab}{\sqrt[4]{a^2 b^3}}$ $(a > 0, b > 0)$

Solution **(a)** $\dfrac{6}{\sqrt[3]{7}} \cdot 1 = \dfrac{6}{\sqrt[3]{7}} \cdot \dfrac{\sqrt[3]{7^2}}{\sqrt[3]{7^2}}$ **(b)** $\dfrac{ab}{\sqrt[4]{a^2 b^3}} \cdot 1 = \dfrac{ab}{\sqrt[4]{a^2 b^3}} \cdot \dfrac{\sqrt[4]{a^2 b}}{\sqrt[4]{a^2 b}}$

$$= \frac{6\sqrt[3]{7^2}}{\sqrt[3]{7^3}} = \frac{6\sqrt[3]{49}}{7} \qquad\qquad = \frac{ab\sqrt[4]{a^2 b}}{\sqrt[4]{a^4 b^4}} = \frac{ab\sqrt[4]{a^2 b}}{ab} = \sqrt[4]{a^2 b}$$ ▲

▼ EXERCISE SET 1.3

A

For Exercises 1–20, simplify the expression. For exercises where negative exponents appear, write the answers without using negative exponents.

1. (a) $x^3 x^{12}$
   (b) $(x^3)^{12}$
   (c) $(x+1)^3(x+1)^{12}$
   (d) $[(x+1)^3]^{12}$

2. (a) $y^2 y^6$
   (b) $(y^2)^6$
   (c) $(y^2 y^6)^3$
   (d) $[(y+1)^2]^6$

3. (a) $\dfrac{a^{15}}{a^9}$
   (b) $\dfrac{(a+1)^{15}}{(a+1)^9}$
   (c) $\dfrac{(a+2)^6(a+1)^{15}}{[(a+2)(a+1)]^5}$

4. (a) $\dfrac{(x^2 a)^5}{(x^3 a^2)^4}$
   (b) $\dfrac{(x^2+1)^5}{(x^2+1)^4}$
   (c) $\dfrac{(x^2+1)^3}{[(x^2+1)^2]^3}$

5. (a) $(64)^0$
   (b) $(64^3)^0$
   (c) $(64^0)^3$

6. (a) $(2^0+3^0)^3$
   (b) $(2+3)^0$
   (c) $[(\tfrac{1}{10})^0]^3$

7. (a) $10^{-1}+10^{-2}$
   (b) $(10^{-1}+10^{-2})^{-1}$
   (c) $(10^{-1})^{-2}$

8. (a) $4^{-2}+4^{-1}$
   (b) $[(\tfrac{1}{4})^{-1}+(\tfrac{1}{4})^{-2}]^{-1}$
   (c) $[(\tfrac{1}{4})^{-1}]^{-2}$

9. (a) $(a^2 b c^0)^{-3}$
   (b) $(a^3 b)^3(a^2 b^4)^{-1}$
   (c) $(a^{-3} b^{-1} c^3)^{-2}$

10. (a) $(x^3 y^{-2} z^{-1})^{-4}$
    (b) $(x^2 y^3)^{-3}(x^2 y^4)^{-2}$
    (c) $\dfrac{(x^{-1} y^{-1})^{-1}}{xy}$

11. $(2^{-2}+2^{-1}+2^0)^{-2}$

12. $(10^{-2}+10^{-1}+10^2)^{-1}$

13. $\left(\dfrac{x^3 y^{-2} z}{xy^2 z^{-3}}\right)^{-3}$

14. $\left(\dfrac{x^4 y^{-8} z^2}{xy^2 z^{-6}}\right)^2$

15. $\left(\dfrac{x^4 y^{-8} z^2}{xy^2 z^{-6}}\right)^{-2}$

16. $\left(\dfrac{a^3 b^{-9} c^2}{a^5 b^2 c^{-4}}\right)^0$

17. $\left(\dfrac{a^{-2} b^{-3} c^{-4}}{a^2 b^3 c^4}\right)^2$

18. $(-2x)^{-3}(-2x^{-2})^{-1}$

19. $\dfrac{x^2}{y^{-3}} \div \dfrac{x^2}{y^3}$

20. $(2x^2)^{-3} - 2(x^2)^{-3}$

In Exercises 21–26, determine whether the statement is true or false.

21. (a) $\sqrt{81} = -9$
    (b) $\sqrt{81} = 9$
    (c) $-\sqrt{81} = -9$

22. (a) $-\sqrt{121} = -11$
    (b) $\sqrt{121} = \pm 11$
    (c) $\sqrt{121} = 11$

23. (a) $\sqrt{9+16} = \sqrt{9}+\sqrt{16}$
    (b) $\sqrt{(9)(16)} = \sqrt{9}\sqrt{16}$

24. (a) $\sqrt{10+6} = \sqrt{10}+\sqrt{6}$
    (b) $\sqrt{10}\sqrt{6} = \sqrt{(10)(6)}$

25. (a) $\sqrt{(-5)^2} = -5$
    (b) $\sqrt{x^2} = x$ $(x<0)$

26. (a) $\sqrt{(-6)^2} = 6$
    (b) $\sqrt{y^2} = |y|$

In Exercises 27–36, evaluate the expression. If the expression is undefined (i.e., does not represent a real number), say so.

27. (a) $\sqrt[3]{-64}$
    (b) $\sqrt[4]{-64}$

28. (a) $\sqrt[5]{32}$
    (b) $\sqrt[5]{-32}$

29. (a) $\sqrt[3]{8/125}$
    (b) $\sqrt[3]{-8/125}$

30. (a) $\sqrt[3]{-1/1000}$
    (b) $\sqrt[6]{-1/1000}$

31. (a) $\sqrt{-16}$
    (b) $\sqrt[4]{-16}$

32. (a) $-\sqrt[4]{16}$
    (b) $-\sqrt[4]{-16}$

33. (a) $\sqrt[4]{256/81}$
    (b) $\sqrt[3]{-27/125}$

34. (a) $\sqrt[6]{64}$
    (b) $\sqrt[6]{-64}$

35. (a) $\sqrt[5]{-32}$
    (b) $-\sqrt[5]{-32}$

36. (a) $\sqrt[4]{(-10)^4}$
    (b) $\sqrt[3]{(-10)^3}$

In Exercises 37–66, simplify the expression. (Unless otherwise specified, assume that all letters in Exercises 53–66 represent positive numbers.)

37. (a) $\sqrt{18}$
    (b) $\sqrt[3]{54}$

38. (a) $\sqrt{150}$
    (b) $\sqrt[3]{375}$

39. (a) $\sqrt{98}$
    (b) $\sqrt[3]{-64}$

40. (a) $\sqrt{27}$
    (b) $\sqrt[3]{-108}$

41. (a) $\sqrt{25/4}$
    (b) $\sqrt[4]{16/625}$

42. (a) $\sqrt{225/49}$
    (b) $\sqrt[5]{-256/243}$

43. (a) $\sqrt{2}+\sqrt{8}$
    (b) $\sqrt[3]{2}+\sqrt[3]{16}$

44. (a) $4\sqrt{3}-2\sqrt{27}$
    (b) $2\sqrt[3]{81}+3\sqrt[3]{24}$

45. (a) $4\sqrt{50}-3\sqrt{128}$
    (b) $\sqrt[4]{32}+\sqrt[4]{162}$

46. (a) $\sqrt{3}-\sqrt{12}+\sqrt{48}$
    (b) $\sqrt[5]{-2}+\sqrt[3]{-64}-\sqrt[5]{486}$

47. (a) $\sqrt{0.09}$
    (b) $\sqrt[3]{0.008}$

48. (a) $\sqrt[3]{-2}+\sqrt[3]{2}$
    (b) $\sqrt{81/121}-\sqrt[3]{-8/1331}$

49. $4\sqrt{24}-8\sqrt{54}+2\sqrt{6}$

50. $\sqrt[3]{192}+\sqrt[3]{-81}+\sqrt{\sqrt[3]{9}}$

51. $\sqrt{\sqrt{64}}$

52. $\sqrt[3]{\sqrt{4096}}$

53. (a) $\sqrt{36x^2}$
    (b) $\sqrt{36y^2}$, $y<0$

54. (a) $\sqrt{225x^4 y^3}$
    (b) $\sqrt[4]{16a^4}$, $a<0$

55. (a) $\sqrt{ab^2}\sqrt{a^2 b}$
    (b) $\sqrt{ab^3}\sqrt{a^3 b}$

56. (a) $\sqrt[3]{125x^6}$
    (b) $\sqrt[4]{64y^4}$, $y<0$

57. $\sqrt{72a^3 b^4 c^5}$

58. $\sqrt{\dfrac{(a+b)^5}{16a^2 b^2}}$

59. $\sqrt[4]{16a^4b^5}, \quad a < 0$

60. $\sqrt[3]{8a^4b^6}$

61. $\sqrt{18a^3b^2}$

62. $\sqrt[5]{64a^6b^{12}}$

63. $\sqrt[3]{\dfrac{16a^{12}b^2}{c^9}}$

64. $\sqrt[4]{ab^3}\,\sqrt[4]{a^3b}$

65. $\sqrt[6]{\dfrac{5a^7}{a^{-5}b^6}}, \quad b < 0$

66. $\sqrt[3]{a^2b}\,\sqrt[3]{ab}\,\sqrt[3]{b^4}$

In Exercises 67–80, rationalize the denominator and simplify where possible. (Assume that all letters represent positive quantities.)

67. **(a)** $\dfrac{4}{\sqrt{7}}$ **(b)** $\dfrac{3}{\sqrt{3}}$ **(c)** $\dfrac{\sqrt{2}}{\sqrt{5}}$

68. **(a)** $\dfrac{3}{\sqrt{2}}$ **(b)** $\dfrac{6}{\sqrt{6}}$ **(c)** $\dfrac{\sqrt{10}}{\sqrt{2}}$

69. **(a)** $\dfrac{1}{1+\sqrt{5}}$ **(b)** $\dfrac{1}{1-\sqrt{5}}$ **(c)** $\dfrac{1+\sqrt{5}}{1-\sqrt{5}}$

70. **(a)** $\dfrac{3}{2-\sqrt{3}}$ **(b)** $\dfrac{3}{2+\sqrt{3}}$ **(c)** $\dfrac{2-\sqrt{3}}{2+\sqrt{3}}$

71. $\dfrac{1}{\sqrt{5}} + 4\sqrt{45}$

72. $\dfrac{3}{\sqrt{8}} - \sqrt{450}$

73. $\dfrac{1}{\sqrt[3]{25}}$

74. $\dfrac{4}{\sqrt[3]{16}}$

75. $\dfrac{3}{\sqrt[4]{3}}$

76. $\dfrac{\sqrt[3]{5}}{\sqrt[3]{6}}$

77. $\dfrac{1}{\sqrt[4]{2ab^5}}$

78. $\dfrac{1}{\sqrt[3]{4a^2b^8}}$

79. $\dfrac{3}{\sqrt[5]{16a^4b^9}}$

80. $\dfrac{3}{\sqrt[4]{27a^5b^{11}}}$

81. $\dfrac{x}{\sqrt{x}-2}$

82. $\dfrac{\sqrt{x}}{5-\sqrt{x}}$

83. $\dfrac{\sqrt{x}-\sqrt{a}}{\sqrt{x}+\sqrt{a}}$

84. $\dfrac{1}{\sqrt{x+h}-\sqrt{x}}$

In Exercises 85–88, rationalize the numerator.

85. $\dfrac{\sqrt{x}-\sqrt{5}}{x-5}$

86. $\dfrac{\sqrt{a}-\sqrt{b}}{a-b}$

87. $\dfrac{\sqrt{2+h}-\sqrt{2}}{h}$

88. $\dfrac{\sqrt{x+h}-\sqrt{x}}{h}$

C **89.** Which number is larger, 10^9 or 9^{10}?

C **90.** **(a)** Which is larger, 11^{12} or 12^{11}?
     (b) Which is larger, 18^{19} or 19^{18}?

B

91. **(a)** **C** Evaluate $\sqrt{8-2\sqrt{7}}$ and $\sqrt{7}-1$. What do you observe?

(b) Prove that $\sqrt{8-2\sqrt{7}} = \sqrt{7}-1$. *Hint:* In view of the definition of a principal square root, you need to check that $(\sqrt{7}-1)^2 = 8-2\sqrt{7}$.

92. Verify that

$$\sqrt{a+b+2\sqrt{ab}} = \sqrt{a}+\sqrt{b} \quad (a \ge 0, b \ge 0)$$

[See the hint in Exercise 91(b).]

C **93.** Use a calculator to provide empirical evidence that both of the following equations may be correct:

$$\dfrac{2-\sqrt{3}}{\sqrt{2-\sqrt{2-\sqrt{3}}}} + \dfrac{2+\sqrt{3}}{\sqrt{2+\sqrt{2+\sqrt{3}}}} = \sqrt{2}$$

$$\sqrt{\sqrt{97.5-\tfrac{1}{11}}} = \pi$$

The point of this exercise is to remind you that as useful as calculators may be, there is still the need for proofs in mathematics. In fact, it can be shown that the first equation is indeed correct, but the second is not.

94. Rationalize the denominator: $\dfrac{1}{1+\sqrt{2}+\sqrt{3}}$.

Hint: First multiply by $\dfrac{1+\sqrt{2}-\sqrt{3}}{1+\sqrt{2}-\sqrt{3}}$.

95. Simplify $\dfrac{\sqrt{a}}{\sqrt{a}+\sqrt{b}} + \dfrac{\sqrt{b}}{\sqrt{a}-\sqrt{b}}$.

Hint: First rationalize the denominators.

96. Simplify $\dfrac{a^{4p+2q}}{a^{3p}a^p(a^q)^2}$.

97. Simplify $\dfrac{x^{3a+2b-c}}{(x^{2a})(x^b)} \cdot x^{3c-a-b}$.

C

98. Let a and b be nonnegative integers, and consider the following two quantities.

$$\sqrt{a}+\sqrt{b} \quad \text{and} \quad \sqrt{a+b}$$

(a) Carry out a few computations to find out which quantity seems, in general, to be the larger.

(b) Write an inequality to summarize your results in part (a).

(c) Prove your statement in part (b). *Hint:* If x and y are nonnegative quantities, the condition $x \le y$ is equivalent to $x^2 \le y^2$.

99. Verify that

$$\sqrt{p+q-r+2\sqrt{q(p-r)}} = \sqrt{p-r}+\sqrt{q}$$

where p, q, and r are positive and $p > r$.

100. Suppose that $p = b^x$, $q = b^y$, and $b^2 = (p^y q^x)^z$, where all the letters denote natural numbers and $b \ne 1$. Show that $xyz = 1$.

1.4 ▼ RATIONAL EXPONENTS

We can use the concept of an nth root to give a meaning to fractional exponents that is useful and, at the same time, consistent with our work in the previous section. First, by way of motivation, suppose that we want to assign a value to $5^{1/3}$. Assuming that the usual properties of exponents continue to apply here, we can write

$$(5^{1/3})^3 = 5^1 = 5$$

That is,

$$(5^{1/3})^3 = 5 \qquad \text{or} \qquad 5^{1/3} = \sqrt[3]{5}$$

By replacing 5 and 3 with b and n, respectively, we can see that we want to define $b^{1/n}$ to mean $\sqrt[n]{b}$. Also, by thinking of $b^{m/n}$ as $(b^{1/n})^m$, we see that the definition for $b^{m/n}$ ought to be $(\sqrt[n]{b})^m$. These definitions are formalized in the box that follows.

DEFINITION

Rational Exponents

EXAMPLES

1. Let b denote a real number and n a natural number. We define $b^{1/n}$ by

$$b^{1/n} = \sqrt[n]{b}$$

(If n is even, we require that $b \geq 0$.)

$4^{1/2} = \sqrt{4} = 2$

$(-8)^{1/3} = \sqrt[3]{-8} = -2$

2. Let m/n be a rational number reduced to lowest terms. Assume that n is positive and that $\sqrt[n]{b}$ exists. Then,

$$b^{m/n} = (\sqrt[n]{b})^m$$

or, equivalently,

$$b^{m/n} = \sqrt[n]{b^m}$$

$8^{2/3} = (\sqrt[3]{8})^2 = 2^2 = 4$

or, equivalently,

$8^{2/3} = \sqrt[3]{8^2} = \sqrt[3]{64} = 4$

It can be shown that the four properties of exponents that we listed in Section 1.3 (on page 11) continue to hold for rational exponents in general. In fact, we'll take this for granted rather than follow the lengthy argument needed for its verification. We will also assume that these properties apply to irrational exponents. So, for instance, we have

$$(2^{\sqrt{5}})^{\sqrt{5}} = 2^5 = 32$$

(The definition of irrational exponents is discussed in Section 4.1.) In the next three examples, we display the basic techniques for working with rational exponents.

EXAMPLE 1 Simplify each expression, writing the answer using positive exponents. If an expression does not represent a real number, say so.

(a) $49^{1/2}$ **(b)** $-49^{1/2}$ **(c)** $(-49)^{1/2}$ **(d)** $49^{-1/2}$

Solution

(a) $49^{1/2} = \sqrt{49} = 7$

(b) $-49^{1/2} = -(49^{1/2}) = -\sqrt{49} = -7$

(c) $(-49)^{1/2}$ does not represent a real number because there is no real number x such that $x^2 = -49$.

(d) $49^{-1/2} = \sqrt{49^{-1}} = \sqrt{\dfrac{1}{49}} = \dfrac{\sqrt{1}}{\sqrt{49}} = \dfrac{1}{7}$

Alternatively, we have

$$49^{-1/2} = (49^{1/2})^{-1} = 7^{-1} = \frac{1}{7}$$

▲

EXAMPLE 2   Simplify each expression, writing the answer using positive exponents. (Assume that $a > 0$.)

(a) $(5a^{2/3})(4a^{3/4})$ **(b)** $\sqrt[5]{\dfrac{16a^{1/3}}{a^{1/4}}}$ **(c)** $(x^2 + 1)^{1/5}(x^2 + 1)^{4/5}$

Solution   **(a)** $(5a^{2/3})(4a^{3/4}) = 20a^{(2/3)+(3/4)}$

$$= 20a^{17/12} \quad \text{because } \frac{2}{3} + \frac{3}{4} = \frac{17}{12}$$

(b) $\sqrt[5]{\dfrac{16a^{1/3}}{a^{1/4}}} = \left(\dfrac{16a^{1/3}}{a^{1/4}}\right)^{1/5}$

$$= (16a^{1/12})^{1/5} \quad \text{because } \frac{1}{3} - \frac{1}{4} = \frac{1}{12}$$

$$= 16^{1/5}a^{1/60}$$

(c) $(x^2 + 1)^{1/5}(x^2 + 1)^{4/5} = (x^2 + 1)^1 = x^2 + 1$

▲

EXAMPLE 3   Simplify: **(a)** $32^{-2/5}$ **(b)** $(-8)^{4/3}$

Solution   **(a)** $32^{-2/5} = (\sqrt[5]{32})^{-2} = 2^{-2} = \dfrac{1}{2^2} = \dfrac{1}{4}$

Alternatively, we have

$$32^{-2/5} = (2^5)^{-2/5} = 2^{-2} = \frac{1}{2^2} = \frac{1}{4}$$

(b) $(-8)^{4/3} = (\sqrt[3]{-8})^4 = (-2)^4 = 16$

Alternatively, we can write

$$(-8)^{4/3} = [(-2)^3]^{4/3} = (-2)^4 = 16$$

▲

Rational exponents can be used to simplify certain expressions containing radicals. For example, one of the properties of nth roots, which was listed but not discussed in the previous section, is $\sqrt[m]{\sqrt[n]{x}} = \sqrt[mn]{x}$. Using exponents, it is easy to verify this property. We have

$$\sqrt[m]{\sqrt[n]{x}} = (x^{1/n})^{1/m}$$
$$= x^{1/mn} = \sqrt[mn]{x}, \qquad \text{as we wished to show.}$$

EXAMPLE 4   Consider the expression $\sqrt{x}\,\sqrt[3]{y^2}$, where x and y are positive.

(a) Rewrite the expression using rational exponents.
(b) Rewrite the expression using only one radical sign.

Solution (a) $\sqrt{x}\,\sqrt[3]{y^2} = x^{1/2}y^{2/3}$

(b) $\sqrt{x}\,\sqrt[3]{y^2} = x^{1/2}y^{2/3}$

$\qquad\qquad = x^{3/6}y^{4/6}$ We've rewritten the fractions using a common denominator.

$\qquad\qquad = \sqrt[6]{x^3}\,\sqrt[6]{y^4}$

$\qquad\qquad = \sqrt[6]{x^3 y^4}$ ▲

EXAMPLE 5   Rewrite the following expression using rational exponents. (Assume that x, y, and z are positive.)

$$\sqrt{\frac{\sqrt[3]{x}\,\sqrt[4]{y^3}}{\sqrt[5]{z^4}}}$$

Solution $\sqrt{\dfrac{\sqrt[3]{x}\,\sqrt[4]{y^3}}{\sqrt[5]{z^4}}} = \left(\dfrac{\sqrt[3]{x}\,\sqrt[4]{y^3}}{\sqrt[5]{z^4}}\right)^{1/2} = \left(\dfrac{x^{1/3}y^{3/4}}{z^{4/5}}\right)^{1/2}$

$\qquad\qquad = \dfrac{x^{1/6}y^{3/8}}{z^{2/5}} = x^{1/6}y^{3/8}z^{-2/5}$ ▲

▼ EXERCISE SET 1.4

A

In Exercises 1–36, evaluate or simplify the expression. Express the answer using positive exponents. If an expression is undefined (i.e., does not represent a real number), say so. (Assume that all letters represent positive numbers.)

1. $16^{1/2}$
2. $100^{1/2}$
3. $(\frac{1}{36})^{1/2}$
4. $0.09^{1/2}$
5. $(-16)^{1/2}$
6. $(-1)^{1/2}$
7. $625^{1/4}$
8. $(\frac{1}{81})^{1/4}$
9. $8^{1/3}$
10. $0.001^{1/3}$
11. $8^{2/3}$
12. $64^{2/3}$
13. $(-32)^{1/5}$
14. $(-\frac{1}{125})^{1/3}$
15. $(-1000)^{1/3}$
16. $(243)^{1/5}$
17. $49^{-1/2}$
18. $121^{-1/2}$
19. $(-49)^{-1/2}$
20. $(-64)^{-1/3}$
21. $36^{-3/2}$
22. $(-0.001)^{-2/3}$
23. $125^{2/3}$
24. $125^{-2/3}$
25. $(-1)^{3/5}$
26. $27^{4/3} + 27^{-4/3} + 27^0$
27. $32^{4/5} - 32^{-4/5}$
28. $64^{1/2} + 64^{-1/2} - 64^{4/3}$
29. $(\frac{9}{16})^{-5/2} - (\frac{1000}{27})^{4/3}$
30. $(256^{-3/4})^{4/3}$
31. $(2a^{1/3})(3a^{1/4})$
32. $\sqrt[5]{3a^4}$
33. $\sqrt[4]{\dfrac{64a^{2/3}}{a^{1/3}}}$
34. $(x^2 + 1)^{2/3}(x^2 + 1)^{4/3}$
35. $\dfrac{(x^2 + 1)^{3/4}}{(x^2 + 1)^{-1/4}}$
36. $\dfrac{(2x^2 + 1)^{-6/5}(2x^2 + 1)^{6/5}(x^2 + 1)^{-1/5}}{(x^2 + 1)^{9/5}}$

For Exercises 37–44, follow Example 4 in the text to rewrite the expression in two ways: (a) using rational exponents; (b) using only one radical sign. (Assume x, y, and z are positive.)

37. $\sqrt{3}\,\sqrt[3]{6}$
38. $\sqrt{5}\,\sqrt[3]{7}$
39. $\sqrt[3]{6}\,\sqrt[4]{2}$
40. $\sqrt[3]{2}\,\sqrt[5]{2}$
41. $\sqrt[3]{x^2}\,\sqrt[5]{y^4}$
42. $\sqrt{x}\,\sqrt[3]{y}\,\sqrt[4]{z}$
43. $\sqrt[4]{x^a}\,\sqrt[3]{x^b}\,\sqrt{x^{a/6}}$
44. $\sqrt[3]{27\sqrt{64x}}$

In Exercises 45–52, rewrite the expression using rational exponents rather than radicals. (Assume that x, y, and z are positive.)

45. $\sqrt[3]{(x + 1)^2}$
46. $\dfrac{1}{\sqrt{x}} + \sqrt{x}$
47. $(\sqrt[3]{x + y})^2$
48. $\sqrt{\sqrt{x}}$
49. $\sqrt[3]{\sqrt{x}} + \sqrt{\sqrt[3]{x}}$
50. $\sqrt[3]{\sqrt{2}}$
51. $\sqrt{\sqrt[3]{x}\,\sqrt[4]{y}}$
52. $\sqrt[5]{\dfrac{\sqrt{x}\,\sqrt[3]{y}}{\sqrt[4]{z^2}}}$

[C] 53. Which is larger, $9^{10/9}$ or $10^{9/10}$?

[C] 54. Which number is closer to 3: $13^{3/7}$ or $6560^{1/8}$?

C **55.** Consider the expression $n^{1/n}$, where n is a natural number. In calculus, it is shown that as n takes on larger and larger values, the resulting value of the expression approaches 1. Confirm this empirically by completing the table. (Round off your results to four decimal places.)

n	2	5	10	100	10^3	10^4	10^5	10^6
$n^{1/n}$								

B

C **56.** In working with roots and radicals, we cannot always tell from the outset that two numbers are equal. (We've already seen examples of this in Exercise Set 1.1, Exercise 67.) In each case, use a calculator to verify that the quantity on the left-hand side of the equation agrees with the quantity on the right-hand side through the first six decimal places. (In each case, it can be shown that the numbers are indeed equal.)

(a) $(176 + 80\sqrt{5})^{1/5} = 1 + \sqrt{5}$

(b) $(7\sqrt[3]{20} - 19)^{1/6} = \sqrt[3]{\frac{5}{3}} - \sqrt[3]{\frac{2}{3}}$

(c) $(2 + \sqrt{5})^{1/3} + (2 - \sqrt{5})^{1/3} = 1$

57. Without using a calculator, decide which number is larger in each case.

(a) $2^{2/3}$ or $2^{3/2}$ (b) $5^{1/2}$ or 5^{-2}

(c) $2^{1/2}$ or $2^{1/3}$ (d) $(\frac{1}{2})^{1/2}$ or $(\frac{1}{2})^{1/3}$

(e) $10^{1/10}$ or $(\frac{1}{10})^{10}$

58. Give an example showing that each statement is false.

(a) $x^{1/m} = \frac{1}{x^m}$ (b) $x^{-1/2} = \frac{1}{x^2}$

(c) $x^{m/n} = \sqrt[m]{x^n}$ (d) $\sqrt{x + y} = \sqrt{x} + \sqrt{y}$

(e) $\sqrt[3]{x + y} = \sqrt[3]{x} + \sqrt[3]{y}$

59. (a) Give an example in which a rational number raised to a rational power is irrational.

(b) Give an example in which an irrational number raised to a rational power is rational.

60. Without using a calculator, decide which number is closer to zero, $(0.5)^{1/3}$ or $(0.5)^{1/4}$. Then use a calculator to check your answer.

61. Without using a calculator, decide which number is smaller, $(-0.5)^{1/3}$ or $(-0.4)^{1/3}$. Then use a calculator to check your answer.

C

62. Simplify: $\sqrt[x]{x^{(x^2 - x)}}$ $(x > 0)$

63. Show that $\frac{a - b}{a + b}\sqrt{\frac{a + b}{a - b}} = \left(\frac{a - b}{a + b}\right)^{1/2}$ (Assume that $a > b > 0$.)

64. Given $a + b = 2c$, evaluate the expression

$$\left[\frac{(2^{a-b})^b (2^{b-c})^{c-a}}{(2^{c+b})^{c-b}}\right]^{1/c}$$

1.5 ▼ POLYNOMIALS AND FACTORING

As background for our work on polynomials, we first review the terms *constant* and *variable*. By way of example, consider the familiar expression for the area of a circle of radius r, namely, πr^2. Here π is a constant; its value never changes throughout the discussion. On the other hand, r is a variable; we can substitute any positive number for r to obtain the area of a particular circle. More generally, by a **constant** we mean either a particular number (e.g., π, -17, or $\sqrt{2}$) or a letter whose value remains fixed (although perhaps unspecified) throughout a given discussion. In contrast, a **variable** is a letter for which we can substitute any number selected from a given set of numbers. The given set of numbers is called the **domain** of the variable.

Some expressions will make sense only for certain values of the variable. For instance, $1/(x - 3)$ will be undefined when x is 3 (for then the denominator is zero). So in this case we would agree that the domain of the variable x consists of all real numbers except $x = 3$. Similarly, throughout this chapter (with the exception of Section 1.9 on complex numbers), we adopt the following convention:

The Domain Convention

> The domain of a variable in a given expression is the set of all real number values of the variable for which the expression is defined.

In algebra it's customary (but not mandatory) to use letters near the end of the alphabet for variables; letters from the beginning of the alphabet are generally used for constants. (This convention was begun by Descartes.) So, for example, in the expression $ax + b$, the letter x is the variable and a and b are constants.

EXAMPLE 1 Specify the variable, the constants, and the domain of the variable for each expression.

(a) $3x + 4$ (b) $\dfrac{1}{(t-1)(t+3)}$ (c) $ay^2 + by + c$ (d) $4x + 3x^{-1}$

Solution

	VARIABLE	CONSTANTS	DOMAIN
(a) $3x + 4$	x	3, 4	The set of all real numbers.
(b) $\dfrac{1}{(t-1)(t+3)}$	t	$1, -1, 3$	The set of all real numbers except $t = 1$ and $t = -3$.
(c) $ay^2 + by + c$	y	a, b, c	The set of all real numbers.
(d) $4x + 3x^{-1}$	x	4, 3	The set of all real numbers except $x = 0$. (Remember, $x^{-1} = 1/x$.) ▲

The expressions in parts (a) and (c) of Example 1 are *polynomials*. By a **polynomial in x**, we mean an expression of the form

$$a_n x^n + a_{n-1} x^{n-1} + \cdots + a_1 x + a_0$$

where n is a nonnegative integer and $a_n \neq 0$. The individual expressions $a_k x^k$ making up the polynomial are called **terms**. In this chapter, the **coefficients** a_k will always be real numbers. For example, the terms of the polynomial $x^2 - 7x + 3$ are x^2, $-7x$, and 3; the coefficients are 1, -7, and 3. In writing a polynomial, it's customary (but not mandatory) to write the terms in order of decreasing powers of x. For instance, we would usually write $x^2 - 7x + 3$ rather than $x^2 + 3 - 7x$. The highest power of x in a polynomial is called the **degree** of the polynomial. For example, the degree of the polynomial $x^2 - 7x + 3$ is 2. In the case of a polynomial consisting only of a nonzero constant a_0, we say that the degree is zero (because $a_0 = a_0 x^0$). No degree is defined for the polynomial whose only term is zero.

Some additional terminology is useful in describing polynomials that have only a few terms. A polynomial with only one term (such as $3x^2$) is a **monomial**; a polynomial with only two terms (such as $2x + 3$) is a **binomial**; and a polynomial with only three terms (such as $5x^2 - 6x + 3$) is a **trinomial**. The table that follows provides examples of the terminology.

		IF POLYNOMIAL		
EXPRESSION	POLYNOMIAL?	DEGREE	TERMS	COEFFICIENTS
$2x^3 - 3x^2 + 4x - 1$	yes	3	$2x^3, -3x^2, 4x, -1$	$2, -3, 4, -1$
$t^2 + 1$	yes	2	$t^2, 1$	1, 1
-12	yes	0	-12	-12
$2x^{-4} + 5$	no			
$\dfrac{1}{2x - 3}$	no			
$\sqrt{4x^2 + 1}$	no			
$\sqrt{2}\,x + 1$	yes	1	$\sqrt{2}\,x, 1$	$\sqrt{2}, 1$

In your previous courses in algebra, you learned to add, subtract, and multiply polynomials. In the multiplication of polynomials, several particular types of products occur so frequently that it is well worth your time to memorize the results. In the box that follows, we display these special products. (Exercises 9 and 10 will ask you to verify these formulas.) As you memorize these formulas, keep in mind that it's the *form*, or *pattern*, that is important—not the specific choice of letters.

PROPERTY SUMMARY

SPECIAL PRODUCTS
1. $(A - B)(A + B) = A^2 - B^2$
2. (a) $(A + B)^2 = A^2 + 2AB + B^2$ (b) $(A - B)^2 = A^2 - 2AB + B^2$
3. (a) $(A + B)^3 = A^3 + 3A^2B + 3AB^2 + B^3$ (b) $(A - B)^3 = A^3 - 3A^2B + 3AB^2 - B^3$
4. (a) $(A + B)(A^2 - AB + B^2) = A^3 + B^3$ (b) $(A - B)(A^2 + AB + B^2) = A^3 - B^3$

EXAMPLE 2   Use the special products to compute each product.

(a) $(5xy - 4)(5xy + 4)$ (b) $(3\sqrt{xy} - z)(3\sqrt{xy} + z)$
(c) $(4x^3 - 3)^2$ (d) $(2x + 5)^3$
(e) $(x - 2)(x^2 + 2x + 4)$

Solution

(a) $(5xy - 4)(5xy + 4) = (5xy)^2 - 4^2 = 25x^2y^2 - 16$ using special product 1

(b) $(3\sqrt{xy} - z)(3\sqrt{xy} + z) = (3\sqrt{xy})^2 - z^2$ using special product 1
$$= 9xy - z^2$$

(c) $(4x^3 - 3)^2 = (4x^3)^2 - 2(4x^3)(3) + 3^2$ special product 2(b)
$$= 16x^6 - 24x^3 + 9$$

(d) $(2x + 5)^3 = (2x)^3 + 3(2x)^2(5) + 3(2x)(5)^2 + (5)^3$ special product 3(a)
$$= 8x^3 + 60x^2 + 150x + 125$$

(e) $(x - 2)(x^2 + 2x + 4) = x^3 - 2^3$ special product 4(b)   ▲
$$= x^3 - 8$$

For many cases in algebra the process of **factoring** simplifies the work at hand. To factor a polynomial means to write it as a product of two or more nonconstant polynomials. For instance, a factorization of $x^2 - 9$ is given by

$$x^2 - 9 = (x - 3)(x + 3)$$

We'll consider several techniques for factoring in this section.

We need to agree on one convention at the outset. If the polynomial or expression that we wish to factor contains only integer coefficients, then the factors (if any) should involve only integer coefficients. For example, according to this convention, we will not consider the following type of factorization in this section:

$$x^2 - 2 = (x - \sqrt{2})(x + \sqrt{2})$$

(note that it involves coefficients that are irrational numbers). We should point out, however, that factorizations such as this are useful at times, particularly in calculus. As it happens, $x^2 - 2$ is an example of a polynomial that cannot be factored using integer coefficients. We say in such a case that the polynomial is **irreducible over the integers**.

Five elementary techniques for factoring are summarized in Table 1. Notice that three of the formulas in the table are just restatements of the corresponding special product formulas. Remember, it is the form, or pattern, in the formula that is important—not the specific choice of letters.

TABLE 1

Basic Factoring Techniques

TECHNIQUE	EXAMPLE OR FORMULA	REMARK
Common factor	$3x^4 + 6x^3 - 12x^2 = 3x^2(x^2 + 2x - 4)$ $4(x^2 + 1) - x(x^2 + 1) = (x^2 + 1)(4 - x)$	In any factoring problem, the first step always is to look for the common factor of highest degree.
Difference of squares	$x^2 - a^2 = (x - a)(x + a)$	There is no corresponding formula for a sum of squares; $x^2 + a^2$ is irreducible over the integers.
Trial and error	$x^2 + 2x - 3 = (x + 3)(x - 1)$	In this example, the only possibilities, or trials, are: **(a)** $(x - 3)(x - 1)$ **(b)** $(x - 3)(x + 1)$ **(c)** $(x + 3)(x - 1)$ **(d)** $(x + 3)(x + 1)$ By inspection or by carrying out the indicated multiplications, we find that only case (c) checks.
Difference of cubes Sum of cubes	$x^3 - a^3 = (x - a)(x^2 + ax + a^2)$ $x^3 + a^3 = (x + a)(x^2 - ax + a^2)$	Verify these formulas for yourself by carrying out the multiplications. Then memorize the formulas.
Grouping	$x^3 - x^2 + x - 1 = (x^3 - x^2) + (x - 1)$ $= x^2(x - 1) + (x - 1) \cdot 1$ $= (x - 1)(x^2 + 1)$	This is actually an application of the common factor technique.

The idea in factoring is to use one or more of these techniques until each of the factors obtained is irreducible. The examples that follow show how this works in practice.

EXAMPLE 3 Factor: **(a)** $x^2 - 49$ **(b)** $2x^3 - 50x$

Solution **(a)** $x^2 - 49 = x^2 - 7^2$

$\qquad = (x - 7)(x + 7)$ difference of squares

(b) $2x^3 - 50x = 2x(x^2 - 25)$ common factor

$\qquad = 2x(x - 5)(x + 5)$ difference of squares ▲

EXAMPLE 4 Factor: $3x^5 - 3x$

Solution $3x^5 - 3x = 3x(x^4 - 1)$ common factor

$\qquad = 3x[(x^2)^2 - 1^2]$

$\qquad = 3x(x^2 - 1)(x^2 + 1)$ difference of squares

$\qquad = 3x(x - 1)(x + 1)(x^2 + 1)$ difference of squares, again ▲

EXAMPLE 5   Factor: **(a)** $x^2 - 4x - 5$ **(b)** $x^2 - 4x + 5$ **(c)** $8x^4 - 24x^3 + 18x^2$

Solution   **(a)** $x^2 - 4x - 5 = (x - 5)(x + 1)$ trial and error
(b) Irreducible   trial and error
(c) The first step is to check for a common factor. It is $2x^2$, and we write

$$8x^4 - 24x^3 + 18x^2 = 2x^2(4x^2 - 12x + 9)$$

Next we consider the expression in parentheses on the right-hand side of this equation. It can be factored either by trial and error or by recalling the special product $(A - B)^2 = A^2 - 2AB + B^2$. In either case, we find that $4x^2 - 12x + 9 = (2x - 3)^2$. Thus, our final factorization is

$$8x^4 - 24x^3 + 18x^2 = 2x^2(2x - 3)^2$$

▲

EXAMPLE 6   Factor: $4(t - b)^4 - (t - b)^2$

Solution $4(t - b)^4 - (t - b)^2 = (t - b)^2[4(t - b)^2 - 1]$ common factor
$$= (t - b)^2[2(t - b) - 1][2(t - b) + 1]$$ difference of squares
$$= (t - b)^2(2t - 2b - 1)(2t - 2b + 1)$$

▲

EXAMPLE 7   Factor: $ax + ay^2 + bx + by^2$

Solution   We factor a from the first two terms and b from the second two, to obtain

$$ax + ay^2 + bx + by^2 = a(x + y^2) + b(x + y^2)$$

Now we recognize the quantity $(x + y^2)$ as a common expression that can be factored out. We then have

$$a(x + y^2) + b(x + y^2) = (x + y^2)(a + b)$$

The required factorization is therefore

$$ax + ay^2 + bx + by^2 = (x + y^2)(a + b)$$

As you may wish to check for yourself, this factorization can also be obtained by trial and error.

▲

EXAMPLE 8   Factor: $x^3 - 6ax^2 + 12a^2x - 8a^3$

Solution   Since this expression involves four terms, our first inclination is to apply the method shown in Example 7. We factor x^2 from the first two terms and $4a^2$ from the second two. This yields

$$x^3 - 6ax^2 + 12a^2x - 8a^3 = x^2(x - 6a) + 4a^2(3x - 2a)$$

But now (as opposed to the corresponding point in the solution of Example 7) no common expression is apparent on the right-hand side. So we need to start over. Again, we will use the method of Example 7, but first we rearrange the terms. We have

$$x^3 - 6ax^2 + 12a^2x - 8a^3 = \underbrace{x^3 - 8a^3}_{} - \underbrace{6ax^2 + 12a^2x}_{}$$

Differences of cubes is applicable.   Common term is $-6ax$.

Thus,

$$x^3 - 6ax^2 + 12a^2x - 8a^3 = (x - 2a)(x^2 + 2ax + 4a^2) - 6ax(x - 2a)$$

Now, on the right-hand side of this last equation we factor out the common expression $x - 2a$ to obtain

$$x^3 - 6ax^2 + 12a^2x - 8a^3 = (x - 2a)[(x^2 + 2ax + 4a^2) - 6ax]$$
$$= (x - 2a)(x^2 - 4ax + 4a^2)$$
$$= (x - 2a)[(x - 2a)(x - 2a)]$$
$$= (x - 2a)^3$$

This is the required factorization.

Note You can check this answer by recalling the special product formula for $(A - B)^3$. Indeed, if you're sufficiently familiar with that formula, the required factorization in this example can be obtained simply by inspection. ▲

For some manipulations in calculus it's helpful to be able to factor an expression involving fractional exponents. For instance, suppose (as in the next example), we want to factor the expression

$$x(2x - 1)^{-1/2} + (2x - 1)^{3/2}$$

The common expression to factor out here is $(2x - 1)^{-1/2}$; the technique is to *choose the expression with the smaller exponent.*

EXAMPLE 9   Factor: $x(2x - 1)^{-1/2} + (2x - 1)^{3/2}$

Solution
$$x(2x - 1)^{-1/2} + (2x - 1)^{3/2} = (2x - 1)^{-1/2}[x + (2x - 1)^{3/2 + 1/2}]$$
$$= (2x - 1)^{-1/2}[x + (2x - 1)^2]$$
$$= (2x - 1)^{-1/2}(4x^2 - 3x + 1)$$ ▲

▼ EXERCISE SET 1.5

A

In Exercises 1–6, specify the domain of the variable.

1. **(a)** $2x^2 - 3x + 5$ **(b)** $2x^{1/2} - 3x + 5$

2. **(a)** $y - 1$ **(b)** $\dfrac{2y}{y - 1}$

3. **(a)** $ax + b$ **(b)** $ax^{1/3} + b$

4. **(a)** $x + x^{-1}$ **(b)** $t^{-1} + 2t^{-2}$

5. **(a)** $4\sqrt{t} + t^2$ **(b)** $\dfrac{4}{(t - 1)\sqrt{t}}$

6. **(a)** $\dfrac{1}{x}$ **(b)** $\dfrac{1}{(x - 1)(x + 2)(x - 3)}$

For Exercises 7 and 8, specify the degree and the (nonzero) coefficients of the polynomial.

7. **(a)** 4 **(b)** $4x^3$ **(c)** $x^6 + 4x^3 - x^2 + 2$
8. **(a)** 0 **(b)** $1 - x + 6x^5$ **(c)** $\sqrt{2}\,x^2 - 2\sqrt{2}$

In Exercises 9 and 10, you are asked to verify the special products in the box on page 24. As an example, here is a verification for special product 4(a).

EXAMPLE

$$(A + B)(A^2 - AB + B^2) = A(A^2 - AB + B^2)$$
$$+ B(A^2 - AB + B^2)$$
$$= A^3 - A^2B + AB^2 + A^2B$$
$$- AB^2 + B^3$$
$$= A^3 + (AB^2 - AB^2)$$
$$+ (A^2B - A^2B) + B^3$$
$$= A^3 + 0 + 0 + B^3 = A^3 + B^3$$

9. Verify special products 1, 2(a), and 2(b), given on page 24.

10. Verify special products 3(a), 3(b), and 4(b), given on page 24.

For Exercises 11–32, use the special products to compute the products.

11. (a) $(x - y)(x + y)$ (b) $(x^2 - 5)(x^2 + 5)$

12. (a) $(3y - a^2)(3y + a^2)$

(b) $(3\sqrt{y} - a^2)(3\sqrt{y} + a^2)$

13. $(A - 4)(A + 4)$

14. $[(a + b) - 4][(a + b) + 4]$

15. $(\sqrt{ab} + \sqrt{c})(\sqrt{ab} - \sqrt{c})$

16. $(a^{1/2} + b^{1/2})(a^{1/2} - b^{1/2})$

17. $(x - 8)^2$ **18.** $(2x^2 - 5)^2$

19. $(2^m + 1)^2$ **20.** $(a^m - a^{-m})^2$

21. $(\sqrt{x} + \sqrt{y})^2$ **22.** $(\sqrt{x + y} - \sqrt{x})^2$

23. $(2x + y)^3$ **24.** $(x - 2y)^3$

25. $(a + 1)^3$ **26.** $(3x^2 - 2a^2)^3$

27. $(x - y)(x^2 + xy + y^2)$

28. $(x^2 - y^2)(x^4 + x^2y^2 + y^4)$

29. $(x + 1)(x^2 - x + 1)$

30. $(x + 2)(x^2 - 2x + 4)$

31. $(x^{1/3} - y^{1/3})(x^{2/3} + x^{1/3}y^{1/3} + y^{2/3})$

32. $(a^{1/3} + 5^{1/3})(a^{2/3} - a^{1/3} \cdot 5^{1/3} + 5^{2/3})$

In Exercises 33–86, factor the polynomial or expression. If a polynomial is irreducible, say so. (In Exercises 33–38, the factoring techniques are specified for you.)

33. (Common factor and difference of squares)

(a) $x^2 - 64$ (b) $7x^4 + 14x^2$

(c) $121z - z^3$ (d) $a^2b^2 - c^2$

34. (Common factor and difference of squares)

(a) $1 - t^4$ (b) $x^6 + x^5 + x^4$

(c) $u^2v^2 - 225$ (d) $81x^4 - x^2$

35. (Trial and error)

(a) $x^2 + 2x - 3$ (b) $x^2 - 2x - 3$

(c) $x^2 - 2x + 3$ (d) $-x^2 + 2x + 3$

36. (Trial and error)

(a) $2x^2 - 7x - 4$ (b) $2x^2 + 7x - 4$

(c) $2x^2 + 7x + 4$ (d) $-2x^2 - 7x + 4$

37. (Sum and difference of cubes)

(a) $x^3 + 1$ (b) $x^3 + 216$

(c) $1000 - 8x^6$ (d) $64a^3x^3 - 125$

38. (Grouping)

(a) $x^4 - 2x^3 + 3x - 6$

(b) $a^2x + bx - a^2z - bz$

39. $2x - 2x^3$ **40.** $3x^4 - 48x^2$

41. $100x^3 - x^5$ **42.** $2x^2 + 5x - 12$

43. $2x^4 + 3x^3 - 9x^2$ **44.** $x^2 - 6x + 1$

45. $4x^3 - 20x^2 + 25x$ **46.** $ab - bc + a^2 - ac$

47. $x^2z^2 + xzt + xyz + yt$ **48.** $x^2 + 32x + 256$

49. $a^2t^2 + b^2t^2 - cb^2 - ca^2$ **50.** $x^4 - 6x^2 + 9$

51. $x^3 - 13x^2 - 90x$ **52.** $a^4 - 4a^2b^2c^2 + 4b^4c^4$

53. $4x^2 - 29xy - 24y^2$

54. $(x + a)^4 - 2(x + a)^2(x + b)^2 + (x + b)^4$

Hint: See special product 2(b).

55. $x^2 + 2x + 16$ **56.** $(x - y)^2 - z^2$

57. $1 - (x + y)^2$

58. $(a + b + 1)^2 - (a - b + 1)^2$

59. $x^8 - 1$ **60.** $x^2 - y^2 + x - y$

61. $x^3 + 3x^2 + 3x + 1$ **62.** $-1 + 6x - 12x^2 + 8x^3$

63. $27x^3 + 108x^2 + 144x + 64$

64. $(a + b)^3 - 1$ **65.** $x^4 - 25x^2 + 144$

66. $x^4 + 25x^2 + 144$ **67.** $x^2 + 16y^2$

68. $81x^4 - 16y^4$ **69.** $x^3 + 2x^2 - 255x$

70. $4(x - a)^3 - (x - a)$ **71.** $x^3 + a^3 + x + a$

72. $x^2 - a^2 + y^2 - 2xy$ **73.** $a^4 - (b + c)^4$

74. $21x^3 + 82x^2 - 39x$

75. $x^3a^2 - 8y^3a^2 - 4x^3b^2 + 32y^3b^2$

76. $12xy + 25 - 4x^2 - 9y^2$

77. $ax^2 + (1 + ab)xy + by^2$

78. $ax^2 + (a + b)x + b$

79. $(5a^2 - 11a + 10)^2 - (4a^2 - 15a + 6)^2$

80. $x^6 - y^6$ *Hint:* $x^6 - y^6 = (x^3)^2 - (y^3)^2$

81. $(x + 1)^{1/2} - (x + 1)^{3/2}$

82. $(x^2 + 1)^{3/2} + (x^2 + 1)^{7/2}$

83. $(x + 1)^{-1/2} - (x + 1)^{-3/2}$

84. $(x^2 + 1)^{-2/3} + (x^2 + 1)^{-5/3}$

85. $x^2(a^2 - x^2)^{-1/2} + \sqrt{a^2 - x^2}$. Write the answer in a form not involving negative exponents.

86. $x^2(x - 2)^{-4} + x(x - 2)^{-3}$. Write the answer in a form not involving negative exponents. *Hint:* As with fractional exponents, factor out the expression with the smaller exponent.

C **87.** The expression $\dfrac{x^2 - 16}{x - 4}$ is undefined when $x = 4$. In this exercise, we investigate the values of the expression when x is very close to 4.

(a) Complete the tables.

x	$\dfrac{x^2 - 16}{x - 4}$
3.9	
3.99	
3.999	
3.9999	
3.99999	

x	$\dfrac{x^2 - 16}{x - 4}$
4.1	
4.01	
4.001	
4.0001	
4.00001	

(b) On the basis of the tables in part (a), what value does the expression $\dfrac{x^2 - 16}{x - 4}$ seems to be approaching as x gets closer and closer to 4? This "target value" is referred to as the *limit* of $\dfrac{x^2 - 16}{x - 4}$ as x approaches 4. (The notation of a limit is made more precise in calculus.)

(c) How could factoring have been used to obtain this limit without the work in part (a)?

88. The expression $\dfrac{x^3 - 8}{x - 2}$ is undefined when $x = 2$. In this exercise, we investigate the values of the expression when x is very close to 2.

(a) Complete the tables.

x	$\dfrac{x^3 - 8}{x - 2}$
1.9	
1.99	
1.999	
1.9999	
1.99999	

x	$\dfrac{x^3 - 8}{x - 2}$
2.1	
2.01	
2.001	
2.0001	
2.00001	

(b) On the basis of the tables in part (a), what "target value," or limit, does the expression $\dfrac{x^3 - 8}{x - 2}$ seem to be approaching as x gets closer and closer to 2?

(c) How could factoring have been used to obtain this limit without the work in part (a)?

C
_____

In Exercises 89–94, factor the expression.

89. $x^4 + 64$ *Hint:* Add and subtract the term $16x^2$.

90. **(a)** $x^4 - 15x^2 + 9$ *Hint:* Add and subtract a term.
(b) $x^4 + x^2y^2 + y^4$

91. $(x + y)^2 + (x + z)^2 - (z + t)^2 - (y + t)^2$

92. $(a - a^2)^3 + (a^2 - 1)^3 + (1 - a)^3$

93. $(b - c)^3 + (c - a)^3 + (a - b)^3$

94. $(a + b + c)^3 - a^3 - b^3 - c^3$

1.6 ▼ QUADRATIC EQUATIONS

The formula for the quadratic equation seems first to have been discovered by the Moslems around 900 A.D., although quadratic equations had been solved by the Babylonians 3000 years earlier.

Charles Robert Hadlock in *Field Theory and Its Classical Problems* (Washington, D.C.: Mathematical Association of America, 1978)

An equation that can be written in the form

$$ax^2 + bx + c = 0 \qquad \text{with } a, b, \text{ and } c \text{ real numbers and } a \neq 0$$

is called a **quadratic equation**. Examples of quadratic equations are

$$x^2 - 8x - 9 = 0 \qquad 2y^2 - 5 = 0 \qquad 3m^2 = 1 - m$$

Recall that a **solution**, or **root**, of an equation is a value for the unknown that makes the equation a true statement. For example, the value $x = -1$ is a solution of the quadratic equation $x^2 - 8x - 9 = 0$, because when x is replaced by -1 in this equation, we have

$$(-1)^2 - 8(-1) - 9 = 0$$
$$1 + 8 - 9 = 0$$
$$0 = 0 \qquad \text{True}$$

One of the simplest techniques for solving quadratic equations involves factoring. This method relies on the following familiar and important property of the real number system.

PROPERTY SUMMARY

ZERO-PRODUCT PROPERTY OF REAL NUMBERS

$pq = 0$ if and only if $p = 0$ or $q = 0$ (or both)

For example, to solve the equation $x^2 - 2x - 3 = 0$ by factoring, we have

$$x^2 - 2x - 3 = 0$$
$$(x - 3)(x + 1) = 0$$
$$x - 3 = 0 \quad | \quad x + 1 = 0$$
$$x = 3 \quad\; | \quad\;\; x = -1$$

As you can easily check, the values $x = 3$ and $x = -1$ both satisfy the given equation.

Unfortunately, not all quadratic expressions can be factored so easily. Consider, for example, the equation $x^2 - 2x - 4 = 0$. Three factorizations with integer coefficients are possible, but none yields the appropriate middle term, $-2x$, when multiplied out:

$$(x - 4)(x + 1) \qquad (x + 4)(x - 1) \qquad (x - 2)(x + 2)$$

Moreover, it is not always obvious whether a particular equation, such as $x^2 + 156x + 5963 = 0$, can be solved by factoring. (We will come back to this equation in Exercise 37.) Clearly, we need a systematic approach to solving those quadratic equations that are not readily solvable through factoring. The technique of **completing the square** provides this approach. We'll demonstrate this technique by solving the equation

$$x^2 - 2x - 4 = 0$$

First, we rewrite the equation in the form

$$x^2 - 2x = 4 \qquad\qquad\qquad\qquad\qquad\qquad\qquad\qquad (1)$$

with the x-terms isolated on the left-hand side of the equation. To complete the square, we follow these two steps:

STEP 1 Take half of the coefficient of x and square it.
STEP 2 Add the number obtained in step 1 to both sides of the equation.

For equation (1), the coefficient of x is -2. Taking half of -2 and then squaring it gives us $(-1)^2$, or 1. Now, as directed in step 2, we add 1 to both sides of equation (1). This yields

$$x^2 - 2x + 1 = 4 + 1$$
$$(x - 1)^2 = 5$$
$$x - 1 = \pm\sqrt{5}$$
$$x = 1 \pm \sqrt{5}$$

We have now obtained the two solutions, $1 + \sqrt{5}$ and $1 - \sqrt{5}$. (Exercise 6 will ask you to verify that these values indeed satisfy the original equation.) In the box that follows, we summarize the technique of completing the square, and we indicate how the process got its name.

ALGEBRAIC PROCEDURE FOR COMPLETING THE SQUARE IN THE EXPRESSION $x^2 + bx$

Add the square of half of the x-coefficient:

$$(x^2 + bx) + \left(\frac{b}{2}\right)^2 = \left(x + \frac{b}{2}\right)^2$$

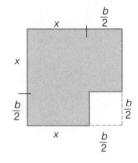

GEOMETRIC INTERPRETATION OF COMPLETING THE SQUARE FOR $x^2 + bx$

The colored region in the figure represents the quantity $x^2 + bx$ $\left(\text{since the area is } x^2 + \frac{b}{2}x + \frac{b}{2}x\right).$ By adding the white square to the colored region, we fill out or "complete" the larger square. The area of the white region that completes the square is

$$\frac{b}{2} \cdot \frac{b}{2} = \frac{b^2}{4}.$$

Rather than repeat the process of completing the square for each new equation to be solved, we can instead use a general formula for solving any quadratic equation. The derivation of this **quadratic formula** runs as follows. We start with the general quadratic equation $ax^2 + bx + c = 0$ ($a \neq 0$). In order to use the procedure for completing the square described in the preceding box, we first divide both sides of the given equation by a, so that the coefficient of x^2 is 1:

$$x^2 + \frac{b}{a}x + \frac{c}{a} = 0$$

Subtracting c/a from both sides yields

$$x^2 + \frac{b}{a}x = -\frac{c}{a}$$

Now, to complete the square, we add $[\frac{1}{2}(b/a)]^2$, or $b^2/4a^2$, to both sides. That gives us

$$x^2 + \frac{b}{a}x + \frac{b^2}{4a^2} = \frac{b^2}{4a^2} - \frac{c}{a}$$

$$\left(x + \frac{b}{2a}\right)^2 = \frac{b^2 - 4ac}{4a^2}$$

$$x + \frac{b}{2a} = \pm\sqrt{\frac{b^2 - 4ac}{4a^2}} = \pm\frac{\sqrt{b^2 - 4ac}}{2|a|}$$

$$= \pm\frac{\sqrt{b^2 - 4ac}}{2a}$$

This last equality follows from the fact that for any real number $a \neq 0$, the expressions $\pm 2|a|$ and $\pm 2a$ both represent the same two numbers. We now conclude that the solutions are

$$x = -\frac{b}{2a} + \frac{\sqrt{b^2 - 4ac}}{2a} \qquad \text{and} \qquad x = -\frac{b}{2a} - \frac{\sqrt{b^2 - 4ac}}{2a}$$

These solutions are usually written in the more compact form displayed in the following box.

The Quadratic Formula

> The solutions of the equation $ax^2 + bx + c = 0 \quad (a \neq 0)$ are given by
>
> $$x = \frac{-b \pm \sqrt{b^2 - 4ac}}{2a}$$

EXAMPLE 1 Use the quadratic formula to solve $2x^2 = 4 - x$.

Solution We first rewrite the equation $2x^2 + x - 4 = 0$, so that it has the form $ax^2 + bx + c = 0$. By comparing these last two equations, we see that $a = 2$, $b = 1$, and $c = -4$. Therefore,

$$x = \frac{-b \pm \sqrt{b^2 - 4ac}}{2a} = \frac{-1 \pm \sqrt{1^2 - 4(2)(-4)}}{2(2)} = \frac{-1 \pm \sqrt{33}}{4}$$

Thus, the two solutions are $\dfrac{-1 + \sqrt{33}}{4}$ and $\dfrac{-1 - \sqrt{33}}{4}$. ▲

FIGURE 1

EXAMPLE 2 Figure 1 shows a right circular cylinder with radius r and height h. The formula for the total surface area S of the cylinder is

$$S = 2\pi r^2 + 2\pi rh \tag{2}$$

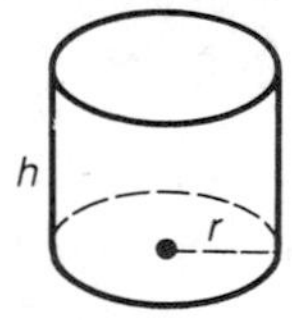

Solve equation (2) for r in terms of h and S.

Solution By rewriting equation (2) in the following form, we can see that the equation is quadratic in r, and we can determine what values to use for a, b, and c in the quadratic formula:

$$(2\pi)r^2 + (2\pi h)r - S = 0$$

We have $a = 2\pi$, $b = 2\pi h$, and $c = -S$ and, therefore,

$$r = \frac{-2\pi h \pm \sqrt{4\pi^2 h^2 - 4(2\pi)(-S)}}{4\pi} = \frac{-2\pi h \pm 2\sqrt{\pi^2 h^2 + 2\pi S}}{4\pi}$$

$$= \frac{-\pi h \pm \sqrt{\pi^2 h^2 + 2\pi S}}{2\pi}$$

This last expression gives us two solutions to equation (2). In the present context, however, the radius r cannot be negative, so we discard the "minus" option within the plus-or-minus sign, since that would lead to a negative value for r.

Thus we have

$$r = \frac{-\pi h + \sqrt{\pi^2 h^2 + 2\pi S}}{2\pi}$$

Question   How do we know that this last expression is indeed positive?   ▲

In Examples 1 and 2, we found that each quadratic equation had two real roots. As the next example indicates, however, it is also possible for quadratic equations to have only one real root, or even no real root. [In Example 3(b), we make use of the complex number system. This topic is reviewed in detail in Section 1.9.]

EXAMPLE 3   Use the quadratic formula to solve each equation.

(a) $4x^2 - 12x + 9 = 0$ (b) $x^2 + x + 2 = 0$

Solution   (a) Here $a = 4$, $b = -12$, $c = 9$, so

$$x = \frac{12 \pm \sqrt{(-12)^2 - 4(4)(9)}}{2(4)} = \frac{12 \pm \sqrt{144 - 144}}{8}$$

$$= \frac{12 \pm 0}{8} = \frac{12}{8} = \frac{3}{2}$$

In this case our only solution is $\frac{3}{2}$. We refer to $\frac{3}{2}$ as a **double root**. (*Note* We've used the quadratic formula here only for an illustration; in this case we can solve the original equation more efficiently by factoring.)

(b) Since $a = 1$, $b = 1$, $c = 2$, we have

$$x = \frac{-1 \pm \sqrt{(1)^2 - 4(1)(2)}}{2(1)} = \frac{-1 \pm \sqrt{-7}}{2}$$

Now, if we confine ourselves to the real number system, the equation has no solutions, because the expression $\sqrt{-7}$ is undefined within the real number system. However, if we take a broader point of view and work within the complex number system, then there are indeed two solutions. As you'll see in Section 1.9 (or as you perhaps already know from a previous course) these two nonreal complex solutions can be written

$$\frac{-1 + i\sqrt{7}}{2} \qquad \text{and} \qquad \frac{-1 - i\sqrt{7}}{2}$$
▲

The quantity $b^2 - 4ac$ that appears under the radical sign in the quadratic formula is called the **discriminant**. In Examples 1 and 2, the discriminant was positive and, consequently, each equation had two real solutions. In Example 3(a), in which the discriminant was zero, we obtained only one (real) solution. In Example 3(b) the discriminant was negative and, consequently, we obtained two solutions involving the imaginary unit i. These observations are generalized in the box that follows.

The Discriminant
$b^2 - 4ac$

Consider the quadratic equation $ax^2 + bx + c = 0$, where a, b, and c are real numbers and $a \neq 0$. The expression $b^2 - 4ac$ is called the **discriminant**.

1. If $b^2 - 4ac > 0$, then the equation has two distinct real roots.
2. If $b^2 - 4ac = 0$, then the equation has exactly one real root.
3. If $b^2 - 4ac < 0$, then the equation has no real root.

EXAMPLE 4

(a) Compute the discriminant to determine how many real solutions there are for the equation $x^2 + x - 1 = 0$.
(b) Find a value for k such that the quadratic equation $x^2 + \sqrt{2}\,x + k = 0$ has exactly one real solution.

Solution

(a) Here $a = 1$, $b = 1$, $c = -1$ and, therefore,

$$b^2 - 4ac = 1^2 - 4(1)(-1) = 5$$

Since the discriminant is positive, the equation has two distinct real solutions.

(b) For the equation to have exactly one real solution, the discriminant must be zero, that is,

$$b^2 - 4ac = 0$$
$$(\sqrt{2})^2 - 4(1)(k) = 0$$
$$2 - 4k = 0$$
$$k = \tfrac{1}{2}$$

The required value for k is $\tfrac{1}{2}$. ▲

The techniques presented in this section can often be used to solve equations that contain radicals. As a simple example, consider the equation

$$x - 2 = \sqrt{x}$$

By squaring both sides, we obtain

$$x^2 - 4x + 4 = x$$
$$x^2 - 5x + 4 = 0$$
$$(x - 4)(x - 1) = 0$$

Therefore,

$$x = 4 \qquad \text{or} \qquad x = 1$$

The value $x = 4$ checks in the original equation, but the value $x = 1$ does not check. (Verify this.) We say in this case that the value $x = 1$ is an **extraneous root** or **extraneous solution** of the original equation. [This extraneous root arises from the fact that if $a^2 = b^2$, it is not necessarily true that $a = b$. For instance, $(-3)^2 = 3^2$, but certainly $-3 \neq 3$.] In summary, whenever we square both sides of an equation (or raise both sides to an even integral power), there is a possibility of introducing extraneous roots. For this reason, it is essential to check all candidates for solutions obtained in this manner.

Our final example in this section involves an equation containing several square root expressions. Before squaring both sides in such cases, it is usually a good idea to see if we can first isolate one of the radical expressions on one side of the equation.

EXAMPLE 5   Solve: $\sqrt{9 + x} + \sqrt{1 + x} - \sqrt{x + 16} = 0$

Solution $\sqrt{9 + x} + \sqrt{1 + x} = \sqrt{x + 16}$ adding $\sqrt{x + 16}$ to both sides in preparation for squaring

$$(\sqrt{9 + x} + \sqrt{1 + x})^2 = (\sqrt{x + 16})^2$$
$$9 + x + 2\sqrt{9 + x}\sqrt{1 + x} + 1 + x = x + 16$$
$$2\sqrt{9 + x}\sqrt{1 + x} = 6 - x$$
$$(2\sqrt{9 + x}\sqrt{1 + x})^2 = (6 - x)^2$$
$$4(9 + 10x + x^2) = 36 - 12x + x^2$$
$$3x^2 + 52x = 0$$
$$x(3x + 52) = 0$$
$$x = 0 \quad \Big| \quad 3x + 52 = 0$$
$$x = -\tfrac{52}{3}$$

We have found two possible solutions, 0 and $-\tfrac{52}{3}$. Since they were obtained by squaring both sides of an equation, we need to check to see if they satisfy the original equation. With $x = 0$, the original equation becomes $\sqrt{9} + \sqrt{1} - \sqrt{16} = 0$, or $3 + 1 - 4 = 0$, which is certainly true. On the other hand, with $x = -\tfrac{52}{3}$, the quantities beneath the radicals in the original equation are negative and, consequently, the (real) square roots are undefined. So $-\tfrac{52}{3}$ is an extraneous solution, and $x = 0$ is the only solution of the given equation.   ▲

▼ EXERCISE SET 1.6

A

In Exercises 1–5, determine if the given value is a solution of the equation.

1. $2x^2 - 6x - 36 = 0; x = -3$
2. $(y - 4)(y - 5) = 0; y = -1$
3. $4x^2 - 1 = 0; x = -\tfrac{1}{4}$
4. $m^2 + m - \tfrac{5}{16} = 0; m = \tfrac{1}{4}$
5. $x^2 - 2x - 6 = 0; x = -1 - \sqrt{7}$
6. Verify that the numbers $1 + \sqrt{5}$ and $1 - \sqrt{5}$ both satisfy the equation $x^2 - 2x - 4 = 0$.

In Exercises 7–22, solve the equation by factoring.

7. $x^2 - 5x - 6 = 0$
8. $x^2 - 5x = -6$
9. $x^2 - 100 = 0$
10. $4y^2 + 4y + 1 = 0$
11. $25x^2 - 60x + 36 = 0$
12. $144 - t^2 = 0$
13. $10z^2 - 13z - 3 = 0$
14. $3t^2 - t - 4 = 0$
15. $(x + 1)^2 - 4 = 0$
16. $x^2 + 3x - 40 = 0$
17. $2x^2 - 15x - 8 = 0$
18. $12x^2 = 12 - 7x$
19. $x(2x - 13) = -6$
20. $x(3x - 23) = 8$
21. $x(x + 1) = 156$
22. $x^2 + (2\sqrt{5})x + 5 = 0$

In Exercises 23–34, use the quadratic formula to solve the equation.

23. $x^2 + 10x + 9 = 0$
24. $x^2 - 10x + 25 = 0$
25. $x^2 - x - 5 = 0$
26. $x^2 - x + 5 = 0$
27. $2x^2 + 3x - 4 = 0$
28. $4x^2 - 3x - 9 = 0$
29. $12x^2 + 32x + 5 = 0$
30. $10x^2 - x - 1 = 0$
31. $2x^2 = x + 5$
32. $3 - 2x = -4x^2$
33. $-6x^2 + 12x = -1$
34. $-\sqrt{2}x^2 + x = -\sqrt{2}$

In Exercises 35 and 36, use the quadratic formula and a calculator to solve for x. Round each answer to two decimal places. (Note: After completing Exercises 35 and 36, compare the two equations and their solutions. These exercises show that a slight change in one of the coefficients can sometimes radically alter the nature of the solutions.)

[C] 35. $x^2 + 3x + 2.249 = 0$
[C] 36. $x^2 + 3x + 2.251 = 0$

In Exercises 37 and 38, use the quadratic formula and a calculator to solve for x. (These exercises provide examples of equations that could be solved by factoring, but for which the quadratic formula is clearly more efficient.)

37. $x^2 + 156x + 5963 = 0$ **38.** $52x^2 - 165x + 108 = 0$

In Exercises 39–52, solve the equations using any method you choose.

39. $(3x - 2)^2 = 3x - 2$

40. $3(x + 1)^2 - 4(x + 1) = 0$

41. $1 - 2x^2 = x$ **42.** $5x^2 - 4x + 1 = 0$

43. $3x^2 + 4x - 3 = 0$ **44.** $13x^2 = 52$

45. $x^2 = 24$ **46.** $x^2 + 16x = 0$

47. $x(x - 1) = 1$ **48.** $x(x - 1) = -4$

49. $\frac{1}{2}x^2 - x - \frac{1}{3} = 0$ **50.** $x^2 + 34x + 288 = 0$

51. $2\sqrt{5}\,x^2 - x - 2\sqrt{5} = 0$ **52.** $\sqrt{2}\,x^2 + x = 10\sqrt{2}$

In Exercises 53–58, solve for x in terms of the other letters.

53. $2y^2x^2 - 3yx + 1 = 0$ $(y > 0)$

54. $(ax + b)^2 - (bx + a)^2 = 0$ $(a \neq \pm b)$

55. $(x - p)^2 + (x - q)^2 = p^2 + q^2$

56. $21x^2 - 2kx - 3k^2 = 0$ $(k > 0)$

57. $12x^2 = ax + 20a^2$ $(a > 0)$

58. $3Ax^2 - 2Ax - 3Bx + 2B = 0$ $(A \neq 0)$

In Exercises 59–64, solve for the indicated letter.

59. $2\pi r^2 + 2\pi rh = 20\pi$;   for r

60. $2\pi y^2 + \pi yx = 12$;   for y

61. $-16t^2 + v_0 t = 0$;   for t

62. $-\frac{1}{2}gt^2 + v_0 t + h_0 = 0$;   for t

63. $x^3 + bx^2 - 2b^2x = 0$

64. $\frac{1}{y^3} + \frac{b}{y^2} - \frac{2b^2}{y} = 0$

In Exercises 65–72, use the discriminant to determine how many real roots the equation has.

65. $x^2 - 12x + 16 = 0$ **66.** $2x^2 - 6x + 5 = 0$

67. $4x^2 - 5x - \frac{1}{2} = 0$ **68.** $4x^2 - 28x + 49 = 0$

69. $x^2 + \sqrt{3}x + \frac{3}{4} = 0$ **70.** $\sqrt{2}x^2 + \sqrt{3}x + 1 = 0$

71. $y^2 - \sqrt{5}y = -1$ **72.** $\frac{m^2}{4} - \frac{4m}{3} + \frac{16}{9} = 0$

In Exercises 73–76, find values for k such that the equation has exactly one real root.

73. $x^2 + 12x + k = 0$ **74.** $3x^2 + (\sqrt{2k})x + 6 = 0$

75. $x^2 + kx + 5 = 0$ **76.** $kx^2 + kx + 1 = 0$

In Exercises 77–90, determine all the real number solutions for the equation.

77. **(a)** $\sqrt{x - 8} = 4$ **(b)** $\sqrt{x - 8} = 2x + 1$

78. **(a)** $\sqrt{3x + 1} = 4$ **(b)** $\sqrt{3x + 1} = 2x - 1$

79. $\sqrt{1 - 3x} = 2$

80. $\sqrt{x^2 + 5x - 2} = 2$

81. $\sqrt{x^4 - 13x^2 + 37} = 1$ *Hint: Let $x^2 = t$ and $x^4 = t^2$*

82. $\sqrt{y + 2} = y - 4$

83. $\sqrt{1 - 2x} + \sqrt{x + 5} = 4$

84. $\sqrt{x - 5} - \sqrt{x + 4} + 1 = 0$

85. $\sqrt{3 + 2t} + \sqrt{-1 + 4t} = 1$

86. $\sqrt{2t + 5} - \sqrt{8t + 25} + \sqrt{2t + 8} = 0$

87. $\sqrt{2y - 3} - \sqrt{3y + 3} + \sqrt{3y - 2} = 0$

88. $\sqrt{2x + 3} + \sqrt{x + 2} = 2$

89. $\sqrt{2x + 1} + \sqrt{x + 4} = 1$

90. $\sqrt{2x + 6} + \sqrt{x + 4} - \sqrt{8x + 9} = 0$

B

In Exercises 91–100, find all the real solutions of the equation. You'll first want to eliminate the fractions by multiplying both sides by the least common denominator. [For example, in Exercise 91, first multiply both sides by $x(x + 5)$.] Be sure to check your answers for these exercises, because multiplying both sides of an equation by an expression involving the variable may introduce extraneous roots.

91. $\frac{3}{x + 5} + \frac{4}{x} = 2$ **92.** $\frac{5}{x + 2} - \frac{2x - 1}{5} = 0$

93. $1 - x - \frac{2}{6x + 1} = 0$ **94.** $\frac{x^2 - 3x}{x + 1} = \frac{4}{x + 1}$

95. $\frac{3x^2 - 6x - 3}{(x + 1)(x - 2)(x - 3)} + \frac{5 - 2x}{x^2 - 5x + 6} = 0$

96. $\frac{6}{x^2 - 1} + \frac{x}{x + 1} = \frac{3}{2}$ **97.** $\frac{2x}{x^2 - 1} - \frac{1}{x + 3} = 0$

98. $\frac{3}{x^2 - x - 2} - \frac{4}{x^2 + x - 6} = \frac{1}{3x + 3}$

99. $\frac{x - 1}{x + 1} - \frac{x + 1}{x + 3} + \frac{4}{x^2 + 4x + 3} = 0$

100. $\frac{x}{x - 2} + \frac{x}{x + 2} = \frac{8}{x^2 - 4}$

In Exercises 101–108, find all the real solutions of the equation.

101. $\sqrt{\sqrt{x + \sqrt{a}} + \sqrt{\sqrt{x - \sqrt{a}}}} = \sqrt{2\sqrt{x} + 2\sqrt{b}}$

102. $x = \sqrt{3x + x^2} - 3\sqrt{3x + x^2}$

103. $\frac{\sqrt{x - a}}{\sqrt{x}} - \frac{\sqrt{x + a}}{\sqrt{x - b}} = 0$ $(a > 0, b > 0)$

   Suggestion: Let $\sqrt{x} = t$.

104. $x - \sqrt{x^2 - x} = \sqrt{x}$ $(x > 0)$

   Suggestion: If $x \neq 0$, then you can divide through by $\sqrt{x}$ to obtain a simpler equation.

105. $\sqrt{x^2 - x - 1} - \dfrac{2}{\sqrt{x^2 - x - 1}} = 1$

Hint: Let $t = x^2 - x - 1$.

106. $\sqrt{x^2 + 3x - 4} - \sqrt{x^2 - 5x + 4} = x - 1 \qquad (x > 4)$

Hint: Factor the expressions beneath the radicals. Then note that $\sqrt{x - 1}$ is a factor of both sides of the equation.

107. $\sqrt{\dfrac{x - a}{x}} + 4\sqrt{\dfrac{x}{x - a}} = 5 \qquad (a \neq 0)$

Hint: Let $t = \dfrac{x - a}{x}$. Then $\dfrac{1}{t} = \dfrac{x}{x - a}$.

108. $\sqrt{p + 4q - 5t} + \sqrt{4p + q - 5t} = 3\sqrt{p + q - 2t}$
(Assume that p and q are constants and $q > p$.)

109. For a certain right circular cylinder, the height is 3 m and the total surface area is 8π m^2. Find the radius of the cylinder.

110. A ball is thrown straight upward. Suppose that the height of the ball at time t is given by $h = -16t^2 + 96t$, where h is in feet and t is in seconds, with $t = 0$ corresponding to the instant that the ball is first tossed.

    (a) How long does it take before the ball lands?

    (b) At what time is the height 80 ft? Why does this question have two answers?

111. During a flu epidemic in a small town, a public health official finds that the total number of people P who have caught the flu after t days is closely approximated by the formula

$$P = -t^2 + 26t + 106 \qquad (1 \leq t \leq 13)$$

    (a) How many have caught the flu after 10 days?

    (b) After approximately how many days will 250 people have caught the flu?

112. The radius of a circle is r units. By how many units should the radius be increased so that the area increases by b square units?

113. A piece of wire L in. long is cut into two pieces. Each piece is then bent to form a square. If the sum of the areas of the two squares is $5L^2/128$, how long is each piece of wire?

114. **(a)** Show that the sum of the roots of the equation $x^2 + px + q = 0$ is $-p$.

    (b) Show that the product of the roots of the equation $x^2 + px + q = 0$ is q.

115. Use the results in Exercise 114 to show that the sum and product of the roots of $ax^2 + bx + c = 0$ are $-b/a$ and c/a, respectively.

C

In Exercises 116–118, find all the real number solutions for the equation.

116. $\sqrt{8 + 2t} + \sqrt{5 + t} = \sqrt{15 + 3t}$

117. $-x = \sqrt{1 - \sqrt{1 + x}}$ *Hint:* After squaring once and rearranging, look for a common factor appearing on both sides of the equation.

118. $\sqrt{x^2 + x} + \dfrac{1}{\sqrt{x^2 + x}} = \dfrac{5}{2}$

Hint: Let $t = x^2 + x$. (Use a calculator to evaluate the final answer. Round to two decimal places.)

119. A piece of wire 16 cm long is cut into two pieces, and each piece is then bent to form a circle. If the sum of the areas of the two circles is 12 cm^2, how long is the shorter piece of wire? (Round your answer to two decimal places.)

1.7 ▼ INEQUALITIES

The fundamental results of mathematics are often inequalities rather than equalities.

E. Beckenbach and R. Bellman in *An Introduction to Inequalities* (New York: Random House, 1961)

If we replace the equal sign in an equation with any one of the four symbols $<$, $\leq$, $>$, or $\geq$, we obtain an **inequality**. As with equations in one variable, a real number is a **solution** of an inequality if we obtain a true statement when the variable is replaced by that real number. For example, the value $x = 5$ is a solution of the inequality $2x - 3 < 8$, because when $x = 5$ we have

$$2(5) - 3 < 8$$
$$7 < 8 \qquad \text{which is true}$$

We also say in this case that the value $x = 5$ **satisfies** the inequality. To **solve** an inequality means to find all of the solutions. The set of all solutions of an inequality is called (naturally enough) the **solution set**.

Recall that two equations are said to be equivalent if they have exactly the same solutions. Similarly, two inequalities are **equivalent** if they have the same solution set. Most of the procedures used for solving inequalities are similar to those for equalities. For example, adding or subtracting the same number on both sides of an inequality produces an equivalent inequality. We need to be careful, however, in multiplying or dividing both sides of an inequality by the same non-zero number. For instance, suppose that we start with the inequality $2 < 3$ and multiply both sides by 5. That yields $10 < 15$, which is certainly true. But if we multiply both sides of the inequality $2 < 3$ by -5, we obtain $-10 < -15$, which is false. Multiplying both sides of an inequality by the same *positive* number preserves the inequality, whereas multiplying by a *negative* number reverses the inequality. In the following box, we list some of the principal properties of inequalities. In general, whenever we use Property 1 or 2 in solving an inequality, we obtain an equivalent inequality. Also, note that each property can be rewritten to reflect the fact that $a < b$ is equivalent to $b > a$. For example, Property 3 can just as well be written this way: If $b > a$ and $c > b$, then $c > a$.

PROPERTY SUMMARY	PROPERTIES OF INEQUALITIES	
	PROPERTY	**EXAMPLE**
	1. If $a < b$, then $a + c < b + c$ and $a - c < b - c$.	If $x - 3 < 0$, then $(x - 3) + 3 < 0 + 3$ and, consequently, $x < 3$.
	2. (a) If $a < b$ and c is positive, then $ac < bc$ and $a/c < b/c$.	If $\frac{1}{2}x < 4$, then $2(\frac{1}{2}x) < 2(4)$ and, consequently, $x < 8$.
	(b) If $a < b$ and c is negative, then $ac > bc$ and $a/c > b/c$.	If $-\dfrac{x}{5} < 6$, then $(-5)\left(-\dfrac{x}{5}\right) > (-5)(6)$ and, consequently, $x > -30$.
	3. The transitive property: If $a < b$ and $b < c$, then $a < c$.	If $a < x$ and $x < 2$, then $a < 2$.

EXAMPLE 1 Solve: $4t + 8 \le 7(1 + t)$

Solution Our work follows the pattern that we would use to solve the equation $4t + 8 = 7(1 + t)$. We have

$$4t + 8 \le 7(1 + t)$$
$$4t + 8 \le 7 + 7t$$
$$-3t \le -1 \quad \text{subtracting } 7t \text{ and } 8 \text{ from both sides}$$
$$t \ge \tfrac{1}{3} \quad \text{dividing by } -3 \text{ (this reverses the inequality)}$$

The solution set is therefore $[\tfrac{1}{3}, \infty)$. See Figure 1. ▲

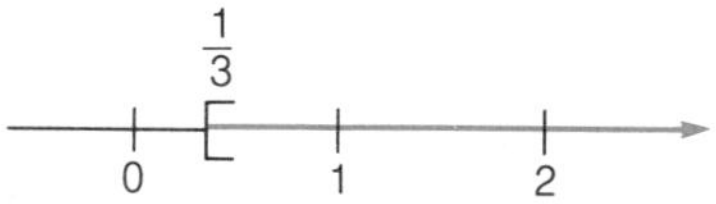

FIGURE 1
$t \ge \tfrac{1}{3}$

In the next example, we solve the inequality

$$-\frac{1}{2} < \frac{3 - x}{-4} < \frac{1}{2}$$

By definition, this is equivalent to the pair of inequalities

$$-\frac{1}{2} < \frac{3-x}{-4} \quad \text{and} \quad \frac{3-x}{-4} < \frac{1}{2}$$

One way to proceed here would be first to determine the solution set for each inequality. Then the set of real numbers common to both solution sets would be the solution set for the original inequality. However, the method shown in Example 2 is more efficient.

EXAMPLE 2   Solve: $-\dfrac{1}{2} < \dfrac{3-x}{-4} < \dfrac{1}{2}$

Solution   We begin by multiplying through by -4. Remember, this will reverse the inequalities:

$$2 > 3 - x > -2$$

Next, with a view toward isolating x, we first subtract 3 to obtain

$$-1 > -x > -5$$

Finally, multiplying through by -1, we have

$$1 < x < 5$$

The solution set is therefore the interval $(1, 5)$.   ▲

In the next example, reference is made to the Celsius and Fahrenheit scales for measuring temperature.* The formula relating the temperature readings on the two scales is

$$F = \frac{9}{5}C + 32$$

EXAMPLE 3   Over the temperature range $32° \le F \le 39.2°$ on the Fahrenheit scale, water contracts (rather than expands) with increasing temperature. What is the corresponding temperature range on the Celsius scale?

Solution

$$32 \le F \le 39.2 \qquad \text{given}$$

$$32 \le \frac{9}{5}C + 32 \le 39.2 \quad \text{substituting } \frac{9}{5}C + 32 \text{ for } F$$

$$0 \le \frac{9}{5}C \le 7.2 \qquad \text{subtracting 32}$$

$$0 \le C \le \frac{5}{9}(7.2) \qquad \text{multiplying by } \frac{5}{9}$$

$$0 \le C \le 4$$

Thus, a range of $32\,°F$–$39.2\,°F$ on the Fahrenheit scale corresponds to $0\,°C$–$4\,°C$ on the Celsius scale.   ▲

---

* The Celsius scale was devised in 1742 by the Swedish astronomer Anders Celsius. The Fahrenheit scale was first used by the German physicist Gabriel Fahrenheit in 1724.

In the next two examples, we solve inequalities that involve absolute values. The following theorem is very useful in this context.

Theorem

> If $a > 0$, then
>
> $|u| < a$ if and only if $-a < u < a$
>
> and
>
> $|u| > a$ if and only if $u < -a$ or $u > a$

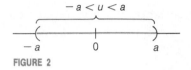

FIGURE 2

You can see why this theorem is valid if you think in terms of distance and position on a number line. The condition $-a < u < a$ means that u lies between $-a$ and a, as indicated in Figure 2. But this is the same as saying that the distance from u to zero is less than a, which in turn can be written $|u| < a$. (The second part of the theorem can be justified in a similar manner.)

EXAMPLE 4   Solve: **(a)** $|x| < 1$ **(b)** $|x| \geq 1$

Solution   **(a)** According to the theorem we just discussed, the condition $|x| < 1$ is equivalent to

$$-1 < x < 1$$

The solution set is therefore the open interval $(-1, 1)$.

(b) In view of the second part of the theorem, the inequality $|x| \geq 1$ is satisfied when x satisfies either of the inequalities

$$x \leq -1 \quad \text{or} \quad x \geq 1$$

The solution sets for these last two inequalities are $(-\infty, -1]$ and $[1, \infty)$, respectively. Consequently, the solution set for the given inequality $|x| \geq 1$ consists of the two intervals $(-\infty, -1]$ and $[1, \infty)$.   ▲

In Example 4(b), we found that the solution set consisted of two intervals on the number line. We have a convenient notation for describing such sets. Given any two sets A and B, we define the set $A \cup B$ (read **A union B**) to be the set of all elements that are in A or in B (or in both). For example, if $A = \{1, 2, 3\}$ and $B = \{4, 5\}$, then $A \cup B = \{1, 2, 3, 4, 5\}$. As another example, the union of the two closed intervals $[3, 5]$ and $[4, 7]$ is given by

$$[3, 5] \cup [4, 7] = [3, 7]$$

because the numbers in the interval $[3, 7]$ are precisely those numbers that are in $[3, 5]$ or $[4, 7]$ (or in both). Using this notation, we can write the solution set for Example 4(b) as

$$(-\infty, -1] \cup [1, \infty)$$

EXAMPLE 5   Solve: $|x - 3| < 1$

Solution   We'll show two methods.

FIRST METHOD   We use the theorem preceding Example 4. With $u = x - 3$ and $a = 1$, the theorem tells us that the given inequality is equivalent to

$$-1 < x - 3 < 1$$

or (by adding 3)

$$2 < x < 4$$

The solution set is therefore the open interval $(2, 4)$.

ALTERNATE METHOD   The given inequality tells us that x must be less than one unit away from 3 on the number line. Looking one unit to either side of 3, then, we see that x must lie strictly between 2 and 4. The solution set is therefore $(2, 4)$, as obtained using the first method.

▲

EXAMPLE 6   Solve: $\left| 1 - \dfrac{t}{2} \right| > 5$

Solution   Referring again to the theorem, we use the fact that $|u| > a$ means that either $u < -a$ or $u > a$. So, in the present example the given inequality means that either

$$1 - \frac{t}{2} < -5 \qquad \text{or} \qquad 1 - \frac{t}{2} > 5$$

$$-\frac{t}{2} < -6 \qquad\qquad -\frac{t}{2} > 4$$

$$t > 12 \qquad\qquad t < -8$$

This tells us that the given inequality is satisfied precisely when t is in either of the intervals $(12, \infty)$ or $(-\infty, -8)$. In other words, the solution set is $(-\infty, -8) \cup (12, \infty)$.

▲

All the inequalities that we have solved so far have involved first-degree polynomials. Now we turn our attention to inequalities involving higher-degree polynomials. Some observations about first-degree polynomials and inequalities will be helpful in introducing the main ideas. Consider, for example the inequality $2x - 5 < 0$, along with its solution set $\left(-\infty, \frac{5}{2}\right)$. Let us call a solution of the corresponding equality $2x - 5 = 0$ a **key number** for the given inequality. So, in this case the only key number is $x = \frac{5}{2}$. As indicated in Figure 3, the key number $\frac{5}{2}$ divides the number line into two intervals, $\left(-\infty, \frac{5}{2}\right)$ and $\left(\frac{5}{2}, \infty\right)$. Notice that on each of the intervals determined by the key number, the algebraic sign of $2x - 5$ is constant. (That is, for $x < \frac{5}{2}$ the value of $2x - 5$ is always negative; for $x > \frac{5}{2}$ the value of $2x - 5$ is always positive.) More generally, it can be shown that this same type of behavior regarding *persistence of sign* occurs with all polynomials, and indeed with quotients of polynomials as well. This important fact, along with the definition of a key number, is presented in the box that follows.

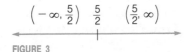

FIGURE 3

Key Numbers and Persistence of Sign

Let P and Q be polynomials with no common factor (other than constants), and consider the following four inequalities:

$$\frac{P}{Q} < 0 \qquad \frac{P}{Q} \leq 0 \qquad \frac{P}{Q} > 0 \qquad \frac{P}{Q} \geq 0$$

The **key numbers** for each of these inequalities are the real numbers for which $P = 0$ or $Q = 0$. It can be proved that the algebraic sign of P/Q is constant on each of the intervals determined by these key numbers.

We'll show how this result is applied by solving the polynomial inequality

$$x^3 - 2x^2 - 3x > 0$$

First, using factoring techniques from Section 1.5, we rewrite the inequality in the equivalent form

$$x(x + 1)(x - 3) > 0$$

The key numbers, then, are the solutions of the equation $x(x + 1)(x - 3) = 0$. That is, the key numbers are $x = 0$, $x = -1$, and $x = 3$. Next, we locate these numbers on a coordinate line. As indicated in Figure 4, this divides the number line into four distinct intervals.

FIGURE 4

Now, according to the result stated in the box just prior to this example, no matter what x-value we choose in the interval $(-\infty, -1)$, the resulting sign of $x^3 - 2x^2 - 3x$ [$= x(x + 1)(x - 3)$] will always be the same. Thus, to see what that sign is, we first choose any convenient *test number* in the interval $(-\infty, -1)$, say, $x = -2$. Then, using $x = -2$, we determine the sign of $x(x + 1)(x - 3)$ simply by considering the sign of each factor, as indicated in Table 1.

TABLE 1

On the interval $(-\infty, -1)$, the sign of $x^3 - 2x^2 - 3x$ [$= x(x + 1)(x - 3)$] is negative because it is the product of three negative factors.

INTERVAL	TEST NUMBER	x	x + 1	x - 3	x(x + 1)(x - 3)
$(-\infty, -1)$	-2	neg.	neg.	neg.	neg.

From this table we conclude that the values of $x^3 - 2x^2 - 3x$ are negative *throughout* the interval $(-\infty, -1)$, and, consequently, no number in this interval satisfies the given inequality. Next, we carry out similar analyses for the remaining three intervals, as shown in the following table. (You should verify for yourself that the entries in the table are correct.)

TABLE 2

On the interval $(-1, 0)$, the product $x(x + 1)(x - 3)$ is positive because it has two negative factors and one positive factor. On $(0, 3)$, the product is negative since it has two positive factors and one negative factor. And on $(3, \infty)$, the product is positive because all the factors are positive.

INTERVAL	TEST NUMBER	x	x + 1	x - 3	x(x + 1)(x - 3)
$(-1, 0)$	$-\frac{1}{2}$	neg.	pos.	neg.	pos.
$(0, 3)$	1	pos.	pos.	neg.	neg.
$(3, \infty)$	4	pos.	pos.	pos.	pos.

Looking at the two tables now, we conclude that the sign of $x(x + 1)(x - 3)$ is positive throughout both of the intervals $(-1, 0)$ and $(3, \infty)$ and, consequently, all the numbers in these intervals satisfy the given inequality. Moreover, our work also shows that the other two intervals that we considered are not part of the solution set. As you can readily check, the key numbers themselves do not satisfy the given inequality for this example. In summary, then, the solution set for the inequality $x^3 - 2x^2 - 3x > 0$ is $(-1, 0) \cup (3, \infty)$. Furthermore, it is important to notice that the work just carried out also provides us with three additional pieces of information:

The solution set for $x^3 - 2x^2 - 3x \geq 0$ is $[-1, 0] \cup [3, \infty)$.

The solution set for $x^3 - 2x^2 - 3x < 0$ is $(-\infty, -1) \cup (0, 3)$.

The solution set for $x^3 - 2x^2 - 3x \leq 0$ is $(-\infty, -1] \cup [0, 3]$.

In the box that follows, we summarize the steps for solving polynomial inequalities.

Steps for Solving Polynomial Inequalities

1. If necessary, rewrite the inequality so that the polynomial is on the left-hand side and zero is on the right-hand side.

2. Find the key numbers for the inequality and locate them on a number line.

3. List the intervals determined by the key numbers.

4. From each interval, choose a convenient test number. Then use the test number to determine the sign of the polynomial throughout the interval.

5. Use the information obtained in the previous step to specify the required solution set. [Don't forget to take into account whether the original inequality is strict ($<$ or $>$) or nonstrict ($\leq$ or $\geq$).]

EXAMPLE 7   Solve: $x^4 \leq 14x^3 - 48x^2$

Solution   First, we rewrite the inequality so that zero is on the right-hand side. Then we factor the left-hand side as follows:

$$x^4 - 14x^3 + 48x^2 \leq 0$$
$$x^2(x^2 - 14x + 48) \leq 0$$
$$x^2(x - 6)(x - 8) \leq 0$$

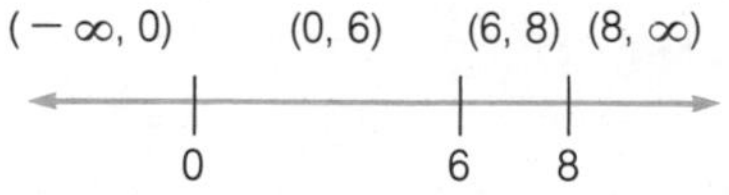

From this last line, we see that the key numbers are $x = 0$, 6, and 8. As indicated in Figure 5, these numbers divide the number line into four distinct intervals. We need to choose a test number from each interval and see whether the polynomial is positive or negative on the interval. This work is carried out in the following table.

$(-\infty, 0)$ $(0, 6)$ $(6, 8)$ $(8, \infty)$

0   6   8

FIGURE 5

INTERVAL	TEST NUMBER	x^2	$x - 6$	$x - 8$	$x^2(x - 6)(x - 8)$
$(-\infty, 0)$	-1	pos.	neg.	neg.	pos.
$(0, 6)$	1	pos.	neg.	neg.	pos.
$(6, 8)$	7	pos.	pos.	neg.	neg.
$(8, \infty)$	9	pos.	pos.	pos.	pos.

The table shows that the quantity $x^2(x - 6)(x - 8)$ is negative only for x-values in the interval $(6, 8)$. Also, as noted at the start, the quantity is equal to zero when $x = 0$, 6, or 8. Thus, the solution set consists of the numbers in the closed interval $[6, 8]$, along with the number 0. We can write this set

$$[6, 8] \cup \{0\}$$

where $\{0\}$ denotes the set whose only member is zero. ▲

EXAMPLE 8 Solve: **(a)** $x^2 - x + 15 > 0$ **(b)** $x^2 - x + 15 < 0$

Solution **(a)** As you can readily check, the equation $x^2 - x + 15 = 0$ has no real solution. So there is no key number, and, consequently, the polynomial $x^2 - x + 15$ never changes sign. To see what that sign is, choose the most convenient test number, namely, $x = 0$, and evaluate the polynomial: $0^2 - 0 + 15 > 0$. So the polynomial is positive for every value of x, and the solution set is $(-\infty, \infty)$, the set of all real numbers.

(b) Our work in part (a) shows that no real number satisfies the inequality $x^2 - x + 15 < 0$. ▲

The technique used in Examples 7 and 8 can also be used to solve inequalities involving quotients of polynomials. For these cases recall that the definition of a key number also includes the x-values for which the denominator is zero. For example, the key numbers for the inequality $\dfrac{x + 3}{x - 4} \geq 0$ are -3 and 4.

EXAMPLE 9 Solve: $\dfrac{x + 3}{x - 4} \geq 0$

$(-\infty, -3)$ $(-3, 4)$ $(4, \infty)$

-3 4

FIGURE 6

Solution The key numbers are -3 and 4. As indicated in Figure 6, these numbers divide the number line into three intervals. In the table that follows, we've chosen a test number from each interval and determined the sign of the quotient $\dfrac{x + 3}{x - 4}$ for each interval.

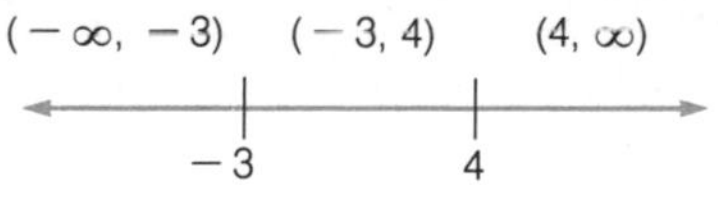

INTERVAL	TEST NUMBER	$x + 3$	$x - 4$	$\dfrac{x + 3}{x - 4}$
$(-\infty, -3)$	-4	neg.	neg.	pos.
$(-3, 4)$	0	pos.	neg.	neg.
$(4, \infty)$	5	pos.	pos.	pos.

From these results, we conclude that the solution set for $\dfrac{x + 3}{x - 4} \geq 0$ contains the two intervals $(-\infty, -3)$ and $(4, \infty)$. However, we still need to consider the two endpoints -3 and 4. As you can easily check, the value $x = -3$ does satisfy the given inequality, but $x = 4$ does not. In summary, then, the solution set is $(-\infty, -3] \cup (4, \infty)$. ▲

EXAMPLE 10 Solve: $\dfrac{2x + 1}{x - 1} - \dfrac{2}{x - 3} < 1$

Solution Our first inclination here might be to multiply through by $(x - 1)(x - 3)$ to eliminate fractions. This strategy is faulty, however, since we don't know whether the quantity $(x - 1)(x - 3)$ is positive or negative. Thus, we begin by rewriting the inequality in an equivalent form, with zero on the right-hand side and a single fraction on the left-hand side.

$$\frac{2x + 1}{x - 1} - \frac{2}{x - 3} - 1 < 0$$

$$\frac{(2x + 1)(x - 3) - 2(x - 1) - 1(x - 1)(x - 3)}{(x - 1)(x - 3)} < 0$$

$$\frac{x^2 - 3x - 4}{(x - 1)(x - 3)} < 0 \quad \text{(Check the algebra!)}$$

$$\frac{(x + 1)(x - 4)}{(x - 1)(x - 3)} < 0 \tag{1}$$

The key numbers are those x-values for which the denominator or the numerator is zero. By inspection then, we see that these numbers are -1, 4, 1, and 3. As Figure 7 indicates, these numbers divide the number line into five distinct intervals. Now, just as in Examples 7 through 9, we choose a test number from each interval and determine the sign of the quotient for that interval. (You should check each entry in the following table for yourself.)

FIGURE 7

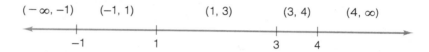

| | | | | |
| $(-\infty, -1)$ | $(-1, 1)$ | $(1, 3)$ | $(3, 4)$ | $(4, \infty)$ |

INTERVAL	TEST NUMBER	$(x + 1)(x - 4)$	$(x - 1)(x - 3)$	$\dfrac{(x + 1)(x - 4)}{(x - 1)(x - 3)}$
$(-\infty, -1)$	-2	pos.	pos.	pos.
$(-1, 1)$	0	neg.	pos.	neg.
$(1, 3)$	2	neg.	neg.	pos.
$(3, 4)$	$\frac{7}{2}$	neg.	pos.	neg.
$(4, \infty)$	5	pos.	pos.	pos.

From these results, we can see that the quotient on the left-hand side of inequality (1) is negative (as required) on the two intervals $(-1, 1)$ and $(3, 4)$. Now we need to check the endpoints of these intervals. When $x = -1$ or $x = 4$, the quotient is zero, and so, in view of the original inequality, we exclude these two x-values from the solution set. Furthermore, the quotient is undefined when $x = 1$ or $x = 3$, so we must also exclude those two values from the solution set. In summary, then, the solution set is $(-1, 1) \cup (3, 4)$. ▲

▼ EXERCISE SET 1.7

A

In Exercises 1–68, solve the inequality and specify the answer using interval notation.

1. $x + 5 < 4$

2. $2x - 7 < 11$

3. $1 - 3x \leq 0$

4. $6 - 4x \leq 22$

5. $4x + 6 < 3(x - 1) - x$

6. $2(t - 1) - 3(t + 1) \leq -5$

7. $1 - 2(t + 3) - t \leq 1 - 2t$

8. $t - 4[1 - (t - 1)] > 7 + 10t$

9. $\dfrac{3x}{5} - \dfrac{x - 1}{3} < 1$

10. $\dfrac{2x + 1}{2} + \dfrac{x - 1}{3} < x + \dfrac{1}{2}$

11. $\dfrac{x-1}{4} - \dfrac{2x+3}{5} \leq x$

12. $\dfrac{x}{2} - \dfrac{8x}{3} + \dfrac{x}{4} > \dfrac{23}{6}$

13. $-2 \leq x - 6 \leq 0$

14. $-3 \leq 2x + 1 \leq 5$

15. $-1 \leq \dfrac{1-4t}{3} \leq 1$

16. $\dfrac{2}{3} \leq \dfrac{5-3t}{-2} \leq \dfrac{3}{4}$

17. $0.99 < \dfrac{x}{2} - 1 < 0.999$

18. $\dfrac{9}{10} < \dfrac{3x-1}{-2} < \dfrac{91}{100}$

19. (a) $|x| \leq \frac{1}{2}$ (b) $|x| > \frac{1}{2}$

20. (a) $|x| > 2$ (b) $|x| \leq 2$

21. (a) $|x| > 0$ (b) $|x| < 0$

22. (a) $|t| \geq 0$ (b) $|t| \leq 0$

23. (a) $|x-2| < 1$ (b) $|x-2| > 1$

24. (a) $|x-4| \geq 4$ (b) $|x-4| \leq 4$

25. (a) $|1-x| \leq 5$ (b) $|1-4x| \leq 5$
     (c) $|1-4x| > 5$

26. (a) $|3x+5| < 17$ (b) $|3x+5| > 17$
     (c) $|3x+5| < 0$

27. $|x-a| < c$

28. $|x-a| + b < c$

29. $\left| \dfrac{x-2}{3} \right| < 4$

30. $\left| \dfrac{4-5x}{2} \right| > 1$

31. $\left| \dfrac{x+1}{2} - \dfrac{x-1}{3} \right| < 1$

32. $\left| \dfrac{3(x-2)}{4} + \dfrac{4(x-1)}{3} \right| \leq 2$

33. (a) $|(x+h)^2 - x^2| < 3h^2$ $(h > 0)$
     (b) $|(x+h)^2 - x^2| < 3h^2$ $(h < 0)$

34. (a) $|3(x+2)^2 - 3x^2| < \frac{1}{10}$
     (b) $|3(x+2)^2 - 3x^2| < \varepsilon$ $(\varepsilon > 0)$

35. $x^2 + x - 6 < 0$

36. $x^2 + 4x - 32 < 0$

37. $x^2 - 11x + 18 > 0$

38. $2x^2 + 7x + 5 > 0$

39. $9x - x^2 \leq 20$

40. $3x^2 + x \leq 4$

41. $x^2 - 16 \geq 0$

42. $24 - x^2 \geq 0$

43. $16x^2 + 24x < -9$

44. $x^4 - 16 < 0$

45. $x^3 + 13x^2 + 42x > 0$

46. $2x^3 - 9x^2 + 4x \geq 0$

47. $2x^2 + 1 \geq 0$

48. $1 + x^2 < 0$

49. $(x-1)(x+3)(x+4) \geq 0$

50. $x^4(x-2)(x-16) \geq 0$

51. $(x-2)^2(3x+1)^3(3x-1) > 0$

52. $(2x-1)^3(2x-3)^5(2x-5) > 0$

53. $x^4 - 25x^2 + 144 \leq 0$

54. $x^4 - 9x^2 + 20 \geq 0$

55. $x^3 + 2x^2 - x - 2 > 0$

56. $2x^4 + x^3 - 16x - 8 > 0$

57. $\dfrac{x-1}{x+1} \leq 0$

58. $\dfrac{x+4}{2x-5} \leq 0$

59. $\dfrac{2-x}{3-2x} \geq 0$

60. $\dfrac{x^2-1}{x^2+8x+15} \geq 0$

61. $\dfrac{2x^3 + 5x^2 - 7x}{3x^2 + 7x + 4} > 0$

62. $\dfrac{x^2 - x - 1}{x^2 + x - 1} > 0$

63. $\dfrac{1}{x-2} - \dfrac{1}{x-1} \geq \dfrac{1}{6}$

64. $\dfrac{2x}{x+5} + \dfrac{x-1}{x-5} < \dfrac{1}{5}$

65. $\dfrac{1+x}{1-x} - \dfrac{1-x}{1+x} < -1$

66. $\dfrac{x+1}{x+2} > \dfrac{x-3}{x+4}$

67. $\dfrac{3-2x}{3+2x} > \dfrac{1}{x}$

68. $\dfrac{x}{x-2} - \dfrac{3}{x+1} \geq 2$

69. Data from the Apollo 11 moon mission in July 1969 showed that temperature readings on the lunar surface vary over the interval $-183° \leq C \leq 112°$ on the Celsius scale. What is the corresponding interval on the Fahrenheit scale? (Round the numbers you obtain to the nearest integer.)

70. From the cloud tops of Venus to the planet's surface, temperature readings range over the interval $-25° \leq C \leq 475°$ on the Celsius scale. What is the corresponding range on the Fahrenheit scale?

71. If an object is projected vertically upward with an initial velocity of v_0 ft/sec, then its velocity v (in ft/sec) after t sec is given by

$$v = -32t + v_0$$

If the initial velocity is 60 ft/sec, during what time interval will the velocity be in the range $50 \geq v \geq 40$? (Leave your answer in fractional form rather than in decimal form.)

72. The width and length of a rectangle are 5 units and x units, respectively.
(a) Show that the perimeter P is given by $P = 2x + 10$.
(b) For which interval of x-values will the perimeter be in the range $100 \leq P \leq 200$?

For Exercises 73 and 74, determine the domain of the variable in the expression. (Recall the domain convention from Section 1.6—the domain is the set of all real number values for the variable for which the expression is defined.)

73. (a) $\sqrt{x^2 - 4x - 5}$ (b) $\sqrt{1/(x^2 - 4x - 5)}$

74. (a) $\sqrt{\dfrac{x+2}{x-4}}$ (b) $\sqrt[3]{\dfrac{x+2}{x-4}}$

B

75. For which values of b will the equation $x^2 + bx + 1 = 0$ have real solutions?

76. The sum of the first n natural numbers is given by

$$1 + 2 + 3 + \cdots + n = \frac{n(n+1)}{2}$$

For which values of n will the sum be less than 1225?

77. For which values of a is $x = 1$ a solution of the inequality
$\dfrac{2a+x}{x-2a} < 1$?

78. Solve: $\dfrac{ax + b}{\sqrt{x}} > 2\sqrt{ab}$ $(a > 0, b > 0)$

79. The two shorter sides in a right triangle have lengths x and $1 - x$ $(x > 0)$. For which values of x will the hypotenuse be less than $\sqrt{17}/5$?

80. A piece of wire 12 cm long is cut into two pieces. Denote the lengths of the two pieces by x and $12 - x$. Both pieces are then bent into squares. For which values of x will the combined areas of the squares exceed 5 cm²?

81. Let V and S denote the volume and total surface area, respectively, for a right circular cylinder of radius r and height 1. For which r-values will the ratio V/S be less than $\frac{1}{3}$?

82. Let V and S denote the volume and total surface area, respectively, for a right circular cone of radius r and height 1. For which r-values will the ratio V/S be less than $\frac{4}{27}$?

C

83. Given two positive numbers a and b, we define the **geometric mean**, the **arithmetic mean**, and the **root mean square** as follows:

$$\text{G.M.} = \sqrt{ab} \qquad \text{A.M.} = \frac{a + b}{2} \qquad \text{R.M.} = \sqrt{\frac{a^2 + b^2}{2}}$$

Complete the table.

85. Let a and b be positive numbers. In Exercise 84 you were asked to show that

$$\sqrt{ab} \le \frac{a + b}{2} \le \sqrt{\frac{a^2 + b^2}{2}}$$

This exercise shows how to establish these important inequalities using geometric, rather than algebraic, methods. In the following figure, $\overline{CF}$ is the diameter of a semicircle with center E. Let $CD = a$ and $DF = b$.

(a) Explain why $CE = EH = \dfrac{a + b}{2}$.

(b) Show that $DG = \sqrt{ab}$ and $DH = \sqrt{\dfrac{a^2 + b^2}{2}}$.

(c) Explain why $DG \le EH \le DH$, and conclude from this that $\sqrt{ab} \le (a + b)/2 \le \sqrt{(a^2 + b^2)/2}$.

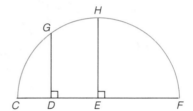

a	b	$\sqrt{ab}$ (G.M)	$\dfrac{a + b}{2}$ (A.M.)	$\sqrt{\dfrac{a^2 + b^2}{2}}$ (R.M.)	WHICH IS LARGEST, G.M., A.M., OR R.M.?	WHICH IS SMALLEST G.M., A.M., OR R.M.?
1	2					
1	3					
1	4					
2	3					
3	4					
9	10					
99	100					
999	1000					

84. Prove that each of the following inequalities is valid for all positive numbers a and b. *Hint:* Use the following property of inequalities: If x and y are positive, then the inequality $x \le y$ is equivalent to $x^2 \le y^2$.

(a) $\sqrt{ab} \le \dfrac{a + b}{2}$ (The geometric mean is less than or equal to the arithmetic mean.)

(b) $\dfrac{a + b}{2} \le \sqrt{\dfrac{a^2 + b^2}{2}}$ (The arithmetic mean is less then or equal to the root mean square.)

86. Let a, b, c, and d be positive real numbers. Use the inequality in Exercise 84(a) to prove each of the following inequalities.

(a) $\sqrt{abcd} \le \dfrac{ab + cd}{2}$

(b) $\sqrt{ab} + \sqrt{cd} \le \sqrt{(a + c)(b + d)}$

87. Let x denote the width of a rectangle whose perimeter is 30 ft.

(a) Show that the area A (in square feet) of the rectangle is given by $A = x(15 - x)$.

(b) Use the inequality in Exercise 84(a) to show that
$A \leq 225/4$.

(c) For which value of x is $A = 225/4$? What are the
dimensions of the rectangle in this case?

88. Let x denote the width of a rectangle whose area is
25 ft^2.

(a) Show that the perimeter P (in feet) of the rectangle
is given by $P = 2x + (50/x)$.

(b) Use the inequality in Exercise 84(a) to show that
$P \geq 20$.

(c) For which value of x is $P = 20$? What are the
dimensions of the rectangle in this case?

89. Find a nonzero value for c so that the solution set for
the inequality

$$x^2 + 2cx - 6c < 0$$

is the open interval $(-3c, c)$.

90. Solve: $(x - a)^2 - (x - b)^2 > \dfrac{(a - b)^2}{4}$

(Assume that $a > b$.)

1.8 ▼ RECTANGULAR COORDINATES

The mathematics I was taught in my
freshman year was algebra (old stuff),
trigonometry (too much of it), and analytical
geometry (a revelation).

Paul R. Halmos in his book *I Want To Be a
Mathematician* (N.Y.: Springer-Verlag, 1985)

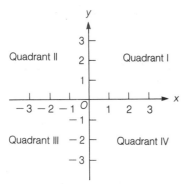

FIGURE 1

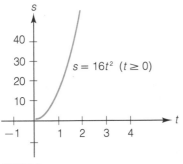

FIGURE 2

A graph of the formula $s = 16t^2$ in a *t-s*
coordinate system. [The formula relates the
distance s (in feet) and the time t (in seconds)
for an object falling in a vacuum.]

In previous courses you learned to work with a rectangular coordinate system such
as that shown in Figure 1. In this section we review some of the most basic formulas
and techniques that are useful here.

The point of intersection of the two perpendicular number lines, or **axes**, is
called the **origin** and is denoted by the letter O. The horizontal and vertical axes
are often labeled the **x-axis** and the **y-axis**, respectively; but any other variables
will do just as well for labeling the axes. For instance, Figure 2 shows a *t-s* coordinate system. (We'll discuss curves, or *graphs*, like the one in the figure in the next
chapter.)

Notice that in Figures 1 and 2 the axes divide the plane into four regions, or
quadrants, labeled I through IV, as shown in Figure 1. Unless indicated otherwise, we assume that the same unit of length is used on both axes. In Figure 1,
the same scales are used on both axes; in Figure 2, the scale used on the *s*-axis is
different than the scale on the *t*-axis.

Now look at the point P in Figure 3(a), on the next page. Starting from the
origin O, one way to reach P is to move three units in the positive *x*-direction
and then two units in the positive *y*-direction. That is, the location of P relative
to the origin and the axes is "right 3, up 2." We say that the **coordinates** of P
are $(3, 2)$. The first number within the parentheses conveys the information "right
3," and the second number conveys the information "up 2." We say that the **x-
coordinate** of P is 3 and the **y-coordinate** of P is 2. Likewise, the coordinates
of point Q in Figure 3(a) are $(-2, 4)$. With this coordinate notation in mind,
observe in Figure 3(b) that $(3, 2)$ and $(2, 3)$ represent different points; that is, the
order in which the two numbers appear within the parentheses affects the location
of the point. Figure 3(c) displays various points with given coordinates; you should
check for yourself that the coordinates correspond correctly to the location of each
point.

Some terminology and notation: The *x-y* coordinate system that we have described is often called a **Cartesian coordinate system**. The term *Cartesian* is
used in honor of René Descartes, the seventeenth-century French philosopher and
mathematician.* The coordinates (x, y) of a point P are referred to as an **ordered**

---

* Professor George F. Simmons in his book *Calculus with Analytic Geometry* (New York: McGraw-Hill,
1985) points out that a careful reading and comparison of the original sources clearly indicates that it
is Pierre de Fermat (1601–1665), not Descartes, who should be credited with the invention of analytic
geometry.

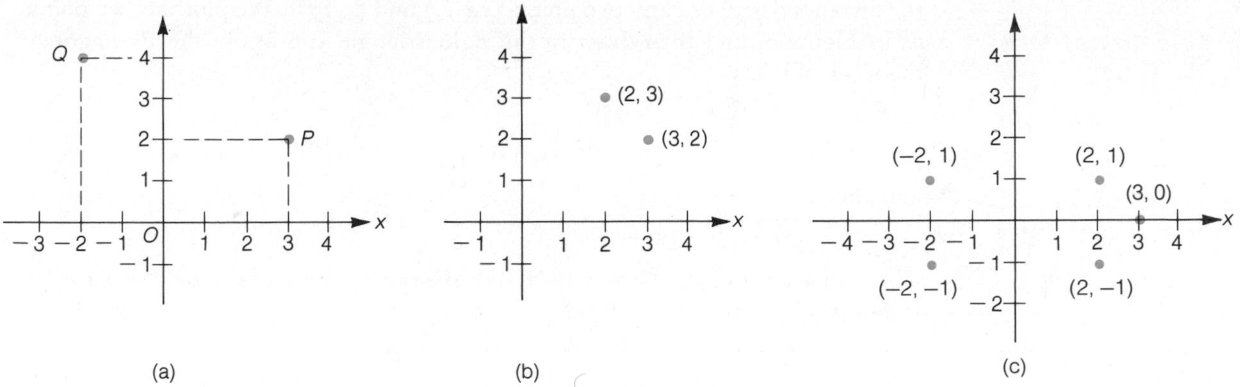

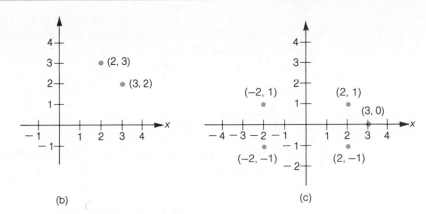

(a)

(b)

(c)

FIGURE 3

pair. Recall, for example, that $(3, 2)$ and $(2, 3)$ represent different points; that is, the order of the numbers matters. The x-coordinate of a point is sometimes referred to as the **abscissa** of the point; the y-coordinate is the **ordinate**. The notation $P(x, y)$ means that P is a point whose coordinates are (x, y). At times, we abbreviate the phrase *the point whose coordinates are* (x, y) to simply *the point* (x, y).

The remainder of our work in this section depends on a key result from elementary geometry, the Pythagorean theorem. For reference, we state this theorem and its converse in the box that follows. (See Exercise 56 for a proof of the theorem.)

The Pythagorean Theorem and Its Converse

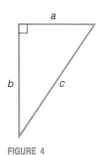

FIGURE 4

1. Pythagorean theorem
   (See Figure 4) In a right triangle, the lengths of the sides are related by the equation

   $$a^2 + b^2 = c^2$$

   where a and b are the lengths of the sides forming the right angle and c is the length of the hypotenuse (the side opposite the right angle).

2. Converse
   If the lengths of the sides of a triangle are related by an equation of the form $a^2 + b^2 = c^2$, then the triangle is a right triangle, and c is the length of the hypotenuse.

EXAMPLE 1 Use the Pythagorean theorem to calculate the distance d between points $(2, 1)$ and $(6, 3)$.

Solution We plot the two given points and draw a line connecting them, as shown in Figure 5. Then we draw in the broken lines as shown, parallel to the axes, and apply the Pythagorean theorem to the right triangle that is formed. The base of the triangle is four units long. You can see this by simply counting spaces or by using absolute value, as discussed in Section 1.2: $|6 - 2| = 4$. The height of the triangle is found to be two units, either by counting spaces or by computing the absolute value: $|3 - 1| = 2$. Thus, we have

$$d^2 = 4^2 + 2^2 = 20$$
$$d = \sqrt{20} = \sqrt{4}\sqrt{5} = 2\sqrt{5}$$

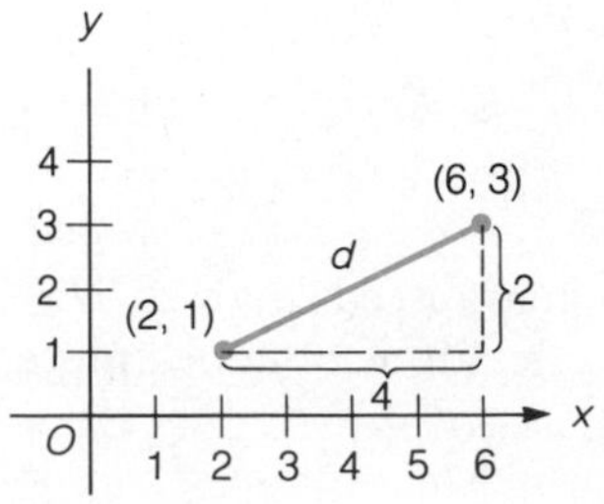

FIGURE 5

The method used in Example 1 can be applied to derive a general formula for

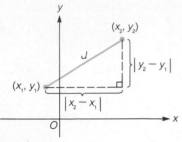

FIGURE 6

The Distance Formula

the distance d between any two points (x_1, y_1) and (x_2, y_2). We plot the two points, as in Figure 6, and then draw in the right triangle and apply the Pythagorean theorem. We have

$$d^2 = |x_2 - x_1|^2 + |y_2 - y_1|^2$$
$$= (x_2 - x_1)^2 + (y_2 - y_1)^2 \qquad \text{(Why?)}$$

and therefore

$$d = \sqrt{(x_2 - x_1)^2 + (y_2 - y_1)^2}$$

This last equation is referred to as the **distance formula**. For reference, we restate it in the box that follows.

> The distance d between the points (x_1, y_1) and (x_2, y_2) is given by
>
> $$d = \sqrt{(x_2 - x_1)^2 + (y_2 - y_1)^2}$$

Examples 2 and 3 demonstrate some simple calculations involving the distance formula.

Note In computing the distance between two given points, it does not matter which point you treat as (x_1, y_1) and which as (x_2, y_2). This is because quantities such as $(x_2 - x_1)$ and $(x_1 - x_2)$ are negatives of each other, and so their squares are equal.

EXAMPLE 2 Calculate the distance between the points $(2, -6)$ and $(5, 3)$.

Solution Substituting $(2, -6)$ for (x_1, y_1) and $(5, 3)$ for (x_2, y_2) in the distance formula, we have

$$d = \sqrt{(5 - 2)^2 + [3 - (-6)]^2}$$
$$= \sqrt{3^2 + 9^2} = \sqrt{90}$$
$$= \sqrt{9}\sqrt{10} = 3\sqrt{10}$$

You should check for yourself that the same answer is obtained using $(2, -6)$ as (x_2, y_2) and $(5, 3)$ as (x_1, y_1). ▲

EXAMPLE 3 Is the triangle with vertices $D(-2, -1)$, $E(4, 1)$, $F(3, 4)$ a right triangle?

Solution First, we sketch the triangle in question; see Figure 7. From the sketch it "appears" that angle E could be a right angle, but certainly this is not a proof. Our strategy is to use the distance formula to calculate the lengths of the three sides, then check whether any relation of the form $a^2 + b^2 = c^2$ holds. The calculations are as follows:

$$DE = \sqrt{[4 - (-2)]^2 + [1 - (-1)]^2} = \sqrt{36 + 4} = \sqrt{40}$$
$$EF = \sqrt{(4 - 3)^2 + (1 - 4)^2} = \sqrt{1 + 9} = \sqrt{10}$$
$$DF = \sqrt{[3 - (-2)]^2 + [4 - (-1)]^2} = \sqrt{25 + 25} = \sqrt{50}$$

Because $(\sqrt{40})^2 + (\sqrt{10})^2 = (\sqrt{50})^2$, we are indeed guaranteed that $\triangle DEF$ is a right triangle. (In Section 2.2, you'll see that this result can be obtained more efficiently using the concept of *slope*.) ▲

We can use the distance formula to obtain the equation of a circle. Figure 8 shows a circle with center (h, k) and radius r. By definition, a point (x, y) is on

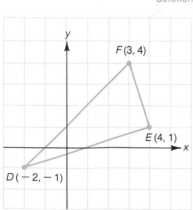

FIGURE 7

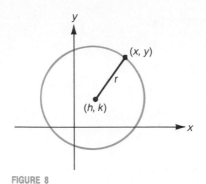

FIGURE 8

this circle if and only if the distance from (x, y) to (h, k) is r. Thus we have

$$\sqrt{(x - h)^2 + (y - k)^2} = r$$

or, equivalently,

$$(x - h)^2 + (y - k)^2 = r^2 \qquad (1)$$

(These last two equations are equivalent because two nonnegative quantities are equal if and only if their squares are equal.)

The work in the previous paragraph shows us two things. First, if a point (x, y) lies on the circle in Figure 8, then x and y together satisfy equation (1). Second, if a pair of numbers x and y satisfies equation (1), then the point (x, y) lies on the circle in Figure 8. (Does it sound to you as if the previous two sentences say the same thing? They don't! Think about it.) Equation (1) is called the **standard form for the equation of a circle**. For reference, we record the result in the box that follows.

The Equation of a Circle in Standard Form

> The equation of a circle with center (h, k) and radius r is
>
> $$(x - h)^2 + (y - k)^2 = r^2$$

EXAMPLE 4   Write the equation of the circle with center $(-2, 5)$ and radius 3.

Solution   In the equation $(x - h)^2 + (y - k)^2 = r^2$, we substitute the given values $h = -2$, $k = 5$, and $r = 3$. This yields

$$[x - (-2)]^2 + (y - 5)^2 = 3^2$$
$$(x + 2)^2 + (y - 5)^2 = 9$$

This is the standard form for the equation of the given circle. An alternative form of this answer is found by carrying out the indicated algebra. We have

$$(x + 2)^2 + (y - 5)^2 = 9$$

Therefore,

$$x^2 + 4x + 4 + y^2 - 10y + 25 = 9$$

and, consequently,

$$x^2 + 4x + y^2 - 10y + 20 = 0$$

The disadvantage to this alternative form is that information regarding the center and radius is no longer readily visible. ▲

When the equation of a circle is not in standard form, the technique of completing the square (discussed in Section 1.6) can be used to convert the equation to standard form. Example 5 shows how this is done.

EXAMPLE 5   Determine the center and radius of the circle given by

$$4x^2 - 24x + 4y^2 + 16y + 51 = 0$$

Solution   First, divide through by 4 so that the coefficients of x^2 and y^2 are both 1:

$$x^2 - 6x + y^2 + 4y + \frac{51}{4} = 0$$

Now, in preparation for completing the squares, we write

$$(x^2 - 6x + \underline{\hspace{1cm}}) + (y^2 + 4y + \underline{\hspace{1cm}}) = -\frac{51}{4}$$

To complete the square in x, we need to add $\left(-\frac{6}{2}\right)^2$, or 9. To complete the square in y, we need to add $\left(\frac{4}{2}\right)^2$, or 4. Thus, we have

$$(x^2 - 6x + 9) + (y^2 + 4y + 4) = -\frac{51}{4} + 9 + 4$$

or

$$(x - 3)^2 + (y + 2)^2 = -\frac{51}{4} + \frac{52}{4} = \frac{1}{4}$$

$$(x - 3)^2 + (y + 2)^2 = \left(\frac{1}{2}\right)^2$$

This is the equation in standard form. The center of the circle is $(3, -2)$; the radius is $\frac{1}{2}$. ▲

In Example 6 we make use of a simple result that you may recall from previous courses: the midpoint formula. This result is summarized in the box that follows. (For a proof of the formula, see Exercise 53.)

The Midpoint Formula

	EXAMPLE
The midpoint of the line segment joining the points $P(x_1, y_1)$ and $Q(x_2, y_2)$ is $$\left(\frac{x_1 + x_2}{2}, \frac{y_1 + y_2}{2}\right)$$	The midpoint of the line segment joining $(2, -15)$ and $(4, 5)$ is $$\left(\frac{2 + 4}{2}, \frac{-15 + 5}{2}\right) = (3, -5)$$

EXAMPLE 6   Points $A(-1, 6)$ and $B(3, -2)$ are the endpoints of a diameter of a circle, as shown in Figure 9. Determine the coordinates of the points P and Q where the circle intersects the y-axis. (The y-coordinates of the points P and Q are called

FIGURE 9
$\overline{AB}$ is a diameter. What are the y-intercepts of the circle?

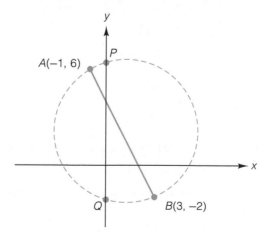

the *y-intercepts* of the circle. Both *y*-intercepts and *x*-intercepts are discussed more generally in Section 2.1.)

Solution

The *x*-coordinates of both *P* and *Q* are zero. So, our strategy here is as follows: If we knew the equation of the circle, then we could find the required *y*-coordinates by setting $x = 0$ in the equation of the circle. To write the equation of the circle, we need to know the center and the radius. We can find these by using the midpoint formula and the distance formula.

The center of the circle is the midpoint of diameter $\overline{AB}$. Therefore, the coordinates of the center are

$$x = \frac{-1 + 3}{2} = 1 \quad \text{and} \quad y = \frac{6 + (-2)}{2} = 2$$

So the center is the point $(1, 2)$. Since the radius *r* is the distance from the center to point $A(-1, 6)$, we have

$$r = \sqrt{[1 - (-1)]^2 + (2 - 6)^2} = \sqrt{4 + 16} = \sqrt{20}$$

Therefore,

$$r^2 = 20$$

It follows now that the equation of the circle is

$$(x - 1)^2 + (y - 2)^2 = 20$$

For the *y*-intercepts, we set $x = 0$ in this last equation to obtain

$$(y - 2)^2 = 19$$
$$y - 2 = \pm\sqrt{19}$$
$$y = 2 \pm \sqrt{19}$$

Thus, the *y*-intercepts are $2 + \sqrt{19}$, which is positive, and $2 - \sqrt{19}$, which is negative. Consequently, the coordinates of *P* and *Q* are $P(0, 2 + \sqrt{19})$ and $Q(0, 2 - \sqrt{19})$. ▲

▼ EXERCISE SET 1.8

A

1. Plot the points $(5, 2)$, $(-4, 5)$, $(-4, 0)$, $(-1, -1)$, and $(5, -2)$.

2. Draw the square *ABCD* whose vertices (corners) are $A(1, 0)$, $B(0, 1)$, $C(-1, 0)$, and $D(0, -1)$.

3. (a) Draw the right triangle *PQR* with vertices $P(1, 0)$, $Q(5, 0)$, and $R(5, 3)$.
   (b) Use the formula for the area of a triangle, $A = \frac{1}{2}bh$, to find the area of triangle *PQR* in part (a).

4. (a) Draw the trapezoid *ABCD* with vertices $A(0, 0)$, $B(7, 0)$, $C(6, 4)$, and $D(4, 4)$.

(b) Compute the area of the trapezoid. (See the inside front cover of this book for the appropriate formula.)

In Exercises 5–10, calculate the distance between the given points.

5. (a) $(0, 0)$ and $(-3, 4)$ (b) $(2, 1)$ and $(7, 13)$

6. (a) $(-1, -3)$ and $(-5, 4)$
   (b) $(6, -2)$ and $(-1, 1)$

7. (a) $(-5, 0)$ and $(5, 0)$ (b) $(0, -8)$ and $(0, 1)$

8. (a) $(-5, -3)$ and $(-9, -6)$
   (b) $(\frac{9}{2}, 3)$ and $(-2\frac{1}{2}, -1)$

9. $(1, \sqrt{3})$ and $(-1, -\sqrt{3})$

10. $(-3, 1)$ and $(374, -335)$

11. Which point is farther from the origin?
  (a) $(3, -2)$ or $(4, \frac{1}{2})$ (b) $(-6, 7)$ or $(9, 0)$

12. Use the distance formula to show that, in each case, the triangle with given vertices is an isosceles triangle.
  (a) $(0, 2), (7, 4), (2, -5)$
  (b) $(-1, -8), (0, -1), (-4, -4)$
  (c) $(-7, 4), (-3, 10), (1, 3)$

13. In each case, determine if the triangle with the given vertices is a right triangle? *Hint:* Find the lengths of the sides and then use the *converse* of the Pythagorean theorem.
  (a) $(7, -1), (-3, 5), (-12, -10)$
  (b) $(4, 5), (-3, 9), (1, 3)$
  (c) $(-8, -2), (1, -1), (10, 19)$

14. (a) Two of the three triangles specified in Exercise 13 are right triangles. Find their areas.
  (b) Calculate the area of the remaining triangle in Exercise 13 by using the following formula for the area A of a triangle with vertices (x_1, y_1), (x_2, y_2), and (x_3, y_3):

$$A = \tfrac{1}{2}|x_1y_2 - x_2y_1 + x_2y_3 - x_3y_2 + x_3y_1 - x_1y_3|$$

  The derivation of this formula is given in Exercise 55.
  (c) Use the formula given in part (b) to check your answers in part (a).

15. Use the formula given in Exercise 14(b) to calculate the area of the triangle with vertices $(1, -4)$, $(5, 3)$, and $(13, 17)$. Conclusion?

16. The coordinates of points A, B, and C are $A(-4, 6)$, $B(-1, 2)$, and $C(2, -2)$.
  (a) Show that $AB = BC$ by using the distance formula.
  (b) Show that $AB + BC = AC$ by using the distance formula.
  (c) What can you conclude from parts (a) and (b)?

In Exercises 17–26, determine the center and the radius for the circle. Also, find the y-coordinates of the points (if any) where the circle intersects the y-axis.

17. $(x - 3)^2 + (y - 1)^2 = 25$

18. $x^2 + (y + 1)^2 = 20$

19. $x^2 + y^2 = \sqrt{2}$

20. $x^2 + y^2 - 10x + 2y + 17 = 0$

21. $x^2 + y^2 + 8x - 6y = -24$

22. $4x^2 - 4x + 4y^2 - 63 = 0$

23. $9x^2 + 54x + 9y^2 - 6y + 64 = 0$

24. $3x^2 + 3y^2 + 5x - 4y = 1$

25. $2x^2 + 2y^2 = 2x + 6y - 3$

26. $x^2 + y^2 - 2\sqrt{2}(x + 2y) + 8 = 0$

In Exercises 27 and 28, determine the equation of the circle in standard form, given the coordinates of the diameter $\overline{PQ}$.

27. $P(-4, -2)$ and $Q(6, 4)$

28. $P(1, -3)$ and $Q(-5, -5)$

In Exercises 29 and 30, find the midpoint of the line segment joining points P and Q.

29. (a) $P(3, 2)$ and $Q(9, 8)$
  (b) $P(-4, 0)$ and $Q(5, -3)$
  (c) $P(3, -6)$ and $Q(-1, -2)$

30. (a) $P(12, 0)$ and $Q(12, 8)$
  (b) $P(\frac{3}{5}, -\frac{2}{3})$ and $Q(0, 0)$
  (c) $P(1, \pi)$ and $Q(3, 3\pi)$

31. The coordinates of A and B are $A(-1, 2)$ and $B(5, -3)$. If B is the midpoint of line segment $\overline{AC}$, what are the coordinates of C?

32. The coordinates of the points S and T are $S(4, 6)$ and $T(10, 2)$. If M is the midpoint of $\overline{ST}$, find the midpoint of $\overline{SM}$.

33. (a) Sketch the parallelogram with vertices $A(-7, -1)$, $B(4, 3)$, $C(7, 8)$, and $D(-4, 4)$.
  (b) Compute the midpoints of the diagonals $\overline{AC}$ and $\overline{BD}$.
  (c) What conclusion can you draw from part (b)?

34. The vertices of $\triangle ABC$ are $A(1, 1)$, $B(9, 3)$, and $C(3, 5)$.
  (a) Find the perimeter of $\triangle ABC$.
  (b) Find the perimeter of the triangle that is formed by joining the midpoints of the three sides of $\triangle ABC$.
  (c) Compute the ratio of the perimeter in part (a) to the perimeter in part (b).
  (d) What theorem from geometry provides the answer for part (c) without using the results in (a) and (b)?

35. This exercise refers to $\triangle ABC$ whose vertices are specified in Exercise 34.
  (a) Compute the sum of the squares of the lengths of the sides of $\triangle ABC$.
  (b) Compute the sum of the squares of the lengths of the three medians. (A **median** is a line segment drawn from a vertex to the midpoint of the opposite side.)
  (c) Check that the ratio of the answer in part (a) to that in part (b) is 4/3. *Remark:* As Exercise 50 asks you to show, this ratio is 4/3 for any triangle.

36. (a) Sketch the circle of radius 1 centered at the origin.
  (b) Write the equation for this circle.

(c) Does the point $(\frac{3}{5}, \frac{4}{5})$ lie on this circle?

(d) Does the point $(-1/2, \sqrt{3}/2)$ lie on this circle?

37. The center of a circle is the point $(3, 2)$. If the point $(-2, -10)$ lies on this circle, find the equation of the circle.

38. Find the equation of the circle tangent to the x-axis and with center $(3, 5)$. *Hint:* First draw a sketch.

39. Find the equation of the circle tangent to the y-axis and with center $(3, 5)$.

40. Find the equation of the circle passing through the origin and with center $(3, 5)$.

41. Use the Pythagorean theorem to find the length a in the figure. Then find b, c, d, e, f, and g.

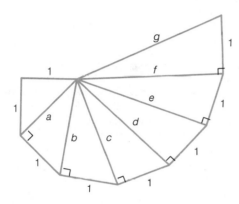

C **42.** (A numerologist's delight) Using the Pythagorean theorem and your calculator, compute the area of a right triangle in which the lengths of the hypotenuse and one leg are 2045 and 693, respectively.

B

In Exercises 43 and 44, determine the equation of the circle that satisfies the given conditions. Write the equation in standard form.

43. The circle passes through the origin and is concentric with the circle $x^2 - 6x + y^2 - 4y + 4 = 0$.

44. The circle passes through $(-4, 1)$, and its center is the midpoint of the line segment joining the centers of the two circles $x^2 + y^2 - 6x - 4y + 12 = 0$ and $x^2 + y^2 - 14x + 47 = 0$.

45. Find a value for t such that points $(0, 2)$ and $(12, t)$ are 13 units apart. *Hint:* By the distance formula, $13 = \sqrt{(12-0)^2 + (t-2)^2}$. Now square both sides and solve for t.

46. **(a)** Find values for t such that the points $(-2, 3)$ and $(t, 1)$ are six units apart.

(b) Can you find a real number t such that the points $(-2, 3$ and $(t, 1)$ are one unit apart? Why or why not?

47. The diagonals of a parallelogram bisect each other. Steps (a), (b), and (c) outline a proof of this theorem.

(a) In the parallelogram $OABC$ shown in the figure, check that the coordinates of B must be $(a + b, c)$.

(b) Use the midpoint formula to calculate the midpoints of diagonals $\overline{OB}$ and $\overline{AC}$.

(c) The two answers in part (b) are identical. This shows that the two diagonals do indeed bisect each other, as we wished to prove.

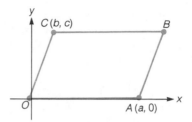

48. Prove that in a parallelogram, the sum of the squares of the lengths of the diagonals equals the sum of the squares of the lengths of the four sides. (Use the figure in Exercise 47.)

49. Use the figure in Exercise 47 to prove the following theorem: If the lengths of the diagonals of a parallelogram are equal, then the parallelogram is a rectangle. *Hint:* Equate the two expressions for the lengths of the diagonals and conclude that $ab = 0$. The case in which $a = 0$ can be discarded.

50. The sum of the squares of the sides of any triangle equals 4/3 the sum of the squares of the medians. Prove this, taking the vertices to be $(0, 0)$, $(2, 0)$, and $(2a, 2b)$. *Note:* The expression *squares of the sides* means *squares of the lengths of the sides.*

51. In $\triangle ABC$, let M be the midpoint of side $\overline{BC}$. Prove that $AB^2 + AC^2 = 2(BM^2 + AM^2)$. *Hint:* Let the coordinates be $A(0, 0)$, $B(2, 0)$, and $C(2a, 2b)$.

52. If the point (x, y) is equidistant from the points $(-3, -3)$ and $(5, 5)$, show that x and y satisfy the equation $x + y = 2$. *Hint:* Use the distance formula.

53. Suppose that the coordinates of points P, Q, and M are

$$P(x_1, y_1) \qquad Q(x_2, y_2) \qquad M\left(\frac{x_1 + x_2}{2}, \frac{y_1 + y_2}{2}\right)$$

Follow steps (a) and (b) to prove that M is the midpoint of the line segment from P to Q.

(a) By computing both of the distances PM and MQ, show that $PM = MQ$. (This shows that M lies on the perpendicular bisector of line segment $\overline{PQ}$, but it does not show that M actually lies *on* $\overline{PQ}$.)

(b) Show that $PM + MQ = PQ$. (This shows that M does lie on $\overline{PQ}$.)

C

54. The point (x, y) is located such that the sum of its distance from $(-1, 0)$ and from $(1, 0)$ is 4. Show that x and y are related by the equation $3x^2 + 4y^2 = 12$.

55. This problem indicates a method for calculating the area of a triangle when the coordinates of the three vertices are given.

(a) Calculate the area of $\triangle ABC$ in the figure. *Hint:* First calculate the area of the rectangle enclosing $\triangle ABC$ and then subtract the areas of the three right triangles.

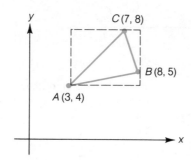

(b) Calculate the area of the triangle with vertices $(1, 3)$, $(4, 1)$, and $(10, 4)$. *Hint:* Work with an enclosing rectangle and three right triangles, as in part (a).

(c) Using the same technique that you used in parts (a) and (b), show that the area of the triangle in the figure is given by

$$A = \tfrac{1}{2}(x_1 y_2 - x_2 y_1 + x_2 y_3 - x_3 y_2 + x_3 y_1 - x_1 y_3)$$

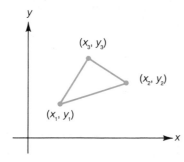

Remark: If we use absolute value signs instead of the parentheses, then the formula in part (c) will

hold regardless of the relative positions or quadrants of the three vertices. Thus, the area of a triangle with vertices (x_1, y_1), (x_2, y_2), (x_3, y_3) is given by

$$A = \tfrac{1}{2}|x_1 y_2 - x_2 y_1 - x_2 y_3 - x_3 y_2 + x_3 y_1 - x_1 y_3|$$

56. This problem outlines one of the shortest proofs of the Pythagorean theorem. In the figure we are given a right triangle ACB with the right angle at C, and we want to prove that $a^2 + b^2 = c^2$. In the figure, $\overline{CD}$ is drawn perpendicular to $\overline{AB}$.

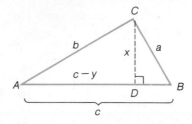

(a) Check that $\angle CAD = \angle DCB$ and that $\triangle BCD$ and $\triangle BAC$ are similar.

(b) Use the result in part (a) to obtain the equation $a/y = c/a$, and conclude that $a^2 = cy$.

(c) Show that $\triangle ACD$ is similar to $\triangle ABC$, and use this to deduce that $b^2 = c^2 - cy$.

(d) Combine the two equaions deduced in parts (b) and (c) to arrive at $a^2 + b^2 = c^2$.

57. Refer to the figure.

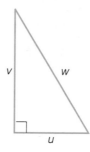

(a) If $u = \dfrac{2(m + n)}{n}$ and $v = \dfrac{4m}{m - n}$, show that

$$w = \frac{2(m^2 + n^2)}{(m - n)n}$$

(b) Show that $\tfrac{1}{2}uv = u + v + w$ (that is, the perimeter is numerically equal to the area).

(c) Give an example of a right triangle in which the length of each side is an integer and the perimeter is numerically equal to the area.

1.9 ▼ THE COMPLEX NUMBER SYSTEM

The very first square root of negative number on record, $\sqrt{81 - 144}$, is in the *Stereometrica of Hero of Alexandria* [second half of the first century A.D.].

David Wells in *The Penguin Dictionary of Curious and Interesting Numbers* (Harmondsworth, Middlesex, England: Penguin Books Ltd., 1986)

In elementary algebra, complex numbers appear as expressions $a + bi$, where a and b are ordinary real numbers and $i^2 = -1$ . . . . Complex numbers are manipulated by the usual rules of algebra, with the convention that i^2 is to be replaced by -1 whenever it occurs.

Ralph Boas in *Invitation to Complex Analysis* (New York: Random House, 1987)

The quotation in the margin from Professor Boas summarizes the basic approach we will follow in this section. Near the end of the section, after you've become accustomed to working with complex numbers, we'll present a formal list of some of the basic definitions and properties that can be used to develop the subject more rigorously, as is required in more advanced courses.

When we solve equations in Chapter 9, you'll see instances in which the real number system proves to be inadequate. In particular, since the square of a real number is never negative, there is no real number x such that $x^2 = -1$. To overcome this inconvenience, mathematicians define the symbol i by the equation

$$i^2 = -1$$

For reasons that are more historical than mathematical, i is referred to as the **imaginary unit**. This name is unfortunate, because to an engineer or a mathematician i is neither less "real" nor less tangible than any real number. Having said this, however, we do have to admit that i does not belong to the real number system.

Algebraically, we operate with the symbol i as if it were any letter in a polynomial expression. However, when we see i^2, we must remember to replace it by -1. Here are some sample calculations involving i.

$$3i + 2i = 5i$$
$$-2i^2 + 6i = -2(-1) + 6i = 2 + 6i$$
$$(-i)^2 = i^2 = -1$$
$$0i = 0$$

An expression of the form $a + bi$, where a and b are real numbers, is called a **complex number**.* Thus, four examples of complex numbers are:

$2 + 3i$

$4 + (-5)i$ (usually written $4 - 5i$)

$1 - \sqrt{2}\,i$ (also written $1 - i\sqrt{2}$)

$\dfrac{1}{2} + \dfrac{3}{2}i$ $\left(\text{also written } \dfrac{1 + 3i}{2}\right)$

Given a complex number $a + bi$, we say that a is the **real part** of $a + bi$, and b is the **imaginary part** of $a + bi$. For example, the real part of $3 - 4i$ is 3 and the imaginary part is -4. Observe that both the real part and the imaginary part of a complex number are themselves real numbers.

We define the notion of *equality* for complex numbers in terms of their real and imaginary parts. Two complex numbers are said to be **equal** if their corresponding real and imaginary parts are equal. We can write this definition symbolically

---

* The phrase *complex number* is attributed to Carl Fredrich Gauss (1777–1855), as is the use of i to denote $\sqrt{-1}$. The term *imaginary number* was originated by René Descartes (1596–1650).

as follows:

$$a + bi = c + di \qquad \text{if and only if} \qquad a = c \quad \text{and} \quad b = d$$

EXAMPLE 1   Determine real numbers c and d such that $10 + 4i = 2c + di$.

Solution   Equating the real parts of the two complex numbers gives us $2c = 10$, and therefore $c = 5$. Similarly, equating the imaginary parts yields $d = 4$. These are the required values for c and d. ▲

As the next example indicates, addition, subtraction, and multiplication of complex numbers are carried out using the usual rules of algebra, with the understanding (as mentioned before) that i^2 is always to be replaced with -1. (We'll discuss division of complex numbers subsequently.)

EXAMPLE 2   Let $z = 2 + 5i$ and $w = 3 - 4i$. Compute each quantity.

(a) $w + z$ **(b)** $3z$ **(c)** $w - 3z$ **(d)** zw **(e)** wz

Solution
(a) $w + z = (3 - 4i) + (2 + 5i) = (3 + 2) + (-4i + 5i) = 5 + i$
(b) $3z = 3(2 + 5i) = 6 + 15i$
(c) $w - 3z = (3 - 4i) - (6 + 15i) = (3 - 6) + (-4i - 15i) = -3 - 19i$
(d) $zw = (2 + 5i)(3 - 4i) = 6 - 8i + 15i - 20i^2$
$$= 6 + 7i - 20(-1) = 26 + 7i$$
(e) $wz = (3 - 4i)(2 + 5i) = 6 + 15i - 8i - 20i^2$
$$= 6 + 7i - 20(-1) = 26 + 7i \qquad ▲$$

If you look over the result of part (a) in Example 2, you can see that the sum is obtained simply by adding the corresponding real and imaginary parts of the given numbers. Likewise, in part (c) the difference is obtained by subtracting the corresponding real and imaginary parts. In part (d), however, notice that the product is *not* obtained in a similar fashion. (Exercise 54 provides some perspective on this.) Finally, notice that the results in parts (d) and (e) are identical. In fact, it can be shown that for any two complex numbers z and w, we always have $zw = wz$. In other words, just as with real numbers, multiplication of complex numbers is commutative. Furthermore, it can be shown that all of the properties of real numbers listed in Section A.2 of the Appendix continue to hold for complex numbers. We'll return to this point at the end of this section and in the exercises.

As background for the discussion of division of complex numbers, we introduce the notion of a *complex conjugate*, or simply a *conjugate*.

DEFINITION

Complex Conjugate

Let $z = a + bi$. The **complex conjugate** of z, denoted by $\bar{z}$, is defined by

$$\bar{z} = a - bi$$

EXAMPLES

If $z = 3 + 4i$, then $\bar{z} = 3 - 4i$.
If $w = 9 - 2i$, then $\bar{w} = 9 + 2i$.

EXAMPLE 3 **(a)** If $z = 6 - 3i$, compute $z\bar{z}$.   **(b)** If $w = a + bi$, compute $w\bar{w}$.

Solution **(a)** $z\bar{z} = (6 - 3i)(6 + 3i)$ **(b)** $w\bar{w} = (a + bi)(a - bi)$
$\qquad = 36 + 18i - 18i - 9i^2$ $\qquad = a^2 - abi + abi - b^2i^2$
$\qquad = 36 + 9 = 45$ $\qquad = a^2 + b^2$

Note The result in part (b) shows that the product of a complex number and its conjugate is always a real number.   ▲

Quotients of complex numbers are easy to compute using conjugates. Suppose, for example, that we wish to compute the quotient

$$\frac{5 - 2i}{3 + 4i}$$

To do this, we take the conjugate of the denominator, namely, $3 - 4i$, and then multiply the given fraction by $\dfrac{3 - 4i}{3 - 4i}$, which equals 1 (assuming that the usual rules of algebra are in force). This yields

$$\frac{5 - 2i}{3 + 4i} = \frac{5 - 2i}{3 + 4i} \cdot \frac{3 - 4i}{3 - 4i}$$

$$= \frac{15 - 26i + 8i^2}{9 - 16i^2}$$

$$= \frac{7 - 26i}{25} \qquad \text{since } 8i^2 = -8 \text{ and } -16i^2 = 16$$

$$= \frac{7}{25} - \frac{26}{25}i$$

In the box that follows, we summarize our procedure for computing quotients. The condition $w \neq 0$ means that w is any complex number other than $0 + 0i$.

Procedure for Computing Quotients

> Let z and w be two complex numbers, and $w \neq 0$. Then z/w is computed as follows:
>
> $$\frac{z}{w} = \frac{z}{w} \cdot \frac{\bar{w}}{\bar{w}}$$

EXAMPLE 4 Let $z = 3 + 4i$ and $w = 1 - 2i$. Compute each quotient.

(a) $\dfrac{1}{z}$ **(b)** $\dfrac{z}{w}$

Solution **(a)** $\dfrac{1}{z} = \dfrac{1}{z} \cdot \dfrac{\bar{z}}{\bar{z}} = \dfrac{1}{3 + 4i} \cdot \dfrac{3 - 4i}{3 - 4i}$ **(b)** $\dfrac{z}{w} = \dfrac{z}{w} \cdot \dfrac{\bar{w}}{\bar{w}} = \dfrac{3 + 4i}{1 - 2i} \cdot \dfrac{1 + 2i}{1 + 2i}$

$\qquad = \dfrac{3 - 4i}{9 - 16i^2} = \dfrac{3 - 4i}{25}$ $\qquad = \dfrac{3 + 10i + 8i^2}{1 - 4i^2}$

$\qquad = \dfrac{3}{25} - \dfrac{4}{25}i$ $\qquad = \dfrac{-5 + 10i}{5}$

$\qquad\qquad\qquad\qquad\qquad\qquad\qquad\qquad = -1 + 2i$ ▲

We began this section by defining i by the equation $i^2 = -1$. This can be rewritten as

$$i = \sqrt{-1}$$

provided that we agree to certain conventions regarding principal square roots and negative numbers. In dealing with the principal square root of a negative real number, say $\sqrt{-5}$, we shall write

$$\sqrt{-5} = \sqrt{(-1)(5)} = \sqrt{-1}\sqrt{5} = i\sqrt{5}$$

In other words, we are allowing the use of the rule $\sqrt{ab} = \sqrt{a}\sqrt{b}$ when a is -1 and b is a positive real number. However, the rule $\sqrt{ab} = \sqrt{a}\sqrt{b}$ *cannot* be used when both a and b are negative. If that were allowed, we could write

$$1 = (-1)(-1)$$

and then

$$\sqrt{1} = \sqrt{(-1)(-1)} = \sqrt{-1}\sqrt{-1} = (i)(i)$$

Consequently, we could write

$$1 = \sqrt{1} = i^2 = -1$$

or

$$1 = -1$$

which is a contradiction. Due to the contradiction that results, the rule $\sqrt{ab} = \sqrt{a}\sqrt{b}$ cannot be applied when both a and b are negative.

EXAMPLE 5   Simplify:

(a) i^4 **(b)** i^{101} **(c)** $\sqrt{-12} + \sqrt{-27}$ **(d)** $\sqrt{-9}\sqrt{-4}$

Solution   **(a)** We make use of the defining equation for i, which is $i^2 = -1$. Thus, we have

$$i^4 = (i^2)^2 = (-1)^2 = 1$$

(The result, $i^4 = 1$, is worth remembering.)

(b) $i^{101} = i^{100}i = (i^4)^{25}i = 1^{25}i = i$

(c) $\sqrt{-12} + \sqrt{-27} = \sqrt{12}\sqrt{-1} + \sqrt{27}\sqrt{-1}$
$$= \sqrt{4}\sqrt{3}i + \sqrt{9}\sqrt{3}i$$
$$= 2\sqrt{3}i + 3\sqrt{3}i = 5\sqrt{3}i$$

(d) $\sqrt{-9}\sqrt{-4} = (3i)(2i) = 6i^2 = -6$

Note $\sqrt{-9}\sqrt{-4} \neq \sqrt{36}$. Why?   ▲

Near the beginning of this section, we mentioned that we'd eventually present a formal list of some of the basic definitions and properties that can be used to develop the subject more rigorously. These are given in the two boxes that follow. As you'll see in some of the exercises, these definitions and properties are indeed consistent with our work in this section and with the properties of real numbers that are listed in Section A.2 of the Appendix.

Addition, Subtraction, Multiplication, and Division for Complex Numbers

Let $z = a + bi$ and $w = c + di$. Then $z + w$, $z - w$, and zw are defined as follows.

1. $z + w = (a + bi) + (c + di) = (a + c) + (b + d)i$

2. $z - w = (a + bi) - (c + di) = (a - c) + (b - d)i$

3. $zw = (a + bi)(c + di) = (ac - bd) + (ad + bc)i$

Furthermore, if $w \neq 0$, then z/w is defined as follows.

4. $\dfrac{z}{w} = \dfrac{a + bi}{c + di} \cdot \dfrac{c - di}{c - di} = \left(\dfrac{ac + bd}{c^2 + d^2}\right) + \left(\dfrac{bc - ad}{c^2 + d^2}\right)i$

Definition and Properties of Complex Conjugates

The **complex conjugate** of $z = a + bi$ is $\bar{z} = a - bi$.
The complex conjugate has the following properties.

1. $\bar{\bar{z}} = z$

2. $z = \bar{z}$ if and only if z is a real number

3. $\overline{z + w} = \bar{z} + \bar{w}$; $\overline{z - w} = \bar{z} - \bar{w}$

4. $\overline{zw} = \bar{z}\bar{w}$; $\dfrac{\bar{z}}{\bar{w}} = \overline{\left(\dfrac{z}{w}\right)}$

5. $(\bar{z})^n = \overline{(z^n)}$ for each natural number n

▼ EXERCISE SET 1.9

A

1. Complete the table.

i^2	i^3	i^4	i^5	i^6	i^7	i^8
-1						

2. Simplify the expression $1 + 3i - 5i^2 + 4 - 2i - i^3$, and write the answer in the form $a + bi$.

For Exercises 3 and 4, specify the real and imaginary parts of each complex number.

3. (a) $4 + 5i$ (b) $4 - 5i$
    (c) $\frac{1}{2} - i$ (d) $16i$

4. (a) $-2 + \sqrt{7}i$ (b) $1 + 5^{1/3}i$
    (c) $-3i$ (d) 0

5. Determine the real numbers c and d such that
$$8 - 3i = 2c + di$$

6. Determine the real numbers a and b such that
$$27 - 64i = a^3 - b^3i$$

7. Simplify each expression.
    (a) $(5 - 6i) + (9 + 2i)$ (b) $(5 - 6i) - (9 + 2i)$

8. If $z = 1 + 4i$, compute $z - 10i$.

In Exercises 9 and 10, compute each product or quotient.

9. (a) $(3 - 4i)(5 + i)$ (b) $(5 + i)(3 - 4i)$
    (c) $\dfrac{3 - 4i}{5 + i}$ (d) $\dfrac{5 + i}{3 - 4i}$

10. (a) $(2 + 7i)(2 - 7i)$ (b) $\dfrac{-1 + 3i}{2 + 7i}$
    (c) $\dfrac{1}{2 + 7i}$ (d) $\dfrac{1}{2 + 7i} \cdot (-1 + 3i)$

In Exercises 11–36, evaluate the expression using the values $z = 2 + 3i$, $w = 9 - 4i$, and $w_1 = -7 - i$.

11. (a) $z + w$ (b) $\bar{z} + w$ (c) $z + \bar{z}$

12. (a) $\bar{z} + \bar{w}$ (b) $\overline{(z + w)}$ (c) $w - \bar{w}$

13. $(z + w) + w_1$ **14.** $z + (w + w_1)$

15. zw **16.** wz

17. $z\bar{z}$ **18.** $w\bar{w}$

19. $z(ww_1)$

20. $(zw)w_1$

21. $z(w + w_1)$

22. $zw + zw_1$

23. $z^2 - w^2$

24. $(z - w)(z + w)$

25. $(zw)^2$

26. z^2w^2

27. z^3

28. z^4

29. $\dfrac{z}{w}$

30. $\dfrac{w}{z}$

31. $\dfrac{\bar{z}}{\bar{w}}$

32. $\overline{\left(\dfrac{z}{w}\right)}$

33. $\dfrac{z}{\bar{z}}$

34. $\dfrac{\bar{z}}{z}$

35. $\dfrac{w - \bar{w}}{2i}$

36. $\dfrac{w + \bar{w}}{2}$

In Exercises 37–40, compute the quotient.

37. $\dfrac{i}{5 + i}$ **38.** $\dfrac{1 - i\sqrt{3}}{1 + i\sqrt{3}}$ **39.** $\dfrac{1}{i}$ **40.** $\dfrac{i + i^2}{i^3 + i^4}$

In Exercises 41–48, simplify the expression.

41. $\sqrt{-49} + \sqrt{-9} + \sqrt{-4}$ **42.** $\sqrt{-25} + i$

43. $\sqrt{-20} - 3\sqrt{-45} + \sqrt{-80}$

44. $\sqrt{-4}\sqrt{-4}$

45. $1 + \sqrt{-36}\sqrt{-36}$ **46.** $i - \sqrt{-100}$

47. $3\sqrt{-128} - 4\sqrt{-18}$ **48.** $64 + \sqrt{-64}\sqrt{-64}$

49. Let $z = a + bi$ and $w = c + di$. Compute each result and then check that it agrees with the definition in the box on page 61.

   (a) $z + w$ **(b)** $z - w$ **(c)** zw **(d)** $\dfrac{z}{w}$

B

50. Show that $\left(\dfrac{-1 + i\sqrt{3}}{2}\right)^2 + \left(\dfrac{-1 - i\sqrt{3}}{2}\right)^2 = -1$.

51. Let $z = \dfrac{-1 + i\sqrt{3}}{2}$ and $w = \dfrac{-1 - i\sqrt{3}}{2}$. Verify each statement.

   (a) $z^3 = 1$ and $w^3 = 1$ **(b)** $zw = 1$
   (c) $z = w^2$ and $w = z^2$
   (d) $(1 - z + z^2)(1 + z - z^2) = 4$

52. Let $z = a + bi$ and $w = c + di$.
   (a) Show that $\bar{\bar{z}} = z$.
   (b) Show that $\overline{(z + w)} = \bar{z} + \bar{w}$.

53. Show that the complex number $0 (= 0 + 0i)$ has the following properties for all complex numbers z.
   (a) $0 + z = z$ and $z + 0 = z$ *Hint:* Let $z = a + bi$.
   (b) $0 \cdot z = 0$ and $z \cdot 0 = 0$

54. This exercise indicates one of the reasons why multiplication of complex numbers is not carried out by simply multiplying the corresponding real and imaginary parts of the numbers. (Recall that addition and subtraction *are* carried out in this manner.) Suppose for the moment that we were to

define multiplication in this seemingly less complicated way:

$$(a + bi)(c + di) = ac + (bd)i \qquad (*)$$

   (a) Compute $(2 + 3i)(5 + 4i)$, assuming that multiplication is defined by $(*)$.
   (b) Still assuming that multiplication is defined by $(*)$, find two complex numbers z and w such that $z \neq 0$, $w \neq 0$, but $zw = 0$ (where 0 denotes the complex number $0 + 0i$).

Notice that the result in part (b) is contrary to our expectation or desire that the product of two non-zero numbers should be nonzero, as is the case for real numbers. It can be shown that when multiplication is carried out as described in the text, then the product of two complex numbers is nonzero if and only if both factors are nonzero.

55. **(a)** Show that addition of complex numbers is commutative. That is, show that $z + w = w + z$ for all complex numbers z and w.   *Hint:* Let $z = a + bi$ and $w = c + di$.
   (b) Show that multiplication of complex numbers is commutative. That is, show that $zw = wz$ for all complex numbers z and w.

56. Let $z = a + bi$.
   (a) Show that $(\bar{z})^2 = \overline{(z^2)}$.
   (b) Show that $(\bar{z})^3 = \overline{(z^3)}$.

C

57. Let a and b be real numbers. Find the real and imaginary parts of the quantity $\dfrac{a + bi}{a - bi} + \dfrac{a - bi}{a + bi}$.

58. Find the real and imaginary parts of the quantity

$$\left(\dfrac{a + bi}{a - bi}\right)^2 - \left(\dfrac{a - bi}{a + bi}\right)^2$$

59. Find the real part of $\dfrac{(a + bi)^2}{a - bi} - \dfrac{(a - bi)^2}{a + bi}$.

60. **(a)** Let

$$\alpha = \left(\dfrac{\sqrt{a^2 + b^2} + a}{2}\right)^{1/2}$$

   and

$$\beta = \left(\dfrac{\sqrt{a^2 + b^2} - a}{2}\right)^{1/2}$$

   Show that the square of the complex number $\alpha + \beta i$ is $a + bi$.
   (b) Use the result in part (a) to find a complex number z such that $z^2 = i$.
   (c) Use the result in part (a) to find a complex number z such that $z^2 = -7 + 24i$.

▼ CHAPTER ONE   SUMMARY OF PRINCIPAL DEFINITIONS AND FORMULAS

TERMS OR NOTATIONS	PAGE REFERENCE	COMMENTS
1. Natural numbers, integers, rational numbers, and irrational numbers	1	The box on page 1 provides both definitions and examples. Also note the theorem in the box that explains how to distinguish between rational and irrational numbers in terms of their decimal representations.
2. $a < b$ $b > a$	3	a is less than b. b is greater than a.
3. (a, b) $[a, b]$	3	The open interval (a, b) consists of all real numbers between a and b, excluding a and b. The closed interval $[a, b]$ consists of all real numbers between a and b, including a and b.
4. (a, ∞)	4	The unbounded interval (a, ∞) consists of all real numbers x such that $x > a$. The infinity sumbol, ∞, does not denote a real number. It is used in the context (a, ∞) to indicate that the interval has no right-hand boundary. For the definitions of the other types of unbounded intervals, see the box on page 4.
5. $|x|$	6, 7	The absolute value of a real number x is the distance of x from the origin. This is equivalent to the following algebraic definition: $$|x| = \begin{cases} x & \text{if } x \geq 0 \\ -x & \text{if } x < 0 \end{cases}$$
6. $a^0 = 1$ $a^{-n} = 1/a^n$ $a^{m/n} = \sqrt[n]{a^m} = (\sqrt[n]{a})^m$	11 12 19	These are the defining equations for zero as an exponent, for negative exponents, and for rational exponents. The definition $a^{m/n}$ presupposes that m/n is in lowest terms and that $\sqrt[n]{a}$ exists. The principal nth root of a, $\sqrt[n]{a}$, is defined in the box on page 14.
7. Constant, variable, and the domain convention	22	A constant is either a particular number (such as -8 or π) or a letter whose value remains fixed (although perhaps unspecified) throughout a given discussion. In contrast, a variable is a letter for which we can substitute any number from a given set of numbers. The given set is called the domain of the variable. According to the *domain convention*, the domain of a variable in a given expression consists of all real numbers for which the expression is defined.
8. Polynomial	23	A polynomial is an expression of the form $a_n x^n + a_{n-1} x^{n-1} + \cdots + a_1 x + a_0$. The individual expressions $a_k x^k$ making up the polynomial are called *terms*. The numbers a_k are called *coefficients*.
9. Degree of a polynomial	23	For a polynomial that is not simply a constant, the degree is the highest power (that is, exponent) of x that appears. The degree of a nonzero constant polynomial is zero. Degree is undefined for the zero polynomial.
10. Special products	24	The box on page 24 lists four basic types of products that you should memorize because they occur so frequently.
11. Factoring techniques	25	The techniques for factoring listed in the box on page 25 are fundamental for simplifying expressions and for solving equations.

TERMS OR NOTATIONS	PAGE REFERENCE	COMMENTS
12. Solution (or root) of an equation	29	A solution, or root, is a number that when substituted for the variable in an equation yields a true statement.
13. Quadratic equation	29	A quadratic equation is an equation that can be written in the form $ax^2 + bx + c = 0$ where $a \neq 0$.
14. Zero-product property	30	This property of real numbers (and of complex numbers in general) can be stated as follows. If $pq = 0$ then $p = 0$ or $q = 0$; conversely, if $p = 0$ or $q = 0$, then $pq = 0$. (We used this property in solving quadratic equations by factoring.)
15. Quadratic formula $$x = \frac{-b \pm \sqrt{b^2 - 4ac}}{2a}$$	32	The quadratic formula provides the solutions of the quadratic equation $ax^2 + bx + c = 0$. The formula is derived on page 31 by completing the square.
16. Discriminant	33	The discriminant of the quadratic equation $ax^2 + bx + c = 0$ is the number $b^2 - 4ac$. As indicated in the box on page 34, the discriminant provides information about the roots of the equation.
17. Extraneous root	34	In solving equations, certain processes (such as squaring both sides) can lead to answers that do not check in the original equation. Such answers are called extraneous roots; they are not solutions of the original equation.
18. Solution set of an inequality	37	This is the set of numbers that satisfy the inequality.
19. $A \cup B$	40	The set $A \cup B$ consists of all elements that are in at least one of the two sets A and B.
20. Key numbers of an inequality	41	Let P and Q denote polynomials and consider the following four inequalities: $$\frac{P}{Q} < 0 \qquad \frac{P}{Q} \leq 0 \qquad \frac{P}{Q} > 0 \qquad \frac{P}{Q} \geq 0$$ The key numbers for each of these inequalities are the real numbers for which $P = 0$ or $Q = 0$. It can be proved that the algebraic sign of P/Q is constant on each of the intervals determined by these key numbers.
21. Pythagorean theorem $a^2 + b^2 = c^2$	49	In a right triangle, the lengths of the sides are related by this equation, where c is the length of the hypotenuse. Conversely, if the lengths of the sides of a triangle are related by an equation of the form $a^2 + b^2 = c^2$, then the triangle is a right triangle, and c is the length of the hypotenuse.
22. $d = \sqrt{(x_2 - x_1)^2 + (y_2 - y_1)^2}$	50	d is the distance between (x_1, y_1) and (x_2, y_2).
23. $(x - h)^2 + (y - k)^2 = r^2$	51	This is the equation of a circle with center at (h, k) and radius r.
24. $\left(\dfrac{x_1 + x_2}{2}, \dfrac{y_1 + y_2}{2} \right)$	52	These are the coordinates of the midpoint of the line segment joining (x_1, y_1) and (x_2, y_2).
25. $i^2 = -1$	57	This is the defining equation for the symbol i. The quantity i is not a real number.

TERMS OR NOTATIONS	PAGE REFERENCE	COMMENTS
26. Complex numbers	57	An expression of the form $a + bi$, where a and b are real numbers, is a complex number. With the exception of division, complex numbers are combined using the usual rules of algebra, along with the convention $i^2 = -1$. Division is carried out using complex conjugates (defined below), as indicated in the box on page 59. All of the properties of real numbers listed in Section A.2 of the Appendix also apply to complex numbers.
27. If $z = c + di$, then $\bar{z} = c - di$.	58	$\bar{z}$ is the *conjugate* of z.

▼ CHAPTER ONE   REVIEW EXERCISES

NOTE   Exercises 1–15 constitute a chapter test on the fundamentals, based on group A problems.

1. Factor each expression.
  (a) $x - 81x^3$
  (b) $(x^2 + a^2)(x^3 + 3b^3 + 1) - (x^2 + a^2)(2b^3 + 1)$
  (c) $2(x^2 + 4)^{1/2} - 2x^2(x^2 + 4)^{-3/2}$

2. Rewrite using absolute values and inequality: The distance between x and 3 is less than 5.

3. Simplify each expression.
  (a) $\sqrt[3]{16} - \sqrt[3]{-250} + \sqrt[3]{-2}$ **(b)** $\left(\dfrac{a^{-3}b^4c^5}{b^0ca^{-1}}\right)^{-3/2}$
  (c) $8^{2/3} + 4^0 + \left(\frac{4}{9}\right)^{-3/2}$

4. Rationalize each denominator:
  (a) $\dfrac{\sqrt{2}}{\sqrt{5}}$ **(b)** $\dfrac{1}{1 + \sqrt{2}}$ **(c)** $\dfrac{4}{\sqrt[3]{4}}$

5. Compute the distance between the points $A(4, -6)$ and $B(-12, -4)$.

6. Determine all the real solutions for each equation.
  (a) $8x^2 - 10x = 3$ **(b)** $3x^2 + x - 1 = 0$
  (c) $2 - \sqrt{4 + 3x} - \sqrt{3 + 2x} = 0$

7. Simplify the expression $|x + 3| - |x + 2|$, given that x lies in the open interval $(-3, -2)$.

8. Write the expression $i^3 - i^4 + 1/i$ in the form $a + bi$, where a and b are real numbers.

9. Evaluate each expression using the values $z = 2 + 3i$ and $w = 1 - 4i$. Write each answer in the form $a + bi$.
  (a) $z^2 + w^2$ **(b)** $\dfrac{w}{z}$ **(c)** $\bar{z} + 3w$

10. Find a positive value for k such that the equation $3x^2 - kx + 5 = 0$ has exactly one real solution.

11. Find the equation of the circle with diameter $\overline{AB}$, given that the coordinates of A and B are $(-6, 1)$ and $(4, -5)$. Write the answer in standard form.

In Exercises 12 and 13, solve the inequality.

12. **(a)** $3 \le \dfrac{2 - 3x}{2} \le 5$ **(b)** $|x - 6| < 1$
  (c) $|2x - 4| \ge 2$

13. **(a)** $6x^3 + 11x^2 - 7x < 0$ **(b)** $\dfrac{x + 3}{x - 4} \ge 0$
  (c) $\dfrac{1}{x} - \dfrac{x}{x - 1} < 3$

14. Express the repeating decimal $5.\overline{7}$ in the form p/q, where p and q are integers.

15. Consider the expression $\sqrt{a}\,\sqrt[4]{b^3}$, where a and b are positive.
  (a) Rewrite the expression using rational exponents.
  (b) Rewrite the expression using only one radical sign.

In Exercises 16–34 factor the expression.

16. $a^2 - 16b^7$ **17.** $x^2 + ax + yx + ay$
18. $8 - (a + 1)^3$ **19.** $x^2 - 18x + 81$
20. $a^2x^3 + 2ax^2b + b^2x$ **21.** $8a^2x^2 + 16a^3$
22. $8x^2 + 6x + 1$ **23.** $12x^2 - 2x - 4$
24. $a^4x^4 - x^8a^8$ **25.** $2x^2 - 2bx + ax - ab$
26. $8 + 12a + 6a^2 + a^3$
27. $(x^2 + 2x - 8)^2 - (2x + 1)^2$
28. $4x^2y^2z^3 - 3xyz^3 - z^3$
29. $(x + y - 1)^2 - (x - y + 1)^2$
30. $1 - x^6$ **31.** $12x^3 + 44x^2 - 16x$

32. $a^2x^2 + 2abx + b^2 - 4a^2b^2x^2$

33. $a^2 - b^2 + ac - bc + a^2b - b^2a$

34. $5(a + 1)^2 + 29(a + 1) - 144$

In Exercises 35–50 simplify the expression. (Assume that the letters represent positive quantities.)

35. $4^{3/2}$

36. $49^{-1/2}$

37. $[(3025)^{1/2}]^0$

38. $(-125)^{2/3}$

39. $8^{-4/3}$

40. $10^{-3} + 10^3$

41. $(-243)^{-2/5}$

42. $\left(\frac{625}{16}\right)^{3/4}$

43. $(a^2b^6c^8)^{1/2}$

44. $\sqrt{25a^4b^{10}}$

45. $\sqrt{a^3b^5}\sqrt{4ab^3}$

46. $\sqrt{28a^3} + a\sqrt{63a}$

47. $\sqrt[3]{16} - \sqrt[3]{-54}$

48. $\sqrt[5]{\frac{-32a^{15}b^{10}}{c^5}}$

49. $\sqrt{24a^2b^3} + ba\sqrt{54b}$

50. $\sqrt{\sqrt{256x^8}}$

In Exercises 51–54 simplify the expression. (Do not assume that the letters represent positive quantities.)

51. $\sqrt{t^2}$

52. $\sqrt[3]{t^3}$

53. $\sqrt[4]{16x^4}$

54. $\sqrt[3]{8x^4}$

In Exercises 55–58, rewrite each expression in two ways:
(a) *using rational exponents;*
(b) *using a single radical.*
Assume that all letters represent nonnegative quantities.

55. $\sqrt[3]{x}\sqrt[4]{x^3}$

56. $\sqrt[4]{\sqrt[3]{x}}$

57. $\sqrt{\sqrt[3]{t}\sqrt[5]{t^4}}$

58. $\sqrt[3]{\frac{\sqrt{a}\sqrt[4]{b^2}}{\sqrt[5]{c^3}}}$

Rationalize the denominator in Exercises 59–64.

59. $\frac{6}{\sqrt{3}}$

60. $\frac{\sqrt{3}}{\sqrt{2}}$

61. $\frac{1}{\sqrt{6} - \sqrt{3}}$

62. $\frac{1}{\sqrt{x + 2} - \sqrt{x}}$

63. $\frac{\sqrt{a^2 + x^2} + \sqrt{a^2 - x^2}}{\sqrt{a^2 + x^2} - \sqrt{a^2 - x^2}}$

64. **(a)** $\frac{1}{\sqrt{a} + \sqrt{b} + \sqrt{c}}$

Hint: First multiply numerator and denominator by $\sqrt{a} + \sqrt{b} - \sqrt{c}$. Then multiply numerator and denominator by $(a + b - c) - 2\sqrt{ab}$.

(b) $\frac{1}{\sqrt{2} + \sqrt{3} + \sqrt{5}}$

In Exercises 65–68, rationalize the numerator.

65. $\frac{\sqrt{x} - 5}{x - 25}$

66. $\frac{\sqrt{t} - a}{t - a^2}$

67. $\frac{\sqrt{a^2 + x^2} - \sqrt{a^2 - x^2}}{\sqrt{a^2 + x^2} + \sqrt{a^2 - x^2}}$

68. $\frac{\sqrt[3]{x} - 2}{x - 8}$

Hint: The denominator can be written $(x^{1/3})^3 - 2^3$.

In Exercises 69–78, use the Special Products in Section 1.5 to compute the product.

69. $(9x - y)(9x + y)$

70. $(3x + 1)^2$

71. $(3x^2 + y^2)^2$

72. $(x^2 - 2)^3$

73. $(1 - 3a)(1 + 3a + 9a^2)$

74. $(x^n + y^n)(x^{2n} - x^ny^n + y^{2n})$

75. $(x^{1/2} - y^{1/2})(x^{1/2} + y^{1/2})$

76. $(\sqrt{x} - \sqrt{3})(\sqrt{x} + \sqrt{3})$

77. $(x^{1/3} + y^{1/3})(x^{2/3} - x^{1/3}y^{1/3} + y^{2/3})$

78. $(x^2 - 3x + 1)^2$ *Hint:* Rewrite the expression as $[x^2 - (3x - 1)]^2$.

For Exercises 79–86, carry out the indicated operations and express the answer in the standard form $a + bi$.

79. $(3 - 2i)(3 + 2i) + (1 + 3i)^2$

80. $2i(1 + i)^2$

81. $(1 + i\sqrt{2})(1 - i\sqrt{2}) + (\sqrt{2} + i)(\sqrt{2} - i)$

82. $\frac{2 + 3i}{1 + i}$

83. $\frac{3 - i\sqrt{3}}{3 + i\sqrt{3}}$

84. $\frac{1 + i}{1 - i} + \frac{1 - i}{1 + i}$

85. $-\sqrt{-2}\sqrt{-9} + \sqrt{-8} - \sqrt{-72}$

86. $\frac{\sqrt{-4} - \sqrt{-3}\sqrt{-3}}{\sqrt{-100}}$

87. **(a)** The **real part** of a complex number z is denoted by $\text{Re}(z)$. For instance, $\text{Re}(2 + 5i) = 2$. Show that for any complex number z, we have $\text{Re}(z) = \frac{1}{2}(z + \bar{z})$.   *Hint:* Let $z = a + bi$.

(b) The **imaginary part** of a complex number z is denoted by $\text{Im}(z)$. For instance, $\text{Im}(2 + 5i) = 5$. Show that for any complex number z, we have

$$\text{Im}(z) = \frac{1}{2i}(z - \bar{z}).$$

88. The **absolute value** of the complex number $a + bi$ is defined by $|a + bi| = \sqrt{a^2 + b^2}$.

(a) Compute $|6 + 2i|$ and $|6 - 2i|$.

(b) As you know, the absolute value of the real number -3 is 3. Now write -3 in the form $-3 + 0i$ and compute its absolute value using the new definition. (Observe that the two results agree.)

(c) Let $z = a + bi$. Show that $z\bar{z} = |z|^2$.

For Exercises 89–93, rewrite the statement using absolute values and an inequality or equality.

89. The distance between x and 6 is 2.

90. The distance between x and a is less than $\frac{1}{2}$.

91. The distance between a and b is 3.

92. The distance between x and -1 is 5.

93. The distance between x and 0 exceeds 10.

94. What can you say about x if $|x - 5| = 0$?

In Exercises 95–100, rewrite the expression in a form that does not contain absolute values.

95. $|\sqrt{6} - 2|$ **96.** $|2 - \sqrt{6}|$ **97.** $|x^4 + x^2 + 1|$

98. (a) $|x - 3|$ if $x < 3$ (b) $|x - 3|$ if $x > 3$

99. $|x - 2| + |x - 3|$ if:
(a) $x < 2$ (b) $2 < x < 3$ (c) $x > 3$

100. $|x + 2| + |x - 1|$ if:
(a) $x < -2$ (b) $-2 < x < 1$ (c) $x > 1$

In Exercises 101 and 102, answer T if the statement is true without exception. Otherwise, answer F.

101. (a) If $x < 3$, then $x + 7 < 10$.
(b) If $x < y$, then $x^2 < y^2$.
(c) If $x \geq -4$, then $x \leq 4$.
(d) If $-x > y$, then $x < -y$.
(e) If $x < 2$, then $\dfrac{1}{x} < \dfrac{1}{2}$ (f) $x \leq x^2$

102. (a) If $\sqrt{x + 1} = 3$, then $x + 1 = 9$.
(b) If $0 < x < 1$, then $\dfrac{1}{x} > x$.
(c) If $x < 2$ and $y < 3$, then $x < y$.
(d) If $x < 2$ and $y > 3$, then $x < y$.
(e) If $a - b \leq a^2 - b^2$, then $1 \leq a + b$.
(f) If $0 < a - b \leq a^2 - b^2$, then $1 \leq a + b$.

103. The four corners of a square $ABCD$ have been cut off to form a regular octagon, as shown in the figure. If each side of the square is 1 cm long, how long is each side of the octagon?

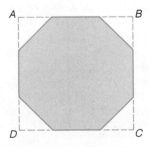

104. Determine p if the larger root of the equation $x^2 + px - 2 = 0$ is p.

105. A piece of wire x cm long is bent into a square. For which values of x will the area be (numerically) greater than the perimeter?

106. A piece of wire $6x$ cm long is bent into an equilateral triangle. For which values of x will the area be (numerically) less than the perimeter?

In Exercises 107–128, solve the equation. In Exercises 119–128, work exclusively within the real number system.

107. $t^2 - 2t - 99 = 0$ **108.** $12x^2 + 2x - 2 = 0$

109. $4y^2 - 21y = 18$ **110.** $\frac{1}{2}x^2 + x - 12 = 0$

111. $\dfrac{1}{1 - x} + \dfrac{4}{2 - x} = \dfrac{11}{6}$ **112.** $\dfrac{2x + 1}{1 - 2x} + \dfrac{1 - 2x}{2x + 1} = 2$

113. $\dfrac{x^2}{(x - 1)(x + 1)} = \dfrac{4}{x + 1} + \dfrac{4}{(x - 1)(x + 1)}$

114. $x^4 - 7x^2 + 12 = 0$ *Hint:* Factor.

115. $y^4 - 8y^2 + 11 = 0$ *Hint:* Let $t = y^2$.

116. $x^3 - 6x^2 + 7x = 0$

117. $1 + 14x^{-1} + 48x^{-2} = 0$ *Hint:* Multiply by x^2.

118. $x^{-2} - x^{-1} - 1 = 0$

119. $\sqrt{4x + 3} = \sqrt{11 - 8x} - 1$

120. $\sqrt{\sqrt{x - 4} + x} = 2$

121. $\sqrt{5 - 2x} - \sqrt{2 - x} - \sqrt{3 - x} = 0$

122. $2 - \sqrt{3\sqrt{2x - 1}} + x = 0$

123. $\sqrt{x + 48} - \sqrt{x} = 4$

124. $\sqrt{x + 6} - \sqrt{x} - \sqrt{2} = 0$

125. $3(4t - 1)^{1/2} - (2t)^{1/2} = (3 + 2t)^{1/2}$

126. $1 + \sqrt{2x^2 + 5x - 9} = \sqrt{2x^2 + 5x - 2}$

127. $\dfrac{2}{\sqrt{x^2 - 36}} + \dfrac{1}{\sqrt{x + 6}} - \dfrac{1}{\sqrt{x - 6}} = 0$

128. $\sqrt{x - \sqrt{1 - x}} + \sqrt{x - 1} = 0$

In Exercises 129–136, solve the equation for x in terms of the other letters.

129. $\frac{1}{4}x^2 + bx - 8b^2 = 0$

130. $\dfrac{1}{a + x} + \dfrac{1}{b + x} = \dfrac{a + b}{ab}$ $\quad (a + b \neq 0)$

131. $4x^2y^2 - 4xy = -1$ $\quad (y \neq 0)$

132. $4x^4y^4 + 2x^2y^2 + \frac{1}{4} = 0$ $\quad (y \neq 0)$

133. $x + \dfrac{1}{a} - \dfrac{1}{b} = \dfrac{2}{a^2x} + \dfrac{2}{abx}$

134. $\dfrac{a}{x - b} + \dfrac{b}{x - a} = \dfrac{x - a}{b} + \dfrac{x - b}{a}$ $\quad (a \neq -b)$

Suggestion: Before clearing fractions, carry out the indicated additions on each side of the equation.

135. $\dfrac{1}{x+a+b} = \dfrac{1}{x} + \dfrac{1}{a} + \dfrac{1}{b}$ $(a+b \neq 0)$

136. $\dfrac{1}{x} + \dfrac{1}{a-x} + \dfrac{1}{x+3a} = 0$ $(a \neq 0)$

In Exercises 137–150, solve the inequality.

137. $x^2 - 21x + 108 \leq 0$ **138.** $x^2 + 3x - 40 < 0$

139. $x^2 \geq 15x$

140. $(x+12)(x-1)(x-8) < 0$

141. $(x-4)^2(x+8)^3 \geq 0$ **142.** $625 - x^4 \leq 0$

143. $\dfrac{x+12}{x-5} > 0$

144. $\dfrac{(x-6)^2(x-8)(x+3)}{(x-3)^2} \leq 0$

145. $\dfrac{x^2 - 10x + 9}{x^3 + 1} \leq 0$ **146.** $\dfrac{3x+1}{x-4} < 1$

147. $\dfrac{1-2x}{1+2x} \leq \dfrac{1}{2}$ **148.** $\dfrac{x}{x-2} + \dfrac{1}{x-1} \leq \dfrac{23}{12}$

149. $x^2 + \dfrac{1}{x^2} > 3$

    Suggestion: Use a calculator to evaluate the key numbers.

150. $\sqrt{x} - \dfrac{5}{\sqrt{x}} \leq 4$

For Exercises 151–154, find the values of k for which the roots of the equation are real numbers.

151. $kx^2 - 6x + 5 = 0$ **152.** $x^2 + x + k^2 = 0$

153. $kx(x+2) = -1$ **154.** $x^2 + (k+1)x + 2k = 0$

155. In the figure, points P and Q trisect the hypotenuse in $\triangle ABC$. Prove that the square of the hypotenuse is equal to $\frac{9}{5}$ the sum of the squares of the distances from the trisection points to the vertex A of the right angle. *Hint:* Let the coordinates of B and C be $(0, 3b)$ and $(3c, 0)$, respectively. Then the coordinates of P and Q are $(c, 2b)$ and $(2c, b)$, respectively.

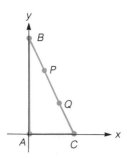

156. Figure (a) shows a triangle with sides of lengths s, t, and u and a median of length m. (A **median** of a

triangle is a line segment drawn from a vertex to the midpoint of the opposite side.) Prove that

$$m^2 = \frac{1}{2}(s^2 + t^2) - \frac{1}{4}u^2$$

Hint: Set up a coordinate system as indicated in Figure (b). Then each of the quantities m^2, s^2, t^2, and u^2 can be computed in terms of a, b, and c.

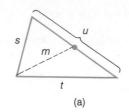

(a)

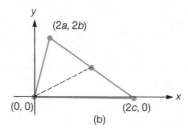

(b)

In Exercises 157 and 158, make use of the following result. If an object is thrown vertically upward from a height of h_0 feet with an initial speed of v_0 ft/sec, then its height h (in feet) after t seconds is given by

$$h = -16t^2 + v_0 t + h_0$$

157. (a) If a ball is thrown vertically upward from ground level with an initial speed of 64 ft/sec, at what times will the height be 15 ft?

    (b) For the ball described in part (a), for how long an interval of time will the height of the ball exceed 63 ft?

    (c) A ball is thrown vertically upward from a height of 50 ft with an initial speed of 40 ft/sec. At the same instant, another ball is thrown vertically upward from a height of 100 ft with an initial speed of 5 ft/sec. Which ball hits the ground first?

158. An object is projected vertically upward. Suppose that its height is H ft at t_1 sec and again at t_2 sec. Express the initial speed in terms of t_1 and t_2.

                                 Answer: $16(t_1 + t_2)$

Exercise 159 appears in the text Introduction to Algebra *by George Chrystal, first published in 1898.*

159. A circle is inscribed in a quadrant of a circle of radius r, as shown in the figure on the next page. Find the radius of the inscribed circle.

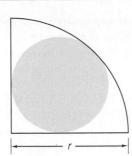

$$\longleftarrow\ r \longrightarrow$$

C **160.** **(a)** Evaluate the expression $\dfrac{\sqrt{3-\sqrt{5}}}{\sqrt{2}+\sqrt{7-3\sqrt{5}}}$.

(b) Evaluate the expression $\sqrt{5}/5$. If you do the calculator work carefully, your results in parts (a) and (b) will suggest (but not prove!) that the two expressions are equal. Follow steps (c), (d), and (e) to prove that the two expressions are indeed equal.

(c) Multiply the expression in part (a) by

$\dfrac{\sqrt{2}-\sqrt{7-3\sqrt{5}}}{\sqrt{2}-\sqrt{7-3\sqrt{5}}}$, which equals 1. After

carrying out the indicated operations, you should obtain

$$\frac{\sqrt{6-2\sqrt{5}}-2\sqrt{9-4\sqrt{5}}}{-5+3\sqrt{5}}$$

(d) Verify that $\sqrt{6-2\sqrt{5}}=\sqrt{5}-1$. Also verify that $\sqrt{9-4\sqrt{5}}=\sqrt{5}-2$.

(e) Use the results in part (d) to simplify the expression in part (c). The final result, after rationalizing the denominator, should be $\sqrt{5}/5$.

Exercise 161 (including the clever solution that is outlined) appears in the book The USSR Olympiad Problem Book *by D. O. Shklarsky, et al. (San Francisco: W. H. Freeman and Co., 1962).*

161. Follow steps (a)–(d) to solve the equation

$$\sqrt{a-\sqrt{a+x}}=x \qquad (a\geq 1)$$

(a) By squaring as usual, obtain the equation

$$x^4-2ax^2-x+a^2-a=0$$

(b) Although the equation obtained in part (a) is a fourth-degree equation in x, it is only a quadratic in a. Solve for a in terms of x to obtain

$$a=x^2+x+1 \quad \text{or} \quad a=x^2-x$$

(c) Solve each of the equations in part (b) for x in terms of a.

(d) Check your results in part (c) to eliminate any extraneous roots.

CHAPTER TWO

FUNCTIONS AND GRAPHS

In the first two sections of this chapter, we summarize and expand upon some of the work you have done on graphing in previous courses. The graphs we look at here will provide an important source of examples for concepts in precalculus and calculus. In addition, familiarity with some simple equations and their graphs will provide you with a strong intuitive background for the crucial topic of functions, beginning in Section 2.3. Sections 2.4 and 2.5 develop techniques that allow us to graph functions with a minimum of calculation. Finally, Sections 2.6 and 2.7 show that two functions can be combined to produce new functions. One way to combine two functions is known as *composition of functions*. As you will see, this is the unifying theme between Sections 2.6 and 2.7.

2.1 ▼ GRAPHS AND SYMMETRY

Symmetry is a working concept. If all the object is symmetrical, then the parts must be halves (or some other rational fraction) and the amount of information necessary to describe the object is halved (etc.).

Alan L. Macay, Department of Crystallography, University of London

One of the most useful ways to obtain information about an equation relating two variables is through a graph. The key definition is as follows.

> The **graph** of an equation in two variables is the set of all points whose coordinates satisfy the equation.

DEFINITION

The Graph of an Equation

TABLE 1 $y = 3x - 2$

x	y
0	−2
1	1
2	4
3	7
−1	−5
−2	−8

Suppose, for example, that we want to graph the equation $y = 3x - 2$. We begin by noting that the domain of the variable x in the expression $3x - 2$ is the set of all real numbers. Now we choose (at will) various values for x and in each case compute the corresponding y-value from the equation $y = 3x - 2$. For example, if x is zero, then $y = 3(0) - 2 = -2$. Table 1 summarizes the results of some of these calculations. The first line in Table 1 tells us that the point with coordinates $(0, -2)$ is on the graph of $y = 3x - 2$. Reading down the table, we see that some other points on the graph are $(1, 1)$, $(2, 4)$, $(3, 7)$, $(-1, -5)$, and $(-2, -8)$.

We now plot (locate) the points we have determined and note, in this example, that they all appear to lie on a straight line. We draw the line indicated; this is the graph of $y = 3x - 2$ (see Figure 1). In Example 1, we ask a simple question that will test your understanding of the process we've just described and of graphing in general.

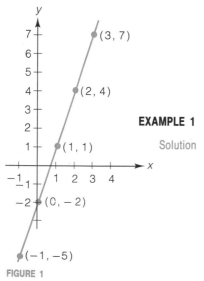

FIGURE 1
$y = 3x - 2$

EXAMPLE 1

Does the point $(1\frac{2}{3}, 2\frac{2}{3})$ lie on the graph of $y = 3x - 2$?

Solution

Looking at the graph in Figure 1, we certainly see that the point $(1\frac{2}{3}, 2\frac{2}{3})$ is very close to the line. But on this visual inspection alone, we cannot be certain that this point actually lies on the line. To settle the question, then, we check to see if the values $x = 1\frac{2}{3} = \frac{5}{3}$ and $y = 2\frac{2}{3} = \frac{8}{3}$ together satisfy the equation $y = 3x - 2$. We have

$$\frac{8}{3} \stackrel{?}{=} 3\left(\frac{5}{3}\right) - 2$$

$$\frac{8}{3} \stackrel{?}{=} 3 \qquad \text{No!}$$

Since the coordinates $(1\frac{2}{3}, 2\frac{2}{3})$ evidently do not satisfy the equation $y = 3x - 2$, we conclude that the point $(1\frac{2}{3}, 2\frac{2}{3})$ is not on the graph. [The calculation further tells us that the point $(1\frac{2}{3}, 3)$ *is* on the graph.] ▲

The next example illustrates a type of graph that often appears in beginning calculus courses (to illustrate the notion of a *limit*).

EXAMPLE 2   Graph the equation: $y = \dfrac{3x^2 - 5x + 2}{x - 1}$

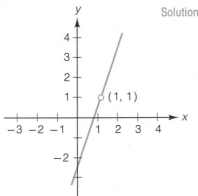

FIGURE 2
$y = \dfrac{3x^2 - 5x + 2}{x - 1}$

Solution   We must do two things here before we set up a table of values. First, note that the domain of the expression on the right-hand side of the given equation consists of all real numbers except $x = 1$. Thus, however the resulting graph may look, it cannot contain a point whose x-coordinate is 1. Next, can we simplify the given expression? (If so, that would save work when we make the calculations for our table.) In this case, we can simplify things as follows:

$$y = \frac{3x^2 - 5x + 2}{x - 1} = \frac{(3x - 2)(x - 1)}{x - 1} = 3x - 2$$

So, to graph the given equation we need only graph the simpler equation $y = 3x - 2$, remembering that the point with an x-coordinate of 1 must be excluded from the graph. We've previously done the work for $y = 3x - 2$—see Figure 1. Thus, the required graph is obtained by deleting the point $(1, 1)$ from the graph in Figure 1. This is shown in Figure 2. [The open circle in Figure 2 is used to mean that the point $(1, 1)$ does not belong to the graph.] ▲

When we graph an equation, it's helpful to know where the curve or line crosses the x- or y-axis. By a **y-intercept** of a graph, we mean the y-coordinate of a point where the graph crosses the y-axis. For instance, returning to Figure 1, the y-intercept is -2. Notice that this y-intercept can be obtained algebraically just by setting x equal to zero in the given equation $y = 3x - 2$. In a similar fashion, an **x-intercept** of a graph is the x-coordinate of a point where the graph crosses the x-axis. In Figure 1 we can see that the x-intercept is a number between 0 and 1. To determine this x-intercept, set y equal to zero in the equation $y = 3x - 2$. That yields $0 = 3x - 2$ and, consequently, $x = \frac{2}{3}$. So, the x-intercept of the graph in Figure 1 is $\frac{2}{3}$.

EXAMPLE 3   Figure 3 shows a computer-generated graph of the equation $y = 2x^3 - 2x^2 - x$. Determine the x-intercepts of the graph.

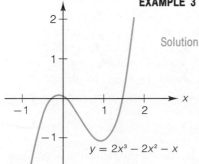

FIGURE 3
$y = 2x^3 - 2x^2 - x$

Solution   At the points where the graph crosses the x-axis, the y-coordinates are zero. Setting y equal to zero in the given equation yields

$$2x^3 - 2x^2 - x = 0$$
$$x(2x^2 - 2x - 1) = 0$$

Thus, $x = 0$ or $2x^2 - 2x - 1 = 0$. This last equation can be solved by using the quadratic formula. As you should check for yourself, the roots are $(1 \pm \sqrt{3})/2$. We've now found the three x-intercepts of the graph in Figure 3: they are 0, $(1 + \sqrt{3})/2$, and $(1 - \sqrt{3})/2$.   *Suggestion:* Use your calculator to check that these last two values are consistent with Figure 3. ▲

In elementary graphing there is always the question of how many points must be plotted before the essential features of a graph are clear. As you'll see throughout this text, there are a number of techniques and concepts that make it unnecessary to plot a large number of points. In this section we introduce three types of *symmetry* that are useful in graphing equations.

Three Types of Symmetry

TYPE OF SYMMETRY	EXAMPLE	DEFINITION
1. Symmetry about the x-axis	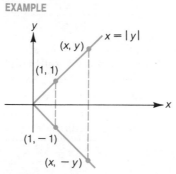 FIGURE 4 Symmetry about the x-axis	For each point (x, y) on the graph, the point $(x, -y)$ is also on the graph. The points (x, y) and $(x, -y)$ are *reflections of one another about* (or *in*) *the x-axis*. In Figure 4, the portions of the graph above and below the x-axis also are said to be reflections of one another about the x-axis.
2. Symmetry about the y-axis	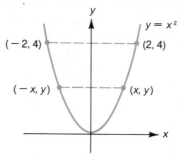 FIGURE 5 Symmetry about the y-axis	For each point (x, y) on the graph, the point $(-x, y)$ is also on the graph. The points (x, y) and $(-x, y)$ are *reflections of one another about* (or *in*) *the y-axis*. In Figure 5, the portions of the graph to the right and left of the y-axis also are said to be reflections of one another about the y-axis.
3. Symmetry about the origin	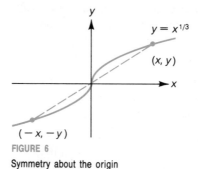 FIGURE 6 Symmetry about the origin	For each point (x, y) on the graph, the point $(-x, -y)$ is also on the graph. The points (x, y) and $(-x, -y)$ are *reflections of one another about the origin*. In Figure 6, the first- and third-quadrant portions of the curve also are said to be reflections of one another about the origin. [In terms of the two previous symmetries, the point $(-x, -y)$ can be obtained from (x, y) as follows: first reflect (x, y) about the y-axis to obtain $(-x, y)$, then reflect $(-x, y)$ about the x-axis to obtain $(-x, -y)$.]

EXAMPLE 4 A line segment $\mathscr{L}$ has endpoints $(1, 2)$ and $(5, 3)$. Sketch the reflection of $\mathscr{L}$ about:

(a) the x-axis.   **(b)** the y-axis.   **(c)** the origin.

FIGURE 7

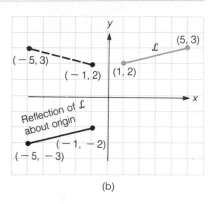

(a)

(b)

Solution (a) First reflect the endpoints $(1, 2)$ and $(5, 3)$ about the x-axis to obtain the new endpoints $(1, -2)$ and $(5, -3)$, respectively. Then join these two points as indicated in Figure 7(a), in quadrant IV.

(b) Reflect the given endpoints about the y-axis to obtain the new endpoints $(-1, 2)$ and $(-5, 3)$; then join these two points as shown in Figure 7(a), in quadrant II.

(c) As described in the box preceding this example, reflection about the origin can be carried out in two steps: first reflect about the y-axis, then reflect about the x-axis. In part (b) we obtained the reflection of $\mathscr{L}$ in the y-axis. So now we need only to reflect the line segment obtained in part (b) about the x-axis. See Figure 7(b). ▲

In the box that follows we list three rules for testing whether the graph of an equation possesses any of the types of symmetry we've been discussing. (The validity of each rule follows directly from the definitions of symmetry.)

Three Tests for Symmetry

> **1.** The graph of an equation is symmetric about the y-axis if replacing x with $-x$ yields an equivalent equation.
>
> **2.** The graph of an equation is symmetric about the x-axis if replacing y with $-y$ yields an equivalent equation.
>
> **3.** The graph of an equation is symmetric about the origin if replacing x and y with $-x$ and $-y$, respectively, yields an equivalent equation.

EXAMPLE 5 (a) Test for symmetry about the y-axis: $y = x^4 - 3x^2 + 1$

(b) Test for symmetry about the x-axis: $x = \dfrac{y^2}{y - 1}$

Solution (a) Replacing x with $-x$ in the original equation gives us

$$y = (-x)^4 - 3(-x)^2 + 1$$

or

$$y = x^4 - 3x^2 + 1$$

Since this last equation is the same as the original equation, we conclude that the graph is symmetric about the y-axis.

(b) Replacing y with $-y$ in the original equation gives us

$$x = \frac{(-y)^2}{(-y) - 1}$$

or

$$x = \frac{y^2}{-y - 1}$$

Since this last equation is not equivalent to the original equation, we conclude that the graph is not symmetric about the x-axis. ▲

EXAMPLE 6   Is the graph of $y = 2x^3 - x$ symmetric about the origin?

Solution   Replacing x and y with $-x$ and $-y$, respectively, gives us

$$-y = 2(-x)^3 - (-x) = -2x^3 + x$$
$$y = 2x^3 - x \quad \text{multiplying through by } -1$$

This last equation is identical to the given equation. Therefore the graph is symmetric about the origin. ▲

In the next two examples we use the notions of symmetry and intercepts as guides for drawing the required graphs.

EXAMPLE 7   Graph the equation: $y = -x^2 + 5$

Solution   The domain of the variable x in the expression $-x^2 + 5$ is the set of all real numbers. However, since the graph must be symmetric about the y-axis (why?), we need only to be able to sketch the graph to the right of the y-axis; the portion to the left will then be the mirror image of this (with the y-axis as the mirror). The x- and y-intercepts (if any) are computed as follows:

x-INTERCEPTS

$$-x^2 + 5 = 0$$
$$x^2 = 5$$
$$x = \pm\sqrt{5} \ (\approx \pm 2.2)$$

y-INTERCEPTS

$$y = -(0)^2 + 5$$
$$y = 5$$

FIGURE 8

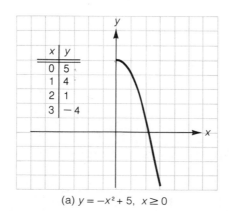

(a) $y = -x^2 + 5$, $x \geq 0$

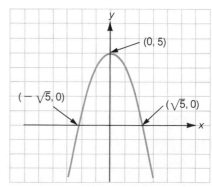

(b) The graph of $y = -x^2 + 5$ is symmetric about the y-axis.

In Figure 8(a) we've set up a short table of values and sketched the graph for $x \geq 0$. The complete graph is then obtained by reflection about the y-axis, as shown in Figure 8(b). [The curve in Figure 8(b) is a *parabola*. This type of curve will be considered in detail in later chapters.] ▲

EXAMPLE 8 Graph: $y = 1/x$

Solution Before doing any calculations, we note that the domain of the variable x consists of all real numbers other than zero. Thus, however the resulting graph may look, it cannot contain a point whose x-coordinate is zero. In other words, the graph cannot cross the y-axis. (Check for yourself that the graph does not cross the x-axis.) Now, since the graph is symmetric about the origin (why?), we need only to be able to sketch the graph for $x > 0$; the portion corresponding to $x < 0$ can then be obtained by reflection about the origin. In Figure 9(a) we've set up a table of values and sketched the graph for $x > 0$. In Figure 9(b) we have reflected the first-quadrant portion of the graph, first about the y-axis, then about the x-axis, to obtain the required reflection about the origin. Figure 9(c) shows our final graph of $y = 1/x$.

FIGURE 9

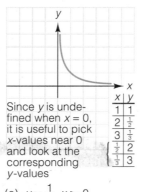

Since y is undefined when $x = 0$, it is useful to pick x-values near 0 and look at the corresponding y-values

x	y
1	1
2	$\frac{1}{2}$
3	$\frac{1}{3}$
$\frac{1}{2}$	2
$\frac{1}{3}$	3

(a) $y = \frac{1}{x}, x > 0$

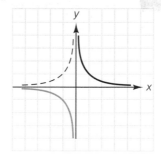

(b) Reflections about the y-axis and then the x-axis yield a reflection about the origin.

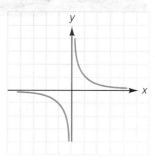

(c) The graph of $y = 1/x$ is symmetric about the origin. ▲

We can use the graph in Figure 9(a) to introduce the idea of an *asymptote*. A line is an **asymptote** for a curve if the distance between the line and the curve approaches zero as we move out farther and farther along the line. So, for the curve in Figure 9(a), both the x-axis and the y-axis are asymptotes. (We'll return to this idea several times later in the text. For other pictures of asymptotes, see Figure 2 on page 191.)

▼ **EXERCISE SET 2.1**

A

In Exercises 1–6, the graph of the equation is a straight line. Graph the equation after finding the x- and y-intercepts.

1. $3x + 4y = 12$

2. $3x - 4y = 12$

3. $y = 2x - 4$

4. $x = 2y - 4$

5. $x + y = 1$

6. $2x - 3y = 6$

In Exercises 7–20, determine any x- or y-intercepts for the graph of the equation. Note: You're not asked to draw the graph.

7. **(a)** $y = x^2 + 3x + 2$ **(b)** $y = x^2 + 2x + 3$

8. **(a)** $y = x^2 - 4x - 12$ **(b)** $y = x^2 - 4x + 12$

9. **(a)** $y = x^2 + x - 1$ **(b)** $y = x^2 + x + 1$

10. **(a)** $y = 6x^3 + 9x^2 + x$ **(b)** $y = 9x^3 + 6x^2 + x$

11. $\dfrac{x}{2} + \dfrac{y}{3} = 1$

12. $y = |x - 1|$

13. $3x - 5y = 10$

14. $xy = 1$

15. $y = x^3 - 8$

16. $y = (x - 8)^3 + 1$

17. (a) $(x - 2)^2 + (y - 3)^2 = 1$
    (b) $(x - 2)^2 + (y - 3)^2 = 4$
    (c) $(x - 2)^2 + (y - 3)^2 = 9$
    (d) $(x - 2)^2 + (y - 3)^2 = 16$

18. (a) $x^2 - 8x + y^2 + 10y = -32$
    (b) $x^2 - 8x + y^2 + 10y = -25$
    (c) $x^2 - 8x + y^2 + 10y = -16$
    (d) $x^2 - 8x + y^2 + 10y = -5$

19. $y = \sqrt{x - 1} - x + 3$

20. $y = 7 + x - \sqrt{5 - x}$

In Exercises 21–24, each figure shows a computer-generated graph of an equation. Find the x- and y-intercepts of the graph. If an intercept involves a radical, give both the radical form of the answer and a calculator approximation rounded to two decimal places. (Check to see that your answer is consistent with the given figure.)

21.

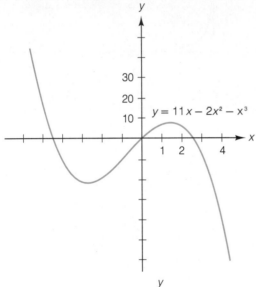

$y = 11x - 2x^2 - x^3$

22.

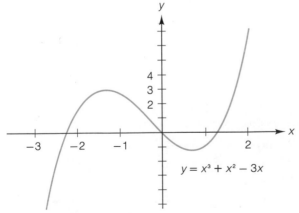

$y = x^3 + x^2 - 3x$

23.

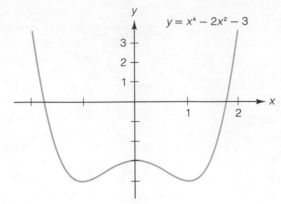

$y = x^4 - 2x^2 - 3$

24.

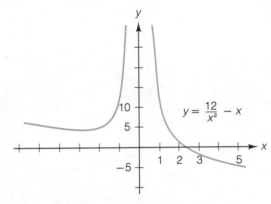

$y = \dfrac{12}{x^2} - x$

In Exercises 25–28, the endpoints of a line segment $\overline{AB}$ are given. Sketch the reflection of $\overline{AB}$ about (a) the x-axis; (b) the y-axis; (c) the origin.

25. $A(1, 4)$ and $B(3, 1)$

26. $A(-1, -2)$ and $B(-5, -2)$

27. $A(-2, -3)$ and $B(2, -1)$

28. $A(-3, -3)$ and $B(-3, -1)$

In Exercises 29–38, test each equation for symmetry about the x-axis, the y-axis, and the origin.

29. (a) $3x^2 + y = 16$ (b) $3x + y^2 = 16$

30. (a) $x^2y + xy^2 = 0$ (b) $x^2y + xy^2 = 1$

31. (a) $y = x^4 - 3$ (b) $y = x^3 - 3$
    (c) $y^2 = x^3 - 3$ (d) $y^3 = x$

32. (a) $y = x^5$ (b) $y = x^5 - 4x^3$
    (c) $y = x^5 - 4x^3 - 1$ (d) $y = x^4 - 4x^2 - 1$

33. $x^2 + y^2 = 16$ **34.** $y = |x - 2|$

35. $y = x^2 + x^3$ **36.** $x = y^2 - 4y^6$

37. $x + y = 1$ **38.** $|x| + |y| = 1$

In Exercises 39–56, graph the equation after determining the x- and y-intercepts and whether the graph possesses any of the three types of symmetry discussed in this section.

39. $y = x^2$ **40.** $y = -x^3$

41. $y = 1/x$

42. $x = y^2 - 1$

43. $y = -x^2$

44. $y = 1/x^2$

45. $y = -1/x^3$

46. $y = |x|$

47. $y = \sqrt{x^2}$

48. $y = x + 1$

49. $y = x^2 - 2x + 1$

50. $x = y^3 - 1$

51. $y^2 = 2x - 4$

52. $|y| = 2x - 4$

53. $y = 2x^2 + x - 4$

54. (a) $y = x^2 - 4$ (b) $y = |x^2 - 4|$

55. (a) $y = 3x - 6$ (b) $y = |3x - 6|$

56. (a) $x + y = 2$ (b) $|x| + y = 2$
   (c) $x + |y| = 2$ (d) $|x + y| = 2$

In Exercises 57–60, specify the domain of the variable x, and then graph the equation (as in Example 2).

57. $y = \dfrac{x^2 + x - 20}{x - 4}$

58. $y = \dfrac{6x^2 + 11x + 4}{3x + 4}$

59. $y = \dfrac{x^2 - 9}{x - 3}$

60. $y = \dfrac{x - x^2}{x}$

B

61. (a) Graph the equation $s = 16t^2$, which gives the distance s (in feet) traveled by a freely falling object in t seconds. (The time $t = 0$ corresponds to the instant the object begins its fall; thus, your graph should lie only in the first quadrant, where $t \geq 0$.)
   (b) Indicate the portion of the s-axis that shows the distance covered during the first second, from $t = 0$ to $t = 1$.
   (c) Indicate the portion of the s-axis that shows the distance covered during the next second, from $t = 1$ to $t = 2$.

62. Suppose that the formula $C = 5 + x$ gives the manufacturer's cost in dollars to produce x units of a certain commodity.
   (a) Graph this equation.
   (b) Indicate on the C-axis the cost corresponding to a production level of $x = 2$ units.
   (c) Indicate on the x-axis the number of units corresponding to a cost of $9.

63. Draw the following graphs on the same set of axes and in each case include only the portion of the graph between $x = 0$ and $x = 1$. Draw the graphs carefully enough to ensure that you can make accurate comparisons among them.
   (a) $y = x$ (b) $y = x^2$
   (c) $y = x^3$ (d) $y = x^4$

What is the pattern here? What would you say that $y = x^{100}$ must look like in this interval?

64. The figure shows the graph of $y = \sqrt{x}$. Use this graph to estimate the following quantities (to one decimal place).
   (a) $\sqrt{2}$ (b) $\sqrt{3}$ (c) $\sqrt{6}$ *Hint:* $\sqrt{ab} = \sqrt{a}\sqrt{b}$

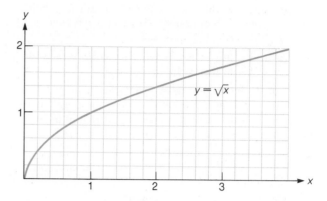

65. (a) On the same set of axes, draw the graphs of $y = x^2$, $y = x^2 + 1$, and $y = x^2 + 2$.
   (b) Based on your results in part (a), how would you say the value of the constant K influences the graph of $y = x^2 + K$?

66. (a) On the same set of axes, draw the graphs of $y = |x|$, $y = |x| + 1$, and $y = |x| + 2$.
   (b) Based on your observations in part (a), draw the graphs of $y = |x| + 3$ and $y = |x| - 1$ without first setting up tables.

67. (a) On the same set of axes, draw the graphs of $y = x^2$, $y = (x - 1)^2$, and $y = (x + 1)^2$.
   (b) Based on your observations in part (a), draw the graph of $y = (x + 2)^2$ without setting up a table.

68. (a) On the same set of axes, draw the graphs of $y = |x|, y = |x - 1|$, and $y = |x + 1|$.
   (b) Based on your observations in part (a), draw the graphs of $y = |x - 2|$ and $y = |x + 2|$ without using tables.

69. In a certain biology experiment, the number N of bacteria increases with time t as indicated in the figure.

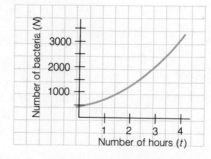

(a) How many bacteria are initially present when $t = 0$?

(b) Approximately how long does it take for the original colony to double in size?

(c) For which value of t is the population approximately 2500?

(d) During which time interval does the population increase more rapidly, between $t = 0$ and $t = 1$ or between $t = 3$ and $t = 4$?

C 70. Determine the intercepts and symmetry for the equation $y = x^{2/3}$, and then graph the equation.

C 71. Find the intercepts and symmetry for $y = x^3 - 27x$. Then use a calculator (as necessary) to set up a table of x- and y-values, with x-values from 0 to 6 in increments of $\frac{1}{2}$. Finally, graph the equation.

C 72. Set up a table of x- and y-values for the equation $y = 4x^2/(1 + x^2)$. Use x-values from 0 to 5 in increments of $\frac{1}{2}$. Now use the results in your table, along with symmetry, to graph the equation.

C

73. Graph the equation $(x^2 + y^2 - 1)(x^2 + y^2 - 4) = 0$.

74. Graph the equation

$$[(x - 1)^2 + y^2 - 1][(x - 3)^2 + y^2 - 1]$$
$$\times [(x - 5)^2 + y^2 - 1] = 0.$$

C 75. Graph the equation $y = x + \dfrac{1}{\sqrt{x + 1}}$.

76. Suppose that the circle $x^2 + 2Ax + y^2 + 2By = C$ has two x-intercepts, x_1 and x_2, and two y-intercepts, y_1 and y_2. Prove each statement.

(a) $\dfrac{x_1 + x_2}{y_1 + y_2} = \dfrac{A}{B}$ **(b)** $x_1 x_2 - y_1 y_2 = 0$

(c) $x_1 x_2 + y_1 y_2 = -2C$

2.2 ▼ EQUATIONS OF LINES

He [Pierre de Fermat (1601–1665)] introduced perpendicular axes and found the general equations of straight lines and circles and the simplest equations of parabolas, ellipses, and hyperbolas; and he further showed in a fairly complete and systematic way that every first- or second-degree equation can be reduced to one of these types.

George F. Simmons in *Calculus with Analytic Geometry* (New York: McGraw-Hill Book Company, 1985)

In this section we take a systematic look at equations of lines and their graphs. We begin by recalling the concept of *slope*, which you've seen in previous courses. The slope of a nonvertical line is a number that measures the slant or direction of the line; it is defined as follows.

DEFINITION

Slope

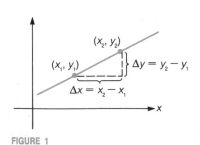

FIGURE 1

The **slope** of a nonvertical line passing through the two points (x_1, y_1) and (x_2, y_2) is the number m defined by

$$m = \frac{y_2 - y_1}{x_2 - x_1}$$

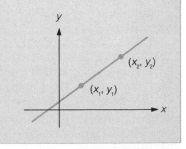

Note that the quantity $x_2 - x_1$ appearing in the definition of slope is the amount by which x changes as we move from (x_1, y_1) to (x_2, y_2) along the line. We denote this change in x by the symbol Δx (read *delta x*). Thus $\Delta x = x_2 - x_1$ (see Figure 1). Similarly, the symbol Δy is defined to mean the change in y: $\Delta y = y_2 - y_1$. Using these ideas, we can rewrite our definition of slope as $m = \Delta y / \Delta x$.

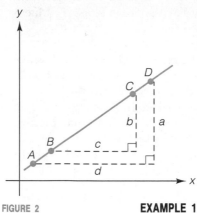

FIGURE 2

The slope of a line does not depend on which two particular points on the line are used in the calculation. To see why this is so, consider Figure 2. The two right triangles are similar (because the corresponding angles are equal). This implies that the corresponding sides of the two triangles are proportional, and so we have

$$\frac{a}{d} = \frac{b}{c}$$

Now notice that the left-hand side of this equation represents the slope $\Delta y/\Delta x$ calculated using the points A and D, and the right-hand side represents the slope calculated using the points B and C. Thus the values we obtain for the slope are indeed equal.

EXAMPLE 1   Compute and compare the slopes of the three lines shown in Figure 3.

Solution   First, we'll calculate the slope of line 1, using the formula $m = (y_2 - y_1)/(x_2 - x_1)$. Which point should we use as (x_1, y_1) and which as (x_2, y_2)? In fact, it does not matter how we label our points. Using $(-3, -1)$ for (x_1, y_1) and $(-2, 4)$ for (x_2, y_2), we have

$$m = \frac{y_2 - y_1}{x_2 - x_1} = \frac{4 - (-1)}{-2 - (-3)} = 5$$

So the slope of line 1 is 5. If, instead, we had used $(-2, 4)$ for (x_1, y_1) and $(-3, -1)$ for (x_2, y_2), then we'd have $m = (-1 - 4)/(-3 + 2) = 5$, the same result. This is not accidental because, in general,

$$\frac{y_2 - y_1}{x_2 - x_1} = \frac{y_1 - y_2}{x_1 - x_2}$$

(Exercise 64 asks you to verify this identity.) Next, we calculate the slopes of lines 2 and 3.

Slope of line 2: $\dfrac{0 - (-1)}{2 - (-1)} = \dfrac{1}{3}$

Slope of line 3: $\dfrac{3 - (-1)}{7 - 9} = -2$

FIGURE 3

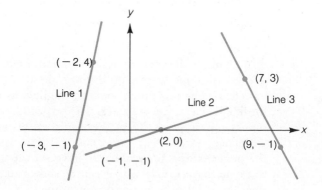

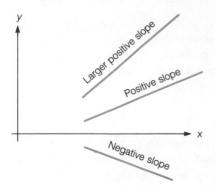

FIGURE 4

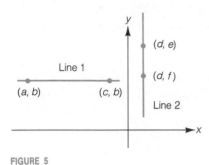

FIGURE 5

FIGURE 6

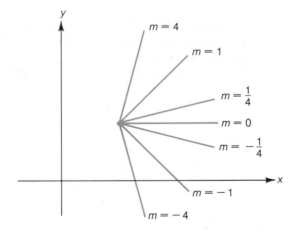

Lines 1 and 2 both have positive slopes and slant upward to the right. Note that line 1 is steeper than line 2 and correspondingly has the larger slope, 5. Line 3 has a negative slope, -2, and slants downward to the right.   ▲

The observations made in Example 1 are true in general. Lines with a positive slope slant upward to the right, the steeper line having the larger slope. Likewise, lines with a negative slope slant downward to the right. See Figure 4.

We have yet to mention slopes for horizontal or vertical lines. In Figure 5, line 1 is horizontal; it passes through the points (a, b) and (c, b). Note that the two y-coordinates must be the same in order for the line to be horizontal. Line 2 in Figure 5 is vertical; it passes through (d, e) and (d, f). Note that the two x-coordinates must be the same in order for the line to be vertical. For the slope of line 1 we have $m = \dfrac{b - b}{c - a} = \dfrac{0}{c - a} = 0$ (provided $a \neq c$). Thus the slope of line 1, a horizontal line, is zero. For the vertical line in Figure 5, the calculation of slope begins with writing $\dfrac{e - f}{d - d}$. But then the denominator is zero, and since division by zero is undefined, we conclude that slope is undefined for vertical lines. We summarize these results in the box that follows. Figure 6 shows for comparison some values of m for various lines.

Horizontal and Vertical Lines

> **1.** The slope of a horizontal line is zero.
>
> **2.** Slope is not defined for vertical lines.

We can use the concept of slope to find the equation of a line. Suppose we have a line with slope m, passing through the point (x_1, y_1) as shown in Figure 7(a).

Let (x, y) be any other point on the line, as in Figure 7(b). Then the slope of the line is given by $m = (y - y_1)/(x - x_1)$ and, therefore, $y - y_1 = m(x - x_1)$. Note that the given point (x_1, y_1) also satisfies this last equation, because in that case we have $y_1 - y_1 = m(x_1 - x_1)$, or $0 = 0$. The equation $y - y_1 = m(x - x_1)$ is called the **point–slope formula**. We have shown that any point on the line satisfies this equation. (Conversely, it can be shown that if a point satisfies this equation, then the point does lie on the given line.)

FIGURE 7

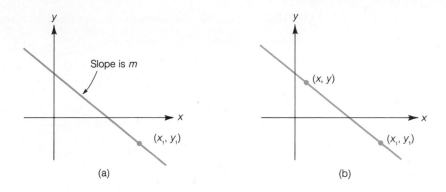

(a)   (b)

The Point–Slope Formula

The equation of a line with slope m passing through the point (x_1, y_1) is

$$y - y_1 = m(x - x_1)$$

EXAMPLE 2

Write the equation of the line passing through $(-3, 1)$ with a slope of -2. Sketch a graph of the line.

Solution

Since the slope and a point are given, we use the point–slope formula:

$$y - y_1 = m(x - x_1)$$
$$y - 1 = -2[x - (-3)]$$
$$y - 1 = -2x - 6$$
$$y = -2x - 5$$

This is the required equation. One way to graph this line is to pick values for x and then compute the corresponding y-values, as done in Section 2.1. The table and graph are displayed in Figure 8. Notice that if we know ahead of time that the graph is a line, a table as brief as that in Figure 8 is sufficient. ▲

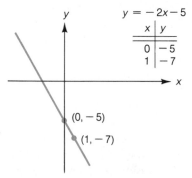

$y = -2x - 5$

x	y
0	−5
1	−7

FIGURE 8
$y = -2x - 5$

There is another way to go about graphing the line $y = -2x - 5$ in Example 2. Since slope is (change in y)/(change in x), a slope of -2 (or $-2/1$) can be interpreted as telling us that if we start at $(-3, 1)$ and let x increase by one unit, then y must decrease by two units to bring us back to the line. Following this path in Figure 9 takes us from $(-3, 1)$ to $(-2, -1)$. We now draw the line through these two points, as shown in Figure 9.

FIGURE 9

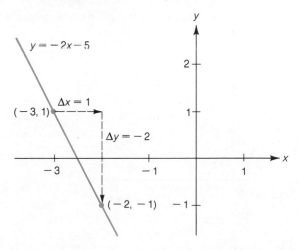

EXAMPLE 3 Find the equation of the line passing through the points $(-2, -3)$ and $(2, 5)$.

Solution The slope of the line is

$$m = \frac{y_2 - y_1}{x_2 - x_1} = \frac{5 - (-3)}{2 - (-2)} = \frac{8}{4} = 2$$

Knowing the slope, we can apply the point–slope formula, making use of either $(-2, -3)$ or $(2, 5)$. Using the point $(2, 5)$ as (x_1, y_1), we have

$$y - y_1 = m(x - x_1)$$
$$y - 5 = 2(x - 2)$$
$$y - 5 = 2x - 4$$
$$y = 2x + 1$$

Thus the required equation is $y = 2x + 1$. You should check for yourself that the same answer is obtained using the point $(-2, -3)$ instead of $(2, 5)$ in the last set of calculations. ▲

EXAMPLE 4 Find the equation of the horizontal line passing through the point $(4, -2)$. See Figure 10.

Solution Since the slope of a horizontal line is zero, we have

$$y - y_1 = m(x - x_1)$$
$$y - (-2) = 0(x - 4)$$
$$y = -2$$

Thus the equation of the horizontal line passing through $(4, -2)$ is $y = -2$. ▲

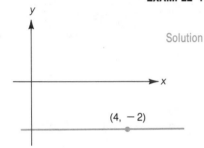

FIGURE 10

By using the point–slope formula exactly as we did in Example 4, we can show more generally that the equation of the horizontal line in Figure 11 is $y = b$. What about the equation of the vertical line in Figure 12 passing through the point (a, b)? Because slope is not defined for vertical lines, the point–slope formula is not applicable this time. However, in Figure 12, note that as you move along the vertical line, the x-coordinate is always a; it is only the y-coordinate that varies. The equation $x = a$ expresses exactly these two facts; it says that x must always be a and it places no restrictions on y.

In the box (at the top of the next page) we summarize our results concerning horizontal and vertical lines.

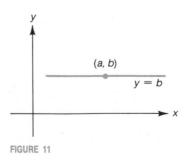

FIGURE 11

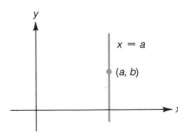

FIGURE 12

Equations of Vertical and Horizontal Lines

1. The equation of a horizontal line through the point (a, b) is $y = b$. (See Figure 11.)
2. The equation of a vertical line through the point (a, b) is $x = a$. (See Figure 12.)

Another basic form for the equation of a line is the *slope–intercept form*. We are given the slope m and y-intercept b, as shown in Figure 13, and we want to find the equation of the line. To say that the line has a y-intercept of b is the same as saying that the line passes through $(0, b)$. The point–slope formula is applicable now, using the slope m and the point $(0, b)$. We have

$$y - y_1 = m(x - x_1)$$
$$y - b = m(x - 0)$$
$$y = mx + b$$

This last equation is called the **slope–intercept formula**.

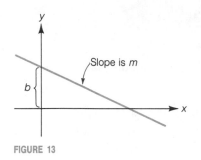

FIGURE 13

The equation of a line with slope m and y-intercept b is

$$y = mx + b$$

The Slope–Intercept Formula

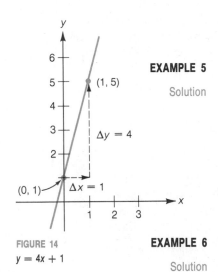

FIGURE 14
$y = 4x + 1$

EXAMPLE 5 Write the equation of a line with slope 4 and y-intercept 1. Graph the line.

Solution Substituting $m = 4$ and $b = 1$ in the equation $y = mx + b$ yields

$$y = 4x + 1$$

This is the required equation. We could draw the graph by first setting up a simple table, but for purposes of emphasis and review we proceed as we did at the end of Example 2. Starting from the point $(0, 1)$, we interpret the slope of 4 as saying that if x increases by 1, then y increases by 4. This takes us from $(0, 1)$ to the point $(1, 5)$ and the line can now be sketched as in Figure 14. ▲

EXAMPLE 6 Find the slope and y-intercept of the line $3x - 5y = 15$.

Solution First we solve for y so that the equation is in the form $y = mx + b$:

$$3x - 5y = 15$$
$$-5y = -3x + 15$$
$$y = \frac{3}{5}x - 3$$

The slope m and y-intercept b can now be read directly from the equation: $m = \frac{3}{5}$ and $b = -3$. ▲

As a consequence of our work up to this point, we can say that the graph of any linear equation

$$Ax + By + C = 0 \qquad \text{where } A \text{ and } B \text{ are not both zero}$$

is a line, since an equation of this type can always be rewritten in one of the following three forms: $y = mx + b$, or $x = a$, or $y = b$. In the box that follows, we summarize our basic results on equations of lines. The box also contains one form we have not considered: the **two-intercept formula**. (The use of this formula is illustrated in Example 7, and Exercise 65 shows you how to derive the formula.)

PROPERTY SUMMARY	EQUATIONS OF LINES
EQUATION	**COMMENT**
$y - y_1 = m(x - x_1)$ (the point–slope formula)	This is the equation of a line with slope m, passing through the point (x_1, y_1).
$y = mx + b$ (the slope–intercept formula)	This is the equation of a line with slope m and y-intercept b.
$\dfrac{x}{a} + \dfrac{y}{b} = 1$ (the two–intercept formula)	This is the equation of a line with x-intercept a and y-intercept b.
$Ax + By + C = 0$	The graph of any equation of this form (where A and B are not both zero) is a line. Special cases: If $A = 0$, then the equation can be written $y = -C/B$, which is the equation of a horizontal line with y-intercept $-C/B$. If $B = 0$, then the equation can be written $x = -C/A$, which represents a vertical line with x-intercept $-C/A$.

EXAMPLE 7 Find the area of the triangle formed by the line $4x + 3y = 12$ and the coordinate axes.

Solution The two-intercept formula is $(x/a) + (y/b) = 1$. We can convert the given equation to this form by dividing both sides by 12. This yields

$$\frac{4x}{12} + \frac{3y}{12} = \frac{12}{12}$$

or

$$\frac{x}{3} + \frac{y}{4} = 1.$$

Thus, the x- and y-intercepts are 3 and 4, respectively, and the line can be sketched as in Figure 15. Looking at Figure 15, we see that the base of the triangle is 3, the height is 4, and the area is

$$\text{area} = \tfrac{1}{2}(\text{base})(\text{height})$$
$$= \tfrac{1}{2}(3)(4) = 6 \text{ square units}$$

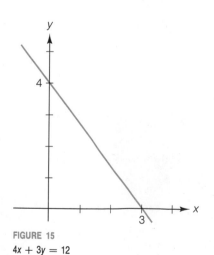

FIGURE 15
$4x + 3y = 12$

We conclude this section by discussing two useful relationships regarding the slopes of parallel lines and the slopes of perpendicular lines. First, nonvertical parallel lines have the same slope. This should seem reasonable if you recall that slope is a number indicating the direction or slant of a line. The relationship concerning slopes of perpendicular lines is not so obvious. The slopes of two non-vertical perpendicular lines are negative reciprocals of each other. That is, if m_1 and m_2 denote the slopes of the two perpendicular lines, then $m_1 = -1/m_2$, or, equivalently, $m_1 m_2 = -1$. For example, if a line has a slope of $\frac{2}{3}$, then any line perpendicular to it must have a slope of $-\frac{3}{2}$. For reference we summarize these facts in the box that follows. (Proofs of these facts are outlined in detail in Exercises 61 and 62.)

Parallel and Perpendicular Lines

Let m_1 and m_2 denote the slopes of two nonvertical lines. Then:

1. The lines are parallel if and only if $m_1 = m_2$;

2. The lines are perpendicular if and only if $m_1 = -1/m_2$.

EXAMPLE 8   Are the two lines $3x - 6y - 8 = 0$ and $2y = x + 1$ parallel?

Solution   By solving each equation for y, we can see what the slopes are:

$$3x - 6y - 8 = 0 \qquad\qquad 2y = x + 1$$
$$-6y = -3x + 8 \qquad\qquad y = \frac{1}{2}x + \frac{1}{2}$$
$$y = \frac{-3}{-6}x + \frac{8}{-6}$$
$$y = \frac{1}{2}x - \frac{4}{3}$$

From this we see that both lines have the same slope, namely, $m = \frac{1}{2}$. It follows therefore that the lines are parallel.   ▲

EXAMPLE 9   Find the equation of the line parallel to $5x + 6y = 30$ and passing through the origin.

Solution   First, find the slope of the line $5x + 6y = 30$:

$$5x + 6y = 30$$
$$6y = -5x + 30$$
$$y = -\frac{5}{6}x + 5$$

The slope is the x-coefficient in the last equation: $m = -\frac{5}{6}$. A parallel line will have the same slope. Since the required line is to pass through $(0, 0)$, we have

$$y - y_1 = m(x - x_1)$$
$$y - 0 = -\frac{5}{6}(x - 0)$$
$$y = -\frac{5}{6}x$$

This is the equation of a line parallel to $5x + 6y = 30$ and passing through the origin, as required. See Figure 16.   ▲

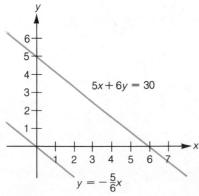

FIGURE 16

FIGURE 17

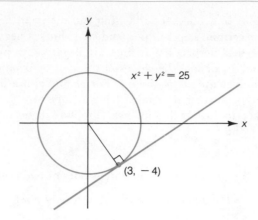

EXAMPLE 10   Find the equation of the line tangent to the circle $x^2 + y^2 = 25$ at the point $(3, -4)$ (see Figure 17). Write the answer in the form $y = mx + b$.

Solution   We have one point on the tangent line, namely, $(3, -4)$. If we knew the slope, then we could apply the point–slope formula. From elementary geometry, we know that the tangent line is perpendicular to the radius. Since the radius passes through $(0, 0)$ and $(3, -4)$, its slope is

$$\frac{-4 - 0}{3 - 0} = -\frac{4}{3}$$

and, consequently, the slope of the tangent is $\frac{3}{4}$. The point–slope formula is now applicable:

$$y - y_1 = m(x - x_1)$$
$$y - (-4) = \frac{3}{4}(x - 3)$$

Check now for yourself that this can be simplified to $y = \frac{3}{4}x - \frac{25}{4}$, which is the required equation of the tangent line. ▲

▼ EXERCISE SET 2.2

A

In Exercises 1–3, compute the slope of the line passing through the two given points. In Exercise 3, include a sketch with your answers.

1. **(a)** $(-3, 2), (1, -6)$ **(b)** $(2, -5), (4, 1)$
   (c) $(-2, 7), (1, 0)$ **(d)** $(4, 5), (5, 8)$

2. **(a)** $(-3, 0), (4, 9)$ **(b)** $(-1, 2), (3, 0)$
   (c) $(\frac{1}{2}, -\frac{3}{5}), (\frac{3}{2}, \frac{3}{4})$ **(d)** $(\frac{17}{3}, -\frac{1}{2}), (-\frac{1}{2}, \frac{17}{3})$

3. **(a)** $(1, 1), (-1, -1)$ **(b)** $(0, 5), (-8, 5)$
   (c) $(-1, 1), (1, -1)$ **(d)** $(a, b), (b, a)$
      (assume $a \neq b$)

4. Compute the slope of the line in the accompanying figure using the two points indicated.

   (a) A and B **(b)** B and C **(c)** A and C

The principle involved here is that no matter which pair of points you choose, the slope is the same.

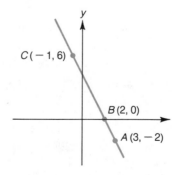

5. The slopes of four lines are indicated in the figure. List the slopes m_1, m_2, m_3, m_4 in order of increasing value.

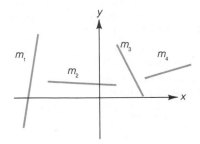

6. Refer to the accompanying figure.

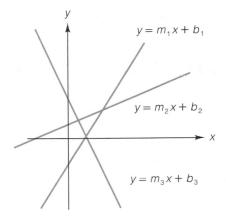

(a) List the slopes m_1, m_2, and m_3 in order of increasing size.
(b) List the numbers b_1, b_2, and b_3 in order of increasing size.

In Exercises 7–9, three points A, B, and C are specified. Determine if A, B, and C are collinear (lie on the same line) by checking to see whether the slope of $\overline{AB}$ equals the slope of $\overline{BC}$.

7. $A(-8, -2)$, $B(2, \frac{1}{2})$, $C(11, -1)$
8. $A(4, -3)$, $B(-1, 0)$, $C(-4, 2)$
9. $A(0, -5)$, $B(3, 4)$, $C(-1, -8)$
10. If the area of the "triangle" formed by three points is zero, then the points must in fact be collinear. Use this observation, along with the formula in Exercise 14(b) of Exercise Set 1.8, to rework Exercise 9.

In Exercises 11 and 12, find the equation for the line having the given slope and passing through the given point. Write your answers in the form $y = mx + b$.

11. (a) $m = -5$; through $(-2, 1)$
    (b) $m = 4$; through $(4, -4)$
    (c) $m = \frac{1}{3}$; through $(-6, -\frac{2}{3})$
    (d) $m = -1$; through $(0, 1)$

12. (a) $m = 22$; through $(0, 0)$
    (b) $m = -222$; through $(0, 0)$
    (c) $m = \sqrt{2}$; through $(0, 0)$

In Exercises 13 and 14, find the equation for the line passing through the two given points. Write your answer in the form $y = mx + b$.

13. (a) $(4, 8)$ and $(-3, -6)$
    (b) $(-2, 0)$ and $(3, -10)$
    (c) $(-3, -2)$ and $(4, -1)$

14. (a) $(7, 9)$ and $(-11, 9)$
    (b) $(\frac{5}{4}, 2)$ and $(\frac{3}{4}, 3)$
    (c) $(12, 13)$ and $(13, 12)$

In Exercises 15 and 16, write the equation of a vertical line passing through the given point. In Exercises 17 and 18, write the equation of a horizontal line passing through the given point.

15. $(-3, 4)$ 16. $(8, 5)$
17. $(-3, 4)$ 18. $(8, 5)$
19. Is the graph of the line $x = 0$ the x-axis or the y-axis?
20. Is the graph of the line $y = 0$ the x-axis or the y-axis?

In Exercises 21 and 22, find the equation of the line with the given slope and y-intercept.

21. (a) slope -4; y-intercept 7
    (b) slope 2; y-intercept $\frac{3}{2}$
    (c) slope $-\frac{4}{3}$, y-intercept 14

22. (a) slope 0; y-intercept 14
    (b) slope 14; y-intercept 0

In Exercises 23–28, find an equation for the line that is described, and sketch the graph. For Exercises 23–25, write the final answer in the form $y = mx + b$; for Exercises 26–28, write the answer in the form $Ax + By + C = 0$.

23. (a) Passing through $(-3, -1)$, with slope 4
    (b) Passing through $(\frac{5}{2}, 0)$, with slope $\frac{1}{2}$
    (c) With x-intercept 6, y-intercept 5
    (d) With x-intercept -2, slope $\frac{3}{4}$
    (e) Passing through $(1, 2)$ and $(2, 6)$

24. (a) Passing through $(-7, -2)$ and $(0, 0)$
    (b) Passing through $(6, -3)$, with y-intercept 8
    (c) Passing through $(0, -1)$ and with the same slope as the line $3x + 4y = 12$
    (d) Passing through $(6, 2)$ and with the same x-intercept as the line $-2x + y = 1$
    (e) With x-intercept -6, y-intercept $\sqrt{2}$

25. With the same x- and y-intercepts as the circle $x^2 + y^2 + 4x - 4y + 4 = 0$

26. Passing through $(3, -5)$ and through the center of the circle $4x^2 + 8x + 4y^2 - 24y + 15 = 0$

27. Passing through $(-3, 4)$ and parallel to the x-axis

28. Passing through $(-3, 4)$ and parallel to the y-axis

In Exercises 29 and 30, rewrite the equation in the two-intercept form $(x/a) + (y/b) = 1$, and then find the area and perimeter of the triangle formed by the given line and the axes.

29. (a) $3x + 5y = 15$ (b) $3x - 5y = 15$

30. (a) $5x + 4y = 40$ (b) $2x + 4y = \sqrt{2}$

31. Determine whether the two lines are parallel, perpendicular, or neither.
(a) $3x - 4y = 12$; $4x - 3y = 12$
(b) $y = 5x - 16$; $y = 5x + 2$
(c) $5x - 6y = 25$; $6x + 5y = 0$
(d) $y = -\frac{2}{3}x - 1$; $y = \frac{3}{2}x - 1$
(e) $-2x - 5y = 1$; $y - \frac{2}{5}x - 4 = 0$
(f) $x = 8y + 3$; $4y - \frac{1}{2}x = 32$

32. Are the lines $y = x + 1$ and $y = 1 - x$ parallel, perpendicular, or neither?

In Exercises 33–38, find an equation for the line that is described. Write the answer in the two forms $y = mx + b$ and $Ax + By + C = 0$.

33. Parallel to $2x - 5y = 10$ and passing through $(-1, 2)$

34. Parallel to $4x + 5y = 20$ and passing through $(0, 0)$

35. Perpendicular to $4y - 3x = 1$ and passing through $(4, 0)$

36. Perpendicular to $x - y + 2 = 0$ and passing through $(3, 1)$

37. Parallel to $3x - 5y = 25$ and with the same y-intercept as $6x - y + 11 = 0$

38. Perpendicular to $9y - 2x = 3$ and with the same x-intercept as the circle $x^2 + y^2 - 12x - 2y = -36$

39. (a) Sketch the circle $x^2 + y^2 = 25$.
(b) Find the equation of the line tangent to this circle at the point $(-4, -3)$. Include a sketch.

40. (a) Tangents are drawn to the circle $x^2 + y^2 = 169$ at the points $(5, 12)$ and $(5, -12)$. Find the equations of the tangents.
(b) Find the coordinates of the point where these two tangents intersect.

B

41. Show that the slope of the line passing through the two points $(3, 9)$ and $(3 + h, (3 + h)^2)$ is $6 + h$.

42. Show that the slope of the line passing through the two points (x, x^2) and $(x + h, (x + h)^2)$ is $2x + h$.

43. Show that the slope of the line passing through the two points (x, x^3) and $(x + h, (x + h)^3)$ is $3x^2 + 3xh + h^2$.

44. Show that the slope of the line passing through the points $(x, \sqrt{x})$ and $(x + h, \sqrt{x + h})$ can be written

$$\frac{1}{\sqrt{x + h} + \sqrt{x}}$$

45. Show that the slope of the line passing through the two points $\left(x, \dfrac{1}{x}\right)$ and $\left(x + h, \dfrac{1}{x + h}\right)$ is $\dfrac{-1}{x(x + h)}$.

46. Why is the letter m used to denote slope? Look up the verb *monter* in a French-to-English dictionary.

C **47.** Sketch the curve $y = x^2$ and indicate the point $T(3, 9)$ on the graph. In each case that follows, you're given the x-coordinate of a point P on the curve. Find the corresponding y-coordinate, and then calculate the slope of the line through P and T. Display your results in a table.

(a) $x = 2.5$ (b) $x = 2.9$ (c) $x = 2.99$
(d) $x = 2.999$ (e) $x = 2.9999$

As P gets closer to T, what numerical value does the slope of $\overline{PT}$ seem to approach? What would you estimate to be the slope of the line that is tangent to the curve $y = x^2$ at point T?

C **48.** Follow the instructions for Exercise 47 using the curve $y = \sqrt{x}$, the point $T(1, 1)$, and the following sequence of x-values.

(a) $x = 1.1$ (b) $x = 1.01$ (c) $x = 1.001$
(d) $x = 1.0001$ (e) $x = 1.00001$
(Round your answers to six decimal places.)

49. Find a point P on the curve $y = x^3$ such that the slope of the line passing through P and $(1, 1)$ is $\frac{3}{4}$.

50. Find a point P on the curve $y = \sqrt{x}$ such that the slope of the line through P and $(1, 1)$ is $\frac{1}{4}$.

51. Find a point P on the curve $y = 1/x$ such that the slope of the line through P and $(2, \frac{1}{2})$ is $-\frac{1}{16}$.

52. Find a point P on the circle $x^2 + y^2 = 20$ such that the slope of the radius drawn from $(0, 0)$ to P is 2. (You will get two answers.)

53. A line with a slope of -5 passes through the point $(3, 6)$. Find the area of the triangle in the first quadrant formed by this line and the coordinate axes.

54. The y-intercept of the line in the figure is 6. Find the slope of the line if the area of the shaded triangle is 72 square units.

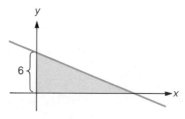

55. (a) Sketch the line $y = \frac{1}{2}x - 5$ and the point $P(1, 3)$. Follow parts (b)–(d) to calculate the perpendicular distance from point $P(1, 3)$ to the line.

(b) Find the equation of the line that passes through $P(1, 3)$ and is perpendicular to the line $y = \frac{1}{2}x - 5$.

(c) Find the coordinates of the point where these two lines intersect.

(d) Use the distance formula to find the perpendicular distance from $P(1, 3)$ to the line $y = \frac{1}{2}x - 5$.

56. (a) Show that the equation of the line tangent to the circle $x^2 + y^2 = a^2$ at the point (x_1, y_1) on the circle is

$$x_1 x + y_1 y = a^2$$

(b) Use the result in part (a) to rework Exercises 39(b) and 40(a).

57. Follow the instructions in parts (a)–(e) to determine the slope m of the line in the figure.

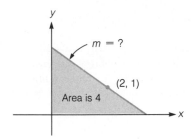

(a) Use the point-slope formula to show that the equation of the line is

$$y - 1 = m(x - 2)$$

(b) Show that the y-intercept for the line is $-2m + 1$. Show that the x-intercept is $\dfrac{2m - 1}{m}$.

(c) Show that the area of the triangle, being one-half base times height, is $\dfrac{1}{2}\left(\dfrac{2m - 1}{m}\right)(-2m + 1)$.

(d) Set the expression for area in part (c) equal to 4, as given in the figure, and simplify the resulting equation to obtain $4m^2 + 4m + 1 = 0$.

(e) Solve the equation for m, either by factoring or by using the quadratic formula. You should obtain $m = -\frac{1}{2}$.

58. A line with a negative slope passes through the point $(3, 1)$. The area of the triangle bounded by this line and the axes is 6 square units. Find the possible slopes of the line. *Hint:* See Exercise 57.

59. By analyzing sales figures, the accountant for College Stereo Company knows that 280 units of a compact disc player can be sold each month when the price is $P = \$195$ per unit. The figures also show that for each $\$15$ hike in the price, 10 fewer units are sold monthly.

(a) Let x denote the number of units sold per month and P the price per unit. Find an equation that

expresses x in terms of P, assuming that this relationship is linear.
Hint: $\Delta x / \Delta P = -10/15 = -2/3$.

(b) Use the equation that you found in part (a) to determine how many units can be sold in a month when the price is $\$270$ per unit.

(c) What should the price be to sell 205 units per month?

60. Imagine that you own a grove of orange trees, and suppose that from past experience you know that when 100 trees are planted, each tree will yield approximately 240 oranges per year. Furthermore, you've noticed that when additional trees are planted in the grove, the yield per tree decreases. Specifically, you have noted that the yield per tree decreases by about 20 oranges for each additional tree planted.

(a) Let y denote the yield per tree when x trees are planted. Find a linear equation relating x and y.

(b) Use the equation in part (a) to determine how many trees should be planted to obtain a yield of 400 oranges per tree.

(c) If the grove contained 95 trees, what yield would you expect from each tree?

61. This exercise outlines a proof of the fact that two nonvertical lines are parallel if and only if their slopes are equal. The proof relies on the following observation for the given figure: The lines $y = m_1 x + b_1$ and $y = m_2 x + b_2$ will be parallel if and only if the two vertical distances AB and CD are equal. (In the figure, the points C and D both have x-coordinate 1.)

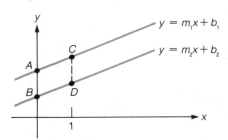

(a) Verify that the coordinates of A, B, C, and D are:

$$A(0, b_1) \quad B(0, b_2) \quad C(1, m_1 + b_1) \quad D(1, m_2 + b_2)$$

(b) Using the coordinates in part (a), check that $AB = b_1 - b_2$ and $CD = (m_1 + b_1) - (m_2 + b_2)$.

(c) Use part (b) to show that the equation $AB = CD$ is equivalent to $m_1 = m_2$.

62. This exercise outlines a proof of the fact that two nonvertical lines with slopes m_1 and m_2 are perpendicular if and only if $m_1 m_2 = -1$. In the figure (on the following page), we've assumed that our two nonvertical lines $y = m_1 x$ and $y = m_2 x$ intersect at the origin. [If they did not intersect there, we could just as well work with lines parallel to these that do intersect at $(0, 0)$, recalling that parallel lines have the same slope.] The proof relies on the following geometric fact:

$$\overline{OA} \perp \overline{OB} \quad \text{if and only if} \quad OA^2 + OB^2 = AB^2$$

(a) Verify that the coordinates of A and B are $A(1, m_1)$ and $B(1, m_2)$.

(b) Show that

$$OA^2 = 1 + m_1^2$$
$$OB^2 = 1 + m_2^2$$
$$AB^2 = m_1^2 - 2m_1 m_2 + m_2^2$$

(c) Use part (b) to show that the equation $OA^2 + OB^2 = AB^2$ is equivalent to $m_1 m_2 = -1$.

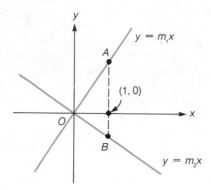

63. Find the equation of the line that is obtained when $y = mx + b$ is reflected about **(a)** the x-axis; **(b)** the y-axis; **(c)** the origin.

64. Verify the identity

$$\frac{y_2 - y_1}{x_2 - x_1} = \frac{y_1 - y_2}{x_1 - x_2}$$

What does this identity tell you about calculating slope?

65. The x- and y-intercepts of a line are a and b, respectively.

(a) What is the slope of the line?

(b) Show that the equation of the line can be written

$$\frac{x}{a} + \frac{y}{b} = 1$$

66. The figure shows two tangent lines drawn from the point (a, b) to the circle $x^2 + y^2 = R^2$. Follow steps (a)–(d) to show that the equation of the line passing through the two points of tangency is

$$ax + by = R^2$$

(This problem, along with the following clever solution, appears in the classic text by Isaac Todhunter, *A Treatise on Plane Co-ordinate Geometry*, first published in 1855.)

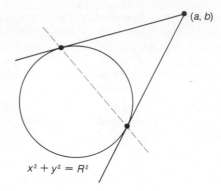

(a) Let (x_1, y_1) be one of the points of tangency. Show that the equation of the tangent line through this point is $x_1 x + y_1 y = R^2$.

(b) Using the result in part (a), explain why $x_1 a + y_1 b = R^2$.

(c) In a similar fashion, explain why $x_2 a + y_2 b = R^2$, where (x_2, y_2) is the other point of tangency.

(d) The equation $ax + by = R^2$ represents a line. Explain why this line must pass through (x_1, y_1) and (x_2, y_2).

C

67. Given two lines L_1 and L_2 passing through the origin as shown, a third line L is said to bisect the area between L_1 and L_2 provided that for each point P on L the area of region A equals the area of region B. (In the figure, each dotted line is parallel to one of the coordinate axes.)

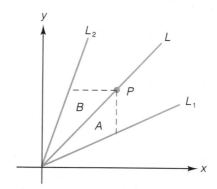

(a) Suppose that the equation of L_1 is $y = \frac{1}{2}x$ and the equation of L_2 is $y = 8x$. Find the equation of L.

(b) Suppose that the equations of L_1 and L_2 are $y = nx$ and $y = mx$, respectively. Find the equation of L.

68. The vertices of a triangle are $A(-4, 0)$, $B(2, 0)$, and $C(0, 6)$. Let M_1, M_2, and M_3 be the midpoints of $\overline{AB}$, $\overline{BC}$, and $\overline{AC}$, respectively. Let H_1, H_2, and H_3 be the feet of the altitudes on sides $\overline{AB}$, $\overline{BC}$, and $\overline{AC}$, respectively.

(a) Find the equation of the circle passing through M_1, M_2, and M_3. _Answer:_ $3x^2 + 3y^2 + 3x - 11y = 0$

(b) Find the equation of the circle passing through H_1, H_2, and H_3. _Answer:_ $3x^2 + 3y^2 + 3x - 11y = 0$

(c) Find the point P at which the three altitudes intersect. _Answer:_ $(0, \frac{4}{3})$

(d) Let $\mathcal{N}_1$, $\mathcal{N}_2$, and $\mathcal{N}_3$ be the midpoints of $\overline{AP}$, $\overline{BP}$, and $\overline{CP}$, respectively. Show that the circle obtained in parts (a) and (b) passes through $\mathcal{N}_1$, $\mathcal{N}_2$, and $\mathcal{N}_3$. _Note:_ This circle is called the _nine-point circle_ of $\triangle ABC$.

(e) The _circumcircle_ of $\triangle ABC$ is the circle passing through the three points A, B, and C. Show that the co-ordinates of the center Q of the circumcircle are $Q(-1, \frac{7}{3})$. _Hint:_ Use the result from geometry that the perpendicular bisectors of the sides of the triangle intersect at the center of the circumcircle.

(f) For $\triangle ABC$, show that the radius of the nine-point circle is one half the radius of the circumcircle.

(g) Show that the midpoint of line segment $\overline{QP}$ is the center of the nine-point circle.

(h) Find the point G where the three medians of $\triangle ABC$ intersect.

(i) Show that G lies on line segment $\overline{QP}$ and that $PG = 2(GQ)$.

2.3 ▼ THE DEFINITION OF A FUNCTION

From the beginning of modern mathematics in the 17th century the concept of function has been at the very center of mathematical thought.

Richard Courant and Fritz John in _Introduction to Calculus and Analysis_ (New York: Wiley-Interscience, 1965)

The word _function_ was introduced into mathematics by Leibniz, who used the term primarily to refer to certain kinds of mathematical formulas. It was later realized that Leibniz's idea of function was much too limited in scope, and the meaning of the word has since undergone many steps of generalization.

Tom M. Apostol in _Calculus_, Second Edition (New York: John Wiley & Sons, 1967)

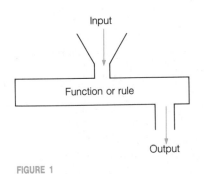

FIGURE 1

TABLE 1

Mercury	0
Venus	0
Earth	1
Mars	2
Jupiter	17
Saturn	22
Uranus	15
Neptune	8
Pluto	1

There are numerous instances in mathematics and its applications in which one quantity corresponds to or depends on another according to some definite rule. Consider, for example, the equation $y = 3x - 2$. Each time that we select an x-value, a corresponding y-value is determined, in this case according to the rule _multiply by 3, then subtract 2_. In this sense, the equation $y = 3x - 2$ is an example of a _function_. It is a _rule_ specifying a y-value corresponding to each x-value. It is useful to think of the x-values as inputs and the corresponding y-values as outputs. The function, or rule, then tells us what output results from a given input. This is indicated schematically in Figure 1. As another example, the area A of a circle depends on the radius r according to the rule or function $A = \pi r^2$. For each value of r there is a corresponding value for A obtained by the rule _square r and multiply the result by π_. In this case the inputs are the values of r and the outputs are the corresponding values of A.

For most of the functions studied in this text (and in beginning calculus), the inputs and outputs are real numbers and the function or rule is specified by means of an equation. This was the case with the two examples we have given; however, this need not always be the case. Consider, for example, the correspondences set up in Table 1, which indicate the number of moons each planet in the solar system was known to have as of 1990. If we think of the inputs as the planets listed in the left-hand column of Table 1 and the outputs as the numbers in the right-hand column, then these correspondences constitute a function. The rule for this function may be stated, _Assign to each planet in the solar system the number of moons it was known to have in 1990._

The following is one definition of the term *function*. As you will see, the definition is broad enough to encompass all the examples we have just looked at.

> Let A and B be two nonempty sets. A **function** from A to B is a rule of correspondence that assigns to each element in A exactly one element in B.

The set A in the definition just given is called the **domain** of the function. Think of the domain as the set of all possible inputs. The set of all outputs, on the other hand, is called the **range** of the function. When a function is defined by means of an equation, the letter representing elements from the domain (that is, the inputs) is called the **independent variable**. For example, in the equation $y = 3x - 2$, the independent variable is x. The letter representing elements from the range (that is, the outputs) is called the **dependent variable**. In the equation $y = 3x - 2$, y is the dependent variable; its value *depends* on x. This is also expressed by saying that y is a function of x.

For functions defined by equations, we'll agree to the following convention regarding the domain: Unless otherwise indicated, the domain is assumed to be the set of all real numbers that lead to unique real-number outputs. Thus, the domain of the function defined by $y = 3x - 2$ is the set of all real numbers, whereas the domain of the function defined by $y = \dfrac{1}{x - 5}$ is the set of all real numbers except 5. (The expression $\dfrac{1}{x - 5}$ is undefined when $x = 5$ because the denominator is then zero.)

EXAMPLE 1   Find the domain of the function defined by each equation.

(a) $y = \sqrt{2x + 6}$ **(b)** $s = \dfrac{1}{t^2 - 6t - 7}$

Solution   **(a)** The quantity under the radical sign must be nonnegative, so we have

$$2x + 6 \geq 0$$
$$2x \geq -6$$
$$x \geq -3$$

The domain is therefore the interval $[-3, \infty)$.

(b) Since division by zero is undefined, the domain of this function consists of all real numbers t except those for which the denominator is zero. Thus to find out which values of t to exclude, we solve the equation $t^2 - 6t - 7 = 0$. We have

$$t^2 - 6t - 7 = 0$$
$$(t - 7)(t + 1) = 0$$
$$t - 7 = 0 \quad \bigg| \quad t + 1 = 0$$
$$t = 7 \quad \bigg| \quad t = -1$$

It follows now that the domain of the function defined by $s = \dfrac{1}{t^2 - 6t - 7}$ is the set of all real numbers except $t = 7$ and $t = -1$.   ▲

EXAMPLE 2   Find the range of the function defined by $y = \dfrac{x + 2}{x - 3}$.

Solution   The range of this function is the set of all outputs y. One way to see what restrictions the given equation imposes on y is to solve the equation for x as follows:

$$y(x - 3) = x + 2 \quad \text{multiplying by } (x - 3)$$
$$xy - 3y = x + 2$$
$$xy - x = 3y + 2$$
$$x(y - 1) = 3y + 2$$
$$x = \frac{3y + 2}{y - 1}$$

From this last equation we see that the value of y cannot be 1. (The denominator is zero when $y = 1$.) The range therefore consists of all real numbers except $y = 1$. ▲

We often use single letters in order to name functions. If f is a function and x is an input for the function, then the resulting output is denoted by $f(x)$. This is read f of x or *the value of f at x*. As an example of this notation, suppose that f is the function defined by

$$f(x) = x^2 - 3x + 1 \tag{1}$$

Then $f(-2)$ denotes the output that results when the input is -2. To calculate this output, just replace x with -2 throughout equation (1). This yields

$$f(-2) = (-2)^2 - 3(-2) + 1$$
$$= 4 + 6 + 1 = 11$$

That is, $f(-2) = 11$. Figure 2 summarizes this result and the notation.

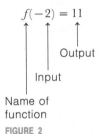

$f(-2) = 11$

Output

Input

Name of function

FIGURE 2

EXAMPLE 3   Let $f(x) = \dfrac{1}{x - 1}$. Compute: **(a)** $f(0)$ **(b)** $f(t)$ **(c)** $f(x - 1)$

Solution   **(a)** $f(0) = \dfrac{1}{0 - 1} = -1$

(b) $f(t) = \dfrac{1}{t - 1}$

(c) Replace x with the quantity $x - 1$ throughout the given equation. This yields

$$f(x - 1) = \frac{1}{(x - 1) - 1} = \frac{1}{x - 2}$$ ▲

The next example points out a faulty assumption that students sometimes make in using function notation. Don't automatically assume that $f(a + b) = f(a) + f(b)$; it is not, in general, true.

EXAMPLE 4   Let $f(x) = x^2$. Show that $f(2 + 3) \neq f(2) + f(3)$.

Solution $f(2 + 3) = f(5) = (5)^2 = 25$; but $f(2) = (2)^2 = 4$ and $f(3) = (3)^2 = 9$. Since $25 \neq 4 + 9$, we have shown that $f(2 + 3) \neq f(2) + f(3)$. ▲

EXAMPLE 5   Let $g(x) = 1 - x^2$. Compute $g(x - 1)$.

Solution   In the equation $g(x) = 1 - x^2$, we substitute the quantity $x - 1$ in place of each occurrence of x. This gives us

$$g(x - 1) = 1 - (x - 1)^2$$
$$= 1 - (x^2 - 2x + 1)$$
$$= -x^2 + 2x$$

Thus, $g(x - 1) = -x^2 + 2x$.   ▲

Here is a slightly different perspective on function notation that we can apply in Example 5 and in similar problems. Instead of writing $g(x) = 1 - x^2$, we can write $g(\ \ ) = 1 - (\ \ )^2$, with the understanding that whatever quantity goes in the parentheses on the left-hand side of the equation must also be placed in the parentheses on the right-hand side. In particular, if we want $g(x - 1)$, we simply write $x - 1$ inside each set of parentheses:

$$g(\ \ ) = 1 - (\ \ )^2$$

Therefore,

$$g(x - 1) = 1 - (x - 1)^2$$

From here on, the algebra is the same as in Example 5. We again obtain $g(x - 1) = -x^2 + 2x$.

The next two examples using function notation involve calculations with a **difference quotient**. This is an expression of the form

$$\frac{f(x + h) - f(x)}{h} \qquad \text{or} \qquad \frac{f(x) - f(a)}{x - a}$$

For now we'll concentrate on the algebraic techniques used in calculating such quantities. (Later you'll see some applications, in Exercise Set 2.4, Exercises 37–39.)

EXAMPLE 6   Let $f(x) = x^2 + 3x$. Compute $\dfrac{f(x) - f(2)}{x - 2}$.

Solution
$$\frac{f(x) - f(2)}{x - 2} = \frac{(x^2 + 3x) - [2^2 + 3(2)]}{x - 2}$$
$$= \frac{x^2 + 3x - 10}{x - 2}$$
$$= \frac{(x - 2)(x + 5)}{x - 2} = x + 5$$

The difference quotient is $x + 5$.   ▲

EXAMPLE 7   Let $G(x) = 2/x$. Find $\dfrac{G(x + h) - G(x)}{h}$.

Solution
$$\frac{G(x + h) - G(x)}{h} = \frac{2/(x + h) - 2/x}{h}$$

An easy way to simplify this last expression is to multiply it by $\dfrac{(x + h)x}{(x + h)x}$, which equals 1. This yields

$$\frac{G(x+h) - G(x)}{h} = \frac{(x+h)x}{(x+h)x} \cdot \frac{\dfrac{2}{x+h} - \dfrac{2}{x}}{h}$$

$$= \frac{2x - 2(x+h)}{h(x+h)x}$$

$$= \frac{2x - 2x - 2h}{h(x+h)x} = \frac{-2h}{h(x+h)x}$$

$$= \frac{-2}{(x+h)x}$$

The difference quotient is therefore $-2/(x+h)x$. ▲

In Examples 1 through 7 we've considered functions that are defined by means of equations. It is important to understand, however, that not all equations and rules define functions. For example, consider the equation $y^2 = x$ and the input $x = 4$. Then we have $y^2 = 4$ and, consequently, $y = \pm 2$. So we have *two* outputs in this case, whereas the definition of a function requires that there be *exactly one* output. Example 8 provides some additional perspective on this situation.

EXAMPLE 8   Let $A = \{b, g\}$ and $B = \{s, t, u, z\}$. Which of the four correspondences in Figure 3 represent functions from A to B? For those correspondences that do represent functions, specify the range in each case.

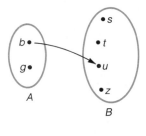
(a)

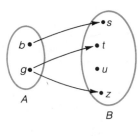
(b)

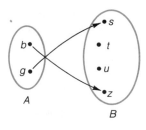
(c)

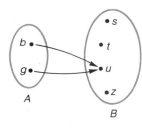
(d)

FIGURE 3

Solution   *Figure 3(a)*   Not a function. The definition requires that *each* element in A be assigned an element in B. The element g in this case has no assignment.

Figure 3(b)   Not a function. The definition requires *exactly one* output for a given input. In this case there are two outputs for the input g.

Figure 3(c) and (d)   Both of these rules qualify as functions from A to B. For each input there is exactly one output. (Regarding the function in Figure 3(d) in particular, notice that nothing in the definition of a function prohibits two different inputs from producing the same output.) For the function in Figure 3(c), the outputs are s and z, and so the range is the set $\{s, z\}$. For the function in Figure 3(d), the only output is u, and consequently the range is the set $\{u\}$. ▲

Because the function concept is so central to the rest of the material in this text, for reference we conclude here with a summary of the terminology introduced in this section.

PROPERTY SUMMARY		TERMINOLOGY FOR FUNCTIONS
TERM	DEFINITION AND COMMENTS	EXAMPLE
Function	Given two nonempty sets A and B, a function f from A to B is a rule that assigns to each element x in A exactly one element $f(x)$ in B. We think of each element x in A as an input and the corresponding element $f(x)$ in B as an output.	The equation $f(x) = x^2$ defines a function. The rule in this case is: *For each real number x, compute its square.* Given the input $x = 3$, for example, the output is $f(3) = 9$.
Domain and range	The domain of a function is the set of all inputs; this is the set A in the definition. The range is the set of all outputs; see Figure 4.	The domain and the range of the function defined by $f(x) = x^2$ are the sets $(-\infty, \infty)$ and $[0, \infty)$, respectively.
Independent and dependent variables	The letter used to represent the elements in the domain of a function (the inputs) is the independent variable. The letter used for elements in the range (the outputs) is the dependent variable.	For the function $y = x^2$, the independent variable is x and the dependent variable is y. The value of y *depends* on x.

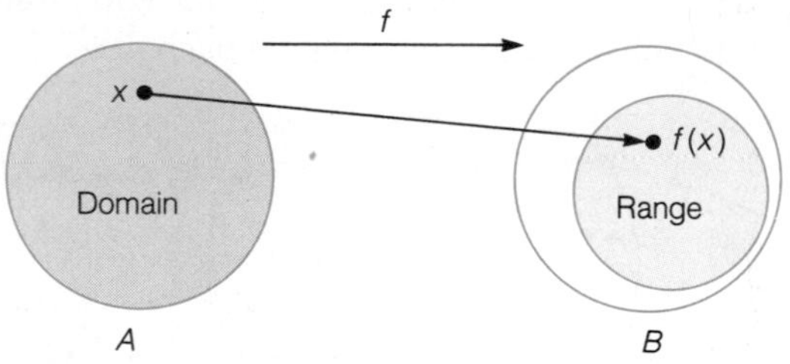

FIGURE 4
A function f from A to B

▼ **EXERCISE SET 2.3**

A

In Exercises 1–6, find the domain of each function.

1. (a) $y = -5x + 1$ (b) $y = \sqrt{-5x + 1}$
     (c) $y = \dfrac{x}{\sqrt{|-5x + 1|}}$

2. (a) $y = x^2 - 3x + 4$
     (b) $y = \sqrt{x^2 - 3x + 4}$ *Hint:* Section 1.7 explains a method for solving inequalities such as $x^2 - 3x + 4 \geq 0$.
     (c) $y = \sqrt{x^2 - 3x - 4}$

3. (a) $y = \dfrac{x + 4}{x - 4}$ (b) $y = \dfrac{x^2 + 4}{x^2 - 4}$ (c) $y = \sqrt{\dfrac{x^2 + 4}{x^2 - 4}}$

4. (a) $y = \dfrac{1}{x^3 - 16x}$ (b) $y = \dfrac{1}{x^3 - 16x^2}$
     (c) $y = \sqrt[3]{x^3 - 16x}$

5. (a) $y = \sqrt{x^2 - 4x - 5}$ (b) $y = \dfrac{2x}{\sqrt{x^2 - 4x - 5}}$
     (c) $y = \sqrt[3]{x^2 - 4x - 5}$

6. (a) $y = x^3 - 14x^2 - 6x - 1$
     (b) $y = \dfrac{1}{x^3 + x^2 - 2x - 2}$ (c) $y = \dfrac{1}{x^3 + x^2 - 2x}$

In Exercises 7–10, find the range of each function.

7. **(a)** $y = \dfrac{x+3}{x-5}$ **(b)** $y = \dfrac{x-5}{x+3}$

8. **(a)** $y = \dfrac{3x-2}{x+4}$ **(b)** $y = \dfrac{1}{x+4}$

9. **(a)** $y = x^2 + 4$ **(b)** $y = x^3 + 4$

10. **(a)** $y = \dfrac{ax+b}{x}$ (*a* and *b* are real numbers and $b \neq 0$)

  (b) $y = cx^3 + d$ $(c \neq 0)$

11. Let $A = \{x, y, z\}$ and $B = \{1, 2, 3\}$. Which of the rules displayed in the figure represent function from A to B?

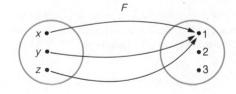

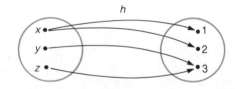

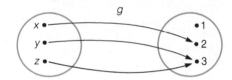

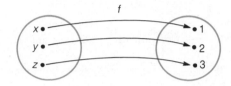

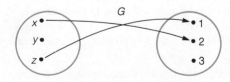

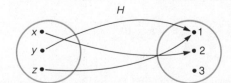

12. Let $D = \{a, b\}$ and $C = \{i, j, k\}$. Which of the rules displayed in the figure represent a function from D to C?

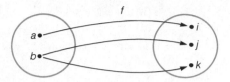

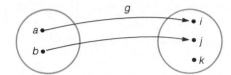

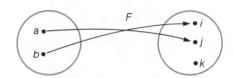

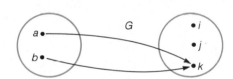

13. **(a)** Specify the range for each rule that represents a function in Exercise 11.
  (b) Specify the range for each rule that represents a function in Exercise 12.

14. **(a)** Suppose that in Exercise 11 all the arrows were reversed. Which, if any, of the new rules would be functions from B to A?
  (b) Suppose that in Exercise 12 all the arrows were reversed. Which, if any, of the new rules would be functions from C to D?

15. Each of the following rules defines a function whose domain is the set of all real numbers. Express each rule by means of an equation.

  EXAMPLE The rule *For each real number, compute its square* can be written $y = x^2$.

  (a) For each real number, subtract 3 and then square the result.
  (b) For each real number, compute its square and then subtract 3 from the result.
  (c) For each real number, multiply it by 3 and then square the result.
  (d) For each real number, compute its square and then multiply the result by 3.

16. Each of the following rules defines a function whose domain is the set of all real numbers. Express each rule in words.

    (a) $y = 2x^3 + 1$ **(b)** $y = 2(x + 1)^3$

    (c) $y = (2x + 1)^3$ **(d)** $y = (2x)^3 + 1$

17. Let $f(x) = x^2 - 3x + 1$. Compute the following.

    (a) $f(1)$ **(b)** $f(0)$ **(c)** $f(-1)$

    (d) $f(\tfrac{3}{2})$ **(e)** $f(z)$ **(f)** $f(x + 1)$

    (g) $f(a + 1)$ **(h)** $f(-x)$ **(i)** $|f(1)|$

    (j) $f(\sqrt{3})$ **(k)** $f(1 + \sqrt{2})$ **(l)** $|1 - f(2)|$

18. Let $H(x) = 1 - x + x^2 - x^3$.

    (a) Which number is larger, $H(0)$ or $H(1)$?

    (b) Find $H(\tfrac{1}{2})$. Does $H(\tfrac{1}{2}) + H(\tfrac{1}{2}) = H(1)$?

19. Let $f(x) = 3x^2$. Find the following.

    (a) $f(2x)$ **(b)** $2f(x)$ **(c)** $f(x^2)$

    (d) $[f(x)]^2$ **(e)** $f(x/2)$ **(f)** $f(x)/2$

For checking: No two answers are the same.

20. Let $f(x) = 4 - 3x$. Find the following.

    (a) $f(2)$ **(b)** $f(-3)$ **(c)** $f(2) + f(-3)$

    (d) $f(2 + 3)$ **(e)** $f(2x)$ **(f)** $2f(x)$

    (g) $f(x^2)$ **(h)** $f(1/x)$ **(i)** $f[f(x)]$

    (j) $x^2 f(x)$ **(k)** $1/f(x)$ **(l)** $f(-x)$

    (m) $-f(x)$ **(n)** $-f(-x)$

21. Let $H(x) = 1 - 2x^2$. Find the following.

    (a) $H(0)$ **(b)** $H(2)$

    (c) $H(\sqrt{2})$ **(d)** $H(\tfrac{5}{6})$

    (e) $H(x + 1)$ **(f)** $H(x + h)$

    (g) $H(x + h) - H(x)$ **(h)** $\dfrac{H(x + h) - H(x)}{h}$

22. **(a)** If $f(x) = 2x + 1$, does $f(3 + 1) = f(3) + f(1)$?

    (b) If $f(x) = 2x$, does $f(3 + 1) = f(3) + f(1)$?

    (c) If $f(x) = \sqrt{x}$, does $f(3 + 1) = f(3) + f(1)$?

23. Let $R(x) = \dfrac{2x - 1}{x - 2}$. Find the following.

    (a) The domain and range of R **(b)** $R(0)$

    (c) $R(\tfrac{1}{2})$ **(d)** $R(-1)$

    (e) $R(x^2)$ **(f)** $R(1/x)$

    (g) $R(a)$ **(h)** $R(x - 1)$

24. Let $g(x) = 2$, for all x. Find the output.

    (a) $g(0)$ **(b)** $g(5)$ **(c)** $g(x + h)$

25. Let $d(t) = -16t^2 + 96t$.

    (a) Compute $d(1)$, $d(\tfrac{3}{2})$, $d(2)$, $d(t_0)$.

    (b) For which values of t is $d(t) = 0$?

    (c) For which values of t is $d(t) = 1$?

26. Let $A(x) = |x^2 - 1|$. Compute $A(2)$, $A(1)$, and $A(0)$.

27. Let $g(t) = |t - 4|$. Find $g(3)$. Find $g(x + 4)$.

28. Let $f(x) = x^2/|x|$.

    (a) What is the domain of f?

    (b) Find $f(2)$, $f(-2)$, $f(20)$, and $f(-20)$.

    (c) What is the range of f? *Hint:* Look over your results in part (b).

29. Compute $\dfrac{f(x + h) - f(x)}{h}$ for the function f specified in each case.

    (a) $f(x) = x^2$ **(b)** $f(x) = 2x^2 - 3x + 1$

    (c) $f(x) = x^3$

30. Compute $\dfrac{f(x) - f(a)}{x - a}$ for the three functions given in Exercise 29.

31. Let $f(x) = \dfrac{x}{x - 1}$. Compute each difference quotient.

    (a) $\dfrac{f(x) - f(a)}{x - a}$ **(b)** $\dfrac{f(x) - f(3)}{x - 3}$

    (c) $\dfrac{f(x + h) - f(x)}{h}$ **(d)** $\dfrac{f(3 + h) - f(3)}{h}$

32. Let $g(t) = \dfrac{1}{3t - 5}$. Compute each difference quotient.

    (a) $\dfrac{g(t) - g(a)}{t - a}$ **(b)** $\dfrac{g(t) - g(2)}{t - 2}$

    (c) $\dfrac{g(t + h) - g(t)}{h}$ **(d)** $\dfrac{g(2 + h) - g(2)}{h}$

33. In each case a pair of functions f and g are given. Find all real numbers x_0 for which $f(x_0) = g(x_0)$.

    (a) $f(x) = 4x - 3$, $g(x) = 8 - x$

    (b) $f(x) = x^2 - 4$, $g(x) = 4 - x^2$

    (c) $f(x) = x^2$, $g(x) = x^3$

    (d) $f(x) = 2x^2 - x$, $g(x) = 3$

C **34.** Let $h(x) = \sqrt{x}$.

    (a) Compute $\dfrac{h(5) - h(1)}{5 - 1}$ and $\dfrac{1}{h(5) + 1}$. Round both answers to three decimal places.

    (b) Why are the two answers in part (a) the same?

C **35.** When \$1000 is deposited in a savings account at an annual rate of 12%, compounded quarterly, the amount in the account after t years is given by

$$A(t) = 1000\left(1 + \frac{0.12}{4}\right)^{4t}$$

    (a) Compute $A(1) - A(0)$. This is the amount by which the account will grow in the first year.

    (b) Compute $A(10) - A(9)$. This is the amount by which the account will grow in the tenth year.

C **36.** Let $f(n) = (1 + 1/n)^n$.

    (a) Complete the table. (Round results to three decimal places.)

n	1	2	5	10	15	20
$f(n)$						

    (b) By trial and error, find the smallest natural number n such that $f(n) > 2.7$.

(c) Using your calculator, can you find a number n such that $f(n) \geq 2.8$?

37. Let $g(n) = n^{1/n}$.

(a) Complete the table. (Round results to four decimal places.)

n	2	3	4	5	6	7	8
$g(n)$							

(b) By trial and error, find the smallest natural number n such that $g(n) < 1.2$.

38. Consider the function f defined by

$$f(x) = x^2 + \frac{2}{x^2}, \qquad x > 0.$$

(a) Complete the table. (Round results to four decimal places.)

x	1	1.05	1.10	1.15	1.20	1.25
$f(x)$						

(b) Which x-value in the table yields the smallest value for $f(x)$?

(c) It can be shown using calculus that the input x yielding the smallest possible output $f(x)$ for this function is $x = 2^{1/4}$. Which x-value in the table is closest to this?

(d) Compute $f(2^{1/4})$. Which value of $f(x)$ in the table is closest to $f(2^{1/4})$?

B

39. Let $f(x) = \dfrac{x-a}{x+a}$.

(a) Find $f(a), f(2a)$, and $f(3a)$. Is it true that $f(3a) = f(a) + f(2a)$?

(b) Show that $f(5a) = 2f(2a)$.

40. Let $M(x) = \dfrac{x-a}{x+u}$. Compute $M\left(\dfrac{1}{x}\right)$.

41. Let $\phi(y) = 2y - 3$. Show that $\phi(y^2) \neq [\phi(y)]^2$.

42. Let $k(x) = 5x^3 + \dfrac{5}{x^3} - x - \dfrac{1}{x}$. Show that $k(x) = k(1/x)$.

43. Let $f(x) = 2x + 3$. Find values for a and b such that the equation $f(ax + b) = x$ is true for all values of x.

44. If $p(x) = 2^x$, verify each identity.

(a) $2p(x) = p(x+1)$ **(b)** $p(a+b) = p(a) \cdot p(b)$

(c) $p(x) \cdot p(-x) - 1 = 0$

45. Let $f(t) = \dfrac{t-x}{t+y}$. Show that

$$f(x+y) + f(x-y) = \frac{-2y^2}{x^2 + 2xy}.$$

46. Let $f(z) = \dfrac{3z-4}{5z-3}$. Find $f\left(\dfrac{3z-4}{5z-3}\right)$.

47. Let $F(x) = \dfrac{ax+b}{cx-a}$. Show that $F\left(\dfrac{ax+b}{cx-a}\right) = x$. (Assume that $a^2 + bc \neq 0$.)

48. If $f(x) = -2x^2 + 6x + k$ and $f(0) = -1$, find k.

49. If $g(x) = x^2 - 3xk - 4$ and $g(1) = -2$, find k.

50. If $f(x) = 1/x^2$, show that

$$\frac{f(x+b) - f(x)}{h} = \frac{-2x - h}{(x+h)^2 x^2}.$$

51. A function doesn't always have to be given by an algebraic formula. For example, let the function L be defined by the following rule: $L(x)$ *is the power to which 2 must be raised to yield x.* (For the moment, we won't concern ourselves with the domain and range.) Then $L(8) = 3$, for example, since the power to which 2 must be raised to yield 8 is 3 $(8 = 2^3)$. Find the following outputs.

(a) $L(1)$ **(b)** $L(2)$ **(c)** $L(4)$ **(d)** $L(64)$

(e) $L(\frac{1}{2})$ **(f)** $L(\frac{1}{4})$ **(g)** $L(\frac{1}{64})$ **(h)** $L(\sqrt{2})$

The function L is called a *logarithm function*. The usual notation for $L(x)$ in this example is $\log_2 x$. Logarithm functions will be studied in Chapter 4.

52. Let $f(x) = ax^2 + bx + c$. Show that

$$\frac{f(x+h) - f(x)}{h} = 2ax + ah + b.$$

53. Let $q(x) = ax^2 + bx + c$. Evaluate

$$q\left(\frac{-b + \sqrt{b^2 - 4ac}}{2a}\right).$$

54. By definition, a **fixed point** for the function f is a number x_0 such that $f(x_0) = x_0$. For instance, to find any fixed points for the function $f(x) = 3x - 2$, we write $3x_0 - 2 = x_0$. On solving this last equation, we find that $x_0 = 1$. Thus, 1 is a fixed point for f. Calculate the fixed points (if any) for each function.

(a) $f(x) = 6x + 10$ **(b)** $g(x) = x^2 - 2x - 4$

(c) $S(t) = t^2$ **(d)** $R(z) = \dfrac{z+1}{z-1}$

55. Let $f(x) = \dfrac{3x-4}{x-3}$.

(a) Find $f[f(x)]$.

(b) Find $f[f(\frac{22}{7})]$. Try not to do it the hard way!

56. Consider the following two rules, F and G, where F is the rule that assigns to each person his or her mother, and G is the rule that assigns to each person his or her aunt. Explain why F is a function but G is not.

*In Exercises 57–59, use this definition: A **prime number** is a positive whole number with no factors other than itself and 1. For example, 2, 13, and 37 are primes, but 24 and 39 are not. By convention, 1 is not considered prime, so the list of the first few primes is as follows:*

2, 3, 5, 7, 11, 13, 17, 19, 23, 29, . . .

57. Let G be the rule that assigns to each positive integer the nearest prime. For example, $G(8) = 7$, since 7 is the prime nearest 8. Explain why G is not a function. How could you alter the definition of G to make it a function?

58. Let f be the function that assigns to each natural number x the number of primes that are less than or equal to x. For example, $f(12) = 5$, because, as you can easily check, five primes are less than or equal to 12. Similarly, $f(3) = 2$, because two primes are less than or equal to 3. Find $f(8), f(10),$ and $f(50)$.

C

59. If $P(x) = x^2 - x + 41$, find $P(1), P(2), P(3),$ and $P(4)$. Can you find a natural number x for which $P(x)$ is not prime?

60. Let $S(x) = \dfrac{3^x - 3^{-x}}{2}$ and $C(x) = \dfrac{3^x + 3^{-x}}{2}$. Show that the functions S and C possess the given properties.
 (a) $S(0) = 0; C(0) = 1; S(1) = \frac{4}{3}; C(1) = \frac{5}{3}$
 (b) $[C(x)]^2 - [S(x)]^2 = 1$
 (c) $S(-x) = -S(x)$ and $C(-x) = C(x)$
 (d) $S(x + y) = S(x)C(y) + C(x)S(y)$
 (e) $C(x + y) = C(x)C(y) + S(x)S(y)$
 (f) $S(2x) = 2S(x)C(x)$ and $C(2x) = [C(x)]^2 + [S(x)]^2$
 (g) $S(3x) = 3S(x) + 4[S(x)]^3$
 (h) $[S(x) + C(x)]^2 = S(2x) + C(2x)$
 (i) $S\left(\dfrac{x}{2}\right) = \pm\sqrt{\dfrac{C(x) - 1}{2}}$
 (j) $C\left(\dfrac{x}{2}\right) = \sqrt{\dfrac{C(x) + 1}{2}}$

2.4 ▼ THE GRAPH OF A FUNCTION

In my own case, I got along fine without knowing the name of the distributive law until my sophomore year in college; meanwhile I had drawn lots of graphs.

Professor Donald E. Knuth (Professor Knuth is judged by many to be the world's preeminent computer scientist.)

When the domain and range of a function are sets of real numbers, the function can be graphed in the same way in which equations were graphed earlier in this chapter. In graphing functions, the usual practice is to reserve the horizontal axis for the independent variable and the vertical axis for the dependent variable. The function or rule then tells you how you must pick your y-coordinate, once you have selected an x-coordinate.

DEFINITION

Graph of a Function

> The **graph** of a function f in the x-y plane consists of those points (x, y) such that x is in the domain of f and $y = f(x)$. See Figure 1.

FIGURE 1

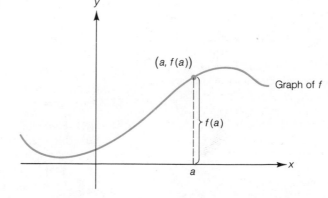

EXAMPLE 1 Graph the functions f and g defined as follows.

(a) $f(x) = x^2 - 1$ (b) $g(x) = |x^2 - 1|$

Solution   (a) The graph of this function is by definition the graph of the equation $y = x^2 - 1$. (Whether we label the vertical axis y or $f(x)$ is immaterial.) As preparation for drawing the graph, we first determine the domain of f, the x- and y-intercepts of the graph, and any symmetries the graph may have. As you can readily check, the results are as follows.

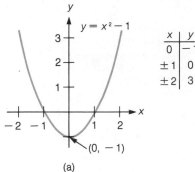

x	y
0	-1
± 1	0
± 2	3

> domain: $(-\infty, \infty)$
>
> x-intercepts: $1, -1$
>
> y-intercept: -1
>
> symmetry: about the y-axis

The required graph is then sketched in Figure 2(a).

(b) The graph of g is shown in Figure 2(b). It is obtained from the graph of f in Figure 2(a) as follows. First, for $x \geq 1$ or $x \leq -1$, the graph of g is identical to that of f because in this case, we have

$$g(x) = |x^2 - 1| = x^2 - 1 = f(x)$$

Second, on the interval $(-1, 1)$, the graph of g is obtained by reflecting the graph of f in the x-axis. This is because on the interval $(-1, 1)$, the quantity $x^2 - 1$ is negative and therefore

$$g(x) = |x^2 - 1| = -(x^2 - 1) = -f(x)$$ ▲

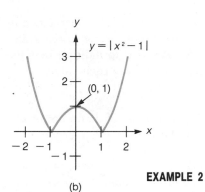

(a)

(b)

FIGURE 2

EXAMPLE 2 Specify the domain and the range of the function g whose graph appears in Figure 3(a).

Solution   The domain of g is just that portion of the x-axis (the inputs) utilized in graphing g. As Figure 3(b) indicates, this amounts to all real numbers x from 1 to 5, inclusive: $1 \leq x \leq 5$. Recall from Section 1.1 that this set of numbers is denoted by $[1, 5]$. To find the range of g, we need to check which part of the y-axis is utilized in graphing g. As Figure 3(b) indicates, this is the set of all real numbers y between 2 and 4, inclusive: $2 \leq y \leq 4$. Our shorthand notation for this interval of numbers is $[2, 4]$.

FIGURE 3

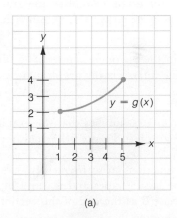

(a)

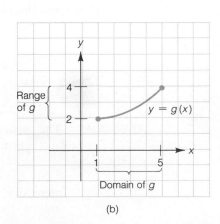

(b)   ▲

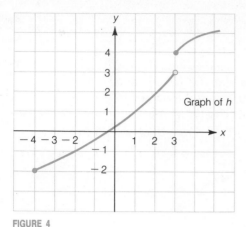

FIGURE 4

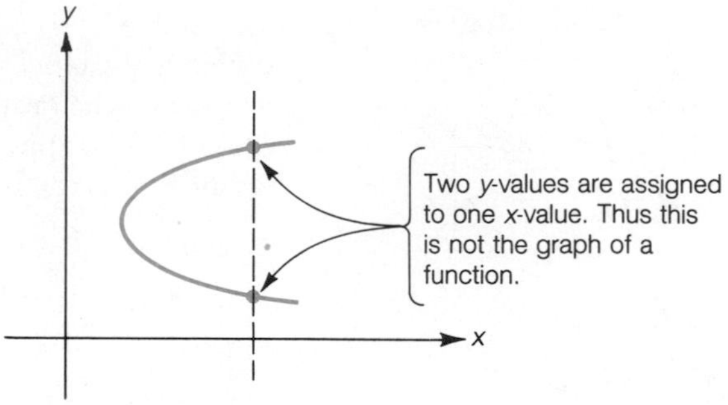

FIGURE 5

Two y-values are assigned to one x-value. Thus this is not the graph of a function.

EXAMPLE 3   A portion of the graph of a function h is shown in Figure 4. The open circle in the figure is used to indicate that the point $(3, 3)$ does not belong to the graph of h. The two filled-in circles, on the other hand, are used to indicate that the points $(-4, -2)$ and $(3, 4)$ do belong to the graph of h. Find each value.

(a) $h(-2)$ (b) $h(3)$ (c) $|h(-4)|$

Solution   (a) The function notation $h(-2)$ stands for the y-coordinate of that point on the graph of h whose x-coordinate is -2. Since the point $(-2, -1)$ is on the graph of h, we conclude $h(-2) = -1$.

(b) We conclude that $h(3) = 4$ because the point $(3, 4)$ lies on the graph of h. Note that h would not be considered a function if the point $(3, 3)$ were also part of the graph. (Why?)

(c) Since the point $(-4, -2)$ lies on the graph of h, we write $h(-4) = -2$. Thus $|h(-4)| = |-2| = 2$.   ▲

Most of the graphs that we looked at in the text and exercises for Section 2.1 are graphs of functions. However, it's important to understand that not every graph represents a function. Consider, for example, the graph in Figure 5.

Figure 5 also shows a vertical line intersecting the graph in two distinct points. The specific coordinates of the two points are unimportant. What the vertical line helps us to see is that two different y-values have been assigned to the same x-value, and therefore the graph cannot be the graph of a function $y = f(x)$.

The preceding remarks can be summarized as follows.

Vertical Line Test

A graph in the x-y plane represents a function $y = f(x)$, provided that any vertical line intersects the graph in at most one point.

EXAMPLE 4   The vertical line test implies that the graph in Figure 6(a) represents a function and that the graph in Figure 6(b) does not.   ▲

Six basic functions arise frequently enough to make it worth your while to memorize the basic shapes and features of their graphs. Figure 7 displays these graphs. Exercise 6 at the end of this section asks you to set up tables and verify

FIGURE 6

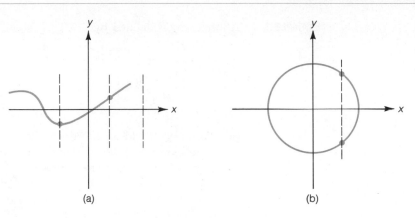

(a)   (b)

for yourself that the graphs in Figure 7 are indeed correct. From now on (with the exception of Exercise 6), if you need to sketch a graph of one of these basic functions, you should do so from memory.

FIGURE 7

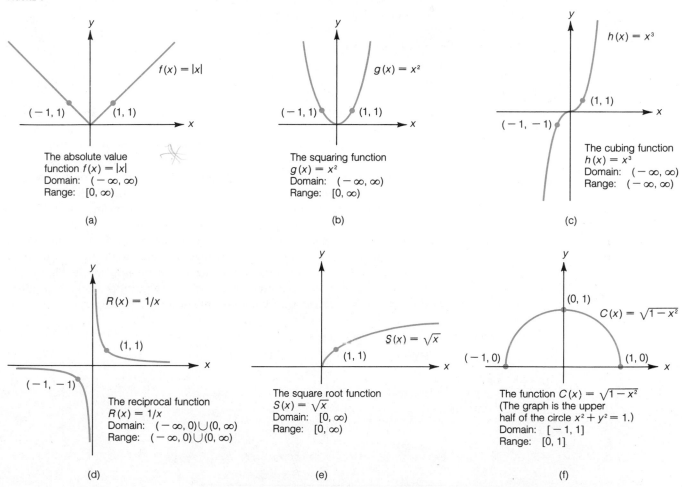

$f(x) = |x|$

$(-1, 1)$ $(1, 1)$

The absolute value
function $f(x) = |x|$
Domain: $(-\infty, \infty)$
Range: $[0, \infty)$

(a)

$g(x) = x^2$

$(-1, 1)$ $(1, 1)$

The squaring function
$g(x) = x^2$
Domain: $(-\infty, \infty)$
Range: $[0, \infty)$

(b)

$h(x) = x^3$

$(1, 1)$

$(-1, -1)$

The cubing function
$h(x) = x^3$
Domain: $(-\infty, \infty)$
Range: $(-\infty, \infty)$

(c)

$R(x) = 1/x$

$(1, 1)$

$(-1, -1)$

The reciprocal function
$R(x) = 1/x$
Domain: $(-\infty, 0) \cup (0, \infty)$
Range: $(-\infty, 0) \cup (0, \infty)$

(d)

$S(x) = \sqrt{x}$

$(1, 1)$

The square root function
$S(x) = \sqrt{x}$
Domain: $[0, \infty)$
Range: $[0, \infty)$

(e)

$(0, 1)$

$C(x) = \sqrt{1 - x^2}$

$(-1, 0)$ $(1, 0)$

The function $C(x) = \sqrt{1 - x^2}$
(The graph is the upper
half of the circle $x^2 + y^2 = 1$.)
Domain: $[-1, 1]$
Range: $[0, 1]$

(f)

All the graphs in Figure 7 arise from functions defined by rather simple equations. In some instances, however, functions may be defined by combinations of equations. The next two examples display instances of this.

EXAMPLE 5 A function g is defined by

$$g(x) = \begin{cases} x^2 & \text{if } x < 2 \\ \dfrac{1}{x} & \text{if } x \ge 2 \end{cases}$$

(a) Find $g(1)$, $g(2)$, and $g(3)$. **(b)** Sketch the graph of g.

Solution

(a) To find $g(1)$, do we substitute the value $x = 1$ in the expression x^2 or in the expression $1/x$? According to the instructions contained in the given definition of g, we should use the expression x^2 whenever the inputs are less than 2. Thus we have $g(1) = 1^2 = 1$, that is, $g(1) = 1$. On the other hand, the definition of g tells us to use the expression $1/x$ whenever the inputs are greater than or equal to 2. So in this case we have $g(2) = \frac{1}{2}$ and $g(3) = \frac{1}{3}$.

(b) For the graph of g, we look back at Figures 7(b) and (d) and choose the appropriate portion of each. The result is displayed in Figure 8. The open circle in the figure is used to indicate that the point $(2, 4)$ does not belong to the graph of g. The filled-in circle, on the other hand, is used to indicate that the point $(2, \frac{1}{2})$ does belong to the graph of g. ▲

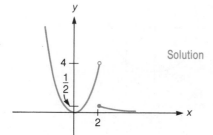

FIGURE 8

A graph of the function

$$g(x) = \begin{cases} x^2 & \text{if } x < 2 \\ \dfrac{1}{x} & \text{if } x \ge 2 \end{cases}$$

EXAMPLE 6 Graph the function f defined by

$$f(x) = \begin{cases} x^2 - 1 & \text{if } x \le 2 \\ 5 - x & \text{if } x > 2 \end{cases}$$

Solution

Given this definition of f, we'll set up two tables. For the first table, the inputs x will all be less than or equal to 2, and the outputs will be computed using the expression $x^2 - 1$. In the second table, the inputs x will be greater than 2, and the outputs will be computed from the expression $5 - x$. These tables and the graph derived from them are shown in Figure 9. [Actually, we could save a few steps by referring to the graph of $y = x^2 - 1$ shown in Figure 2(a).]

We make two observations here. First, notice that, in setting up a table for $y = 5 - x$ with $x > 2$, only two points are needed, since the graph is a portion of a straight line. Second, notice that in this example, as opposed to the function in Example 5, the two portions of the graph together form a graph with no break or gap in it. We say that the function f (whose graph appears in Figure 9) is *continuous* at $x = 2$ but that the function g (whose graph appears in Figure 8) is *discontinuous* at $x = 2$. A rigorous definition of continuity is properly a subject for

FIGURE 9

$$f(x) = \begin{cases} x^2 - 1, & x \le 2 \\ 5 - x, & x > 2 \end{cases}$$

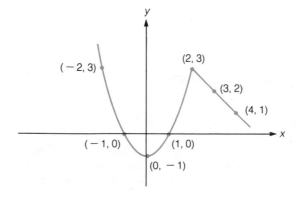

$y = x^2 - 1, \quad x \le 2$

x	± 2	± 1	0
y	3	0	-1

$y = 5 - x, \quad x > 2$

x	3	4
y	2	1

calculus. However, even at the intuitive level at which we've presented the idea here, you'll find that the concept is useful in helping you to organize your thoughts about the graph of a function. ▲

We conclude this section with some terminology that is helpful in analyzing functions. To introduce this terminology we'll use the graphs in Figures 10 and 11.

The points P and Q in Figures 10 and 11 are called *turning points*. At a **turning point**, the graph changes from rising to falling, or vice versa. In Figure 10, the highest point on the graph of the function G is $P(2, 102)$. We say that the *maximum value* of the function G is 102 and that this maximum value occurs at $t = 2$. More generally, we say that $f(x_0)$ is the **maximum value** of a function f if the inequality $f(x_0) \geq f(x)$ holds for every x in the domain of f. Minimum values are defined similarly: $f(x_0)$ is the **minimum value** for a function f if the inequality $f(x_0) \leq f(x)$ holds for every x in the domain of f. Assuming that the domain of the function in Figure 10 is $[0, 6]$, the minimum value of the function G is 99, and it occurs when $t = 6$. In Figure 11, the minimum value of the function is -1, and this minimum occurs when $x = 3$. Not every function has a maximum or minimum value. For example, the function $y = 1/x$ possesses neither a maximum nor a minimum value; you can see this by looking back at the graph of this function in Figure 7(d).

The function G in Figure 10 is said to be *increasing* on the open interval $(0, 2)$, and *decreasing* on the interval $(2, 6)$. In terms of the temperature interpretation for Figure 10, the patient's temperature is rising between noon and two o'clock and falling between two and six o'clock. In Figure 11, the function is decreasing on the interval $(-\infty, 3)$ and increasing on the interval $(3, \infty)$. For theoretical work, the terms increasing and decreasing can be defined in terms of inequalities. A function f is **increasing** on an interval if the following condition holds: If x_1 and x_2 are in the interval and $x_1 < x_2$, then $f(x_1) < f(x_2)$. Similarly, a function f is said to be **decreasing** on an interval if the following condition holds: If x_1 and x_2 are in the interval and $x_1 < x_2$, then $f(x_1) > f(x_2)$.

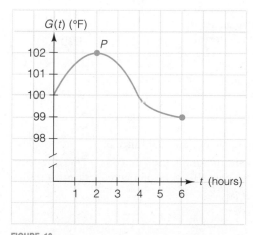

FIGURE 10

The graph of the function G is a fever graph. $G(t)$ is a patient's temperature, t hours after 12 noon, $0 \leq t \leq 6$.

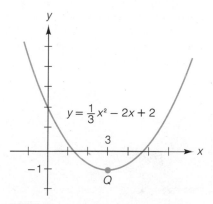

FIGURE 11

The point Q is a turning point. The minimum value of the function is -1. The function is decreasing on $(-\infty, 3)$ and increasing on $(3, \infty)$.

▼ EXERCISE SET 2.4

A

1. The figure displays the graph of a function f.
 (a) Is $f(0)$ positive or negative?
 (b) Find $f(-2), f(1), f(2)$, and $f(3)$.
 (c) Which is larger, $f(2)$ or $f(4)$?
 (d) Find $f(4) - f(1)$.
 (e) Find $|f(4) - f(1)|$.
 (f) Write the domain and range of f using the interval notation $[a, b]$.

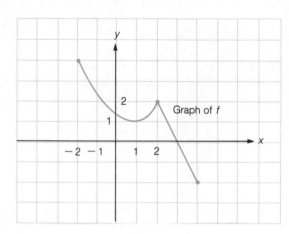

Graph of f

2. The figure shows the graph of a function h.
 (a) Find $h(a), h(b), h(c)$, and $h(d)$.
 (b) Is $h(0)$ positive or negative?
 (c) For which values of x does $h(x) = 0$?
 (d) Which is larger, $h(b)$ or $h(0)$?
 (e) As x increases from c to d, do the corresponding values of $h(x)$ increase or decrease?
 (f) As x increases from a to b, do the corresponding values of $h(x)$ increase or decrease?

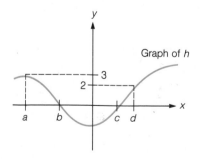

Graph of h

In Exercises 3–5, refer to the graphs of the functions f and g in the figure. Assume that the domain of each function is $[-3, 3]$ and that the axes are marked off in one-unit intervals.

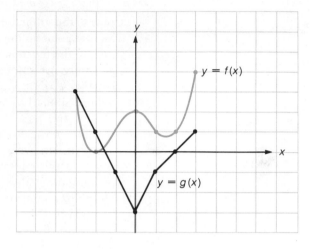

3. (a) Which is larger, $f(-2)$ or $g(-2)$?
 (b) Compute $f(0) - g(0)$.
 (c) Which among the following three quantities is the smallest?

$$f(1) - g(1), \qquad f(2) - g(2), \qquad f(3) - g(3)$$

 (d) For which value(s) of x does $g(x) = f(1)$?
 (e) Is the number 4 in the range of f or in the range of g?

4. (a) For the interval $[0, 3]$, is the quantity $g(x) - f(x)$ positive or negative?
 (b) For the interval $(-3, -2)$, is the quantity $g(x) - f(x)$ positive or negative?
 (c) Compute $\dfrac{f(x) - f(2)}{x - 2}$ when $x = 3$.
 (d) Compute $\dfrac{g(x) - g(-2)}{x + 2}$ when $x = -3$.

5. Specify the range of f and the range of g.

6. Set up a table and graph each function. (The symmetry tests from Section 2.1 are helpful here.)
 (a) $f(x) = |x|$
 (b) $g(x) = x^2$
 (c) $h(x) = x^3$
 (d) $R(x) = 1/x$
 (e) $S(x) = \sqrt{x}$
 (f) $C(x) = \sqrt{1 - x^2}$

7. Complete the following table.

FUNCTION	$|x|$	x^2	x^3
TURNING POINT			
MAXIMUM VALUE			
MINIMUM VALUE			
INTERVAL(S) WHERE INCREASING			
INTERVAL(S) WHERE DECREASING			

8. Set up and complete a table as in Exercise 7 for the three functions $1/x$, $\sqrt{x}$, and $\sqrt{1-x^2}$.

In Exercises 9 and 10, use the vertical line test to determine if each graph represents a function $y = f(x)$.

9.

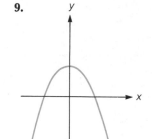

(a)

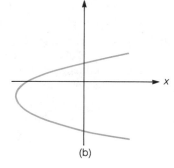

(b)

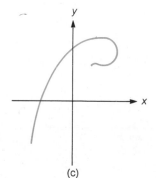

(c)

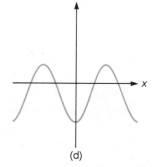

(d)

10.

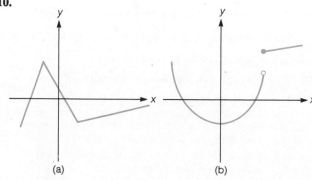

(a)   (b)

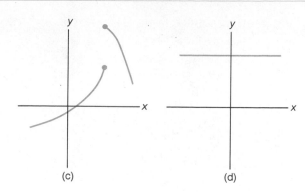

(c)   (d)

11. Let $f(x) = x^2$. Find the slope of the straight line that passes through the points $(3, f(3))$ and $(4, f(4))$. Include a sketch.

12. Let $G(x) = x^3$. Find the slope of the straight line that passes through the points $(0, G(0))$ and $(-2, G(-2))$. Include a sketch.

13. Let $T(x) = \sqrt{x}$. Which line has the larger slope, the line that passes through $(1, T(1))$ and $(4, T(4))$ or the line that passes through $(4, T(4))$ and $(9, T(9))$?

14. In Exercises 14 and 15, sketch a graph of each function for the given domain.
  (a) $p(x) = x^2$ with domain $[-1, 2]$
  (b) $q(x) = x^3$ with domain $[0, 1]$
  (c) $r(x) = 1/x$ with domain $(0, \infty)$
  (d) $s(x) = \sqrt{x}$ with domain $[1, 4]$

15. **(a)** $k(x) = \sqrt{1-x^2}$ with domain $[0, 1]$
  (b) $m(x) = \sqrt{1-x^2}$ with domain $(0, 1)$
  (c) $n(x) = |x|$ with domain consisting only of the numbers 0, 1, 2, 3, and 4
  (d) $z(x) = x^2$ with domain consisting of the numbers -2, -1, 0, 1, and 4 as well as those numbers in the open interval $(2, 3)$

16. What are the domain and range of the function whose graph is the horizontal line $y = 3$?

In Exercises 17–30, graph the function defined by the given rules.

17. $f(x) = \begin{cases} |x| & \text{if } x \le 0 \\ x^2 & \text{if } x > 0 \end{cases}$

18. **(a)** $g(x) = \begin{cases} |x| & \text{if } x \le 0 \\ x+1 & \text{if } x > 0 \end{cases}$

  (b) $F(x) = \begin{cases} |x| & \text{if } x < 0 \\ x+1 & \text{if } x \ge 0 \end{cases}$

19. $A(x) = \begin{cases} x^3 & \text{if } -2 \le x \le -1 \\ x^2 & \text{if } x > -1 \end{cases}$

20. $B(x) = \begin{cases} \sqrt{1-x^2} & \text{if } -1 \le x < 1 \\ 1/x & \text{if } x \ge 1 \end{cases}$

21. $C(x) = \begin{cases} x^3 & \text{if } x < 1 \\ \sqrt{x} & \text{if } x > 1 \end{cases}$

22. (a) $f(x) = \begin{cases} x^2/|x| & \text{if } x \neq 0 \\ 0 & \text{if } x = 0 \end{cases}$

    (b) $F(x) = \begin{cases} x^2/|x| & \text{if } x \neq 0 \\ 1 & \text{if } x = 0 \end{cases}$

23. (a) $g(x) = \begin{cases} x/|x| & \text{if } x \neq 0 \\ 0 & \text{if } x = 0 \end{cases}$

    (b) $G(x) = \begin{cases} x/|x| & \text{if } x \neq 0 \\ 1 & \text{if } x = 0 \end{cases}$

24. $U(x) = \begin{cases} 1 & \text{if } x \leq -2 \\ -1 & \text{if } x > -2 \end{cases}$

25. $V(x) = \begin{cases} x & \text{if } x \leq -2 \\ 1 & \text{if } x > -2 \end{cases}$

26. (a) $W(x) = \begin{cases} x & \text{if } x \leq -2 \\ -2 & \text{if } x > -2 \end{cases}$

    (b) $D(x) = \begin{cases} x & \text{if } x < -2 \\ -2 & \text{if } x > -2 \end{cases}$

27. $f(x) = \begin{cases} 1/x & \text{if } x < -1 \\ x & \text{if } -1 \leq x \leq 1 \\ 1/x & \text{if } x > 1 \end{cases}$

28. $g(x) = \begin{cases} 1/x & \text{if } x < -\frac{1}{2} \\ 1 & \text{if } -\frac{1}{2} \leq x \leq 1 \\ x^3 & \text{if } x > 1 \end{cases}$

29. $y = \begin{cases} x & \text{if } 0 \leq x < 1 \\ x - 1 & \text{if } 1 \leq x < 2 \\ x - 2 & \text{if } 2 \leq x < 3 \end{cases}$

30. $T(x) = \begin{cases} 2x + 2 & \text{if } -1 \leq x < 0 \\ -2x + 2 & \text{if } 0 \leq x < 1 \\ 2x - 2 & \text{if } 1 \leq x < 2 \\ -2x + 6 & \text{if } 2 \leq x \leq 3 \end{cases}$

B

31. (a) Sketch the graph of $f(x) = \dfrac{x^2 - 25}{x + 5}$. What is the domain of f?

    (b) Graph the function F defined by the rule

$$F(x) = \begin{cases} \dfrac{x^2 - 25}{x + 5} & \text{if } x \neq -5 \\ -10 & \text{if } x = -5 \end{cases}$$

32. In the figure, find the coordinates of the point T if the slope of the straight line passing through the two points P and T is

    (a) 5   *Hint:* Let the coordinates of T be (x, x^2).

    (b) 1000   **(c)** k

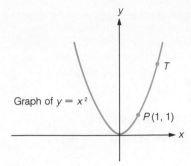

Graph of $y = x^2$

T

$P(1, 1)$

33. Let $f(x) = 1/x$. Find a number t such that the slope of the straight line passing through the two points $(1, 1)$ and $(t, f(t))$ on the graph of f is $-\frac{1}{5}$. Include a sketch.

34. Find the coordinates of a point P on the graph of $f(x) = \sqrt{x}$ if the slope of the straight line passing through P and the point $(1, 1)$ is $\frac{1}{7}$. Include a sketch.

35. The graph of a function f is a straight line. As the figure shows, A and B are two points on the line and the x-coordinates of A and B are h units apart. The coordinates of A are $(x, f(x))$. What are the coordinates of B? Show that the slope of the line is $\dfrac{f(x + h) - f(x)}{h}$.

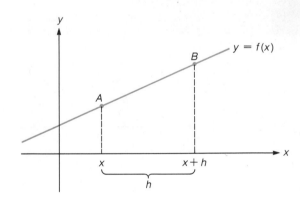

$y = f(x)$

B

A

x $x + h$

h

36. Let $F(x) = 2x + 3$. Using the values $x = 1$ and $h = \frac{1}{2}$, compute the value of $\dfrac{F(x + h) - F(x)}{h}$. Include a sketch of the type shown in Exercise 35.

37. The **average rate of change** of a function f between $x = a$ and $x = b$ is defined to be the quantity $\dfrac{f(b) - f(a)}{b - a}$.

    (a) By means of a sketch, indicate why the average rate of change represents the slope of the line joining the two points on the graph of f whose x-coordinates are a and b.

(b) Which is larger, the average rate of change of $f(x) = x^2$ between 0 and 1 or between 10 and 11?

(c) Which of the following functions has the largest average rate of change between 3 and 4?

$$f(x) = x^2 \qquad g(x) = x^3 \qquad h(x) = \frac{1}{x}$$

(d) Find the average rate of change for the function $f(x) = x^2$ over the following intervals:
  (i) $a = 1$ to $b = 1.1$
  (ii) $a = 1$ to $b = 1.01$
  (iii) $a = 1$ to $b = 1.001$
  As b gets closer and closer to 1, what value does your average rate of change seem to be approaching? This "target value" is sometimes called the **instantaneous rate of change**.

38. Suppose that during the first 4 hours of a laboratory experiment, the temperature of a certain substance is given by the formula $f(t) = t^3 - 6t^2 + 9t$, where t is measured in hours with $t = 0$ corresponding to the time the experiment begins, and where $f(t)$ is the temperature (degrees Fahrenheit) of the substance at time t. Calculate the average rate of change of temperature between the following times:
  (a) $t = 0$ and $t = 1$
  (b) $t = 1$ and $t = 2$
  (c) $t = 0$ and $t = 3$
  (d) $t = 0$ and $t = 4$
  What are the units associated with your answers?

39. Assume that the distance covered by a falling object is given by the formula $f(t) = 16t^2$, where t is measured in seconds, $f(t)$ is in feet, and $t = 0$ corresponds to the instant the object first begins to fall. Thus, after 1 sec, the object has fallen $16(1)^2$, or 16 ft; after 2 sec, the object has fallen a total of $16(2)^2$, or 64 ft; and so on.
  (a) Calculate $f(2) - f(1)$, the distance the object has fallen during the time interval from $t = 1$ to $t = 2$ (i.e., during the second second). Include a sketch of the graph of f. On the t-axis, indicate the time interval from $t = 1$ to $t = 2$; on the y- or $f(t)$-axis, indicate the interval corresponding to the distance $f(2) - f(1)$.
  (b) Calculate $f(3) - f(2)$, the distance the object has fallen during the time interval from $t = 2$ to $t = 3$ (i.e., during the third second). Include a sketch as you did in part (a).
  (c) The **average velocity** during the time interval from $t = a$ to $t = b$ is defined to be the quantity $\dfrac{f(b) - f(a)}{b - a}$. Note that this is just distance divided by elapsed time and that the units are feet per second. Calculate the average velocity from $t = 1$ to $t = 3$. On a graph of f, locate the two points

$(1, f(1))$ and $(3, f(3))$. What is the slope of the line joining these two points?

(d) Calculate the average velocity over each of the following intervals:
  (i) $a = 1$ to $b = 1.1$
  (ii) $a = 1$ to $b = 1.001$
  (iii) $a = 1$ to $b = 1.00001$
  As b comes closer and closer to 1, what value does your average velocity seem to be approaching? This "target value" is called the **instantaneous velocity** (or just **velocity**) at $t = 1$.

40. The notation $[x]$ denotes the greatest integer that is less than or equal to x. For instance, $[3] = 3$, $[4\frac{1}{2}] = 4$, and $[-\frac{4}{3}] = -2$.
  (a) Complete the following table.

x	0	0.1	0.5	0.9	1.0
$[x]$					

  (b) Graph the **greatest integer function** $y = [x]$ on the interval $-2 \le x < 3$.
  (c) The domain of the greatest integer function is $(-\infty, \infty)$. What is the range?

41. A bug travels counterclockwise around the square in the figure, beginning from the point $(1, 0)$. As indicated in the figure, t denotes the distance the bug has traveled, and $P(t)$ denotes the bug's location when it has traveled the distance t. For example, when $t = 0$, $P(t)$ is the point $(1, 0)$, and when $t = 1$, $P(t)$ is the point $(1, 1)$.
  (a) A function S is defined as follows: $S(t)$ is the y-coordinate of the point $P(t)$. For example, if $t = \frac{1}{2}$, then $P(t)$ is the point $(1, \frac{1}{2})$ and, therefore, $S(t) = \frac{1}{2}$. Set up a table showing the values of $S(t)$ with t running from 0 to 3 in increments of $\frac{1}{4}$. Then graph the function S on the interval $0 \le t \le 3$.

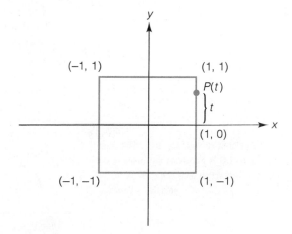

(b) Set up a table showing the values of $S(t)$ with t running from 3 to 5 in increments of $\frac{1}{4}$. Then graph the function S on the interval $3 \leq t \leq 5$.

(c) Graph the function S for $0 \leq t \leq 8$. This corresponds to one complete trip around the square by the bug.

(d) Graph the function S for $8 \leq t \leq 16$. How does your result relate to the graph in part (c)?

42. Let $P(t)$ be defined as in Exercise 41. Define a function C as follows: $C(t)$ is the x-coordinate of the point $P(t)$. Graph the function C for $0 \leq t \leq 8$.

2.5 ▼ TECHNIQUES IN GRAPHING

One good picture is worth a thousand words. For us, in our study of functions, this means draw graphs. Even more, cultivate the habit of thinking graphically, to the point where it becomes almost second nature.

George F. Simmons in *Calculus with Analytic Geometry* (New York: McGraw-Hill, 1985)

. . . geometrical figures are graphic formulas.

David Hilbert (1862–1943)

The simple geometric concepts of reflection and translation can be used to great advantage in graphing. We discussed the idea of reflection in the x-axis and in the y-axis in Section 2.1. (As background for the present section, you should review the first two definitions in the box on page 72, as well as Example 4 on pages 72–73.) By a **translation** of a graph, we mean a shift or movement in its location that does not change the size or shape of the graph.

EXAMPLE 1

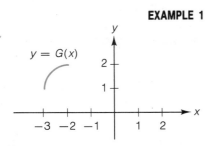

$y = G(x)$

FIGURE 1

The graph of a function G is shown in Figure 1. In each case, sketch the resulting graph after the following operations are carried out on the graph of G:

(a) a translation of four units in the positive x-direction followed by a translation of one unit in the negative y-direction;

(b) a translation of three units in the positive x-direction followed by a reflection in the y-axis;

(c) a reflection in the y-axis followed by a translation of three units in the positive x-direction. [Note that these are the same two operations that are used in part (b), but here the order is reversed.]

Solution

(a) Figure 2 shows the result of translating the graph of G four units in the positive x-direction and then one unit in the negative y-direction. (As you can check by drawing a sketch, the same end result is obtained if we first translate one unit in the negative y-direction and then four units in the positive x-direction. *Fact:* If several translations are to be carried out in succession, the end result will be the same, no matter in what order the individual translations are carried out.)

FIGURE 2

The colored curve is the end result when the graph of G is translated four units in the positive x-direction and then one unit in the negative y-direction.

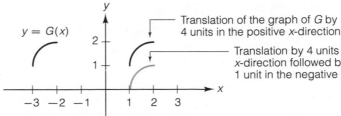

$y = G(x)$

Translation of the graph of G by 4 units in the positive x-direction

Translation by 4 units in the positive x-direction followed by a translation of 1 unit in the negative y-direction

FIGURE 3

Figure (a) shows the result of translating the graph of G three units in the positive x-direction. When this is followed by a reflection in the y-axis, we obtain the colored curve in figure (b).

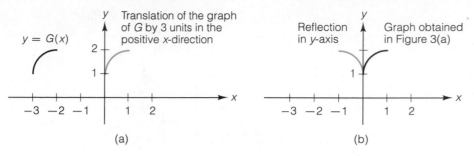

(a)   (b)

FIGURE 4

The reflection of the graph of G in the y-axis followed by a translation of three units in the positive x-direction.

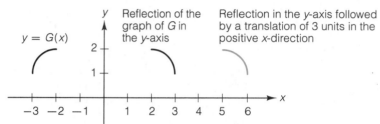

(**b**) Figure 3(a) shows the graph of G after a translation of three units in the positive x-direction. When this is followed by a reflection in the y-axis, we obtain the result shown in Figure 3(b).

(**c**) Figure 4 shows what happens to the graph of G if we use the same two operations that were used in part (b), but we reverse the order in which they are carried out. Note that the end result is quite different than that obtained in part (b).   ▲

There are some very basic, very useful connections between the geometric ideas that we have just been discussing and the equation of a graph. We'll explore these connections using some familiar graphs. In Figure 5(a), we begin with the line $y = -2x$. When we translate this line to the right four units and up three units, we obtain the line $y - 3 = -2(x - 4)$, as shown in Figure 5(a). (The point–slope formula tells us that this is indeed the correct equation for the translated line.) The pertinent observation here is this: *In the equation $y = -2x$, the effect of replacing x and y with $x - 4$ and $y - 3$, respectively, is to translate the graph of $y = -2x$ to the right four units and up three units.*

Now let's see what happens if we translate in the opposite direction; that is, suppose that we translate the graph of $y = -2x$ four units to the left and three

FIGURE 5

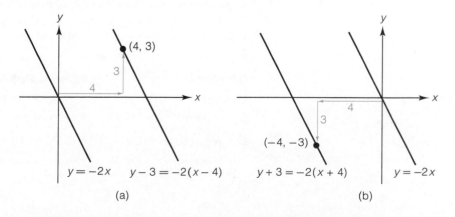

(a)   (b)

FIGURE 6

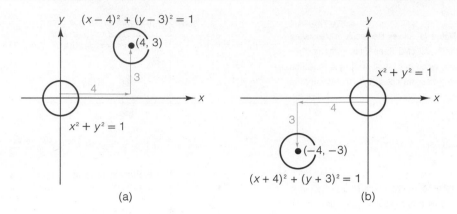

(a)

(b)

units down. As indicated in Figure 5(b) on the previous page, the line now passes through the point $(-4, -3)$ and, therefore (according to the point–slope formula), the equation of the translated line is

$$y - (-3) = -2[x - (-4)]$$

or

$$y + 3 = -2(x + 4)$$

In summary, then: *In the equation $y = -2x$, the effect of replacing x and y with $x + 4$ and $y + 3$, respectively, is to translate the graph of $y = -2x$ to the left four units and down three units.*

As a second example of this same pattern, consider the circles in Figure 6. Looking at the figure in part (a), we make this observation: The effect of replacing x with $x - 4$ and y with $y - 3$ in the equation $x^2 + y^2 = 1$ is to translate the graph of $x^2 + y^2 = 1$ to the right four units and up three units. Similarly, Figure 6(b) shows that the effect of replacing x with $x + 4$ and y with $y + 3$ in the equation $x^2 + y^2 = 1$ is to translate the graph to the left four units and down three units.

The examples that we have been considering are specific instances of the following basic result. The result is valid whether or not the given equation and graph represent a function.

PROPERTY SUMMARY

TRANSLATION AND COORDINATES

Suppose that we have an equation that determines a graph in the x-y plane, and let h and k denote positive numbers. Then, replacing x with $x - h$ or $x + h$, or replacing y with $y - k$ or $y + k$, has the following effects on the graph of the original equation.

REPLACEMENT	RESULTING TRANSLATION
1. x replaced with $x - h$	h units in the positive x-direction
2. y replaced with $y - k$	k units in the positive y-direction
3. x replaced with $x + h$	h units in the negative x-direction
4. y replaced with $y + k$	k units in the negative y-direction

EXAMPLE 2 The graph of the equation $|x| + |y| = 1$ is shown in Figure 7. Sketch the graph of each of the following equations. (*Question for review* Is the graph in Figure 7 the graph of a function?)

(**a**) $|x - 4| + |y| = 1$ (**b**) $|x + 4| + |y| = 1$

(**c**) $|x| + |y - 4| = 1$ (**d**) $|x| + |y + 4| = 1$

(**e**) $|x - 4| + |y + 4| = 1$

FIGURE 7

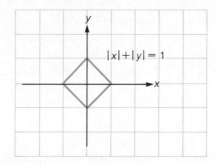

Solution

EQUATION	HOW TO GRAPH
(**a**) $|x - 4| + |y| = 1$	Translate the graph of $|x| + |y| = 1$ four units to the right. See Figure 8(a).
(**b**) $|x + 4| + |y| = 1$	Translate the graph of $|x| + |y| = 1$ four units to the left. See Figure 8(a).
(**c**) $|x| + |y - 4| = 1$	Translate the graph of $|x| + |y| = 1$ four units up (i.e., four units in the positive y-direction). See Figure 8(b).
(**d**) $|x| + |y + 4| = 1$	Translate the graph of $|x| + |y| = 1$ four units down (i.e., four units in the negative y-direction). See Figure 8(b).
(**e**) $|x - 4| + |y + 4| = 1$	Translate the graph of $|x| + |y| = 1$ four units to the right and four units down. See Figure 8(c).

FIGURE 8

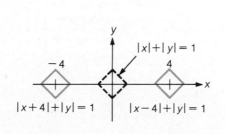

(a)

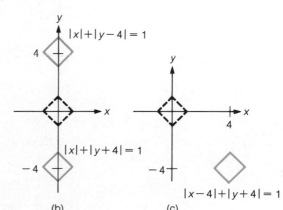

(b)   (c)

FIGURE 9
The graph of $y = x^2 - 4$ is obtained by translating the graph of $y = x^2$ four units in the negative y-direction.

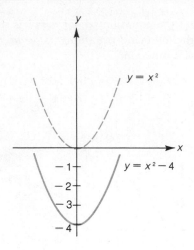

EXAMPLE 3 Graph the function $y = x^2 - 4$.

Solution For the moment, think of the given equation as

$$y + 4 = x^2$$

This last equation is obtained from $y = x^2$ by replacing y with $y + 4$. Therefore, the graph of $y + 4 = x^2$ is obtained by translating the graph of $y = x^2$ four units in the negative y-direction. See Figure 9. ▲

EXAMPLE 4 Graph the functions defined by $y = |x + 2|$ and $y = |x + 2| - 1$.

Solution The equation $y = |x + 2|$ is obtained from $y = |x|$ by replacing x with $x + 2$. Therefore, the graph of $y = |x + 2|$ is obtained by translating the graph of $y = |x|$ two units in the negative x-direction, as indicated in Figures 10(a) and 10(b).

FIGURE 10

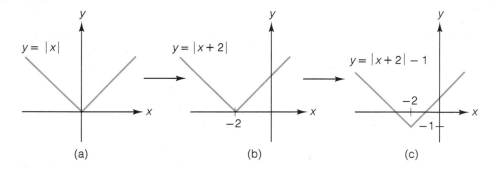

Next, what about the graph of $y = |x + 2| - 1$? If we think of this last equation as

$$y + 1 = |x + 2|$$

we see that it is obtained from $y = |x + 2|$ by replacing y with $y + 1$. Therefore the required graph can be obtained by translating the graph of $y = |x + 2|$ in Figure 10(b) one unit in the negative y-direction. See Figure 10(c). ▲

FIGURE 11

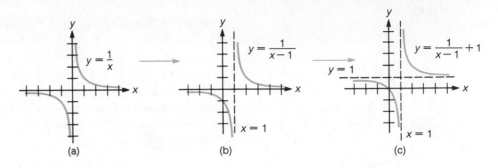

(a)   (b)   (c)

EXAMPLE 5   Graph the functions defined by $y = \dfrac{1}{x-1}$ and $y = \dfrac{1}{x-1} + 1$.

Solution   Begin with the graph of $y = 1/x$ in Figure 11(a). As we explained in Section 2.1, the x- and y-axes are asymptotes for this graph. Moving this graph to the right one unit yields the graph of $y = \dfrac{1}{x-1}$, shown in Figure 11(b). (Note that the vertical asymptote moves one unit to the right also, but the horizontal asymptote is unchanged.) Next, we move the graph in Figure 11(b) up one unit (why?) to obtain the graph of $y = \dfrac{1}{x-1} + 1$, as shown in Figure 11(c). (You should verify for yourself that the y-intercepts in Figures 11(b) and 11(c) are -1 and 0 respectively.)   ▲

FIGURE 12

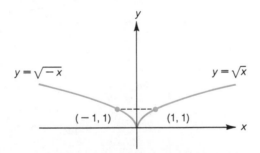

In addition to using translation as an aid in graphing, we can also use the idea of reflection, which we discussed in Section 2.1. Consider the graphs of $y = \sqrt{x}$ and $y = \sqrt{-x}$ in Figure 12. Notice that each point on the graph of $y = \sqrt{-x}$ is the reflection about the y-axis of a corresponding point on $y = \sqrt{x}$. For example, the point $(-1, 1)$ on $y = \sqrt{-x}$ is the reflection of the point $(1, 1)$ on $y = \sqrt{x}$. More generally, each point on $y = \sqrt{-x}$ is of the form $(-x, y)$, where (x, y) is a point on $y = \sqrt{x}$. For the purpose of obtaining a working rule that will be easy to apply, we rephrase our observations here as follows. The effect of replacing x with $-x$ in the equation $y = \sqrt{x}$ is to reflect the graph of $y = \sqrt{x}$ in the y-axis.

In just the same way that the graphs of $y = \sqrt{x}$ and $y = \sqrt{-x}$ are reflections of each other in the y-axis, the graphs of $y = \sqrt{x}$ and $y = -\sqrt{x}$ are reflections of one another in the x-axis; see Figure 13. For each point (x, y) on the graph of $y = \sqrt{x}$ in the figure, there is a corresponding point $(x, -y)$ on the graph of $y = -\sqrt{x}$. For instance, $(1, 1)$ is on the graph of $y = \sqrt{x}$ and $(1, -1)$ is on the graph of $y = -\sqrt{x}$.

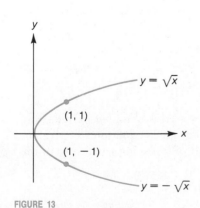

FIGURE 13

We've just seen that the graph of $y = -\sqrt{x}$ can be obtained by reflecting the graph of $y = \sqrt{x}$ in the x-axis. If, for the moment, we rewrite the equation $y = -\sqrt{x}$ in the equivalent form $-y = \sqrt{x}$, then we can rephrase the observations of the previous paragraph this way: The effect of replacing y with $-y$ in the equation $y = \sqrt{x}$ is to reflect the graph of $y = \sqrt{x}$ in the x-axis. This observation, along with our earlier remarks about reflection in the y-axis, are specific instances of the following general result (which is valid whether or not the given equation represents a function).

PROPERTY SUMMARY

REFLECTION AND COORDINATES

Suppose that we have an equation that determines a graph in the x-y plane. Then replacing x with $-x$ or y with $-y$ has the following effects on the graph of the original equation.

REPLACEMENT	RESULT
1. x replaced with $-x$	Reflection about the y-axis
2. y replaced with $-y$	Reflection about the x-axis

EXAMPLE 6   The graph of a function f is a line segment joining the points $(-3, 1)$ and $(2, 4)$, as shown in Figure 14. Graph each of the following functions.

(a) $y = f(-x)$ **(b)** $y = -f(x)$ **(c)** $y = -f(-x)$
(d) $y = f(x + 1)$ **(e)** $y = -f(x + 1)$ **(f)** $y = f(1 - x)$

FIGURE 14

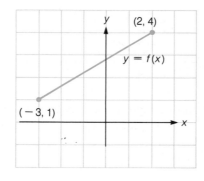

Solution   See Figure 15.

EXAMPLE 7   Graph each equation: **(a)** $y = -|x|$ **(b)** $y = -|x - 2| + 3$

Solution   **(a)** As indicated in Figure 16(a), the graph of $y = -|x|$ is obtained by reflecting the graph of $y = |x|$ about the x-axis. (*Reason:* The given equation is equivalent to $-y = |x|$. But this last equation is the result of replacing y with $-y$ in the equation $y = |x|$.)

(b) For the moment, think of the given equation as

$$y - 3 = -|x - 2|$$

(*Continued on page 118.*)

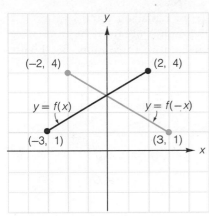

(a) Reflect the graph of $y = f(x)$ in the y-axis to obtain the graph of $y = f(-x)$.

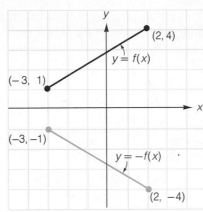

(b) Reflect the graph of $y = f(x)$ in the x-axis to obtain the graph of $y = -f(x)$. [Reason: $y = -f(x)$ is equivalent to $-y = f(x)$, which is obtained from $y = f(x)$ by replacing y with $-y$.]

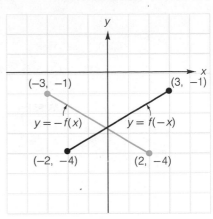

(c) Reflect the graph of $y = -f(x)$ in the y-axis to obtain the graph of $y = -f(-x)$. [Or, reflect the graph of $y = f(-x)$ in the x-axis.]

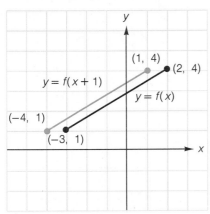

(d) Translate the graph of $y = f(x)$ one unit to the left to obtain the graph of $y = f(x + 1)$.

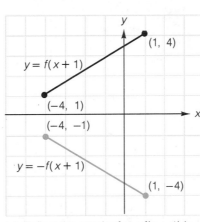

(e) Reflect the graph of $y = f(x + 1)$ in the x-axis to obtain the graph of $y = -f(x + 1)$.

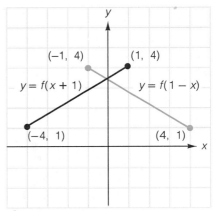

(f) Reflect the graph of $y = f(x + 1)$ in Figure 15(d) in the y-axis to obtain the graph of $y = f(1 - x)$. [Reason: Replacing x with $-x$ in $y = f(x + 1)$ yields $y = f(1 - x)$.]

FIGURE 15

FIGURE 16

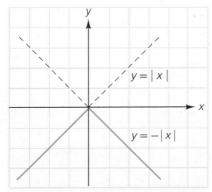

(a) Reflecting the graph of $y = |x|$ about the x-axis yields the graph of $y = -|x|$.

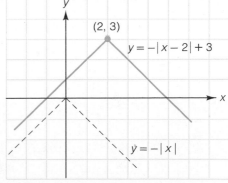

(b) Translating the graph of $y = -|x|$ two units to the right and three units up yields the graph of $y = -|x - 2| + 3$.

This last equation is obtained from $y = -|x|$ by replacing x and y with $x - 2$ and $y - 3$, respectively. Thus, the required graph is obtained by translating the graph of $y = -|x|$ to the right two units and up three units. See Figure 16(b). ▲

EXAMPLE 8   Figure 17(a) shows the graph of $y = G(x)$. What equations describe the graphs in parts (b), (c), and (d)?

Solution

GRAPH IN GIVEN FIGURE	HOW OBTAINED	EQUATION
17(b)	Reflection in the x-axis of graph of G.	$y = -G(x)$
17(c)	Reflection in the y-axis of the graph of $y = -G(x)$	$y = -G(-x)$
17(d)	Translation of the graph of $y = -G(-x)$ by three units in the positive y-direction	$y - 3 = -G(-x)$ or $y = -G(-x) + 3$

FIGURE 17

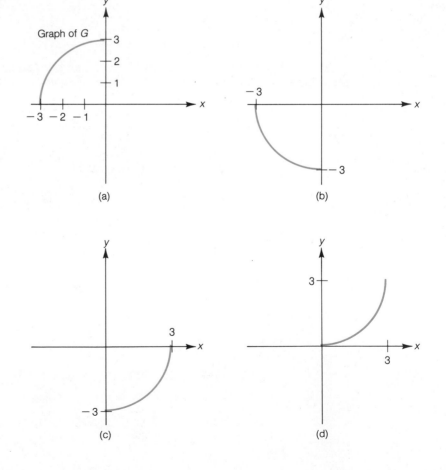

(a)

(b)

(c)

(d)   ▲

▼ EXERCISE SET 2.5

A

In Exercises 1 and 2, the right-hand column contains instructions for translating and/or reflecting the graph of $y = f(x)$. With each equation in the left-hand column, match an appropriate set of instructions in the right-hand column.

1. (a) $y = f(x - 1)$ (**A**) Translate left 1 unit.

    (b) $y = f(x) - 1$ (**B**) Reflect in the x-axis, then translate left 1 unit.

    (c) $y = f(x) + 1$ (**C**) Translate right 1 unit.

    (d) $y = f(x + 1)$ (**D**) Reflect in the x-axis, then translate up 1 unit.

    (e) $y = f(-x) + 1$ (**E**) Reflect in the x-axis, then translate down 1 unit.

    (f) $y = f(-x) - 1$ (**F**) Translate down 1 unit.

    (g) $y = -f(x) + 1$ (**G**) Reflect in the x-axis, reflect in the y-axis, then translate up 1 unit.

    (h) $y = -f(x + 1)$ (**H**) Translate left 1 unit, then reflect in the y-axis, then translate up 1 unit.

    (i) $y = -f(x) - 1$ (**I**) Translate up 1 unit.

    (j) $y = f(1 - x) + 1$ (**J**) Reflect in the y-axis, then translate up 1 unit.
    Hint: See Example 6, part (f)

    (k) $y = -f(-x) + 1$ (**K**) Reflect in the y-axis, then translate down 1 unit.

2. (a) $y = f(x + 2) + 3$ (**A**) Translate left 2 units, then translate down 2 units.

    (b) $y = f(x + 3) + 2$ (**B**) Translate left 3 units, then translate up 2 units.

    (c) $y = f(x - 2) + 3$ (**C**) Translate right 3 units, then translate up 2 units.

    (d) $y = f(x - 2) - 3$ (**D**) Translate left 3 units, then translate down 2 units.

    (e) $y = f(x + 2) - 3$ (**E**) Translate right 3 units, then translate down 2 units.

    (f) $y = f(x - 3) + 2$ (**F**) Reflect in the y-axis, then translate up 2 units.

    (g) $y = f(x - 3) - 2$ (**G**) Reflect in the x-axis, then translate right 2 units.

    (h) $y = f(x + 3) - 2$ (**H**) Reflect in the x-axis, then translate left 2 units.

    (i) $y = -f(x + 2)$ (**I**) Translate left 2 units, then reflect in the y-axis.

    (j) $y = -f(x - 2)$ (**J**) Translate right 2 units, then translate up 3 units.

    (k) $y = f(2 - x)$ (**K**) Translate left 2 units, then translate up 3 units.
    Hint: See Example 6, part (f)

    (l) $y = f(-x) + 2$ (**L**) Translate right 2 units, then translate down 3 units.

In Exercises 3–6, graph each equation, using the following two graphs for reference. (The graph of the equation $4x^2 + 9y^2 = 36$ is an ellipse; this curve will be studied in a later chapter).

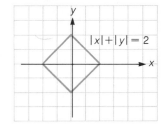 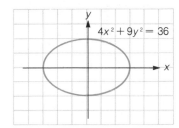

3. (a) $|x - 2| + |y| = 2$ (b) $|x + 2| + |y| = 2$

    (c) $|x| + |y - 2| = 2$ (d) $|x| + |y + 2| = 2$

    (e) $|x - 2| + |y + 2| = 2$

    (f) $|x + 2| + |y + 2| = 2$

4. (a) $|x - 3| + |y - 4| = 2$

    (b) $|x + 3| + |y - 4| = 2$

    (c) $|x + 3| + |y + 4| = 2$

    (d) $|x - 3| + |y + 4| = 2$

    (e) $|x - 3| + |y| = 2$

    (f) $|x| + |y + 4| = 2$

5. (a) $4(x - 1)^2 + 9y^2 = 36$

    (b) $4x^2 + 9(y - 1)^2 = 36$

    (c) $4(x + 1)^2 + 9(y + 1)^2 = 36$

    (d) $4(x - 1)^2 + 9(y - 1)^2 = 36$

6. (a) $4(x - 3)^2 + 9y^2 = 36$

    (b) $4x^2 + 9(y + 2)^2 = 36$

    (c) $4(x + 3)^2 + 9(y - 1)^2 = 36$

    (d) $4(x + 6)^2 + 9y^2 = 36$

In Exercises 7–28, sketch the graph of the function.

7. $y = x^3 - 3$

8. $y = x^2 + 3$

9. $y = (x + 4)^2$

10. $y = (x + 4)^2 - 3$

11. $y = (x - 4)^2$

12. $y = (x - 4)^2 + 1$

13. $y = -x^2$

14. $y = -x^2 - 3$

15. $y = -(x - 3)^2$

16. $y = -(x - 3)^2 - 3$

17. $y = \sqrt{x - 3}$

18. $y = \sqrt{x - 3} + 1$

19. $y = -\sqrt{x + 1}$

20. $y = -\sqrt{x + 1} + 1$

21. $y = \dfrac{1}{x + 2} + 2$

22. $y = \dfrac{1}{x - 3} - 1$

23. $y = (x - 2)^3$

24. $y = (x - 2)^3 + 1$

25. $y = -x^3 + 4$

26. $y = -(x - 1)^3 + 4$

27. (a) $y = |x + 4|$ (b) $y = |4 - x|$

     (c) $y = -|4 - x| + 1$

28. (a) $y = \sqrt{x + 2}$ (b) $y = \sqrt{2 - x}$

     (c) $y = -\sqrt{2 - x}$

In Exercises 29–44, sketch the graph of the function, given that f, F, and g are defined as follows:

$$f(x) = |x| \qquad F(x) = \frac{1}{x} \qquad g(x) = \sqrt{1 - x^2}$$

29. $y = f(x - 5)$

30. $y = -f(x - 5)$

31. $y = f(5 - x)$

32. $y = -f(5 - x)$

     Hint: See Example 6, part (f)

33. $y = 1 - f(x - 5)$

34. $y = f(-x)$

35. $y = F(x + 3)$

36. $y = F(x) + 3$

37. $y = -F(x + 3)$

38. $y = F(-x) + 3$

39. $y = g(x - 2)$

40. $y = -g(x - 2)$

41. $y = 1 - g(x - 2)$

42. $y = g(-x)$

43. $y = g(2 - x)$

44. $y = -g(2 - x)$

In Exercises 45 and 46, refer to the graph of the function f shown in the figure to graph each function.

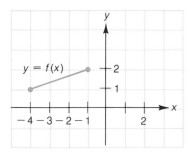

45. (a) $y = -f(x)$ (b) $y = f(-x)$

     (c) $y = -f(-x)$

46. (a) $y = f(x) - 2$ (b) $y = f(x - 2)$

     (c) $y = f(x - 2) - 2$

In Exercises 47 and 48, refer to the graph of the function g shown in the figure in order to graph each function.

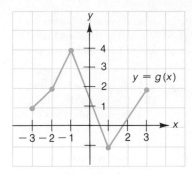

47. (a) $y = g(-x)$ (b) $y = -g(x)$

     (c) $y = -g(-x)$

48. $y = -g(x - 3) - 1$

49. The figure shows the graph of the function $y = 10^x$. Sketch the graphs of the following functions.

     (a) $y = 10^{-x}$ (b) $y = -10^x$

     (c) $y = -10^{-x}$ (d) $y = 10^{x-1}$

     (e) $y = 10^x + 1$ (f) $y = -10^{x-1} - 1$

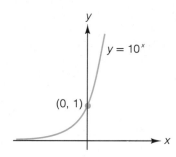

50. Let $f(x) = x^2$. In each case, sketch the graph of the function and determine its x- and y-intercepts, if any.

     (a) $y = f(x) + 4$ (b) $y = f(x) - 3$

     (c) $y = f(x + 1)$ (d) $y = f(x - 4) + 3$

     (e) $y = -f(x - 4) + 2$ (f) $y = f(x - 100)$

B

For Exercises 51 and 52, assume that (a, b) is a point on the graph of $y = f(x)$, and specify the corresponding point on the graph of each equation. [For example, the point that corresponds to (a, b) on the graph of $y = f(x - 1)$ is $(a + 1, b)$.]

51. (a) $y = f(x - 3)$ (b) $y = f(x) - 3$

     (c) $y = f(x - 3) - 3$ (d) $y = -f(x)$

     (e) $y = f(-x)$ (f) $y = -f(-x)$

     (g) $y = f(3 - x)$ (h) $y = -f(3 - x) + 1$

52. (a) $y = f(-x) + 2$ (b) $y = -f(-x) + 2$

     (c) $y = -f(x - 3)$ (d) $y = 1 - f(x + 1)$

     (e) $y = f(1 - x)$ (f) $y = -f(1 - x) + 1$

53. Show that $\dfrac{x}{x-1} = \dfrac{1}{x-1} + 1$, provided that $x \neq 1$. Then use this fact to graph the function g defined by

$$g(x) = \frac{x}{x-1}.$$

54. Verify that $-x^2 + 2x + 2 = -(x-1)^2 + 3$. Then use this fact to graph the function h defined by $h(x) = -x^2 + 2x + 2$.

55. Show that $x^3 - 6x^2 + 12x - 9 = (x-2)^3 - 1$. Then use this fact to sketch the graph of the equation $y = x^3 - 6x^2 + 12x - 9$.

56. Show that $\dfrac{2-x}{1-\sqrt{x-1}} = 1 + \sqrt{x-1}$, provided that $x \geq 1$ and $x \neq 2$. Then use this information to graph the function f defined by $f(x) = \dfrac{2-x}{1-\sqrt{x-1}}$.

57. What is the range of the function f defined by

$$f(x) = \frac{x^2}{1 - \sqrt{1-x^2}}?$$ *Hint:* The function is easy to graph after rationalizing the denominator.

58. (a) A function f is said to be **even** if the equation $f(-x) = f(x)$ is satisfied by all values of x in the domain of f. Explain why the graph of an even function must be symmetric about the y-axis.

(b) Show that each function is even by computing $f(-x)$ and then noting that $f(x)$ and $f(-x)$ are equal.

    (i) $f(x) = x^2$ **(ii)** $f(x) = 2x^4 - 6$

    (iii) $f(x) = 3x^6 - \dfrac{4}{x^2} + 1$

59. (a) A function f is said to be **odd** if the equation $f(-x) = -f(x)$ is satisfied by all values of x in the domain of f. Show that if (x, y) is a point on the graph of an odd function f, then the point $(-x, -y)$ is also on the graph. (This implies that the graph of an odd function must be symmetric about the origin.)

(b) Show that each function is odd by computing $f(-x)$ as well as $-f(x)$ and then noting that the two expressions obtained are equal.

    (i) $f(x) = x^3$ **(ii)** $f(x) = -2x^5 + 4x^3 - x$

    (iii) $f(x) = \dfrac{|x|}{x+x^7}$

60. Is each function odd, even, or neither? (See Exercises 58 and 59 for definitions.)

    (a) $f(x) = \dfrac{1-x^2}{2+x^2}$ **(b)** $g(x) = \dfrac{x-x^3}{2x+x^3}$

    (c) $h(x) = x^2 + x$ **(d)** $F(x) = (x^2+x)^2$

    (e) $G(x) = \begin{cases} 1 & \text{if } x > 0 \\ 0 & \text{if } x = 0 \\ -1 & \text{if } x < 0 \end{cases}$

    Suggestion for part (e): Look at the graph.

61. Suppose that the function f is increasing on the interval $(2, 4)$ and decreasing on the intervals $(-\infty, 2)$ and $(4, \infty)$. On what interval(s) is each function increasing?

    (a) $y = f(-x)$ **(b)** $y = -f(x)$

62. Suppose that the function g is decreasing on $(-\infty, 1)$ and increasing on $(1, \infty)$. On what interval(s) is the function $y = -f(-x)$ increasing?

2.6 ▼ METHODS OF COMBINING FUNCTIONS

Two given numbers a and b can be combined in various ways to produce a third number. For instance, we can form the sum $a + b$ or the difference $a - b$ or the product ab. Also, if $b \neq 0$, we can form the quotient a/b. Similarly, two functions can be combined in various ways to produce a third function. Suppose, for example, that we start with the two functions $y = x^2$ and $y = x^3$. It would seem natural to define their sum, difference, product, and quotient as follows:

$$\text{sum:} \quad y = x^2 + x^3$$

$$\text{difference:} \quad y = x^2 - x^3$$

$$\text{product:} \quad y = x^2 x^3 = x^5$$

$$\text{quotient:} \quad y = \frac{x^2}{x^3} = \frac{1}{x}$$

Indeed, this is the idea behind the formal definitions we now give.

DEFINITIONS

Let f and g be two functions. Then the **sum** $f + g$, the **difference** $f - g$, the **product** fg, and the **quotient** f/g are functions defined by the following equations:

$$(f + g)(x) = f(x) + g(x) \tag{1}$$
$$(f - g)(x) = f(x) - g(x) \tag{2}$$
$$(fg)(x) = f(x) \cdot g(x) \tag{3}$$
$$(f/g)(x) = f(x)/g(x) \qquad \text{provided } g(x) \neq 0 \tag{4}$$

For the functions defined by equations (1), (2), and (3), the domain is the set of all inputs x belonging to both the domain of f and the domain of g. For the quotient function in equation (4), we impose the additional restriction that the domain exclude all inputs x for which $g(x) = 0$.

EXAMPLE 1   Let $f(x) = 3x + 1$ and $g(x) = x - 1$. Compute each of the following functions.

(a) $(f + g)(x)$ **(b)** $(f - g)(x)$ **(c)** $(fg)(x)$ **(d)** $(f/g)(x)$

Solution
(a) $(f + g)(x) = f(x) + g(x) = (3x + 1) + (x - 1) = 4x$
(b) $(f - g)(x) = f(x) - g(x) = (3x + 1) - (x - 1) = 2x + 2$
(c) $(fg)(x) = [f(x)][g(x)] = (3x + 1)(x - 1) = 3x^2 - 2x - 1$

(d) $(f/g)(x) = \dfrac{f(x)}{g(x)} = \dfrac{3x + 1}{x - 1} \qquad (x \neq 1)$

▲

For the remainder of this section, we are going to discuss a method for combining functions known as **composition of functions**. As you will see, this method is based on the familiar algebraic process of substitution. Suppose, for example, that f and g are two functions defined by

$$f(x) = x^2 \qquad g(x) = 3x + 1$$

Choose any number in the domain of g, say $x = -2$. We can compute $g(-2)$:

$$g(-2) = 3(-2) + 1 = -5$$

Now let's use the output -5 that g has produced as an *input* for f. We obtain

$$f(-5) = (-5)^2 = 25$$

and, consequently,

$$f[g(-2)] = 25$$

So, beginning with the input -2, we've successively applied g and then f to obtain the output 25. Similarly, we could carry out this same procedure for any other number in the domain of g. Here is a summary of the procedure:

1. Start with an input x and calculate $g(x)$.
2. Use $g(x)$ as an input for f; that is, calculate $f[g(x)]$.

We use the notation $f \circ g$ to denote the function, or rule, that tells us to assign the output $f[g(x)]$ to the initial input x. In other words, $f \circ g$ denotes the rule

FIGURE 1
Diagram for the function $f \circ g$

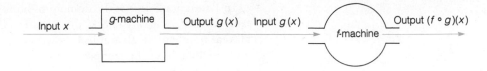

consisting of two steps: *First apply g; then apply f.* We read the notation $f \circ g$ as *f circle g* or *f composed with g.* In Figure 1, we summarize these ideas.

When we write $g(x)$, we assume that x is in the domain of the function g. Likewise, for the notation $f[g(x)]$ to make sense, the outputs $g(x)$ must themselves be acceptable inputs for the function f. Our formal definition, then, for the composite function $f \circ g$ is as follows.

DEFINITION

Composition of Functions: $f \circ g$

Given two functions f and g, the function $f \circ g$ is defined by

$$(f \circ g)(x) = f[g(x)]$$

The domain of $f \circ g$ consists of those inputs x (in the domain of g) for which $g(x)$ is in the domain of f.

EXAMPLE 2 Let $f(x) = x^2$ and $g(x) = 3x + 1$. Compute $(f \circ g)(x)$ and $(g \circ f)(x)$.

Solution

$$\begin{aligned}
(f \circ g)(x) &= f[g(x)] & \text{definition of } f \circ g \\
&= f(3x + 1) & \text{definition of } g \\
&= (3x + 1)^2 & \text{definition of } f \\
&= 9x^2 + 6x + 1
\end{aligned}$$

$$\begin{aligned}
(g \circ f)(x) &= g[f(x)] & \text{definition of } g \circ f \\
&= g(x^2) & \text{definition of } f \\
&= 3(x^2) + 1 & \text{definition of } g \\
&= 3x^2 + 1
\end{aligned}$$

▲

Notice that the two results obtained in Example 2 are not the same. This shows that, in general, $f \circ g$ and $g \circ f$ represent different functions.

EXAMPLE 3 Let f and g be defined as in Example 2: $f(x) = x^2$ and $g(x) = 3x + 1$. Compute $(f \circ g)(-2)$.

Solution We will show two methods.

FIRST METHOD Using the formula for $(f \circ g)(x)$ developed in Example 2, we have

$$(f \circ g)(x) = 9x^2 + 6x + 1$$

and therefore,

$$\begin{aligned}
(f \circ g)(-2) &= 9(-2)^2 + 6(-2) + 1 \\
&= 36 - 12 + 1 \\
&= 25
\end{aligned}$$

ALTERNATIVE METHOD Working directly from the definition of $f \circ g$, we have

$$\begin{aligned}
(f \circ g)(-2) &= f[g(-2)] \\
&= f[3(-2) + 1] \\
&= f(-5) \\
&= (-5)^2 = 25
\end{aligned}$$

▲

In Examples 2 and 3, the domain of both f and g is the set of all real numbers. And as you can easily check, the domain of both $f \circ g$ and $g \circ f$ is also the set of all real numbers. In Example 4, however, some care needs to be taken in describing the domain of the composite function.

EXAMPLE 4 Let f and g be defined as follows:

$$f(x) = x^2 + 1 \qquad g(x) = \sqrt{x}$$

Compute $(f \circ g)(x)$. Find the domain of $f \circ g$ and sketch its graph.

Solution

$$(f \circ g)(x) = f[g(x)]$$
$$= f(\sqrt{x}) = (\sqrt{x})^2 + 1 = x + 1$$

So we have $(f \circ g)(x) = x + 1$. Now what about the domain of $f \circ g$? Our first inclination may be to say (incorrectly!) that the domain is the set of all real numbers, since any real number can be used as an input in the expression $x + 1$. However, the definition of $f \circ g$ on page 123 tells us that the inputs for $f \circ g$ must first of all be acceptable inputs for g. Given the definition of g, then, we must require that x be nonnegative. On the other hand, for any nonnegative input x, the number $g(x)$ will be an acceptable input for f. (Why?) In summary, then, the domain of $f \circ g$ is the interval $[0, \infty)$. The graph of $f \circ g$ is shown in Figure 2. ▲

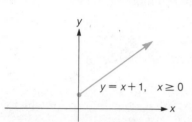

FIGURE 2
Graph of the function $f \circ g$ in Example 4

$y = x + 1, \quad x \geq 0$

One reason for studying the composition of functions is that it lets us express a given function in terms of simpler functions. This is often useful in calculus. Suppose, for example, that we wish to express the function C defined by

$$C(x) = (2x^3 - 5)^2$$

as a composition of simpler functions. That is, we want to come up with two functions f and g so that the equation

$$C(x) = (f \circ g)(x)$$

holds for every x in the domain of C.

We begin by thinking what we would do to compute $(2x^3 - 5)^2$ for a given value of x. First, we would compute the quantity $2x^3 - 5$, then we would square the result. Therefore, recalling that the rule $f \circ g$ tells us to do g *first*, we let $g(x) = 2x^3 - 5$. And then, since the next step is squaring, we let $f(x) = x^2$. Now let us see if these choices for f and g are correct; that is, let us calculate $(f \circ g)(x)$ and see if it really is the same as $C(x)$.

Using $f(x) = x^2$ and $g(x) = 2x^3 - 5$, we have

$$(f \circ g)(x) = f[g(x)] = f(2x^3 - 5)$$
$$= (2x^3 - 5)^2 = C(x)$$

This shows that our choices for f and g were indeed correct, and we have expressed C as a composition of two simpler functions.*

Note that in expressing C as $f \circ g$, we chose g to be the "inner" function, that is, the quantity inside parentheses: $g(x) = 2x^3 - 5$. This observation is used in Example 5.

---

* Other answers are possible, too. For instance, if $F(x) = (x - 5)^2$ and $G(x) = 2x^3$, then (as you should verify for yourself) $C(x) = F[G(x)]$.

EXAMPLE 5 Let $s(x) = \sqrt{1 + x^4}$. Express the function s as a composition of two simpler functions f and g.

Solution Let g be the "inner" function; that is, let $g(x)$ be the quantity inside the radical:

$$g(x) = 1 + x^4$$

And let us take f to be the square root function:

$$f(x) = \sqrt{x}$$

Now we need to verify that these are appropriate choices for f and g; that is, we need to check that the equation $(f \circ g)(x) = s(x)$ is true for every x in the domain of s. We have

$$(f \circ g)(x) = f[g(x)] = f(1 + x^4)$$
$$= \sqrt{1 + x^4} = s(x)$$

Thus, $(f \circ g)(x) = s(x)$, as required. ▲

▼ **EXERCISE SET 2.6**

A

In Exercises 1–10, compute each expression, given that the functions f, g, h, k, and m are defined as follows:

$$f(x) = 2x - 1 \qquad k(x) = 2, \quad \text{for all } x$$
$$g(x) = x^2 - 3x - 6 \qquad m(x) = x^2 - 9$$
$$h(x) = x^3$$

1. **(a)** $(f + g)(x)$ **(b)** $(f - g)(x)$ **(c)** $(f - g)(0)$
2. **(a)** $(fh)(x)$ **(b)** $(h/f)(x)$ **(c)** $(f/h)(1)$
3. **(a)** $(m - f)(x)$ **(b)** $(f - m)(x)$
4. **(a)** $\langle fg \rangle(x)$ **(b)** $(fg)(\frac{1}{2})$
5. **(a)** $(fk)(x)$ **(b)** $(kf)(x)$ **(c)** $(fk)(1) - (kf)(2)$
6. **(a)** $(g + m)(x)$ **(b)** $(g + m)(x) - (g - m)(x)$
7. **(a)** $(f/m)(x) - (m/f)(x)$ **(b)** $(f/m)(0) - (m/f)(0)$
8. **(a)** $[h \cdot (f + m)](x)$ *Note:* h and $(f + m)$ are two functions; the notation $h \cdot (f + m)$ denotes the product function.
   (b) $(hf)(x) + (hm)(x)$
9. **(a)** $[m \cdot (k - h)](x)$ **(b)** $(mk)(x) - (mh)(x)$
   (c) $(mk)(-1) - (mh)(-1)$
10. **(a)** $(g + g)(x)$ **(b)** $(g - g)(x)$
    (c) $(kg)(x)$ **(d)** $(g + g)(-3) - (kg)(-3)$
11. Let $f(x) = 3x + 1$ and $g(x) = -2x - 5$. Compute the following.
    (a) $(f \circ g)(x)$ **(b)** $(f \circ g)(10)$
    (c) $(g \circ f)(x)$ **(d)** $(g \circ f)(10)$

12. Let $f(x) = 1 - 2x^2$ and $g(x) = x + 1$. Compute the following.
    (a) $(f \circ g)(x)$ **(b)** $(f \circ g)(-1)$
    (c) $(g \circ f)(x)$ **(d)** $(g \circ f)(-1)$
    (e) $(f \circ f)(x)$ **(f)** $(g \circ g)(-1)$

13. Compute $(f \circ g)(x)$, $(f \circ g)(-2)$, $(g \circ f)(x)$, and $(g \circ f)(-2)$ for each pair of functions.
    (a) $f(x) = 1 - x$, $g(x) = 1 + x$
    (b) $f(x) = x^2 - 3x - 4$, $g(x) = 2 - 3x$
    (c) $f(x) = x/3$, $g(x) = 1 - x^4$
    (d) $f(x) = 2^x$, $g(x) = x^2 + 1$
    (e) $f(x) = x$, $g(x) = 3x^5 - 4x^2$
    (f) $f(x) = 3x - 4$, $g(x) = \frac{1}{3}(x + 4)$

14. Let $h(x) = 4x^2 - 5x + 1$, $k(x) = x$, and $m(x) = 7$ for all x. Compute the following.
    (a) $h[k(x)]$ **(b)** $k[h(x)]$ **(c)** $h[m(x)]$
    (d) $m[h(x)]$ **(e)** $k[m(x)]$ **(f)** $m[k(x)]$

15. Let $F(x) = \dfrac{3x - 4}{3x + 3}$ and $G(x) = \dfrac{x + 1}{x - 1}$. Compute the following.
    (a) $(F \circ G)(x)$ **(b)** $F[G(t)]$ **(c)** $(F \circ G)(2)$
    (d) $(G \circ F)(x)$ **(e)** $G[F(y)]$ **(f)** $(G \circ F)(2)$

16. Let $f(x) = \dfrac{1}{x^2} + 1$ and $g(x) = \dfrac{1}{x - 1}$.
    (a) Compute $(f \circ g)(x)$.
    (b) What is the domain of $f \circ g$?
    (c) Graph the function $f \circ g$.

17. Let $M(x) = \dfrac{2x - 1}{x - 2}$.

   (a) Compute $M(7)$ and then $M[M(7)]$.

   (b) Compute $(M \circ M)(x)$.

   (c) Compute $(M \circ M)(7)$, using the formula you obtained in part (b). Check that your answer agrees with that obtained in part (a).

18. Let $F(x) = (x + 1)^5$, $f(x) = x^5$, and $g(x) = x + 1$. Which of the following is true for all x?

   $(f \circ g)(x) = F(x)$ or $(g \circ f)(x) = F(x)$

19. Refer to the graphs of the functions f, g, and h to compute the required quantities. Assume that all the axes are marked off in one-unit intervals.

   (a) $f[g(3)]$ **(b)** $g[f(3)]$ **(c)** $f[h(3)]$

   (d) $(h \circ g)(2)$ **(e)** $h\{f[g(3)]\}$

   (f) $(g \circ f \circ h \circ f)(2)$ *Note:* This means first do f, then h, then f, then g.

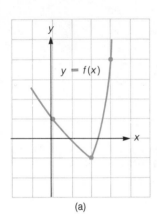

(a)

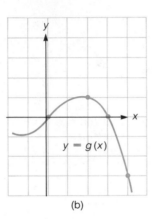

(b)

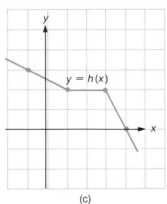

(c)

20. Let $f(x) = 2x - 1$. Find $f[f(x)]$. If $z = f[f(x)]$, find $f(z)$.

21. **(a)** Let $T(x) = 4x^3 - 3x^2 + 6x - 1$ and $I(x) = x$. Find $(T \circ I)(x)$ and $(I \circ T)(x)$.

   (b) Let $G(x) = ax^2 + bx + c$ and $I(x) = x$. Find $(G \circ I)(x)$ and $(I \circ G)(x)$.

   (c) What general conclusion do you arrive at from the results of parts (a) and (b)?

22. The figure shows the graphs of two functions, F and G. Use the graphs to compute each quantity.

   (a) $(G \circ F)(1)$ **(b)** $(F \circ G)(1)$

   (c) $(F \circ F)(1)$ **(d)** $(G \circ G)(-3)$

   (e) $(G \circ F)(5)$ **(f)** $(F \circ F)(5)$

   (g) $(F \circ G)(-1)$ **(h)** $(G \circ F)(2)$

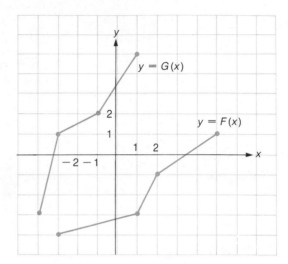

23. The domain of a function f consists of the numbers $-1, 0, 1, 2,$ and 3. The following table shows the output f assigns to each input.

x	-1	0	1	2	3
$f(x)$	2	2	0	3	1

The domain of a function g consists of the numbers $0, 1, 2, 3,$ and 4. The following table shows the output g assigns to each input.

x	0	1	2	3	4
$g(x)$	3	2	0	4	-1

Use this information to complete the tables for $f \circ g$ and $g \circ f$.   *Note:* Two of the entries will be undefined.

x	0	1	2	3	4
$(f \circ g)(x)$					

x	-1	0	1	2	3	4
$(g \circ f)(x)$						

24. The following two tables show certain pairs of inputs and outputs for functions f and g.

x	0	$\dfrac{\pi}{6}$	$\dfrac{\pi}{4}$	$\dfrac{\pi}{3}$	$\dfrac{\pi}{2}$
$f(x)$	0	$\dfrac{1}{2}$	$\dfrac{\sqrt{2}}{2}$	$\dfrac{\sqrt{3}}{2}$	1

y	0	$\dfrac{1}{4}$	$\dfrac{\sqrt{2}}{4}$	$\dfrac{1}{2}$	$\dfrac{\sqrt{2}}{2}$	$\dfrac{3}{4}$	$\dfrac{\sqrt{3}}{2}$	1
$g(y)$	$\dfrac{\pi}{2}$	π	0	$\dfrac{\pi}{3}$	$\dfrac{\pi}{4}$	0	$\dfrac{\pi}{6}$	0

Use this information to complete the following table of values for $(g \circ f)(x)$.

x	0	$\dfrac{\pi}{6}$	$\dfrac{\pi}{4}$	$\dfrac{\pi}{3}$	$\dfrac{\pi}{2}$
$(g \circ f)(x)$					

25. Let $f(x) = 2x + 1$ and $g(x) = 3x - 4$.
  (a) Find $(f \circ g)(x)$ and graph the function $f \circ g$.
  (b) Find $(g \circ f)(x)$ and graph the function $g \circ f$.

26. Let $F(x) = x^2$ and $G(x) = x - 4$.
  (a) Find $F[G(x)]$ and graph the function $F \circ G$.
  (b) Find $G[F(x)]$ and graph the function $G \circ F$.

27. Let $g(x) = \sqrt{x} - 3$ and $f(x) = x - 1$.
  (a) Sketch a graph of g. Specify the domain and range.
  (b) Sketch a graph of f. Specify the domain and range.
  (c) Compute $(f \circ g)(x)$. Graph the function $f \circ g$ and specify its domain and range.
  (d) Find a formula for $g[f(x)]$. Which values of x are acceptable inputs here? That is, what is the domain of $g \circ f$?
  (e) Use the results of part (d) to sketch a graph of the function $g \circ f$.

28. Let $F(x) = -x^2$ and $G(x) = \sqrt{x}$. Determine the domains of $F \circ G$ and $G \circ F$.

29. Let $C(x) = (3x - 1)^4$. Express C as a composition of two simpler functions.

30. Let $C(x) = (1 + x^2)^3$. Find functions f and g so that $C(x) = (f \circ g)(x)$ is true for all values of x.

31. Express each function as a composition of two functions.
  (a) $F(x) = \sqrt[3]{3x + 4}$ **(b)** $G(x) = |2x - 3|$
  (c) $H(x) = (ax + b)^5$ **(d)** $T(x) = 1/\sqrt{x}$

32. Let $a(x) = x^2$, $b(x) = |x|$, $c(x) = 3x - 1$. Express each function as a composition of two of the given functions.
  (a) $f(x) = (3x - 1)^2$ **(b)** $g(x) = |3x - 1|$
  (c) $h(x) = 3x^2 - 1$

33. Let $a(x) = 1/x$, $b(x) = \sqrt[3]{x}$, $c(x) = 2x + 1$, and $d(x) = x^2$. Express each function as a composition of two of the given functions.
  (a) $f(x) = \sqrt[3]{2x + 1}$ **(b)** $g(x) = 1/x^2$
  (c) $h(x) = 2x^2 + 1$ **(d)** $K(x) = 2\sqrt[3]{x} + 1$
  (e) $l(x) = \dfrac{2}{x} + 1$ **(f)** $m(x) = \dfrac{1}{2x + 1}$
  (g) $n(x) = x^{2/3}$

34. Express $G(x) = 1/(1 + x^4)$ as a composition of two simpler functions, one of which is $f(x) = 1/x$.

B

35. The circumference of a circle with radius r is given by $C(r) = 2\pi r$. Suppose that the circle is shrinking in size and that the radius at time t is given by

$$r = f(t) = \frac{1}{t^2 + 1}$$

Assume that t is measured in minutes and that r, or $f(t)$, is in feet. Find $(C \circ f)(t)$ and use this to find the circumference when $t = 3$ min.

36. A spherical weather balloon is being inflated in such a way that the radius is given by

$$r = g(t) = \tfrac{1}{2}t + 2$$

Assume that r is in meters and t is in seconds, with $t = 0$ corresponding to the time that inflation begins. If the volume of a sphere of radius r is given by

$$V(r) = \tfrac{4}{3}\pi r^3$$

compute $V[g(t)]$ and use this to find the time at which the volume of the balloon is $36\pi \text{ m}^3$.

37. Suppose that a manufacturer knows that the daily production cost to build x bicycles is given by the function C, where

$$C(x) = 100 + 90x - x^2 \qquad (0 \le x \le 40)$$

That is, $C(x)$ represents the cost in dollars of building x bicycles. Furthermore, suppose that the number of bicycles that can be built in t hr is given by the function f, where

$$x = f(t) = 5t \qquad (0 \le t \le 8)$$

  (a) Compute $(C \circ f)(t)$.
  (b) Compute the production cost on a day that the factory operates for $t = 3$ hr.
  (c) If the factory runs for 6 hr instead of 3 hr, is the cost twice as much?

38. Suppose that in a certain biology lab experiment, the number of bacteria is related to the temperature of the environment by the function

$$N(T) = -2T^2 + 240T - 5400 \qquad (40 \le T \le 90)$$

Here, $N(T)$ represents the number of bacteria present when the temperature is T degrees Fahrenheit. Also, suppose that t hr after the experiment begins, the temperature is given by

$$T(t) = 10t + 40 \qquad (0 \le t \le 5)$$

(a) Compute $N[T(t)]$.
(b) How many bacteria are present when $t = 0$? When $t = 2$ hr? When $t = 5$ hr?

39. Let $g(x) = 4x - 1$. Find $f(x)$, given that the equation $(g \circ f)(x) = x + 5$ is true for all values of x.

40. Let $g(x) = 2x + 1$. Find $f(x)$, given that $(g \circ f)(x) = 10x - 7$.

41. Let $f(x) = -2x + 1$ and $g(x) = ax + b$. Find values for a and b so the equation $f[g(x)] = x$ holds for all values of x.

42. Let $f(x) = \dfrac{3x - 4}{x - 3}$.

(a) Compute $(f \circ f)(x)$.
(b) Find $f[f(\frac{113}{355})]$. (Try not to do it the hard way.)

43. Let $f(x) = x^2$ and $g(x) = 2x - 1$.

(a) Compute $\dfrac{f[g(x)] - f[g(a)]}{g(x) - g(a)}$.

(b) Compute $\dfrac{f[g(x)] - f[g(a)]}{x - a}$.

44. Suppose that $y = f(x)$ and $g(y) = x$. Show that $(g \circ f)(x) - (f \circ g)(y) = x - y$.

45. Given three functions f, g, and h, we define the function $f \circ g \circ h$ by the formula

$$(f \circ g \circ h)(x) = f\{g[h(x)]\}$$

For example, if $f(x) = x^2$, $g(x) = x + 1$, and $h(x) = x/2$, we evaluate $(f \circ g \circ h)(x)$ as

$$(f \circ g \circ h)(x) = f\{g[h(x)]\}$$

$$= f\left[g\left(\frac{x}{2}\right)\right] = f\left(\frac{x}{2} + 1\right)$$

$$= \left(\frac{x}{2} + 1\right)^2 = \frac{x^2}{4} + x + 1$$

Using $f(x) = x^2$, $g(x) = x + 1$, and $h(x) = x/2$, compute each composition.

(a) $(g \circ h \circ f)(x)$ (b) $(h \circ f \circ g)(x)$
(c) $(g \circ f \circ h)(x)$ (d) $(f \circ h \circ g)(x)$
(e) $(h \circ g \circ f)(x)$

Hint for checking: No two answers are the same.

46. Let $F(x) = 2x - 1$, $G(x) = 4x$, and $H(x) = 1 + x^2$. Compute each composition.

(a) $(F \circ G \circ H)(x)$ (b) $(G \circ F \circ H)(x)$
(c) $(F \circ H \circ G)(x)$ (d) $(H \circ F \circ G)(x)$
(e) $(H \circ G \circ F)(x)$ (f) $(G \circ H \circ F)(x)$

47. Let $f(x) = x^2$, $g(x) = 1 - x$, and $h(x) = 3x$. Express each function as a composition of f, g, and h.
(a) $p(x) = 1 - 9x^2$ (b) $q(x) = 3 - 3x^2$
(c) $r(x) = 1 - 6x + 9x^2$ (d) $s(x) = 3 - 6x + 3x^2$

48. Suppose that three functions f', g, and g' are defined as follows (f' is read f *prime*):

$$f'(x) = 3x^2 \qquad g(x) = 2x^2 + 5 \qquad g'(x) = 4x$$

(a) Find $f'[g(x)] \cdot g'(x)$. (The dot denotes multiplication, not composition.)
(b) Evaluate $f'[g(-1)] \cdot g'(-1)$.
(c) Find $f'[g(t)] \cdot g'(t)$.

49. Suppose that three functions F', G, and G' are defined by

$$F'(x) = \frac{1}{2\sqrt{x}} \qquad G(x) = x^2 + 2x + 2 \qquad G'(x) = 2x + 2$$

(a) Find $F'[G(x)] \cdot G'(x)$. (The dot denotes multiplication.)
(b) Find $F'[G(9)] \cdot G'(9)$.

C

50. Let p and q be two numbers whose sum is 1. Define the function F by

$$F(x) = p - \frac{1}{x + q}$$

Show that $F\{F[F(x)]\} = x$.

51. Let $i(x) = x$, $a(x) = -x$, $b(x) = 1/x$, and $c(x) = -1/x$. Assume that the domain of each function is $(0, \infty)$.
(a) Complete the following table, where the operation is composition of functions. For example, c is the proper entry in the second row, third column, since if you compute $(a \circ b)(x)$, you'll get $-1/x$, which is $c(x)$.

$\circ$	i	a	b	c
i				
a			c	
b				
c				

(b) According to your table, is the operation commutative; that is, does $a \circ b = b \circ a$?
(c) Use your table to find a^2 (meaning $a \circ a$), b^2, c^2, and c^3.
(d) Use your table to find out if $(a \circ b) \circ c = a \circ (b \circ c)$.
(e) Let G denote the set consisting of the four functions you've worked with in this exercise. Now define a function f, with domain G, as follows: $f(t) = t^2$ for each element t in the set G. [For example, $f(a) = a^2 = a \circ a = i$.] What is the range of f?
(f) Verify that $f(a \circ b) = [f(a)] \circ [f(b)]$.
(g) Find out if $f(a \circ b \circ c) = [f(a)] \circ [f(b)] \circ [f(c)]$.

2.7 ▼ INVERSE FUNCTIONS

If anybody ever told me why the graph of $y = x^{1/2}$ is the reflection of the graph of $y = x^2$ in a 45° line, it didn't sink in. To this day, there are textbooks that expect students to think that it is so obvious as to need no explanation. This is a pity, if only because [in calculus] it is such a common practice to define the natural logarithm first and then define the exponential function as its inverse.

Professor Ralph P. Boas, recalling his student days in the article "Inverse Functions" in *The College Mathematics Journal* 16(1985): 42.

PROBLEM
SOLUTION

© 1981 Scott Kim. From Inversions
(W. H. Freeman, 1989).

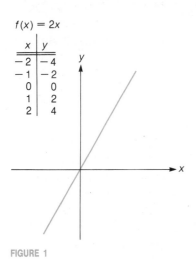

$f(x) = 2x$

x	y
−2	−4
−1	−2
0	0
1	2
2	4

FIGURE 1

We shall introduce the idea of inverse functions through an easy example. Consider the function f defined by

$$f(x) = 2x$$

A short table of values for f and the graph of f are displayed in Figure 1.

Now we ask this question: What happens if we take the table for f and interchange the entries in the x and y columns to obtain a new table that looks like the following?

x	y
−4	−2
−2	−1
0	0
2	1
4	2

← New table formed by interchanging the inputs and outputs for f

Several observations are in order. Our new table does itself represent a function (in this case); for each input x there is exactly one output y. Let us call this new function g. Note that each output is half the corresponding input. So the function g can be described by the formula $g(x) = x/2$. In summary, we began with the "doubling function," $f(x) = 2x$; then, by interchanging the inputs with the outputs, we obtained the "halving function," $g(x) = x/2$.

The doubling function f and the halving function g are in a certain sense opposites; each reverses, or undoes, the effect of the other. That is, if you begin with x and then calculate $g[f(x)]$, you get x again; similarly, $f[g(x)]$ is also equal to x. (Verify these last two statements for yourself.)

The two functions f and g that we've been discussing are an example of a pair of *inverse functions*. The general definition for inverse functions is given in the box that follows.

DEFINITION

Inverse Functions

Two functions f and g are **inverses** of one another provided that

$$f[g(x)] = x \qquad \text{for each } x \text{ in the domain of } g$$

and

$$g[f(x)] = x \qquad \text{for each } x \text{ in the domain of } f$$

EXAMPLE 1   Verify that the functions f and g are inverses, where

$$f(x) = \frac{1}{3}x + 2 \qquad \text{and} \qquad g(x) = 3x - 6$$

Solution   In view of the definition, we must check that $f[g(x)] = x$ and $g[f(x)] = x$. We have

$$f[g(x)] = f(3x - 6)$$
$$= \frac{1}{3}(3x - 6) + 2 = x \qquad \text{(Check the algebra.)}$$

Thus, $f[g(x)] = x$. Now we still need to check that $g[f(x)] = x$. We have

$$g[f(x)] = g\left(\frac{1}{3}x + 2\right)$$
$$= 3\left(\frac{1}{3}x + 2\right) - 6 = x \qquad \text{(Again, check the algebra.)}$$

Having shown that $f[g(x)] = x$ and $g[f(x)] = x$, we conclude that f and g are indeed inverse functions.   ▲

EXAMPLE 2   Suppose that f and g are a pair of inverse functions. If $f(2) = 3$, what is $g(3)$?

Solution   We can use either of two methods.

FIRST METHOD   The quantities 3 and $f(2)$ are declared to be equal. Thus, whether we use 3 or $f(2)$ as an input for g, the result must be the same. We then have

$$3 = f(2)$$
$$g(3) = g[f(2)]$$
$$g(3) = 2 \quad \text{using the fact that } g[f(x)] = x \text{ for any } x \text{ in the domain of } f$$

Thus $g(3) = 2$.

ALTERNATIVE METHOD   Think of a table of values for the function f. Since $f(2) = 3$, one entry in the table must look like this:

x		2	
y		3	

Now, g reverses the roles of x and y, so in the table for g, one entry must look like this:

x		3	
y		2	

Thus, $g(3) = 2$, as obtained previously. In Figure 2, we show a graphical interpretation of this method.

FIGURE 2

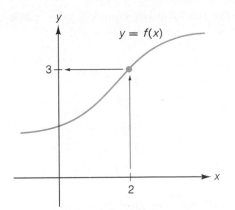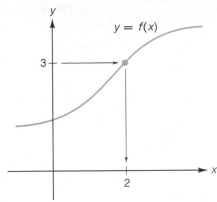

(a) The function *f* assigns the output 3 to the input 2. (Note the direction of the arrows.)

(b) The inverse of the function *f* assigns the output 2 to the input 3. (Note the direction of the arrows.) ▲

It is customary to use the notation f^{-1} (read *f inverse*) for the function that is the inverse of *f*. So in Example 1, for instance, we have

$$f(x) = \frac{1}{3}x + 2 \quad \text{and} \quad f^{-1}(x) = 3x - 6$$

Note In the context of functions, f^{-1} does *not* in general mean $1/f$. (For an example of this, see Exercise 4 at the end of this section.) In the following box, we rewrite the defining equations for inverse functions, using the f^{-1} notation rather than *g*. Figure 3 provides a graphic summary of these equations.

$$f[f^{-1}(x)] = x \quad \text{for each } x \text{ in the domain of } f^{-1}$$

and

$$f^{-1}[f(x)] = x \quad \text{for each } x \text{ in the domain of } f$$

FIGURE 3

The action of inverse functions: The set *A* in the figure is both the domain of *f* and the range of f^{-1}. The set *B* is both the domain of f^{-1} and the range of *f*.

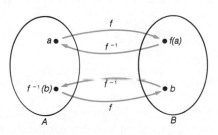

EXAMPLE 3   Suppose $f(x) = 4x^3 + 7$. Find $f[f^{-1}(5)]$. (Assume that f^{-1} exists and that 5 is in its domain.)

Solution   By definition, $f[f^{-1}(x)] = x$. Substituting $x = 5$ on both sides of this identity directly gives us

$$f[f^{-1}(5)] = 5, \quad \text{as required}$$

Note that we didn't need to make use of the formula $f(x) = 4x^3 + 7$, nor did we need to find a formula for $f^{-1}(x)$. ▲

EXAMPLE 4 Solve the following equation for x, given that the domain of both f and f^{-1} is $(-\infty, \infty)$ and that $f(1) = -2$.

$$3 + f^{-1}(x - 1) = 4$$

Solution

$$3 + f^{-1}(x - 1) = 4$$
$$f^{-1}(x - 1) = 1$$
$$f[f^{-1}(x - 1)] = f(1) \quad \text{applying } f \text{ to both sides}$$
$$x - 1 = f(1) \quad \text{definition of inverse function}$$
$$x = 1 + f(1) = 1 + (-2) = -1$$

▲

We'll postpone until the end of this section a discussion of which functions have inverses and which do not. For functions that do have inverses, however, there's a simple method that can often be used to determine those inverses. This method is illustrated in Examples 5 and 6.

EXAMPLE 5 Let $f(x) = 3x - 4$. Find $f^{-1}(x)$.

Solution

We begin by rewriting the given equation as

$$y = 3x - 4$$

We know that f^{-1} interchanges the inputs and outputs of f. So to determine f^{-1}, we only need to switch the x's and y's in the equation $y = 3x - 4$. That gives us $x = 3y - 4$. Now we solve for y as follows:

$$x = 3y - 4$$
$$x + 4 = 3y$$
$$\frac{x + 4}{3} = y$$

Thus, the inverse function is $y = \dfrac{x + 4}{3}$. We can also write this as

$$f^{-1}(x) = \frac{x + 4}{3}$$

Actually, we should call this result just a *candidate* for f^{-1}, since certain technical matters remain to be discussed. However, if you compute $f[f^{-1}(x)]$ and $f^{-1}[f(x)]$ and find that they both do equal x, then the matter is settled, by definition. Exercise 13 at the end of this section asks you to carry out those calculations.

▲

Here is a summary of our procedure for calculating $f^{-1}(x)$.

To Find $f^{-1}(x)$ for the Function $y = f(x)$

1. Interchange x and y in the equation $y = f(x)$.
2. Solve the resulting equation for y.

EXAMPLE 6   Let $f(x) = \dfrac{2x - 1}{3x + 5}$. Find $f^{-1}(x)$.

Solution

STEP 1 Write the given function as $y = \dfrac{2x - 1}{3x + 5}$. Interchange x and y to get

$$x = \frac{2y - 1}{3y + 5}$$

STEP 2 Solve the resulting equation for y.

$$x = \frac{2y - 1}{3y + 5}$$

$$3xy + 5x = 2y - 1 \qquad \text{multiplying by } 3y + 5$$

$$3xy - 2y = -5x - 1 \qquad \text{rearranging}$$

$$y(3x - 2) = -5x - 1$$

$$y = \frac{-5x - 1}{3x - 2}$$

Thus, the inverse function is given by

$$f^{-1}(x) = \frac{-5x - 1}{3x - 2}$$

▲

A certain symmetry always exists between the graph of a function and the graph of its inverse. As background for this discussion, we need the following definition of **symmetry about a line**.

DEFINITION

Symmetry about a Line

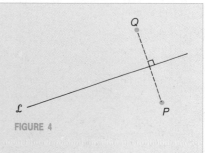

Refer to Figure 4. Two points P and Q are **symmetric about the line** $\mathscr{L}$ provided that

$\overline{PQ}$ is perpendicular to $\mathscr{L}$

and

points P and Q are equidistant from $\mathscr{L}$

FIGURE 4

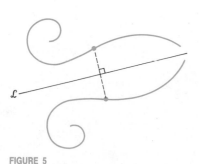

FIGURE 5

In other words, P and Q are symmetric about the line $\mathscr{L}$ if $\mathscr{L}$ is the perpendicular bisector of line segment $\overline{PQ}$. We say that $\mathscr{L}$ is the **axis of symmetry** and that P and Q are mirror images, or **reflections**, of each other through $\mathscr{L}$. (Actually, you already encountered two instances of this type of symmetry in Section 2.1, where we discussed symmetry about the x- and y-axes.) In addition, we say that two curves are symmetric about a line if each point on one curve is the reflection of a corresponding point on the other curve, and vice versa (see Figure 5).

EXAMPLE 7 Verify that the points $P(4, 1)$ and $Q(1, 4)$ are symmetric about the line $y = x$. See Figure 6.

Solution We have to show that the line $y = x$ is the perpendicular bisector of the line segment $\overline{PQ}$. Now, the slope of $\overline{PQ}$ is

$$m = \frac{4 - 1}{1 - 4} = -1$$

We know that the slope of the line $y = x$ is 1. Since these two slopes are negative reciprocals, we conclude that the line $y = x$ is perpendicular to $\overline{PQ}$. Next, we must show that the line $y = x$ passes through the midpoint of $\overline{PQ}$. But (as you should check for yourself), the midpoint of $\overline{PQ}$ is $(\frac{5}{2}, \frac{5}{2})$, which does lie on the line $y = x$. In summary, then, we've shown that the line $y = x$ passes through the midpoint of $\overline{PQ}$ and is perpendicular to $\overline{PQ}$. Thus P and Q are symmetric about the line $y = x$, as we wished to show. ▲

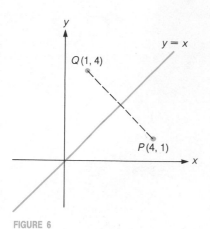

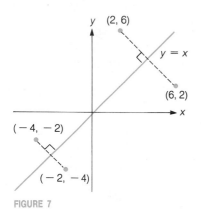

FIGURE 6

FIGURE 7

Just as we've shown that $(4, 1)$ and $(1, 4)$ are symmetric about the line $y = x$, we can also show in general that (a, b) and (b, a) are always symmetric about the line $y = x$. Figure 7 displays two examples of this, and Exercise 41 at the end of this section asks you to supply a proof for the general situation.

Now let us see how these ideas about symmetry relate to inverse functions. By way of example, consider the functions f and f^{-1} from Example 5:

$$f(x) = 3x - 4 \qquad f^{-1}(x) = \frac{1}{3}x + \frac{4}{3}$$

Figure 8 shows the graphs of f, f^{-1}, and the line $y = x$. Note that the graphs of f and f^{-1} are symmetric about the line $y = x$. This kind of symmetry, in fact, always occurs for the graphs of any pair of inverse functions.

Why are the graphs of f and f^{-1} always mirror images of each other about the line $y = x$? First, recall that the function f^{-1} switches the inputs and outputs of f. Thus, (a, b) is on the graph of f if and only if (b, a) is on the graph of f^{-1}. But as we have stated, the points (a, b) and (b, a) are mirror images in the line $y = x$. It follows that the graphs of f and f^{-1} are reflections of each other about the line $y = x$. For reference, we restate this useful fact in the box that follows.

FIGURE 8

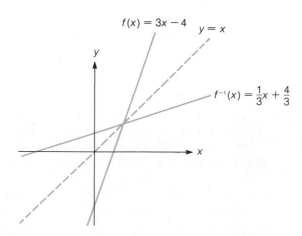

PROPERTY SUMMARY

THE GRAPHS OF f AND f^{-1}

The graphs of f and f^{-1} are symmetric about the line $y = x$.

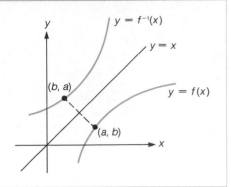

EXAMPLE 8   The graph of a function f consists of the line segment joining the points $(-2, -3)$ and $(-1, 4)$, as shown in Figure 9. Sketch a graph of f^{-1}.

Solution   The graph of f^{-1} is obtained by reflecting the graph of f about the line $y = x$. The reflection of $(-2, -3)$ is $(-3, -2)$, and the reflection of $(-1, 4)$ is $(4, -1)$. To graph f^{-1}, then, we plot the reflected points $(-3, -2)$ and $(4, -1)$ and connect them with a line segment, as shown in Figure 10. For reference, Figure 10 also shows the graphs of f and $y = x$.

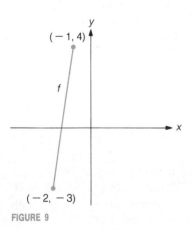

FIGURE 9

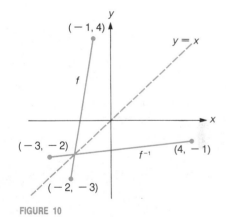

FIGURE 10

EXAMPLE 9   Let $g(x) = x^3$. Find $g^{-1}(x)$ and then, on the same set of axes, sketch the graphs of g, g^{-1}, and $y = x$.

Solution   We begin by writing $y = x^3$, then switching x and y and solving for y. We first have

$$x = y^3$$

To solve this equation for y, we take the cube root of both sides to obtain

$$\sqrt[3]{x} = y$$

So the inverse function is $g^{-1}(x) = \sqrt[3]{x}$. We could graph the function g^{-1} by plotting points. But the easier way is to reflect the graph of $g(x) = x^3$ about the line $y = x$. See Figure 11.

FIGURE 11

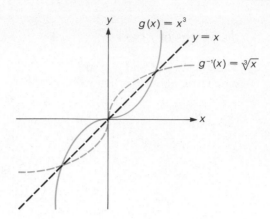

Earlier in this section, we mentioned that not every function has an inverse function. For instance, Tables 1(a) and (b) at the bottom of this page indicate what happens if we begin with the function defined by $F(x) = x^2$ and interchange the inputs and outputs. The resulting table, Table 1(b), does not define a function. For instance, the input 4 in Table 1(b) produces two distinct outputs, 2 and -2, but the definition of a function requires exactly one output for a given input.

Why didn't this difficulty arise when we looked for the inverse of $f(x) = 2x$ at the beginning of this section? The answer is that there is an essential difference between the functions defined by $f(x) = 2x$ and $F(x) = x^2$. With $f(x) = 2x$, distinct inputs never yield the same output. That is, if $x_1 \neq x_2$, then $f(x_1) \neq f(x_2)$. This condition guarantees that interchanging the inputs and outputs of f will yield a function. With $F(x) = x^2$, however, distinct inputs can yield the same output. For instance,

$$2 \neq -2$$

but

$$F(2) = F(-2) \quad \text{because } 2^2 = 4 = (-2)^2$$

It's useful to have a name for the type of function in which distinct inputs always yield distinct outputs. We call such a function a **one-to-one function**. So in view of the discussion in the previous paragraph, the function defined by $f(x) = 2x$ is one-to-one, but $F(x) = x^2$ is not. Using graphs, there is an easy way

TABLE 1

(b) The table formed by reversing the inputs and outputs for $F(x) = x^2$

(a) $F(x) = x^2$

x	y		x	y
1	1		1	1
2	4		4	2
3	9		9	3
-1	1		1	-1
-2	4		4	-2
-3	9		9	-3

Reverse inputs and outputs →

to tell which functions are one-to-one:

Horizontal Line Test

> A function f is one-to-one if and only if each horizontal line intersects the graph of $y = f(x)$ in at most one point.

EXAMPLE 10 Determine if each function is one-to-one.

(a) $F(x) = 2x$ **(b)** $H(x) = x^2$ **(c)** $h(x) = (\sqrt{x})^4$

Solution We sketch the graphs and apply the horizontal line test, as in Figure 12. The horizontal line test tells us that F and h are one-to-one, but H is not.

FIGURE 12

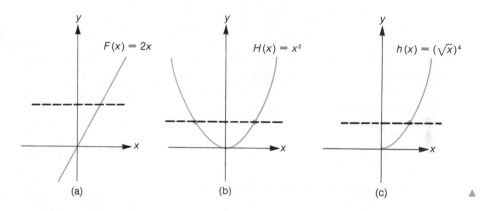

(a)   (b)   (c)

The following theorem tells us that the functions with inverses are precisely the one-to-one functions. (The proof of the theorem isn't difficult, but we'll omit giving it here.)

Theorem

> A function f has an inverse function if and only if f is one-to-one.

EXAMPLE 11 Which of the three functions in Example 10 have inverse functions?

Solution According to our theorem, the functions with inverses are those that are one-to-one. In Example 10, we found that F and h were one-to-one, but H was not. So both F and h each have inverses, but H does not.

▼ **EXERCISE SET 2.7**

A

1. Verify that the given pairs of functions are inverse functions.
   (a) $f(x) = 3x$; $g(x) = x/3$
   (b) $f(x) = 4x - 1$; $g(x) = \frac{1}{4}x + \frac{1}{4}$
   (c) $g(x) = \sqrt{x}$; $h(x) = x^2$ [Assume that the domain of both g and h is $[0, \infty)$.]

2. Which pairs of functions are inverses?
   (a) $f(x) = -3x + 2$; $g(x) = \frac{2}{3} - \frac{1}{3}x$
   (b) $F(x) = 2x + 1$; $G(x) = \frac{1}{2}x - 1$
   (c) $G(x) = x^3$; $H(x) = 1 - x^3$
   (d) $f(t) = t^3$; $g(t) = \sqrt[3]{t}$

3. Let $f(x) = x^3 + 2x + 1$ and assume that the domain of f^{-1} is $(-\infty, \infty)$. Simplify the expression.
   (a) $f[f^{-1}(4)]$
   (b) $f^{-1}[f(-1)]$
   (c) $(f \circ f^{-1})(\sqrt{2})$
   (d) $f[f^{-1}(t + 1)]$
   (e) $f(0)$; use your answer to find $f^{-1}(1)$.
   (f) $f(-1)$; use your answer to find $f^{-1}(-2)$.

4. Let $f(x) = 2x + 1$.
   (a) Find $f^{-1}(x)$.
   (b) Calculate $f^{-1}(5)$ and $1/f(5)$. Are your answers the same?

5. Let $f(x) = 3x - 1$.
   (a) Compute $f^{-1}(x)$.
   (b) Verify that $f[f^{-1}(x)] = x$ and that $f^{-1}[f(x)] = x$.
   (c) On the same set of axes, sketch the graphs of f, f^{-1}, and the line $y = x$. Note that the graphs of f and f^{-1} are symmetric about the line $y = x$.

6. Follow Exercise 5, but use $f(x) = \frac{1}{3}x - 2$.

7. Follow Exercise 5, but use $f(x) = \sqrt{x - 1}$. [The domain of f^{-1} will be $[0, \infty)$.]

8. Follow Exercise 5, but use $f(x) = 1/x$.

9. Let $f(x) = \dfrac{x + 2}{x - 3}$.
   (a) Find the domain and range of the function f.
   (b) Find $f^{-1}(x)$.
   (c) Find the domain and range of the function f^{-1}. What do you observe?

10. Let $f(x) = \dfrac{2x - 3}{x + 4}$. Find $f^{-1}(x)$. Find the domain and range for f and f^{-1}. What do you observe?

11. Let $f(x) = 2x^3 + 1$. Find $f^{-1}(x)$.

12. In our preliminary discussion at the beginning of this section, we considered the functions $f(x) = 2x$ and $g(x) = x/2$. Verify that $f[g(x)] = x$ and that $g[f(x)] = x$. On the same set of axes, sketch the graphs of f, g, and $y = x$.

13. This exercise refers to the comments made at the end of Example 5. Compute $f[f^{-1}(x)]$ and $f^{-1}[f(x)]$ using the functions $f(x) = 3x - 4$ and $f^{-1}(x) = \frac{1}{3}x + \frac{4}{3}$. By actually carrying out the calculations, you can see in each case that the answer is indeed x. Then, on the same set of axes, sketch the graphs of f, f^{-1}, and the line $y = x$.

14. Let $f(x) = (x - 3)^2$ and take the domain to be $[3, \infty)$.
   (a) Find $f^{-1}(x)$. Give the domain of f^{-1}.
   (b) On the same set of axes, sketch the graphs of f and f^{-1}.

15. Let $f(x) = (x - 3)^3 - 1$.
   (a) Compute $f^{-1}(x)$.
   (b) On the same set of axes, sketch the graphs of f, f^{-1}, and the line $y = x$. *Hint:* Sketch f using the ideas of Section 2.5. Then reflect f to get f^{-1}.

16. The figure shows the graph of a function f. (The axes are marked off in one-unit intervals.) Graph each of the following functions.
   (a) $y = f^{-1}(x)$
   (b) $y = f(x - 2)$
   (c) $y = f(x) - 2$
   (d) $y = f(-x)$
   (e) $y = -f(x)$
   (f) $y = f^{-1}(x - 2)$
   (g) $y = f^{-1}(x) - 2$
   (h) $y = f^{-1}(x - 2) - 2$
   (i) $y = f^{-1}(-x)$
   (j) $y = -f^{-1}(x)$

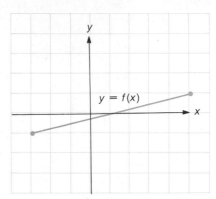

17. The figure shows the graph of a function g. (The axes are marked off in one-unit intervals.) Graph each of the following functions.
   (a) $y = g^{-1}(x)$
   (b) $y = g^{-1}(x) - 1$
   (c) $y = g^{-1}(x - 1)$
   (d) $y = g^{-1}(-x)$
   (e) $y = -g^{-1}(x)$
   (f) $y = -g^{-1}(-x)$

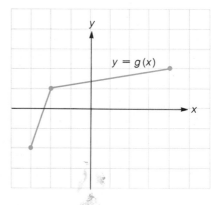

18. The figure on the next page shows the graph of a function $y = h(x)$.
   (a) Sketch a graph that is the reflection of h about the y-axis. What is the equation of the graph you obtain?
   (b) On the same set of axes, sketch graphs of h, h^{-1}, and the line $y = x$.
   (c) Let k be the function whose graph is the reflection of h in the x-axis. Sketch a graph of the function k^{-1}.

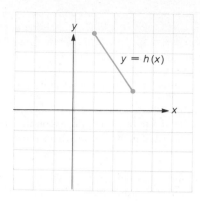

In Exercises 19–30, use the horizontal line test to determine if the function is one-to-one (and therefore has an inverse).

19. $y = x^2 + 1$

20. $y = \sqrt{x}$

21. $f(x) = 1/x$

22. $g(x) = |x|$

23. $y = x^3$

24. $y = 1 - x^3$

25. $y = \sqrt{1 - x^2}$

26. $y = 3x$

27. $g(x) = 5$ (for all x)

28. $y = mx + b$ ($m \neq 0$)

29. $f(x) = \begin{cases} x^2 & \text{if } -1 \leq x \leq 0 \\ x^2 + 1 & \text{if } x > 0 \end{cases}$

30. $g(x) = \begin{cases} x^2 & \text{if } -1 \leq x < 0 \\ x^2 + 1 & \text{if } x \geq 0 \end{cases}$

31. Let $f(x) = \dfrac{3x - 2}{5x - 3}$. Show that $f[f(x)] = x$. This says that f is its own inverse.

32. Let $f(x) = \dfrac{ax + b}{cx - a}$. Show that f is its own inverse. (Assume that $a^2 + bc \neq 0$.)

In Exercises 33–36, assume that the domain of f and f^{-1} is $(-\infty, \infty)$. Solve the equation for x or for t (whichever is appropriate) using the given information.

33. (a) $7 + f^{-1}(x - 1) = 9$; $f(2) = 6$
(b) $4 + f(x + 3) = -3$; $f^{-1}(-7) = 0$

34. (a) $f^{-1}(2x + 3) = 5$; $f(5) = 13$
(b) $f(1 - 2x) = -4$; $f^{-1}(-4) = -5$

35. $f^{-1}\left(\dfrac{t + 1}{t - 2}\right) = 12$; $f(12) = 13$

36. $f\left(\dfrac{1 - 2t}{1 + 2t}\right) = 7$; $f^{-1}(7) = -3$

B

37. Let $f(x) = \sqrt{x}$.
(a) Find $f^{-1}(x)$. What is the domain of f^{-1}? [The domain is not $(-\infty, \infty)$.]

(b) In each case, determine whether the given point lies on the graph of f or f^{-1}.
(i) $(4, 2)$ (ii) $(2, 4)$
(iii) $(5, \sqrt{5})$ (iv) $(\sqrt{5}, 5)$
(v) $(a, f(a))$ (Assume $a \geq 0$.)
(vi) $(f(a), a)$
(vii) $(b, f^{-1}(b))$ (Assume $b \geq 0$.)
(viii) $(f^{-1}(b), b)$

38. Let $F(x) = x^3 + 7x - 5$. Find $F^{-1}(-5)$. [Assume that the domain of F^{-1} is $(-\infty, \infty)$.]

39. Reread the definition of symmetry about a line given on page 133. Then use the method shown in Example 7 to show that the points $(7, -1)$ and $(-1, 7)$ are symmetric about the line $y = x$.

40. Use the method shown in Example 7 to show that $(5, 2)$ and $(2, 5)$ are symmetric about $y = x$.

41. Use the method of Example 7 to show that the two points (a, b) and (b, a) are symmetric about the line $y = x$.

42. Points P and Q are both reflected in the line $y = x$ to obtain points P' and Q', respectively. Does the distance from P to Q equal the distance from P' and Q'?

43. Pick any three points on the line $y = \frac{1}{2}x + 3$. Reflect each point in the line $y = x$. Show that the three reflected points all lie on one line. What is the equation of that line?

C

44. Determine values for m and b so that the points $(8, 2)$ and $(4, 8)$ are symmetric about the line $y = mx + b$.

45. Let $f(x) = 2x^3 - 5$. Define a function F by $F(x) = f(x - 1)$. Find $F[f^{-1}(x) + 1]$.

46. Let f be a function whose domain is $(-\infty, \infty)$ and suppose that f^{-1} exists. Let F be the function defined by $F(x) = f(x - 1)$. Show that $F^{-1}(x) = f^{-1}(x) + 1$.

47. Reflect the point $(2, 1)$ in the line $y = 3x$. What are the coordinates of the point you obtain?

48. (a) If the point (a, b) is reflected in the line $y = 3x$, show that the coordinates of the reflected point are $\left(\dfrac{3b - 4a}{5}, \dfrac{3a + 4b}{5}\right)$.
(b) If the point (a, b) is reflected in the line $y = mx$, what are the coordinates of the reflected point?

49. (a) Find the reflection of the point (a, b) in the line $y = -x$.   *Answer:* $(-b, -a)$
(b) Begin with the point (a, b). Reflect this point about the x-axis to obtain a point Q. Then reflect Q about the y-axis to obtain a point R. Then reflect R about the line $y = x$ to obtain a point S. Show that S is the reflection of (a, b) in the line $y = -x$.

▼ **CHAPTER TWO   SUMMARY OF PRINCIPAL TERMS AND FORMULAS**

TERMS OR FORMULAS	PAGE REFERENCE	COMMENTS
1. Graph of an equation	70	The graph of an equation in the variables x and y is the set of all points (x, y) whose coordinates satisfy the equation.
2. x-intercept and y-intercept	71	An x-intercept of a graph is the x-coordinate of a point where the graph intersects the x-axis. Similarly, a y-intercept is the y-coordinate of a point where the graph intersects the y-axis.
3. Symmetry about the x-axis	72	A graph is symmetric about the x-axis if, for each point (x, y) on the graph, the point $(x, -y)$ is also on the graph. The points (x, y) and $(x, -y)$ are *reflections* of each other about (or in) the x-axis.
4. Symmetry about the y-axis	72	A graph is symmetric about the y-axis if, for each point (x, y) on the graph, the point $(-x, y)$ is also on the graph. The points (x, y) and $(-x, y)$ are *reflections* of each other about (or in) the y-axis.
5. Symmetry about the origin	72	A graph is symmetric about the origin if, for each point (x, y) on the graph, the point $(-x, -y)$ is also on the graph. The points (x, y) and $(-x, -y)$ are *reflections* of each other about the origin.
6. Symmetry tests	73	**(i)** The graph of an equation is symmetric about the y-axis if replacing x with $-x$ yields an equivalent equation. **(ii)** The graph of an equation is symmetric about the x-axis if replacing y with $-y$ yields an equivalent equation. **(iii)** The graph of an equation is symmetric about the origin if replacing x and y with $-x$ and $-y$, respectively, yields an equivalent equation.
7. $m = \dfrac{y_2 - y_1}{x_2 - x_1}$	78	m is the slope of a nonvertical line passing through the points (x_1, y_1) and (x_2, y_2).
8. $y - y_1 = m(x - x_1)$	80	This is the point–slope form for the equation of a line passing through the point (x_1, y_1) with slope m.
9. $y = mx + b$	83	This is the slope–intercept form for the equation of a line with slope m and y-intercept b.
10. $\dfrac{x}{a} + \dfrac{y}{b} = 1$	84	This is the two-intercept form for the equation of a line with x- and y-intercepts of a and b, respectively.
11. $m_1 = m_2$	85	Nonvertical parallel lines have the same slope.
12. $m_1 = \dfrac{-1}{m_2}$	85	Two nonvertical lines are perpendicular if and only if the slopes are negative reciprocals of each other.
13. Function	92	Given two nonempty sets A and B, a function from A to B is a rule of correspondence that assigns to each element of A exactly one element in B.
14. Domain	92	The domain of a function is the set of all inputs for that function. If f is a function from A to B, then the domain is the set A.
15. Range	92	The range of a function from A to B is the set of all elements in B that are actually used as outputs.
16. $f(x)$	93	Given a function f, the notation $f(x)$ denotes the output that results from the input x.

TERMS OR FORMULAS	PAGE REFERENCE	COMMENTS
17. Graph of a function	100	The graph of a function f consists of those points (x, y) such that x is in the domain of f and $y = f(x)$.
18. Vertical line test	102	A graph in the x-y plane represents a function $y = f(x)$ provided that any vertical line intersects the graph in at most one point.
19. Turning point	105	A turning point on a graph is a point where the graph changes from rising to falling, or vice versa. See, for example, Figures 10 and 11 on page 105.
20. Maximum value and minimum value	105	An output $f(a)$ is a maximum value of the function f if $f(a) \geq f(x)$ for every x in the domain of f. An output $f(b)$ is a minimum value if $f(b) \leq f(x)$ for every x in the domain of f.
21. Increasing function and decreasing function	105	A function f is increasing on an interval if the following condition holds: If a and b are in the interval and $a < b$, then $f(a) < f(b)$. Geometrically, this means that the graph is rising as we move in the positive x-direction. A function f is decreasing on an interval if the following condition holds: If a and b are in the interval and $a < b$, then $f(a) > f(b)$. Geometrically, this means that the graph is falling as we move in the positive x-direction.
22. Translation, reflection, and coordinates	112, 116	If we have an equation that determines a graph in the x-y plane, and h and k are positive numbers, then: **(i)** Replacing x with $x - h$ translates the graph h units in the positive x-direction; **(ii)** Replacing y with $y - k$ translates the graph k units in the positive y-direction; **(iii)** Replacing x with $x + h$ translates the graph h units in the negative x-direction; **(iv)** Replacing y with $y + k$ translates the graph k units in the negative y-direction; **(v)** Replacing x with $-x$ reflects the graph in the y-axis; **(vi)** Replacing y with $-y$ reflects the graph in the x-axis. For comprehensive examples, see Example 2 on page 113 and Example 6 on page 116.
23. $f \circ g$	123	A composition of functions f and g: $(f \circ g)(x) = f[g(x)]$
24. Inverse functions	129	Two functions f and g are inverses of one another provided that the following two conditions are met. First, $f[g(x)] = x$ for each x in the domain of g. Second, $g[f(x)] = x$ for each x in the domain of f.
25. f^{-1}	131	f^{-1} denotes the inverse function for f. *Note:* In this context, f^{-1} does not mean $1/f$. For an example, see Exercise 4 on page 138.
26. Symmetry about a line	133	Two points P and Q are symmetric about a line if that line is the perpendicular bisector of $\overline{PQ}$.
27. One-to-one function	136	A function f is said to be one-to-one provided that distinct inputs always yield distinct outputs. That is, if x_1 and x_2 are in the domain of f and $x_1 \neq x_2$, then $f(x_1) \neq f(x_2)$. The relationship between inverse functions and the notion of *one-to-one* is given by the following theorem: A function f has an inverse if and only if f is one-to-one.
28. Horizontal line test	137	A function f is one-to-one if and only if each horizontal line intersects the graph of $y = f(x)$ in at most one point.

▼ CHAPTER TWO   REVIEW EXERCISES

NOTE   Exercises 1–16 constitute a chapter test on the fundamentals, based on group A problems.

1. **(a)** Find the domain of the function defined by
    $$f(x) = \sqrt{15 - 5x}.$$
    (b) Find the range of the function defined by
    $$g(x) = \frac{3 + x}{2x - 5}.$$

2. Let $f(x) = 3x^2 - 4x$ and $g(x) = 2x + 1$. Compute each of the following.
    (a) $(f - g)(x)$ **(b)** $(f \circ g)(x)$ **(c)** $f[g(-1)]$

3. Find the equation of the line that is tangent to the circle $x^2 + 4x + y^2 - 2y - 8 = 0$ at the point $(1, 3)$. Write your answer in the form $y = mx + b$.

4. Test for symmetry with respect to the x-axis, the y-axis, and the origin: $y = 2x^3 - x^5$

5. **(a)** Compute $\dfrac{F(x) - F(a)}{x - a}$ given that $F(x) = 1/x$.

    (b) Compute $\dfrac{g(x + h) - g(x)}{h}$ for the function defined by $g(x) = x - 2x^2$.

6. Find the y-intercept and slope of the line $3x - 5y = 15$. Graph the line.

7. **(a)** Find $g^{-1}(x)$ given that $g(x) = \dfrac{1 - 5x}{3x}$.

    (b) The figure displays the graph of a function f. Sketch the graph of f^{-1}.

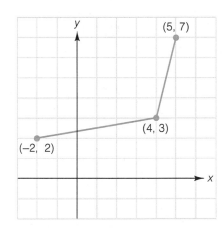

8. Find the equation of a line passing through the origin and parallel to the line $5x - 6y = 30$. Write the answer in the form $y = mx + b$.

9. Graph each function and specify the intercepts.
    (a) $y = |x + 2| - 3$ **(b)** $y = \dfrac{1}{x + 2} - 1$

10. Refer to the graph of $y = g(x)$ in the figure. The domain of g is $[-5, 2]$.
    (a) What are the coordinates of the turning point(s)?
    (b) What is the maximum value of g?
    (c) Which input yields a minimum value for g?
    (d) On which interval(s) is g increasing?

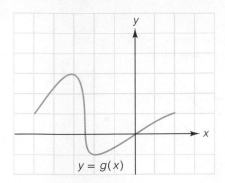

11. Let $f(x) = 3x^2 - 2x$.
    (a) Find $f(-1)$.   **(b)** Find $f(1 - \sqrt{2})$.

12. Graph the function G defined by
    $$G(x) = \begin{cases} \sqrt{1 - x^2} & \text{if } -1 \leq x < 0 \\ \sqrt{x} & \text{if } x \geq 0 \end{cases}$$

13. Find the slope of a line passing through the two points $(5, 25)$ and $(5 + h, (5 + h)^2)$. Express your answer in terms of h.

14. Graph the equation $y = |9 - x^2|$. Specify symmetry and intercepts.

15. The figure shows the graph of a function $y = f(x)$. Sketch the graph of $y = f(-x)$.

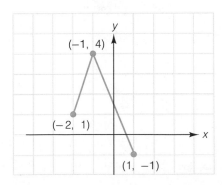

16. Given that the domains of f and f^{-1} are both $(-\infty, \infty)$ and that $f^{-1}(1) = -4$, solve the following equation for t:
    $$2 + f(3t + 5) = 3$$

In Exercises 17–34, find an equation for the line satisfying the given conditions. Write your answer in the form $y = mx + b$.

17. Passing through $(-4, 2)$ and $(-6, 6)$

18. $m = -2$; y-intercept 5

19. $m = \frac{1}{4}$; passing through $(-2, -3)$

20. $m = \frac{1}{3}$; x-intercept -1

21. x-intercept -4; y-intercept 8

22. $m = -10$; y-intercept 0

23. y-intercept -2; parallel to the x-axis

24. Passing through $(0, 0)$ and parallel to $6x - 3y = 5$

25. Passing through $(1, 2)$ and perpendicular to the line $x + y + 1 = 0$

26. Passing through $(1, 1)$ and through the center of the circle $x^2 - 4x + y^2 - 8y + 16 = 0$

27. Passing through the centers of the circles $x^2 + 4x + y^2 + 2y = 0$ and $x^2 - 4x + y^2 - 16y = 0$

28. $m = 3$; the same x-intercept as the line $3x - 8y = 12$

29. Passing through the origin and the midpoint of the line segment joining the points $(-2, -3)$ and $(6, -5)$

30. Tangent to the circle $x^2 + y^2 = 20$ at the point $(-2, 4)$

31. Tangent to the circle $x^2 - 6x + y^2 + 8y = 0$ at the point $(0, 0)$

32. Passing through $(2, 4)$; the y-intercept is twice the x-intercept

33. Passing through $(2, -1)$; the sum of the x- and y-intercepts is 2 (There are two answers.)

34. Passing through $(4, 5)$; no x-intercept

35. Suppose that (a, b) is a point on the graph of $y = f(x)$. Match the functions defined in the left-hand column with the points in the right-hand column. For example, the appropriate match for (a) in the left-hand column is determined as follows. The graph of $y = f(x) + 1$ is obtained by translating the graph of $y = f(x)$ up one unit. Thus, the point (a, b) moves up to $(a, b + 1)$ and, consequently, (E) is the appropriate match for (a).

(**a**) $y = f(x) + 1$ (**A**) $(-a, b)$

(**b**) $y = f(x + 1)$ (**B**) (b, a)

(**c**) $y = f(x - 1) + 1$ (**C**) $(a - 1, b)$

(**d**) $y = f(-x)$ (**D**) $(-b, a + 1)$

(**e**) $y = -f(x)$ (**E**) $(a, b + 1)$

(**f**) $y = f(-x)$ (**F**) $(1 - a, b)$

(**g**) $y = f^{-1}(x)$ (**G**) $(-a, -b)$

(**h**) $y = f^{-1}(x) + 1$ (**H**) $(-b, 1 - a)$

(**i**) $y = f^{-1}(x - 1)$ (**I**) $(b, -a)$

(**j**) $y = f^{-1}(-x) + 1$ (**J**) $(a, -b)$

(**k**) $y = -f^{-1}(x)$ (**K**) $(b + 1, a)$

(**l**) $y = -f^{-1}(-x) + 1$ (**L**) $(a + 1, b + 1)$

(**m**) $y = 1 - f^{-1}(x)$ (**M**) $(b, a + 1)$

(**n**) $y = f(1 - x)$ (**N**) $(b, 1 - a)$

In Exercises 36–58, sketch the graph and specify any x- or y-intercept.

36. $y = 4 - x^2$

37. $y = -(x - 1)^2 + 2$

38. $y = \dfrac{1}{x} + 1$

39. $f(x) = \dfrac{1}{x + 1}$

40. $y = \dfrac{1}{x + 1} + 1$

41. $y = |x + 3|$

42. $g(x) = -\sqrt{x - 4}$

43. $h(x) = \sqrt{1 - x^2}$

44. $f(x) = \sqrt{1 - x^2} + 1$

45. $y = 1 - (x + 1)^3$

46. $y = 4 - \sqrt{-x}$

47. $y = (\sqrt{x})^2$

48. $f \circ g$, where $g(x) = x + 3$ and $f(x) = x^2$

49. $f \circ g$, where $g(x) = \sqrt{x - 1}$ and $f(x) = -x^2$

50. $f(x) = \begin{cases} \sqrt{1 - x^2} & \text{if } -1 \le x \le 0 \\ \sqrt{x + 1} & \text{if } x > 0 \end{cases}$

51. $F(x) = \begin{cases} -\sqrt{1 - x^2} & \text{if } -1 \le x < 0 \\ \sqrt{x} & \text{if } x \ge 0 \end{cases}$

52. $y = \begin{cases} |x - 1| & \text{if } 0 \le x \le 2 \\ |x - 3| & \text{if } 2 < x \le 4 \end{cases}$

53. $y = \begin{cases} 1/x & \text{if } 0 < x \le 1 \\ 1/(x - 1) & \text{if } 1 < x \le 2 \end{cases}$

54. $(y + |x| - 1)(y - |x| + 1) = 0$

55. f^{-1}, where $f(x) = \frac{1}{2}(x + 1)$

56. g^{-1}, where $g(x) = \sqrt[3]{x + 2}$

57. $f \circ f^{-1}$, where $f(x) = \sqrt{x - 2}$

58. $f^{-1} \circ f$, where $f(x) = \sqrt{x - 2}$

For Exercises 59–64, determine if the graph is symmetric in the x-axis, the y-axis, or the origin.

59.

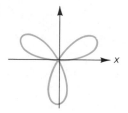

60.

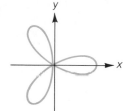

61.

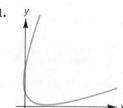

62.

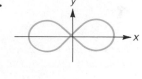

63.

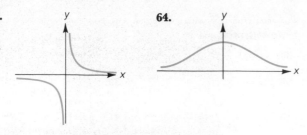

64.

In Exercises 65–74, find the domain of the function.

65. $y = \dfrac{1}{x^2 - 9}$

66. $y = x^3 - x^2$

67. $y = \sqrt{8 - 2x}$

68. $y = \dfrac{x}{6x^2 + 7x - 3}$

69. $y = \sqrt{|2 - 5x|}$

70. $y = \dfrac{25 - x^2}{\sqrt{x^2 + 1}}$

71. $y = \sqrt{x^2 - 2x - 3}$

72. $y = \sqrt{5 - x^2}$

73. $y = x + \dfrac{1}{x}$

74. $z = \dfrac{1}{t - \sqrt{t + 2}}$

In Exercises 75–80, determine the range of the function.

75. $y = \dfrac{x + 4}{3x - 1}$

76. $y = \dfrac{2x - 3}{x - 2}$

77. $f \circ g$, where $f(x) = 1/x$ and $g(x) = 3x + 4$

78. $g \circ f$, where $f(x) = \dfrac{x + 2}{x - 1}$ and $g(x) = \dfrac{x + 1}{x + 4}$

79. f^{-1}, where $f(x) = \dfrac{x}{3x - 6}$

80. f^{-1}, where $f(x) = \dfrac{5 - x}{1 + x}$

In Exercises 81–88, express the function as a composition of two or more of the following functions:

$$f(x) = \frac{1}{x} \qquad g(x) = x - 1 \qquad F(x) = |x| \qquad G(x) = \sqrt{x}$$

81. $a(x) = \dfrac{1}{x - 1}$

82. $b(x) = \dfrac{1}{x} - 1$

83. $c(x) = \sqrt{x - 1}$

84. $d(x) = \sqrt{x} - 1$

85. $A(x) = \dfrac{1}{\sqrt{x}} - 1$

86. $B(x) = |x - 2|$

87. $C(x) = \sqrt[4]{x} - 1$

88. $D(x) = \dfrac{1}{\sqrt{x - 3}}$

For Exercises 89–122, compute the indicated quantity using the functions f, g, and F defined as follows:

$$f(x) = x^2 - x \qquad g(x) = 1 - 2x \qquad F(x) = \frac{x - 3}{x + 4}$$

89. $f(-3)$

90. $f(1 + \sqrt{2})$

91. $F\left(\tfrac{3}{4}\right)$

92. $f(t)$

93. $f(-t)$

94. $g(2x)$

95. $f(x - 2)$

96. $g(x + h)$

97. $g(2) - g(0)$

98. $f(x) - g(x)$

99. $|f(1) - f(3)|$

100. $f(x + h) - f(x)$

101. $f(x^2)$

102. $f(x)/x \quad (x \neq 0)$

103. $[f(x)][g(x)]$

104. $f[f(x)]$

105. $f[g(x)]$

106. $g[f(3)]$

107. $(g \circ f)(x)$

108. $(g \circ f)(x) - (f \circ g)(x)$

109. $(F \circ g)(x)$

110. $\dfrac{g(x + h) - g(x)}{h}$

111. $\dfrac{f(x + h) - f(x)}{h}$

112. $\dfrac{F(x) - F(a)}{x - a}$

113. $F^{-1}(x)$

114. $F[F^{-1}(x)]$

115. $F^{-1}[F(x)]$

116. $F^{-1}(0) - \dfrac{1}{F(0)}$

117. $(g \circ g^{-1})(x)$

118. $g^{-1}(x)$

119. $g^{-1}(-x)$

120. $\dfrac{g^{-1}(x + h) - g^{-1}(x)}{h}$

121. $F^{-1}[F(\tfrac{22}{7})]$

122. $T^{-1}(x)$, where $T(x) = f(x)/x \quad (x \neq 0)$

In Exercises 123–136, refer to the graph of the function f in the figure. (The axes are marked off in one-unit intervals.)

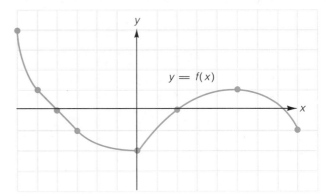

$y = f(x)$

123. Is $f(0)$ positive or negative?

124. Specify the domain and range of f.

125. Find $f(-3)$.

126. Which is larger, $f(-\tfrac{5}{2})$ or $f(-\tfrac{1}{2})$?

127. Compute $f(0) - f(8)$.

128. Compute $|f(0) - f(8)|$.

129. Specify the coordinates of the turning points.

130. What are the minimum and the maximum values of f?

131. On which interval(s) is f decreasing?

132. For which x-values is it true that $1 \leq f(x) \leq 4$?

133. What is the largest value of $f(x)$ when $|x| \leq 2$?

134. Is f a one-to-one function?

135. Does f possess an inverse function?

136. Compute $f[f(-4)]$.

For Exercises 137–152, refer to the graphs of the functions f and g in the figure. Assume that the domain of each function is [0, 10].

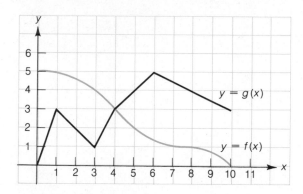

137. For which x-value is $f(x) = g(x)$?

138. For which x-values is it true that $g(x) \leq f(x)$?

139. **(a)** For which x-value is $f(x) = 0$?
  (b) For which x-value is $g(x) = 0$?

140. Compute $f(0) + g(0)$.

141. Compute each of the following:
  (a) $(f + g)(8)$ **(b)** $(f - g)(8)$
  (c) $(fg)(8)$ **(d)** $(f/g)(8)$

142. Compute each of the following:
  (a) $g[f(5)]$ **(b)** $f[g(5)]$
  (c) $(g \circ f)(5)$ **(d)** $(f \circ g)(5)$

143. Which is larger, $(f \circ f)(10)$ or $(g \circ g)(10)$?

144. Compute $g[f(10)] - f[g(10)]$.

145. For which x-values is it true that $f(x) \geq 3$?

146. For which x-values is it true that $|f(x) - 3| \leq 1$?

147. What is the largest number in the range of g?

148. Specify the coordinates of the highest point on the graph of each of the following equations.
  (a) $y = g(-x)$ **(b)** $y = -g(x)$
  (c) $y = g(x - 1)$ **(d)** $y = f(-x)$
  (e) $y = -f(x)$ **(f)** $y = -f(-x)$

149. On which intervals is the function g decreasing?

150. What are the coordinates of the turning points of g?

151. For which values of x in the interval $(4, 7)$ is the quantity $\dfrac{f(x) - f(5)}{x - 5}$ negative?

152. For which values of x in the interval $(0, 5)$ is the quantity $\dfrac{f(x) - f(2)}{x - 2}$ positive?

153. Let $P_1(x_1, y_1)$ and $P_2(x_2, y_2)$ be two given points. Let Q be the point

$$(\tfrac{1}{3}x_1 + \tfrac{2}{3}x_2, \tfrac{1}{3}y_1 + \tfrac{2}{3}y_2)$$

  (a) Show that the points P_1, Q, and P_2 are collinear. *Hint:* Compute the slope of $\overline{P_1Q}$ and the slope of $\overline{QP_2}$.

  (b) Show that $P_1Q = \tfrac{2}{3}P_1P_2$. (In other words, Q is on the line segment $\overline{P_1P_2}$ and two thirds of the way from P_1 to P_2.)

154. **(a)** Let the vertices of $\triangle ABC$ be $A(-5, 3)$, $B(7, 7)$, and $C(3, 1)$. Find the point on each median that is two thirds of the way from the vertex to the midpoint of the opposite side. (Recall that a median of a triangle is a line segment drawn from a vertex to the midpoint of the opposite side.) What do you observe? *Hint:* Use the result in Exercise 153.

  (b) Follow part (a), but take the vertices to be $A(0, 0)$, $B(2a, 0)$, and $C(2b, 2c)$. What do you observe? What does this prove?

155. In this exercise you'll derive a useful formula for the (perpendicular) distance d from the point (x_0, y_0) to the line $y = mx + b$. The formula is

$$d = \frac{|y_0 - mx_0 - b|}{\sqrt{1 + m^2}}$$

  (a) Refer to the figure. Use similar triangles to show that

$$\frac{d}{AB} = \frac{1}{\sqrt{1 + m^2}}.$$

  Therefore, $d = AB / \sqrt{1 + m^2}$.

  (b) Check that $AB = AC - BC = y_0 - mx_0 - b$.

  (c) Conclude from parts (a) and (b) that

$$d = \frac{y_0 - mx_0 - b}{\sqrt{1 + m^2}}.$$

  For the general case (in which the point and line may not be situated as in our figure), we need to use the absolute value of the quantity in the numerator to assure that AB and d are nonnegative.

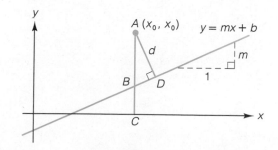

156. A circle with center $(4, 1)$ is tangent to the line $y = -\frac{1}{2}x + 8$. Find the equation of the circle in standard form. *Hint:* Use the formula in Exercise 155 to find the radius of the circle.

157. In the figure, the circle is tangent to the x-axis, to the y-axis, and to the line $3x + 4y = 12$.

(a) Find the equation of the circle. *Suggestion:* First decide what the relationships must be between h, k, and r in the equation $(x - h)^2 + (y - k)^2 = r^2$. Then find a way to apply the formula given in Exercise 155.

(b) Let S, T, and U denote the points where the circle touches the x-axis, the line $3x + 4y = 12$, and the y-axis, respectively. Find the equation of the line through A and T; through B and U; through S and C.

(c) Where do the line segments $\overline{AT}$ and $\overline{CS}$ intersect? Where do the line segments $\overline{AT}$ and $\overline{BU}$ intersect?

What do you observe? *Remark:* The point determined here is called the *Gergonne point* of triangle ABC, so named in honor of its discoverer, French mathematician Joseph-Diez Gergonne (1771–1859).

CHAPTER THREE

POLYNOMIAL AND RATIONAL FUNCTIONS. APPLICATIONS TO OPTIMIZATION

INTRODUCTION In Chapter 2, we studied some rather general rules for operating with functions and their graphs. Now, in Chapters 3 through 6, we will focus on several particular types of functions. In Section 3.1 we will discuss linear functions and their applications. As you'll see, these applications, although very elementary, are quite diverse, ranging from business and economics to physics and to statistics. Section 3.2 contains a discussion of quadratic functions. Section 3.3 could aptly be titled "Translating English into Algebra." In that section we will develop the background skills for solving the maximum and minimum problems we will encounter in Section 3.4. (If a friend of yours is taking calculus, you will find that you can solve some of your friend's maximum and minimum problems without using calculus.) The last two sections of this chapter, 3.5 and 3.6, present a number of techniques for graphing polynomial and rational functions.

3.1 ▼ LINEAR FUNCTIONS

All decent functions are practically linear.

Professor Andrew Gleason

By a **linear function**, we mean a function defined by an equation of the form

$$f(x) = Ax + B$$

where A and B are constants. In this chapter, the constants A and B will always be real numbers. From our work in Chapter 2, we know that the graph of $y = Ax + B$ is a straight line.

EXAMPLE 1 Suppose f is a linear function. If $f(1) = 0$ and $f(2) = 3$, find an equation defining f.

Solution From the statement of the problem, we know that the graph of f is a straight line passing through the points $(1, 0)$ and $(2, 3)$. Thus the slope of the line is

$$m = \frac{y_2 - y_1}{x_2 - x_1} = \frac{3 - 0}{2 - 1} = 3$$

Now we can use the point–slope formula to find the required equation. We have

$$y - y_1 = m(x - x_1)$$
$$y - 0 = 3(x - 1)$$
$$y = 3x - 3$$

This is the equation defining f. If we wish, we can rewrite it using function notation: $f(x) = 3x - 3$. ▲

One basic application of linear functions that occurs in business and economics is *linear* or *straight-line depreciation*. In this situation we assume that the value V of an asset (such as a machine or an apartment building) decreases linearly over time t; that is, $V = mt + b$, where the slope m is negative.

EXAMPLE 2

A factory owner buys a new machine for $8000. After ten years, the machine has a salvage value of $500.

(a) Assuming linear depreciation (as indicated in Figure 1), find a formula for the value V of the machine after t years, where $0 \le t \le 10$.

(b) Use the formula derived in part (a) to find the value of the machine after five years.

Solution

(a) We need to determine m and b in the equation $V = mt + b$. From Figure 1 we see that the V-intercept of the line segment is $b = 8000$. Furthermore, since the line segment passes through the two points $(10, 500)$ and $(0, 8000)$, the slope m is $\dfrac{8000 - 500}{0 - 10}$, or -750. So we have $m = -750$ and $b = 8000$, and the required equation is

$$V = -750t + 8000 \qquad (0 \le t \le 10)$$

(b) Substituting $t = 5$ in the equation $V = -750t + 8000$ yields

$$V = -750(5) + 8000$$
$$= -3750 + 8000 = 4250$$

Thus the value of the machine after five years is $4250. ▲

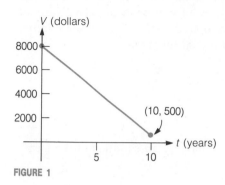

FIGURE 1

It is instructive to keep track of the units associated with slope in Example 2. Repeating the slope calculation and keeping track of the units, we have

$$m = \frac{\$8000 - \$500}{0 \text{ years} - 10 \text{ years}} = -\$750/\text{year}$$

Thus, slope is a *rate of change*. In this example, the slope represents the rate of change of the value of the machine.

As a second example of slope as a rate of change, let us suppose that a small manufacturer of handmade running shoes knows that her total cost, $C(x)$, in dollars, for producing x pairs of shoes each business day is given by the linear function

$$C(x) = 10x + 50$$

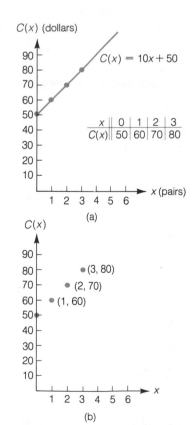

Figure 2(a) displays a table and graph for this function. Actually, since x represents the daily number of pairs of shoes, x can assume only whole-number values. So, technically, the graph that should be given is the one in Figure 2(b). However, it turns out to be useful in practice to make use of the graph in Figure 2(a), and we shall follow this convention.

The slope of the line $C(x) = 10x + 50$ in Figure 2(a) is 10, the coefficient of x. But to understand the units involved, let's calculate the slope using two of the points in Figure 2(a). Using the points $(0, 50)$ and $(1, 60)$ and keeping track of the units, we have

$$m = \frac{\$60 - \$50}{1 \text{ pair} - 0 \text{ pair}} = \frac{\$10}{1 \text{ pair}} = \$10/\text{pair}$$

Again the slope is a rate. In this case, the slope represents the rate of increase of cost; each additional pair of shoes produced costs the manufacturer $10.

FIGURE 2

In the study of economics, an equation that gives the cost $C(x)$ for producing x units of a commodity is called a **cost equation** or **cost function**. When the graph of the cost equation is a line, we define the **marginal cost** as the additional cost to produce one more unit. Thus, in the preceding example, the marginal cost is $10 per pair, and we see that the slope of the line in Figure 2(a) represents this marginal cost.

EXAMPLE 3 Suppose that the cost $C(x)$ in dollars of producing x bicycles is given by

$$C(x) = 625 + 45x$$

(a) Find the cost of producing ten bicycles.
(b) What is the marginal cost?
(c) Use the answers in parts (a) and (b) to find the cost of producing 11 bicycles. Then check the answer by evaluating $C(11)$.

Solution **(a)** Using $x = 10$ in the cost equation, we have

$$C(10) = 625 + 45(10) = 1075$$

Thus, the cost of producing ten bicycles is $1075.

(b) Since C is a linear function, the marginal cost is the slope and we have

$$\text{marginal cost} = \$45 \text{ per bicycle}$$

(c) According to the result in part (b), each additional bicycle costs $45. Therefore, we can compute the cost of 11 bicycles by adding $45 to the cost for 10 bicycles:

$$\text{cost of 11 bicycles} = \text{cost of 10 bicycles} + \text{marginal cost}$$
$$= \$1075 + \$45$$
$$= \$1120$$

So, the cost of producing 11 bicycles is $1120. We can check this result by using the cost function $C(x) = 625 + 45x$ to compute $C(11)$ directly:

$$C(11) = 625 + 45(11) = 625 + 495$$
$$= \$1120 \qquad \text{as obtained previously} \qquad \blacktriangle$$

Interpreting slope as a rate of change is not restricted to applications in business or economics. Suppose, for example, that you are driving a car at a steady rate of 50 mph. Using the distance formula from elementary mathematics,

$$\text{distance} = \text{rate} \times \text{time}$$

or

$$d = rt$$

we have, in this case, $d = 50t$, where d represents the distance traveled (in miles) in t hours. In Figure 3, we show a graph of the linear function $d = 50t$. The slope of this line is 50, the coefficient of t. But 50 is also the given rate of speed, in miles per hour. So again, slope is a rate of change. In this case, slope is the **velocity**, or *rate of change of distance with respect to time*.

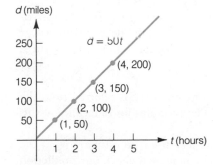

FIGURE 3

The slope of the line is the rate of change of distance with respect to time.

x (YEAR)	y (TIME)	
1911	4:15.4	(John Paul Jones, U.S.)
1913	4:14.6	(John Paul Jones, U.S.)
1915	4:12.6	(Norman Taber, U.S.)
1923	4:10.4	(Paavo Nurmi, Finland)
1931	4:09.2	(Jules Ladoumegue, France)
1933	4:07.6	(Jack Lovelock, New Zealand)
1934	4:06.8	(Glen Cunningham, U.S.)
1937	4:06.4	(Sidney Wooderson, Great Britain)
1942	4:06.2	(Gunder Haegg, Sweden)
1942	4:06.2	(Arne Andersson, Sweden)
1942	4:04.6	(Gunder Haegg, Sweden)
1943	4:02.6	(Arne Andersson, Sweden)
1944	4:01.6	(Arne Andersson, Sweden)
1945	4:01.4	(Gunder Haegg, Sweden)
1954	3:59.4	(Roger Bannister, Great Britain)
1954	3:58.0	(John Landy, Australia)
1957	3:57.2	(Derek Ibbotson, Great Britain)
1958	3:54.5	(Herb Elliott, Australia)
1962	3:54.4	(Peter Snell, New Zealand)
1964	3:54.1	(Peter Snell, New Zealand)
1965	3:53.6	(Michel Jazy, France)
1966	3:51.3	(Jim Ryun, U.S.)
1967	3:51.1	(Jim Ryun, U.S.)
1975	3:51.0	(Filbert Bayi, Tanzania)
1975	3:49.4	(John Walker, New Zealand)
1979	3:49.0	(Sebastian Coe, Great Britain)
1980	3:48.8	(Steve Ovett, Great Britain)

We conclude this section with an example showing one way that linear functions are used in statistical applications. Table 1 shows the evolution of the record for the one-mile run during the years 1911–1980. (In case you're wondering why the table begins with 1911, John Paul Jones was the first twentieth-century runner to break the previous century's record of 4:15.6, set in 1895.) In Figure 4(a) we've plotted the (x, y) pairs given in the table. The resulting plot is called a **scatter diagram**.

A striking feature of the scatter diagram in Figure 4(a) is that the records do not appear to be leveling off; rather, they seem to be decreasing in an approximately linear fashion. Using the *least-squares technique* from the field of statistics, it can be shown that the linear function that best fits this particular set of data is

$$f(x) = -0.41x + 1040.94 \tag{1}$$

The graph of this line is shown in Figure 4(b). The line itself is referred to as the **regression line**, or the **least-squares line**. (The formulas for determining this line are given in Exercise 33.)

As an intuitive check on the least-squares line, let's use it to estimate what the record for the mile run might have been in 1985 (then we'll check our prediction against the actual record). With $x = 1985$, we have

$$f(1985) = -0.41(1985) + 1040.94$$
$$= 227.09 \text{ seconds}$$

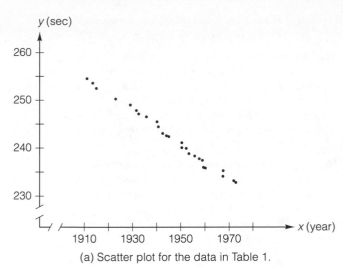

(a) Scatter plot for the data in Table 1.

(b) Scatter plot and regression line for the data in Table 1.

FIGURE 4

Thus (after converting to minutes and rounding off), our least-squares estimate for the 1985 record is 3:47.1. This agrees favorably with the actual record of 3:46.31 set by Steve Cram of Great Britain.

EXAMPLE 4 The June 1976 issue of *Scientific American* contained an article entitled "Future Performance in Footracing" by H. W. Ryder, H. J. Carr, and P. Herget. According to the authors of the article, "It appears likely that within 50 years the record [for the mile] will be down to 3:30." Use the regression line defined in equation (1) to make a projection for the mile time in the year 2026 (which will be 50 years after the article appeared). Is your projection close to the one given in the *Scientific American* article?

Solution The given equation for the regression line is

$$f(x) = -0.41x + 1040.94$$

Setting $x = 2026$ and using a calculator, we have

$$f(2026) = -0.41(2026) + 1040.94$$
$$= 210.28 \text{ seconds}$$

Converting this result to minutes and then rounding off, our projection for the year 2026 is 3:30.3, which is essentially the same as the projection in *Scientific American*. ▲

▼ EXERCISE SET 3.1

A

In Exercises 1–10, find the linear functions satisfying the given conditions.

1. $f(-1) = 0$ and $f(5) = 4$

2. $f(3) = 2$ and $f(-3) = -4$

3. $g(0) = 0$ and $g(1) = \sqrt{2}$

4. The graph passes through the points $(2, 4)$ and $(3, 9)$.

5. $f(\frac{1}{2}) = -3$ and the graph of f is a line parallel to the line $x - y = 1$.

6. $g(2) = 1$ and the graph of g is perpendicular to the line $6x - 3y = 2$.

7. The graph of f is a horizontal line that passes through the larger of the two y-intercepts of the circle $x^2 - 2x + y^2 - 3 = 0$.

8. The x- and y-intercepts of the graph of g are 1 and 4, respectively.

9. The graph of the inverse function passes through the points $(-1, 2)$ and $(0, 4)$.

10. The x- and y-intercepts of the inverse function are 5 and -1, respectively.

11. Let $f(x) = 3x - 4$ and $g(x) = 1 - 2x$. Determine if the function $f \circ g$ is linear.

12. Explain why there is no linear function with a graph that passes through all three of the points $(-3, 2)$, $(1, 1)$, and $(5, 2)$.

13. A factory owner buys a new machine for $20,000. After eight years, the machine has a salvage value of $1000. Find a formula for the value of the machine after t years, where $0 \le t \le 8$.

14. A manufacturer buys a new machine costing $120,000. It is estimated that the machine has a useful lifetime of ten years and a salvage value of $4000 at that time.
 (a) Find a formula for the value of the machine after t years, where $0 \le t \le 10$.
 (b) Find the value of the machine after eight years.

15. A factory owner installs a new machine costing $60,000. Its expected lifetime is five years, and at the end of that time the machine has no salvage value.
 (a) Find a formula for the value of the machine after t years, where $0 \le t \le 5$.
 (b) Complete the following depreciation schedule.

END OF YEAR	YEARLY DEPRECIATION	ACCUMULATED DEPRECIATION	VALUE V
0	0	0	60,000
1			
2			
3			
4			
5		60,000	0

16. Let x denote the temperature on the Celsius scale, and let y denote the corresponding temperature on the Fahrenheit scale.
 (a) Find a linear function relating x and y; use the facts that $32\,°F$ corresponds to $0\,°C$, and $212\,°F$ corresponds to $100\,°C$. Write your answer in the form $y = Ax + B$.

(b) What Celsius temperature corresponds to $98.6\,°F$?
(c) Find a number z for which $z\,°F = z\,°C$.

17. Suppose that the cost $C(x)$ in dollars of producing x electric fans is given by $C(x) = 450 + 8x$.
 (a) Find the cost to produce 10 fans.
 (b) Find the cost to produce 11 fans.
 (c) Use your answers in parts (a) and (b) to find the marginal cost. (Then check that your answer is the slope of the line.)

18. Suppose that the cost to a manufacturer of producing x units of a certain motorcycle is given by $C(x) = 220x + 4000$, where $C(x)$ is in dollars.
 (a) Find the marginal cost.
 (b) Find the cost of producing 500 motorcycles.
 (c) Use your answers in parts (a) and (b) to find the cost of producing 501 motorcycles.

19. Suppose that the cost $C(x)$, in dollars, of producing x units of a certain cassette tape player is given by $C(x) = 400 + 50x$.
 (a) Compute $C(n + 1) - C(n)$.
 (b) What is the marginal cost?
 (c) What is the relationship between the answers in parts (a) and (b)?

20. Suppose that the cost $C(x)$, in dollars, of producing x cassette tapes is given by $C(x) = 0.5x + 500$.
 (a) Graph the given equation.
 (b) Compute $C(150)$.
 (c) If you add the marginal cost to the answer in part (b), you obtain a certain dollar amount. What does this amount represent?

21. The following graphs each relate distance and time for a moving object. Determine the velocity in each case.

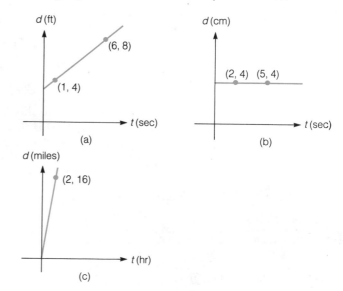

22. The distance d, in feet, covered by a particular object in t sec is given by the equation $d = 5t$.
    (a) Find the velocity of the object.
    (b) Find the distance covered in 15 sec.
    (c) Use your answers in parts (a) and (b) to find the distance covered in 16 sec. Check your answer by substituting $t = 16$ in the given distance formula.

23. Two points A and B move along the x-axis. After t sec, their positions are given by the equations

    A: $x = 3t + 100$

    B: $x = 20t - 36$

    (a) Which point is traveling faster, A or B?
    (b) Which point is farther to the right when $t = 0$?
    (c) At what time t do A and B have the same x-coordinate?

24. A point moves along the x-axis, and its x-coordinate after t seconds is $x = 4t + 10$. (Assume that x is in cm.)
    (a) What is the velocity?
    (b) What is the x-coordinate when $t = 2$ sec?
    (c) Use your answers in parts (a) and (b) to find the x-coordinate when $t = 3$ sec. Check your answer by letting $t = 3$ in the given equation.

25. (a) On graph paper, plot the following points: $(1, 0)$, $(2, 3)$, $(3, 6)$, $(4, 7)$.
    (b) In your scatter diagram from part (a), sketch a line that, you feel, seems to best fit the data. Estimate the slope and y-intercept of the line.
    (c) The actual regression line in this case is $y = 2.4x - 2$. Add the graph of this line to your sketch from parts (a) and (b).

26. (a) On graph paper, plot the following points: $(1, 2)$, $(3, 2)$, $(5, 4)$, $(8, 5)$, $(9, 6)$.
    (b) In your scatter diagram from part (a), sketch a line that, you feel, seems to best fit the data.
    (c) Using your sketch from part (b), estimate the y-intercept and the slope of the regression line.
    (d) The actual regression line is $y = 0.518x + 1.107$. Check to see if your estimates in part (c) are consistent with the actual y-intercept and slope. Graph this line along with the points given in part (a). (In sketching the line, use the approximation $y = 0.5x + 1.1$.)

27. The following table shows the population of the city of Los Angeles over the years 1930–1988. The accompanying graph shows the corresponding scatter plot and regression line. The equation of the regression line (after some rounding off) is

    $y = 37,025x - 70,226,566$

(a) Predict whether the population of Los Angeles will reach 4 million by the year 2000.
(b) Make a projection for the population of Los Angeles in the year 2010. Round off your answer to the nearest 5,000.

Year	Population of Los Angeles
1930	1238048
1940	1504277
1950	1970358
1960	2479015
1970	2816061
1980	2966850
1988	3361500

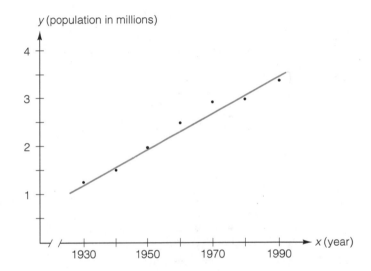

28. The following table (on the next page) shows expenditures on pollution control and abatement in the United States during the period 1975–1985. (The figures include both government and private-sector expenditures.) The accompanying graph displays the scatter plot and regression line for the data. The equation of the regression line (after rounding off) is

    $f(x) = (4198.8)x - 8262216$

    (a) Use the regression line (and a calculator, of course) to make a projection for the expenditures in the year 1995.
    (b) Find $f^{-1}(x)$.
    (c) Use your answer in part (b) to find the year when the expenditures might reach 125 billion dollars.

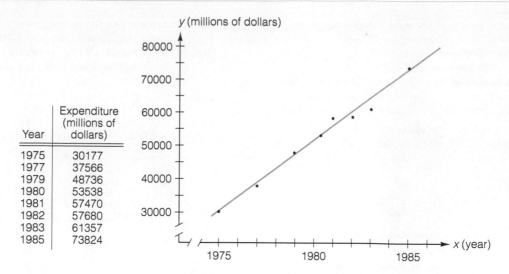

Year	Expenditure (millions of dollars)
1975	30177
1977	37566
1979	48736
1980	53538
1981	57470
1982	57680
1983	61357
1985	73824

B

29. Find a linear function $f(x) = mx + b$ such that m is positive and $(f \circ f)(x) = 9x + 4$.

30. Let $f(x) = Ax + B$. Find $\dfrac{f(x + h) - f(x)}{h}$.

31. (a) Let f be a linear function. Show that

$$f\left(\frac{x_1 + x_2}{2}\right) = \frac{f(x_1) + f(x_2)}{2}$$

(In words: The output of the average is the average of the outputs.)

(b) Show that the equation in part (a) does not hold for the function $f(x) = x^2$.

32. Let f be a linear function such that

$$f(a + b) = f(a) + f(b)$$

for all real numbers a and b. Show that the graph of f passes through the origin. *Hint:* Let $f(x) = Ax + B$.

33. This exercise shows how to compute the slope and the y-intercept for the regression line. As an example, we'll work with the simple data set given in Exercise 25:

x	1	2	3	4
y	0	3	6	7

(a) Let $\sum x$ denote the sum of the x-coordinates in the data set, and let $\sum y$ denote the sum of the y-coordinates. Check that $\sum x = 10$ and $\sum y = 16$.

(b) Let $\sum x^2$ denote the sum of the squares of the x-coordinates, and let $\sum xy$ denote the sum of the products of the corresponding x- and y-coordinates. Check that $\sum x^2 = 30$ and $\sum xy = 52$.

(c) The slope m and the y-intercept b of the regression line satisfy the following pair of simultaneous equations [in the first equation, n denotes the number of points (x, y) in the data set]:

$$\begin{cases} nb + (\sum x)m = \sum y \\ (\sum x)b + (\sum x^2)m = \sum xy \end{cases}$$

So, in the present example, these equations become

$$\begin{cases} 4b + 10m = 16 \\ 10b + 30m = 52 \end{cases}$$

Solve this pair of equations for m and b, and check that your answers agree with the values in Exercise 25(c).

In Exercises 34–38, use the method described in Exercise 33 to find the equation of the regression line for the given data sets.

34.

x	2	4	8	10
y	-7	-5	-2	-1

35.

x	1	2	3	4	5
y	2	3	9	9	11

36.

x	1	2	3	4	5
y	16	13.1	10.5	7.5	2

37.

x	520	740	560	610	650
y	81	98	83	88	95

3.2 ▼ QUADRATIC FUNCTIONS

After the linear functions, the next simplest are the **quadratic functions**. These are functions defined by equations of the form

$$f(x) = ax^2 + bx + c \qquad (a \neq 0)$$

where a, b, and c are constants and a is not zero. In this chapter the constants a, b, and c will always be real numbers. What we will see is that the graph of any quadratic function is a curve called a **parabola**, which is similar or identical in shape to the graph of $y = x^2$. Figure 1 displays the graphs of two typical parabolas.

FIGURE 1

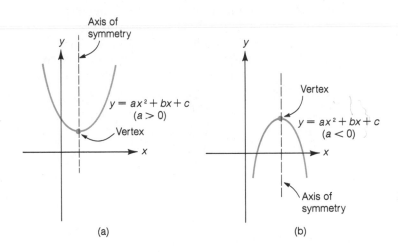

(a)   (b)

Subsequent examples will demonstrate that the parabola $y = ax^2 + bx + c$ opens upward when $a > 0$ and downward when $a < 0$. As Figure 1 indicates, the turning point on the parabola is called the **vertex**. The **axis of symmetry** of the parabola $y = ax^2 + bx + c$ is the vertical line passing through the vertex.

The methods that we will use for dealing with quadratic functions have already been developed in Chapters 1 and 2. In particular, the following two topics are prerequisites for understanding the rest of the examples in this section:

1. Completing the square; for a review, see Section 1.6.
2. Translations and reflections; for a review, see Section 2.5.

EXAMPLE 1   Graph the function $y = x^2 - 2x + 3$.

Solution   The idea here is to use the technique of completing the square; this will then enable us to obtain the required graph simply by moving the basic $y = x^2$ graph around in the plane. In preparation for completing the square, we rewrite the given equation

$$y - 3 = x^2 - 2x$$

To complete the square for the x-terms, we need to add 1. (Check this.) Then, to keep the equation in balance, we need to add 1 on the left-hand side, also. So

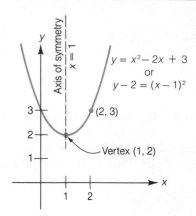

FIGURE 2

we have

$$y - 2 = x^2 - 2x + 1 \qquad \text{adding 1 to both sides}$$
$$y - 2 = (x - 1)^2 \qquad\qquad (1)$$

or

$$y = (x - 1)^2 + 2$$

Now, from our work on translation in Section 2.5, we know that the graph of equation (1) is obtained by moving the parabola $y = x^2$ one unit to the right and two units up. See Figure 2.

Note As a guide before sketching the graph yourself, you'll want to know the y-intercept. To find the y-intercept, substitute $x = 0$ in the given equation to obtain $y = 3$. Then, given the vertex $(1, 2)$ and the point $(0, 3)$, a reasonably accurate graph can be quickly sketched. [Actually, once you find that $(0, 3)$ is on the graph, you also know that the reflection of this point about the axis of symmetry is on the graph. This is why the point $(2, 3)$ is shown in Figure 2.] ▲

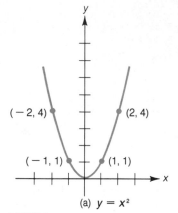

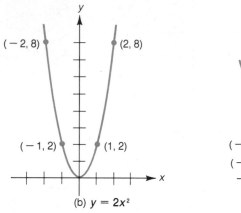

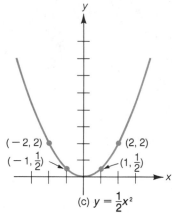

FIGURE 3

Now we want to compare the graphs of $y = x^2$, $y = 2x^2$, and $y = \frac{1}{2}x^2$. The last two equations were not discussed in Chapter 2, so for the moment you can think about graphing them by first setting up tables. Figure 3 displays all three graphs. They are all parabolas that open upward, but notice that the shapes are not identical. The parabola $y = 2x^2$ is narrower than $y = x^2$, while $y = \frac{1}{2}x^2$ is wider than $y = x^2$. These same observations about shape would also apply to $y = -x^2$, $y = -2x^2$, and $y = -\frac{1}{2}x^2$. In these cases, though, the parabolas open downward rather than upward. The observations that we have just made are generalized in the box that follows.

PROPERTY SUMMARY

THE GRAPH OF $y = ax^2$

1. The graph of $y = ax^2$ is a parabola with vertex at the origin. It is similar in shape to $y = x^2$.

2. The parabola $y = ax^2$ opens upward if $a > 0$, downward if $a < 0$.

3. The parabola $y = ax^2$ is narrower than $y = x^2$ if $|a| > 1$, wider than $y = x^2$ if $|a| < 1$.

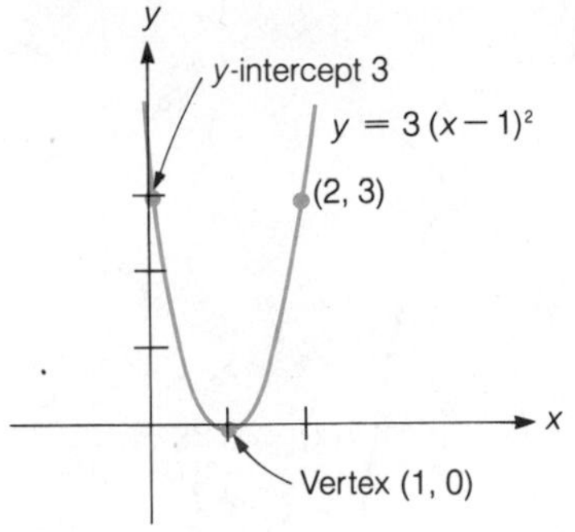

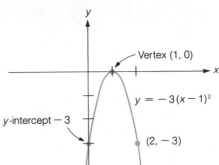

FIGURE 4

FIGURE 5

EXAMPLE 2 Sketch the graph of $y = 3(x - 1)^2$.

Solution Because of the factor $x - 1$, we shift the basic parabola $y = x^2$ one unit to the right. The factor of 3 in the given equation tells us that we want to draw a parabola that is narrower than $y = x^2$. To see exactly how narrow to draw $y = 3(x - 1)^2$, we would like to know another point on the graph other than the vertex $(1, 0)$. An easy point to obtain is the y-intercept. Setting $x = 0$ in the equation yields

$$y = 3(0 - 1)^2 = 3$$

Now that we know the vertex, $(1, 0)$, and the y-intercept, 3, a reasonably accurate graph can be sketched. See Figure 4. ▲

EXAMPLE 3 Sketch the graph of $y = -3(x - 1)^2$.

Solution In Example 2 we sketched the graph of $y = 3(x - 1)^2$. By reflecting that graph about the x-axis, we obtain the graph of $y = -3(x - 1)^2$. See Figure 5. ▲

EXAMPLE 4 Graph the function $f(x) = -2x^2 + 12x - 16$ and specify the vertex, axis of symmetry, maximum or minimum value of f, and x- and y-intercepts.

Solution The idea is to complete the square, as in Example 1. We have

$$y + 16 = -2x^2 + 12x$$
$$y + 16 = -2(x^2 - 6x)$$
$$y + 16 - 18 = -2(x^2 - 6x + 9)$$
$$y - 2 = -2(x - 3)^2 \qquad (2)$$

or

$$f(x) = -2(x - 3)^2 + 2$$

From equation (2), we see that the required graph is obtained simply by shifting the graph of $y = -2x^2$ "right 3, up 2," so that the vertex is $(3, 2)$. As a guide for actually sketching the graph, we want to compute the intercepts. The y-intercept is -16. (Why?) For the x-intercepts, we replace y with 0 in equation (2) to obtain

$$-2(x - 3)^2 = -2$$
$$(x - 3)^2 = 1$$
$$x - 3 = \pm 1$$
$$x = 4 \quad \text{or} \quad 2$$

FIGURE 6

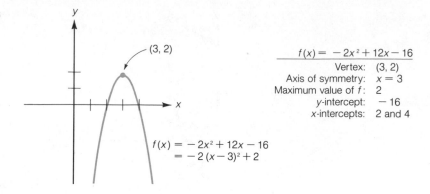

$$f(x) = -2x^2 + 12x - 16$$

Vertex:	$(3, 2)$
Axis of symmetry:	$x = 3$
Maximum value of f:	2
y-intercept:	-16
x-intercepts:	2 and 4

$$f(x) = -2x^2 + 12x - 16$$
$$= -2(x-3)^2 + 2$$

Thus the x-intercepts are $x = 4$ and $x = 2$. Knowing these intercepts and the vertex, we can sketch the graph as in Figure 6. You should check for yourself that the information accompanying Figure 6 is correct. ▲

We conclude this section with several examples involving maximum and minimum values of functions. As background for this, we first summarize our basic technique for graphing parabolas.

PROPERTY SUMMARY

THE GRAPH OF THE PARABOLA $y = ax^2 + bx + c$

By completing the square, the equation of the parabola $y = ax^2 + bx + c$ can always be rewritten in the form

$$y = a(x - h)^2 + k$$

In this form, the vertex of the parabola is (h, k) and the axis of symmetry is the line $x = h$. The parabola opens upward if $a > 0$ and downward if $a < 0$.

EXAMPLE 5 Among all possible inputs for the function

$$g(x) = 3x^2 - 2x - 6$$

which yields the smallest output? What is this minimum output?

Solution Think in terms of a graph. Since the graph of this function is a parabola opening upward, the input we want is the x-coordinate of the vertex. By completing the square (as shown in Example 4), we find that the given equation can be rewritten

$$g(x) = 3\left(x - \frac{1}{3}\right)^2 - \frac{19}{3}$$

(You should verify this for yourself, using Example 4 as a model.) Thus, the vertex of the parabola is $(\frac{1}{3}, -\frac{19}{3})$. From this, and from the fact that the parabola opens upward, we conclude that the input $x = \frac{1}{3}$ produces the smallest output, and this minimum output is $g(\frac{1}{3}) = -\frac{19}{3}$. ▲

For the previous two examples, keep in mind that we were able to find the maximum or minimum easily only because the functions were quadratics. By way of contrast, we cannot expect to find the minimum of $y = x^4 - 8x$ using the method of Examples 4 and 5, for this is not a quadratic function. In general, the techniques of calculus are required to find maxima and minima for functions other than quadratics. There are some cases, however, in which our present method can be adapted to functions that are closely related to quadratics. The next two examples show instances of this.

EXAMPLE 6 Let $D = \sqrt[3]{x^2 - x + 1}$. For which value of x is D minimum?

Solution The same x-value that makes D as small as possible will also cause the quantity $D^3 = x^2 - x + 1$ to be its smallest. That is, we only need to find the x-value that minimizes the quantity $x^2 - x + 1$. Call this quantity y, and complete the square as follows:

$$y = x^2 - x + 1$$
$$y - 1 = x^2 - x$$
$$y - 1 + \tfrac{1}{4} = x^2 - x + \tfrac{1}{4}$$
$$y - \tfrac{3}{4} = (x - \tfrac{1}{2})^2$$

This shows that the vertex of the parabola $y = x^2 - x + 1$ is $(\tfrac{1}{2}, \tfrac{3}{4})$. Since this parabola opens upward, we conclude that the input $x = \tfrac{1}{2}$ will produce the minimum value $\tfrac{3}{4}$ for the quantity $x^2 - x + 1$. And, according to our initial remark, this same value, $x = \tfrac{1}{2}$, also minimizes the quantity $D = \sqrt[3]{x^2 - x + 1}$. (Our work also shows that the minimum value of D is $\sqrt[3]{\tfrac{3}{4}}$.) ▲

EXAMPLE 7 Find the maximum value of the function $y = 16t^2 - 4t^4$.

Solution Although this is not a quadratic function, we can complete the square as follows:

$$y = -4(t^4 - 4t^2 \quad)$$
$$= -4(t^4 - 4t^2 + 4) + 16$$
$$= -4(t^2 - 2)^2 + 16$$
$$= 16 - 4(t^2 - 2)^2$$

From this last equation we see that y never exceeds 16 (because the quantity being subtracted from 16 is nonnegative). Furthermore, y does attain the value 16 when $t^2 = 2$ or $t = \pm\sqrt{2}$. It now follows that the maximum value of the given function is 16. ▲

▼ EXERCISE SET 3.2

A

In Exercises 1–18, graph the quadratic function. Specify the vertex, axis of symmetry, maximum or minimum value, and intercepts.

1. $y = (x + 2)^2$

2. $y = -(x + 2)^2$

3. $y = 2(x + 2)^2$

4. $y = 2(x + 2)^2 + 4$

5. $y = -2(x + 2)^2 + 4$

6. $y = x^2 + 6x - 1$

7. $f(x) = x^2 - 4x$

8. $F(x) = x^2 - 3x + 4$

9. $g(x) = 1 - x^2$

10. $y = 2x^2 + \sqrt{2}\,x$

11. $y = x^2 - 2x - 3$

12. $y = 2x^2 + 3x - 2$

13. $y = -x^2 + 6x + 2$

14. $y = -3x^2 + 12x$

15. $s = 16t^2$

16. $s = -\tfrac{1}{2}t^2 - t$

17. $s = 2 + 3t - 9t^2$

18. $s = -\tfrac{1}{4}t^2 + t - 1$

For Exercises 19–24, determine the input that produces the largest or smallest output (whichever is appropriate). State whether the output is largest or smallest.

19. $y = 2x^2 - 4x + 11$ **20.** $f(x) = 8x^2 + x - 5$

21. $g(x) = -6x^2 + 18x$ **22.** $s = -16t^2 + 196t + 80$

23. $f(x) = x^2 - 10$ **24.** $h(x) = x^2 - 10x$

In Exercises 25–30, find the maximum or minimum value for each function (whichever is appropriate). State whether the value is a maximum or minimum.

25. $y = x^2 - 8x + 3$ **26.** $y = \frac{1}{2}x^2 + x + 1$

27. $y = -2x^2 - 3x + 2$ **28.** $y = -\frac{1}{3}x^2 - 2x$

29. $f(t) = -12t^2 + 1000$ **30.** $g(t) = 400t^2$

31. How far from the origin is the vertex of the parabola $y = x^2 - 6x + 13$?

32. Find the distance between the vertices of the parabolas $y = -\frac{1}{2}x^2 + 4x$ and $y = 2x^2 - 8x - 1$.

For Exercises 33–38, the functions f, g, and h are defined as follows.

$$f(x) = 2x - 3 \qquad g(x) = x^2 + 4x + 1 \qquad h(x) = 1 - 2x^2$$

In each exercise, classify the function as linear, quadratic, or neither.

33. $f \circ g$ **34.** $g \circ f$ **35.** $g \circ h$

36. $h \circ g$ **37.** $f \circ f$ **38.** $h \circ h$

In Exercises 39 and 40, determine the inputs that yield the minimum values for each function. Compute the minimum value in each case.

39. (a) $f(x) = \sqrt{x^2 - 6x + 73}$

   (b) $g(x) = \sqrt[3]{x^2 - 6x + 73}$

   (c) $h(x) = x^4 - 6x^2 + 73$

40. (a) $F(x) = (4x^2 - 4x + 109)^{1/2}$

   (b) $G(x) = (4x^2 - 4x + 109)^{1/3}$

   (c) $H(x) = 4x^4 - 4x^2 + 109$

In Exercises 41 and 42, determine the maximum value for each function.

41. (a) $f(x) = \sqrt{-x^2 + 4x + 12}$

   (b) $g(x) = \sqrt[3]{-x^2 + 4x + 12}$

   (c) $h(x) = -x^4 + 4x^2 + 12$

42. (a) $F(x) = (27x - x^2)^{1/2}$

   (b) $G(x) = (27x - x^2)^{1/3}$

   (c) $H(x) = 27x^2 - x^4$

B

In Exercises 43–46, find quadratic functions satisfying the given conditions.

43. The graph passes through the origin and the vertex is (2, 2).

44. The graph is obtained by translating $y = x^2$ four units in the negative x-direction and three units in the positive y-direction.

45. The vertex is $(3, -1)$ and one x-intercept is 1.

46. The axis of symmetry is the line $x = 1$. The y-intercept is 1. There is only one x-intercept.

47. By completing the square, show that the coordinates of the vertex of the parabola $y = ax^2 + bx + c$ are

$$\left(\frac{-b}{2a}, \frac{4ac - b^2}{4a} \right)$$

48. Find the equation of the circle that passes through the origin and that is centered at the vertex of the parabola $y = 2x^2 + 12x + 14$.

49. For which value of c will the minimum value of the function $f(x) = x^2 + 2x + c$ be $\sqrt{2}$?

50. Explain why the graph of $y = -x^2 - 10x - \frac{51}{2}$ has no x-intercepts by finding the vertex and noting whether the parabola opens up or down.

51. Find the x-coordinate of the vertex of the parabola $y = (x - a)(x - b)$, where a and b are constants.

52. Compute the average of the two x-intercepts of the graph of $y = ax^2 + bx + c$. (Assume $b^2 - 4ac > 0$.) How does your answer relate to the result in Exercise 47?

C

53. Consider the quadratic function $y = px^2 + px + r$, where $p \neq 0$.

   (a) Show that if the vertex lies on the x-axis, then $p = 4r$.

   (b) Show that if $p = 4r$, then the vertex lies on the x-axis.

54. Let $g(x) = x^2 + 2(a + b)x + 2(a^2 + b^2)$, where a and b are constants.

   (a) Show that the coordinates of the vertex are

$$(-(a + b), (a - b)^2)$$

   (b) Use the result in part (a) to explain why the graph of g has no x-intercepts unless $a = b$, in which case there is only one x-intercept.

55. As you have seen in this section, if the quadratic equation $f(x) = x^2 + Bx + C = 0$ has two real roots, there is a geometric interpretation: these roots are the x-intercepts for the graph of f. Less well known is the fact that there is a geometric interpretation when the quadratic equation $f(x) = x^2 + Bx + C = 0$ has nonreal complex roots. If these roots are $a + bi$ and $a - bi$, where a and b are real numbers, show that the coordinates of the vertex for the graph of f are (a, b^2). *Hint:* If r_1 and r_2 are the roots of the equation $x^2 + Bx + C = 0$, then $x^2 + Bx + C$ can be written $(x - r_1)(x - r_2)$.

3.3 ▼ APPLIED FUNCTIONS: SETTING UP EQUATIONS

I hope that I shall shock a few people in asserting that the most important single task of mathematical instruction in the secondary schools is to teach the setting up of equations to solve word problems.

George Polya (1887–1985)

Each problem that I solved became a rule which afterwards served to solve other problems.

René Descartes (1596–1650)

One of the first steps in problem solving often involves defining a function. The function then serves to describe or summarize the given situation in a way that is both concise and (we hope) revealing. In this section, we want to practice setting up equations that define such functions. Then, in Section 3.4, we will use this skill in solving a certain class of applied problems involving quadratic functions. For many of the examples we look at, we'll rely on the following four-step procedure to set up the required equation. You may eventually want to modify this procedure to fit your own style. The point, however, is that it is possible to approach these problems in a systematic manner. A word of advice: You're accustomed to working mathematics problems in which the answers are numbers. In this section, the answers are functions (or, more precisely, equations defining functions); you'll need to get used to this.

Steps for Setting Up Equations

> **STEP 1** After reading the problem carefully, draw a picture that conveys the given information.
>
> **STEP 2** State in your own words, as specifically as you can, what the problem is asking for. (This usually requires rereading the problem.) Now, assuming that the problem asks you to find a particular quantity (or to find a formula for a particular quantity), assign a variable to denote that key quantity.
>
> **STEP 3** Label any other quantities in your figure that appear relevant. Are there equations relating these quantities?
>
> **STEP 4** Find an equation involving the key variable that you identified in step 2. (Some people prefer to do this right after step 2.) Now, as necessary, substitute in this equation using the auxiliary equations from step 3 to obtain an equation involving only the required variables.

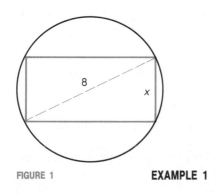

FIGURE 1

EXAMPLE 1 A rectangle is inscribed in a circle of diameter 8 cm. Express the perimeter of the rectangle as a function of its width x.

Solution Let's follow our four-step procedure.

STEP 1 See Figure 1. Note that the diagonal of the rectangle is a diameter of the circle.

STEP 2 The problem asks us to come up with a formula or function that gives the perimeter of the rectangle in terms of x, the width. Let P denote the perimeter.

STEP 3 Let L denote the length of the rectangle, as shown in Figure 2. Then, by the Pythagorean theorem, we have

$$L^2 + x^2 = 8^2 = 64$$

This equation relates the length L and the width x. Rather than leaving the equation in this form, however, we'll solve for L in terms of x (because the instructions in the problem mention x, not L):

$$L^2 = 64 - x^2$$
$$L = \sqrt{64 - x^2} \tag{1}$$

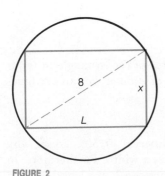

FIGURE 2

STEP 4 The perimeter of the rectangle is the sum of the lengths of the four sides. Thus,

$$P = x + x + L + L = 2x + 2L \tag{2}$$

This expresses P in terms of x and L. However, the problem asks for P in terms of just x. Using equation (1) to substitute for L in equation (2), we have

$$P = 2x + 2\sqrt{64 - x^2} \tag{3}$$

This is the required equation. It expresses the perimeter P as a function of the width x, so if you know x, you can calculate P. To emphasize this dependence of P on x, we can employ function notation to rewrite equation (3):

$$P(x) = 2x + 2\sqrt{64 - x^2} \tag{4}$$

Before leaving this example, we need to specify the domain of the perimeter function in equation (4). An easy way to do this is to look back at Figure 1. Since x represents a width, we certainly want $x > 0$. Furthermore, Figure 1 tells us that $x < 8$, because in any right triangle, a leg is always shorter than the hypotenuse. Putting these observations together, we conclude that the domain of the perimeter function in equation (4) is the open interval $(0, 8)$.

Note Although it would make sense *algebraically* to use an input such as $x = -1$ in equation (4), it does not make sense in our *geometric* context, where x denotes the width. ▲

EXAMPLE 2 The perimeter of a rectangle is 100 cm. Express the area of the rectangle in terms of the width x.

Solution **STEP 1** See Figure 3.

STEP 2 We want to express the area of the rectangle in terms of x, the width. Let A stand for the area of the rectangle.

STEP 3 Call the length of the rectangle L, as indicated in Figure 4. Then, since the perimeter is given as 100 cm, we have

$$2x + 2L = 100$$
$$x + L = 50$$
$$L = 50 - x \tag{5}$$

STEP 4 Area of a rectangle equals width times length:

$$A = x \cdot L$$
$$= x(50 - x) \quad \text{substituting for } L \text{ using equation (5)}$$
$$= 50x - x^2 \tag{6}$$

This is the required equation expressing the area of the rectangle in terms of the width x. To emphasize this dependence of A on x, we can use function notation to rewrite equation (6):

$$A(x) = 50x - x^2$$

The domain of this area function is the open interval $(0, 50)$. To see why this is so, first note that $x > 0$ because x denotes a width. Furthermore, in view of equation (5), we must have $x < 50$ (otherwise L, the length, would be zero or negative). ▲

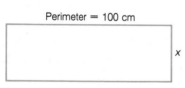
Perimeter = 100 cm

FIGURE 3

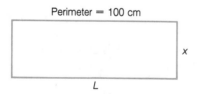
Perimeter = 100 cm

FIGURE 4

EXAMPLE 3   Let $P(x, y)$ be a point on the curve $y = \sqrt{x}$. Express the distance from P to the point $(1, 0)$ as a function of x.

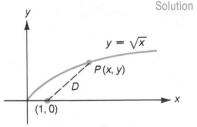

FIGURE 5

Solution   **STEP 1** See Figure 5.

STEP 2 We want to express the length of the broken line in Figure 5 in terms of x. Call this length D.

STEP 3 There are no other quantities in Figure 5 that need labeling. But don't forget that we are given

$$y = \sqrt{x} \tag{7}$$

STEP 4 By the distance formula, we have

$$\begin{aligned} D &= \sqrt{(x - 1)^2 + (y - 0)^2} \\ &= \sqrt{x^2 - 2x + 1 + y^2} \end{aligned} \tag{8}$$

Now we can use equation (7) to eliminate y in equation (8):

$$\begin{aligned} D(x) &= \sqrt{x^2 - 2x + 1 + (\sqrt{x})^2} \\ &= \sqrt{x^2 - 2x + 1 + x} \\ &= \sqrt{x^2 - x + 1} \end{aligned} \tag{9}$$

Equation (9) expresses the distance as a function of x, as required. What about the domain of this distance function? Since the x-coordinate of a point on the curve $y = \sqrt{x}$ can be any nonnegative number, the domain of the distance function is $[0, \infty)$.   ▲

EXAMPLE 4   A point $P(x, y)$ lies in the first quadrant on the parabola $y = 16 - x^2$, as indicated in Figure 6. Express the area of the triangular region in Figure 6 as a function of x.

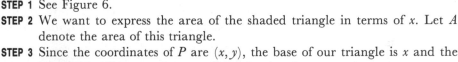

FIGURE 6

Solution   **STEP 1** See Figure 6.

STEP 2 We want to express the area of the shaded triangle in terms of x. Let A denote the area of this triangle.

STEP 3 Since the coordinates of P are (x, y), the base of our triangle is x and the height is y. Also, x and y are related by the given equation

$$y = 16 - x^2 \tag{10}$$

STEP 4 The area of a triangle equals $\frac{1}{2}$(base)(height):

$$A = \tfrac{1}{2}(x)(y)$$

so

$$\begin{aligned} A(x) &= \tfrac{1}{2}(x)(16 - x^2) \quad \text{substituting for } y \text{ using equation (10)} \\ &= 8x - \tfrac{1}{2}x^3 \end{aligned}$$

This last equation expresses the area of the triangle as a function of x, as required. (Exercise 46 at the end of this section will ask you to specify the domain of this area function.)   ▲

EXAMPLE 5   A piece of wire x in. long is bent into the shape of a circle. Express the area of the circle in terms of x.

Solution   **STEP 1** See Figure 7.

STEP 2 We are supposed to express the area of the circle in terms of x, the circumference. Let A denote the area.

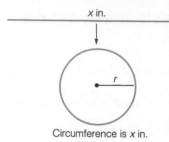

Circumference is x in.

FIGURE 7

STEP 3 The general formula for circumference in terms of radius is

$$C = 2\pi r$$

Since in our case the circumference is given as x, our equation becomes

$$x = 2\pi r$$

or

$$r = \frac{x}{2\pi} \qquad (11)$$

STEP 4 The general formula for the area of a circle in terms of the radius is

$$A = \pi r^2 \qquad (12)$$

This expresses A in terms of r, but we want A in terms of x. So we replace r in equation (12) by the quantity given in equation (11). This yields

$$A(x) = \pi\left(\frac{x}{2\pi}\right)^2 = \frac{\pi x^2}{4\pi^2} = \frac{x^2}{4\pi}$$

Thus, we have

$$A(x) = \frac{x^2}{4\pi}$$

This is the required equation. It expresses the area of the circle in terms of the circumference x. In other words, if we know the length of the piece of wire that is to be bent into a circle, from that length we can calculate what the area of the circle will be. ▲

EXAMPLE 6 Figure 8 displays a right circular cylinder, along with the formulas for the volume V and the total surface area S. Given that the volume is 10 cm³, express the surface area S as a function of r, the radius of the base.

Solution **STEP 1** See Figure 8.

STEP 2 We are given a formula that expresses the surface area S in terms of both r and h. We want to express S in terms of just r.

STEP 3 We are given that $V = 10$ and also that $V = \pi r^2 h$. Thus,

$$\pi r^2 h = 10$$

and, consequently, expressing h in terms of r,

$$h = \frac{10}{\pi r^2} \qquad (13)$$

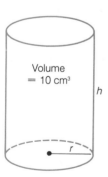

Volume
$= 10$ cm³

$V = \pi r^2 h$
$S = 2\pi r^2 + 2\pi rh$

FIGURE 8

STEP 4 We take the given formula for S, namely,

$$S = 2\pi r^2 + 2\pi rh$$

and replace h using the quantity given in equation (13). We get

$$S(r) = 2\pi r^2 + 2\pi r\left(\frac{10}{\pi r^2}\right)$$

$$= 2\pi r^2 + \frac{20}{r}$$

This is the required equation. It expresses the total surface area in terms of the radius r. Since the only restriction on r is that it be positive, the domain of the area function is $(0, \infty)$. ▲

We have followed the same four-step procedure in Examples 1 through 6. Of course, no single method can cover all possible cases. As usual, common sense and experience are often necessary. Also, you should not feel compelled to follow this procedure at any cost. Keep this in mind as you study the last three examples in this section.

EXAMPLE 7 Two numbers add up to 8. Express the product P of these two numbers in terms of a single variable.

Solution If we call the two numbers x and $8 - x$, then their product P is given by

$$P(x) = x(8 - x) = 8x - x^2$$

That's it. This last equation expresses the product as a function of the variable x. Since there are no restrictions on x (other than it's being a real number), the domain of this function is $(-\infty, \infty)$. ▲

EXAMPLE 8 Express the radius of a circle as a function of the area of the circle.

Solution We already know a formula relating these two quantities:

$$A = \pi r^2$$

In essence, the problem asks us to solve this equation for r instead of A. We then have

$$\pi r^2 = A$$
$$r^2 = \frac{A}{\pi}$$
$$r(A) = \sqrt{A/\pi}$$

This is the required equation. It tells us how to calculate the radius of the circle if the area is known.

Question Are there any restrictions on A in this last equation? ▲

EXAMPLE 9 In economics, the revenue R generated by selling x units at a price of p dollars per unit is given by

$$R = x \cdot p$$

price per unit
number of units

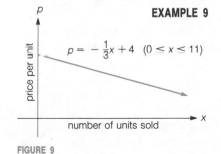

$p = -\frac{1}{3}x + 4$ $(0 < x < 11)$

price per unit

number of units sold

FIGURE 9

In Figure 9, we are given a hypothetical function relating the selling price of a certain item to the number of units sold. Such a function is called a **demand function**. Express the revenue as a function of x.

Solution More than anything else, this problem is an exercise in reading. After reading the problem several times, we find that it comes down to this:

given: $R = x \cdot p$ and $p = -\dfrac{1}{3}x + 4$ $(0 \le x \le 11)$

find: an equation expressing R in terms of x

In view of this, we write

$$R = x \cdot p$$

$$R(x) = x\left(-\frac{1}{3}x + 4\right)$$

Thus, we have

$$R(x) = -\frac{1}{3}x^2 + 4x \qquad (0 \le x \le 11)$$

This is the required function. It allows us to calculate the revenue when we know the number of units sold. Note that this revenue function is a quadratic function.

Question for review and also preview: How would you find the maximum revenue in this case?

▲

▼ EXERCISE SET 3.3

A

1. A rectangle is inscribed in a circle of diameter 12 in.
   (a) Express the perimeter of the rectangle as a function of its width x. *Suggestion:* First reread Example 1.
   (b) Express the area of the rectangle as a function of its width x.

2. (a) The perimeter of a rectangle is 16 cm. Express the area of the rectangle in terms of the width x. *Suggestion:* First reread Example 2.
   (b) The area of a rectangle is 85 cm^2. Express the perimeter as a function of the width x.

3. A point $P(x, y)$ is on the curve $y = x^2 + 1$.
   (a) Express the distance from P to the origin as a function of x. *Suggestion:* First reread Example 3.
   (b) In part (a), you expressed the length of a certain line segment as a function of x. Now express the slope of that line segment in terms of x.

4. A point $P(x, y)$ lies on the curve $y = \sqrt{x}$, as shown in the figure.
   (a) Express the area of the shaded triangle in terms of x. *Suggestion:* First reread Example 4.
   (b) Express the perimeter of the shaded triangle in terms of x.

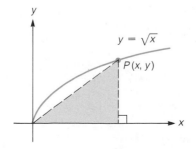

5. A piece of wire πy in. long is bent into a circle.
   (a) Express the area of the circle as a function of y. *Suggestion:* First reread Example 5.
   (b) If the original piece of wire were bent into a square instead of a circle, how would you express the area in terms of y?

6. The volume of a certain right circular cylinder is 20 in.^{3} Express the total surface area of the cylinder as a function of r, the radius of the base. *Suggestion:* First reread Example 6

7. Two numbers add to 16.
   (a) Express the product of the two numbers in terms of a single variable. *Suggestion:* First reread Example 7.
   (b) Express the sum of the squares of the two numbers in terms of a single variable.
   (c) Express the difference of the cubes of the two numbers in terms of a single variable. (There are two answers.)
   (d) What happens when you try to express the average of the two numbers in terms of one variable?

8. (a) Express the radius of a circle as a function of the circumference. *Suggestion:* Reread Example 8.
   (b) Express the diameter of a circle as a function of the circumference.

9. Given a demand function $p = -\frac{1}{4}x + 8$, express the revenue as a function of x. (See Example 9 for terminology and definitions.)

10. In Example 2, we considered a rectangle with perimeter 100 cm. We found that the area of such a rectangle is given by $A(x) = 50x - x^2$, where x is the width of the rectangle. Compute the numbers $A(1)$, $A(10)$, $A(20)$, $A(25)$, and $A(35)$. Which width x seems to yield the largest area $A(x)$?

C **11.** In Example 1, we considered a rectangle inscribed in a circle of diameter 8 cm. We found that the perimeter of such a rectangle is given by

$$P(x) = 2x + 2\sqrt{64 - x^2}$$

where x is the width of the rectangle.

(a) Use a calculator to complete the following table. Round off each answer to two decimal places.

x	1	2	3	4	5	6	7
$P(x)$							

(b) In your table, what is the largest value for $P(x)$? What is the width x in this case?

(c) Using calculus, it can be shown that among all possible widths x, the width $x = 4\sqrt{2}$ cm yields the largest possible perimeter. Use a calculator to compute the perimeter in this case, and check to see that the value you obtain is indeed larger than all of the values obtained in part (a).

12. In economics, the demand function for a given commodity tells us how the unit price p is related to the number of units x that are sold. Suppose we are given a demand function

$$p = 5 - \frac{x}{4} \qquad (p \text{ in dollars})$$

(a) Graph this demand function.

(b) How many units can be sold when the unit price is \$3? Locate the point on the graph of the demand function that conveys this information.

(c) To sell 12 items, how should the unit price be set? Locate the point on the graph of the demand function that conveys this information.

(d) Find the revenue function corresponding to the given demand function. (Use the formula $R = x \cdot p$.) Graph the revenue function.

(e) Find the revenue when $x = 2$, when $x = 8$, and when $x = 14$.

(f) According to your graph in part (d), which x-value yields the greatest revenue? What is that revenue? What is the corresponding unit price?

13. Let $2s$ denote the length of the side of an equilateral triangle.

(a) Express the height of the triangle as a function of s.

(b) Express the area of the triangle as a function of s.

(c) Use the function you found in part (a) to determine the height of an equilateral triangle, each side of which is 8 cm long.

(d) Use the function you found in part (b) to determine the area of an equilateral triangle, each side of which is 5 in. long.

14. If x denotes the length of a side of an equilateral triangle, express the area of the triangle as a function of x.

15. In a certain right circular cylinder, the height is twice the radius. Express the volume as a function of the radius.

16. Using the information given in Exercise 15, express the radius as a function of the volume.

17. The volume of a right circular cylinder is 12π in.^{3}

(a) Express the height as a function of the radius.

(b) Express the total surface area as a function of the radius.

18. In a certain right circular cylinder, the total surface area is 14 in.^{2} Express the volume as a function of the radius.

19. The volume V and the surface area S of a sphere of radius r are given by the formulas $V = \frac{4}{3}\pi r^3$ and $S = 4\pi r^2$. Express V as a function of S.

20. The base of a rectangle lies on the x-axis, while the upper two vertices lie on the parabola $y = 10 - x^2$. Suppose that the coordinates of the upper right vertex of the rectangle are (x, y). Express the area of the rectangle as a function of x.

21. The hypotenuse of a right triangle is 20 cm. Express the area of the triangle as a function of the length x of one of the legs.

22. (a) Express the area of the shaded triangle in the accompanying figure as a function of x.

(b) Express the perimeter of the shaded triangle as a function of x.

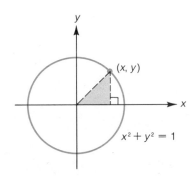

23. For the following figure, express the length of $\overline{AB}$ as a function of x. *Hint:* Note the similar triangles.

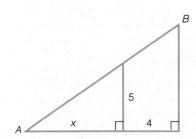

24. Five hundred feet of fencing are available to enclose a rectangular pasture alongside a river, which serves as one side of the rectangle (so only three sides require fencing). Express the area of the rectangular pasture as a function of x. See the figure.

C **25.** After rereading Example 2, complete the given table. Which x-value in the table yields the largest area A? What is the corresponding value of the length L in that case?

x	5	10	20	24	24.8	24.9	25	25.1	25.2	45
$A(x)$										

C **26.** After rereading Example 3, complete the following three tables. For each table, specify the x-value that yields the smallest distance D. (For Table 2, round off the values of D to two significant digits. For Table 3, round off the values of D to seven significant digits.)

TABLE 1

x	1	2	3	4	5
D					

TABLE 2

x	0.25	0.50	0.75
D			

TABLE 3

x	0.498	0.499	0.500	0.501	0.502
D					

C **27.** **(a)** After rereading Example 4, complete the following three tables. For each table, specify the x-value that yields the largest area A. [For Tables 2 and 3, round off the values of A to six significant digits. In part (b), you'll see why we are asking for that many digits.]
(b) Using calculus, it can be shown that the x-value yielding the largest area A is $x = 4\sqrt{3}/3$. Which x-value in the tables is closest to this x-value? To six significant digits, what is the area A when $x = 4\sqrt{3}/3$?

TABLE 1

x	1	2	3	4
A				

TABLE 2

x	1.75	2.00	2.25	2.50	2.75
A					

TABLE 3

x	2.15	2.20	2.25	2.30	2.35
A					

B

In Exercises 28–31, refer to the following figure, which displays a right circular cone along with the formulas for the volume V and the lateral surface area S.

$V = \frac{1}{3}\pi r^2 h$

$S = \pi r \sqrt{r^2 + h^2}$

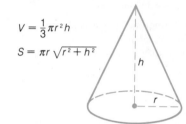

28. The volume of a right circular cone is 12π cm^3.
(a) Express the height as a function of the radius.
(b) Express the radius as a function of the height.

29. Suppose that, for a right circular cone, the height and radius are related by the equation $h = \sqrt{3}\, r$.
(a) Express the volume as a function of r.
(b) Express the lateral surface area as a function of r.

30. The volume of a right circular cone is 2 ft^3. Show that the lateral surface area as a function of r is given by

$$S = \frac{\sqrt{\pi^2 r^6 + 36}}{r}$$

31. In a certain right circular cone, the volume is numerically equal to the lateral surface area.
(a) Express the radius as a function of the height.
(b) Express the height as a function of the radius.

32. A line is drawn from the origin O to a point $P(x, y)$ in the first quadrant on the graph of $y = 1/x$. From point P, a line is drawn perpendicular to the x-axis, meeting the x-axis at B.
(a) Draw a figure of the situation described.
(b) Express the perimeter of triangle OPB as a function of x.
(c) Try to express the area of triangle OPB as a function of x. What happens?

33. A piece of wire 14 in. long is cut into two pieces. The first piece is bent into a circle, the second into a square. Express the combined total area of the circle and the square as a function of x, where x denotes the length of the wire that is used for the circle.

34. A wire of length L is cut into two pieces. The first piece is bent into a square, the second into an equilateral triangle. Express the combined total area of the square and triangle as a function of x, where x denotes the length of wire used for the triangle. (L is to be considered a constant here, not another variable.)

35. An athletic field with a perimeter of $\frac{1}{4}$ mile consists of a rectangle with a semicircle at each end, as shown in the figure. Express the area of the field as a function of r, the radius of the semicircle.

36. A square of side x is inscribed in a circle. Express the area of the circle as a function of x.

37. An equilateral triangle of side x is inscribed in a circle. Express the area of the circle as a function of x.

38. Consider an object of mass m moving at a velocity v. The kinetic energy K is defined as $K = \frac{1}{2}mv^2$. The momentum p is defined as $p = mv$. Express the kinetic energy in terms of mass and momentum (but not velocity).

39. An isosceles triangle is inscribed in a circle of radius R, where R is a constant. Express the area within the circle but outside the triangle as a function of h, where h denotes the height of the triangle.

40. Refer to the accompanying figure. An offshore oil rig is located at point A, which is 10 miles out to sea. An oil pipeline is to be constructed from A to a point C on the shore and then to an oil refinery at point D, farther up the coast. If it costs \$8000 per mile to lay the pipeline in the sea and \$2000 per mile on land, express the cost of laying the pipeline in terms of x, where x is the distance from B to C.

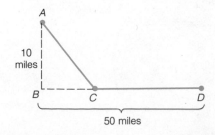

41. An open-top box is constructed from a 6-by-8-in. rectangular sheet of tin by cutting out equal squares at each corner and then folding up the flaps, as shown in the figure. Express the volume of the box as a function of x, the length of the side of each cut-out square.

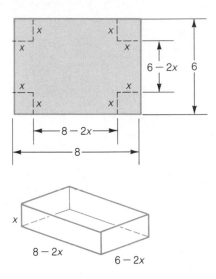

42. Follow Exercise 41, but assume that the original piece of tin is a square, 12 in. on each side.

43. A Norman window is in the shape of a rectangle surmounted by a semicircle, as shown in the next figure. Assume that the perimeter of the window is 32 ft.

(a) Express the area of the window as a function of r, the radius of the semicircle.

(b) The function you were asked to find in part (a) is a quadratic function, so its graph is a parabola. Does the parabola open upward or downward? Does it pass through the origin? Show that the vertex of the parabola is

$$\left(\frac{32}{\pi + 4}, \frac{512}{\pi + 4} \right)$$

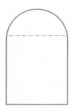

44. Refer to the following figure. Express the lengths of $\overline{CB}$, $\overline{CD}$, $\overline{BD}$, and $\overline{AB}$ in terms of x. *Hint:* Recall the theorem from geometry stating that in a 30°–60°–90° right triangle, the side opposite the 30° angle is half the hypotenuse.

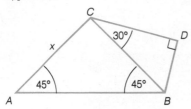

45. Refer to the following figure. Let s denote the ratio of y to z.
   (a) Express y as a function of s.
   (b) Express s as a function of y.
   (c) Express z as a function of s.
   (d) Express s as a function of z.

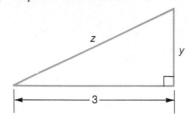

46. Refer to Example 4.
   (a) What is the x-intercept of the curve $y = 16 - x^2$ in Figure 6?
   (b) What is the domain of the area function in Example 4, assuming that the point P does not lie on the x- or y-axis?

47. The following figure shows the parabola $y = x^2$ and a line segment $\overline{AP}$ drawn from the point $A(0, -1)$ to the point $P(a, a^2)$ on the parabola.
   (a) Express the slope of $\overline{AP}$ in terms of a.
   (b) Show that the area of the shaded triangle in the figure is given by

$$\text{area} = \frac{a^5}{2(a^2 + 1)}$$

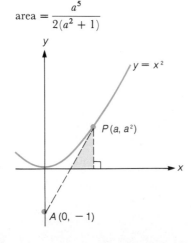

48. A rancher who wishes to fence off a rectangular area finds that the fencing in the east-west direction will require extra reinforcement owing to strong prevailing winds. Because of this, fencing in the east-west direction will be \$12 per (linear) yard, as opposed to a cost of \$8 per yard for fencing in the north-south direction. Given that the rancher wants to spend \$4800 on fencing, express the area of the rectangle as a function of x, its width. The required function is in fact a quadratic, so its graph is a parabola. Does the parabola open upward or downward? By considering this graph, find which width x yields the rectangle of largest area. What is this maximum area?

C

49. The following figure shows two concentric squares. Express the area of the shaded triangle as a function of x.

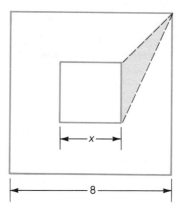

The following figure shows two concentric circles of radii r and R. Let A denote the area within the larger circle but outside the smaller one. Express A as a function of x (where x is defined in the figure).

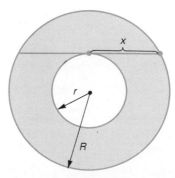

51. A straight line with slope m $(m < 0)$ passes through the point $(1, 2)$ and intersects the line $y = 4x$ at a point in the first quadrant. Let A denote the area of the triangle bounded by $y = 4x$, the x-axis, and the given line of slope m. Express A as a function of m.

52. A line with slope m $(m < 0)$ passes through the point (a, b) in the first quadrant and intersects the line $y = Mx$ $(M > 0)$ at another point in the first quadrant. Let A denote the area of the triangle bounded by $y = Mx$, the x-axis, and the given line with slope m. Express A in terms of m, M, a, and b.

Answer: $A = \dfrac{M(am - b)^2}{2m(m - M)}$

53. A line with slope m $(m < 0)$ passes through the point (a, b) in the first quadrant. Express the area of the triangle bounded by this line and the axes in terms of m.

54. One corner of a page of width a is folded over and just reaches the opposite side, as indicated in the following figure. Express L, the length of the crease, in terms of x and a.

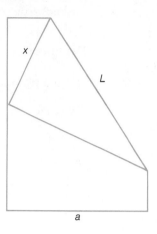

3.4 ▼ MAXIMUM AND MINIMUM PROBLEMS

. . . problems on maxima and minima, although new features in an English textbook, stand so little in need of apology with the scientific public that I offer none.

G. Chrystal, in his preface to *Textbook of Algebra* (1886)

You have already met several examples of maximum and minimum problems in the second section of this chapter. In particular, before reading further in the present section, you should first look back and review Examples 5 through 7 on pages 158–159.

Actually, we begin this section's discussion of maximum and minimum problems at a more intuitive level. Consider, for example, the following question: If two numbers add to 9, what is the largest possible value of their product? To gain some insight here, we carry out a few preliminary calculations (see Table 1).

We have circled the 20 in the right-hand column of Table 1, for that "appears" to be the largest product. We say *appears* because our table is incomplete. For instance, what if we allowed x- and y-values that are not whole numbers? Might we get a product exceeding 20? Table 2 shows the results of some additional calculations along these lines.

As you can see from Table 2, there is a product exceeding 20, namely, 20.25. Now the question is, if we further expand our tables, can we find yet another product, this time exceeding 20.25? And here we have come about as far as we want to go using this approach involving arithmetic and tables. For no matter what candidate we come up with for the largest product, there will always be the question of whether we might do still better using a larger table.

Nevertheless, this approach was useful, for it showed us what is really at the heart of a typical maximum or minimum problem. Essentially, we are trying to sort through an infinite number of possible cases and then pick out the required extreme case. For instance, in the example at hand, there are infinitely many pairs

TABLE 1

x	y	$x + y$	xy
-2	11	9	-22
-1	10	9	-10
0	9	9	0
1	8	9	8
2	7	9	14
3	6	9	18
4	5	9	⟨20⟩

TABLE 2

x	y	$x + y$	xy
1	8	9	8
1.5	7.5	9	11.25
2	7	9	14
2.5	6.5	9	16.25
3	6	9	18
3.5	5.5	9	19.25
4	5	9	20
4.5	4.5	9	20.25

of numbers x and y adding to 9. Among all of these pairs, we are asked to look at the products and see which (if any!) is the largest. Example 1 shows how to apply our knowledge of quadratic functions to solve this problem in a definitive manner.

EXAMPLE 1   Two numbers add to 9. What is the largest possible value for their product?

Solution   Call the two numbers x and $9 - x$. Then their product P is given by

$$P = x(9 - x) = 9x - x^2$$

The graph of this quadratic function is the parabola in Figure 1. Note the accompanying calculation for the vertex. As Figure 1 and the accompanying calculations show, the largest value of the product P is 20.25. This is the required solution. (Note from the graph that there is no smallest value of P.)

FIGURE 1

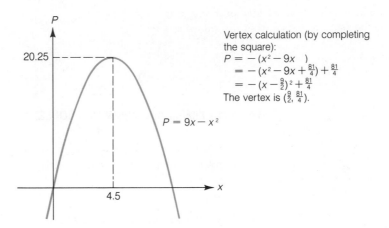

Vertex calculation (by completing the square):
$$P = -(x^2 - 9x \quad)$$
$$= -(x^2 - 9x + \tfrac{81}{4}) + \tfrac{81}{4}$$
$$= -(x - \tfrac{9}{2})^2 + \tfrac{81}{4}$$
The vertex is $(\tfrac{9}{2}, \tfrac{81}{4})$.

$$P = 9x - x^2$$

Example 1 illustrates the general strategy we will follow for solving the maximum and minimum problems in this section.

Strategy for Solving the Maximum and Minimum Problems in This Section

1. Express the quantity to be maximized or minimized in terms of a single variable. For instance, in Example 1, we found $P = 9x - x^2$. In trying to set up such functions, you'll want to keep in mind the four-step procedure used in Section 3.3.

2. Assuming that the function you have determined is a quadratic, note whether its graph, a parabola, opens upward or downward. Check whether this is consistent with the requirements of the problem. For instance, the parabola in Figure 1 opens downward; so it makes sense to look for a *largest*, not smallest, value of P. Now complete the square to locate the vertex. (If the function is not a quadratic but is closely related to a quadratic, these ideas may still apply. See, for instance, Examples 6 and 7 in Section 3.2.)

3. After you have determined the vertex, you must relate that information to the original question. In Example 1, for instance, we were asked for the product P, not for x.

EXAMPLE 2   Among all rectangles having a perimeter of 10 ft, find the dimensions (length and width) of the one with the greatest area.

Solution   First we want to set up a function that expresses the area of the rectangle in terms of a single variable. In doing this, we'll be guided by the four-step procedure that we used in Section 3.3. Figure 2(a) displays the given information. Our problem is to determine the dimensions of the rectangle that has the greatest area. As Figure 2(b) indicates, we can label the dimensions x and y. Since the perimeter is given as 10 ft, we have

$$2x + 2y = 10$$
$$x + y = 5$$
$$y = 5 - x \tag{1}$$

Letting A denote the area of the rectangle, we can write

$$A = xy$$
$$A(x) = x(5 - x) \quad \text{substituting for } y \text{ using equation (1)}$$
$$= 5x - x^2$$

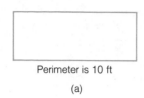

Perimeter is 10 ft

(a)

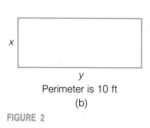

x

y

Perimeter is 10 ft

(b)

FIGURE 2

This expresses the area of the rectangle in terms of the width x. Since the graph of this quadratic function is a parabola that opens downward, it does make sense to talk about the maximum. We can find the vertex of the parabola by completing the square:

$$A(x) = -(x^2 - 5x \qquad)$$
$$= -\left(x^2 - 5x + \frac{25}{4}\right) + \frac{25}{4} \quad \text{adding 0}$$
$$= -\left(x - \frac{5}{2}\right)^2 + \frac{25}{4}$$

So the vertex is $\left(\frac{5}{2}, \frac{25}{4}\right)$ and, in particular, the width $x = \frac{5}{2}$ ft yields the maximum area. Now, the problem asks us for the length and width of this rectangle, not its area. To compute the length y, we again use equation (1):

$$y = 5 - x = 5 - \frac{5}{2} = \frac{5}{2}$$

Thus, among all rectangles having a perimeter of 10 ft, the one with the greatest area is actually the square with dimensions of $2\frac{1}{2}$ ft by $2\frac{1}{2}$ ft.   ▲

EXAMPLE 3   Suppose that a baseball is tossed straight up and that its height as a function of time is given by the formula

$$h = -16t^2 + 64t + 6$$

In this formula, h is measured in feet and t in seconds, with $t = 0$ corresponding to the instant that the ball is released. What is the maximum height of the ball? When does the ball reach this height?

Solution   The given function tells us how the height h of the ball depends on the time t. We want to know the largest possible value for h. Since the graph of the given function is a parabola opening downward, we can determine the largest value of h just by finding the vertex of the parabola. After completing the square, we find that the original equation can be rewritten

$$h = -16(t - 2)^2 + 70 \qquad (2)$$

[You should verify this for yourself. (If you get stuck, follow the model for completing the square in Example 4 of Section 3.2.)] From equation (2) we see that the vertex of the parabola $h = -16t^2 + 64t + 6$ is $(2, 70)$. Therefore the maximum height of the ball is 70 ft, and the ball reaches this height at $t = 2$ sec. ▲

EXAMPLE 4   Which point on the curve $y = \sqrt{x}$ is closest to the point $(1, 0)$?

Solution   In Figure 3, we let D denote the distance from a point (x, y) on the curve to the point $(1, 0)$. We are asked to find out exactly which point (x, y) will make the distance D as small as possible. Using the distance formula, we have

$$D = \sqrt{(x - 1)^2 + (y - 0)^2}$$
$$= \sqrt{x^2 - 2x + 1 + y^2} \qquad (3)$$

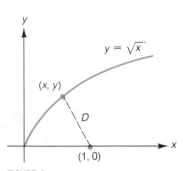

$y = \sqrt{x}$

(x, y)

D

$(1, 0)$

FIGURE 3

This expresses D in terms of both x and y. To express D in terms of x alone, we use the given equation $y = \sqrt{x}$ to substitute for y in equation (3). This yields

$$D(x) = \sqrt{x^2 - 2x + 1 + (\sqrt{x})^2}$$
$$= \sqrt{x^2 - x + 1}$$

Since the same x-value that makes $D(x)$ its smallest will also cause the quantity $[D(x)]^2 = x^2 - x + 1$ to be its smallest, we need only find the x-value that minimizes the quantity $x^2 - x + 1$. Denoting this quantity by $f(x)$ for the moment (so that we have an equation with a graph that is a parabola), we have

$$f(x) = x^2 - x + 1$$
$$= \left(x - \frac{1}{2}\right)^2 + \frac{3}{4} \quad \text{completing the square}$$

From this last equation we conclude that the quantity $x^2 - x + 1$ will be smallest when $x = \frac{1}{2}$.

Now that we have x, we want to calculate y. (The problem did not ask for D; we needn't calculate it.) We know that $y = \sqrt{x}$ and $x = \frac{1}{2}$. Therefore,

$$y = \sqrt{\frac{1}{2}} = \frac{1}{\sqrt{2}} \cdot \frac{\sqrt{2}}{\sqrt{2}} = \frac{\sqrt{2}}{2}$$

Thus, among all points (x, y) on the curve $y = \sqrt{x}$, the point closest to $(1, 0)$ is $(\frac{1}{2}, \sqrt{2}/2)$. This is the required solution.

Question   Why would it have made no sense to ask instead for the point farthest from $(1, 0)$? ▲

EXAMPLE 5   Suppose that you have 600 m of fencing with which to build two adjacent rectangular corrals. The two corrals are to share a common fence on one side, as shown in Figure 4. Find the dimensions x and y so that the total enclosed area is as large as possible.

Solution   You have 600 m of fencing to be set up as shown in Figure 4. The question is how you should choose x and y so that the total area is a maximum. Since the

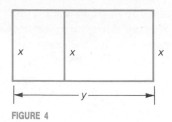

FIGURE 4

total length of fencing is 600 m, we can relate x and y by writing

$$x + x + x + y + y = 600$$
$$3x + 2y = 600$$
$$2y = 600 - 3x$$
$$y = 300 - \frac{3}{2}x \qquad (4)$$

Letting A denote the total area, we have

$$A = xy$$
$$A(x) = x\left(300 - \frac{3}{2}x\right) \qquad \text{using equation (4) to substitute for } y$$
$$= 300x - \frac{3}{2}x^2$$

This last equation expresses the area as a function of x. Note that the graph of this function is a parabola opening downward, so it does make sense to talk about a maximum. We complete the square, as usual, to locate the vertex. As you can check, the result is

$$A(x) = -\frac{3}{2}(x - 100)^2 + 15000$$

Thus the x-coordinate of the vertex is 100, and this is the x-value that maximizes the area. The corresponding y-value can now be calculated using equation (4):

$$y = 300 - \frac{3}{2}x = 300 - \frac{3}{2}(100)$$
$$= 300 - 3(50) = 150$$

Thus, by choosing x to be 100 m and y to be 150 m, the total area in Figure 4 will be as large as possible. Incidentally, note that the exact location of the fence dividing the two corrals does not influence our work or final answer. ▲

EXAMPLE 6 Suppose that the following function relates the selling price, p, of an item to the quantity sold, x:

$$p = -\frac{1}{3}x + 40 \qquad (p \text{ in dollars})$$

For which value of x will the corresponding revenue be a maximum? What are the maximum revenue and the unit price in this case?

Solution First, let us recall the formula for revenue given on page 165:

$$R = \text{number of units} \times \text{price per unit}$$

In view of this, we have

$$R = x \cdot p$$
$$R(x) = x\left(-\frac{1}{3}x + 40\right)$$
$$= -\frac{1}{3}x^2 + 40x \qquad (5)$$

We want to know which value of x yields the largest revenue R. Since the graph of the revenue function in equation (5) is a parabola opening downward, the required x-value is the x-coordinate of the vertex of the parabola. We complete the square to locate the vertex:

$$R(x) = -\frac{1}{3}(x^2 - 120x \qquad\quad)$$

$$= -\frac{1}{3}(x^2 - 120x + 3600) + 1200 \quad \text{adding } 0 \ (= -1200 + 1200)$$

$$= -\frac{1}{3}(x - 60)^2 + 1200$$

The vertex is therefore (60, 1200). Consequently, the revenue is a maximum when $x = 60$ items, and this maximum revenue is \$1200. For the corresponding unit price, we have

$$p(x) = -\frac{1}{3}x + 40$$

$$p(60) = -\frac{1}{3}(60) + 40 = 20$$

Thus the unit price corresponding to the maximum revenue is \$20.   ▲

EXAMPLE 7   Figure 5 shows a computer-generated graph of the function

$$f(x) = x^4 - 3x^2$$

From the graph, it is clear that the minimum of the function is slightly less than -2. Find the exact value of this minimum and the corresponding x-values at which it occurs.

Solution   Although this is not a quadratic function, we can nevertheless use the technique of completing the square in this particular case. (You have already seen one instance of this in Example 7 of Section 3.2.) We have

$$f(x) = x^4 - 3x^2 + \frac{9}{4} - \frac{9}{4}$$

$$= \left(x^2 - \frac{3}{2}\right)^2 - \frac{9}{4}$$

$$= -\frac{9}{4} + \left(x^2 - \frac{3}{2}\right)^2$$

From this last equation we see that $f(x)$ is never less than $-\frac{9}{4}$ (because the quantity being added to $-\frac{9}{4}$ is nonnegative). Furthermore, $f(x)$ does attain the value $-\frac{9}{4}$ when $x^2 = \frac{3}{2}$; that is, when $x = \pm\sqrt{\frac{3}{2}}$. Thus the minimum value of f is $-\frac{9}{4}$ and the corresponding x-values are

$$x = \pm\frac{\sqrt{3}}{\sqrt{2}} = \pm\frac{\sqrt{6}}{2} \quad \text{rationalizing the denominator}$$

Note   Use your calculator to approximate these x-values, and then check that they are consistent with the graph in Figure 5.   ▲

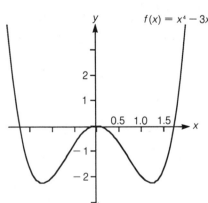

$f(x) = x^4 - 3x^2$

FIGURE 5

▼ EXERCISE SET 3.4

A

1. Two numbers add to 5. What is the largest possible value of their product?

2. Find two numbers adding to 20 such that the sum of their squares is as small as possible.

3. The difference of two numbers is 1. What is the smallest possible value for the sum of their squares?

4. For each quadratic function, state whether it would make sense to look for a highest or a lowest point on the graph. Then determine the coordinates of that point.
   (a) $y = 2x^2 - 8x + 1$ (b) $y = -3x^2 - 4x - 9$
   (c) $h = -16t^2 + 256t$ (d) $f(x) = 1 - (x + 1)^2$
   (e) $g(t) = t^2 + 1$
   (f) $f(x) = 1000x^2 - x + 100$

5. Among all rectangles having a perimeter of 25 m, find the dimensions of the one with the largest area.

6. What is the largest possible area for a rectangle whose perimeter is 80 cm?

7. What is the largest possible area for a right triangle in which the sum of the lengths of the two shorter sides is 100 in.?

8. The perimeter of a rectangle is 12 m. Find the dimensions for which the diagonal is as short as possible.

9. Two numbers add to 6.
   (a) Let T denote the sum of the squares of the two numbers. What is the smallest possible value for T?
   (b) Let S denote the sum of the first number and the square of the second. What is the smallest possible value for S?
   (c) Let U denote the sum of the first number and twice the square of the second number. What is the smallest possible value for U?
   (d) Let V denote the sum of the first number and the square of twice the second number. What is the smallest possible value for V?

10. Suppose that the height of an object shot straight up is given by $h = 512t - 16t^2$ (here h is in feet and t is in seconds). Find the maximum height and the time at which the object hits the ground.

11. A baseball is thrown straight up, and its height as a function of time is given by the formula $h = -16t^2 + 32t$ (where h is in feet and t is in seconds).
   (a) Find the height of the ball when $t = 1$ and when $t = \frac{3}{2}$.
   (b) Find the maximum height of the ball and the time at which that height is attained.
   (c) At what times is the height 7 ft?

12. Find the point on the curve $y = \sqrt{x}$ that is nearest to the point $(3, 0)$.

13. Which point on the curve $y = \sqrt{x - 2} + 1$ is closest to the point $(4, 1)$? What is this minimum distance?

14. Find the coordinates of the point on the line $y = 3x + 1$ closest to $(4, 0)$.

15. (a) What number exceeds its square by the greatest amount?
    (b) What number exceeds twice its square by the greatest amount?

16. Suppose that you have 1800 m of fencing with which to build three adjacent rectangular corrals, as shown in the figure. Find the dimensions so that the total enclosed area is as large as possible.

17. Five hundred feet of fencing is available for a rectangular pasture alongside a river, the river serving as one side of the rectangle (so only three sides require fencing). Find the dimensions yielding the greatest area.

18. Let $A = 3x^2 + 4x - 5$ and $B = x^2 - 4x - 1$. Find the minimum value of $A - B$.

19. Let $R = 0.4x^2 + 10x + 5$ and $C = 0.5x^2 + 2x + 101$. For which value of x is $R - C$ a maximum?

20. Suppose that the revenue generated by selling x units of a certain commodity is given by $R = -\frac{1}{5}x^2 + 200x$. Assume that R is in dollars. What is the maximum revenue possible in this situation?

21. Suppose that the function $p = -\frac{1}{4}x + 30$ relates the selling price p of an item to the number of units x that are sold. Assume that p is in dollars. For which value of x will the corresponding revenue be a maximum? What is this maximum revenue and what is the unit price?

22. The action of sunlight on automobile exhaust produces air pollutants known as *photochemical oxidants*. In a study of cross-country runners in Los Angeles, it was shown that running performances can be adversely affected when the oxidant level reaches 0.03 part per million. Suppose that on a given day, the oxidant level L is approximated by the formula

    $$L = 0.059t^2 - 0.354t + 0.557 \qquad (0 \le t \le 7)$$

    where t is measured in hours, with $t = 0$ corresponding to 12 noon, and L is in parts per million. At what time is the oxidant level L a minimum? At this time, is the oxidant level high enough to affect a runner's performance?

23. **(a)** Find the smallest possible value of the quantity $x^2 + y^2$ under the restriction that $2x + 3y = 6$.

(b) Find the radius of the circle whose center is at the origin and that is tangent to the line $2x + 3y = 6$. How does this answer relate to your answer in part (a)?

B

24. Through a type of chemical reaction known as *autocatalysis*, the human body produces the enzyme trypsin from the enzyme trypsinogen. (Trypsin then breaks down proteins into amino acids, which the body needs for growth.) Let r denote the rate of this chemical reaction in which trypsin is formed from trypsinogen. It has been shown experimentally that $r = kx(a - x)$, where k is a positive constant, a is the initial amount of trypsinogen, and x is the amount of trypsin produced (so x increases as the reaction proceeds). Show that the reaction rate r is a maximum when $x = a/2$. In other words, the speed of the reaction is greatest when the amount of trypsin formed is half the original amount of trypsinogen.

25. **(a)** Let $x + y = 15$. Find the minimum value of the quantity $x^2 + y^2$.

(b) Let C be a constant and $x + y = C$. Show that the minimum value of $x^2 + y^2$ is $C^2/2$. Then use this result to check your answer in part (a).

26. Suppose that A, B, and C are positive constants and that $x + y = C$. Show that the minimum value of $Ax^2 + By^2$ occurs when $x = BC/(A + B)$ and $y = AC/(A + B)$.

27. The following figure shows a square inscribed within a unit square. For which value of x is the area of the inscribed square a minimum? What is the minimum area? *Hint:* Denote the lengths of the two segments that make up the base of the unit square by t and $1 - t$. Now use the Pythagorean Theorem and congruent triangles to express x in terms of t.

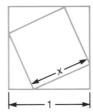

28. **(a)** Find the coordinates of the point on the line $y = mx + b$ that is closest to the origin.

$$\textit{Answer:} \left(\frac{-mb}{1 + m^2}, \frac{b}{1 + m^2} \right)$$

(b) Show that the perpendicular distance from the origin to the line $y = mx + b$ is $|b|/\sqrt{1 + m^2}$. *Suggestion:* Use the result in part (a).

(c) Use part (b) to show that the perpendicular distance from the origin to the line $Ax + By + C = 0$ is $|C|/\sqrt{A^2 + B^2}$.

29. The point P lies in the first quadrant on the graph of the line $y = 7 - 3x$. From the point P, perpendiculars are drawn to both the x-axis and the y-axis. What is the largest possible area for the rectangle thus formed?

30. Show that the largest possible area for the shaded rectangle shown in the figure is $-b^2/4m$. Then use this to check your answer in Exercise 29.

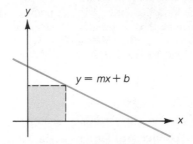

31. Show that the maximum possible area for a rectangle inscribed in a circle of radius R is $2R^2$. *Hint:* Maximize the square of the area.

32. An athletic field with a perimeter of $\frac{1}{4}$ mile consists of a rectangle with a semicircle at each end, as shown in the figure. Find the dimensions x and r that yield the greatest possible area for the rectangular region.

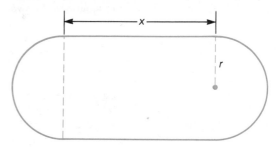

33. A rancher who wishes to fence off a rectangular area finds that the fencing in the east-west direction will require extra reinforcement owing to strong prevailing winds. Because of this, the cost of fencing in the east-west direction will be $12 per (linear) yard, as opposed to a cost of $8 per yard for fencing in the north-south direction. Find the dimensions of the largest possible rectangular area that can be fenced for $4800.

34. Let $f(x) = (x-a)^2 + (x-b)^2 + (x-c)^2$, where a, b, and c are constants. Show that $f(x)$ will be a minimum when x is the average of a, b, and c.

35. Let $y = a_1(x-x_1)^2 + a_2(x-x_2)^2$, where a_1, a_2, x_1, and x_2 are all constants. In addition, suppose that a_1 and a_2 are both positive. Show that the minimum of this function occurs when

$$x = \frac{a_1 x_1 + a_2 x_2}{a_1 + a_2}$$

36. Among all rectangles with a given perimeter P, find the dimensions of the one with the shortest diagonal.

37. By analyzing sales figures, the economist for a stereo manufacturer knows that 150 units of a tape player can be sold each month when the price is set at $p = \$200$ per unit. The figures also show that for each $10 hike in price, five fewer units are sold each month.
  (a) Let x denote the number of units sold per month and let p denote the price per unit. Find a linear function relating p and x.
  Hint: $\Delta p / \Delta x = 10/(-5) = -2$.
  (b) The revenue R is given by $R = xp$. What is the maximum revenue? At what level should the price be set to achieve this maximum revenue?

38. Let $f(x) = x^2 + px + q$, and suppose that the minimum value of this function is 0. Show that $q = p^2/4$.

39. For which numbers t will the value of the function $f(t) = t^2 - t^4$ be as large as possible?

40. Find the minimum value of the function $f(t) = t^4 - 8t^2$.

41. Among all possible inputs for the function $f(t) = -t^4 + 6t^2 - 6$, which ones yield the largest output?

42. Let $f(x) = x - 3$ and $g(x) = x^2 - 4x + 1$.
  (a) Find the minimum value of $g \circ f$.
  (b) Find the minimum value of $f \circ g$.
  (c) Are the results in parts (a) and (b) the same?

43. A piece of wire 16 in. long is to be cut into two pieces. Let x denote the length of the first piece and $16 - x$ the length of the second. The first piece is to be bent into a circle and the second piece into a square.
  (a) Express the total combined area A of the circle and the square as a function of x.
  (b) For which value of x is the area A a minimum?
  (c) Using the x-value that you found in part (b), find the ratio of the lengths of the shorter to the longer piece of wire. *Answer:* $\pi/4$.

44. A 30-in. piece of string is to be cut into two pieces. The first piece will be formed into the shape of an equilateral triangle and the second piece into a square. Find the length of the first piece if the combined area of the triangle and the square is to be as small as possible.

C

45. Repeat Exercise 43, but assume that the length of the wire is L in.

46. Let $f(x) = x^2 + bx + 1$. Find a positive value for b such that the distance from the origin to the vertex of the parabola is as small as possible.

47. The next figure shows a rectangle inscribed in a given triangle of base b and height h. Find the ratio of the area of the triangle to the area of the rectangle when the area of the rectangle is a maximum.

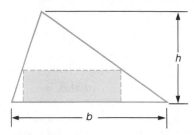

48. A Norman window is in the shape of a rectangle surmounted by a semicircle, as shown in the figure. Assume that the perimeter of the window is P, a constant. Show that the area of the window is a maximum when both x and r are equal to $P/(\pi + 4)$. Show that this maximum area is $\frac{1}{2}P^2/(\pi + 4)$.

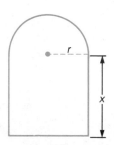

49. A triangle is inscribed in a semicircle of diameter $2R$, as shown in the figure. Show that the smallest possible value for the area of the shaded region is $\frac{1}{2}(\pi - 2)R^2$. *Hint:* The area of the shaded region is a minimum when the area of the triangle is a maximum. Find the value of x that maximizes the *square* of the area of the triangle. This will be the same x that maximizes the area of the triangle.

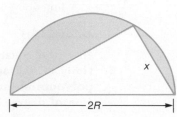

50. **(a)** Complete the following table. Which x-y pair in the table yields the smallest sum $x + y$?

x	0.5	1	1.5	2	2.5	3	3.5
y							
xy	12	12	12	12	12	12	12
$x + y$							

(b) Find two positive numbers with a product of 12 and as small a sum as possible. *Hint:* The quantity that you need to minimize is $x + (12/x)$, where $x > 0$. But

$$x + \frac{12}{x} = \left(\sqrt{x} - \sqrt{\frac{12}{x}}\right)^2 + 2\sqrt{12}$$

This last expression is minimized when the quantity within parentheses is zero. Why?

(c) Use a calculator to verify that the two numbers obtained in part (b) produce a sum that is smaller than any of the sums obtained in part (a).

51. What is the smallest possible value for the sum of a positive number and its reciprocal? *Hint:* After setting up the appropriate function, adapt the hint given in Exercise 50(b).

52. Let $f(x) = \dfrac{2x^2 - 4x + 1}{2x}$, and assume that $x > 0$.

(a) Complete the following two tables. In both cases, specify the x-value that yields the smallest output. (Round off the outputs in the tables to six decimal places.)

x	0.4	0.5	0.6	0.7	0.8	0.9
$f(x)$						

x	0.68	0.69	0.70	0.71	0.72	0.73
$f(x)$						

(b) Among all positive inputs for the given function f, find the one for which $f(x)$ is a minimum.
Hint: Write $f(x)$ as $\left(x + \dfrac{1}{2x}\right) - 2$, and then adapt the hint given in Exercise 50(b).

(c) Use a calculator to verify that the x-value obtained in part (b) produces an output that is smaller than any of the outputs obtained in part (a).

53. Let $G(x) = (a + x)(b + x)/x$, where a and b are positive constants and $x > 0$. Show that the minimum value of G is $(\sqrt{a} + \sqrt{b})^2$ and that this minimum value occurs when $x = \sqrt{ab}$. *Hint:* Show that $G(x) = (a + b) + \left(\dfrac{ab}{x} + x\right)$. Then adapt the hint in Exercise 50(b).

54. A line with a positive slope m passes through the point $(-2, 1)$.
(a) Show that the area of the triangle bounded by this line and the axes is $\left(2m + \dfrac{1}{2m}\right) + 2$.

(b) For which value of m will the area be a minimum? *Hint:* Adapt the hint given in Exercise 50(b).

55. Prove that for any value of x we always have

$$\frac{1}{2x^2 - x + 1} \leq \frac{8}{7}$$

56. Find the largest value of the function

$$f(x) = \frac{1}{x^4 + 2x^2 + 1}$$

3.5 ▼ POLYNOMIAL FUNCTIONS

We can rephrase the definitions of linear and quadratic functions using the terminology for polynomials given in Section 1.5.

A function f is **linear** if $f(x)$ is a **polynomial of degree 1**:

$$f(x) = a_1 x + a_0 \tag{1}$$

A function f is **quadratic** if $f(x)$ is a **polynomial of degree 2**:

$$f(x) = a_2 x^2 + a_1 x + a_0 \tag{2}$$

In view of equation (1), linear functions are sometimes called **polynomial functions of degree 1**. Similarly, in view of equation (2), quadratic functions are **polynomial functions of degree 2**. More generally, by a **polynomial function of degree n**, we mean a function defined by an equation of the form

$$f(x) = a_n x^n + a_{n-1} x^{n-1} + \cdots + a_1 x + a_0 \tag{3}$$

FIGURE 1

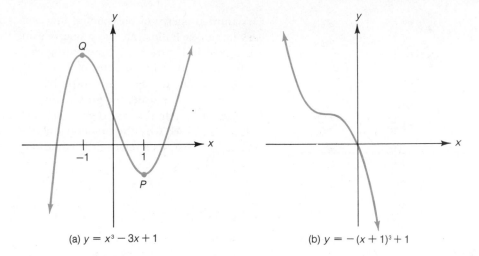

(a) $y = x^3 - 3x + 1$ (b) $y = -(x + 1)^3 + 1$

where n is a nonnegative integer and $a_n \neq 0$. Throughout the remainder of this chapter, the coefficients a_k will always be real numbers.

In principle, we can obtain the graph of any polynomial function by setting up a table and plotting a sufficient number of points. Indeed, this is just the way a computer equipped with a curve plotter operates. However, to *understand* why the graphs look as they do, we want to discuss some additional methods for graphing polynomial functions.

There are three facts that we shall need. By way of example, look at the graphs of the polynomial functions in Figure 1. First, notice that both graphs are unbroken, smooth curves, with no "corners." As is shown in calculus, this is true for the graph of every polynomial function. By way of contrast, the graphs in Figures 2 and 3 cannot represent polynomial functions. The graph in Figure 2 has a break in it, and the graph in Figure 3 has a **cusp**.

Now look back at the graph in Figure 1(a). Recall (from Section 2.4) that points such as P and Q are called **turning points**. These are points where the graph changes from rising to falling, or vice versa. It is a fact (proved in calculus) that the graph of a polynomial function of degree n has *at most* $n - 1$ turning points. For instance, in Figure 1(a), there are two turning points, while the degree of the polynomial is 3. However, as Figure 1(b) indicates, we needn't have any turning points at all.

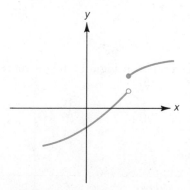

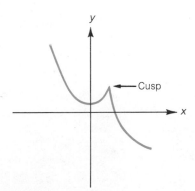

FIGURE 2

Since the graph has a break, it cannot represent a polynomial function.

FIGURE 3

Since the graph has a cusp, it cannot represent a polynomial function.

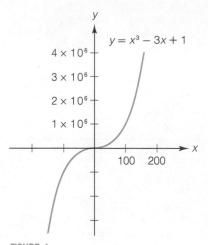

FIGURE 4
When $|x|$ is very large, the graph of
$y = x^3 - 3x + 1$ appears indistinguishable
from that of $y = x^3$.

A third property of polynomial functions concerns their behavior when $|x|$ is very large. We'll illustrate the property using the function $y = x^3 - 3x + 1$, which is graphed in Figure 1(a). Now, in Figure 1(a), the x-values are relatively small; for instance, the x-coordinates of P and Q are 1 and -1, respectively. In Figure 4, however, we show the graph of this same function using units of 100 on the x-axis. On this scale, the graph appears indistinguishable from that of $y = x^3$. In particular, note that as $|x|$ gets very large, $|y|$ grows very large.

It's easy to see why the function $y = x^3 - 3x + 1$ resembles $y = x^3$ when $|x|$ is very large. First, let's rewrite the equation $y = x^3 - 3x + 1$ as

$$y = x^3\left(1 - \frac{3}{x^2} + \frac{1}{x^3}\right)$$

Now, when $|x|$ is very large, both $3/x^2$ and $1/x^3$ are close to zero. So we have

$$y \approx x^3(1 - 0 + 0)$$
$$\approx x^3 \qquad \text{when } |x| \text{ is very large}$$

The same technique that we've just used in analyzing $y = x^3 - 3x + 1$ can be applied to any (nonconstant) polynomial function. The result is summarized in item 3 in the box that follows.

PROPERTY SUMMARY

GRAPHS OF POLYNOMIAL FUNCTIONS

1. The graph of a polynomial function of degree 2 or greater is an unbroken smooth curve. (For degrees 1 and 0, the graph is a line.)

2. The graph of a polynomial function of degree n has at most $n - 1$ turning points.

3. For the graph of any polynomial function (other than a constant function), as $|x|$ gets very large, $|y|$ grows very large. If

$$f(x) = a_n x^n + a_{n-1} x^{n-1} + \cdots + a_1 x + a_0 \qquad (a_n \neq 0)$$

then

$$f(x) \approx a_n x^n \quad \text{when } |x| \text{ is very large.}$$

EXAMPLE 1   A function f is defined by

$$f(x) = -x^3 + x^2 + 9x + 9$$

which of the graphs in Figure 5 might represent this function?

FIGURE 5

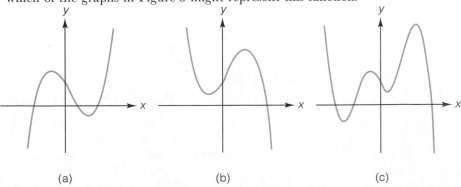

(a)   (b)   (c)

Solution   When $|x|$ is very large, $f(x) \approx -x^3$. This rules out the graph in Figure 5(a). The graph in Figure 5(c) can also be ruled out, but for a different reason. That graph has four turning points, whereas the graph of the cubic function f can have at most two turning points. The graph in Figure 5(b), on the other hand, does have two turning points; furthermore, that graph does behave like $y = -x^3$ when $|x|$ is very large. The graph in Figure 5(b) might be (in fact, it is) the graph of the given function f.   ▲

The simplest polynomial functions to graph are those of the form $y = x^n$. From our work in Chapter 2, we already know about the graphs of $y = x$, $y = x^2$, and $y = x^3$. For reference, these are shown in Figure 6.

FIGURE 6

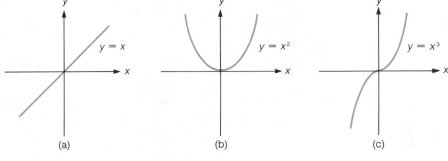

(a)   (b)   (c)

The graph of $y = x^n$, when n is greater than 3, resembles the graph of $y = x^2$ or $y = x^3$, depending on whether n is even or odd. Consider, for instance, the graph of $y = x^4$, shown in Figure 7 along with the graph of $y = x^2$. Just as with $y = x^2$, the graph of $y = x^4$ is a symmetric, U-shaped curve passing through the three points $(-1, 1)$, $(1, 1)$, and $(0, 0)$. However, in the interval $(-1, 1)$, the graph of $y = x^4$ is flatter than that of $y = x^2$. Similarly, the graph of $y = x^6$ in this interval would be flatter still. The data in Table 1 show why this is true. Figure 8 displays the graphs of $y = x^2$, $y = x^4$, and $y = x^6$ for the interval $0 \le x \le 1$.

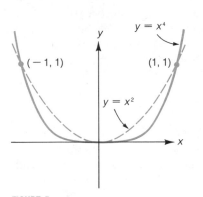

FIGURE 7

TABLE 1

x	0.2	0.4	0.6	0.8	1.0
x^2	0.04	0.16	0.36	0.64	1.0
x^4	0.0016	0.0256	0.1296	0.4096	1.0
x^6	0.000064	0.004096	0.046656	0.262144	1.0

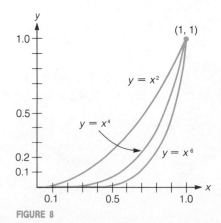

FIGURE 8

Incidentally, Figure 8 indicates one of the practical difficulties you may encounter in trying to draw an accurate graph of $y = x^n$. Suppose, for instance, that you want to graph $y = x^6$, and the lines you draw are 0.01 cm thick. Also, suppose that you use the same scale on both axes, taking the common unit to be 1 cm. Then, in the first quadrant, your graph of $y = x^6$ will be indistinguishable from the x-axis when $x^6 < 0.01$, or $x < \sqrt[6]{0.01} \approx 0.46$ cm (using a calculator). This explains why sections of the graphs in Figure 8 appear horizontal.

For $|x| > 1$, the graph of $y = x^4$ rises more rapidly than that of $y = x^2$. Similarly, the graph of $y = x^6$ rises still more rapidly. This is shown in Figures 9(a) and (b). (Note the different scales used on the y-axes in the two figures.)

As mentioned earlier, when n is odd, the graph of $y = x^n$ resembles that of $y = x^3$. In Figure 10 we compare the graphs of $y = x^3$ and $y = x^5$. Notice that both curves pass through $(0, 0)$, $(1, 1)$, and $(-1, -1)$. For reasons similar to those explained

FIGURE 9

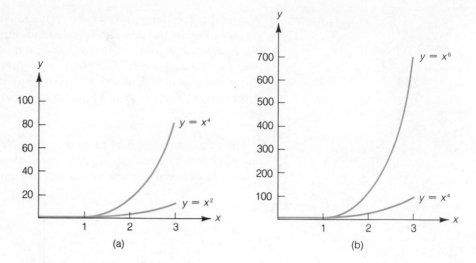

(a)

(b)

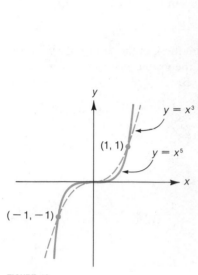

FIGURE 10

for even n, the graph of $y = x^5$ is flatter than that of $y = x^3$ in the interval $-1 < x < 1$, and the graph of $y = x^7$ would be flatter still. For $|x| > 1$, the graph of $y = x^5$ is steeper than that of $y = x^3$, and $y = x^7$ would be steeper still.

In Section 3.2, we observed the effect of the constant a on the graph of $y = ax^2$. Those same comments apply to the graph of $y = ax^n$. For instance, the graph of $y = \frac{1}{2}x^4$ is wider than that of $y = x^4$, while the graph of $y = -\frac{1}{2}x^4$ is obtained by reflecting $y = \frac{1}{2}x^4$ in the x-axis.

EXAMPLE 2   Sketch the graph of $y = (x + 2)^5$ and specify the y-intercept.

Solution   The graph of $y = (x + 2)^5$ is obtained by moving the graph of $y = x^5$ two units to the left. As a guide to drawing the curve, we recall that $y = x^5$ passes through the points $(1, 1)$ and $(-1, -1)$. Thus, $y = (x + 2)^5$ must pass through $(-1, 1)$ and $(-3, -1)$, as shown in Figure 11. Although the curve rises and falls very sharply, it is important to realize that it is never really vertical. For instance, the curve eventually crosses the y-axis. To find the y-intercept, we set $x = 0$ to obtain $y = 2^5 = 32$. Thus, the y-intercept is 32.

FIGURE 11

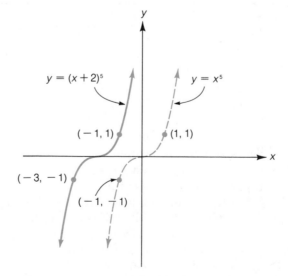

FIGURE 12

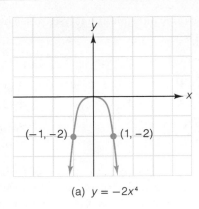

(a) $y = -2x^4$

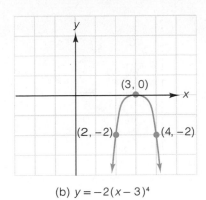

(b) $y = -2(x - 3)^4$

EXAMPLE 3   Graph the function $y = -2(x - 3)^4$.

Solution   We begin with the graph of $y = -2x^4$ in Figure 12(a). The points $(1, -2)$ and $(-1, -2)$ are obtained by substituting $x = 1$ and $x = -1$, respectively, in the equation $y = -2x^4$. Now if we replace x with $x - 3$ in the equation $y = -2x^4$, we have $y = -2(x - 3)^4$, whose graph is obtained by translating the graph in Figure 12(a) three units to the right. See Figure 12(b).   ▲

In general, the most efficient ways to graph a given polynomial function involve methods of calculus. However, if the polynomial can be written as a product of linear factors with real coefficients, then a fairly accurate graph can be obtained without the use of calculus. For example, let us graph the function

$$f(x) = (x + 2)(x - 1)(x - 3)$$

By inspection, we see that $f(x) = 0$ when $x = -2$, $x = 1$, and $x = 3$. These are the x-intercepts of the graph. Also, note that the y-intercept is 6. (Why?) Next, we want to find out what the graph looks like in the immediate vicinity of each x-intercept. First, suppose that x is very close to -2. We have

$$f(x) = (x + 2)\underbrace{(x - 1)}_{}\underbrace{(x - 3)}_{}$$

If x is close to -2, this is close to -3.   If x is close to -2, this is close to -5.

Thus, if x is very close to -2, we have

$$f(x) \approx (x + 2)(-3)(-5)$$
$$\approx 15(x + 2)$$
$$\approx 15x + 30$$

(Notice the technique used to obtain this approximation: We retained the factor corresponding to the intercept -2 and we estimated the remaining factors.) We conclude from this that when x is very close to -2, the graph of f resembles the line $y = 15x + 30$. In particular, the graph is extremely steep and rises to the right (see Figure 13).

FIGURE 13

Near $x = -2$, the graph of $f(x) = (x + 2)(x - 1)(x - 3)$ is nearly vertical; it resembles the line $y = 15x + 30$.

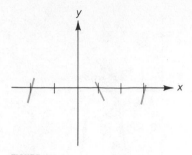

FIGURE 14
The graph of f near its x-intercepts.

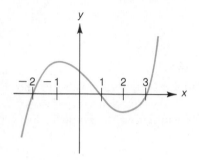

FIGURE 15
A rough sketch of
$f(x) = (x + 2)(x - 1)(x - 3)$.

FIGURE 16
$f(x) = (x + 2)(x - 1)(x - 3)$.

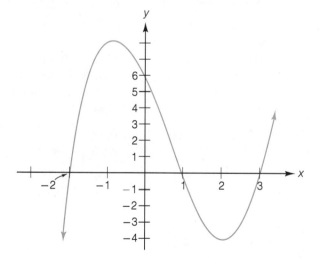

Next, we carry out similar analyses to see how the graph of f looks near its other x-intercepts:

x near 1

$$f(x) = (x + 2)(x - 1)(x - 3)$$
$$\approx (1 + 2)(x - 1)(1 - 3)$$
$$\approx 3(x - 1)(-2)$$
$$\approx -6x + 6$$

x near 3

$$f(x) = (x + 2)(x - 1)(x - 3)$$
$$\approx (3 + 2)(3 - 1)(x - 3)$$
$$\approx 5(2)(x - 3)$$
$$\approx 10x - 30$$

Thus, for x very close to 1, the graph of f resembles the line $y = -6x + 6$, while for x very close to 3, the graph resembles the line $y = 10x - 30$. Figure 14 summarizes our information about the behavior of the graph of f near its x-intercepts. With this as a guide, we can then draw a rough sketch of the graph of f, as shown in Figure 15.

Notice that to draw a smooth curve satisfying the conditions of Figure 14, we need at least two turning points: one between $x = -2$ and $x = 1$, and one between $x = 1$ and $x = 3$. On the other hand, since the degree of $f(x)$ is 3, there can be no more than two turning points. This is why Figure 15 shows exactly two turning points. While the precise location of the turning points is a matter for calculus, we can nevertheless improve upon the sketch in Figure 15 by computing $f(x)$ for several specific values of x. Some reasonable choices here are $x = -1$ and $x = 2$. (We've already noted that when $x = 0$, we have $y = 6$.) As you can check, the resulting calculations show that $(-1, 8)$ and $(2, -4)$ are on the graph. We can now sketch the graph as shown in Figure 16.

The next example further illustrates our technique for graphing those polynomial functions that can be expressed as a product of linear factors with real coefficients.

EXAMPLE 4   Sketch the graph of $y = x^3(x + 1)(x - 2)$.

Solution   From the given equation, we see that the y-intercept is 0, and the x-intercepts are $x = 0$, $x = -1$, and $x = 2$. Now, just as in the previous example, we want to see how the graph looks in the immediate vicinity of the x-intercepts.

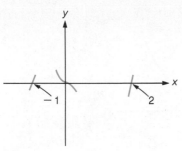

FIGURE 17

The graph of $y = x^3(x + 1)(x - 2)$ near its x-intercepts.

x near 0

$y \approx x^3(0 + 1)(0 - 2)$

$\approx -2x^3$

x near −1

$y \approx (-1)^3(x + 1)(-1 - 2)$

$\approx 3(x + 1)$

$\approx 3x + 3$

x near 2

$y \approx 2^3(2 + 1)(x - 2)$

$\approx 24(x - 2)$

$\approx 24x - 48$

These results are displayed in Figure 17. With this as a guide, we can draw a rough sketch of the required graph, as shown in Figure 18. To obtain a slightly more accurate sketch, we compute the y-values corresponding to $x = -\frac{1}{2}$ and $x = 1$. As you can check, the curve passes through the points $(-\frac{1}{2}, \frac{5}{32})$ and $(1, -2)$. This additional information is used to draw the graph shown in Figure 19.

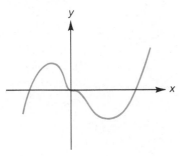

FIGURE 18

A rough sketch of $y = x^3(x + 1)(x - 2)$.

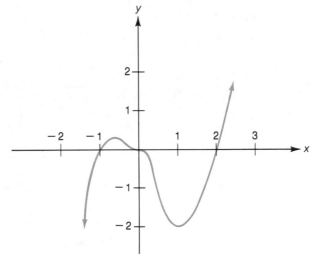

FIGURE 19

$y = x^3(x + 1)(x - 2)$

▼ **EXERCISE SET 3.5**

A

In Exercises 1 1, give a reason (as in Example 1) why each graph cannot represent a polynomial function of degree 3.

1.

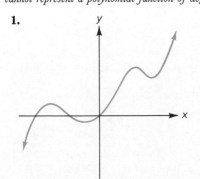

2.

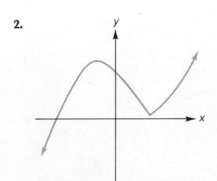

3.

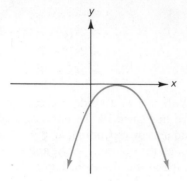

7.

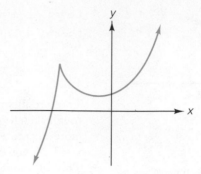

4.

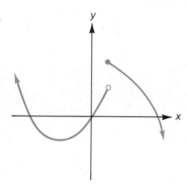

8.

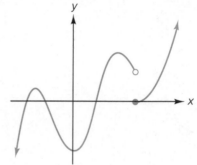

In Exercises 9–20, sketch the graph of each function and specify all x- and y-intercepts.

9. $y = (x - 2)^2 + 1$ **10.** $y = -3x^4$

11. $y = -(x - 1)^4$ **12.** $y = -(x + 2)^3$

13. $y = (x - 4)^3 - 2$ **14.** $y = -(x - 4)^3 - 2$

15. $y = -2(x + 5)^4$ **16.** $y = -2x^4 + 5$

17. $y = \frac{1}{2}(x + 1)^5$ **18.** $y = \frac{1}{2}x^5 + 1$

19. $y = -(x - 1)^3 - 1$ **20.** $y = x^8$

In Exercises 5–8, give a reason why each graph cannot represent a polynomial function that has highest-degree term $2x^5$.

5.

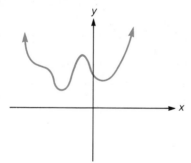

In Exercises 21–34, first sketch the graph in the immediate vicinity of each of the x-intercepts, then sketch the graph.

21. $y = (x - 2)(x - 1)(x + 1)$

22. $y = (x - 3)(x + 2)(x + 1)$

23. $y = 2x(x + 1)(x + 3)$

24. $y = -x^2(x + 2)$

25. $y = x^3(x + 2)$

26. $y = (x - 1)(x - 4)^2$

27. $y = 2(x - 1)(x - 4)^3$

28. $y = (x - 1)^2(x - 4)^2$

29. $y = (x + 1)^2(x - 1)(x - 3)$

30. $y = x^2(x - 4)(x + 2)$

31. $y = -x^3(x - 4)(x + 2)$

32. $y = 4(x - 2)^2(x + 2)^3$

33. $y = -4x(x - 2)^2(x + 2)^3$

34. $y = -3x^3(x + 1)^4$

6.

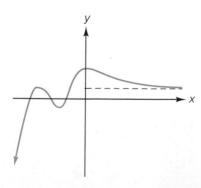

B

In Exercises 35–40, six functions are defined as follows:

$$f(x) = x \qquad g(x) = x^2 \qquad h(x) = x^3$$
$$F(x) = x^4 \qquad G(x) = x^5 \qquad H(x) = x^6$$

Refer also to the following figure.

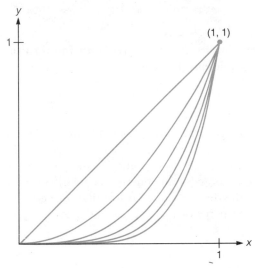

35. The six graphs in the figure are the graphs of the six given functions for the interval [0, 1], but the graphs are not labeled. Which is which?

36. For which x-values in [0, 1] will the graph of g lie strictly below the horizontal line $y = 0.1$? Use a calculator to evaluate your answer. Round off the result to two significant figures. [C]

37. Follow Exercise 36, using the function H instead of g. [C]

38. Find a number t in [0, 1] such that the vertical distance between $f(t)$ and $g(t)$ is $\frac{1}{4}$.

39. Is there a number t in [0, 1] such that the vertical distance between $g(t)$ and $F(t)$ is 0.26?

40. Find all numbers t in [0, 1] such that $F(t) = G(t) + H(t)$.

41. Do the graphs of $y = x$ and $y = \frac{1}{100}x^2$ intersect anywhere other than at the origin? (See figure.)

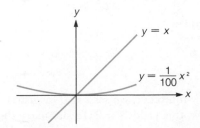

In Exercises 42–47, graph each function. (Use the factoring techniques from Section 1.5 to factor the right-hand side of each equation.)

42. $y = x^3 - 2x^2$

43. $y = 2x^3 - x^4$

44. $y = x^3 - 9x$

45. $y = x^3 + x^2 - 2x$

46. $y = x^3 + 3x^2 + 3x + 1$

47. $y = 4x^2 - x^4$

48. Find the coordinates of the turning points for the graph in Exercise 47. *Hint:* See Example 7 in Section 3.2.

49. **(a)** Graph the function $D(x) = x^2 - x^4$.
   (b) Find the turning points of the graph. *Hint:* See Example 7 in Section 3.2.
   (c) On the same set of axes, sketch the graphs of $y = x^2$ and $y = x^4$ for $0 \le x \le 1$. What is the maximum vertical distance between the graphs?

50. **(a)** An open-top box is to be constructed from a 6-in.-by-8-in. rectangular sheet of tin by cutting out equal squares at each corner and then folding up the resulting flaps. Let x denote the length of the side of each cut-out square. Show that the volume $V(x)$ is

$$V(x) = x(6 - 2x)(8 - 2x)$$

   (b) What is the domain of the volume function in part (a)? [The answer is *not* $(-\infty, \infty)$.]
   (c) Graph the volume function.
   (d) Use your graph to estimate the maximum possible volume for the box.

51. A cylinder is inscribed in a sphere of radius 6 cm. Let r and h denote the radius and height of the cylinder, as shown in the figure. We are going to estimate the maximum possible volume for the cylinder.

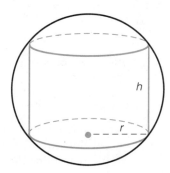

   (a) Show that $h = 2\sqrt{36 - r^2}$.
   (b) Let V denote the volume of the cylinder. Check that

$$V = 2\pi r^2 \sqrt{36 - r^2}$$

   (c) What is the domain of the volume function in part (b)?
   (d) The expression for V in part (b) is not a polynomial, but the square of the expression is. So, we'll work with the function $f(r) = 4\pi^2 r^4(36 - r^2)$. At the end, we can convert back to V by taking the square root of $f(r)$. Graph the function $f(r) = 4\pi^2 r^4(36 - r^2)$. (To fine-tune your graph, use a calculator to compute $f(r)$ for r running from 0 to 6 in increments of 0.5.) From the graph, estimate the maximum value of $f(r)$. Now estimate the maximum possible volume of the cylinder.

3.6 ▼ GRAPHS OF RATIONAL FUNCTIONS

We shall not attempt to explain the numerous situations in life sciences where the study of such graphs is important nor the chemical reactions which give rise to the rational functions $r(x)$ .... Let it suffice to mention that, to the experimental biochemist, theoretical results concerning the shapes of rational functions are of considerable interest

W. G. Bardsley and R. M. W. Wood in the article, "Critical Points and Sigmoidicity of Positive Rational Functions," in *The American Mathematical Monthly*, **92** (1985) 37–42.

After the polynomial functions, the next simplest functions are the **rational functions**. These are functions defined by equations of the form

$$y = \frac{f(x)}{g(x)}$$

where $f(x)$ and $g(x)$ are polynomials. In general, throughout this section, when we write a function such as $y = f(x)/g(x)$, we assume that $f(x)$ and $g(x)$ contain no common factors (other than constants). [Exercises 36 and 37 ask you to consider several cases in which $f(x)$ and $g(x)$ do contain common factors.] Also, for each of the examples that we discuss, the degree of $f(x)$ is less than or equal to the degree of $g(x)$. [A case in which the degree of $f(x)$ exceeds the degree of $g(x)$ is developed in the exercises.]

EXAMPLE 1　Specify the domain of the rational function defined by

$$y = \frac{3x - 2}{x^2 - 1}$$

Also, find the x- and y-intercepts (if any) for the graph of this function.

Solution　By factoring the denominator, we can rewrite the given equation

$$y = \frac{3x - 2}{(x - 1)(x + 1)}$$

Since the denominator is zero when $x = 1$ and when $x = -1$, it follows that the domain of the given function consists of all real numbers except 1 and -1. To determine the x-intercepts of the graph, we set $y = 0$ (as usual) in the given equation to obtain

$$\frac{3x - 2}{x^2 - 1} = 0 \qquad (x \neq 1, x \neq -1)$$

$$3x - 2 = 0 \quad \text{multiplying both sides of the equation by the nonzero quantity } x^2 - 1, \text{ since } x \neq \pm 1$$

$$x = \frac{2}{3}$$

Thus, the only x-intercept is $x = \frac{2}{3}$. For the y-intercept, we set $x = 0$ (as usual) in the given equation. As you can readily check, this yields $y = 2$. In summary then, the x- and y-intercepts are $\frac{2}{3}$ and 2, respectively.　▲

In the box that follows, we summarize the ideas used in Example 1 for determining the domain and x-intercepts of a rational function.

Let R be a rational function defined by

$$R(x) = \frac{f(x)}{g(x)}$$

where $f(x)$ and $g(x)$ are polynomials with no common factors (other than constants). Then the domain of R consists of all real numbers for which $g(x) \neq 0$; the x-intercepts (if any) are the real solutions of the equation $f(x) = 0$.

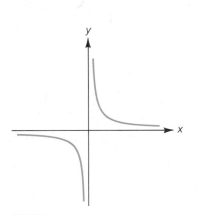

FIGURE 1

The graph of $y = \dfrac{1}{x}$

You already know about the graph of one rational function, because in Chapter 2 we graphed $y = 1/x$. For convenience, the graph is shown again in Figure 1. The graph in Figure 1 differs from the graph of every polynomial function in two important aspects. First, the graph has a break in it; it is composed of two distinct pieces or **branches**. (Recall from Section 3.5 that the graph of a polynomial function never has a break in it.) In general, the graph of a rational function has one more branch than the number of distinct real values for which the denominator is zero. For instance, referring to the function in Example 1, we can expect the graph of $y = \dfrac{3x - 2}{x^2 - 1}$ to have three branches.

The second way that the graph in Figure 1 differs from that of a polynomial function has to do with the *asymptotes*. Recall that a line is an **asymptote** for a curve if the distance between the line and the curve approaches zero as we move out farther and farther along the line. Thus, the x-axis is a horizontal asymptote for the graph in Figure 1, while the y-axis is a vertical asymptote. It can be shown that the graph of a polynomial function never has an asymptote. Figure 2 displays additional examples of curves with asymptotes.

FIGURE 2

Curves with asymptotes

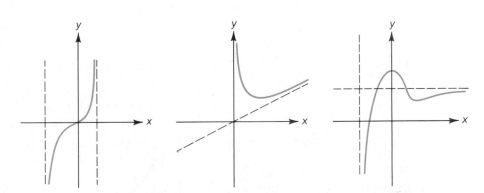

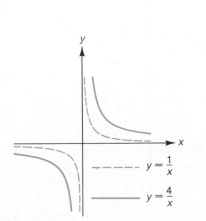

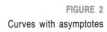

$y = \dfrac{1}{x}$

$y = \dfrac{4}{x}$

FIGURE 3

If k is a positive constant, the graph of $y = k/x$ resembles that of $y = 1/x$. Figure 3 shows the graphs of $y = 1/x$ and $y = 4/x$. Once we know about the graph of $y = k/x$, we can graph any rational function of the form

$$y = \frac{ax + b}{cx + d}$$

The next three examples show how this is done. These examples make use of the techniques of translation and reflection developed in Chapter 2.

EXAMPLE 2 Graph $y = \dfrac{1}{x-1}$.

Solution By translating the graph of $y = 1/x$ one unit to the right, we obtain the graph of $y = 1/(x-1)$ shown in Figure 4. Notice that the horizontal asymptote is unchanged by the translation. However, the vertical asymptote moves one unit to the right. (The y-intercept is -1; why?)

FIGURE 4
$y = \dfrac{1}{x-1}$

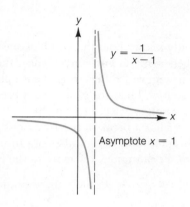

$y = \dfrac{1}{x-1}$

Asymptote $x = 1$

EXAMPLE 3 Graph $y = \dfrac{2}{x-1}$ and $y = \dfrac{-2}{x-1}$.

Solution The graph of $y = 2/(x-1)$ has the same basic shape and location as the graph of $y = 1/(x-1)$ shown in Figure 4. As a further guide to sketching $y = 2/(x-1)$, we can pick several convenient x-values on either side of the asymptote $x = 1$ and then compute the corresponding y-values. After doing this, we obtain the graph shown in Figure 5(a). By reflecting this graph about the x-axis, we obtain the graph of $y = -2/(x-1)$, which is shown in Figure 5(b).

FIGURE 5

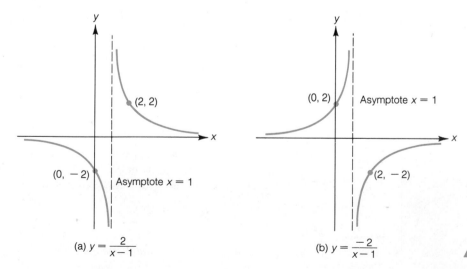

(a) $y = \dfrac{2}{x-1}$

(b) $y = \dfrac{-2}{x-1}$

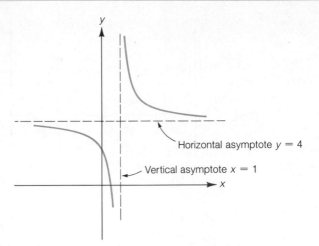

FIGURE 6

$$y = \frac{4x - 2}{x - 1}$$

Horizontal asymptote $y = 4$

Vertical asymptote $x = 1$

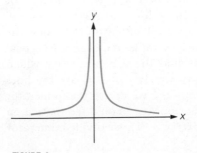

(a)

(b)

FIGURE 7

FIGURE 8

$$y = \frac{1}{x^2}$$

EXAMPLE 4 Graph $y = \dfrac{4x - 2}{x - 1}$.

Solution First, as you can readily check, the x- and y-intercepts are $\frac{1}{2}$ and 2, respectively. Next, using long division, we find that

$$\frac{4x - 2}{x - 1} = 4 + \frac{2}{x - 1}$$

We conclude that the required graph can be obtained by moving up the graph of $y = 2/(x - 1)$ four units in the y-direction (see Figure 6). Notice that the vertical asymptote is still $x = 1$, but the horizontal asymptote is now $y = 4$ instead of the x-axis. ▲

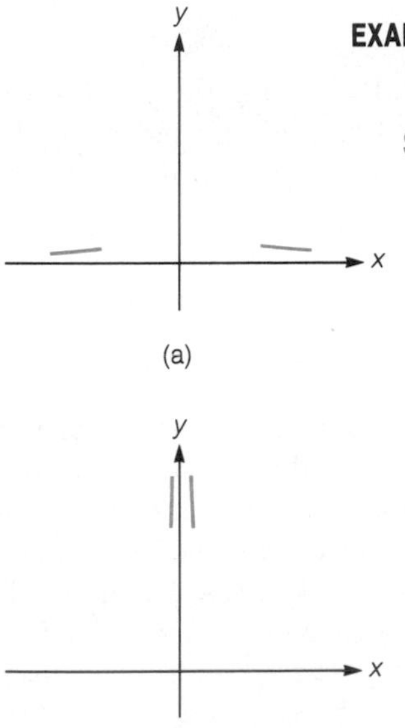

Now let us look at rational functions of the form $y = 1/x^n$. First we'll consider $y = 1/x^2$. As with $y = 1/x$, the domain consists of all real numbers except $x = 0$. For $x \neq 0$, the quantity $1/x^2$ is always positive. This means that the graph will always lie above the x-axis. As $|x|$ becomes very large, the quantity $1/x^2$ approaches zero; this is true whether x itself is positive or negative. So when $|x|$ is very large, we expect the graph of $y = 1/x^2$ to look as shown in Figure 7(a). On the other hand, when x is a very small fraction close to zero, either negative or positive, the quantity $1/x^2$ is very large. For instance, if $x = \frac{1}{10}$, we find that $y = 100$, and if $x = \frac{1}{100}$, we find that $y = 10,000$. Thus, as x approaches zero, from the right or from the left, the graph of $y = 1/x^2$ must look as shown in Figure 7(b). Now, by plotting several points and taking Figures 7(a) and (b) into account, we obtain the graph of $y = 1/x^2$, shown in Figure 8. Also, by following a similar line of reasoning, we find that the graph of $y = 1/x^3$ looks as shown in Figure 9. Note that $y = 1/x^2$ is symmetric about the y-axis, and $y = 1/x^3$ is symmetric about the origin.

In general, when n is an even integer greater than 2, the graph of $y = 1/x^n$ resembles that of $y = 1/x^2$. When n is an odd integer greater than 3, the graph of $y = 1/x^n$ resembles that of $y = 1/x^3$.

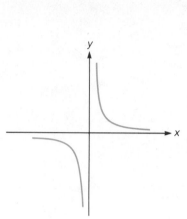

FIGURE 9
$y = \dfrac{1}{x^3}$

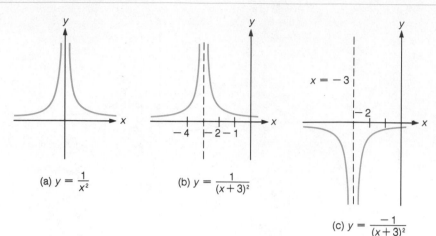

(a) $y = \dfrac{1}{x^2}$

(b) $y = \dfrac{1}{(x+3)^2}$

(c) $y = \dfrac{-1}{(x+3)^2}$

FIGURE 10

EXAMPLE 5   Graph $y = \dfrac{-1}{(x+3)^2}$.

Solution   Refer to Figure 10. Begin with the graph of $y = 1/x^2$. By moving the graph three units to the left, we obtain the graph of $y = 1/(x+3)^2$. Then by reflecting the graph of $y = 1/(x+3)^2$ about the x-axis, we get the graph of $y = -1/(x+3)^2$.

Question   What are the y-intercepts in Figures 10(b) and (c)? ▲

EXAMPLE 6   Graph $y = \dfrac{4}{(x-2)^3}$.

Solution   Moving the graph of $y = 1/x^3$ two units to the right gives us the graph of $y = 1/(x-2)^3$. The graph of $y = 4/(x-2)^3$ will have the same basic shape and location. As a further guide to sketching the required graph, we can pick several convenient x-values near the asymptote $x = 2$ and compute the corresponding y-values. Using $x = 0$, $x = 1$, and $x = 3$, we find that the points $(0, -\frac{1}{2})$, $(1, -4)$, and $(3, 4)$ are on the graph. With this information, the graph can be sketched as in Figure 11. ▲

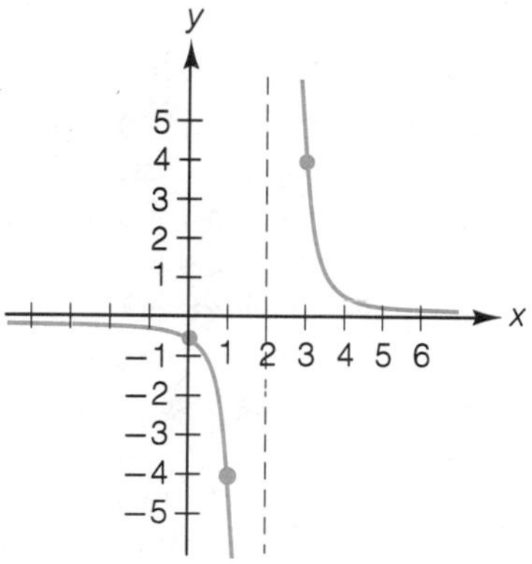

FIGURE 11
$y = \dfrac{4}{(x-2)^3}$

In Section 3.5, you learned a technique for graphing those polynomial functions that can be expressed as a product of linear factors. That technique can also be used to graph rational functions in which both the numerator and denominator can be expressed as products of linear factors. In addition to investigating the behavior of the function near its x-intercepts, now we will also want to see how the function looks near its asymptotes. As the previous examples indicate, the vertical asymptotes of $y = f(x)/g(x)$ are known once we find the values of x for which $g(x) = 0$. For instance, the vertical asymptotes of $y = 3x^2/(x^2 - 1)$ are the lines $x = 1$ and $x = -1$. To find the horizontal asymptotes, we divide the numerator and denominator by the highest power of x that appears. (You'll see the reason for doing this in a moment.) In the case of $y = 3x^2/(x^2 - 1)$, we divide numerator and denominator by x^2 to obtain

$$y = \dfrac{3}{1 - \dfrac{1}{x^2}}$$

Now, when $|x|$ grows very large, $1/x^2$ approaches zero. We therefore have

$$y \approx \frac{3}{1-0} \qquad \text{when } |x| \text{ is very large}$$

This tells us that the line $y = 3$ is a horizontal asymptote for the graph of $y = 3x^2/(x^2 - 1)$. You'll see another example of this type of calculation in Example 7.

EXAMPLE 7   Graph the function $y = \dfrac{(x-3)(x+2)}{(x+1)(x-2)}$.

Solution   By inspection, we see that the x-intercepts are $x = 3$ and $x = -2$, the y-intercept is 3, and the vertical asymptotes are the lines $x = -1$ and $x = 2$. To find the horizontal asymptote, we write

$$y = \frac{x^2 - x - 6}{x^2 - x - 2}$$

$$= \frac{1 - \dfrac{1}{x} - \dfrac{6}{x^2}}{1 - \dfrac{1}{x} - \dfrac{2}{x^2}} \approx \frac{1 - 0 - 0}{1 - 0 - 0} \qquad \text{when } |x| \text{ is large}$$

The horizontal asymptote is therefore the line $y = 1$.

Now we want to see how the graph looks in the immediate vicinity of the x-intercepts and the vertical asymptotes. To do this, we use the approximation technique explained in the previous section for polynomial functions. Let's start with the x-intercept $x = 3$. We have, for x near 3,

$$y = \frac{(x-3)(x+2)}{(x+1)(x-2)} \approx \frac{(x-3)(3+2)}{(3+1)(3-2)} = \frac{5}{4}(x-3)$$

So, in the immediate vicinity of the x-intercept $x = 3$, the required graph will closely resemble the graph of the line $y = \frac{5}{4}x - \frac{15}{4}$. The remaining calculations for approximating the graph near the other x-intercept and near the two vertical asymptotes are carried out in a similar manner. As Exercise 35 will ask you to verify, the results are as follows:

$$\begin{array}{ll} x \text{ near } 3: & y \approx \dfrac{5}{4}x - \dfrac{15}{4} \\[3mm] x \text{ near } -2: & y \approx -\dfrac{5}{4}x - \dfrac{5}{2} \\[3mm] x \text{ near } -1: & y \approx \dfrac{\frac{4}{3}}{x+1} \\[3mm] x \text{ near } 2: & y \approx \dfrac{-\frac{4}{3}}{x-2} \end{array}$$

We can summarize these results as follows. As the graph passes through the points $(3, 0)$ and $(-2, 0)$, it resembles the lines $y = \frac{5}{4}x - \frac{15}{4}$ and $y = -\frac{5}{4}x - \frac{5}{2}$, respectively. Near the vertical asymptote $x = -1$, the graph has the same basic shape as $y = 1/(x + 1)$. Near the vertical asymptote $x = 2$, the graph has the same basic

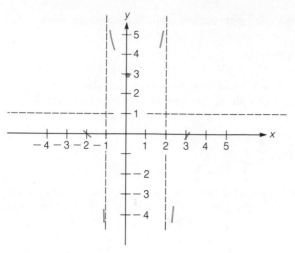

FIGURE 12

The graph of $y = \dfrac{(x-3)(x+2)}{(x+1)(x-2)}$ near its

x-intercepts and vertical asymptotes.

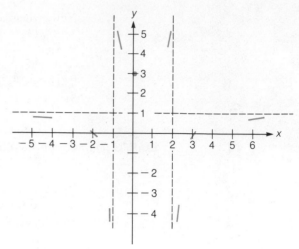

FIGURE 13

The graph of $y = \dfrac{(x-3)(x+2)}{(x+1)(x-2)}$ near its

x-intercepts and horizontal and vertical asymptotes.

shape as $y = -1/(x-2)$ (see Figure 12). Next we need to see how the graph approaches the horizontal asymptote $y = 1$. When $|x|$ is large and $x > 0$, we find that y is less than 1. (Check this for yourself using $x = 10$ or $x = 100$.) This means that to the right of the origin, the graph approaches the asymptote $y = 1$ from below. Similarly, when $|x|$ is large and $x < 0$, we find that y is again less than 1. This means that to the left of the origin, the graph approaches the asymptote $y = 1$ from below. In Figure 13, we summarize what we have discovered up to this point about the graph. Then, using Figure 13 as a guide, we can sketch the required graph as shown in Figure 14.

FIGURE 14

$y = \dfrac{(x-3)(x+2)}{(x+1)(x-2)}$

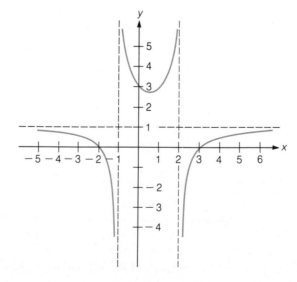

Note A general method for finding the coordinates of the lowest point on the middle branch of the graph is studied in calculus. Those coordinates can also be found, however, by applying algebraic techniques. (See Exercise 38.) ▲

▼ EXERCISE SET 3.6

A

In Exercises 1–6, find the domain and the x- and y-intercepts for each rational function.

1. $y = \dfrac{3x + 15}{4x - 12}$

2. $y = \dfrac{(x + 6)(x + 4)}{(x - 1)^2}$

3. $y = \dfrac{x^2 - 8x - 9}{x^2 - x - 6}$

4. $y = \dfrac{2x^2 + x - 5}{x^2 + 1}$

5. $y = \dfrac{(x^2 - 4)(x^3 - 1)}{x^6}$

6. $y = \dfrac{x^5 - 2x^4 - 9x + 18}{8x^3 + 2x^2 - 3x}$

In Exercises 7–28, sketch the graph of each rational function. Specify the intercepts and the asymptotes.

7. $y = \dfrac{1}{x + 4}$

8. $y = \dfrac{-1}{x + 4}$

9. $y = \dfrac{3}{x + 2}$

10. $y = \dfrac{-3}{x + 2}$

11. $y = \dfrac{-2}{x - 3}$

12. $y = \dfrac{x - 1}{x + 1}$

13. $y = \dfrac{x - 3}{x - 1}$

14. $y = \dfrac{2x}{x + 3}$

15. $y = \dfrac{4x - 2}{2x + 1}$

16. $y = \dfrac{3x + 2}{x - 3}$

17. $y = \dfrac{1}{(x - 2)^2}$

18. $y = \dfrac{-1}{(x - 2)^2}$

19. $y = \dfrac{3}{(x + 1)^2}$

20. $y = \dfrac{-3}{(x + 1)^2}$

21. $y = \dfrac{1}{(x + 2)^3}$

22. $y = \dfrac{-1}{(x + 2)^3}$

23. $y = \dfrac{-4}{(x + 5)^3}$

24. $y = \dfrac{x}{(x + 1)(x - 3)}$

25. $y = \dfrac{-x}{(x + 2)(x - 2)}$

26. $y = \dfrac{2x}{(x + 1)^2}$

27. $y = \dfrac{x}{(x - 1)(x + 3)}$

28. $y = \dfrac{x^2 + 1}{x^2 - 1}$

29. (a) $f(x) = \dfrac{(x - 2)(x - 4)}{x(x - 1)}$

   (b) $g(x) = \dfrac{(x - 2)(x - 4)}{x(x - 3)}$

[Compare the graphs you obtain in parts (a) and (b). Notice how a change in only one constant can radically alter the nature of the graph.]

30. (a) $f(x) = \dfrac{(x - 1)(x + 2.75)}{(x + 1)(x + 3)}$

   (b) $g(x) = \dfrac{(x - 1)(x + 3.25)}{(x + 1)(x + 3)}$

[Compare the graphs you obtain in parts (a) and (b). Notice how a relatively small change in one of the constants can radically alter the graph.]

B

In Exercises 31–34, graph the functions. *Note: Each graph intersects its horizontal asymptote once. Find the intersection point before sketching your final graph.*

31. $y = \dfrac{(x - 4)(x + 2)}{(x - 1)(x - 3)}$

32. $y = \dfrac{(x - 1)(x - 3)}{(x + 1)^2}$

33. $y = \dfrac{(x + 1)^2}{(x - 1)(x - 3)}$

34. $y = \dfrac{2x^2 - 3x - 2}{x^2 - 3x - 4}$

35. (This exercise refers to Example 7.) Let

$$y = \dfrac{(x - 3)(x + 2)}{(x + 1)(x - 2)}$$

Verify each of the following approximations.

(a) When x is close to -2, then $y \approx -\frac{5}{4}x - \frac{5}{2}$.

(b) When x is close to -1, then $y \approx \dfrac{\frac{4}{3}}{x + 1}$.

(c) When x is close to 2, then $y \approx \dfrac{-\frac{4}{3}}{x - 2}$.

In Exercises 36 and 37, graph the functions. Notice in each case that the numerator and denominator contain at least one common factor. Thus, you can simplify each quotient; but don't lose track of the domain of the function as it was initially defined.

36. (a) $y = \dfrac{x + 2}{x + 2}$

   (b) $y = \dfrac{x^2 - 4}{x - 2}$

   (c) $y = \dfrac{x - 1}{(x - 1)(x - 2)}$

37. (a) $y = \dfrac{x^2 - 9}{x + 3}$

   (b) $y = \dfrac{x^2 - 5x + 6}{x^2 - 2x - 3}$

   (c) $y = \dfrac{(x - 1)(x - 2)(x - 3)}{(x - 1)(x - 2)(x - 3)(x - 4)}$

38. This exercise shows you how to determine the coordinates of the lowest point on the middle branch of the curve in Figure 14. The basic idea is as follows. Suppose that the required coordinates are (h, k). Then the horizontal line $y = k$ is the unique horizontal line intersecting the curve in one and only one point. (Any other horizontal line intersects the curve either in two points or not at all.) In steps (a) through (c) that follow, we use these observations to determine the point (h, k).

(a) Given any horizontal line $y = k$, its intersection with the curve in Figure 14 is determined by solving the following pair of simultaneous equations:

$$\begin{cases} y = k \\ y = \dfrac{x^2 - x - 6}{x^2 - x - 2} \end{cases}$$

In the second equation of the system, replace y with k and show that the resulting equation can be written

$$(k - 1)x^2 - (k - 1)x + (6 - 2k) = 0 \qquad (1)$$

(b) If k is indeed the required y-coordinate, then equation (1) must have exactly one real solution. Set the discriminant of the quadratic equation equal to zero to obtain

$$(k - 1)^2 - 4(k - 1)(6 - 2k) = 0$$

and deduce from this that $k = 1$ or $k = \frac{25}{9}$. The solution $k = 1$ can be discarded. (To see why, look at Figure 14.)

(c) Using the value $y = \frac{25}{9}$, show that the corresponding x-coordinate is $\frac{1}{2}$. Thus, the required point is $(\frac{1}{2}, \frac{25}{9})$.

39. Graph the function $y = x/(x - 3)^2$. Use the technique explained in Exercise 38 to find the coordinates of any turning points on the graph.

40. Graph the function $y = 2/(x - x^2)$. Use the technique explained in Exercise 38 to find the coordinates of any turning points on the graph.

An asymptote that is neither horizontal nor vertical is called a **slant**, *or* **oblique**, **asymptote**. *For example, as indicated in the following figure, the line $y = x$ is a slant asymptote for the graph of $y = (x^2 + 1)/x$. To understand why the line $y = x$ is an asymptote, we carry out the indicated division and write the function in the form*

$$y = x + \frac{1}{x} \qquad (2)$$

From equation (2), we see that if $|x|$ is very large, then $y \approx x + 0$; that is, $y \approx x$, as we wished to show. Equation (2) actually tells us more than this. When $|x|$ is very close to zero, equation (2) yields $y \approx 0 + (1/x)$. In other words, as we approach the y-axis, the curve looks more and more like the graph of $y = 1/x$.

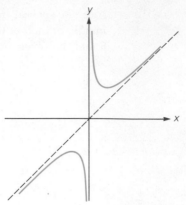

The curve $y = (x^2 + 1)/x$ with slant asymptote $y = x$

In general, if we have a rational function $f(x)/g(x)$ in which the degree of $f(x)$ is 1 greater than the degree of $g(x)$, then the graph has a slant asymptote that is obtained as follows. Divide $f(x)$ by $g(x)$ to obtain

$$\frac{f(x)}{g(x)} = (mx + b) + \frac{h(x)}{g(x)}$$

where the degree of $h(x)$ is less than the degree of $g(x)$. Then the equation of the slant asymptote is $y = mx + b$. [For instance, using the example in the previous paragraph, we have $mx + b = x$ and $h(x)/g(x) = 1/x$.] In Exercises 41–43, you are asked to graph functions that have slant asymptotes.

41. Let $y = F(x) = \dfrac{x^2 + x - 6}{x - 3}$.

(a) Use long division to show that

$$\frac{x^2 + x - 6}{x - 3} = (x + 4) + \frac{6}{x - 3}$$

(b) $\boxed{\text{C}}$ The result in part (a) shows that the line $y = x + 4$ is a slant asymptote for the graph of the function F. Verify this fact empirically by completing the following two tables.

x	$x + 4$	$\dfrac{x^2 + x - 6}{x - 3}$
10		
100		
1000		

x	$x + 4$	$\dfrac{x^2 + x - 6}{x - 3}$
-10		
-100		
-1000		

(c) Determine the vertical asymptote and the x- and y-intercepts for the graph of F.

(d) Graph the function F. (Make use of the techniques in this section along with the fact that $y = x + 4$ is a slant asymptote.)

(e) Use the technique explained in Exercise 38 to find the coordinates of the two turning points on the graph of F.

42. (a) Show that the line $y = x - 2$ is a slant asymptote for the graph of $F(x) = x^2/(x + 2)$.

(b) Sketch the graph of F.

43. Show that the line $y = -x$ is a slant asymptote for the graph of $y = (1 - x^2)/x$. Then sketch the graph of this function.

▼ **CHAPTER THREE   SUMMARY OF PRINCIPAL TERMS**

TERMS	PAGE REFERENCE	COMMENTS
1. Linear function	147	A linear function is a function of the form $f(x) = Ax + B$. The graph of a linear function is a straight line. An important idea that arose in several of the examples is that the slope of a line can be interpreted as a rate of change. Two instances of this are marginal cost and velocity.
2. Quadratic function	155	A quadratic function is a function of the form $f(x) = ax^2 + bx + c$. The graph of a quadratic function is a parabola. See the boxes on pages 156 and 158.
3. Polynomial function	180	A polynomial function of degree n is a function of the form $$f(x) = a_n x^n + a_{n-1} x^{n-1} + \cdots + a_1 x + a_0$$ where n is a nonnegative integer and $a_n \neq 0$. Three basic properties of polynomial functions are summarized in the box on page 182.
4. Rational function	190	A rational function is a function of the form $y = f(x)/g(x)$, where $f(x)$ and $g(x)$ are polynomials.
5. Asymptote	191	A line is said to be an asymptote for a curve if the distance between the line and the curve approaches zero as we move out farther and farther along the line. See, for example, Figure 2 on page 191, in which the dashed lines are asymptotes.

▼ **CHAPTER THREE   REVIEW EXERCISES**

NOTE   Exercises 1–14 constitute a chapter test on the fundamentals, based on group A problems.

1. Find $G(0)$ if G is a linear function such that $G(1) = -2$ and $G(-2) = -11$.

2. (a) Let $f(x) = 3x^2 + 6x - 10$. For which input x is the value of the function a minimum? What is that minimum value?

(b) Let $g(t) = 6t^2 - t^4$. For which input t is the value of the function a maximum?

3. Suppose the function $p = -\frac{1}{8}x + 100$ $(0 \leq x \leq 12)$ relates the selling price p of an item to the quantity x that is sold. Assume that p is in dollars. What is the maximum revenue possible in this situation?

4. Graph the function $y = -1/(x + 1)^3$.

5. Graph the function $y = (x - 4)(x - 1)(x + 1)$.

6. Graph the function $f(x) = x^2 + 4x - 5$. Specify the vertex, the x- and y-intercepts, and the axis of symmetry.

7. A factory owner buys a new machine for $1000. After five years, the machine has a salvage value of $100. Assuming linear depreciation, find a formula for the value V of the machine after t years, where $0 \leq t \leq 5$.

8. Graph the function $y = 2(x - 3)^4$. Does the graph cross the y-axis? If so, where?

9. Graph the function $y = \dfrac{3x + 5}{x + 2}$. Specify all intercepts and asymptotes.

10. What is the largest area possible for a right triangle in which the sum of the lengths of the two shorter sides is 12 cm?

11. Graph the function $x/[(x + 2)(x - 4)]$.

12. Let $P(x, y)$ be a point [other than $(-1, -1)$] on the graph of $f(x) = x^3$. Express the slope of the line passing through the points P and $(-1, -1)$ as a function of x. Simplify your answer as much as possible.

13. A rectangle is inscribed in a circle. The circumference of the circle is 12 cm. Express the perimeter of the rectangle as a function of its width w.

14. Give a reason why each of the following two graphs cannot represent a polynomial function with highest-degree term $-\frac{1}{3}x^3$.

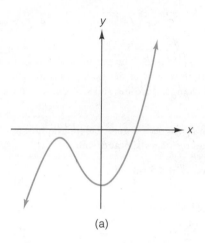

(a)

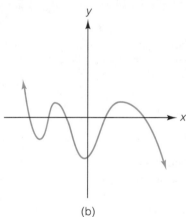

(b)

In Exercises 15–20, find equations for the linear functions satisfying the given conditions. Write each answer in the form $f(x) = mx + b$.

15. $f(3) = 5$ and $f(-2) = 0$

16. $f(1) = 5$ and the graph of f passes through the origin

17. $f(4) = -1$ and the graph of f is parallel to the line $3x - 8y = 16$

18. The graph passes through $(6, 1)$ and the x-intercept is twice the y-intercept

19. $f(-3) = 5$, and the graph of the inverse function passes through $(2, 1)$

20. The graph of f passes through the vertices of the two parabolas $y = x^2 + 4x + 1$ and $y = \frac{1}{2}x^2 + 9x + \frac{81}{2}$

In Exercises 21–26, graph the quadratic functions. In each case, specify the vertex and the x- and y-intercepts.

21. $y = x^2 + 2x - 3$

22. $f(x) = x^2 - 2x - 15$

23. $y = -x^2 + 2\sqrt{3}x + 3$

24. $f(x) = 2x^2 - 2x + 1$

25. $y = -3x^2 + 12x$

26. $f(x) = -4x^2 + 16x$

27. Find the distance between the vertices of the two parabolas $y = x^2 - 4x + 6$ and $y = -x^2 - 4x - 5$.

28. Find the value of a, given that the maximum value of the function $f(x) = ax^2 + 3x - 4$ is 5.

29. The sum of two numbers is $\sqrt{3}$. Find the largest possible value for their product.

30. The sum of two numbers is $\frac{2}{3}$. What is the smallest possible value for the sum of their squares?

31. Suppose that an object is thrown vertically upward (from ground level) with an initial velocity of v_0 ft/sec. It can be shown that the height h (in feet) after t sec is given by the formula $h = v_0 t - 16t^2$.
   (a) At what time does the object reach its maximum height? What is that maximum height?
   (b) At what time does the object strike the ground?

32. Let $f(x) = 4x^2 - x + 1$ and $g(x) = (x - 3)/2$.
   (a) For which input will the value of the function $f \circ g$ be a minimum?
   (b) For which input will the value of $g \circ f$ be a minimum?

33. **(a)** Let P be a point on the parabola $y = x^2$. Express the distance from P to the point $(0, 2)$ in terms of x.
   (b) Which point in the second quadrant on the parabola $y = x^2$ is closest to the point $(0, 2)$?

34. Find the coordinates of the point on the line $y = 2x - 1$ closest to $(-5, 0)$.

35. Find all the values of b such that the minimum distance from the point $(2, 0)$ to the line $y = \frac{4}{3}x + b$ is 5.

36. What number exceeds one-half its square by the greatest amount?

37. Suppose that $x + y = \sqrt{2}$. Find the minimum value of the quantity $x^2 + y^2$.

38. For which numbers t will the value of $9t^2 - t^4$ be as large as possible?

39. Find the maximum area possible for a right triangle with a hypotenuse of 15 cm. *Hint:* Let x denote the length of one leg. Show that the area is $A = \frac{1}{2}x\sqrt{225 - x^2}$. Now work with A^2.

40. For which point (x, y) on the curve $y = 1 - x^2$ is the sum $x + y$ a maximum?

41. Let $f(x) = x^2 - (a^2 + 2a)x + 2a^3$, where $0 < a < 2$. For which value of a will the distance between the x-intercepts of f be a maximum?

42. Suppose that the revenue R (in dollars) generated by selling x units of a certain product is given by $R = 300x - \frac{1}{4}x^2$ $(0 \le x \le 1200)$. How many units should be sold to maximize the revenue? What is that maximum revenue?

43. Suppose that the function $p = 160 - \frac{1}{5}x$ relates the selling price p of an item to the quantity x that is sold. Assume that p is in dollars. For which value of x will the revenue R $(= xp)$ be a maximum? What is the selling price p in this case?

44. A piece of wire 16 cm long is cut into two pieces. Let x denote the length of the first piece and $16 - x$ the length of the second. The first piece is formed into a rectangle in which the length is twice the width. The second piece of wire is also formed into a rectangle, but with the length three times the width. For which value of x is the total area of the two rectangles a minimum?

In Exercises 45–60, graph each function and specify the x- and y-intercepts and asymptotes, if any.

45. $y = (x + 4)(x - 2)$ **46.** $y = (x + 4)(x - 2)^2$

47. $y = -(x + 5)^3$ **48.** $y = -x(x + 1)$

49. $y = -x^2(x + 1)$ **50.** $y = -x^3(x + 1)$

51. $y = x(x - 2)(x + 2)$

52. $y = (x - 3)(x + 1)(x + 5)$

53. $y = \dfrac{3x + 1}{x}$ **54.** $y = \dfrac{1 - 2x}{x}$

55. $y = \dfrac{-1}{(x - 1)^2}$ **56.** $y = \dfrac{x + 1}{x + 2}$

57. $y = \dfrac{x - 2}{x - 3}$ **58.** $y = \dfrac{x}{(x - 2)(x + 4)}$

59. $y = \dfrac{x^2 - 2x + 1}{x^2 - 4x + 4}$ **60.** $y = \dfrac{x(x - 2)}{(x - 4)(x + 4)}$

61. Let $f(x) = x^2 + 2bx + 1$.
  (a) If $b = 1$, find the distance from the vertex of the parabola to the origin.
  (b) If $b = 2$, find the distance from the vertex of the parabola to the origin.
  (c) For which real numbers b will the distance from the vertex to the origin be as small as possible?

62. Find the range of the function $f(x) = -2x^2 + 12x - 5$. *Hint:* Look at the graph.

63. The range of the function $y = x^2 - 2x + k$ is the interval $[5, \infty)$. Find the value of k.

64. Find a value for b such that the range of the function $f(x) = x^2 + bx + b$ is the interval $[-15, \infty)$.

65. Find the range of the function $y = \dfrac{(x - 1)(x - 3)}{x - 4}$.

Hint: Solve the equation for x in terms of y using the quadratic formula. If you're careful with the algebra, you will find that the expression under the resulting

radical sign is $y^2 - 8y + 4$. The range of the given function can then be found by solving the inequality $y^2 - 8y + 4 \ge 0$.

66. In the figure, triangle OAB is equilateral and $\overline{AB}$ is parallel to the x-axis. Find the length of a side and the area of the triangle OAB.

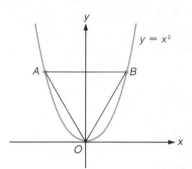

67. Let A denote the area of the right triangle in the first quadrant that is formed by the y-axis and the lines $y = mx$ and $y = m$. (Assume $m > 0$.) Express the area of the triangle as a function of m.

68. (a) Express the distance from the origin to the vertex of the parabola $y = x^2 - 2bx$ $(b > 0)$ as a function of b.
  (b) Let V denote the vertex of the parabola in part (a), and let A and B denote the points where the curve meets the x-axis. Express the area of $\triangle VAB$ as a function of b.
  (c) A circle is drawn with center V and passing through A and B. Then, through the smaller y-intercept of the circle, a tangent line is drawn. Express the x-intercept of this tangent line as a function of b.

69. In the accompanying figure, the radius of the circle is $OC = 1$. Express the area of $\triangle ABC$ as a function of x.

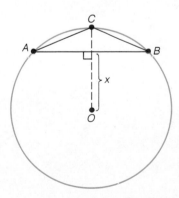

70. (a) Factor the expression $x^3 - 3x^2 + 4$. *Hint:* Add and subtract 1, then factor by grouping.
  (b) Use the factorization from part (a) to graph the function $y = x^3 - 3x^2 + 4$.

71. (a) Draw a scatter diagram (i.e., plot the points) for the data in the following table.

x (YEAR)	y (NUMBER OF REFUGEES)
1983	12083
1984	10285
1985	9350
1986	8713
1987	8606
1988	7818

Refugees admitted into the United States from Eastern Europe during the (fiscal) years 1983–1988. Source: U.S. State Department

(b) In your scatter diagram, draw the line that you feel seems best to fit the data. Estimate the slope and the y-intercept of the line.

(c) Using the formulas in Exercise 33 of Section 3.1, it can be shown that the equation of the regression line is

$$f(x) = (-771.4)x + 1,541,090$$

Use this equation (and your calculator, of course) to estimate to the nearest 100 the number of refugees from Eastern Europe that will arrive in the United States in 1995, assuming that the current trend continues.

(d) Find $f^{-1}(x)$.

(e) Use your answer in part (d) to make a projection for the year when the number of refugees will dip below 1000.

72. (a) Let $f(x) = ax^2 + bx + c$ $(a \neq 0)$. Find a constant x_0 such that the following equation is an identity.

$$f(x_0 + x) = f(x_0 - x)$$

(b) What is the geometric significance of your answer in part (a)?

73. In Section 3.2, we saw how the sign of the constant a influences the graph of the parabola $f(x) = ax^2 + bx + c$. In this exercise you'll see how the sign of b affects the graph.

(a) On the same set of axes, graph the parabolas $y = x^2 + 4x + 1$ and $y = x^2 - 4x + 1$. What type of symmetry do you observe?

(b) Let $f(x) = ax^2 + bx + c$. Compute $f(-x)$.

(c) Use the result in part (b) to describe what happens to the graph of $f(x) = ax^2 + bx + c$ when the sign of b is reversed.

CHAPTER FOUR

EXPONENTIAL AND LOGARITHMIC FUNCTIONS

The exponential growth of population and its attendant assault on the environment is so recent that it is difficult for people to appreciate how much damage is being done.

Nathan Keyfitz in his article "The Growing Human Population," in *Scientific American*, **261** (September 1989) 119–126.

INTRODUCTION   In this chapter we study exponential functions and their inverses. Figure 1 introduces one of the key properties of many of these exponential functions: they grow or increase very rapidly. Section 4.1 begins with an example involving perhaps the simplest exponential function, $y = 2^x$. In this section we discuss in detail exponential functions and their graphs. Section 4.2 introduces the constant e and the corresponding exponential function $y = e^x$. In Sections 4.3 and 4.4 we study the inverses of the exponential functions; these inverse functions are called logarithmic functions. Section 4.5 is devoted to some of the applications of exponential and logarithmic functions. As you'll see, these applications are quite diverse, ranging from interest rates in banking to population growth and to nuclear energy.

FIGURE 1

Drawing by Professor Ann Jones, University of Colorado, Boulder. From the cover of *The Physics Teacher*, Vol. 14, No. 7 (October 1976).

4.1 ▼ EXPONENTIAL FUNCTIONS

We begin with an example. Suppose that your mathematics instructor, in an effort to improve classroom attendance, offers to pay you each day for attending class! Say you are to receive 2¢ on the first day you attend class, 4¢ the second day, 8¢ the third day, and so on, as indicated in Table 1. How much money will you receive for attending class on the 30th day?

As you can see by looking at Table 1, the amount y earned on day x is given by the rule, or *exponential function*,

$$y = 2^x$$

Thus on the 30th day (when $x = 30$), you will receive

$$y = 2^{30} \text{ cents}$$

If you use a calculator, you will find this amount to be well over 10 million dollars. The point here is simply this: Although we begin with a small amount, $y = 2$¢,

TABLE 1

x (day number)	y (amount earned that day)
1	2¢ $(= 2^1)$
2	4¢ $(= 2^2)$
3	8¢ $(= 2^3)$
4	16¢ $(= 2^4)$
5	32¢ $(= 2^5)$
⋮	⋮
x	2^x

repeated doubling quickly leads to a very large amount. Put in other terms, the exponential function grows very rapidly.

Before leaving this example, we mention a simple method for quickly estimating numbers such as 2^{30} (or any power of two) in terms of the more familiar powers of ten. Begin by observing that

$$2^{10} \approx 10^3 \qquad \text{(a useful coincidence, worth remembering)}$$

Now just cube both sides to obtain

$$(2^{10})^3 \approx (10^3)^3 \qquad \text{or} \qquad 2^{30} \approx 10^9$$

Thus 2^{30} is about one billion. To convert this number of cents to dollars, we divide by 100 or 10^2 to obtain

$$\frac{10^9}{10^2} = 10^7 \text{ dollars}$$

which is 10 million dollars, as mentioned before.

EXAMPLE 1   Estimate 2^{40} in terms of a power of 10.

Solution   Take the basic approximation $2^{10} \approx 10^3$ and raise both sides to the fourth power. This yields

$$(2^{10})^4 \approx (10^3)^4$$

or

$$2^{40} \approx 10^{12}, \qquad \text{as required} \qquad \blacktriangle$$

EXAMPLE 2   Estimate the power to which 10 must be raised to yield 2.

Solution   We begin with our approximation

$$10^3 \approx 2^{10}$$

Raising both sides to the power $\frac{1}{10}$ yields

$$(10^3)^{1/10} \approx (2^{10})^{1/10}$$

and, consequently,

$$10^{3/10} \approx 2$$

Thus the power to which 10 must be raised to yield 2 is approximately $\frac{3}{10}$. $\blacktriangle$

In Section 1.4 we defined the expression b^x, where x is a rational number. We also stated that if x is irrational, then b^x can be defined so that the usual properties of exponents continue to hold. Although a rigorous definition of irrational exponents requires concepts from calculus, we can nevertheless indicate the basic idea by means of an example. (We need to do this before we give the general definition for exponential functions.)

How shall we assign a meaning to $2^{\sqrt{2}}$, for example? The basic idea is to evaluate successively the expression 2^x using rational numbers x that are closer and closer to $\sqrt{2}$. The following table displays the results of some calculations along these lines.

TABLE 2
Values of 2^x for Rational Numbers x
Approaching $\sqrt{2}$ (= 1.41421356 . . .)

x	1.4	1.41	1.414	1.4142	1.41421	1.414213
2^x	2.6 . . .	2.65 . . .	2.664 . . .	2.6651 . . .	2.66514 . . .	2.665143 . . .

The data in the table suggest that as x approaches $\sqrt{2}$, the corresponding values of 2^x approach a unique real number, call it t, whose decimal expansion begins as 2.665. Furthermore, by continuing this process we could obtain (in theory, at least) as many places in the decimal expansion of t as we wished. The value of the expression $2^{\sqrt{2}}$ is then defined to be this number t. The following results (stated here without proof) serve to summarize this discussion and also to pave the way for the definition we will give for exponential functions.

PROPERTY SUMMARY

REAL NUMBER EXPONENTS

Let b denote an arbitrary positive real number. Then:

1. For each real number x, the quantity b^x is a unique real number.

2. When x is irrational, we can approximate b^x as closely as we wish by evaluating b^r, where r is a rational number sufficiently close to the number x.

3. The properties of rational exponents continue to hold for irrational exponents.

4. If $b^x = b^y$ and $b \neq 1$, then $x = y$.

EXAMPLE 3 Solve the equation $4^x = 8$.

Solution First, let's estimate x, just to get a feeling for what kind of answer to expect. Since $4^1 = 4$, which is less than 8, and $4^2 = 16$, which is more than 8, we know that our final answer should be a number between 1 and 2. To obtain this answer, we take advantage of the fact that both 4 and 8 are powers of 2. Using this fact, we can write the given equation as

$$(2^2)^x = 2^3$$
$$2^{2x} = 2^3$$
$$2x = 3 \quad \text{using Property 4 in the box}$$
$$x = \frac{3}{2}$$

Note that the answer $x = \frac{3}{2}$ is indeed between 1 and 2, as we had estimated. ▲

For the remainder of this section, b denotes an arbitrary positive constant other than 1. In the box that follows, we define the exponential function with base b.

DEFINITION

The Exponential Function with Base b

Let b denote an arbitrary positive constant other than 1. The **exponential function with base b** is defined by the equation

$$y = b^x$$

EXAMPLES

1. The equations $y = 2^x$ and $y = 3^x$ define the exponential functions with bases 2 and 3, respectively.

2. The equation $y = (1/2)^x$ defines the exponential function with base 1/2.

3. The equations $y = x^2$ and $y = x^3$ do not define exponential functions.

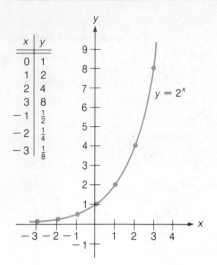

x	y
0	1
1	2
2	4
3	8
−1	$\frac{1}{2}$
−2	$\frac{1}{4}$
−3	$\frac{1}{8}$

FIGURE 2

To help with our analysis of exponential functions, let's set up a table and use it to graph the exponential function $y = 2^x$. This is done in Figure 2. (In drawing a smooth and unbroken curve, we are actually relying on the results in the Property Summary on page 205. The key features of the exponential function $y = 2^x$ and its graph are these:

1. The domain of $y = 2^x$ is the set of all real numbers. The range is the set of all *positive* real numbers.
2. The y-intercept of the graph is 1. The graph has no x-intercept.
3. For $x > 0$, the function increases or grows very rapidly. For $x < 0$, the graph rapidly approaches the x-axis; the x-axis is a horizontal asymptote for the graph. (Recall from Section 2.1 that a line is an **asymptote** for a curve if the separation distance between the curve and the line approaches zero as we move out farther and farther along the line.)

You should memorize the basic shape and features of the graph of $y = 2^x$ so that you can sketch it as needed without first setting up a table. The next example shows why this is useful.

EXAMPLE 4   Graph each of the following functions. In each case specify the domain, range, intercept(s), and asymptote.

(a) $y = -2^x$ **(b)** $y = 2^{-x}$ **(c)** $y = (\frac{1}{2})^x$ **(d)** $y = 2^{-x} - 2$

Solution   **(a)** Recall that -2^x means $-(2^x)$, not $(-2)^x$. The graph of $y = -2^x$ is obtained by reflecting the graph of $y = 2^x$ in the x-axis. See Figure 3(a).
(b) Similarly, the graph of $y = 2^{-x}$ is obtained from the graph of $y = 2^x$ by reflection in the y-axis. See Figure 3(b).

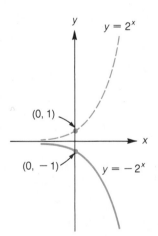

$y = -2^x$

Domain: $(-\infty, \infty)$
Range: $(-\infty, 0)$
y-intercept: -1
x-intercept: none
Asymptote: x-axis

(a)

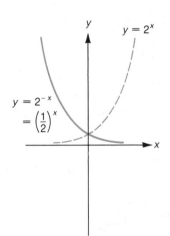

$y = 2^{-x}$ (also $y = (\frac{1}{2})^x$)

Domain: $(-\infty, \infty)$
Range: $(0, \infty)$
y-intercept: 1
x-intercept: none
Asymptote: x-axis

(b)

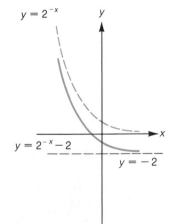

$y = 2^{-x} - 2$

Domain: $(-\infty, \infty)$
Range: $(-2, \infty)$
y-intercept: -1
x-intercept: -1
Asymptote: $y = -2$

(c)

FIGURE 3

(c) Next, regarding $y = (\frac{1}{2})^x$, observe that

$$\left(\frac{1}{2}\right)^x = \frac{1^x}{2^x} = \frac{1}{2^x} = 2^{-x}$$

In other words, $y = (\frac{1}{2})^x$ is really the same function as $y = 2^{-x}$, and this we already graphed in Figure 3(b).

(d) Finally, to graph $y = 2^{-x} - 2$, take the graph of $y = 2^{-x}$ in Figure 3(b) and move it two units in the negative y-direction, as shown in Figure 3(c). Note that the asymptote and y-intercept will also move down two units. To find the x-intercept, we set $y = 0$ in the given equation to obtain

$$2^{-x} - 2 = 0$$
$$2^{-x} = 2^1$$
$$-x = 1 \quad \text{using Property 4 on page 205}$$
$$x = -1$$

Thus, the x-intercept is -1. ▲

In the next example, we apply our knowledge about the graph of $y = 2^x$ to solve an equation. In particular, we use the fact that the graph of $y = 2^x$ always lies above the x-axis; for no value of x is 2^x ever zero.

EXAMPLE 5 Solve the equation $x^2 2^x - 2^x = 0$.

Solution First, we factor the left-hand side of the equation; the common term is 2^x. This gives us

$$2^x(x^2 - 1) = 0$$

Since 2^x is never zero, we may now divide both sides of this last equation by 2^x (without losing a solution) to obtain

$$x^2 - 1 = 0$$
$$(x - 1)(x + 1) = 0$$

and, consequently,

$$x = 1 \quad \text{or} \quad x = -1$$

Thus the required solutions are $x = 1$, $x = -1$. (You should check for yourself that each of these values satisfies the original equation.) ▲

Now what about exponential functions with bases other than 2? As Figure 4 indicates, the graphs are similar to $y = 2^x$. As you would expect, the graph of $y = 4^x$ rises more rapidly than $y = 2^x$ when x is positive. For negative x-values, the graph of $y = 4^x$ is below that of $y = 2^x$. You can see why this happens by taking $x = -1$, for example, and comparing the values of 4^x and 2^x. If $x = -1$, then

$$2^x = 2^{-1} = \frac{1}{2} \quad \text{but} \quad 4^x = 4^{-1} = \frac{1}{4}$$

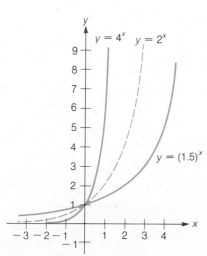

FIGURE 4

Therefore $4^x < 2^x$ when $x = -1$. Notice also in Figure 4 that all three graphs have the same y-intercept of 1. This follows from the fact that $b^0 = 1$ for any positive number b.

The exponential functions in Figure 4 all have bases larger than 1. To see examples in which the bases are in the interval $0 < b < 1$, we need only reflect those graphs in Figure 4 in the y-axis. For instance, the reflection of $y = 4^x$ in the y-axis gives us the graph of $y = (\frac{1}{4})^x$. (We discussed the idea behind this in Example 4; see Figure 3(b), for instance.)

In the box that follows, we summarize what we've learned up to this point regarding the exponential function $y = b^x$.

PROPERTY SUMMARY

THE EXPONENTIAL FUNCTION $y = b^x$

Domain: $(-\infty, \infty)$
Range: $(0, \infty)$
y-intercept: 1
x-intercept: none
Asymptote: x-axis

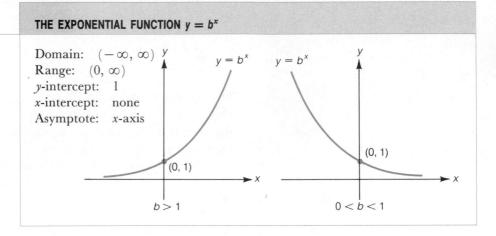

EXAMPLE 6 Graph the function $y = -3^{-x} + 1$.

Solution The required graph is obtained by reflecting and translating the graph of $y = 3^x$, as shown in Figure 5.

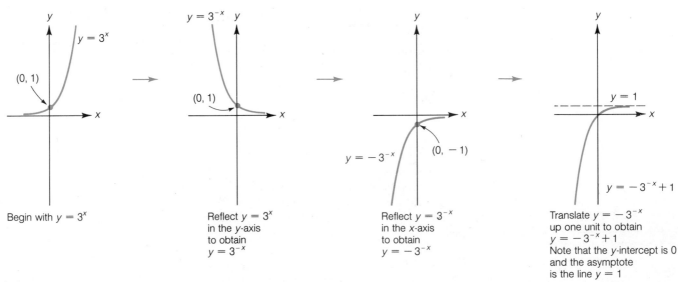

FIGURE 5

▼ EXERCISE SET 4.1

A

In Exercises 1 and 2, estimate each quantity in terms of powers of 10, as in Example 1.

1. **(a)** 2^{30} **(b)** 2^{50}

2. **(a)** 2^{90} **(b)** 4^{20}

In Exercises 3 and 4, solve each equation, as in Example 3.

3. **(a)** $3^x = 27$ **(b)** $9^t = 27$
   (c) $3^{1-2y} = \sqrt{3}$ **(d)** $3^z = 9\sqrt{3}$

4. **(a)** $2^x = 32$ **(b)** $2^t = \frac{1}{4}$
   (c) $2^{3y+1} = \sqrt{2}$ **(d)** $8^{z+1} = 32\sqrt{2}$

In Exercises 5–8, specify the domain of the function.

5. $y = 2^x$ 6. $y = \dfrac{1}{2^x}$

7. $y = \dfrac{1}{2^{x-1}}$ 8. $y = \dfrac{1}{2^x - 1}$

In Exercises 9–16, graph the pair of functions on the same set of axes.

9. $y = 2^x, y = 2^{-x}$ 10. $y = 3^x, y = 3^{-x}$

11. $y = 3^x, y = -3^x$ 12. $y = 4^x, y = -4^x$

13. $y = 2^x, y = 3^x$ 14. $y = (\frac{1}{3})^x, y = 3^x$

15. $y = (\frac{1}{2})^x, y = (\frac{1}{3})^x$ 16. $y = (\frac{1}{2})^{-x}, y = (\frac{1}{3})^{-x}$

For Exercises 17–24, graph the function and specify the domain, range, intercept(s), and asymptote.

17. $y = -2^x + 1$ 18. $y = -3^x + 3$

19. $y = 3^{-x} + 1$ 20. $y = 3^{-x} - 3$

21. $y = 2^{x-1}$ 22. $y = 2^{x-1} - 1$

23. $y = 3^{x+1} + 1$ 24. $y = 1 - 3^{x-1}$

In Exercises 25–28, solve the equation.

25. $3x(10^x) + 10^x = 0$ 26. $4x^2(2^x) - 9(2^x) = 0$

27. $3(3^x) \quad 5x(3^x) + 2x^2(3^x) - 0$

28. $\dfrac{(x+4)10^x}{x-3} = 2x(10^x)$

B

29. Let $f(x) = 2^x$. Show that
$$\frac{f(x+h) - f(x)}{h} = 2^x\left(\frac{2^h - 1}{h}\right)$$

30. Let $\phi(t) = 1 + a^t$. Show that
$$\frac{1}{\phi(t)} + \frac{1}{\phi(-t)} = 1$$

31. Let $f(x) = 2^x$ and let g denote the function that is the inverse of f.
    (a) On the same set of axes, sketch the graphs of f, g, and the line $y = x$.
    (b) Using the graph you obtained in part (a), specify the domain, range, intercept, and asymptote for the function g.

32. Let $S(x) = \dfrac{2^x - 2^{-x}}{2}$ and $C(x) = \dfrac{2^x + 2^{-x}}{2}$. Compute $[C(x)]^2 - [S(x)]^2$.

33. Use the figure to estimate each quantity.
    (a) $\sqrt{2}$ **(b)** $\sqrt[5]{2}$ **(c)** $\sqrt[5]{8}$

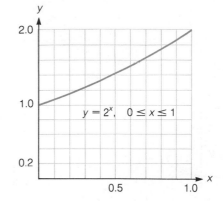

C

34. The functions f, g, h, S, T, and C are defined as follows:

$$f(x) = |x| \qquad g(x) = x^2 \qquad h(x) = x^3$$
$$T(x) = 2^x \qquad S(x) = \sqrt{x} \qquad C(x) = \sqrt{1 - x^2}$$

Match the given functions (a)–(i) with the appropriate graphs (A)–(I).

(a) $T \circ g$ **(b)** $g \circ T$ **(c)** $T \circ f$
(d) $g \circ f$ **(e)** $T \circ S \circ f$ **(f)** $S \circ T$
(g) $T \circ h$ **(h)** $T \circ C$ **(i)** $C \circ T$

(A)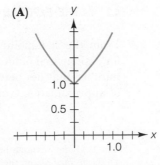

(B)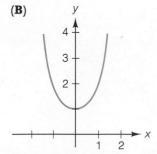

(C)

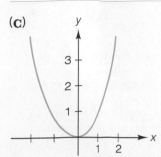

(D)

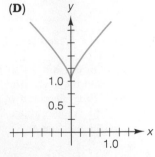

(E)

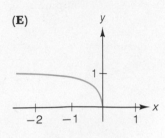

(F)

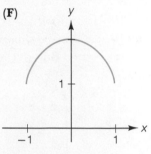

(G)

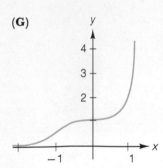

(H)

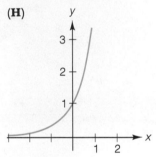

(I)

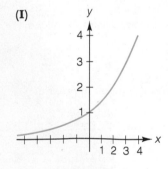

35. This exercise serves as a preview for the work on logarithms in Section 4.3. Follow steps (a)–(f) to complete the table. (Notice that one entry in the table is already filled in. Reread Example 2 in the text to see how that entry was obtained.)

(a) Fill in the entries in the right-hand column corresponding to $x = 1$ and $x = 10$.

(b) Note that 4 and 8 are powers of 2. Use this information along with the approximation $10^{0.3} \approx 2$ to find the entries in the table corresponding to $x = 4$ and $x = 8$.

(c) Find the entry corresponding to $x = 5$.

Hint: $5 = \dfrac{10}{2} \approx \dfrac{10}{10^{0.3}}$.

(d) Find the entry corresponding to $x = 7$.

Hint: $7^2 \approx 50 = 5 \times 10$. Now make use of your answer in part (c).

(e) Find the entry corresponding to $x = 3$.

Hint: $3^4 \approx 80 = 8 \times 10$.

(f) Find the entries corresponding to $x = 6$ and $x = 9$.

Hint: $6 = 3 \times 2$ and $9 = 3^2$.

Remark   This table of powers of 10 is called a *table of logarithms to the base* 10. We say, for example, that the logarithm of 2 to the base 10 is (about) 0.3. We write this symbolically as $\log_{10} 2 \approx 0.3$.

x	POWER TO WHICH 10 MUST BE RAISED TO YIELD x
1	
2	≈ 0.3
3	
4	
5	
6	
7	
8	
9	
10	

4.2 ▼ THE EXPONENTIAL FUNCTION $y = e^x$

This is undoubtedly the most important function in mathematics.

Walter Rudin in *Real and Complex Analysis* (New York: McGraw-Hill, 1966)

From the standpoint of calculus and scientific applications, one particular base for exponential functions is by far the most useful. This base is a certain irrational number that lies between 2 and 3 and is denoted by the letter e.* For purposes of approximation, you'll need to know that

$$e \approx 2.7$$

---

* In 1741 the Swiss mathematician Leonhard Euler introduced the letter e to denote this number. To six decimal places, the value of e is 2.718281 . . . .

FIGURE 1

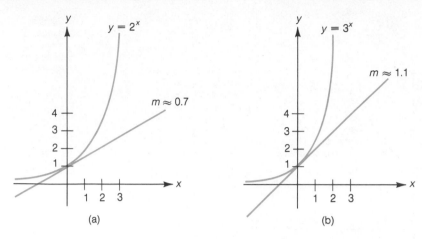

(a)   (b)

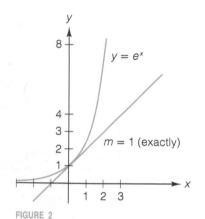

FIGURE 2
The slope of the tangent to the curve
$y = e^x$ at the point (0, 1) is $m = 1$.

At the precalculus level, it's hard to escape the feeling that $y = 2^x$ or $y = 10^x$ is by far more simple and more natural than $y = e^x$. So as we work with e and e^x in this chapter, you will need to take it on faith that, in the long run, the number e and the function $y = e^x$ make life simpler, not more complex.

Here is one way to define the number e. (For another approach, see Exercise 31; also see Table 3 in Section 4.5.) You know that the graph of each exponential function $y = b^x$ passes through the point (0, 1). Figure 1(a) shows the exponential function $y = 2^x$ along with a line that is tangent to the curve at the point (0, 1). By carefully measuring rise and run, it can be shown that the slope of this tangent line is about 0.7. Figure 1(b) shows a similar situation with the curve $y = 3^x$. Here the slope of the tangent line through (0, 1) is approximately 1.1. Now, since the slope of the tangent to $y = 2^x$ is a bit less than 1, while that for $y = 3^x$ is a bit more than 1, it seems reasonable to suppose that there is a number between 2 and 3, call it e, with the property that the slope of the tangent to $y = e^x$ through (0, 1) is exactly 1. See Figure 2 and the summary box that follows.

PROPERTY SUMMARY

THE EXPONENTIAL FUNCTION $y = e^x$

Domain: $(-\infty, \infty)$
Range: $(0, \infty)$
y-intercept: 1
Asymptote: x-axis

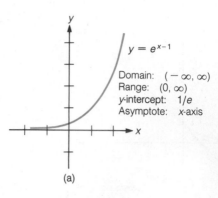

$y = e^{x-1}$

Domain: $(-\infty, \infty)$
Range: $(0, \infty)$
y-intercept: $1/e$
Asymptote: x-axis

(a)

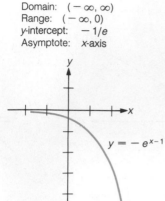

Domain: $(-\infty, \infty)$
Range: $(-\infty, 0)$
y-intercept: $-1/e$
Asymptote: x-axis

$y = -e^{x-1}$

(b)

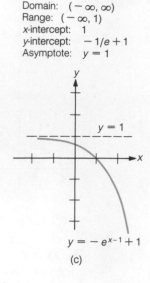

Domain: $(-\infty, \infty)$
Range: $(-\infty, 1)$
x-intercept: 1
y-intercept: $-1/e + 1$
Asymptote: $y = 1$

$y = 1$

$y = -e^{x-1} + 1$

(c)

FIGURE 3

EXAMPLE 1   Graph each of the following functions. Specify the domain, range, intercept, and asymptote.

(a) $y = e^{x-1}$ (b) $y = -e^{x-1}$ (c) $y = -e^{x-1} + 1$

Solution   (a) We begin with the graph of $y = e^x$ in the Property Summary figure. Moving the graph to the right one unit yields the graph of $y = e^{x-1}$ shown in Figure 3(a). The x-axis is still an asymptote for this translated graph, but the y-intercept will no longer be 1. For the y-intercept, we replace x with 0 in the given equation to obtain

$$y = e^{-1} = \frac{1}{e} \approx 0.37$$

(b) Reflecting the graph from part (a) in the x-axis yields the graph of $y = -e^{x-1}$ [see Figure 3(b)]. Note that under this reflection, the y-intercept moves from $1/e$ to $-1/e$.

(c) Translating the graph in Figure 3(b) up one unit produces the graph of $y = -e^{x-1} + 1$, shown in Figure 3(c). Under this translation, the asymptote moves from $y = 0$ (the x-axis) to $y = 1$. Also, the y-intercept moves from $\frac{-1}{e}$ to

$\frac{-1}{e} + 1$ (≈ 0.63). The x-intercept in Figure 3(c) is obtained by setting $y = 0$ in the equation $y = -e^{x-1} + 1$. This yields

$$0 = -e^{x-1} + 1$$
$$e^{x-1} = 1 = e^0$$
$$x - 1 = 0 \qquad \text{(Why?)}$$
$$x = 1$$

In some of the applications that we will discuss here and in Section 4.5, we will need to work with equations of the form

$$e^{ax} = b \qquad (1)$$

where a and b are constants. In Section 4.3 we'll see how to solve this equation. (It will require logarithms.) For now, however, it will be sufficient to isolate the quantity e^x in equation (1). To do this, we just raise both sides of equation (1) to the power $1/a$:

$$(e^{ax})^{1/a} = b^{1/a}$$
$$e^x = b^{1/a}$$

EXAMPLE 2   Solve the equation $e^{4k} = 3$ for the quantity e^k.

Solution   To isolate e^k, we raise both sides of the given equation to the power $\frac{1}{4}$. This yields

$$(e^{4k})^{1/4} = 3^{1/4}$$
$$e^k = 3^{1/4}$$

Thus, the value of e^k is $3^{1/4}$. As you can check using a calculator, this is approximately 1.3.   ▲

We now turn to some applications involving the number e. In Section 4.5 we'll look at these and other applications in greater detail. (Thus your immediate goals in studying the remainder of this section are these: Begin to get used to seeing formulas involving e^x; master the algebra that you will be seeing in Examples 3 and 4.)

Under ideal conditions involving unlimited food and space, a colony of bacteria increases according to the *growth law*:

$$N = N_0 e^{kt}$$

In this formula, N is the population at time t and k is a positive constant related to the growth rate of the population. The number N_0 is also a constant; it represents the size of the population at time $t = 0$. Sometimes, to emphasize the fact that N depends on (is a function of) t, we employ function notation to rewrite the growth law as

$$N(t) = N_0 e^{kt}$$

EXAMPLE 3   Suppose that at the start of an experiment in a biology laboratory, 1200 bacteria are present in a colony. Four hours later, the population is found to be 3600. How many bacteria were there three hours after the experiment began?

Solution   We begin with the growth law $N = N_0 e^{kt}$. Our strategy will be to evaluate the quantity e^k; then we can determine the required value of N. (*Note:* In Section 4.3 we'll see how to find k itself, rather than e^k.) We are given that $N_0 = 1200$. Thus we can write

$$N = 1200 e^{kt} \qquad (2)$$

We are also given that $N = 3600$ when $t = 4$. Substituting these values in equation (2) gives us

$$3600 = 1200 e^{4k}$$

Now we divide both sides of this last equation by 1200 to obtain

$$3 = e^{4k} \qquad (3)$$

To isolate e^k, we raise both sides of the equation to the power $\frac{1}{4}$ (as in Example 2).

$$3^{1/4} = (e^{4k})^{1/4}$$

Therefore

$$3^{1/4} = e^k \tag{4}$$

Now we can use this value of e^k to determine the size of the colony after three hours, as required. We take equation (2) and replace t by three:

$$\begin{aligned}
N &= 1200e^{3k} = 1200(e^k)^3 \\
&= 1200(3^{1/4})^3 \quad \text{using equation (4) to substitute for } e^k \\
&= 1200(3^{3/4})
\end{aligned}$$

Using a calculator, we find that $1200(3^{3/4}) \approx 2735$. Rather than claim that there are precisely 2735 bacteria after three hours, we follow common sense and round our answer to the nearest hundred. Thus we say that after three hours there are about 2700 bacteria in the colony. ▲

In Example 3 we used the function $N = N_0 e^{kt}$ to describe population growth. It is a remarkable fact, shown in calculus, that the same general equation also governs radioactive decay, but in that case the constant k is negative, not positive. Thus we will assume here the following *decay law* for radioactive substances:

$$N = N_0 e^{kt}$$

where N_0 is the original amount present at time $t = 0$, N is the amount present at time t, and k is a negative constant related to the rate of decay of the substance. In discussing radioactive decay, it is convenient to introduce the term *half-life*.

DEFINITION

Half-life

EXAMPLE

The **half-life** of a radioactive substance is the time required for half of a given sample to disintegrate. The half-life is an intrinsic property of the substance; it does not depend on the given sample size.

Iodine-131 is a radioactive substance with a half-life of 8 days. Suppose that 2 g are present initially. Then:

at $t = 0$, 2 g are present;

at $t = 8$ days, 1 g is left;

at $t = 16$ days, $\frac{1}{2}$ g is left;

at $t = 24$ days, $\frac{1}{4}$ g is left;

at $t = 32$ days, $\frac{1}{8}$ g is left.

EXAMPLE 4   Hospitals utilize the radioactive substance iodine-131 (with a half-life of 8 days) in the diagnosis of the thyroid gland. If a hospital acquires 2 g of iodine-131, how much of this sample will remain after 30 days?

Solution   First, let's estimate the answer to get a feeling for the situation. We noted just prior to this example that, after 24 days, $\frac{1}{4}$ g will remain, while after 32 days, $\frac{1}{8}$ g will remain. Since 30 is between 24 and 32, it follows that, after 30 days, the amount remaining will be something between $\frac{1}{4}$ g and $\frac{1}{8}$ g.

Our actual calculations for the answer begin with the decay law

$$N = N_0 e^{kt} \tag{5}$$

We are asked to find N when $t = 30$. As in Example 3, we first determine the quantity e^k. To do this, we use the half-life information. This says that when $t = 8$, the value of N is $\frac{1}{2}N_0$. Using these values in equation (5), we find that

$$\frac{1}{2}N_0 = N_0 e^{8k}$$
$$\frac{1}{2} = e^{8k}$$
$$\left(\frac{1}{2}\right)^{1/8} = \left(e^{8k}\right)^{1/8}$$
$$\left(\frac{1}{2}\right)^{1/8} = e^k \tag{6}$$

Returning to equation (5) now, we substitute $N_0 = 2$ and $t = 30$ to obtain

$$N = 2e^{30k}$$
$$= 2(e^k)^{30}$$
$$= 2\left[\left(\frac{1}{2}\right)^{1/8}\right]^{30} \quad \text{using equation (6) to substitute for } e^k$$

Using the properties for exponents, this can be simplified to

$$N = 2^{-11/4}$$

(Exercise 32 at the end of this section asks you to verify this.) Using a calculator now, we obtain

$$N = 0.148\ldots$$

Thus, after 30 days, approximately 0.15 g of the iodine-131 remains. As you can check, this figure is indeed between $\frac{1}{4}$ g and $\frac{1}{8}$ g, as we first estimated.   ▲

▼ EXERCISE SET 4.2

A

In Exercises 1–12, graph the function and specify the domain, range, intercept(s), and asymptote.

1. $y = e^x$

2. $y = e^{-x}$

3. $y = -e^x$

4. $y = -e^{-x}$

5. $y = e^x + 1$

6. $y = e^{x+1}$

7. $y = e^{x+1} + 1$

8. $y = e^{x-1} - 1$

9. $y = -e^{x-2}$

10. $y = -e^{x-2} - 2$

11. $y = e - e^x$

12. $y = e^{-x} - e$

13. On the same set of axes, graph the functions $y = 2^x$, $y = e^x$, and $y = 3^x$.

14. On the same set of axes, graph the functions $y = 2^{-x}$, $y = e^{-x}$, and $y = 3^{-x}$.

In Exercises 15–18, solve the given equation for the quantity indicated.

15. $e^{5k} = 32$; e^k

16. $e^{2k} = 225$; e^k

17. $e^{3t} = 4$; e^t

18. $e^{0.2t} = 10.3$; e^t

19. At the start of an experiment, 2000 bacteria are present in a colony. In 2 hours, the population has tripled. Assume that the growth law $N = N_0 e^{kt}$ applies.
  (a) Determine e^k.
  (b) Determine the population 10 hours after the start of the experiment.

20. At the start of an experiment 2×10^4 bacteria are present in a colony. After 8 hours the population is 3×10^4. Assume that the growth law $N = N_0 e^{kt}$ applies.
  (a) Determine the population 1 hour after it is 3×10^4.
  (b) Determine the population 24 hours after the start of the experiment.

21. A colony of bacteria grows according to the law $N = N_0 e^{kt}$. It is known that the value of e^k is $\sqrt[5]{2}$. If the initial population is 3200, what will the population be 5 hours later?

22. The half-life of a certain radioactive substance is 5 days. The initial sample is 8 g. What is the amount remaining after **(a)** 5 days? **(b)** 10 days? **(c)** 15 days?
  (d) 20 days? **(e)** 50 days?

23. The half-life of iodine-131 is 8 days. How much of a 1-g sample will remain after 7 days?

24. The half-life of strontium-90 is 28 years. How much of a 10-g sample will remain after **(a)** 1 year? **(b)** 10 years?

25. Plutonium-239 is a product of nuclear reactors. The half-life of plutonium-239 is about 24,000 years. What percentage of a given sample will remain after 1000 years?

26. Krypton-91 has a half-life of 10 sec. What percentage of a given sample will remain after 5 min?

B

27. If $E(x) = e^x$, show that

$$\frac{E(x+h) - E(x)}{h} = e^x\left(\frac{e^h - 1}{h}\right)$$

In Exercises 28 and 29, set up a table and graph the function.

28. $S(x) = \dfrac{e^x - e^{-x}}{2}$ **29.** $C(x) = \dfrac{e^x + e^{-x}}{2}$

30. Refer to the graph of $y = e^x$ in Figure 2.
  (a) What is the equation of the tangent line shown in the figure?
  (b) Provided that x is close to zero, e^x may be approximated by the quantity $x + 1$. Use part (a) to explain why.
  (c) Complete the following table. What do you observe?

x	$x + 1$	e^x
1.0		
0.5		
0.1		
0.01		
0.001		
0.0001		

31. Complete the following table. It can be shown that if the table is continued indefinitely, the numbers in the right-hand column approach the value e. (In fact, this is one way of defining e in calculus.)

n	$\left(1 + \dfrac{1}{n}\right)$	$\left(1 + \dfrac{1}{n}\right)^n$
1		
10		
100		
1000		
10000		
100000		

32. Carry out the simplification referred to in Example 4 to show that $2[(\frac{1}{2})^{1/8}]^{30} = 2^{-11/4}$. *Hint:* First replace $\frac{1}{2}$ with 2^{-1}.

33. Assume that the growth law $N = N_0 e^{kt}$, which describes bacterial growth, also approximates the Earth's human population.
  (a) Use the following data to estimate the population for the year 1985. In 1850 (call this $t = 0$), the population was estimated to be one billion. In 1930, it was 2 billion.
  (b) Statistics from the United Nations say that the actual 1985 population was about 4.8 billion. How does your estimate from part (a) compare with this? Does the formula $N = N_0 e^{kt}$ predict too high or too low a figure?

34. Let $f(x) = e^x$ and $g(x) = x - 1$. Graph the functions F and G defined as follows, and specify any intercepts or asymptotes.

$$F(x) = (f \circ g)(x)$$
$$G(x) = (g \circ f)(x)$$

35. Let $f(x) = e^x$. Let L denote the function that is the inverse of f.
  (a) On the same set of axes, sketch the graphs of f and L.
  (b) Specify the domain, range, intercept, and asymptote for the function L and its graph.
  (c) Graph each of the following functions. Specify the intercept and asymptote in each case.
    (i) $y = -L(x)$
    (ii) $y = L(-x)$
    (iii) $y = L(x - 1)$

C

36. Let S and C denote the functions that were defined in Exercises 28 and 29, respectively. (The function S is called the *hyperbolic sine*; C is the *hyperbolic cosine*.) Prove each of the following identities.
  (a) $[C(x)]^2 - [S(x)]^2 = 1$
  (b) $S(-x) = -S(x)$
  (c) $C(-x) = C(x)$
  (d) $S(x + y) = S(x)C(y) + C(x)S(y)$
  (e) $C(x + y) = C(x)C(y) + S(x)S(y)$
  (f) $S(2x) = 2S(x)C(x)$
  (g) $C(2x) = [C(x)]^2 + [S(x)]^2$

4.3 ▼ LOGARITHMIC FUNCTIONS

[John Napier] hath set my head and hands a work with his new and admirable logarithms. I hope to see him this summer, if it please God, for I never saw book that pleased me better, or made me more wonder.

Henry Briggs, March 10, 1615

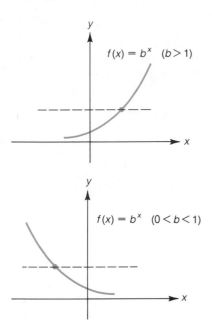

FIGURE 1
The exponential function $y = b^x$ is one-to-one.

In the previous two sections we studied exponential functions. Now we consider functions that are inverses of exponential functions. These inverse functions are called *logarithmic functions*.

Having said this, let's back up for a moment to review briefly some of the basic ideas behind inverse functions (as discussed in Section 2.7). We start with a given function F, say $F(x) = 3x$, for example, that is one-to-one. (That is, for each output there is but one input.) Then, by interchanging the inputs and outputs, we obtain a new function, the so-called inverse function. In the case of $F(x) = 3x$, it's easy to find an equation defining the inverse function. We just interchange x and y in the equation $y = 3x$ to obtain $x = 3y$. Solving for y in this last equation then gives us $y = \frac{1}{3}x$, which defines the inverse function. Using function notation, we can summarize the situation by writing $F(x) = 3x$ and $F^{-1}(x) = \frac{1}{3}x$.

In the preceding paragraph we saw that a particular linear function had an inverse, and we found a formula for that inverse. Now let's repeat that same reasoning beginning with an exponential function. First, we must make sure the exponential function f defined by $f(x) = b^x$ is one-to-one. We can see this by applying the horizontal line test, as indicated in Figure 1. Next, since $f(x) = b^x$ is one-to-one, it has an inverse function. Let's study this inverse.

We begin by writing the exponential function $f(x) = b^x$ in the form

$$y = b^x \tag{1}$$

Then, to obtain an equation for f^{-1}, we interchange x and y in equation (1). This gives us

$$x = b^y \tag{2}$$

The crucial step now is to express equation (2) in words:

$$y \text{ is the power to which } b \text{ must be raised to yield } x \tag{3}$$

Statement (3) defines the function that is the inverse of $y = b^x$. Now we introduce a notation that will allow us to write this statement in a more compact form.

DEFINITION

$\log_b x$

We define the expression **$\log_b x$** to mean "the power to which b must be raised to yield x." ($\log_b x$ is read *log base b of x* or *the logarithm of x to the base b*.)

EXAMPLES

(a) $\log_2 8 = 3$, since 3 is the power to which 2 must be raised to yield 8.

(b) $\log_{10} \frac{1}{10} = -1$, since -1 is the power to which 10 must be raised to yield $\frac{1}{10}$.

(c) $\log_5 1 = 0$, since 0 is the power to which 5 must be raised to yield 1.

Using this notation, statement (3) becomes

$$y = \log_b x$$

Since equations (2) and (3) are equivalent, we have the following important relationship.

TABLE 1

EXPONENTIAL FORM OF EQUATION	LOGARITHMIC FORM OF EQUATION
$8 = 2^3$	$\log_2 8 = 3$
$\dfrac{1}{9} = 3^{-2}$	$\log_3 \dfrac{1}{9} = -2$
$1 = e^0$	$\log_e 1 = 0$
$a = b^c$	$\log_b a = c$

$$y = \log_b x \qquad \text{is equivalent to} \qquad x = b^y.$$

We say that the equation $y = \log_b x$ is in **logarithmic form** and that the equivalent equation $x = b^y$ is in **exponential form**. Table 1 displays some examples.

Let us now summarize our discussion up to this point.

1. According to the horizontal line test, the function $f(x) = b^x$ is one-to-one and therefore possesses an inverse. This inverse function is written

$$f^{-1}(x) = \log_b x$$

2. $\log_b a = c$ means that $a = b^c$.

To graph the function $y = \log_b x$, we recall from Chapter 2 that the graph of a function and its inverse are reflections of one another about the line $y = x$. Thus, to graph $y = \log_b x$, we need only reflect the curve $y = b^x$ about the line $y = x$. This is shown in Figure 2.

FIGURE 2

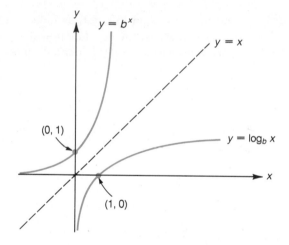

Note For the rest of this chapter we assume that the base b is greater than 1 when we use the expression $\log_b x$.

With the aid of Figure 2, we can make the following observations about the function $y = \log_b x$.

PROPERTY SUMMARY

THE LOGARITHMIC FUNCTION $y = \log_b x$ $(b > 1)$

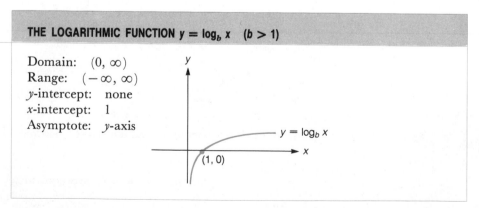

Domain: $(0, \infty)$
Range: $(-\infty, \infty)$
y-intercept: none
x-intercept: 1
Asymptote: y-axis

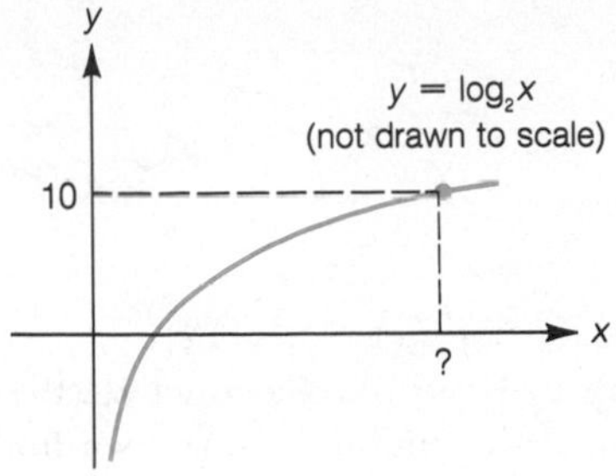

FIGURE 3

One aspect of the function $y = \log_b x$ may not be immediately apparent to you from Figure 2. The function grows or increases *very* slowly. Consider $y = \log_2 x$, for example. Let us ask how large x must be before the curve reaches the height $y = 10$. See Figure 3.

To answer this question, we substitute $y = 10$ in the equation $y = \log_2 x$. This gives us

$$10 = \log_2 x$$

Writing this equation in exponential form yields

$$x = 2^{10} = 1024$$

In other words, we must go out beyond 1000 on the x-axis before the curve $y = \log_2 x$ reaches a height of 10 units. Exercise 47 at the end of this section asks you to show that the graph of $y = \log_2 x$ doesn't reach a height of 100 until x is greater than 10^{30}. (Numbers as large as 10^{30} rarely occur in any of the sciences. For instance, the distance in inches to the Andromeda galaxy is less than 10^{24}.) The point we are emphasizing here is this: the graph of $y = \log_2 x$ (or $\log_b x$) is always rising, but very slowly.

We conclude this section with a number of examples involving logarithms and logarithmic functions. In one way or another, each example makes use of the key fact that the equation $\log_b a = c$ is equivalent to $b^c = a$.

EXAMPLE 1　Which quantity is larger: $\log_3 10$ or $\log_7 40$?

Solution　First we estimate $\log_3 10$. This quantity represents the power to which 3 must be raised to yield 10. Since $3^2 = 9$ (less than 10), but $3^3 = 27$ (more than 10), we conclude that the quantity $\log_3 10$ lies between 2 and 3. In a similar way we can estimate $\log_7 40$; this quantity represents the power to which 7 must be raised to yield 40. Since $7^1 = 7$ (less than 40), while $7^2 = 49$ (more than 40), we conclude that the quantity $\log_7 40$ lies between 1 and 2. It now follows from these two estimates that $\log_3 10$ is larger than $\log_7 40$.　▲

EXAMPLE 2　Evaluate $\log_4 32$.

Solution　Let $y = \log_4 32$. The exponential form of this equation is $4^y = 32$. Now, since both 4 and 32 are powers of 2, we can rewrite the equation $4^y = 32$ using the same base on both sides:

$$(2^2)^y = 2^5$$
$$2^{2y} = 2^5$$
$$2y = 5 \quad \text{using Property 4 on page 205}$$
$$y = \frac{5}{2} \quad \text{as required}$$
　▲

EXAMPLE 3　Solve $\log_{10}(x^2 + 3x + 12) = 1$ for x.

Solution　Writing the given equation in exponential form yields

$$x^2 + 3x + 12 = 10^1$$
$$x^2 + 3x + 2 = 0$$
$$(x + 2)(x + 1) = 0$$

Consequently, we have

$$x + 2 = 0 \qquad \text{or} \qquad x + 1 = 0$$

and, therefore,

$$x = -2 \qquad \text{or} \qquad x = -1$$

Check With $x = -2$ the given equation reads $\log_{10}(4 - 6 + 12) = 1$, or $\log_{10} 10 = 1$. Since this last equation is equivalent to $10^1 = 10$, we conclude that the value $x = -2$ does satisfy the given equation. In a similar manner, we find that $x = -1$ also satisfies the equation. In summary, then, the equation has two solutions, $x = -2$ and $x = -1$. ▲

EXAMPLE 4 Graph the equation: **(a)** $y = \log_{10} x$ **(b)** $y = -\log_{10} x$

Solution

(a) The function $y = \log_{10} x$ is the inverse function for the exponential function $y = 10^x$. Thus we obtain the graph of $y = \log_{10} x$ by reflecting the graph of $y = 10^x$ in the line $y = x$. See Figure 4(a).

(b) To graph $y = -\log_{10} x$, we reflect $y = \log_{10} x$ in the *x*-axis. See Figure 4(b).

FIGURE 4

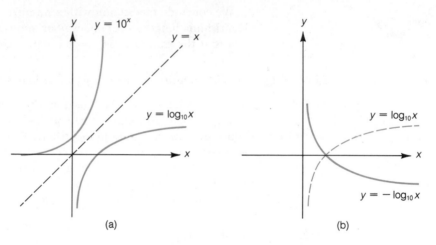

(a)   (b)   ▲

EXAMPLE 5 Find the domain of the function defined by $y = \log_2(12 - 4x)$.

Solution

As you can see by looking back at the figure in the box on page 218, the inputs for the logarithmic function must be positive. So, in the case at hand, we require that the quantity $12 - 4x$ be positive. Consequently, we have

$$12 - 4x > 0$$
$$-4x > -12$$
$$x < 3$$

Therefore the domain of the function defined by $y = \log_2(12 - 4x)$ is the interval $(-\infty, 3)$. ▲

The next example concerns the exponential function $y = e^x$ and its inverse function $y = \log_e x$. Many books, as well as calculators, abbreviate the expression $\log_e x$ by $\ln x$, read *natural log of x*. For reference and emphasis, we repeat this in the box that follows. (Incidentally, on calculators the expression *log* is an abbreviation for $\log_{10}$.)

DEFINITION

$$\ln x \quad \text{means} \quad \log_e x.$$

EXAMPLE 6   Graph the equation:

(**a**) $y = \ln x$ (**b**) $y = \ln(x - 1)$

Solution   (**a**) The function $y = \ln x \ (= \log_e x)$ is the inverse of $y = e^x$. Thus its graph is obtained by reflecting $y = e^x$ in the line $y = x$, as in Figure 5(a).

(**b**) To graph $y = \ln(x - 1)$, we take the graph of $y = \ln x$ and move it one unit in the positive x-direction. See Figure 5(b).

FIGURE 5

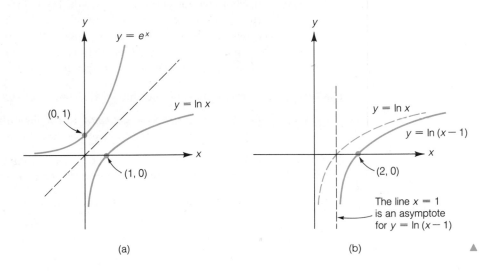

(a)   (b)

EXAMPLE 7   Simplify the expression:

(**a**) $\ln e$ (**b**) $\ln 1$

Solution   (**a**) The expression $\ln e$ denotes the power to which e must be raised to yield e. Clearly this power is 1. Thus $\ln e = 1$.

(**b**) Similarly, $\ln 1$ denotes the power to which e must be raised to yield 1. Since $e^0 = 1$, zero is the required power. Thus $\ln 1 = 0$.

EXAMPLE 8   Solve the equations: (**a**) $10^{2x} = 200$ (**b**) $e^{3t-1} = 2$

Solution   (**a**) We write the equation $10^{2x} = 200$ in logarithmic form to obtain

$$2x = \log_{10} 200$$

$$x = \frac{\log_{10} 200}{2}$$

Without a calculator or tables, we leave the answer in this form. Alternatively, using a calculator we obtain

$$x \approx 1.15$$

(**b**) To solve $e^{3t-1} = 2$, we write the equation in logarithmic form:

$$3t - 1 = \log_e 2$$

that is,

$$3t - 1 = \ln 2$$

$$3t = 1 + \ln 2$$

$$t = \frac{1 + \ln 2}{3}$$

Without a calculator or tables, this is the final answer. On the other hand, using a calculator we obtain

$$t \approx 0.56$$ ▲

In the previous section, we used the equation $N = N_0 e^{kt}$ to describe radioactive decay as well as population growth. If you look back at the examples and problems there, you'll see that every question was of the same type: Given t, find N. In other words, we were always asked "how much" or "how many," but never "when." Now that we have defined logarithms, we can solve a much wider range of problems, as the next example indicates.

EXAMPLE 9 Suppose that the half-life of a certain radioactive substance is 4 days.

(a) Compute the *decay constant* k in the formula $N = N_0 e^{kt}$.

(b) If you begin with a 2-g sample, how long will it be until only 0.01 g remains?

Solution **(a)** The half-life is 4 days. This means that when t is 4, the value of N is $\frac{1}{2}N_0$. Using these values for t and N in the formula $N = N_0 e^{kt}$, we obtain

$$\frac{1}{2} N_0 = N_0 e^{k4}$$

and, therefore,

$$\frac{1}{2} = e^{4k}$$

We solve for k by writing this last equation in logarithmic form:

$$4k = \ln \frac{1}{2}$$

$$k = \frac{\ln \frac{1}{2}}{4} = \frac{\ln 0.5}{4}$$

If a numerical value for k is required, we can use tables or a calculator to obtain

$$k \approx -0.17$$

(b) We are given that $N_0 = 2$ g and we want to find the time t at which N is 0.01 g. Substituting the values $N_0 = 2$ and $N = 0.01$ into the decay law $N = N_0 e^{kt}$ yields

$$0.01 = 2e^{kt} \qquad \text{where } k \text{ has the value determined in part (a)}$$

or, dividing by 2,

$$0.005 = e^{kt}$$

Converting this last equation into its equivalent logarithmic form gives us

$$kt = \ln(0.005)$$

and, therefore,

$$t = \frac{\ln(0.005)}{k}$$

Finally, upon replacing k by the expression $\frac{1}{4}\ln 0.5$, we obtain

$$t = \frac{\ln(0.005)}{\frac{1}{4}\ln 0.5} = \frac{4\ln(0.005)}{\ln 0.5}$$

$$\approx 30.58 \quad \text{(using a calculator)}$$

Thus after about $30\frac{1}{2}$ days, only 0.01 g remains of the original 2-g sample. ▲

▼ EXERCISE SET 4.3

A

Exercises 1–6 are review exercises dealing with inverse functions.

1. Which of the following functions is one-to-one and therefore has an inverse?
    (a) $y = x^2 + 1$ **(b)** $y = 3x$ **(c)** $y = (x + 1)^3$

2. Does each of the following functions have an inverse?
    (a) $f(x) = \begin{cases} x^2 & \text{if } -1 \le x \le 0 \\ x^2 + 1 & \text{if } x > 0 \end{cases}$

    (b) $g(x) = \begin{cases} x^2 & \text{if } -1 \le x < 0 \\ x^2 + 1 & \text{if } x \ge 0 \end{cases}$

3. Let $f(x) = \dfrac{2x - 1}{3x + 4}$. Find each quantity.
    (a) $f^{-1}(x)$ **(b)** $1/f(x)$
    (c) $f^{-1}(0)$ **(d)** $1/f(0)$

4. Let $f(x) = x^3 + 2x + 1$. Evaluate $f[f^{-1}(5)]$. [Assume that the domain of f^{-1} is $(-\infty, \infty)$.]

5. The graph of $y = f(x)$ is a line segment joining the two points $(3, -2)$ and $(-1, 5)$. What are the corresponding endpoints for the graph of $y = f^{-1}(x - 1)$?

6. Which (if either) of the following two conditions tells us that a function is one-to-one?
    (a) For each input there is exactly one output.
    (b) For each output there is exactly one input.

In Exercises 7 and 8, write each equation in logarithmic form.

7. **(a)** $9 = 3^2$ **(b)** $1000 = 10^3$
    (c) $7^3 = 343$ **(d)** $\sqrt{2} = 2^{1/2}$

8. **(a)** $\frac{1}{125} = 5^{-3}$ **(b)** $e^0 = 1$
    (c) $5^x = 6$ **(d)** $e^{3t} = 8$

9. Write each equation in exponential form.
    (a) $\log_2 32 = 5$ **(b)** $\log_{10} 1 = 0$
    (c) $\log_e \sqrt{e} = \frac{1}{2}$ **(d)** $\log_3 \frac{1}{81} = -4$
    (e) $\log_t u = v$

10. Complete the tables.

(a)

x	1	10	10^2	10^3	10^4	10^{-1}	10^{-2}	10^{-3}
$\log_{10} x$								

(b)

x	1	e	e^2	e^3	e^4	e^{-1}	e^{-2}	e^{-3}
$\ln x$								

11. Which quantity is larger, $\log_5 30$ or $\log_8 60$?

12. Which quantity is larger?
    (a) $\log_{10} 90$ or $\log_e e^5$ **(b)** $\log_2 3$ or $\log_3 2$

In Exercises 13 and 14, evaluate each expression.

13. **(a)** $\log_9 27$ **(b)** $\log_4 \frac{1}{32}$ **(c)** $\log_5 5\sqrt{5}$
14. **(a)** $\log_{25} \frac{1}{625}$ **(b)** $\log_{16} \frac{1}{64}$
     (c) $\log_{10} 10$ **(d)** $\log_2 8\sqrt{2}$

In Exercises 15 and 16, solve each equation for x.

15. **(a)** $\log_x 256 = 8$ **(b)** $\log_5 x = -1$
16. **(a)** $\log_{10}(x^2 + 36) = 2$ **(b)** $\log_2(x^2 - 8x + 1) = 0$
     (c) $\log_{10}(x^2 - 5x + 14) = 1$

In Exercises 17 and 18, find the domain of each function.

17. **(a)** $y = \log_4 5x$ **(b)** $y = \log_{10}(3 - 4x)$
     (c) $y = \ln(x^2)$ **(d)** $y = (\ln x)^2$
     (e) $y = \ln(x^2 - 25)$

18. **(a)** $y = \ln(2 - x - x^2)$ **(b)** $y = \log_{10} \dfrac{2x + 3}{x - 5}$

19. Find the domain and the range of the function defined by $y = 1/(1 - \ln x)$.

In Exercises 20–24, graph the function and specify the domain, range, intercept(s), and asymptote.

20. $y = \log_2(x - 3)$
21. **(a)** $y = \log_{10}(x + 1)$ **(b)** $y = -\log_{10}(x + 1)$

22. $y = 1 + \log_2 x$

23. (a) $y = \ln x$ (b) $y = \ln(-x)$
(c) $y = -1 + \ln(-x)$

24. $y = 1 - \log_2(x - 1)$

In Exercises 25 and 26, simplify each expression.

25. (a) $\ln e^4$ (b) $\ln(1/e)$ (c) $\ln \sqrt{e}$

26. (a) $\ln e$ (b) $\ln e^{-2}$ (c) $(\ln e)^{-2}$

In Exercises 27–34, find all the real-number solutions for each equation. In each case, give both the exact value of the answer and a calculator approximation rounded off to two decimal places.

27. $10^x = 25$ **28.** $10^x = 145$

29. $10^{x^2} = 40$ **30.** $(10^x)^2 = 40$

31. $e^{2t+3} = 10$ **32.** $e^{t-1} = 16$

33. $e^{1-4t} = 12.405$ **34.** $e^{3x^2} = 112$

35. The half-life of uranium-238 is 4.5×10^9 years.
(a) Find the decay constant k.
(b) What percentage of a given amount N_0 remains after 1000 years?

36. Compute the half-life of carbon-14 if the value of the decay constant is $k = -0.00012$. (Assume that t is in years.)

37. The half-life of a certain radioactive substance is 1 year.
(a) Find the decay constant k.
(b) Find the time required for 90% of a given 4-g sample to decay.

38. The initial population of a bacteria culture is 1.5×10^5. Three hours later, the population is found to be 2×10^5. Assume that the growth law $N = N_0 e^{kt}$ applies.
(a) Find the growth constant k.
(b) How long does it take for the initial population to double?

39. Initially a bacteria culture has a population of 2×10^7. Two hours later, the population is 3×10^8.
(a) How long does it take for the culture to double its original size?
(b) When does the population reach one billion?

40. The half-life of radium-226 is 1620 years.
(a) Find the decay constant k.
(b) How much of a 0.1-g sample remains after 100 years?

B

41. Let $f(x) = e^{x+1}$. Find $f^{-1}(x)$ and sketch its graph. Specify any intercept or asymptote.

42. Let $g(t) = \ln(t - 1)$. Find $g^{-1}(t)$ and draw its graph. Specify any intercept or asymptote.

43. Sketch the region bounded by $y = e^x$, $y = e^{-x}$, the x-axis, and the vertical lines $x = \pm 1$. Why must the area of this region be less than two square units?

44. Solve $2^{2x} - 2^{x+1} - 15 = 0$ for x. *Hint:* First show that the equation is equivalent to
$$(2^x)^2 - 2(2^x) - 15 = 0.$$

45. Solve $e^{2x} - 5e^x - 6 = 0$ for x.

46. Solve $e^x - 3e^{-x} = 2$ for x. *Hint:* Multiply both sides by e^x.

47. Estimate a value for x such that $\log_2 x = 100$. Use the approximation $10^3 \approx 2^{10}$ to express your answer as a power of 10.   *Answer:* 10^{30}

48. (a) How large must x be before the graph of $y = \ln x$ reaches a height of $y = 100$?
(b) How large must x be before the graph of $y = e^x$ reaches a height of (i) $y = 100$?   (ii) $y = 10^6$?

49. Suppose that $A = A_0 e^{k_1 t}$ and $B = B_0 e^{k_2 t}$. Find the value of t for which $A = B$.   *Answer:* $t = \dfrac{\ln(A_0/B_0)}{k_2 - k_1}$

How would you interpret this question and answer in terms of radioactive decay?

50. Chemists define pH by the formula
$$pH = -\log_{10}[H^+]$$
where $[H^+]$ is the hydrogen ion concentration, measured in moles per liter. For example, if $[H^+] = 10^{-5}$, then pH = 5. Solutions with a pH of 7 are said to be *neutral*; a pH below 7 indicates an *acid* and a pH above 7 indicates a *base*.
(a) For some fruit juices, $[H^+] = 3 \times 10^{-4}$. Determine the pH and classify as acid or base.
(b) For sulfuric acid, $[H^+] = 1$. Find the pH.
(c) An unknown substance has a hydrogen ion concentration of 3.5×10^{-9}. Classify the substance as acid or base.

C   **51.** This exercise indicates one of the ways the natural logarithm function is used in the study of prime numbers. Recall that a prime number is a natural number greater than 1 with no factors other than itself and 1. For example, the first ten prime numbers are 2, 3, 5, 7, 11, 13, 17, 19, 23, and 29.
(a) Let $P(x)$ denote the number of prime numbers that do not exceed x. For instance, $P(6) = 3$, since there are three prime numbers (2, 3, and 5) that do not exceed 6. Compute $P(10)$, $P(18)$, and $P(19)$.
(b) According to the **prime number theorem**, $P(x)$ can be approximated by $x/\ln x$ when x is large, and in fact, the ratio $\dfrac{P(x)}{x/\ln x}$ approaches 1 as x grows larger and larger. Verify this empirically by completing the following table. Round your results to three decimal places. (These facts were discovered by Carl Friedrich Gauss in 1792, when he was 15 years old. It was more than 100 years later, however, in 1896, before the prime number

theorem was formally proved by the French mathematician J. Hadamard and also by the Belgian mathematician C. J. de la Vallée-Poussin).

x	$P(x)$	$\dfrac{x}{\ln x}$	$\dfrac{P(x)}{x/\ln x}$
10^2	25		
10^4	1229		
10^6	78498		
10^8	5761455		
10^9	50847534		
10^{10}	455052512		

(c) In 1808, A. M. Legendre found that he could improve upon Gauss's approximation for $P(x)$ by using the expression $\dfrac{x}{\ln x - 1.08366}$ rather than $\dfrac{x}{\ln x}$. Complete the following table to see how well Legendre's expression approximates $P(x)$. Round off your results to four decimal places.

x	$P(x)$	$\dfrac{x}{\ln x - 1.08366}$	$\dfrac{P(x)}{x/(\ln x - 1.08366)}$
10^2	25		
10^4	1229		
10^6	78498		
10^8	5761455		
10^9	50847534		
10^{10}	455052512		

52. The figure shows a portion of the graph of $y = e^x$ for $1.5 \leq x \leq 1.7$. Using the figure, estimate the quantity $\ln 5$ to two decimal places. Then use a calculator to determine the percent error in your approximation. *Hint:* Write the equation $x = \ln 5$ in exponential form.

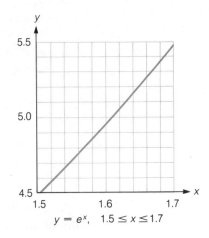

$y = e^x$, $1.5 \leq x \leq 1.7$

53. Using, as usual, x for the inputs and y for the outputs, graph the function defined by the equation $\ln y = x$.

C

54. (a) Find the domain of the function f defined by $f(x) = \ln(\ln x)$.
   (b) Find $f^{-1}(x)$ for the function f in part (a).
   (c) Find the domain of the function g defined by $g(x) = \ln[\ln(\ln x)]$.

4.4 ▼ PROPERTIES OF LOGARITHMS

he [John Napier] invented the word "logarithms," using two Greek words,   arithmos, "number" and *logos*, "ratio." It is impossible to say exactly what he had in mind when making up this word.

Alfred Hooper in *Makers of Mathematics* (New York: Random House, Inc., 1948)

The teaching of logarithms as a computing device is vanishing from the schools . . .
   The logarithmic function, however, will never die, for the simple reason that logarithmic and exponential variations are a vital part of nature and of analysis.

Howard Eves in *An Introduction to the History of Mathematics*, 5th ed. (Philadelphia: Saunders College Publishing, 1983)

A few basic properties of logarithms are used repeatedly. Our procedure in this section will be to state these properties, discuss their proofs, and then look at examples.

Properties of Logarithms

> 1. (a) $\log_b b = 1$ (b) $\log_b 1 = 0$
>
> 2. $\log_b PQ = \log_b P + \log_b Q$
> *The log of a product is the sum of the logs of the factors.*
>
> 3. $\log_b (P/Q) = \log_b P - \log_b Q$
> *The log of a quotient is the log of the numerator minus the log of the denominator.*
> As a useful particular case, we have $\log_b(1/Q) = -\log_b Q$.
>
> 4. $\log_b P^n = n \log_b P$
>
> 5. $b^{\log_b P} = P$
>
> *Note* P and Q are assumed to be positive in Properties 2–5.

In essence, each of the properties follows from the equivalence of the two equations $y = \log_b x$ and $b^y = x$. For instance, the equivalent exponential forms for properties 1(a) and 1(b) are $b^1 = b$ and $b^0 = 1$, respectively, both of which are certainly valid.

To prove Property 2, we begin by letting $x = \log_b P$. The equivalent exponential form of this equation is

$$P = b^x \tag{1}$$

Similarly, we let $y = \log_b Q$. The exponential form of this equation is

$$Q = b^y \tag{2}$$

If we multiply equation (1) by equation (2), we get

$$PQ = b^x b^y$$

and, therefore,

$$PQ = b^{x+y} \tag{3}$$

Next we write equation (3) in its equivalent logarithmic form. This yields

$$\log_b PQ = x + y$$

But, using the definitions of x and y, this last equation is equivalent to

$$\log_b PQ = \log_b P + \log_b Q$$

That completes the proof of Property 2.

The proof of Property 3 is quite similar to the proof given for Property 2. Exercise 76 asks you to carry out the proof of Property 3.

We turn now to the proof of Property 4. We begin by letting $x = \log_b P$. In exponential form, this last equation becomes

$$b^x = P \tag{4}$$

Now we raise both sides of equation (4) to the power n. This yields

$$(b^x)^n = b^{nx} = P^n$$

The logarithmic form of this last equation is

$$\log_b P^n = nx$$

or (from the definition of x)

$$\log_b P^n = n \log_b P$$

The proof of Property 4 is now complete.

Property 5 is again just a restatement of the meaning of the expression $\log_b P$. To derive this property, let $x = \log_b P$. Therefore,

$$b^x = P$$

Now, in this last equation, we simply replace x with $\log_b P$. The result is

$$b^{\log_b P} = P \qquad \text{as required.}$$

Now let's see how these properties are used. To begin with, we display some simple numerical examples in the box that follows.

PROPERTY SUMMARY

PROPERTIES OF LOGARITHMS

PROPERTY

$$\log_b P + \log_b Q = \log_b PQ$$

EXAMPLE

Simplify: $\log_{10} 50 + \log_{10} 2$

Solution: $\log_{10} 50 + \log_{10} 2 = \log_{10} (50 \cdot 2)$
$$= \log_{10} 100$$
$$= 2$$

$$\log_b P - \log_b Q = \log_b \frac{P}{Q}$$

Simplify: $\log_8 56 - \log_8 7$

Solution: $\log_8 56 - \log_8 7 = \log_8 \frac{56}{7}$
$$= \log_8 8$$
$$= 1$$

$$\log_b P^n = n \log_b P$$

Simplify: $\log_2 \sqrt[5]{16}$

Solution: $\log_2 \sqrt[5]{16} = \log_2 (16^{1/5})$
$$= \frac{1}{5} \log_2 16$$
$$= \frac{1}{5} \cdot 4 = \frac{4}{5}$$

$$b^{\log_b P} = P$$

Simplify: $3^{\log_3 7}$

Solution: $3^{\log_3 7} = 7$

The preceding examples showed how we can simplify or shorten certain expressions involving logarithms. The next example is also of this type.

EXAMPLE 1 Express as a single logarithm with a coefficient of 1:

$$\tfrac{1}{2} \log_b x - \log_b (1 + x^2)$$

Solution

$$\tfrac{1}{2} \log_b x - \log_b (1 + x^2) = \log_b x^{1/2} - \log_b (1 + x^2) \quad \text{using Property 4 on page 226}$$

$$= \log_b \frac{x^{1/2}}{1 + x^2} \quad \text{using Property 3}$$

This last expression is the required answer.

Property 2 says that the logarithm of a product of two factors is equal to the sum of the logarithms of the two factors. This can be generalized to any number of factors. For instance, with three factors, we have

$$\log_b(ABC) = \log_b[A(BC)]$$
$$= \log_b A + \log_b BC \qquad \text{using Property 2}$$
$$= \log_b A + \log_b B + \log_b C \qquad \text{using Property 2, again}$$

The next example makes use of this idea.

EXAMPLE 2 Express as a single logarithm with a coefficient of 1:

$$\ln(x^2 - 9) + 2\ln\frac{1}{x+3} + 4\ln x \qquad (x > 3)$$

Solution

$$\ln(x^2 - 9) + 2\ln\frac{1}{x+3} + 4\ln x = \ln(x^2 - 9) + \ln\left[\left(\frac{1}{x+3}\right)^2\right] + \ln x^4$$

$$= \ln(x^2 - 9) + \ln\frac{1}{(x+3)^2} + \ln x^4$$

$$= \ln\left[(x^2 - 9) \cdot \frac{1}{(x+3)^2} \cdot x^4\right]$$

$$= \ln\frac{(x^2 - 9)x^4}{(x+3)^2}$$

This last expression can be simplified still further by writing $x^2 - 9$ as $(x - 3)(x + 3)$. Then a factor of $x + 3$ can be divided out of the numerator and denominator of the fraction. The result is

$$\ln(x^2 - 9) + 2\ln\frac{1}{x+3} + 4\ln x = \ln\frac{(x-3)x^4}{x+3}$$

$$= \ln\frac{x^5 - 3x^4}{x+3} \qquad \blacktriangle$$

In the examples considered so far, we've used the properties of logarithms to shorten given expressions. We can also use these properties to expand an expression. (This is useful in calculus.)

EXAMPLE 3 Write each of the following quantities as sums and differences of simpler logarithmic expressions. Express each answer in such a way that no logarithm of products, quotients, or powers appears.

(a) $\log_{10}\sqrt{3x}$ **(b)** $\log_{10}\sqrt[3]{\dfrac{2x}{3x^2 + 1}}$ **(c)** $\ln\dfrac{x^2\sqrt{2x - 1}}{(2x + 1)^{3/2}}$

Solution

(a) $\log_{10}\sqrt{3x} = \log_{10}(3x)^{1/2} = \frac{1}{2}\log_{10}(3x)$
$$= \frac{1}{2}(\log_{10} 3 + \log_{10} x)$$

(b) $\log_{10}\sqrt[3]{\dfrac{2x}{3x^2 + 1}} = \log_{10}\left[\left(\dfrac{2x}{3x^2 + 1}\right)^{1/3}\right]$

$$= \frac{1}{3}\log_{10}\frac{2x}{3x^2 + 1}$$

$$= \frac{1}{3}[\log_{10} 2x - \log_{10}(3x^2 + 1)]$$

$$= \frac{1}{3}[\log_{10} 2 + \log_{10} x - \log_{10}(3x^2 + 1)]$$

(c) $\ln \dfrac{x^2 \sqrt{2x - 1}}{(2x + 1)^{3/2}} = \ln \dfrac{x^2 (2x - 1)^{1/2}}{(2x + 1)^{3/2}}$

$$= \ln x^2 + \ln(2x - 1)^{1/2} - \ln(2x + 1)^{3/2}$$

$$= 2 \ln x + \tfrac{1}{2} \ln(2x - 1) - \tfrac{3}{2} \ln(2x + 1)$$

EXAMPLE 4 Given that $\log_{10} A = a$, $\log_{10} B = b$, and $\log_{10} C = c$, express $\log_{10} \dfrac{A^3}{B^4 \sqrt{C}}$ in terms of a, b, and c.

Solution $\log_{10} \dfrac{A^3}{B^4 \sqrt{C}} = \log_{10} \dfrac{A^3}{B^4 C^{1/2}}$

$$= \log_{10} A^3 - \log_{10} B^4 C^{1/2}$$

$$= \log_{10} A^3 - (\log_{10} B^4 + \log_{10} C^{1/2})$$

$$= 3 \log_{10} A - (4 \log_{10} B + \tfrac{1}{2} \log_{10} C)$$

$$= 3 \log_{10} A - 4 \log_{10} B - \tfrac{1}{2} \log_{10} C$$

$$= 3a - 4b - \tfrac{1}{2}c$$

The properties of logarithms are also useful in solving equations, as the next two examples demonstrate.

EXAMPLE 5 Use logarithms to the base e to solve the equation $10 = 3e^{1 - 2x}$ for x.

Solution We take the natural (base e) logarithm of both sides of the given equation to obtain

$$\ln 10 = \ln(3e^{1 - 2x})$$

$$= \ln 3 + \ln e^{1 - 2x} \quad \text{using Property 2}$$

$$= \ln 3 + (1 - 2x) \quad \text{using the definition of ln}$$

$$2x = \ln 3 - \ln 10 + 1$$

$$x = \frac{\ln 3 - \ln 10 + 1}{2} \approx -0.102$$

This is the required solution. If we wish, we can use Property 3 to rewrite this solution as

$$x = \frac{\ln \tfrac{3}{10} + 1}{2}$$

(There are yet other ways to write this answer. See Exercise 77.)

EXAMPLE 6 Solve for x: $\log_3 x + \log_3(x + 2) = 1$

Solution Using Property 2, we can write the given equation as

$$\log_3 [x(x + 2)] = 1$$

or

$$\log_3 (x^2 + 2x) = 1$$

Writing this last equation in exponential form yields

$$x^2 + 2x = 3^1$$

$$x^2 + 2x - 3 = 0$$

$$(x + 3)(x - 1) = 0$$

Thus, we have

$$x + 3 = 0 \quad \text{or} \quad x - 1 = 0$$

and, consequently,

$$x = -3 \quad \text{or} \quad x = 1$$

Now let us check these values in the original equation to see if they are indeed solutions.

If $x = -3$, the equation becomes	If $x = 1$, the equation becomes
$\underbrace{\log_3(-3)} + \underbrace{\log_3(-1)} \overset{?}{=} 1$	$\log_3 1 + \log_3 3 \overset{?}{=} 1$
	$0 + 1 \overset{?}{=} 1 \qquad$ True

Neither expression is defined, since the domain of the logarithm function does not contain negative numbers.

Thus, the value $x = 1$ is a solution of the original equation, but $x = -3$ is not. Why is an extraneous solution ($x = -3$) generated along with the correct solution $x = 1$? The reason is that in the second line of our solution we used the property $\log_b PQ = \log_b P + \log_b Q$. This property is valid only when P and Q are both positive. ▲

It is sometimes necessary to convert logarithms in one base to logarithms in another base. After the next example, we will state a formula for this. However, as the next example indicates, it is easy to work this type of problem from the basics, without relying on a formula.

EXAMPLE 7 Express the quantity $\log_2 5$ in terms of base 10 logarithms.

Solution Let $z = \log_2 5$. The exponential form of this equation is

$$2^z = 5$$

We now take the base 10 logarithm of each side of this equation to obtain

$$\log_{10} 2^z = \log_{10} 5$$
$$z \log_{10} 2 = \log_{10} 5 \quad \text{using property 4}$$
$$z = \frac{\log_{10} 5}{\log_{10} 2}$$

Given our definition of z, this last equation can be written

$$\log_2 5 = \frac{\log_{10} 5}{\log_{10} 2}$$

This is the required answer. ▲

The method shown in Example 7 can be used to convert between any two bases. Exercise 76(b) at the end of this section asks you to follow this method, using letters rather than numbers, to arrive at the following general formula.

Change of Base Formula

$$\log_a x = \frac{\log_b x}{\log_b a}$$

Examples 1–7 have dealt with applications of the five properties of logarithms. However, you also need to understand what the properties *don't* say. For instance, Property 3 does not apply to an expression such as $\dfrac{\log_{10} 5}{\log_{10} 2}$. (Property 3 *would* apply if the expression were $\log_{10} \frac{5}{2}$.)

EXAMPLE 8 How does the statement $(\log_b P)^n = n \log_b P$ differ from Property 4? Give an example showing that this equation is not an identity.

Solution In Property 4, the exponent n on the left-hand side of the equation applies only to P, not to the quantity $\log_b P$. To provide the required example, we choose (almost any) convenient values for b, P, and n for which both sides of the equation can be easily evaluated. Using $b = 10$, $P = 10$, and $n = 2$, the equation reads

$$(\log_{10} 10)^2 \overset{?}{=} 2 \log_{10} 10$$
$$1^2 \overset{?}{=} (2)(1)$$
$$1 \overset{?}{=} 2 \quad \text{No!}$$

Thus, the given statement is not always true. The point here is that you must learn not only what the properties say, but also what they don't say. ▲

▼ EXERCISE SET 4.4

A

In Exercises 1–10, simplify the expression by using the definition and properties for logarithms.

1. $\log_{10} 70 - \log_{10} 7$ **2.** $\log_{10} 40 + \log_{10} \frac{5}{2}$

3. $\log_7 \sqrt{7}$ **4.** $\log_9 25 - \log_9 75$

5. $\log_3 108 + \log_3 \frac{3}{4}$ **6.** $\ln e^3 - \ln e$

7. $-\frac{1}{2} + \ln \sqrt{e}$ **8.** $e^{\ln 3} + e^{\ln 2} - e^{\ln e}$

9. $2^{\log_2 b} - 3 \log_5 \sqrt[3]{5}$ **10.** $\log_b b^b$

In Exercises 11–19, write the expression as a single logarithm with a coefficient of 1.

11. $\log_{10} 30 + \log_{10} 2$ **12.** $2 \log_{10} x - 3 \log_{10} y$

13. $\log_5 6 + \log_5 \frac{1}{3} + \log_5 10$

14. $p \log_b A - q \log_b B + r \log_b C$

15. (a) $\ln 3 - 2 \ln 4 + \ln 32$
(b) $\ln 3 - 2(\ln 4 + \ln 32)$

16. (a) $\log_{10}(x^2 - 16) - 3 \log_{10}(x + 4) + 2 \log_{10} x$
(b) $\log_{10}(x^2 - 16) - 3[\log_{10}(x + 4) + 2 \log_{10} x]$

17. $\log_b 4 + 3[\log_b(1 + x) - \frac{1}{2} \log_b(1 - x)]$

18. $\ln(x^3 - 1) - \ln(x^2 + x + 1)$

19. $4 \log_{10} 3 - 6 \log_{10}(x^2 + 1) + \frac{1}{2}[\log_{10}(x + 1) - 2 \log_{10} 3]$

In Exercises 20–26, write the quantity using sums and differences of simpler logarithmic expressions. Express the answer so that logarithms of products, quotients, and powers do not appear.

20. (a) $\log_{10} \sqrt{(x + 1)(x + 2)}$

(b) $\ln \sqrt{\dfrac{(x + 1)(x + 2)}{(x - 1)(x - 2)}}$

21. (a) $\log_{10} \dfrac{x^2}{1 + x^2}$ (b) $\ln \dfrac{x^2}{\sqrt{1 + x^2}}$

22. (a) $\log_b \dfrac{\sqrt{1 - x^2}}{x}$ (b) $\ln \dfrac{x \sqrt[3]{4x + 1}}{\sqrt{2x - 1}}$

23. (a) $\log_{10} \sqrt{9 - x^2}$ (b) $\ln \dfrac{\sqrt{4 - x^2}}{(x - 1)(x + 1)^{3/2}}$

24. (a) $\log_b \sqrt[3]{\dfrac{x + 3}{x}}$ (b) $\ln \dfrac{1}{\sqrt{x^2 + x + 1}}$

25. (a) $\log_b \sqrt{x/b}$
(b) $2 \ln \sqrt{(1 + x^2)(1 + x^4)(1 + x^6)}$

26. **(a)** $\log_b \sqrt[3]{\dfrac{(x-1)^2(x-2)}{(x+2)^2(x+1)}}$ **(b)** $\ln\left(\dfrac{e-1}{e+1}\right)^{3/2}$

27. Suppose that $\log_{10} A = a$, $\log_{10} B = b$, and $\log_{10} C = c$. Express each logarithm in terms of a, b, and c.

(a) $\log_{10} AB^2C^3$ **(b)** $\log_{10}\sqrt{10ABC}$
(c) $\log_{10}(10A/\sqrt{BC})$ **(d)** $\log_{10}(100A^2/B^4\sqrt[3]{C})$
(e) $\log_{10}[(AB)^5/C]$

28. If $\ln x = t$ and $\ln y = u$, write each expression in terms of t and u.

(a) $\ln xy - \ln x^2$ **(b)** $\ln \sqrt{xy} + \ln(x/e)$
(c) $\ln(e^2 x^3 \sqrt{y})$

In Exercises 29–35, solve the equation. Express the answer in terms of base e logarithms.

29. $5 = 2e^{2x-1}$ **30.** $100 = 3e^x$ **31.** $3e^{1+t} = 2$
32. $4e^{1-2t} = 7$ **33.** $2^x = 9$ **34.** $5^{3x-1} = 27$
35. $10 \cdot 2^x = 5^x$

In Exercises 36–45, solve the equation for x; check the answer to remove any extraneous roots.

36. $\log_6 x + \log_6(x+1) = 1$
37. $\log_9(x+1) = \frac{1}{2} + \log_9 x$
38. $\log_2(x+4) = 2 - \log_2(x+1)$
39. $\log_{10}(2x+4) + \log_{10}(x-2) = 1$
40. $\ln x + \ln(x+1) = \ln 12$
41. $\log_{10}(x+3) - \log_{10}(x-2) = 2$
42. $\ln(x+1) = 2 + \ln(x-1)$
43. $\log_b(x+1) = 2\log_b(x-1)$
44. $\log_2(2x^2+4) = 5$
45. $\log_{10}(x-6) + \log_{10}(x+3) = 1$
46. Solve for x in terms of a:

$\log_2(x+a) - \log_2(x-a) = 1$

47. Solve for x in terms of y.
(a) $\log_{10} x - y = \log_{10}(3x-1)$
(b) $\log_{10}(x-y) = \log_{10}(3x-1)$
48. Solve for x in terms of b: $\log_b(1-3x) = 3 + \log_b x$.

In Exercises 49–54, express the quantity in terms of base 10 logarithms.

49. $\log_2 5$ **50.** $\log_5 10$ **51.** $\ln 3$
52. $\ln 10$ **53.** $\log_b 2$ **54.** $\log_2 b$

In Exercises 55–59, express the quantity in terms of natural logarithms.

55. $\log_{10} 6$ **56.** $\log_2 10$ **57.** $\log_{10} e$
58. $\log_b 2$, where $b = e^2$ **59.** $\log_{10}(\log_{10} x)$
60. Give specific examples showing that each statement is *false*.

(a) $\log(x+y) = \log x + \log y$
(b) $(\log x)/(\log y) = \log x - \log y$
(c) $(\log x)(\log y) = \log x + \log y$
(d) $(\log x)^k = k \log x$

61. True or false?

(a) $\log_{10} A + \log_{10} B - \frac{1}{2}\log_{10} C = \log_{10}(AB/\sqrt{C})$
(b) $\log_e \sqrt{e} = \frac{1}{2}$ **(c)** $\ln \sqrt{e} = \frac{1}{2}$
(d) $\ln x^3 = \ln 3x$ **(e)** $\ln x^3 = 3 \ln x$
(f) $\ln 2x^3 = 3 \ln 2x$ **(g)** $\log_a c = b$ means $a^b = c$.
(h) $\log_5 24$ is between 5^1 and 5^2.
(i) $\log_5 24$ is between 1 and 2.
(j) $\log_5 24$ is closer to 1 than to 2.
(k) The domain of $g(x) = \ln x$ is the set of all real numbers.
(l) The range of $g(x) = \ln x$ is the set of all real numbers.
(m) The function $g(x) = \ln x$ is one-to-one.

C *Use a calculator for Exercises 62–75.*

62. Check Property 2 using the values $b = 10$, $P = \pi$, and $Q = \sqrt{2}$.
63. Let $P = 3$ and $Q = 4$. Show that $\ln(P + Q) \neq \ln P + \ln Q$.
64. Check Property 3 using the values $b = 10$, $P = 2$, and $Q = 3$.
65. If $P = 10$ and $Q = 20$, show that $\ln(PQ) \neq (\ln P)(\ln Q)$.
66. Check Property 3 using natural logarithms and the values $P = 19$ and $Q = 89$.
67. Show that $(\log_{10} 19)/(\log_{10} 89) \neq \log_{10} 19 - \log_{10} 89$.
68. Show that $(\ln 19)/(\ln 89) \neq \ln 19 - \ln 89$.
69. Check Property 4 using the values $b = 10$, $P = \pi$, and $n = 7$.
70. Using the values given for b, P, and n in Exercise 69, show that $\log_b P^n \neq (\log_b P)^n$.
71. Verify Property 5 using the values $b = 10$ and $P = 1776$.
72. Verify that $\ln 2 + \ln 3 + \ln 4 = \ln 24$.
73. Verify that $\log_{10} A + \log_{10} B + \log_{10} C = \log_{10}(ABC)$, using the values $A = 11$, $B = 12$, and $C = 13$.
74. Let $f(x) = e^x$ and $g(x) = \ln x$. Compute $f[g(2345.6)]$.
75. Let $f(x) = 10^x$ and $g(x) = \log_{10} x$. Compute $g[f(0.123456)]$.

B

76. **(a)** Prove that $\log_b(P/Q) = \log_b P - \log_b Q$.
Hint: Study the proof of Property 2 in the text.
(b) Prove the **change of base formula**:

$$\log_a x = \frac{\log_b x}{\log_b a}$$

Hint: Use the method of Example 7 in the text.

77. In Example 5 we solved the equation $10 = 3e^{1-2x}$ and found that $x = (\ln 0.3 + 1)/2$.

  (a) Show that this x-value indeed satisfies the given equation.

  (b) Show that an equivalent form for x is $\frac{1}{2} + \ln \sqrt{0.3}$.

  (c) Show that another equivalent form for x is $\ln \sqrt{3e/10}$.

78. Solve $\log_b(2x + 1) = 2 + \log_b x$ for x (in terms of b). What restrictions does your solution impose on b?

79. Show that $\log_b \dfrac{\sqrt{3} + \sqrt{2}}{\sqrt{3} - \sqrt{2}} = 2 \log_b(\sqrt{3} + \sqrt{2})$.

80. **(a)** Show that $\log_b(P/Q) + \log_b(Q/P) = 0$.

  (b) Simplify: $\log_a x + \log_{1/a} x$

81. Simplify: $b^{3 \log_b x}$

82. Let $\log_{10} 2 = a$ and $\log_{10} 3 = b$. Express each logarithm in terms of a and/or b.

  (a) $\log_{10} 4$ **(b)** $\log_{10} 8$ **(c)** $\log_{10} 5$

  (d) $\log_{10} 6$ **(e)** $\log_{10} \frac{5}{64}$ **(f)** $\log_{10} 108$

  (g) $\log_{10} \sqrt[3]{12}$ **(h)** $\log_{10} 0.0027$

83. Simplify: $\log_2 \sqrt[5]{4\sqrt{2}}$

In Exercises 84 and 85, solve for x in terms of α and β.

84. $\alpha \ln x + \ln \beta = 0$ **85.** $3 \ln x = \alpha + 3 \ln \beta$

86. Is there a constant k such that the equation $e^x = 2^{kx}$ holds for all values of x?

87. Prove that $\log_b a = 1/(\log_a b)$.

88. Simplify: $(\log_2 3)(\log_3 4)(\log_4 5)$

C

89. Prove that $(\log_a x)/(\log_{ab} x) = 1 + \log_a b$.

90. Simplify a^x when $x = [\log_b(\log_b a)]/\log_b a$.

91. If $a^2 + b^2 = 7ab$, and a and b are positive, show that

$$\log[\tfrac{1}{3}(a + b)] = \tfrac{1}{2}(\log a + \log b)$$

no matter which base is used for the logarithms. (But it is understood that the same base is used throughout.)

92. Solve for x (assuming $a > b > 0$):

$$(a^4 - 2a^2b^2 + b^4)^{x-1} = (a - b)^{2x}(a + b)^{-2}$$

Answer: $x = \dfrac{\ln(a - b)}{\ln(a + b)}$

93. Let $f(x) = \ln(x + \sqrt{x^2 + 1})$. Find $f^{-1}(x)$.

94. Suppose that $\log_{10} 2 = a$ and $\log_{10} 3 = b$. Solve for x in terms of a and b:

$$6^x = \frac{10}{3} - 6^{-x}$$

C **95.** Let $y = \ln[\ln(\ln x)]$. First, complete the table using a calculator. After doing that, disregard the evidence in your table and prove (without a calculator, of course) that the range of the given function is actually the set of all real numbers.

x	100	1000	10^6	10^{20}	10^{50}	10^{99}
y						

4.5 ▼ APPLICATIONS

$S = Pe^{rt}$. This result is remarkable both because of its simplicity and the occurrence of e. (Who would expect that number to pop up in finance theory?)

Philip Gillett in *Calculus and Analytic Geometry*, 3rd ed. (Lexington, Mass., D.C. Heath & Company, 1988)

We begin this section by considering how money accumulates in a savings account. Eventually, this will lead us back to the number e and the growth law $N = N_0 e^{kt}$, both of which we used in different contexts in Section 4.2.

The following idea from arithmetic is a prerequisite for our discussion. To increase a given quantity by, say, 15%, we multiply the quantity by 1.15. For instance, suppose that we want to increase $100 by 15%. The calculations can be written

$$100 + 0.15(100) = 100(1 + 0.15) = 100(1.15)$$

Similarly, to increase a quantity by 30%, we would multiply by 1.30; and so on. The next example displays some calculations involving percentage increase. The results may surprise you unless you're already familiar with this topic.

EXAMPLE 1 An amount of $100 is increased by 15% and then the new amount is increased by 15%. Is this the same as an overall increase of 30%?

Solution   To increase $100 by 15%, we multiply by 1.15 to obtain $100(1.15). Now to increase this new amount by 15%, we multiply it by 1.15 to obtain

$$[(\$100)(1.15)](1.15) = \$100(1.15)^2 = \$132.25$$

Alternatively, if we increase the original $100 by 30%, we obtain

$100(1.30) = $130

Comparing, we see that the result of two successive 15% increases is greater than the result of a single 30% increase.   ▲

Now let's look at another example and use it to introduce some terminology. Suppose that you place $1000 in a savings account at 10% interest *compounded annually*. This means that at the end of each year, the bank contributes to your account 10% of the amount that is in the account at that time. Interest compounded in this manner is called **compound interest**. The original deposit of $1000 is called the **principal**, denoted P. The interest rate, expressed as a decimal, is denoted by r. Thus, $r = 0.10$ in this example. The variable A is used to denote the **amount** in the account at any given time. The calculations displayed in Table 1 show how the account grows.

TABLE 1

TIME PERIOD	ALGEBRA	ARITHMETIC
After 1 year	$A = P(1 + r)$	$A = 1000(1.10)$
		$= \$1100$
After 2 years	$A = [P(1 + r)](1 + r)$	$A = 1000(1.10)^2$
	$= P(1 + r)^2$	$= \$1210$
After 3 years	$A = [P(1 + r)^2](1 + r)$	$A = 1000(1.10)^3$
	$= P(1 + r)^3$	$= \$1331$

We can learn several things from Table 1. First, consider how much interest is paid each year.

Interest paid for first year: $1100 − $1000 = $100
Interest paid for second year: $1210 − $1100 = $110
Interest paid for third year: $1331 − $1210 = $121

Thus the interest earned each year is not a constant; it increases each year. However, notice that the *ratio* of interest earned in successive years is constant:

$$\frac{110}{100} = 1.1$$

$$\frac{121}{110} = 1.1$$

This is because the given growth rate of 10% per year is constant.

If you look at the algebra in Table 1, you can see what the general formula should be for the amount after t years.

Compound Interest Formula
(Interest compounded annually)

Suppose that a principal of P dollars is invested at an annual rate r that is compounded annually. Then the amount A after t years is given by

$$A = P(1 + r)^t$$

EXAMPLE 2   Suppose that $2000 is invested at $7\frac{1}{2}\%$ interest compounded annually. How many years will it take for the money to double?

Solution   In the formula $A = P(1 + r)^t$, we use the given values $P = \$2000$ and $r = 0.075$. We want to find how long it will take for the money to double; that is, we want to find t when $A = \$4000$. Making these substitutions in the formula, we obtain

$$4000 = 2000(1 + 0.075)^t$$

and, therefore,

$$2 = 1.075^t \quad \text{dividing by 2000}$$

We can solve this exponential equation by taking the logarithm of both sides. We will use base e logarithms. (Base 10 would be just as convenient here.) This yields

$$\ln 2 = \ln 1.075^t$$

and, consequently,

$$\ln 2 = t \ln 1.075 \qquad \text{(Why?)}$$

To isolate t, we divide both sides of this last equation by $\ln 1.075$. This yields

$$t = \frac{\ln 2}{\ln 1.075}$$

or, using a calculator

$$t \approx 9.6 \text{ years}$$

Now, assuming that the bank computes the compound interest only at the end of the year, we must round the preliminary answer of 9.6 years and say that when $t = 10$ years, the initial $2000 will have *more than* doubled. Table 2 adds some perspective to this. The table shows that after 9 years, something less than $4000 is in the account; whereas after 10 years, the amount exceeds $4000.   ▲

TABLE 2
$A = 2000(1.075)^t$

t (years)	A (dollars)
9	3834.48
10	4122.06

In Example 2, the interest was compounded annually. In practice, though, the interest is usually computed more often. For instance, a bank may advertise a rate of 10% per year compounded semiannually. This means that after half a year, the interest is compounded at 5%, and then after another half year, the interest is again compounded at 5%. If you review Example 1, you'll see that one compounding at a rate of r is not the same as two compoundings, each at a rate of $r/2$. The formula $A = P(1 + r)^t$ can be generalized to cover such cases where interest is compounded more than once each year.

Compound Interest Formula
(Interest compounded n times per year)

Suppose that a principal of P dollars is invested at an annual rate r that is compounded n times per year. Then the amount A after t years is given by

$$A = P\left(1 + \frac{r}{n}\right)^{nt}$$

EXAMPLE 3   Suppose that $1000 is placed in a savings account at 10% per annum. How much is in the account at the end of one year if the interest is **(a)** compounded once each year $(n = 1)$? **(b)** compounded quarterly $(n = 4)$?

Solution   We use the formula $A = P(1 + r/n)^{nt}$.

(a) For $n = 1$, we obtain   **(b)** For $n = 4$, we obtain

$$A = 1000\left(1 + \frac{0.10}{1}\right)^{(1)(1)} \qquad\qquad A = 1000\left(1 + \frac{0.10}{4}\right)^{4(1)}$$

$$= 1000(1.1) \qquad\qquad\qquad\qquad = 1000(1.025)^4$$

$$= \$1100 \qquad\qquad\qquad\qquad\quad = \$1103.81 \quad \text{using a calculator}$$

Notice here that compounding the interest quarterly rather than annually yields the greater amount. This is in agreement with our observations in Example 1.   ▲

    The results in Example 3 will serve to illustrate some additional terminology used by financial institutions. In that example, the interest for the year under quarterly compounding was

$1103.81 - $1000 = $103.81

Now, $103.81 is 10.381% of $1000. We say in this case that the **effective rate** of interest is 10.381%. The given rate of 10% per annum compounded once a year is called the **nominal rate**.* The next example further illustrates these ideas.

EXAMPLE 4   A bank offers a nominal interest rate of 12% per annum for certain accounts. Compute the effective rate if interest is compounded monthly.

Solution   Let P denote the principal earning 12% $(r = 0.12)$ compounded monthly $(n = 12)$. Then with $t = 1$, our formula yields

$$A = P\left(1 + \frac{0.12}{12}\right)^{12(1)} = P(1.01)^{12}$$

Using a calculator to approximate the quantity $(1.01)^{12}$, we obtain

$$A \approx P(1.12683)$$

This shows that the effective interest rate is about 12.68%.   ▲

    There are two rather natural questions to ask when one first encounters compound interest calculations:

QUESTION 1   For a fixed period of time (say one year), does more and more frequent compounding of interest continue to yield greater and greater amounts?

QUESTION 2   Is there a limit on how much money can accumulate in a year when interest is compounded more and more frequently?

The answer to both of these questions is yes. If you look back over Example 1, you'll see evidence for the affirmative answer to question 1. For additional evi-

---

* The nominal rate and the effective rate are also referred to as the *annual rate* and the *annual yield*, respectively.

dence, and for the answer to question 2, let's do some calculations. To keep things as simple as possible, suppose a principal of $1 is invested for 1 year at the nominal rate of 100% per annum. (More realistic figures could be used here, but the algebra becomes more cluttered.) With these data, our formula becomes

$$A = 1\left(1 + \frac{1}{n}\right)^{n(1)}$$

or

$$A = \left(1 + \frac{1}{n}\right)^{n}$$

Table 3 shows the results of compounding the interest more and more frequently.

TABLE 3

NUMBER OF COMPOUNDINGS, n	AMOUNT, $\left(1 + \frac{1}{n}\right)^{n}$
$n = 1$ (annually)	$\left(1 + \frac{1}{1}\right)^{1} = 2$
$n = 2$ (semiannually)	$\left(1 + \frac{1}{2}\right)^{2} = 2.25$
$n = 4$ (quarterly)	$\left(1 + \frac{1}{4}\right)^{4} \approx 2.44$
$n = 12$ (monthly)	$\left(1 + \frac{1}{12}\right)^{12} \approx 2.61$
$n = 365$ (daily)	$\left(1 + \frac{1}{365}\right)^{365} \approx 2.7146$
$n = 8760$ (hourly)	$\left(1 + \frac{1}{8760}\right)^{8760} \approx 2.7181$
$n = 525,600$ (each minute)	$\left(1 + \frac{1}{525,600}\right)^{525,600} \approx 2.71827$
$n = 31,536,000$ (each second)	$\left(1 + \frac{1}{31,536,000}\right)^{31,536,000} \approx 2.71828$

Table 3 shows that the amount does increase with the number of compoundings. But assuming that the bank rounds off to the nearest penny, Table 3 also shows that there is no difference between compounding hourly, compounding each minute, and compounding each second. In each case, the amount, when rounded off, is $2.72.

The data in Table 3 suggest that the quantity $(1 + 1/n)^n$ gets closer and closer to the number e as n becomes larger and larger. In symbols,

$$\left(1 + \frac{1}{n}\right)^{n} \approx e \qquad \text{when } n \text{ is large}$$

Indeed, in many calculus books, the number e is defined as the *limiting value* or *limit* of the quantity $(1 + 1/n)^n$ as n grows ever larger. Admittedly, we have not defined here the meaning of limiting value or limit. That is a topic for calculus. Nevertheless, Table 3 should give you a reasonable, if intuitive, appreciation of the idea.

Some banks advertise interest compounded not monthly, daily, or even hourly, but *continuously*—that is, at each instant. The formula for the amount earned under continuous compounding of interest is as follows.

**Compound Interest Formula
(Interest compounded continuously)**

> Suppose that a principal of P dollars is invested at an annual rate r that is compounded continuously. Then the amount A after t years is given by
>
> $$A = Pe^{rt}$$

EXAMPLE 5   A sum of $100 is placed in a savings account at 5% per annum compounded continuously. Assuming no subsequent withdrawals or deposits, when will the balance reach $150?

Solution   Substitute the values $A = 150$, $P = 100$, and $r = 0.05$ in the formula $A = Pe^{rt}$ to obtain

$$150 = 100e^{0.05t}$$

and, therefore,

$$1.5 = e^{0.05t}$$

To solve this last equation for t, we rewrite it in its equivalent exponential form:

$$0.05t = \ln 1.5$$
$$t = \frac{\ln 1.5}{0.05} \text{ years}$$

Using a calculator, we find that $t \approx 8.1$ years. In other words, it will take slightly more than 8 years 1 month for the balance to reach $150.   ▲

In the next example, we compare the nominal rate with the effective rate under continuous compounding of interest.

EXAMPLE 6   **(a)** Given a nominal rate of 8% per annum compounded continuously, compute the effective interest rate.
(b) Given an effective rate of 8% per annum, compute the nominal rate.

Solution   **(a)** With the values $r = 0.08$ and $t = 1$, the formula $A = Pe^{rt}$ yields

$$A = Pe^{0.08(1)}$$

$$A \approx P(1.08329) \quad \text{using a calculator}$$

This shows that the effective interest rate is approximately 8.33% per year.
(b) We now wish to compute the nominal rate r, given an effective rate of 8% per year. An effective rate of 8% means that the initial principal P grows to $P(1.08)$ by the end of the year. Thus, in the formula $A = Pe^{rt}$, we make the substitutions $A = P(1.08)$ and $t = 1$. This yields

$$P(1.08) = Pe^{r(1)}$$

Dividing both sides of this last equation by P, we have

$$1.08 = e^{r}$$

To solve this equation for r, we rewrite it in its equivalent logarithmic form:

$$r = \ln(1.08)$$

$r \approx 0.07696$ using a calculator

Thus, a nominal rate of about 7.70% per annum yields an effective rate of 8%. Table 4 summarizes these results. ▲

Now we come to one of the remarkable and characteristic features of growth governed by the formula $A = Pe^{rt}$. By the **doubling time** we mean, as the name implies, the amount of time required for a given principal to double. The surprising fact here is that the doubling time does not depend on the principal P. To see why this is so, we begin with the formula

$$A = Pe^{rt}$$

We are interested in the time t at which $A = 2P$. Replacing A by $2P$ in the formula yields

$$2P = Pe^{rt}$$
$$2 = e^{rt}$$
$$rt = \ln 2$$
$$t = \frac{\ln 2}{r}$$

Denoting the doubling time by T_2, we have the following formula.

$$\text{Doubling time} = T_2 = \frac{\ln 2}{r}$$

As you can see, the formula for the doubling time T_2 does not involve P, but only r. Thus, at a given rate under continuous compounding, $2 and $2000 would both take the same amount of time to double. (This idea takes some getting used to.)

EXAMPLE 7   Compute the doubling time T_2 when a sum is invested at an interest rate of 4% per annum compounded continuously.

Solution $T_2 = \dfrac{\ln 2}{r} = \dfrac{\ln 2}{0.04} \approx 17.3$ years   using a calculator

▲

There is a convenient approximation that allows us easily to estimate doubling times. Using a calculator, we see that

$$\ln 2 \approx 0.7$$

Using this approximation, we have the following rule of thumb for estimating doubling time.

$$T_2 \approx \frac{0.7}{r}$$

Let's use this rule to rework Example 7. With $r = 0.04$, we obtain

$$T_2 \approx \frac{0.7}{0.04} = \frac{70}{4} = 17.5 \text{ years}$$

Notice that this estimation is quite close to the actual doubling time obtained in Example 7.

The idea of doubling time is useful in graphing the function $A = Pe^{rt}$. In this discussion, we assume that P and r are constants, so that the amount A is a function of the time t. Suppose, for example, that a principal of $1000 is invested at 10% per annum compounded continuously. Then the function we wish to graph is

$$A = 1000e^{0.1t}$$

Now, the doubling time in this situation is

$$T_2 \approx \frac{0.7}{r} = \frac{0.7}{0.1} = 7 \text{ years}$$

Table 5 shows the results of doubling a principal of $1000 every 7 years.

We'll use the data in Table 5 to graph the function $A = 1000e^{0.1t}$. We'll mark off units on the t-axis in multiples of 7; on the A-axis, we'll use multiples of 2000. Figure 1 shows the result of plotting the points from Table 5 and then joining them with a smooth curve. Notice that the domain in this context is $[0, \infty)$.

You may have already noticed that aside from the arbitrary choice of letters, there is no difference between the functions $A = Pe^{rt}$ and $N = N_0e^{kt}$. Both formulas have the same form. Thus, this single function serves as a model for phenomena as diverse as continuous compounding of interest, population growth, and radioactive decay.

In newspapers and in everyday speech, the term *exponential growth* is used rather loosely to describe any situation involving rapid growth. In the sciences, however, **exponential growth** refers specifically to growth governed by functions of the form $N = N_0e^{kt}$ with $k > 0$. And since the function $A = Pe^{rt}$ has this form, we say that money grows exponentially under continuous compounding of interest. Simi-

TABLE 5

t (years)	A (dollars)
0	1000
7	2000
14	4000
21	8000
28	16000

FIGURE 1

Exponential growth function

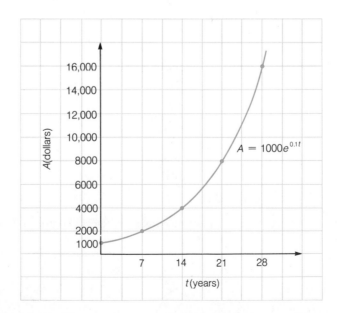

larly, in the sciences, **exponential decay** refers specifically to decay governed by functions of the form $N = N_0 e^{kt}$ with $k < 0$.

In the next example we predict the population of the world in the year 2000, assuming that the population grows exponentially. In this context, it can be shown that the constant k in the formula $N = N_0 e^{kt}$ represents the **relative growth rate** of the population N.* Algebraically, we will work with k just as we did in the calculations with compound interest, where k represented an interest rate.

EXAMPLE 8 Statistical projections from the United Nations indicate that the relative growth rate for the world's population during the years 1985–2000 will be approximately 1.6% per year. (This is down from the all-time high of 2% per year in 1970.) Use the formula $N = N_0 e^{kt}$ to predict the world population in the year 2000, given that the population in 1985 was 4.837 billion.

Solution Let $t = 0$ correspond to the year 1985. Then the year 2000 corresponds to the value $t = 15$. Our given data therefore are $k = 0.016$, $t = 15$, and $N_0 = 4.837$ (in units of one billion). Using these values in the formula $N = N_0 e^{kt}$, we obtain

$$N = 4.837 e^{(0.016)15}$$

$$N \approx 6.149$$

Thus, our prediction for the world's population in the year 2000 is 6.149 billion.

▲

In previous sections, we used the function $N = N_0 e^{kt}$ with $k < 0$ to describe radioactive decay. Let's return to that idea now. Just as we used the idea of a doubling time to graph an exponential growth function, we can use the half-life concept, introduced in Section 4.2, to graph an exponential decay function. Consider, for example, the radioactive element iodine-131, which has a half-life of about 8 days. Table 6 shows what fraction of an initial amount remains at 8-day intervals. Using the data in this table, we can draw the graph for the decay function $N = N_0 e^{kt}$ for iodine-131 (see Figure 2). Note that we are able to construct this graph without specifically evaluating the decay constant k.

TABLE 6

t (days)	N (amount)
0	N_0
8	$\frac{1}{2}N_0$
16	$\frac{1}{4}N_0$
24	$\frac{1}{8}N_0$
32	$\frac{1}{16}N_0$

FIGURE 2
Exponential decay function

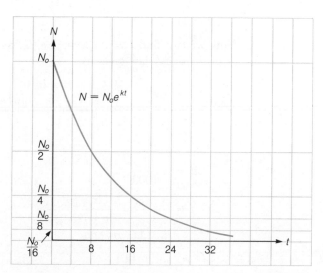

---

* The *relative growth rate* is $(\Delta N/N)/(\Delta t)$; it is the fractional change in N per unit time. This is distinct from (but sometimes confused with) the *average growth rate*, $\Delta N/\Delta t$.

The next example deals with the subject of nuclear energy. The purpose of the example is not to present an argument for or against the use of nuclear power. Rather, the purpose is to show you that with an understanding of exponential decay, you will be better equipped to read about and to evaluate the issues.

EXAMPLE 9   An article on nuclear energy appeared in the January 1976 issue of *Scientific American*. The author of the article is physicist Hans Bethe, a Nobel Prize winner. At one point in the article, Professor Bethe discusses the disposal (through burial) of radioactive waste material from a nuclear reactor. The particular waste product under discussion is plutonium-239.

> . . . Plutonium-239 has a half-life of nearly 25,000 years, and 10 half-lives are required to cut the radioactivity by a factor of 1000. Thus, the buried wastes must be kept out of the biosphere for 250,000 years.

(a) Supply the detailed calculations to support the statement that 10 half-lives are required before the radioactivity is reduced by a factor of 1000.

(b) Show how the figure of 10 half-lives can be obtained by estimation, as opposed to detailed calculation.

Solution   (a) Let N_0 denote the initial amount of plutonium-239 at time $t = 0$. Then the amount N present at time t is given by $N = N_0 e^{kt}$. We wish to determine t when $N = \frac{1}{1000} N_0$. First, we determine the value of k by using the half-life information. To say that the half-life is 25,000 years is to say that $N = \frac{1}{2} N_0$ when $t = 25,000$. Substituting these values into the formula $N = N_0 e^{kt}$ yields

$$\tfrac{1}{2} N_0 = N_0 e^{k(25,000)}$$
$$\tfrac{1}{2} = e^{25,000k} \qquad \text{dividing by } N_0$$

Writing this last equation in its equivalent logarithmic form, we obtain

$$25,000k = \ln \tfrac{1}{2}$$

and, therefore,

$$k = \frac{\ln \tfrac{1}{2}}{25,000}$$

Now we can find t when $N = \frac{1}{1000} N_0$, as required. Substituting $\frac{1}{1000} N_0$ for N in the decay law gives us

$$\frac{1}{1000} N_0 = N_0 e^{kt} \qquad \left(\text{where } k \text{ is } \frac{\ln \tfrac{1}{2}}{25,000} \right)$$

$$\frac{1}{1000} = e^{kt}$$

$$kt = \ln \frac{1}{1000}$$

$$t = \frac{\ln \frac{1}{1000}}{k} = \frac{\ln \frac{1}{1000}}{(\ln \tfrac{1}{2})/25,000}$$

$$t = \frac{(\ln \frac{1}{1000})(25,000)}{\ln \tfrac{1}{2}}$$

Using a calculator, we obtain the value 249,144.6 years; however, given the time scale involved, it would be ludicrous to announce the answer in this form. Instead, we round off the answer to the nearest thousand years and say that after 249,000 years, the radioactivity will have decreased by a factor of 1000. Notice that this result confirms Professor Bethe's ballpark estimate of 10 half-lives, or 250,000 years.

(b) After 1 half-life: $N = \dfrac{N_0}{2}$

After 2 half-lives: $N = \dfrac{1}{2}\left(\dfrac{N_0}{2}\right) = \dfrac{N_0}{2^2}$

After 3 half-lives: $N = \dfrac{1}{2}\left(\dfrac{N_0}{2^2}\right) = \dfrac{N_0}{2^3}$

Following this pattern, we see that after 10 half-lives, we should have

$$N = \dfrac{N_0}{2^{10}}$$

But as we noted in the first section of this chapter, 2^{10} is approximately 1000. Therefore, we have

$$N \approx \dfrac{N_0}{1000} \qquad \text{after 10 half-lives}$$

This is in agreement with Professor Bethe's statement.   ▲

▼ EXERCISE SET 4.5

1. You invest $800 at 6% interest, compounded annually. How much is in the account after 4 years, assuming that you make no subsequent withdrawal or deposit?

2. A sum of $1000 is invested at an interest rate of $5\frac{1}{2}\%$ compounded annually. How many years will it take before the sum exceeds $2500? (First find out when the amount equals $2500; then round off as in Example 2.)

3. At what interest rate compounded annually will a sum of $4000 grow to $6000 in 5 years?

4. A bank pays 7% interest compounded annually. What principal will grow to $10,000 in 10 years?

5. You place $500 in a savings account at 5% compounded annually. After 4 years you withdraw all your money and take it to a different bank, which advertises a rate of 6% compounded annually. What is the balance in this new account after 4 more years? (As usual, assume that no subsequent withdrawal or deposit is made.)

6. A sum of $3000 is placed in a savings account at 6% per annum. How much is in the account after 1 year if the interest is compounded **(a)** annually?   **(b)** semiannually? **(c)** daily?

7. A sum of $1000 is placed in a savings account at 7% per annum. How much is in the account after 20 years if the interest is compounded **(a)** annually?   **(b)** quarterly?

8. Your friend invests $2000 at $5\frac{1}{4}\%$ per annum compounded semiannually. You invest an equal amount at the same yearly rate, but compounded daily. How much larger is your account than your friend's after 8 years?

9. You invest $100 at 6% per annum compounded quarterly. How long will it take for your balance to exceed $120? (Round off your answer to the next quarter.)

10. A bank offers an interest rate of 7% per annum compounded daily. What is the effective rate?

11. What principal should you deposit at $5\frac{1}{2}\%$ per annum compounded semiannually so as to have $6000 after 10 years?

12. You place a sum of $800 in a savings account at 6% per annum compounded continuously. Assuming that you make no subsequent withdrawal or deposit, how much is in the account after 1 year? When will the balance reach $1000?

13. A bank offers an interest rate of $6\frac{1}{2}\%$ per annum compounded continuously. What principal will grow to $5000 in 10 years under these conditions?

14. Given a nominal rate of 6% per annum, compute the effective rate under continuous compounding of interest.

15. Suppose that under continuous compounding of interest, the effective rate is 6% per annum. Compute the nominal rate.

16. You have two savings accounts, each with an initial principal of $1000. The nominal rate on both accounts is $5\frac{1}{4}\%$ per annum. In the first account, interest is compounded semiannually. In the second account, interest is compounded continuously. How much more is in the second account after 12 years?

17. You want to invest $10,000 for 5 years and you have a choice between two accounts. The first pays 6% per annum compounded annually. The second pays 5% per annum compounded continuously. Which is the better investment?

18. Suppose that a certain principal is invested at 6% per annum compounded continuously.
   (a) Use the rule $T_2 \approx 0.7/r$ to estimate the doubling time.
   (b) Compute the doubling time using the formula $T_2 = (\ln 2)/r$.
   (c) Do your answers in (a) and (b) differ by more than 2 months?

19. A sum of $1500 is invested at 5% per annum compounded continuously.
   (a) Estimate the doubling time.
   (b) Compute the actual doubling time.
   (c) Let d_1 and d_2 denote the actual and estimated doubling times, respectively. Define d by

   $$d = |d_1 - d_2|$$

   What percent is d of the actual doubling time?

20. A sum of $5000 is invested at 10% per annum compounded continuously.
   (a) Estimate the doubling time.
   (b) Estimate the time required for the $5000 to grow to $40,000.

21. After carrying out the calculations in this problem, you'll see one of the reasons why the government imposes inheritance taxes and why laws are passed to prohibit savings accounts from being passed from generation to generation without restriction. Suppose that a family invests $1000 at 8% per annum compounded continuously. If this account were to remain intact, being passed from generation to generation, for 300 years, how much would be in the account at the end of those 300 years?

22. A principal of $500 is invested at 7% per annum compounded continuously.
   (a) Estimate the doubling time.
   (b) Sketch a graph similar to the one in Figure 1, showing how the amount increases with time.

23. A principal of $7000 is invested at 5% per annum, compounded continuously.
   (a) Estimate the doubling time.
   (b) Sketch a graph showing how the amount increases with time.

24. In one savings account, a principal of $1000 is deposited at 5% per annum. In a second account, a principal of $500 is deposited at 10% per annum. Both accounts compound interest continuously.
   (a) Estimate the doubling time for each account.
   (b) On the same set of axes, sketch graphs showing the amount of money in each account over time. Give the (approximate) coordinates of the point where the two curves meet. In financial terms, what is the significance of this point? (In working this problem, assume that the initial deposits in each account were made at the same time.)
   (c) During what period of time does the first account have the larger balance?

The statistics presented in Exercises 25–27 are taken from the Global 2000 Report to the President. *The following statement by then President Jimmy Carter provides some background on the report.*

I am directing the Council on Environmental Quality and the Department of State, working in cooperation with . . . other appropriate agencies, to make a one-year study of the probable changes in the world's population, natural resources, and environment through the end of the century.

President Jimmy Carter, May 23, 1977

25. Complete the following table on world population projections, assuming that the populations grow exponentially and that the indicated relative growth rates are valid through the year 2000.

REGION	1975 POPULATION (billions)	PERCENT OF POPULATION IN 1975	RELATIVE GROWTH RATE (percent per year)	YEAR 2000 POPULATION (billions)	PERCENT OF WORLD POPULATION IN 2000
World	4.090	100	1.8	?	?
More developed regions	1.131	?	0.6	?	?
Less developed regions	2.959	?	2.1	?	?

26. Complete the following table. Assume that the growth is exponential and that the given relative growth rates are in effect through the year 2000.

COUNTRY	1975 POPULATION (billions)	RELATIVE GROWTH RATE (percent per year)	YEAR 2000 POPULATION (billions)	PERCENT INCREASE IN POPULATION
United States	0.214	0.6	?	?
People's Republic of China	0.935	1.4	?	?
Mexico	0.060	3.1	?	?

27. The *Global 2000 Report* actually contains three sets of world population estimates and projections, as shown in the following table. Complete the table.

PROJECTION	1975 POPULATION (billions)	RELATIVE GROWTH RATE (percent per year)	YEAR 2000 POPULATION (billions)	PERCENT INCREASE IN POPULATION
Low	4.043	1.5	?	?
Medium	4.090	1.8	?	?
High	4.134	2.0	?	?

28. According to the United States Bureau of the Census, the population of the United States grew most rapidly during the period 1800–1810 and least rapidly during the period 1930–1940. Use the following data to compute the relative growth rate (percent per year) for each of these two periods. In 1800, the population was 5,308,483; in 1810, it was 7,239,881. In 1930, the population was 123,202,624; in 1940, it was 132,164,569.

29. According to the U.S. Bureau of the Census, the population of the United States in the year 1850 was 23,191,876; in 1900, the population was 62,947,714.
(a) Assuming that the population grew exponentially during this period, compute the growth constant k.
(b) Assuming continued growth at the same rate, predict the 1950 population.
(c) The actual population for 1950, according to the Bureau of the Census, was 150,697,361. How does this compare to your prediction in part (b)? Was the actual growth over the period 1900–1950 faster or slower than exponential growth with the growth constant k determined in part (a)?

In Exercises 30–32, use the population statistics given in the following table, provided by the U.S. Bureau of the Census.

STATE	1930	1940	1950
California	5,677,251	6,907,387	10,586,223
New York	12,588,066	13,479,142	14,830,192
North Dakota	680,845	641,935	619,636

30. (a) Assume that the population of California grew exponentially over the period 1930–1940. Compute the growth constant k and express it as a percent (per year).
(b) Assume that the population of California grew exponentially over the period 1940–1950. Compute the growth constant k and express it as a percent.
(c) What would the 1980 California population have been if the relative growth rate obtained in part (b) has remained in effect throughout the period 1950–1980?   *Hint:* Let $t = 0$ correspond to 1950. Then find N when $t = 30$.
(d) How does your prediction in part (c) compare to the actual 1980 population of 23,668,562?

31. Repeat Exercise 30 for New York.   *Note:* The 1980 population of New York was 17,557,288.

32. Repeat Exercise 30 for North Dakota.   *Note:* The 1980 population of North Dakota was 652,695.

33. (a) The half-life of radium-226 is 1620 years. Draw a graph of the decay function for radium-226, similar to that shown in Figure 2.
(b) The half-life of radium-A is 3 min. Draw a graph of the decay function for radium-A.

34. (a) The half-life of thorium-232 is 1.4×10^{10} years. Draw a graph of the decay function.
(b) The half-life of thorium-A is 0.16 sec. Draw a graph of the decay function.

35. The half-life of plutonium-241 is 13 years.
   (a) Compute the decay constant k.
   (b) What fraction of an initial sample $\mathcal{N}_0$ would remain after 10 years? after 100 years?

36. The half-life of thorium-229 is 7340 years.
   (a) Compute the time required for a given sample to be reduced by a factor of 1000. Show detailed calculations, as in Example 9(a).
   (b) Express your answer in part (a) in terms of half-lives.
   (c) As in Example 9(b), estimate the time required for a given sample of thorium-229 to be reduced by a factor of 1000. Compare your answer with that obtained in part (b).

37. Strontium-90, with a half-life of 28 years, is one of the waste products from nuclear fission reactors. One of the reasons great care is taken in the storage and disposal of this substance stems from the fact that strontium-90 is, in some chemical respects, similar to ordinary calcium. Thus, strontium-90 in the biosphere, entering the food chain via plants or animals, would eventually be absorbed into our bones. (In fact, everyone already has a measurable amount of strontium-90 in their bones as a result of fallout from atmospheric nuclear tests.)
   (a) Compute the decay constant k for strontium-90.
   (b) Compute the time required if a given quantity of strontium-90 is to be stored until the radioactivity is reduced by a factor of 1000.
   (c) Using half-lives, estimate the time required for a given sample to be reduced by a factor of 1000. Compare your answer with that obtained in (b).

38. **(a)** Suppose that a certain country violates the ban against above-ground nuclear testing and, as a result, an island is contaminated with debris containing the radioactive substance iodine-131. A team of scientists from the United Nations wants to visit the island to look for clues in determining which country was involved. However, the level of radioactivity from the iodine-131 is estimated to be 30,000 times the safe level. Approximately how long must the team wait before it is safe to visit the island? The half-life of iodine-131 is 8 days.
   (b) Rework part (a), assuming instead that the radioactive substance is strontium-90 rather than iodine-131. The half-life of strontium-90 is 28 years. Assume, as before, that the initial level of radioactivity is 30,000 times the safe level. (This exercise underscores the difference between a half-life of 8 days and one of 28 years.)

39. In 1969 the United States National Academy of Sciences issued a report entitled *Resources and Man.* One conclusion in the report is that a world population of 10 billion "is close to (if not above) the maximum that an intensively managed world might hope to support with some degree of comfort and individual choice." (The figure

"10 billion" is sometimes referred to as the *carrying capacity* of the Earth.)
   (a) When the report was issued in 1969, the world population was about 3.6 billion, with a relative growth rate of 2% per year. Assuming continued exponential growth at this rate, estimate the year in which the Earth's carrying capacity of 10 billion might be reached.
   (b) The data in the following table, from the *Global 2000 Report*, show the world population estimates for 1975. Use the "low" figures to estimate the year in which the Earth's carrying capacity of 10 billion might be reached.
   (c) Use the "high" figures to estimate the year in which the Earth's carrying capacity of 10 billion might be reached.

PROJECTION	POPULATION (billions)	RELATIVE GROWTH RATE (percent per year)
Low	4.043	1.5
High	4.134	2.0

40. *Depletion of Nonrenewable Energy Resources.* Suppose that the world population grows exponentially. Then, as a first approximation, it is reasonable to assume that the use of nonrenewable energy resources, such as petroleum and coal, also grows exponentially. Under these conditions, the following formula can be derived (using calculus):

$$A = \frac{A_0}{k}\left(e^{kT} - 1\right)$$

where A is the amount of the resource consumed from time $t = 0$ to $t = T$, the quantity A_0 is the amount of the resource consumed during the year $t = 0$, and k is the relative growth rate of annual consumption.
   (a) Show that solving the formula for T yields

$$T = \frac{\ln[(Ak/A_0) + 1)]}{k}$$

This formula gives the "life expectancy" T for a given resource. In the formula, A_0 and k are as previously defined, and A represents the total amount of the resource available.
   (b) In the *Global 2000 Report to the President*, it is estimated that the 1976 worldwide consumption of oil was 21.7 billion barrels and that the total remaining oil resources ultimately available was 1661 billion barrels. Compute the "life expectancy" for oil if **(i)** the relative growth rate $k = 1\%$/year; **(ii)** $k = 2\%$/year; **(iii)** $k = 3\%$/year. (This last is, in fact, the predicted rate for the next decade.)
   (c) In the *Global 2000 Report*, it is estimated that the 1976 worldwide consumption of natural gas was

approximately 50 tcf (trillions of cubic feet), while the total remaining gas ultimately available was 8493 tcf. Compute the "life expectancy" for natural gas if $k = 2\%$/year.

41. In this exercise, the term *nonrenewable energy resources* refers collectively to petroleum, natural gas, coal, shale oil, and uranium. In the *Global 2000 Report*, it was estimated that in 1976, the total worldwide consumption of these nonrenewable resources amounted to 250 quadrillion Btu. It was also estimated that in 1976, the total nonrenewable resources ultimately available were 161,241 quadrillion Btu. Use the formula for T in Exercise 40 to compute the life expectancy for these nonrenewable energy resources if **(a)** $k = 1\%$ per year; **(b)** $k = 2\%$ per year.

42. The data in the following table, from the *Global 2000 Report*, show the world reserves for selected minerals. The term *world reserves* refers to known resources available with current technology. Use the formula for T derived in Exercise 40 to estimate the "depletion date" for each mineral under these conditions (depletion date $= T + 1976$).

44. The age of some rocks can be estimated by measuring the ratio of the amounts of certain chemical elements within the rock. The method known as the *rubidium-strontium method* will be discussed here. This method has been used in dating the moon rocks brought back on the Apollo missions.

Rubidium-87 is a radioactive substance with a half-life of 4.7×10^{10} years. Rubidium-87 decays into strontium-87, which is stable (nonradioactive). We are going to derive the following formula for the age of a rock:

$$T = \frac{\ln(N_s/N_r + 1)}{-k}$$

where T is the age of the rock, k is the decay constant for rubidium-87, N_s is the number of atoms of strontium-87 now present in the rock, and N_r is the number of atoms of rubidium-87 now present in the rock.

(a) Assume that initially, when the rock was formed, there were N_0 atoms of rubidium-87 and none of strontium-87. Then as time goes by, some of the rubidium atoms decay into strontium atoms, but

MINERAL	1976 WORLD RESERVES	1976 CONSUMPTION	RELATIVE GROWTH RATE (percent per year)
Fluorine (million short tons)	37	2.1	4.58
Silver (million troy ounces)	6,100	305	2.33
Tin (thousand metric tons)	10,000	241	2.05
Copper (million short tons)	503	8.0	2.94
Phosphate rock (million metric tons)	25,732	107	5.17
Aluminum in bauxite (million short tons)	5,610	18	4.29

43. In 1956, scientists B. Gutenberg and C. F. Richter developed a formula to estimate the amount of energy E released in an earthquake. The formula is

$$\log_{10} E = 11.4 + 1.5M$$

where E is the energy in units of ergs and M is the Richter magnitude.

(a) Solve the formula for E.

(b) The table shows the Richter magnitudes for two large earthquakes that occurred in 1985. If E_1 denotes the energy of the quake in Indonesia, and E_2 denotes the energy of the quake in Mexico, compute the ratio E_2/E_1 to compare the energies of the two earthquakes.

DATE	LOCATION	M (magnitude)
April 13, 1985	Indonesia	6.8
September 19, 1985	Mexico	7.8

the total number of atoms must still be N_0. Thus, after T years, we have

$$N_0 = N_r + N_s$$

or, equivalently,

$$N_s = N_0 - N_r \tag{1}$$

However, according to the law of exponential decay for the rubidium-87, we must have $N_r = N_0 e^{kT}$. Solve this equation for N_0 and then use the result to eliminate N_0 from equation (1). Show that the result can be written

$$N_s = N_r e^{-kT} - N_r \tag{2}$$

(b) Solve equation (2) for T to obtain the formula given at the beginning of this exercise.

45. (Continuation of Exercise 44) The half-life of rubidium-87 is 4.7×10^{10} years. Compute the decay constant k.

46. (Continuation of Exercise 44) Analysis of lunar rock samples taken on the Apollo 11 mission showed the strontium-rubidium ratio to be

$$\frac{N_s}{N_r} = 0.0588$$

Estimate the age of these lunar rocks.

47. (Continuation of Exercise 44) Analysis of the so-called genesis rock sample taken on the Apollo 15 mission revealed a strontium-rubidium ratio of 0.0636. Estimate the age of this rock.

48. *Radiocarbon Dating.* Because rubidium-87 decays so slowly, the technique of rubidium-strontium dating is generally considered effective only for objects older than 10 million years. By way of contrast, archeologists and geologists rely on the method of *radiocarbon dating* in assigning ages ranging from 500 to 50,000 years.

Two types of carbon occur naturally in our environment—carbon-12, which is nonradioactive, and carbon-14, which has a half-life of 5730 years. All living plant and animal tissue contains both types of carbon, always in the same ratio. (The ratio is one part carbon-14 to 10^{12} parts carbon-12.) As long as the plant or animal is living, this ratio is maintained. When the organism dies, however, no new carbon-14 is absorbed, and the amount of carbon-14 begins to decrease exponentially. Since the amount of carbon-14 decreases exponentially, it follows that the level of radioactivity also must decrease exponentially. The formula describing this situation is

$$N = N_0 e^{kT}$$

where T is the age of the sample, N is the present level of radioactivity (in units of distintegrations per hour per gram of carbon), and N_0 is the level of radioactivity T years ago, when the organism was alive. Given that the half-life of carbon-14 is 5730 years and that $N_0 = 920$ disintegrations per hour per gram, show that the age T of a sample is given by

$$T = \frac{5730 \ln(N/920)}{\ln \frac{1}{2}}$$

In Exercises 49–52, use the formula derived in Exercise 48 to estimate the age of each sample. Note: *Some technical complications arise in interpreting such results. Studies have shown that the ratio of carbon-12 to carbon-14 in the air, and therefore in living matter, has not in fact been constant over time. For instance, air pollution from factory smokestacks tends to increase the level of carbon-12. In the other direction, nuclear bomb testing increases the level of carbon-14.*

49. Prehistoric cave paintings were discovered in the Lascaux cave in France. Charcoal from the site was analyzed and the level of radioactivity was found to be $N = 141$ disintegrations per hour per gram. Estimate the age of the paintings.

50. Before radiocarbon dating was used, historians estimated that the tomb of Vizier Hemaka, in Egypt, was constructed about 4900 years ago. After radiocarbon dating became available, wood samples from the tomb were analyzed, and it was determined that the radioactivity level was 510 disintegrations per hour per gram. Estimate the age of the tomb on the basis of this reading and compare your answer to the figure already mentioned.

51. Analyses of the oldest campsites in the Western Hemisphere reveal a carbon-14 radioactivity level of $N = 226$ disintegrations per hour per gram. Show that this implies an age of 11,500 years, to the nearest 500 years. (This would correspond to the last Ice Age. At that time, the sea level was significantly lower than today. It is believed that at the time of the last Ice Age, land stretched across what is now the Bering Strait, and humans first entered the Western Hemisphere by this route.)

52. The Dead Sea Scrolls are a collection of ancient manuscripts discovered in caves along the west bank of the Dead Sea. (The discovery occurred by accident when an Arab herdsman of the Taamireh tribe was searching for a stray goat.) When the linen wrappings on the scrolls were analyzed, the carbon-14 radioactivity level was found to be 723 disintegrations per hour per gram. Estimate the age of the scrolls using this information. Historical evidence suggests that some of the scrolls date back somewhere between 150 B.C. and A.D. 40. How do these dates compare with the estimate derived using radiocarbon dating?

▼ **CHAPTER FOUR SUMMARY OF PRINCIPAL TERMS AND FORMULAS**

TERMS OR FORMULAS	PAGE REFERENCE	COMMENTS
1. Exponential function with base b	205	The exponential function with base b is defined by the equation $y = b^x$. It is understood here that the base b is a positive number other than 1.
2. Asymptote	206	A line is an asymptote for a curve if the separation distance between the curve and the line approaches zero as we move out farther and farther along the line

TERMS OR FORMULAS	PAGE REFERENCE	COMMENTS
3. The number e	211	The irrational number e is one of the basic constants in mathematics, as is the irrational number π. To five decimal places, the value of e is 2.71828. In calculus e is the base most commonly used for exponential functions. The graph of the exponential function $y = e^x$ is shown in the box on page 211.
4. $N = N_0 e^{kt}$ $(k > 0)$	213	In many instances throughout this chapter, we studied situations in which a quantity N grows or increases according to this formula, called the *growth law*. In this formula, the quantity N is a function of the time t, and N_0 denotes the value of N at time $t = 0$. The number k is a constant, called the *growth constant*, that is related to the rate at which N increases.
5. $N = N_0 e^{kt}$ $(k < 0)$	214	This is called the *decay law* for radioactive substances. In the formula, N denotes the amount of the substance present at time t, and N_0 is the amount of the substance present at time $t = 0$. The negative number k is a constant, called the *decay constant*, that is related to the rate at which the decay takes place.
6. Half-life	214	The half-life of a radioactive substance is the time required for half of a given sample to disintegrate. Notice that the half-life is an amount of time, not an amount of the substance.
7. $\log_b x$	217	The expression $\log_b x$ denotes the power to which b must be raised to yield x. The equation $\log_b x = y$ is equivalent to $b^y = x$.
8. $\ln x$	220	The expression $\ln x$ means $\log_e x$. Logarithms to the base e are known as *natural logarithms*. For the graph of $y = \ln x$, see Figure 5 on page 221.
9. $\log_a x = \dfrac{\log_b x}{\log_b a}$	231	This is the change of base formula for converting logarithms from one base to another.
10. $A = P\left(1 + \dfrac{r}{n}\right)^{nt}$	235	This formula gives the amount A that accumulates after t years when a principal of P dollars is invested at an annual rate r that is compounded n times per year.
11. $A = Pe^{rt}$	238	This formula gives the amount A that accumulates after t years when a principal of P dollars is invested at an annual rate r that is compounded continuously.

▼ CHAPTER FOUR REVIEW EXERCISES

NOTE Exercises 1–20 constitute a chapter test on the fundamentals, based on group A problems.

1. Which is larger, $\log_5 126$ or $\log_{10} 999$?

2. Graph the function $y = 3^{-x} - 3$. Specify the domain, range, intercept(s), and asymptote.

3. Suppose that the population of a colony of bacteria increases exponentially. If the population at the start of an experiment is 8000, and 4 hours later it is 10,000, how long (from the start of the experiment) will it take for the population to reach 12,000? (Express the answer in terms of base e logarithms.)

4. Express $\log_{10} 2$ in terms of base e logarithms.

5. Let f be the function defined by
$$f(x) = \begin{cases} 2^{-x} & \text{if } x < 0 \\ x^2 & \text{if } x \geq 0 \end{cases}$$
Sketch the graph of f and then use the horizontal line test to determine whether f is one-to-one.

6. Estimate 2^{60} in terms of an integral power of 10.

7. Solve for x: $\ln(x + 1) - 1 = \ln(x - 1)$

8. Suppose that \$5000 is invested at 8% interest compounded annually. How many years will it take for

the money to double? Make use of the approximations $\ln 2 \approx 0.7$ and $\ln 1.08 \approx 0.08$ to obtain a numerical answer.

9. On the same set of axes, sketch the graphs of $y = e^x$ and $y = \ln x$. Specify the domain and range for each function.

10. Solve for x: $\quad xe^x - 2e^x = 0$

11. Simplify: $\quad \log_9 \frac{1}{27}$

12. Given that $\ln A = a$, $\ln B = b$, and $\ln C = c$, express $\ln[(A^2 \sqrt{B})/C^3]$ in terms of a, b, and c.

13. The half-life of plutonium-241 is 13 years. What is the decay constant? Use the approximation $\ln \frac{1}{2} \approx -0.7$ to obtain a numerical answer.

14. Express as a single logarithm with a coefficient of 1: $3 \log_{10} x - \log_{10}(1 - x)$

15. Solve for x, leaving your answer in terms of base e logarithms: $\quad 5e^{2-x} = 12$

16. Let $f(x) = e^{x+1}$. Find a formula for $f^{-1}(x)$ and specify the domain of f^{-1}.

17. Suppose that in 1980 the population of a certain country was 2 million and increasing with a relative growth rate of 2% per year. Estimate the year in which the population will reach 3 million. (Use the approximation $\ln 1.5 \approx 0.40$ to obtain a numerical answer.)

18. Simplify $\ln e + \ln \sqrt{e} + \ln 1 + \ln e^{\ln 10}$

19. A principal of $1000 is deposited at 10% per annum compounded continuously. Estimate the doubling time and then sketch a graph showing how the amount increases with time.

20. Simplify: $\quad \ln(\log_8 56 - \log_8 7)$

In Exercises 21–32, graph the function and specify the asymptote(s) or intercept(s).

21. $y = e^x$

22. $y = -e^{-x}$

23. $y = \ln x$

24. $y = \ln(x + 2)$

25. $y = 2^{x+1} + 1$

26. $y = \log_{10}(-x)$

27. $y = (1/e)^x$

28. $y = (\frac{1}{2})^{-x}$

29. $y = e^{x+1} + 1$

30. $y = -\log_2(x + 1)$

31. $y = \ln(e^x)$

32. $y = e^{\ln x}$

In Exercises 33–49, solve the equation for x. (When logarithms appear in your answer, leave the answer in that form, rather than using a calculator.)

33. $\log_4 x + \log_4(x - 3) = 1$

34. $\log_3 x + \log_3(2x + 5) = 1$

35. $\ln x + \ln(x + 2) = \ln 15$

36. $\log_6 \dfrac{x + 4}{x - 1} = 1$

37. $\log_2 x + \log_2(3x + 10) - 3 = 0$

38. $2 \ln x - 1 = 0$

39. $3 \log_9 x = \frac{1}{2}$

40. $e^{2x} = 6$

41. $e^{1-5x} = 3\sqrt{e}$

42. $2^x = 100$

43. $\log_{10} x - 2 = \log_{10}(x - 2)$

44. $\log_{10}(x^2 - x - 10) = 1$

45. $\ln(x + 2) = \ln x + \ln 2$

46. $\ln(2x) = \ln 2 + \ln x$

47. $\ln x^4 = 4 \ln x$

48. $(\ln x)^4 = 4 \ln x$

49. $\log_{10} x = \ln x$

50. Solve for x: $\quad (\ln x)/(\ln 3) = \ln x - \ln 3$.
Use a calculator to evaluate your result and round off to the nearest integer. *Answer:* 206765

In Exercises 51–66, simplify the expression without using a calculator.

51. $\log_{10} \sqrt{10}$

52. $\log_7 1$

53. $\ln \sqrt[5]{e}$

54. $\log_3 54 - \log_3 2$

55. $\log_{10} \pi - \log_{10} 10\pi$

56. $\log_2 2$

57. 10^t, where $t = \log_{10} 16$

58. $e^{\ln 5}$

59. $\ln(e^4)$

60. $\log_{10}(10^{\sqrt{2}})$

61. $\log_{12} 2 + \log_{12} 18 + \log_{12} 4$

62. $(\log_{10} 8)/(\log_{10} 2)$

63. $(\ln 100)/(\ln 10)$

64. $\log_5 2 + \log_{1/5} 2$

65. $\log_2 \sqrt[7]{16 \sqrt[3]{2\sqrt{2}}}$

66. $\log_{1/8}(\frac{1}{16}) + \log_5 0.02$

In Exercises 67–70, express the quantity in terms of a, b, and c, where $a = \log_{10} A$, $b = \log_{10} B$, and $c = \log_{10} C$.

67. $\log_{10} A^2 B^3 \sqrt{C}$

68. $\log_{10} \sqrt[3]{AC/B}$

69. $16 \log_{10} \sqrt{A} \sqrt[4]{B}$

70. $6 \log_{10}[B^{1/3}/(A\sqrt{C})]$

In Exercises 71–76 find consecutive integers n and $n + 1$ such that the given expression lies between n and $n + 1$. Do not use a calculator.

71. $\log_{10} 209$

72. $\ln 2$

73. $\log_6 100$

74. $\log_{10} \frac{1}{12}$

75. $\log_{10} 0.003$

76. $\log_3 244$

77. **(a)** On the same set of axes, graph the curves $y = \ln(x + 2)$ and $y = \ln(-x) - 1$. According to your graph, in which quadrant do the two curves intersect?

    (b) Find the x-coordinate of the intersection point.

78. The curve $y = ae^{bt}$ passes through the point $(3, 4)$ and has a y-intercept of 2. Find a and b.

79. A certain radioactive substance has a half-life of T years. Find the decay constant (in terms of T).

80. Find the half-life of a certain radioactive substance if it takes T years for one third of a given sample to disintegrate. (Your answer will be in terms of T.)

81. A radioactive substance has a half-life of M min. What percentage of a given sample will remain after $4M$ min?

82. At the start of an experiment, the initial population is a bacteria in a colony. After b hours, c bacteria are present. Determine the population d hours after the start of the experiment.

83. The half-life of a radioactive substance is d days. If you begin with a sample weighing b g, how long will it be until c g remain?

84. Find the half-life of a radioactive substance if it takes D days for P percent of a given sample to disintegrate.

In Exercises 85–90, write the expression as a single logarithm with a coefficient of 1.

85. $\log_{10} 8 + \log_{10} 3 - \log_{10} 12$

86. $4 \log_{10} x - 2 \log_{10} y$

87. $\ln 5 - 3 \ln 2 + \ln 16$

88. $\ln(x^4 - 1) - \ln(x^2 - 1)$

89. $a \ln x + b \ln y$

90. $\ln(x^3 + 8) - \ln(x + 2) - 2 \ln(x^2 - 2x + 4)$

In Exercises 91–98, write the quantity using sums and differences of simpler logarithmic expressions. Express the answer so that logarithms of products, quotients, and powers do not appear.

91. $\ln \sqrt{(x - 3)(x + 4)}$

92. $\log_{10} \dfrac{x^2 - 4}{x + 3}$

93. $\log_{10} \dfrac{3}{\sqrt{1 + x}}$

94. $\ln \dfrac{x^2 \sqrt{2x + 1}}{2x - 1}$

95. $\log_{10} \sqrt[3]{x/100}$

96. $\log_{10} \sqrt{\dfrac{x^2 + 8}{(x^2 + 9)(x + 1)}}$

97. $\ln\left(\dfrac{1 + 2e}{1 - 2e}\right)^3$

98. $\log_{10} \sqrt[3]{x \sqrt{1 + y^2}}$

99. Suppose that A dollars are invested at $R\%$ compounded annually. How many years will it take for the money to double?

100. A sum of $2800 is placed in a savings account at 9% per annum. How much is in the account after two years if the interest is compounded quarterly?

101. A bank offers an interest rate of 9.5% per annum compounded monthly. Compute the effective interest rate.

102. A sum of D dollars is placed in an account at $R\%$ per annum compounded continuously. When will the balance reach E dollars?

103. Compute the doubling time for a sum of D dollars invested at $R\%$ per annum compounded continuously.

104. Your friend invests D dollars at $R\%$ per annum compounded semiannually. You invest an equal amount at the same yearly rate, but compounded daily. How much larger is your account than your friend's after T years?

105. **(a)** You invest $660 at 5.5% per annum compounded quarterly. How long will it take for your balance to reach $1000? Round off your answer to the next quarter of a year.

   (b) You invest D dollars at $R\%$ per annum, compounded quarterly. How long will it take for your balance to reach nD dollars? (Assume that $n > 1$.)

In Exercises 106–110, find the domain of each function.

106. **(a)** $y = \ln x$

   (b) $y = e^x$

107. **(a)** $y = \log_{10} \sqrt{x}$

   (b) $y = \sqrt{\log_{10} x}$

108. **(a)** $f \circ g$ where $f(x) = \ln x$ and $g(x) = e^x$

   (b) $g \circ f$ where f and g are as in part (a)

109. **(a)** $y = \log_{10} |x^2 - 2x - 15|$

   (b) $y = \log_{10}(x^2 - 2x - 15)$

110. $y = \dfrac{2 + \ln x}{2 - \ln x}$

111. Find the range of the function defined by $y = \dfrac{e^x + 1}{e^x - 1}$.

In Exercises 112–122, use the graph and the properties of logarithms to estimate the quantity to the nearest tenth.

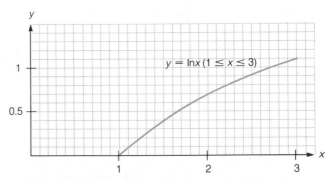

112. $\ln 2$ **113.** $\ln 0.5$ **114.** $\ln 3$

115. $\ln \frac{1}{9}$ **116.** $\ln 6$ **117.** $\ln 72$

118. $\ln 40.5$ *Hint:* $40.5 = \frac{81}{2}$

119. e *Hint:* If $x = e$, then $\ln x = 1$.

120. $e^{1.1}$ *Hint:* If $x = e^{1.1}$, then $\ln x = 1.1$.

121. $\log_2 3$ **122.** $\log_3 6$

123. Lester R. Brown, in his book *State of the World 1985* (New York: W. W. Norton and Company, 1985), makes the following statement:

The projected [*population*] growth for North America, all of Europe, and the Soviet Union is less than the additions expected in either Bangladesh or Nigeria.

In this exercise, you are asked to carry out the type of calculations that could be used to support Lester Brown's projection for the period 1990–2025. The source of the data in the following table is the Population Division of the United Nations.

(a) Complete the table, assuming that the populations grow exponentially and that the indicated growth rates are valid over the period 1990–2025.

(b) According to the projections in part (a), what will be the net increase in Nigeria's population over the period 1990–2025?

REGION	1990 POPULATION (millions)	PROJECTED RELATIVE GROWTH RATE (percent per year)	2025 PROJECTED POPULATION (millions)
North America	275.2	0.7	?
Soviet Union	291.3	0.7	?
Europe	499.5	0.2	?
Nigeria	113.3	3.1	?

(c) According to the projections in part (a), what will be the net increase in the combined populations of North America, the Soviet Union, and Europe over the period 1990–2025?

(b) Compare your answers in parts (b) and (c). Do your results support or contradict Lester Brown's projection?

124. As of 1989, ten countries in the world had populations exceeding 100 million. These countries are listed (in order of population, from largest to smallest) in the table. According to *World Population Profile: 1987*, published by the United States Bureau of the Census, "The latest projections suggest that India's population may surpass China's in less than 60 years, or before today's youngsters in both countries reach old age." Using the data in the table, estimate the populations of the ten countries in the year 2050. (Assume that the indicated growth rates are valid over the period 1989–2050.) Tabulate your final answers, listing the countries in order of population, from largest to smallest. Do your results for China and India support the projection by the Census Bureau?

COUNTRY	1989 POPULATION (millions)	RELATIVE GROWTH RATE (percent per year)
1. China	1103.9	1.4
2. India	835.0	2.2
3. Soviet Union	289.0	1.0
4. United States	248.8	0.7
5. Indonesia	184.6	2.0
6. Brazil	147.4	2.0
7. Japan	123.2	0.5
8. Nigeria	115.3	2.9
9. Bangladesh	114.7	2.8
10. Pakistan	110.4	2.9

Source: *1989 World Population Data Sheet*, published by Population Reference Bureau, Inc., Washington, D.C.

125. The figure shows a portion of the graph of $y = e^x$, for $1.0970 \leq x \leq 1.1000$.

(a) Use the graph to estimate $\ln 3$ to four decimal places.

(b) Using your calculator, compute the percent error in the approximation in part (a).

(c) Using the approximation in part (a), estimate the quantities $\ln \sqrt{3}$, $\ln 9$, and $\ln \frac{1}{3}$.

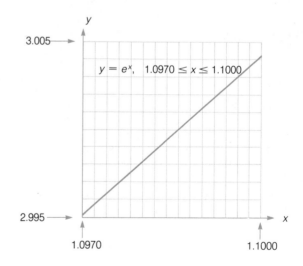

The subject of trigonometry is an excellent example of a branch of mathematics . . . which was motivated by both practical and intellectual interests—surveying, map-making, and navigation on the one hand, and curiosity about the size of the universe on the other. With it the Alexandrian mathematicians triangulated the universe and rendered precise their knowledge about the Earth and the heavens.

Morris Klein in *Mathematics in Western Culture* (New York: Oxford University Press, 1953)

INTRODUCTION   There are two general approaches to the study of trigonometry at the precalculus level. The first approach centers around the study of triangles. Indeed, the word "trigonometry" is derived from two Greek words, *trigonon*, meaning triangle, and *metria*, meaning measurement. This is the approach that we shall follow in this chapter. The second approach, which is in a sense a generalization of the first, we will take up in Chapter 6. Before getting down to specifics, we offer the following overview of Chapters 5 and 6 in terms of functions.

	INPUTS FOR THE FUNCTIONS	OUTPUTS	EMPHASIS
Chapter 5	Angles	Real numbers	The trigonometric functions are used to study triangles.
Chapter 6	Real numbers	Real numbers	The objects of study are the trigonometric functions themselves.

5.1 ▼ TRIGONOMETRIC FUNCTIONS OF ACUTE ANGLES

Most of the functions that we have considered up to this point were defined by means of equations that involved only the basic operations of algebra. Examples of such functions are

$$f(x) = x^2 \qquad g(x) = mx + b \qquad h(x) = \sqrt{x}$$

In this and the next two chapters, we will study the so-called trigonometric functions. The definitions of these functions involve some geometry, as well as algebra.

Let θ denote an acute angle in a right triangle, as indicated in Figure 1.* We define the six trigonometric functions of the acute angle θ in the box that follows.

FIGURE 1

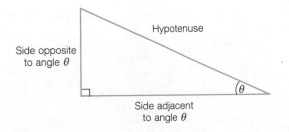

---

* The symbol "θ" is the lowercase Greek letter *theta*. Greek letters are often used to represent angles. See the back endpapers for a listing of the complete Greek alphabet.

Notice that $\sin\theta$ and $\csc\theta$ are reciprocals of one another. Similarly, $\cos\theta$ and $\sec\theta$ are reciprocals, as are $\tan\theta$ and $\cot\theta$.

NAME OF FUNCTION	ABBREVIATION	DEFINITION
sine	sin	$\sin\theta = \dfrac{\text{length of side opposite angle } \theta}{\text{length of hypotenuse}}$
cosine	cos	$\cos\theta = \dfrac{\text{length of side adjacent angle } \theta}{\text{length of hypotenuse}}$
tangent	tan	$\tan\theta = \dfrac{\text{length of side opposite angle } \theta}{\text{length of side adjacent angle } \theta}$
cosecant	csc	$\csc\theta = \dfrac{\text{length of hypotenuse}}{\text{length of side opposite angle } \theta}$
secant	sec	$\sec\theta = \dfrac{\text{length of hypotenuse}}{\text{length of side adjacent angle } \theta}$
cotangent	cot	$\cot\theta = \dfrac{\text{length of side adjacent angle } \theta}{\text{length of side opposite angle } \theta}$

EXAMPLE 1   Figure 2 shows two right triangles. The first right triangle has sides 5, 12, and 13. The second right triangle is similar to the first (the angles are the same), but each side is 10 times longer than the corresponding side in the first triangle. Calculate and compare the values of $\sin\theta$, $\cos\theta$, and $\tan\theta$ for both triangles.

FIGURE 2

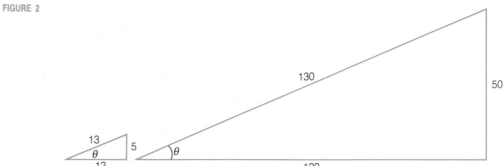

Solution   SMALL TRIANGLE

$$\sin\theta = \frac{\text{opposite}}{\text{hypotenuse}} = \frac{5}{13}$$

$$\cos\theta = \frac{\text{adjacent}}{\text{hypotenuse}} = \frac{12}{13}$$

$$\tan\theta = \frac{\text{opposite}}{\text{adjacent}} = \frac{5}{12}$$

LARGE TRIANGLE

$$\sin\theta = \frac{50}{130} = \frac{5}{13}$$

$$\cos\theta = \frac{120}{130} = \frac{12}{13}$$

$$\tan\theta = \frac{50}{120} = \frac{5}{12}$$

Observation   The corresponding values for $\sin\theta$, $\cos\theta$, and $\tan\theta$ are the same for both triangles.   ▲

The point of Example 1 is this: For a given acute angle θ, the values of the trigonometric functions depend on the *ratios* of the lengths of the sides but not on the size of the particular right triangle in which θ resides.

EXAMPLE 2 Let θ be the acute angle indicated in Figure 3. Determine the six quantities $\sin\theta$, $\cos\theta$, $\tan\theta$, $\csc\theta$, $\sec\theta$, and $\cot\theta$.

Solution We use the definitions:

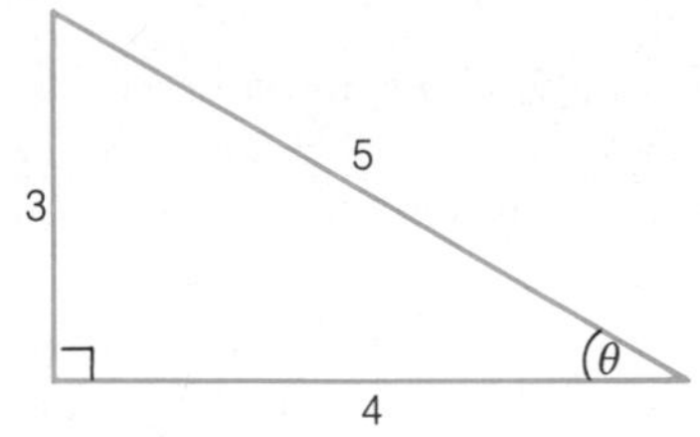

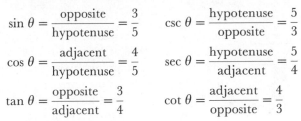

$$\sin\theta = \frac{\text{opposite}}{\text{hypotenuse}} = \frac{3}{5} \qquad \csc\theta = \frac{\text{hypotenuse}}{\text{opposite}} = \frac{5}{3}$$

$$\cos\theta = \frac{\text{adjacent}}{\text{hypotenuse}} = \frac{4}{5} \qquad \sec\theta = \frac{\text{hypotenuse}}{\text{adjacent}} = \frac{5}{4}$$

$$\tan\theta = \frac{\text{opposite}}{\text{adjacent}} = \frac{3}{4} \qquad \cot\theta = \frac{\text{adjacent}}{\text{opposite}} = \frac{4}{3}$$

FIGURE 3

Note the pairs of answers that are reciprocals. (This helps in memorizing the definitions.) ▲

EXAMPLE 3 Let β be the acute angle indicated in Figure 4. Find $\sin\beta$ and $\cos\beta$.

Solution In view of the definitions, we need to know the length of the hypotenuse in Figure 4. If we call this length h, then by the Pythagorean theorem we have

$$h^2 = 3^2 + 1^2 = 10$$
$$h = \sqrt{10}$$

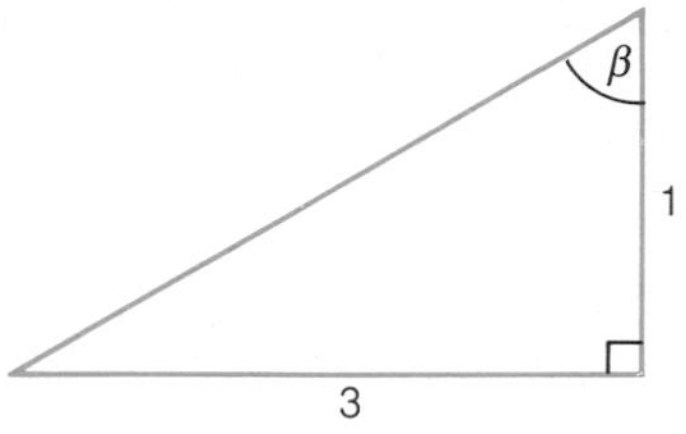

Therefore

$$\sin\beta = \frac{\text{opposite}}{\text{hypotenuse}} = \frac{3}{\sqrt{10}} = \frac{3\sqrt{10}}{10}$$

FIGURE 4

and

$$\cos\beta = \frac{\text{adjacent}}{\text{hypotenuse}} = \frac{1}{\sqrt{10}} = \frac{\sqrt{10}}{10}$$ ▲

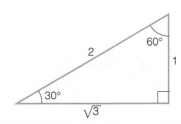

FIGURE 5

There are certain angles for which we can evaluate the trigonometric functions without using a calculator or tables. Three such angles are $30°$, $45°$, and $60°$. To do this, we rely on the following two basic results from elementary geometry

PROPERTY SUMMARY

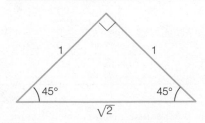

FIGURE 6

30°–60° AND 45°–45° RIGHT TRIANGLES
1. (Refer to Figure 5.) In a 30°–60° right triangle, the length of the side opposite the 30° angle is one-half the length of the hypotenuse. The lengths of the three sides are therefore in the proportions 1–2–$\sqrt{3}$.
2. (Refer to Figure 6.) In a 45°–45° right triangle, the lengths of the two legs are equal and the lengths of the three sides are in the proportions 1–1–$\sqrt{2}$.

Using Figures 5 and 6, we can evaluate the trigonometric functions of 30°, 45°, and 60° by inspection. For instance, using Figure 5, we have

$$\sin 30° = \frac{\text{opposite}}{\text{hypotenuse}} = \frac{1}{2}$$

As another example, this time using Figure 6, we have

$$\cos 45° = \frac{\text{adjacent}}{\text{hypotenuse}} = \frac{1}{\sqrt{2}} = \frac{\sqrt{2}}{2}$$

In Table 1, we list the results that are obtained in this manner for sine, cosine, and tangent. (We've listed only the values for sine, cosine, and tangent in the table since the remaining three values are just the reciprocals of these.) The results in Table 1 should be memorized.

There are numerous identities involving the trigonometric functions. For now, in the context of right-triangle trigonometry we'll consider only the three most basic ones. In Section 5.4 and in Chapter 6, we will take up this subject in greater detail and from a broader perspective. Recall that an **identity** is an equation that is satisfied by all relevant values of the variables concerned. Two examples of identities are $(x - y)(x + y) = x^2 - y^2$ and $x^3 = x^4/x$. The first of these is true no matter what real numbers are used for x and y; the second is true for all real numbers except $x = 0$. In the box that follows, the notations

$$\sin^2 \theta \qquad \text{and} \qquad \cos^2 \theta$$

stand for

$$(\sin \theta)^2 \qquad \text{and} \qquad (\cos \theta)^2$$

respectively. (We will say more about this common notational convention in Section 5.4.)

TABLE 1

θ	$\sin \theta$	$\cos \theta$	$\tan \theta$
30°	$\dfrac{1}{2}$	$\dfrac{\sqrt{3}}{2}$	$\dfrac{\sqrt{3}}{3}$
45°	$\dfrac{\sqrt{2}}{2}$	$\dfrac{\sqrt{2}}{2}$	1
60°	$\dfrac{\sqrt{3}}{2}$	$\dfrac{1}{2}$	$\sqrt{3}$

PROPERTY SUMMARY   **BASIC IDENTITIES FOR SINE AND COSINE** $(0° < \theta < 90°)$

IDENTITY	EXAMPLES
1. $\sin^2 \theta + \cos^2 \theta = 1$	1. $\sin^2 30° + \cos^2 30° = 1$ $\sin^2 19° + \cos^2 19° = 1$
2. $\dfrac{\sin \theta}{\cos \theta} = \tan \theta$	2. $\dfrac{\sin 60°}{\cos 60°} = \tan 60°$
3. $\sin(90° - \theta) = \cos \theta$ and $\cos(90° - \theta) = \sin \theta$	3. $\sin 70° = \cos 20°$ $\sin 20° = \cos 70°$

All three of these identities can be proved by referring to Figure 7.

PROOF THAT $\sin^2 \theta + \cos^2 \theta = 1$ Looking at Figure 7, we have $\sin \theta = a/c$ and $\cos \theta = b/c$. Thus

$$\sin^2 \theta + \cos^2 \theta = \left(\frac{a}{c}\right)^2 + \left(\frac{b}{c}\right)^2 = \frac{a^2}{c^2} + \frac{b^2}{c^2} = \frac{a^2 + b^2}{c^2}$$

$$= \frac{c^2}{c^2} \qquad \text{using } a^2 + b^2 = c^2, \text{ the Pythagorean theorem}$$

$$= 1 \qquad \text{as required}$$

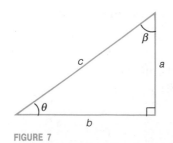

FIGURE 7

PROOF THAT $(\sin\theta)/(\cos\theta) = \tan\theta$ Again, with reference to Figure 7, we have, by definition, $\sin\theta = a/c$, $\cos\theta = b/c$, and $\tan\theta = a/b$. Therefore

$$\frac{\sin\theta}{\cos\theta} = \frac{a/c}{b/c}$$

$$= \frac{a}{c} \times \frac{c}{b} = \frac{a}{b}$$

$$= \tan\theta \qquad \text{as required}$$

PROOF THAT $\sin(90° - \theta) = \cos\theta$ First of all, since the sum of the angles in any triangle is 180°, we have

$$\theta + \beta + 90° = 180°$$

$$\beta = 90° - \theta$$

Then $\sin(90° - \theta) = \sin\beta = b/c$. But also $\cos\theta = b/c$. Thus

$$\sin(90° - \theta) = \cos\theta$$

since both expressions equal b/c. This is what we wanted to prove.

The proof that $\cos(90° - \theta) = \sin\theta$ is entirely similar and we shall omit it. We can conveniently summarize these last two results by recalling the notion of complementary angles. Two acute angles are said to be **complementary** provided their sum is 90°. Thus, the two angles θ and $90° - \theta$ are complementary. In view of this, we can restate the last two results as follows.

> If two angles are complementary, then the sine of (either) one equals the cosine of the other.

Incidentally, this result gives us an insight into the origin of the term *cosine*. *Cosine* is a shortened form of the phrase *complement's sine*.

EXAMPLE 4   Suppose that B is an acute angle and $\cos B = \frac{2}{5}$. Find $\sin B$ and $\tan B$.

Solution   We'll show two methods. The first uses the identities $\sin^2 B + \cos^2 B = 1$ and $(\sin B)/(\cos B) = \tan B$. The second makes direct use of the Pythagorean theorem.

FIRST METHOD   Replace $\cos B$ with $\frac{2}{5}$ in the identity $\sin^2 B + \cos^2 B = 1$. This yields

$$\sin^2 B + \left(\frac{2}{5}\right)^2 = 1$$

$$\sin^2 B = 1 - \frac{4}{25} = \frac{21}{25}$$

$$\sin B = \sqrt{\frac{21}{25}} = \frac{\sqrt{21}}{5}$$

Notice that we've chosen the positive square root here. This is because the values of the trigonometric functions of an acute angle are by definition positive.

Caution   In Section 5.3 we'll extend the definitions of the trigonometric functions to include angles that are not acute. In those cases, some of the trigonometric values will not be positive; we'll then need to pay more attention to the matter of signs.

For tan B, we have

$$\tan B = \frac{\sin B}{\cos B} = \frac{\sqrt{21}/5}{2/5} = \frac{\sqrt{21}}{2}$$

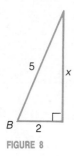

FIGURE 8

ALTERNATE METHOD Since $\cos B = \frac{2}{5} = $ adjacent/hypotenuse, we can work with a right triangle labeled as in Figure 8. Using the Pythagorean theorem, we have $2^2 + x^2 = 5^2$, from which it follows that $x = \sqrt{21}$. Consequently,

$$\sin B = \frac{\text{opposite}}{\text{hypotenuse}} = \frac{x}{5} = \frac{\sqrt{21}}{5}$$

and

$$\tan B = \frac{\text{opposite}}{\text{adjacent}} = \frac{x}{2} = \frac{\sqrt{21}}{2} \qquad \text{as obtained previously}$$

▲

EXAMPLE 5 If θ is an acute angle and $\sin \theta = y$, express the other five trigonometric values as functions of y.

Solution We can use either of the methods shown in Example 4. We'll demonstrate the first method here. Replacing $\sin \theta$ with y in the identity $\sin^2 \theta + \cos^2 \theta = 1$ yields

$$y^2 + \cos^2 \theta = 1$$
$$\cos^2 \theta = 1 - y^2$$
$$\cos \theta = \sqrt{1 - y^2} \qquad \text{(Why is the positive root appropriate in this context?)}$$

For $\tan \theta$, now, we have

$$\tan \theta = \frac{\sin \theta}{\cos \theta} = \frac{y}{\sqrt{1 - y^2}}$$

The values of $\csc \theta$, $\sec \theta$, and $\cot \theta$ are, by definition, the reciprocals of $\sin \theta$, $\cos \theta$, and $\tan \theta$, respectively. So we have

$$\csc \theta = \frac{1}{y} \qquad \sec \theta = \frac{1}{\sqrt{1 - y^2}} \qquad \cot \theta = \frac{\sqrt{1 - y^2}}{y}$$

▲

▼ EXERCISE SET 5.1

A

1. **(a)** Check that $8^2 + 15^2 = 17^2$. This shows that the triangle in the accompanying figure is indeed a right triangle.

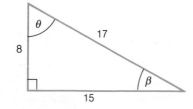

(b) Find $\sin \theta$ and $\cos \beta$.
(c) Find $\cos \theta$ and $\sin \beta$.
(d) Find $\tan \theta$, $\csc \theta$, $\sec \theta$, and $\cot \theta$.
(e) Find $\tan \beta$, $\csc \beta$, $\sec \beta$, and $\cot \beta$.

2. **(a)** Find x in the accompanying figure.
(b) Evaluate the six trigonometric functions of θ.
(c) Evaluate the six trigonometric functions of β.

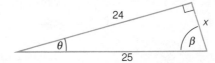

In Exercises 3–10, refer to the following figure. (Each problem, however, is independent of the others.)

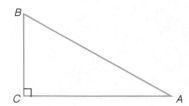

3. Suppose $AC = 3$ and $BC = 2$.
    (a) Find $\sin A$, $\cos A$, and $\tan A$.
    (b) Using the values in part (a), verify that $\sin^2 A + \cos^2 A = 1$ and $(\sin A)/(\cos A) = \tan A$.
    (c) Find $\cos B$. Then check that $\sin^2 A + \cos^2 B \neq 1$.

4. Suppose $AC = 6$ and $BC = 2$.
    (a) Compute $\sin B$, $\cos B$, and $\tan B$.
    (b) Using the values in part (a), verify that $\sin^2 B + \cos^2 B = 1$ and $(\sin B)/(\cos B) = \tan B$.
    (c) Find $\sin A$. Then check that $\sin^2 A + \cos^2 B \neq 1$.

5. If $AB = 13$ and $BC = 5$, compute the values of the six trigonometric functions of angle A.

6. If $AB = 29$ and $AC = 21$, compute the values of the six trigonometric functions of angle B.

7. Suppose $AC = 1$ and $BC = \frac{3}{4}$.
    (a) Compute $\sin B$ and $\cos A$.
    (b) Compute $\cos B$ and $\sin A$.
    (c) Compute the product $(\tan A)(\tan B)$.

8. Suppose $AC = BC = 1$.
    (a) Compute $\csc A$, $\sec A$, and $\cot A$.
    (b) Compute $\csc B$, $\sec B$, and $\cot B$.

9. Suppose $AC = \sqrt{11}$ and $BC = \sqrt{5}$.
    (a) Compute $\sin A$ and $\cos A$. Then verify that $\sin^2 A + \cos^2 A = 1$.
    (b) Compute $\tan B$, $\sin B$, and $\cos B$. Then verify that $\tan B = (\sin B)/(\cos B)$.

10. Suppose $AC = \sqrt{2} + 1$ and $BC = \sqrt{2} - 1$. Compute the values of the six trigonometric functions of angle B. For each answer, rationalize the denominator.

In Exercises 11–22, verify that each equation is correct. (The purpose of Exercises 11–22 is twofold. First, doing the problems will help you to memorize the values in Table 1. Second, the exercises serve as an algebra review.)

11. $\cos 60° = \cos^2 30° - \sin^2 30°$
12. $\cos 60° = 1 - 2 \sin^2 30°$
13. $\sin^2 30° + \sin^2 45° + \sin^2 60° = \frac{3}{2}$
14. $\sin 30° \cos 60° + \cos 30° \sin 60° = 1$
15. $2 \sin 30° \cos 30° = \sin 60°$

16. $2 \sin 45° \cos 45° = 1$
17. $\sin 30° = \sqrt{\dfrac{1 - \cos 60°}{2}}$
18. $\cos 30° = \sqrt{\dfrac{1 + \cos 60°}{2}}$
19. $\tan 30° = \dfrac{\sin 60°}{1 + \cos 60°}$
20. $\tan 30° = \dfrac{1 - \cos 60°}{\sin 60°}$
21. $1 + \tan^2 45° = \sec^2 45°$
22. $1 + \cot^2 60° = \csc^2 60°$

For Exercises 23–26, refer to the following figures. In each case, express the indicated trigonometric values as functions of x. (Rationalize any denominators containing radicals.)

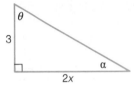

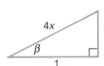

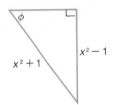

23. **(a)** $\sin \theta$, $\cos \theta$, $\tan \theta$ **(b)** $\sin^2 \theta$, $\cos^2 \theta$, $\tan^2 \theta$
     (c) $\sin(90° - \theta)$, $\cos(90° - \theta)$, $\tan(90° - \theta)$
24. **(a)** $\csc \alpha$, $\sec \alpha$, $\cot \alpha$ **(b)** $\sin^2 \alpha + \cos^2 \alpha + \tan^2 \alpha$
25. **(a)** $\sin \beta$, $\cos \beta$, $\tan \beta$ **(b)** $\csc \beta$, $\sec \beta$, $\cot \beta$
     (c) $\sin(90° - \beta)$, $\cos(90° - \beta)$, $\tan(90° - \beta)$
26. **(a)** $\sin \phi$, $\cos \phi$, $\tan \phi$ **(b)** $(\csc \phi)(\sec \phi)(\cot \phi)$
     (c) $\sin(90° - \phi)$, $\cos(90° - \phi)$, $\tan(90° - \phi)$

In Exercises 27–32, use the given information to determine the values of the remaining five trigonometric functions. (The angles are assumed to be acute angles.)

27. $\cos B = \frac{4}{7}$
28. $\cos B = \frac{3}{8}$
29. $\sin \theta = 2\sqrt{3}/5$
30. $\sin \theta = \frac{7}{25}$
31. $\tan A = \dfrac{\sqrt{2} - 1}{\sqrt{2} + 1}$
32. $\tan A = \dfrac{2 - \sqrt{3}}{2 + \sqrt{3}}$

In Exercises 33–36, use the given information to express the remaining five trigonometric values of the acute angle θ in terms of x. (Rationalize any denominators containing radicals.)

33. $\sin \theta = x/2$
34. $\sin \theta = 3x/5$
35. $\cos \theta = x^2$
36. $\cos \theta = (x - 1)/(x + 1)$

[c] *Exercises 37–40 are calculator exercises. The goal is simply to familiarize you with the use of your calculator for computing trigonometric values. Round off each of the final answers to five decimal places. Note: Check to see that your calculator is set in*

the "degree mode" rather than the "radian mode." (Radian measure for angles is discussed in the next chapter.)

37. **(a)** Evaluate cos 30°, cos 45°, and sin 60°.
(b) In Table 1, there are expressions (involving radicals) for cos 30°, cos 45°, and sin 60°. Evaluate these expressions, and check to see that your results agree with those in part (a).

38. **(a)** Evaluate tan 30° and tan 60°.
(b) Evaluate the expressions for tan 30° and tan 60° given in Table 1. Check to see that the results agree with those in part (a).

39. Evaluate csc 25°, sec 25°, and cot 25°. (To evaluate csc 25°, for example, first evaluate sin 25°, then use the $\boxed{1/x}$ key.)

40. **(a)** Evaluate (sin 59°)/(cos 59°).
(b) Evaluate tan 59°, and check that your answer agrees with that in part (a).
(c) Use your calculator to evaluate $\sin^2 59° + \cos^2 59°$.

B

For Exercises 41 and 42, refer to the following figure. In each case, say which of the two given quantities is larger. If the quantities are equal, say so.

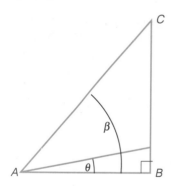

41. **(a)** sin θ, sin β
(b) csc θ, csc β
(c) cos θ, cos β

42. **(a)** tan θ, tan β
(b) cot θ, cot β
(c) sec θ, sec β

43. The following figure shows a 30°–60° right triangle, △ABC. Prove that AC = 2AB. *Suggestion:* Construct

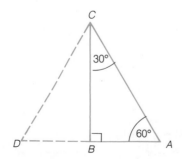

△DBC as shown, congruent to △ABC. Then note that △ADC is equilateral.

44. Suppose that β is an acute angle and

$$\sin \beta = \frac{m^2 - n^2}{m^2 + n^2} \qquad (m > n > 0)$$

Show that

$$\cos \beta = \frac{2mn}{m^2 + n^2} \qquad \text{and} \qquad \tan \beta = \frac{m^2 - n^2}{2mn}$$

45. This exercise shows how to obtain radical expressions for sin 15° and cos 15°. In the figure, assume that AB = BD = 2.

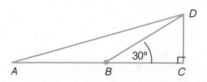

(a) In the right triangle BCD, note that DC = 1 because $\overline{DC}$ is opposite the 30° angle and BD = 2. Use the Pythagorean theorem to show that $BC = \sqrt{3}$.
(b) Use the Pythagorean theorem to show that $AD = 2\sqrt{2 + \sqrt{3}}$.
(c) Show that the expression for AD in part (b) is equal to $\sqrt{6} + \sqrt{2}$. *Hint:* Two nonnegative quantities are equal if and only if their squares are equal.
(d) Explain why ∠BAD = ∠BDA.
(e) According to a theorem from geometry, an exterior angle of a triangle is equal to the sum of the two nonadjacent interior angles. Apply this to △ABD with exterior angle DBC = 30°, and show that ∠BAD = 15°.
(f) Using the figure and the values that you have obtained for the lengths, conclude that

$$\sin 15° = \frac{1}{\sqrt{6} + \sqrt{2}} \qquad \cos 15° = \frac{2 + \sqrt{3}}{\sqrt{6} + \sqrt{2}}$$

(g) Rationalize the denominators in part (f) to obtain

$$\sin 15° = \frac{\sqrt{6} - \sqrt{2}}{4} \qquad \cos 15° = \frac{\sqrt{6} + \sqrt{2}}{4}$$

(h) Use your calculator to check the results in part (g).

46. If an angle θ is an integral multiple of 3°, then the real number sin θ is either rational or expressible in terms of radicals. You've already seen examples of this with the angles 30°, 45°, and 60°. The accompanying table gives the values of sin θ for some multiples of 3°. Use your calculator to check the entries in the table. In Exercise Set 6.6, you will see how these results are derived. *Remark:* If θ is not an integral multiple of $3°$, then the real number $\sin \theta$ cannot be expressed in terms of radicals within the real number system.

θ	$\sin \theta$
$3°$	$\frac{1}{16}[(\sqrt{6} + \sqrt{2})(\sqrt{5} - 1) - 2(\sqrt{3} - 1)\sqrt{5 + \sqrt{5}}]$
$6°$	$\frac{1}{8}(\sqrt{30 - 6\sqrt{5}} - \sqrt{5} - 1)$
$9°$	$\frac{1}{8}(\sqrt{10} + \sqrt{2} - 2\sqrt{5 - \sqrt{5}})$
$12°$	$\frac{1}{8}(\sqrt{10 + 2\sqrt{5}} - \sqrt{15} + \sqrt{3})$
$15°$	$\frac{1}{4}(\sqrt{6} - \sqrt{2})$
$18°$	$\frac{1}{4}(\sqrt{5} - 1)$

47. This exercise shows how to obtain radical expressions for $\sin 18°$ and $\cos 18°$, using the following figure.
(a) Find $\angle B$, $\angle BDC$, and $\angle ADC$.
(b) Why does $AC = BC = BD$?
For the rest of this problem, assume $AD = 1$.
(c) Why does $CD = 1$?
(d) Let x denote the common lengths of $\overline{AC}$, $\overline{BC}$, and $\overline{BD}$. Use similar triangles to deduce that $x/(1 + x) = 1/x$. Then show that $x = \frac{1}{2} + \frac{1}{2}\sqrt{5}$.

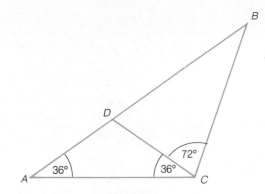

(e) In $\triangle BDC$, draw an altitude from B to $\overline{DC}$, meeting $\overline{DC}$ at F. Use right triangle BFC to conclude that $\sin 18° = 1/(1 + \sqrt{5})$.
(f) Rationalize the denominator in part (e) to obtain $\sin 18° = \frac{1}{4}(\sqrt{5} - 1)$.
(g) Use the result in part (f), along with the identity $\sin^2 \theta + \cos^2 \theta = 1$, to show that

$$\cos 18° = \tfrac{1}{4}\sqrt{10 + 2\sqrt{5}}$$

5.2 ▼ RIGHT-TRIANGLE APPLICATIONS

We continue the work we began in the previous section on right-triangle trigonometry. As prerequisites for this section, you'll need to have memorized the definitions of the trigonometric functions, given in the box on page 254, and the table of trigonometric values for $30°$, $45°$, and $60°$ on page 256.

EXAMPLE 1 Use one of the trigonometric functions to find x in Figure 1.

Solution Relative to the given $30°$ angle, x is the adjacent side. The length of the hypotenuse is 100 cm. Since the adjacent side and the hypotenuse are involved, we use the cosine function here:

$$\cos 30° = \frac{\text{adjacent}}{\text{hypotenuse}} = \frac{x}{100}$$

Consequently

$$x = 100 \cos 30° = 100 \cdot \left(\frac{\sqrt{3}}{2}\right) = 50\sqrt{3} \text{ cm}$$

This is the result we are looking for. ▲

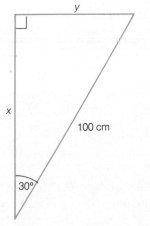

FIGURE 1

We used the cosine function in Example 1 because the adjacent side and the hypotenuse were involved. We could instead use the secant. In that case, again with reference to Figure 1, the calculations look like this:

$$\sec 30° = \frac{\text{hypotenuse}}{\text{adjacent}} = \frac{100}{x}$$

$$\frac{2}{\sqrt{3}} = \frac{100}{x}$$

$$2x = 100\sqrt{3}$$

$$x = \frac{100\sqrt{3}}{2} = 50\sqrt{3} \text{ cm} \qquad \text{as obtained previously}$$

EXAMPLE 2   Find y in Figure 1 (on the previous page).

Solution   As you can see from Figure 1, the side of length y is opposite the 30° angle. Furthermore, we are given the length of the hypotenuse. Since the opposite side and the hypotenuse are involved, we employ the sine function. This yields

$$\sin 30° = \frac{\text{opposite}}{\text{hypotenuse}} = \frac{y}{100}$$

That is,

$$\sin 30° = \frac{y}{100}$$

and therefore

$$y = 100 \sin 30° = 100 \cdot \tfrac{1}{2} = 50 \text{ cm}$$

This is the required answer. Actually, we could have obtained this particular result much faster by recalling that in the 30°–60° right triangle, the side opposite the 30° angle, namely y, is half the hypotenuse. That is, $y = \frac{100}{2} = 50$ cm, as we obtained previously.   ▲

EXAMPLE 3   A ladder, which is leaning against the side of a building, forms an angle of 50° with the ground. If the foot of the ladder is 12 ft from the base of the building, how far up the side of the building does the ladder reach? See Figure 2.

Solution   In Figure 2 we have used y to denote the required distance. Notice that y is opposite the 50° angle, while the given side is adjacent to that angle. Since the opposite and adjacent sides are involved, we employ the tangent function. (The cotangent function could also be used.) We have

$$\tan 50° = \frac{y}{12}$$

and therefore

$$y = 12 \tan 50°$$

Without the use of a calculator or tables, this is our final answer. On the other hand, using a calculator we find that tan 50° is approximately 1.19. Thus we have

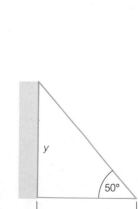

FIGURE 2

$$y \approx (12)(1.19) = 14 \text{ ft} \quad \text{to the nearest foot} \qquad ▲$$

In the next example we derive a useful formula for the area of a triangle. The formula can be used when we are given two sides and the included angle, but not the height.

EXAMPLE 4   Show that the area of the triangle in Figure 3 is given by $A = \frac{1}{2}ab \sin \theta$.

Solution   We draw an altitude, as shown in Figure 3, and we call the length of this altitude h. Then we have

$$\sin \theta = \frac{h}{a}$$

and therefore

$$h = a \sin \theta$$

This value for h can now be used in the usual formula for the area of a triangle:

$$A = \frac{1}{2}bh = \frac{1}{2}b(a \sin \theta) = \frac{1}{2}ab \sin \theta \qquad \blacktriangle$$

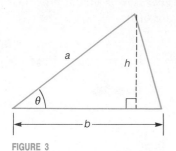

FIGURE 3

The formula that we just derived in Example 4 is worth remembering. In words, the formula states that *the area of a triangle is equal to half the product of the lengths of two of the sides times the sine of the included acute angle.* In Section 5.3, we will see that the formula remains valid even when the included angle is not an acute angle.

EXAMPLE 5   Find the area of the triangle in Figure 4.

Solution   In the formula $A = \frac{1}{2}ab \sin \theta$, we let $a = 3$ cm, $b = 12$ cm, and $\theta = 60°$. Then

$$A = \frac{1}{2}(3)(12) \sin 60° = (18)\frac{\sqrt{3}}{2}$$

$$= 9\sqrt{3} \text{ cm}^2$$

This is the required area. $\qquad \blacktriangle$

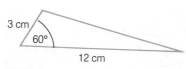

3 cm
60°
12 cm

FIGURE 4

EXAMPLE 6   Figure 5 shows a regular pentagon inscribed in a circle of radius 2 in. (*Regular* means that all of the sides are equal and all of the angles are equal.) Find the area of the pentagon.

Solution   The idea here is to find first the area of triangle BOA using the area formula from Example 4. Then, since the pentagon is composed of five such identical triangles, the area of the pentagon will be five times the area of triangle BOA. We will make use of the result from geometry that, in a regular n-sided polygon, the central angle is $360°/n$. In our case, we therefore have

$$\angle BOA = \frac{360°}{5} = 72°$$

We can now find the area of triangle BOA:

$$\text{area} = \frac{1}{2}ab \sin \theta$$

$$= \frac{1}{2}(2)(2) \sin 72°$$

$$= 2 \sin 72° \text{ in.}^2$$

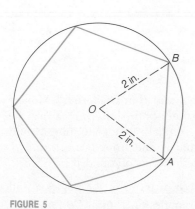

O
B
A
2 in.
2 in.

FIGURE 5

The area of the pentagon is five times this, or $10 \sin 72°$ in.^{2} (Using a calculator, this is about 9.51 in.^{2}) ▲

Now we introduce some terminology that will be used in the next two examples. Suppose that a surveyor sights an object at a point above the horizontal, as indicated in Figure 6(a). Then the angle between the line of sight and the horizontal is called the **angle of elevation**. The **angle of depression** is similarly defined for an object below the horizontal, as shown in Figure 6(b).

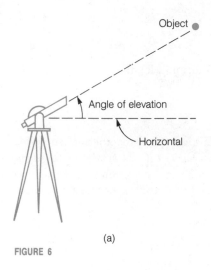

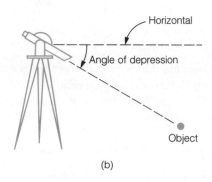

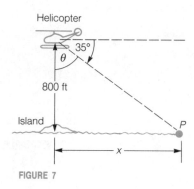

(a)

(b)

FIGURE 6

FIGURE 7

EXAMPLE 7   A helicopter hovers 800 ft directly above a small island that is off the California coast. From the helicopter, the pilot takes a sighting to a point P directly ashore on the mainland, at the water's edge. If the angle of depression is 35°, how far off the coast is the island? See Figure 7.

Solution   Let x denote the distance from the island to the mainland. Then, as you can see from Figure 7, we have $\theta + 35° = 90°$, from which it follows that $\theta = 55°$. Now we can write

$$\tan 55° = \frac{x}{800}$$

or

$$x = 800 \tan 55° \approx 1150 \text{ ft} \quad \text{using a calculator and} \atop \text{rounding to the nearest 50 feet}$$

▲

EXAMPLE 8   Two satellite tracking stations, located at points A and B in California's Mojave Desert, are 200 miles apart. At a prearranged time, both stations measure the angle of elevation of a satellite as it crosses the vertical plane containing A and B. (See Figure 8.) If the angles of elevation from A and from B are α and β, respectively, express the altitude h of the satellite in terms of α and β.

Solution   We want to express the length $h = SC$ in Figure 8 in terms of the angles α and β. Note that $\overline{SC}$ is a side of both of the right triangles SCA and SCB. Working

FIGURE 8

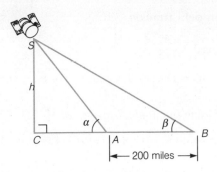

first in right triangle SCB, we have

$$\cot \beta = \frac{CA + 200}{h}$$

or

$$h \cot \beta = CA + 200 \qquad (1)$$

We can eliminate CA from equation (1) as follows. Looking at right triangle SCA, we have

$$\cot \alpha = \frac{CA}{h} \qquad \text{and thus} \qquad CA = h \cot \alpha$$

Using this last equation to substitute for CA in equation (1), we obtain

$$h \cot \beta = h \cot \alpha + 200$$
$$h \cot \beta - h \cot \alpha = 200$$
$$h(\cot \beta - \cot \alpha) = 200$$
$$h = \frac{200}{\cot \beta - \cot \alpha} \text{ miles} \qquad \text{as required}$$

▲

EXAMPLE 9   The arc shown in Figure 9 is a portion of the unit circle, $x^2 + y^2 = 1$. Express the following quantities in terms of θ.

(a) OA **(b)** AB **(c)** Area of $\triangle OAB$ **(d)** OC

Solution   **(a)** In right triangle OAB:

$$\cos \theta = \frac{OA}{OB} = \frac{OA}{1}$$
$$OA = \cos \theta$$

(b) In right triangle OAB:

$$\sin \theta = \frac{AB}{OB} = \frac{AB}{1}$$
$$AB = \sin \theta$$

(c) Area $\triangle OAB = \frac{1}{2}(OA)(OB) \sin \theta$
$$= \frac{1}{2}(\cos \theta)(1)(\sin \theta)$$
$$= \frac{1}{2} \cos \theta \sin \theta$$

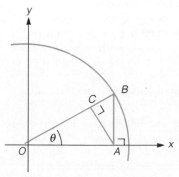

FIGURE 9

(d) In right triangle OAC:

$$\cos \theta = \frac{OC}{OA}$$

$$OC = OA \cos \theta = (\cos \theta)(\cos \theta) = \cos^2 \theta$$ ▲

EXAMPLE 10   Determine the angle θ in Figure 10.

Solution   By inspection, $\tan \theta = \frac{3}{4} = 0.75$. From here, we have three choices.

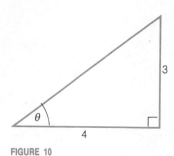

FIGURE 10

CHOICE 1   We can take the easy way out and say, "θ is the acute angle whose tangent is $\frac{3}{4}$." True, this doesn't say what θ really is in degrees; nevertheless, this form of the answer will often suffice in calculus.

CHOICE 2   Using a table of the trigonometric functions, we can hunt until we find an angle whose tangent is as close to 0.75 as possible.

CHOICE 3   We can use a calculator to do all the work. For instance, with a Texas Instruments calculator, we push the buttons

0.75 | INV | | TAN |

The result in this case is

$$\theta \approx 36.8°$$ ▲

Note   When performing the calculations in Example 10, check to see that your calculator is set in the "degree mode," rather than the "radian mode." (Radian measure for angles is discussed in the next chapter, as are the inverse trigonometric functions.)

▼ EXERCISE SET 5.2

A

For Exercises 1–6, refer to the following figure. (However, each problem is independent of the others.)

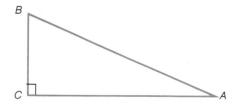

1. If $\angle A = 30°$ and $AB = 60$ cm, find AC and BC.

2. If $\angle A = 60°$ and $AB = 12$ cm, find AC and BC.

3. If $\angle B = 60°$ and $AC = 16$ cm, find BC and AB.

4. If $\angle B = 45°$ and $AC = 9$ cm, find BC and AB.

5. If $\angle B = 50°$ and $AB = 15$ cm, find BC and AC. (Round off your answers to one decimal place.)

6. If $\angle A = 25°$ and $AC = 100$ cm, find BC and AB. (Round off your answers to one decimal place.)

7. A ladder 18 ft long leans against a building. The ladder forms an angle of 60° with the ground.
  (a) How high up the side of the building does the ladder reach? [Here and in part (b), give two forms for your answers: one with radicals and one (using a calculator) with decimals, rounded off to two places.]
  (b) Find the horizontal distance from the foot of the ladder to the base of the building.

8. From a point level with and 1000 ft away from the base of the Washington Monument, the angle of elevation to the top of the monument is 29.05°. Determine the height of the monument to the nearest half foot.

9. Refer to the figure. At certain times, the planets Earth and Mercury line up in such a way that $\angle EMS$ is a right angle, as shown in the figure. At such times, $\angle SEM$ is found to be 21.16°. Use this information to estimate the distance MS of Mercury from the Sun. Assume that the distance from the Earth to the Sun is 93 million miles. (Round off your answer to the nearest million miles. Because Mercury's orbit is not exactly circular, the actual distance of Mercury from the Sun varies from about 28 million miles to 43 million miles.)

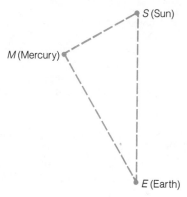

10. Determine the distance AB across the lake shown in the figure, using the following data: $AC = 400$ m, $\angle C = 90°$, and $\angle CAB = 40°$. Round off the answer to the nearest meter.

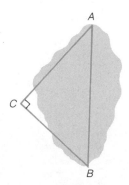

11. Find the area of each triangle. In part (b) use a calculator and round off the final answer to two decimal places.

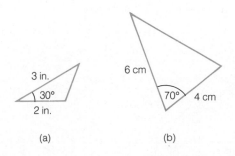

12. Show that the perimeter of the pentagon in Example 6 is 20 sin 36°. *Hint:* In Figure 5, draw a perpendicular from O to $\overline{AB}$.

13. A regular decagon (ten sides) is inscribed in a circle of radius 20 cm. Find the area of the decagon. Round off the answer to one decimal place.

14. In triangle OAB, lengths $OA = OB = 6$ in. and $\angle AOB = 72°$. Find AB. *Hint:* Draw a perpendicular from O to $\overline{AB}$. Round off the answer to one decimal place.

15. The accompanying figure shows two ships at points P and Q, which are in the same vertical plane with an airplane at point R. When the height of the airplane is 3500 ft, the angle of depression to P is 48° and that to Q is 25°. Find the distance between the two ships. Round off the answer to the nearest 10 feet.

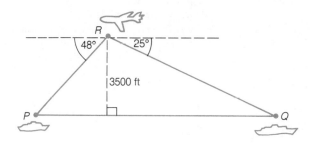

16. An observer in a lighthouse is 66 ft above the surface of the water. The observer sees a ship and finds the angle of depression to be 0.7°. Estimate the distance of the ship from the base of the lighthouse. Round off the answer to the nearest 5 feet.

17. From a point on ground level, you measure the angle of elevation to the top of a mountain to be 38°. Then you walk 200 meters farther away from the mountain and find that the angle of elevation is now 20°. Find the height of the mountain. Round off the answer to the nearest meter.

18. A surveyor stands 30 yd from the base of a building. On top of the building is a vertical radio antenna. Let α denote the angle of elevation when the surveyor sights to the top of the building. Let β denote the angle of elevation when the surveyor sights to the top of the antenna. Express the length of the antenna in terms of the angles α and β.

19. In $\triangle ACD$, you are given $\angle C = 90°$, $\angle A = 60°$, and $AC = 18$ cm. If B is a point on $\overline{CD}$ and $\angle BAC = 45°$, find BD. Express the answer in terms of a radical (rather than using a calculator).

20. The radius of the circle in the following figure is 1 unit. Express the lengths OA, AB, and DC in terms of α.

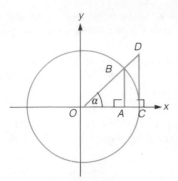

FIGURE FOR EXERCISE 20

21. The arc in the next figure is a portion of the unit circle, $x^2 + y^2 = 1$.
  (a) Express the following angles in terms of θ:
    $\angle BOA$, $\angle OAB$, $\angle BAP$, $\angle BPA$.
  (b) Express the following lengths in terms of $\sin \theta$ and $\cos \theta$: AO, AP, OB, BP.

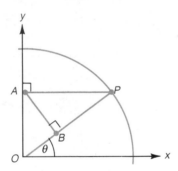

22. In the accompanying figure, suppose that $AD = 1$. Find the length of each of the other line segments in the figure. When radicals appear in an answer, leave the answer in that form, rather than using a calculator.

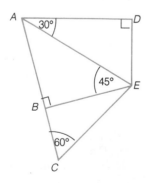

23. (First look over Example 10.) Use a calculator to determine the angle θ in each figure. Round off each answer to one decimal place.

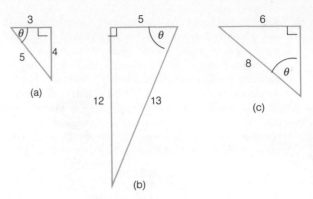

24. First, use the Pythagorean theorem to find the lengths OA, OB, OC, OD, and OE shown in the figure. Then use a calculator or tables to estimate the angles α, β, γ, δ, and ε. (Actually, you don't need any aid for two of the angles.)

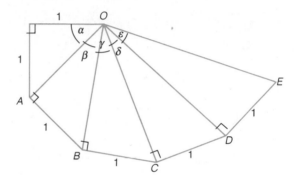

B

25. Express AC as a function of θ.

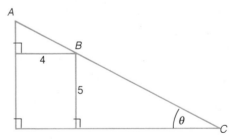

26. If $AB = 8$ in., express x as a function of θ.

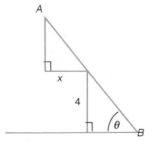

27. In the figure, line segment $\overline{BA}$ is tangent to the circle at A. Also, $\overline{CF}$ is tangent to the circle at F. Express the following lengths in terms of θ.

    (a) DE **(b)** OE **(c)** CF

    (d) OC **(e)** AB **(f)** OB

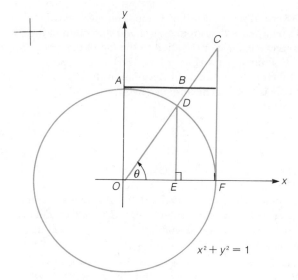

$$x^2 + y^2 = 1$$

28. At point P on the earth's surface, the moon is observed to be directly overhead, while at the same time at point T, the moon is just visible. See the figure.

    (a) Show that $MP = \dfrac{OT}{\cos \theta} - OP$.

    (b) [C] Use a calculator and the following data to estimate the distance MP from the earth to the moon: $\theta = 89.05°$ and $OT = OP = 4000$ miles. Round off your answer to the nearest thousand miles. (Because the moon's orbit is not really circular, the actual distance varies from about 216,400 miles to 247,000 miles.)

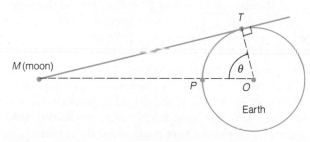

29. Refer to the figure. Let r denote the radius of the moon.

    (a) Show that $r = \left(\dfrac{\sin \theta}{1 - \sin \theta} \right) PS$.

    (b) [C] Use a calculator and the following data to estimate the radius r of the moon: $PS = 238{,}857$ mi. and $\theta = 0.257°$. Round off your answer to the nearest 10 miles.

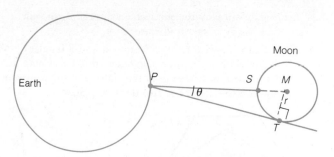

30. Figure A shows a regular hexagon inscribed in a circle of radius 1. Figure B shows a regular heptagon (seven-sided polygon) inscribed in a circle of radius 1. In Figure A, a line segment drawn from the center of the circle perpendicular to one of the sides is called an **apothem** of the polygon.

    (a) Show that the length of the apothem in Figure A is $\sqrt{3}/2$.

    (b) Show that the length of one side of the heptagon in Figure B is $2 \sin(180°/7)$.

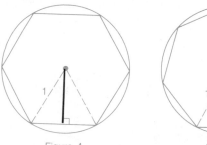

Figure A Figure B

    (c) [C] Use a calculator to evaluate the expressions in parts (a) and (b). Round off each answer to four decimal places, and note how close the two values are. Approximately two thousand years ago, Heron of Alexandria made use of this coincidence when he used the length of the apothem of the hexagon to approximate the length of the side of the heptagon. (The apothem of the hexagon can be constructed with ruler and compass; the side of the regular heptagon cannot.)

31. **(a)** Show that the area of a regular n-gon inscribed in a circle of radius 1 unit is given by

$$A_n = \frac{n}{2} \sin \frac{360°}{n}$$

    (b) [C] Use a calculator and the formula in part (a) to complete the following table.

n	5	10	50	100	1000	5000	10,000
A_n							

(c) Explain why the successive values of A_n in your table get closer and closer to π.

32. The figure shows a regular pentagon and a regular hexagon, with a common side of length 2 cm. Compute the area within the hexagon but outside of the pentagon. Round off the answer to two decimal places.

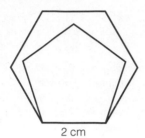

2 cm

C

33. In the accompanying figure, the smaller circle is tangent to the larger circle. Ray PQ is a common tangent and

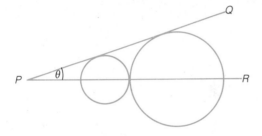

ray PR passes through the centers of both circles. If the radius of the smaller circle is a and the radius of the larger circle is b, show that $\sin \theta = (b - a)/(a + b)$. Then, using the identity $\sin^2 \theta + \cos^2 \theta = 1$, show that $\cos \theta = 2\sqrt{ab}/(a + b)$.

34. A vertical tower of height h stands on level ground. From a point P at ground level and due south of the tower, the angle of elevation to the top of the tower is θ. From a point Q at ground level and due west of the tower, the angle of elevation to the top of the tower is β. If d is the distance between P and Q, show that

$$h = \frac{d}{\sqrt{\cot^2 \theta + \cot^2 \beta}}$$

35. The following problem is taken from *A Treatise on Plane Trigonometry*, 7th ed., by E. W. Hobson (New York: Cambridge University Press, 1928): $ABCD$ is the rectangular floor of a room whose length AB is a feet. Find its height, which at C subtends at A an angle α, and at B an angle β.

36. The following problem is taken from *An Elementary Treatise on Plane Trigonometry*, 8th ed., by R. D. Beasley (London: Macmillan and Co., 1884): The [angle of] elevation of a tower standing on a horizontal plane is observed; a feet nearer it is found to be 45°; b feet nearer still it is the complement of what it was at the first station; shew that the height of the tower is $ab/(a - b)$ feet.

5.3 ▼ TRIGONOMETRIC FUNCTIONS OF GENERAL ANGLES

There is nothing strange in the circle being the origin of any and every marvel.

Aristotle

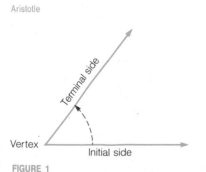

FIGURE 1

Recall that the definitions of the trigonometric functions given in Section 5.1 apply only to acute angles. That is, as defined on page 254, the domains of trigonometric functions consist of all angles θ such that $0° < \theta < 90°$. In this section, we want to expand our definitions of the trigonometric functions to include angles of any size. To do this, we will need to be more explicit than we have been about angles and their size or measure.

An angle is formed by two rays, or half-lines, issuing from a common point called the **vertex**. For our purposes, it is useful to think of the two rays as originally coincident. Then, while one ray is held fixed, the other is rotated to create the given angle. As Figure 1 indicates, the fixed ray is called the **initial side** of the angle, while the rotated ray is called the **terminal side**. By convention, we take the measure of an angle to be **positive** if the rotation is counterclockwise and **negative** if the rotation is clockwise. Figure 2 shows some examples of this. We say that an angle is in **standard position** provided:

(a) the vertex of the angle coincides with the origin in a rectangular coordinate system;

(b) the initial side of the angle lies along the positive horizontal axis.

Figure 3 displays examples of angles in standard position.

FIGURE 2

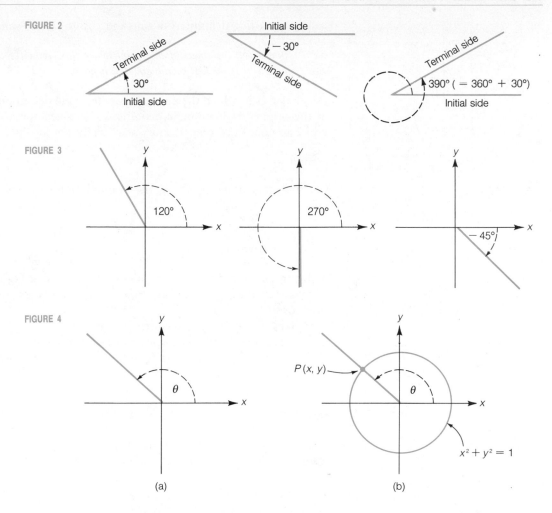

FIGURE 3

FIGURE 4

(a)   (b)

Now, to extend our definitions of the trigonometric functions to accommodate any angle θ, we begin by placing the angle θ in standard position; see Figure 4(a). Next we draw the unit circle $x^2 + y^2 = 1$, as shown in Figure 4(b). (Recall from Chapter 1 that the equation $x^2 + y^2 = 1$ represents a circle of radius 1, with center at the origin.) As Figure 4(b) indicates, $P(x, y)$ denotes the point where the terminal side of angle θ intersects the unit circle. Then with this notation, we state the following definitions.

Trigonometric Functions of General Angles

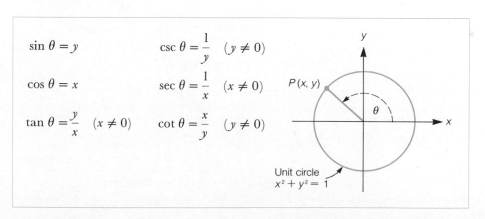

$$\sin \theta = y \qquad \csc \theta = \frac{1}{y} \quad (y \neq 0)$$

$$\cos \theta = x \qquad \sec \theta = \frac{1}{x} \quad (x \neq 0)$$

$$\tan \theta = \frac{y}{x} \quad (x \neq 0) \qquad \cot \theta = \frac{x}{y} \quad (y \neq 0)$$

Unit circle
$x^2 + y^2 = 1$

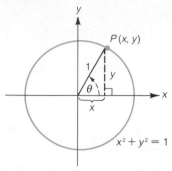

FIGURE 5

The preceding definitions of sine and cosine can be summarized this way:

sin θ *is the second coordinate of that point where the terminal side of angle θ intersects the unit circle;* cos θ *is the first coordinate of that point.*

For the cases in which the angle θ is acute, these definitions are really equivalent to the original right-triangle definitions. Consider, for instance, the acute angle θ in Figure 5. According to our "new" definition of sine, we have

$$\sin \theta = y$$

On the other hand, applying the original right-triangle definition in Figure 5 yields

$$\sin \theta = \frac{\text{opposite}}{\text{hypotenuse}} = \frac{y}{1} = y$$

Thus, in both cases we obtain the same result. Using Figure 5, you should check for yourself that the same agreement also occurs for the other trigonometric functions.

In fact, these definitions agree in other ways, too, with our previous work. Notice that the definitions imply that tan θ = (sin θ)/(cos θ). Additionally, from the definitions we see that sin θ and csc θ are reciprocals, as are cos θ and sec θ and tan θ and cot θ.

EXAMPLE 1 Compute sin 90°, cos 90°, tan 90°, csc 90°, sec 90°, and cot 90°.

Solution We place the angle $\theta = 90°$ in standard position. Then, as Figure 6 indicates, the terminal side of the angle meets the unit circle at the point $(0, 1)$. Now we apply the definitions.

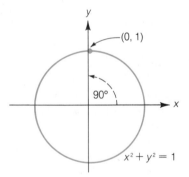

FIGURE 6

$$(0, 1)$$

By definition, cos 90° ⟶ ⟵ By definition, sin 90°
is this number.   is this number.

Thus,

$$\sin 90° = 1 \qquad \text{and} \qquad \cos 90° = 0$$

For the remaining trigonometric functions of 90°, we have

$$\tan 90° = \frac{y}{x} = \frac{1}{0}, \qquad \text{undefined}$$

$$\csc 90° = \frac{1}{y} = \frac{1}{1} = 1$$

$$\sec 90° = \frac{1}{x} = \frac{1}{0}, \qquad \text{undefined}$$

$$\cot 90° = \frac{x}{y} = \frac{0}{1} = 0$$

These are the required results. ▲

EXAMPLE 2   Evaluate the trigonometric functions of $-180°$.

Solution   The results can be read off from Figure 7.

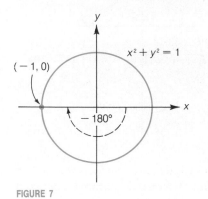

$$\sin(-180°) = y = 0 \qquad\qquad \csc(-180°) = \frac{1}{y} = \frac{1}{0}, \quad \text{Undefined}$$

$$\cos(-180°) = x = -1$$

$$\tan(-180°) = \frac{y}{x} = \frac{0}{-1} = 0 \qquad \sec(-180°) = \frac{1}{x} = \frac{1}{-1} = -1$$

$$\cot(-180°) = \frac{x}{y} = \frac{-1}{0}, \quad \text{Undefined}$$

FIGURE 7

In the two examples just concluded, we evaluated the trigonometric functions for $\theta = 90°$ and $\theta = -180°$. In the same manner, the trigonometric functions can be evaluated just as easily for any angle that is an integral multiple of $90°$. Table 1 shows the results of such calculations. Exercise 6 at the end of this section asks you to make these calculations for yourself.

TABLE 1

θ	$\sin\theta$	$\cos\theta$	$\tan\theta$	$\csc\theta$	$\sec\theta$	$\cot\theta$
0°	0	1	0	undefined	1	undefined
90°	1	0	undefined	1	undefined	0
180°	0	−1	0	undefined	−1	undefined
270°	−1	0	undefined	−1	undefined	0
360°	0	1	0	undefined	1	undefined

To evaluate the trigonometric functions for angles that are not multiples of $90°$, we introduce the notion of a *reference angle*. Given an angle θ that is not a multiple of $90°$, the **reference angle** associated with θ is the acute angle formed by the x-axis and the terminal side of θ. Figure 8 displays examples of some angles and their respective reference angles.

FIGURE 8

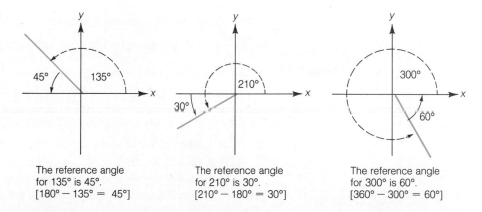

The reference angle for 135° is 45°.
[180° − 135° = 45°]

The reference angle for 210° is 30°.
[210° − 180° = 30°]

The reference angle for 300° is 60°.
[360° − 300° = 60°]

Now let's look at an example to see how reference angles are used to evaluate the trigonometric functions. Suppose we want to evaluate $\cos 150°$. In Figure 9(a) we've placed the angle $\theta = 150°$ in standard position. As you can see, the reference angle for $150°$ is $30°$. By definition, the value of $\cos 150°$ is the x-coordinate of the point P in Figure 9(a). To determine this x-coordinate, we'll use Figure 9(b), in

FIGURE 9

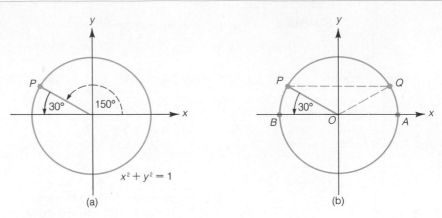

(a)   (b)

which the line segment $\overline{PQ}$ is parallel to the x-axis. There are two facts that follow from the symmetry of the situation in Figure 9(b). (Exercise 35 outlines a detailed verification of these two facts.)

FIRST FACT   The x-coordinate of P is the negative of the x-coordinate of Q.

SECOND FACT   The reference angle POB is equal to angle QOA, and therefore angle QOA is 30°.

Now we apply these two facts. Since angle QOA is 30°, the x-coordinate of Q is by definition cos 30°, or $\sqrt{3}/2$. The x-coordinate of P is then the negative of this. That is, the x-coordinate of P is $-\sqrt{3}/2$. It follows now, again by definition, that the value of cos 150° is $-\sqrt{3}/2$.

The same method that we have just used to evaluate cos 150° can be used to evaluate any of the trigonometric functions when the angles are not multiples of 90°. The following three steps summarize this method.

STEP 1   Determine the reference angle associated with the given angle.
STEP 2   Evaluate the given trigonometric function using the reference angle for the input.
STEP 3   Affix the appropriate sign to the number found in Step 2.

The next three examples illustrate this procedure.

EXAMPLE 3   Evaluate the following quantities.

(a) sin 135°   **(b)** cos 135°   **(c)** tan 135°

Solution   As Figure 10 indicates, the reference angle associated with 135° is 45°.

(a) STEP 1   The reference angle is 45°.
STEP 2 $\sin 45° = \sqrt{2}/2$
STEP 3 $\sin \theta$ is the y-coordinate. In the second quadrant y-coordinates are positive. Thus sin 135° is positive, since the terminal side of $\theta = 135°$ lies in the second quadrant. We have therefore

$$\sin 135° = \frac{\sqrt{2}}{2}$$

(b) STEP 1   The reference angle is 45°.
STEP 2 $\cos 45° = \sqrt{2}/2$

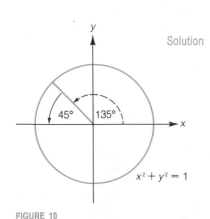

FIGURE 10

STEP 3 $\cos \theta$ is the x-coordinate. In the second quadrant x-coordinates are negative. Thus $\cos 135°$ is negative, since the terminal side of $\theta = 135°$ lies in the second quadrant. We have therefore

$$\cos 135° = -\frac{\sqrt{2}}{2}$$

(c)   STEP 1   The reference angle is $45°$.

STEP 2 $\tan 45° = 1$

STEP 3   By definition, $\tan \theta = y/x$. Now, the terminal side of $\theta = 135°$ lies in the second quadrant, in which y is positive and x is negative. Thus $\tan 135°$ is negative. We have therefore

$$\tan 135° = -1$$ ▲

EXAMPLE 4   Evaluate each of the following quantities.

(a) $\cos(-120°)$ **(b)** $\cot(-120°)$

Solution   As Figure 11 shows, the reference angle for $-120°$ is $60°$.

(a)   STEP 1   The reference angle for $-120°$ is $60°$.

STEP 2 $\cos 60° = \frac{1}{2}$

STEP 3 $\cos \theta$ is the x-coordinate. In the third quadrant, the x-coordinates are negative. Thus, $\cos(-120°)$ is negative, since the terminal side of $\theta = -120°$ lies in the third quadrant. It now follows that

$$\cos(-120°) = -\frac{1}{2}$$

(b)   STEP 1   The reference angle for $-120°$ is $60°$.

STEP 2 $\cot 60° = \sqrt{3}/3$

STEP 3   By definition, $\cot \theta = x/y$. Now, the terminal side of $\theta = -120°$ lies in the third quadrant, in which both x and y are negative. So x/y is positive and we have

$$\cot(-120°) = \frac{\sqrt{3}}{3}$$ ▲

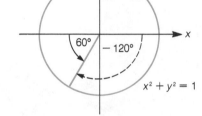

FIGURE 11

In Section 5.1, we showed that the identity $\sin^2 \theta + \cos^2 \theta = 1$ is valid for all acute angles θ. Now we're in a position to show that this identity holds for *all* angles, not just acute angles. We are also going to prove two useful results concerning supplementary angles. Recall that two angles are **supplementary** pro vided their sum is $180°$. Our theorem is as follows.

Theorem

1. For any angle θ, we have

$$\sin^2 \theta + \cos^2 \theta = 1$$

2. If $0° < \theta < 180°$, then

$$\cos(180° - \theta) = -\cos \theta$$

and

$$\sin(180° - \theta) = \sin \theta$$

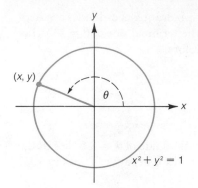

FIGURE 12

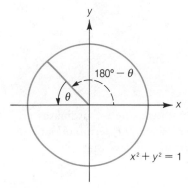

FIGURE 13

To prove that $\sin^2 \theta + \cos^2 \theta = 1$, we need to use only the unit circle definitions of sine and cosine (see Figure 12). Since the point (x, y) lies on the unit circle, we have

$$x^2 + y^2 = 1$$

Replacing x by $\cos \theta$ and y by $\sin \theta$ in this equation then yields

$$(\cos \theta)^2 + (\sin \theta)^2 = 1$$

That is,

$$\sin^2 \theta + \cos^2 \theta = 1 \qquad \text{as required}$$

We will prove the second part of our theorem for the case in which θ is an acute angle. (See Exercises 36 and 37 at the end of this section for the cases in which θ is an angle between $90°$ and $180°$.) Now, referring to Figure 13, notice that the reference angle for $180° - \theta$ is θ itself. Thus, we have

$$\cos(180° - \theta) = \underbrace{-\cos \theta}$$

reference angle for $180° - \theta$

negative, since the terminal side of $180° - \theta$ lies in the second quadrant

and

$$\sin(180° - \theta) = \underbrace{\sin \theta}$$

reference angle for $180° - \theta$

positive, since the terminal side of $180° - \theta$ lies in the second quadrant

These observations complete the proof of the second part of our theorem.

EXAMPLE 5 Given that $\sin \theta = \frac{2}{3}$ and $90° < \theta < 180°$, find $\cos \theta$ and $\tan \theta$.

Solution Substituting $\sin \theta = \frac{2}{3}$ into the identity $\sin^2 \theta + \cos^2 \theta = 1$ yields

$$\left(\frac{2}{3}\right)^2 + \cos^2 \theta = 1$$

$$\cos^2 \theta = 1 - \frac{4}{9} = \frac{5}{9}$$

$$\cos \theta = \pm\sqrt{\frac{5}{9}} = \frac{\pm\sqrt{5}}{3}$$

To decide whether to choose the positive or negative value, note that the given inequality $90° < \theta < 180°$ tells us that the terminal side of θ lies in the second quadrant. Since in the second quadrant x-coordinates are negative, we choose the negative value here. Thus

$$\cos \theta = \frac{-\sqrt{5}}{3}$$

For $\tan \theta$, we have

$$\tan \theta = \frac{y}{x} = \frac{\sin \theta}{\cos \theta} = \frac{2/3}{-\sqrt{5}/3}$$

$$= -\frac{2}{\sqrt{5}} = -\frac{2\sqrt{5}}{5}$$

▲

EXAMPLE 6 Show that the area A of the triangle in Figure 14(a) is given by

$$A = \frac{1}{2} ab \sin \theta$$

Note In Section 5.2 we proved this formula for the case in which θ is an acute angle.

FIGURE 14

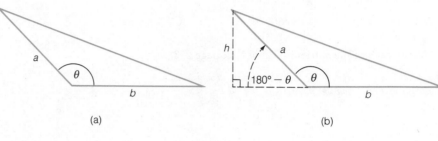

(a)   (b)

Solution The area A of any triangle is given by $A = \frac{1}{2}$(base)(height). Thus, referring to Figure 14(b), we have

$$A = \tfrac{1}{2}bh \tag{1}$$

Also from Figure 14(b), we have

$$\sin(180° - \theta) = \frac{\text{opposite}}{\text{hypotenuse}} = \frac{h}{a}$$

$$h = a \sin(180° - \theta)$$

$$h = a \sin \theta \quad \text{using part 2 of the theorem on page 275} \tag{2}$$

Now we can use equation (2) to substitute for h in equation (1). This yields

$$A = \tfrac{1}{2}b(a \sin \theta) = \tfrac{1}{2}ab \sin \theta$$ ▲

The formula for the area of a triangle that we derived in Example 6 is a useful one. We will use it in later work to prove the *law of sines*. For reference, we summarize the result from Example 6 in the box that follows.

Formula for the Area of a Triangle

If a and b are the lengths of two sides of a triangle and θ is the angle included between those two sides, then the area of the triangle is given by

$$\text{area} = \tfrac{1}{2}ab \sin \theta$$

In words The area of a triangle equals half the product of the lengths of two sides times the sine of the included angle.

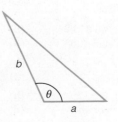

▼ EXERCISE SET 5.3

A

In Exercises 1 and 2, sketch each angle in standard position, and specify the reference angle.

1. (a) $110°$ (b) $240°$ (c) $60°$ (d) $-60°$
2. (a) $300°$ (b) $1000°$ (c) $-15°$ (d) $15°$

In Exercises 3–5, evaluate each expression. (If the value is not defined, say so.)

3. (a) $\cos 0°$ (b) $\sin 450°$
   (c) $\sin 270°$ (d) $\sin(-630°)$
4. (a) $\sin(-270°)$ (b) $\tan 180°$
   (c) $\sec(-90°)$ (d) $\sec 90°$
5. (a) $\csc(-90°)$ (b) $\cot 720°$
   (c) $\cos(-540°)$ (d) $\sin 810°$
6. Verify each of the entries in Table 1 on page 273.

In Exercises 7 and 8, refer to the following figure, in which the x-coordinate of P is $\frac{3}{4}$ and the y-coordinate of Q is $\frac{1}{5}$.

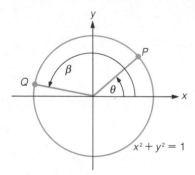

7. (a) Evaluate $\cos \theta$.
   (b) Evaluate $\sin \theta$ and $\tan \theta$.
   (c) Which is larger: $\cos(\beta - 90°)$ or $\cos(\theta + 90°)$?
       Hint: Sketch both angles in standard position.
8. (a) Evaluate $\sin \beta$.
   (b) Evaluate $\cos \beta$ and $\cot \beta$.
   (c) Which is larger: $\cos(\beta + 180°)$ or $\sin(\theta + 180°)$?
       Hint: Sketch both angles in standard position.

In Exercises 9–12, use the method shown in Example 3 to evaluate each expression.

9. (a) $\sin 315°$ (b) $\cos 300°$
   (c) $\tan 330°$ (d) $\sin(-315°)$
10. (a) $\cos 210°$ (b) $\sec 210°$
    (c) $\csc 225°$ (d) $\cos(-210°)$
11. (a) $\tan 135°$ (b) $\cot 120°$
    (c) $\cot 480°$ (d) $\tan(-135°)$
12. (a) $\cos 120°$ (b) $\csc 240°$
    (c) $\sec 600°$ (d) $\cos(-120°)$

13. Evaluate the six trigonometric functions of $\theta = -30°$.
14. Evaluate the six trigonometric functions of $\theta = -45°$.
15. Complete the following two tables.

(a)

θ	$\sin \theta$	$\cos \theta$
$0°$		
$90°$		
$180°$		
$270°$		
$360°$		
$450°$		
$540°$		
$630°$		
$720°$		

(b)

θ	$\sin \theta$
$30°$	
$60°$	
$90°$	
$120°$	
$150°$	
$180°$	
$210°$	
$240°$	
$270°$	
$300°$	
$330°$	
$360°$	

16. Complete the following tables.

(a)

θ	$\sin \theta$	$\cos \theta$	$\tan \theta$
$0°$			
$30°$			
$45°$			
$60°$			
$90°$			
$120°$			
$135°$			
$150°$			
$180°$			

(b)

θ	$\sin \theta$	$\cos \theta$	$\tan \theta$
$180°$			
$210°$			
$225°$			
$240°$			
$270°$			
$300°$			
$315°$			
$330°$			
$360°$			

In Exercises 17–20, use the given information to determine the remaining five trigonometric values.

17. $\sin\theta = \frac{1}{5}$, $90° < \theta < 180°$

18. $\sin\theta = -\frac{24}{25}$, $180° < \theta < 270°$

19. $\cos\theta = -\frac{3}{5}$, $180° < \theta < 270°$

20. $\cos\theta = \frac{1}{4}$, $270° < \theta < 360°$

In Exercises 21–24, use the given information to determine the area of each triangle.

21. Two of the sides are 5 cm and 7 cm, and the angle between those sides is 120°.

22. Two of the sides are 3 m and 6 m, and the included angle is 150°.

23. Two of the sides are 21.4 cm and 28.6 cm, and the included angle is 98.5°. (Round off the final answer to one decimal place.)

24. Two of the sides are 5.98 cm and 8.05 cm, and the included angle is 107.1°. (Round off the answer to one decimal place.)

25. List three angles for which the sine of each is equal to 1.

26. List three angles for which the cosine of each is $-\frac{1}{2}$.

27. Use the figure to estimate the following quantities.
 (a) $\sin 10°$; $\cos 10°$ **(b)** $\sin 20°$; $\cos 20°$
 (c) $\sin 30°$; $\cos 30°$ **(d)** $\sin 40°$; $\cos 40°$
 (e) $\sin 50°$; $\cos 50°$
 Hint: $\sin\theta = \cos(90° - \theta)$
 (f) $\sin 70°$; $\cos 70°$
 (g) $\sin 80°$; $\cos 80°$
 (h) $\sin 100°$; $\cos 100°$
 (i) $\sin 130°$; $\cos 130°$

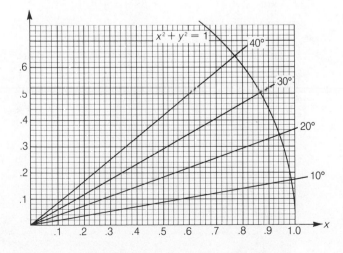

28. **(a)** Complete the following table, using the words "positive" or "negative" as appropriate.

	TERMINAL SIDE OF ANGLE θ LIES IN			
	Quadrant I	Quadrant II	Quadrant III	Quadrant IV
$\sin\theta$	positive	positive		
$\cos\theta$				
$\tan\theta$				

(b) The mnemonic (memory device) ASTC (*all students take calculus*) is sometimes used to recall the signs of the trigonometric values in each quadrant:

A . . . . . All are positive in the first quadrant.
S . . . . . Sine is positive in the second quadrant.
T . . . . . Tangent is positive in the third quadrant.
C . . . . . Cosine is positive in the fourth quadrant.

Check the validity of this mnemonic against your chart in part (a).

B

29. Given that $\sin 15° = \frac{1}{4}(\sqrt{6} - \sqrt{2})$ and $\cos 18° = \frac{1}{4}\sqrt{10 + 2\sqrt{5}}$, evaluate the following.
 (a) $\sin 195°$ **(b)** $\cos 162°$ **(c)** $\tan 345°$
 (d) $\sin(-15°)$ **(e)** $\cos(-18°)$ **(f)** $\cos 918°$

30. Suppose that A, B, and C are the three angles of a triangle. Prove the following identities.
 (a) $\sin A = \sin(B + C)$
 (b) $\cos A = -\cos(B + C)$
 (c) $\cos\dfrac{B + C}{2} = \sin\dfrac{A}{2}$
 (d) $\sin\dfrac{B + C}{2} = \cos\dfrac{A}{2}$
 (e) $\tan A = -\tan(B + C)$

31. Refer to the accompanying figure.
 (a) What are the coordinates of P? What are the coordinates of Q?
 (b) Show that the length of line segment $\overline{PQ}$ is
$$\sqrt{2}\sqrt{1 - \cos\theta\cos\phi - \sin\theta\sin\phi}$$

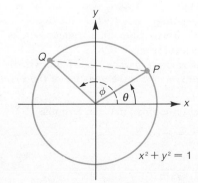

32. Use the figure to prove the following theorem: The area of a quadrilateral is equal to half the product of the diagonals times the sine of the included angle. *Hint:* Find the areas of each of the four triangles that make up the quadrilateral. Show that the sum of those areas is $\frac{1}{2}(\sin\theta)[qs + rq + rp + ps]$. Then factor the quantity within the brackets.

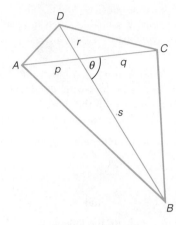

33. Prove the following identity for $0° < \theta < 180°$:

$$2\log_{10}\sin\theta = \log_{10}(1 - \cos\theta) + \log_{10}(1 + \cos\theta)$$

Why is the restriction $0° < \theta < 180°$ necessary?

34. Use the identity $\sin^2\theta + \cos^2\theta = 1$ to explain why, for any angle θ, we have

$$|\sin\theta| \le 1 \quad \text{and} \quad |\cos\theta| \le 1$$

35. In this exercise we are going to verify two statements that were made in the text concerning Figure 9(b). In particular, with reference to that figure, we'll show that angle QOA is 30°, and also that the x-coordinate of P is the negative of the x-coordinate of Q.

(a) Since $\overline{PQ}$ is parallel to $\overline{BA}$, the two angles BOP and QPO are equal. Thus angle QPO is 30°. Now explain why angle PQO is 30°. *Hint:* The base angles of an isosceles triangle are equal.

(b) Since angle PQO is 30°, it follows that angle QOA is 30°. Why?

(c) Let C denote the point where the line segment $\overline{PQ}$ meets the y-axis. Show that triangle OPC is congruent to triangle OQC.

(d) Use the result in part (c) to explain why the x-coordinate of P is the negative of the x-coordinate of Q.

36. In this exercise we prove that the formulas

$$\cos(180° - \theta) = -\cos\theta$$

and

$$\sin(180° - \theta) = \sin\theta$$

are valid when $90° < \theta < 180°$. (Recall that the proof in the text assumes θ is an acute angle.)

The statements that follow refer to the figure below. Supply the reason or reasons for each statement.

(a) The coordinates of P are $(\cos\theta, \sin\theta)$.

(b) Draw $\overline{PQ}$ parallel to the x-axis, as shown in the figure. Then the coordinates of Q are $(-\cos\theta, \sin\theta)$.

(c) $\angle QOA = 180° - \theta$

(d) The coordinates of Q are $(\cos(180° - \theta), \sin(180° - \theta))$.

(e) $\cos(180° - \theta) = -\cos\theta$ and $\sin(180° - \theta) = \sin\theta$.

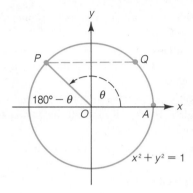

37. Check that the two formulas in Exercise 36 (e) are valid in the cases $\theta = 0°$, $\theta = 90°$, and $\theta = 180°$.

38. *Formula for* $\sin(\alpha + \beta)$ In the following figure, $\overline{AD} \perp \overline{BC}$ and $AD = 1$.

(a) Show that $AC = \sec\alpha$ and $AB = \sec\beta$.

(b) Show that

$$\text{area } \triangle ADC = \tfrac{1}{2}\sec\alpha\sin\alpha$$
$$\text{area } \triangle ADB = \tfrac{1}{2}\sec\beta\sin\beta$$
$$\text{area } \triangle ABC = \tfrac{1}{2}\sec\alpha\sec\beta\sin(\alpha + \beta)$$

(c) The sum of the areas of the two smaller triangles in part (b) equals the area of $\triangle ABC$. Use this fact and the expressions given in part (b) to show that

$$\sin(\alpha + \beta) = \sin\alpha\cos\beta + \cos\alpha\sin\beta$$

(d) Find $\sin 75°$. *Hint:* $75° = 30° + 45°$

(e) Show that $\sin 75° \ne \sin 30° + \sin 45°$.

(f) Compute $\sin 105°$ and then check that $\sin 105° \ne \sin 45° + \sin 60°$.

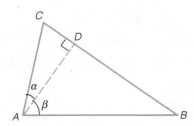

5.4 ▼ ALGEBRA AND THE TRIGONOMETRIC FUNCTIONS

It was Robert of Chester's translation from the Arabic that resulted in our word "sine." The Hindus had given the name *jiva* to the half chord in trigonometry, and the Arabs had taken this over as *jiba*. In the Arabic language there is also a word *jaib* meaning "bay" or "inlet." When Robert of Chester came to translate the technical word *jiba*, he seems to have confused this with the word *jaib* (perhaps because vowels were omitted); hence he used the word *sinus*, the Latin word for "bay" or "inlet."

Carl B. Boyer in *A History of Mathematics* (New York: Wiley, 1968)

In this section, we are going to concentrate on the algebra involved in manipulating trigonometric expressions. This will help to pave the way for some of the more analytical parts of trigonometry that are developed in the next chapter. For reference, in the box that follows we summarize some common notational conventions that have been introduced in the previous sections.

Some Notational Conventions

1. The quantity $(\sin \theta)^n$ is usually written $\sin^n \theta$. For example, $(\sin \theta)^2$ is written $\sin^2 \theta$. The same convention also applies to the other five trigonometric functions.*

2. Parentheses are often omitted in multiplication. For example, the product $(\sin \theta)(\cos \theta)$ is usually written $\sin \theta \cos \theta$. Similarly, $2(\sin \theta)$ is written $2 \sin \theta$.

3. An expression such as $\sin \theta$ really means $\sin(\theta)$, where sin or sine is the name of the function and θ is an input. It is for historical rather than mathematical reasons that the parentheses are suppressed. An exception to this, however, occurs in expressions such as $\sin(A + B)$; here the parentheses are necessary.

EXAMPLE 1   Multiply $(1 - \sin \theta)^2$.

Solution   We do this the same way we would expand $(1 - s)^2$. Since

$$(1 - s)^2 = 1 - 2s + s^2$$

it follows that

$$(1 - \sin \theta)^2 = 1 - 2 \sin \theta + \sin^2 \theta \qquad \blacktriangle$$

EXAMPLE 2   Add $\sin \theta + \dfrac{1}{\cos \theta}$.

Solution   The least common denominator is the quantity $\cos \theta$. We have

$$\sin \theta + \frac{1}{\cos \theta} = \frac{\sin \theta}{1} + \frac{1}{\cos \theta} = \frac{\sin \theta}{1} \cdot \frac{\cos \theta}{\cos \theta} + \frac{1}{\cos \theta}$$

$$= \frac{\sin \theta \cos \theta}{\cos \theta} + \frac{1}{\cos \theta} = \frac{\sin \theta \cos \theta + 1}{\cos \theta}$$

This is the required result. $\qquad \blacktriangle$

---

* A single exception to this convention is the case $n = -1$. The meaning of expressions such as $\sin^{-1} x$ will be explained in Section 8 of Chapter 6.

EXAMPLE 3   Show that the statement $\cos A + \cos B = \cos(A + B)$ is not true in general.

Solution   Let's back up for a minute. How would we convince a beginning algebra student that the statement $(x + y)^2 = x^2 + y^2$ is not true in general? One way would be simply to pick specific values for x and y and then show that the equation fails in that case. For instance, using $x = 1$ and $y = 1$ would give

$$(1 + 1)^2 \overset{?}{=} 1^2 + 1^2$$
$$4 \overset{?}{=} 2 \qquad \text{No!}$$

We can do the same thing with the problem at hand. Let's choose $A = 30°$ and $B = 30°$. Then we have

$$\cos A + \cos B \overset{?}{=} \cos(A + B)$$
$$\cos 30° + \cos 30° \overset{?}{=} \cos(30° + 30°)$$
$$\frac{\sqrt{3}}{2} + \frac{\sqrt{3}}{2} \overset{?}{=} \cos 60°$$
$$\sqrt{3} \overset{?}{=} \frac{1}{2} \qquad \text{No!}$$

We conclude that the statement $\cos A + \cos B = \cos(A + B)$ is not true in general. [To see what $\cos(A + B)$ does equal, see Exercise 48 at the end of this section.]

▲

EXAMPLE 4   Factor: $\tan^2 A + 5 \tan A + 6$

Solution   **PRELIMINARY SOLUTION**   To help you focus on the algebra that is actually involved, let's replace each occurrence of the quantity $\tan A$ by the letter T. Then

$$T^2 + 5T + 6 = (T + 3)(T + 2)$$

ACTUAL SOLUTION

$$\tan^2 A + 5 \tan A + 6 = (\tan A + 3)(\tan A + 2)$$

Note   After you are accustomed to working with trigonometric expressions, you will be able to eliminate the preliminary step.   ▲

EXAMPLE 5   Simplify: $\dfrac{\dfrac{\sin \theta + 1}{\sin \theta} + 1}{\dfrac{\sin \theta - 1}{\sin \theta} - 1}$

Solution   **PRELIMINARY SOLUTION**

$$\frac{\dfrac{S + 1}{S} + 1}{\dfrac{S - 1}{S} - 1} = \frac{S\left(\dfrac{S + 1}{S} + 1\right)}{S\left(\dfrac{S - 1}{S} - 1\right)}$$

$$= \frac{S + 1 + S}{S - 1 - S}$$

$$= \frac{2S + 1}{-1}$$

$$= -2S - 1$$

ACTUAL SOLUTION

$$\frac{\dfrac{\sin \theta + 1}{\sin \theta} + 1}{\dfrac{\sin \theta - 1}{\sin \theta} - 1} = \frac{\sin \theta\left(\dfrac{\sin \theta + 1}{\sin \theta} + 1\right)}{\sin \theta\left(\dfrac{\sin \theta - 1}{\sin \theta} - 1\right)}$$

$$= \frac{\sin \theta + 1 + \sin \theta}{\sin \theta - 1 - \sin \theta}$$

$$= \frac{2 \sin \theta + 1}{-1}$$

$$= -2 \sin \theta - 1$$

▲

EXAMPLE 6   Combine and simplify: $\dfrac{\sin A}{\cos A} + \dfrac{\cos A}{\sin A}$

Solution   The common denominator is $\cos A \sin A$. Therefore we have

$$\frac{\sin A}{\cos A} + \frac{\cos A}{\sin A} = \frac{\sin A}{\cos A} \cdot \frac{\sin A}{\sin A} + \frac{\cos A}{\sin A} \cdot \frac{\cos A}{\cos A}$$

$$= \frac{\sin^2 A}{\cos A \sin A} + \frac{\cos^2 A}{\sin A \cos A}$$

$$= \frac{\sin^2 A + \cos^2 A}{\cos A \sin A} = \frac{1}{\cos A \sin A}$$

This is the required result. If we wish, we can write this answer in an alternative form that doesn't involve fractions:

$$\frac{1}{\cos A \sin A} = \frac{1}{\cos A} \cdot \frac{1}{\sin A} = \sec A \csc A \qquad \blacktriangle$$

One of the most useful techniques for simplifying a trigonometric expression is first to rewrite it in terms of sines and cosines and then to carry out the usual algebraic simplifications. The next example shows how this works.

EXAMPLE 7   Simplify the expression $\dfrac{\sec A + 1}{\sin A + \tan A}$.

Solution

$$\frac{\sec A + 1}{\sin A + \tan A} = \frac{\dfrac{1}{\cos A} + 1}{\sin A + \dfrac{\sin A}{\cos A}}$$

$$= \frac{\cos A\left(\dfrac{1}{\cos A} + 1\right)}{\cos A\left(\sin A + \dfrac{\sin A}{\cos A}\right)} \qquad \text{multiplying numerator and denominator by } \cos A$$

$$= \frac{1 + \cos A}{\cos A \sin A + \sin A}$$

$$= \frac{\overset{1}{\cancel{1 + \cos A}}}{\sin A \underset{1}{(\cancel{\cos A + 1})}} = \frac{1}{\sin A}$$

This last expression can also be written $\csc A$. $\blacktriangle$

In the next four examples, we are asked to show that certain trigonometric equations are, in fact, identities. This type of problem is similar to Example 7, which we just completed, except now we are given an answer to work toward. The identities in these examples should not be memorized; they are too specialized. Instead, you should concentrate on the proofs themselves, noting where the basic definitions are used, and where the fundamental identities (such as $\sin^2 \theta + \cos^2 \theta = 1$) come into play.

EXAMPLE 8   Prove that the equation $\csc A \tan A \cos A = 1$ is an identity.

Solution   We begin with the left-hand side and express each factor in terms of sines or cosines.

$$\csc A \tan A \cos A = \frac{1}{\underset{1}{\cancel{\sin A}}} \cdot \frac{\overset{1}{\cancel{\sin A}}}{\underset{1}{\cancel{\cos A}}} \cdot \overset{1}{\cancel{\cos A}} = 1 \qquad \text{as required}$$

▲

EXAMPLE 9   Prove that $\cos^2 B - \sin^2 B = \dfrac{1 - \tan^2 B}{1 + \tan^2 B}$.

Solution   We begin with the right-hand side this time; it is the more complicated expression. As in the previous examples, we express everything in terms of sines and cosines.

$$\frac{1 - \tan^2 B}{1 + \tan^2 B} = \frac{1 - \dfrac{\sin^2 B}{\cos^2 B}}{1 + \dfrac{\sin^2 B}{\cos^2 B}}$$

$$= \frac{\cos^2 B\left(1 - \dfrac{\sin^2 B}{\cos^2 B}\right)}{\cos^2 B\left(1 + \dfrac{\sin^2 B}{\cos^2 B}\right)}$$

$$= \frac{\cos^2 B - \sin^2 B}{\cos^2 B + \sin^2 B} \qquad \text{(Check the algebra.)}$$

$$= \frac{\cos^2 B - \sin^2 B}{1} = \cos^2 B - \sin^2 B \qquad \text{as required}$$

▲

EXAMPLE 10   Prove that $\dfrac{\cos \theta}{1 - \sin \theta} = \dfrac{1 + \sin \theta}{\cos \theta}$.

Solution   The suggestions given in the previous examples are not applicable here. Everything is already in terms of sines and cosines. Furthermore, neither side appears more complicated than the other. A technique that does work here is to begin with the left-hand side and multiply numerator *and* denominator by the same quantity, namely, $1 + \sin \theta$. Doing so gives us

$$\frac{\cos \theta}{1 - \sin \theta} = \frac{\cos \theta}{1 - \sin \theta} \cdot \frac{1 + \sin \theta}{1 + \sin \theta}$$

$$= \frac{\cos \theta(1 + \sin \theta)}{1 - \sin^2 \theta}$$

$$= \frac{\cos \theta(1 + \sin \theta)}{\cos^2 \theta} = \frac{1 + \sin \theta}{\cos \theta} \qquad \text{as required}$$

▲

The general strategy for each of the proofs in Examples 8 through 10 was the same. In each case we worked with one side of the given equation, and we transformed it (into equivalent expressions) until it was identical to the other side of the equation. This is not the only strategy that can be used. For instance, an alternate way to establish the identity in Example 10 is as follows. The given equation is equivalent to

$$\frac{\cos \theta}{1 - \sin \theta} - \frac{1 + \sin \theta}{\cos \theta} = 0 \qquad (1)$$

Now we show that equation (1) is an identity. To do this, we combine the two fractions on the left-hand side, using the common denominator $\cos \theta(1 - \sin \theta)$. We have

$$\frac{\cos \theta}{1 - \sin \theta} - \frac{1 + \sin \theta}{\cos \theta} = \frac{\cos^2 \theta - (1 + \sin \theta)(1 - \sin \theta)}{\cos \theta(1 - \sin \theta)}$$

$$= \frac{\cos^2 \theta - (1 - \sin^2 \theta)}{\cos \theta(1 - \sin \theta)}$$

$$= \frac{\cos^2 \theta + \sin^2 \theta - 1}{\cos \theta(1 - \sin \theta)} = \frac{1 - 1}{\cos \theta(1 - \sin \theta)}$$

$$= 0$$

This shows that equation (1) is an identity, and thus the equation given in Example 10 is an identity.

Another strategy that can be used in establishing trigonometric identities is to work independently with each side of the given equation until a common expression is obtained. This is the strategy used in Example 11.

EXAMPLE 11 Prove that $\dfrac{\cos(180° - A) + \sin(180° - A)}{\cos(180° - A) - \sin(180° - A)} = \dfrac{1 - \tan A}{1 + \tan A}.$

Solution

$$\text{Left-hand side} = \frac{-\cos A + \sin A}{-\cos A - \sin A} \qquad \qquad \text{Right hand side} = \frac{1 - \dfrac{\sin A}{\cos A}}{1 + \dfrac{\sin A}{\cos A}} \cdot \frac{\cos A}{\cos A}$$

$$= \frac{\cos A - \sin A}{\cos A + \sin A} \qquad \qquad \qquad = \frac{\cos A - \sin A}{\cos A + \sin A}$$

We've now established the required identity by showing that both sides are equal to the same quantity. ▲

▼ EXERCISE SET 5.4

A

In Exercises 1 and 2, carry out the indicated operations and simplify where possible.

1. (a) $(1 - \cos \theta)^2$
   (b) $(\sin \theta + \cos \theta)^2$
   (c) $\cos \theta + \dfrac{1}{\sin \theta}$

2. (a) $(\sin A + \cos A)(\csc A + \sec A)$
   (b) $(\sin^2 B + 1)(\cos^2 B + 1)$
   (c) $(\sin \theta \csc \theta + \cos \theta \sec \theta)^2$

In Exercises 3 and 4, factor each expression.

3. (a) $\tan^2 \theta - 5 \tan \theta - 6$ (b) $\sin^2 B - \cos^2 B$
   (c) $\cos^2 A + 2 \cos A + 1$

4. (a) $3 \sin \theta \cos^2 \theta + 6 \sin \theta$ (b) $\csc^2 \alpha - 2 \csc \alpha - 3$
   (c) $6 \sin^2 \theta + 7 \sin \theta + 2$

In Exercises 5–24, simplify each expression.

5. $\dfrac{\dfrac{\cos \theta + 1}{\cos \theta} + 1}{\dfrac{\cos \theta - 1}{\cos \theta} - 1}$

6. $\dfrac{\dfrac{\cot \beta + 1}{\cot \beta} + 1}{\dfrac{\cot \beta - 1}{\cot \beta} - 1}$

7. $\dfrac{1 - \tan \theta}{\dfrac{\sin \theta}{\cos \theta} - 1}$

8. $\dfrac{2 + \dfrac{1}{\cos \theta}}{\dfrac{1}{\cos^2 \theta}}$ *Hint:* Multiply numerator and denominator by $\cos^2 \theta$.

9. $\dfrac{\sin^2 A - \cos^2 A}{\sin A - \cos A}$ *Hint:* Factor the numerator.

10. $\dfrac{\sin^4 A - \cos^4 A}{\cos A - \sin A}$

11. $\sin^2 \theta \cos \theta \csc^3 \theta \sec \theta$

12. $\sin \theta \csc \theta \tan \theta$

13. $\cot B \sin^2 B \cot B$

14. $\dfrac{3 \sin \theta + 6}{\sin^2 \theta - 4}$

15. $\dfrac{\cos^2 A + \cos A - 12}{\cos A - 3}$

16. **(a)** $\dfrac{x - y}{y - x}$ **(b)** $\dfrac{\sin \theta - \cos \theta}{\cos \theta - \sin \theta}$

17. $\dfrac{\cos A - 2 \sin A \cos A}{\cos^2 A - \sin^2 A + \sin A - 1}$

18. $\dfrac{\tan \theta}{\sec \theta - 1} + \dfrac{\tan \theta}{\sec \theta + 1}$

19. $\cot \theta + \dfrac{1 - 2 \cos^2 \theta}{\sin \theta \cos \theta}$

20. $\sec A \csc A - \tan A - \cot A$

21. $\dfrac{\sec \theta - 1}{\sec \theta + 1} - \dfrac{\tan \theta - \sin \theta}{\tan \theta + \sin \theta}$

22. $(\sec A + \tan A)(\sec A - \tan A)$

23. $\dfrac{\cot^2 \theta}{\csc^2 \theta} + \dfrac{\tan^2 \theta}{\sec^2 \theta}$

24. $\dfrac{\tan \theta + \tan \theta \sin \theta - \cos \theta \sin \theta}{\sin \theta \tan \theta}$

In Exercises 25–43, prove that the equations are identities.

25. $\sin \theta \cos \theta \sec \theta \csc \theta = 1$

26. $\tan^2 A + 1 = \sec^2 A$

27. $(\sin \theta \sec \theta)/(\tan \theta) = 1$

28. $\tan \beta \sin \beta = \sec \beta - \cos \beta$

29. $(1 - 5 \sin x)/\cos x = \sec x - 5 \tan x$

30. $\dfrac{1}{\sin \theta} - \sin \theta = \cot \theta \cos \theta$

31. $\cos A(\sec A - \cos A) = \sin^2 A$

32. $\dfrac{\sin \theta}{\csc \theta} + \dfrac{\cos \theta}{\sec \theta} = 1$

33. $(1 - \sin \theta)(\sec \theta + \tan \theta) = \cos \theta$

34. $(\cos \theta - \sin \theta)^2 + 2 \sin \theta \cos \theta = 1$

35. $(\sec \alpha - \tan \alpha)^2 = \dfrac{1 - \sin \alpha}{1 + \sin \alpha}$

36. $\dfrac{\sin B}{1 + \cos B} + \dfrac{1 + \cos B}{\sin B} = 2 \csc B$

37. $\sin A + \cos A = \dfrac{\sin A}{1 - \cot A} - \dfrac{\cos A}{\tan A - 1}$

38. $(1 - \cos C)(1 + \sec C) = \tan C \sin C$

39. $\csc^2 \theta + \sec^2 \theta = \csc^2 \theta \sec^2 \theta$

40. $1 - \sin(90° - \theta) \cos \theta = \sin^2 \theta$

41. $\dfrac{2 \sin^3 \beta}{1 - \cos \beta} = 2 \sin \beta + 2 \sin \beta \cos \beta$
Hint: Write $\sin^3 \beta$ as $(\sin \beta)(\sin^2 \beta)$.

42. $\dfrac{\cos B \tan B}{\tan(90° - B)} = \sin^2 B \sec B$

43. $\dfrac{\sin^3 \theta + \cos^3 \theta}{\sin \theta + \cos \theta} = 1 - \sin \theta \cos \theta$

44. Prove the identity $\dfrac{\sin \theta}{1 - \cos \theta} = \dfrac{1 + \cos \theta}{\sin \theta}$ in two different ways.
   (a) Adapt the method of Example 10.
   (b) Begin with the left-hand side and multiply numerator and denominator by $\sin \theta$.

B

45. Prove the identity $\dfrac{\sec \theta - \csc \theta}{\sec \theta + \csc \theta} = \dfrac{\tan \theta - 1}{\tan \theta + 1}$.

46. **(a)** Factor the expression $\cos^3 \theta - \sin^3 \theta$.
   (b) Prove the identity.

$$\dfrac{\cos \phi \cot \phi - \sin \phi \tan \phi}{\csc \phi - \sec \phi} = 1 + \sin \phi \cos \phi$$

47. Prove the identity.

$$(r \sin \theta \cos \phi)^2 + (r \sin \theta \sin \phi)^2 + (r \cos \theta)^2 = r^2$$

48. *Formula for* $\cos(\alpha + \beta)$ The figure shown is constructed as follows. Begin with right triangle ABC in which $AC = 1$ and $\alpha + \beta = \angle A$. Extend line segment $\overline{AP}$ and draw $\overline{CE}$ perpendicular to it. Draw $\overline{ED}$ perpendicular to $\overline{BC}$. Finally, from E, draw a perpendicular to $\overline{AB}$ extended, meeting this extension at F:

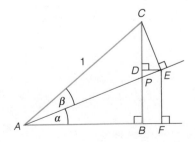

   (a) Show that $CE = \sin \beta$. (Look at triangle AEC.)
   (b) Show that $\angle DCE = \alpha$.

(c) Show that $DE = \sin\alpha\sin\beta$. (Look at triangle CDE.)
(d) Show that $AE = \cos\beta$.
(e) Show that $AF = \cos\alpha\cos\beta$.
(f) Show that $\cos(\alpha + \beta) = AF - DE$.
  Hint: $DE = BF$.
(g) On the basis of steps (f), (e), and (c), conclude that

$$\cos(\alpha + \beta) = \cos\alpha\cos\beta - \sin\alpha\sin\beta$$

(h) Find $\cos 75°$.   *Hint:* $75° = 30° + 45°$
(i) Show that $\cos 75° \neq \cos 30° + \cos 45°$.

49. Suppose that

$$A\sin\theta + \cos\theta = 1 \quad \text{and} \quad B\sin\theta - \cos\theta = 1$$

Show that $AB = 1$.   *Hint:* Solve the first equation for A, the second for B, and then compute AB.

50. (a) Make the substitution $u = a\sin\theta$ in the expression $\sqrt{a^2 - u^2}$ and show that the expression becomes $a\cos\theta$. (Assume $a > 0$ and $0° < \theta < 90°$.)
(b) Make the substitution $u = a\tan\theta$ in the expression $\sqrt{a^2 + u^2}$ and show that the result is $a\sec\theta$. (Assume $a > 0$ and $0° < \theta < 90°$.)
(c) Make the substitution $x = 2\sin\theta$ in the expression $x^2/(4 - x^2)^{3/2}$ and show that the result is $\frac{1}{2}\tan^2\theta\sec\theta$.

51. Show that the equation of line L in the following figure is $x\cos\theta + y\sin\theta = d$.

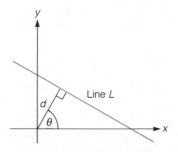

52. Suppose that $\tan\theta + \sin\theta = a$ and $\tan\theta - \sin\theta = b$. Show that $a^2 - b^2 = 4\sqrt{ab}$.

53. If $\sin\alpha + \cos\alpha = a$ and $\sin\alpha - \cos\alpha = b$, show that $\tan\alpha = \dfrac{a+b}{a-b}$.

54. If $a\sin^2\theta + b\cos^2\theta = 1$, show that

$$\sin^2\theta = \frac{1-b}{a-b} \quad \text{and} \quad \tan^2\theta = \frac{b-1}{1-a}$$

55. Refer to the figure. Express m as a function of θ.

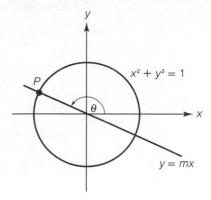

C

56. In the figure, triangle ABC is a right triangle with the right angle at C. The remainder of the figure is constructed as follows. From A a line is drawn perpendicular to $\overline{AB}$, meeting $\overline{BC}$ (extended) at E. Similarly, from B a line is drawn perpendicular to $\overline{AB}$, meeting $\overline{AC}$ (extended) at D. Finally, segment $\overline{DE}$ is drawn.
(a) Show that $\tan\theta = \tan^3\beta$.
(b) Find the ratio of the area of triangle ABC to the area of triangle CED.

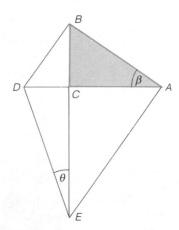

57. Prove the identity $\dfrac{\tan\theta + \sec\theta - 1}{\tan\theta - \sec\theta + 1} = \dfrac{1 + \sin\theta}{\cos\theta}$.

58. Given that $\dfrac{\sin\alpha}{\sin\beta} = p$ and $\dfrac{\cos\alpha}{\cos\beta} = q$, express $\tan\alpha$ and $\tan\beta$ in terms of p and q. (Assume that α and β are acute angles.)

5.5 ▼ THE LAW OF SINES AND THE LAW OF COSINES

The law of sines and the law of cosines are, in essence, formulas relating the sides and angles in any triangle. These formulas can be used to determine an unknown side or angle using given information about the triangle. As you will see, which formula to apply in a particular case depends on what data are initially given. To help you keep track of this, we'll use the following notation in our examples.

NOTATION	EXPLANATION
SAA	One side and two angles are given.
SSA	Two sides and an angle opposite one of those sides are given.
SAS	Two sides and the included angle are given.
SSS	Three sides of the triangle are given.

Also, we will often follow the convention of denoting the angles of a triangle by A, B, and C and the lengths of the corresponding opposite sides by a, b, and c (see Figure 1). With this notation, we are ready to state the law of sines.

Law of Sines

In any triangle, the sines of the angles are proportional to the lengths of the opposite sides:

$$\frac{\sin A}{a} = \frac{\sin B}{b} = \frac{\sin C}{c}$$

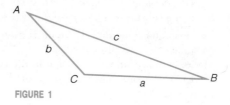

FIGURE 1

The proof of the law of sines is easy. We use the following result from Section 5.3: The area of any triangle is equal to half the product of two sides times the sine of the included angle. Thus, with reference to Figure 1, we have

$$\tfrac{1}{2}bc \sin A = \tfrac{1}{2}ac \sin B = \tfrac{1}{2}ab \sin C$$

since each of these three expressions equals the area of triangle ABC. Now we just multiply through by the quantity $2/abc$ to obtain

$$\frac{\sin A}{a} = \frac{\sin B}{b} = \frac{\sin C}{c}$$

which completes the proof.

EXAMPLE 1   (SAA) Find the length x in Figure 2.

Solution   We can determine x directly by applying the law of sines. We have

$$\frac{\sin 30°}{20} = \frac{\sin 135°}{x}$$

length of side opposite the 30° angle ⟶   ⟵ length of side opposite the 135° angle

$$x \sin 30° = 20 \sin 135°$$
$$x\left(\frac{1}{2}\right) = 20\left(\frac{\sqrt{2}}{2}\right)$$
$$x = 20\sqrt{2} \text{ cm}$$

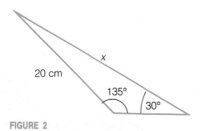

FIGURE 2

EXAMPLE 2   (SAA) Find the length y in Figure 3.

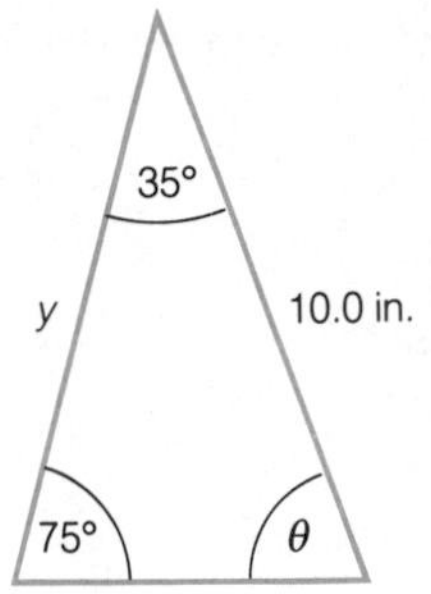

Solution   To determine y using the law of sines, it is necessary to know the opposite angle, denoted by θ in Figure 3. Since the sum of the angles in any triangle is 180°, we have

$$\theta + 75° + 35° = 180°$$

and, consequently,

$$\theta = 180° - 75° - 35° = 70°$$

Then, by the law of sines, we obtain

$$\frac{\sin 75°}{10.0} = \frac{\sin 70°}{y}$$

$$y \sin 75° = 10 \sin 70°$$

$$y = \frac{10 \sin 70°}{\sin 75°} \text{ in.}$$

Without the use of a calculator or tables, this is the final form of the answer. On the other hand (as you should check for yourself), a calculator yields

$$y \approx 9.7 \text{ in.} \quad\blacktriangle$$

EXAMPLE 3   (SSA) In triangle ABC, we are given $\angle C = 45°$, $b = 4\sqrt{2}$ ft, and $c = 8$ ft. Determine the remaining side and angles.

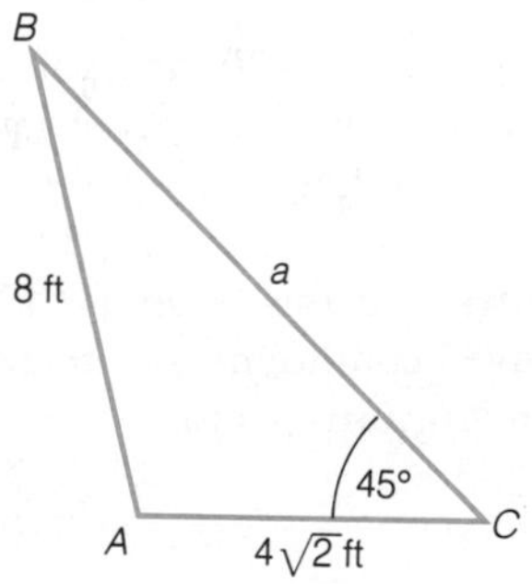

Solution   First let's draw a preliminary sketch conveying the given data; see Figure 4. (The sketch must be considered tentative. At the outset, we don't know whether the other angles are acute or even whether the given data are compatible.) To find angle B, we have (according to the law of sines)

$$\frac{\sin B}{4\sqrt{2}} = \frac{\sin 45°}{8}$$

and therefore,

$$8 \sin B = 4\sqrt{2} \sin 45° = 4\sqrt{2}\left(\frac{\sqrt{2}}{2}\right) = 4$$

$$\sin B = \frac{1}{2}$$

From our previous work, we know that one possibility for B is 30°, since $\sin 30° = \frac{1}{2}$.* However, there is another possibility. Since the reference angle for 150° is 30°, we know that $\sin 150°$ is also equal to $\frac{1}{2}$. Which angle do we want? For the problem at hand, this is easy to answer. Since angle C is given as 45°, angle B cannot equal 150°, for the sum of 45° and 150° exceeds 180°. We conclude that

$$\angle B = 30°$$

---

* If the value were not so familiar, say, $\sin B = \frac{2}{3}$, we could use a calculator or tables, as in Example 10 of Section 5.2.

FIGURE 3

FIGURE 4

Next, since $\angle B = 30°$ and $\angle C = 45°$, we have

$$\angle A = 180° - 30° - 45° = 105°$$

Finally, since angle A is opposite to side $\overline{BC}$, the length of $\overline{BC}$ can be determined using the law of sines:

$$\frac{\sin C}{c} = \frac{\sin A}{a}$$

$$\frac{\sin 45°}{8} = \frac{\sin 105°}{a}$$

$$a \sin 45° = 8 \sin 105°$$

$$a = \frac{8 \sin 105°}{\sin 45°} = \frac{8 \sin 105°}{\sqrt{2}/2}$$

$$= \frac{16}{\sqrt{2}} \sin 105° = \frac{16\sqrt{2}}{2} \sin 105°$$

$$= 8\sqrt{2} \sin 105° \text{ ft}$$

Using a calculator, this expression for a can be evaluated directly. The result is $a = 11$ ft to the nearest foot. On the other hand, using tables, it would first be necessary to use the identity $\sin(180° - \theta) = \sin\theta$ to write

$$\sin 105° = \sin(180° - 105°) = \sin 75°$$

This is because trigonometric tables generally do not include values of angles beyond 90°.   ▲

In the preceding example, two possibilities arose for angle B: both 30° and 150°. However, it turned out that the value 150° was incompatible with the given information in the problem. In Exercise 11 at the end of this section, you will see a case in which both of two possibilities are compatible with the given data. This results in two distinct solutions to the problem. In contrast to this, Exercise 10 shows a case in which there is no triangle fulfilling the given conditions. For these reasons, the case SSA is sometimes referred to as the **ambiguous case**.

Now we turn to the law of cosines.

Law of Cosines

In any triangle, the square of the length of any side equals the sum of the squares of the lengths of the other two sides minus twice the product of the lengths of those other two sides times the cosine of their included angle.

$$a^2 = b^2 + c^2 - 2bc \cos A$$
$$b^2 = c^2 + a^2 - 2ca \cos B$$
$$c^2 = a^2 + b^2 - 2ab \cos C$$

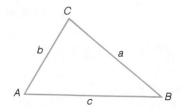

Before looking at a proof of this law, we make two preliminary comments. First, it is important to understand that the three equations in the box all follow the same pattern. For example, look at the first equation.

$$\overbrace{a^2 = b^2 + c^2 - 2bc \cos A}$$

side and opposite angle

sides that include this angle

Now check for yourself that the other two equations also follow this pattern. It is the pattern that is important here; after all, not every triangle will be labeled *ABC*.

The second observation is that the law of cosines is a generalization of the Pythagorean theorem. In fact, look what happens to the equation

$$a^2 = b^2 + c^2 - 2bc \cos A$$

when angle *A* is a right angle:

$$a^2 = b^2 + c^2 - 2bc \underbrace{\cos 90°}_{0}$$

$$a^2 = b^2 + c^2 \qquad \text{which is the Pythagorean theorem}$$

Now let us prove the law of cosines:

$$a^2 = b^2 + c^2 - 2bc \cos A$$

(The other two equations can be proved in the same way. Indeed, just relabeling the figure would suffice.) The proof that we give uses coordinate geometry in a very nice way to complement the trigonometry. We begin by placing angle *A* in standard position, as indicated in Figure 5. (So, in the figure, angle *A* is then identified with angle *CAB*.) Then if *u* and *v* denote the lengths indicated in Figure 5, the coordinates of *C* are (u, v) and we have

$$\cos A = \frac{\text{adjacent}}{\text{hypotenuse}} = \frac{u}{b}$$

and therefore,

$$u = b \cos A$$

Similarly, we have

$$\sin A = \frac{\text{opposite}}{\text{hypotenuse}} = \frac{v}{b}$$

and therefore,

$$v = b \sin A$$

Thus, the coordinates of *C* are

$$(b \cos A, b \sin A)$$

(Exercise 38 at the end of this section asks you to check that these represent the coordinates of *C* even when angle *A* is not acute.) Now we use the distance formula,

$$d = \sqrt{(x_2 - x_1)^2 + (y_2 - y_1)^2}$$

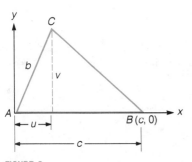

FIGURE 5

to compute the required distance a between the points $C(b \cos A, b \sin A)$ and $B(c, 0)$. We have

$$a = \sqrt{(b \cos A - c)^2 + (b \sin A - 0)^2}$$
$$= \sqrt{b^2 \cos^2 A - 2bc \cos A + c^2 + b^2 \sin^2 A}$$
$$= \sqrt{b^2 \underbrace{(\cos^2 A + \sin^2 A)}_{1} - 2bc \cos A + c^2}$$
$$= \sqrt{b^2 + c^2 - 2bc \cos A}$$

Squaring both sides of this last equation gives us the law of cosines, as we set out to prove.

EXAMPLE 4   (SAS) Compute the length x in Figure 6.

Solution   The law of cosines is directly applicable. We have

$$x^2 = 7^2 + 8^2 - 2(7)(8) \cos 120°$$
$$= 49 + 64 - 112(-\tfrac{1}{2}) = 169$$
$$x = \sqrt{169} = 13 \text{ cm}$$

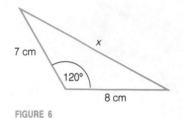

FIGURE 6

If the equation $a^2 = b^2 + c^2 - 2bc \cos A$ is solved for $\cos A$, the result is

$$\cos A = \frac{b^2 + c^2 - a^2}{2bc}$$

This expresses the cosine of an angle in a triangle in terms of the lengths of the sides. In a similar fashion, we obtain the corresponding formulas

$$\cos B = \frac{c^2 + a^2 - b^2}{2ca}$$

and

$$\cos C = \frac{a^2 + b^2 - c^2}{2ab}$$

These alternative forms for the law of cosines are used in the next example.

EXAMPLE 5   (SSS) In triangle ABC, the sides are $a = 3$ units, $b = 5$ units, and $c = 7$ units. Find the angles.

Solution   Figure 7 summarizes the given data. We now have

$$\cos A = \frac{b^2 + c^2 - a^2}{2bc}$$
$$= \frac{5^2 + 7^2 - 3^2}{2(5)(7)} = \frac{65}{70} = \frac{13}{14}$$

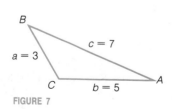

FIGURE 7

Using a calculator (set in the degree mode), we obtain $\angle A \approx 21.8°$. For example, the sequence of keystrokes on a Texas Instruments calculator is as follows:

$$13 \boxed{\div} 14 \boxed{=} \boxed{\text{INV}} \boxed{\text{COS}}$$

In a similar manner, we have

$$\cos B = \frac{c^2 + a^2 - b^2}{2ca}$$

$$= \frac{7^2 + 3^2 - 5^2}{2(7)(3)} = \frac{33}{42} = \frac{11}{14}$$

Thus, $\cos B = \frac{11}{14}$ and, using a calculator, we find that $\angle B \approx 38.2°$.

Finally, working in the same manner to compute angle C, we have

$$\cos C = \frac{a^2 + b^2 - c^2}{2ab}$$

$$= \frac{3^2 + 5^2 - 7^2}{2(3)(5)} = \frac{-15}{30} = -\frac{1}{2}$$

Thus, $\cos C = -\frac{1}{2}$. A calculator is not needed in this case. On the basis of our previous work, we conclude that $\angle C = 120°$. The required angles are thus

$$\angle A \approx 21.8°, \qquad \angle B \approx 38.2°, \qquad \angle C = 120° \qquad \blacktriangle$$

▼ EXERCISE SET 5.5

A

In Exercises 1–8, assume that the vertices and the lengths of the sides of a triangle are labeled as in Figure 1 in the text. For Exercises 1–4, leave your answers in terms of radicals or the trigonometric functions; that is, don't use a calculator. In Exercises 5–8, use a calculator and round off your final answers to one decimal place.

1. If $\angle A = 60°$, $\angle B = 45°$, and $BC = 12$ cm, find AC.

2. If $\angle A = 30°$, $\angle B = 135°$, and $BC = 4$ cm, find AC.

3. If $\angle B = 100°$, $\angle C = 30°$, and $AB = 10$ cm, find BC.

4. If $\angle A = \angle B = 35°$, and $AB = 16$ cm, find AC and BC.

5. If $\angle A = 36°$, $\angle B = 50°$, and $b = 12.61$ cm, find a and c.

6. If $\angle B = 81°$, $\angle C = 55°$, and $b = 6.24$ cm, find c and a.

7. If $a = 29.45$ cm, $b = 30.12$ cm, and $\angle B = 66°$, find the remaining side and angles of the triangle.

8. If $a = 52.15$ cm, $c = 42.90$ cm, and $\angle A = 125°$, find the remaining side and angles of the triangle.

9. (a) In triangle ABC, $\sin B = \sqrt{2}/2$. What are the possible values for angle B?
   (b) In triangle ABC, if $\cos A = -\sqrt{3}/2$, what are the possible values for angle A?
   (c) In triangle CDE, $\sin D = \frac{1}{4}$. Using a calculator, determine the possible values for angle D.

10. Show that there is no triangle for which $a = 23$, $b = 50$, and $\angle A = 30°$.   *Hint:* Try computing $\sin B$ using the law of sines.

11. Let $b = 1$, $a = \sqrt{2}$, and $\angle B = 30°$.
    (a) Use the law of sines to show that $\sin A = \sqrt{2}/2$.
    (b) Conclude that $A = 45°$ or $A = 135°$.
    (c) Assuming that $A = 45°$, determine the remaining parts of triangle ABC.
    (d) Assuming that $A = 135°$, determine the remaining parts of triangle ABC.

12. Let $a = 30$, $b = 36$, and $\angle A = 20°$. Find the remaining parts of triangle ABC. (There are two distinct triangles possible.)

13. Find the lengths a, b, c, and d in the following figure. Leave your answers in terms of trigonometric functions (rather than decimals) as necessary.

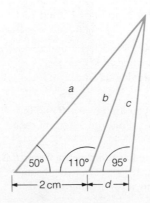

14. Find the area of triangle ABC if $c = 7$ in., $\angle A = 40°$, and $\angle B = 85°$.

15. Use the law of cosines to determine x in each of the figures. For part (b), use a calculator or tables and round off to one decimal place.

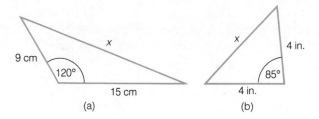

(a)   (b)

16. In applying the law of cosines to the following figure, a student incorrectly writes

$$x^2 = 5^2 + 11^2 - 2(5)(11) \cos 95°$$

Why is this incorrect? What is the correct equation?

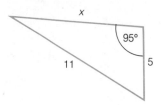

In Exercises 17 and 18, use the given information to find the cosine of each angle in △ABC.

17. $a = 6$ cm, $b = 7$ cm, $c = 10$ cm

18. $a = 17$ cm, $b = 8$ cm, $c = 15$ cm (For this particular triangle, you can check your answers because there is an alternate method of solution that does not require the law of cosines.)

In Exercises 19–22, round off each answer to one decimal place.

19. A regular pentagon is inscribed in a circle of unit radius. Find the perimeter of the pentagon. *Hint:* First find the length of a side using the law of cosines.

20. In triangle ABC, $a = 2$, $b = 3$, and $c = 4$. Find each angle.

21. In triangle ABC, $\angle A = 40°$, $b = 6.1$ cm, and $c = 3.2$ cm.
(a) Find a using the law of cosines.
(b) Find angle C using the law of sines.
(c) Find angle B.

22. In parallelogram $ABCD$, you are given $AB = 6$ in., $AD = 4$ in., and $\angle A = 40°$. Find the length of each diagonal.

23. In square $ABCD$, each side has unit length. Point P is on diagonal AC such that $AP = 1$. Find the distance from P to D. (Express your answer using radicals.)

24. Two points P and Q are on opposite sides of a river (see the sketch). From P to another point R on the same side is 300 ft. Angles PRQ and RPQ are found to be 20° and 120°, respectively. Compute the distance from P to Q, across the river. (Round off your answer to the nearest foot.)

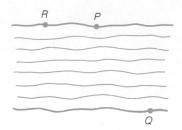

B

25. (a) Let m and n be positive numbers, with $m > n$. Furthermore, suppose that in triangle ABC the lengths a, b, and c are given by

$$a = 2mn + n^2 \qquad b = m^2 - n^2$$
$$c = m^2 + n^2 + mn$$

Show that $\cos C = -\frac{1}{2}$ and conclude that $\angle C = 120°$.
(b) Give an example of a triangle in which the lengths of the sides are whole numbers and one of the angles is 120°. (Specify the three sides; you needn't find the other angles.)

26. If the lengths of two adjacent sides of a parallelogram are a and b, and if the acute angle formed by these two sides is θ, show that the product of the lengths of the two diagonals is given by the expression

$$\sqrt{(a^2 + b^2)^2 - 4a^2b^2 \cos^2 \theta}$$

27. Compute the length x in the following figure. Round off the answer to the nearest one-half foot. *Suggestion:* Draw the line segment joining the vertices of the 80° and 50° angles.

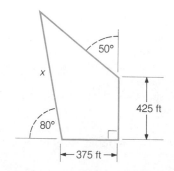

28. In $\triangle ABC$, angle B is a right angle, and D is a point on the hypotenuse such that $\overline{BD}$ bisects angle B. Show that

$$\frac{\sin A}{\sin C} = \frac{DC}{DA}$$

29. In $\triangle ABC$, angle $A = 60°$, $c = 1/(\sqrt{6} + \sqrt{2})$, and $b = 1/(\sqrt{6} - \sqrt{2})$.
   (a) Show that $a = \sqrt{3}/2$.
   (b) Find $\sin B$ and $\sin C$.

30. In $\triangle ABC$, $a = 3 + \sqrt{3}$, $b = \sqrt{6}$, and $c = 2\sqrt{3}$.
   (a) Find $\cos B$ and $\cos C$.
   (b) Find the angles of the triangle.

31. Suppose that you are riding in a hot air balloon directly above a straight road. At a certain instant you see two consecutive mileposts on the road and measure the angles of depression to be $60°$ and $45°$. Compute the height of the balloon, in feet. Round off your answer to the nearest 5 feet.

32. At the foot of a hill is a vertical flagpole. The angle of inclination of the hill (to the horizontal) is θ. When you walk a ft up the hill, you find that the flagpole subtends an angle β. Find the height of the flagpole in terms of θ, β, and a. Simplify the answer using the identity $\sin(90° - A) = \cos A$.

33. Two trains leave the railroad station at noon. The first train travels along a straight track at 90 mph. The second train travels at 75 mph along another straight track that makes an angle of $130°$ with the first track. At what time are the trains 400 miles apart? Round off your answer to the nearest minute.

34. The figure shows three mutually tangent circles with centers A, B, C and radii of 2 cm, 3 cm, and 4 cm, respectively. Determine the angles in triangle ABC. Round off your answers to the nearest tenth of a degree.

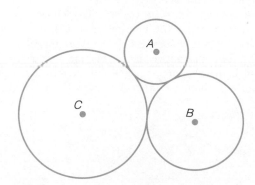

Exercises 35 and 36 use the term bearing *in specifying the location of one point relative to another. The following figure and accompanying statements illustrate the meaning of this term.*

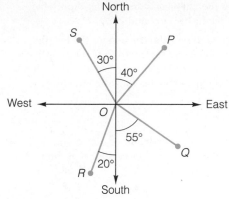

The bearing of P from O is N40°E (read "north, 40° east").
The bearing of Q from O is S55°E.
The bearing of R from O is S20°W.
The bearing of S from O is N30°W.
(In each case, note that the angle is an acute angle measured from the north–south line.)

35. An airplane crashes in a lake and is spotted by observers at lighthouses A and B along the coast. Lighthouse B is 1.50 miles due east of lighthouse A. The bearing of the airplane from lighthouse A is S20°E; the bearing of the plane from lighthouse B is S42°W. Find the distance from each lighthouse to the crashsite. (Round off your final answer to two decimal places.)

36. (Continuation of Exercise 35) A rescue boat is in the lake, three-fourths of a mile from lighthouse B, and at a bearing of S35°E from lighthouse B.
   (a) Find the distance from the rescue boat to the airplane. Express your answer using miles and feet, with the portion in feet rounded off to the nearest ten feet.
   (b) Find the bearing of the plane from the rescue boat. (Your answer should have the form Sθ°W. Round off θ to two decimal places.)

37. In the following figure, $\overline{AD}$ bisects angle A in triangle ABC. Show that

$$\frac{n}{m} = \frac{c}{b}$$

In words, this says that the bisector of an angle in a triangle divides the opposite side into segments that are

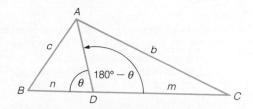

proportional to the adjacent sides.   *Hint:* Apply the law of sines in triangles *ABD* and *ACD*.

38. Let the positive numbers *u* and *v* denote the lengths indicated in the following figure, so that the coordinates of *C* are $(-u, v)$. Show that $u = -b \cos A$ and $v = b \sin A$. Conclude from this that the coordinates of *C* are

$(b \cos A, b \sin A)$

Hint: Use the right-triangle definitions for cosine and sine along with the formulas for $\cos(180° - \theta)$ and $\sin(180° - \theta)$.

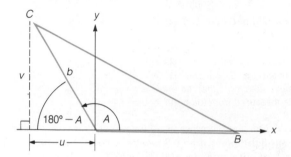

39. *Heron's formula*   Approximately 2000 years ago, Heron of Alexandria derived a formula for the area of a triangle in terms of the lengths of the sides. A more modern derivation of Heron's formula using the law of cosines is indicated in the steps that follow.

(a) Solve the equation $c^2 = a^2 + b^2 - 2ab \cos C$ for $\cos C$ to obtain

$$\cos C = \frac{a^2 + b^2 - c^2}{2ab}$$

(b) Show that $1 - \cos C = \frac{(c + a - b)(c - a + b)}{2ab}$.

Suggestion: Work with $1 - \cos C$. You can factor the expression $c^2 - (a - b)^2$ that arises by using the difference of squares technique.

(c) Using the result in part (a), show that

$$1 + \cos C = \frac{(a + b + c)(a + b - c)}{2ab}.$$

(d) Show that $\sin^2 C =$

$$\frac{(a + b + c)(-a + b + c)(a - b + c)(a + b - c)}{4a^2b^2}$$

(e) Using the formula area $\triangle ABC = \frac{1}{2}ab \sin C$, show that area $\triangle ABC =$

$$\frac{1}{4}\sqrt{(a + b + c)(-a + b + c)(a - b + c)(a + b - c)}$$

(f) The formula in part (e) expresses the area in terms of *a*, *b*, and *c*. We can simplify this formula somewhat by introducing the following notation.

   Let *s* denote one-half the perimeter of triangle *ABC*. That is, let $s = \frac{1}{2}(a + b + c)$. Using this notation, check that

   (i) $a + b + c = 2s$

(ii) $-a + b + c = 2(s - a)$
(iii) $a - b + c = 2(s - b)$
(iv) $a + b - c = 2(s - c)$

Then show that, with this notation, we obtain

area $\triangle ABC = \sqrt{s(s - a)(s - b)(s - c)}$

This is Heron's formula. [For historical background and a purely geometric proof, see *An Introduction to the History of Mathematics*, 5th ed., by Howard Eves (Philadelphia: Saunders, 1983), pp. 133 and 147.]

40. The accompanying figure shows a quadrilateral, with sides *a*, *b*, *c*, and *d*, inscribed in a circle.

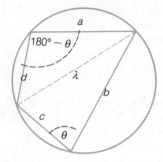

If λ denotes the length of the diagonal indicated in the figure, prove that

$$\lambda^2 = \frac{(ab + cd)(ac + bd)}{bc + ad}$$

This result is known as *Brahmagupta's Theorem*. It is named after its discoverer, a seventh-century Hindu mathematician.   *Hint:* Assume as given the theorem from geometry stating that when a quadrilateral is inscribed in a circle, the opposite angles are supplementary. Apply the law of cosines in both of the triangles in the figure to obtain expressions for λ^2. Then eliminate $\cos \theta$ from one equation.

41. *Alternate derivation for the law of sines*   The following figure shows triangle *ABC* inscribed in a circle of radius *r*.

(a) Show that $\angle AOD = \angle C$.   *Hint:* Use the following theorem from geometry: The measure of an angle inscribed in a circle is half the measure of the corresponding central angle. (In the figure, angle

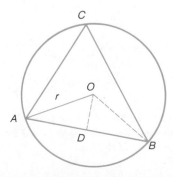

C is the inscribed angle and angle AOB is the central angle.)

(b) Show that $r = c/(2 \sin C)$.

(c) Using the reasoning in parts (a) and (b), conclude that

$$r = \frac{a}{2 \sin A} \quad \text{and} \quad r = \frac{b}{2 \sin B}$$

Then show that

$$\frac{\sin A}{a} = \frac{\sin B}{b} = \frac{\sin C}{c}$$

(d) In the figure we used, the center of the circle falls within triangle ABC. Draw a figure in which the center lies outside of triangle ABC and prove that the equation $r = c/(2 \sin C)$ is true in this case, too.

C

42. Suppose that triangle ABC is a right triangle with the right angle at C. Use the law of cosines to prove the following statements.

(a) The square of the distance from C to the midpoint of the hypotenuse is equal to one-fourth the square of the hypotenuse.

(b) The sum of the squares of the distances from C to the two points that trisect the hypotenuse is equal to five-ninths the square of the hypotenuse.

(c) Let P, Q, and R be points on the hypotenuse such that $AP = PQ = QR = RB$. Derive a result similar to your results in parts (a) and (b) for the sum of the squares of the distances from C to P, Q, and R.

43. The perimeter and the area of the triangle in the figure are 20 cm and $10\sqrt{3}$ cm^2, respectively. Find a and b. (Assume $a < b$.)

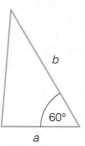

44. In triangle ABC, $\angle B - \angle A = 90°$. Show that $c^2 = (b^2 - a^2)^2/(a^2 + b^2)$.   *Hint:* Let P be the point on $\overline{AC}$ such that $\angle CBP = 90°$ and $\angle PBA = \angle A$. Show that $CP = (a^2 + b^2)/(2b)$.

45. In triangle ABC, $\angle B - \angle A = 90°$. Show that

$$\frac{2}{c^2} = \frac{1}{(b+a)^2} + \frac{1}{(b-a)^2}$$

Hint: Use the result in Exercise 44.

FORMULAS OR NOTATIONS	PAGE REFERENCE	COMMENT
1. $\sin \theta$ $\cos \theta$ $\tan \theta$ $\csc \theta$ $\sec \theta$ $\cot \theta$	254, 271	When θ is an acute angle in a right triangle, the six trigonometric functions of θ can be defined as follows: $\sin \theta = \dfrac{\text{opposite}}{\text{hypotenuse}} \qquad \csc \theta = \dfrac{\text{hypotenuse}}{\text{opposite}}$ $\cos \theta = \dfrac{\text{adjacent}}{\text{hypotenuse}} \qquad \sec \theta = \dfrac{\text{hypotenuse}}{\text{adjacent}}$ $\tan \theta = \dfrac{\text{opposite}}{\text{adjacent}} \qquad \cot \theta = \dfrac{\text{adjacent}}{\text{opposite}}$ More generally, if θ denotes any angle in standard position, and $P(x, y)$ is the point where the terminal side of the angle meets the unit circle, then the six trigonometric functions of θ can be defined as follows: $\sin \theta = y \qquad \csc \theta = \dfrac{1}{y} \quad (y \neq 0)$ $\cos \theta = x \qquad \sec \theta = \dfrac{1}{x} \quad (x \neq 0)$ $\tan \theta = \dfrac{y}{x} \quad (x \neq 0) \qquad \cot \theta = \dfrac{x}{y} \quad (y \neq 0)$

FORMULAS OR NOTATIONS	PAGE REFERENCE	COMMENT
2. $\sin(90° - \theta) = \cos\theta$ $\cos(90° - \theta) = \sin\theta$	256	If two angles are complementary, then the sine of (either) one equals the cosine of the other.
3. $\tan\theta = \dfrac{\sin\theta}{\cos\theta}$	256	This equation is an identity. It holds for all angles θ except those for which $\cos\theta$ is zero.
4. $\sin^2\theta + \cos^2\theta = 1$	256, 275	This equation is an identity. It holds for all angles θ.
5. $\cos(180° - \theta) = -\cos\theta$ $\sin(180° - \theta) = \sin\theta$	275	In Section 5.3, we saw that these two identities are valid for all angles θ such that $0° < \theta < 180°$. In the next chapter, we will see that these two identities actually hold for all angles θ.
6. $A = \frac{1}{2}ab\sin\theta$	263, 277	The area of a triangle equals half the product of the lengths of two sides times the sine of the included angle.
7. $\sin^n\theta$	281	$\sin^n\theta$ means $(\sin\theta)^n$. The same convention also applies to the other five trigonometric functions.
8. $\dfrac{\sin A}{a} = \dfrac{\sin B}{b} = \dfrac{\sin C}{c}$	288	This is the law of sines. In words, it states that in any triangle, the sines of the angles are proportional to the lengths of the opposite sides.
9. $a^2 = b^2 + c^2 - 2bc\cos A$	290	This is the law of cosines. We can use it to calculate the third side of a triangle when we are given two sides and the included angle. The formula can also be used to calculate the angles of a triangle when we know the three sides, as in Example 5 on page 292.

▼ CHAPTER FIVE   REVIEW EXERCISES

NOTE   Exercises 1–20 constitute a chapter test on the fundamentals, based on group A problems.

1. Specify a value for each of the following expressions: $\cos 30°$, $\tan 60°$, and $\sin^2 7° + \cos^2 7°$.

2. Evaluate each of the following.
(a) $\sin(-270°)$ **(b)** $\cos 540°$ **(c)** $\cot 450°$

3. Factor $2\cos^2\theta + 11\cos\theta + 12$.

4. Find the area of a triangle in which the lengths of two sides are 5 cm and 6 cm, and the angle included between these sides is 135°.

5. In triangle ABC, $A = 120°$, $b = 5$ cm, and $c = 3$ cm. Find a.

6. The sides of a triangle are 2 cm, 3 cm, and 4 cm. Determine the cosine of the angle opposite the longest side. On the basis of your answer, explain whether the angle opposite the longest side is an acute angle.

7. Evaluate each of the following.
(a) $\cos 330°$
(b) $\tan(-135°)$

8. If $\sin\theta = -\frac{1}{3}$ and $270° < \theta < 360°$, determine $\cos\theta$ and $\cot\theta$.

9. Compute $\tan\theta$ in the figure shown.

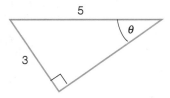

10. Simplify: $\dfrac{\dfrac{\tan\theta + 1}{\tan\theta} + 1}{\dfrac{\tan\theta - 1}{\tan\theta} - 1}$

11. Two of the angles in a triangle are 30° and 45°. If the side opposite the 45° angle is $20\sqrt{2}$ cm, find the side opposite the 30° angle.

12. A 10-ft ladder, which is leaning against the side of a building, makes an angle of 60° with the ground, as shown on the next page. How far up the building does the ladder reach?

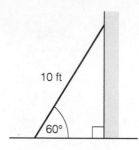

10 ft

60°

13. Prove that the following equation is an identity:

$$\frac{\cos\theta + 1}{\csc\theta + \cot\theta} = \sin\theta$$

14. The arc in the following figure is a portion of the unit circle. Express AB and BP in terms of θ.

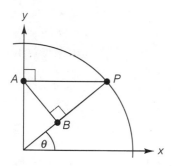

15. Prove that the following equation is an identity:

$$\frac{\sin\theta}{1 + \cos(180° - \theta)} = \frac{1 + \cos\theta}{\sin(180° - \theta)}$$

16. Simplify: $\dfrac{(\cos\theta + \sin\theta)^2}{1 + 2\sin\theta\cos\theta}$.

17. If $\cos 15° = \frac{1}{4}(\sqrt{6} + \sqrt{2})$, find $\sin 15°$.

18. Each side of the square $STUV$ is 8 cm long. The point P is on diagonal $\overline{SU}$ such that $SP = 2$ cm. Find the distance from P to V.

19. In the following figure, $\angle DAB = 25°$, $\angle CAB = 55°$, and $AB = 50$ cm. Find CD. (Leave your answer in terms of the trigonometric functions rather than decimals.)

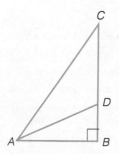

20. A regular nine-sided polygon is inscribed in a circle of radius 2 m. Find the area of the polygon. (Leave your answer in terms of the trigonometric functions rather than decimals.)

In Exercises 21–36, the lengths of the three sides of a triangle are denoted by a, b, and c; the angles opposite these sides are A, B, and C, respectively. In each exercise, use the given information to find the required quantities.

	GIVEN	FIND
21.	$B = 90°$, $\sin A = \frac{2}{5}$, $a = 7$	b
22.	$B = 90°$, $\sec C = 4$, $c = \sqrt{2}$	b
23.	$B = 90°$, $\cos A = \frac{3}{8}$	$\sin A$ and $\cot A$
24.	$B = 90°$, $b = 1$, $\tan C = \sqrt{5}$	a
25.	$b = 4$, $c = 5$, $A = 150°$	a and area of $\triangle ABC$
26.	$A = 120°$, $b = 8$, area of $\triangle ABC = 12\sqrt{3}$	c
27.	$a = 6$, $b = 3$, $c = 5$	$\cos A$ and $\sin A$
28.	$a = 1$, $c = \sqrt{2}$, $B = 60°$	b
29.	$c = 4$, $a = 2$, $B = 90°$	$\sin^2 A + \cos^2 B$
30.	$B = 90°$, $2a = b$	A
31.	$B = 45°$, $C = 30°$, $b = 3\sqrt{2}$	c
32.	$A = 30°$, $B = 120°$, $b = 16$	a
33.	$a = 3$, $c = 12$, $C = 135°$	$\sin A$, $\cos A$, and B
34.	$a = 7$, $b = 8$, $\sin C = \frac{1}{4}$	area of $\triangle ABC$
35.	$b = 4$, $a = 5$, $\cos C = \frac{1}{8}$	c
36.	$a = b = 5$, $\sin(C/2) = \frac{9}{10}$	C

For Exercises 37–52, evaluate each expression. (Don't use a calculator.)

37. $\sin 135°$	**38.** $\cos(-60°)$	**39.** $\tan(-240°)$
40. $\sin 450°$	**41.** $\csc 210°$	**42.** $\sec 225°$
43. $\sin 270°$	**44.** $\cot(-330°)$	**45.** $\cos(-315°)$
46. $\cos 180°$	**47.** $\cos 1800°$	**48.** $\sec 120°$
49. $\csc 240°$	**50.** $\csc(-45°)$	**51.** $\sec 780°$
52. $\sin^2 33° + \sin^2 57°$		

53. For the following figure, show that $y = x[\tan(\alpha + \beta) - \tan\beta]$.

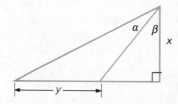

54. For the following figure, show that $y = \dfrac{x}{\cot\alpha - \cot\beta}$.

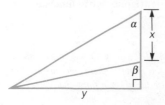

55. For the following figure, show that $\cot\theta = \dfrac{a}{b} + \cot\alpha$.

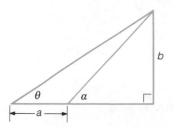

56. This problem is adapted from the text *An Elementary Treatise on Plane Trigonometry* by R. D. Beasley, first published in 1884.
   (a) Prove the following identity, which will be used in part (b): $1 + \tan^2\alpha = \sec^2\alpha$.
   (b) Suppose that α and θ are acute angles and $\tan\theta = \dfrac{1 + \tan\alpha}{1 - \tan\alpha}$. Express $\sin\theta$ and $\cos\theta$ in terms of α. *Hint:* Draw a right triangle, label one of the angles θ, and let the lengths of the sides opposite and adjacent to θ be $1 + \tan\alpha$ and $1 - \tan\alpha$, respectively.

   Answer: $\sin\theta = \dfrac{1}{\sqrt{2}}(\cos\alpha + \sin\alpha)$,

   $\cos\theta = \dfrac{1}{\sqrt{2}}(\cos\alpha - \sin\alpha)$.

57. Suppose that θ is an acute angle in a right triangle and $\sin\theta = \dfrac{2p^2q^2}{p^4 + q^4}$. Find $\cos\theta$ and $\tan\theta$.

58. Evaluate each expression in terms of a, where $a = \cos 20°$.
   (a) $\sin 20°$ **(b)** $\tan 20°$ **(c)** $\cos 70°$
   (d) $\sin 70°$ **(e)** $\sin(-20°)$ **(f)** $\cos(-20°)$
   (g) $\cos 160°$ **(h)** $\cos 340°$ **(i)** $\cos 200°$

59. Suppose that α and β are acute angles such that $(\sin\alpha)/(\sin\beta) = \sqrt{2}$ and $(\tan\alpha)/(\tan\beta) = \sqrt{3}$. Find α and β.

60. In the following figure, $\overline{AC}$ is tangent to the unit circle at P. Express the area of triangle AOC as a function of θ.

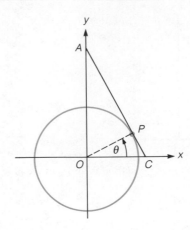

In Exercises 61–70, use the given information to find the required quantities.

	GIVEN	FIND
61.	$\cos\theta = \frac{3}{5}, 0° < \theta < 90°$	$\sin\theta$ and $\tan\theta$
62.	$\sin\theta = -\frac{12}{13}, 180° < \theta < 270°$	$\cos\theta$
63.	$\sec\theta = \frac{25}{7}, 270° < \theta < 360°$	$\tan\theta$
64.	$\cot\theta = \frac{4}{3}, 0° < \theta < 90°$	$\sin\theta$
65.	$\csc\theta = \frac{13}{12}, 0° < \theta < 90°$	$\cot\theta$
66.	$\sin\theta = \frac{1}{5}, 0° < \theta < 90°$	$\cos(90° - \theta)$
67.	$\cos\theta = -\frac{3}{4}, 90° < \theta < 180°$	$\tan(90° - \theta)$
68.	$\sin\theta = -\sqrt{5}/6, 180° < \theta < 270°$	$\sec\theta$
69.	$\cos\theta = \frac{7}{9}, 0° < \theta < 90°$	$\cos(180° - \theta)$
70.	$\cos\theta = -\frac{1}{4}, 90° < \theta < 180°$	$\cos(180° - \theta)$

71. Suppose θ is an acute angle and $\tan\theta + \cot\theta = 2$. Show that $\sin\theta + \cos\theta = \sqrt{2}$.

72. A 100-ft vertical antenna is on the roof of a building. From a point on the ground, the angles of elevation to the top and the bottom of the antenna are 51° and 37°, respectively. Find the height of the building.

73. In an isosceles triangle, the two base angles are each 35°, and the length of the base is 120 cm. Find the area of the triangle.

74. Find the perimeter and the area of a regular pentagon inscribed in a circle of radius 9 cm.

75. In triangle ABC, let h denote the length of the altitude from A to $\overline{BC}$. Show that $h = a/(\cot B + \cot C)$.

76. The length of each side of an equilateral triangle is $2a$. Show that the radius of the inscribed circle is $a/\sqrt{3}$ and the radius of the circumscribed circle is $2a/\sqrt{3}$.

77. From a helicopter h ft above the sea, the angle of depression to the pilot's horizon is θ. Show that $\cot\theta = R/\sqrt{2Rh + h^2}$, where R is the radius of the Earth.

78. In triangle ABC, angle C is a right angle. Show that
$$\frac{\sin A - \sin B}{\sin A + \sin B} = \frac{a - b}{a + b}.$$

79. In this exercise, we prove that the equation given in Exercise 78 holds for any triangle, not just a right triangle.
   (a) First, verify that if $\dfrac{a}{b} = \dfrac{c}{d}$, then $\dfrac{a-b}{a+b} = \dfrac{c-d}{c+d}$.
   (b) From the law of sines, $a/b = (\sin A)/(\sin B)$. Now apply the result from part (a).

80. In triangle ABC, suppose $BC = t(4 + t)$, $AC = 4 - t^2$, and $AB = t^2 + 2t + 4$. Compute $\cos C$. What is angle C? (Assume that $0 < t < 2$.)

81. The radius of the circle in the following figure is 1 unit. The line segment $\overline{PT}$ is tangent to the circle at P, and $\overline{PN}$ is perpendicular to $\overline{OT}$. Express each of the following quantities in terms of θ.
   (a) PN **(b)** ON **(c)** PT
   (d) OT **(e)** NA **(f)** NT

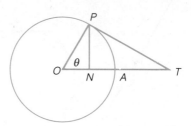

82. Refer to the figure accompanying Exercise 81.
   (a) Show that the ratio of the area of triangle ONP to the area of triangle NPT is $\cot^2 \theta$.
   (b) Show that the ratio of the area of triangle NPT to the area of triangle OPT is $\sin^2 \theta$.

In Exercises 83–108, show that each equation is an identity.

83. $\sin^2(90° - \theta) \csc \theta - \tan^2(90° - \theta) \sin \theta = 0$

84. $\dfrac{1 - \sin \theta \cos \theta}{\cos \theta (\sec \theta - \csc \theta)} \times \dfrac{\sin^2 \theta - \cos^2 \theta}{\sin^3 \theta + \cos^3 \theta} = \sin \theta$

85. $\cos^2 \theta - \sin^2 \theta = 2 \cos^2 \theta - 1$

86. $\cos^2 \theta - \sin^2 \theta = 1 - 2 \sin^2 \theta$

87. $\sin A \tan A = \dfrac{1 - \cos^2 A}{\cos A}$

88. $\dfrac{\cot A - 1}{\cot A + 1} = \dfrac{1 - \tan A}{1 + \tan A}$

89. $\cot^2 A + \csc^2 A = -\cot^4 A + \csc^4 A$

90. $\sin^3 A + \cos^3 A = (\sin A + \cos A)(1 - \sin A \cos A)$

91. $\dfrac{\cot^2 A - \tan^2 A}{(\cot A + \tan A)^2} = 2 \cos^2 A - 1$

92. $\dfrac{\sin A - \cos A}{\sin A} + \dfrac{\cos A - \sin A}{\cos A} = 2 - \sec A \csc A$

93. $\dfrac{\cos A - \sin A}{\cos A + \sin A} = \dfrac{1 - \tan A}{1 + \tan A}$

94. $\dfrac{\cos A - \sin A}{\sec A \cot A - \csc A \tan A} = \cos A \sin A$

95. $\dfrac{1 - \sin A}{\cos A} - \dfrac{1}{\sec A + \tan A} = 0$

96. $\dfrac{\cos A}{1 - \tan A} + \dfrac{\sin A}{1 - \cot A} = \cos A + \sin A$

97. $\tan A \tan B = \dfrac{\tan A + \tan B}{\cot A + \cot B}$

98. $\dfrac{\sin A}{1 + \cos A} + \dfrac{1 + \cos A}{\sin A} = 2 \csc A$

99. $\tan A - \dfrac{\sec A \sin^3 A}{1 + \cos A} = \sin A$

100. $\dfrac{1}{1 - \cos A} + \dfrac{1}{1 + \cos A} = 2 + 2 \cot^2 A$

101. $\dfrac{1}{\csc A - \cot A} - \dfrac{1}{\csc A + \cot A} = 2 \cot A$

102. $\dfrac{\sin^3 A}{\cos A - \cos^3 A} = \tan A$

103. $\dfrac{1}{1 + \sin^2 A} + \dfrac{1}{1 + \cot^2 A} + \dfrac{1}{1 + \sec^2 A} + \dfrac{1}{1 + \csc^2 A} = 2$

104. $\dfrac{\sec A - \tan A}{\sec A + \tan A} = \left(\dfrac{\cos A}{1 + \sin A}\right)^2$

105. $\dfrac{\sec A - \csc A}{\sec A + \csc A} = \dfrac{\sin A - \cos A}{\sin A + \cos A}$

106. $\sin^6 A + \cos^6 A = 1 - 3 \sin^2 A \cos^2 A$

107. $\sin A \cos A = \dfrac{\tan A}{1 + \tan^2 A}$

108. $\dfrac{2 \tan A}{1 - \tan^2 A} + \dfrac{1}{\cos^2 A - \sin^2 A} = \dfrac{\cos A + \sin A}{\cos A - \sin A}$

109. In the accompanying figure, the radius of the circle is $OA = 1$.

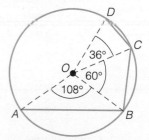

(a) Show that
$$AB = \sqrt{2 + 2 \cos 72°}$$
$$BC = 1$$
$$CD = \sqrt{2 - 2 \cos 36°}$$

(b) Use your calculator to verify that

$$AB = BC + CD$$

(You'll be able to prove this after studying Section 6.6.)

In Exercises 110–120, convert each expression into one involving only sines and cosines and then simplify. (Leave your answers in terms of sines and/or cosines.)

110. $\dfrac{\sin A + \cos A}{\sec A + \csc A}$

111. $\dfrac{\csc A \sec A}{\sec^2 A + \csc^2 A}$

112. $\dfrac{\sin A \sec A}{\tan A + \cot A}$

113. $\cos A + \tan A \sin A$

114. $\dfrac{\cos A}{1 - \tan A} + \dfrac{\sin A}{1 - \cot A}$

115. $\dfrac{1}{\sec A - 1} \div \dfrac{1}{\sec A + 1}$

116. $(\sec A + \csc A)^{-1}[(\sec A)^{-1} + (\csc A)^{-1}]$

117. $\dfrac{\dfrac{\tan^2 A - 1}{\tan^3 A + \tan A}}{\dfrac{\tan A + 1}{\tan^2 A + 1}}$

118. $\dfrac{\dfrac{\sin A + \cos A}{\sin A - \cos A} - \dfrac{\sin A - \cos A}{\sin A + \cos A}}{\dfrac{\sin A + \cos A}{\sin A - \cos A} + \dfrac{\sin A - \cos A}{\sin A + \cos A}}$

119. $\dfrac{\sin A \tan A - \cos A \cot A}{\sec A - \csc A}$

120. $\dfrac{\cos \theta + 1}{\csc \theta + \cot \theta} - \dfrac{1}{\sin \theta + \cot \theta \cos \theta}$

121. The equations

$$a = b \cos C + c \cos B$$
$$b = c \cos A + a \cos C$$
$$c = a \cos B + b \cos A$$

are called the *projection laws* for a triangle. Use the law of cosines to give a proof of the projection laws as follows. To obtain the formula

$$a = b \cos C + c \cos B$$

add the two equations

$$b^2 = a^2 + c^2 - 2ac \cos B$$

and

$$c^2 = a^2 + b^2 - 2ab \cos C$$

In the resulting equation, combine like terms and solve for a. The other projection laws are obtained in a similar manner.

122. **(a)** In triangle ABC, prove that

$$\sin C = \sin A \cos B + \cos A \sin B$$

Hint: In the equation $c = a \cos B + b \cos A$ (obtained in Exercise 121), multiply the term c by $(\sin C)/c$; multiply the term $a \cos B$ by $(\sin A)/a$; multiply the term $b \cos A$ by $(\sin B)/b$.

(b) Use the formula in part (a) and the following figure to show that $\sin 15° = \frac{1}{4}(\sqrt{6} - \sqrt{2})$.

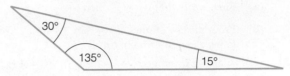

123. The two figures show the same circle of diameter D inscribed first in an equilateral triangle and then in a regular hexagon. Let a denote the length of a side of the triangle and let b denote the length of a side of the hexagon. Show that $ab = D^2$.

124. In $\triangle ABC$, let h denote the length of the altitude drawn from vertex B to side $\overline{AC}$. Show that $h = c \sin A = a \sin C$.

125. The figure shows a circle inscribed in $\triangle ABC$. Follow steps (a), (b), and (c) to show that the radius r of the inscribed circle is given by

$$r = \frac{2S}{a + b + c}$$

where S denotes the area of $\triangle ABC$ and a, b, and c denote (as usual) the lengths of the sides.

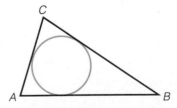

(a) Let O be the center of the circle. Why is the area of $\triangle AOB$ equal to $\frac{1}{2}cr$?

(b) Find expressions similar to the one in part (a) for the areas of triangles AOC and BOC.

(c) The area of $\triangle ABC$ is the sum of the areas of triangles AOB, AOC, and BOC. Use this fact along with the results in parts (a) and (b) to show that $\frac{1}{2}r(a + b + c) = S$. Thus $r = 2S/(a + b + c)$, as required.

126. This exercise shows you how to prove the following remarkable fact. In any triangle, the sum of the reciprocals of the lengths of the altitudes equals the reciprocal of the radius of the inscribed circle. In the following steps, we use the results and the notation from Exercises 124 and 125.

(a) Using the result in Exercise 125, show that

$$\frac{1}{r} = \frac{a}{ab \sin C} + \frac{b}{bc \sin A} + \frac{c}{ca \sin B}$$

$$= \frac{1}{b \sin C} + \frac{1}{c \sin A} + \frac{1}{a \sin B}$$

(b) Use the result in Exercise 124 to check that the expression on the right-hand side of the last equation in part (a) is equal to the sum of the reciprocals of the lengths of the altitudes.

CHAPTER SIX

TRIGONOMETRIC FUNCTIONS OF REAL NUMBERS

INTRODUCTION   In the previous chapter we measured angles using the familiar units of degrees. However, there is another unit for measuring angles, called the *radian*, that has proven to be more useful in calculus and in many of the sciences. In Section 6.1 we define radian measure. Then, in Section 6.2, we use radian measure to restate the definitions of the trigonometric functions in such a way that the domains are sets of real numbers. Among other advantages, this allows us to analyze the trigonometric functions using the graphing techniques developed in Chapter 2. With this in mind, we discuss the graphs of the trigonometric functions in Sections 6.3 and 6.4. In the two sections after this, Sections 6.5 and 6.6, we develop the basic trigonometric identities. Then in Section 6.7 we apply some of these identities in solving trigonometric equations. Finally, in Section 6.8 we consider the inverse trigonometric functions.

6.1 ▼ RADIAN MEASURE

. . . I wrote to him [*to Alexander J. Ellis, in 1874*], and he declared at once for the form "radian," on the ground that it could be viewed as a contraction for "radial angle. . . ."

Thomas Muir, in a letter appearing in the April 7, 1910, issue of *Nature*

I shall be very pleased to send Dr. Muir a copy of my father's examination questions of June, 1873, containing the word *radian*. . . . It thus appears that *radian* was thought of independently by Dr. Muir and my father, and, what is really more important than the exact form of the name, they both independently thought of the necessity of giving a name to the unit-angle.

James Thomson, in a letter appearing in the June 16, 1910, issue of *Nature*

For the portion of trigonometry dealing with triangles, the units of degrees are quite suitable for measuring angles. For the more analytical portions of trigonometry, however, radian measure is used. The radian measure of an angle is defined as follows.

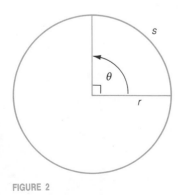

FIGURE 1

DEFINITION

Radian Measure of an Angle

Place the vertex of the angle at the center of a circle of radius r. Let s denote the length of the arc intercepted by the angle, as indicated in Figure 1. Then the **radian measure** θ is the ratio of the arc length s to the radius r. In symbols,

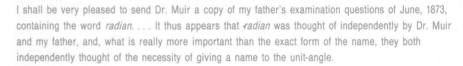

$$\theta = \frac{s}{r}$$

At first it may appear to you that the radian measure depends on the radius of the particular circle that we use. But as you will see, this is not the case.

To gain some experience in working with the definition of radian measure, let us calculate the radian measure of the right angle in Figure 2. We begin with the formula

$$\theta = \frac{s}{r} \tag{1}$$

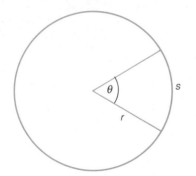

FIGURE 2

Now, since the angle is a right angle, the arc length s is one-quarter of the entire circumference. Thus,

$$s = \frac{1}{4}(2\pi r) = \frac{\pi r}{2} \tag{2}$$

Using equation (2) to substitute for s in equation (1), we get

$$\theta = \frac{\pi r/2}{r} = \frac{\pi r}{2} \times \frac{1}{r} = \frac{\pi}{2}$$

In other words,

$$90° = \frac{\pi}{2} \text{ radians} \tag{3}$$

Notice that the radius r does not appear in our answer.

For practical reasons, we would like to be able to convert rapidly between degree and radian measure. Multiplying both sides of equation (3) by 2 yields

$$180° = \pi \text{ radians} \tag{4}$$

Equation (4) is useful and should be memorized. For instance, dividing both sides of equation (4) by 6 yields

$$\frac{180°}{6} = \frac{\pi}{6}$$

or

$$30° = \frac{\pi}{6} \text{ radians}$$

Similarly, dividing both sides of equation (4) by 4 and 3, respectively, yields

$$45° = \frac{\pi}{4} \text{ radians} \qquad \text{and} \qquad 60° = \frac{\pi}{3} \text{ radians}$$

And, multiplying both sides of equation (4) by 2 gives us

$$360° = 2\pi \text{ radians}$$

Figure 3 summarizes some of these results.

FIGURE 3

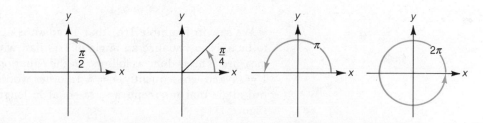

EXAMPLE 1　(a) Express 1° in radian measure.

(b) Express 1 radian in terms of degrees.

Solution　(a) Dividing both sides of equation (4) by 180 yields

$$1° = \frac{\pi}{180} \text{ radian}$$

(b) Dividing both sides of equation (4) by π yields

$$\frac{180°}{\pi} = 1 \text{ radian}$$

In other words, 1 radian is approximately 180°/3.14, or 57.3°.　▲

From the results in Example 1, we have the following rules for converting between radians and degrees.

> To convert from degrees to radians, multiply by $\frac{\pi}{180°}$. To convert from radians to degrees, multiply by $\frac{180°}{\pi}$.

EXAMPLE 2　Convert 150° to radians.

Solution　$$150°\left(\frac{\pi}{180°}\right) = \frac{5\pi}{6} \quad \text{reducing the fraction}$$

Thus,

$$150° = \frac{5\pi}{6} \text{ radians}$$

　　▲

EXAMPLE 3　Convert $11\pi/6$ radians to degrees.

Solution　$$\frac{11\pi}{6}\left(\frac{180°}{\pi}\right) = \frac{(11)(180°)}{6} = 11(30°) = 330°$$

Thus,

$$\frac{11\pi}{6} \text{ radians} = 330°$$

　　▲

We saw in Example 1(b) that 1 radian is approximately 57°. It is also useful to be able to visualize an angle of 1 radian without thinking in terms of degree measure. This is done as follows. In the equation $\theta = s/r$, we let $\theta = 1$. This yields $1 = s/r$ and, consequently, $r = s$. In other words, in a circle, 1 radian is the central angle that intercepts an arc equal in length to the radius of the circle (see Figure 4).

PROPERTY SUMMARY

RADIAN MEASURE

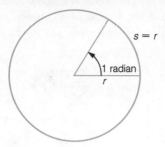

FIGURE 4

In a circle, 1 radian is the measure of the central angle that intercepts an arc equal in length to the radius of the circle.

From now on, when we specify the measure of an angle, we will assume that the units are radians unless the degree symbol is explicitly used. (This convention is also used in calculus.) For instance, the equation $\theta = 2$ means that θ is 2 radians. Similarly, the expression $\sin \pi/6$ refers to the sine of $\pi/6$ radians.

EXAMPLE 4   Evaluate the following quantities.

(a) $\sin \dfrac{\pi}{6}$ **(b)** $\cos 2\pi$

Solution   **(a)** $\dfrac{\pi}{6}$ radians $= 30°$

Therefore,

$$\sin \frac{\pi}{6} = \sin 30° = \frac{1}{2} \quad \text{because } \sin 30° = \frac{1}{2}$$

(b) 2π radians $= 360°$

Therefore,

$$\cos 2\pi = \cos 360° = 1 \quad \text{because } \cos 360° = 1 \qquad \blacktriangle$$

EXAMPLE 5   Without using a calculator, determine whether the following values are positive or negative.

(a) $\cos 3$ **(b)** $\sin 1$ **(c)** $\tan 6$

Solution   **(a)** Since π radians is 180°, we estimate that 3 radians is slightly less than 180° (because 3 is slightly less than π). Thus, in standard position, the terminal side of an angle of 3 radians lies in the second quadrant. Consequently, $\cos 3$ is negative.

(b) Since 1 radian is approximately 60° [as we saw in Example 1(b) and in Figure 4], $\sin 1$ is positive.

(c) We know that 6 radians is slightly less than 2π radians, which is one revolution, or 360°. Thus, in standard position, the terminal side of an angle of 6 radians lies in the fourth quadrant. Consequently, $\tan 6$ is negative. $\blacktriangle$

Let us now return to the defining equation for radian measure, $\theta = s/r$, and rewrite it as $s = r\theta$. This useful equation expresses arc length on a circle in terms of the radius r and the central angle θ subtended by the arc. Caution: In applying the formula $s = r\theta$, θ must be expressed in radians (because the formula was obtained from the defining equation for radians).

Formula for Arc Length

> In a circle of radius r, the length s of an arc that subtends a central angle of radian measure θ is given by
>
> $$s = r\theta$$

EXAMPLE 6   Find the arc length s in Figure 5.

Solution   We have seen previously that $30° = \pi/6$ radians. Thus

$$s = r\theta = (10)\left(\frac{\pi}{6}\right)$$

$$= \frac{5\pi}{3} \text{ cm} \qquad \text{(This is approximately 5 cm. Why?)}$$

Note that the units for s and for r are the same.   ▲

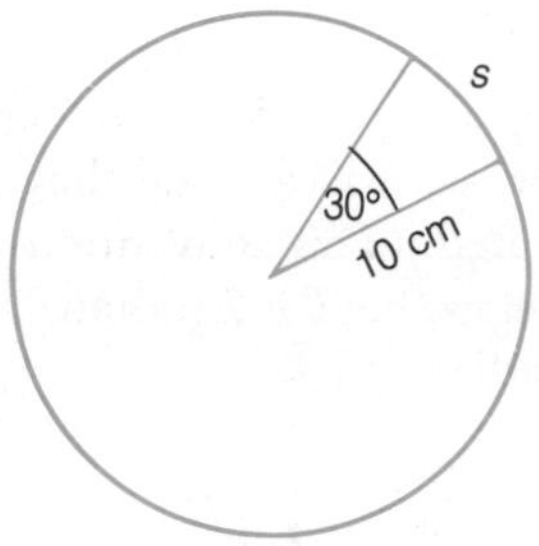

The next example indicates how radian measure is used in the study of rotating objects. As a prerequisite for this example, we first define the terms *angular speed* and *linear speed*.

FIGURE 5

DEFINITION

Angular Speed and Linear Speed

> Suppose that a wheel rotates about its axis at a constant rate.
>
> 1. [Refer to Figure 6(a)] If a radial line turns through an angle θ in time t, then the **angular speed** of the wheel (denoted by the Greek letter ω) is defined to be
>
> $$\text{angular speed} = \omega = \frac{\theta}{t}$$
>
> 2. [Refer to Figure 6(b)] If a point P on the rotating wheel travels a distance d in time t, then the **linear speed** of P, denoted by v, is defined to be
>
> $$\text{linear speed} = v = \frac{d}{t}$$

FIGURE 6

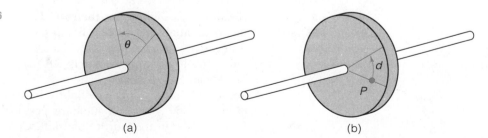

(a)   (b)

EXAMPLE 7 A circular gear in a motor rotates at the rate of 100 rpm (revolutions per minute).

(a) What is the angular speed of the gear in radians per minute?
(b) Find the linear speed of a point on the gear 4 cm from the center.

Solution (a) Each revolution of the gear is 2π radians. So in 100 revolutions, there are

$$\theta = 100(2\pi) = 200\pi \text{ radians}$$

Consequently, we have

$$\omega = \frac{\theta}{t} = \frac{200\pi \text{ radians}}{1 \text{ min}} = 200\pi \text{ radians/min}$$

(b) We can use the formula $s = r\theta$ to find the distance traveled by the point in 1 min. Using $r = 4$ cm and $\theta = 200\pi$, we obtain

$$s = r\theta = 4(200\pi) = 800\pi \text{ cm}$$

The linear speed therefore is

$$v = \frac{d}{t} = \frac{800\pi \text{ cm}}{1 \text{ min}} = 800\pi \text{ cm/min}$$ ▲

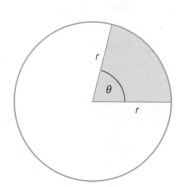

FIGURE 7

One of the advantages of using radian measure is that many formulas then take on particularly simple forms. The formula $s = r\theta$ is one example of this. As another example, let us derive a formula for the area of a *sector* of a circle. (In Figure 7, the shaded region is the sector.) From geometry, we know that the area A of the sector is directly proportional to the measure θ of its central angle. In symbols,

$$A = k\theta \qquad (5)$$

To determine the constant k, we use the fact that when $\theta = 2\pi$, we have a complete circle. Thus, when $\theta = 2\pi$, the area must be πr^2. Substituting the values $\theta = 2\pi$ and $A = \pi r^2$ in equation (5) yields

$$\pi r^2 = k(2\pi)$$

and therefore,

$$\tfrac{1}{2}r^2 = k$$

Using this value for k in equation (5) gives us the following result:

FIGURE 8

Formula for the Area of a Sector

> In a circle of radius r, the area of a sector with central angle of radian measure θ is given by
>
> $$A = \frac{1}{2}r^2\theta$$

EXAMPLE 8 Compute the area of the sector in Figure 8.

Solution We first convert 120° to radians:

$$120°\left(\frac{\pi}{180°}\right) = \frac{2\pi}{3} \text{ radians}$$

We then have

$$A = \frac{1}{2} r^2 \theta$$

$$= \frac{1}{2} (5^2) \frac{2\pi}{3} = \frac{25\pi}{3} \text{ cm}^2$$

▲

EXAMPLE 9 **(a)** In Figure 9, the radian measure of $\angle AOB$ is θ. Express the area of the shaded region as a function of θ.
(b) Compute the area of the shaded region assuming that $\angle AOB = 75°$.

Solution **(a)** Let S denote the area of the shaded region in Figure 9. Then we have

$$S = (\text{area of sector } OAB) - (\text{area of } \triangle OAB)$$

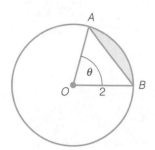

↑ Use the formula area $= \frac{1}{2} r^2 \theta$ to find this.

↑ Use the result from Chapter 5 stating that the area of a triangle equals half the product of two sides times the sine of the included angle.

Thus,

$$S = \frac{1}{2}(2^2)\theta - \frac{1}{2}(2)(2) \sin \theta$$

$$= 2\theta - 2 \sin \theta, \quad \text{where } \theta \text{ is in radians.}$$

FIGURE 9

(b) We'll use the formula obtained in part (a). Since that was obtained using the sector-area formula, we first need to convert 75° to radian measure. As you can check, the result is $\theta = 5\pi/12$. So we have

$$S = 2\theta - 2 \sin \theta$$

$$= 2\left(\frac{5\pi}{12}\right) - 2 \sin\left(\frac{5\pi}{12}\right) = \frac{5\pi}{6} - 2 \sin\left(\frac{5\pi}{12}\right)$$

$$\approx 2.57 \text{ cm}^2 \qquad (\text{Use your calculator to check this.})$$

▲

In Example 9(b), we first needed to convert 75° to radian measure because the sector-area formula is valid only when radian measure is used. It is instructive to observe what would occur had we mistakenly been using degree measure. Had we been using degree measure, we would have obtained $S = (75° - 2 \sin 75°) \text{ cm}^2$ or, using a calculator, $S \approx (75° - 1.93) \text{ cm}^2$. But this last expression for S does not make sense because the two quantities in the parentheses do not have the same units, and so the subtraction cannot be carried out.

▼ **EXERCISE SET 6.1**

A

In Exercises 1 and 2, convert to radian measure.

1. **(a)** 60° **(b)** 225° **(c)** 36°
  (d) 450° **(e)** 0°

2. **(a)** 35° **(b)** 22.5° **(c)** 2° **(d)** 100°

In Exercises 3 and 4, convert the radian measures to degrees.

3. **(a)** $\pi/12$ **(b)** $3\pi/2$ **(c)** 6π
  (d) $\pi/10$ **(e)** $\pi/2$ **(f)** 3

4. **(a)** $11\pi/6$ **(b)** $3\pi/7$ **(c)** 2π
  (d) 2 **(e)** $1/\pi$ **(f)** π^2

5. Suppose that $\theta = \frac{3}{2}$. Without using a calculator or tables, decide whether θ is larger or smaller than a right angle.

In Exercises 6–8, complete each table.

6.

θ	0	$\pi/2$	π	$3\pi/2$	2π
$\sin \theta$					

7.

θ	0	$\pi/2$	π	$3\pi/2$	2π
$\cos \theta$					

8.

θ	0	$\pi/6$	$\pi/4$	$\pi/3$	$\pi/2$
$\tan \theta$					

For Exercises 9 and 10, refer to the following figure, which shows all angles from 0° to 360° that are multiples of 30° or 45°.

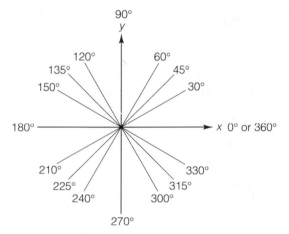

9. In the figure, relabel the angles in quadrants I and II using radian measure.

10. In the figure, relabel the angles in quadrants III and IV using radian measure.

In Exercises 11–14, determine whether the given quantities are positive or negative; do not use a calculator.

11. **(a)** $\sin 2$ **(b)** $\sin 3$
12. **(a)** $\cos 2$ **(b)** $\tan 2$
13. **(a)** $\sin 3.16$ **(b)** $\tan 3.16$
14. **(a)** $\sec 6.3$ **(b)** $\sec(-6.3)$

In Exercises 15–18, find the arc length s in each case.

15.

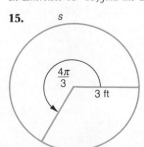

16.

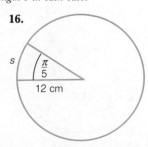

17.

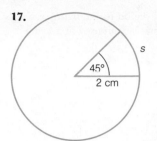

18.

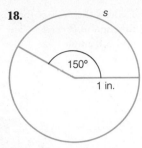

19. In a circle of radius 6 cm, an arc of length 1 cm subtends an angle θ at the center of the circle. What is the radian measure of θ?

20. In a circle of radius 4 in., an arc of length 4 in. subtends an angle θ at the center of the circle. What is the radian measure of θ?

In Exercises 21–26, you are given the rate of rotation of a wheel and its radius. In each case, determine the following:
(a) the angular speed, in units of radians/sec;
(b) the linear speed, in units of cm/sec, of a point on the circumference of the wheel;
(c) the linear speed, in cm/sec, of a point halfway between the center of the wheel and the circumference.

21. 6 revolutions/sec; $r = 12$ cm
22. 15 revolutions/sec; $r = 20$ cm
23. 1080°/sec; $r = 25$ cm 24. 2160°/sec; $r = 60$ cm
25. 500 rpm; $r = 45$ cm 26. 1250 rpm; $r = 10$ cm
27. Find the area of the shaded sectors.

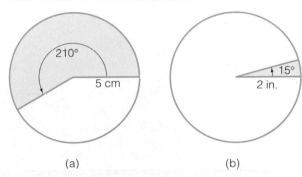

(a) (b)

28. In a circle of radius 1 cm, the area of a certain sector is $\pi/5$ cm². Find the central angle of this sector.

29. Find the area of the shaded region in the following figure.

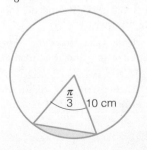

30. Express the area of the shaded region in the figure as a function of θ. *Hint:* The region is composed of a sector and a triangle.

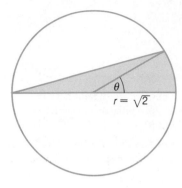

B

31. Suppose that you have two sticks and a piece of wire, each of length 1 ft, fastened at the ends to form an equilateral triangle. See Figure (a). If side $\overline{BC}$ is bent out to form an arc of a circle with center A, then the angle at A will decrease from 60° to something less. See Figure (b). What is the measure of this new angle at A?

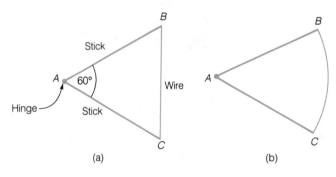

(a)   (b)

32. **(a)** When a clock reads 4:00, what is the radian measure of the (smaller) angle between the hour hand and minute hand?
(b) When a clock reads 5:30, what is the radian measure of the (smaller) angle between the hour hand and minute hand?

33. A wheel 3 ft in diameter makes x revolutions. Find x, given that the distance traveled by a point on the circumference of the wheel is 22619 ft. (Round off your answer to the nearest whole number.)

34. A point P is on a rotating wheel, 12 cm from the center. If the linear speed of P is 204 cm/sec, find the angular speed of the wheel (in radians/sec).

35. For this problem, assume that the earth is a sphere with a radius of 3960 miles and a rotation rate of 1 revolution per 24 hours.

(a) Find the angular speed. Express your answer in units of radians/sec, and round off to two significant digits.
(b) Find the linear speed of a point on the equator. Express the answer in units of miles per hour, and round off to the nearest 10 mph.

36. The accompanying figure shows a circular sector with radius r cm and central angle 2θ (radian measure). The perimeter of the sector is 12 cm.
(a) Express r as a function of θ.
(b) Express the area A of the sector as a function of θ.

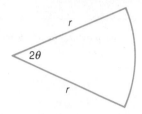

37. The following figure shows a semicircle of radius 1 unit and two adjacent sectors, AOC and COB.
(a) Show that the product P of the areas of the two sectors is given by

$$P = \frac{\pi\theta}{4} - \frac{\theta^2}{4}$$

Is this a quadratic function?
(b) For what value of θ is P a maximum?

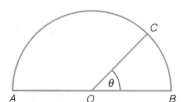

C

38. Refer to the accompanying figure. From a point R, two tangent lines are drawn to a circle of radius 1. These tangents meet the circle at the points P and Q. At the

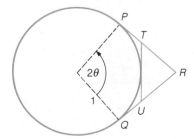

midpoint of the circular arc PQ, a third tangent line is drawn, meeting the other two tangents at T and U, as shown. Express the area of triangle RTU as a function of θ (θ is defined in the figure).

39. The following problem is taken from *A Treatise on Plane Trigonometry*, 7th ed., by E. W. Hobson (New York: Cambridge University Press, 1928): Two circles, the sum of whose radii is a, are placed in the same plane, with their centers at a distance $2a$, and an endless string, quite stretched, partly surrounds the circles and crosses itself between them. Show that the length of the string is

$$\left(\frac{4\pi}{3} + 2\sqrt{3}\right)a.$$

6.2 ▼ TRIGONOMETRIC FUNCTIONS OF REAL NUMBERS

The definitions of the trigonometric functions in Section 5.3 are based on the unit circle. Let us look at radian measure in this context. By definition, the radian measure θ is given by $\theta = s/r$. So when $r = 1$, we obtain

$$\theta = \frac{s}{1} = s$$

Thus, in the unit circle, the arc length *is* the radian measure of the angle (see Figure 1).

In view of this observation, we can now define the trigonometric functions in such a way that the domains are sets of real numbers, rather than angles. There are two reasons for doing this. First, for analytical work, the domains of the trigonometric functions are sets of real numbers. Second, if the domains are sets of real numbers, then the techniques for graphing that we developed in Chapter 2 can be used to analyze these functions.

In the definitions that follow, you may think of θ as either the measure of an arc length or the radian measure of an angle. But in both cases—and this is the point—θ denotes a real number. The conventions regarding the measurement of arc length on the unit circle are the same as those used previously for angles: We measure from the point $(1, 0)$ and we assume that the positive direction is counterclockwise.

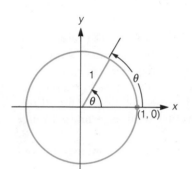

FIGURE 1

In the unit circle, the length of the intercepted arc is the radian measure of the central angle.

Trigonometric Functions of Real Numbers

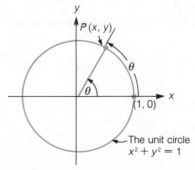

FIGURE 2

(Refer to Figure 2.) Let $P(x, y)$ denote the point on the unit circle whose arc length from $(1, 0)$ is θ. Or, equivalently, let $P(x, y)$ denote the point where the terminal side of the angle with radian measure θ intersects the unit circle. Then the six trigonometric functions are defined as follows.

$$\sin \theta = y \qquad\qquad \csc \theta = \frac{1}{y} \quad (y \neq 0)$$

$$\cos \theta = x \qquad\qquad \sec \theta = \frac{1}{x} \quad (x \neq 0)$$

$$\tan \theta = \frac{y}{x} \quad (x \neq 0) \qquad \cot \theta = \frac{x}{y} \quad (y \neq 0)$$

Let us immediately note that there is nothing new here as far as evaluating the trigonometric functions. For instance, $\sin \pi/6$ is still equal to $\frac{1}{2}$. [If this seems unfamiliar, review Example 4(a) on page 307.] Also, whether we are working in radians or degrees, our three-step procedure for evaluating the trigonometric functions is applicable. For instance, in Example 3(b) on page 274, we computed cos 135° (cos $3\pi/4$). Here is how that same procedure looks when radian measure is used. You should carefully compare each of the corresponding steps.

STEP 1 The reference angle for $3\pi/4$ is $\pi/4$.* ($\pi - 3\pi/4 = \pi/4$)

STEP 2 $\cos \pi/4 = \sqrt{2}/2$

STEP 3 Cos θ is the x-coordinate. In the second quadrant, x-coordinates are negative. Thus, cos $3\pi/4$ is negative, since the terminal side of $\theta = 3\pi/4$ lies in the second quadrant. (Equivalently, the arc of length $3\pi/4$ terminates in the second quadrant.) Thus we have

$$\cos \frac{3\pi}{4} = -\frac{\sqrt{2}}{2}$$

There are many identities involving trigonometric functions. (In fact, that is one reason why these functions are so useful.) We will be concerned with these identities, in one form or another, throughout the remainder of this chapter. Some of the identities will probably seem familiar to you from the text and exercises in Chapter 5. However, keep in mind that now the identities will be true for all real numbers θ for which the expressions are defined, rather than for a limited range of angles.

First of all, there are five trigonometric identities that are immediate consequences of the definitions. These are as follows.

$$\csc \theta = \frac{1}{\sin \theta} \qquad \tan \theta = \frac{\sin \theta}{\cos \theta}$$

$$\sec \theta = \frac{1}{\cos \theta} \qquad \cot \theta = \frac{\cos \theta}{\sin \theta}$$

$$\cot \theta = \frac{1}{\tan \theta}$$

The next identities that we consider are the three *Pythagorean identities*.

The Pythagorean Identities

$$\sin^2 \theta + \cos^2 \theta = 1$$
$$\tan^2 \theta + 1 = \sec^2 \theta$$
$$\cot^2 \theta + 1 = \csc^2 \theta$$

To prove the identity $\sin^2 \theta + \cos^2 \theta = 1$, we refer back to Figure 2. Since (x, y) is a point on the unit circle, we have

$$x^2 + y^2 = 1$$

Replacing x by cos θ and y by sin θ then gives us

$$\cos^2 \theta + \sin^2 \theta = 1$$

---

* Some texts refer to this as the reference *number*, instead of reference angle, to emphasize the fact that $3\pi/4$ is a real number.

which is essentially what we wished to show. Incidentally, you should also become familar with the equivalent forms of this identity:

$$\cos^2 \theta = 1 - \sin^2 \theta \qquad \text{and} \qquad \sin^2 \theta = 1 - \cos^2 \theta$$

To prove the second of the Pythagorean identities, we begin with $\sin^2 \theta + \cos^2 \theta = 1$ and divide both sides by the quantity $\cos^2 \theta$ to obtain

$$\frac{\sin^2 \theta}{\cos^2 \theta} + \frac{\cos^2 \theta}{\cos^2 \theta} = \frac{1}{\cos^2 \theta} \qquad (\text{assuming } \cos \theta \neq 0)$$

and, consequently,

$$\tan^2 \theta + 1 = \sec^2 \theta \qquad \text{as required}$$

Since the proof of the third Pythagorean identity is similar, we omit it here.

EXAMPLE 1   If $\sin \theta = \frac{2}{3}$ and $\pi/2 < \theta < \pi$, compute $\cos \theta$ and $\tan \theta$.

Solution
$$\cos^2 \theta = 1 - \sin^2 \theta \quad \text{using the first Pythagorean identity}$$
$$= 1 - \left(\frac{2}{3}\right)^2 \quad \text{substituting}$$
$$= 1 - \frac{4}{9} = \frac{5}{9}$$

Consequently,

$$\cos \theta = \frac{\sqrt{5}}{3} \qquad \text{or} \qquad \cos \theta = \frac{-\sqrt{5}}{3}$$

Now, since $\pi/2 < \theta < \pi$, it follows that $\cos \theta$ is negative. (Why?) Thus,

$$\cos \theta = \frac{-\sqrt{5}}{3} \qquad \text{as required}$$

To compute $\tan \theta$, we use the identity $\tan \theta = \sin \theta / \cos \theta$ to obtain

$$\tan \theta = \frac{2/3}{-\sqrt{5}/3} = \frac{2}{3} \times \frac{3}{-\sqrt{5}}$$
$$= -\frac{2}{\sqrt{5}}$$

If required, we can rationalize the denominator (by multiplying by $\sqrt{5}/\sqrt{5}$) to obtain

$$\tan \theta = -\frac{2\sqrt{5}}{5}$$

▲

EXAMPLE 2   If $\sec \theta = -\frac{5}{3}$ and $\pi < \theta < 3\pi/2$, compute $\cos \theta$ and $\tan \theta$.

Solution
Since $\cos \theta$ is the reciprocal of $\sec \theta$, we have $\cos \theta = -\frac{3}{5}$. We can compute $\tan \theta$ using the second Pythagorean identity as follows.

$$\tan^2 \theta = \sec^2 \theta - 1$$
$$= \left(-\frac{5}{3}\right)^2 - 1 = \frac{25}{9} - \frac{9}{9} = \frac{16}{9}$$

Therefore,

$$\tan \theta = \frac{4}{3} \quad \text{or} \quad \tan \theta = -\frac{4}{3}$$

Since θ is between π and $3\pi/2$, the quantity $\tan \theta$ is positive. Thus,

$$\tan \theta = \frac{4}{3}$$

▲

The next example shows how certain expressions containing radicals can be simplified through an appropriate trigonometric substitution. (This is often useful in calculus.)

EXAMPLE 3   In the expression $u/\sqrt{u^2 - 1}$, make the substitution $u = \sec \theta$ and show that the resulting expression is equal to $\csc \theta$. (Assume that $0 < \theta < \pi/2$.)

Solution   Replacing u by $\sec \theta$ in the given expression yields

$$\frac{u}{\sqrt{u^2 - 1}} = \frac{\sec \theta}{\sqrt{\sec^2 \theta - 1}}$$

$$= \frac{\sec \theta}{\sqrt{\tan^2 \theta}} \quad \text{using the second Pythagorean identity}$$

$$= \frac{\sec \theta}{\tan \theta} = \frac{1/\cos \theta}{\sin \theta/\cos \theta} = \frac{1}{\cos \theta} \times \frac{\cos \theta}{\sin \theta}$$

$$= \frac{1}{\sin \theta} = \csc \theta$$

Question   Where did we use the condition $0 < \theta < \pi/2$?   ▲

The next three identities indicate the effects of replacing θ by $-\theta$ in the expressions $\sin \theta$, $\cos \theta$, and $\tan \theta$.

$$\cos(-\theta) = \cos \theta$$
$$\sin(-\theta) = -\sin \theta$$
$$\tan(-\theta) = -\tan \theta$$

To see why the first two of these identities are true, consider Figure 3. (Although Figure 3 shows an arc of length θ terminating in the first quadrant, the same kind of argument we use will work when θ terminates in any quadrant.) By definition, the coordinates of P and Q are as follows:

P: $(\cos \theta, \sin \theta)$

Q: $(\cos(-\theta), \sin(-\theta))$

However, as you can see by looking at Figure 3, the x-coordinates of P and Q are the same, while the y-coordinates are negatives of each other. Thus,

$$\cos(-\theta) = \cos \theta$$

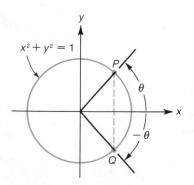

FIGURE 3

and

$$\sin(-\theta) = -\sin\theta \qquad \text{as we wished to show}$$

Now we can establish the third identity, involving $\tan(-\theta)$, as follows:

$$\tan(-\theta) = \frac{\sin(-\theta)}{\cos(-\theta)} = \frac{-\sin\theta}{\cos\theta}$$

$$= -\frac{\sin\theta}{\cos\theta} = -\tan\theta$$

The last identities we are going to discuss in this section are simply consequences of the fact that the circumference C of the unit circle is 2π.* Thus, if we begin at any point P on the unit circle and travel a distance of 2π units along the circle, we return to the same point P. In other words, arc lengths of θ and $\theta + 2\pi$ [measured from $(1, 0)$, as usual] yield the same terminal point P on the unit circle. Since the trigonometric functions are defined in terms of the coordinates of that point P, we conclude that

$$\sin(\theta + 2\pi) = \sin\theta$$
$$\cos(\theta + 2\pi) = \cos\theta$$

These two identities are true for all real numbers θ. Similar identities hold for the other trigonometric functions in their respective domains:

$$\tan(\theta + 2\pi) = \tan\theta \qquad \csc(\theta + 2\pi) = \csc\theta$$
$$\cot(\theta + 2\pi) = \cot\theta \qquad \sec(\theta + 2\pi) = \sec\theta$$

EXAMPLE 4   Evaluate $\sin(5\pi/2)$.

Solution   First, simply as a matter of arithmetic, we observe that $5\pi/2 = \pi/2 + 2\pi$. Thus, in view of our earlier remarks, we have

$$\sin\frac{5\pi}{2} = \sin\frac{\pi}{2} = 1$$

▲

The preceding set of identities can be generalized as follows. If we start at a point P on the unit circle and make two complete revolutions, the arc length we travel is

$$2\pi + 2\pi = 4\pi$$

Similarly, the arc length covered in three complete revolutions is

$$2\pi + 2\pi + 2\pi = 6\pi$$

And the arc length for k complete revolutions is $2k\pi$. (When k is positive, the revolutions are counterclockwise; when k is negative, the revolutions are clockwise.) It follows that the arc lengths θ and $\theta + 2\pi k$ yield the same terminal point P. We summarize this as follows.

---

* This is because $C = 2\pi r$. Thus, when $r = 1$ we obtain $C = 2\pi$.

> For any integer k, the following identities hold.
>
> $$\sin(\theta + 2\pi k) = \sin \theta$$
> $$\cos(\theta + 2\pi k) = \cos \theta$$

EXAMPLE 5   Evaluate $\cos(-17\pi)$.

Solution $\cos(-17\pi) = \cos(\pi - 18\pi) = \cos \pi = -1$ ▲

▼ EXERCISE SET 6.2

A

1. Complete the following table.

θ	$\sin \theta$	$\cos \theta$	$\tan \theta$	$\csc \theta$	$\sec \theta$	$\cot \theta$
0						
$\pi/6$						
$\pi/4$						
$\pi/3$						
$\pi/2$						
$2\pi/3$						
$3\pi/4$						
$5\pi/6$						
π						

2. Complete the following table.

θ	$\sin \theta$	$\cos \theta$	$\tan \theta$	$\csc \theta$	$\sec \theta$	$\cot \theta$
π						
$7\pi/6$						
$5\pi/4$						
$4\pi/3$						
$3\pi/2$						
$5\pi/3$						
$7\pi/4$						
$11\pi/6$						

3. If $\sin \theta = -\frac{3}{5}$ and $\pi < \theta < 3\pi/2$, compute $\cos \theta$ and $\tan \theta$.

4. If $\cos \theta = \frac{5}{13}$ and $3\pi/2 < \theta < 2\pi$, compute $\sin \theta$ and $\cot \theta$.

5. If $\sin t = \sqrt{3}/4$ and $\pi/2 < t < \pi$, compute $\tan t$.

6. If $\sec \theta = -\sqrt{13}/2$ and $\sin \theta > 0$, compute $\tan \theta$.

7. If $\tan \alpha = \frac{12}{5}$ and $\cos \alpha > 0$, compute $\sec \alpha$, $\cos \alpha$, and $\sin \alpha$.

8. If $\cot \theta = -1/\sqrt{3}$ and $\cos \theta < 0$, compute $\csc \theta$ and $\sin \theta$.

9. In the expression $\sqrt{9 - x^2}$, make the substitution $x = 3 \sin \theta$ $(0 < \theta < \pi/2)$ and show that the result is $3 \cos \theta$.

10. Make the substitution $u = 2 \cos \theta$ in the expression $1/\sqrt{4 - u^2}$ and simplify the result. (Assume that $0 < \theta < \pi$.)

11. In the expression $1/(u^2 - 25)^{3/2}$, make the substitution $u = 5 \sec \theta$ $(0 < \theta < \pi/2)$ and show that the result is $(\cot^3 \theta)/125$.

12. In the expression $1/(x^2 + 5)^2$, replace x by $\sqrt{5} \tan \theta$ and show that the result is $(\cos^4 \theta)/25$.

13. In the expression $1/\sqrt{u^2 + 7}$, let $u = \sqrt{7} \tan \theta$ $(0 < \theta < \pi/2)$ and simplify the result.

14. In the expression $\sqrt{x^2 - a^2}/x$ $(a > 0)$, let $x = a \sec \theta$ $(0 < \theta < \pi/2)$ and simplify the result.

15. (a) If $\sin \theta = \frac{2}{3}$, find $\sin(-\theta)$.
  (b) If $\sin \phi = -\frac{1}{4}$, find $\sin(-\phi)$.
  (c) If $\cos \alpha = \frac{1}{5}$, find $\cos(-\alpha)$.
  (d) If $\cos \beta = -\frac{1}{5}$, find $\cos(-\beta)$.

16. (a) If $\sin \theta = 0.35$, find $\sin(-\theta)$.
  (b) If $\sin \phi = -0.47$, find $\sin(-\phi)$.
  (c) If $\cos \alpha = 0.21$, find $\cos(-\alpha)$.
  (d) If $\cos \beta = -0.56$, find $\cos(-\beta)$.

17. If $\cos \theta = -\frac{1}{3}$ $(\pi/2 < \theta < \pi)$, compute the following.
  (a) $\sin(-\theta) + \cos(-\theta)$ (b) $\sin^2(-\theta) + \cos^2(-\theta)$

18. If $\sin(-\theta) = \frac{3}{5}$ $(\pi < \theta < 3\pi/2)$, compute the following.
  (a) $\sin \theta$ (b) $\cos(-\theta)$
  (c) $\cos \theta$ (d) $\tan \theta + \tan(-\theta)$

19. Evaluate each of the following using the identities $\cos(\theta + 2\pi k) = \cos \theta$ and $\sin(\theta + 2\pi k) = \sin \theta$.

  (a) $\cos\left(\dfrac{\pi}{4} + 2\pi\right)$ (b) $\sin\left(\dfrac{\pi}{3} + 2\pi\right)$

  (c) $\sin\left(\dfrac{\pi}{2} - 6\pi\right)$

20. Evaluate each of the following using the identities $\cos(\theta + 2\pi k) = \cos \theta$ and $\sin(\theta + 2\pi k) = \sin \theta$.

(a) $\sin \dfrac{17\pi}{4}$ **(b)** $\sin\left(-\dfrac{17\pi}{4}\right)$

(c) $\cos 11\pi$ **(d)** $\cos \dfrac{53\pi}{4}$

(e) $\tan\left(\dfrac{-7\pi}{4}\right)$ **(f)** $\cos \dfrac{7\pi}{4}$

(g) $\sec\left(\dfrac{11\pi}{6} + 2\pi\right)$ **(h)** $\csc\left(2\pi - \dfrac{\pi}{3}\right)$

In Exercises 21–24, use the Pythagorean identities to simplify the given expressions.

21. $\dfrac{\sin^2 \theta + \cos^2 \theta}{\tan^2 \theta + 1}$ **22.** $\dfrac{\sec^2 \theta - 1}{\tan^2 \theta}$

23. $\dfrac{\sec^2 \theta - \tan^2 \theta}{1 + \cot^2 \theta}$ **24.** $\dfrac{\csc^4 \theta - \cot^4 \theta}{\csc^2 \theta + \cot^2 \theta}$

In Exercises 25–32, prove that the equations are identities.

25. $\csc \theta = \sin \theta + \cot \theta \cos \theta$

26. $\sin^2 \theta - \cos^2 \theta = \dfrac{1 - \cot^2 \theta}{1 + \cot^2 \theta}$

27. $\dfrac{1}{1 + \sec \theta} + \dfrac{1}{1 - \sec \theta} = -2 \cot^2 \theta$

28. $\dfrac{1 + \tan \theta}{1 - \tan \theta} = \dfrac{\sec^2 \theta + 2 \tan \theta}{2 - \sec^2 \theta}$

29. $\cot \theta + \tan \theta + 1 = \dfrac{\cot \theta}{1 - \tan \theta} + \dfrac{\tan \theta}{1 - \cot \theta}$

30. $\dfrac{\sec s + \cot s \csc s}{\cos s} = \csc^2 s \sec^2 s$

31. $(\tan \theta)(1 - \cot^2 \theta) + (\cot \theta)(1 - \tan^2 \theta) = 0$

32. $(\cos \alpha \cos \beta - \sin \alpha \sin \beta) \times (\cos \alpha \cos \beta + \sin \alpha \sin \beta)$
$$= \cos^2 \alpha - \sin^2 \beta$$

33. If $\sec t = \tfrac{13}{5}$ and $3\pi/2 < t < 2\pi$, evaluate
$$\dfrac{2 \sin t - 3 \cos t}{4 \sin t - 9 \cos t}$$

34. If $\sec t = (b^2 + 1)/2b$ and $\pi < t < 3\pi/2$, find $\tan t$ and $\sin t$.

35. Use the accompanying figure to explain why the following four identities are valid. (The identities can be used to provide an algebraic foundation for the "reference-angle technique" that we've used to evaluate the trigonometric functions.)

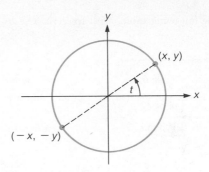

(i) $\sin(t + \pi) = -\sin t$
(ii) $\sin(t - \pi) = -\sin t$
(iii) $\cos(t + \pi) = -\cos t$
(iv) $\cos(t - \pi) = -\cos t$

36. Suppose that $\tan \theta = (b^2 - a^2)/2ab$, where a and b are positive and $\pi/2 < \theta < \pi$. Show that
$$\cos \theta = \dfrac{-2ab}{a^2 + b^2}$$

37. In the equation $x^4 + 6x^2y^2 + y^4 = 32$, make the substitutions
$$x = X \cos \frac{\pi}{4} - Y \sin \frac{\pi}{4} \quad \text{and} \quad y = X \sin \frac{\pi}{4} + Y \cos \frac{\pi}{4}$$
and show that the resulting equation simplifies to
$$X^4 + Y^4 = 16$$

38. Suppose that $\tan \theta = 2$ and $0 < \theta < \pi/2$.
(a) Compute $\sin \theta$ and $\cos \theta$
(b) Using the values obtained in part (a), make the substitutions
$$x = X \cos \theta - Y \sin \theta \quad \text{and} \quad y = X \sin \theta + Y \cos \theta$$
in the expression $7x^2 - 8xy + y^2$, and simplify the result.

In Section 6.1, we pointed out that one of the advantages in using radian measure is that many formulas then take on particularly simple forms. Another reason for using radian measure is that the trigonometric functions can be closely approximated by very simple polynomial functions.

Note *The approximating polynomials in Exercises 39 and 40 are known as* Taylor polynomials, *after the English mathematician Brook Taylor (1685–1731). The theory of Taylor polynomials is developed in calculus.*

39. Complete the following table. The values for $\cos \theta$ in the table were obtained using a calculator.

θ	$1 - \dfrac{\theta^2}{2}$	$\cos \theta$
0.1		0.995004 ...
0.2		0.980066 ...
0.3		0.955336 ...

40. Complete the following table, as in Exercise 39.

θ	$\theta - \dfrac{\theta^3}{6}$	$\sin \theta$
0.1		0.099833 . . .
0.2		0.198669 . . .
0.3		0.295520 . . .

41. Let $a = \sin^2 \theta + \csc^2 \theta$, $b = \cos^2 \theta + \sec^2 \theta$, and $c = \tan^2 \theta + \cot^2 \theta$. Show that $a + b - c = 3$.

42. **(a)** [C] Use a calculator to complete the following table. (Set your calculator in the radian mode.)

θ	$\sin \theta$	**WHICH IS LARGER, θ OR $\sin \theta$?**
0.1		
0.2		
0.3		
0.4		
0.5		

  (b) From the following figure, explain why $PQ < PR < \theta$, and use this to show that if $0 < \theta < \pi/2$, then

$$\sin \theta < \theta$$

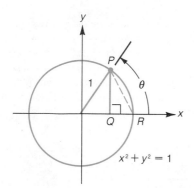

43. **(a)** List four distinct real numbers θ for which $\sin \theta = 0$.
  (b) What is the domain of the function $f(\theta) = 1/\sin \theta$?
  (c) What is the domain of the function $g(\theta) = 1/\sin \pi\theta$?

44. **(a)** List four distinct real numbers t for which $\cos t = 0$. Include at least one negative number in your response.
  (b) What is the domain of the function $f(t) = 1/\cos t$?
  (c) What is the domain of the function $g(t) = 1/\cos \pi t$?

45. In this exercise, we are going to find the minimum value of the function

$$f(t) = \tan^2 t + 9 \cot^2 t \qquad (0 < t < \pi/2)$$

  (a) Set your calculator to the radian mode and complete the following table. (Round off the results to two decimal places.)

t	0.2	0.4	0.6	0.8	1.0	1.2	1.4
$f(t)$							

  (b) Of the seven outputs you calculated in part (a), which is the smallest? What is the corresponding input?
  (c) Prove that $\tan^2 t + 9 \cot^2 t = (\tan t - 3 \cot t)^2 + 6$.
  (d) Use the identity in part (c) to explain why

$$\tan^2 t + 9 \cot^2 t \geq 6$$

  (e) The inequality in part (d) tells us that $f(t)$ is never less than 6. Furthermore, in view of part (c), the value of $f(t)$ will equal 6 when $\tan t - 3 \cot t = 0$. From this last equation, show that $\tan^2 t = 3$, and conclude that $t = \pi/3$. In summary, the minimum value of $f(t)$ is 6, and this occurs when $t = \pi/3$. How do these values compare with your answers in part (b)?

46. Let $f(\theta) = \sin \theta \cos \theta$ $(0 \leq \theta \leq \pi/2)$.
  (a) [C] Set your calculator in the radian mode and complete the following table. (Round off the results to two decimal places.)

θ	0	$\dfrac{\pi}{10}$	$\dfrac{\pi}{5}$	$\dfrac{\pi}{4}$	$\dfrac{3\pi}{10}$	$\dfrac{2\pi}{5}$	$\dfrac{\pi}{2}$
$f(\theta)$							

  (b) What is the largest value of $f(\theta)$ in your table in part (a)?
  (c) Show that $\sin \theta \cos \theta \leq \frac{1}{2}$ for all real numbers θ in the interval $0 \leq \theta \leq \pi/2$. *Hint:* Use the inequality $\sqrt{ab} \leq \frac{1}{2}(a + b)$ [given in Exercise Set 1.7, Exercise 84(a)], with $a = \sin \theta$ and $b = \cos \theta$.
  (d) Does the inequality $\sin \theta \cos \theta \leq \frac{1}{2}$ hold for all real numbers θ?

6.3 ▼ GRAPHS OF THE SINE AND COSINE FUNCTIONS

In this section and the next, we consider the graphs of the trigonometric functions. Our focus in this section is on the sine and cosine functions. As preparation for the discussion, we want to understand what is meant by the term *periodic function*. By way of example, all of the functions in Figure 1 are **periodic**. That is, their graphs display patterns that repeat themselves at regular intervals.

In Figure 1(a), the graph of the function *f* repeats itself every six units. We say that *the period of f is 6*. Similarly, the period of *g* in Figure 1(b) is 2, while the period of *h* in Figure 1(c) is 2π. Notice that in each case, the period represents the minimum number of units that we must travel along the horizontal axis before the graph begins to repeat itself. Because of this, we can state the definition of a periodic function as follows.

DEFINITION

A Periodic Function and Its Period

A function *f* is said to be **periodic** provided there is a number $p > 0$ such that

$$f(x + p) = f(x)$$

for all *x* in the domain of *f*. The smallest such number *p* is called the **period** of *f*.

We also want to define the term **amplitude** as it applies to periodic functions. For a function such as *h* in Figure 1(c), in which the graph is centered about the horizontal axis, the amplitude is simply the maximum height of the graph above that horizontal axis. Thus, the amplitude of *h* is 4. More generally, we define the amplitude of any periodic function as follows (on the next page).

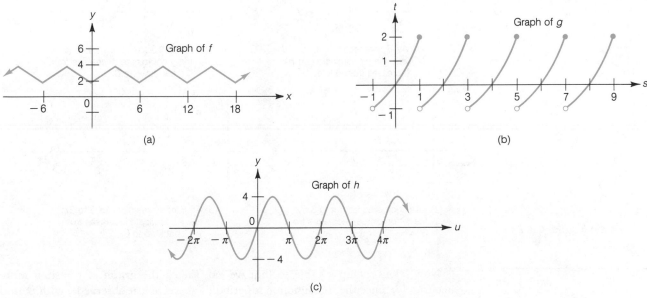

(a)

(b)

(c)

FIGURE 1

DEFINITION

Amplitude

Let f be a periodic function and let m and M denote, respectively, the smallest and largest values of the function. Then, by definition, the **amplitude** of f is the number

$$\frac{M - m}{2}$$

For the function h in Figure 1(c), this definition tells us that the amplitude is $\frac{4 - (-4)}{2} = 4$, which agrees with our previous value. Check for yourself that the amplitudes of the functions f and g in Figure 1(a) and (b) are, respectively, 1 and $\frac{3}{2}$.

Let us now graph the sine function $f(\theta) = \sin\theta$. We are assuming that θ is in radians, so that the domain of the sine function is the set of all real numbers. Even before making any calculations, we can gain a strong intuitive insight into how the graph must look by carrying out the following experiment. After drawing the unit circle, $x^2 + y^2 = 1$, place your fingertip at the point $(1, 0)$ and then move your finger counterclockwise around the circle. As you do this, keep track of what happens to the y-coordinate of your fingertip. (The y-coordinate is $\sin\theta$.) Figure 2 indicates the general results.

In summary, as θ increases from 0 to 2π, the graph of $y = \sin\theta$ should display the general features shown in Figure 3. Furthermore, since at $\theta = 2\pi$ we've returned to our starting point, $(1, 0)$, additional counterclockwise trips around the unit circle will just result in repetitions of the pattern established in Figure 3. In other words, the period of $y = \sin\theta$ is 2π.

FIGURE 2

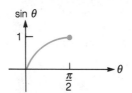

(a) As θ increases from 0 to $\pi/2$, the y-coordinate ($\sin\theta$) increases from 0 to 1.

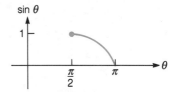

(b) As θ increases from $\pi/2$ to π, the y-coordinate decreases from 1 back down to 0.

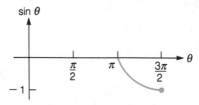

(c) As θ increases from π to $3\pi/2$, the y-coordinate decreases from 0 to -1.

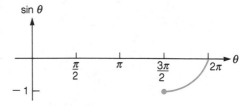

(d) As θ increases from $3\pi/2$ to 2π, the y-coordinate increases from -1 back up to 0.

Now let us set up a table so that we can sketch the graph of $y = \sin\theta$ more accurately. Since the sine function is periodic (as we've just observed), with period 2π, our table needs only to contain values of θ between 0 and 2π, as shown in Table 1. This will establish the basic pattern for the graph.

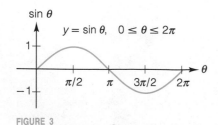

FIGURE 3

TABLE 1

θ	0	$\dfrac{\pi}{6}$	$\dfrac{\pi}{3}$	$\dfrac{\pi}{2}$	$\dfrac{2\pi}{3}$	$\dfrac{5\pi}{6}$	π	$\dfrac{7\pi}{6}$	$\dfrac{4\pi}{3}$	$\dfrac{3\pi}{2}$	$\dfrac{5\pi}{3}$	$\dfrac{11\pi}{6}$	2π
$\sin\theta$	0	$\dfrac{1}{2}$	$\dfrac{\sqrt{3}}{2}$	1	$\dfrac{\sqrt{3}}{2}$	$\dfrac{1}{2}$	0	$-\dfrac{1}{2}$	$-\dfrac{\sqrt{3}}{2}$	-1	$-\dfrac{\sqrt{3}}{2}$	$-\dfrac{1}{2}$	0

FIGURE 4

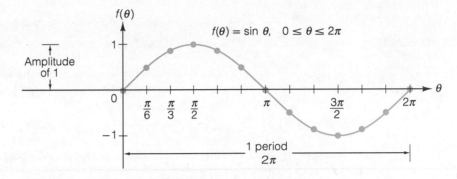

In plotting the points obtained from this table, we use the approximation $\sqrt{3}/2 \approx 0.87$. Rather than approximating π, however, we mark off units on the horizontal axis in terms of π. The resulting graph is shown in Figure 4. By continuing this same pattern to the left and right, we obtain the complete graph of $f(\theta) = \sin \theta$, as indicated in Figure 5.

Before going on to analyze the sine function or to study the graphs of the other trigonometric functions, we are going to make a slight change in the notation we have been using. To conform with common usage, we will use x instead of θ on the horizontal axis, and we will use y for the vertical axis. The sine function is then written simply as $y = \sin x$, where the real number x denotes the radian measure of an angle or, equivalently, the length of the corresponding arc on the unit circle. For reference, we redraw Figures 4 and 5 using this familiar x-y notation, as shown in Figure 6.

FIGURE 5

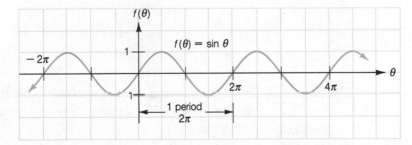

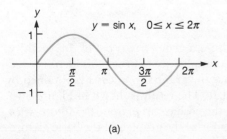

(a)

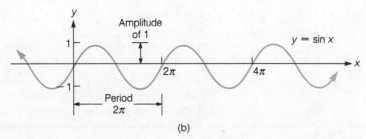

(b)

FIGURE 6

You should memorize the graph of the sine curve in Figure 6(b), so that you can sketch it without first setting up a table. Of course, once you know the shape of the basic wave in Figure 6(a), you automatically know the graph of the full sine curve in Figure 6(b).

We can use the graphs in Figure 6 to help us list some of the key properties of the sine function.

PROPERTY SUMMARY

THE SINE FUNCTION: $y = \sin x$

1. The domain of the sine function is the set of all real numbers. The range of the sine function is the closed interval $[-1, 1]$. Another way to express this last fact is

    $-1 \leq \sin x \leq 1,$ for all x

2. The sine function is periodic, with period 2π. The amplitude is 1.

3. (Refer to Figure 7.) The basic sine wave crosses the x-axis at the beginning, midpoint, and end of the period. The curve reaches its highest point after one-quarter of the period and its lowest point after three-quarters of the period.

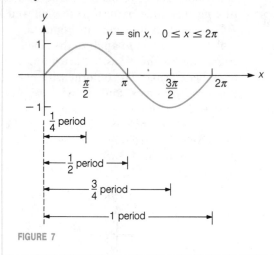

FIGURE 7

We could obtain the graph of the cosine function by setting up a table, just as we did with the sine function. A more interesting way to proceed, however, is to use the identity

$$\cos x = \sin\left(x + \frac{\pi}{2}\right) \tag{1}$$

(Exercise 47 at the end of this section outlines a geometric proof of this identity. Also, after studying Section 6.5, you'll see a simple way to prove this identity algebraically.) From equation (1), it follows that the graph of $y = \cos x$ is obtained by translating the sine curve $\pi/2$ units to the left. The result is shown in Figure 8(a). Figure 8(b) displays one complete cycle of the cosine curve, from $x = 0$ to $x = 2\pi$.

As with the sine function, you should memorize the graph and the basic features of the cosine function. The key features of the cosine are summarized in the box that follows.

FIGURE 8

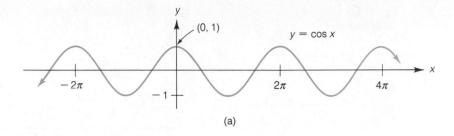

(a)

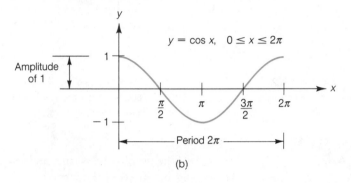

(b)

PROPERTY SUMMARY	THE COSINE FUNCTION: *y* = cos *x*

1. The domain of the cosine function is the set of all real numbers. The range of the cosine function is the closed interval $[-1, 1]$. Another way to express this last fact is

   $$-1 \leq \cos x \leq 1, \qquad \text{for all } x$$

2. The cosine function is periodic, with period 2π. The amplitude is 1.

3. (Refer to Figure 9.) At the beginning and at the end of the period, the basic cosine wave reaches its highest point. At the midpoint of the period, the curve reaches its lowest point. The x-intercepts occur one-quarter and three-quarters of the way through the period.

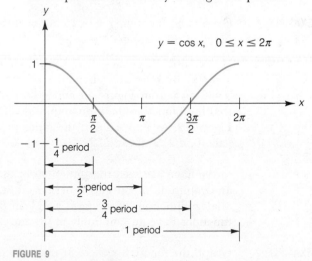

FIGURE 9

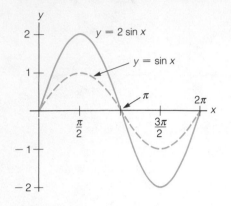

FIGURE 10

The amplitude of $y = 2 \sin x$ is 2.
Both $y = \sin x$ and $y = 2 \sin x$ have
a period of 2π.

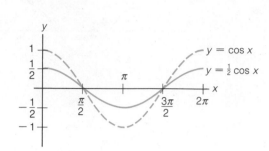

FIGURE 11

The amplitude of $y = \frac{1}{2} \cos x$ is $\frac{1}{2}$.
Both $y = \cos x$ and $y = \frac{1}{2} \cos x$
have a period of 2π.

The graphs of $y = \sin x$ and $y = \cos x$ are the building blocks we need for graphing functions of the form

$$y = A \sin(Bx - C) \qquad \text{and} \qquad y = A \cos(Bx - C)$$

where A, B, and C are constants and neither A nor B is zero. These functions are used throughout the sciences to analyze a variety of periodic phenomena ranging from the vibrations of an electron to the variations in the size of an animal population as it interacts with its environment.

As a first example, consider $y = 2 \sin x$. To obtain the graph of $y = 2 \sin x$ from that of $y = \sin x$, we multiply each y-coordinate on the graph of $y = \sin x$ by 2. As indicated in Figure 10, this changes the amplitude from 1 to 2, but it does not affect the period, which remains 2π. As a second example, Figure 11 shows the graphs of $y = \cos x$ and $y = \frac{1}{2} \cos x$. Note that the amplitude of $y = \frac{1}{2} \cos x$ is $\frac{1}{2}$ and the period is, again, 2π. More generally, graphs of functions of the form $y = A \sin x$ and $y = A \cos x$ always have an amplitude of $|A|$ and a period of 2π.

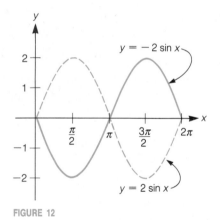

FIGURE 12

EXAMPLE 1 Graph the function $y = -2 \sin x$ over one period. On which interval(s) is the function decreasing?

Solution In Chapter 2 we saw that the graph of $y = -f(x)$ is obtained from that of $y = f(x)$ by reflection about the x-axis. Thus, we need only take the graph of $y = 2 \sin x$ from Figure 10 and reflect it about the x-axis. See Figure 12. Note that both functions have an amplitude of 2 and a period of 2π. From the graph in Figure 12, we can see that the function $y = -2 \sin x$ is decreasing on the intervals $(0, \pi/2)$ and $(3\pi/2, 2\pi)$. ▲

We have just seen that functions of the form $y = A \sin x$ and $y = A \cos x$ have an amplitude of $|A|$ and a period of 2π. The next two examples show how to analyze functions of the form $y = A \sin Bx$ and $y = A \cos Bx$. As you'll see, these functions have an amplitude of $|A|$ and a period of $2\pi/B$.

EXAMPLE 2 Graph the function $y = \cos 3x$ over one period.

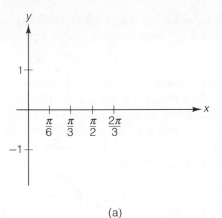

(a)

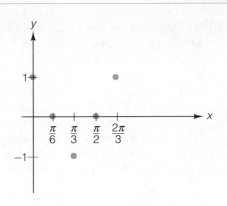

(b)

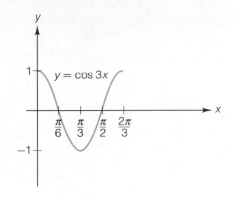

(c) The period of $y = \cos 3x$ is $2\pi/3$; the amplitude is 1.

FIGURE 13

Solution   We know that the cosine curve $y = \cos x$ displays its basic pattern beginning with $x = 0$ and completes that pattern when $x = 2\pi$. Thus, $y = \cos 3x$ will begin its basic pattern when $3x = 0$ and will complete that pattern when $3x = 2\pi$. From the equation $3x = 0$, we conclude that $x = 0$ and, from the equation $3x = 2\pi$, we conclude that $x = 2\pi/3$. Thus, the graph of $y = \cos 3x$ begins its basic pattern at $x = 0$ and completes the pattern at $x = 2\pi/3$. This tells us that the period is $2\pi/3$. Next, in preparation for drawing the graph, we divide the period into quarters, as shown in Figure 13(a). In Figure 13(b) we've plotted the points that have the x-coordinates shown in Figure 13(a). (We've also plotted the point on the curve corresponding to $x = 0$, where the basic pattern begins.) From Figure 13(b), you can see that the amplitude is going to be 1. Now, with the points in Figure 13(b) as a guide, we can sketch one cycle of $y = \cos 3x$, as shown in Figure 13(c).　　　▲

EXAMPLE 3   Graph each function over one period.

(a) $y = \frac{1}{2} \cos 3x$　　(b) $y = -\frac{1}{2} \cos 3x$

Solution   (a) In Example 2 we graphed $y = \cos 3x$. To obtain the graph of $y = \frac{1}{2} \cos 3x$ from that of $y = \cos 3x$, we multiply each y-coordinate on the graph of $y = \cos 3x$ by $\frac{1}{2}$. As indicated in Figure 14(a), this changes the amplitude from 1 to $\frac{1}{2}$, but it does not affect the period, which is still $2\pi/3$.

FIGURE 14

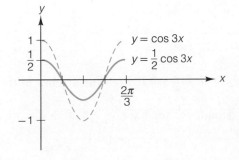

(a)

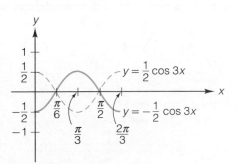

(b) The period of $y = -\frac{1}{2} \cos 3x$ is $2\pi/3$; the amplitude is $\frac{1}{2}$.

(b) The graph of $y = -\frac{1}{2} \cos 3x$ is obtained by reflecting the graph of $y = \frac{1}{2} \cos 3x$ about the x-axis, as indicated in Figure 14(b). Both functions have a period of $2\pi/3$ and an amplitude of $\frac{1}{2}$. ▲

Before looking at more examples, let's take a moment to summarize where we are. Our work in Examples 2 and 3(a) shows how to graph $y = \frac{1}{2} \cos 3x$. The same technique we used in those examples can be applied to any function of the form $y = A \cos Bx$ or $y = A \sin Bx$. As indicated in the box that follows, the amplitude is $|A|$ and the period is $2\pi/B$ for both functions. (Exercise 50 asks you to use the method of Example 2 to show that the period is indeed $2\pi/B$.)

PROPERTY SUMMARY **THE GRAPHS OF** $y = A \sin Bx$ **AND** $y = A \cos Bx$ **($B > 0$)**

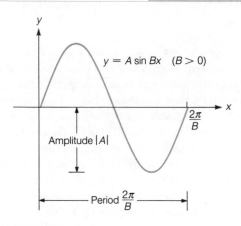

FIGURE 15
The period of $y = A \sin Bx$ is $2\pi/B$; the amplitude is $|A|$.

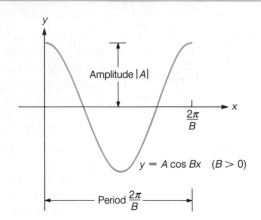

FIGURE 16
The period of $y = A \cos Bx$ is $2\pi/B$; the amplitude is $|A|$.

EXAMPLE 4 Figure 17 is the graph of a function of the form $y = A \sin Bx$ for one period. Determine the values of A and B.

Solution From the figure, we see that the amplitude is 4. Also from the figure, we know that three fourths of the period is 9, so

$$\frac{3}{4}\left(\frac{2\pi}{B}\right) = 9$$

$$\frac{\pi}{2B} = 3$$

$$B = \frac{\pi}{6} \qquad \text{(Check the algebra in the last two lines.)}$$

In summary, we have $A = 4$ and $B = \pi/6$; the equation of the curve is $y = 4 \sin(\pi x/6)$. ▲

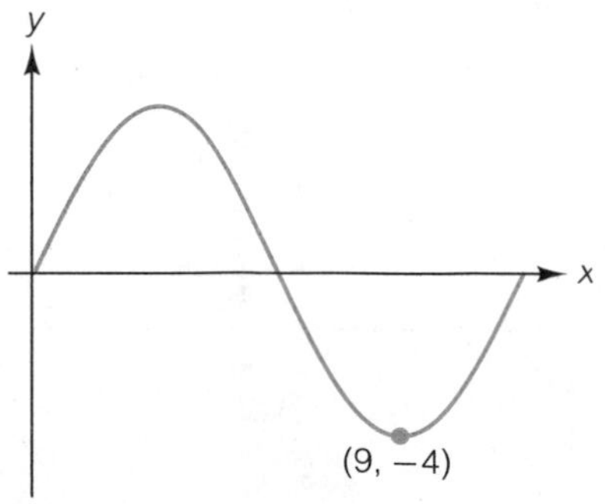

FIGURE 17

$(9, -4)$

EXAMPLE 5 Graph the function $y = 4 \sin\left(2x - \frac{2\pi}{3}\right)$ over one period.

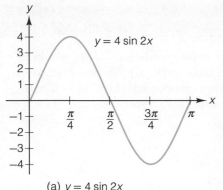

(a) $y = 4 \sin 2x$
Period: $2\pi/B = \pi$
Amplitude: $|A| = 4$

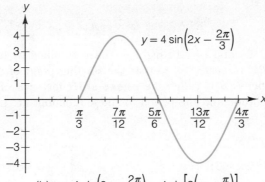

(b) $y = 4 \sin\left(2x - \frac{2\pi}{3}\right) = 4 \sin\left[2\left(x - \frac{\pi}{3}\right)\right]$

The graph is obtained by translating the
graph of $y = 4 \sin 2x$ a distance of $\pi/3$
units to the right. The period and
amplitude are still π and 4, respectively.

FIGURE 18

Solution   The technique here is to factor the quantity within parentheses so that the coefficient of x is 1. We'll then be able to graph the function using a simple translation,
as in Chapter 2. We have

$$y = 4 \sin\left(2x - \frac{2\pi}{3}\right)$$

$$= 4 \sin\left[2\left(x - \frac{\pi}{3}\right)\right] \tag{2}$$

Now, note that equation (2) is obtained from $y = 4 \sin 2x$ by replacing x with
$x - \pi/3$. Thus, the graph of equation (2) is obtained by translating the graph of
$y = 4 \sin 2x$ a distance of $\pi/3$ units to the right. Figure 18(a) shows the graph of
$y = 4 \sin 2x$ over one period. By translating this graph $\pi/3$ units to the right, we
obtain the required graph, as shown in Figure 18(b). Note that the translated
graph has the same amplitude and period as the original graph. Also, as a matter
of arithmetic, you should check for yourself that each of the x-coordinates shown
in Figure 18(b) is obtained simply by adding $\pi/3$ to a corresponding x-coordinate
in Figure 18(a). For example, in Figure 18(a), the cycle ends at $x = \pi$; in Figure
18(b) the cycle ends at $\pi + \pi/3 = 4\pi/3$.   ▲

In the example just completed, we used translation to graph $y = 4 \sin\left(2x - \frac{2\pi}{3}\right)$.
In particular, we translated the graph of $y = 4 \sin 2x$ so that the starting point of
the basic cycle was shifted from $x = 0$ to $x = \pi/3$. The number $\pi/3$ in this case is
called the *phase shift* of the function $y = 4 \sin\left(2x - \frac{2\pi}{3}\right)$. In the box that follows,
we define phase shift and we generalize the results of the graphing technique used
in Example 5.

PROPERTY SUMMARY   THE GRAPHS OF $y = A \sin(Bx - C)$ AND $y = A \cos(Bx - C)$ $(B > 0, C \neq 0)$

The graphs of $y = A \sin(Bx - C)$ and $y = A \cos(Bx - C)$ are obtained by horizontal translations of the graphs of $y = A \sin Bx$ and $y = A \cos Bx$, respectively, so that the starting point of the basic cycle of each is shifted from $x = 0$ to $x = C/B$. The number C/B is called the **phase shift** for each of the functions $y = A \sin(Bx - C)$ and $y = A \cos(Bx - C)$. The amplitude and the period for these functions are $|A|$ and $2\pi/B$, respectively.

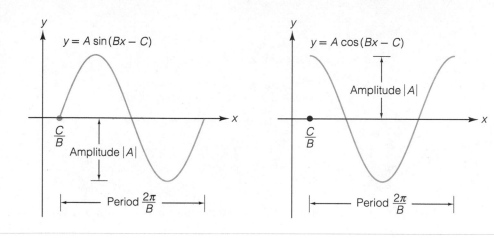

EXAMPLE 6   Specify the amplitude, period, and phase shift for each function.

(a) $f(x) = 3 \cos(4x - 5)$ **(b)** $g(x) = -2 \cos\left(\pi x + \dfrac{2\pi}{3}\right)$

Solution   **(a)** By comparing the given equation with $y = A \cos(Bx - C)$, we see that $A = 3$, $B = 4$, and $C = 5$. Consequently, we have

$$\text{amplitude} = |A| = 3$$

$$\text{period} = \frac{2\pi}{B} = \frac{2\pi}{4} = \frac{\pi}{2}$$

$$\text{phase shift} = \frac{C}{B} = \frac{5}{4}$$

For purposes of review, let's also calculate the phase shift without explicitly relying on the expression C/B. In the equation $f(x) = 3 \cos(4x - 5)$, we can factor a 4 out of the parentheses to obtain

$$f(x) = 3 \cos\left[4\left(x - \frac{5}{4}\right)\right]$$

This last equation tells us that we can obtain the graph of f by translating the graph of $y = 3 \cos 4x$. In particular, the translation would shift the starting point of the basic cycle from $x = 0$ to $x = \frac{5}{4}$. The number $\frac{5}{4}$ is the phase shift, as obtained previously.

(b) We have $A = -2$, $B = \pi$, and $C = -2\pi/3$, therefore

$$\text{amplitude} = |A| = 2$$

$$\text{period} = \frac{2\pi}{B} = \frac{2\pi}{\pi} = 2$$

$$\text{phase shift} = \frac{C}{B} = \frac{-2\pi/3}{\pi} = -\frac{2}{3}$$

▲

EXAMPLE 7   Graph the following function over one period.

$$g(x) = -2\cos\left(\pi x + \frac{2\pi}{3}\right)$$

Solution   Our strategy is first to obtain the graph of $y = 2\cos\left(\pi x + \dfrac{2\pi}{3}\right)$. We can then obtain the graph of g by reflection about the x-axis. We have

$$y = 2\cos\left(\pi x + \frac{2\pi}{3}\right)$$

$$= 2\cos\left[\pi\left(x + \frac{2}{3}\right)\right]$$

Now, the graph of this last equation is obtained by translating the graph of $y = 2\cos \pi x$ a distance of $\frac{2}{3}$ unit to the left. Figures 19(a) and (b) show the graphs of $y = 2\cos \pi x$ and $y = 2\cos[\pi(x + \frac{2}{3})]$, respectively. By reflecting the second graph about the x-axis, we obtain the graph of g, as required. See Figure 19(c).

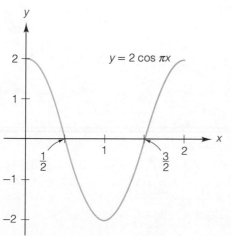

(a) $y = 2\cos \pi x$
Amplitude: $|A| = 2$
Period: $2\pi/B = 2$

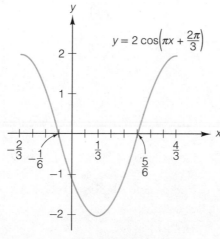

(b) $y = 2\cos\left(\pi x + \frac{2\pi}{3}\right)$
Amplitude: 2
Period: 2
Phase shift: $-2/3$

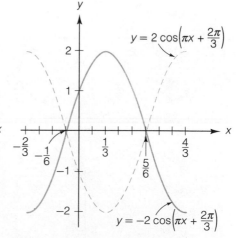

(c) $y = -2\cos\left(\pi x + \frac{2\pi}{3}\right)$
Amplitude: 2
Period: 2
Phase shift: $-2/3$

FIGURE 19

▲

▼ EXERCISE SET 6.3

A

In Exercises 1–8, specify the period and amplitude for each function.

1.

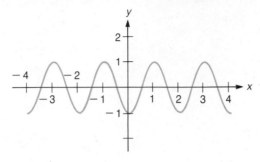

2.

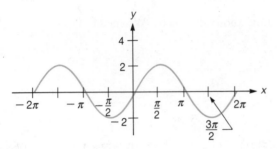

3.

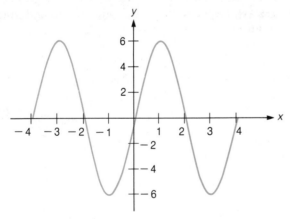

4.

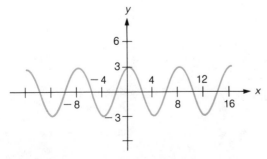

5.

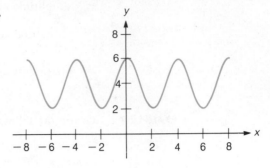

6.

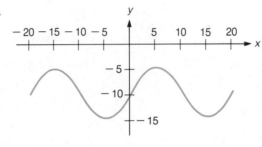

7.

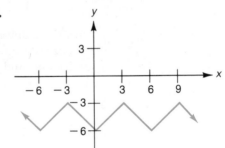

8.

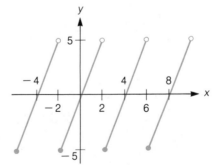

In Exercises 9–20, graph the functions for one period. In each case, specify the amplitude, period, x-intercepts, and interval(s) on which the function is increasing.

9. (a) $y = 2 \sin x$
    (b) $y = -\sin 2x$

10. (a) $y = 3 \sin x$
    (b) $y = \sin 3x$

11. **(a)** $y = \cos 2x$
    (b) $y = 2 \cos 2x$

12. **(a)** $y = \cos(x/2)$
    (b) $y = -\frac{1}{2}\cos(x/2)$

13. **(a)** $y = 3\sin(\pi x/2)$
    (b) $y = -3\sin(\pi x/2)$

14. **(a)** $y = 2\sin \pi x$
    (b) $y = -2\sin \pi x$

15. **(a)** $y = \cos 2\pi x$
    (b) $y = -4\cos 2\pi x$

16. **(a)** $y = -2\cos(x/4)$
    (b) $y = -2\cos(\pi x/4)$

17. $y = 1 + \sin 2x$

18. $y = \sin(x/2) - 2$

19. $y = 1 - \cos(\pi x/3)$

20. $y = -2 - 2\cos 3\pi x$

In Exercises 21–36, determine the amplitude, period, and phase shift for the given function. Graph the function over one period. Indicate the x-intercepts and the coordinates of the highest and lowest points on the graph.

21. $f(x) = \sin\left(x - \frac{\pi}{6}\right)$

22. $g(x) = \cos\left(x + \frac{\pi}{3}\right)$

23. $F(x) = -\cos\left(x + \frac{\pi}{4}\right)$

24. $G(x) = -\sin(x + 2)$

25. $y = \sin\left(2x - \frac{\pi}{2}\right)$

26. $y = \sin\left(3x + \frac{\pi}{2}\right)$

27. $y = \cos(2x - \pi)$

28. $y = \cos\left(x - \frac{\pi}{2}\right)$

29. $y = 3\sin\left(\frac{1}{2}x + \frac{\pi}{6}\right)$

30. $y = -2\sin(\pi x + \pi)$

31. $y = 4\cos\left(3x - \frac{\pi}{4}\right)$

32. $y = \cos(x + 1)$

33. $y = \frac{1}{2}\sin\left(\frac{\pi x}{2} - \pi^2\right)$

34. $y = \cos\left(2x - \frac{\pi}{3}\right) + 1$

35. $y = 1 - \cos\left(2x - \frac{\pi}{3}\right)$

36. $y = 3\cos\left(\frac{2x}{3} + \frac{\pi}{6}\right)$

In Exercises 37–42, a function of the form $y = A\sin Bx$ or $y = A\cos Bx$ is graphed for one period. Find the values of A and B in each case.

37.
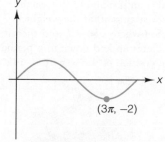
$(3\pi, -2)$

38.
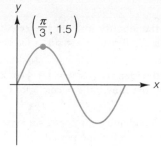
$\left(\frac{\pi}{3}, 1.5\right)$

39.
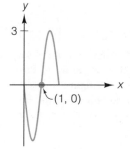
3
$(1, 0)$

40.
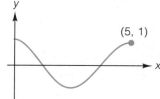
$(5, 1)$

41.
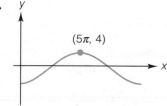
$(5\pi, 4)$

42.
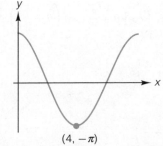
$(4, -\pi)$

B

43. In Section 6.6 you'll see the identity $\sin^2 x = \frac{1}{2} - \frac{1}{2} \cos 2x$. Use this identity to graph the function $y = \sin^2 x$ for one period.

44. In Section 6.6 you'll see the identity $\cos^2 x = \frac{1}{2} + \frac{1}{2} \cos 2x$. Use this identity to graph the function $y = \cos^2 x$ for one period.

45. In Section 6.6 we derive the identity $\sin 2x = 2 \sin x \cos x$. Use this to graph $y = \sin x \cos x$ for one period.

46. In Section 6.6 we derive the identity $\cos 2x = \cos^2 x - \sin^2 x$. Use this to graph $y = \cos^2 x - \sin^2 x$ for one period.

47. In the text we used the identity $\cos \theta = \sin\left(\theta + \dfrac{\pi}{2}\right)$ in obtaining the graph of $y = \cos x$ from that of $y = \sin x$. In this exercise you'll derive this identity. Refer to the accompanying figure. (Although the figure shows the angle (or arc length) of radian measure θ in the first quadrant, the proof can be easily carried over for the other quadrants as well.)

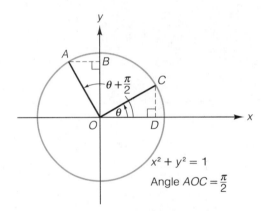

(a) What are the coordinates of C?
(b) Show that the triangles AOB and COD are congruent.
(c) Use the answers in parts (a) and (b) to show that the coordinates of A are $(-\sin \theta, \cos \theta)$.
(d) Since angle $DOA = \theta + \dfrac{\pi}{2}$, the coordinates of A (by definition) are $\left(\cos\left(\theta + \dfrac{\pi}{2}\right), \sin\left(\theta + \dfrac{\pi}{2}\right)\right)$. Now explain why.

$$\cos \theta = \sin\left(\theta + \frac{\pi}{2}\right)$$

and

$$-\sin \theta = \cos\left(\theta + \frac{\pi}{2}\right)$$

48. The figure shows a portion of the graph of the function $y = \sin(e^x)$. Determine each of the following quantities. Specify both exact values and calculator approximations, rounded off to two decimal places.
(a) The y-intercept
(b) The x-intercept at Q
(c) The coordinates of the turning points P and R

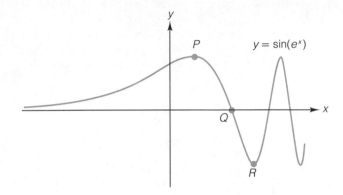

49. The figure shows a portion of the graph of the periodic function $y = e^{\sin x}$. What are the coordinates of the turning points P and Q?

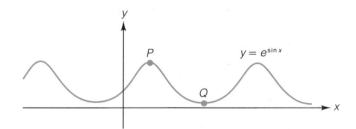

50. In Example 2 we showed that the period of $y = \cos 3x$ is $2\pi/3$. Use the same method to show that the period of $y = A \cos Bx$ is $2\pi/B$.

51. A weight is attached to a spring hung from the ceiling, as shown in the figure. The coordinate system has been chosen so that the equilibrium position of the weight corresponds to $s = 0$. Now suppose that the weight is pulled down to the position where $s = -3$, and then it is released, so that it oscillates up and down in a periodic fashion. Assume that the position of the weight at time t is given by the function

$$s = -3 \cos \frac{\pi t}{3}$$

where s is in feet and t is in seconds.

3 +
2 +
1 +
0 +
−1 +
−2 +
−3 +

s-axis

(a) Find the amplitude and period of this function, and sketch its graph over the interval $0 \leq t \leq 12$.

(b) Use your graph to determine the times in this interval at which the weight is farthest from the origin.

(c) When during this interval of time is the weight passing through the equilibrium position? (Specify the values of t.)

(d) Now assume that under these conditions, the velocity v of the weight is given by the function

$$v = \pi \sin \frac{\pi t}{3}$$

where t is in seconds and v is in feet per second. Graph this velocity function over the interval $0 \leq t \leq 12$. Specify the amplitude and period.

(e) Use your graph to find the times during this interval when the velocity is zero. At these times, what is the position (s-coordinate) of the weight?

(f) At what times during this interval is the weight moving downward?

(g) Use your graph to find the times when the velocity is greatest and least. What is the numerical value of the velocity in those cases? (Use $\pi \approx 3.14$.) Where is the weight located at those times when the velocity is greatest and least?

(h) On the same set of axes, graph the velocity function and the position function over the interval $0 \leq t \leq 12$.

52. The following figure shows a simple pendulum consisting of a string with a weight attached at one end, while the other end is suspended from a fixed point. It can be shown that if the pendulum is displaced 0.1 radian from the equilibrium (vertical) position and then released, the size of the angle θ at time t is very closely approximated by $\theta = 0.1 \cos(t\sqrt{g/L})$. In this formula, t is measured in seconds, L is the length of the pendulum, and g is a constant (the acceleration due to gravity). Show that under these conditions, the time T required for one complete oscillation of the pendulum is given by $T = 2\pi\sqrt{L/g}$. [There are two conclusions that follow from this formula: (1) We can adjust a pendulum clock by altering the length L of the pendulum. (2) On the other hand, altering the amplitude slightly will have no effect on the period, since the value of T does not depend on the amplitude.]

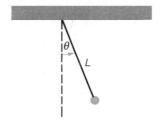

6.4 ▼ GRAPHS OF THE TANGENT AND THE RECIPROCAL FUNCTIONS

A third . . . function, the tangent of θ, or tan θ, is of secondary importance, in that it is not associated with wave phenomena. Nevertheless, it enters into the body of analysis so prominently that we cannot ignore it.

Samuel E. Urner and William B. Orange in *Elements of Mathematical Analysis* (Boston: Ginn and Company, 1950)

We have seen in the previous section that the sine and cosine functions are periodic. The remaining four trigonometric functions also are periodic, but their graphs differ significantly from those of sine and cosine. The graphs of $y = \tan x$, cot x, sec x, and csc x all possess asymptotes.

We'll obtain the graph of the tangent function by a combination of point plotting and symmetry considerations. Table 1 (on the next page) displays a list of values for $y = \tan x$ using x-values in the interval $[0, \pi/2)$.

As indicated in Table 1, tan x is undefined when $x = \pi/2$. This follows from the identity

$$\tan x = \frac{\sin x}{\cos x} \tag{1}$$

TABLE 1

As x increases from 0 toward $\pi/2$, the values of tan x increase slowly at first, then more and more rapidly.

x	0	$\dfrac{\pi}{6}$	$\dfrac{\pi}{4}$	$\dfrac{\pi}{3}$	$\dfrac{5\pi}{12}\ (=75°)$	$\dfrac{17\pi}{36}\ (=85°)$	$\dfrac{89\pi}{180}\ (=89°)$	$\dfrac{\pi}{2}$
tan x	0	$\dfrac{\sqrt{3}}{3}\approx0.58$	1	$\sqrt{3}\approx1.73$	3.73	11.43	57.29	undefined

When $x = \pi/2$, the denominator in this identity is zero. Indeed, when x is equal to any odd integral multiple of $\pi/2$ (e.g., $\pm3\pi/2$, $\pm5\pi/2$), the denominator in equation (1) will be zero and, consequently, tan x will be undefined.

Because tan x is undefined when $x = \pi/2$, we want to see how the graph behaves as x gets closer and closer to $\pi/2$. This is why the x-values $5\pi/12$, $17\pi/36$, and $89\pi/180$ are used in Table 1. In Figure 1, we've used the data in Table 1 to draw the graph of $y = \tan x$ for $0 \le x < \pi/2$. As the figure indicates, the vertical line $x = \pi/2$ is an asymptote for the graph.

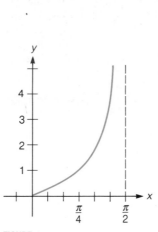

FIGURE 1
$y = \tan x, \quad 0 \le x < \pi/2$

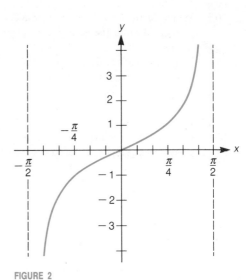

FIGURE 2
$y = \tan x, \quad -\pi/2 < x < \pi/2$
The graph is symmetric about the origin.
The lines $x = \pm\pi/2$ are asymptotes.

The graph of $y = \tan x$ can now be completed without further need for tables or a calculator. First, the identity $\tan(-x) = -\tan x$ (from Section 6.2) tells us that the graph of $y = \tan x$ is symmetric about the origin. So, after reflecting the graph in Figure 1 about the origin, we can draw the graph of $y = \tan x$ on the interval $(-\pi/2, \pi/2)$, as shown in Figure 2.

Now to complete the graph of $y = \tan x$, we'll use the identity

$$\tan(s + \pi) = \tan s \tag{2}$$

Looking at Figure 3, we can see why this identity is valid. By definition, the coordinates of P and Q are

$$P(\cos s, \sin s) \qquad \text{and} \qquad Q(\cos(s + \pi), \sin(s + \pi))$$

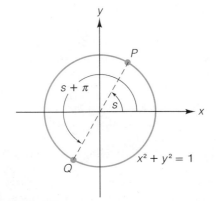

FIGURE 3

On the other hand, the points P and Q are symmetric about the origin and therefore the coordinates of Q are just the negatives of the coordinates of P. That is,

$$\cos(s + \pi) = -\cos s \qquad \text{and} \qquad \sin(s + \pi) = -\sin s$$

Consequently, we have

$$\tan(s + \pi) = \frac{\sin(s + \pi)}{\cos(s + \pi)} = \frac{-\sin s}{-\cos s}$$

$$= \frac{\sin s}{\cos s} = \tan s \qquad \text{as required}$$

[Although Figure 3 shows the angle (or arc) with radian measure s terminating in the first quadrant, our proof is valid for the other quadrants as well. (Draw a figure for yourself and verify this.)]

Identity (2) tells us that the graph of $y = \tan x$ must repeat itself at intervals of length π. Taking this fact into account, along with Figure 2, we conclude that the period of $y = \tan x$ is exactly π. Our final graph of $y = \tan x$ is shown in the summary box that follows.

PROPERTY SUMMARY   **THE TANGENT FUNCTION:** $y = \tan x$

Domain: The set of all real numbers other than
$\pm\pi/2, \pm 3\pi/2, \pm 5\pi/2, \ldots$

Range: $(-\infty, \infty)$

Period: π

Asymptotes: $x = \pm\pi/2, \; x = \pm 3\pi/2, \; x = \pm 5\pi/2, \ldots$

x-intercepts: $0, \pm\pi, \pm 2\pi, \pm 3\pi, \ldots$

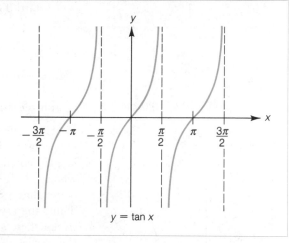

$y = \tan x$

EXAMPLE 1   Graph the following functions for one period. In each case, specify the period, the asymptotes, and the intercepts.

(a) $y = \tan\left(x - \dfrac{\pi}{4}\right)$ **(b)** $y = -\tan\left(x - \dfrac{\pi}{4}\right)$

Solution   We begin with the graph of one period of $y = \tan x$, as shown in Figure 4(a). By translating this graph $\pi/4$ units to the right, we obtain the graph of $y = \tan\left(x - \dfrac{\pi}{4}\right)$, in Figure 4(b). With this translation, note that the left asymptote shifts from $x = -\dfrac{\pi}{2}$ to $x = -\dfrac{\pi}{2} + \dfrac{\pi}{4} = -\dfrac{\pi}{4}$; the right asymptote shifts from $x = \dfrac{\pi}{2}$ to $x = \dfrac{\pi}{2} + \dfrac{\pi}{4} = \dfrac{3\pi}{4}$; and the x-intercept shifts from 0 to $\dfrac{\pi}{4}$. For the y-intercept of

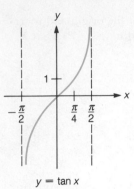

$y = \tan x$

Period: π
Asymptotes: $x = \pm \pi/2$
x-intercept: 0
y-intercept: 0

(a)

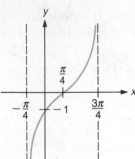

$y = \tan(x - \frac{\pi}{4})$

Period: π
Asymptotes: $x = -\pi/4, \; x = 3\pi/4$
x-intercept: $\pi/4$
y-intercept: -1

(b)

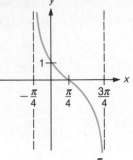

$y = -\tan(x - \frac{\pi}{4})$

Period: π
Asymptotes: $x = -\pi/4, \; x = 3\pi/4$
x-intercept: $\pi/4$
y-intercept: 1

(c)

FIGURE 4

$y = \tan(x - \pi/4)$, replace x with 0 in the equation. This yields

$$y = \tan\left(0 - \frac{\pi}{4}\right) = \tan\left(-\frac{\pi}{4}\right) = -\tan\frac{\pi}{4} = -1$$

Finally, for the graph of $y = -\tan\left(x - \frac{\pi}{4}\right)$, we need only reflect the graph in Figure 4(b) about the x-axis; see Figure 4(c). ▲

EXAMPLE 2   Graph the function $y = \tan(x/2)$ for one period.

Solution   First we refer back to Figure 2, which shows the basic pattern for one period of $y = \tan x$. In this basic pattern, the asymptotes occur when the angle (or arc length) equals $-\pi/2$ or $\pi/2$. Consequently, for $y = \tan(x/2)$, the asymptotes occur when $x/2 = -\pi/2$ and when $x/2 = \pi/2$. From the equation $x/2 = -\pi/2$, we conclude that $x = -\pi$, and from the equation $x/2 = \pi/2$, we conclude that $x = \pi$. Thus, the asymptotes for $y = \tan(x/2)$ are $x = -\pi$ and $x = \pi$. The distance between these asymptotes, namely, 2π, is the period of $y = \tan(x/2)$. This is twice the period of $y = \tan x$. So basically, we want to draw a curve with the same general shape as $y = \tan x$ but twice as wide. See Figure 5. Note that the graph in Figure 5 passes through the origin, since when $x = 0$, we have

$$y = \tan\frac{0}{2} = \tan 0 = 0$$

▲

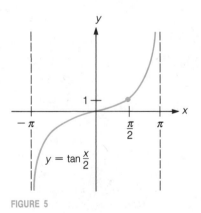

$y = \tan\frac{x}{2}$

FIGURE 5

The graph of the cotangent function can be obtained from that of the tangent function by means of the identity

$$\cot x = -\tan\left(x - \frac{\pi}{2}\right) \tag{3}$$

(Exercise 33 shows how to derive this identity.) According to identity (3), the graph of $y = \cot x$ can be obtained by first translating the graph of $y = \tan x$ (in the box on page 337) to the right $\pi/2$ units and then reflecting the translated graph about the *x*-axis. When this is done, we obtain the graph shown in the box that follows.

PROPERTY SUMMARY | **THE COTANGENT FUNCTION:** $y = \cot x$

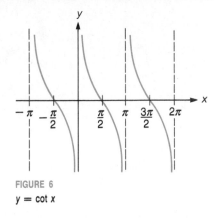

Domain: The set of all real numbers other than $0, \pm\pi, \pm 2\pi, \ldots$

Range: $(-\infty, \infty)$

Period: π (See Figure 6.)

Asymptotes: $x = 0, x = \pm\pi, x = \pm 2\pi, \ldots$

x-intercepts: $\pm\dfrac{\pi}{2}, \pm\dfrac{3\pi}{2}, \pm\dfrac{5\pi}{2}, \ldots$

FIGURE 6
$y = \cot x$

EXAMPLE 3   Graph each of the following functions for one period.

   (a) $y = \cot \pi x$ **(b)** $y = \frac{1}{2}\cot \pi x$ **(c)** $y = -\frac{1}{2}\cot \pi x$

Solution   **(a)** Looking at the graph of $y = \cot x$ in Figure 6, we see that one complete pattern or cycle of the graph is bounded by the asymptotes $x = 0$ and $x = \pi$. Now, for the function we are given, x has been replaced by πx. Thus, the corresponding asymptotes occur when $\pi x = 0$ and when $\pi x = \pi$; in other words, when $x = 0$ and when $x = 1$. See Figure 7(a).

   (b) The graph of $y = \frac{1}{2}\cot \pi x$ will have the same general shape as that of $y = \cot \pi x$, but each *y*-coordinate on $y = \frac{1}{2}\cot \pi x$ will be half of the corresponding coordinate on $y = \cot \pi x$. See Figure 7(b).

   (c) The graph of $y = -\frac{1}{2}\cot \pi x$ is obtained by reflecting the graph of $y = \frac{1}{2}\cot \pi x$ about the *x*-axis. See Figure 7(c).   ▲

FIGURE 7

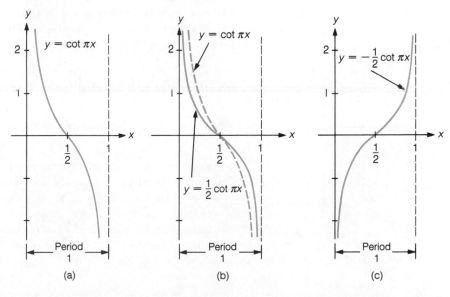

(a)   (b)   (c)

FIGURE 8

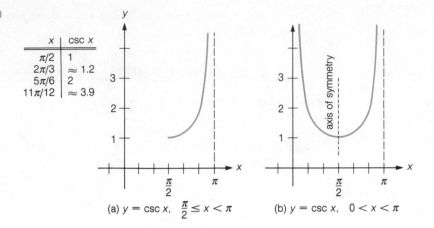

x	csc x
π/2	1
2π/3	≈ 1.2
5π/6	2
11π/12	≈ 3.9

(a) $y = \csc x, \ \frac{\pi}{2} \le x < \pi$

(b) $y = \csc x, \ 0 < x < \pi$

We conclude this section with a discussion of the graphs of $y = \csc x$ and $y = \sec x$. We will obtain these graphs in a series of easy steps, relying on the ideas of symmetry and translation. First consider the function $y = \csc x$. In Figure 8(a), we've set up a table and used it to sketch the graph of $y = \csc x$ on the interval $[\pi/2, \pi)$. Note that csc x is undefined when $x = \pi$. (Why?) As Figure 8(a) indicates, the vertical line $x = \pi$ is an asymptote for the graph. The graph in Figure 8(a) can be extended to the interval $(0, \pi)$ by means of the identity

$$\csc\left(\frac{\pi}{2} + s\right) = \csc\left(\frac{\pi}{2} - s\right)$$

(Exercise 34 shows you how to verify this identity.) This identity tells us that, starting at $x = \pi/2$, whether we travel a distance s to the right or a distance s to the left, the value of $y = \csc x$ is the same. In other words, the graph of $y = \csc x$ is symmetric about the line $x = \pi/2$. In view of this symmetry, we can sketch the graph of $y = \csc x$ on the interval $(0, \pi)$, as shown in Figure 8(b).

The next step in obtaining the graph of $y = \csc x$ is to use the fact that the graph is symmetric about the origin. To verify this, we need to check that $\csc(-x) = -\csc x$. We have

$$\csc(-x) = \frac{1}{\sin(-x)}$$

$$= \frac{1}{-\sin x} = -\csc x \qquad \text{as required}$$

Now, taking into account this symmetry about the origin and the portion of the graph that we've already obtained in Figure 8(b), we can sketch the graph of $y = \csc x$ over the interval $(-\pi, \pi)$, as shown in Figure 9.

To complete the graph of $y = \csc x$, we observe that the values of csc x must repeat themselves at intervals of 2π. This is because csc $x = 1/(\sin x)$, and the sine function has a period of 2π. In view of this, we can draw the graph of $y = \csc x$ as shown in the box that follows.

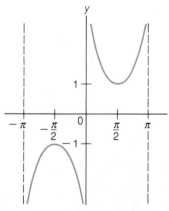

FIGURE 9

$y = \csc x, \quad -\pi < x < \pi$

The graph is symmetric about the origin.

PROPERTY SUMMARY   **THE COSECANT FUNCTION:** $y = \csc x$

Domain: All real numbers other than
$\qquad$ $0, \pm\pi, \pm 2\pi, \ldots$

Range: $(-\infty, -1] \cup [1, \infty)$

Period: 2π (See Figure 10.)

Asymptotes: $x = 0$, $x = \pm\pi$, $x = \pm 2\pi, \ldots$

Intercepts: None

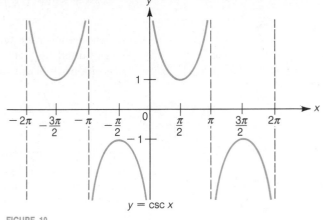

FIGURE 10

EXAMPLE 4   Graph the function $y = \csc \dfrac{x}{3}$ for one period.

Solution   Since $\csc(x/3) = 1/\sin(x/3)$, it will be helpful first to graph one period of $y = \sin(x/3)$. This is done in Figure 11(a) (using the techniques of the previous section). Note that the period of $y = \sin(x/3)$ is 6π. This is also the period of $y = \csc(x/3)$, because $\csc(x/3)$ and $\sin(x/3)$ are just reciprocals. The asymptotes for $y = \csc(x/3)$ occur when $\sin(x/3) = 0$. From Figure 11(a), we see that $\sin(x/3) = 0$ when $x = 0$, when $x = 3\pi$, and when $x = 6\pi$. These asymptotes are sketched in Figure 11(b). The colored points in Figure 11(b) indicate where the value of $\sin(x/3)$ is 1 or -1. The graph of $y = \csc(x/3)$ must pass through these points. (Why?) Finally, using the points and the asymptotes in Figure 11(b), we can sketch the graph of $y = \csc(x/3)$, as shown in Figure 11(c).   ▲

FIGURE 11

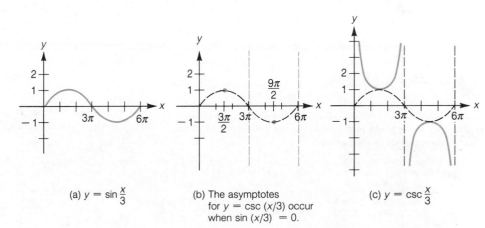

(a) $y = \sin \dfrac{x}{3}$

(b) The asymptotes
for $y = \csc (x/3)$ occur
when $\sin (x/3) = 0$.

(c) $y = \csc \dfrac{x}{3}$

In the previous section we used the identity

$$\cos x = \sin\left(x + \frac{\pi}{2}\right) \tag{4}$$

to graph the cosine function. This identity tells us that the graph of $y = \cos x$ can be obtained by translating the graph of $y = \sin x$ to the left by $\pi/2$ units. Now, from identity (4), it follows that

$$\sec x = \csc\left(x + \frac{\pi}{2}\right)$$

This last identity tells us that the graph of $y = \sec x$ can also be obtained by a translation. In this case, it is the graph of $y = \csc x$ that we translate; by translating $y = \csc x$ a distance of $\pi/2$ units to the left, we obtain the graph of $y = \sec x$, as shown in the box that follows.

PROPERTY SUMMARY   **THE SECANT FUNCTION:** $y = \sec x$

Domain: All real numbers other than $\pm\pi/2$, $\pm 3\pi/2$, $\pm 5\pi/2$, ...

Range: $(-\infty, -1] \cup [1, \infty)$

Period: 2π (See Figure 12.)

Asymptotes: $x = \pm\pi/2$, $x = \pm 3\pi/2$,
               $x = \pm 5\pi/2$, ...

y-intercept: 1

x-intercept: None

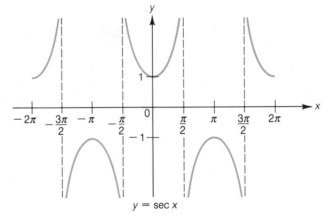

$y = \sec x$

FIGURE 12

▼ **EXERCISE SET 6.4**

A

In Exercises 1–24, graph each function for one period, and show (or specify) the intercepts and asymptotes.

1. (a) $y = \tan\left(x + \dfrac{\pi}{4}\right)$

    (b) $y = -\tan\left(x + \dfrac{\pi}{4}\right)$

2. (a) $y = \tan\left(x - \dfrac{\pi}{3}\right)$

    (b) $y = -\tan\left(x - \dfrac{\pi}{3}\right)$

3. (a) $y = \tan(x/3)$

    (b) $y = -\tan(x/3)$

4. (a) $y = 2\tan \pi x$

    (b) $y = -2\tan \pi x$

5. $y = \frac{1}{2}\tan(\pi x/2)$

6. $y = -\frac{1}{2}\tan 2\pi x$

7. $y = \cot(\pi x/2)$

8. $y = \cot 2\pi x$

9. $y = -\cot\left(x - \dfrac{\pi}{4}\right)$

10. $y = \cot\left(x + \dfrac{\pi}{6}\right)$

11. $y = \frac{1}{2}\cot 2x$

12. $y = -\frac{1}{2}\cot(x/2)$

13. $y = \csc\left(x - \dfrac{\pi}{4}\right)$

14. $y = \csc\left(x - \dfrac{\pi}{6}\right)$

15. $y = -\csc(x/2)$

16. $y = 2\csc x$

17. $y = \frac{1}{3}\csc \pi x$

18. $y = -\frac{1}{2}\csc 2\pi x$

19. $y = -\sec x$

20. $y = -2\sec x$

21. $y = \sec(x - \pi)$

22. $y = \sec(x + 1)$

23. $y = 3\sec(\pi x/2)$

24. $y = -2\sec(\pi x/3)$

B

For Exercises 25–28, six functions are defined as follows.

$$f(x) = \sin x \qquad g(x) = \csc x \qquad h(x) = \pi x - \dfrac{\pi}{6}$$

$$F(x) = \cos x \qquad G(x) = \sec x \qquad H(x) = \pi x + \dfrac{\pi}{4}$$

In each case, graph the indicated function over one period.

25. (a) $f \circ h$
    (b) $g \circ h$

26. (a) $F \circ H$
    (b) $G \circ H$

27. (a) $f \circ H$
    (b) $g \circ H$

28. (a) $F \circ h$
    (b) $G \circ h$

For Exercises 29–32, four functions are defined as follows.

$$f(x) = \csc x \qquad T(x) = \tan x$$

$$g(x) = \sec x \qquad A(x) = |x|$$

In each case, graph the indicated function over the interval $[-2\pi, 2\pi]$.

29. $A \circ T$

30. $A \circ g$

31. $A \circ f$

32. $f \circ A$

33. In the text, we used the identity

$$\cot s = -\tan\left(s - \dfrac{\pi}{2}\right)$$

to obtain the graph of $y = \cot x$ from that of $y = \tan x$. The following steps show one way to derive this identity. [Although the accompanying figure shows the angle with

radian measure s terminating in the first quadrant, the proof is valid no matter in which quadrant the angle terminates. In the figure $\overline{PO} \perp \overline{QO}$.

(a) Why are the coordinates of P and Q as follows?

$$P(\cos s, \sin s) \quad \text{and} \quad Q\left(\cos\left(s - \dfrac{\pi}{2}\right), \sin\left(s - \dfrac{\pi}{2}\right)\right)$$

(b) Using congruent triangles [and without reference to part (a)], explain why the y-coordinate of Q is the negative of the x-coordinate of P, and the x-coordinate of Q equals the y-coordinate of P.

(c) Use the results in parts (a) and (b) to conclude that

$$\sin\left(s - \dfrac{\pi}{2}\right) = -\cos s \quad \text{and} \quad \cos\left(s - \dfrac{\pi}{2}\right) = \sin s$$

(d) Use the result in part (c) to show that

$$\cot s = -\tan\left(s - \dfrac{\pi}{2}\right)$$

34. In this exercise we verify the identity $\csc[(\pi/2) + s] = \csc[(\pi/2) - s]$. Refer to the following figure, in which $\angle AOB = \angle BOC = s$ radians. [Although the figure shows s in the interval $0 < s < \pi/2$, a similar proof will work for other intervals as well. (A proof that does not depend upon a picture can be given using the formula for $\sin(s + t)$ that is developed in Section 6.5.)]

(a) Show that the triangles AOE and COD are congruent and, consequently, $AE = CD$.

(b) Explain why the y-coordinates of the points C and A are $\sin[(\pi/2) - s]$ and $\sin[(\pi/2) + s]$, respectively.

(c) Use parts (a) and (b) to conclude that $\sin[(\pi/2) + s] = \sin[(\pi/2) - s]$. It follows from this that $\csc[(\pi/2) + s] = \csc[(\pi/2) - s]$, as required.

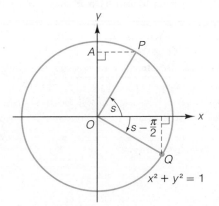

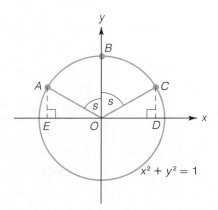

6.5 ▼ THE ADDITION FORMULAS

It has long been recognized that the addition formulas are the heart of trigonometry. Indeed, Professor Rademacher and others have shown that the entire body of trigonometry can be derived from the assumption that there exist functions S and C such that

(1) $S(x - y) = S(x)C(y) - C(x)S(y)$

(2) $C(x - y) = C(x)C(y) + S(x)S(y)$

(3) $\lim\limits_{x \to 0^+} \dfrac{S(x)}{x} = 1$

Professor Frederick H. Young in his article, "The Addition Formulas," in *The Mathematics Teacher*, Vol. L (1957), pp. 45–48

For any real numbers r, s, and t, it is always true that $r(s + t) = rs + rt$. This is the so-called **distributive law** for real numbers. If f is a function, however, it is not true in general that $f(s + t) = f(s) + f(t)$. By way of example, consider the cosine function. It is not true in general that $\cos(s + t) = \cos s + \cos t$. For instance, with $s = \pi/6$ and $t = \pi/3$, we have

$$\cos\left(\frac{\pi}{6} + \frac{\pi}{3}\right) \overset{?}{=} \cos\frac{\pi}{6} + \cos\frac{\pi}{3}$$

$$\cos\frac{\pi}{2} \overset{?}{=} \cos\frac{\pi}{6} + \cos\frac{\pi}{3}$$

$$0 \overset{?}{=} \frac{\sqrt{3}}{2} + \frac{1}{2} \qquad \text{No!}$$

In this section, we will see just what $\cos(s + t)$ does equal. The correct formula for $\cos(s + t)$ is one of a group of four important trigonometric identities called the **addition formulas**.

The Addition Formulas

$$\sin(s + t) = \sin s \cos t + \cos s \sin t$$
$$\sin(s - t) = \sin s \cos t - \cos s \sin t$$
$$\cos(s + t) = \cos s \cos t - \sin s \sin t$$
$$\cos(s - t) = \cos s \cos t + \sin s \sin t$$

Our strategy for deriving these formulas will be as follows. First we'll prove the fourth formula—this takes some effort. The other formulas are then relatively easy to derive from the fourth one.

To prove the formula $\cos(s - t) = \cos s \cos t + \sin s \sin t$, we make use of Figure 1. The idea behind the proof is simply this. There are two distinct ways to calculate the quantity PQ^2 in Figure 1: We can use the law of cosines in triangle QOP, or we can use the distance formula $d = \sqrt{(x_2 - x_1)^2 + (y_2 - y_1)^2}$. Of course, the results of these two calculations must agree because they'll both represent PQ^2. It is by equating the two results that we will arrive at the required formula for $\cos(s - t)$.

Applying the law of cosines in triangle QOP, we have

$$PQ^2 = OP^2 + OQ^2 - 2OP \cdot OQ \cos(s - t)$$
$$= 1^2 + 1^2 - 2(1)(1)\cos(s - t)$$
$$= 2 - 2\cos(s - t) \tag{1}$$

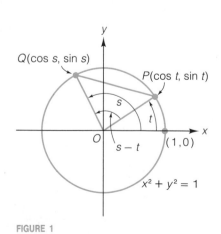

FIGURE 1

On the other hand, using the distance formula, we find that

$$PQ = \sqrt{(\cos s - \cos t)^2 + (\sin s - \sin t)^2}$$

and therefore

$$PQ^2 = (\cos s - \cos t)^2 + (\sin s - \sin t)^2$$

These two terms add to 1.

$$= \cos^2 s - 2 \cos s \cos t + \cos^2 t + \sin^2 s - 2 \sin s \sin t + \sin^2 t$$

These two terms add to 1.

$$= 2 - 2 \cos s \cos t - 2 \sin s \sin t \qquad (2)$$

Now we equate the two expressions for PQ^2 given in equations (1) and (2). This yields

$$2 - 2 \cos(s - t) = 2 - 2 \cos s \cos t - 2 \sin s \sin t$$
$$-2 \cos(s - t) = -2 \cos s \cos t - 2 \sin s \sin t$$
$$\cos(s - t) = \cos s \cos t + \sin s \sin t$$

This completes the proof of the fourth addition formula. Before deriving the other three, let's take a look at some applications of this one.

EXAMPLE 1 Simplify the expression $\cos 2\theta \cos \theta + \sin 2\theta \sin \theta$.

Solution According to the identity that we just proved, we have

$$\cos 2\theta \cos \theta + \sin 2\theta \sin \theta = \cos(2\theta - \theta)$$
$$= \cos \theta$$

Thus, the given expression is equal to $\cos \theta$. ▲

EXAMPLE 2 Simplify $\cos(\theta - \pi)$.

Solution We use the formula for $\cos(s - t)$ with s and t replaced by θ and π, respectively. This yields

$$\cos(s - t) = \cos s \cos t + \sin s \sin t$$
$$\cos(\theta - \pi) = \cos \theta \cos \pi + \sin \theta \sin \pi$$
$$= (\cos \theta)(-1) + (\sin \theta)(0)$$
$$= -\cos \theta$$

Thus, the required simplification is $\cos(\theta - \pi) = -\cos \theta$. ▲

In Example 2, we found that $\cos(\theta - \pi) = -\cos \theta$. This type of identity is often referred to as a **reduction formula**. The next example develops two basic reduction formulas that we will need to use later in this section.

EXAMPLE 3 Prove the following identities.

(a) $\cos\left(\dfrac{\pi}{2} - \alpha\right) = \sin \alpha$ **(b)** $\sin\left(\dfrac{\pi}{2} - \beta\right) = \cos \beta$

Solution **(a)** $\cos\left(\dfrac{\pi}{2} - \alpha\right) = \cos \dfrac{\pi}{2} \cos \alpha + \sin \dfrac{\pi}{2} \sin \alpha$

$$= (0) \cos \alpha + (1) \sin \alpha$$
$$= \sin \alpha$$

This proves the identity. Incidentally, if we use degree measure instead of radian measure, this identity states that

$$\cos(90° − α) = \sin α$$

as we saw earlier, in Chapter 5.

(b) Since the identity $\cos\left(\dfrac{\pi}{2} − α\right) = \sin α$ holds for all values of $α$, we can simply replace $α$ by the quantity $\dfrac{\pi}{2} − β$ to obtain

$$\cos\left[\frac{\pi}{2} − \left(\frac{\pi}{2} − β\right)\right] = \sin\left(\frac{\pi}{2} − β\right)$$

$$\cos\left(\frac{\pi}{2} − \frac{\pi}{2} + β\right) = \sin\left(\frac{\pi}{2} − β\right)$$

$$\cos β = \sin\left(\frac{\pi}{2} − β\right)$$

This proves the identity. ▲

The two identities in Example 3 are worth memorizing. For simplicity, we replace both $α$ and $β$ by the single letter $θ$ to get

$$\cos\left(\frac{\pi}{2} − θ\right) = \sin θ$$

$$\sin\left(\frac{\pi}{2} − θ\right) = \cos θ$$

We developed the fourth addition formula under the assumption that the angles were measured in radians. However, the formula is still valid when the angles are measured in degrees. In the next example, we apply the formula in just such a case.

EXAMPLE 4   Use the formula for $\cos(s − t)$ to determine the exact value of $\cos 15°$.

Solution   First observe that $15° = 45° − 30°$. Then we have

$$\cos 15° = \cos(45° − 30°)$$
$$= \cos 45° \cos 30° + \sin 45° \sin 30° \quad \text{using the formula for } \cos(s − t) \text{ with } s = 45° \text{ and } t = 30°$$

$$= \left(\frac{\sqrt{2}}{2}\right)\left(\frac{\sqrt{3}}{2}\right) + \left(\frac{\sqrt{2}}{2}\right)\left(\frac{1}{2}\right)$$

$$= \frac{\sqrt{6}}{4} + \frac{\sqrt{2}}{4} = \frac{\sqrt{6} + \sqrt{2}}{4}$$

Thus, the exact value of $\cos 15°$ is $\frac{1}{4}(\sqrt{6} + \sqrt{2})$. ▲

Now let us return to our derivations of the addition formulas. Using the fourth addition formula, we can easily derive the third formula as follows. In the formula

$$\cos(s − t) = \cos s \cos t + \sin s \sin t$$

we replace t by the quantity $-t$. This is permissible because the formula holds for all real numbers. We obtain

$$\cos[s - (-t)] = \cos s \cos(-t) + \sin s \sin(-t)$$

On the right-hand side of this equation, we can use the identities developed for $\cos(-t)$ and $\sin(-t)$ in Section 6.2. Doing this yields

$$\cos(s + t) = (\cos s)(\cos t) + (\sin s)(-\sin t)$$

which is equivalent to

$$\cos(s + t) = \cos s \cos t - \sin s \sin t$$

This is the third addition formula, as we wished to prove.

Next we derive the formula for $\sin(s + t)$. We have

$$
\begin{aligned}
\sin(s + t) &= \cos\left[\frac{\pi}{2} - (s + t)\right] \quad &&\text{replacing } \theta \text{ by } s + t \text{ in the identity} \\
&&&\sin \theta = \cos[(\pi/2) - \theta] \\
&= \cos\left[\left(\frac{\pi}{2} - s\right) - t\right] \\
&= \cos\left(\frac{\pi}{2} - s\right) \cos t + \sin\left(\frac{\pi}{2} - s\right) \sin t \quad &&\text{(Why?)} \\
&= \sin s \cos t + \cos s \sin t \quad &&\text{(Why?)}
\end{aligned}
$$

This proves the first addition formula.

Finally, we can use the first addition formula to prove the second as follows:

$$
\begin{aligned}
\sin(s - t) &= \sin[s + (-t)] \\
&= \sin s \cos(-t) + \cos s \sin(-t) \\
&= (\sin s)(\cos t) + (\cos s)(-\sin t) \\
&= \sin s \cos t - \cos s \sin t
\end{aligned}
$$

This completes the proofs of the four addition formulas.

EXAMPLE 5   If $\sin s = \frac{3}{5}$, $0 < s < \pi/2$, and $\sin t = -\sqrt{3}/4$, $\pi < t < 3\pi/2$, compute $\sin(s - t)$.

Solution $$\sin(s - t) = \underset{\substack{\uparrow \\ \text{given as } \frac{3}{5}}}{\sin s} \cos t - \cos s \underset{\substack{\uparrow \\ \text{given as } \frac{-\sqrt{3}}{4}}}{\sin t} \qquad (3)$$

In view of equation (3), we need to find only $\cos t$ and $\cos s$. These can be determined using the Pythagorean identity $\cos^2 \theta = 1 - \sin^2 \theta$. We have

$$
\begin{aligned}
\cos^2 t &= 1 - \sin^2 t \\
&= 1 - \left(\frac{-\sqrt{3}}{4}\right)^2 = 1 - \frac{3}{16} = \frac{13}{16}
\end{aligned}
$$

Therefore,

$$\cos t = \frac{\sqrt{13}}{4} \qquad \text{or} \qquad \cos t = \frac{-\sqrt{13}}{4}$$

We choose the negative value here for cosine, since it is given that $\pi < t < 3\pi/2$. Thus,

$$\cos t = \frac{-\sqrt{13}}{4}$$

Similarly, to find cos s, we have

$$\cos^2 s = 1 - \sin^2 s$$
$$= 1 - \left(\frac{3}{5}\right)^2 = \frac{16}{25}$$

Therefore,

$$\cos s = \frac{4}{5} \quad \text{cos } s \text{ is positive, since } 0 < s < \pi/2.$$

Finally, we substitute the values we've obtained for cos t and cos s, along with the given data, back into equation (3). This yields

$$\sin(s - t) = \left(\frac{3}{5}\right)\left(\frac{-\sqrt{13}}{4}\right) - \left(\frac{4}{5}\right)\left(\frac{-\sqrt{3}}{4}\right)$$
$$= \frac{-3\sqrt{13}}{20} + \frac{4\sqrt{3}}{20} = \frac{-3\sqrt{13} + 4\sqrt{3}}{20}$$

▼ **EXERCISE SET 6.5**

A

In Exercises 1–6, use the addition formulas to simplify the expressions.

1. (a) $\sin\theta\cos 2\theta + \cos\theta\sin 2\theta$

   (b) $\sin\dfrac{\pi}{6}\cos\dfrac{\pi}{3} + \cos\dfrac{\pi}{6}\sin\dfrac{\pi}{3}$

2. (a) $\sin 3\theta\cos\theta - \cos 3\theta\sin\theta$
   (b) $\sin 110°\cos 20° - \cos 110°\sin 20°$

3. (a) $\cos 2u\cos 3u - \sin 2u\sin 3u$
   (b) $\cos 2u\cos 3u + \sin 2u\sin 3u$

4. (a) $\cos\dfrac{3\pi}{10}\cos\dfrac{\pi}{5} - \sin\dfrac{3\pi}{10}\sin\dfrac{\pi}{5}$

   (b) $\cos\dfrac{2\pi}{9}\cos\dfrac{\pi}{18} + \sin\dfrac{2\pi}{9}\sin\dfrac{\pi}{18}$

5. $\sin(A + B)\cos A - \cos(A + B)\sin A$
6. $\cos(s - t)\cos t - \sin(s - t)\sin t$

In Exercises 7–10, simplify each term (as we did in Example 2).

7. $\sin\left(\theta - \dfrac{3\pi}{2}\right)$

8. $\cos\left(\dfrac{3\pi}{2} + \theta\right)$

9. $\cos(\theta + \pi)$

10. $\sin(\theta - \pi)$

11. Expand $\sin(\theta + 2\pi)$ using the appropriate addition formula, and check to see that your answer agrees with the formula for $\sin(\theta + 2\pi)$ given on page 317.

12. Follow the directions in Exercise 11, but use $\cos(\theta + 2\pi)$.

13. Use the formula for $\cos(s + t)$ to compute the exact value for cos 75°.

14. Use the formula for $\sin(s - t)$ to compute the exact value of $\sin(\pi/12)$.

15. Use the formula for $\sin(s + t)$ to find $\sin(7\pi/12)$.

16. If $\sin s = \frac{4}{5}$, $0 < s < \pi/2$, and $\sin t = -\frac{1}{4}$, $\pi < t < 3\pi/2$, compute $\sin(s - t)$.

In Exercises 17–20, use the addition formulas to simplify each expression.

17. $\sin\left(\dfrac{\pi}{4} + s\right) - \sin\left(\dfrac{\pi}{4} - s\right)$

18. $\sin\left(t + \dfrac{\pi}{6}\right) - \sin\left(t - \dfrac{\pi}{6}\right)$

19. $\cos\left(\dfrac{\pi}{3} - \theta\right) - \cos\left(\dfrac{\pi}{3} + \theta\right)$

20. $\cos\left(\theta - \dfrac{\pi}{4}\right) + \cos\left(\theta + \dfrac{\pi}{4}\right)$

21. If $\sin\alpha = \frac{12}{13}$, $\pi/2 < \alpha < \pi$, and $\cos\beta = \frac{3}{5}$, $3\pi/2 < \beta < 2\pi$, compute the following.
    (a) $\sin(\alpha + \beta)$ (b) $\sin(\alpha - \beta)$
    (c) $\cos(\alpha + \beta)$ (d) $\cos(\alpha - \beta)$

22. Suppose that $\sin\theta = \frac{1}{5}$ and $0 < \theta < \pi/2$.
    (a) Compute $\cos\theta$.
    (b) Compute $\sin 2\theta$. *Hint:* $\sin 2\theta = \sin(\theta + \theta)$

23. Suppose that $\cos\theta = \frac{12}{13}$ and $3\pi/2 < \theta < 2\pi$.
    (a) Compute $\sin\theta$.
    (b) Compute $\cos 2\theta$. *Hint:* $\cos 2\theta = \cos(\theta + \theta)$

24. Given $\tan\theta = -\frac{2}{3}$, $\pi/2 < \theta < \pi$, and $\csc\beta = 2$, $0 < \beta < \pi/2$, find $\sin(\theta + \beta)$ and $\cos(\beta - \theta)$.

25. Given $\sec s = \frac{5}{4}$, $\sin s < 0$, and $\cot t = -1$, $\pi/2 < t < \pi$, find $\sin(s - t)$ and $\cos(s + t)$.

26. In the following figure, $AB = CD = 1$ and $BC = 2$. Find $\alpha + \beta$ and express your answer using radian measure. *Hint:* Find $\cos(\alpha + \beta)$.

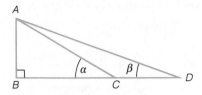

B

In Exercises 27–34, prove that each equation is an identity.

27. $\dfrac{\sin(s + t)}{\cos s \cos t} = \tan s + \tan t$

28. $\dfrac{\cos(s - t)}{\cos s \sin t} = \cot t + \tan s$

29. $\cos(A - B) - \cos(A + B) = 2 \sin A \sin B$

30. $\sin(A - B) + \sin(A + B) = 2 \sin A \cos B$

31. $\cos(A + B) \cos(A - B) = \cos^2 A - \sin^2 B$

32. $\sin(A + B) \sin(A - B) = \cos^2 B - \cos^2 A$

33. $\cos(\alpha + \beta) \cos \beta + \sin(\alpha + \beta) \sin \beta = \cos \alpha$

34. $\cos\left(\theta + \dfrac{\pi}{4}\right) + \sin\left(\theta - \dfrac{\pi}{4}\right) = 0$

35. Let $f(t) = \cos^2 t + \cos^2\left(t + \dfrac{2\pi}{3}\right) + \cos^2\left(t - \dfrac{2\pi}{3}\right)$.

    (a) [C] Complete the table.

t	1	2	3	4
$f(t)$				

    (b) On the basis of your results in part (a), make a conjecture about the function f. Prove that your conjecture is correct.

36. Let A, B, and C be the angles of a triangle.
    (a) Show that $\sin(A + B) = \sin C$.
    (b) Show that $\cos(A + B) = -\cos C$.
    (c) Show that $\tan(A + B) = -\tan C$.

37. Let f be the function defined by $f(\theta) = \sin \theta$. Prove that

$$\frac{f(\theta + h) - f(\theta)}{h} = (\sin \theta)\left(\frac{\cos h - 1}{h}\right) + (\cos \theta)\left(\frac{\sin h}{h}\right)$$

38. Prove the following identities.
    (a) $\sin 2\theta = 2 \sin \theta \cos \theta$ *Hint:* $\sin 2\theta = \sin(\theta + \theta)$
    (b) $\cos 2\theta = \cos^2 \theta - \sin^2 \theta$

39. Prove that

$$\frac{\sin(\alpha - \beta)}{\cos \alpha \cos \beta} + \frac{\sin(\beta - \gamma)}{\cos \beta \cos \gamma} + \frac{\sin(\gamma - \alpha)}{\cos \gamma \cos \alpha} = 0$$

40. Suppose that A, B, and C are the angles of a triangle. Show that

$$\cos^2 A + \cos^2 B + \cos^2 C + 2 \cos A \cos B \cos C = 1$$

41. Suppose that $a^2 + b^2 = 1$ and $c^2 + d^2 = 1$. Prove that $|ac + bd| \le 1$. *Hint:* Let $a = \cos \theta$, $b = \sin \theta$, $c = \cos \phi$, and $d = \sin \phi$.

Exercises 42 and 43 outline simple geometric derivations of the formulas for $\sin(\alpha + \beta)$ and $\cos(\alpha + \beta)$ in the case where α and β are acute angles, with $\alpha + \beta < 90°$. The exercises rely on the accompanying figures, which are constructed as follows. Begin, in Figure A, with $\alpha = \angle GAD$, $\beta = \angle HAG$, and $\overline{AH} = 1$. Then, from H, draw perpendiculars to $\overline{AD}$ and to $\overline{AG}$, as shown in Figure B. Finally, draw $\overline{FE} \perp \overline{BH}$ and $\overline{FC} \perp \overline{AD}$.

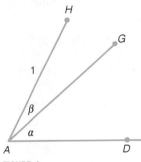

FIGURE A

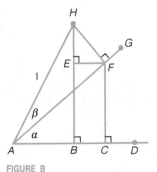

FIGURE B

42. *Formula for $\sin(\alpha + \beta)$* Supply the reasons or steps behind each statement.
    (a) $BH = \sin(\alpha + \beta)$ **(b)** $FH = \sin \beta$
    (c) $\angle BHF = \alpha$
    (d) $EH = \cos \alpha \sin \beta$ *Hint:* Use $\triangle EFH$ and the result in part (b).
    (e) $AF = \cos \beta$ **(f)** $CF = \sin \alpha \cos \beta$
    (g) $\sin(\alpha + \beta) = \sin \alpha \cos \beta + \cos \alpha \sin \beta$
        Hint: $\sin(\alpha + \beta) = BH = EH + CF$

43. *Formula for $\cos(\alpha + \beta)$* Supply the reasons or steps behind each statement.
    (a) $\cos(\alpha + \beta) = AB$
    (b) $AC = \cos \alpha \cos \beta$ *Hint:* Use $\triangle ACF$ and the result in Exercise 42(e).
    (c) $EF = \sin \alpha \sin \beta$
    (d) $\cos(\alpha + \beta) = \cos \alpha \cos \beta - \sin \alpha \sin \beta$
        Hint: $AB = AC - BC$

C

44. Prove the following addition formulas for the tangent function.

    (a) $\tan(\alpha + \beta) = \dfrac{\tan \alpha + \tan \beta}{1 - \tan \alpha \tan \beta}$

    (b) $\tan(\alpha - \beta) = \dfrac{\tan \alpha - \tan \beta}{1 + \tan \alpha \tan \beta}$

45. (a) The angles of a triangle are $A = 20°$, $B = 50°$, and $C = 110°$. Use your calculator to compute the two quantities $\tan A + \tan B + \tan C$ and $\tan A \tan B \tan C$. What do you observe?

(b) The angles of a triangle are $\alpha = \pi/10$, $\beta = 3\pi/10$, and $\gamma = 3\pi/5$. Use your calculator to compute $\tan \alpha + \tan \beta + \tan \gamma$ and $\tan \alpha \tan \beta \tan \gamma$.

(c) If triangle ABC is not a right triangle, prove that

$$\tan A + \tan B + \tan C = \tan A \tan B \tan C$$

46. If triangle ABC is not a right triangle, and $\cos A = \cos B \cos C$, show that $\tan B \tan C = 2$.

6.6 ▼ FURTHER IDENTITIES

There are a number of basic identities that follow from the addition formulas for sine and cosine. In the box, we summarize some of the most useful of these.

Addition Formulas for Tangent

(a) $\tan(s + t) = \dfrac{\tan s + \tan t}{1 - \tan s \tan t}$

(b) $\tan(s - t) = \dfrac{\tan s - \tan t}{1 + \tan s \tan t}$

Double-Angle Formulas

(a) $\sin 2\theta = 2 \sin \theta \cos \theta$
(b) $\cos 2\theta = \cos^2 \theta - \sin^2 \theta$

(c) $\tan 2\theta = \dfrac{2 \tan \theta}{1 - \tan^2 \theta}$

Half-Angle Formulas

(a) $\sin \dfrac{s}{2} = \pm \sqrt{\dfrac{1 - \cos s}{2}}$

(b) $\cos \dfrac{s}{2} = \pm \sqrt{\dfrac{1 + \cos s}{2}}$

(c) $\tan \dfrac{s}{2} = \dfrac{\sin s}{1 + \cos s}$

To prove the formula for $\tan(s + t)$, we begin with

$$\tan(s + t) = \frac{\sin(s + t)}{\cos(s + t)}$$

$$= \frac{\sin s \cos t + \cos s \sin t}{\cos s \cos t - \sin s \sin t} \tag{1}$$

Now we divide both numerator and denominator on the right-hand side of equation (1) by the quantity $\cos s \cos t$. This yields

$$\tan(s + t) = \frac{\dfrac{\sin s \cos t}{\cos s \cos t} + \dfrac{\cos s \sin t}{\cos s \cos t}}{\dfrac{\cos s \cos t}{\cos s \cos t} - \dfrac{\sin s \sin t}{\cos s \cos t}}$$

$$= \frac{\tan s + \tan t}{1 - \tan s \tan t}$$

This proves the formula for $\tan(s + t)$. The formula for $\tan(s - t)$ can be deduced from this with the aid of the identity $\tan(-t) = -\tan t$, which was derived in Section 6.2. We have

$$\tan(s - t) = \tan[s + (-t)]$$

$$= \frac{\tan s + \tan(-t)}{1 - \tan s \tan(-t)} = \frac{\tan s + (-\tan t)}{1 - (\tan s)(-\tan t)}$$

$$= \frac{\tan s - \tan t}{1 + \tan s \tan t} \qquad \text{as required}$$

EXAMPLE 1   Compute $\tan \dfrac{7\pi}{12}$, using the fact that $\dfrac{7\pi}{12} = \dfrac{\pi}{4} + \dfrac{\pi}{3}$.

Solution

$$\tan \frac{7\pi}{12} = \tan\left(\frac{\pi}{4} + \frac{\pi}{3}\right)$$

$$= \frac{\tan(\pi/4) + \tan(\pi/3)}{1 - \tan(\pi/4)\,\tan(\pi/3)} \qquad \text{using the formula for } \tan(s + t)$$
$$\text{with } s = \pi/4 \text{ and } t = \pi/3$$

$$= \frac{1 + \sqrt{3}}{1 - (1)(\sqrt{3})}$$

Thus, $\tan \dfrac{7\pi}{12} = \dfrac{1 + \sqrt{3}}{1 - \sqrt{3}}$. We can write this answer in a more compact form by rationalizing the denominator. As you can check, the result is

$$\tan \frac{7\pi}{12} = -2 - \sqrt{3} \qquad\qquad\qquad\qquad\qquad\quad \blacktriangle$$

We turn now to the identities for $\sin 2\theta$, $\cos 2\theta$, and $\tan 2\theta$. All of these are derived in the same way: we replace 2θ by $(\theta + \theta)$ and use the appropriate addition formula. For instance, for $\sin 2\theta$ we have

$$\sin 2\theta = \sin(\theta + \theta) = \sin \theta \cos \theta + \cos \theta \sin \theta$$
$$= 2 \sin \theta \cos \theta$$

This establishes the formula for $\sin 2\theta$. (Exercise 29 asks you to carry out the corresponding derivations for $\cos 2\theta$ and $\tan 2\theta$.)

EXAMPLE 2   If $\sin \theta = \frac{4}{5}$ and $\pi/2 < \theta < \pi$, find $\cos \theta$, $\sin 2\theta$, and $\cos 2\theta$.

Solution   We have

$$\cos^2 \theta = 1 - \sin^2 \theta$$

$$= 1 - \left(\frac{4}{5}\right)^2 = \frac{9}{25}$$

Consequently,

$$\cos \theta = \frac{3}{5} \qquad \text{or} \qquad \cos \theta = -\frac{3}{5}$$

We want the negative value for the cosine here, since $\pi/2 < \theta < \pi$. Thus,

$$\cos \theta = -\frac{3}{5}$$

Now that we know the values of $\cos\theta$ and $\sin\theta$, the double-angle formulas can be used to determine $\sin 2\theta$ and $\cos 2\theta$. We have

$$\sin 2\theta = 2\sin\theta\cos\theta$$
$$= 2\left(\frac{4}{5}\right)\left(-\frac{3}{5}\right) = -\frac{24}{25}$$

and

$$\cos 2\theta = \cos^2\theta - \sin^2\theta$$
$$= \left(-\frac{3}{5}\right)^2 - \left(\frac{4}{5}\right)^2 = \frac{9}{25} - \frac{16}{25} = -\frac{7}{25}$$

The required values are therefore $\cos\theta = -\frac{3}{5}$, $\sin 2\theta = -\frac{24}{25}$, and $\cos 2\theta = -\frac{7}{25}$.

▲

EXAMPLE 3 If $x = 4\sin\theta$, $0 < \theta < \pi/2$, express $\sin 2\theta$ in terms of x.

Solution The given equation is equivalent to $\sin\theta = x/4$, so we have

$$\sin 2\theta = 2\sin\theta\cos\theta = 2\left(\frac{x}{4}\right)\cos\theta$$

$$= \frac{x}{2}\sqrt{1 - \sin^2\theta} \qquad \text{(Why is the positive root appropriate?)}$$

$$= \frac{x}{2}\sqrt{1 - \frac{x^2}{16}} = \frac{x}{2}\sqrt{\frac{16 - x^2}{16}}$$

$$= \frac{x\sqrt{16 - x^2}}{8}$$

▲

EXAMPLE 4 Prove the following identity:

$$\cos 3\theta = 4\cos^3\theta - 3\cos\theta$$

Solution $\cos 3\theta = \cos(2\theta + \theta)$

$$= \underbrace{\cos 2\theta}\cos\theta - \underbrace{\sin 2\theta}\sin\theta$$

$$\uparrow\uparrow$$

$$(\cos^2\theta - \sin^2\theta)\ (2\sin\theta\cos\theta)$$

$$= (\cos^2\theta - \sin^2\theta)\cos\theta - (2\sin\theta\cos\theta)\sin\theta$$

$$= \cos^3\theta - \sin^2\theta\cos\theta - 2\sin^2\theta\cos\theta$$

Collecting like terms now gives us

$$\cos 3\theta = \cos^3\theta - 3\sin^2\theta\cos\theta$$

Finally, we replace $\sin^2\theta$ by the quantity $1 - \cos^2\theta$. This yields

$$\cos 3\theta = \cos^3\theta - 3(1 - \cos^2\theta)\cos\theta$$

$$= \cos^3\theta - 3\cos\theta + 3\cos^3\theta$$

$$= 4\cos^3\theta - 3\cos\theta \qquad \text{as required}$$

▲

In the box that follows, we list several alternate ways of writing the formula for $\cos 2\theta$. Formulas (c) and (d) are quite useful in calculus.

Equivalent Forms of the Formula
$$\cos 2\theta = \cos^2 \theta - \sin^2 \theta$$

(a) $\cos 2\theta = 2 \cos^2 \theta - 1$ **(b)** $\cos 2\theta = 1 - 2 \sin^2 \theta$

(c) $\cos^2 \theta = \dfrac{1 + \cos 2\theta}{2}$ **(d)** $\sin^2 \theta = \dfrac{1 - \cos 2\theta}{2}$

One way to prove identity (a) is as follows:

$$\begin{aligned}
\cos 2\theta &= \cos^2 \theta - \sin^2 \theta \\
&= \cos^2 \theta - (1 - \cos^2 \theta) \\
&= \cos^2 \theta - 1 + \cos^2 \theta \\
&= 2 \cos^2 \theta - 1 \qquad \text{as required}
\end{aligned}$$

If, in this last equation, we add 1 to both sides and then divide by 2, the result is identity (c). (Verify this.) The proofs for (b) and (d) are similar; see Exercise 30.

EXAMPLE 5 Express $\cos^4 t$ in a form that does not involve powers of the trigonometric functions.

Solution
$$\cos^4 t = (\cos^2 t)^2 = \left(\frac{1 + \cos 2t}{2} \right)^2 \qquad \text{using the formula for } \cos^2 \theta$$

$$= \frac{1 + 2 \cos 2t + \cos^2 2t}{4}$$

$$= \frac{1 + 2 \cos 2t + \frac{1}{2}(1 + \cos 4t)}{4} \qquad \begin{array}{l} \text{using the formula} \\ \text{for } \cos^2 \theta \text{ with } \theta = 2t \end{array}$$

An easy way to simplify this last expression is to multiply both the numerator and the denominator by 2. As you should check for yourself, the final result is

$$\cos^4 t = \frac{3 + 4 \cos 2t + \cos 4t}{8}$$

The last three formulas we are going to prove in this section are the **half-angle formulas**:

$$\cos \frac{s}{2} = \pm \sqrt{\frac{1 + \cos s}{2}}$$

$$\sin \frac{s}{2} = + \sqrt{\frac{1 - \cos s}{2}}$$

$$\tan \frac{s}{2} = \frac{\sin s}{1 + \cos s}$$

To derive the formula for $\cos(s/2)$, we begin with one of the alternate forms of the cosine double-angle formula:

$$\cos^2 \theta = \frac{1 + \cos 2\theta}{2}$$

or

$$\cos \theta = \pm \sqrt{\frac{1 + \cos 2\theta}{2}}$$

Since this identity holds for all values of θ, we may replace θ by $s/2$ to obtain

$$\cos\frac{s}{2} = \pm\sqrt{\frac{1 + \cos 2(s/2)}{2}} = \pm\sqrt{\frac{1 + \cos s}{2}}$$

This is the required formula for $\cos(s/2)$. To derive the formula for $\sin(s/2)$, we follow exactly the same procedure, except that we begin with the identity $\sin^2\theta = \frac{1 - \cos 2\theta}{2}$. [Exercise 30(c) asks you to complete the proof.] In both formulas, the sign before the radical is determined by the quadrant in which the arc (or angle) $s/2$ terminates.

EXAMPLE 6   Evaluate $\cos 105°$ using the half-angle formula.

Solution

$$\cos 105° = \cos\frac{210°}{2} = \pm\sqrt{\frac{1 + \cos 210°}{2}} \qquad \text{using the formula for } \cos(s/2), \text{ with } s = 210°$$

$$= \pm\sqrt{\frac{1 + (-\sqrt{3}/2)}{2}}$$

$$= \pm\sqrt{\frac{1 - (\sqrt{3}/2)}{2} \cdot \frac{2}{2}} = \pm\sqrt{\frac{2 - \sqrt{3}}{4}}$$

$$= \frac{\pm\sqrt{2 - \sqrt{3}}}{2}$$

We want to choose the negative value here, since the terminal side of $105°$ lies in the second quadrant. Thus, we finally obtain

$$\cos 105° = \frac{-\sqrt{2 - \sqrt{3}}}{2}$$

▲

Our last task is to establish the formula for $\tan(s/2)$. To do this, we first prove the equivalent identity $\tan\theta = \dfrac{\sin 2\theta}{1 + \cos 2\theta}$.

PROOF THAT $\tan\theta = \dfrac{\sin 2\theta}{1 + \cos 2\theta}$

$$\frac{\sin 2\theta}{1 + \cos 2\theta} = \frac{2\sin\theta\cos\theta}{2\cos^2\theta} \qquad \text{using the identity } \cos 2\theta = 2\cos^2\theta - 1 \text{ in the denominator}$$

$$= \frac{\sin\theta}{\cos\theta} = \tan\theta$$

If we now replace θ by $s/2$ in the identity $\tan\theta = \dfrac{\sin 2\theta}{1 + \cos 2\theta}$, the result is

$$\tan\frac{s}{2} = \frac{\sin s}{1 + \cos s}$$

This is the half-angle formula for the tangent.

We conclude this section with a summary of the principal trigonometric identities developed in this and the previous section. For completeness, the list also includes two sets of trigonometric identities that we did not discuss in the text but that are occasionally useful. These are called **product-to-sum formulas** and **sum-to-product formulas**. Proofs and applications of these formulas are included in the exercises.

PROPERTY SUMMARY	PRINCIPAL TRIGONOMETRIC IDENTITIES

I. Consequences of the Definitions

(a) $\csc \theta = \dfrac{1}{\sin \theta}$ (b) $\sec \theta = \dfrac{1}{\cos \theta}$ (c) $\cot \theta = \dfrac{1}{\tan \theta}$

(d) $\tan \theta = \dfrac{\sin \theta}{\cos \theta}$ (e) $\cot \theta = \dfrac{\cos \theta}{\sin \theta}$

II. Pythagorean Identities

(a) $\sin^2 \theta + \cos^2 \theta = 1$ (b) $\tan^2 \theta + 1 = \sec^2 \theta$ (c) $\cot^2 \theta + 1 = \csc^2 \theta$

III. Opposite-Angle Formulas

(a) $\sin(-\theta) = -\sin \theta$ (b) $\cos(-\theta) = \cos \theta$ (c) $\tan(-\theta) = -\tan \theta$

IV. Reduction Formulas

(a) $\sin(\theta + 2\pi k) = \sin \theta$ (b) $\cos(\theta + 2\pi k) = \cos \theta$

(c) $\sin\left(\dfrac{\pi}{2} - \theta\right) = \cos \theta$ (d) $\cos\left(\dfrac{\pi}{2} - \theta\right) = \sin \theta$

V. Addition Formulas

(a) $\sin(s + t) = \sin s \cos t + \cos s \sin t$ (b) $\sin(s - t) = \sin s \cos t - \cos s \sin t$
(c) $\cos(s + t) = \cos s \cos t - \sin s \sin t$ (d) $\cos(s - t) = \cos s \cos t + \sin s \sin t$

(e) $\tan(s + t) = \dfrac{\tan s + \tan t}{1 - \tan s \tan t}$ (f) $\tan(s - t) = \dfrac{\tan s - \tan t}{1 + \tan s \tan t}$

VI. Double-Angle Formulas

(a) $\sin 2\theta = 2 \sin \theta \cos \theta$ (b) $\cos 2\theta = \cos^2 \theta - \sin^2 \theta$ (c) $\tan 2\theta = \dfrac{2 \tan \theta}{1 - \tan^2 \theta}$

VII. Half-Angle Formulas

(a) $\sin \dfrac{\theta}{2} = \pm\sqrt{\dfrac{1 - \cos \theta}{2}}$ (b) $\cos \dfrac{\theta}{2} = \pm\sqrt{\dfrac{1 + \cos \theta}{2}}$ (c) $\tan \dfrac{\theta}{2} = \dfrac{\sin \theta}{1 + \cos \theta}$

VIII. Product-to-Sum Formulas

(a) $\sin A \sin B = \frac{1}{2}[\cos(A - B) - \cos(A + B)]$
(b) $\sin A \cos B = \frac{1}{2}[\sin(A + B) + \sin(A - B)]$
(c) $\cos A \cos B = \frac{1}{2}[\cos(A + B) + \cos(A - B)]$

IX. Sum-to-Product Formulas

(a) $\sin \alpha + \sin \beta = 2 \sin \dfrac{\alpha + \beta}{2} \cos \dfrac{\alpha - \beta}{2}$ (b) $\sin \alpha - \sin \beta = 2 \cos \dfrac{\alpha + \beta}{2} \sin \dfrac{\alpha - \beta}{2}$

(c) $\cos \alpha + \cos \beta = 2 \cos \dfrac{\alpha + \beta}{2} \cos \dfrac{\alpha - \beta}{2}$ (d) $\cos \alpha - \cos \beta = -2 \sin \dfrac{\alpha + \beta}{2} \sin \dfrac{\alpha - \beta}{2}$

▼ EXERCISE SET 6.6

A

In Exercises 1–12, refer to the two triangles and compute the quantities indicated.

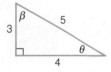
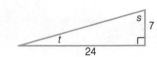

1. (a) $\tan(\theta + t)$ (b) $\tan(\theta - t)$
2. (a) $\tan(s + \beta)$ (b) $\tan(s - \beta)$
3. (a) $\tan\left(\theta + \dfrac{\pi}{4}\right)$ (b) $\tan\left(\theta - \dfrac{\pi}{4}\right)$
4. (a) $\tan\left(\beta + \dfrac{\pi}{3}\right)$ (b) $\tan\left(\beta - \dfrac{\pi}{3}\right)$
5. (a) $\sin 2\theta$ (b) $\cos 2\theta$ (c) $\tan 2\theta$
6. (a) $\sin 2t$ (b) $\cos 2t$ (c) $\tan 2t$
7. (a) $\sin 2\beta$ (b) $\cos 2\beta$ (c) $\tan 2\beta$
8. (a) $\sin 2s$ (b) $\cos 2s$ (c) $\tan 2s$
9. (a) $\sin(\theta/2)$ (b) $\cos(\theta/2)$ (c) $\tan(\theta/2)$
10. (a) $\sin(s/2)$ (b) $\cos(s/2)$ (c) $\tan(s/2)$
11. (a) $\sin(\beta/2)$ (b) $\cos(\beta/2)$ (c) $\tan(\beta/2)$
12. (a) $\sin(t/2)$ (b) $\cos(t/2)$ (c) $\tan(t/2)$

In Exercises 13–16, use the given information to compute each of the following: (a) $\sin 2\theta$;   (b) $\cos 2\theta$;   (c) $\sin(\theta/2)$;   (d) $\cos(\theta/2)$.

13. $\sin \theta = \frac{3}{4}$ and $\pi/2 < \theta < \pi$
14. $\cos \theta = \frac{2}{5}$ and $3\pi/2 < \theta < 2\pi$
15. $\cos \theta = -\frac{1}{3}$ and $\pi < \theta < 3\pi/2$
16. $\sin \theta = -\frac{1}{10}$ and $3\pi/2 < \theta < 2\pi$

In Exercises 17–20, use an appropriate half-angle formula to evaluate each quantity,

17. (a) $\sin(\pi/12)$ (b) $\cos(\pi/12)$ (c) $\tan(\pi/12)$
18. (a) $\sin(\pi/8)$ (b) $\cos(\pi/8)$ (c) $\tan(\pi/8)$
19. (a) $\sin 105°$ (b) $\cos 105°$ (c) $\tan 105°$
20. (a) $\sin 165°$ (b) $\cos 165°$ (c) $\tan 165°$

In Exercises 21–24, use the given information to express $\sin 2\theta$ and $\cos 2\theta$ in terms of x.

21. $x = 5 \sin \theta, 0 < \theta < \pi/2$
22. $x = \sqrt{2} \cos \theta, 0 < \theta < \pi/2$
23. $x - 1 = 2 \sin \theta, 0 < \theta < \pi/2$
24. $x + 1 = 3 \sin \theta, \pi/2 < \theta < \pi$

In Exercises 25–28, express each quantity in a form that does not involve powers of the trigonometric functions (as in Example 5).

25. $\sin^4 \theta$ 26. $\cos^6 \theta$
27. $\sin^4(\theta/2)$ 28. $\sin^6(\theta/4)$
29. Prove each of the following double-angle formulas. *Hint:* As in the text, replace 2θ with $\theta + \theta$, and use an appropriate addition formula.
    (a) $\cos 2\theta = \cos^2 \theta - \sin^2 \theta$
    (b) $\tan 2\theta = \dfrac{2 \tan \theta}{1 - \tan^2 \theta}$

30. (a) Beginning with the identity $\cos 2\theta = \cos^2 \theta - \sin^2 \theta$, prove that $\cos 2\theta = 1 - 2 \sin^2 \theta$.
    (b) Using the result in part (a), prove that $\sin^2 \theta = \frac{1}{2}(1 - \cos 2\theta)$.
    (c) Derive the formula for $\sin(s/2)$ as follows. Using the identity in part (b), replace θ with $s/2$, and then take square roots.

B

In Exercises 31–46, prove that the given equations are identities.

31. $\cos 2s = \dfrac{1 - \tan^2 s}{1 + \tan^2 s}$ 32. $1 + \cos 2t = \cot t \sin 2t$

33. $\cos \theta = 2 \cos^2(\theta/2) - 1$ 34. $\dfrac{\sin 2\theta}{\sin \theta} - \dfrac{\cos 2\theta}{\cos \theta} = \sec \theta$

35. $\sin^4 \theta = \dfrac{3 - 4 \cos 2\theta + \cos 4\theta}{8}$

36. $\sin 3\theta = 3 \sin \theta - 4 \sin^3 \theta$

37. $\sin 2\theta = \dfrac{2 \tan \theta}{1 + \tan^2 \theta}$ 38. $2 \csc 2\theta = \dfrac{\csc^2 \theta}{\cot \theta}$

39. $\sin 2\theta = 2 \sin^3 \theta \cos \theta + 2 \sin \theta \cos^3 \theta$

40. $\cot \theta = \dfrac{1 + \cos 2\theta}{\sin 2\theta}$

41. $\dfrac{1 + \tan(\theta/2)}{1 - \tan(\theta/2)} = \tan \theta + \sec \theta$

42. $\tan \theta + \cot \theta = 2 \csc 2\theta$

43. $2 \sin^2(45° - \theta) = 1 - \sin 2\theta$

44. $(\sin \theta - \cos \theta)^2 = 1 - \sin 2\theta$

45. $1 + \tan \theta \tan 2\theta = \tan 2\theta \cot \theta - 1$

46. $\tan\left(\dfrac{\pi}{4} + \theta\right) - \tan\left(\dfrac{\pi}{4} - \theta\right) = 2 \tan 2\theta$

47. Let triangle ABC be a right triangle with $\angle C = 90°$. Prove the following statements. [In part (c), the letters a, b, and c denote the lengths of the sides opposite angles A, B, and C, respectively.]
    (a) $\sin 2A = \sin 2B$ (b) $\cos 2A + \cos 2B = 0$

(c) $\sin 3A = \dfrac{3ab^2 - a^3}{c^3}$

48. Prove that $\tan \dfrac{\theta}{2} = \dfrac{1 - \cos \theta}{\sin \theta}$.

49. If $\cos(\alpha + \beta) = 0$, show that $\sin(\alpha + 2\beta) = \sin \alpha$.

50. If $\tan \alpha = \frac{1}{11}$ and $\tan \beta = \frac{5}{6}$, find $\alpha + \beta$, given that $0 < \alpha < \pi/2$ and $0 < \beta < \pi/2$.

51. If $\sin \theta = \dfrac{a^2 - b^2}{a^2 + b^2}$, show that $\tan \dfrac{\theta}{2} = \dfrac{a - b}{a + b}$. (Assume that θ is an acute angle and a and b are positive.)

52. Let $z = \tan \theta$. Show that

$$\cos 2\theta = \frac{1 - z^2}{1 + z^2} \qquad \text{and} \qquad \sin 2\theta = \frac{2z}{1 + z^2}$$

53. Prove the following three formulas, known as the product-to-sum formulas.
(a) $\sin A \sin B = \frac{1}{2}[\cos(A - B) - \cos(A + B)]$
(b) $\sin A \cos B = \frac{1}{2}[\sin(A + B) + \sin(A - B)]$
(c) $\cos A \cos B = \frac{1}{2}[\cos(A + B) + \cos(A - B)]$

54. Express each product as a sum or difference of sines or cosines.
(a) $4 \sin 6\theta \cos 2\theta$ **(b)** $2 \sin A \sin 3A$
(c) $\cos 4\theta \cos 2\theta$

55. **(a)** Show that $\sin(\pi/4) \cos(\pi/12) = \frac{1}{4}(\sqrt{3} + 1)$.
(b) Evaluate $2(\sin 82.5°)(\cos 37.5°)$ without using a calculator.

56. Prove the following identities.
(a) $\sin(A + B) \sin(A - B) = \sin^2 A - \sin^2 B$
(b) $\cos(A + B) \cos(A - B) = \cos^2 A - \sin^2 B$

57. In the formula (from Exercise 53)

$$\sin A \cos B = \frac{1}{2}[\sin(A + B) + \sin(A - B)]$$

make the substitutions $A + B = \alpha$ and $A - B = \beta$ and deduce that

$$\sin \alpha + \sin \beta = 2 \sin \frac{\alpha + \beta}{2} \cos \frac{\alpha - \beta}{2}$$

This is one of the sum-to-product formulas mentioned in the text.

58. Prove the following sum-to-product identities. *Hint:* See the previous exercise.

(a) $\sin \alpha - \sin \beta = 2 \cos \dfrac{\alpha + \beta}{2} \sin \dfrac{\alpha - \beta}{2}$

(b) $\cos \alpha + \cos \beta = 2 \cos \dfrac{\alpha + \beta}{2} \cos \dfrac{\alpha - \beta}{2}$

(c) $\cos \alpha - \cos \beta = -2 \sin \dfrac{\alpha + \beta}{2} \sin \dfrac{\alpha - \beta}{2}$

59. **(a)** Show that $\cos \dfrac{2\pi}{9} + \cos \dfrac{\pi}{9} = \sqrt{3} \cos \dfrac{\pi}{18}$.

(b) Show that $\sin 105° + \sin 15° = \sqrt{6}/2$.

60. Prove that $\dfrac{\sin \theta + \sin 3\theta}{\cos \theta + \cos 3\theta} = \tan 2\theta$. *Hint:* Use the sum-to-product formulas.

61. Prove that $\dfrac{\sin 7\theta - \sin 5\theta}{\cos 7\theta + \cos 5\theta} = \tan \theta$.

62. Prove that $\tan(x + y) = \dfrac{\sin 2x + \sin 2y}{\cos 2x + \cos 2y}$.

63. The following figure shows a semicircle with radius $AO = 1$.
(a) Use the figure to derive the formula

$$\tan \frac{\theta}{2} = \frac{\sin \theta}{1 + \cos \theta}$$

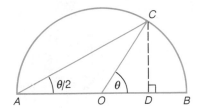

Hint: Show that $CD = \sin \theta$ and $OD = \cos \theta$. Then look at right triangle ADC to find $\tan(\theta/2)$.
(b) Use the formula developed in part (a) to show that

(i) $\tan 15° = \dfrac{1}{2 + \sqrt{3}} = 2 - \sqrt{3}$;

(ii) $\tan(\pi/8) = \sqrt{2} - 1$.

64. This exercise outlines geometric derivations of the formulas for $\sin 2\theta$ and $\cos 2\theta$. Refer to the following figure, which shows a semicircle with diameter $\overline{SV}$ and radius $UV = 1$. In each case, supply the reasons for the given statement.
(a) $TU = \cos 2\theta$ **(b)** $TV = 1 + \cos 2\theta$
(c) $WT = \sin 2\theta$
(d) $WV = 2 \cos \theta$ *Hint:* Consider right triangle SWV.
(e) $SW = 2 \sin \theta$

(f) $\sin \theta = \dfrac{\sin 2\theta}{2 \cos \theta}$ *Hint:* Use $\triangle WTV$ and the results in parts (c) and (d).

(g) $\sin 2\theta = 2 \sin \theta \cos \theta$

(h) $\cos \theta = \dfrac{1 + \cos 2\theta}{2 \cos \theta}$ *Hint:* Use $\triangle WTV$.

(i) $\cos 2\theta = 2 \cos^2 \theta - 1$

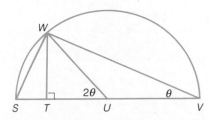

C

65. **(a)** Prove that

$$\tan \theta \tan(60° - \theta) \tan(60° + \theta) = \tan 3\theta.$$

(b) Prove that $\tan 20° \tan 40° \tan 80° = \sqrt{3}$.

66. *Calculation of* $\sin 18°$, $\cos 18°$, *and* $\sin 3°$
   (a) Prove that $\cos 3\theta = 4 \cos^3 \theta - 3 \cos \theta$.
   (b) Supply a reason for each statement.
      (i) $\sin 36° = \cos 54°$
      (ii) $2 \sin 18° \cos 18° = 4 \cos^3 18° - 3 \cos 18°$
      (iii) $2 \sin 18° = 4 \cos^2 18° - 3$
   (c) In equation (iii), replace $\cos^2 18°$ by $1 - \sin^2 18°$ and then solve the resulting equation for $\sin 18°$. Thus, show that

$$\sin 18° = \tfrac{1}{4}(\sqrt{5} - 1)$$

   (d) Show that

$$\cos 18° = \tfrac{1}{4}\sqrt{10 + 2\sqrt{5}}$$

   (e) Show that

$$\sin 3° = \tfrac{1}{16}[(\sqrt{5} - 1)(\sqrt{6} + \sqrt{2})$$
$$- 2(\sqrt{3} - 1)\sqrt{5 + \sqrt{5}}]$$

   Hint: $3° = 18° - 15°$
   (f) $\boxed{\text{C}}$ Use your calculator to check the results in parts (c), (d), and (e).

67. In this exercise, you will derive a formula for the length of an angle bisector in a triangle. (The formula will be needed in Exercise 68.) Let f denote the length of the bisector of angle C in triangle ABC, as shown in the following figure.

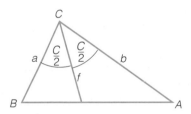

(a) Explain why

$$\frac{1}{2} af \sin \frac{C}{2} + \frac{1}{2} bf \sin \frac{C}{2} = \frac{1}{2} ab \sin C$$

Hint: Use areas.

(b) Show that $f = \dfrac{2ab \cos(C/2)}{a + b}$.

(c) By the cosine law, $\cos C = \dfrac{a^2 + b^2 - c^2}{2ab}$. Use this to show that

$$\cos \frac{C}{2} = \frac{1}{2}\sqrt{\frac{(a + b - c)(a + b + c)}{ab}}$$

(d) Show that the length of the angle bisector in terms of the sides is given by

$$f = \frac{\sqrt{ab}}{a + b} \sqrt{(a + b - c)(a + b + c)}$$

68. In triangle XYZ (in the accompanying figure), $\overline{SX}$ bisects angle ZXY and $\overline{TY}$ bisects angle ZYX. In this exercise, you are going to prove the following theorem, known as the Steiner–Lehmus theorem: If the lengths of the angle bisectors $\overline{SX}$ and $\overline{TY}$ are equal, then triangle XYZ is isosceles (with $XZ = YZ$).

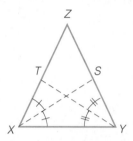

(a) Let $x, y,$ and z denote the lengths of the sides $\overline{YZ}$, $\overline{XZ}$, and $\overline{XY}$, respectively. Use the formula in Exercise 67(d) to show that the equation $TY = SX$ is equivalent to

$$\frac{\sqrt{xz}}{x + z} \sqrt{(x + z - y)(x + z + y)}$$
$$= \frac{\sqrt{yz}}{y + z} \sqrt{(y + z - x)(y + z + x)} \quad (1)$$

(b) What common factors do you see on both sides of equation (1)? Divide both sides of equation (1) by those common factors. You should obtain

$$\frac{\sqrt{x}}{x + z} \sqrt{x + z - y} = \frac{\sqrt{y}}{y + z} \sqrt{y + z - x} \quad (2)$$

(c) Clear equation (2) of fractions, and then square both sides. After combining like terms and then grouping, the equation can be written

$$(3x^2yz - 3xy^2z) + (x^3y - xy^3)$$
$$+ (x^2z^2 - y^2z^2) + (xz^3 - yz^3) = 0 \quad (3)$$

(d) Show that equation (3) can be written

$$(x - y)[3xyz + xy(x + y) + x^2(x + y) + z^3] = 0$$

Now notice that the quantity in brackets in this last equation must be positive. (Why?) Consequently, $x - y = 0$, and so $x = y$, as required.
Remark: This theorem has a fascinating history, beginning in 1840 when C. L. Lehmus first proposed the theorem to the great Swiss geometer Jacob Steiner (1796–1863). For background (and much shorter proofs!), see either of the following references: *Scientific American* **204** (1961): 166–168; *American Mathematical Monthly* **70** (1963): 79–80.

6.7 ▼ TRIGONOMETRIC EQUATIONS

In this section, we consider some techniques for solving equations involving the trigonometric functions. As usual, by a **solution** of an equation, we mean a value of the variable for which the equation becomes a true statement.

EXAMPLE 1   Consider the trigonometric equation $\sin x + \cos x = 1$. Is $x = \pi/4$ a solution? Is $x = \pi/2$ a solution?

Solution   To see if the value $x = \pi/4$ satisfies the given equation, we write

$$\sin \frac{\pi}{4} + \cos \frac{\pi}{4} \overset{?}{=} 1$$

$$\frac{\sqrt{2}}{2} + \frac{\sqrt{2}}{2} \overset{?}{=} 1$$

$$\sqrt{2} \overset{?}{=} 1 \qquad \text{No!}$$

Thus, $x = \pi/4$ is not a solution. In a similar fashion, we can check to see if $x = \pi/2$ is a solution:

$$\sin \frac{\pi}{2} + \cos \frac{\pi}{2} \overset{?}{=} 1$$

$$1 + 0 \overset{?}{=} 1 \qquad \text{Yes!}$$

Thus, $x = \pi/2$ is a solution.   ▲

The example that we have just concluded serves to remind us of the difference between a *conditional equation* and an *identity*. An identity is true for all values of the variable in its domain. For example, the equation $\sin^2 t + \cos^2 t = 1$ is an identity; it is true for every real number t. A conditional equation, on the other hand, is true for only some (or perhaps even none) of the values of the variable. The equation in Example 1 is a conditional equation; we saw that it is false when $x = \pi/4$ and true when $x = \pi/2$. The equation $\sin t = 2$ is an example of a conditional equation that has no solution (because the maximum value of the sine function is 1). The equations that we are going to solve in this section are conditional equations that involve the trigonometric functions. In general, there is no single technique that can be used to solve every trigonometric equation. In the examples that follow, we illustrate some of the more common approaches to solving trigonometric equations.

EXAMPLE 2   Solve the equation $\sin \theta = \frac{1}{2}$.

Solution   First of all, we know that $\sin(\pi/6) = \frac{1}{2}$. Thus, one solution is certainly $\theta = \pi/6$. To find another solution, we note that $\sin \theta$ is positive in the second quadrant, as well as in the first. In the second quadrant, the angle with a reference angle of $\pi/6$ is $5\pi/6$ (150°). Thus, $\sin(5\pi/6) = \frac{1}{2}$, and we conclude that $\theta = 5\pi/6$ is also a solution to the given equation. (See Figure 1.)

Since $\sin \theta$ is negative in the third and fourth quadrants, we needn't look there for solutions. Nevertheless, there are other solutions, because the sine function is periodic, with period 2π. Therefore, since $\theta = \pi/6$ is a solution, we know that

$$\theta = \frac{\pi}{6} + 2\pi$$

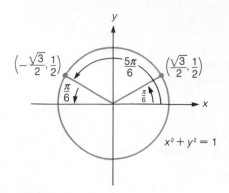

FIGURE 1

$$\sin \frac{\pi}{6} = \sin \frac{5\pi}{6} = \frac{1}{2}$$

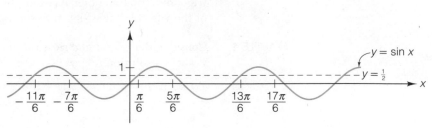

FIGURE 2

There are infinitely many real numbers x for which $\sin x = \frac{1}{2}$.

is another solution. So is $\theta = (\pi/6) + 4\pi$ a solution, and so on. Similarly, adding multiples of 2π to $\theta = 5\pi/6$ produces additional solutions. Therefore, the solutions of the equation $\sin \theta = \frac{1}{2}$ are

$$\frac{\pi}{6} + 2\pi k \qquad \text{and} \qquad \frac{5\pi}{6} + 2\pi k$$

where k is any integer. (See Figure 2.) If we wish, we may express these solutions in degrees:

$$30° + 360k° \qquad \text{and} \qquad 150° + 360k°$$

▲

Conceptually, the next example is similar to the example we just completed, but now we'll need to use a calculator. As background for this, you should review Example 10 in Section 5.2.

EXAMPLE 3 Find all solutions of the equation

$$\cos x = 0.351$$

in the open interval $(0, 2\pi)$. Round off the answers to three decimal places.

Solution As indicated in Figure 3, there are two numbers, x_1 and x_2, in the interval $(0, 2\pi)$ that satisfy the equation $\cos x = 0.351$. We can determine x_1 by using a calculator (set in the radian mode). On a Texas Instruments calculator, for example, the keystrokes are

0.351 $\boxed{\text{INV}}$ $\boxed{\text{COS}}$

FIGURE 3

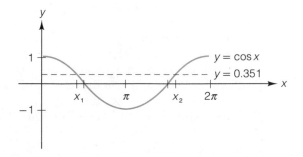

Again, for example, on a Casio calculator, the keystrokes are

0.351 $\boxed{\text{COS}^{-1}}$

In any case, as you should check for yourself, the result is

$x_1 = 1.212$

For the value of x_2, we can reason as follows. From the graph in Figure 3 we see that x_2 lies in the range $3\pi/2 < x_2 < 2\pi$. Thinking in terms of angles, x_2 is the radian measure of an angle with a terminal side in quadrant IV. Furthermore, x_1 is the measure of the reference angle for x_2. (How do we know, without relying on a calculator, that $x_1 < \pi/2$?) Therefore

$x_2 = 2\pi - x_1$

$\quad = 5.071$ using a calculator and rounding off

In summary now, the two solutions in the interval $(0, 2\pi)$ are 1.212 and 5.071.

▲

The next example shows how factoring can be used to solve a trigonometric equation.

EXAMPLE 4   Solve the equation $\cos^2 x + \cos x - 2 = 0$.

Solution   By factoring the expression on the left-hand side of the equation, we obtain

$(\cos x + 2)(\cos x - 1) = 0$

Therefore,

$\cos x = -2 \quad$ or $\quad \cos x = 1$

We discard the result $\cos x = -2$, since the value of $\cos x$ is never less than -1. From the equation $\cos x = 1$, we conclude that $x = 0$ is one solution. After $x = 0$, the next time we have $\cos x = 1$ is when $x = 2\pi$. (You can see this by looking at the graph of $y = \cos x$ or by considering the unit circle definition of the cosine.) In general, then, the solutions of the equation $\cos^2 x + \cos x - 2 = 0$ are given by $x = 2\pi k$, where k is an integer. ▲

In some equations, more than one trigonometric function is present. A common approach here is to express the various functions in terms of a single one. The next example demonstrates this technique.

EXAMPLE 5   Find all solutions of the equation $3 \tan^2 x - \sec^2 x - 5 = 0$.

Solution   We use the Pythagorean identity

$\tan^2 x + 1 = \sec^2 x$

to substitute for $\sec^2 x$ in the given equation. This gives us

$3 \tan^2 x - (\tan^2 x + 1) - 5 = 0$

$2 \tan^2 x - 6 = 0$

$2 \tan^2 x = 6$

$\tan^2 x = 3$

$\tan x = \pm\sqrt{3}$

Since the period of the tangent function is π, we need only find values of x between 0 and π that satisfy $\tan x = \pm\sqrt{3}$. The other solutions will then be obtained by adding multiples of π to these solutions. Now, in the first quadrant, the tangent is positive and we know that $\tan x = \sqrt{3}$ when $x = \pi/3$. In the second quadrant, $\tan x$ is negative and we know that $\tan 2\pi/3 = -\sqrt{3}$, since the reference angle for $2\pi/3$ is $\pi/3$. Thus, the solutions between 0 and π are $x = \pi/3$ and $x = 2\pi/3$. It follows that all of the solutions to the equation $3\tan^2 x - \sec^2 x - 5 = 0$ are given by

$$x = \frac{\pi}{3} + \pi k \qquad \text{and} \qquad x = \frac{2\pi}{3} + \pi k$$

where k is an integer. ▲

The technique used in Example 5, of expressing the various functions in terms of a single function, is most useful when it does not involve introducing a radical expression. For instance, consider the equation $\sin s + \cos s = 1$. Although we could begin by replacing $\cos s$ by the expression $\pm\sqrt{1 - \sin^2 s}$, it turns out to be easier in this situation to begin by squaring both sides of the given equation. This is done in the next example.

EXAMPLE 6   Find all solutions of the equation $\sin s + \cos s = 1$ satisfying $0° \le s < 360°$.

Solution   Squaring both sides of the equation yields

$$(\sin s + \cos s)^2 = 1^2$$

$$\sin^2 s + 2\sin s \cos s + \cos^2 s = 1$$

These add to 1

Consequently, we have

$$2\sin s \cos s = 0$$

From this last equation, we conclude that $\sin s = 0$ or $\cos s = 0$. When $\sin s = 0$, we know that $s = 0°$ or $s = 180°$. And when $\cos s = 0$, we know that $s = 90°$ or $s = 270°$. Now we must go back and check which (if any) of these values is a solution to the *original* equation. This must be done whenever we square both sides in the process of solving an equation.

$s = 0°$:	$\sin 0° + \cos 0° \overset{?}{=} 1$	
	$0 + 1 \overset{?}{=} 1$	True
$s = 90°$:	$\sin 90° + \cos 90° \overset{?}{=} 1$	
	$1 + 0 \overset{?}{=} 1$	True
$s = 180°$:	$\sin 180° + \cos 180° \overset{?}{=} 1$	
	$0 + (-1) \overset{?}{=} 1$	False
$s = 270°$:	$\sin 270° + \cos 270° \overset{?}{=} 1$	
	$-1 + 0 \overset{?}{=} 1$	False

We conclude that the only solutions of the equation $\sin s + \cos s = 1$ on the interval $0° \le s < 360°$ are $s = 0°$ and $s = 90°$. ▲

In the example that follows, we consider an equation that involves a multiple of the unknown angle. (Methods for solving other types of equations involving multiple angles are indicated in the exercises.)

EXAMPLE 7 Solve the equation $\sin 3x = 1$ on the interval $0 \le x \le 2\pi$.

Solution We know that $\sin(\pi/2) = 1$. Thus, one solution can be found by writing $3x = \pi/2$, from which we conclude that $x = \pi/6$. We can look for other solutions in the required interval by writing, more generally,

$$3x = \frac{\pi}{2} + 2\pi k$$

from which it follows that

$$x = \frac{\pi}{6} + \frac{2\pi k}{3}$$

Thus, when $k = 1$, we obtain

$$x = \frac{\pi}{6} + \frac{2\pi}{3} = \frac{\pi}{6} + \frac{4\pi}{6} = \frac{5\pi}{6}$$

When $k = 2$, we obtain

$$x = \frac{\pi}{6} + \frac{2\pi(2)}{3} = \frac{\pi}{6} + \frac{8\pi}{6} = \frac{3\pi}{2}$$

When $k = 3$, we obtain

$$x = \frac{\pi}{6} + \frac{2\pi(3)}{3}$$

$$= \frac{\pi}{6} + 2\pi \quad \text{which is greater than } 2\pi$$

We conclude that the solutions of $\sin 3x = 1$ on the interval $0 \le x \le 2\pi$ are $\pi/6$, $5\pi/6$, and $3\pi/2$. ▲

In Example 7, notice that we did not need to make use of a formula for $\sin 3x$, even though the expression $\sin 3x$ did appear in the given equation. In the next example, however, we do make use of the identity $\sin 2x = 2 \sin x \cos x$.

EXAMPLE 8 Solve the equation $\sin x \cos x = 1$.

Solution We could begin by squaring both sides. (Exercise 57 will ask you to use this approach.) However, with the double-angle formula for sine in mind, we can proceed instead as follows. We multiply both sides of the given equation by 2. This yields

$$2 \sin x \cos x = 2$$

and, consequently,

$$\sin 2x = 2 \quad \text{using the double-angle formula}$$

This last equation has no solution, since the value of the sine function never exceeds 1. Thus, the equation $\sin x \cos x = 1$ has no solution. ▲

For the last example in this section, we look at another case in which tables or a calculator are required.

EXAMPLE 9   Find all angles θ between $0°$ and $360°$ satisfying the equation $\sin \theta = 2 \cos \theta$.

Solution   We first want to rewrite the given equation using a single function. The easiest way to do this is to divide both sides by $\cos \theta$. (Nothing is lost here in assuming that $\cos \theta \neq 0$. If $\cos \theta$ were 0, then θ would be $90°$ or $270°$; but neither of those angles is a solution of the given equation.) Dividing through by $\cos \theta$, we obtain

$$\frac{\sin \theta}{\cos \theta} = \frac{2 \cos \theta}{\cos \theta} \qquad \text{or} \qquad \tan \theta = 2$$

From experience, we know that none of the angles with which we are familiar (the multiples of $30°$ and $45°$) has a value of 2 for their tangent. Thus, a calculator or table is required at this point. Using a Texas Instruments calculator (set in the degree mode), we find an angle whose tangent is 2 by using the following keystrokes:

2 $\boxed{\text{INV}}$ $\boxed{\text{TAN}}$

By so doing, we find that

$$\theta \approx 63.4°$$

(Even if we were to retain all the decimals displayed by the calculator, we would still have only an approximate solution.) To find another value of θ, we note that the tangent is also positive in the third quadrant. Thus, a second (approximate) solution is given by

$$\theta = 180° + 63.4° = 243.4°$$

We conclude that the solutions between $0°$ and $360°$ to the equation $\sin \theta = 2 \cos \theta$ are approximately $63.4°$ and $243.4°$. Note that the calculator provided only one value for θ; we still needed to work out the second value ourselves using the reference-angle concept.   ▲

▼ EXERCISE SET 6.7

A

1. Is $\theta = \pi/2$ a solution of the following equation?

   $2 \cos^2 \theta - 3 \cos \theta = 0$

2. Is $x = 15°$ a solution of $(\sqrt{3}/3) \cos 2x + \sin 2x = 1$?

3. Is $x = 3\pi/4$ a solution of $\tan^2 x - 3 \tan x + 2 = 0$?

4. Is $t = 2\pi/3$ a solution of $2 \sin t + 2 \cos t = \sqrt{3} - 1$?

In Exercises 5–22, determine all solutions of the given equations. Express your answers using radian measure.

5. $\sin \theta = \sqrt{3}/2$

6. $\sin \theta = \sqrt{2}/2$

7. $\sin \theta = -\frac{1}{2}$

8. $\sin \theta + \dfrac{\sqrt{2}}{2} = 0$

9. $\cos \theta = -1$

10. $\cos \theta = \frac{1}{2}$

11. $\tan \theta = \sqrt{3}$

12. $\tan \theta + \dfrac{\sqrt{3}}{3} = 0$

13. $\tan x = 0$

14. $2 \sin^2 x - 3 \sin x + 1 = 0$

15. $2 \cos^2 \theta + \cos \theta = 0$

16. $\sin^2 x - \sin x - 6 = 0$

17. $\cos^2 t \sin t - \sin t = 0$

18. $\cos \theta + 2 \sec \theta = -3$

19. $2 \cos^2 x - \sin x - 1 = 0$

20. $2 \cot^2 x + \csc^2 x - 2 = 0$

21. $\sqrt{3} \sin t - \sqrt{1 + \sin^2 t} = 0$

22. $\sec \alpha + \tan \alpha = \sqrt{3}$

In Exercises 23–30, determine all solutions in the interval $0° \leq \theta < 360°$.

23. $\cos 3\theta = 1$

24. $\tan 2\theta = -1$

25. $\sin 3\theta = -\sqrt{2}/2$ **26.** $\sin(\theta/2) = \frac{1}{2}$

27. $\sin \theta = \cos(\theta/2)$ *Hint:* $\sin \theta = 2 \sin(\theta/2) \cos(\theta/2)$

28. $2 \sin^2 \theta - \cos 2\theta = 0$

29. $\sin 2\theta = \sqrt{3} \cos 2\theta$ *Hint:* Divide by $\cos 2\theta$.

30. $\sin 2\theta = -2 \cos \theta$

[c] *In Exercises 31–36, use a calculator where necessary to find all solutions in the interval $0° \leq \theta < 360°$.*

31. $\sin \theta = \frac{1}{4}$ **32.** $3 \cos \theta = 5$

33. $2 \tan \theta = -4$

34. $3 \sin^2 \theta - 2 \sin \theta - 1 = 0$

35. $\cos^2 x - \cos x - 1 = 0$ **36.** $2 \sin x = 3$

[c] *In Exercises 37–42, use a calculator to find all solutions in the interval $(0, 2\pi)$. Round off the answers to two decimal places.*

37. $\cos x = 0.184$ **38.** $\cos t = -0.567$

39. $\sin x = 1/\sqrt{5}$ **40.** $\sin t = -0.301$

41. $\sin t = 5 \cos t$ **42.** $\sin x \cos x = 0.035$

B

43. Find all solutions of the equation $\tan 3x - \tan x = 0$ in the interval $0 \leq x < 2\pi$.   *Hint:* Write $\tan 3x = \tan(2x + x)$ and use the addition formula for tangent.

[c] **44.** Find all solutions of the equation $2 \sin x = 1 - \cos x$ in the interval $0° \leq x < 360°$.

45. Find all solutions of the equation $\cos(x/2) = 1 + \cos x$ in the interval $0 \leq x < 2\pi$.

46. Find all solutions of the equation
$\sin 3x \cos x + \cos 3x \sin x = \sqrt{3}/2$ in the interval
$0 < x < 2\pi$.

47. Find all real numbers θ for which $\sec 4\theta + 2 \sin 4\theta = 0$.

48. Consider the equation $\sin^2 x - \cos^2 x = \frac{7}{25}$.
   (a) Solve the equation for $\cos x$.
   (b) [c] Find all solutions of the equation satisfying
      $0 < x < \pi$.

Each of the equations in Exercises 49–54 can be solved with one of the sum-to-product identities listed on page 355. Find all solutions of the equations in Exercises 49–54 on the interval $0 \leq x \leq 2\pi$.

49. $\sin 5x = \sin 3x$ **50.** $\cos 3x = \cos 2x$

51. $\sin 3x = \cos 2x$ *Hint:* $\cos 2x = \sin\left(\dfrac{\pi}{2} - 2x\right)$

52. $\sin 5x - \sin 3x + \sin x = 0$ *Hint:* First subtract $\sin x$ from both sides.

53. $\sin\left(x + \dfrac{\pi}{18}\right) = \cos\left(x - \dfrac{2\pi}{9}\right)$

54. $\cos 3x + \cos 2x + \cos x = 0$

[c] **55.** Find a solution of the equation $4 \sin \theta - 3 \cos \theta = 2$ in the interval $0° < \theta < 90°$.   *Hint:* Add $3 \cos \theta$ to both sides, and then square.

56. Find all solutions of the equation
$\sin^3 \theta \cos \theta - \sin \theta \cos^3 \theta = -\frac{1}{4}$ in the interval
$0 < \theta < \pi$.   *Hint:* Factor the left-hand side, and then use the double-angle formulas.

57. Consider the equation $\sin x \cos x = 1$.
   (a) Square both sides, and then replace $\cos^2 x$ by $1 - \sin^2 x$. Show that the resulting equation can be written $\sin^4 x - \sin^2 x + 1 = 0$.
   (b) Show that the equation $\sin^4 x - \sin^2 x + 1 = 0$ has no solutions. Conclude from this that the original equation has no real-number solutions.

[c] **58.** The figure shows a portion of the graph of the periodic function
$$f(x) = \sin(1 + \sin x)$$
Find the x-coordinates of the turning points P, Q, and R. Round off the answers to three decimal places.
Hint: For Q, the x-coordinate is halfway between the x-coordinates of P and R.

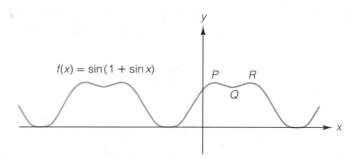

[c] **59.** **(a)** Find the smallest solution of $\cos x = 0.412$ in the interval $(1000, \infty)$. (Round off the answer to three decimal places.)
   (b) Find the smallest solution of $\cos x = -0.412$ in the interval $(1000, \infty)$.

C

60. In this exercise, you will see how certain cubic equations can be solved by using the following identity (which we proved in Example 4 in Section 6.6):
$$4 \cos^3 \theta - 3 \cos \theta = \cos 3\theta \tag{1}$$
For example, suppose that we wish to solve the equation
$$8x^3 - 6x - 1 = 0 \tag{2}$$
To transform this equation into a form where the stated identity is useful, we make the substitution $x = a \cos \theta$, where a is a constant to be determined. With this substitution, equation (2) can be written
$$8a^3 \cos^3 \theta - 6a \cos \theta = 1 \tag{3}$$

In equation (3), the coefficient of $\cos^3 \theta$ is $8a^3$. Since we want this coefficient to be 4 [as it is in equation (1)], we divide both sides of equation (3) by $2a^3$ to obtain

$$4 \cos^3 \theta - \frac{3}{a^2} \cos \theta = \frac{1}{2a^3} \qquad (4)$$

Next, a comparison of equations (4) and (1) leads us to require that $3/a^2 = 3$. Thus, $a = \pm 1$. For convenience, we choose $a = 1$; equation (4) then becomes

$$4 \cos^3 \theta - 3 \cos \theta = \tfrac{1}{2} \qquad (5)$$

Comparing equation (5) with the identity in (1) leads us to the equation

$$\cos 3\theta = \tfrac{1}{2}$$

As you can check, the solutions here are of the form

$$\theta = 20° + 120k° \qquad \text{and} \qquad \theta = 100° + 120k°$$

Thus,

$$x = \cos(20° + 120k°) \qquad \text{and} \qquad x = \cos(100° + 120k°)$$

Now, however, as you can again check, only three of the angles yield distinct values for $\cos \theta$, namely, $\theta = 20°$, $\theta = 140°$, and $\theta = 260°$. Thus, the solutions of the equation $8x^3 - 6x - 1 = 0$ are given by $x = \cos 20°$, $x = \cos 140°$, and $x = \cos 260°$.

Use the method just described to solve the following equations.

(a) $x^3 - 3x + 1 = 0$
    Answers: $2 \cos 40°$, $-2 \cos 20°$, $2 \cos 80°$
(b) $x^3 - 36x - 72 = 0$
(c) $x^3 - 6x + 4 = 0$ *Answers:* 2, $-1 \pm \sqrt{3}$
(d) [C] $x^3 - 7x - 7 = 0$ (Round off your answers to three decimal places.)

6.8 ▼ THE INVERSE TRIGONOMETRIC FUNCTIONS

This notation $\cos^{-1} e$ must not be understood to signify $\dfrac{1}{\cos e}$.

John Herschel in *Philosophical Transactions of London* (1813)

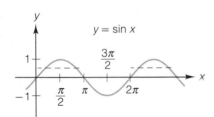

FIGURE 1
$y = \sin x$ is not one-to-one.

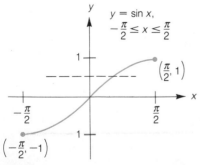

FIGURE 2
The restricted sine function is one-to-one.

According to the horizontal line test (page 137), the function $y = \sin x$ is not one-to-one (see Figure 1). Since the function is not one-to-one, there is no inverse function. However, let us consider the **restricted sine function**:

$$y = \sin x \qquad \left(-\frac{\pi}{2} \le x \le \frac{\pi}{2} \right)$$

As Figure 2 indicates, this restricted sine function is one-to-one according to the horizontal line test.

Since the restricted sine function is one-to-one, the inverse function does exist. There are two notations commonly used to denote this inverse:

$$y = \sin^{-1} x \qquad \text{and} \qquad y = \arcsin x$$

Initially, at least, we will use the notation $y = \sin^{-1} x$. The graph of $y = \sin^{-1} x$ is easily obtained using the fact that the graph of a function and its inverse are reflections of one another about the line $y = x$. Figure 3(a) shows the graph of the restricted sine function and its inverse, $y = \sin^{-1} x$. Figure 3(b) shows the graph of $y = \sin^{-1} x$ alone. From Figure 3(b), we see that the domain of $y = \sin^{-1} x$ is the interval $[-1, 1]$, while the range is $[-\pi/2, \pi/2]$.

Values of $\sin^{-1} x$ are computed according to the following rule.

$\sin^{-1} x$ is that number in the interval $[-\pi/2, \pi/2]$ whose sine is x.

To see why this is so, we begin with the restricted sine function:

$$y = \sin x \qquad \left(-\frac{\pi}{2} \le x \le \frac{\pi}{2} \right)$$

FIGURE 3

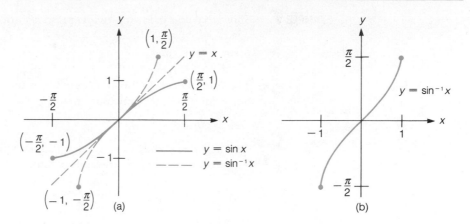

(a)

(b)

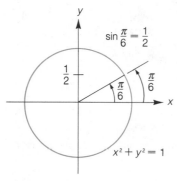

FIGURE 4

As explained in Chapter 2, the inverse function is obtained by interchanging x and y (the inputs and the outputs). So, for the inverse function, we have

$$x = \sin y \qquad \left(-\frac{\pi}{2} \leq y \leq \frac{\pi}{2} \right) \tag{1}$$

Equation (1) tells us that y is the number in the interval $[-\pi/2, \pi/2]$ whose sine is x. This is what we wished to show.

EXAMPLE 1   Evaluate each of the following.

(**a**) $\sin^{-1}(\frac{1}{2})$ (**b**) $\sin^{-1}(-\frac{1}{2})$

Solution   (**a**) $\sin^{-1}(\frac{1}{2})$ is that number in the interval $[-\pi/2, \pi/2]$ whose sine is $\frac{1}{2}$. Since $\sin(\pi/6) = \frac{1}{2}$, we conclude that $\sin^{-1}(\frac{1}{2}) = \pi/6$.

(**b**) Since $\sin(-\pi/6) = -\frac{1}{2}$ and $-\pi/6$ belongs to the interval $[-\pi/2, \pi/2]$, we conclude that $\sin^{-1}(-\frac{1}{2}) = -\pi/6$.   ▲

Using the arcsin notation, we can write the result in Example 1(a) in the form

$$\arcsin \frac{1}{2} = \frac{\pi}{6}$$

Now, as Figure 4 indicates, the arc length with a sine of $\frac{1}{2}$ is $\pi/6$. This is the idea behind the arcsin notation.

EXAMPLE 2   Evaluate $\arcsin \frac{3}{4}$.

Solution   The quantity $\arcsin \frac{3}{4}$ is that number (or arc length, or angle in radians) in the interval $[-\pi/2, \pi/2]$ whose sine is $\frac{3}{4}$. Since we are not familiar with an angle with a sine of $\frac{3}{4}$, we use a calculator. For instance, using a Texas Instruments calculator, we set the calculator to the radian mode and then use the keystrokes

0.75 INV SIN

The result is

$$\arcsin \frac{3}{4} \approx 0.85$$

▲

EXAMPLE 3   Show that the following two expressions are not equal:

$$\sin^{-1} 0 \qquad \text{and} \qquad \frac{1}{\sin 0}$$

Solution   The quantity $\sin^{-1} 0$ is that number in the interval $[-\pi/2, \pi/2]$ whose sine is 0. Since $\sin 0 = 0$, we conclude that

$$\sin^{-1} 0 = 0$$

On the other hand, since $\sin 0 = 0$, the expression $1/(\sin 0)$ is not even defined. Thus, the two given expressions certainly are not equal.   ▲

If f and f^{-1} are any pair of inverse functions, then by definition,

$$f[f^{-1}(x)] = x \qquad \text{for every } x \text{ in the domain of } f^{-1}$$

and

$$f^{-1}[f(x)] = x \qquad \text{for every } x \text{ in the domain of } f$$

Applying these facts to the restricted sine function, $y = \sin x$, and its inverse, $y = \sin^{-1} x$, we obtain the following two identities.

$$\sin(\sin^{-1} x) = x \qquad (-1 \le x \le 1)$$

$$\sin^{-1}(\sin x) = x \qquad \left(-\frac{\pi}{2} \le x \le \frac{\pi}{2}\right)$$

The following example indicates that the domain restrictions accompanying these two identities cannot be disregarded.

EXAMPLE 4   Compute each of the following quantities that is defined.

(a) $\sin^{-1}\left(\sin \dfrac{\pi}{4}\right)$ **(b)** $\sin^{-1}(\sin \pi)$

(c) $\sin(\sin^{-1} 2)$ **(d)** $\sin\left[\sin^{-1}\left(-\dfrac{1}{\sqrt{5}}\right)\right]$

Solution   **(a)** Since $\pi/4$ lies in the domain of the restricted sine function, the identity $\sin^{-1}(\sin x) = x$ is applicable here. Thus,

$$\sin^{-1}\left(\sin \frac{\pi}{4}\right) = \frac{\pi}{4}$$

Check $\sin(\pi/4) = \sqrt{2}/2$, therefore $\sin^{-1}(\sin \pi/4) = \sin^{-1}(\sqrt{2}/2) = \pi/4$.

(b) The number π is not in the domain of the restricted sine function, so the identity $\sin^{-1}(\sin x) = x$ does not apply in this case. However, since $\sin \pi = 0$, we have

$$\sin^{-1}(\sin \pi) = \sin^{-1} 0 = 0$$

Thus, $\sin^{-1}(\sin \pi)$ is equal to 0, not π.

(**c**) The number 2 is not in the domain of the inverse sine function. Thus, the expression $\sin(\sin^{-1} 2)$ is undefined.

(**d**) The identity $\sin(\sin^{-1} x) = x$ is applicable here. (Why?) Thus,

$$\sin\left[\sin^{-1}\left(-\frac{1}{\sqrt{5}}\right)\right] = -\frac{1}{\sqrt{5}}$$

▲

FIGURE 5

To define the inverse cosine function, we begin by defining the **restricted cosine function**:

$$y = \cos x \qquad (0 \leq x \leq \pi)$$

As indicated by the horizontal line test in Figure 5, this function is one-to-one. Since the restricted cosine function is one-to-one, the inverse function exists. We denote this inverse by

$$y = \cos^{-1} x \qquad \text{or} \qquad y = \arccos x$$

The graph of $y = \cos^{-1} x$ is obtained by reflecting the restricted cosine function about the line $y = x$. Figure 6(a) displays the graph of the restricted cosine function along with $y = \cos^{-1} x$. Figure 6(b) shows the graph of $y = \cos^{-1} x$ alone. From Figure 6(b), we see that the domain of $y = \cos^{-1} x$ is the interval $[-1, 1]$, while the range is $[0, \pi]$. By reasoning the way we did for the inverse sine, we obtain the following results for the inverse cosine.

$\cos^{-1} x$ is that number in the interval $[0, \pi]$ whose cosine is x.

$$\cos(\cos^{-1} x) = x \qquad (-1 \leq x \leq 1)$$
$$\cos^{-1}(\cos x) = x \qquad (0 \leq x \leq \pi)$$

FIGURE 6

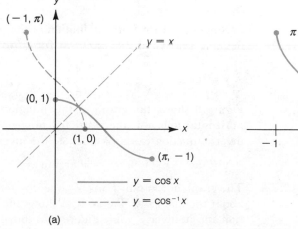

(a)

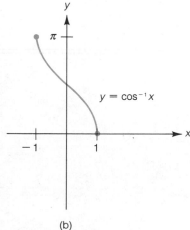

(b)

EXAMPLE 5   Compute each of the following.

(a) $\cos^{-1}(0)$ **(b)** $\arccos(\cos\frac{2}{5})$

Solution   **(a)** $\cos^{-1}(0)$ is that number in the interval $[0, \pi]$ whose cosine is 0. Since $\cos(\pi/2) = 0$, we have

$$\cos^{-1}(0) = \frac{\pi}{2}$$

(b) Since the number $\frac{2}{5}$ is in the domain of the restricted cosine function, the identity $\arccos(\cos x) = x$ is applicable. Thus,

$$\arccos\left(\cos\frac{2}{5}\right) = \frac{2}{5}$$

▲

EXAMPLE 6   Show that $\sin(\cos^{-1} x) = \sqrt{1 - x^2}$ for $-1 \le x \le 1$.

Solution   We use the identity

$$\sin y = \sqrt{1 - \cos^2 y}, \qquad \text{which is valid for } 0 \le y \le \pi$$

Substituting $\cos^{-1} x$ for y in this identity, we obtain

$$\sin(\cos^{-1} x) = \sqrt{1 - [\cos(\cos^{-1} x)]^2}$$

or

$$\sin(\cos^{-1} x) = \sqrt{1 - x^2} \qquad \text{as required}$$

Before leaving Example 6, we point out an alternate method of solution that is useful when the restriction on x is $0 < x < 1$. In this case, we let $\theta = \cos^{-1} x$. Then θ is the radian measure of the acute angle whose cosine is x and we can sketch θ as shown in Figure 7. The sides of the triangle in Figure 7 have been labeled in such a way that $\cos \theta = x$. Then by the Pythagorean theorem we find that the third side of the triangle is $\sqrt{1 - x^2}$. We therefore have

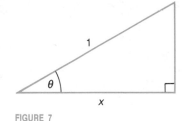

FIGURE 7

$$\sin \theta = \frac{\text{opposite}}{\text{hypotenuse}} = \frac{\sqrt{1 - x^2}}{1} = \sqrt{1 - x^2}$$

▲

Now let us turn to the definition of the inverse tangent function. We begin by defining the **restricted tangent function**:

$$y = \tan x \qquad \left(-\frac{\pi}{2} < x < \frac{\pi}{2}\right)$$

Figure 8 shows the graph of this function. As you can check by applying the horizontal line test, the restricted tangent function is one-to-one; therefore the inverse function exists. We denote the inverse tangent function by

$$y = \tan^{-1} x \qquad \text{or} \qquad y = \arctan x$$

The graph of $y = \tan^{-1} x$ is obtained by reflecting the restricted tangent function about the line $y = x$. Figure 9(a) shows the graphs of the restricted tangent function and its inverse, while Figure 9(b) shows the graph of $y = \tan^{-1} x$ alone. From Figure 9(b), we see that the domain of $y = \tan^{-1} x$ is the set of all real numbers,

FIGURE 8

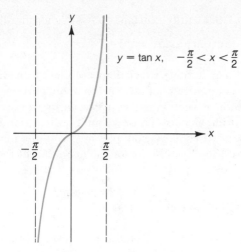

FIGURE 8

$$y = \tan x, \quad -\frac{\pi}{2} < x < \frac{\pi}{2}$$

FIGURE 9

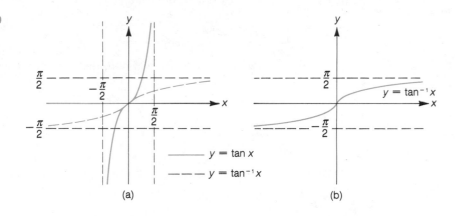

$$y = \tan x$$

$$y = \tan^{-1} x$$

(a)

$$y = \tan^{-1} x$$

(b)

and the range is the open interval $(-\pi/2, \pi/2)$. Applying the same type of reasoning used for the inverse sine, we obtain the results in the box that follows.

$\tan^{-1} x$ is that number in the interval $(-\pi/2, \pi/2)$ whose tangent is x.

$$\tan(\tan^{-1} x) = x \qquad \text{(for all } x\text{)}$$

$$\tan^{-1}(\tan x) = x \qquad \left(-\frac{\pi}{2} < x < \frac{\pi}{2}\right)$$

EXAMPLE 7   Compute the following quantities.

(a) $\tan^{-1}(-1)$ **(b)** $\tan(\tan^{-1} \sqrt{5})$

Solution   **(a)** The quantity $\tan^{-1}(-1)$ is that number in the interval $(-\pi/2, \pi/2)$ whose tangent is -1. Since $\tan(-\pi/4) = -1$, we have

$$\tan^{-1}(-1) = -\frac{\pi}{4}$$

(b) The identity $\tan(\tan^{-1} x) = x$ holds for all real numbers x. We therefore have

$$\tan(\tan^{-1} \sqrt{5}) = \sqrt{5} \qquad \blacktriangle$$

When suitable restrictions are placed on the domains of the secant, cosecant, and cotangent functions, corresponding inverse functions can be defined. However, with the exception of the inverse secant, these functions are rarely, if ever, encountered in calculus. Thus, we omit a discussion of these functions here; instead, we look at three more examples involving the inverse sine and inverse tangent functions. (See Exercise 56 for the definition of the inverse secant.)

EXAMPLE 8　Evaluate $\cos\left(\sin^{-1} \dfrac{3}{5}\right)$.

Solution　Let $\theta = \sin^{-1}(\frac{3}{5})$. Then θ is the measure of the acute angle whose sine is $\frac{3}{5}$, as indicated in Figure 10. Using the Pythagorean theorem, we find that the length of the third side of the triangle is 4. Thus,

$$\cos \theta = \frac{\text{adjacent}}{\text{hypotenuse}} = \frac{4}{5}$$

In view of the definition of θ, this last equation states that

$$\cos\left(\sin^{-1} \frac{3}{5}\right) = \frac{4}{5}$$

This is the required answer. 　　　　　　　　　　　　　　　　　　　　$\blacktriangle$

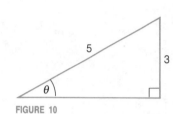

FIGURE 10

EXAMPLE 9　If $\sin \theta = x/3$ and $0 < \theta < \pi/2$, express the following quantity as a function of x:

$$\theta - \sin 2\theta$$

Solution　The given conditions tell us that $\theta = \sin^{-1}(x/3)$. That expresses θ in terms of x. To express $\sin 2\theta$ in terms of x, we have

$$\sin 2\theta = 2 \sin \theta \cos \theta = 2 \sin \theta \sqrt{1 - \sin^2 \theta}$$
$$= 2\left(\frac{x}{3}\right) \sqrt{1 - \frac{x^2}{9}} = \frac{2x}{3} \sqrt{\frac{9 - x^2}{9}}$$
$$= \frac{2x\sqrt{9 - x^2}}{9}$$

Now, combining the results, we have

$$\theta - \sin 2\theta = \sin^{-1} \frac{x}{3} - \frac{2x\sqrt{9 - x^2}}{9} \qquad \blacktriangle$$

EXAMPLE 10　Show that $\pi = 4[\tan^{-1}(\frac{1}{2}) + \tan^{-1}(\frac{1}{5}) + \tan^{-1}(\frac{1}{8})]$.

Solution　The given equation is equivalent to

$$\underbrace{\frac{\pi}{4} - \tan^{-1} \frac{1}{2}}_{A} = \underbrace{\tan^{-1} \frac{1}{5} + \tan^{-1} \frac{1}{8}}_{B}$$

Then, using the letters A and B as indicated, we wish to show that $A = B$. We will accomplish this if we can show that $\tan A = \tan B$, since both A and B lie between 0 and $\pi/2$. (Why?) Using the formula

$$\tan(x - y) = \frac{\tan x - \tan y}{1 + \tan x \tan y}$$

we have

$$\tan A = \tan\left(\frac{\pi}{4} - \tan^{-1}\frac{1}{2}\right) = \frac{\tan\dfrac{\pi}{4} - \tan\left(\tan^{-1}\dfrac{1}{2}\right)}{1 + \tan\dfrac{\pi}{4}\tan\left(\tan^{-1}\dfrac{1}{2}\right)}$$

$$= \frac{1 - \frac{1}{2}}{1 + 1\left(\frac{1}{2}\right)} = \frac{\frac{1}{2}}{\frac{3}{2}} = \frac{1}{3}$$

Similarly, using the formula

$$\tan(x + y) = \frac{\tan x + \tan y}{1 - \tan x \tan y}$$

we have

$$\tan B = \tan\left(\tan^{-1}\frac{1}{5} + \tan^{-1}\frac{1}{8}\right)$$

$$= \frac{\tan(\tan^{-1}\frac{1}{5}) + \tan(\tan^{-1}\frac{1}{8})}{1 - [\tan(\tan^{-1}\frac{1}{5})][\tan(\tan^{-1}\frac{1}{8})]}$$

$$= \frac{\frac{1}{5} + \frac{1}{8}}{1 - \left(\frac{1}{5}\right)\left(\frac{1}{8}\right)} = \frac{1}{3} \qquad \text{(Check the arithmetic.)}$$

We have now shown that $\tan A = \tan B$, since both quantities are equal to $\frac{1}{3}$. It follows that $A = B$, as we wished to show. ▲

▼ EXERCISE SET 6.8

A

In Exercises 1–20, evaluate each of the quantities that is defined, without using a calculator or tables. If a quantity is undefined, say so.

1. $\sin^{-1}(\sqrt{3}/2)$ **2.** $\cos^{-1}(-1)$

3. $\tan^{-1}\sqrt{3}$ **4.** $\arccos(-\sqrt{2}/2)$

5. $\arctan(-1/\sqrt{3})$ **6.** $\arcsin(-1)$

7. $\tan^{-1}1$ **8.** $\sin^{-1}0$

9. $\cos^{-1}2\pi$ **10.** $\arctan 0$

11. $\sin[\sin^{-1}(\frac{1}{4})]$ **12.** $\cos[\cos^{-1}(\frac{4}{3})]$

13. $\cos[\cos^{-1}(\frac{3}{4})]$ **14.** $\tan(\arctan 3\pi)$

15. $\arctan[\tan(-\pi/7)]$ **16.** $\sin(\arcsin 2)$

17. $\arcsin[\sin(\pi/2)]$ **18.** $\arccos[\cos(\pi/8)]$

19. $\arccos(\cos 2\pi)$ **20.** $\sin^{-1}[\sin(3\pi/2)]$

In Exercises 21–30, evaluate the given quantities without using a calculator or tables.

21. $\tan[\sin^{-1}(\frac{4}{5})]$ **22.** $\cos(\arcsin\frac{2}{7})$

23. $\sin(\tan^{-1}1)$ **24.** $\sin[\tan^{-1}(-1)]$

25. $\tan(\arccos\frac{5}{13})$ **26.** $\cos[\sin^{-1}(\frac{2}{3})]$

27. $\cos(\arctan\sqrt{3})$ **28.** $\sin[\cos^{-1}(\frac{1}{3})]$

29. $\sin[\arccos(-\frac{1}{3})]$ **30.** $\tan(\arcsin\frac{20}{29})$

[C] **31.** Use a calculator to evaluate each of the following quantities. Express your answers to two decimal places; don't round off.

   (a) $\sin^{-1}(\frac{3}{4})$ **(b)** $\cos^{-1}(\frac{2}{3})$

   (c) $\tan^{-1}\pi$ **(d)** $\tan^{-1}(\tan^{-1}\pi)$

In Exercises 32–34, evaluate the given expressions without using a calculator or tables.

32. $\csc[\sin^{-1}(\frac{1}{2}) - \cos^{-1}(\frac{1}{2})]$

33. $\sec[\cos^{-1}(\sqrt{2}/2) + \sin^{-1}(-1)]$

34. $\cot[\cos^{-1}(-1/2) + \cos^{-1}(0) + \tan^{-1}(1/\sqrt{3})]$

35. Show that $\cos(\sin^{-1} x) = \sqrt{1 - x^2}$ for $-1 \le x \le 1$. *Suggestion:* Use the method of Example 6 in the text.

36. If $\sin\theta = 2x$ and $0 < \theta < \pi/2$, express $\theta + \cos 2\theta$ as a function of x.

37. If $\sin\theta = 3x/2$ and $0 < \theta < \pi/2$, express $(\theta/4) - \sin 2\theta$ as a function of x.

38. If $\cos\theta = x - 1$ and $0 < \theta < \pi/2$, express $2\theta - \cos 2\theta$ as a function of x.

39. If $\tan\theta = \frac{1}{2}(x - 1)$ and $0 < \theta < \pi/2$, express $\theta - \cos\theta$ as a function of x.

40. If $\tan\theta = \frac{1}{3}(x + 1)$ and $0 < \theta < \pi/2$, express $2\theta + \tan 2\theta$ as a function of x.

B

41. Evaluate $\sin(2 \tan^{-1} 4)$.   *Hint:* $\sin 2\theta = 2 \sin\theta \cos\theta$.

42. Evaluate $\cos[2 \sin^{-1}(\frac{5}{13})]$.

43. Evaluate $\sin(\arccos \frac{3}{5} - \arctan \frac{7}{13})$.   *Suggestion:* Use the formula for $\sin(x - y)$.

44. Show that $\sin\left(\sin^{-1}\frac{1}{3} + \sin^{-1}\frac{1}{4}\right) = \dfrac{\sqrt{15} + 2\sqrt{2}}{12}$.

45. Evaluate $\cos(\tan^{-1} 2 + \tan^{-1} 3)$.

46. **(a)** Explain why $0 < \tan^{-1}(\frac{1}{3}) + \tan^{-1}(\frac{1}{2}) < \pi/2$.
  (b) Show that $\pi/4 = \tan^{-1}(\frac{1}{3}) + \tan^{-1}(\frac{1}{2})$.

47. Show that $\tan^{-1}(\frac{4}{3}) - \tan^{-1}(\frac{1}{7}) = \pi/4$.

48. Show that $\sin^{-1} x + \cos^{-1} x = \pi/2$.   *Suggestion:* Let $\theta = \sin^{-1} x$ and $\beta = \cos^{-1} x$. First explain why $-\pi/2 \le \theta + \beta \le 3\pi/2$. Then show $\sin(\theta + \beta) = 1$.

49. If $-1 < x < 1$, show that $\sin^{-1} x = \tan^{-1}(x/\sqrt{1 - x^2})$.

50. Show that $\arctan x + \arctan y = \arctan \dfrac{x + y}{1 - xy}$ when x and y are positive and $xy < 1$.

51. Show that $\tan^{-1}(\frac{1}{7}) + 2 \tan^{-1}(\frac{1}{3}) = \pi/4$.

52. Evaluate $\arcsin \frac{4}{5} + \arctan \frac{3}{4}$.

53. Show that $\sin(2 \sin^{-1} x) = 2x\sqrt{1 - x^2}$ for $-1 \le x \le 1$.

54. Let $f(x) = 2 \tan^{-1}\left(\dfrac{1 + x}{1 - x}\right) + \sin^{-1}\left(\dfrac{1 - x^2}{1 + x^2}\right)$ $(0 < x < 1)$.
  (a) [C] Set your calculator in the radian mode and complete the table. (Record your answers using the same number of decimal places as is shown on your calculator.)

x	0.1	0.2	0.3	0.4	0.5
$f(x)$					

  (b) On the basis of your results in part (a), make a conjecture about the function f. Prove that your conjecture is correct.   *Hint:* For the proof, compute $\sin[f(x)]$.

55. Let $f(x) = \sin^{-1} x + \cos^{-1} x$.
  (a) What is the domain of f?
  (b) [C] Compute $f(x)$ for $x = 0.1$, 0.2, and 0.3.
  (c) Compute $f(x)$ for $x = 1/2$, $\sqrt{2}/2$, and $\sqrt{3}/2$.
  (d) On the basis of your results in parts (b) and (c), make a conjecture about the value of $f(x)$; then prove that your conjecture is valid.

56. **(a)** The domain of the **restricted secant function** is $[0, \pi/2) \cup [\pi, 3\pi/2)$. Sketch the restricted secant function and note that it is one-to-one.
  (b) The inverse secant function can be defined as follows:

  $\sec^{-1} x$ is the unique number in the set $[0, \pi/2) \cup [\pi, 3\pi/2)$ whose secant is x.

  Using the fact that this function is the inverse of the restricted secant function, sketch the graph of $y = \sec^{-1} x$.
  (c) Evaluate $\sec^{-1}(2/\sqrt{3})$ and $\sec^{-1}(-2/\sqrt{3})$.
  (d) Evaluate $\sec^{-1}(\sqrt{2})$ and $\sec^{-1}(-\sqrt{2})$.
  (e) Evaluate $\sec(\sec^{-1} 2)$ and $\sec^{-1}(\sec 0)$.

57. Evaluate each of the following quantities. (The inverse secant function is defined in the previous exercise.)
  (a) $\cos(\sec^{-1} \frac{4}{3})$ **(b)** $\sin(\sec^{-1} 4)$
  (c) $\tan[\sec^{-1}(-3)]$ **(d)** $\cot[\sec^{-1} \frac{13}{12}]$

In Exercises 58–60, solve the equations.

58. $\cos^{-1} t = \sin^{-1} t$ *Hint:* Compute the cosine of both sides.

59. $\sin^{-1}(3t - 2) = \sin^{-1} t - \cos^{-1} t$

60. $2 \tan^{-1} \sqrt{t - t^2} = \tan^{-1} t + \tan^{-1}(1 - t)$

C

61. Show that $\tan^{-1} 1 + \tan^{-1} 2 + \tan^{-1} 3 = \pi$.

62. Suppose that $\arccos A + \arccos B + \arccos C = \pi$. Show that $A^2 + B^2 + C^2 + 2ABC = 1$.

63. **(a)** Show that $2 \tan^{-1} x = \tan^{-1}[2x/(1 - x^2)]$ for $0 \le x < 1$.
  (b) Show that $4 \tan^{-1}(\frac{1}{5}) = \tan^{-1}(\frac{120}{119})$.
  (c) Show that $4 \tan^{-1}(\frac{1}{5}) - \tan^{-1}(\frac{1}{239}) = \pi/4$. (In 1706, John Machin made use of this formula in calculating the value of π to 100 decimal places.)

▼   **CHAPTER SIX   SUMMARY OF PRINCIPAL FORMULAS AND TERMS**

FORMULAS OR TERMS	PAGE REFERENCE	COMMENT
1. $\theta = \dfrac{s}{r}$	304	This equation defines the *radian measure* θ of an angle. We assume here that the vertex of the angle is placed at the center of a circle of radius r and that s is the length of the intercepted arc. See Figure 1 in Section 6.1.
2. $s = r\theta$	308	This formula expresses the arc length s on a circle in terms of the radius r and the radian measure θ of the central angle subtended by the arc.
3. $A = \frac{1}{2}r^2\theta$	309	This formula expresses the area of a sector of a circle in terms of the radius r and the radian measure θ of the central angle.
4. $\sin\theta$ $\cos\theta$ $\tan\theta$ $\csc\theta$ $\sec\theta$ $\cot\theta$	313	Let θ denote the radian measure of an angle in standard position, and let $P(x, y)$ denote the point where the terminal side of the angle intersects the unit circle. Or, equivalently, let θ denote the measure of the corresponding arc length on the unit circle. Then the values of θ are real numbers, and we define the six trigonometric functions of θ as follows: $$\sin\theta = y \qquad \csc\theta = \frac{1}{y} \quad (y \neq 0)$$ $$\cos\theta = x \qquad \sec\theta = \frac{1}{x} \quad (x \neq 0)$$ $$\tan\theta = \frac{y}{x} \quad (x \neq 0) \qquad \cot\theta = \frac{x}{y} \quad (y \neq 0)$$
5. $\csc\theta = \dfrac{1}{\sin\theta}$ $\sec\theta = \dfrac{1}{\cos\theta}$ $\cot\theta = \dfrac{1}{\tan\theta}$ $\tan\theta = \dfrac{\sin\theta}{\cos\theta}$ $\cot\theta = \dfrac{\cos\theta}{\sin\theta}$	314	These five identities are direct consequences of the definitions of the trigonometric functions.
6. $\sin^2\theta + \cos^2\theta = 1$ $\tan^2\theta + 1 = \sec^2\theta$ $\cot^2\theta + 1 = \csc^2\theta$	314	These are the three *Pythagorean identities*.
7. $\cos(-\theta) = \cos\theta$ $\sin(-\theta) = -\sin\theta$ $\tan(-\theta) = -\tan\theta$	316	These three identities are sometimes referred to as the *opposite angle identities*. In each case, a trigonometric function of $-\theta$ is expressed in terms of that same function of θ.
8. Periodic function	321	A function f is periodic if there is a positive number p such that the equation $f(x + p) = f(x)$ holds for all x in the domain of f. The smallest such number p is called the *period* of f. Important examples: The period of $y = \sin x$ is 2π; the period of $y = \cos x$ is 2π; the period of $y = \tan x$ is π.

FORMULAS OR TERMS	PAGE REFERENCE	COMMENT
9. Amplitude	322	Let m and M denote the smallest and the largest values, respectively, of the periodic function f. Then the amplitude of f is defined to be the number $\frac{1}{2}(M - m)$. Important examples: The amplitude for both $y = \sin x$ and $y = \cos x$ is 1; amplitude is not defined for the function $y = \tan x$.
10. $\text{Period} = \dfrac{2\pi}{B}$	328	This formula gives the period for the functions $y = A \sin(Bx - C)$ and $y = A \cos(Bx - C)$.
11. $\text{Phase shift} = \dfrac{C}{B}$	330	The phase shift serves as a guide in graphing functions of the form $y = A \sin(Bx - C)$ or $y = A \cos(Bx - C)$. For instance, to graph one complete cycle of $y = A \sin(Bx - C)$, first sketch one complete cycle of $y = A \sin Bx$, beginning at $x = 0$. Then draw a curve with exactly the same shape, but beginning at $x = C/B$ rather than $x = 0$. This will represent one cycle of $y = A \sin(Bx - C)$.
12. $\sin(s + t) = \sin s \cos t + \cos s \sin t$ $\sin(s - t) = \sin s \cos t - \cos s \sin t$ $\cos(s + t) = \cos s \cos t - \sin s \sin t$ $\cos(s - t) = \cos s \cos t + \sin s \sin t$	344	These identities are referred to as the *addition formulas* for sine and cosine.
13. $\cos\left(\dfrac{\pi}{2} - \theta\right) = \sin \theta$ $\sin\left(\dfrac{\pi}{2} - \theta\right) = \cos \theta$	346	These two identities, called *reduction formulas*, hold for all real numbers θ. In terms of angles, the identities state that the sine of an angle is equal to the cosine of its complement.
14. $\tan(s + t) = \dfrac{\tan s + \tan t}{1 - \tan s \tan t}$ $\tan(s - t) = \dfrac{\tan s - \tan t}{1 + \tan s \tan t}$	350	These two identities are referred to as the *addition formulas* for tangent.
15. $\sin 2\theta = 2 \sin \theta \cos \theta$ $\cos 2\theta = \cos^2 \theta - \sin^2 \theta$ $\tan 2\theta = \dfrac{2 \tan \theta}{1 - \tan^2 \theta}$	350	These identities are known as the *double-angle formulas*. There are four other forms of the double-angle formula for cosine. They appear in the box on page 353.
16. $\sin \dfrac{s}{2} = \pm\sqrt{\dfrac{1 - \cos s}{2}}$ $\cos \dfrac{s}{2} = \pm\sqrt{\dfrac{1 + \cos s}{2}}$ $\tan \dfrac{s}{2} = \dfrac{\sin s}{1 + \cos s}$	350	These three identities are referred to as the *half-angle formulas*. In the case of the half-angle formulas for sine and cosine, the sign before the radical is determined by the quadrant in which the angle or arc terminates.
17. Restricted sine function	366	The domain of the function $y = \sin x$ is the set of all real numbers. By allowing inputs only from the closed interval $[-\pi/2, \pi/2]$, we obtain the restricted sine function.
18. $\sin^{-1} x$	366	$\sin^{-1} x$ is that number in the interval $[-\pi/2, \pi/2]$ whose sine is x. An alternate, but entirely equivalent, form of this expression is arcsin x. The function defined by $y = \sin^{-1} x$ is the inverse function for the restricted sine function.

FORMULAS OR TERMS	PAGE REFERENCE	COMMENT
19. Restricted cosine function	369	The domain of the function $y = \cos x$ is the set of all real numbers. By allowing only inputs from the closed interval $[0, \pi]$, we obtain the restricted cosine function.
20. $\cos^{-1} x$	369	$\cos^{-1} x$ is that number in the closed interval $[0, \pi]$ whose cosine is x. An alternate, but entirely equivalent, expression is arccos x. The function defined by $y = \cos^{-1} x$ is the inverse function for the restricted cosine function.
21. Restricted tangent function	370	The restricted tangent function is defined by the equation $y = \tan x$ for $-\pi/2 < x < \pi/2$.
22. $\tan^{-1} x$	370	$\tan^{-1} x$ is that number in the open interval $(-\pi/2, \pi/2)$ whose tangent is x. An alternate, but entirely equivalent, form of this expression is arctan x. The function defined by $y = \tan^{-1} x$ is the inverse function for the restricted tangent function.

▼ CHAPTER SIX   REVIEW EXERCISES

NOTE Exercises 1–20 constitute a chapter test on the fundamentals, based on group A problems.

1. Evaluate the following.

(a) $\sin \dfrac{5\pi}{3}$ (b) $\cot \dfrac{11\pi}{6}$

2. Graph the function $y = 3 \cos 3\pi x$ over the interval $-\frac{1}{3} \leq x \leq \frac{1}{3}$. Specify the x-intercepts and the coordinates of the highest point on the graph.

3. Use an appropriate addition formula to simplify the expression $\sin(\theta + 3\pi/2)$.

4. Referring to the accompanying figure, compute the following.

(a) $\cos 2\beta$ (b) $\tan \dfrac{\alpha}{2}$

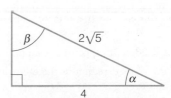

5. Suppose B and C are two points on a circle with radius $\sqrt{5}$ cm and center at A, such that angle BAC is 75°.
   (a) Find the length of the (shorter) arc of the circle joining B to C.
   (b) Find the area of the (smaller) sector determined by angle BAC.

6. In the expression $1/\sqrt{4 - t^2}$, make the substitution $t = 2 \cos x$ and simplify the result. Assume that $0 < x < \pi$.

7. Is an angle of $\pi^2/180$ radian larger or smaller than an angle of 3°? Give reasons for your answer. (A calculator is not needed.)

8. Find all solutions of the equation

$$2 \sin^2 x + 7 \sin x + 3 = 0$$

on the interval $0 \leq x \leq 2\pi$.

9. If $\cos \alpha = 2/\sqrt{5}$ $(3\pi/2 < \alpha < 2\pi)$ and $\sin \beta = \frac{4}{5}$ $(\pi/2 < \beta < \pi)$, compute $\sin(\beta - \alpha)$.

10. Prove that the following equation is an identity:

$$\cos 4\theta = 8 \cos^4 \theta - 8 \cos^2 \theta + 1$$

11. If $\sec t = -\frac{5}{3}$ and $\pi < t < 3\pi/2$, compute $\cot t$.

12. Find all solutions of the equation
$\sin(x + 30°) = \sqrt{3} \sin x$ on the interval $0° < x < 90°$.

13. Graph the function $y = -\sin(2x - \pi)$ for one complete period. Specify the amplitude, the period, and the phase shift.

14. If $\csc \theta = -3$ and $\pi < \theta < 3\pi/2$, compute $\sin(\theta/2)$.

15. Graph the function $y = -\frac{1}{2} \tan(\pi x/3)$ for one period.

16. In the following figure, P is the center of the circle. The radius of the circle is $\sqrt{2}$ units. If the radian measure of angle BPA is θ, express the area of the shaded region in terms of θ. Simplify your answer as much as possible.

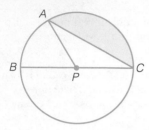

17. On the same set of axes, sketch the graphs of the restricted sine function and the function $y = \sin^{-1} x$. Specify the domain and the range for each function.

18. Compute each of the following.
(a) $\sin^{-1}[\sin(\pi/10)]$ **(b)** $\sin^{-1}(\sin 2\pi)$

19. Compute $\cos(\arcsin \frac{3}{4})$.

20. Prove that the following equation is an identity:
$$\tan\left(\frac{\pi}{4} + \frac{\theta}{2}\right) = \frac{1 + \cos\theta + \sin\theta}{1 + \cos\theta - \sin\theta}$$

In Exercises 21–28, convert the radian measures to degrees.

21. $3\pi/4$	**22.** $\pi/15$	**23.** 5π
24. $\pi/3$	**25.** $5\pi/6$	**26.** 1
27. 2	**28.** $180/\pi$	

In Exercises 29–36, convert the degree measures to radians.

29. $360°$	**30.** $36°$	**31.** $1°$
32. $330°$	**33.** $7.5°$	**34.** $\pi°$
35. $(1/\pi)°$	**36.** $(\pi/180)°$	

In Exercises 37–48, evaluate each expression without using a calculator or tables.

37. $\cos\pi$	**38.** $\sin(-3\pi/2)$	**39.** $\csc(2\pi/3)$
40. $\tan(\pi/3)$	**41.** $\cot(11\pi/6)$	**42.** $\cos 0$
43. $\sin(\pi/6)$	**44.** $\sec(3\pi/4)$	
45. $\cot(5\pi/4)$	**46.** $\tan(-7\pi/4)$	
47. $\csc(-5\pi/6)$	**48.** $\sin^2(\pi/7) + \cos^2(\pi/7)$	

In Exercises 49–56, use the arc length formula $s = r\theta$ and the sector area formula $A = \frac{1}{2}r^2\theta$ to find the required quantities.

	GIVEN	FIND
49.	$\theta = \pi/8$, $r = 16$ cm	s and A
50.	$\theta = 120°$, $r = 12$ cm	s and A
51.	$r = s = 1$ cm	A
52.	$r = s = \sqrt{3}$ cm	θ
53.	$s = 4$ cm, $\theta = 36°$	r and A
54.	$\theta = \pi/10$, $A = 200\pi$ cm²	r
55.	$s = 12$ cm, $\theta = r + 1$	r and θ
56.	$\theta = 50°$, $A = 20\pi$ cm²	s

57. Two angles in a triangle are 40° and 70°. What is the radian measure of the third angle?

58. The radian measures of two angles in a triangle are $\frac{1}{6}$ and $\frac{5}{12}$. What is the radian measure of the third angle?

C *Exercises 59–68 are calculator exercises. (Set your calculator to the radian mode.) In Exercises 59–64, where numerical answers are required, round off your results to three decimal places.*

59. Evaluate $\sin 1$. **60.** Evaluate $\cos 2$.

61. Evaluate $\sin(3\pi/2)$. **62.** Evaluate $\sin(0.78)$.

63. Evaluate $\sin(\sin 0.0123)$.

64. Evaluate $\sin[\sin(\sin 0.0123)]$.

65. Verify that $\sin^2 1986 + \cos^2 1986 = 1$.

66. Verify that $\sin 14 = 2\sin 7\cos 7$.

67. Verify that $\cos(0.5) = \cos^2(0.25) - \sin^2(0.25)$.

68. Verify that $\cos(0.3) = [\frac{1}{2}(1 + \cos 0.6)]^{1/2}$.

69. In the expression $\sqrt{25 - x^2}$, make the substitution $x = 5\sin\theta$ $(0 < \theta < \pi/2)$ and simplify the result.

70. In the expression $(49 + x^2)^{1/2}$, make the substitution $x = 7\tan\theta$ $(0 < \theta < \pi/2)$ and simplify the result.

71. In the expression $(x^2 - 100)^{1/2}$, make the substitution $x = 10\sec\theta$ $(0 < \theta < \pi/2)$ and simplify the result.

72. In the expression $(x^2 - 4)^{-3/2}$, make the substitution $x = 2\sec\theta$ $(0 < \theta < \pi/2)$ and simplify the result.

73. In the expression $(x^2 + 5)^{-1/2}$, make the substitution $x = \sqrt{5}\tan\theta$ $(0 < \theta < \pi/2)$ and simplify the result.

74. If $\sin\theta = -\frac{5}{13}$ and $\pi < \theta < 3\pi/2$, compute $\cos\theta$.

75. If $\sin\theta = \frac{3}{5}$, $\tan\theta$ is positive, and $0 < \theta < 2\pi$, compute $\sin(\theta/2)$.

76. If $\cos\theta = \frac{8}{17}$, $\sin\theta$ is negative, and $0 < \theta < 2\pi$, compute $\tan\theta$.

77. If $\tan\theta = \frac{7}{24}$, $\cos\theta$ is positive, and $0 < \theta < 2\pi$, compute $\cos 2\theta$.

78. If $\sec\theta = -\frac{25}{7}$ and $\pi < \theta < 3\pi/2$, compute $\cot\theta$.

79. If $\csc\theta = \frac{29}{20}$, $\cos\theta$ is negative, and $0 < \theta < 2\pi$, compute $\cos(\theta/2)$.

80. If $\cos\theta = \frac{3}{5}$, $\sin\theta$ is positive, and $0 < \theta < 2\pi$, compute
(a) $\sin 3\theta$ **(b)** $\sin(\theta/4)$.

81. If $\tan 2\theta = \frac{7}{24}$ and $\pi < 2\theta < 3\pi/2$, compute $\sin\theta$.

82. Express the area of the shaded region in the following figure in terms of r and θ.

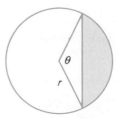

83. In the following figure, $ABCD$ is a square, each side of which is 1 cm. The two arcs are portions of circles with radii of 1 cm and with centers A and C, respectively. Find the area of the shaded region.   *Hint:* Draw $\overline{BD}$ and use your result from Exercise 82.

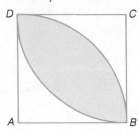

84. In the following figure, the three circles are mutually tangent, and each circle has a radius of 2 cm. Compute the area of the shaded region.

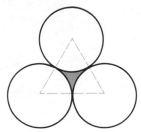

C 85. The following identity is known as *Euler's formula:*

$$\cos(36° + A) + \cos(36° - A)$$
$$= \cos A + \sin(18° + A) + \sin(18° - A)$$

Use a calculator to evaluate each side of the formula for the following values of A. In each case, round off your final results to three decimal places.
  (a) 1°   (b) 4°   (c) 6°

86. In Euler's formula (given in Exercise 85), let $A = 0°$. Then use one of the double-angle formulas for cosine to transform the equation you've obtained into an equation that is quadratic in the quantity $\sin 18°$. Solve this equation to obtain the exact value of $\sin 18°$ in radical form.

In Exercises 87–93, prove that the given equations are identities.

87. $(\sin x + \cos x)^2 + (\sin x - \cos x)^2 = 2$

88. $(\sin x + \cos x)^3 - (\sin x - \cos x)^3$
$$= 6 \cos x - 4 \cos^3 x$$

89. $(\cos x + \sin x - 1)(\cos x + \sin x + 1) = 2 \sin x \cos x$

90. $\dfrac{1}{1 - \sin x} + \dfrac{1}{1 + \sin x} + \dfrac{\sin x}{1 + \cos x} + \dfrac{1 + \cos x}{\sin x}$
$$= 2(\csc x + \sec^2 x)$$

91. $(9 - 4 \sin x - \cos^2 x)(9 + 4 \sin x - \cos^2 x)$
$$= \sin^4 x + 64$$

92. $(\sin x - \cos x)(1 + \tan x + \cot x)$
$$= (\sin^2 x \cos^2 x)(\sec^3 x - \csc^3 x)$$

93. $(a \sin x - b \cos x)^2 + (a \cos x + b \sin x)^2$
$$= a^2 + b^2$$

94. If A and B are acute angles such that $\tan A = \frac{12}{5}$ and $\tan B = \frac{3}{4}$, compute $\sin(A + B)$.

95. Using the data in Exercise 94, compute $\cos(A + B)$.

In Exercises 96–105, use the addition formulas to simplify the expressions.

96. $\sin\left(x + \dfrac{3\pi}{2}\right)$ 97. $\sin(\pi - x)$ 98. $\cos(\pi - x)$

99. $\sin 10° \cos 80° + \cos 10° \sin 80°$

100. $\cos 175° \cos 25° + \sin 175° \sin 25°$

101. $\cos \dfrac{2\pi}{5} \cos \dfrac{\pi}{10} - \sin \dfrac{2\pi}{5} \sin \dfrac{\pi}{10}$

102. $\cos\left(x - \dfrac{2\pi}{3}\right) - \cos\left(x + \dfrac{2\pi}{3}\right)$

103. $\sin\left(x + \dfrac{\pi}{6}\right) - \sin\left(x - \dfrac{\pi}{6}\right)$

104. $\tan(x + 45°) \tan(x - 45°)$

105. $\tan x \tan y + \tan y \tan z + \tan z \tan x$, where $y = x + 60°$ and $z = x - 60°$

106. (a) Show that $\tan(\pi/12) = 2 - \sqrt{3}$.
       Hint: $\pi/12 = (\pi/4) - (\pi/6)$
   (b) Show that $\cot(\pi/12) = 2 + \sqrt{3}$.

107. If $x, y,$ and z are acute angles with sines $\frac{3}{5}, \frac{5}{13},$ and $\frac{8}{17},$ respectively, compute $\sin(x + y + z)$.

In Exercises 108–112, a function of the form $y = A \sin Bx$ or $y = A \cos Bx$ is graphed for one period. Specify the equation in each case.

108.

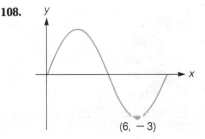

109.

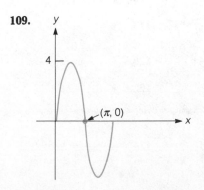

110.

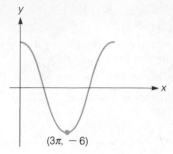

$(3\pi, -6)$

111.

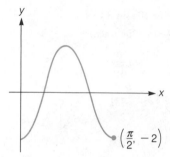

$\left(\dfrac{\pi}{2}, -2\right)$

112.

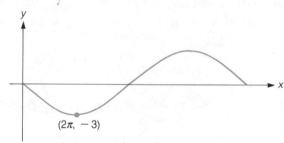

$(2\pi, -3)$

113. Find the amplitude of the function $y = A \sin\left(\dfrac{3\pi x}{2} - \pi\right)$, given that the graph passes through the point $(1, \sqrt{2})$.

114. Find the amplitude of the function $y = -3 \sin(2x - 3)\cos(2x - 3)$. *Hint:* Make use of the formula for $\sin 2\theta$.

In Exercises 115–122, sketch the graph of each function for one complete cycle. In each case, specify the x-intercepts and the coordinates of the highest and lowest points on the graph.

115. $y = -3 \cos 4x$

116. $y = 2 \sin(3\pi x/4)$

117. $y = 2 \sin\left(\dfrac{\pi x}{2} - \dfrac{\pi}{4}\right)$

118. $y = -\sin(x - 1)$

119. $y = 3 \cos\left(\dfrac{\pi x}{3} - \dfrac{\pi}{3}\right)$

120. $y = -2 \cos(x + \pi)$

121. $y = -\cos 2x \sin 2x$

122. $y = 1 - 2 \sin^2 x$

In Exercises 123–128, sketch the graph of each function for one period.

123. (a) $y = \tan(\pi x/4)$
(b) $y = \cot(\pi x/4)$

124. (a) $y = 2 \cot 2x$
(b) $y = 2 \tan 2x$

125. (a) $y = 3 \sec(x/4)$
(b) $y = 3 \csc(x/4)$

126. (a) $y = \sec \pi x$
(b) $y = \csc \pi x$

127. $y = \tan(x - 2)$

128. $y = \dfrac{\sin 2x}{1 + \cos 2x}$

In Exercises 129–154, prove that the equations are identities.

129. $\cot(x + y) = \dfrac{\cot x \cot y - 1}{\cot x + \cot y}$

130. $\cos 2x = \dfrac{1 - \tan^2 x}{1 + \tan^2 x}$

131. $\sin 2x = \dfrac{2 \tan x}{1 + \tan^2 x}$

132. $\sin^2 x - \sin^2 y = \sin(x + y)\sin(x - y)$

133. $\tan^2 x - \tan^2 y = \dfrac{\sin(x + y)\sin(x - y)}{\cos^2 x \cos^2 y}$

134. $2 \csc 2x = \sec x \csc x$

135. $(\sin x)\left(\tan \dfrac{x}{2} + \cot \dfrac{x}{2}\right) = 2$

136. $\tan\left(x + \dfrac{\pi}{4}\right) = \dfrac{1 + \tan x}{1 - \tan x}$

137. $\tan\left(\dfrac{\pi}{4} + x\right) - \tan\left(\dfrac{\pi}{4} - x\right) = 2 \tan 2x$

138. $\dfrac{\cot x - 1}{\cot x + 1} = \dfrac{1 - \sin 2x}{\cos 2x}$

139. $2 \sin\left(\dfrac{\pi}{4} - \dfrac{x}{2}\right)\cos\left(\dfrac{\pi}{4} - \dfrac{x}{2}\right) = \cos x$

140. $\dfrac{\tan(x + y) - \tan y}{1 + \tan(x + y) \tan y} = \tan x$

141. $\tan 2x + \sec 2x = \dfrac{\cos x + \sin x}{\cos x - \sin x}$

142. $\cos^4 x - \sin^4 x = \cos 2x$

143. $2 \sin x + \sin 2x = \dfrac{2 \sin^3 x}{1 - \cos x}$

144. $1 + \tan x \tan(x/2) = \sec x$

145. $\tan \dfrac{x}{2} = \dfrac{1 - \cos x + \sin x}{1 + \cos x + \sin x}$

146. $\dfrac{\sin 3x}{\sin x} - \dfrac{\cos 3x}{\cos x} = 2$

147. $\sin(x + y) \cos y - \cos(x + y) \sin y = \sin x$

148. $\dfrac{\sin x + \sin 2x}{\cos x - \cos 2x} = \cot \dfrac{x}{2}$

149. $\dfrac{1 - \tan^2(x/2)}{1 + \tan^2(x/2)} = \cos x$

150. $4 \sin \dfrac{x}{4} \cos \dfrac{x}{4} \cos \dfrac{x}{2} = \sin x$

151. $\sin 4x = 4 \sin x \cos x - 8 \sin^3 x \cos x$

152. $\cos 4x = 8 \cos^4 x - 8 \cos^2 x + 1$

153. $\sin 5x = 16 \sin^5 x - 20 \sin^3 x + 5 \sin x$

154. $\cos 5x = 16 \cos^5 x - 20 \cos^3 x + 5 \cos x$

155. Is it possible to find a triangle with the dimensions indicated in the following figure?   *Hint:* Apply the law of sines, followed by the formula for $\sin 2\theta$.

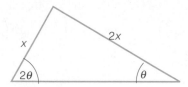

156. In this exercise, we'll use the accompanying figure to prove the following identities.

$$\cos 2\theta = 2 \cos^2 \theta - 1$$
$$\sin 2\theta = 2 \sin \theta \cos \theta$$
$$\cos 3\theta = 4 \cos^3 \theta - 3 \cos \theta$$
$$\sin 3\theta = 3 \sin \theta - 4 \sin^3 \theta$$

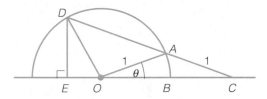

[The figure and technique in this exercise are adapted from the article by Wayne Dancer, "Geometric Proofs of Multiple Angle Formulas," in *American Mathematical Monthly* **44** (1937) 366–367.] The figure is constructed as follows. We start with $\angle AOB = \theta$ in standard position in the unit circle, as shown. The point C is chosen on the extended diameter such that $CA = 1$. Then $\overline{CA}$ is extended to meet the circle at D, and radius $\overline{DO}$ is drawn. Finally, from D, a perpendicular is drawn to the diameter, as shown.

Supply the reason or reasons that justify each of the following statements.

(a) $\angle ACO = \theta$

(b) $\angle DAO = 2\theta$ *Hint:* $\angle DAO$ is an exterior angle to $\triangle AOC$.

(c) $\angle ODA = 2\theta$

(d) $\angle DOE = 3\theta$

(e) From O, draw a perpendicular to $\overline{CD}$, meeting $\overline{CD}$ at F. From A, draw a perpendicular to $\overline{OC}$, meeting $\overline{OC}$ at G. Then $GC = OG = \cos \theta$ and $FA = DF = \cos 2\theta$.

(f) $\cos \theta = \dfrac{1 + \cos 2\theta}{2 \cos \theta}$ *Hint:* Right triangle CFO

(g) $\cos 2\theta = 2 \cos^2 \theta - 1$ *Hint:* In the equation in part (f), solve for $\cos 2\theta$.

(h) $OF = \sin 2\theta$

(i) $\sin 2\theta = 2 \sin \theta \cos \theta$ *Hint:* Find $\sin \theta$ in $\triangle CFO$, and then solve the resulting equation for $\sin 2\theta$.

(j) $DC = 1 + 2 \cos 2\theta$ and $EO = \cos 3\theta$

(k) $\cos \theta = \dfrac{2 \cos \theta + \cos 3\theta}{1 + 2 \cos 2\theta}$ *Hint:* Compute $\cos \theta$ in $\triangle CDE$ and then use part (j).

(l) $\cos 3\theta = 4 \cos^3 \theta - 3 \cos \theta$ *Hint:* Use the results in parts (k) and (g).

(m) $DE = \sin 3\theta$

(n) $\sin 3\theta = 3 \sin \theta - 4 \sin^3 \theta$ *Hint:* Compute $\sin \theta$ in $\triangle CDE$.

In Exercises 157–166, establish the identities by applying the sum-to-product formulas.

157. $\sin 80° - \sin 20° = \cos 50°$

158. $\sin 65° + \sin 25° = \sqrt{2} \cos 20°$

159. $\dfrac{\cos x - \cos 3x}{\sin x + \sin 3x} = \tan x$

160. $\dfrac{\sin 3° + \sin 33°}{\cos 3° + \cos 33°} = \tan 18°$

161. $\sin(5\pi/12) + \sin(\pi/12) = \sqrt{6}/2$

162. $\cos 10° - \sin 10° = \sqrt{2} \sin 35°$

163. $\dfrac{\cos 3y + \cos(2x - 3y)}{\sin 3y + \sin(2x - 3y)} = \cot x$

164. $\dfrac{\sin 10° - \sin 50°}{\cos 50° - \cos 10°} = \sqrt{3}$

165. $\dfrac{\sin 40° - \sin 20°}{\cos 20° - \cos 40°} = \dfrac{\sin 10° - \sin 50°}{\cos 50° - \cos 10°}$

166. Suppose that in $\triangle ABC$, $\sin A + \sin B = \cos A + \cos B$. Show that angle C is a right angle.   *Suggestion:* Begin with the sum-to-product formulas.

167. Angles A and B are acute angles in right triangle ABC.
(a) [C] Complete the following table. Which triangle yields the largest value for the quantity $\cos A + \cos B$?

a	b	c	cos A + cos B
3	4	5	
5	12	13	
7	24	25	
20	21	29	
8	15	17	
1	$\sqrt{3}$	2	
1	1	$\sqrt{2}$	
$\sqrt{2}$	$\sqrt{3}$	$\sqrt{5}$	
696	697	985	

(b) Using the sum-to-product formula, prove that

$$\cos A + \cos B = \sqrt{2} \cos \frac{A - B}{2}$$

(c) Use the result obtained in part (b) to deduce that

$$\cos A + \cos B \le \sqrt{2}$$

with equality occurring when $A = B = \pi/4$.

In Exercises 168–180, find all solutions of each equation in the range $0 \le x < 2\pi$.

168. $\tan^2 x - 3 = 0$

169. $\cot^2 x - \cot x = 0$

170. $1 + \sin x = \cos x$

171. $2 \sin 3x - \sqrt{3} = 0$

172. $\sin x - \cos 2x + 1 = 0$

173. $\sin x + \sin 2x = 0$

174. $3 \csc x - 4 \sin x = 0$

175. $2 \sin^2 x + \sin x - 1 = 0$

176. $2 \sin^4 x - 3 \sin^2 x + 1 = 0$

177. $\sec^2 x - \sec x - 2 = 0$

178. $\sin^4 x + \cos^4 x = \frac{5}{8}$

179. $4 \sin^2 2x + \cos 2x - 2 \cos^2 x - 2 = 0$

180. $\cot x + \csc x + \sec x = \tan x$ *Suggestion:* Using sines and cosines, the given equation becomes $\cos^2 x - \sin^2 x + \cos x + \sin x = 0$, which can be solved by factoring.

181. If A and B both are solutions of the equation $a \cos x + b \sin x = c$, show that $\tan[\frac{1}{2}(A + B)] = b/a$. *Hint:* The given information yields two equations. After subtracting one of those equations from the other and rearranging, you will have $\dfrac{\cos A - \cos B}{\sin A - \sin B} = -(b/a)$. Now use the sum-to-product formulas.

182. If $a \sin x + b \cos x = a \csc x + b \sec x$, show that $\tan x = -\sqrt[3]{a/b}$.

183. Evaluate $\cos \tan^{-1} \sin \tan^{-1}(\sqrt{2}/2)$.

In Exercises 184–205, evaluate each expression (without using a calculator or tables).

184. $\cos^{-1}(-\sqrt{2}/2)$

185. $\arctan(\sqrt{3}/3)$

186. $\sin^{-1} 0$

187. $\arcsin \frac{1}{2}$

188. $\arctan \sqrt{3}$

189. $\cos^{-1}(\frac{1}{2})$

190. $\tan^{-1}(-1)$

191. $\cos^{-1}(-\frac{1}{2})$

192. $\sin(\sin^{-1} 1)$

193. $\cos[\cos^{-1}(\frac{2}{7})]$

194. $\sin[\arccos(-\frac{1}{2})]$

195. $\sin[\tan^{-1}(-1)]$

196. $\cot[\cos^{-1}(\frac{1}{2})]$

197. $\sec[\cos^{-1}(\sqrt{2}/3)]$

198. $\sin\left(\dfrac{3\pi}{2} + \arccos \dfrac{3}{5}\right)$

199. $\tan\left[\dfrac{\pi}{4} + \sin^{-1}\left(\dfrac{5}{13}\right)\right]$

200. $\sin[2 \sin^{-1}(\frac{4}{5})]$

201. $\tan(2 \tan^{-1} 2)$

202. $\sin^{-1}[\sin(\pi/7)]$

203. $\cos[\frac{1}{2} \cos^{-1}(\frac{4}{5})]$

204. $\sin(\arctan \frac{1}{2} + \arctan \frac{1}{3})$

205. $\tan[\frac{1}{2}(\arcsin \frac{1}{3} + \arccos \frac{1}{2})]$

In Exercises 206–213, show that each equation is an identity.

206. $\tan(\tan^{-1} x + \tan^{-1} y) = (x + y)/(1 - xy)$

207. $\tan^{-1}(x/\sqrt{1 - x^2}) = \sin^{-1} x$

208. $\sin(2 \arctan x) = 2x/(1 + x^2)$

209. $\cos(2 \cos^{-1} x) = 2x^2 - 1$

210. $\sin[\frac{1}{2} \sin^{-1}(x^2)] = \sqrt{\frac{1}{2} - \frac{1}{2}\sqrt{1 - x^4}}$

211. $\tan^{-1}(\frac{1}{3}) + \tan^{-1}(\frac{1}{5}) = \tan^{-1}(\frac{4}{7})$

212. $\arcsin(4\sqrt{41}/41) + \arcsin(\sqrt{82}/82) = \pi/4$

213. $\tan\left(\sin^{-1} \dfrac{1}{3} + \cos^{-1} \dfrac{1}{2}\right) = \dfrac{8\sqrt{2} + 9\sqrt{3}}{5}$

214. (a) [C] Use a calculator to compute the quantity $\cos 20° \cos 40° \cos 60° \cos 80°$. Give your answer to as many decimal places as is shown on your calculator.

(b) Now use a product-to-sum formula to *prove* that the display on your calculator is the exact value of the given expression, not an approximation.

CHAPTER SEVEN

SYSTEMS OF EQUATIONS

INTRODUCTION In this chapter we consider systems of equations. Roughly speaking, a system of equations is just a collection of equations with a common set of unknowns. In solving such systems, we try to find values for the unknowns that simultaneously satisfy each of the equations in the system. In Section 7.1 we review two techniques that are often presented in intermediate algebra for solving systems involving two linear equations in two unknowns. An important technique for solving larger systems of equations is developed in Section 7.2: this technique is known as *Gaussian elimination*. After that, in Section 7.3, we introduce matrices as a tool for reducing the amount of bookkeeping involved in Gaussian elimination. Two additional techniques for solving systems of linear equations are explained in Sections 7.4 and 7.5. The technique in Section 7.4 uses the idea of an inverse matrix; the technique in Section 7.5 involves determinants and Cramer's rule. Section 7.6 contains a brief discussion of nonlinear systems of equations. In the last section of this chapter, Section 7.7, we consider systems of inequalities.

7.1 ▼ SYSTEMS OF TWO LINEAR EQUATIONS IN TWO UNKNOWNS

Many problems in a variety of disciplines give rise to . . . [linear] systems. For example, in physics, in order to find the currents in an electrical circuit containing known voltages and resistances, a system of linear equations must be solved. In chemistry, the balancing of chemical equations requires the solution of a system of linear equations. And in economics, the Leontief input-output model reduces problems concerning the production and consumption of goods to systems of linear equations.

Leslie Hogben in her text *Elementary Linear Algebra* (St. Paul: West Publishing Company, 1987)

Both in theory and in applications, it's often necessary to solve two equations in two unknowns. You may have been introduced to the idea of simultaneous equations in a previous course in algebra; however, to put matters on a firm foundation, we begin here with the basic definitions. By a **linear equation in two variables** we mean an equation of the form

$$ax + by = c$$

where the constants a and b are not both zero. The two variables needn't always be denoted by the letters x and y, of course; it is the *form* of the equation that matters. Table 1 displays some examples.

An ordered pair of numbers (x_0, y_0) is said to be a **solution of the linear equation** $ax + by = c$ provided we obtain a true statement when we replace x and y in the equation by x_0 and y_0, respectively. For example, the ordered pair $(3, 2)$ is a solution of the equation $x - y = 1$, since $3 - 2 = 1$. On the other hand, $(2, 3)$ is not a solution of $x - y = 1$, since $2 - 3 \neq 1$.

Now consider a **system** of two linear equations in two unknowns:

$$\begin{cases} ax + by = c \\ dx + ey = f \end{cases}$$

If we can find an ordered pair that is a solution to both equations, then we say that ordered pair is a **solution of the system**. Sometimes, to emphasize the fact that a solution must satisfy both equations, we refer to the system as a pair of

TABLE 1

EQUATION IN TWO VARIABLES	IS IT LINEAR?	
	Yes	No
$3x - 8y = 12$	✓	
$-s + 4t = 0$	✓	
$2x - 3y^2 = 1$		✓
$y = 4 - 2x$	✓	
$\dfrac{4}{u} + \dfrac{5}{v} = 3$		✓

simultaneous equations. A system that has at least one solution is said to be **consistent**. If there are no solutions, the system is **inconsistent**.

EXAMPLE 1   Consider the system

$$\begin{cases} x + y = 2 \\ 2x - 3y = 9 \end{cases}$$

(a) Is $(1, 1)$ a solution of the system?
(b) Is $(3, -1)$ a solution of the system?

Solution   **(a)** Although $(1, 1)$ is a solution of the first equation, it is not a solution of the system because it does not satisfy the second equation. (Check this for yourself.)
(b) $(3, -1)$ satisfies the first equation:

$$3 + (-1) \overset{?}{=} 2 \qquad \text{True}$$

$(3, -1)$ satisfies the second equation:

$$2(3) - 3(-1) \overset{?}{=} 9$$
$$6 + 3 \overset{?}{=} 9 \qquad \text{True}$$

Since $(3, -1)$ satisfies both equations, it is a solution of the system.   ▲

We can gain an important perspective on systems of linear equations by looking at Example 1 in graphical terms. Table 2 shows how each of the statements in that example can be rephrased using geometric ideas with which we are already familiar.

TABLE 2

ALGEBRAIC IDEA	CORRESPONDING GEOMETRIC IDEA
1. The ordered pair $(1, 1)$ is a solution of the equation $x + y = 2$.	**1.** The point $(1, 1)$ lies on the line $x + y = 2$. See Figure 1.
2. The ordered pair $(1, 1)$ is not a solution of the equation $2x - 3y = 9$.	**2.** The point $(1, 1)$ does not lie on the line $2x - 3y = 9$. See Figure 1.
3. The ordered pair $(3, -1)$ is a solution of the system $\begin{cases} x + y = 2 \\ 2x - 3y = 9 \end{cases}$	**3.** The point $(3, -1)$ lies on both of the lines $x + y = 2$ and $2x - 3y = 9$. See Figure 1.

In Example 1, we verified that $(3, -1)$ is a solution of the system

$$\begin{cases} x + y = 2 \\ 2x - 3y = 9 \end{cases}$$

Are there any other solutions of this particular system? No; Figure 1 shows us that there are no other solutions, since $(3, -1)$ is clearly the only point common to both lines. In a moment, we'll look at two important methods for solving systems of linear equations in two unknowns. But even before we consider these methods, we can say something about the solutions of linear systems.

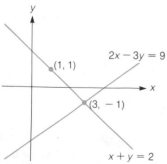

FIGURE 1

PROPERTY SUMMARY	POSSIBILITIES FOR SOLUTIONS OF LINEAR SYSTEMS

Given a system of two linear equations in two unknowns, exactly one of the following cases must occur.

CASE 1   The graphs of the two linear equations intersect in exactly one point. Thus, there is exactly one solution to the system. See Figure 2.

CASE 2   The graphs of the two linear equations are parallel lines. Therefore, the lines do not intersect, and the system has no solution. See Figure 3.

CASE 3   The two equations actually represent the same line. Thus, there are infinitely many points of inter-section and correspondingly infinitely many solutions. See Figure 4.

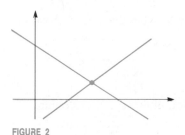

FIGURE 2
A consistent system with exactly one solution

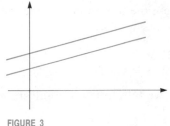

FIGURE 3
An inconsistent system has no solution

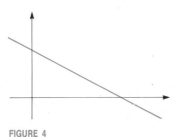

FIGURE 4
A consistent system with infinitely many solutions

We are going to review two methods from basic algebra for solving systems of two linear equations in two unknowns. These methods are the **substitution method** and the **addition–subtraction method**. We'll begin by demonstrating the substitution method. Consider the system

$$\begin{cases} 3x + 2y = 17 & (1) \\ 4x - 5y = -8 & (2) \end{cases}$$

We first choose one of the two equations and then use it to express one of the variables in terms of the other. In the case at hand, neither equation appears particularly simpler than the other, so let's just start with the first equation and solve for x in terms of y. We have

$$3x = 17 - 2y$$
$$x = \tfrac{1}{3}(17 - 2y) \qquad (3)$$

Now we use equation (3) to substitute for x in the equation that we have not used yet, namely, equation (2). This yields

$$4[\tfrac{1}{3}(17 - 2y)] - 5y = -8$$
$$4(17 - 2y) - 15y = -24 \quad \text{multiplying by 3}$$
$$-23y = -92$$
$$y = 4$$

The value $y = 4$ that we have just obtained can be used in equation (3) to find x. Replacing y with 4 in equation (3) yields

$$x = \tfrac{1}{3}[17 - 2(4)] = \tfrac{1}{3}(9) = 3$$

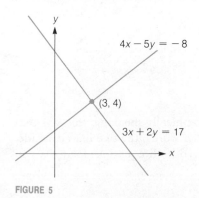

FIGURE 5

We have now found that $x = 3$ and $y = 4$. As you can easily check, this pair of values indeed satisfies both of the original equations. We write our solution as the ordered pair $(3, 4)$. Figure 5 (on the previous page) summarizes the situation. The figure shows that the system is consistent and that $(3, 4)$ is the only solution.

Generally speaking, it is not necessary to graph the equations in a given system in order to decide whether the system is consistent. Rather, this information will emerge as you attempt to follow an algebraic method of solution. Examples 2 and 3 will illustrate this.

EXAMPLE 2   Solve the system

$$\begin{cases} \frac{3}{2}x - 3y = -9 \\ x - 2y = 4 \end{cases}$$

Solution   We use the substitution method. Since it is easy to solve the second equation for x, we begin there:

$$x - 2y = 4$$
$$x = 4 + 2y$$

Now we substitute this result in the first equation of our system to obtain

$$\frac{3}{2}(4 + 2y) - 3y = -9$$
$$6 + 3y - 3y = -9$$
$$6 = -9 \qquad \text{False}$$

Since the substitution process leads us to this obviously false statement, we conclude that the given system has no solution; that is, the system is inconsistent. *Question:* What can you say about the graphs of the two given equations?   ▲

EXAMPLE 3   Solve the system

$$\begin{cases} 3x + 4y = 12 \\ 2y = 6 - \frac{3}{2}x \end{cases}$$

Solution   We use the method of substitution. Since it is easy to solve the second equation for y, we begin there:

$$2y = 6 - \frac{3}{2}x$$
$$y = 3 - \frac{3}{4}x$$

Now we use this result to substitute for y in the first equation of the original system. The result is

$$3x + 4(3 - \frac{3}{4}x) = 12$$
$$3x + 12 - 3x = 12$$
$$3x - 3x = 12 - 12$$
$$0 = 0 \qquad \text{Always true}$$

This last identity imposes no restrictions on x. Graphically speaking, this says that our two lines intersect for every value of x. In other words, the two lines coincide. We could have foreseen this initially had we solved both equations for y. As you can verify, the result in both cases is

$$y = -\frac{3}{4}x + 3$$

Every point on this line yields a solution to our system of equations. In summary then, our system is consistent and the solutions to the system have the form $(x, -\frac{3}{4}x + 3)$, where x can be any real number. For instance, when $x = 0$, we obtain the solution $(0, 3)$. When $x = 1$, we obtain the solution $(1, \frac{9}{4})$. The idea here is that for *each* value of x, we obtain a solution; thus, there are infinitely many solutions. ▲

Now let us turn to the addition–subtraction method of solving systems of equations. By way of example, consider the system

$$\begin{cases} 2x + 3y = 5 \\ 4x - 3y = -1 \end{cases}$$

Notice that if we add these two equations, the result will be an equation involving only the unknown x. Carrying this out gives us

$$6x = 4$$
$$x = \frac{4}{6} = \frac{2}{3}$$

There are now several ways in which the corresponding value of y may be obtained. As you can easily check, substituting the value $x = \frac{2}{3}$ in either of the original equations leads to the result $y = \frac{11}{9}$.

Another way to find y is by multiplying both sides of the first equation by -2. (You'll see why in a moment.) We display the work this way:

$$2x + 3y = 5 \xrightarrow{\text{Multiply by } -2} -4x - 6y = -10$$

$$4x - 3y = -1 \xrightarrow{\text{No change}} 4x - 3y = -1$$

Adding the last two equations then gives us

$$-9y = -11$$
$$y = \frac{11}{9}$$

The required solution is therefore $(\frac{2}{3}, \frac{11}{9})$.

In the previous example, we were able to find x directly by adding the two equations. As the next example shows, it may be necessary first to multiply both sides of each equation by an appropriate constant.

EXAMPLE 4　Solve the system

$$\begin{cases} 5x - 3y = 4 \\ 2x + 4y = 1 \end{cases}$$

Solution　To eliminate x, we could multiply the second equation by $\frac{5}{2}$ and then subtract the resulting equation from the first equation. However, to avoid working with fractions, we proceed as follows.

$$5x - 3y = 4 \xrightarrow{\text{Multiply by 2}} 10x - 6y = 8 \tag{4}$$

$$2x + 4y = 1 \xrightarrow{\text{Multiply by 5}} 10x + 20y = 5 \tag{5}$$

Subtracting equation (5) from equation (4) then yields

$$-26y = 3$$
$$y = -\frac{3}{26}$$

To find x, we return to the original system and work in a similar manner:

$$5x - 3y = 4 \xrightarrow{\text{Multiply by 4}} 20x - 12y = 16$$

$$2x + 4y = 1 \xrightarrow{\text{Multiply by 3}} 6x + 12y = 3$$

Upon adding the last two equations, we obtain

$$26x = 19$$
$$x = \tfrac{19}{26}$$

The solution of the given system of equations is therefore $(\tfrac{19}{26}, -\tfrac{3}{26})$. ▲

We conclude this section with some problems that can be solved using simultaneous equations.

EXAMPLE 5   Determine constants b and c so that the parabola $y = x^2 + bx + c$ passes through the points $(-3, 1)$ and $(1, -2)$.

Solution   Since the point $(-3, 1)$ lies on the curve $y = x^2 + bx + c$, the coordinates must satisfy the equation. Thus, we have

$$1 = (-3)^2 + b(-3) + c$$
$$-8 = -3b + c \tag{6}$$

This gives us one equation in two unknowns. We need another equation involving b and c. Since the point $(1, -2)$ also lies on the graph of $y = x^2 + bx + c$, we must have

$$-2 = 1^2 + b(1) + c$$
$$-3 = b + c \tag{7}$$

Rewriting equations (6) and (7), we have the system

$$\begin{cases} -3b + c = -8 & (8) \\ b + c = -3 & (9) \end{cases}$$

Subtracting equation (9) from (8) then yields

$$-4b = -5$$
$$b = \tfrac{5}{4}$$

One way to obtain the corresponding value of c is to replace b by $\tfrac{5}{4}$ in equation (9). This yields

$$\tfrac{5}{4} + c = -3$$
$$c = -3 - \tfrac{5}{4} = -\tfrac{17}{14}$$

The required values of b and c are therefore

$$b = \tfrac{5}{4}, \qquad c = -\tfrac{17}{4}$$

(Exercise 54 at the end of this section asks you to check that the parabola $y = x^2 + \tfrac{5}{4}x - \tfrac{17}{4}$ indeed passes through the given points.) ▲

EXAMPLE 6   In Figure 6, find the area of the triangular region bounded by the lines $y = 3x$, $y = -\tfrac{1}{2}x + 7$, and the x-axis.

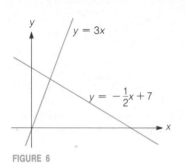

FIGURE 6

Solution

To compute the area of the triangle, we will need to know the base and the height. To determine the base, we first compute the x-intercept of the line $y = -\frac{1}{2}x + 7$:

$$0 = -\frac{1}{2}x + 7$$
$$\frac{1}{2}x = 7$$
$$x = 14$$

The height of the triangle is the y-coordinate of the intersection point of the lines $y = 3x$ and $y = -\frac{1}{2}x + 7$ (see Figure 6). The substitution method can be used to solve this pair of simultaneous equations for y. From the equation $y = 3x$, we obtain $x = \frac{1}{3}y$. Then substituting this in the equation $y = -\frac{1}{2}x + 7$ yields

$$y = -\frac{1}{2}(\tfrac{1}{3}y) + 7$$
$$6y = -y + 42$$
$$7y = 42$$
$$y = 6$$

Now that we know that the base of the triangle is 14 and the height is 6, we can compute the required area:

$$A = \tfrac{1}{2}bh$$
$$= \tfrac{1}{2}(14)(6) = 42 \text{ square units}$$

In the next example, we solve a mixture problem using a system of two equations in two unknowns.

EXAMPLE 7

Suppose that a chemistry student can obtain two acid solutions from the stockroom. The first solution is 20% acid and the second solution is 45% acid. (The percentages are by volume.) How many milliliters of each solution should the student mix together to obtain 100 ml of a 30% acid solution?

Solution

We begin by assigning letters to denote the required quantities.

Let x denote the number of milliliters of the 20% solution to be used.

Let y denote the number of milliliters of the 45% solution to be used.

Then the data can be summarized as in Table 3. Since the final mixture must total 100 ml, we have the equation

$$x + y = 100 \tag{10}$$

This gives us one equation in two unknowns. However, we need a second equation. Looking at the data in the right-hand column of Table 3, we can write

$$\underbrace{0.20x}_{\substack{\text{Amount of acid} \\ \text{in } x \text{ ml of the} \\ 20\% \text{ solution}}} + \underbrace{0.45y}_{\substack{\text{Amount of acid} \\ \text{in } y \text{ ml of the} \\ 45\% \text{ solution}}} = \underbrace{(0.30)(x+y)}_{\substack{\text{Amount of acid in} \\ \text{the final mixture}}}$$

TABLE 3

TYPE OF SOLUTION	NUMBER OF ML	PERCENT OF ACID	TOTAL ACID (ML)
First solution (20% acid)	x	20	$(0.20)x$
Second solution (45% acid)	y	45	$(0.45)y$
Mixture	$x + y$	30	$(0.30)(x+y)$

Thus,

$$0.20x + 0.45y = 0.30(x + y)$$
$$20x + 45y = 30(x + y)$$
$$4x + 9y = 6(x + y) = 6x + 6y$$
$$-2x + 3y = 0 \qquad (11)$$

Equations (10) and (11) can be solved by either the substitution method or the addition method. As Exercise 55 at the end of this section asks you to show, the results are

$$x = 60 \text{ ml} \qquad \text{and} \qquad y = 40 \text{ ml} \qquad \blacktriangle$$

▼ EXERCISE SET 7.1

A

1. Which of the following are linear equations in two variables?
   (a) $3x + 3y = 10$ (b) $2x + 4xy + 3y = 1$
   (c) $u - v = 1$ (d) $x = 2y + 6$

2. Which of the following are linear equations in two variables?
   (a) $y = x$ (b) $y = x^2$
   (c) $\dfrac{4}{x} - \dfrac{3}{y} = -1$ (d) $2w + 8z = -4w + 3$

3. Is $(5, 1)$ a solution of the following system?
$$\begin{cases} 2x - 8y = 2 \\ 3x + 7y = 22 \end{cases}$$

4. Is $(14, -2)$ a solution of the following system?
$$\begin{cases} x + y = 12 \\ x - y = 4 \end{cases}$$

5. Is $(0, -4)$ a solution of the following system?
$$\begin{cases} \frac{1}{6}x + \frac{1}{2}y = -2 \\ \frac{2}{3}x + \frac{3}{4}y = 2 \end{cases}$$

6. Is $(12, -8)$ a solution of the system in Exercise 5?

7. Is $(3, -2)$ a solution of the following system?
$$\begin{cases} \frac{2}{7}x - \frac{1}{5}y = \frac{44}{35} \\ \frac{1}{3}x - \frac{5}{4}y = \frac{7}{2} \end{cases}$$

In Exercises 8–18, use the substitution method to find all solutions of each system.

8. $\begin{cases} 4x - y = 7 \\ -2x + 3y = 9 \end{cases}$

9. $\begin{cases} 3x - 2y = -19 \\ x + 4y = -4 \end{cases}$

10. $\begin{cases} 6x - 2y = -3 \\ 5x + 3y = 4 \end{cases}$

11. $\begin{cases} 4x + 2y = 3 \\ 10x + 4y = 1 \end{cases}$

12. $\begin{cases} \frac{3}{2}x - 5y = 1 \\ x + \frac{3}{4}y = -1 \end{cases}$

13. $\begin{cases} 13x - 8y = -3 \\ -7x + 2y = 0 \end{cases}$

14. $\begin{cases} 4x + 6y = 3 \\ -6x - 9y = -\frac{9}{2} \end{cases}$

15. $\begin{cases} -\frac{2}{5}x + \frac{1}{4}y = 3 \\ \frac{1}{4}x - \frac{2}{5}y = -3 \end{cases}$

16. $\begin{cases} 0.02x - 0.03y = 1.06 \\ 0.75x + 0.50y = -0.01 \end{cases}$

17. $\begin{cases} \sqrt{2}x - \sqrt{3}y = \sqrt{3} \\ \sqrt{3}x - \sqrt{8}y = \sqrt{2} \end{cases}$

18. $\begin{cases} 7x - 3y = -12 \\ \frac{14}{3}x - 2y = 2 \end{cases}$

In Exercises 19–28, use the addition–subtraction method to find all solutions of each system of equations.

19. $\begin{cases} 5x + 6y = 4 \\ 2x - 3y = -3 \end{cases}$

20. $\begin{cases} -8x + y = -2 \\ 4x - 3y = 1 \end{cases}$

21. $\begin{cases} 4x + 13y = -5 \\ 2x - 54y = -1 \end{cases}$

22. $\begin{cases} 16x - 3y = 100 \\ 16x + 10y = 10 \end{cases}$

23. $\begin{cases} \frac{1}{4}x - \frac{1}{3}y = 4 \\ \frac{2}{7}x - \frac{1}{7}y = \frac{1}{10} \end{cases}$

Suggestion: First clear both equations of fractions.

24. $\begin{cases} 2.1x - 3.5y = 1.2 \\ 1.4x + 2.6y = 1.1 \end{cases}$

25. $\begin{cases} 8x + 16y = 5 \\ 2x + 5y = \frac{5}{4} \end{cases}$

26. $\begin{cases} 8x + 16y = 5 \\ 2x + 4y = 1 \end{cases}$

27. $\begin{cases} 125x - 40y = 45 \\ \frac{1}{10}x + \frac{1}{10}y = \frac{3}{10} \end{cases}$

28. $\begin{cases} \sqrt{6}x - \sqrt{3}y = 3\sqrt{2} - \sqrt{3} \\ \sqrt{2}x - \sqrt{5}y = \sqrt{6} + \sqrt{5} \end{cases}$

29. Find b and c, given that the parabola $y = x^2 + bx + c$ passes through $(0, 4)$ and $(2, 14)$.

30. Determine the constants a and b, given that the parabola $y = ax^2 + bx + 1$ passes through $(-1, 11)$ and $(3, 1)$.

31. Determine the constants A and B, given that the line $Ax + By = 2$ passes through the points $(-4, 5)$ and $(7, -9)$.

32. Find the area of the triangular region in the first quadrant bounded by the lines $y = 5x$ and $y = -3x + 6$ and the x-axis.

33. Find the area of the triangular region in the first quadrant bounded by the x-axis and the lines $y = 2x - 5$ and $y = -\frac{1}{2}x + 3$.

34. Find the area of the triangular region in the first quadrant bounded by the y-axis and the lines $y = 2x + 2$ and $y = -\frac{3}{2}x + 9$.

35. A student in a chemistry laboratory has access to two acid solutions. The first solution is 10% acid and the second is 35% acid. (The percentages are by volume.) How many cubic centimeters of each should she mix together to obtain 200 cm³ of a 25% acid solution?

36. One salt solution is 15% salt and another is 20% salt. How many cubic centimeters of each solution must be mixed to obtain 50 cm³ of a 16% salt solution?

37. A shopkeeper has two types of coffee beans on hand. One type sells for \$5.20/lb, the other for \$5.80/lb. How many pounds of each type must be mixed to produce 16 lb of a blend that sells for \$5.50/lb?

38. A certain alloy contains 10% tin and 30% copper. (The percentages are by weight.) How many pounds of tin and how many pounds of copper must be melted with 1000 lb of the given alloy to yield a new alloy containing 20% tin and 35% copper?

B

39. Find x and y in terms of a and b:

$$\begin{cases} \dfrac{x}{a} + \dfrac{y}{b} = 1 \\ \dfrac{x}{b} + \dfrac{y}{a} = 1 \end{cases}$$

Does your solution impose any conditions on a and b?

40. Solve the following system for x and y in terms of a and b, where $a \neq b$:

$$\begin{cases} ax + by = \dfrac{1}{a} \\ b^2x + a^2y = 1 \end{cases}$$

41. Solve the following system for x and y in terms of a and b, where $a \neq b$:

$$\begin{cases} ax + a^2y = 1 \\ bx + b^2y = 1 \end{cases}$$

Does your solution impose any additional conditions on a and b?

42. Solve the following system for s and t:

$$\begin{cases} \dfrac{3}{s} - \dfrac{4}{t} = 2 \\ \dfrac{5}{s} + \dfrac{1}{t} = -3 \end{cases}$$

Hint: Make the substitutions $1/s = x$ and $1/t = y$ in order to obtain a system of two linear equations.

43. Solve the following system for s and t:

$$\begin{cases} \dfrac{1}{2s} - \dfrac{1}{2t} = -10 \\ \dfrac{2}{s} + \dfrac{3}{t} = 5 \end{cases}$$

(Use the hint in Exercise 42.)

In Exercises 44–49, find all solutions of the given systems. For Exercises 47–49, round off the final answers to two decimal places.

44. $\begin{cases} 0.5x - 0.8y = 0.3 \\ 0.4x - 0.1y = 0.9 \end{cases}$

45. $\begin{cases} \dfrac{2w-1}{3} + \dfrac{z+2}{4} = 4 \\ \dfrac{w+3}{2} - \dfrac{w-z}{3} = 3 \end{cases}$

46. $\dfrac{x-y}{2} = \dfrac{x+y}{3} = 1$

47. $\begin{cases} 1.03x - 2.54y = 5.47 \\ 3.85x + 4.29y = -1.84 \end{cases}$

48. $\begin{cases} 2.39x + 8.16y = -2.83 \\ 1.01x + 2.98y = 4.41 \end{cases}$

49. $\begin{cases} \sqrt{5}x - 4\sqrt{3}y = 6 \\ \sqrt{2}x + 3\sqrt{3}y = 8 \end{cases}$

50. The sum of two numbers is 64. Twice the larger number plus five times the smaller number is 20. Find the two numbers. (Let x denote the larger number and let y denote the smaller number.)

51. In a two-digit number, the sum of the digits is 14. Twice the tens digit exceeds the units digit by one. Find the number.

52. The sum of the digits in a two-digit number is 14. Furthermore, the number itself is 2 greater than 11 times the tens digit. Find the number.

53. The perimeter of a rectangle is 34 in. If the length is 2 in. more than twice the width, find the length and the width.

54. Verify that the parabola $y = x^2 + \frac{5}{4}x - \frac{17}{4}$ indeed passes through $(-3, 1)$ and $(1, -2)$.

55. Consider the following system from Example 7.

$$\begin{cases} x + y = 100 \\ -2x + 3y = 0 \end{cases}$$

(a) Solve this system using the method of substitution.

Answer: $(60, 40)$

(b) Solve the system using the addition–subtraction method.

56. Consider the following system:

$$\begin{cases} 2x - 5y = -2 \\ 3x + 4y = 5 \end{cases}$$

(a) Solve the system using the substitution method.
(b) Solve the system using the addition–subtraction method. *Answer:* $\left(\frac{17}{23}, \frac{16}{23}\right)$
(c) Verify that your solution indeed satisfies both equations of the system.

57. Solve the following system for x and y:

$$\begin{cases} by = x + ab \\ cy = x + ac \end{cases}$$

(Assume a, b, and c are constants and that $b \neq c$.)

58. Solve for x and y in terms of a, b, c, d, e, and f:

$$\begin{cases} ax + by = c \\ dx + ey = f \end{cases}$$

(Assume that $ae - bd \neq 0$.)

In Exercises 59 and 60, solve the system of equations. Give the exact values for the unknowns as well as calculator approximations rounded to two decimal places.

59. $\begin{cases} 3 \ln s + 4 \ln t = 1 \\ 5 \ln s + 3 \ln t = 9 \end{cases}$

Hint: Use the substitutions $x = \ln s$ and $y = \ln t$ to obtain a linear system.

60. $\begin{cases} e^s + e^t = 6 \\ 2e^s - 3e^t = 7 \end{cases}$ (Adapt the hint in Exercise 59.)

61. Solve the following system for x and y in terms of a and b, where $a \neq b$:

$$\begin{cases} \dfrac{a}{bx} + \dfrac{b}{ay} = a + b \\ \dfrac{b}{x} + \dfrac{a}{y} = a^2 + b^2 \end{cases}$$

C

62. Solve the following system for x and y in terms of a and b, where $ab \neq -1$:

$$\begin{cases} \dfrac{x + y - 1}{x - y + 1} = a \\ \dfrac{y - x + 1}{x - y + 1} = ab \end{cases}$$

Answer: $\left(\dfrac{a + 1}{ab + 1}, \dfrac{a(b + 1)}{ab + 1} \right)$

63. Given that the lines $7x + 5y = 4$, $x + ky = 3$, and $5x + y + k = 0$ are concurrent (pass through a common point), what are the possible values for k?

64. Solve the following system for x and y in terms of a and b:

$$\begin{cases} (a + b)x + (a^2 + b^2)y = a^3 + b^3 \\ (a - b)x + (a^2 - b^2)y = a^3 - b^3 \end{cases}$$

(Assume that a and b are nonzero and that $a \neq \pm b$.)

65. The vertices of triangle ABC are $A(2a, 0)$, $B(2b, 0)$, and $C(0, 2)$.

(a) Show that the equations of the sides are $x + by = 2b$, $x + ay = 2a$, and $y = 0$.
(b) Show that the equations of the medians are $x + (2a - b)y = 2a$, $x + (2b - a)y = 2b$, and $2x + (a + b)y = 2(a + b)$.
(c) Show that all three medians intersect at the point $\left(\frac{2}{3}(a + b), \frac{2}{3}\right)$.
(d) Show that the equations of the altitudes are $bx - y = 2ab$, $ax - y = 2ab$, and $x = 0$.
(e) Show that all three altitudes intersect at the point $(0, -2ab)$.
(f) Show that the equations of the perpendicular bisectors of the sides are $x = a + b$, $bx - y = b^2 - 1$, and $ax - y = a^2 - 1$.
(g) Show that all three perpendicular bisectors intersect at the point $(a + b, ab + 1)$.
(h) Show that the three points found in parts (c), (e), and (g) all lie on a line. This is the **Euler line** of the triangle.

7.2 ▼ GAUSSIAN ELIMINATION

A method of solution is perfect if we can foresee from the start, and even prove, that following that method we shall obtain our aim.

Gottfried Wilhelm von Leibniz (1646–1716)

In the previous section, we solved systems of linear equations in two unknowns. In this section, we introduce the technique known as Gaussian elimination for solving systems of linear equations in which there are more than two unknowns.*

———————

* The technique is named in honor of Carl Friedrich Gauss (1777–1885), however, the technique itself was in existence before Gauss's time. Indeed, the essentials of the method appear in a Chinese text, "Nine Chapters on the Mathematical Art" (*Chiu Chang Suan Shu*), written approximately 2000 years ago.

As a first example, consider the following system of three linear equations in the three unknowns x, y, and z:

$$\begin{cases} 3x + 2y - z = -3 \\ 5y - 2z = 2 \\ 5z = 20 \end{cases}$$

This system is easy to solve using the process of *back-substitution*. Dividing both sides of the third equation by 5 yields $z = 4$. Then substituting $z = 4$ back into the second equation gives us

$$5y - 2(4) = 2$$
$$5y = 10$$
$$y = 2$$

Finally, substituting the values $z = 4$ and $y = 2$ back into the first equation yields

$$3x + 2(2) - 4 = -3$$
$$3x = -3$$
$$x = -1$$

We have now found that $x = -1$, $y = 2$, and $z = 4$. If you go back and check, you will find that these values indeed satisfy all three equations in the given system. Furthermore, the algebra we've just carried out shows that these are the only possible values for x, y, and z satisfying all three equations. We summarize by saying that the **ordered triple** $(-1, 2, 4)$ is the solution of the given system.

The system that we just considered was easy to solve (using back-substitution) because of the special form in which it was written. This form is called **upper-triangular form**. Although the following definition of upper-triangular form refers to systems with three unknowns, the same type of definition can be given for systems with any number of unknowns. Table 1 displays examples of systems in upper-triangular form.

Upper-Triangular Form (Three Variables)

A system of linear equations in x, y, and z is said to be in **upper-triangular** form provided x appears in no equation after the first and y appears in no equation after the second. (It is possible that y may not even appear in the second equation.)

TABLE 1
Examples of Systems in Upper-Triangular Form

2 UNKNOWNS: x, y	3 UNKNOWNS: x, y, z	4 UNKNOWNS: x, y, z, t
$\begin{cases} 3x + 5y = 7 \\ 8y = 5 \end{cases}$	$\begin{cases} 4x - 3y + 2z = -5 \\ 7y + z = 9 \\ -4z = 3 \end{cases}$	$\begin{cases} x - y + z - 4t = 1 \\ 3y - 2z + t = -1 \\ 3z - 5t = 4 \\ 6t = 7 \end{cases}$
	$\begin{cases} 15x - 2y + z = 1 \\ 3z = -8 \end{cases}$	$\begin{cases} 2x + y + 2z - t = -3 \\ 4z + 3t = 1 \\ 5t = 6 \end{cases}$
		$\begin{cases} 8x + 3y - z + t = 2 \\ 2y + z - 4t = 1 \end{cases}$

When we were solving linear systems of equations with two unknowns in the previous section, we observed that there were three possibilities: a unique solution; infinitely many solutions; and no solution. As the next three examples indicate, the situation is similar when dealing with larger systems.

EXAMPLE 1   Find all solutions of the system

$$\begin{cases} x + y + 2z = 2 \\ \quad\quad 3y - 4z = -5 \\ \quad\quad\quad\quad 6z = 3 \end{cases}$$

Solution   The system is in upper-triangular form, and we can use back-substitution. Dividing the third equation by 6 yields $z = \frac{1}{2}$. Substituting this value for z back into the second equation then yields

$$3y - 4(\tfrac{1}{2}) = -5$$
$$3y = -3$$
$$y = -1$$

Now, substituting the values $z = \frac{1}{2}$ and $y = -1$ back in the first equation, we obtain

$$x + (-1) + 2(\tfrac{1}{2}) = 2$$
$$x = 2$$

As you can easily check, the values $x = 2$, $y = -1$, and $z = \frac{1}{2}$ indeed satisfy all three equations. Furthermore, the algebra we've just carried out shows that these are the only possible values for $x, y,$ and z satisfying all three equations. We summarize by saying that the unique solution to our system is the ordered triple $(2, -1, \frac{1}{2})$.   ▲

EXAMPLE 2   Find all solutions of the system

$$\begin{cases} -2x + y + 3z = 6 \\ \quad\quad\quad\quad 2z = 10 \end{cases}$$

Solution   Again, the system is in upper-triangular form, and we use back-substitution. Solving the second equation for z yields $z = 5$. Then replacing z by 5 in the first equation gives us

$$-2x + y + 3(5) = 6$$
$$-2x + y = -9$$
$$y = 2x - 9$$

At this point, we've made use of both equations in the given system. There is no third equation to provide additional restrictions on $x, y,$ or z. We know from the previous section that the equation $y = 2x - 9$ has infinitely many solutions, all of the form

$(x, 2x - 9)$ where x is a real number

It follows, then, that there are infinitely many solutions to the given system and they may be written

$(x, 2x - 9, 5)$ where x is a real number

For instance, choosing in succession $x = 0$, $x = 1$, and $x = 2$ yields the solutions $(0, -9, 5)$, $(1, -7, 5)$, and $(2, -5, 5)$. (We remark in passing that any linear system in upper-triangular form in which the number of unknowns exceeds the number of equations will always have infinitely many solutions.) ▲

EXAMPLE 3   Find all solutions of the system

$$\begin{cases} 4x - 7y + 3z = 1 \\ 3x + y - 2z = 4 \\ 4x - 7y + 3z = 6 \end{cases}$$

(Note that the system is not in upper-triangular form.)

Solution   Look at the left-hand sides of the first and third equations: they are identical. Thus, if there were values for x, y, and z that satisfied both equations, it would follow that $1 = 6$, which is clearly impossible. We conclude that the given system has no solutions. ▲

As Examples 1 and 2 have demonstrated, systems in upper-triangular form can be readily solved. In view of this, it would be useful to have a technique for converting a given system into an equivalent system in upper-triangular form. (The expression **equivalent system** means a system with exactly the same set of solutions as the original system.) **Gaussian elimination** is one such technique. We will demonstrate this technique in Examples 4, 5, and 6. In using Gaussian elimination, we will rely on what are called the three **elementary operations**, listed in the box that follows. These are operations that can be performed on an equation in a system without altering the set of solutions of the system.

The Elementary Operations

1. Multiply both sides of an equation by a nonzero constant.
2. Interchange the order in which two equations of a system are listed.
3. To one equation add a multiple of another equation in the system.

EXAMPLE 4   Find all solutions of the system

$$\begin{cases} x + 2y + z = 3 \\ 2x + y + z = 16 \\ x + y + 2z = 9 \end{cases}$$

Solution   First we want to eliminate x from the second and third equations. To eliminate x from the second equation, we add to it -2 times the first equation. The result is the equivalent system

$$\begin{cases} x + 2y + z = 3 \\ -3y - z = 10 \\ x + y + 2z = 9 \end{cases}$$

To eliminate x from the third equation, we add to it -1 times the first equation. The result is the equivalent system

$$\begin{cases} x + 2y + z = 3 \\ -3y - z = 10 \\ -y + z = 6 \end{cases}$$

Now to bring the system into upper-triangular form, we need to eliminate y from the third equation. We could do this by adding $-\frac{1}{3}$ times the second equation to the third equation. However, to avoid working with fractions as long as possible, we proceed instead to interchange the second and third equations to obtain the equivalent system

$$\begin{cases} x + 2y + z = 3 \\ -y + z = 6 \\ -3y - z = 10 \end{cases}$$

Now we add -3 times the second equation to the last equation to obtain the equivalent system

$$\begin{cases} x + 2y + z = 3 \\ -y + z = 6 \\ -4z = -8 \end{cases}$$

The system is now in upper-triangular form, and back-substitution yields, in turn, $z = 2$, $y = -4$, and $x = 9$. (Check this for yourself.) The required solution is therefore $(9, -4, 2)$. ▲

In Table 2 we list some convenient abbreviations that are used in describing the elementary operations. You should plan on using these abbreviations yourself: they'll make it simpler for you (and your instructor) to check your work. In Table 2, the notation E_i stands for the ith equation in a system. For instance, for the initial system in Example 4, the symbol E_1 denotes the first equation: $x + 2y + z = 3$.

TABLE 2
Abbreviations for the Elementary Operations

ABBREVIATION	EXPLANATION
1. cE_i	Multiply both sides of the ith equation by c.
2. $E_i \leftrightarrow E_j$	Interchange the ith and jth equations.
3. $cE_i + E_j$	To the jth equation, add c times the ith equation.

EXAMPLE 5 Solve the system

$$\begin{cases} 4x - 3y + 2z = 40 \\ 5x + 9y - 7z = 47 \\ 9x + 8y - 3z = 97 \end{cases}$$

Solution

$$\begin{cases} 4x - 3y + 2z = 40 \\ 5x + 9y - 7z = 47 \\ 9x + 8y - 3z = 97 \end{cases} \xrightarrow{(-1)E_2 + E_1} \begin{cases} -x - 12y + 9z = -7 \\ 5x + 9y - 7z = 47 \\ 9x + 8y - 3z = 97 \end{cases}$$

adding -1 times the second equation to the first equation

$$\xrightarrow[9E_1 + E_3]{5E_1 + E_2} \begin{cases} -x - 12y + 9z = -7 \\ -51y + 38z = 12 \\ -100y + 78z = 34 \end{cases}$$

$$\xrightarrow{\frac{1}{2}E_3} \begin{cases} -x - 12y + 9z = -7 \\ -51y + 38z = 12 \\ -50y + 39z = 17 \end{cases}$$

$$\xrightarrow{(-1)E_3 + E_2} \begin{cases} -x - 12y + 9z = -7 \\ -y - z = -5 \\ -50y + 39z = 17 \end{cases}$$

to allow working with smaller but integral coefficients

$$\xrightarrow{-50E_2 + E_3} \begin{cases} -x - 12y + 9z = -7 \\ -y - z = -5 \\ 89z = 267 \end{cases}$$

The system is now in upper-triangular form. Solving the third equation, we obtain $z = 3$. Substituting this value back into the second equation yields $y = 2$. (Check this for yourself.) Finally, substituting $z = 3$ and $y = 2$ back into the first equation yields $x = 10$. (Again, check this for yourself.) The solution to the system is therefore $(10, 2, 3)$. ▲

EXAMPLE 6 Solve the system

$$\begin{cases} x + 2y + 4z = 0 \\ x + 3y + 9z = 0 \end{cases}$$

Solution This system is similar to the one in Example 2 in that there are fewer equations than there are unknowns. By subtracting the first equation from the second, we readily obtain an equivalent system in upper-triangular form:

$$\begin{cases} x + 2y + 4z = 0 \\ y + 5z = 0 \end{cases}$$

Although the system is now in upper-triangular form, notice that the second equation does not determine y or z uniquely; that is, there are infinitely many number pairs (y, z) satisfying the second equation. We can solve the second equation for y in terms of z; the result is $y = -5z$. Now we replace y with $-5z$ in the first equation to obtain

$$x + 2(-5z) + 4z = 0 \quad \text{or} \quad x = 6z$$

At this point, we've used both of the equations in the system to express x and y in terms of z. Furthermore, there is no third equation in the system to provide additional restrictions on $x, y,$ or z. We therefore conclude that the given system has infinitely many solutions. These solutions have the form

$$(6z, -5z, z) \qquad \text{where } z \text{ is any real number} \qquad \blacktriangle$$

EXAMPLE 7 Solve the system

$$\begin{cases} x - 4y + z = 3 \\ 3x + 5y - 2z = -1 \\ 7x + 6y - 3z = 2 \end{cases}$$

Solution

$$\begin{cases} x - 4y + z = 3 \\ 3x + 5y - 2z = -1 \\ 7x + 6y - 3z = 2 \end{cases} \quad \xrightarrow[\substack{-3E_1 + E_2 \\ -7E_1 + E_3}]{} \quad \begin{cases} x - 4y + z = 3 \\ 17y - 5z = -10 \\ 34y - 10z = -19 \end{cases}$$

$$\xrightarrow[\quad -2E_2 + E_3 \quad]{} \quad \begin{cases} x - 4y + z = 3 \\ 17y - 5z = -10 \\ 0 = 1 \end{cases}$$

From the third equation in this last system, we conclude that this system, and consequently the original system, has no solution. (Reason: If there *were* values for $x, y,$ and z satisfying the original system, then it would follow that $0 = 1$, which is clearly impossible.) $\blacktriangle$

In the next two examples, we introduce the subject of **partial fractions**. The basic goal here is to write a given fractional expression as a sum or difference of two or more *simpler* fractions. For instance, in Example 8, we will be given the fraction $1/(x - 1)(x + 1)$ and we will be asked to find constants A and B so that

$$\frac{1}{(x - 1)(x + 1)} = \frac{A}{x - 1} + \frac{B}{x + 1}$$

When A and B are determined, we refer to the right-hand side of the preceding equation as the **partial fraction decomposition** of the given fraction. Detailed techniques for working with partial fractions are developed as the need arises in calculus courses. The two examples and the exercises that follow will provide you with ample background for that later work.

EXAMPLE 8 Determine constants A and B so that the following equation is an identity (i.e., the equation is to hold for all values of x for which the denominators are not zero):

$$\frac{1}{(x - 1)(x + 1)} = \frac{A}{x - 1} + \frac{B}{x + 1} \tag{1}$$

Solution First, to clear equation (1) of fractions, we multiply both sides of the equation by the quantity $(x - 1)(x + 1)$. This yields

$$1 = (x + 1)A + (x - 1)B = Ax + A + Bx - B$$
$$1 = (A + B)x + (A - B) \tag{2}$$

Since this last equation is to be an identity, we now reason as follows:

$$\begin{bmatrix} \text{coefficient of } x \text{ on the left-} \\ \text{hand side of equation (2)} \end{bmatrix} = \begin{bmatrix} \text{coefficient of } x \text{ on the right-} \\ \text{hand side of equation (2)} \end{bmatrix}$$

Therefore,

$$0 = A + B \tag{3}$$

Similarly, we have

$$\begin{bmatrix} \text{constant term on the left-} \\ \text{hand side of equation (2)} \end{bmatrix} = \begin{bmatrix} \text{constant term on the right-} \\ \text{hand side of equation (2)} \end{bmatrix}$$

Therefore,

$$1 = A - B \tag{4}$$

Rewriting equations (3) and (4) together gives us a system of two linear equations in the unknowns A and B:

$$\begin{cases} A + B = 0 \\ A - B = 1 \end{cases}$$

Adding these two equations yields

$$2A = 1$$
$$A = \tfrac{1}{2}$$

Subtracting the two equations gives us

$$B - (-B) = 0 - 1$$
$$2B = -1$$
$$B = -\tfrac{1}{2}$$

We have now found that $A = \tfrac{1}{2}$ and $B = -\tfrac{1}{2}$, as required. The partial fraction decomposition of $1/(x - 1)(x + 1)$ is therefore

$$\frac{1}{(x-1)(x+1)} = \frac{\tfrac{1}{2}}{x-1} + \frac{-\tfrac{1}{2}}{x+1}$$

$$= \frac{1}{2(x-1)} - \frac{1}{2(x+1)}$$

You should check for yourself that the result of combining the two fractions on the right-hand side of this last equation is indeed $1/(x - 1)(x + 1)$. ▲

In Example 8, we were able to find the partial fraction decomposition by solving a system of two linear equations. There is a shortcut that is sometimes useful in such problems. As before, we multiply both sides of equation (1) by $(x - 1)(x + 1)$ to obtain

$$1 = (x + 1)A + (x - 1)B \tag{5}$$

Now, equation (5) is to hold for all values of x; in particular, it must hold when x is -1. Replacing x with -1 in the equation then yields

$$1 = -2B$$
$$B = -\tfrac{1}{2} \quad \text{as obtained previously}$$

The value of A can be obtained in a similar way. We replace x by 1 in equation (5). As you can check, this readily yields $A = \tfrac{1}{2}$, which again agrees with the result we obtained previously.

EXAMPLE 9   Determine the constants A, B, and C so that the following equation is an identity:

$$\frac{2x + 1}{(x + 2)(x - 3)^2} = \frac{A}{x + 2} + \frac{B}{(x - 3)} + \frac{C}{(x - 3)^2}$$

Solution   First, to clear the given equation of fractions, we multiply both sides by the quantity $(x + 2)(x - 3)^2$. After dividing out the common factors, we have

$$\begin{aligned} 2x + 1 &= A(x - 3)^2 + B(x + 2)(x - 3) + C(x + 2) \\ &= A(x^2 - 6x + 9) + B(x^2 - x - 6) + C(x + 2) \\ &= Ax^2 - 6Ax + 9A + Bx^2 - Bx - 6B + Cx + 2C \\ 2x + 1 &= (A + B)x^2 + (-6A - B + C)x + 9A - 6B + 2C \end{aligned}$$

Equating the coefficients of x^2 on both sides of this identity yields

$$A + B = 0$$

Similarly, equating the x-coefficients on both sides yields

$$-6A - B + C = 2$$

Finally, equating the constant terms on both sides gives us

$$9A - 6B + 2C = 1$$

We now have a system of three equations in the three unknowns A, B, and C:

$$\begin{cases} A + B & = 0 \\ -6A - B + C = 2 \\ 9A - 6B + 2C = 1 \end{cases}$$

Exercise 12 at the end of this section asks you to verify that the required values here are $A = -\frac{3}{25}$, $B = \frac{3}{25}$, and $C = \frac{7}{5}$. The partial fraction decomposition is therefore

$$\frac{2x + 1}{(x + 2)(x - 3)^2} = \frac{-3}{25(x + 2)} + \frac{3}{25(x - 3)} + \frac{7}{5(x - 3)^2}$$ ▲

Here is an alternate method for working Example 9, which relies on the shortcut mentioned earlier. Again, we first multiply the given equation by $(x + 2)(x - 3)^2$ to obtain

$$2x + 1 = A(x - 3)^2 + B(x + 2)(x - 3) + C(x + 2) \qquad (6)$$

Now we just replace x by 3 in this equation. This yields

$$7 = 5C \quad \text{or} \quad C = \tfrac{7}{5} \quad \text{as obtained previously}$$

The value of A can be obtained in a similar manner. We replace x by -2 in equation (6). As you can check, this readily yields $A = -\frac{3}{25}$, which again agrees with the value obtained previously. We've now computed A and C using the short method. But as you'll see if you try, the value of B cannot be obtained in the same way. One way to find B here is by substituting the values found for A and C in equation (6) to obtain

$$2x + 1 = -\tfrac{3}{25}(x - 3)^2 + B(x + 2)(x - 3) + \tfrac{7}{5}(x + 2)$$

This equation is to hold for all values of x; in particular, it must hold for the convenient value $x = 0$. Substituting $x = 0$ in this last equation yields

$$1 = -\tfrac{3}{25}(-3)^2 + B(2)(-3) + \tfrac{7}{5}(2)$$

$$25 = -27 - 150B + 70 \quad \text{multiplying by 25}$$

$$150B = 18 \quad\quad\quad\quad\quad \text{collecting like terms}$$

$$B = \tfrac{18}{150} = \tfrac{3}{25} \quad\quad\quad \text{dividing by 150 and simplifying}$$

We have now found that $B = \tfrac{3}{25}$, in agreement with the value obtained previously.

▼ EXERCISE SET 7.2

A

In Exercises 1–10, the systems of linear equations are in upper-triangular form. Find all solutions of each system.

1. $\begin{cases} 2x + y + z = -9 \\ 3y - 2z = -4 \\ 8z = -8 \end{cases}$

2. $\begin{cases} -3x + 7y + 2z = -19 \\ y + z = 1 \\ -2z = -2 \end{cases}$

3. $\begin{cases} 8x + 5y + 3z = 1 \\ 3y + 4z = 2 \\ 5z = 3 \end{cases}$

4. $\begin{cases} 2x + 7z = -4 \\ 5y - 3z = 6 \\ 6z = 18 \end{cases}$

5. $\begin{cases} -4x + 5y = 0 \\ 3y + 2z = 1 \\ 3z = -1 \end{cases}$

6. $\begin{cases} 3x - 2y + z = 4 \\ 3z = 9 \end{cases}$

7. $\begin{cases} -x + 8y + 3z = 0 \\ 2z = 0 \end{cases}$

8. $\begin{cases} -x + y + z + w = 9 \\ 2y - z - w = 9 \\ 3z + 2w = 1 \\ 11w = 22 \end{cases}$

9. $\begin{cases} 2x + 3y + z + w = -6 \\ y + 3z - 4w = 23 \\ 6z - 5w = 31 \\ - 2w = 10 \end{cases}$

10. $\begin{cases} 7x - y - z + w = 3 \\ 2y - 3z - 4w = -2 \\ 3w = 6 \end{cases}$

In Exercises 11–30, find all solutions of each system.

11. $\begin{cases} x + y + z = 12 \\ 2x - y - z = -1 \\ 3x + 2y + z = 22 \end{cases}$

12. $\begin{cases} A + B = 0 \\ -6A - B + C = 2 \\ 9A - 6B + 2C = 1 \end{cases}$
Answer: $(-\tfrac{3}{25}, \tfrac{3}{25}, \tfrac{7}{5})$

13. $\begin{cases} 2x - 3y + 2z = 4 \\ 4x + 2y + 3z = 7 \\ 5x + 4y + 2z = 7 \end{cases}$

14. $\begin{cases} x + 2z = 5 \\ y - 30z = -16 \\ x - 2y + 4z = 8 \end{cases}$

15. $\begin{cases} 3x + 3y - 2z = 13 \\ 6x + 2y - 5z = 13 \\ 7x + 5y - 3z = 26 \end{cases}$

16. $\begin{cases} 2x + 5y - 3z = 4 \\ 4x - 3y + 2z = 9 \\ 5x + 6y - 2z = 18 \end{cases}$

17. $\begin{cases} x + y + z = 1 \\ -2x + y + z = -2 \\ 3x + 6y + 6z = 5 \end{cases}$

18. $\begin{cases} 7x + 5y - 7z = -10 \\ 2x + y + z = 7 \\ x + y - 3z = -8 \end{cases}$

19. $\begin{cases} 2x - y + z = -1 \\ x + 3y - 2z = 2 \\ -5x + 6y - 5z = 5 \end{cases}$

20. $\begin{cases} -2x + 2y - z = 0 \\ 3x - 4y + z = 1 \\ 5x - 8y + z = 4 \end{cases}$

21. $\begin{cases} 2x - y + z = 4 \\ x + 3y + 2z = -1 \\ 7x + 5z = 11 \end{cases}$

22. $\begin{cases} 3x + y - z = 10 \\ 8x - y - 6z = -3 \\ 5x - 2y - 5z = 1 \end{cases}$

23. $\begin{cases} x + y + z + w = 4 \\ x - 2y - z - w = 3 \\ 2x - y + z - w = 2 \\ x - y + 2z - 2w = -7 \end{cases}$

24. $\begin{cases} x + y - 3z + 2w = 0 \\ -2x - 2y + 6z + w = -5 \\ -x + 3y + 3z + 3w = -5 \\ 2x + y - 3z - w = 4 \end{cases}$

25. $\begin{cases} 2x + 3y + 2z = 5 \\ x + 4y - 3z = 1 \end{cases}$

26. $\begin{cases} 4x - y - 3z = 2 \\ 6x + 5y - z = 0 \end{cases}$

27. $\begin{cases} x - 2y - 2z + 2w = -10 \\ 3x + 4y - z - 3w = 11 \\ -4x - 3y - 3z + 8w = -21 \end{cases}$

28. $\begin{cases} 2x + y + z + w = 1 \\ x + 3y - 3z - 3w = 0 \\ -3x - 4y + 2z + 2w = -1 \end{cases}$

29. $\begin{cases} 4x - 2y + 3z = -2 \\ 6y - 4z = 6 \end{cases}$

30. $\begin{cases} 6x - 2y - 5z + w = 3 \\ 5x - y + z + 5w = 4 \\ -4y - 7z - 9w = -5 \end{cases}$

In Exercises 31–40, determine the constants (denoted by capital letters) so that each equation is an identity.

31. **(a)** $\dfrac{1}{(x-2)(x+2)} = \dfrac{A}{x-2} + \dfrac{B}{x+2}$

(b) $\dfrac{x}{(x-2)(x+2)} = \dfrac{A}{x-2} + \dfrac{B}{x+2}$

32. **(a)** $\dfrac{2}{(x+3)(x+2)} = \dfrac{A}{x+3} + \dfrac{B}{x+2}$

(b) $\dfrac{x-9}{(x+3)(x+2)} = \dfrac{A}{x+3} + \dfrac{B}{x+2}$

33. $\dfrac{3x}{(x+4)(x-5)} = \dfrac{A}{x+4} + \dfrac{B}{x-5}$

34. $\dfrac{4x+26}{(x-1)(x+5)} = \dfrac{A}{x-1} + \dfrac{B}{x+5}$

35. **(a)** $\dfrac{3}{(x-4)(x+4)} = \dfrac{A}{x-4} + \dfrac{B}{x+4}$

(b) $\dfrac{3}{(x-4)(x+4)^2} = \dfrac{A}{x-4} + \dfrac{B}{x+4} + \dfrac{C}{(x+4)^2}$

36. **(a)** $\dfrac{1}{(x+1)(x+6)} = \dfrac{A}{x+1} + \dfrac{B}{x+6}$

(b) $\dfrac{1}{(x+1)(x+6)^2} = \dfrac{A}{x+1} + \dfrac{B}{x+6} + \dfrac{C}{(x+6)^2}$

37. $\dfrac{1}{(x+1)(x^2-x+1)} = \dfrac{A}{x+1} + \dfrac{Bx+C}{x^2-x+1}$

38. $\dfrac{1}{(x+1)^2(x^2-x+1)} = \dfrac{A}{x+1} + \dfrac{B}{(x+1)^2} + \dfrac{Cx+D}{x^2-x+1}$

39. $\dfrac{4}{x(1-x)} = \dfrac{A}{x} + \dfrac{B}{1-x}$

40. $\dfrac{4}{x^2(1-x)} = \dfrac{A}{x} + \dfrac{B}{x^2} + \dfrac{C}{1-x}$

B

41. Determine the constants A, B, C, and D so that the following equation is an identity:

$$\frac{1}{(x^2+x+1)(x^2-x+1)} = \frac{Ax+B}{x^2+x+1} + \frac{Cx+D}{x^2-x+1}$$

42. Suppose that the height of an object as a function of time is given by $f(t) = at^2 + bt + c$, where t is time in seconds, $f(t)$ is the height in feet at time t, and a, b, and c are certain constants. If after 1, 2, and 3 seconds the corresponding heights are 184 ft, 136 ft, and 56 ft, find the time at which the object is at ground level (height = 0 ft).

43. Find the equation of the circle passing through the points $(-1, -5)$, $(1, 2)$, and $(-2, 0)$. Write your answer in the form $Ax^2 + Ay^2 + Bx + Cy + D = 0$.

44. Find the equation of a circle that passes through the origin and the points $(8, -4)$ and $(7, -1)$. Specify the center and radius.

45. Solve the following system for x, y, and z:

$$\begin{cases} e^x + e^y - 2e^z = 2a \\ e^x + 2e^y - 4e^z = 3a \\ \tfrac{1}{2}e^x - 3e^y + e^z = -5a \end{cases}$$

(Assume that $a > 0$.)   *Hint:* Let $A = e^x$, $B = e^y$, and $C = e^z$. Solve the resulting linear system for A, B, and C.

46. Solve the following system for α, β, and γ, given that each of these numbers lies in the closed interval $[0, \pi]$.

$$\begin{cases} \cos\alpha - \cos\beta - \cos\gamma = 2 \\ 3\cos\alpha + 5\cos\beta - 2\cos\gamma = 1 \\ 2\cos\alpha - 4\cos\beta + \cos\gamma = 2 \end{cases}$$

Hint: First make the substitutions $x = \cos\alpha$, $y = \cos\beta$, and $z = \cos\gamma$; then solve the resulting linear system for x, y, and z.

47. The following figure displays three circles that are mutually tangent. The line segments joining the centers have lengths a, b, and c, as shown. Let r_1, r_2, and r_3 denote the radii of the circles, as indicated in the figure. Show that

$$r_1 = \frac{a+c-b}{2}$$

$$r_2 = \frac{a+b-c}{2}$$

$$r_3 = \frac{b+c-a}{2}$$

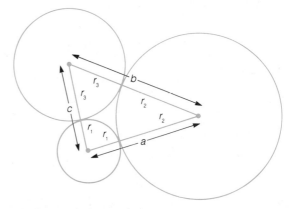

48. Consider the system

$$\begin{cases} \lambda x + y + z = a \\ x + \lambda y + z = b \\ x + y + \lambda z = c \end{cases}$$

(a) Assuming that the value of the constant λ is neither 1 nor -2, find all solutions of the system.

(b) If $\lambda = -2$, how must a, b, and c be related for the system to have solutions? Find these solutions.

(c) If $\lambda = 1$, how must a, b, and c be related for the system to have solutions? Find these solutions.

In Exercises 49 and 50, the lowercase letters a, b, p, and q denote given constants. In each case, determine the values of A and B so that the equation is an identity.

49. (a) $\dfrac{1}{(x-a)(x-b)} = \dfrac{A}{x-a} + \dfrac{B}{x-b}$ $(a \neq b)$

(b) $\dfrac{px+q}{(x-a)(x-b)} = \dfrac{A}{x-a} + \dfrac{B}{x-b}$ $(a \neq b)$

50. (a) $\dfrac{1}{(x-a)(x+a)} = \dfrac{A}{x-a} + \dfrac{B}{x+a}$ $(a \neq 0)$

(b) $\dfrac{px+q}{(x-a)(x+a)} = \dfrac{A}{x-a} + \dfrac{B}{x+a}$ $(a \neq 0)$

C

51. Express A, B, and C in terms of a, b, c, p, and q so that the following is an identity. (Assume a, b, and c are all unequal.)

$$\frac{x^2 + px + q}{(x-a)(x-b)(x-c)} = \frac{A}{x-a} + \frac{B}{x-b} + \frac{C}{x-c}$$

52. Express A, B, and C in terms of a, b, and c so that the following equation is an identity. (Assume a, b, and c are all unequal.)

$$\frac{1}{(1-ax)(1-bx)(1-cx)} = \frac{A}{1-ax} + \frac{B}{1-bx} + \frac{C}{1-cx}$$

The following exercise appears in Algebra for Colleges and Schools *by H. S. Hall and S. R. Knight, revised by F. L. Sevenoak (New York: The Macmillan Company, 1906).*

53. A, B, and C are three towns forming a triangle. A man has to walk from one to the next, ride thence to the next, and drive thence to his starting point. He can walk, ride, and drive a mile in a, b, and c minutes, respectively. If he starts from B he takes $a + c - b$ hours, if he starts from C he takes $b + a - c$ hours, and if he starts from A he takes $c + b - a$ hours. Find the length of the circuit. [Assume that the circuit from A to B to C is counterclockwise.]

7.3 ▼ MATRICES

Arthur Cayley (1821–1895) and James Joseph Sylvester (1814–1897), two English mathematicians, invented the matrix . . . in the 1850s. . . . The operations of addition and multiplication of matrices were later defined and the algebra of matrices was then developed. In 1925, Werner Heisenberg, a German physicist, used matrices in developing his theory of quantum mechanics, extending the role of matrices from algebra to the area of applied mathematics.

John K. Luedeman and Stanley M. Lukawecki in their text *Elementary Linear Algebra* (St. Paul: West Publishing Company, 1986)

As you know from the previous section, there can be a good deal of bookkeeping involved in using Gaussian elimination to solve systems of equations. We can lighten the load somewhat by using matrices, where a **matrix** is simply a rectangular array of numbers, enclosed in parentheses or brackets. Here are three examples:

$$\begin{pmatrix} 2 & 3 \\ -5 & 4 \end{pmatrix} \quad \begin{pmatrix} -6 & 0 & 1 & \frac{1}{4} \\ \frac{2}{3} & 1 & 5 & 8 \end{pmatrix} \quad \begin{pmatrix} \pi & 0 & 0 \\ 0 & 1 & 9 \\ -1 & -2 & 3 \\ -4 & 8 & 6 \end{pmatrix}$$

The particular numbers constituting a matrix are called its **entries** or **elements**. For instance, the entries in the matrix $\begin{pmatrix} 2 & 3 \\ -5 & 4 \end{pmatrix}$ are the four numbers, 2, 3, -5, and 4. In this section, the entries will always be real numbers. However, it is also possible to consider matrices in which some or all of the entries are nonreal complex numbers.

It is convenient to agree on a standard system for labeling the rows and columns of a matrix. The rows are numbered from top to bottom and the columns from left to right, as indicated in the following example:

$$\begin{array}{c} \text{column 1} \quad \text{column 2} \\ \begin{array}{c} \text{row 1} \longrightarrow \\ \text{row 2} \longrightarrow \\ \text{row 3} \longrightarrow \end{array} \begin{pmatrix} 5 & -3 \\ 0 & 2 \\ -1 & 16 \end{pmatrix} \end{array}$$

We express the **size**, or **dimension**, of a matrix by specifying the number of rows and the number of columns, in that order. For instance, we would say that the matrix

$$\begin{pmatrix} 5 & -3 \\ 0 & 2 \\ -1 & 16 \end{pmatrix}$$

is a 3×2 (read "3 by 2") matrix, not 2×3. The following example will help fix in your mind the terminology that we have introduced.

EXAMPLE 1   Consider the matrix

$$\begin{pmatrix} 1 & 3 & 5 \\ 7 & 9 & 11 \end{pmatrix}$$

(**a**) List the entries.
(**b**) What is the size of the matrix?
(**c**) Which element is in the second row, third column?

Solution   (**a**) The entries are 1, 3, 5, 7, 9, and 11.
(**b**) Since there are two rows and three columns, this is a 2×3 matrix.
(**c**) To locate the element in the second row, third column, draw lines through the second row and the third column and see where they intersect:

$$\begin{pmatrix} 1 & 3 & 5 \\ 7 & 9 & 11 \end{pmatrix}$$

Thus, the entry in row 2, column 3 is 11.

There is a natural way to use matrices to describe and solve systems of linear equations. Consider, for example, the following system of linear equations in **standard form** (with the x, y, and z terms lined up on the left-hand side and the constant terms on the right-hand side of each equation):

$$\begin{cases} x + 2y - 3z = 4 \\ 3x + z = 5 \\ -x - 3y + 4z = 0 \end{cases}$$

The **coefficient matrix** of this system is the matrix

$$\begin{pmatrix} 1 & 2 & -3 \\ 3 & 0 & 1 \\ -1 & -3 & 4 \end{pmatrix}$$

As the name implies, the coefficient matrix of the system is the matrix whose entries are the coefficients of x, y, and z, written in the same relative positions as they appear in the system. Notice the zero appearing in the second row, second column of the matrix. It is there because the coefficient of y in the second equation is in fact zero. The **augmented matrix** of the system of equations considered here is

$$\begin{pmatrix} 1 & 2 & -3 & 4 \\ 3 & 0 & 1 & 5 \\ -1 & -3 & 4 & 0 \end{pmatrix}$$

As you can see, the augmented matrix is formed by *augmenting* the coefficient matrix with the column of constant terms taken from the right-hand side of the given system of equations. To help relate the augmented matrix to the original system of equations, we will write the augmented matrix this way:

$$\begin{pmatrix} 1 & 2 & -3 & | & 4 \\ 3 & 0 & 1 & | & 5 \\ -1 & -3 & 4 & | & 0 \end{pmatrix}$$

EXAMPLE 2   Write the coefficient matrix and the augmented matrix for the system

$$\begin{cases} 8x - 2y + z = 1 \\ 3x - 4z + y = 2 \\ 12y - 3z - 6 = 0 \end{cases}$$

Solution   First write the system in standard form, with the $x, y,$ and z terms lined up and the constant terms on the right. This yields

$$\begin{cases} 8x - 2y + z = 1 \\ 3x + y - 4z = 2 \\ 12y - 3z = 6 \end{cases}$$

The coefficient matrix is then

$$\begin{pmatrix} 8 & -2 & 1 \\ 3 & 1 & -4 \\ 0 & 12 & -3 \end{pmatrix}$$

while the augmented matrix is

$$\begin{pmatrix} 8 & -2 & 1 & | & 1 \\ 3 & 1 & -4 & | & 2 \\ 0 & 12 & -3 & | & 6 \end{pmatrix}$$ ▲

In the previous section we used the three elementary operations in solving systems of linear equations. In Table 1 we express these operations in the language of matrices. The matrix operations are called the **elementary row operations**.

Table 2 displays an example of each elementary row operation. In the table, notice that the notation for describing these operations is essentially the same as that introduced in the previous section. For example, in the previous section, $10E_1$ indicated that the first equation in a system was multiplied by 10. Now, $10R_1$ indicates that each entry in the first *row* of a matrix is multiplied by 10.

We are now ready to use matrices to solve systems of equations. In the example that follows, we'll use the same system of equations used in Example 5 of the

TABLE 1

ELEMENTARY OPERATIONS FOR A SYSTEM OF LINEAR EQUATIONS	CORRESPONDING ELEMENTARY ROW OPERATION FOR A MATRIX
1. Multiply both sides of an equation by a nonzero constant.	**1′.** Multiply each entry in a given row by a nonzero constant.
2. Interchange two equations.	**2′.** Interchange two rows.
3. To one equation, add a multiple of another equation.	**3′.** To one row, add a multiple of another row.

TABLE 2

EXAMPLES OF THE ELEMENTARY ROW OPERATIONS	COMMENTS
$\begin{pmatrix} 1 & 2 & 3 \\ 4 & 5 & 6 \\ 7 & 8 & 9 \end{pmatrix} \xrightarrow{\ 10R_1\ } \begin{pmatrix} 10 & 20 & 30 \\ 4 & 5 & 6 \\ 7 & 8 & 9 \end{pmatrix}$	Multiply each entry in row 1 by 10.
$\begin{pmatrix} 1 & 2 & 3 \\ 4 & 5 & 6 \\ 7 & 8 & 9 \end{pmatrix} \xrightarrow{\ R_2 \leftrightarrow R_3\ } \begin{pmatrix} 1 & 2 & 3 \\ 7 & 8 & 9 \\ 4 & 5 & 6 \end{pmatrix}$	Interchange rows 2 and 3.
$\begin{pmatrix} 1 & 2 & 3 \\ 4 & 5 & 6 \\ 7 & 8 & 9 \end{pmatrix} \xrightarrow{\ -4R_1 + R_2\ } \begin{pmatrix} 1 & 2 & 3 \\ 0 & -3 & -6 \\ 7 & 8 & 9 \end{pmatrix}$	To each entry in row 2, add -4 times the corresponding entry in row 1.

previous section. *Suggestion:* After reading the next example, carefully compare each step with the corresponding one taken in Example 5 of the previous section.

EXAMPLE 3 Solve the system

$$\begin{cases} 4x - 3y + 2z = 40 \\ 5x + 9y - 7z = 47 \\ 9x + 8y - 3z = 97 \end{cases}$$

Solution

$$\begin{pmatrix} 4 & -3 & 2 & | & 40 \\ 5 & 9 & -7 & | & 47 \\ 9 & 8 & -3 & | & 97 \end{pmatrix} \xrightarrow{(-1)R_2 + R_1} \begin{pmatrix} -1 & -12 & 9 & | & -7 \\ 5 & 9 & -7 & | & 47 \\ 9 & 8 & -3 & | & 97 \end{pmatrix}$$

$$\xrightarrow[9R_1 + R_3]{5R_1 + R_2} \begin{pmatrix} -1 & -12 & 9 & | & -7 \\ 0 & -51 & 38 & | & 12 \\ 0 & -100 & 78 & | & 34 \end{pmatrix}$$

$$\xrightarrow{\frac{1}{2}R_3} \begin{pmatrix} -1 & -12 & 9 & | & -7 \\ 0 & -51 & 38 & | & 12 \\ 0 & -50 & 39 & | & 17 \end{pmatrix}$$

$$\xrightarrow{(-1)R_3 + R_2} \begin{pmatrix} -1 & -12 & 9 & | & -7 \\ 0 & -1 & -1 & | & -5 \\ 0 & -50 & 39 & | & 17 \end{pmatrix}$$

$$\xrightarrow{-50R_2 + R_3} \begin{pmatrix} -1 & -12 & 9 & | & -7 \\ 0 & -1 & -1 & | & -5 \\ 0 & 0 & 89 & | & 267 \end{pmatrix}$$

This last augmented matrix represents a system of equations in upper-triangular form:

$$\begin{cases} -x - 12y + 9z = -7 \\ \quad\ -y - \ z = -5 \\ \qquad\qquad 89z = 267 \end{cases}$$

As you should now check for yourself, this yields the values $z = 3$, then $y = 2$, then $x = 10$. The solution of the original system is therefore $(10, 2, 3)$. ▲

For the remainder of this section (and in part of the next) we are going to study matrices without reference to systems of equations. As motivation for this, we point out that matrices are essential tools in many fields of study. For example, a knowledge of matrices and their properties is needed for work in computer graphics. To begin, we need to say what it means for two matrices to be equal.

DEFINITION

Equality of Matrices

EXAMPLES

Two matrices are equal provided they have the same size (same number of rows, same number of columns) and the corresponding entries are equal.

$$\begin{pmatrix} 2 & 3 \\ 4 & 5 \end{pmatrix} = \begin{pmatrix} 2 & 3 \\ 4 & 5 \end{pmatrix}$$

$$\begin{pmatrix} 2 & 3 & 0 \\ 4 & 5 & 0 \end{pmatrix} \neq \begin{pmatrix} 2 & 3 \\ 4 & 5 \end{pmatrix}$$

$$\begin{pmatrix} 2 & 3 \\ 4 & 5 \end{pmatrix} \neq \begin{pmatrix} 2 & 3 \\ 5 & 4 \end{pmatrix}$$

Now we can define matrix addition and subtraction. These operations are defined only between matrices of the same size.

DEFINITION

Matrix Addition

EXAMPLES

To add (or subtract) two matrices of the same size, add (or subtract) the corresponding entries.

$$\begin{pmatrix} 2 & 3 \\ -1 & 4 \end{pmatrix} + \begin{pmatrix} 6 & 1 \\ 0 & -4 \end{pmatrix} = \begin{pmatrix} 8 & 4 \\ -1 & 0 \end{pmatrix}$$

$$\begin{pmatrix} 2 & 3 \\ -1 & 4 \\ 9 & 10 \end{pmatrix} - \begin{pmatrix} 6 & 1 \\ 0 & -4 \\ 3 & 2 \end{pmatrix} = \begin{pmatrix} -4 & 2 \\ -1 & 8 \\ 6 & 8 \end{pmatrix}$$

Many properties of the real numbers also apply to matrices. For instance, matrix addition is *commutative*:

$$A + B = B + A \qquad \text{where } A \text{ and } B \text{ are matrices of the same size}$$

Matrix addition is also *associative*:

$$A + (B + C) = (A + B) + C \qquad \text{where } A, B, \text{ and } C \text{ are matrices of the same size}$$

In the next example, we verify these properties in two specific instances.

EXAMPLE 4 Let $A = \begin{pmatrix} 1 & 2 \\ 3 & 4 \end{pmatrix}$, $B = \begin{pmatrix} 0 & -5 \\ 8 & -1 \end{pmatrix}$, and $C = \begin{pmatrix} 6 & 7 \\ 8 & 9 \end{pmatrix}$.

(a) Show that $A + C = C + A$.
(b) Show that $A + (B + C) = (A + B) + C$.

Solution **(a)** $A + C = \begin{pmatrix} 1 & 2 \\ 3 & 4 \end{pmatrix} + \begin{pmatrix} 6 & 7 \\ 8 & 9 \end{pmatrix} = \begin{pmatrix} 7 & 9 \\ 11 & 13 \end{pmatrix}$

$C + A = \begin{pmatrix} 6 & 7 \\ 8 & 9 \end{pmatrix} + \begin{pmatrix} 1 & 2 \\ 3 & 4 \end{pmatrix} = \begin{pmatrix} 7 & 9 \\ 11 & 13 \end{pmatrix}$

This shows that $A + C = C + A$, since both $A + C$ and $C + A$ represent the matrix $\begin{pmatrix} 7 & 9 \\ 11 & 13 \end{pmatrix}$.

(b) First we compute $A + (B + C)$:

$$A + (B + C) = \begin{pmatrix} 1 & 2 \\ 3 & 4 \end{pmatrix} + \left[\begin{pmatrix} 0 & -5 \\ 8 & -1 \end{pmatrix} + \begin{pmatrix} 6 & 7 \\ 8 & 9 \end{pmatrix} \right]$$

$$= \begin{pmatrix} 1 & 2 \\ 3 & 4 \end{pmatrix} + \begin{pmatrix} 6 & 2 \\ 16 & 8 \end{pmatrix}$$

$$= \begin{pmatrix} 7 & 4 \\ 19 & 12 \end{pmatrix}$$

Next we compute $(A + B) + C$:

$$(A + B) + C = \left[\begin{pmatrix} 1 & 2 \\ 3 & 4 \end{pmatrix} + \begin{pmatrix} 0 & -5 \\ 8 & -1 \end{pmatrix} \right] + \begin{pmatrix} 6 & 7 \\ 8 & 9 \end{pmatrix}$$

$$= \begin{pmatrix} 1 & -3 \\ 11 & 3 \end{pmatrix} + \begin{pmatrix} 6 & 7 \\ 8 & 9 \end{pmatrix}$$

$$= \begin{pmatrix} 7 & 4 \\ 19 & 12 \end{pmatrix}$$

We conclude from these calculations that $A + (B + C) = (A + B) + C$, since both sides of that equation represent the matrix $\begin{pmatrix} 7 & 4 \\ 19 & 12 \end{pmatrix}$. ▲

We will now define an operation on matrices called **scalar multiplication**. First of all, the word *scalar* here just means *real number*. So we are talking about multiplying a matrix by a real number. (In more advanced work, nonreal complex scalars are also considered.)

DEFINITION

Scalar Multiplication

EXAMPLES

To multiply a matrix by a scalar, multiply each entry in the matrix by that scalar.

$$2 \begin{pmatrix} 5 & 9 & 0 \\ -1 & 2 & 3 \end{pmatrix} = \begin{pmatrix} 10 & 18 & 0 \\ -2 & 4 & 6 \end{pmatrix}$$

$$1 \begin{pmatrix} 1 & 2 \\ 3 & 4 \end{pmatrix} = \begin{pmatrix} 1 & 2 \\ 3 & 4 \end{pmatrix}$$

There are two simple but useful properties of scalar multiplication that are worth noting at this point. We'll omit the proofs of these two properties; however, Example 5 does ask us to verify them for a particular case

PROPERTY SUMMARY

PROPERTIES OF SCALAR MULTIPLICATION

1. $c(kM) = (ck)M$ for all scalars c and k and any matrix M

2. $c(M + N) = cM + cN$,   where c is any scalar and M and N are any matrices of the same size

EXAMPLE 5   Let $c = 2$, $k = 3$, $M = \begin{pmatrix} 1 & 2 \\ 3 & 4 \end{pmatrix}$, and $N = \begin{pmatrix} 5 & 6 \\ 7 & 8 \end{pmatrix}$.

(**a**) Show that $c(kM) = (ck)M$.

(**b**) Show that $c(M + N) = cM + cN$.

Solution

(**a**) $c(kM) = 2\left[3\begin{pmatrix} 1 & 2 \\ 3 & 4 \end{pmatrix}\right] = 2\begin{pmatrix} 3 & 6 \\ 9 & 12 \end{pmatrix} = \begin{pmatrix} 6 & 12 \\ 18 & 24 \end{pmatrix}$

$(ck)M = (2 \cdot 3)\begin{pmatrix} 1 & 2 \\ 3 & 4 \end{pmatrix} = 6\begin{pmatrix} 1 & 2 \\ 3 & 4 \end{pmatrix} = \begin{pmatrix} 6 & 12 \\ 18 & 24 \end{pmatrix}$

Thus, $c(kM) = (ck)M$, since in both cases the result is $\begin{pmatrix} 6 & 12 \\ 18 & 24 \end{pmatrix}$.

(**b**) $c(M + N) = 2\left[\begin{pmatrix} 1 & 2 \\ 3 & 4 \end{pmatrix} + \begin{pmatrix} 5 & 6 \\ 7 & 8 \end{pmatrix}\right]$

$= 2\begin{pmatrix} 6 & 8 \\ 10 & 12 \end{pmatrix} = \begin{pmatrix} 12 & 16 \\ 20 & 24 \end{pmatrix}$

$cM + cN = 2\begin{pmatrix} 1 & 2 \\ 3 & 4 \end{pmatrix} + 2\begin{pmatrix} 5 & 6 \\ 7 & 8 \end{pmatrix}$

$= \begin{pmatrix} 2 & 4 \\ 6 & 8 \end{pmatrix} + \begin{pmatrix} 10 & 12 \\ 14 & 16 \end{pmatrix} = \begin{pmatrix} 12 & 16 \\ 20 & 24 \end{pmatrix}$

Thus, $c(M + N) = cM + cN$, since both sides equal $\begin{pmatrix} 12 & 16 \\ 20 & 24 \end{pmatrix}$. ▲

A matrix with zeros for all of its entries plays the same role in matrix addition as does the number zero in ordinary addition of real numbers. For instance, in the case of 2×2 matrices, we have

$$\begin{pmatrix} a & b \\ c & d \end{pmatrix} + \begin{pmatrix} 0 & 0 \\ 0 & 0 \end{pmatrix} = \begin{pmatrix} a & b \\ c & d \end{pmatrix}$$

and

$$\begin{pmatrix} 0 & 0 \\ 0 & 0 \end{pmatrix} + \begin{pmatrix} a & b \\ c & d \end{pmatrix} = \begin{pmatrix} a & b \\ c & d \end{pmatrix}$$

for all real numbers a, b, c, and d. The matrix $\begin{pmatrix} 0 & 0 \\ 0 & 0 \end{pmatrix}$ is called the **additive identity** for 2×2 matrices. Similarly, any matrix with all zero entries is the additive identity for matrices of that size. It is sometimes convenient to denote an additive identity matrix by a boldface zero: **0**. With this notation, we can write

$A + \mathbf{0} = \mathbf{0} + A = A$ for any matrix A

For the preceding matrix equation, it is understood that the size of the matrix **0** is the same as the size of A. With this notation, we also have

$A - A = \mathbf{0}$ for any matrix A

Our last topic in this section is matrix multiplication. We will begin with the simplest case and then work up to the more general situation. By convention, a matrix with only one row is called a **row vector**. Examples of row vectors are

$(2 \quad 13), \quad (-1 \quad 4 \quad 3), \quad \text{and} \quad (0 \quad 0 \quad 0 \quad 1)$

Similarly, a matrix with only one column is called a **column vector**. Examples are

$$\begin{pmatrix} 2 \\ 13 \end{pmatrix}, \qquad \begin{pmatrix} -1 \\ 4 \\ 3 \end{pmatrix}, \qquad \text{and} \qquad \begin{pmatrix} 0 \\ 0 \\ 0 \\ 1 \end{pmatrix}$$

The following definition tells us how to multiply a row vector and a column vector when they have the same number of entries.

The Inner Product of a Row Vector and a Column Vector

Let A be a row vector and B a column vector, and assume that the number of columns in A is the same as the number of rows in B. Then the **inner product** $A \cdot B$ is defined to be the number obtained by multiplying the corresponding entries and then adding the products.

EXAMPLES

$$(1 \quad 2 \quad 3) \cdot \begin{pmatrix} 4 \\ 5 \\ 6 \end{pmatrix} = 1 \cdot 4 + 2 \cdot 5 + 3 \cdot 6$$

$$= 32$$

$$(1 \quad 2) \cdot \begin{pmatrix} 4 \\ 5 \\ 6 \end{pmatrix} \quad \text{is not defined}$$

$$(1 \quad 2 \quad 3) \cdot \begin{pmatrix} 4 \\ 5 \end{pmatrix} \quad \text{is not defined}$$

An important observation here is that the end result of taking the inner product is always just a number. The definition of matrix product that we now give depends on this observation.

The Product of Two Matrices

Let A and B be two matrices, and assume that the number of columns in A is the same as the number of rows in B. Then the product matrix AB is computed according to the following rule.

The entry in the ith row and the jth column of AB is the inner product of the ith row of A with the jth column of B.

The matrix AB will have as many rows as A and as many columns as B.

As an example of matrix multiplication, we will compute the product AB, where $A = \begin{pmatrix} 1 & 2 \\ 3 & 4 \end{pmatrix}$ and $B = \begin{pmatrix} 5 & 6 & 0 \\ 7 & 8 & 1 \end{pmatrix}$. In other words, we will compute $\begin{pmatrix} 1 & 2 \\ 3 & 4 \end{pmatrix}\begin{pmatrix} 5 & 6 & 0 \\ 7 & 8 & 1 \end{pmatrix}$. Before we attempt to carry out the calculations of any matrix multiplication, however, we should check on two points.

1. Is the product defined? That is, does the number of columns in A equal the number of rows in B? In this case, yes; the common number is 2.
2. What is the size of the product? According to the definition, the product AB will have as many rows as A and as many columns as B. Thus, the size of AB will be 2×3.

Schematically, then, the situation looks like this:

$$\begin{pmatrix} 1 & 2 \\ 3 & 4 \end{pmatrix}\begin{pmatrix} 5 & 6 & 0 \\ 7 & 8 & 1 \end{pmatrix} = \begin{pmatrix} ? & ? & ? \\ ? & ? & ? \end{pmatrix}$$

We have six positions to fill. The computations are presented in Table 3.

TABLE 3

POSITION	HOW TO COMPUTE	COMPUTATION
row 1, column 1	inner product of row 1 and column 1 $$\begin{pmatrix} 1 & 2 \\ 3 & 4 \end{pmatrix}\begin{pmatrix} 5 & 6 & 0 \\ 7 & 8 & 1 \end{pmatrix}$$	$1 \cdot 5 + 2 \cdot 7 = 19$
row 1, column 2	inner product of row 1 and column 2 $$\begin{pmatrix} 1 & 2 \\ 3 & 4 \end{pmatrix}\begin{pmatrix} 5 & 6 & 0 \\ 7 & 8 & 1 \end{pmatrix}$$	$1 \cdot 6 + 2 \cdot 8 = 22$
row 1, column 3	inner product of row 1 and column 3 $$\begin{pmatrix} 1 & 2 \\ 3 & 4 \end{pmatrix}\begin{pmatrix} 5 & 6 & 0 \\ 7 & 8 & 1 \end{pmatrix}$$	$1 \cdot 0 + 2 \cdot 1 = 2$
row 2, column 1	inner product of row 2 and column 1 $$\begin{pmatrix} 1 & 2 \\ 3 & 4 \end{pmatrix}\begin{pmatrix} 5 & 6 & 0 \\ 7 & 8 & 1 \end{pmatrix}$$	$3 \cdot 5 + 4 \cdot 7 = 43$
row 2, column 2	inner product of row 2 and column 2 $$\begin{pmatrix} 1 & 2 \\ 3 & 4 \end{pmatrix}\begin{pmatrix} 5 & 6 & 0 \\ 7 & 8 & 1 \end{pmatrix}$$	$3 \cdot 6 + 4 \cdot 8 = 50$
row 2, column 3	inner product of row 2 and column 3 $$\begin{pmatrix} 1 & 2 \\ 3 & 4 \end{pmatrix}\begin{pmatrix} 5 & 6 & 0 \\ 7 & 8 & 1 \end{pmatrix}$$	$3 \cdot 0 + 4 \cdot 1 = 4$

Reading from the table, we see that our result is

$$AB = \begin{pmatrix} 1 & 2 \\ 3 & 4 \end{pmatrix}\begin{pmatrix} 5 & 6 & 0 \\ 7 & 8 & 1 \end{pmatrix} = \begin{pmatrix} 19 & 22 & 2 \\ 43 & 50 & 4 \end{pmatrix}$$

EXAMPLE 6 Let $A = \begin{pmatrix} 1 & 2 \\ 3 & 4 \end{pmatrix}$ and $B = \begin{pmatrix} 5 & 6 \\ 7 & 8 \end{pmatrix}$. By computing AB and then BA, show that $AB \neq BA$. This shows that, in general, matrix multiplication is not commutative.

Solution

$$AB = \begin{pmatrix} 1 & 2 \\ 3 & 4 \end{pmatrix}\begin{pmatrix} 5 & 6 \\ 7 & 8 \end{pmatrix} = \begin{pmatrix} 1 \cdot 5 + 2 \cdot 7 & 1 \cdot 6 + 2 \cdot 8 \\ 3 \cdot 5 + 4 \cdot 7 & 3 \cdot 6 + 4 \cdot 8 \end{pmatrix} = \begin{pmatrix} 19 & 22 \\ 43 & 50 \end{pmatrix}$$

$$BA = \begin{pmatrix} 5 & 6 \\ 7 & 8 \end{pmatrix}\begin{pmatrix} 1 & 2 \\ 3 & 4 \end{pmatrix} = \begin{pmatrix} 5 \cdot 1 + 6 \cdot 3 & 5 \cdot 2 + 6 \cdot 4 \\ 7 \cdot 1 + 8 \cdot 3 & 7 \cdot 2 + 8 \cdot 4 \end{pmatrix} = \begin{pmatrix} 23 & 34 \\ 31 & 46 \end{pmatrix}$$

Comparing the two matrices AB and BA, we conclude that $AB \neq BA$. ▲

▼ EXERCISE SET 7.3

A

In Exercises 1–4, specify the size of each matrix.

1. (a) $\begin{pmatrix} -4 & 0 & 5 \\ 2 & 8 & -1 \end{pmatrix}$ (b) $\begin{pmatrix} 7 & 1 \\ 4 & -3 \\ 0 & 0 \end{pmatrix}$

2. (a) $\begin{pmatrix} 1 & 0 \\ 0 & -1 \end{pmatrix}$ (b) $\begin{pmatrix} 1 \\ 6 \\ 8 \\ 1 \end{pmatrix}$

3. $\begin{pmatrix} 1 & a & b & c \\ a & 1 & 0 & a \\ b & 0 & 1 & b \\ c & a & b & 1 \\ 0 & 0 & 0 & 1 \end{pmatrix}$ **4.** $(-3 \quad 1 \quad 6 \quad 0)$

In Exercises 5–8, write the coefficient matrix and the augmented matrix for each system.

5. $\begin{cases} 2x + 3y + 4z = 10 \\ 5x + 6y + 7z = 9 \\ 8x + 9y + 10z = 8 \end{cases}$ **6.** $\begin{cases} 5x - y + z = 0 \\ 4y + 2z = 1 \\ 3x + y + z = -1 \end{cases}$

7. $\begin{cases} x + z + w = -1 \\ x + y + 2w = 0 \\ y + z + w = 1 \\ 2x - y - z = 2 \end{cases}$ **8.** $\begin{cases} 8x - 8y = 5 \\ x - y + z = 1 \end{cases}$

In Exercises 9–22, use matrices to solve each system of equations.

9. $\begin{cases} x - y + 2z = 7 \\ 3x + 2y - z = -10 \\ -x + 3y + z = -2 \end{cases}$ **10.** $\begin{cases} 2x - 3y + 4z = 14 \\ 3x - 2y + 2z = 12 \\ 4x + 5y - 5z = 16 \end{cases}$

11. $\begin{cases} x + z = -2 \\ -3x + 2y = 17 \\ x - y - z = -9 \end{cases}$ **12.** $\begin{cases} 5x + y + 10z = 23 \\ 4x + 2y - 10z = 76 \\ 3x - 4y = 18 \end{cases}$

13. $\begin{cases} x + y + z = -4 \\ 2x - 3y + z = -1 \\ 4x + 2y - 3z = 33 \end{cases}$ **14.** $\begin{cases} 2x + 3y - 4z = 7 \\ x - y + z = -\frac{3}{2} \\ 6x - 5y - 2z = -7 \end{cases}$

15. $\begin{cases} 3x - 2y + 6z = 0 \\ x + 3y + 20z = 15 \\ 10x - 11y - 10z = -9 \end{cases}$ **16.** $\begin{cases} 3A - 3B + C = 4 \\ 6A + 9B - 3C = -7 \\ A - 2B - 2C = -3 \end{cases}$

17. $\begin{cases} 4x - 3y + 3z = 2 \\ 5x + y - 4z = 1 \\ 9x - 2y - z = 3 \end{cases}$ **18.** $\begin{cases} 6x + y - z = -1 \\ -3x + 2y + 2z = 2 \\ 5y + 3z = 1 \end{cases}$

19. $\begin{cases} x - y + z + w = 6 \\ x + y - z + w = 4 \\ x + y + z - w = -2 \\ -x + y + z + w = 0 \end{cases}$

20. $\begin{cases} x + 2y - z - 2w = 5 \\ 2x + y + 2z + w = -7 \\ -2x - y - 3z - 2w = 10 \\ z + w = -3 \end{cases}$

21. $\begin{cases} 15A + 14B + 26C = 1 \\ 18A + 17B + 32C = -1 \\ 21A + 20B + 38C = 0 \end{cases}$

22. $\begin{cases} A + B + C + D + E = -1 \\ 3A - 2B - 2C + 3D + 2E = 13 \\ 3C + 4D - 4E = 7 \\ 5A - 4B + E = 30 \\ C - 2E = 3 \end{cases}$

In Exercises 23–50, the matrices A, B, C, D, E, F, and G are defined as follows:

$$A = \begin{pmatrix} 2 & 3 \\ -1 & 4 \end{pmatrix} \quad B = \begin{pmatrix} 1 & -1 \\ 3 & 0 \end{pmatrix} \quad C = \begin{pmatrix} 1 & 0 \\ 0 & 1 \end{pmatrix}$$

$$D = \begin{pmatrix} -1 & 2 & 3 \\ 4 & 0 & 5 \end{pmatrix} \quad E = \begin{pmatrix} 2 & 1 \\ 8 & -1 \\ 6 & 5 \end{pmatrix}$$

$$F = \begin{pmatrix} 5 & -1 \\ -4 & 0 \\ 2 & 3 \end{pmatrix} \quad G = \begin{pmatrix} 0 & 0 \\ 0 & 0 \\ 0 & 0 \end{pmatrix}$$

In each exercise, carry out the indicated matrix operations if they are defined. If an operation is not defined, say so.

23. $A + B$	**24.** $A - B$	**25.** $2A + 2B$
26. $2(A + B)$	**27.** AB	**28.** BA
29. AC	**30.** CA	**31.** $3D + E$
32. $E + F$	**33.** $2F - 3G$	**34.** DE
35. ED	**36.** DF	**37.** FD
38. $A + D$	**39.** $G + A$	**40.** DG
41. GD	**42.** $(A + B) + C$	**43.** $A + (B + C)$
44. CD	**45.** DC	**46.** $5E - 3F$
47. $A^2 (= AA)$	**48.** A^2A	**49.** AA^2
50. C^2		

51. Let

$$A = \begin{pmatrix} -1 & 3 & 4 \\ 3 & 2 & -3 \\ 9 & 1 & 6 \end{pmatrix} \quad B = \begin{pmatrix} 7 & 0 & 1 \\ 0 & 0 & 3 \\ -1 & 2 & 4 \end{pmatrix}$$

$$C = \begin{pmatrix} 4 & 6 & 1 \\ 2 & 1 & 3 \\ -1 & -1 & 2 \end{pmatrix}$$

(a) Compute $A(B + C)$.   **(b)** Compute $AB + AC$.
(c) Compute $(AB)C$.   **(d)** Compute $A(BC)$.

B

52. Let $A = \begin{pmatrix} 1 & 2 \\ 3 & 4 \end{pmatrix}$ and $B = \begin{pmatrix} 5 & 6 \\ 7 & 8 \end{pmatrix}$. Let A^2 and B^2 denote the matrix products AA and BB, respectively. Compute each of the following.
(a) $(A + B)(A + B)$ **(b)** $A^2 + 2AB + B^2$
(c) $A^2 + AB + BA + B^2$

53. Let $A = \begin{pmatrix} 3 & 5 \\ 7 & 9 \end{pmatrix}$ and $B = \begin{pmatrix} 2 & 4 \\ 6 & 8 \end{pmatrix}$. Compute each of the following.
(a) $A^2 - B^2$ **(b)** $(A - B)(A + B)$
(c) $(A + B)(A - B)$ **(d)** $A^2 + AB - BA - B^2$

54. Let

$$A = \begin{pmatrix} 1 & 0 \\ 0 & 1 \end{pmatrix} \qquad B = \begin{pmatrix} 1 & 0 \\ 0 & -1 \end{pmatrix}$$

$$C = \begin{pmatrix} -1 & 0 \\ 0 & 1 \end{pmatrix} \qquad D = \begin{pmatrix} -1 & 0 \\ 0 & -1 \end{pmatrix}$$

Complete the following multiplication table.

	A	B	C	D
A				
B			D	
C				
D				

Hint: In the second row, third column, D is the proper entry because (as you can check) $BC = D$.

55. In this exercise, let us agree to write the coordinates (x, y) of a point in the plane as the 2×1 matrix $\begin{pmatrix} x \\ y \end{pmatrix}$.
(a) Let $A = \begin{pmatrix} 1 & 0 \\ 0 & -1 \end{pmatrix}$ and $Z = \begin{pmatrix} x \\ y \end{pmatrix}$. Compute the matrix AZ. After computing AZ, observe that it represents the point obtained by reflecting $\begin{pmatrix} x \\ y \end{pmatrix}$ about the x-axis.
(b) Let $B = \begin{pmatrix} -1 & 0 \\ 0 & 1 \end{pmatrix}$ and $Z = \begin{pmatrix} x \\ y \end{pmatrix}$. Compute the matrix BZ. After computing BZ, observe that it represents the point obtained by reflecting $\begin{pmatrix} x \\ y \end{pmatrix}$ about the y-axis.

(c) Let A, B, and Z represent the matrices defined in parts (a) and (b). Compute the matrix $(AB)Z$, and then interpret it in terms of reflection about the axes.

56. In this exercise, we continue to explore some of the connections between matrices and geometry. As in Exercise 55, we will use 2×1 matrices to specify the coordinates of points in the plane. Let P, S, and T be the matrices defined as follows:

$$P = \begin{pmatrix} \cos x \\ \sin x \end{pmatrix} \qquad S = \begin{pmatrix} \cos \theta & -\sin \theta \\ \sin \theta & \cos \theta \end{pmatrix}$$

$$T = \begin{pmatrix} \cos \beta & -\sin \beta \\ \sin \beta & \cos \beta \end{pmatrix}$$

Notice that the point P lies on the unit circle.
(a) Compute the matrix SP. After computing SP, observe that it represents the point on the unit circle obtained by rotating P (about the origin) through an angle θ.
(b) Show that

$$ST = TS = \begin{pmatrix} \cos(\theta + \beta) & -\sin(\theta + \beta) \\ \sin(\theta + \beta) & \cos(\theta + \beta) \end{pmatrix}$$

(c) Compute $(ST)P$. What is the angle through which P is rotated?

57. A function f is defined as follows. The domain of f is the set of all 2×2 matrices (with real entries). If $A = \begin{pmatrix} a & b \\ c & d \end{pmatrix}$, then $f(A) = ad - bc$.
(a) Let $A = \begin{pmatrix} 1 & 2 \\ 3 & 4 \end{pmatrix}$ and $B = \begin{pmatrix} 3 & -1 \\ 5 & 8 \end{pmatrix}$. Compute $f(A), f(B)$, and $f(AB)$. Is it true, in this case, that $f(A) \cdot f(B) = f(AB)$?
(b) Let $A = \begin{pmatrix} a & b \\ c & d \end{pmatrix}$ and $B = \begin{pmatrix} e & f \\ g & h \end{pmatrix}$. Show that $f(A) \cdot f(B) = f(AB)$.

58. The **trace** of a 2×2 matrix $\begin{pmatrix} a & b \\ c & d \end{pmatrix}$ is defined by

$$\mathrm{tr}\begin{pmatrix} a & b \\ c & d \end{pmatrix} = a + d$$

(a) If $A = \begin{pmatrix} 1 & 2 \\ 3 & 4 \end{pmatrix}$ and $B = \begin{pmatrix} 5 & 6 \\ 7 & 8 \end{pmatrix}$, verify that $\mathrm{tr}(A + B) = \mathrm{tr}\, A + \mathrm{tr}\, B$.
(b) If $A = \begin{pmatrix} a & b \\ c & d \end{pmatrix}$ and $B = \begin{pmatrix} e & f \\ g & h \end{pmatrix}$, show that $\mathrm{tr}(A + B) = \mathrm{tr}\, A + \mathrm{tr}\, B$.

59. Let $A = \begin{pmatrix} a & b \\ c & d \end{pmatrix}$. The **transpose** of A is the matrix denoted by A^T and defined by

$$A^T = \begin{pmatrix} a & c \\ b & d \end{pmatrix}$$

In other words, A^T is obtained by switching the columns and rows of A. Show that the following equations hold for all 2×2 matrices A and B.

(a) $(A + B)^T = A^T + B^T$ **(b)** $(A^T)^T = A$
(c) $(AB)^T = B^T A^T$

60. Find an example of two 2×2 matrices A and B for which $AB = \mathbf{0}$ but neither A nor B is $\mathbf{0}$.

7.4 ▼ THE INVERSE OF A SQUARE MATRIX

A matrix in which there are as many rows as there are columns is called a **square matrix**. So, two examples of square matrices are

$$A = \begin{pmatrix} 1 & 2 \\ 3 & 4 \end{pmatrix} \quad \text{and} \quad B = \begin{pmatrix} -5 & 6 & 7 \\ \frac{1}{2} & 0 & 1 \\ 8 & 4 & -3 \end{pmatrix}$$

The matrix A is said to be a square matrix of **order two** (or, more simply, a 2×2 matrix); B is a square matrix of **order three** (that is, a 3×3 matrix). Initially in this section, we will present the concepts and techniques in terms of square matrices of order two. After that, we'll show how the ideas carry over to larger square matrices.

We begin by defining a special matrix I_2:

$$I_2 = \begin{pmatrix} 1 & 0 \\ 0 & 1 \end{pmatrix}$$

The matrix I_2 has the following property. For every 2×2 matrix $A = \begin{pmatrix} a & b \\ c & d \end{pmatrix}$, we have (as you can easily verify)

$$\begin{pmatrix} a & b \\ c & d \end{pmatrix}\begin{pmatrix} 1 & 0 \\ 0 & 1 \end{pmatrix} = \begin{pmatrix} a & b \\ c & d \end{pmatrix} \quad \text{and} \quad \begin{pmatrix} 1 & 0 \\ 0 & 1 \end{pmatrix}\begin{pmatrix} a & b \\ c & d \end{pmatrix} = \begin{pmatrix} a & b \\ c & d \end{pmatrix}$$

In other words,

$$AI_2 = A \quad \text{and} \quad I_2 A = A$$

This shows that the matrix I_2 plays the same role in the multiplication of 2×2 matrices as does the number 1 in the multiplication of real numbers. For this reason, I_2 is called the **multiplicative identity** for square matrices of order two.

If we have two real numbers a and b such that $ab = 1$, then we say that a and b are (multiplicative) inverses. In the box that follows we apply this same terminology to matrices.

DEFINITION

The Inverse of a 2 × 2 Matrix

> If A and B are 2×2 matrices such that
>
> $$AB = I_2 \quad \text{and} \quad BA = I_2$$
>
> then A and B are said to be **inverses** of one another.

EXAMPLE 1   Find the inverse of $A = \begin{pmatrix} 1 & 2 \\ 3 & 4 \end{pmatrix}$.

Solution   We need to find numbers a, b, c, and d such that the following two matrix equations are valid.

$$\begin{pmatrix} 1 & 2 \\ 3 & 4 \end{pmatrix}\begin{pmatrix} a & b \\ c & d \end{pmatrix} = \begin{pmatrix} 1 & 0 \\ 0 & 1 \end{pmatrix} \tag{1}$$

$$\begin{pmatrix} a & b \\ c & d \end{pmatrix}\begin{pmatrix} 1 & 2 \\ 3 & 4 \end{pmatrix} = \begin{pmatrix} 1 & 0 \\ 0 & 1 \end{pmatrix} \tag{2}$$

From equation (1), we have

$$\begin{pmatrix} a + 2c & b + 2d \\ 3a + 4c & 3b + 4d \end{pmatrix} = \begin{pmatrix} 1 & 0 \\ 0 & 1 \end{pmatrix}$$

For this last equation to be valid, the corresponding entries in the two matrices must be equal. (Why?) Consequently, we obtain four equations, two involving a and c, and two involving b and d:

$$\begin{cases} a + 2c = 1 \\ 3a + 4c = 0 \end{cases} \qquad \begin{cases} b + 2d = 0 \\ 3b + 4d = 1 \end{cases}$$

As Exercise 32(a) asks you to show, the solution to the first system is $a = -2$ and $c = \frac{3}{2}$; and for the second system, $b = 1$ and $d = -\frac{1}{2}$. Furthermore (as you can check), these same values are obtained for a, b, c, and d if we begin with matrix equation (2) rather than (1). Thus, the required inverse matrix is

$$\begin{pmatrix} -2 & 1 \\ \frac{3}{2} & -\frac{1}{2} \end{pmatrix}$$

Exercise 32(b) asks you to carry out the matrix multiplication to confirm that we indeed have

$$\begin{pmatrix} 1 & 2 \\ 3 & 4 \end{pmatrix}\begin{pmatrix} -2 & 1 \\ \frac{3}{2} & -\frac{1}{2} \end{pmatrix} = \begin{pmatrix} -2 & 1 \\ \frac{3}{2} & -\frac{1}{2} \end{pmatrix}\begin{pmatrix} 1 & 2 \\ 3 & 4 \end{pmatrix}$$
$$= \begin{pmatrix} 1 & 0 \\ 0 & 1 \end{pmatrix} \qquad \blacktriangle$$

In Example 1 we found the inverse of a square matrix by solving systems of equations. However, as you've seen in previous sections, there are systems of equations that do not have solutions. In the present context, this implies that there are matrices that do not have inverses. For example (as Exercise 33 asks you to show), the matrix $\begin{pmatrix} 2 & 5 \\ 6 & 15 \end{pmatrix}$ does not have an inverse. A matrix that does not have an inverse is called a **singular matrix**. If a matrix does possess an inverse, we say that the matrix is **nonsingular**. So, according to the result in Example 1, the matrix $\begin{pmatrix} 1 & 2 \\ 3 & 4 \end{pmatrix}$ is nonsingular.

It can be proven that if a square matrix has an inverse, then that inverse is unique. (In other words, you can never find two different inverses for the same matrix.) The inverse of a nonsingular matrix A is denoted by A^{-1} (read A *inverse*).

There are several methods that can be used to compute the inverse of a matrix (or to show that there is no inverse). One method involves working with systems of equations, as in Example 1. A more efficient method, however, is the following. (We'll demonstrate the method with an example, but we won't give a formal proof of its validity.)

Suppose that we want to compute the inverse (if it exists) for the matrix

$$A = \begin{pmatrix} 6 & 2 \\ 13 & 4 \end{pmatrix}$$

We begin by writing a larger matrix, formed by merging A and I_2 as follows.

$$\begin{pmatrix} 6 & 2 & | & 1 & 0 \\ 13 & 4 & | & 0 & 1 \end{pmatrix}$$

$$A \longrightarrow \qquad \qquad I_2$$

A convenient abbreviation for this matrix is $(A \,|\, I_2)$. Now we carry out elementary row operations on this matrix until one of the following two situations occurs:

1. Either the matrix takes on the form

$$\begin{pmatrix} 1 & 0 & | & a & b \\ 0 & 1 & | & c & d \end{pmatrix}$$

$$I_2 \longrightarrow$$

In this case, $A^{-1} = \begin{pmatrix} a & b \\ c & d \end{pmatrix}$.

2. Or, one of the rows to the left of the dashed line consists entirely of zeros. In this case, A^{-1} does not exist.

The calculations for our example then run as follows.

$$\begin{pmatrix} 6 & 2 & | & 1 & 0 \\ 13 & 4 & | & 0 & 1 \end{pmatrix} \xrightarrow{\;-2R_1 + R_2\;} \begin{pmatrix} 6 & 2 & | & 1 & 0 \\ 1 & 0 & | & -2 & 1 \end{pmatrix}$$

$$\xrightarrow{\;R_1 \leftrightarrow R_2\;} \begin{pmatrix} 1 & 0 & | & -2 & 1 \\ 6 & 2 & | & 1 & 0 \end{pmatrix}$$

$$\xrightarrow{\;-6R_1 + R_2\;} \begin{pmatrix} 1 & 0 & | & -2 & 1 \\ 0 & 2 & | & 13 & -6 \end{pmatrix}$$

$$\xrightarrow{\;\frac{1}{2}R_2\;} \begin{pmatrix} 1 & 0 & | & -2 & 1 \\ 0 & 1 & | & \frac{13}{2} & -3 \end{pmatrix}$$

The inverse matrix can now be read off:

$$A^{-1} = \begin{pmatrix} -2 & 1 \\ \frac{13}{2} & -3 \end{pmatrix}$$

(You should verify for yourself that we indeed have $AA^{-1} = I_2$ and $A^{-1}A = I_2$.)

Inverse matrices can be used in solving certain systems of equations in which the number of unknowns is the same as the number of equations. Before explaining how this works, we describe how a system of equations can be written in matrix form. Consider the system

$$\begin{cases} x + 2y = 8 \\ 3x + 4y = 6 \end{cases} \tag{3}$$

As defined in the previous section, the coefficient matrix A for this system is

$$A = \begin{pmatrix} 1 & 2 \\ 3 & 4 \end{pmatrix}$$

Now we define two matrices X and B:

$$X = \begin{pmatrix} x \\ y \end{pmatrix} \qquad B = \begin{pmatrix} 8 \\ 6 \end{pmatrix}$$

Then system (3) can be written as a single matrix equation:

$$AX = B \tag{4}$$

To see why this is so, we expand equation (4) to obtain

$$\begin{pmatrix} 1 & 2 \\ 3 & 4 \end{pmatrix}\begin{pmatrix} x \\ y \end{pmatrix} = \begin{pmatrix} 8 \\ 6 \end{pmatrix} \qquad \text{using the definitions of } A, X, \text{ and } B$$

$$\begin{pmatrix} x + 2y \\ 3x + 4y \end{pmatrix} = \begin{pmatrix} 8 \\ 6 \end{pmatrix} \qquad \text{carrying out the matrix multiplication}$$

By equating the corresponding entries of the matrices in this last equation, we obtain $x + 2y = 8$ and $3x + 4y = 6$, as given initially in system (3).

EXAMPLE 2   Use an inverse matrix to solve the system

$$\begin{cases} x + 2y = 8 \\ 3x + 4y = 6 \end{cases}$$

Solution   As explained just prior to this example, the matrix form for this system is

$$AX = B \tag{5}$$

where

$$A - \begin{pmatrix} 1 & 2 \\ 3 & 4 \end{pmatrix} \qquad X - \begin{pmatrix} x \\ y \end{pmatrix} \qquad B - \begin{pmatrix} 8 \\ 6 \end{pmatrix}$$

From Example 1, we know that

$$A^{-1} = \begin{pmatrix} -2 & 1 \\ \frac{3}{2} & -\frac{1}{2} \end{pmatrix}$$

Now we multiply both sides of equation (5) by A^{-1} to obtain

$$A^{-1}(AX) = A^{-1}B$$
$$(A^{-1}A)X = A^{-1}B \qquad \text{Matrix multiplication is associative.}$$
$$I_2 X = A^{-1}B$$
$$X = A^{-1}B \qquad \text{(Why?)}$$

Substituting the actual matrices into this last equation, we have

$$\begin{pmatrix} x \\ y \end{pmatrix} = \begin{pmatrix} -2 & 1 \\ \frac{3}{2} & -\frac{1}{2} \end{pmatrix}\begin{pmatrix} 8 \\ 6 \end{pmatrix} = \begin{pmatrix} -10 \\ 9 \end{pmatrix}$$

Therefore $x = -10$ and $y = 9$, as required. ▲

All of the ideas we have discussed for square matrices of order two can be carried over rather directly to larger square matrices. It is easy to check that the following matrices, I_3 and I_4, are the multiplicative identities for 3×3 and 4×4 matrices, respectively.

$$I_3 = \begin{pmatrix} 1 & 0 & 0 \\ 0 & 1 & 0 \\ 0 & 0 & 1 \end{pmatrix} \qquad I_4 = \begin{pmatrix} 1 & 0 & 0 & 0 \\ 0 & 1 & 0 & 0 \\ 0 & 0 & 1 & 0 \\ 0 & 0 & 0 & 1 \end{pmatrix}$$

Thses identity matrices are described by saying that they have ones down the **main diagonal** and zeros everywhere else. (Larger identity matrices can be defined following this same pattern.) Sometimes, when it is clear from the context, or as a matter of convenience, we'll omit the subscript and denote the appropriately sized identity matrix simply by I. (This is done in the box that follows.)

PROPERTY SUMMARY

THE INVERSE OF A SQUARE MATRIX

1. Suppose that A and B are square matrices of the same size, and let I denote the identity matrix of the same size. Then A and B are said to be **inverses** of one another provided

$$AB = I \qquad \text{and} \qquad BA = I$$

2. It can be shown that every square matrix has at most one inverse.

3. A square matrix that has an inverse is said to be **nonsingular** (or **invertible**). A square matrix that does not have an inverse is called **singular**.

4. If the matrix A is nonsingular, then the inverse of A is denoted by A^{-1}. In this case, we have

$$AA^{-1} = I \qquad \text{and} \qquad A^{-1}A = I$$

EXAMPLE 3   Let $A = \begin{pmatrix} 5 & 0 & 2 \\ 2 & 2 & 1 \\ -3 & 1 & -1 \end{pmatrix}$. Use the elementary row operations to compute A^{-1}, if it exists.

Solution   Following the method that we described for the 2×2 case, we first write down the matrix $(A \mid I_3)$ formed by merging A and I_3:

$$\begin{pmatrix} 5 & 0 & 2 & | & 1 & 0 & 0 \\ 2 & 2 & 1 & | & 0 & 1 & 0 \\ -3 & 1 & -1 & | & 0 & 0 & 1 \end{pmatrix}$$

Now we carry out the elementary row operations, trying to obtain I_3 to the left of the dashed line. We have

$$\left(\begin{array}{rrr|rrr} 5 & 0 & 2 & 1 & 0 & 0 \\ 2 & 2 & 1 & 0 & 1 & 0 \\ -3 & 1 & -1 & 0 & 0 & 1 \end{array}\right) \xrightarrow{-2R_2 + R_1} \left(\begin{array}{rrr|rrr} 1 & -4 & 0 & 1 & -2 & 0 \\ 2 & 2 & 1 & 0 & 1 & 0 \\ -3 & 1 & -1 & 0 & 0 & 1 \end{array}\right)$$

$$\xrightarrow[3R_1 + R_3]{-2R_1 + R_2} \left(\begin{array}{rrr|rrr} 1 & -4 & 0 & 1 & -2 & 0 \\ 0 & 10 & 1 & -2 & 5 & 0 \\ 0 & -11 & -1 & 3 & -6 & 1 \end{array}\right)$$

$$\xrightarrow{1R_3 + R_2} \left(\begin{array}{rrr|rrr} 1 & -4 & 0 & 1 & -2 & 0 \\ 0 & -1 & 0 & 1 & -1 & 1 \\ 0 & -11 & -1 & 3 & -6 & 1 \end{array}\right) \xrightarrow{(-)R_2} \left(\begin{array}{rrr|rrr} 1 & -4 & 0 & 1 & -2 & 0 \\ 0 & 1 & 0 & -1 & 1 & -1 \\ 0 & -11 & -1 & 3 & -6 & 1 \end{array}\right)$$

$$\xrightarrow[11R_2 + R_3]{4R_2 + R_1} \left(\begin{array}{rrr|rrr} 1 & 0 & 0 & -3 & 2 & -4 \\ 0 & 1 & 0 & -1 & 1 & -1 \\ 0 & 0 & -1 & -8 & 5 & -10 \end{array}\right) \xrightarrow{(-1)R_3} \left(\begin{array}{rrr|rrr} 1 & 0 & 0 & -3 & 2 & -4 \\ 0 & 1 & 0 & -1 & 1 & -1 \\ 0 & 0 & 1 & 8 & -5 & 10 \end{array}\right)$$

The required inverse is therefore

$$A^{-1} = \left(\begin{array}{rrr} -3 & 2 & -4 \\ -1 & 1 & -1 \\ 8 & -5 & 10 \end{array}\right)$$

▲

▼ EXERCISE SET 7.4

A

In Exercises 1–4, the matrices A, B, C, and D are defined as follows.

$$A = \begin{pmatrix} 4 & -1 \\ -5 & 2 \end{pmatrix} \qquad B = \begin{pmatrix} \frac{1}{2} & 5 \\ 3 & 1 \end{pmatrix}$$

$$C = \begin{pmatrix} 3 & 0 & -2 \\ 0 & 5 & 6 \\ 1 & 4 & -7 \end{pmatrix} \qquad D = \begin{pmatrix} 1 & 2 & 3 \\ 4 & 5 & 6 \\ 7 & 8 & 9 \end{pmatrix}$$

1. Compute AI_2 and I_2A to verify that $AI_2 = I_2A = A$.
2. Compute BI_2 and I_2B to verify that $BI_2 = I_2B = B$.
3. Compute CI_3 and I_3C to verify that $CI_3 = I_3C = C$.
4. Compute DI_3 and I_3D to verify that $DI_3 = I_3D = D$.

In Exercises 5–12, compute A^{-1}, if it exists, using the method of Example 1.

5. $A = \begin{pmatrix} 7 & 9 \\ 4 & 5 \end{pmatrix}$

6. $A = \begin{pmatrix} 3 & -8 \\ 2 & -5 \end{pmatrix}$

7. $A = \begin{pmatrix} -3 & 1 \\ 5 & 6 \end{pmatrix}$

8. $A = \begin{pmatrix} -4 & 0 \\ 9 & 3 \end{pmatrix}$

9. $A = \begin{pmatrix} -2 & 3 \\ -4 & 6 \end{pmatrix}$

10. $A = \begin{pmatrix} \frac{5}{3} & -2 \\ -\frac{2}{3} & 1 \end{pmatrix}$

11. $A = \begin{pmatrix} \frac{1}{3} & \frac{1}{3} \\ -\frac{1}{9} & \frac{2}{9} \end{pmatrix}$

12. $A = \begin{pmatrix} -3 & 7 \\ 12 & -28 \end{pmatrix}$

In Exercises 13–26, compute the inverse matrix, if it exists, using elementary row operations (as shown in Example 3).

13. $\begin{pmatrix} 2 & 1 \\ 3 & 2 \end{pmatrix}$

14. $\begin{pmatrix} -6 & 5 \\ 18 & -15 \end{pmatrix}$

15. $\begin{pmatrix} 0 & -11 \\ 1 & 6 \end{pmatrix}$

16. $\begin{pmatrix} -2 & 13 \\ -4 & 25 \end{pmatrix}$

17. $\begin{pmatrix} \frac{2}{3} & -\frac{1}{4} \\ -8 & 3 \end{pmatrix}$

18. $\begin{pmatrix} -\frac{2}{5} & \frac{1}{3} \\ -6 & 5 \end{pmatrix}$

19. $\begin{pmatrix} -5 & 4 & -3 \\ 10 & -7 & 6 \\ 8 & -6 & 5 \end{pmatrix}$

20. $\begin{pmatrix} 1 & 0 & -2 \\ -3 & -1 & 6 \\ 2 & 1 & -5 \end{pmatrix}$

21. $\begin{pmatrix} 1 & 2 & -1 \\ 0 & 3 & 0 \\ -4 & 0 & 5 \end{pmatrix}$

22. $\begin{pmatrix} 1 & -4 & -8 \\ 1 & 2 & 5 \\ 1 & 1 & 3 \end{pmatrix}$

23. $\begin{pmatrix} -7 & 5 & 3 \\ 3 & -2 & -2 \\ 3 & -2 & -1 \end{pmatrix}$

24. $\begin{pmatrix} 2 & -1 & -1 \\ 1 & 0 & -1 \\ -2 & 1 & 2 \end{pmatrix}$

25. $\begin{pmatrix} 1 & 2 & 3 \\ 4 & 5 & 6 \\ 7 & 8 & 9 \end{pmatrix}$

26. $\begin{pmatrix} 2 & 1 & 3 \\ 4 & 5 & -7 \\ 2 & 1 & 3 \end{pmatrix}$

27. If $A = \begin{pmatrix} 3 & 8 \\ 4 & 11 \end{pmatrix}$, then $A^{-1} = \begin{pmatrix} 11 & -8 \\ -4 & 3 \end{pmatrix}$. Use this fact and the method of Example 2 to solve the following systems.

(a) $\begin{cases} 3x + 8y = 5 \\ 4x + 11y = 7 \end{cases}$ **(b)** $\begin{cases} 3x + 8y = -12 \\ 4x + 11y = 0 \end{cases}$

28. If $A = \begin{pmatrix} 3 & -7 \\ 4 & -9 \end{pmatrix}$, then $A^{-1} = \begin{pmatrix} -9 & 7 \\ -4 & 3 \end{pmatrix}$. Use this fact and the method of Example 2 to solve the following systems.

(a) $\begin{cases} 3x - 7y = 30 \\ 4x - 9y = 39 \end{cases}$ **(b)** $\begin{cases} 3x - 7y = -45 \\ 4x - 9y = -71 \end{cases}$

29. The inverse of the matrix $A = \begin{pmatrix} 3 & 2 & 6 \\ 1 & 1 & 2 \\ 2 & 2 & 5 \end{pmatrix}$ is

$A^{-1} = \begin{pmatrix} 1 & 2 & -2 \\ -1 & 3 & 0 \\ 0 & -2 & 1 \end{pmatrix}$. Use this fact to solve the following systems.

(a) $\begin{cases} 3x + 2y + 6z = 28 \\ x + y + 2z = 9 \\ 2x + 2y + 5z = 22 \end{cases}$ **(b)** $\begin{cases} 3x + 2y + 6z = -7 \\ x + y + 2z = -2 \\ 2x + 2y + 5z = -6 \end{cases}$

30. The inverse of the matrix $A = \begin{pmatrix} 1 & -1 & 1 \\ 2 & -3 & 2 \\ -4 & 6 & 1 \end{pmatrix}$ is

$A^{-1} = \begin{pmatrix} 3 & -\frac{7}{5} & -\frac{1}{5} \\ 2 & -1 & 0 \\ 0 & \frac{2}{5} & \frac{1}{5} \end{pmatrix}$. Use this fact to solve the following systems.

(a) $\begin{cases} x - y + z = 5 \\ 2x - 3y + 2z = -15 \\ -4x + 6y + z = 25 \end{cases}$

(b) $\begin{cases} x - y + z = 1 \\ 2x - 3y + 2z = -2 \\ -4x + 6y + z = 0 \end{cases}$

B

31. Let $A = \begin{pmatrix} 1 & -6 & 3 \\ 2 & -7 & 3 \\ 4 & -12 & 5 \end{pmatrix}$.

(a) Compute the matrix product AA. What do you observe?

(b) Use the result in part (a) to solve the following system.

$$\begin{cases} x - 6y + 3z = \frac{19}{2} \\ 2x - 7y + 3z = 11 \\ 4x - 12y + 5z = 19 \end{cases}$$

32. (a) Solve the following two systems and then check to see that your results agree with those given in Example 1.

$$\begin{cases} a + 2c = 1 \\ 3a + 4c = 0 \end{cases} \qquad \begin{cases} b + 2d = 0 \\ 3b + 4d = 1 \end{cases}$$

(b) At the end of Example 1, it is asserted that

$$\begin{pmatrix} 1 & 2 \\ 3 & 4 \end{pmatrix}\begin{pmatrix} -2 & 1 \\ \frac{3}{2} & -\frac{1}{2} \end{pmatrix} = \begin{pmatrix} -2 & 1 \\ \frac{3}{2} & -\frac{1}{2} \end{pmatrix}\begin{pmatrix} 1 & 2 \\ 3 & 4 \end{pmatrix}$$

$$= \begin{pmatrix} 1 & 0 \\ 0 & 1 \end{pmatrix}$$

Carry out the indicated matrix multiplications to verify that these equations are valid.

33. Use the technique in Example 1 to show that the matrix $\begin{pmatrix} 2 & 5 \\ 6 & 15 \end{pmatrix}$ does not have an inverse.

34. Use the elementary row operations (as in Example 3) to find the inverse of the following matrix.

$$\begin{pmatrix} 1 & 1 & 1 & 1 \\ 1 & 2 & 3 & 4 \\ 1 & 3 & 6 & 10 \\ 1 & 4 & 10 & 20 \end{pmatrix}$$

35. Let $A = \begin{pmatrix} 2 & 3 \\ 4 & 5 \end{pmatrix}$ and $B = \begin{pmatrix} 7 & 8 \\ 6 & 7 \end{pmatrix}$.

(a) Compute A^{-1}, B^{-1}, and $B^{-1}A^{-1}$.

(b) Compute $(AB)^{-1}$. What do you observe?

36. Let $A = \begin{pmatrix} a & b \\ c & d \end{pmatrix}$. Compute A^{-1}. (Assume that $ad - bc \neq 0$.)

7.5 ▼ DETERMINANTS AND CRAMER'S RULE

As you saw in the previous section, square matrices and their inverses can be used to solve certain systems of equations. In this section, we are going to associate a number with each square matrix. This number is called the **determinant** of the matrix. As you'll see, this too has an application in solving systems of equations.

The determinant of a matrix A may be denoted by det A or $|A|$. Determinants are also denoted simply by replacing the parentheses of matrix notation with vertical lines. Thus, three examples of determinants are

$$\begin{vmatrix} 1 & 2 \\ 3 & 4 \end{vmatrix} \qquad \begin{vmatrix} 1 & 2 & 3 \\ 4 & 5 & 6 \\ 7 & 8 & 9 \end{vmatrix} \qquad \begin{vmatrix} 3 & 7 & 8 & 9 \\ 5 & 6 & 4 & 3 \\ -9 & 9 & 0 & 1 \\ 1 & 3 & -2 & 1 \end{vmatrix}$$

A determinant with n rows and n columns is said to be an **nth-order determinant**. Therefore, the determinants we've just written are, respectively, second-, third-, and fourth-order determinants. As with matrices, we speak of the numbers in a determinant as its **entries**. We also number the rows and the columns of a determinant as we do with matrices. However, unlike matrices, each determinant has a definite value. The value of a second-order determinant is defined as follows.

DEFINITION

2 × 2 Determinant

$$\begin{vmatrix} a & b \\ c & d \end{vmatrix} = ad - bc$$

Table 1 illustrates how this definition is used to evaluate, or *expand*, a second-order determinant.

In general, the value of an nth-order determinant ($n > 2$) is defined in terms of certain determinants of order $n - 1$. For instance, the value of a third-order determinant is defined in terms of second-order determinants. We'll use the following example to introduce the necessary terminology here:

$$\begin{vmatrix} 8 & 3 & 5 \\ 2 & 4 & 6 \\ 9 & 1 & 7 \end{vmatrix}$$

Pick a given entry—say, 8—and imagine crossing out all entries occupying the same row or the same column as 8.

$$\begin{vmatrix} 8 & 3 & 5 \\ 2 & 4 & 6 \\ 9 & 1 & 7 \end{vmatrix}$$

Then we are left with the second-order determinant $\begin{vmatrix} 4 & 6 \\ 1 & 7 \end{vmatrix}$. The second-order determinant is called the **minor** of the entry 8. Similarly, to find the minor of

TABLE 1

DETERMINANT	VALUE OF DETERMINANT
$\begin{vmatrix} 3 & 7 \\ 5 & 10 \end{vmatrix}$	$\begin{vmatrix} 3 & 7 \\ 5 & 10 \end{vmatrix} = 3(10) - 7(5) = 30 - 35 = -5$
$\begin{vmatrix} 3 & -7 \\ 5 & 10 \end{vmatrix}$	$\begin{vmatrix} 3 & -7 \\ 5 & 10 \end{vmatrix} = 3(10) - (-7)(5) = 30 + 35 = 65$
$\begin{vmatrix} a & a^3 \\ 1 & a^2 \end{vmatrix}$	$\begin{vmatrix} a & a^3 \\ 1 & a^2 \end{vmatrix} = a(a^2) - a^3(1) = a^3 - a^3 = 0$

the entry 6 in the original determinant, imagine crossing out all entries that occupy the same row or the same column as 6.

$$\begin{vmatrix} 8 & 3 & 5 \\ 2 & 4 & 6 \\ 9 & 1 & 7 \end{vmatrix}$$

We are left with the second-order determinant $\begin{vmatrix} 8 & 3 \\ 9 & 1 \end{vmatrix}$, which by definition is the minor of the entry 6. In the same manner, *the minor of any element is the determinant obtained by crossing out the entries occupying the same row or column as the given element.*

Closely related to the minor of an entry in a determinant is the cofactor of that entry. The **cofactor** of an entry is defined as the minor multiplied by $+1$ or -1, according to the scheme displayed in Figure 1.

After looking at an example, we'll give a more formal rule for computing cofactors, one that will not rely on a figure and that will also apply to larger determinants.

$$\begin{vmatrix} + & - & + \\ - & + & - \\ + & - & + \end{vmatrix}$$

FIGURE 1

EXAMPLE 1   Consider the determinant

$$\begin{vmatrix} 1 & 2 & 3 \\ 4 & 5 & 6 \\ 7 & 8 & 9 \end{vmatrix}$$

Compute the minor and the cofactor of the entry 4.

Solution   By definition, we have

$$\text{minor of } 4 = \begin{vmatrix} 2 & 3 \\ 8 & 9 \end{vmatrix} = 18 - 24 = -6$$

Thus, the minor of the entry 4 is -6. To compute the cofactor of 4, we first notice that 4 is located in the second row and first column of the given determinant. Upon checking the corresponding position in Figure 1, we see a negative sign. Therefore, we have by definition

$$\text{cofactor of } 4 = (-1)(\text{minor of } 4)$$
$$= (-1)(-6) = 6$$

The cofactor of 4 is therefore 6.   ▲

The following rule tells us how cofactors can be computed without relying on Figure 1. For reference, we also restate the definition of a minor. (You should verify for yourself that this rule yields results that are consistent with Figure 1.)

DEFINITION

Minors and Cofactors

The **minor** of an entry b in a determinant is the determinant formed by suppressing the entries in the row and in the column in which b appears.

Suppose that the entry b is in the ith row and the jth column. Then the **cofactor** of b is given by the expression

$$(-1)^{i+j}(\text{minor of } b)$$

We are now prepared to state the definition that tells us how to evaluate a third-order determinant. Actually, as you'll see later, the definition is quite general and may be applied to determinants of any size.

Multiply each entry in the first row of the determinant by its cofactor and then add the results. The value of the determinant is defined to be this sum.

To see how this definition is used, let us evaluate the determinant

$$\begin{vmatrix} 1 & 2 & 3 \\ 4 & 5 & 6 \\ 7 & 8 & 9 \end{vmatrix}$$

The definition tells us to multiply each entry in the first row by its cofactor and then add the results. Carrying out this procedure, we have

$$\begin{vmatrix} 1 & 2 & 3 \\ 4 & 5 & 6 \\ 7 & 8 & 9 \end{vmatrix} = 1\begin{vmatrix} 5 & 6 \\ 8 & 9 \end{vmatrix} - 2\begin{vmatrix} 4 & 6 \\ 7 & 9 \end{vmatrix} + 3\begin{vmatrix} 4 & 5 \\ 7 & 8 \end{vmatrix}$$

$$= 1(45 - 48) - 2(36 - 42) + 3(32 - 35)$$

$$= 0 \qquad \text{(Check the arithmetic!)}$$

So the value of this particular determinant is zero. The procedure we've used here is referred to as **expanding the determinant along its first row**. The following theorem (stated here without proof) tells us that the value of a determinant can be obtained by expanding along any row or along any column; the results are the same in all cases.

Theorem

Select any row or any column in a determinant and multiply each element in that row or column by its cofactor. Then add the results. The number obtained will be the value of the determinant. (In other words, the number obtained will be the same as that obtained by expanding the determinant along its first row.)

According to this theorem, we could have evaluated the determinant

$$\begin{vmatrix} 1 & 2 & 3 \\ 4 & 5 & 6 \\ 7 & 8 & 9 \end{vmatrix}$$

by expanding it along any row or any column. Let's expand it along the second column and check to see that the result agrees with the value obtained earlier. Expanding along the second column, we have

$$\begin{vmatrix} 1 & 2 & 3 \\ 4 & 5 & 6 \\ 7 & 8 & 9 \end{vmatrix} = -2\begin{vmatrix} 4 & 6 \\ 7 & 9 \end{vmatrix} + 5\begin{vmatrix} 1 & 3 \\ 7 & 9 \end{vmatrix} - 8\begin{vmatrix} 1 & 3 \\ 4 & 6 \end{vmatrix}$$

$$= -2(36 - 42) + 5(9 - 21) - 8(6 - 12)$$

$$= -2(-6) + 5(-12) - 8(-6)$$

$$= 12 - 60 + 48 = 0 \qquad \text{as obtained previously}$$

We would have obtained the same result had we chosen to begin with any other row or column. (Exercise 13 at the end of this section asks you to verify this.)

There are three basic rules that make it easier to evaluate determinants. These are summarized in the box that follows. (Suggestions for proving these can be found in the exercises.)

PROPERTY SUMMARY   **RULES FOR MANIPULATING DETERMINANTS**

EXAMPLES

1. If each entry in a given row is multiplied by the constant k, then the value of the determinant is multiplied by k. This is true for columns, also.

$$10 \begin{vmatrix} 1 & 3 & 4 \\ 1 & 2 & 3 \\ 4 & 5 & 6 \end{vmatrix} = \begin{vmatrix} 10 & 30 & 40 \\ 1 & 2 & 3 \\ 4 & 5 & 6 \end{vmatrix}$$

$$k \begin{vmatrix} a & b & c \\ d & e & f \\ g & h & i \end{vmatrix} = \begin{vmatrix} a & kb & c \\ d & ke & f \\ g & kh & i \end{vmatrix}$$

2. If a multiple of one row is added to another row, the value of the determinant is not changed. This applies to columns, also.

$$\begin{vmatrix} a & b & c \\ d & e & f \\ g & h & i \end{vmatrix} = \begin{vmatrix} a & b & c \\ d+ka & e+kb & f+kc \\ g & h & i \end{vmatrix}$$

3. If two rows are interchanged, then the value of the determinant is multiplied by -1. This applies to columns, also.

$$\begin{vmatrix} 1 & 2 & 3 \\ 4 & 5 & 6 \\ a & b & c \end{vmatrix} = - \begin{vmatrix} 4 & 5 & 6 \\ 1 & 2 & 3 \\ a & b & c \end{vmatrix}$$

EXAMPLE 2   Evaluate the determinant $\begin{vmatrix} 15 & 14 & 26 \\ 18 & 17 & 32 \\ 21 & 20 & 42 \end{vmatrix}$.

Solution

$$\begin{vmatrix} 15 & 14 & 26 \\ 18 & 17 & 32 \\ 21 & 20 & 42 \end{vmatrix} = (3 \times 2) \begin{vmatrix} 5 & 14 & 13 \\ 6 & 17 & 16 \\ 7 & 20 & 21 \end{vmatrix}$$

using rule 1 to factor 3 from the first column and 2 from the third column

$$= 6 \begin{vmatrix} 5 & 1 & 13 \\ 6 & 1 & 16 \\ 7 & -1 & 21 \end{vmatrix}$$

using rule 2 to subtract the third column from the second column

$$= 6 \begin{vmatrix} 12 & 0 & 34 \\ 13 & 0 & 37 \\ 7 & -1 & 21 \end{vmatrix}$$

using rule 2 to add the third row to the first and second rows

$$= 6 \left[-\left(-1 \begin{vmatrix} 12 & 34 \\ 13 & 37 \end{vmatrix} \right) \right]$$

expanding the determinant along the second column

$$= 6 \begin{vmatrix} 12 & 34 \\ 1 & 3 \end{vmatrix}$$

using rule 2 to subtract the first row from the second row

$$= 6(36 - 34) = 12$$

The value of the given determinant is therefore 12. Notice the general strategy. We use rules 1 and 2 until one column (or one row) contains two zeros. At that point, it is a simple matter to expand the determinant along that column (or row). ▲

EXAMPLE 3 Show that

$$\begin{vmatrix} 1 & 1 & 1 \\ a & b & c \\ a^2 & b^2 & c^2 \end{vmatrix} = (b-a)(c-a)(c-b)$$

Solution

$$\begin{vmatrix} 1 & 1 & 1 \\ a & b & c \\ a^2 & b^2 & c^2 \end{vmatrix} = \begin{vmatrix} 1 & 0 & 0 \\ a & b-a & c-a \\ a^2 & b^2-a^2 & c^2-a^2 \end{vmatrix}$$

using rule 2 to subtract the first column from the second and the third columns

$$= \begin{vmatrix} 1 & 0 & 0 \\ a & b-a & c-a \\ a^2 & (b-a)(b+a) & (c-a)(c+a) \end{vmatrix}$$

$$= (b-a)(c-a)\begin{vmatrix} 1 & 0 & 0 \\ a & 1 & 1 \\ a^2 & b+a & c+a \end{vmatrix}$$

using rule 1 to factor $(b-a)$ from the second column and $(c-a)$ from the third column

$$= (b-a)(c-a)[(c+a)-(b+a)]$$

expanding the determinant along the first row

$$= (b-a)(c-a)(c-b)$$ ▲

The definition that we gave for third-order determinants can also be applied to define fourth-order (or larger) determinants. Consider, for example, the fourth-order determinant A given by

$$A = \begin{vmatrix} 25 & 40 & 5 & 10 \\ 9 & 0 & 3 & 6 \\ -2 & 3 & 11 & -17 \\ -3 & 4 & 7 & 2 \end{vmatrix}$$

By definition, then, we could evaluate this determinant by selecting the first row, multiplying each entry by its cofactor, and then adding the results. This yields

$$A = 25\begin{vmatrix} 0 & 3 & 6 \\ 3 & 11 & -17 \\ 4 & 7 & 2 \end{vmatrix} - 40\begin{vmatrix} 9 & 3 & 6 \\ -2 & 11 & -17 \\ -3 & 7 & 2 \end{vmatrix}$$

$$+ 5\begin{vmatrix} 9 & 0 & 6 \\ -2 & 3 & -17 \\ -3 & 4 & 2 \end{vmatrix} - 10\begin{vmatrix} 9 & 0 & 3 \\ -2 & 3 & 11 \\ -3 & 4 & 7 \end{vmatrix}$$

The problem is now reduced to evaluating four third-order determinants. If we had instead expanded down the second column, the ensuing work would be somewhat less because of the zero in that column. Nevertheless, there would still be three third-order determinants to evaluate.

Because of the amount of computation involved, we in fact rarely evaluate fourth-order determinants directly from the definition. Instead, we use the three rules (given just before Example 2) to simplify matters first. In the example that follows, we show how this is done. (We will accept the fact that the three rules are applicable in evaluating determinants of any size.)

EXAMPLE 4 Evaluate the determinant A given by

$$A = \begin{vmatrix} 25 & 40 & 5 & 10 \\ 9 & 0 & 3 & 6 \\ -2 & 3 & 11 & -17 \\ -3 & 4 & 7 & 2 \end{vmatrix}$$

Solution

$$A = 5 \times 3 \begin{vmatrix} 5 & 8 & 1 & 2 \\ 3 & 0 & 1 & 2 \\ -2 & 3 & 11 & -17 \\ -3 & 4 & 7 & 2 \end{vmatrix} = 15 \begin{vmatrix} 2 & 8 & 0 & 0 \\ 3 & 0 & 1 & 2 \\ -35 & 3 & 0 & -39 \\ -24 & 4 & 0 & -12 \end{vmatrix}$$

Factor 5 from the first row and 3 from the second row.

Subtract the second row from the first. Subtract 11 times the second row from the third. Subtract 7 times the second row from the fourth.

$$= 15 \times 2 \times 4 \begin{vmatrix} 1 & 4 & 0 & 0 \\ 3 & 0 & 1 & 2 \\ -35 & 3 & 0 & -39 \\ -6 & 1 & 0 & -3 \end{vmatrix} = 120 \cdot (-1) \begin{vmatrix} 1 & 4 & 0 \\ -35 & 3 & -39 \\ -6 & 1 & -3 \end{vmatrix}$$

Factor 2 from the first row and 4 from the fourth row.

Expand the determinant along the third column.

$$= -120 \cdot (-3) \begin{vmatrix} 1 & 4 & 0 \\ -35 & 3 & 13 \\ -6 & 1 & 1 \end{vmatrix} = 360 \begin{vmatrix} 1 & 0 & 0 \\ -35 & 143 & 13 \\ -6 & 25 & 1 \end{vmatrix}$$

Factor (-3) from the third column.

Subtract 4 times the first column from the second.

$$= 360 \begin{vmatrix} 143 & 13 \\ 25 & 1 \end{vmatrix} = 360 \begin{vmatrix} -7 & 7 \\ 25 & 1 \end{vmatrix}$$

Expand the determinant along the first row.

Subtract 6 times the second row from the first row.

$$= 360(-7 - 175) = -65520$$

Do the arithmetic!

The value of the given fourth-order determinant is therefore -65520. ▲

Determinants can be used to solve certain systems of linear equations in which there are as many unknowns as there are equations. In the box that follows, we state **Cramer's rule** for solving a system of three linear equations in three unknowns.* A more general, but entirely similar, version of Cramer's rule holds for n equations in n unknowns.

---

* The rule is named after one of its discoverers, the Swiss mathematician Gabriel Cramer (1704–1752).

Cramer's Rule

Consider the system

$$\begin{cases} a_1 x + b_1 y + c_1 z = d_1 \\ a_2 x + b_2 y + c_2 z = d_2 \\ a_3 x + b_3 y + c_3 z = d_3 \end{cases}$$

Let the four determinants D, D_x, D_y, and D_z be defined as follows:

$$D = \begin{vmatrix} a_1 & b_1 & c_1 \\ a_2 & b_2 & c_2 \\ a_3 & b_3 & c_3 \end{vmatrix} \qquad D_x = \begin{vmatrix} d_1 & b_1 & c_1 \\ d_2 & b_2 & c_2 \\ d_3 & b_3 & c_3 \end{vmatrix}$$

$$D_y = \begin{vmatrix} a_1 & d_1 & c_1 \\ a_2 & d_2 & c_2 \\ a_3 & d_3 & c_3 \end{vmatrix} \qquad D_z = \begin{vmatrix} a_1 & b_1 & d_1 \\ a_2 & b_2 & d_2 \\ a_3 & b_3 & d_3 \end{vmatrix}$$

Then if $D \neq 0$, the unique values of x, y, and z satisfying the system are given by

$$x = \frac{D_x}{D} \qquad y = \frac{D_y}{D} \qquad z = \frac{D_z}{D}$$

[If $D = 0$, the solutions (if any) can be found using Gaussian elimination.]

Notice that the determinant D in Cramer's rule is just the determinant of the coefficient matrix of the given system. If you replace the first column of D with the column of numbers on the right side of the given system, you obtain D_x. The determinants D_y and D_z are obtained similarly.

Before actually proving Cramer's rule, let's take a look at how the rule is applied.

EXAMPLE 5   Use Cramer's rule to find all solutions of the following system of equations:

$$\begin{cases} 2x + 2y - 3z = -20 \\ x - 4y + z = 6 \\ 4x - y + 2z = -1 \end{cases}$$

Solution   First, we list the determinants D, D_x, D_y, and D_z:

$$D = \begin{vmatrix} 2 & 2 & -3 \\ 1 & -4 & 1 \\ 4 & -1 & 2 \end{vmatrix} \qquad D_x = \begin{vmatrix} -20 & 2 & -3 \\ 6 & -4 & 1 \\ -1 & -1 & 2 \end{vmatrix}$$

$$D_y = \begin{vmatrix} 2 & -20 & -3 \\ 1 & 6 & 1 \\ 4 & -1 & 2 \end{vmatrix} \qquad D_z = \begin{vmatrix} 2 & 2 & -20 \\ 1 & -4 & 6 \\ 4 & -1 & -1 \end{vmatrix}$$

The calculations for evaluating D begin as follows:

$$D = \begin{vmatrix} 2 & 2 & -3 \\ 1 & -4 & 1 \\ 4 & -1 & 2 \end{vmatrix} = \begin{vmatrix} 0 & 10 & -5 \\ 1 & -4 & 1 \\ 0 & 15 & -2 \end{vmatrix}$$

Subtract twice the second row from the first.
Subtract 4 times the second row from the third.

Since we now have two zeros in the first column, it is an easy matter to expand D along that column to obtain

$$D = -1 \begin{vmatrix} 10 & -5 \\ 15 & -2 \end{vmatrix}$$
$$= -1[-20 - (-75)] = -1(55) = -55$$

The value of D is therefore -55. (Since this value is nonzero, Cramer's rule does apply.) As Exercise 31 at the end of this section asks you to verify, the values of the other three determinants are

$$D_x = 144 \qquad D_y = 61 \qquad D_z = -230$$

By Cramer's rule, then, the unique values of x, y, and z that satisfy the system are

$$x = \frac{D_x}{D} = \frac{144}{-55} = -\frac{144}{55} \qquad y = \frac{D_y}{D} = \frac{61}{-55} = -\frac{61}{55}$$

$$z = \frac{D_z}{D} = \frac{-230}{-55} = \frac{46}{11}$$

▲

One way we can prove Cramer's rule is to use Gaussian elimination to solve the system

$$\begin{cases} a_1x + b_1y + c_1z = d_1 \\ a_2x + b_2y + c_2z = d_2 \\ a_3x + b_3y + c_3z = d_3 \end{cases} \tag{1}$$

A much shorter and simpler proof, however, has been found by D. E. Whitford and M. S. Klamkin.* This is the proof we give here; it makes effective use of the rules employed in this section for manipulating determinants.

Consider the system of equations (1) and assume that $D \neq 0$. We will show that *if* x, y, and z satisfy the system, then in fact $x = D_x/D$, with similar equations giving y and z. (Exercise 62 at the end of this section then shows how to check that these values indeed satisfy the given system.) We have

$$D_x = \begin{vmatrix} d_1 & b_1 & c_1 \\ d_2 & b_2 & c_2 \\ d_3 & b_3 & c_3 \end{vmatrix} \qquad \text{by definition}$$

$$= \begin{vmatrix} (a_1x + b_1y + c_1z) & b_1 & c_1 \\ (a_2x + b_2y + c_2z) & b_2 & c_2 \\ (a_3x + b_3y + c_3z) & b_3 & c_3 \end{vmatrix} \qquad \begin{array}{l} \text{using the equations in (1)} \\ \text{to substitute for } d_1, d_2, \text{ and } d_3 \end{array}$$

$$= \begin{vmatrix} a_1x & b_1 & c_1 \\ a_2x & b_2 & c_2 \\ a_3x & b_3 & c_3 \end{vmatrix} \qquad \begin{array}{l} \text{from the first column, subtracting} \\ y \text{ times the second column as well} \\ \text{as } z \text{ times the third column} \end{array}$$

$$= x \begin{vmatrix} a_1 & b_1 & c_1 \\ a_2 & b_2 & c_2 \\ a_3 & b_3 & c_3 \end{vmatrix} \qquad \begin{array}{l} \text{factoring } x \text{ out of} \\ \text{the first column} \end{array}$$

$$= xD \qquad \text{by definition}$$

---

* The proof was published in the *American Mathematical Monthly*, **60** (1953) 186–187.

We now have $D_x = xD$, which is equivalent to $x = D_x/D$, as required. The formulas for y and z are obtained similarly.

▼ EXERCISE SET 7.5

A

In Exercises 1–6, evaluate the determinants.

1. (a) $\begin{vmatrix} 2 & -17 \\ 1 & 6 \end{vmatrix}$ (b) $\begin{vmatrix} 1 & 6 \\ 2 & -17 \end{vmatrix}$

2. (a) $\begin{vmatrix} 1 & 0 \\ 0 & 1 \end{vmatrix}$ (b) $\begin{vmatrix} 0 & 1 \\ 0 & 1 \end{vmatrix}$

3. (a) $\begin{vmatrix} 5 & 7 \\ 500 & 700 \end{vmatrix}$ (b) $\begin{vmatrix} 5 & 500 \\ 7 & 700 \end{vmatrix}$

4. (a) $\begin{vmatrix} -8 & -3 \\ 4 & -5 \end{vmatrix}$ (b) $\begin{vmatrix} -3 & -8 \\ -5 & 4 \end{vmatrix}$

5. $\begin{vmatrix} \sqrt{2}-1 & \sqrt{2} \\ \sqrt{2} & \sqrt{2}+1 \end{vmatrix}$

6. $\begin{vmatrix} \sqrt{3}+\sqrt{2} & 1+\sqrt{5} \\ 1-\sqrt{5} & \sqrt{3}-\sqrt{2} \end{vmatrix}$

In Exercises 7–12, refer to the following determinant:

$$\begin{vmatrix} -6 & 3 & 8 \\ 5 & -4 & 1 \\ 10 & 9 & -10 \end{vmatrix}$$

7. Evaluate the minor of the entry 3.

8. Evaluate the cofactor of the entry 3.

9. Evaluate the minor of -10.

10. Evaluate the cofactor of -10.

11. (a) Multiply each entry in the first row by its minor and find the sum of the results.
    (b) Multiply each entry in the first row by its cofactor and find the sum of the results.
    (c) Does the answer in part (a) or part (b) give you the value of the determinant?

12. (a) Multiply each entry in the first column by its cofactor and find the sum of the results.
    (b) Follow the same instructions as in part (a), but use the second column.
    (c) Follow the same instructions as in part (a), but use the third column.

13. Let $A = \begin{vmatrix} 1 & 2 & 3 \\ 4 & 5 & 6 \\ 7 & 8 & 9 \end{vmatrix}$. Evaluate A by expanding it along the indicated row or column.
    (a) second row   (b) third row
    (c) first column   (d) third column

In Exercises 14–21, evaluate the determinants.

14. $\begin{vmatrix} 1 & 2 & -1 \\ 2 & -1 & 1 \\ 4 & 0 & 2 \end{vmatrix}$ **15.** $\begin{vmatrix} 5 & 10 & 15 \\ 1 & 2 & 3 \\ -9 & 11 & 7 \end{vmatrix}$

16. $\begin{vmatrix} 8 & 4 & 2 \\ 3 & 9 & 3 \\ -2 & 8 & 6 \end{vmatrix}$ **17.** $\begin{vmatrix} 1 & 2 & -3 \\ 4 & 5 & -9 \\ 0 & 0 & 1 \end{vmatrix}$

18. $\begin{vmatrix} 3 & 0 & 0 \\ 0 & 19 & 0 \\ 0 & 0 & 10 \end{vmatrix}$ **19.** $\begin{vmatrix} -6 & -8 & 18 \\ 25 & 12 & 15 \\ -9 & 4 & 13 \end{vmatrix}$

20. $\begin{vmatrix} 23 & 0 & 47 \\ -37 & 0 & 18 \\ 14 & 0 & 25 \end{vmatrix}$ **21.** $\begin{vmatrix} 16 & 0 & -64 \\ -8 & 15 & -12 \\ 30 & -20 & 10 \end{vmatrix}$

22. Use the method illustrated in Example 3 in the text to show that

$$\begin{vmatrix} 1 & 1 & 1 \\ a & b & c \\ a^3 & b^3 & c^3 \end{vmatrix} = (b-a)(c-a)(c^2 - b^2 + ac - ab)$$

$$= (b-a)(c-a)(c-b)(a+b+c)$$

23. Use the method shown in Example 3 to express the determinant $\begin{vmatrix} 1 & x & x^2 \\ 1 & y & y^2 \\ 1 & z & z^2 \end{vmatrix}$ as a product.

24. Show that

$$\begin{vmatrix} 1 & 1 & 1 \\ a^2 & b^2 & c^2 \\ a^3 & b^3 & c^3 \end{vmatrix} = (b-a)(c-a)(bc^2 - b^2c + ac^2 - ab^2)$$

$$= (b-a)(c-a)(c-b)(bc+ac+ab)$$

25. Simplify the determinant $\begin{vmatrix} 1 & 1 & 1 \\ 1 & 1+x & 1 \\ 1 & 1 & 1+y \end{vmatrix}$.

26. Use the method shown in Example 3 to express the following determinant as a product of three factors:

$$\begin{vmatrix} 1 & 1 & 1 \\ a & b & c \\ bc & ca & ab \end{vmatrix}$$

In Exercises 27–30, evaluate the determinants.

27. $\begin{vmatrix} 1 & -1 & 0 & 2 \\ 0 & 1 & -1 & 0 \\ 2 & 1 & 0 & -1 \\ -2 & 2 & 1 & 1 \end{vmatrix}$ **28.** $\begin{vmatrix} 3 & -2 & 3 & 4 \\ 1 & 4 & -3 & 2 \\ 6 & 3 & -6 & -3 \\ -1 & 0 & 1 & 5 \end{vmatrix}$

29. $\begin{vmatrix} 2 & 0 & 0 & 0 \\ 0 & 3 & 0 & 0 \\ 0 & 0 & 4 & 0 \\ 0 & 0 & 0 & 5 \end{vmatrix}$ **30.** $\begin{vmatrix} 7 & -8 & 1 & 2 \\ 21 & 4 & 3 & -1 \\ -35 & 8 & 3 & -2 \\ 14 & 16 & 0 & 1 \end{vmatrix}$

31. Verify the following statements (from Example 5).

(a) $\begin{vmatrix} -20 & 2 & -3 \\ 6 & -4 & 1 \\ -1 & -1 & 2 \end{vmatrix} = 144$

(b) $\begin{vmatrix} 2 & -20 & -3 \\ 1 & 6 & 1 \\ 4 & -1 & 2 \end{vmatrix} = 61$

(c) $\begin{vmatrix} 2 & 2 & -20 \\ 1 & -4 & 6 \\ 4 & -1 & -1 \end{vmatrix} = -230$

32. Consider the following system:

$$\begin{cases} x + y + 3z = 0 \\ x + 2y + 5z = 0 \\ x - 4y - 8z = 0 \end{cases}$$

(a) Without doing any calculations, find one obvious solution of this system.
(b) Calculate the determinant D.
(c) List all solutions of this system.

In Exercises 33–42, use Cramer's rule to solve those systems for which $D \neq 0$. In cases where $D = 0$, use Gaussian elimination or matrix methods.

33. $\begin{cases} 3x + 4y - z = 5 \\ x - 3y + 2z = 2 \\ 5x - 6z = -7 \end{cases}$

34. $\begin{cases} 3A - B - 4C = 3 \\ A + 2B - 3C = 9 \\ 2A - B + 2C = -8 \end{cases}$

35. $\begin{cases} 3x + 2y - z = -6 \\ 2x - 3y - 4z = -11 \\ x + y + z = 5 \end{cases}$

36. $\begin{cases} 5x - 3y - z = 16 \\ 2x + y - 3z = 5 \\ 3x - 2y + 2z = 5 \end{cases}$

37. $\begin{cases} 2x + 5y + 2z = 0 \\ 3x - y - 4z = 0 \\ x + 2y - 3z = 0 \end{cases}$

38. $\begin{cases} 4u + 3v - 2w = 14 \\ u + 2v - 3w = 6 \\ 2u - v + 4w = 2 \end{cases}$

39. $\begin{cases} 12x - 11z = 13 \\ 6x + 6y - 4z = 26 \\ 6x + 2y - 5z = 13 \end{cases}$

40. $\begin{cases} 3x + 4y + 2z = 1 \\ 4x + 6y + 2z = 7 \\ 2x + 3y + z = 11 \end{cases}$

41. $\begin{cases} x + y + z + w = -7 \\ x - y + z - w = -11 \\ 2x - 2y - 3z - 3w = 26 \\ 3x + 2y + z - w = -9 \end{cases}$

42. $\begin{cases} 2A - B - 3C + 2D = -2 \\ A - 2B + C - 3D = 4 \\ 3A - 4B + 2C - 4D = 12 \\ 2A + 3B - C - 2D = -4 \end{cases}$

43. Find all values of x for which

$$\begin{vmatrix} x - 4 & 0 & 0 \\ 0 & x + 4 & 0 \\ 0 & 0 & x + 1 \end{vmatrix} = 0$$

44. Find all values of x for which

$$\begin{vmatrix} 1 & x & x^2 \\ 1 & 1 & 1 \\ 4 & 5 & 0 \end{vmatrix} = 0$$

45. By expanding the determinant $\begin{vmatrix} a & b & c \\ a & b & c \\ d & e & f \end{vmatrix}$ down the first column, show that its value is zero.

46. By expanding the determinant $\begin{vmatrix} ka & kb & kc \\ d & e & f \\ g & h & i \end{vmatrix}$ along its first row, show that it is equal to

$$k \begin{vmatrix} a & b & c \\ d & e & f \\ g & h & i \end{vmatrix}$$

B

47. Show that

$$\begin{vmatrix} a_1 + A_1 & b_1 & c_1 \\ a_2 + A_2 & b_2 & c_2 \\ a_3 + A_3 & b_3 & c_3 \end{vmatrix} = \begin{vmatrix} a_1 & b_1 & c_1 \\ a_2 & b_2 & c_2 \\ a_3 & b_3 & c_3 \end{vmatrix} + \begin{vmatrix} A_1 & b_1 & c_1 \\ A_2 & b_2 & c_2 \\ A_3 & b_3 & c_3 \end{vmatrix}$$

48. Show that

$$\begin{vmatrix} a_1 & b_1 & c_1 \\ a_2 & b_2 & c_2 \\ a_3 & b_3 & c_3 \end{vmatrix} = \begin{vmatrix} a_1 + kb_1 & b_1 & c_1 \\ a_2 + kb_2 & b_2 & c_2 \\ a_3 + kb_3 & b_3 & c_3 \end{vmatrix}$$

49. By expanding each determinant along a row or column, show that

$$\begin{vmatrix} a_1 & b_1 & c_1 \\ a_2 & b_2 & c_2 \\ a_3 & b_3 & c_3 \end{vmatrix} = - \begin{vmatrix} a_2 & b_2 & c_2 \\ a_1 & b_1 & c_1 \\ a_3 & b_3 & c_3 \end{vmatrix}$$

50. Solve for x in terms of a, b, and c:

$$\begin{vmatrix} a & a & x \\ c & c & c \\ b & x & b \end{vmatrix} = 0 \qquad (c \neq 0)$$

51. Show that

$$\begin{vmatrix} 1 + a & 1 & 1 & 1 \\ 1 & 1 + b & 1 & 1 \\ 1 & 1 & 1 + c & 1 \\ 1 & 1 & 1 & 1 + d \end{vmatrix}$$

$$= abcd\left(\frac{1}{a} + \frac{1}{b} + \frac{1}{c} + \frac{1}{d} + 1\right)$$

52. Show that

$$\begin{vmatrix} 1 & a & a^2 \\ a^2 & 1 & a \\ a & a^2 & 1 \end{vmatrix} = (a^3 - 1)^2$$

53. Consider the determinant $D = \begin{vmatrix} a_1 & b_1 & c_1 \\ a_2 & b_2 & c_2 \\ a_3 & b_3 & c_3 \end{vmatrix}$. Let A_1 denote the cofactor of a_1, let B_1 denote the cofactor of b_1, and so on. Prove that

$$\begin{vmatrix} B_2 & C_2 \\ B_3 & C_3 \end{vmatrix} = a_1 D$$

54. Show that

$$\begin{vmatrix} 1 & bc & b + c \\ 1 & ca & c + a \\ 1 & ab & a + b \end{vmatrix} = (b - c)(c - a)(a - b)$$

55. Find the value of each determinant.

(a) $\begin{vmatrix} 1 & 0 & 0 \\ x & 1 & 0 \\ x & y & 1 \end{vmatrix}$ **(b)** $\begin{vmatrix} 1 & 0 & 0 & 0 \\ x & 1 & 0 & 0 \\ x & y & 1 & 0 \\ x & y & z & 1 \end{vmatrix}$

56. Evaluate the determinant.

$$\begin{vmatrix} \sin\theta & \cos\theta & 0 \\ -\cos\theta & \sin\theta & 0 \\ 1 & 0 & 1 \end{vmatrix}$$

57. Show that

$$\begin{vmatrix} 1 & a & a & a \\ 1 & b & a & a \\ 1 & a & b & a \\ 1 & a & a & b \end{vmatrix} = (b - a)^3$$

58. Show that

$$\begin{vmatrix} a & 1 & 1 & 1 \\ 1 & a & 1 & 1 \\ 1 & 1 & a & 1 \\ 1 & 1 & 1 & a \end{vmatrix} = (a - 1)^3(a + 3)$$

59. Solve the following system for x, y, and z:

$$\begin{cases} ax + by + cz = k \\ a^2x + b^2y + c^2z = k^2 \\ a^3x + b^3y + c^3z = k^3 \end{cases}$$

(Assume that the values of a, b, and c are all distinct and all nonzero.)

60. Show that

$$\begin{vmatrix} x + a & a & a & a \\ b & x + b & b & b \\ c & c & x + c & c \\ d & d & d & x + d \end{vmatrix} = x^3(a + b + c + d + x)$$

Hint: To the first row add each of the three other rows. For convenience, let $t = a + b + c + d + x$.

C

61. Show that the area of the triangle in Figure A is $\frac{1}{2}\begin{vmatrix} a & b \\ c & d \end{vmatrix}$.

Hint: Figure B indicates how the required area may be found using a rectangle and three right triangles.

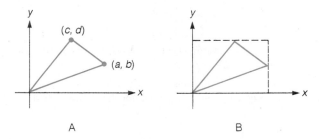

A   B

62. This exercise completes the derivation of Cramer's rule given in the text. Using the same notation, and assuming $D \neq 0$, we need to show that the values $x = D_x/D$, $y = D_y/D$, and $z = D_z/D$ satisfy the equations in (1) on page 428. We will show that these values satisfy the first equation in (1), the verification for the other equations being entirely similar.

(a) Check that substituting the values $x = D_x/D$, $y = D_y/D$, and $z = D_z/D$ in the first equation of (1) yields an equation equivalent to

$$a_1 D_x + b_1 D_y + c_1 D_z - d_1 D = 0$$

(b) Show that the equation in part (a) may be written

$$a_1 \begin{vmatrix} b_1 & c_1 & d_1 \\ b_2 & c_2 & d_2 \\ b_3 & c_3 & d_3 \end{vmatrix} - b_1 \begin{vmatrix} a_1 & c_1 & d_1 \\ a_2 & c_2 & d_2 \\ a_3 & c_3 & d_3 \end{vmatrix}$$

$$+ c_1 \begin{vmatrix} a_1 & b_1 & d_1 \\ a_2 & b_2 & d_2 \\ a_3 & b_3 & d_3 \end{vmatrix} - d_1 \begin{vmatrix} a_1 & b_1 & c_1 \\ a_2 & b_2 & c_2 \\ a_3 & b_3 & c_3 \end{vmatrix} = 0$$

(c) Show that the equation in part (b) may be written

$$\begin{vmatrix} a_1 & b_1 & c_1 & d_1 \\ a_1 & b_1 & c_1 & d_1 \\ a_2 & b_2 & c_2 & d_2 \\ a_3 & b_3 & c_3 & d_3 \end{vmatrix} = 0$$

(d) Now explain why the equation in (c) indeed holds.

63. Let D denote the determinant of the matrix $\begin{pmatrix} a & b \\ c & d \end{pmatrix}$.

(a) Show that the inverse of this matrix is
$\dfrac{1}{D}\begin{pmatrix} d & -b \\ -c & a \end{pmatrix}$. Assume that $ad - bc \neq 0$.

(b) Use the result in part (a) to find the inverse of the
matrix $\begin{pmatrix} -6 & 7 \\ 1 & 9 \end{pmatrix}$.

64. Assuming that $p + q + r = 0$, solve the following
equation for x:

$$\begin{vmatrix} p-x & r & q \\ r & q-x & p \\ q & p & r-x \end{vmatrix} = 0$$

Hint: After adding certain rows or columns, the quantity
$p + q + r$ will appear.

65. The following figure shows a parabola that passes
through the origin and through the two points (x_1, y_1)
and (x_2, y_2). Show that the x-coordinate of the vertex is
given by

$$x = \frac{\begin{vmatrix} x_1^2 & y_1 \\ x_2^2 & y_2 \end{vmatrix}}{2\begin{vmatrix} x_1 & y_1 \\ x_2 & y_2 \end{vmatrix}}$$

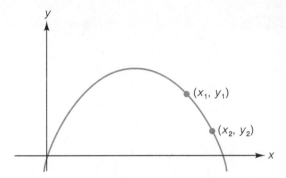

7.6 ▼ NONLINEAR SYSTEMS OF EQUATIONS

In the previous sections of this chapter, we looked at several techniques for solving
systems of linear equations. In particular, we saw that if solutions exist, they can
always be found by Gaussian elimination. In the present section, we consider
nonlinear systems of equations, that is, systems in which at least one of the
equations is not linear. There is no single technique that serves to solve all nonlinear systems. However, simple substitution will often suffice. The work in this
section focuses on examples showing some of the more common approaches. In
all the examples and in the exercises, we will be concerned exclusively with solutions (x, y) in which both x and y are real numbers. In Example 1, the system
consists of one linear equation and one quadratic equation. Such a system can
always be solved by substitution.

EXAMPLE 1 Find all solutions (x, y) of the following system, where x and y are real numbers:

$$\begin{cases} 2x + y = 1 \\ y = 4 - x^2 \end{cases}$$

Solution We use the second equation to substitute for y in the first equation. This will
yield an equation with only one unknown:

$$2x + (4 - x^2) = 1$$
$$-x^2 + 2x + 3 = 0$$
$$x^2 - 2x - 3 = 0$$
$$(x - 3)(x + 1) = 0$$
$$x - 3 = 0 \quad \bigm| \quad x + 1 = 0$$
$$x = 3 \quad \bigm| \quad x = -1$$

The values $x = 3$ and $x = -1$ can now be substituted back into either of the original equations. Substituting $x = 3$ in the equation $y = 4 - x^2$ yields $y = -5$. Similarly, substituting $x = -1$ in the equation $y = 4 - x^2$ gives us $y = 3$. We have now determined the two ordered pairs $(3, -5)$ and $(-1, 3)$. As you can easily check, both of these are solutions of the given system. Figure 1 displays the graphical interpretation of this result. The line $2x + y = 1$ intersects the parabola $y = 4 - x^2$ at the points $(-1, 3)$ and $(3, -5)$. ▲

EXAMPLE 2   Where do the graphs of the parabola $y = x^2$ and the circle $x^2 + y^2 = 1$ intersect? See Figure 2.

Solution   The system that we wish to solve is

$$\begin{cases} y = x^2 & (1) \\ x^2 + y^2 = 1 & (2) \end{cases}$$

In view of equation (1), we can replace the x^2-term of equation (2) by y. Doing this yields

$$y + y^2 = 1$$

or

$$y^2 + y - 1 = 0$$

This last equation can be solved by using the quadratic formula with $a = 1$, $b = 1$, and $c = -1$. As you can check, the results are

$$y = \frac{-1 + \sqrt{5}}{2} \quad \text{and} \quad y = \frac{-1 - \sqrt{5}}{2}$$

However, from Figure 2, it is clear that the y-coordinate at each intersection point is positive. Therefore, we discard the negative number $y = -\frac{1}{2}(1 + \sqrt{5})$ from further consideration in this context. Substituting the positive number $y = \frac{1}{2}(-1 + \sqrt{5})$ back in the equation $y = x^2$ then gives us

$$x^2 = \tfrac{1}{2}(-1 + \sqrt{5})$$

Therefore,

$$x = \pm\sqrt{\tfrac{1}{2}(-1 + \sqrt{5})}$$

By choosing the positive square root, we obtain the x-coordinate for the intersection point in the first quadrant. That point is therefore

$$\left(\sqrt{\tfrac{1}{2}(-1 + \sqrt{5})}, \tfrac{1}{2}(-1 + \sqrt{5})\right) \approx (0.79, 0.62) \quad \text{using a calculator}$$

Similarly, the negative square root yields the x-coordinate for the intersection point in the second quadrant. That point is

$$\left(-\sqrt{\tfrac{1}{2}(-1 + \sqrt{5})}, \tfrac{1}{2}(-1 + \sqrt{5})\right) \approx (-0.79, 0.62) \quad \text{using a calculator}$$

We have now found the two intersection points, as required. ▲

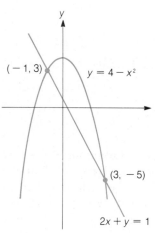

$(-1, 3)$

$y = 4 - x^2$

$(3, -5)$

$2x + y = 1$

FIGURE 1

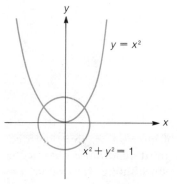

$y = x^2$

$x^2 + y^2 = 1$

FIGURE 2

EXAMPLE 3 Find all solutions (x, y) of the following system, where x and y are real numbers:

$$\begin{cases} xy = 1 & (3) \\ y = 3x + 1 & (4) \end{cases}$$

Solution

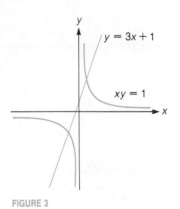

FIGURE 3

Since these equations are easy to graph, we do so, because that will tell us something about the required solutions. As Figure 3 indicates, there are two intersection points, one in the first quadrant, the other in the third quadrant. One way to begin now would be to solve equation (3) for one unknown in terms of the other. However, to avoid introducing fractions at the outset, let us use equation (4) to substitute for y in equation (3). This yields

$$x(3x + 1) = 1$$

or

$$3x^2 + x - 1 = 0$$

This last equation can be solved by using the quadratic formula. As you should verify, the solutions are

$$x = \frac{-1 + \sqrt{13}}{6} \qquad \text{and} \qquad x = \frac{-1 - \sqrt{13}}{6}$$

The corresponding y-values can now be obtained by substituting for x in either of the given equations. We will substitute in equation (4). (Exercise 23 at the end of this section asks you to substitute in equation (3) as well and then to show that the very different-looking answers obtained in that way are in fact equal to those found here.) Substituting $x = \frac{1}{6}(-1 + \sqrt{13})$ in the equation $y = 3x + 1$ gives us

$$y = 3\left(\frac{-1 + \sqrt{13}}{6}\right) + 1$$

$$= \frac{-1 + \sqrt{13}}{2} + \frac{2}{2} = \frac{1 + \sqrt{13}}{2}$$

Thus, one of the intersection points is

$$\left(\tfrac{1}{6}(-1 + \sqrt{13}), \tfrac{1}{2}(1 + \sqrt{13})\right)$$

Notice that this must be the first-quadrant point of intersection, since both coordinates are positive. The intersection point in the third quadrant is obtained in exactly the same manner. As you can check, substituting the value $x = -\frac{1}{6}(1 + \sqrt{13})$ in the equation $y = 3x + 1$ yields $y = \frac{1}{2}(1 - \sqrt{13})$. Thus, the other intersection point is

$$\left(-\tfrac{1}{6}(1 + \sqrt{13}), \tfrac{1}{2}(1 - \sqrt{13})\right)$$

As Figure 3 indicates, there are no other solutions. ▲

EXAMPLE 4 Find all real numbers x and y satisfying the system of equations

$$\begin{cases} y = \sqrt{x} \\ (x + 2)^2 + y^2 = 1 \end{cases}$$

Solution

We use the first equation to substitute for y in the second equation. This yields

$$(x + 2)^2 + (\sqrt{x})^2 = 1$$
$$x^2 + 4x + 4 + x = 1$$
$$x^2 + 5x + 3 = 0$$

Using the quadratic formula to solve this last equation, we have

$$x = \frac{-5 \pm \sqrt{5^2 - 4(1)(3)}}{2(1)} = \frac{-5 \pm \sqrt{13}}{2}$$

The two values of x are thus

$$-\tfrac{1}{2}(5 - \sqrt{13}) \quad \text{and} \quad -\tfrac{1}{2}(5 + \sqrt{13})$$

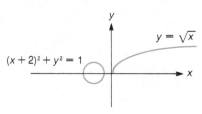

$y = \sqrt{x}$

$(x + 2)^2 + y^2 = 1$

FIGURE 4

However, notice that both of these quantities are negative. (The second is obviously negative; without using a calculator, can you explain why the first is negative?) Thus, neither of these quantities is an appropriate x-input in the equation $y = \sqrt{x}$, since we are looking for y-values that are real numbers. We conclude from this that there are no pairs of real numbers x and y satisfying the given system. In geometric terms, this means that the two graphs do not intersect. See Figure 4.

▲

In the next example, we look at a system that can be reduced to a linear system through appropriate substitutions.

EXAMPLE 5

Solve the system

$$\begin{cases} \dfrac{2}{x^2} - \dfrac{3}{y^2} = -6 \\ \dfrac{3}{x^2} + \dfrac{4}{y^2} = 59 \end{cases}$$

Solution

Let $u = 1/x^2$ and $v = 1/y^2$, so that the system becomes

$$\begin{cases} 2u - 3v = -6 \\ 3u + 4v = 59 \end{cases}$$

This is now a linear system. As Exercise 44 asks you to show, the solution is $u = 9$, $v = 8$. In view of the definitions of u and v, then, we have

$$\frac{1}{x^2} = 9 \qquad \qquad \frac{1}{y^2} = 8$$
$$x^2 = 1/9 \qquad \qquad y^2 = 1/8$$
$$x = \pm 1/3 \qquad \qquad y = \pm 1/\sqrt{8} = \pm 1/(2\sqrt{2})$$
$$= \pm \sqrt{2}/4$$

This gives us four possible solutions for the original system:

$$\left(\frac{1}{3}, \frac{\sqrt{2}}{4}\right) \quad \left(\frac{1}{3}, -\frac{\sqrt{2}}{4}\right) \quad \left(-\frac{1}{3}, \frac{\sqrt{2}}{4}\right) \quad \left(-\frac{1}{3}, -\frac{\sqrt{2}}{4}\right)$$

As you can check, all four of these pairs satisfy the given system.

▲

EXAMPLE 6 Determine all solutions (x, y) of the following system, where x and y are real numbers:

$$\begin{cases} y = 3^x & (5) \\ y = 3^{2x} - 2 & (6) \end{cases}$$

Solution We'll use the substitution method. First we rewrite equation (6) as

$$y = (3^x)^2 - 2 \qquad (7)$$

Now, in view of equation (5), we can replace 3^x with y in equation (7) to obtain

$$\begin{aligned} y &= y^2 - 2 \\ 0 &= y^2 - y - 2 \\ &= (y + 1)(y - 2) \end{aligned}$$

From this last equation, we see that $y = -1$ or $y = 2$. With $y = -1$, equation (5) becomes $-1 = 3^x$, contrary to the fact that 3^x is positive for all real numbers x. Thus, we discard the case where $y = -1$. On the other hand, if $y = 2$, equation (5) becomes

$$2 = 3^x$$

We can solve this exponential equation by taking the logarithm of both sides. Using base e logarithms, we have

$$\ln 2 = \ln 3^x = x \ln 3$$

and, consequently,

$$x = \frac{\ln 2}{\ln 3} \qquad Caution: \frac{\ln 2}{\ln 3} \neq \ln 2 - \ln 3$$

We've now found that $x = (\ln 2)/(\ln 3)$ and $y = 2$. Figure 5 displays a graphical interpretation of this result. (Using a calculator for the x-coordinate, we find that $x \approx 0.6$, which is consistent with Figure 5.) ▲

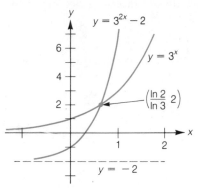

FIGURE 5

▼ EXERCISE SET 7.6

A

In Exercises 1–22, find all solutions (x, y) of the given systems, where x and y are real numbers.

1. $\begin{cases} y = 3x \\ y = x^2 \end{cases}$

2. $\begin{cases} y = x + 3 \\ y = 9 - x^2 \end{cases}$

3. $\begin{cases} x^2 + y^2 = 25 \\ 24y = x^2 \end{cases}$

4. $\begin{cases} 3x + 4y = 12 \\ x^2 - y + 1 = 0 \end{cases}$

5. $\begin{cases} xy = 1 \\ y = -x^2 \end{cases}$

6. $\begin{cases} x + 2y = 0 \\ xy = -2 \end{cases}$

7. $\begin{cases} 2x^2 + y^2 = 17 \\ x^2 + 2y^2 = 22 \end{cases}$

8. $\begin{cases} x - 2y = 1 \\ y^2 - x^2 = 3 \end{cases}$

9. $\begin{cases} y = 1 - x^2 \\ y = x^2 - 1 \end{cases}$

10. $\begin{cases} xy = 4 \\ y = 4x \end{cases}$

11. $\begin{cases} xy = 4 \\ y = 4x + 1 \end{cases}$

12. $\begin{cases} \dfrac{2}{x^2} + \dfrac{5}{y^2} = 3 \\ \dfrac{3}{x^2} - \dfrac{2}{y^2} = 1 \end{cases}$

13. $\begin{cases} \dfrac{1}{x^2} - \dfrac{3}{y^2} = 14 \\ \dfrac{2}{x^2} + \dfrac{1}{y^2} = 35 \end{cases}$

14. $\begin{cases} y = -\sqrt{x} \\ (x - 3)^2 + y^2 = 4 \end{cases}$

15. $\begin{cases} y = -\sqrt{x - 1} \\ (x - 3)^2 + y^2 = 4 \end{cases}$

16. $\begin{cases} y = -\sqrt{x - 6} \\ (x - 3)^2 + y^2 = 4 \end{cases}$

17. $\begin{cases} y = 2^x \\ y = 2^{2x} - 12 \end{cases}$

18. $\begin{cases} y = e^{4x} \\ y = e^{2x} + 6 \end{cases}$

19. $\begin{cases} 2(\log_{10} x)^2 - (\log_{10} y)^2 = -1 \\ 4(\log_{10} x)^2 - 3(\log_{10} y)^2 = -11 \end{cases}$

20. $\begin{cases} y = \log_2(x + 1) \\ y = 5 - \log_2(x - 3) \end{cases}$

21. $\begin{cases} 2^x \cdot 3^y = 4 \\ x + y = 5 \end{cases}$

22. $\begin{cases} a^{2x} + a^{2y} = 10 \\ a^{x+y} = 4 \end{cases}$ $(a > 0)$

Hint: Use the substitutions $a^x = t$ and $a^y = u$.

23. Let $x = \frac{1}{6}(-1 + \sqrt{13})$. Using this x-value, show that the equations $y = 3x + 1$ and $y = 1/x$ yield the same y-value.

24. A sketch shows that the line $y = 100x$ intersects the parabola $y = x^2$ at the origin. Are there any other intersection points? If so, find them. If not, explain why not.

25. Solve the following system for x and y:

$\begin{cases} ax + by = 2 \\ abxy = 1 \end{cases}$

(Assume that neither a nor b is zero.)

B
___

26. Let a, b, and c be constants (with $a \neq 0$), and consider the system

$\begin{cases} y = ax^2 + bx + c \\ y = k \end{cases}$

For which value of k (in terms of a, b, and c) will the system have exactly one solution? What is that solution? What is the relationship between the solution you've found and the graph of $y = ax^2 + bx + c$?

27. Find all solutions of the system

$\begin{cases} x^3 + y^3 = 3473 \\ x + y = 23 \end{cases}$

28. Solve the following system for x, y, and z:

$\begin{cases} yz = p^2 \\ zx = q^2 \\ xy = r^2 \end{cases}$

(Assume that p, q, and r are positive constants.)

29. If the diagonal of a rectangle has length d and the perimeter of the rectangle is $2p$, express the lengths of the sides in terms of d and p.

30. Solve the following system for x and y:

$\begin{cases} xy + pq = 2px \\ x^2y^2 + p^2q^2 = 2q^2y^2 \end{cases}$ $(pq \neq 0)$

Hint: Square the first equation, then substract the second. This results in an equation that can be written $(2px + qy)(px - qy) = 0$.

31. Solve the following system for u and v:

$\begin{cases} u^2 - v^2 = 9 \\ \sqrt{u + v} + \sqrt{u - v} = 4 \end{cases}$

Hint: Square the second equation.

32. Solve the given system for x and y in terms of p and q:

$\begin{cases} q^{\ln x} = p^{\ln y} \\ (px)^{\ln a} = (qy)^{\ln b} \end{cases}$

(Assume all constants and variables are positive.)

33. Solve the following system for x, y, and z in terms of p, q, and r:

$\begin{cases} x(x + y + z) = p^2 \\ y(x + y + z) = q^2 \\ z(x + y + z) = r^2 \end{cases}$

(Assume that p, q, and r are nonzero.) *Hint:* Denote $x + y + z$ by w; then add the three equations.

34. (a) Find the points where the line $y = -2x - 2$ intersects the parabola $y = \frac{1}{2}x^2$.

(b) On the same set of axes, sketch the line $y = -2x - 2$ and the parabola $y = \frac{1}{2}x^2$. Be certain that your sketch is consistent with the results obtained in part (a).

35. The area of a right triangle is 180 cm^2, and the hypotenuse is 41 cm. Find the lengths of the two legs.

36. The sum of two numbers is 8, while their product is -128. What are the two numbers?

37. The perimeter of a rectangle is 46 cm, and the area is 60 cm^2. Find the length and the width.

38. Find all right triangles for which the perimeter is 24 units and the area is 24 square units.

39. Solve the following system for x and y using the substitution method:

$\begin{cases} x^2 + y^2 = 5 & (1) \\ xy = 2 & (2) \end{cases}$

40. The substitution method in Exercise 39 leads to a quadratic equation. Here is an alternative approach to solving that system; this approach leads to linear equations. Multiply equation (2) by 2 and add the resulting equation to equation (1). Now take square roots to conclude that $x + y = \pm 3$. Next multiply equation (2) by 2 and subtract the resulting equation from equation (1). Take square roots to conclude that $x - y = \pm 1$. You now have the following four linear systems, each of which can be solved (with almost no work) by the addition–subtraction method:

$\begin{cases} x + y = 3 \\ x - y = 1 \end{cases}$ $\begin{cases} x + y = 3 \\ x - y = -1 \end{cases}$

$\begin{cases} x + y = -3 \\ x - y = 1 \end{cases}$ $\begin{cases} x + y = -3 \\ x - y = -1 \end{cases}$

Solve these systems and compare your results with those obtained in Exercise 39.

41. Solve the following system using the method explained in Exercise 40:

$$\begin{cases} x^2 + y^2 = 7 \\ xy = 3 \end{cases}$$

42. Solve the following system using the substitution method:

$$\begin{cases} 3xy - 4x^2 = 2 \\ -5x^2 + 3y^2 = 7 \end{cases}$$

(Begin by solving the first equation for y.)

43. Here is an alternative approach for solving the system in Exercise 42. Let $y = mx$, where m is a constant to be determined. Replace y with mx in both equations of the system to obtain the following pair of equations:

$$x^2(3m - 4) = 2 \qquad \qquad (1)$$
$$x^2(-5 + 3m^2) = 7 \qquad \qquad (2)$$

Now divide equation (2) by equation (1). After clearing fractions and simplifying, you can write the resulting equation as $2m^2 - 7m + 6 = 0$. Solve this last equation by factoring. The values of m can then be used in equation (1) to determine values for x. In each case, the corresponding y-values are determined by the equation $y = mx$.

44. Solve the following system. (You should obtain $u = 9$ and $v = 8$, as stated in Example 5.)

$$\begin{cases} 2u - 3v = -6 \\ 3u + 4v = 59 \end{cases}$$

45. Solve the following system for x and y.

$$xy + pq = 2px$$
$$x^2y^2 + p^2q^2 = 2q^2y^2$$

Hint: Square the first equation, then subtract the second. This results in an equation that can be factored.

46. Solve the following system for x and y:

$$\begin{cases} \dfrac{1}{x^2} + \dfrac{1}{xy} = \dfrac{1}{a^2} \\ \dfrac{1}{y^2} + \dfrac{1}{xy} = \dfrac{1}{b^2} \end{cases}$$

(Assume that a and b are positive.)

47. Solve the following system for x and y:

$$\begin{cases} x^4 = y^6 \\ \ln \dfrac{x}{y} = \dfrac{\ln x}{\ln y} \end{cases}$$

48. Find two acute angles such that the sum of their sines is 1 and the sum of their cosines is $\frac{3}{2}$. Express the answers both in degrees and in radians. (Round to two decimal places.)

7.7 ▼ SYSTEMS OF INEQUALITIES

In the first section of this chapter, we solved systems of equations in two unknowns. Now let us consider systems of inequalities in two unknowns. The techniques we develop are used in calculus when discussing functions of two variables, and in business and economics in the study of linear programming.

Let a, b, and c denote real numbers, and assume that a and b are not both zero. Then all of the following are called **linear inequalities**:

$$ax + by + c < 0 \qquad ax + by + c > 0$$
$$ax + by + c \le 0 \qquad ax + by + c \ge 0$$

An ordered pair of numbers (x_0, y_0) is said to be a **solution** of a given inequality (linear or not) provided we obtain a true statement upon substituting x_0 and y_0 for x and y, respectively. For instance, $(-1, 1)$ is a solution of the inequality $2x + 3y - 6 < 0$, since substitution yields

$$2(-1) + 3(1) - 6 < 0$$
$$-5 < 0 \qquad \text{True}$$

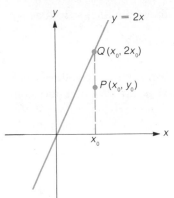

FIGURE 1

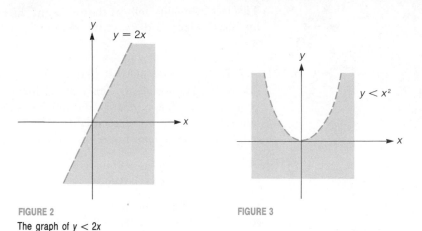

FIGURE 2
The graph of $y < 2x$

FIGURE 3

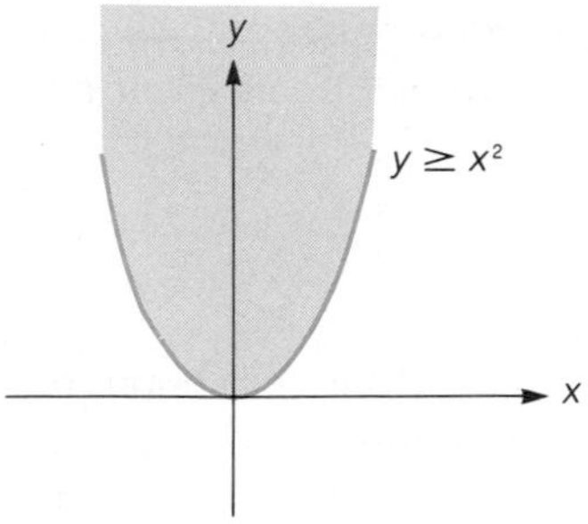

FIGURE 4

As with a linear equation in two unknowns, a linear inequality has infinitely many solutions. For this reason, we often represent the solutions graphically. When we do this, we say that we are **graphing the inequality**.

For example, let us graph the inequality $y < 2x$. First we observe that the coordinates of the points on the line $y = 2x$ do not, by definition, satisfy this inequality. So it remains to consider points above the line and points below the line. We will in fact show that the required graph consists of all points that lie below the line. Take any point $P(x_0, y_0)$ not on the line $y = 2x$. Let $Q(x_0, 2x_0)$ be the point on $y = 2x$ with the same first coordinate as P. Then, as indicated in Figure 1, the point P lies below the line if and only if the y-coordinate of P is less than the y-coordinate of Q. In other words, (x_0, y_0) lies below the line if and only if $y_0 < 2x_0$. This last statement is equivalent to saying that (x_0, y_0) satisfies the inequality $y < 2x$. This shows that the graph of $y < 2x$ consists of all points below the line $y = 2x$ (see Figure 2). The broken line in Figure 2 indicates that the points on $y = 2x$ are not part of the required graph. If the original inequality had been $y \leq 2x$, then we would use a solid line rather than a broken one, indicating that the line is included in the graph. And if the original inequality had been $y > 2x$, the graph would be the region above the line.

Just as the graph of $y < 2x$ is the region below the line $y = 2x$, it is true in general that the graph of $y < f(x)$ is the region below the graph of the function f. For example, Figures 3 and 4 display the graphs of $y < x^2$ and $y \geq x^2$.

Example 1 summarizes the technique developed so far for graphing an inequality. Following this example, we will point out a useful alternative method.

EXAMPLE 1 Graph the inequality $4x - 3y \leq 12$.

Solution The graph will include the line $4x - 3y = 12$ and either the region above or the region below the line. To decide which region, solve the inequality for y:

$$4x - 3y \leq 12$$
$$-3y \leq -4x + 12$$
$$y \geq \frac{4}{3}x - 4 \qquad \text{multiplying or dividing by a negative number reverses an inequality}$$

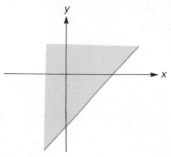

FIGURE 5

The graph of $4x - 3y \leq 12$

This last inequality tells us that we want the region above the line, as well as the line itself. Figure 5 displays the required graph. ▲

There is another method that we can use in Example 1 to determine which side of the line we want. This method involves a **test point**. We pick any convenient point that is not on the line $4x - 3y = 12$. Then we test to see if this point satisfies the given inequality. For example, let us pick the point $(0, 0)$. Substituting these coordinates in the inequality $4x - 3y \leq 12$ yields

$$4(0) - 3(0) \overset{?}{\leq} 12$$
$$0 \overset{?}{\leq} 12 \qquad \text{True}$$

We conclude from this that the required side of the line is that on which the point $(0, 0)$ resides. In agreement with Figure 5, we see this is the region above the line.

Next, we discuss systems of inequalities in two unknowns. As with systems of equations, a **solution** of a system of inequalities is an ordered pair (x_0, y_0) that satisfies all of the inequalities in the system. As a first example, let us graph the points that satisfy the following nonlinear system:

$$\begin{cases} y - x^2 \geq 0 \\ x^2 + y^2 < 1 \end{cases}$$

By writing the first inequality as $y \geq x^2$, we see that it describes the set of points on or above the parabola $y = x^2$. For the second inequality, we must decide whether it describes the points within or outside the circle $x^2 + y^2 = 1$. One way to do this is to choose $(0, 0)$ as a test point. Substituting the values $x = 0$ and $y = 0$ in the inequality $x^2 + y^2 < 1$ yields the true statement $0^2 + 0^2 < 1$. Since $(0, 0)$ lies within the circle and satisfies the inequality, we conclude that $x^2 + y^2 < 1$ describes the set of all points within the circle. Now we put our information together. We wish to graph the points that lie on or above the parabola $y = x^2$ but within the circle $x^2 + y^2 = 1$. This is the shaded region shown in Figure 6.

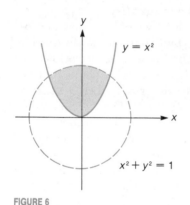

FIGURE 6

EXAMPLE 2   Graph the system

$$\begin{cases} -x + 3y \leq 12 \\ x + y \leq 8 \\ x \geq 0 \\ y \geq 0 \end{cases}$$

Solution   Solving the first inequality for y, we have

$$3y \leq x + 12$$
$$y \leq \tfrac{1}{3}x + 4$$

Thus, the first inequality is satisfied by the points on or below the line $-x + 3y = 12$. Similarly, by solving the second inequality for y, we see that it describes the set of points on or below the line $x + y = 8$. The third inequality, $x \geq 0$, describes the points on or to the right of the y-axis. Similarly, the fourth inequality, $y \geq 0$, describes the points on or above the x-axis. We summarize these statements in Figure 7(a). The arrows indicate which side of each line we wish to consider.

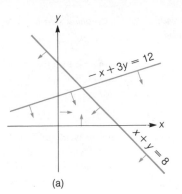

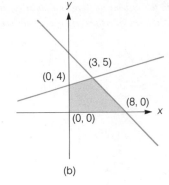

FIGURE 7

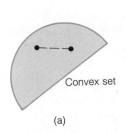

Convex set

(a)

Not a convex set

(b)

FIGURE 8

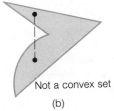

The first quadrant region between the curve $y = 1/x$ and the axes is not bounded.

FIGURE 9

Finally, Figure 7(b) shows the graph of the given system. The coordinates of each point in the shaded region satisfy all four of the given inequalities. (Exercise 25 at the end of this section asks you to show that the coordinates listed in the figure are correct.) ▲

We can use Figure 7(b) to introduce some terminology that is useful in describing sets of points in the plane. The **vertices** of a region are the corners, or points, where the adjacent bounding sides meet. Thus, the vertices of the shaded region in Figure 7(b) are the four points $(0, 0)$, $(8, 0)$, $(3, 5)$, and $(0, 4)$. The shaded region in Figure 7(b) is **convex**. This means that given any two points in that region, the straight line segment joining these two points lies wholly within the region. Figure 8(a) displays another example of a convex set, while Figure 8(b) shows a set that is not convex. The shaded region in Figure 8(b) is also an example of a **bounded region**. By this we mean that the region can be wholly contained within some (sufficiently large) circle. Perhaps the simplest example of a region that is not bounded is the entire x-y plane itself. Figure 9 shows another example of an unbounded region.

▼ EXERCISE SET 7.7

A

In Exercises 1 and 2, decide whether or not the ordered pairs are solutions of the given inequality.

1. $4x - 6y + 3 \geq 0$ **(a)** $(1, 2)$ **(b)** $(0, \frac{1}{2})$

2. $5x + 2y < 1$ **(a)** $(-1, 3)$ **(b)** $(0, 0)$

In Exercises 3–16, graph the given inequalities.

3. $2x - 3y > 6$ **4.** $2x - 3y < 6$

5. $2x - 3y \geq 6$ **6.** $2x + 3y \leq 6$

7. $x - y < 0$ **8.** $y \leq \frac{1}{2}x - 1$

9. $x \geq 1$ **10.** $y < 0$

11. $x > 0$ **12.** $y \leq \sqrt{x}$

13. $y > x^3 + 1$ **14.** $y \leq |x - 2|$

15. $x^2 + y^2 \geq 25$ **16.** $y \geq e^x - 1$

In Exercises 17–22, graph the systems of inequalities.

17. $\begin{cases} y \leq x^2 \\ x^2 + y^2 \leq 1 \end{cases}$ **18.** $\begin{cases} y < x \\ x^2 + y^2 < 1 \end{cases}$

19. $\begin{cases} y \geq 1 \\ y \leq |x| \end{cases}$ **20.** $\begin{cases} x \geq 0 \\ y \geq 0 \\ y < \sqrt{x} \\ x \leq 4 \end{cases}$

21. $\begin{cases} x \geq 0 \\ y \geq 0 \\ y \leq 1 - x^2 \end{cases}$ **22.** $\begin{cases} y < 2x \\ y > \frac{1}{2}x \end{cases}$

In Exercises 23–34, graph the systems of linear inequalities. In each case specify the vertices. Is the region convex? Is the region bounded?

23. $\begin{cases} y \leq x + 5 \\ y \leq -2x + 14 \\ x \geq 0 \\ y \geq 0 \end{cases}$

24. $\begin{cases} y \geq x + 5 \\ y \geq -2x + 14 \\ x \geq 0 \\ y \geq 0 \end{cases}$

25. $\begin{cases} -x + 3y \leq 12 \\ x + y \leq 8 \\ x \geq 0 \\ y \geq 0 \end{cases}$

26. $\begin{cases} y \geq 2x \\ y \geq -x + 6 \end{cases}$

27. $\begin{cases} 0 \leq 2x - y + 3 \\ x + 3y \leq 23 \\ 5x + y \leq 45 \end{cases}$

28. $\begin{cases} 0 \leq 2x - y + 3 \\ x + 3y \leq 23 \\ 5x + y \leq 45 \\ x \geq 0 \\ y \geq 0 \end{cases}$

29. $\begin{cases} 5x + 6y < 30 \\ y > 0 \end{cases}$

30. $\begin{cases} 5x + 6y < 30 \\ x > 0 \end{cases}$

31. $\begin{cases} 5x + 6y < 30 \\ x > 0 \\ y > 0 \end{cases}$

32. $\begin{cases} 2x + 3y \geq 6 \\ 2x + 3y \leq 12 \end{cases}$

33. $\begin{cases} x \geq 0 \\ y \geq 0 \\ 20 - x \geq 0 \\ 30 - y \geq 0 \\ x + y \leq 40 \\ x + y \geq 35 \end{cases}$

34. $\begin{cases} x \geq 0 \\ y \geq 0 \\ 3x - y + 1 \geq 0 \\ 0 \leq x - y + 3 \\ y \leq 5 \\ x \leq \frac{1}{3}(17 - y) \\ x \leq \frac{1}{2}(y + 8) \end{cases}$

B

In Exercises 35 and 36, graph the system of inequalities and specify the vertices.

35. $\begin{cases} -e \leq x \leq e \\ y \geq 0 \\ y \leq e^x \\ y \leq e^{-x} \end{cases}$

36. $\begin{cases} x \geq 0 \\ y \geq 0 \\ y \leq \cos x \\ y \geq \sin x \\ x \leq \pi \end{cases}$

A formula such as

$$f(x, y) = \sqrt{2x - y + 1} \qquad (1)$$

defines a function of two variables. The inputs for such a function are ordered pairs (x, y) of real numbers. For example, using the ordered pair $(3, 5)$ as an input, we have

$$f(3, 5) = \sqrt{2(3) - 5 + 1} = \sqrt{2}$$

So the input $(3, 5)$ yields an output of $\sqrt{2}$. We define the domain for this function just as we did in Chapter 2: The domain is the set of all inputs that yield real-number outputs. For instance, the ordered pair $(1, 4)$ is not in the domain of the function we have been discussing, because (as you should check for yourself) $f(1, 4) = \sqrt{-1}$, which is not a real number. We can determine the domain of the function in equation (1) by requiring that the quantity under the radical sign be nonnegative. Thus, we require that $2x - y + 1 \geq 0$ and, consequently, $y \leq 2x + 1$. (Check this.) The following figure shows the graph of this inequality; the domain of our function is the set of ordered pairs making up the graph. In Exercises 37–42, follow a similar procedure and sketch the domain of the given function.

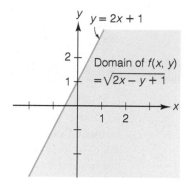

37. $f(x, y) = \sqrt{x + y + 2}$

38. $f(x, y) = \sqrt{x^2 + y^2 - 1}$

39. $g(x, y) = \sqrt{25 - x^2 - y^2}$

40. $g(x, y) = \ln(x^2 - y)$

41. $h(x, y) = \ln(xy)$

42. $h(x, y) = \sqrt{x} + \sqrt{y}$

▼ **CHAPTER SEVEN   SUMMARY OF PRINCIPAL TERMS**

TERMS	PAGE REFERENCE	COMMENT
1. Linear equation in two variables	383	A linear equation in two variables is an equation of the form $ax + by = c$, where a, b, and c are constants and x and y are variables or unknowns. Similarly, a linear equation in three variables is an equation of the form $ax + by + cz = d$.
2. Solution of a linear equation	383	A solution of the linear equation $ax + by = c$ is an ordered pair of numbers (x_0, y_0) such that $ax_0 + by_0 = c$. Similarly, a solution of the linear equation $ax + by + cz = d$ is an ordered triple of numbers (x_0, y_0, z_0) such that $ax_0 + by_0 + cz_0 = d$.

TERMS	PAGE REFERENCE	COMMENT
3. Consistent system; inconsistent system	384	A system of equations is *consistent* if it has at least one solution; otherwise, the system is *inconsistent*. See Figures 2, 3, and 4 in Section 7.1 for a geometric interpretation of these terms.
4. Upper-triangular form (three variables)	393	A system of linear equations in x, y, and z is said to be in *upper-triangular form* if x appears in no equation after the first, and y appears in no equation after the second. This definition can be extended to include systems with any number of unknowns. See Table 1 in Section 7.2 for examples of systems that are in upper-triangular form. When a system is in upper-triangular form, it is a simple matter to obtain the solutions; see, for instance, Examples 1 and 2 in Section 7.2.
5. Elementary operations	395	These are operations that can be performed on an equation in a system without altering the solution set. See the box on page 395 for a list of these operations.
6. Gaussian elimination	395	This is a technique for converting a system of equations to upper-triangular form. See Examples 4 and 5 in Section 7.2 for demonstrations of this technique.
7. Matrix	403	A *matrix* is a rectangular array of numbers, enclosed in parentheses or brackets. The numbers constituting the rectangular array are called the *entries* or the *elements* in the matrix. The *size or dimension* of a matrix is expressed by specifying the number of rows and the number of columns, in that order. For examples of this terminology, see Example 1 in Section 7.3.
8. Matrix equality	407	Two matrices are said to be *equal* provided they are the same size and their corresponding entries are equal.
9. Matrix addition and subtraction	407	To add two matrices of the same size, add the corresponding entries. Similarly, to subtract two matrices of the same size, subtract the corresponding entries.
10. Matrix multiplication	410	Let A and B be two matrices. The matrix product AB is defined only when the number of columns in A is the same as the number of rows in B. In this case, the matrix AB will have as many rows as A and as many columns as B. The entry in the ith row and jth column of AB is the number formed as follows: Multiply the corresponding entries in the ith row of A and the jth column of B; then add the results.
11. Square matrix	414	A matrix in which there are as many rows as there are columns is called a *square matrix*. An $n \times n$ square matrix is said to be an nth-order square matrix.
12. Multiplicative identity matrix	414, 418	For the set of $n \times n$ matrices, the multiplicative identity matrix I_n is the $n \times n$ matrix with ones down the main diagonal (upper left corner to bottom right corner) and zeros everywhere else. For example, $$I_2 = \begin{pmatrix} 1 & 0 \\ 0 & 1 \end{pmatrix} \quad \text{and} \quad I_3 = \begin{pmatrix} 1 & 0 & 0 \\ 0 & 1 & 0 \\ 0 & 0 & 1 \end{pmatrix}$$ For every $n \times n$ matrix A, we have $AI_n = A$ and $I_nA = A$.

TERMS	PAGE REFERENCE	COMMENT
13. Inverse matrix	414, 418	Given an $n \times n$ matrix A, the inverse matrix (if it exists) is denoted by A^{-1}. The defining equations for A^{-1} are $AA^{-1} = I$ and $A^{-1}A = I$. Here, I stands for the multiplicative identity matrix that is the same size as A. If a square matrix has an inverse, it is said to be *nonsingular*; if there is no inverse, then the matrix is *singular*.
14. Minor	421	The *minor* of an entry b in a determinant is the determinant formed by suppressing the entries in the row and column in which b appears.
15. Cofactor	422	If an entry b appears in the ith row and the jth column of a determinant, then the *cofactor* of b is computed by multiplying the minor of b times the number $(-1)^{i+j}$.
16. Determinant	421, 423	The value of the 2×2 determinant $\begin{vmatrix} a & b \\ c & d \end{vmatrix}$ is defined to be $ad - bc$. For larger determinants, the value can be found as follows. Select any row or column and multiply each entry in that row or column by its cofactor. Then add the results. The resulting sum is the value of the determinant. It can be shown that this value is independent of the particular row or column that is chosen.
17. Cramer's rule	427	This rule yields the solutions of certain systems of linear equations in terms of determinants. For a statement of the rule, see page 427. Example 5 on pages 427–428 shows how the rule is applied. The proof of Cramer's rule begins on page 428.
18. Linear inequality in two variables	438	A linear equality in two variables is any one of the four types of inequalities that result when the equal sign in the equation $ax + by = c$ is replaced by one of the four symbols $<, \leq, >, \geq$. For any of these linear inequalities, a *solution* is an ordered pair of numbers (x_0, y_0) with the property that a true statement is obtained when x and y (in the inequality) are replaced by x_0 and y_0, respectively.
19. Vertices	441	The *vertices* of a region in the x-y plane are the corners or points where the adjacent bounding sides meet.
20. Convex region	441	A region in the x-y plane is *convex* if, given any two points in the region, the line segment joining those points lies wholly within the region.
21. Bounded region	441	A region in the x-y plane is *bounded* if it can be wholly contained within some (sufficiently large) circle.

▼ CHAPTER SEVEN   REVIEW EXERCISES

NOTE   Exercises 1–20 constitute a chapter test on the fundamentals, based on group A problems.

1. Determine all solutions of the system
$$\begin{cases} 3x + 4y = 12 \\ y = x^2 + 2x + 3 \end{cases}$$

2. Find all solutions of the system
$$\begin{cases} x - 2y = 13 \\ 3x + 5y = -16 \end{cases}$$

3. (a) Find all solutions of the following system using Gaussian elimination:

$$\begin{cases} x + 4y - \ z = 0 \\ 3x + \ y + \ z = -1 \\ 4x - 4y + 5z = -7 \end{cases}$$

(b) Compute D, D_x, D_y, and D_z for the system in part (a). [Then check your answer in part (a) using Cramer's rule.]

4. Suppose that the matrices A and B are defined as follows:

$$A = \begin{pmatrix} 1 & -3 \\ 2 & -1 \end{pmatrix} \qquad B = \begin{pmatrix} 0 & 4 \\ 1 & 3 \end{pmatrix}$$

(a) Compute $2A - B$ **(b)** Compute BA.

5. Determine the area of the triangular region in the first quadrant that is bounded by the x-axis and the lines $y = 2x$ and $y = -x + 6$.

6. Find all solutions of the system

$$\begin{cases} \dfrac{1}{2x} + \dfrac{1}{3y} = 10 \\ -\dfrac{5}{x} - \dfrac{4}{y} = -4 \end{cases}$$

7. Specify the coefficient matrix for the system

$$\begin{cases} x + \ y - \ z = -1 \\ 2x - \ y + 2z = 11 \\ x - 2y + \ z = 10 \end{cases}$$

Also specify the augmented matrix for this system.

8. Use matrix methods to find all solutions of the system displayed in the previous problem.

9. Find the equation of a line that passes through the point of intersection of the lines $x + y = 11$ and $3x + 2y = 7$ and that is perpendicular to the line $2x - 4y = 7$.

10. Determine the constants A, B, and C such that the following equation is an identity:

$$\frac{x - 2}{(x + 1)(x - 1)^2} = \frac{A}{x + 1} + \frac{B}{x - 1} + \frac{C}{(x - 1)^2}$$

11. Consider the determinant

$$\begin{vmatrix} 2 & 3 & -1 \\ 0 & 1 & 4 \\ 5 & -2 & 6 \end{vmatrix}$$

(a) What is the minor of the entry in the third row, second column?
(b) What is the cofactor of the entry in the third row, second column?

12. Evaluate the determinant

$$\begin{vmatrix} 4 & -5 & 0 \\ -8 & 10 & 7 \\ 16 & 20 & 14 \end{vmatrix}$$

13. Find all solutions of the system

$$\begin{cases} x^2 + y^2 = 15 \\ xy = 5 \end{cases}$$

14. Find the solutions of the system

$$\begin{cases} A + 2B + 3C = 1 \\ 2A - \ B - \ C = 2 \end{cases}$$

15. (a) Determine the inverse of the following matrix.

$$\begin{pmatrix} 10 & -2 & 5 \\ 6 & -1 & 4 \\ 1 & 0 & 1 \end{pmatrix}$$

(b) Use the inverse matrix determined in part (a) to solve the following system.

$$\begin{cases} 10u - 2v + 5w = -1 \\ 6u - \ v + 4w = -2 \\ u \quad\ \ + \ w = 3 \end{cases}$$

16. Graph the inequality $5x - 6y \geq 30$.

17. Determine the constants P and Q, given that the parabola $y = Px^2 + Qx - 5$ passes through the two points $(-2, -1)$ and $(-1, -2)$.

18. Graph the inequality $x^2 - 4x + y^2 + 3 > 0$. Is the solution set bounded? Is it convex?

19. Graph the following system of inequalities and specify the vertices:

$$\begin{cases} x \geq 0 \\ y \geq 0 \\ 2y - x \leq 14 \\ x + 3y \leq 36 \\ 9x + y \leq 99 \end{cases}$$

20. Determine the constant k, given that the three lines $kx + 3y = -4$, $x - 2y = -3$, and $y = x$ are concurrent (pass through a common point).

In Exercises 21–58, solve each system of equations. If there are no solutions in a particular case, say so. In cases where there are literal (rather than numerical) coefficients, specify any restrictions that your solutions impose on those coefficients.

21. $\begin{cases} x + y = -2 \\ x - y = 8 \end{cases}$ **22.** $\begin{cases} x - y = 1 \\ x + y = 5 \end{cases}$

23. $\begin{cases} 2x + y = 2 \\ x + 2y = 7 \end{cases}$

24. $\begin{cases} 3x + 2y = 6 \\ 5x + 4y = 4 \end{cases}$

25. $\begin{cases} 7x + 2y = 9 \\ 4x + 5y = 63 \end{cases}$

26. $\begin{cases} \dfrac{x}{2} + \dfrac{y}{3} = 9 \\ \dfrac{x}{5} - \dfrac{y}{2} = -4 \end{cases}$

27. $\begin{cases} 2x - \dfrac{y}{2} = -8 \\ \dfrac{x}{3} + \dfrac{y}{8} = -1 \end{cases}$

28. $\begin{cases} 3x - 14y - 1 = 0 \\ -6x + 28y - 3 = 0 \end{cases}$

29. $\begin{cases} 3x + 5y - 1 = 0 \\ 9x - 10y - 8 = 0 \end{cases}$

30. $\begin{cases} 9x + 15y - 1 = 0 \\ 6x + 10y + 1 = 0 \end{cases}$

31. $\begin{cases} \frac{2}{3}x = -\frac{1}{2}y - 12 \\ \dfrac{x}{2} = y + 2 \end{cases}$

32. $\begin{cases} 0.1x + 0.2y = -5 \\ -0.2x - 0.5y = 13 \end{cases}$

33. $\begin{cases} \dfrac{1}{x} + \dfrac{1}{y} = -1 \\ \dfrac{2}{x} + \dfrac{5}{y} = -14 \end{cases}$

34. $\begin{cases} \dfrac{2}{x} + \dfrac{15}{y} = -9 \\ \dfrac{1}{x} + \dfrac{10}{y} = -2 \end{cases}$

35. $\begin{cases} ax + (1 - a)y = 1 \\ (1 - a)x + y = 0 \end{cases}$

36. $\begin{cases} ax - by - 1 = 0 \\ (a - 1)x + by + 2 = 0 \end{cases}$

37. $\begin{cases} 2x - y = 3a^2 - 1 \\ 2y + x = 2 - a^2 \end{cases}$

38. $\begin{cases} 3ax + 2by = 3a^2 - ab + 2b^2 \\ 3bx + 2ay = 2a^2 + 5ab - 3b^2 \end{cases}$

39. $\begin{cases} 5x - y = 4a^2 - 6b^2 \\ 2x + 3y = 5a^2 + b^2 \end{cases}$

40. $\begin{cases} \dfrac{2b}{x} - \dfrac{3}{y} = 7ab \\ \dfrac{4a}{x} + \dfrac{5a}{by} = 3a^2 \end{cases}$

41. $\begin{cases} px - qy = q^2 \\ qx + py = p^2 \end{cases}$

42. $\begin{cases} x - y = \dfrac{a - b}{a + b} \\ x + y = 1 \end{cases}$

43. $\begin{cases} \dfrac{4a}{x} - \dfrac{3b}{y} = a - 7b \\ \dfrac{3a^2}{x} - \dfrac{2b^2}{y} = (3a + b)(a - 2b) \end{cases}$

44. $\begin{cases} 6b^2x^{-1/2} - 5a^2y^{-1/2} - a^2b^2 = 0 \\ 2bx^{-1/2} + a^2y^{-1/2} - 3a^2b = 0 \end{cases}$
Hint: Let $u = x^{-1/2}$ and $v = y^{-1/2}$.

45. $\begin{cases} x + y + z = 9 \\ x - y - z = -5 \\ 2x + y - 2z = -1 \end{cases}$

46. $\begin{cases} x - 4y + 2z = 9 \\ 2x + y + z = 3 \\ 3x - 2y - 3z = -18 \end{cases}$

47. $\begin{cases} 4x - 4y + z = 4 \\ 2x + 3y + 3z = -8 \\ x + y + z = -3 \end{cases}$

48. $\begin{cases} x - 8y + z = 1 \\ 5x + 16y + 3z = 3 \\ 4x - 4y - 4z = -4 \end{cases}$

49. $\begin{cases} -2x + y + z = 1 \\ x - 2y + z = -2 \\ x + y - 2z = 4 \end{cases}$

50. $\begin{cases} -x + y + z = 1 \\ x - y + z = -1 \\ x + y - z = 1 \end{cases}$

51. $\begin{cases} 4x + 2y - 3z = 15 \\ 2x + y + 3z = 3 \end{cases}$

52. $\begin{cases} 3x + 2y + 17z = 1 \\ x + 2y + 3z = 3 \end{cases}$

53. $\begin{cases} x + 2y - 3z = -2 \\ 2x - y + z = 1 \\ 3x - 4y + 5z = 1 \end{cases}$

54. $\begin{cases} 9x + y + z = 0 \\ -3x + y - z = 0 \\ 3x - 5y + 3z = 0 \end{cases}$

55. $\begin{cases} x + y + z = a + b \\ 2x - y + 2z = -a + 5b \\ x - 2y + z = -2a + 4b \end{cases}$

56. $\begin{cases} ax + by - 2az = 4ab + 2b^2 \\ x + y + z = 4a + 2b \\ bx + ay + 4az = 5a^2 + b^2 \end{cases}$

57. $\begin{cases} x + y + z + w = 8 \\ 3x + 3y - z - w = 20 \\ 4x - y - z + 2w = 18 \\ 2x + 5y + 5z - 5w = 8 \end{cases}$

58. $\begin{cases} x - 2y + 3z + w = 1 \\ x + y + z + w = 5 \\ 2x + 3y + 2z - w = 3 \\ 3x + y - z + 2w = 4 \end{cases}$

For Exercises 59–70, determine the constants (denoted by capital letters) so that each equation is an identity.

59. $\dfrac{1}{(x - 10)(x + 10)} = \dfrac{A}{x - 10} + \dfrac{B}{x + 10}$

60. $\dfrac{x}{(x - 5)(x + 5)} = \dfrac{A}{x - 5} + \dfrac{B}{x + 5}$

61. $\dfrac{2x}{(x + 1)^2} = \dfrac{A}{x + 1} + \dfrac{B}{(x + 1)^2}$

62. $\dfrac{x + 2}{(x - 1)^2} = \dfrac{A}{x - 1} + \dfrac{B}{(x - 1)^2}$

63. $\dfrac{5}{x(x - 4)} = \dfrac{A}{x} + \dfrac{B}{x - 4}$

64. $\dfrac{6x}{(x + 2)(x + 3)} = \dfrac{A}{x + 2} + \dfrac{B}{x + 3}$

65. $\dfrac{1}{(x - 1)(x + 3)^2} = \dfrac{A}{x - 1} + \dfrac{B}{x + 3} + \dfrac{C}{(x + 3)^2}$

66. $\dfrac{x^2}{(x + 2)(x - 2)^2} = \dfrac{A}{x + 2} + \dfrac{B}{x - 2} + \dfrac{C}{(x - 2)^2}$

67. $\dfrac{4x^2 + 2x + 15}{(x - 1)(x^2 + x + 5)} = \dfrac{A}{x - 1} + \dfrac{Bx + C}{x^2 + x + 5}$

68. $\dfrac{1}{x(x^2 + 16)} = \dfrac{A}{x} + \dfrac{Bx + C}{x^2 + 16}$

69. $\dfrac{1}{x^3 + 64} = \dfrac{A}{x + 4} + \dfrac{Bx + C}{x^2 - 4x + 16}$

70. $\dfrac{1}{x^3 + 3x^2 + 3x + 1} = \dfrac{A}{x + 1} + \dfrac{B}{(x + 1)^2} + \dfrac{C}{(x + 1)^3}$

In Exercises 71–74, the lowercase letters a and b denote nonzero constants, with $a \neq b$. In each case, determine the values of A and B (and C, if applicable) so that the equation is an identity.

71. $\dfrac{x}{(x - a)^3} = \dfrac{A}{x - a} + \dfrac{B}{(x - a)^2} + \dfrac{C}{(x - a)^3}$

72. $\dfrac{ax(1 - a)}{(x - a)(x + a^2)} = \dfrac{A}{x - a} + \dfrac{B}{x + a^2}$

73. $\dfrac{(a - b)(a + b - x)}{(x - a)(x - b)} = \dfrac{A}{x - a} + \dfrac{B}{x - b}$

74. $\dfrac{4a^2 x + 6a^3}{(x + a)(x + 2a)(x + 3a)} = \dfrac{A}{x + a} + \dfrac{B}{x + 2a} + \dfrac{C}{x + 3a}$

In Exercises 75–82, evaluate each of the determinants.

75. $\begin{vmatrix} 1 & 5 \\ -6 & 4 \end{vmatrix}$

76. $\begin{vmatrix} \frac{1}{6} & 1 \\ 2 & 12 \end{vmatrix}$

77. $\begin{vmatrix} 4 & 0 & 3 \\ -2 & 1 & 5 \\ 0 & 2 & -1 \end{vmatrix}$

78. $\begin{vmatrix} 2 & 6 & 4 \\ 6 & 18 & 24 \\ 15 & 5 & -10 \end{vmatrix}$

79. $\begin{vmatrix} 1 & 5 & 7 \\ 1 & 5 & 7 \\ 17 & 19 & 21 \end{vmatrix}$

80. $\begin{vmatrix} 0 & 2 & 4 & 0 \\ 4 & 0 & 6 & 2 \\ 0 & 0 & 1 & 1 \\ 14 & 7 & 1 & 0 \end{vmatrix}$

81. $\begin{vmatrix} 1 & 0 & 0 & 0 \\ 0 & 2 & 0 & 0 \\ 0 & 0 & 3 & 0 \\ 0 & 0 & 0 & 4 \end{vmatrix}$

82. $\begin{vmatrix} 1 & a & b & c \\ 0 & 2 & d & e \\ 0 & 0 & 3 & f \\ 0 & 0 & 0 & 4 \end{vmatrix}$

83. Show that $\begin{vmatrix} a & b & c \\ b & c & a \\ c & a & b \end{vmatrix} = 3abc - a^3 - b^3 - c^3$.

84. Show that $\begin{vmatrix} 1 & 1 & 1 \\ 1 & 1 + x & 1 \\ 1 & 1 & 1 + x^2 \end{vmatrix} = x^3$.

85. Show that

$\begin{vmatrix} a^2 + x & b & c & d \\ -b & 1 & 0 & 0 \\ -c & 0 & 1 & 0 \\ -d & 0 & 0 & 1 \end{vmatrix} = a^2 + b^2 + c^2 + d^2 + x$

86. In a two-digit number, the sum of the digits is 11. Four times the units digit exceeds the tens digit by 4. Find the number.

87. Determine constants a and b so that the parabola $y = ax^2 + bx - 1$ passes through the points $(-2, 5)$ and $(2, 9)$.

88. Find two numbers whose sum and difference are 52 and 10, respectively.

89. The vertices of triangle ABC are $A(-2, 0)$, $B(4, 0)$, and $C(0, 6)$. Let A_1, B_1, and C_1 denote the midpoints of sides $\overline{BC}$, $\overline{AC}$, and $\overline{AB}$, respectively.
   (a) Find the point where the line segments $\overline{AA_1}$ and $\overline{BB_1}$ intersect.   *Answer:* $(\frac{2}{3}, 2)$
   (b) Follow the instructions given in part (a) using $\overline{BB_1}$ and $\overline{CC_1}$.
   (c) Follow the instructions given in part (a) using $\overline{AA_1}$ and $\overline{CC_1}$.
   (d) Let P denote the point $(\frac{2}{3}, 2)$ that you found in part (a). Compute each of the following ratios: AP/PA_1, BP/PB_1, CP/PC_1. What do you observe?

90. An *altitude* of a triangle is a line segment drawn from a vertex perpendicular to the opposite side. Suppose that the vertices of triangle ABC are as given in Exercise 89. Find the intersection point for each pair of altitudes. What do you observe about the three answers?

91. This exercise appears in *Plane and Solid Analytic Geometry*, by W. F. Osgood and W. G. Graustein (New York: Macmillan, 1920): Let P be any point (a, a) of the line $x - y = 0$, other than the origin. Through P draw two lines, of arbitrary slopes m_1 and m_2, intersecting the x-axis in A_1 and A_2, and the y-axis in B_1 and B_2, respectively. Prove that the lines $\overline{A_1 B_2}$ and $\overline{A_2 B_1}$ will, in general, meet on the line $x + y = 0$.

92. Determine constants h, k, and r so that the circle $(x - h)^2 + (y - k)^2 = r^2$ passes through the three points $(0, 0)$, $(0, 1)$, and $(1, 0)$.

93. The vertices of a triangle are the points of intersection of the lines $y = x - 1$, $y = -x - 2$, and $y = 2x + 3$. Find the equation of the circle passing through these three intersection points.

94. The vertices of triangle ABC are $A(0, 0)$, $B(3, 0)$, and $C(0, 4)$.
   (a) Find the center and the radius of the circle that passes through A, B, and C. This circle is called the **circumcircle** for triangle ABC.
   (b) The figure shows the **inscribed circle** for triangle ABC; this is the circle that is tangent to all three sides of the triangle. Find the center and the radius of this circle using the following two facts.
   (i) The center of the inscribed circle is the common intersection point of the three angle bisectors of the triangle.
   (ii) The line that bisects angle B has slope $-\frac{1}{2}$.

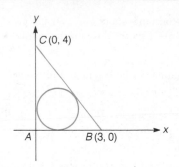

(c) Verify the following statement for triangle ABC. (The statement actually holds for any triangle; the result is known as *Euler's theorem*.) Let R and r denote the radii of the circles in parts (a) and (b), respectively. Then the distance d between the centers of the circles satisfies the equation

$$d^2 = R^2 - 2rR$$

(d) Verify the following statement for triangle ABC. (The statement actually holds for any triangle.) The area of triangle ABC is equal to one-half the product of the perimeter and the radius of the inscribed circle.

(e) Verify the following statement for triangle ABC. (The statement actually holds for any triangle.) The sum of the reciprocals of the lengths of the altitudes is equal to the reciprocal of the radius of the inscribed circle.

In Exercises 95–118, the matrices A, B, C, D, E, and F are defined as follows.

$$A = \begin{pmatrix} 3 & -2 \\ 1 & 5 \end{pmatrix} \qquad B = \begin{pmatrix} 2 & 1 \\ 1 & 8 \end{pmatrix} \qquad C = \begin{pmatrix} -1 & 0 \\ 0 & -1 \end{pmatrix}$$

$$D = \begin{pmatrix} -4 & 0 & 6 \\ 1 & 3 & 2 \end{pmatrix} \qquad E = \begin{pmatrix} 3 & -1 \\ 4 & 1 \\ -5 & 9 \end{pmatrix} \qquad F = \begin{pmatrix} 2 & 6 \\ 5 & 3 \\ 5 & 8 \end{pmatrix}$$

In each exercise, carry out the indicated matrix operations if they are defined. If an operation is not defined, say so.

95. $2A + 2B$

96. $2(A + B)$

97. $4B$

98. $B + 4$

99. AB

100. BA

101. $AB - BA$

102. $B + E$

103. $B + C$

104. $A(B + C)$

105. $AB + AC$

106. $(B + C)A$

107. $BA + CA$

108. $D + E$

109. DE

110. $(EE)D$

111. $E(ED)$

112. $E + F$

113. EF

114. $3E - 2F$

115. $(A + B) + C$

116. $A + (B + C)$

117. $(AB)C$

118. $A(BC)$

119. For a square matrix A, the notation A^2 means AA. Similarly, A^3 means AAA. If $A = \begin{pmatrix} 1 & 1 \\ 0 & 1 \end{pmatrix}$, verify that

$$A^2 = \begin{pmatrix} 1 & 2 \\ 0 & 1 \end{pmatrix} \qquad \text{and} \qquad A^3 = \begin{pmatrix} 1 & 3 \\ 0 & 1 \end{pmatrix}$$

120. Let A be the matrix $\begin{pmatrix} 0 & 0 & 0 \\ a & 0 & 0 \\ b & c & 0 \end{pmatrix}$. Compute A^2 and A^3.

For Exercises 121–124, in each case compute the inverse of the matrix in part (a), and then use that inverse to solve the system of equations in part (b).

121. **(a)** $\begin{pmatrix} 1 & 5 \\ 2 & 9 \end{pmatrix}$ **(b)** $\begin{cases} x + 5y = 3 \\ 2x + 9y = -4 \end{cases}$

122. **(a)** $\begin{pmatrix} 5 & -4 \\ 14 & -11 \end{pmatrix}$ **(b)** $\begin{cases} 5x - 4y = 2 \\ 14x - 11y = 5 \end{cases}$

123. **(a)** $\begin{pmatrix} 1 & -2 & 3 \\ 2 & -5 & 10 \\ -1 & 2 & -2 \end{pmatrix}$

(b) $\begin{cases} x - 2y + 3z = -2 \\ 2x - 5y + 10z = -3 \\ -x + 2y - 2z = 6 \end{cases}$

124. **(a)** $\begin{pmatrix} -1 & 4 & 2 \\ -3 & 10 & 5 \\ 0 & 3 & 1 \end{pmatrix}$

(b) $\begin{cases} -x + 4y + 2z = 8 \\ -3x + 10y + 5z = 0 \\ 3y + z = -12 \end{cases}$

In Exercises 125 and 126, find the inverse of each matrix.

125. $\begin{pmatrix} 5 & 3 & 6 & -7 \\ 3 & -4 & 0 & -9 \\ 0 & 1 & -1 & -1 \\ 2 & 2 & 3 & -2 \end{pmatrix}$

126. $\begin{pmatrix} 1 & -1 & 0 & -3 \\ 5 & -2 & 3 & -11 \\ 2 & 3 & 2 & -3 \\ 4 & 5 & 5 & -5 \end{pmatrix}$

In Exercises 127–134, compute D, D_x, D_y, D_z (and D_w where appropriate) for each system of equations. Use Cramer's rule to solve the systems in which $D \neq 0$. If $D = 0$, solve the system using Gaussian elimination or matrix methods.

127. $\begin{cases} 2x - y + z = 1 \\ 3x + 2y + 2z = 0 \\ x - 5y - 3z = -2 \end{cases}$ **128.** $\begin{cases} x + 2y - z = -1 \\ 2x - 3y + 3z = 3 \\ 2x + 3y + z = 1 \end{cases}$

129. $\begin{cases} x + 2y + 3z = -1 \\ 4x + 5y + 6z = 2 \\ 7x + 8y + 9z = -3 \end{cases}$

130. $\begin{cases} 3x + 2y - 2z = 0 \\ 2x + 3y - z = 0 \\ 8x + 7y - 5z = 0 \end{cases}$

131. $\begin{cases} 3x + 2y - 2z = 1 \\ 2x + 3y - z = -2 \\ 8x + 7y - 5z = 0 \end{cases}$

132. $\begin{cases} x + y + z + w = 5 \\ x - y - z + w = 3 \\ 2x + 3y + 3z + 2w = 21 \\ 4z - 3w = -7 \end{cases}$

133. $\begin{cases} 2x - y + z + 3w = 15 \\ x + 2y + 2w = 12 \\ 3y + 3z + 4w = 12 \\ -4x + y - 4z = -11 \end{cases}$

134. $\begin{cases} x + y + z = (a + b)^2 \\ \dfrac{bx}{a} + \dfrac{ay}{b} - z = 0 \\ x + y - z = (a - b)^2 \end{cases}$

In Exercises 135–148, find all solutions (x, y) for each system, where x and y are real numbers.

135. $\begin{cases} y = 6x \\ y = x^2 \end{cases}$

136. $\begin{cases} y = 4x \\ y = x^3 \end{cases}$

137. $\begin{cases} y = 9 - x^2 \\ y = x^2 - 9 \end{cases}$

138. $\begin{cases} x^3 - y = 0 \\ xy - 16 = 0 \end{cases}$

139. $\begin{cases} x^2 - y^2 = 9 \\ x^2 + y^2 = 16 \end{cases}$

140. $\begin{cases} 2x + 3y = 6 \\ y = \sqrt{x + 1} \end{cases}$

141. $\begin{cases} x^2 + y^2 = 1 \\ y = \sqrt{x} \end{cases}$

142. $\begin{cases} \dfrac{x}{11} + \dfrac{y}{12} = 2 \\ \dfrac{xy}{132} = 1 \end{cases}$

143. $\begin{cases} x^2 + y^2 = 1 \\ y = 2x^2 \end{cases}$

144. $\begin{cases} x^2 - 3xy + y^2 = -11 \\ 2x^2 + xy - y^2 = 8 \end{cases}$

145. $\begin{cases} x^2 + 2xy + 3y^2 = 68 \\ 3x^2 - xy + y^2 = 18 \end{cases}$

146. $\begin{cases} \dfrac{x^2}{a^2} + \dfrac{y^2}{b^2} = 1 \\ \dfrac{x^2}{b^2} + \dfrac{y^2}{a^2} = 1 \end{cases}$ $(a > b > 0)$

147. $\begin{cases} 2(x - 3)^2 - (y + 1)^2 = -1 \\ -3(x - 3)^2 + 2(y + 1)^2 = 6 \end{cases}$
Hint: Let $u = x - 3$ and $v = y + 1$.

148. $\begin{cases} x^4 = y - 1 \\ y - 3x^2 + 1 = 0 \end{cases}$

Exercises 149–154 appear (in German) in an algebra text by Leonhard Euler, first published in 1770. The English versions given here are taken from the translated version, Elements of Algebra, *5th ed., by Leonhard Euler (London: Longman, Orme, and Co., 1840). [This, in turn, has been reprinted by Springer-Verlag (New York, 1984).]*

149. Required two numbers, whose sum may be s, and their proportion as a to b.

Answer: $\dfrac{as}{a + b}$ and $\dfrac{bs}{a + b}$

150. The sum $2a$, and the sum of the squares $2b$, of two numbers being given; to find the numbers.

Answer: $a - \sqrt{b - a^2}$ and $a + \sqrt{b - a^2}$

151. To find three numbers, so that [the sum of] one-half of the first, one-third of the second, and one-quarter of the third, shall be equal to 62; one-third of the first, one-quarter of the second, and one-fifth of the third, equal to 47; and one-quarter of the first, one-fifth of the second, and one-sixth of the third, equal to 38.

Answer: 24, 60, 120

152. Required two numbers, whose product may be 105, and whose squares [when added] may together make 274.

153. Required two numbers, whose product may be m, and the sum of the squares n $(n \geq 2m)$.

154. Required two numbers such that their sum, their product, and the difference of their squares may all be equal.

In Exercises 155–160, graph each system of inequalities and specify whether the region is convex or bounded.

155. $\begin{cases} x^2 + y^2 \geq 1 \\ y - 4x \leq 0 \\ y - x \geq 0 \\ x \geq 0 \\ y \geq 0 \end{cases}$

156. $\begin{cases} x^2 + y^2 \leq 1 \\ x \geq 0 \\ y \geq 0 \end{cases}$

157. $\begin{cases} y - \sqrt{x} \leq 0 \\ y \geq 0 \\ x \geq 1 \\ x - 4 \leq 0 \end{cases}$

158. $\begin{cases} y - |x| \leq 0 \\ x + 1 \geq 0 \\ x - 1 \leq 0 \\ y + 1 \geq 0 \end{cases}$

159. $\begin{cases} y \leq 1/x \\ y \geq 0 \\ x \geq 1 \end{cases}$

160. $\begin{cases} y - 100x \leq 0 \\ y - x^2 \geq 0 \\ x \geq 0 \end{cases}$

In Exercises 161–164, sketch the domains of the functions. (For background, see the discussion preceding Exercise 37 in Exercise Set 7.7.)

161. $f(x, y) = \sqrt{2x + y}$

162. $f(x, y) = \sqrt{9 - x^2 - y^2}$

163. $g(x, y) = \sqrt{9 - x^2 - y^2} + \sqrt{x^2 + y^2 - 4}$

164. $h(x, y) = 1/(1 - e^{xy})$.

In Exercises 165 and 166, find a solution (x, y) for the given system.

165. $\begin{cases} y \cos x = a \\ y \sin x = b \end{cases}$ **166.** $\begin{cases} y \cos(x + a) = p \\ y \sin(x + b) = q \end{cases}$

Hint: (for Exercise 166) Use an addition formula to expand the left-hand side of each equation. In the resulting system, make the substitutions $s = y \sin x$ and $t = y \cos x$. Solve for s and t, and then use the result in Exercise 165.

CHAPTER EIGHT

ANALYTIC GEOMETRY

The Greeks knew the properties of the curves given by cutting a cone with a plane—the ellipse, the parabola and hyperbola. Kepler discovered by analysis of astronomical observations, and Newton proved mathematically . . . that the planets move in ellipses. The geometry of Ancient Greece thus became the cornerstone of modern astronomy.

John Lighton Synge (1897–1987)

INTRODUCTION In this chapter, we continue the work we began in Chapters 1 and 2 on coordinates and graphs. We begin in Section 8.1 by first listing the key formulas with which you need to be familiar. Then we develop two additional results that have to do with slope and distance in the plane. In Sections 8.2 through 8.4, we study the parabola, the ellipse, and the hyperbola. These are the three curves referred to in the opening quotation. In the next section (8.5), we study rotation of axes, a useful tool in graphing equations (and in computer graphics). Finally, in Section 8.6, we introduce polar coordinates.

8.1 ▼ THE BASIC EQUATIONS

As background for the work in this chapter, you need to be familiar with the following material from Chapters 1 and 2.

1. The distance formula: $d = \sqrt{(x_2 - x_1)^2 + (y_2 - y_1)^2}$

2. The equation for a circle: $(x - h)^2 + (y - k)^2 = r^2$

3. The definition of slope: $m = \dfrac{y_2 - y_1}{x_2 - x_1}$

4. The point–slope formula: $y - y_1 = m(x - x_1)$

5. The slope–intercept formula: $y = mx + b$

6. The two-intercept formula: $\dfrac{x}{a} + \dfrac{y}{b} = 1$

7. The condition for two nonvertical lines to be parallel: $m_1 = m_2$

8. The condition for two nonvertical lines to be perpendicular: $m_1 m_2 = -1$

9. The midpoint formula: $(x_0, y_0) = \left(\dfrac{x_1 + x_2}{2}, \dfrac{y_1 + y_2}{2} \right)$

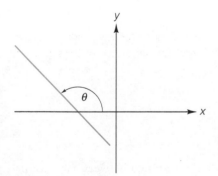

FIGURE 1
The angle of inclination

In this section we are going to develop two additional results to supplement those we just listed. The first of these results concerns the slope of a line. We begin by defining the **angle of inclination** (or, simply, the **inclination**) of a line to be the angle θ measured counterclockwise from the positive side or positive direction of the x-axis to the line (see Figure 1). If θ denotes the angle of inclination, we always have $0° \leq \theta < 180°$ if θ is measured in degrees and $0 \leq \theta < \pi$ if θ is in radians. (Notice that when $\theta = 0°$, the line is horizontal.)

As you might suspect, there is a simple relationship between the angle of inclination and the slope of a line. We state this relationship in the box that follows.

PROPERTY SUMMARY

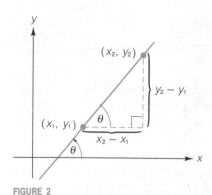

FIGURE 2

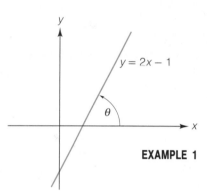

FIGURE 3

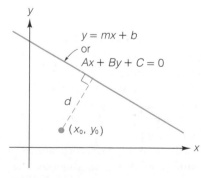

FIGURE 4

SLOPE AND THE ANGLE OF INCLINATION

The slope m of a line and the angle of inclination θ are related by the equation

$$m = \tan \theta$$

The formula $m = \tan \theta$ provides a useful connection between elementary coordinate geometry and trigonometry. To derive the formula, we consider three cases:

Case 1: $\theta = 0°$

Case 2: $0° < \theta < 90°$

Case 3: $90° < \theta < 180°$

The case where $\theta = 90°$ needn't be considered, for then the line is vertical and both m and $\tan \theta$ are undefined. The proofs for cases 1 and 2 run as follows. (Case 3 is covered in Exercise 48 at the end of this section.)

CASE 1: $\theta = 0°$ When $\theta = 0°$, the line is horizontal and both m and $\tan 0°$ are zero. Thus, $m = \tan \theta$ in this case.

CASE 2: $0° < \theta < 90°$ See Figure 2. We have

$$\tan \theta = \frac{\text{opposite}}{\text{adjacent}} = \frac{y_2 - y_1}{x_2 - x_1}$$

$$= m \quad \text{by definition}$$

Thus, $\tan \theta = m$.

EXAMPLE 1 Determine the acute angle θ between the x-axis and the line $y = 2x - 1$. (See Figure 3.)

Solution From the equation $y = 2x - 1$, we read directly that $m = 2$. Then, since $\tan \theta = m$, we conclude that

$$\tan \theta = 2 \quad \text{or} \quad \theta = \tan^{-1} 2$$

If required, we can obtain an appropximate numerical value for θ with a calculator. For instance, using a Texas Instruments calculator, the sequence of keystrokes is

2 $\boxed{\text{INV}}$ $\boxed{\text{TAN}}$

When expressed in degrees, the result in this case is

$$\theta \approx 63.4°$$ ▲

In Chapter 1, we reviewed the formula for the distance between two points. Now we consider another kind of distance formula, one that gives the distance d from a point to a line. As indicated in Figure 4, distance in this context means the *shortest* distance, which is the perpendicular distance. In the box that follows, we show two equivalent forms for this formula. Although the second form is more widely known, the first is just as useful and somewhat simpler to derive.

PROPERTY SUMMARY

DISTANCE FROM A POINT TO A LINE
1. The distance d from the point (x_0, y_0) to the line $y = mx + b$ is given by $$d = \frac{|mx_0 + b - y_0|}{\sqrt{1 + m^2}}$$
2. The distance d from the point (x_0, y_0) to the line $Ax + By + C = 0$ is given by $$d = \frac{|Ax_0 + By_0 + C|}{\sqrt{A^2 + B^2}}$$

We will derive formula (1) in the box with the aid of Figure 5. The broken lines in the figure are parallel to the x- and y-axes.

FIGURE 5

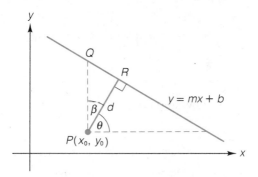

In right triangle PRQ, we have $\cos \beta = d/PQ$, and therefore

$$d = PQ \cos \beta \tag{1}$$

In view of equation (1), the distance d is determined once we find the quantities PQ and $\cos \beta$. For PQ, we have

$$PQ = (y\text{-coordinate of } Q) - (y\text{-coordinate of } P)$$
$$= (mx_0 + b) - y_0$$

Thus, the length PQ in Figure 5 is $mx_0 + b - y_0$. However, to allow for the possibility that the figure might just as well show $P(x_0, y_0)$ above rather than below the line $y = mx + b$, we write

$$PQ = |mx_0 + b - y_0| \tag{2}$$

Having determined PQ, we turn our attention to $\cos \beta$. From Figure 5, we have $\beta = 90° - \theta$, and so

$$\cos \beta = \cos(90° - \theta) = \sin \theta$$

That is,

$$\cos \beta = \sin \theta$$

But since θ is the angle of inclination of the line segment $\overline{PR}$, we have

$$\tan \theta = \text{slope of } \overline{PR}$$

and consequently

$$\tan \theta = -\frac{1}{m} \quad \text{(Why?)}$$

Now that we know $\tan \theta$, we can determine $\sin \theta$ (and consequently $\cos \beta$) as follows:

$$\csc^2 \theta = 1 + \cot^2 \theta$$
$$= 1 + (-m)^2 = 1 + m^2$$

Thus,

$$\sin^2 \theta = \frac{1}{1 + m^2}$$

or

$$\sin \theta = \frac{1}{\sqrt{1 + m^2}}$$

(Note that the positive square root is appropriate, since $0° \le \theta < 180°$.) Therefore,

$$\cos \beta = \frac{1}{\sqrt{1 + m^2}} \tag{3}$$

In view of equations (1), (2), and (3), we conclude that

$$d = \frac{|mx_0 + b - y_0|}{\sqrt{1 + m^2}}$$

which is what we wanted to prove.

EXAMPLE 2   (a) Find the distance from the point $(-3, 1)$ to the line $y = -2x + 7$. See Figure 6.

(b) Find the equation of the circle that has center $(-3, 1)$ and that is tangent to the line $y = -2x + 7$.

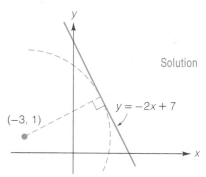

FIGURE 6

Solution   (a) To find the distance from the point $(-3, 1)$ to the line $y = -2x + 7$, we use the formula $d = |mx_0 + b - y_0| / \sqrt{1 + m^2}$ with $x_0 = -3, y_0 = 1, m = -2$, and $b = 7$. This yields

$$d = \frac{|(-2)(-3) + 7 - 1|}{\sqrt{1 + (-2)^2}} = \frac{12}{\sqrt{5}} = \frac{12\sqrt{5}}{5} \text{ units}$$

(b) The equation of a circle with center (h, k) and radius of length r is $(x - h)^2 + (y - k)^2 = r^2$. We are given here that (h, k) is $(-3, 1)$. Furthermore, the distance determined in part (a) is the length of the radius. (A theorem from geometry tells us that the radius drawn to the point of tangency is perpendicular to the tangent.) Thus, the equation of the required circle is

$$[x - (-3)]^2 + (y - 1)^2 = \left(\frac{12\sqrt{5}}{5} \right)^2$$

or

$$(x + 3)^2 + (y - 1)^2 = \frac{144}{5}$$

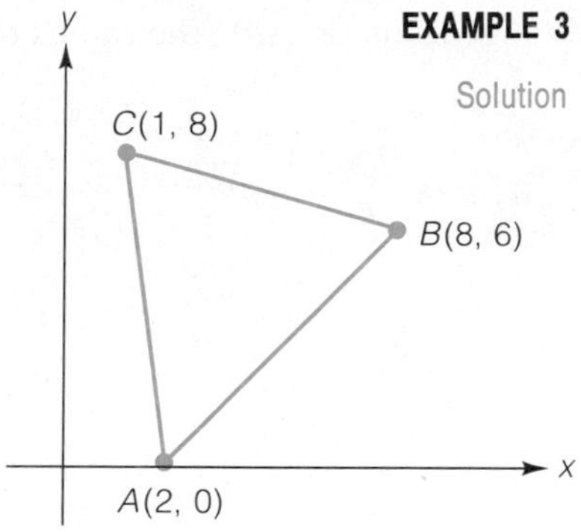

EXAMPLE 3

Find the area of triangle ABC in Figure 7.

Solution

The area of any triangle is one-half the product of the base and the height. Let us view $\overline{AB}$ as the base. Then, using the formula for the distance between two points, we have

$$
\begin{aligned}
AB &= \sqrt{(x_2 - x_1)^2 + (y_2 - y_1)^2} \\
&= \sqrt{(8 - 2)^2 + (6 - 0)^2} \\
&= \sqrt{36 + 36} = \sqrt{2 \times 36} = 6\sqrt{2}
\end{aligned}
$$

With $\overline{AB}$ as the base, the height of the triangle is the perpendicular distance from C to $\overline{AB}$. To compute that distance, we first need the equation of the line through A and B. The slope of the line is

$$
m = \frac{y_2 - y_1}{x_2 - x_1} = \frac{6 - 0}{8 - 2} = \frac{6}{6} = 1
$$

Then, using the point $(2, 0)$ and the slope $m = 1$, we have

$$
\begin{aligned}
y - y_1 &= m(x - x_1) \\
y - 0 &= 1(x - 2) \\
y &= x - 2
\end{aligned}
$$

We can now compute the height of the triangle by finding the perpendicular distance from $C(1, 8)$ to $y = x - 2$:

$$
\begin{aligned}
\text{height} &= \frac{|mx_0 + b - y_0|}{\sqrt{1 + m^2}} \\
&= \frac{|1(1) + (-2) - 8|}{\sqrt{1 + 1^2}} = \frac{|-9|}{\sqrt{2}} = \frac{9}{\sqrt{2}}
\end{aligned}
$$

Now we're ready to compute the area A of the triangle, since we know the base and the height. We have

$$
A = \frac{1}{2}(6\sqrt{2})\,\frac{9}{\sqrt{2}} = 27 \text{ square units}
$$

▲

EXAMPLE 4

From the point $(8, 1)$, a line is drawn tangent to the circle $x^2 + y^2 = 20$, as shown in Figure 8. Find the slope of this tangent line.

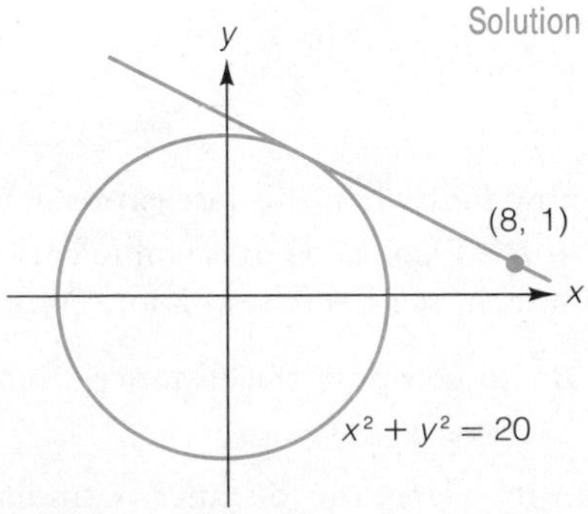

Solution

This is a problem in which the direct approach is not the simplest. The direct approach would be to first determine the coordinates of the point of tangency. Then, using those coordinates along with $(8, 1)$, the required slope could be computed. As it turns out, however, the coordinates of the point of tangency are not very easy to determine. In fact, one of the advantages of the following method is that those coordinates need not be found.

In Figure 8, let m denote the slope of the tangent line. Because the tangent line passes through $(8, 1)$, we can write its equation

$$
\begin{aligned}
y - y_1 &= m(x - x_1) \\
y - 1 &= m(x - 8) \\
y &= mx - 8m + 1
\end{aligned}
$$

FIGURE 7

FIGURE 8

Now, since the distance from the origin to the tangent line is $\sqrt{20}$ (the radius of the circle), we have

$$\sqrt{20} = \frac{|mx_0 + b - y_0|}{\sqrt{1 + m^2}}$$

$$= \frac{|m(0) - 8m + 1 - 0|}{\sqrt{1 + m^2}} \quad \text{using } x_0 = 0 = y_0 \text{ and } b = -8m + 1$$

$$= \frac{|-8m + 1|}{\sqrt{1 + m^2}}$$

To solve this equation for m, we square both sides to obtain

$$20 = \frac{64m^2 - 16m + 1}{1 + m^2}$$

or

$$20(1 + m^2) = 64m^2 - 16m + 1$$
$$0 = 44m^2 - 16m - 19$$
$$0 = (22m - 19)(2m + 1)$$

From this we see that the two roots of the equation are $\frac{19}{22}$ and $-\frac{1}{2}$. Because the slope of the tangent line specified in Figure 8 is negative, we choose the value $m = -\frac{1}{2}$; this is the required slope. ▲

The last formula that we consider in this section is

$$d = \frac{|Ax_0 + By_0 + C|}{\sqrt{A^2 + B^2}}$$

As we pointed out earlier, this formula gives the distance from the point (x_0, y_0) to the line $Ax + By + C = 0$. To derive the formula, first note that the slope and the y-intercept of the line $Ax + By + C = 0$ are

$$m = -\frac{A}{B} \quad \text{and} \quad b = -\frac{C}{B}$$

So we have

$$d = \frac{|mx_0 + b - y_0|}{\sqrt{1 + m^2}}$$

$$= \frac{|(-A/B)x_0 + (-C/B) - y_0|}{\sqrt{1 + (-A/B)^2}}$$

Now, to complete the derivation, we need to show that when this last expression is simplified, the result is $|Ax_0 + By_0 + C|/\sqrt{A^2 + B^2}$. Exercise 47 asks you to carry out the details. (For a completely different approach, see Exercises 32 and 33.)

EXAMPLE 5 Use the formula $d = |Ax_0 + By_0 + C|/\sqrt{A^2 + B^2}$ to compute the distance from the point $(-3, 1)$ to the line $y = -2x + 7$.

Note In Example 2, we computed this quantity using the distance formula $d = |mx_0 + b - y_0|/\sqrt{1 + m^2}$.

Solution   First, we write the equation $y = -2x + 7$ in the form $Ax + By + C = 0$:

$$2x + y - 7 = 0$$

From this we see that $A = 2$, $B = 1$, and $C = -7$. Thus, we have

$$d = \frac{|Ax_0 + By_0 + C|}{\sqrt{A^2 + B^2}}$$

$$= \frac{|2(-3) + 1(1) + (-7)|}{\sqrt{2^2 + 1^2}} = \frac{12}{\sqrt{5}} = \frac{12\sqrt{5}}{5}$$

The required distance is therefore $12\sqrt{5}/5$ units, as was obtained previously in Example 2(a).   ▲

▼ EXERCISE SET 8.1

A

Exercises 1–12 are review exercises. To solve these problems, you will need to utilize the nine formulas listed at the beginning of this section.

1. Find the distance between the points $(-5, -6)$ and $(3, -1)$.

2. Find the equation of the line that passes through $(2, -4)$ and is parallel to the line $3x - y = 1$. Write your answer in the form $y = mx + b$.

3. Find the equation of a line that is perpendicular to the line $4x - 5y - 20 = 0$ and has the same y-intercept as the line $x - y + 1 = 0$. Write your answer in the form $Ax + By + C = 0$.

4. Find the equation of the line passing through the points $(6, 3)$ and $(1, 0)$. Write your answer in the form $y = mx + b$.

5. Find the equation of the line that is the perpendicular bisector of the line segment joining the points $(2, 1)$ and $(6, 7)$. Write your answer in the form $Ax + By + C = 0$.

6. Find the area of the circle $(x - 12)^2 + (y + \sqrt{5})^2 = 49$.

7. Find the x- and the y-intercepts of the circle whose center is $(1, 0)$ and whose radius is 5.

8. Find the equation of the line with a positive slope that is tangent to the circle $(x - 1)^2 + (y - 1)^2 = 4$ at one of its y-intercepts. Write your answer in the form $y = mx + b$.

9. Suppose that the coordinates of A, B, and C are $A(1, 2)$, $B(6, 1)$, and $C(7, 8)$. Find the equation of the line passing through C and through the midpoint of the line segment $\overline{AB}$. Write your answer in the form $ax + by + c = 0$.

10. Find the equation of the line passing through the point $(-4, 0)$ and through the point of intersection of the lines $2x - y + 1 = 0$ and $3x + y - 16 = 0$. Write your answer in the form $y = mx + b$.

11. Find the perimeter of triangle ABC in the following figure.

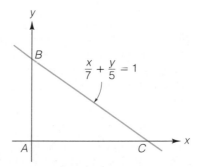

12. Find the sum of the x- and y-intercepts of the line

$$(\csc^2 \alpha)x + (\sec^2 \alpha)y = 1$$

(Assume that α is a constant.)

In each of Exercises 13–16, determine the angle of inclination of the line. Express the answer both in radians and degrees. In cases where a calculator is necessary, round the answer to two decimal places.

13. $y = \sqrt{3}x + 4$

14. $x + \sqrt{2}y - 2 = 0$

15. (a) $y = 5x + 1$
    (b) $y = -5x + 1$

16. (a) $3x - y - 3 = 0$
    (b) $3x + y - 3 = 0$

In each of Exercises 17–20, find the distance from the point to the line using (a) the formula $d = |mx_0 + b - y_0|/\sqrt{1 + m^2}$; (b) the formula $d = |Ax_0 + By_0 + C|/\sqrt{A^2 + B^2}$

17. $(1, 4)$; $y = x - 2$

18. $(-2, -3)$; $y = -4x + 1$

19. $(-3, 5)$; $4x + 5y + 6 = 0$

20. $(0, -3)$; $3x - 2y = 1$

21. (a) Find the equation of the circle with center $(-2, -3)$ that is tangent to the line $2x + 3y = 6$.
    (b) Find the radius of the circle with center $(1, 3)$ that is tangent to the line $y = \frac{1}{2}x + 5$.

22. Find the area of the triangle whose vertices are $(3, 1)$, $(-2, 7)$, and $(6, 2)$. (Use the method shown in Example 3.)

23. Find the area of the quadrilateral $ABCD$ with vertices $A(0, 0)$, $B(8, 2)$, $C(4, 7)$, and $D(1, 6)$.   *Suggestion:* Draw a diagonal and use the method shown in Example 3 for the two resulting triangles.

24. From the point $(7, -1)$, tangent lines are drawn to the circle $(x - 4)^2 + (y - 3)^2 = 4$. Find the slopes of these lines.

25. From the point $(0, -5)$, tangent lines are drawn to the circle $(x - 3)^2 + y^2 = 4$. Find the slope of each tangent.

26. Find the distance between the two parallel lines $y = 2x - 1$ and $y = 2x + 4$.   *Hint:* Draw a sketch; then find the distance from the origin to each line.

27. Find the distance between the two parallel lines $3x + 4y = 12$ and $3x + 4y = 24$.

28. Find the equation of the line that passes through $(3, 2)$ and whose x- and y-intercepts are equal. (There are two answers.)

29. Find the equation of the line that passes through the point $(2, 6)$ in such a way that the segment of the line cut off between the axes is bisected by the point $(2, 6)$.

30. Find the equation of the line whose angle of inclination is $60°$ and whose distance from the origin is four units. (There are two answers.)

B

31. Find the equation of the angle bisector in the accompanying figure.   *Hint:* Let (x, y) be a point on the angle bisector. Then (x, y) is equidistant from the two given lines.

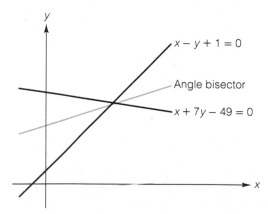

32. In this exercise, you'll use the techniques developed in Section 3.4 to show that the distance from the origin to the line $y = mx + b$ is given by the formula

$$d = \frac{|b|}{\sqrt{1 + m^2}}$$

(a) Show that the distance from the origin to a point (x, y) on the line $y = mx + b$ is given by

$$d = \sqrt{(1 + m^2)x^2 + 2mbx + b^2}$$

(b) Let $f(x) = (1 + m^2)x^2 + 2mbx + b^2$. [This is the expression under the radical sign in part (a).] By completing the square, show that the vertex of this parabola is

$$\left(\frac{-mb}{1 + m^2}, \frac{b^2}{1 + m^2} \right)$$

(c) Use the result in part (b) to show that the minimum value of d is $|b|/\sqrt{1 + m^2}$. (This is the perpendicular distance from the origin to $y = mx + b$, as required.)

33. In this exercise, you are going to use the result in Exercise 32 along with the ideas about translation in Section 2.5 to derive the formula for the distance d from the point (x_0, y_0) to the line $Ax + By + C = 0$. The basic idea is this. If we translate both the point (x_0, y_0) and the line $Ax + By + C = 0$ in the same direction and by the same amounts, then the distance from the new point to the new line is the same as the distance d that we want to compute. In particular, let us translate the point (x_0, y_0) horizontally and vertically so that it coincides with the origin. The same translations applied to the line $Ax + By + C = 0$ give us a new line with equation

$$A(x + x_0) + B(y + y_0) + C = 0 \qquad (1)$$

Our task now is to compute the distance from the origin to the line given by equation (1).

(a) Find the slope and the y-intercept of the line given by equation (1).

(b) Use the formula derived in Exercise 32 along with the results in part (a) of this exercise to show that the distance d from the point (x_0, y_0) to the line $Ax + By + C = 0$ is given by

$$d = \frac{|Ax_0 + By_0 + C|}{\sqrt{A^2 + B^2}}$$

34. Find the center and the radius of the circle that passes through the points $(-2, 7)$, $(0, 1)$, and $(2, -1)$.

35. **(a)** Find the center and the radius of the circle passing through the points $A(-12, 1)$, $B(2, 1)$, and $C(0, 7)$.

(b) Let R denote the radius of the circle in part (a), and let a, b, and c be the lengths of the sides opposite angles A, B, and C, respectively. Show that the area of triangle ABC is equal to $abc/4R$.
Note: It can be shown that this result and the formula in Exercise 36(b) are valid for all triangles.

36. (Continuation of Exercises 35.)
    (a) Let H denote the point where the altitudes of triangle ABC intersect. Find the coordinates of H.
    (b) Let d denote the distance from H to the center of the circle in Exercise 35(a). Show that

$$d^2 = 9R^2 - (a^2 + b^2 + c^2)$$

37. Suppose the line $x - 7y + 44 = 0$ intersects the circle $x^2 - 4x + y^2 - 6y = 12$ at points P and Q. Find the length of chord $\overline{PQ}$.

38. The point $(1, -2)$ is the midpoint of a chord of the circle $x^2 - 4x + y^2 + 2y = 15$. Find the length of the chord.

39. Show that the product of the distances from the point $(0, c)$ to the lines $ax + y = 0$ and $x + by = 0$ is

$$\frac{|bc^2|}{\sqrt{a^2 + a^2b^2 + b^2 + 1}}$$

40. Suppose that the point (x_0, y_0) lies on the circle $x^2 + y^2 = a^2$. Show that the equation of the line tangent to the circle at (x_0, y_0) is

$$x_0x + y_0y = a^2$$

41. The vertices of triangle ABC are $A(0, 0)$, $B(8, 0)$, and $C(8, 6)$.
    (a) Find the equations of the three lines that bisect the angles in triangle ABC. *Hint:* Make use of the identity $\tan(\theta/2) = (\sin \theta)/(1 + \cos \theta)$.
    (b) Find the points where each pair of angle bisectors intersect. What do you observe?

42. Show that the equations of the lines with slope m that are tangent to the circle $x^2 + y^2 = a^2$ are

$$y = mx + a\sqrt{1 + m^2} \quad \text{and} \quad y = mx - a\sqrt{1 + m^2}$$

43. Let P and Q be two points on the circle $x^2 + y^2 = a^2$. Show that the distance from P to the tangent line through Q is equal to the distance from Q to the tangent line through P. *Suggestion:* See Exercise 40.

44. Find the slope of a line that is three units from the origin and that passes through the point $(4, 1)$. (There are two answers.)

45. The point (x, y) is equidistant from the point $(0, \frac{1}{4})$ and the line $y = -\frac{1}{4}$. Show that x and y satisfy the equation $y = x^2$.

46. The point (x_0, y_0) is equidistant from the line $x + 2y = 0$ and the point $(3, 1)$. Find (and simplify) an equation relating x_0 and y_0.

47. **(a)** Find the slope m and the y-intercept b of the line $Ax + By + C = 0$.
    (b) Use the formula $d = |mx_0 + b - y_0|/\sqrt{1 + m^2}$ to show that the distance from the point (x_0, y_0) to the

line $Ax + By + C = 0$ is given by

$$d = \frac{|Ax_0 + By_0 + C|}{\sqrt{A^2 + B^2}}$$

48. Suppose that the angle of inclination θ for the line $y = mx + b$ is such that $90° < \theta < 180°$. We are going to verify that the formula $m = \tan \theta$ is valid in this case.
    (a) Explain why the lines $y = mx + b$ and $y = mx$ have equal angles of inclination.

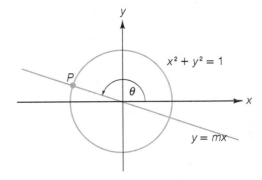

    (b) As indicated in the figure, let P denote the point in the second quadrant where the line $y = mx$ intersects the unit circle $x^2 + y^2 = 1$. Why are the coordinates of P given by $(\cos \theta, \sin \theta)$?
    (c) Compute the slope of $y = mx$ using the two points $(\cos \theta, \sin \theta)$ and $(0, 0)$. Observe that the result is equal to $\tan \theta$.

C

49. Show that the distance of the point (x_1, y_1) from the line passing through the two points (x_2, y_2) and (x_3, y_3) is given by $d = |D|/\sqrt{(x_2 - x_3)^2 + (y_2 - y_3)^2}$, where

$$D = \begin{vmatrix} x_1 & y_1 & 1 \\ x_2 & y_2 & 1 \\ x_3 & y_3 & 1 \end{vmatrix}$$

50. Let (a_1, b_1), (a_2, b_2), and (a_3, b_3) be three noncollinear points. Show that the equation of the circle passing through these three points is

$$\begin{vmatrix} x^2 + y^2 & x & y & 1 \\ a_1^2 + b_1^2 & a_1 & b_1 & 1 \\ a_2^2 + b_2^2 & a_2 & b_2 & 1 \\ a_3^2 + b_3^2 & a_3 & b_3 & 1 \end{vmatrix} = 0$$

51. Find the equation of the circle that passes through the points $(6, 3)$ and $(-4, -3)$ and that has its center on the line $y = 2x - 7$.

52. Let a be a positive number and suppose that the coordinates of points P and Q are $P(a\cos\theta, a\sin\theta)$ and $Q(a\cos\beta, a\sin\beta)$. Show that the distance from the origin to the line passing through P and Q is

$$a\left|\cos\left(\frac{\theta - \beta}{2}\right)\right|$$

53. Find the equation of a circle that has a radius of 5 and that is tangent to the line $2x + 3y = 26$ at the point $(4, 6)$. Write your answer in standard form. (There are two answers.)

54. Find the equation of the circle passing through $(2, -1)$ and tangent to the line $y = 2x + 1$ at $(1, 3)$. Write your answer in standard form.

8.2 ▼ THE PARABOLA

In Section 3.2, we saw that the graph of a quadratic function $y = ax^2 + bx + c$ is a symmetric U-shaped curve called a parabola. In this section, we give a more general definition of the parabola, a definition that emphasizes the geometric properties of the curve.

The Parabola

> A **parabola** is the set of all points in the plane equally distant from a fixed line and a fixed point not on the line. The fixed line is called the **directrix**. The fixed point is called the **focus**.

Let us initially suppose that the focus of the parabola is the point $(0, p)$ and the directrix is the line $y = -p$. We will assume throughout this section that p is positive. To understand the geometric content of the definition of the parabola, the special graph paper displayed in Figure 1(a) is useful. The common center of the concentric circles in Figure 1(a) is the focus $(0, p)$. Thus, all the points on a given circle are at a fixed distance from the focus. The radii of the circles increase in increments of p units. Similarly, the broken horizontal lines in the figure are drawn at intervals that are multiples of p units from the directrix $y = -p$. By considering the points where the circles intersect the horizontal lines, we can find

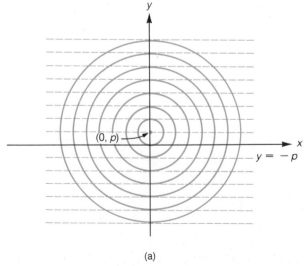

(a)

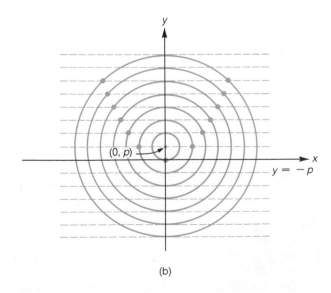

(b)

FIGURE 1

a number of points equally distant from the focus $(0, p)$ and the directrix $y = -p$ [see Figure 1(b)].

The graphical method just described lets us locate many points equally distant from the focus $(0, p)$ and the directrix $y = -p$. Figure 1(b) shows that the points on the parabola are symmetric about a line, in this case the y-axis. Also, by studying the figure, you should be able to convince youself that in this case there can be no points below the x-axis that satisfy the stated condition. However, to completely describe the required set of points, and to show that our new definition is consistent with the old, Figure 1(b) is inadequate. We need to bring algebraic methods to bear on the problem. Thus, let d_1 denote the distance from the point $P(x, y)$ to the focus $(0, p)$ and let d_2 denote the distance from $P(x, y)$ to the directrix $y = -p$, as shown in Figure 2. The distance d_1 is then

$$d_1 = \sqrt{(x - 0)^2 + (y - p)^2} = \sqrt{x^2 + y^2 - 2py + p^2}$$

FIGURE 2

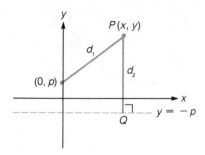

The distance d_2 in Figure 2 is just the vertical distance between the points P and Q. Thus,

$$d_2 = (y\text{-coordinate of } P) - (y\text{-coordinate of } Q)$$
$$= y - (-p) = y + p$$

(Absolute value signs are unnecessary here for, as noted earlier, P cannot lie below the x-axis.) The condition that P be equally distant from the focus $(0, p)$ and the directrix $y = -p$ can be expressed by the equation

$$d_1 = d_2$$

Using the expressions we've found for d_1 and d_2, this last equation becomes

$$\sqrt{x^2 + y^2 - 2py + p^2} = y + p$$

We can obtain a simpler but equivalent equation by squaring both sides. (Two nonnegative quantities are equal if and only if their squares are equal.) Thus,

$$x^2 + y^2 - 2py + p^2 = y^2 + 2py + p^2$$

After combining like terms, this equation becomes

$$x^2 = 4py$$

This is the equation of a parabola with focus $(0, p)$ and directrix $y = -p$. In the box that follows, we summarize the properties of the parabola $x^2 = 4py$. As Figure 3(b) indicates, the terminology we've introduced applies equally well to an arbitrary parabola for which the axis of symmetry is not necessarily parallel to one of the coordinate axes and the vertex is not necessarily the origin.

PROPERTY SUMMARY	THE PARABOLA

1. A **parabola** is the set of points equidistant from a fixed line called the **directrix** and a fixed point, not on the line, called the **focus**.

2. The **axis** of a parabola is the line drawn through the focus and perpendicular to the directrix.

3. The **vertex** of a parabola is the point where the parabola intersects its axis. The vertex is located halfway between the focus and the directrix. See Figure 3.

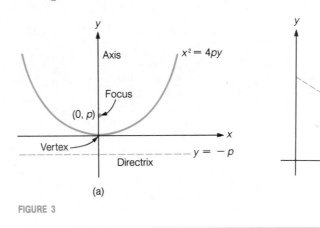

FIGURE 3

EXAMPLE 1 Determine the equation of the parabola in Figure 4, given that the curve passes through the point $(3, 5)$. Specify the focus and the directrix.

Solution We have just seen that the general equation for a parabola in this position is $x^2 = 4py$. Since the point $(3, 5)$ lies on the curve, its coordinates must satisfy the equation $x^2 = 4py$. Thus,

$$3^2 = 4p(5)$$
$$9 = 20p$$
$$\frac{9}{20} = p$$

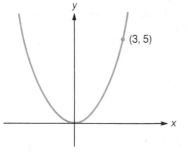

FIGURE 4

With $p = \frac{9}{20}$, the equation $x^2 = 4py$ becomes

$$x^2 = 4\left(\frac{9}{20}\right)y = \frac{9}{5}y \qquad \text{or} \qquad x^2 = \frac{9}{5}y \qquad \text{as required}$$

Furthermore, since $p = \frac{9}{20}$, the focus is $(0, \frac{9}{20})$ and the directrix is $y = -\frac{9}{20}$. ▲

We have seen that the equation of a parabola with focus $(0, p)$ and directrix $y = -p$ is $x^2 = 4py$. By following the same method, we can obtain general equations for parabolas with other orientations. The basic results are summarized in Figure 5.

| PROPERTY SUMMARY | BASIC EQUATIONS FOR THE PARABOLA |

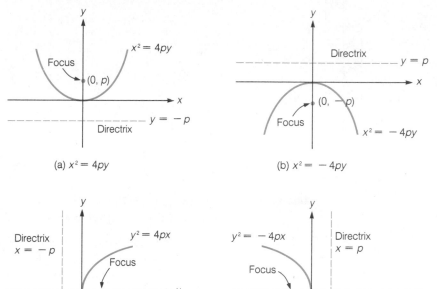

(a) $x^2 = 4py$

(b) $x^2 = -4py$

(c) $y^2 = 4px$

(d) $y^2 = -4px$

FIGURE 5

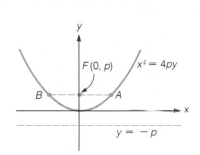

FIGURE 6
Focal width = $AB = 4p$

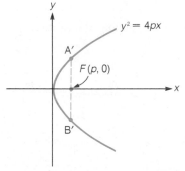

FIGURE 7
Focal width = $A'B' = 4p$

By a **chord** of a parabola, we mean a straight line segment joining any two points on the curve. If the chord passes through the focus, it is called a **focal chord**. For purposes of graphing, it is useful to know the length of the focal chord perpendicular to the axis of the parabola. This is the length of the horizontal line segment $\overline{AB}$ in Figure 6 and the vertical line segment $\overline{A'B'}$ in Figure 7. We will call this length the **focal width**.

In Figure 6, the distance from A to F is the same as the distance from A to the line $y = -p$. (Why?) But the distance from A to the line $y = -p$ is $2p$. Therefore, $AF = 2p$ and AB is twice this, or $4p$. We have shown that the focal width of the parabola is $4p$. In other words, given a parabola $x^2 = 4py$, the width at its focus is $4p$, the coefficient of y. In the same way, the length of the focal chord $\overline{A'B'}$ in Figure 7 is also $4p$, the coefficient of x in that case.

EXAMPLE 2   Find the focus and directrix of the parabola $y^2 = -4x$, and sketch the graph.

Solution   Comparing the basic equation $y^2 = -4px$ [in Figure 5(d)] with the equation at hand, we see that $4p = 4$ and thus $p = 1$. The focus is therefore $(-1, 0)$, and the directrix is $x = 1$. The basic form of the graph will be as in Figure 5(d). For purposes of graphing, we note that the focal width is 4 (the absolute value of the coefficient of x). This, along with the fact that the vertex is $(0, 0)$, is enough information to draw the graph. See Figure 8.   ▲

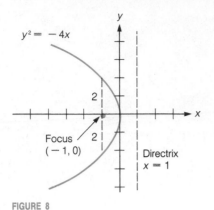

FIGURE 8

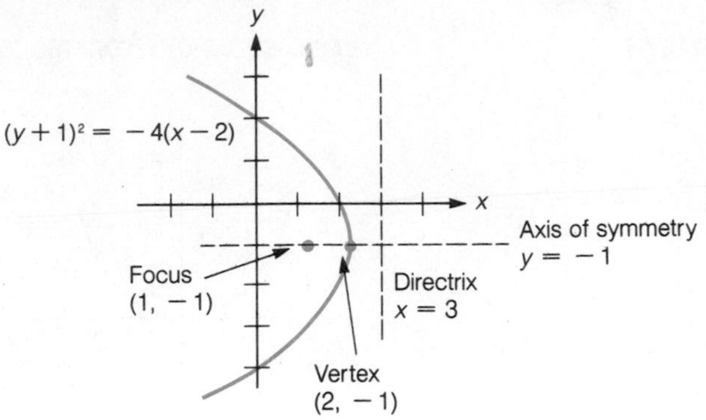

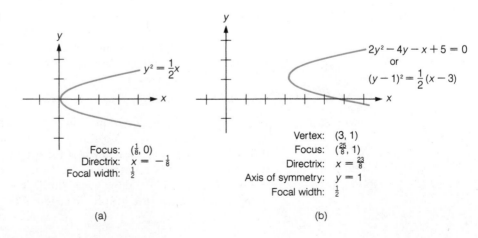

FIGURE 9

EXAMPLE 3 Graph the parabola $(y + 1)^2 = -4(x - 2)$. Specify the vertex, the focus, the directrix, and the axis of symmetry.

Solution The given equation is obtained from $y^2 = -4x$ (which we graphed in the previous example) by replacing x and y with $x - 2$ and $y + 1$, respectively. So (as we saw in Section 2.5 when we discussed translation) the required graph is obtained by translating the parabola in Figure 8 two units to the right and one unit down. In particular, this means that the vertex moves from $(0, 0)$ to $(2, -1)$; the focus moves from $(-1, 0)$ to $(1, -1)$; the directrix moves from $x = 1$ to $x = 3$; and the axis of symmetry moves from $y = 0$ (which is the x-axis) to $y = -1$. The required graph is shown in Figure 9. ▲

EXAMPLE 4 Graph the parabola $2y^2 - 4y - x + 5 = 0$, and specify each of the following: vertex, focus, directrix, axis of symmetry, and focal width.

Solution Just as we did in Section 3.2, we use the technique of completing the square:

$$2(y^2 - 2y \quad) = x - 5$$
$$2(y^2 - 2y + 1) = x - 5 + 2 \quad \text{adding 2 to both sides}$$
$$(y - 1)^2 = \frac{1}{2}(x - 3)$$

FIGURE 10

The graph of this last equation is obtained by translating the graph of $y^2 = \frac{1}{2}x$ "right 3, up 1." This moves the vertex from $(0, 0)$ to $(3, 1)$. Now, for $y^2 = \frac{1}{2}x$, the focus and directrix are determined by setting $4p = \frac{1}{2}$. Therefore $p = \frac{1}{8}$, and consequently the focus and directrix of $y^2 = \frac{1}{2}x$ are $(\frac{1}{8}, 0)$ and $x = -\frac{1}{8}$, respectively. Thus, the focus and directrix of the translated curve are $(3\frac{1}{8}, 1)$ and $x = 2\frac{7}{8}$, respectively. Figure 10(a) shows the graph of $y^2 = \frac{1}{2}x$, and Figure 10(b) shows the translated graph. (You should check for yourself that the information accompanying Figure 10(b) is correct.)   ▲

There are numerous applications of the parabola in the sciences. Parabolic reflectors are used in communications systems, in telescope mirrors, and in automobile headlights. In the case of the telescope mirror, an incoming ray of light that is parallel to the axis of the parabola will be reflected through the focus, as indicated in Figure 11.* Indeed, the very word *focus* comes from a Latin word meaning "fireplace."

Many of the more important properties of the parabola relate to the tangent to the curve. For instance, there is a close connection between the optical property illustrated in Figure 11 and the tangents to the parabola. (See Exercise 72 at the end of this section for details.) In general, the techniques of calculus are required to deal with tangents to curves. However, for curves with equations and graphs as simple as the parabola, the methods of algebra are often adequate. The following discussion shows how we can determine the tangent to a parabola without using calculus.

To begin with, we need a definition for the tangent to a parabola. As motivation, recall from geometry that a tangent line to a circle is defined to be a line that intersects the circle in exactly one point (see Figure 12). However, this definition is not quite adequate for the parabola. For instance, the y-axis intersects the parabola $y = x^2$ in exactly one point, but surely it is not tangent to the curve. With this in mind, we adopt the definition in the following box for the tangent to a parabola. Along with this definition, we make the assumption that through each point P on a parabola, there is only one tangent line that can be drawn.

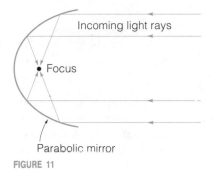

Incoming light rays

Focus

Parabolic mirror

FIGURE 11

FIGURE 12

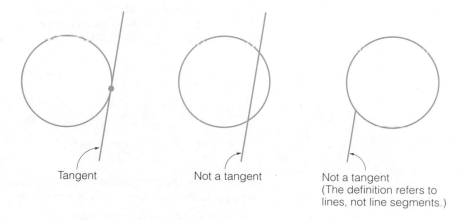

Tangent

Not a tangent

Not a tangent
(The definition refers to lines, not line segments.)

---

* For a more complete description of the uses of the parabola, see the article "The Standup Conic Presents: The Parabola and Applications," by Lee Whitt, in *The Journal of Undergraduate Mathematics and its Applications*, Vol. III no. 3 (1982).

Tangent to a Parabola

Let P be a point on a parabola. Then a line through P is said to be **tangent** to the parabola at P provided that the line intersects the parabola only at P and the line is not parallel to the axis of the parabola.

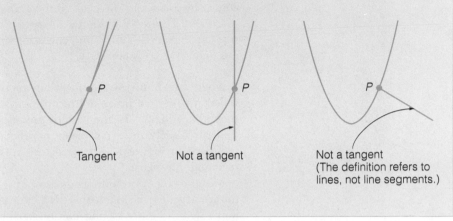

Tangent   Not a tangent   Not a tangent
(The definition refers to lines, not line segments.)

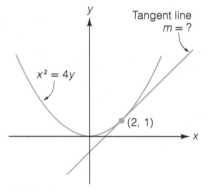

FIGURE 13

We'll demonstrate how to find tangents to parabolas by means of an example. The method used here can be used for any parabola. Suppose that we want to find the equation of the tangent to the parabola $x^2 = 4y$ at the point $(2, 1)$ on the curve (see Figure 13). Let m denote the slope of the tangent line. Since the line must pass through the point $(2, 1)$, its equation is

$$y - 1 = m(x - 2)$$

This is the tangent line, so it intersects the parabola in only one point, namely, $(2, 1)$. Algebraically, this means that the ordered pair $(2, 1)$ is the only solution of the following system of equations:

$$\begin{cases} y - 1 = m(x - 2) & (1) \\ x^2 = 4y & (2) \end{cases}$$

The strategy now is to solve this system of equations in terms of m. Then we'll reconcile the results with the fact that $(2, 1)$ is known to be the unique solution. This will allow us to determine m. From equation (2), we have $y = x^2/4$. Then, substituting for y in equation (1), we obtain

$$\frac{x^2}{4} - 1 = m(x - 2)$$

$$x^2 - 4 = 4m(x - 2) \qquad \text{multiplying by 4}$$

$$(x - 2)(x + 2) = 4m(x - 2) \qquad \text{factoring}$$

$$(x - 2)(x + 2) - 4m(x - 2) = 0 \qquad \text{subtracting } 4m(x - 2) \text{ from both sides}$$

Observe now that the quantity $x - 2$ appears twice on the left-hand side of the last equation, so we can factor it out to obtain

$$(x - 2)[(x + 2) - 4m] = 0$$

$$x - 2 = 0 \quad \bigg| \quad x + 2 - 4m = 0$$

$$x = 2 \quad \bigg| \qquad\quad x = 4m - 2$$

Now look at these two x-values that we've just determined. The first value, $x = 2$, yields no new information, since we knew from the start that this was the x-coordinate of the intersection point. However, the second value, $x = 4m - 2$, must also equal 2, since the system has but one solution. Thus, we have

$$4m - 2 = 2$$
$$4m = 4$$
$$m = 1$$

Therefore, the slope of the tangent line is 1. Substituting this value of m in equation (1) yields

$$y - 1 = 1(x - 2)$$

or

$$y = x - 1 \qquad \text{as required}$$

We conclude this section with some remarks concerning **conic sections**, which are the curves that are formed when a plane intersects the surface of a right circular cone. As Figure 14 indicates, these curves are the circle, the ellipse, the hyperbola, and the parabola. In the case of the parabola in Figure 14, the cutting plane is parallel to the line L.*

The study of the conic sections dates back over 2000 years to Ancient Greece. Apollonius of Perga (262–190 B.C.) wrote an eight-volume treatise on the subject. As the philosopher and mathematician Alfred North Whitehead (1881–1947) pointed out, it is a remarkable fact that from the time of Apollonius until the seventeenth century, the conic sections were studied only as a portion of pure (as opposed to applied) mathematics. Then in the seventeenth century, it was discovered that the conic sections were crucial in expressing some of the most important laws of nature. This is essentially the same observation made in the quotation at the beginning of this chapter.

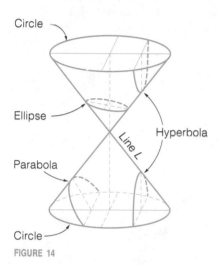

FIGURE 14

▼ EXERCISE SET 8.2

A

In Exercises 1–22, graph the parabolas. In each case, specify the focus, the directrix, and the focal width. For Exercises 13–22, also specify the vertex.

1. $x^2 = 4y$

2. $x^2 = 16y$

3. $y^2 = -8x$

4. $y^2 = 12x$

5. $x^2 = -20y$

6. $x^2 - y = 0$

7. $y^2 + 28x = 0$

8. $4y^2 + x = 0$

9. $x^2 = 6y$

10. $y^2 = -10x$

11. $4x^2 = 7y$

12. $3y^2 = 4x$

13. $y^2 - 6y - 4x + 17 = 0$

14. $y^2 + 2y + 8x + 17 = 0$

15. $x^2 - 8x - y + 18 = 0$

16. $x^2 + 6y + 18 = 0$

17. $y^2 + 2y - x + 1 = 0$

18. $2y^2 - x + 1 = 0$

19. $2x^2 - 12x - y + 18 = 0$

20. $y + \sqrt{2} = (x - 2\sqrt{2})^2$

21. $2x^2 - 16x - y + 33 = 0$

22. $\frac{1}{4}y^2 - y - x + 1 = 0$

---

* For a proof that this definition of the parabola is equivalent to our focus–directrix definition, see *A Calculus Notebook*, by C. Stanley Ogilvy (Boston, Mass.: Prindle, Weber & Schmidt, 1968). The very readable proof presented there requires only a knowledge of high school geometry; no calculus is used in the presentation.

For Exercises 23 and 24, refer to the following figure.

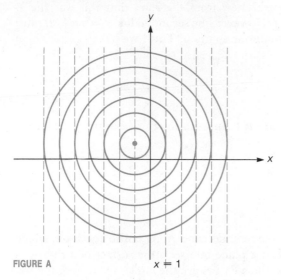

FIGURE A $x = 1$

23. Make a photocopy of Figure *A*. Then, in your copy of Figure *A*, indicate (by drawing dots) eleven points that are equidistant from the point $(-1, 1)$ and the line $x = 1$. What is the equation of the line of symmetry for the set of dots?

24. The eleven dots that you located in Exercise 23 are part of a parabola. Find the equation of that parabola and sketch its graph. Specify the vertex, the focus and directrix, and the focal width.

In Exercises 25–32, find the equation of the parabola satisfying the given conditions. In each case, assume that the vertex is at the origin.

25. The focus is $(0, 3)$.

26. The directrix is $y - 8 = 0$.

27. The directrix is $x + 32 = 0$.

28. The focus lies on the *y*-axis, and the parabola passes through the point $(7, -10)$.

29. The parabola passes through the points $(5, 6)$ and $(5, -6)$.

30. The focus lies on the line $3x - 4y + 12 = 0$, and the axis of symmetry is the *x*-axis.

31. The parabola is symmetric about the *x*-axis, the *x*-coordinate of the focus is negative, and the length of the focal chord perpendicular to the *x*-axis is 9.

32. The focus is the smaller of the two *x*-intercepts of the circle $x^2 - 8x + y^2 - 6y + 9 = 0$.

33. $\overline{PQ}$ is a focal chord of the parabola $x^2 = 4y$. If the coordinates of *P* are $(2, 1)$, what are the coordinates of *Q*?

34. $\overline{PQ}$ is a focal chord of the parabola $y^2 = -16x$. If the coordinates of *P* are $(-4, 8)$, what are the coordinates of *Q*?

35. $\overline{PQ}$ is a focal chord of the parabola $y = x^2$. The coordinates of *P* are $(-3, 9)$. Find the length of $\overline{PQ}$.

36. Let *P* denote the point $(8, 8)$ on the parabola $x^2 = 8y$, and let $\overline{PQ}$ be a focal chord.
  (a) Find the coordinates of *Q*.
  (b) Find the length of $\overline{PQ}$.
  (c) Find the equation of the circle with this focal chord as a diameter.
  (d) Show that the circle determined in part (c) intersects the directrix of the parabola in only one point. Conclude from this that the directrix is tangent to the circle. Draw a sketch of the situation.

37. Let *P* denote the point $(2, 4)$ on the parabola $y^2 = 8x$, and let $\overline{PQ}$ be a focal chord. By following the method in Exercise 36, show that the circle with this focal chord as a diameter is tangent to the directrix of the parabola. Include a sketch with your work.

38. The following figure shows a parabolic cross section of a reflecting mirror. Find the distance from the vertex to the focus. *Hint:* Choose a convenient coordinate system.

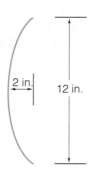

39. An arch is in the shape of a parabola with a vertical axis. The arch is 15 ft high at the center and 40 ft wide at the base. At what height above the base is the width 20 ft? *Hint:* Choose a convenient coordinate system in which the vertex of the parabola is at the origin.

40. The following figure depicts the cable of a suspension bridge. The cable is in the form of a parabola with the axis vertical. The horizontal highway is 300 ft long. The longest of the vertical supporting wires is 100 ft; the shortest is 40 ft. Find the length of a supporting wire that is 50 ft from the middle.

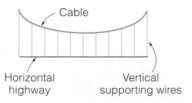

41. The segments $\overline{AA'}$ and $\overline{BB'}$ are focal chords of the parabola $x^2 = 2y$. The coordinates of A and B are $(4, 8)$ and $(-2, 2)$, respectively.
  (a) Find the equation of the line through A and B'.
  (b) Find the equation of the line through B and A'.
  (c) Show that the two lines you have found intersect at a point on the directrix.

42. Let $\overline{PQ}$ be the horizontal focal chord of the parabola $x^2 = 8y$. Let R denote the point where the directrix of the parabola meets the y-axis. Show that $\overline{PR}$ is perpendicular to $\overline{QR}$.

43. Suppose $\overline{PQ}$ is a focal chord of the parabola $y = x^2$ such that the coordinates of P are $(2, 4)$.
  (a) Find the coordinates of Q.
  (b) Find the coordinates of M, the midpoint of $\overline{PQ}$.
  (c) A perpendicular is drawn from M to the y-axis, meeting the y-axis at S. Also, a line perpendicular to the focal chord is drawn through M, meeting the y-axis at T. Find ST and verify that it is equal to one-half the focal width of the parabola.

44. Let F be the focus of the parabola $x^2 = 8y$, and let P denote the point on the parabola with coordinates $(8, 8)$. Let $\overline{PQ}$ be a focal chord. If V denotes the vertex of the parabola, verify that

$$PF \cdot FQ = VF \cdot PQ$$

In Exercises 45–50, use the method shown in the text to find the equation of the line that is tangent to the parabola at the given point. In each case, include a sketch with your answer.

45. $x^2 = 8y$; $(4, 2)$ **46.** $x^2 = 12y$; $(6, 3)$
47. $x^2 = -y$; $(-3, -9)$ **48.** $y^2 = 4x$; $(1, 2)$
49. $y^2 = -8x$; $(-2, 4)$ **50.** $x^2 = -6y$; $(\sqrt{6}, -1)$

B

51. In the following figure, triangle OAB is equilateral and $\overline{AB}$ is parallel to the x-axis. Find the length of a side and the area of triangle OAB.

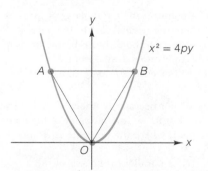

52. Let $\overline{AB}$ be the vertical focal chord of the parabola $y^2 = 4px$. Let C be the point where the directrix meets

the axis of the parabola. Show that $\overline{AC}$ is perpendicular to $\overline{BC}$.

53. Consider the parabola $x^2 = 4y$, and let (x_0, y_0) denote a point on the parabola in the first quadrant.
  (a) Find the y-intercept of the line tangent to the parabola at the point on the parabola where $y_0 = 1$.
  (b) Repeat part (a) using $y_0 = 2$.
  (c) Repeat part (a) using $y_0 = 3$.
  (d) On the basis of your results in parts (a)–(c), make a conjecture about the y-intercept of the line that is tangent to the parabola at the point (x_0, y_0) on the curve. Verify your conjecture by computing this y-intercept.

54. Let V and F denote the vertex and the focus, respectively, of the parabola $x^2 = 4py$. Suppose that P and Q are points on the parabola such that $\overline{PV}$ is perpendicular to $\overline{QV}$. If G denotes the point where $\overline{PQ}$ meets the axis of the parabola, show that $VG = 4p$.

In Exercises 55–57, find the slope of the tangent to the curve at the indicated point. (Use the method shown in the text for parabolas.)

55. $y = \sqrt{x}$; $(4, 2)$ *Hint:* $x - 4 = (\sqrt{x} - 2)(\sqrt{x} + 2)$
56. $y = 1/x$; $(3, \frac{1}{3})$ **57.** $y = x^3$; $(2, 8)$

C

58. If $\overline{PQ}$ is a focal chord of the parabola $x^2 = 4py$ and the coordinates of P are (x_0, y_0), show that the coordinates of Q are

$$\left(\frac{-4p^2}{x_0}, \frac{p^2}{y_0} \right)$$

59. If $\overline{PQ}$ is a focal chord of the parabola $y^2 = 4px$ and the coordinates of P are (x_0, y_0), show that the coordinates of Q are

$$\left(\frac{p^2}{x_0}, \frac{-4p^2}{y_0} \right)$$

60. Let F and V denote the focus and the vertex, respectively, of the parabola $x^2 = 4py$. If $\overline{PQ}$ is a focal chord of the parabola, show that

$$PF \cdot FQ = VF \cdot PQ$$

61. Let $\overline{PQ}$ be a focal chord of the parabola $y^2 = 4px$, and let M be the midpoint of $\overline{PQ}$. A perpendicular is drawn from M to the x-axis, meeting the x-axis at S. Also from M, a line segment is drawn that is perpendicular to $\overline{PQ}$ and that meets the x-axis at T. Show that the length of $\overline{ST}$ is one-half the focal width of the parabola.

62. Let $\overline{AB}$ be a chord (not necessarily a focal chord) of the parabola $y^2 = 4px$, and suppose that $\overline{AB}$ subtends a right angle at the vertex. (In other words, $\angle AOB = 90°$, where O is the origin in this case.) Find the x-intercept

of the segment $\overline{AB}$. What is surprising about the result?
Hint: Begin by writing the coordinates of A and B as

$$A\left(\frac{a^2}{4p}, a\right) \text{ and } B\left(\frac{b^2}{4p}, b\right).$$

63. Two points A and B are in the first quadrant and lie on the parabola $x^2 = 4py$. Let $\overline{AA'}$ and $\overline{BB'}$ be focal chords of the parabola. Show that the line through A and B intersects the line through A' and B' at a point on the directrix. *Hint:* Use the result in Exercise 58.

64. Suppose $\overline{PQ}$ is a focal chord of the parabola $x^2 = 4py$. From P and from Q, line segments are drawn perpendicular to the directrix, meeting the directrix at P' and Q', respectively.
  (a) Show that the line through P and Q' meets the line through Q and P' at the vertex of the parabola.
  (b) If F denotes the focus of the parabola, show that the angle $P'FQ'$ is a right angle.

65. Let $\overline{PQ}$ be a focal chord of the parabola $x^2 = 4py$. Complete the following steps to prove that the circle with $\overline{PQ}$ as a diameter is tangent to the directrix of the parabola. Let the coordinates of P be (x_0, y_0).
  (a) Show that the coordinates of Q are

  $$\left(\frac{-4p^2}{x_0}, \frac{p^2}{y_0}\right).$$

  (b) Show that the midpoint of $\overline{PQ}$ is

  $$\left(\frac{x_0^2 - 4p^2}{2x_0}, \frac{y_0^2 + p^2}{2y_0}\right).$$

  (c) Show that the length of $\overline{PQ}$ is

  $$\frac{(y_0 + p)^2}{y_0}.$$

  Suggestion: This can be done using the formula for the distance between two points, but the following is simpler. Let F be the focus. Then $PQ = PF + FQ$. Now, both PF and FQ can be determined by using the definition of the parabola rather than the distance formula.

  (d) Show that the distance from the center of the circle to the directrix equals the radius of the circle. How does this complete the proof?

66. Let (x_0, y_0) be a point on the parabola $x^2 = 4py$. Using the method explained in the text, show that the equation of the line tangent to the parabola at (x_0, y_0) is

$$y = \frac{x_0}{2p}x - y_0.$$

Thus, the y-intercept of the line tangent to $x^2 = 4py$ at (x_0, y_0) is just $-y_0$.

67. Let (x_0, y_0) be a point on the parabola $y^2 = 4px$. Show

that the equation of the line tangent of the parabola at (x_0, y_0) is

$$y = \frac{2p}{y_0}x + \frac{y_0}{2}.$$

Show that the x-intercept of this line is $-x_0$.

Exercises 68–75 contain results about tangent lines to parabolas. In working these problems, you will find it convenient to use the following facts that were developed in Exercises 66 and 67.

  (i) Let (x_0, y_0) be a point (other than the vertex) on the parabola $x^2 = 4py$. Then the tangent to the parabola at (x_0, y_0) is the line passing through (x_0, y_0) and $(0, -y_0)$. The equation of this tangent is $y = \frac{x_0}{2p}x - y_0$.

  (ii) Let (x_0, y_0) be a point (other than the vertex) on the parabola $y^2 = 4px$. Then the tangent to the parabola at (x_0, y_0) is the line passing through (x_0, y_0) and $(-x_0, 0)$. The equation of this tangent is $y = \frac{2p}{y_0}x + \frac{y_0}{2}$.

In some of the problems, reference is made to the normal line. *The* normal line *or* normal *to a parabola at the point* (x_0, y_0) *on the parabola is defined as the line through* (x_0, y_0) *that is perpendicular to the tangent at* (x_0, y_0).

68. Let $\overline{PQ}$ be a focal chord of the parabola $x^2 = 4py$.
  (a) Show that the tangents to the parabola at P and Q are perpendicular to each other.
  (b) Show that the tangents to the parabola at P and Q intersect at a point on the directrix.
  (c) Let D be the intersection point of the tangents at P and Q. Show that the line segment from D to the focus is perpendicular to $\overline{PQ}$.

69. Let $P(x_0, y_0)$ be a point [other than $(0, 0)$] on the parabola $y^2 = 4px$. Let A be the point where the normal line to the parabola at P meets the axis of the parabola. Let B be the point where the line drawn from P perpendicular to the axis of the parabola meets the axis. Show that $AB = 2p$.

70. Let $P(x_0, y_0)$ be a point on the parabola $y^2 = 4px$. Let A be the point where the normal line to the parabola at P meets the axis of the parabola. Let F be the focus of the parabola. If a line is now drawn from A perpendicular to $\overline{FP}$, meeting $\overline{FP}$ at Z, show that $ZP = 2p$.

71. Let $\overline{AB}$ be a chord (not necessarily a focal chord) of the parabola $x^2 = 4py$. Let M be the midpoint of the chord and let C be the point where the tangents at A and B intersect.
  (a) Show that $\overline{MC}$ is parallel to the axis of the parabola.
  (b) If D is the point where the parabola meets $\overline{MC}$, show that $CD = DM$.

72. In this exercise, we prove the *reflection property* of parabolas. The following figure shows a line tangent to the parabola $y^2 = 4px$ at $P(x_0, y_0)$. The line through H and P is parallel to the axis of the parabola. We wish to prove that $\alpha = \beta$. That is, we wish to prove the reflection property: the angle of incidence equals the angle of reflection.

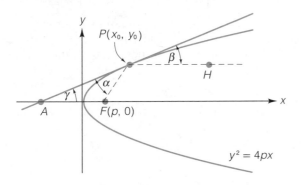

$y^2 = 4px$

(a) Show that $FA = x_0 + p = FP$.   *Hint:* Regarding FP, it is easier to rely on the definition of a parabola than on the distance formula.

(b) Why does $\alpha = \gamma$? Why does $\gamma = \beta$?

(c) Conclude that $\alpha = \beta$, as required.

73. From a point $P(x_0, y_0)$ on the parabola $x^2 = 4py$, a tangent line is drawn meeting the axis of the parabola at A. From the focus F, a line is drawn perpendicular to $\overline{AP}$, meeting $\overline{AP}$ at B. Finally, from P, a line is drawn perpendicular to the directrix, meeting the directrix at C.

(a) Show that B lies on the line that is tangent to the parabola at the vertex.

(b) Show that B is the midpoint of $\overline{FC}$.

74. Verify that the point $A(3, 2)$ lies on the line that is tangent to the parabola $x^2 = 4y$ at $P(4, 4)$. Let F be the focus of the parabola. A perpendicular is drawn from A to $\overline{FP}$, meeting $\overline{FP}$ at B. Also from A, a perpendicular is drawn to the directrix, meeting the directrix at C. Show that $FB = FC$. (This result is known as *Adams' theorem*; it holds for any parabola and any point on a tangent line.)

75. The segment $\overline{AB}$ is a focal chord of the parabola $y^2 = 4x$ and the coordinates of A are $(4, 4)$. Normals are drawn through A and B that meet the parabola again at A' and B', respectively. Prove that $\overline{A'B'}$ is three times as long as $\overline{AB}$.

8.3 ▼ THE ELLIPSE

"The heavenly motions are nothing but a continuous song for several voices, to be perceived by the intellect, not by the ear."

Johannes Kepler (1571–1630)

In this section, we discuss the symmetric oval-shaped curve known as the *ellipse*. As Kepler discovered, and Newton later proved, this is the curve described by the planets in their motions around the sun.

The Ellipse

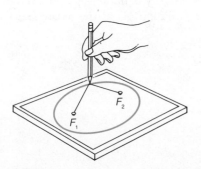

FIGURE 1

An **ellipse** is the set of all points in the plane, the sum of whose distances from two fixed points is constant. Each fixed point is called a **focus** (plural: **foci**).

Subsequently, we will derive an equation describing the ellipse, just as we found an equation for the parabola in the previous section. But first let us consider some rather immediate consequences of the definition. In fact, we can learn a great deal about the ellipse even before we derive its equation.

There is a simple mechanical method for constructing an ellipse that arises directly from the definition of the curve. Mark the given foci, say F_1 and F_2, on a drawing board and insert thumbtacks at those points. Now take a piece of string whose length is greater than the distance from F_1 to F_2 and tie the ends of the string to the tacks. Next pull the string taut with a pencil point. Then if you move the pencil while keeping the string taut, the curve traced out will be an ellipse, as indicated in Figure 1. The reason the curve is an ellipse is that for each

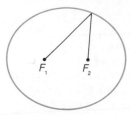

FIGURE 2
When the distance between the foci is small compared to the length of the string, the ellipse resembles a circle.

FIGURE 3
When the distance between the foci is nearly equal to the length of the string, the ellipse is relatively flat.

point on the curve, the sum of the distances from the foci is constant, the constant being the length of the string. By actually carrying out this construction for yourself several times, each time varying the distance between the foci or the length of the string, you can learn a great deal about the ellipse. For instance, when the distance between the foci is small compared to the length of the string, the ellipse begins to resemble a circle, as in Figure 2. On the other hand, when the distance between the foci is nearly equal to the length of the string, the ellipse becomes relatively flat, as in Figure 3.

We now derive one of the standard forms for the equation of an ellipse. As indicated in Figure 4, we assume that the foci are $F_1(-c, 0)$ and $F_2(c, 0)$ and that the sum of the distances from a point $P(x, y)$ on the ellipse to the foci is $2a$. Since the point P lies on the ellipse, we have

$$F_1P + F_2P = 2a$$

and therefore

$$\sqrt{(x + c)^2 + (y - 0)^2} + \sqrt{(x - c)^2 + (y - 0)^2} = 2a$$

or

$$\sqrt{(x - c)^2 + y^2} = 2a - \sqrt{(x + c)^2 + y^2} \qquad (1)$$

Now, by following a straightforward but lengthy process of squaring and simplifying (as outlined in detail in Exercise 40), we find that equation (1) becomes

$$(a^2 - c^2)x^2 + a^2y^2 = a^2(a^2 - c^2) \qquad (2)$$

To write equation (2) in a more symmetric form, we define the positive number b by the equation

$$b^2 = a^2 - c^2 \qquad (3)$$

Note   For this definition to make sense, we need to know that the right-hand side of equation (3) is positive. (See Exercise 41 at the end of this section for details.)

Finally, using equation (3) to substitute for $a^2 - c^2$ in equation (2), we obtain

$$b^2x^2 + a^2y^2 = a^2b^2$$

which can be written

$$\frac{x^2}{a^2} + \frac{y^2}{b^2} = 1 \qquad (a > b) \qquad (4)$$

We have now shown that the coordinates of each point on the ellipse satisfy equation (4). Conversely, it can be shown that if x and y satisfy equation (4), then the point (x, y) indeed lies on the ellipse. We refer to equation (4) as the **standard**

FIGURE 4

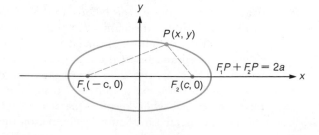

form for the equation of an ellipse with foci $(-c, 0)$ and $(c, 0)$. For an ellipse in this form, it will always be the case that a is greater than b; this follows from equation (3).

For purposes of graphing, we want to know the intercepts of the ellipse. To find the x-intercepts, we set $y = 0$ in equation (4) to obtain

$$\frac{x^2}{a^2} = 1$$

$$x^2 = a^2$$

$$x = \pm a$$

The x-intercepts are therefore a and $-a$. In a similar fashion, you can check that the y-intercepts are b and $-b$. Also (according to the symmetry tests in Section 2.1), note that the graph of equation (4) must be symmetric about both coordinate axes. Figure 5 shows the graph of the ellipse $(x^2/a^2) + (y^2/b^2) = 1$. (Calculator exercises at the end of this section will help convince you that the general shape of the curve in Figure 5 is correct.)

In the box that follows, we define several terms that are useful in describing and analyzing the ellipse.

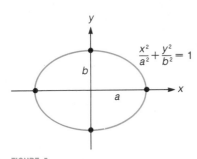

FIGURE 5

The intercepts of the ellipse $(x^2/a^2) + (y^2/b^2) = 1$ are $x = \pm a$ and $y = \pm b$

DEFINITION

Terminology for the Ellipse

1. The **focal axis** is the line passing through the foci of the ellipse.
2. The **center** is the point midway between the foci. This is the point C in Figure 6.

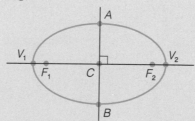

FIGURE 6

3. The **vertices** (singular: **vertex**) are the two points where the focal axis meets the ellipse. These are the points V_1 and V_2 in Figure 6.
4. The **major axis** is the line segment joining the vertices. This is the line segment $\overline{V_1 V_2}$ in Figure 6.
5. The **minor axis** is the line segment through the center of the ellipse, perpendicular to the major axis and with endpoints on the ellipse. This is the segment $\overline{AB}$ in Figure 6.
6. The **eccentricity** is the ratio $\dfrac{c}{a} = \dfrac{\sqrt{a^2 - b^2}}{a}$.

The eccentricity (as defined in the box) provides a numerical measure of how much the ellipse deviates from being a circle. As Figure 7 indicates, the closer the eccentricity is to zero, the more the ellipse resembles a circle. In the other direction, as the eccentricity approaches 1, the ellipse becomes increasingly flat.

In the box that follows, we summarize our discussion up to this point. (In the box, note that we've used the letter e to denote the eccentricity. This is the conventional choice, even though the same letter is used with a very different meaning in connection with exponential functions and logarithms.)

FIGURE 7

Eccentricity is a number between 0 and 1 that describes the shape of an ellipse. The narrowest ellipse in this figure has the same proportions as the orbit of Halley's comet. By way of contrast, the eccentricity of Earth's orbit is 0.0017; if an ellipse with this eccentricity were included in Figure 7, it would appear indistinguishable from a circle.

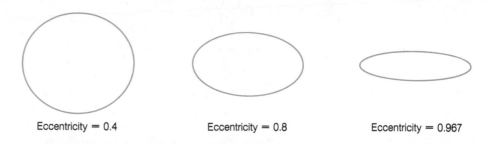

Eccentricity = 0.4   Eccentricity = 0.8   Eccentricity = 0.967

PROPERTY SUMMARY

THE ELLIPSE $\dfrac{x^2}{a^2} + \dfrac{y^2}{b^2} = 1$ $(a > b)$

Foci: $(\pm c, 0)$, where $c^2 = a^2 - b^2$

Center: $(0, 0)$

Vertices: $(\pm a, 0)$

Length of major axis: $2a$

Length of minor axis: $2b$

Eccentricity: $e = \dfrac{c}{a} = \dfrac{\sqrt{a^2 - b^2}}{a}$

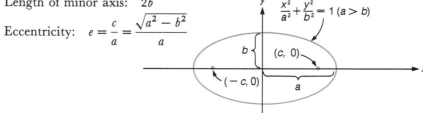

EXAMPLE 1   Find the lengths of the major and minor axes of the ellipse $9x^2 + 16y^2 = 144$. Specify the coordinates of the foci and the eccentricity. Graph the ellipse.

Solution   To convert the equation $9x^2 + 16y^2 = 144$ to standard form, divide both sides by 144. This yields

$$\frac{9x^2}{144} + \frac{16y^2}{144} = \frac{144}{144}$$

$$\frac{x^2}{16} + \frac{y^2}{9} = 1$$

$$\frac{x^2}{4^2} + \frac{y^2}{3^2} = 1$$

This is the standard form. Comparing this equation with $(x^2/a^2) + (y^2/b^2) = 1$, we see that $a = 4$ and $b = 3$. Thus, the major and minor axes are 8 and 6 units, respectively. Next, to determine the foci, we use the equation $c^2 = a^2 - b^2$. We have

$$c^2 = 4^2 - 3^2 = 7$$
$$c = \sqrt{7}$$

(We choose the positive square root because $c > 0$.) It follows that the coordinates of the foci are $(-\sqrt{7}, 0)$ and $(\sqrt{7}, 0)$. We now calculate the eccentricity using the formula $e = c/a$. This yields $e = \sqrt{7}/4$. Figure 8 shows the graph along with the required information.

FIGURE 8

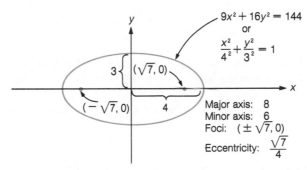

EXAMPLE 2 The foci of an ellipse are $(-1, 0)$ and $(1, 0)$. The eccentricity is $\frac{1}{3}$. Find the equation of the ellipse (in standard form), and specify the lengths of the major and minor axes.

Solution Since the foci are $(\pm 1, 0)$, we have $c = 1$. Using the equation $e = c/a$ with $e = \frac{1}{3}$ and $c = 1$, we obtain

$$\frac{1}{3} = \frac{1}{a} \qquad \text{and therefore} \qquad a = 3$$

Recall now that b^2 is defined by the equation $b^2 = a^2 - c^2$. In view of this, we have

$$b^2 = 3^2 - 1^2 = 8$$
$$b = \sqrt{8} = 2\sqrt{2}$$

The equation of the ellipse in standard form is therefore

$$\frac{x^2}{3^2} + \frac{y^2}{(2\sqrt{2})^2} = 1$$

Furthermore, since $a = 3$ and $b = 2\sqrt{2}$, the lengths of the major and minor axes are 6 and $4\sqrt{2}$ units, respectively.

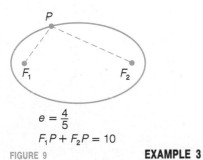

$$e = \frac{4}{5}$$
$$F_1P + F_2P = 10$$

FIGURE 9

EXAMPLE 3 The eccentricity of an ellipse is $\frac{4}{5}$, and the sum of the distances from a point P on the ellipse to the foci is 10 units. Compute the distance between the foci F_1 and F_2. See Figure 9.

Solution We are required to find the distance between the foci F_1 and F_2. By definition, this is the quantity $2c$. Since the sum of the distances from a point on the ellipse to the foci is 10 units, we have

$$2a = 10 \qquad \text{by definition of } 2a$$
$$a = 5$$

Now we substitute the values $a = 5$ and $e = \frac{4}{5}$ in the formula $e = c/a$:

$$\frac{4}{5} = \frac{c}{5} \qquad \text{and therefore} \qquad c = 4$$

It now follows that the required distance $2c$ is 8 units. ▲

Suppose now that we translate the graph of $(x^2/a^2) + (y^2/b^2) = 1$ by h units in the x-direction and k units in the y-direction, as shown in Figure 10. Then (according to our work on translation in Section 2.5) the equation of the translated ellipse is

$$\frac{(x - h)^2}{a^2} + \frac{(y - k)^2}{b^2} = 1 \tag{5}$$

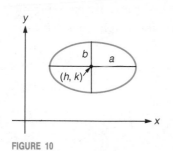

FIGURE 10

Equation (5) is another **standard form** for the equation of an ellipse. As indicated in Figure 10, the center of this ellipse is the point (h, k). In the next example, we use the technique of completing the square to convert an equation for an ellipse to standard form. Once the equation is in standard form, the graph is readily obtained.

EXAMPLE 4   Determine the center, foci, and eccentricity of the ellipse

$$4x^2 + 9y^2 - 8x - 54y + 49 = 0$$

Graph the ellipse.

Solution   We will convert the given equation to standard form by using the technique of completing the square. We have

$$4x^2 - 8x + 9y^2 - 54y = -49$$
$$4(x^2 - 2x) + 9(y^2 - 6y) = -49$$
$$4(x^2 - 2x + 1) + 9(y^2 - 6y + 9) = -49 + (4)(1) + 9(9)$$
$$4(x - 1)^2 + 9(y - 3)^2 = 36$$

Dividing this last equation by 36, we obtain

$$\frac{(x - 1)^2}{9} + \frac{(y - 3)^2}{4} = 1$$

or

$$\frac{(x - 1)^2}{3^2} + \frac{(y - 3)^2}{2^2} = 1$$

This last equation represents an ellipse with center at $(1, 3)$ and with $a = 3$ and $b = 2$. We can calculate c using the formula $c^2 = a^2 - b^2$:

$$c^2 = 3^2 - 2^2 = 5$$
$$c = \sqrt{5}$$

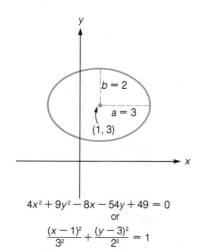

$$4x^2 + 9y^2 - 8x - 54y + 49 = 0$$
or
$$\frac{(x - 1)^2}{3^2} + \frac{(y - 3)^2}{2^2} = 1$$

FIGURE 11

Since the center of this ellipse is $(1, 3)$, the foci are therefore $(1 + \sqrt{5}, 3)$ and $(1 - \sqrt{5}, 3)$. Finally the eccentricity is c/a, which is $\sqrt{5}/3$. Figure 11 shows the graph of this ellipse. ▲

In developing the equation $(x^2/a^2) + (y^2/b^2) = 1$, we assumed that the foci of the ellipse were located on the x-axis at the points $(-c, 0)$ and $(c, 0)$. If, instead, the foci are located on the y-axis at the points $(0, c)$ and $(0, -c)$, then the same method we used in the previous case will show the equation of the ellipse to be

$$\frac{x^2}{b^2} + \frac{y^2}{a^2} = 1 \qquad (a > b)$$

We still assume that $2a$ represents the sum of the distances from a point on the ellipse to the foci. In the box that follows, we summarize the situation for the ellipse with foci $(0, c)$ and $(0, -c)$.

PROPERTY SUMMARY

THE ELLIPSE $\dfrac{x^2}{b^2} + \dfrac{y^2}{a^2} = 1$ $(a > b)$

Foci: $(0, \pm c)$, where $c^2 = a^2 - b^2$

Center: $(0, 0)$

Vertices: $(0, \pm a)$

Length of major axis: $2a$

Length of minor axis: $2b$

Eccentricity: $e = \dfrac{c}{a} = \dfrac{\sqrt{a^2 - b^2}}{a}$

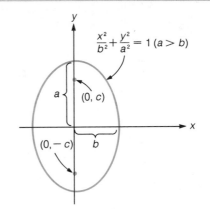

We now have two standard forms for the ellipse:

Foci on the x-axis at $(\pm c, 0)$: $\dfrac{x^2}{a^2} + \dfrac{y^2}{b^2} = 1$ $(a > b)$

Foci on the y-axis at $(0, \pm c)$: $\dfrac{x^2}{b^2} + \dfrac{y^2}{a^2} = 1$ $(a > b)$

EXAMPLE 5   Because a is greater than b in both standard forms, it is always easy to determine by inspection whether the foci lie on the x-axis or the y-axis. Consider, for instance, the equation $(x^2/5^2) + (y^2/6^2) = 1$. In this case, since $6 > 5$, we have $a = 6$. And since 6^2 appears under y^2, we conclude that the foci lie on the y-axis.

The point $(5, 3)$ lies on an ellipse with vertices $(0, \pm 2\sqrt{21})$. Find the equation of the ellipse. Write the answer in both standard form and the form $Ax^2 + By^2 = C$.

Solution   Since the vertices are $(0, \pm 2\sqrt{21})$, the standard form in this case is $(x^2/b^2) + (y^2/a^2) = 1$. Furthermore, in view of the coordinates of the vertices, we have $a = 2\sqrt{21}$. Therefore,

$$\frac{x^2}{b^2} + \frac{y^2}{84} = 1$$

Now, since the point (5, 3) lies on the ellipse, its coordinates must satisfy this last equation. We then have

$$\frac{5^2}{b^2} + \frac{3^2}{84} = 1$$

$$\frac{25}{b^2} + \frac{3}{28} = 1$$

$$\frac{25}{b^2} = \frac{25}{28}$$

From this last equation, we see that $b^2 = 28$ and therefore

$$b = \sqrt{28} = 2\sqrt{7}$$

Now that we've determined a and b, we can write the equation of the ellipse in standard form. It is

$$\frac{x^2}{(2\sqrt{7})^2} + \frac{y^2}{(2\sqrt{21})^2} = 1$$

As you should verify for yourself, this equation can also be written in the equivalent form

$$3x^2 + y^2 = 84 \qquad \blacktriangle$$

As with the parabola, many of the interesting properties of the ellipse are related to tangent lines. We define a **tangent to an ellipse** as a line that intersects the ellipse in exactly one point. Figure 12 shows a line tangent to an ellipse at an arbitrary point P on the curve. Line segments $\overline{F_1P}$ and $\overline{F_2P}$ are drawn from the foci to the point of tangency. These two segments are called **focal radii**. One of the most basic properties of the ellipse is the **reflection property**. According to this property, the focal radii drawn to the point of tangency make equal angles with the tangent. (The steps for proving this fact are outlined in detail in Exercise 65 at the end of this section. For a similar result concerning the parabola, see Exercise 72 in the previous section.)

If $P(x_1, y_1)$ is a point on the ellipse $(x^2/a^2) + (y^2/b^2) = 1$, then the equation of the line tangent to the ellipse at $P(x_1, y_1)$ is

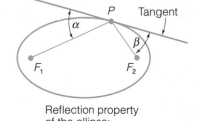

Reflection property
of the ellipse:
$\alpha = \beta$

FIGURE 12

$$\frac{x_1 x}{a^2} + \frac{y_1 y}{b^2} = 1$$

This equation is easy to remember because it so closely resembles the equation of the ellipse itself. However, do notice that the equation $(x_1 x/a^2) + (y_1 y/b^2) = 1$ is indeed linear, since x_1, y_1, a, and b all denote constants. This equation for the tangent to an ellipse can be derived using the same technique we employed with the parabola in the previous section. (See Exercise 64 at the end of this section for the details.) The two examples that follow show how this equation is used.

EXAMPLE 6   Find the equation of the line that is tangent to the ellipse $x^2 + 3y^2 = 57$ at the point $(3, 4)$ on the ellipse. See Figure 13. Write the answer in the form $y = mx + b$.

Solution   The equation of the line tangent to the ellipse at the point (x_1, y_1) is $(x_1 x/a^2) + (y_1 y/b^2) = 1$. We are given that $x_1 = 3$ and $y_1 = 4$. Thus, we need to determine

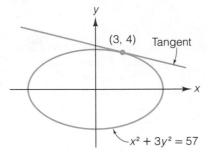

FIGURE 13

a^2 and b^2. To do that, we convert the equation $x^2 + 3y^2 = 57$ to standard form by dividing through by 57. This yields

$$\frac{x^2}{57} + \frac{y^2}{19} = 1$$

It follows that $a^2 = 57$ and $b^2 = 19$. The equation of the tangent line then becomes

$$\frac{3x}{57} + \frac{4y}{19} = 1$$

$$\frac{x}{19} + \frac{4y}{19} = 1$$

$$x + 4y = 19$$

Solving for y, we obtain

$$y = -\frac{1}{4}x + \frac{19}{4}$$

▲

EXAMPLE 7 Show that the slope of the tangent line in Figure 14 is equal to the eccentricity of the ellipse, $e = c/a$.

Solution We first determine the coordinates of the point P in Figure 14. Evidently, the x-coordinate of P is $-c$. Substituting this value for x in the equation of the ellipse yields

$$\frac{c^2}{a^2} + \frac{y^2}{b^2} = 1$$

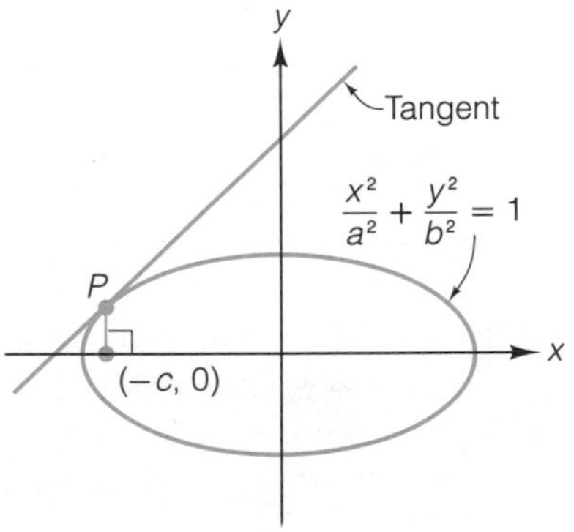

To solve this equation for y, let us first clear the equation of fractions by multiplying by a^2b^2. This yields

$$b^2c^2 + a^2y^2 = a^2b^2$$

$$a^2y^2 = a^2b^2 - b^2c^2 = b^2(a^2 - c^2)$$

$$= b^2(b^2) = b^4 \quad \text{by definition of } b^2$$

$$y^2 = \frac{b^4}{a^2}$$

$$y = \frac{b^2}{a}$$

The coordinates of P are therefore $(-c, b^2/a)$. Substituting these coordinates for x_1 and y_1 in the equation $(x_1 x/a^2) + (y_1 y/b^2) = 1$, we obtain

$$-\frac{cx}{a^2} + \frac{(b^2/a)y}{b^2} = 1$$

$$-\frac{cx}{a^2} + \frac{y}{a} = 1$$

Therefore,

$$\frac{y}{a} = \frac{c}{a^2}x + 1 \qquad \text{and} \qquad y = \frac{c}{a}x + a$$

The slope of the line is c/a, the coefficient of x. This is what we wanted to show.

▲

FIGURE 14

▼ EXERCISE SET 8.3

A

In Exercises 1–24, graph the ellipses. In each case, specify the lengths of the major and minor axes, the foci, and the eccentricity. For Exercises 13–24, also specify the center of the ellipse.

1. $4x^2 + 9y^2 = 36$
2. $4x^2 + 25y^2 = 100$
3. $x^2 + 16y^2 = 16$
4. $9x^2 + 25y^2 = 225$
5. $x^2 + 2y^2 = 2$
6. $2x^2 + 3y^2 = 3$
7. $16x^2 + 9y^2 = 144$
8. $25x^2 + y^2 = 25$
9. $15x^2 + 3y^2 = 5$
10. $9x^2 + y^2 = 4$
11. $2x^2 + y^2 = 4$
12. $36x^2 + 25y^2 = 400$
13. $\dfrac{(x-5)^2}{5^2} + \dfrac{(y+1)^2}{3^2} = 1$
14. $\dfrac{(x-1)^2}{2^2} + \dfrac{(y+4)^2}{3^2} = 1$
15. $\dfrac{(x-1)^2}{1^2} + \dfrac{(y-2)^2}{2^2} = 1$
16. $\dfrac{x^2}{4^2} + \dfrac{(y-3)^2}{2^2} = 1$
17. $\dfrac{(x+3)^2}{3^2} + \dfrac{y^2}{1^2} = 1$
18. $\dfrac{(x-2)^2}{2^2} + \dfrac{(y-2)^2}{2^2} = 1$
19. $3x^2 + 4y^2 - 6x + 16y + 7 = 0$
20. $16x^2 + 64x + 9y^2 - 54y + 1 = 0$
21. $5x^2 + 3y^2 - 40x - 36y + 188 = 0$
22. $x^2 + 16y^2 - 160y + 384 = 0$
23. $16x^2 + 25y^2 - 64x - 100y + 564 = 0$
24. $4x^2 + 4y^2 - 32x + 32y + 127 = 0$

In Exercises 25–32, find the equation of the ellipse satisfying the given conditions. Write the answer both in standard form and in the form $Ax^2 + By^2 = C$.

25. Foci $(\pm 3, 0)$; vertices $(\pm 5, 0)$
26. Foci $(0, \pm 1)$; vertices $(0, \pm 4)$
27. Vertices $(\pm 4, 0)$; eccentricity $\frac{1}{4}$
28. Foci $(0, \pm 2)$; endpoints of minor axis $(\pm 5, 0)$
29. Foci $(0, \pm 2)$; endpoints of major axis $(0, \pm 5)$
30. Endpoints of major axis $(\pm 10, 0)$; endpoints of minor axis $(0, \pm 4)$
31. Center at the origin; vertices on the x-axis; length of major axis twice the length of minor axis; $(1, \sqrt{2})$ lies on ellipse
32. Eccentricity $\frac{3}{5}$; one endpoint of minor axis $(-8, 0)$; center at the origin
33. Find the equation of the tangent to the ellipse $x^2 + 3y^2 = 76$ at each of the following points. Write your answers in the form $y = mx + b$.
    (a) $(8, 2)$
    (b) $(-7, 3)$
    (c) $(1, -5)$

34. (a) Find the equation of the line tangent to the ellipse $x^2 + 3y^2 = 84$ at the point $(3, 5)$ on the ellipse. Write your answer in the form $y = mx + b$.
    (b) Repeat part (a), but at the point $(-3, -5)$ on the ellipse.
    (c) Are the lines determined in (a) and (b) parallel?

35. A line is drawn tangent to the ellipse $3x^2 + y^2 = 52$ at the point $(4, 2)$ on the ellipse.
    (a) Find the equation of this tangent line.
    (b) Find the area of the first-quadrant triangle bounded by the axes and this tangent line.

36. Tangent lines are drawn to the ellipse $x^2 + 3y^2 = 12$ at the points $(3, -1)$ and $(-3, -1)$ on the ellipse.
    (a) Find the equation of each tangent line. Write your answers in the form $y = mx + b$.
    (b) Find the point where the tangent lines intersect.

37. Let F_1 and F_2 be the foci of the ellipse $9x^2 + 25y^2 = 225$. Let P be the point $(1, 6\sqrt{6}/5)$.
    (a) Show that P lies on the ellipse.
    (b) Find the equation of the line that is tangent to the ellipse at P. Write your answer in the form $Ax + By + C = 0$.
    (c) Let d_1 denote the distance from F_1 to the tangent line determined in part (b). Similarly let d_2 denote the distance from F_2 to this tangent line. Compute d_1 and d_2.
    (d) Verify that $d_1 d_2 = b^2$, where b is half the length of the minor axis.

[C] 38. Solve the equation $(x^2/3^2) + (y^2/2^2) = 1$ for y to obtain $y = \pm\frac{1}{3}\sqrt{36 - 4x^2}$. Then complete the following table and use the results to graph the given equation. (Use the fact that the graph must be symmetric about the y-axis.)

x	0	0.5	1.0	1.5	2.0	2.5	3
y	± 2						0

[C] 39. Solve the equation $(x^2/1^2) + (y^2/4^2) = 1$ for y to obtain $y = \pm 4\sqrt{1 - x^2}$. Then complete the following table and use the results to graph the given equation. (Use the fact that the graph must be symmetric about the y-axis.)

x	0	0.1	0.2	0.3	0.4	0.5	0.6
y	± 4						

x	0.7	0.8	0.9	1.0
y				0

B

40. This exercise outlines the steps needed to complete the derivation of the equation $(x^2/a^2) + (y^2/b^2) = 1$.

  (a) Square both sides of equation (1) on page 472. After simplifying, you should obtain

$$a\sqrt{(x+c)^2 + y^2} = a^2 + xc$$

  (b) Square both sides of the equation in part (a). Show that the result can be written

$$a^2x^2 - x^2c^2 + a^2y^2 = a^4 - a^2c^2$$

  (c) Verify that the equation in part (b) is equivalent to

$$(a^2 - c^2)x^2 + a^2y^2 = a^2(a^2 - c^2)$$

  (d) Using the equation in part (c), replace the quantity $a^2 - c^2$ with b^2. Then show that the resulting equation can be rewritten $(x^2/a^2) + (y^2/b^2) = 1$, as required.

41. In the text we defined the positive number b^2 by the equation $b^2 = a^2 - c^2$. For this definition to make sense, we need to show that the quantity $a^2 - c^2$ is indeed positive. This can be done as follows. First, recall that in any triangle, the sum of the lengths of two sides is always greater than the length of the third side. Now apply this fact to triangle F_1PF_2 in Figure 4 to show that $2a > 2c$. Conclude from this that $a^2 - c^2 > 0$, as required.

42. A line is drawn tangent to the ellipse $(x^2/a^2) + (y^2/b^2) = 1$ at the point (x_1, y_1) on the ellipse. Let P and Q denote the points where the tangent meets the y- and x-axis, respectively. Show that the midpoint of PQ is $(a^2/2x_1, b^2/2y_1)$.

43. A *normal* to an ellipse is a line drawn perpendicular to the tangent at the point of tangency. Show that the equation of the normal to the ellipse $(x^2/a^2) + (y^2/b^2) = 1$ at (x_1, y_1) can be written

$$a^2y_1x - b^2x_1y = (a^2 - b^2)x_1y_1$$

44. Let (x_1, y_1) be any point on the ellipse

$$\frac{x^2}{a^2} + \frac{y^2}{b^2} = 1 \qquad (a > b)$$

other than one of the endpoints of the major or minor axis. Show that the normal at (x_1, y_1) does not pass through the origin.   *Hint:* Find the y-intercept of the normal.

45. Find the points of intersection of the ellipses

$$\frac{x^2}{a^2} + \frac{y^2}{b^2} = 1 \qquad \text{and} \qquad \frac{x^2}{b^2} + \frac{y^2}{a^2} = 1 \qquad (a > b)$$

Include a sketch with your answer.

46. Let F_1 and F_2 denote the foci of the ellipse $(x^2/a^2) + (y^2/b^2) = 1$, $a > b$. Suppose that P is one of the endpoints of the minor axis and angle F_1PF_2 is a right angle. Compute the eccentricity of the ellipse.

47. Let $P(x_1, y_1)$ be a point on the ellipse $(x^2/a^2) + (y^2/b^2) = 1$. Let N be the point where the normal through P meets the x-axis, and let F be the focus $(-c, 0)$. Show that $FN/FP = e$, where e denotes the eccentricity.

48. Let $P(x_1, y_1)$ be a point on the ellipse $b^2x^2 + a^2y^2 = a^2b^2$. Suppose that the tangent to the ellipse at P meets the y-axis at A and the x-axis at B. If $AP = PB$, what are x_1 and y_1 (in terms of a and b)?

49. **(a)** Verify that the points $A(5, 1)$, $B(4, -2)$, and $C(-1, 3)$ all lie on the ellipse $x^2 + 3y^2 = 28$.

  (b) Find a point D on the ellipse such that $\overline{CD}$ is parallel to $\overline{AB}$.

  (c) If O denotes the center of the ellipse, show that the triangles OAC and OBD have equal areas.
  Suggestion: In computing the areas, the formula given in Exercise 55(c) of Exercise Set 1.8 is useful.

50. Find the points of intersection of the parabola $y = x^2$ and the ellipse $b^2x^2 + a^2y^2 = a^2b^2$.

51. Recall that the two line segments joining a point on the ellipse to the foci are called *focal radii*. These are the segments $\overline{F_1P}$ and $\overline{F_2P}$ in the following figure.

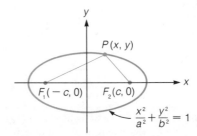

  (a) Show that $F_1P = a + ex$.   *Hint:* If you try to do this from scratch, it can involve a rather lengthy calculation. Begin instead with the equation $a\sqrt{(x+c)^2 + y^2} = a^2 + xc$ [from Exercise 40(a)], and divide both side by a.

  (b) Show that $F_2P = a - ex$.   *Hint:* Make use of the result in part (a), along with the fact that the sum of F_1P and F_2P is, by definition, $2a$.

52. Find the coordinates of a point P in the first quadrant on the ellipse $9x^2 + 25y^2 = 225$ such that angle F_2PF_1 is a right angle.

53. The accompanying figure shows the two tangent lines drawn from the point (h, k) to the ellipse $b^2x^2 + a^2y^2 = a^2b^2$. Follow steps (a) through (d) to show that the equation of the line passing through the two points of tangency is

$$b^2hx + a^2ky = a^2b^2$$

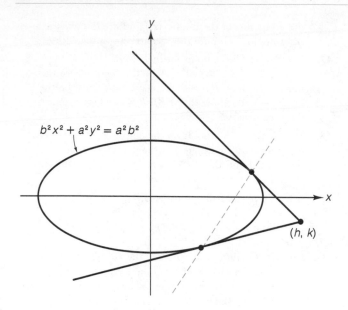

$b^2x^2 + a^2y^2 = a^2b^2$

(h, k)

(a) Let (x_1, y_1) be one of the points of tangency. Check that the equation of the tangent line through this point is $b^2x_1x + a^2y_1y = a^2b^2$.

(b) Using the result in part (a), explain why $b^2x_1h + a^2y_1k = a^2b^2$.

(c) In a similar fashion, show that $b^2x_2h + a^2y_2k = a^2b^2$, where (x_2, y_2) is the other point of tangency.

(d) The equation $b^2hx + a^2ky = a^2b^2$ represents a line. Explain why this line must pass through the points (x_1, y_1) and (x_2, y_2).

C

54. The normal to the ellipse $b^2x^2 + a^2y^2 = a^2b^2$ at $P(x_1, y_1)$ meets the x-axis at A and the y-axis at B. Show that

$$PA \cdot PB = F_1P \cdot F_2P$$

where, as usual, F_1 and F_2 are the foci.

55. In the following figure, $P(x, y)$ denotes an arbitrary point on the ellipse $(x^2/a^2) + (y^2/b^2) = 1$, and R is the foot of the perpendicular drawn from P to the line $x = a/e$. Show that $F_2P/PR = e$. The line $x = a/e$ is called a *directrix* of the ellipse. *Hint:* This result is easy to prove if you make use of the formula for F_2P in Exercise 51(b).

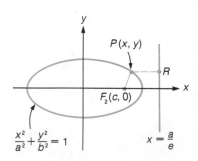

$P(x, y)$

R

$F_2(c, 0)$

$\dfrac{x^2}{a^2} + \dfrac{y^2}{b^2} = 1$

$x = \dfrac{a}{e}$

56. Let $P(x, y)$ be a point on the ellipse $(x^2/a^2) + (y^2/b^2) = 1$. Let Q be the foot of the perpendicular drawn from P to the line $x = -a/e$. Let F_1 be the focus located at $(-c, 0)$. Show that $F_1P/PQ = e$. The line $x = -a/e$ is called a *directrix* of the ellipse.

57. Let D denote the distance from the center of the ellipse $(x^2/a^2) + (y^2/b^2) = 1$ to the directrix $x = a/e$, and let d denote the distance from the focus $F_2(c, 0)$ to the directrix. Show that $D/d = a^2/b^2$.

58. At the point $P(x_1, y_1)$ on the ellipse $(x^2/a^2) + (y^2/b^2) = 1$, a tangent line is drawn that meets the directrix $x = a/e$ at Q. If F denotes the focus located at $(c, 0)$, show that angle PFQ is a right angle.

59. (a) Show that the point $(c, b^2/a)$ lies on the ellipse $(x^2/a^2) + (y^2/b^2) = 1$, where c has its usual meaning.

(b) A tangent is drawn to the ellipse $(x^2/a^2) + (y^2/b^2) = 1$ at the point $(c, b^2/a)$. Find the point where this tangent intersects the directrix $x = a/e$.

60. The *auxiliary circle* of an ellipse is the circle centered at the center of the ellipse and with radius half the length of the major axis of the ellipse.

(a) Find the equation of the auxiliary circle of the ellipse $x^2 + 3y^2 = 12$. Sketch the circle and the ellipse.

(b) Verify that the point $P(3, 1)$ lies on the ellipse.

(c) Find the equation of the line tangent to the ellipse at P.

(d) Let T be the point where the perpendicular drawn from a focus to the tangent meets the tangent. Show that T lies on the auxiliary circle.

61. The points P and Q have the same x-coordinate. The point P lies on the circle $x^2 + y^2 = a^2$, and Q lies on the ellipse $(x^2/a^2) + (y^2/b^2) = 1$. Show that the tangent to the circle at P meets the tangent to the ellipse at Q at a point on the x-axis. (Assume that P and Q are distinct, and that neither point is an endpoint of the major or minor axis of the ellipse.)

62. A line segment that passes through the center of an ellipse is called a *diameter*.

(a) If (x_1, y_1) is an endpoint of a diameter of the ellipse $(x^2/a^2) + (y^2/b^2) = 1$, show that the other endpoint is $(-x_1, -y_1)$.

(b) If (x_1, y_1) is an endpoint of a diameter of the ellipse $(x^2/a^2) + (y^2/b^2) = 1$, show that the length of the diameter is $2\sqrt{(a^2 - b^2)x_1^2 + a^2b^2}/a$.

63. Let $\overline{PQ}$ be a diameter of the ellipse $(x^2/a^2) + (y^2/b^2) = 1$. Show that the tangents to the ellipse at P and Q are parallel.

64. This exercise outlines the steps required to show that the equation of the tangent to the ellipse $(x^2/a^2) + (y^2/b^2) = 1$ at the point (x_1, y_1) on the ellipse is $(x_1x/a^2) + (y_1y/b^2) = 1$.

(a) Show that the equation $(x^2/a^2) + (y^2/b^2) = 1$ is equivalent to

$$b^2x^2 + a^2y^2 = a^2b^2 \qquad (1)$$

Conclude that (x_1, y_1) lies on the ellipse if and only if

$$b^2x_1^2 + a^2y_1^2 = a^2b^2 \qquad (2)$$

(b) Subtract equation (2) from (1) to show that

$$b^2(x^2 - x_1^2) + a^2(y^2 - y_1^2) = 0 \qquad (3)$$

Equation (3) is equivalent to equation (1) provided only that (x_1, y_1) lies on the ellipse. In the following steps, we will find the algebra much simpler if we use equation (3) to represent the ellipse, rather than the equivalent and perhaps more familiar equation (1).

(c) Let the equation of the line tangent to the ellipse at (x_1, y_1) be

$$y - y_1 = m(x - x_1)$$

Explain why the following system of equations must have exactly one solution, namely, (x_1, y_1):

$$\begin{cases} b^2(x^2 - x_1^2) + a^2(y^2 - y_1^2) = 0 & (4) \\ y - y_1 = m(x - x_1) & (5) \end{cases}$$

(d) Solve equation (5) for y and then substitute for y in equation (4) to obtain

$$b^2(x^2 - x_1^2) + a^2m^2(x - x_1)^2 + 2a^2my_1(x - x_1) = 0 \qquad (6)$$

(e) Show that equation (6) can be written

$$(x - x_1)[b^2(x + x_1) + a^2m^2(x - x_1) + 2a^2my_1] = 0 \qquad (7)$$

(f) Equation (7) is a quadratic equation in x, but as pointed out earlier, $x = x_1$ must be the only solution. (That is, $x = x_1$ is a double root.) Thus, the factor in brackets must equal zero when x is replaced by x_1. Use this observation to show that

$$m = -\frac{b^2x_1}{a^2y_1}$$

This represents the slope of the line tangent to the ellipse at (x_1, y_1).

(g) Using this value for m, show that equation (5) becomes

$$b^2x_1x + a^2y_1y = b^2x_1^2 + a^2y_1^2 \qquad (8)$$

(h) Use equation (2) to show that equation (8) can be written

$$\frac{x_1x}{a^2} + \frac{y_1y}{b^2} = 1$$

which is what we set out to show.

65. This exercise outlines a proof of the reflection property of the ellipse. We want to prove that right triangles F_1AP and F_2BP in the accompanying figure are

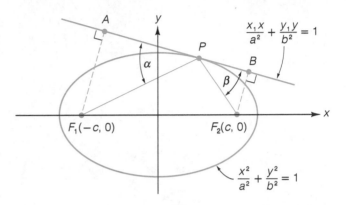

similar. This will imply that $\alpha = \beta$, as required, since corresponding angles in similar triangles are equal. To show that the right triangles F_1AP and F_2BP are similar, it is enough to demonstrate that

$$\frac{F_1A}{F_1P} = \frac{F_2B}{F_2P}$$

(a) The quantity F_1A is the distance from the point $(-c, 0)$ to the line $(x_1x/a^2) + (y_1y/b^2) = 1$. Verify that this distance is given by

$$F_1A = \frac{\left| \dfrac{x_1c}{a^2} + 1 \right|}{\sqrt{A^2 + B^2}}$$

where $A = x_1/a^2$ and $B = y_1/b^2$.

(b) In the preceding expression for F_1A, replace c by ae. Then, by making use of the expression for F_1P obtained in Exercise 51, show that

$$F_1A = \frac{F_1P}{a\sqrt{A^2 + B^2}}$$

(c) Show similarly that

$$F_2B = \frac{F_2P}{a\sqrt{A^2 + B^2}}$$

(d) Now use the expressions for F_1A and F_2B obtained in parts (b) and (c) to verify that

$$\frac{F_1A}{F_1P} = \frac{F_2B}{F_2P}$$

This implies that right triangles F_1AP and F_2BP are similar and, consequently, $\alpha = \beta$, as required.

FIGURE 1

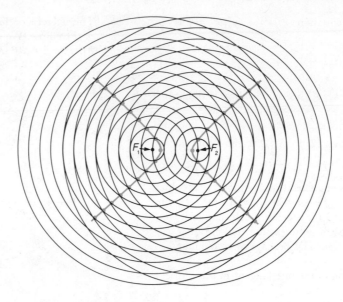

8.4 ▼ THE HYPERBOLA

In the previous section, we defined an ellipse as the set of points P such that the sum of the distances from P to two fixed points is constant. By considering the difference instead of the sum, we are led to the definition of the hyperbola.

The Hyperbola

> A **hyperbola** is the set of all points in the plane, the difference of whose distances from two fixed points is a positive constant. Each fixed point is called a **focus**.

As with the ellipse, we label the foci F_1 and F_2. Before obtaining an equation for the hyperbola, we can see the general features of the curve by using two sets of concentric circles, with centers F_1 and F_2, to locate points satisfying the definition of a hyperbola. In Figure 1, we've plotted a number of points P such that either $F_1P - F_2P = 3$ or $F_2P - F_1P = 3$. By joining these points, we obtain the graph of the hyperbola shown in Figure 1.

Unlike the parabola or the ellipse, the hyperbola is composed of two distinct parts, or **branches**. As you can check, the left branch in Figure 1 corresponds to the equation $F_2P - F_1P = 3$, while the right branch corresponds to the equation $F_1P - F_2P = 3$. Figure 1 also reveals that the hyperbola possesses two types of symmetry. First, it is symmetric about the line passing through the two foci F_1 and F_2; this line is referred to as the **focal axis** of the hyperbola. Second, the hyperbola is symmetric about the line that is the perpendicular bisector of the segment $\overline{F_1F_2}$.

To derive an equation for the hyperbola, let us initially assume that the foci are located at the points $F_1(-c, 0)$ and $F_2(c, 0)$, as indicated in Figure 2. We will use $2a$ to denote the positive constant referred to in the definition of the hyperbola. By definition, then, $P(x, y)$ lies on the hyperbola if and only if

$$|F_1P - F_2P| = 2a$$

or, equivalently,

$$F_1P - F_2P = \pm 2a$$

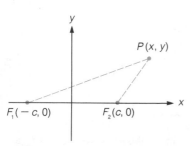

FIGURE 2

If we use the formula for the distance between two points, this last equation becomes

$$\sqrt{(x + c)^2 + (y - 0)^2} - \sqrt{(x - c)^2 + (y - 0)^2} = \pm 2a$$

or

$$\sqrt{(x + c)^2 + y^2} - \sqrt{(x - c)^2 + y^2} = \pm 2a$$

We can simplify this equation by carrying out the same procedure that we used for the ellipse in the previous section. As Exercise 42 asks you to verify, the resulting equation is

$$(c^2 - a^2)x^2 - a^2y^2 = a^2(c^2 - a^2) \tag{1}$$

Before further simplifying equation (1), we point out that the quantity $c^2 - a^2$ [which appears twice in equation (1)] is positive. To see why this is so, refer back to Figure 2. In triangle F_1F_2P (as in any triangle), the length of any side is less than the sum of the lengths of the other two sides. Thus

$$F_1P < F_1F_2 + F_2P \qquad \text{and} \qquad F_2P < F_1F_2 + F_1P$$

and therefore

$$F_1P - F_2P < F_1F_2 \qquad \text{and} \qquad F_2P - F_1P < F_1F_2$$

These last two equations tell us that

$$|F_1P - F_2P| < F_1F_2$$

Therefore, in view of the definitions of $2a$ and $2c$, we have

$$0 < 2a < 2c \qquad \text{or} \qquad 0 < a < c$$

From this last inequality we conclude that $c^2 - a^2$ is positive, as we wished to show.

Now, since $c^2 - a^2$ is positive, we may define the positive number b by the equation

$$b^2 = c^2 - a^2$$

With this notation, equation (1) becomes

$$b^2x^2 - a^2y^2 = a^2b^2$$

Dividing by a^2b^2, we obtain

$$\frac{x^2}{a^2} - \frac{y^2}{b^2} = 1 \tag{2}$$

We have now shown that the coordinates of every point on the hyperbola satisfy equation (2). Conversely, it can be shown that if the coordinates of a point satisfy equation (2), then the point satisfies the original definition of the hyperbola. Equation (2) is the **standard form** for the equation of a hyperbola with foci $F_1(-c, 0)$ and $F_2(c, 0)$.

The intercepts of the hyperbola are readily obtained from equation (2). To find the x-intercepts, we set y equal to zero to obtain

$$\frac{x^2}{a^2} = 1$$

$$x^2 = a^2$$

$$x = \pm a$$

Thus, the hyperbola crosses the x-axis at the points $(-a, 0)$ and $(a, 0)$. On the other hand, the curve does not cross the y-axis, for if we set x equal to zero in equation (2), we obtain $-y^2/b^2 = 1$ or

$$y^2 = -b^2 \qquad (b > 0)$$

Since the square of any real number y is nonnegative, this last equation has no solution. Therefore, the graph does not cross the y-axis. Finally, let us note that (according to the symmetry tests in Section 2.1) the graph of equation (2) must be symmetric about both coordinate axes.

Before graphing the hyperbola $(x^2/a^2) - (y^2/b^2) = 1$, we point out the important fact that the two lines $y = (b/a)x$ and $y = -(b/a)x$ are asymptotes for the curve. We can see why as follows. First we solve equation (2) for y:

$$\frac{x^2}{a^2} - \frac{y^2}{b^2} = 1$$

$$b^2x^2 - a^2y^2 = a^2b^2 \quad \text{multiplying by } a^2b^2$$

$$-a^2y^2 = a^2b^2 - b^2x^2$$

$$y^2 = \frac{b^2x^2 - a^2b^2}{a^2} = \frac{b^2(x^2 - a^2)}{a^2}$$

$$y = \pm\frac{b}{a}\sqrt{x^2 - a^2} \tag{3}$$

Now, as x grows arbitrarily large, the value of the quantity $\sqrt{x^2 - a^2}$ becomes closer and closer to x itself. Table 1 provides some empirical evidence for this statement in the case when $a = 5$. (A formal proof of the statement properly belongs to calculus.) In summary, then, we have the approximation $\sqrt{x^2 - a^2} \approx x$ as x grows arbitrarily large. So, in view of equation (3), we have

$$y = \pm\frac{b}{a}\sqrt{x^2 - a^2} \approx \pm\frac{b}{a}x \qquad \text{as } x \text{ grows arbitrarily large}$$

In other words, the two lines $y = \pm(b/a)x$ are asymptotes for the hyperbola.

A simple way to sketch the two asymptotes and then graph the hyperbola is as follows. First draw the rectangle with vertices (a, b), $(-a, b)$, $(-a, -b)$, and $(a, -b)$, as indicated in Figure 3(a). The slopes of the diagonals in this rectangle are b/a and $-b/a$. Thus by extending these diagonals as in Figure 3(b), we obtain the two asymptotes $y = \pm(b/a)x$. Now, since the x-intercepts of the hyperbola are a and $-a$, we can sketch the curve as shown in Figure 3(c).

For the hyperbola in Figure 3(c), the two points $(\pm a, 0)$, where the curve meets the x-axis, are referred to as **vertices**. The midpoint of the line segment

TABLE 1

x	$\sqrt{x^2 - 5^2}$
100	99.875
1000	999.987
10000	9999.999

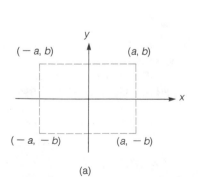

(a)

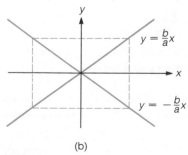

(b)

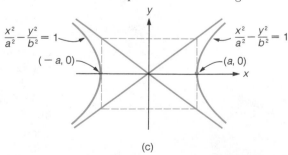

(c)

FIGURE 3

Steps in graphing the hyperbola $(x^2/a^2) - (y^2/b^2) = 1$ and its asymptotes

joining the two vertices is called the **center** of the hyperbola. (Equivalently, we can define the center as the point of intersection of the two asymptotes.) For the hyperbola in Figure 3(c), the center coincides with the origin. The line segment joining the vertices of a hyperbola is the **transverse axis** of the hyperbola. For reference, in the box that follows we summarize our work up to this point on the hyperbola. (Note: several new terms describing the hyperbola are also given in the box.)

PROPERTY SUMMARY

THE HYPERBOLA $\dfrac{x^2}{a^2} - \dfrac{y^2}{b^2} = 1$

1. The **foci** are the points $F_1(-c, 0)$ and $F_2(c, 0)$. The hyperbola is the set of points P such that $|F_1P - F_2P| = 2a$.

2. The **focal axis** is the line passing through the foci.

3. The **vertices** are the points at which the hyperbola intersects its focal axis. In Figure 4, these are the two points $V_1(-a, 0)$ and $V_2(a, 0)$.

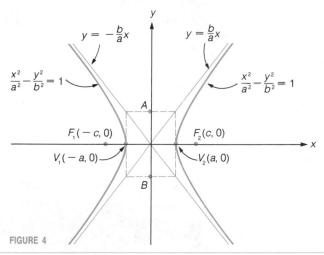

FIGURE 4

4. The **center** is the point on the focal axis midway between the foci. The center of the hyperbola in Figure 4 is the origin.

5. The **transverse axis** is the line segment joining the two vertices. In Figure 4, the length of the transverse axis $\overline{V_1V_2}$ is $2a$.

6. The **conjugate axis** is the line segment perpendicular to the transverse axis, passing through the center and extending a distance b on either side of the center. In Figure 4, this is the segment $\overline{AB}$.

7. The **eccentricity** e is defined by $e = c/a$, where $c^2 = a^2 + b^2$.

In general, if A, B, and C are positive numbers, then the graph of an equation of the form

$$Ax^2 - By^2 = C$$

will be a hyperbola of the type shown in Figure 4. Example 1 shows why this is so.

EXAMPLE 1 Graph the hyperbola $16x^2 - 9y^2 = 144$ after determining the following: vertices, foci, eccentricity, lengths of the transverse and conjugate axes, and asymptotes.

Solution First convert the given equation to standard form by dividing through by 144. This yields

$$\frac{x^2}{9} - \frac{y^2}{16} = 1 \qquad \text{or} \qquad \frac{x^2}{3^2} - \frac{y^2}{4^2} = 1$$

By comparing this with the equation $(x^2/a^2) - (y^2/b^2) = 1$, we see that $a = 3$ and $b = 4$. The value of c can be determined by using the equation $c^2 = a^2 + b^2$. We have

$$c^2 = 3^2 + 4^2 = 25$$
$$c = 5$$

Now that we know the values of a, b, and c, we can list the required information:

Vertices: $(\pm 3, 0)$ Length of transverse axis $(= 2a)$:   6
Foci: $(\pm 5, 0)$ Length of conjugate axis $(= 2b)$:   8

Eccentricity: $\dfrac{5}{3}$ Asymptotes: $y = \pm \dfrac{4}{3}x$

The graph of the hyperbola is shown in Figure 5. ▲

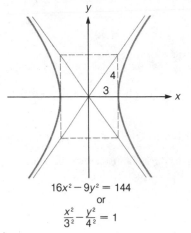

$16x^2 - 9y^2 = 144$
or
$\dfrac{x^2}{3^2} - \dfrac{y^2}{4^2} = 1$

FIGURE 5

The same method that is used to derive the equation for the hyperbola with foci $(\pm c, 0)$ can be used when the foci are instead located on the y-axis at the points $(0, \pm c)$. The equation of the hyperbola in this case is

$$\frac{y^2}{a^2} - \frac{x^2}{b^2} = 1$$

We summarize the basic properties of this hyperbola in the following box.

PROPERTY SUMMARY

THE HYPERBOLA $\dfrac{y^2}{a^2} - \dfrac{x^2}{b^2} = 1$

Foci: $(0, \pm c)$, where $c^2 = a^2 + b^2$
Vertices: $(0, \pm a)$
Asymptotes: $y = \pm (a/b)x$
Length of transverse axis: $2a$
Length of conjugate axis: $2b$
Eccentricity: $e = c/a$

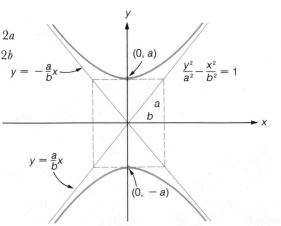

EXAMPLE 2   Use the technique of completing the square to show that the graph of the following equation is a hyperbola.

$$9y^2 - 54y - 25x^2 + 200x - 544 = 0$$

Graph the hyperbola and specify the center, the vertices, the foci, the length of the transverse axis, and the equations of the asymptotes.

Solution

$$9(y^2 - 6y \qquad) - 25(x^2 - 8x \qquad) = 544 \qquad \text{factoring}$$
$$9(y^2 - 6y + 9) - 25(x^2 - 8x + 16) = 544 + 81 - 400 \quad \text{completing the squares}$$
$$9(y - 3)^2 - 25(x - 4)^2 = 225$$
$$\frac{(y - 3)^2}{5^2} - \frac{(x - 4)^2}{3^2} = 1 \qquad \text{dividing by 225}$$

Now, the graph of this last equation is obtained by translating the graph of the equation

$$\frac{y^2}{5^2} - \frac{x^2}{3^2} = 1 \tag{4}$$

four units to the right and three units up. So first we analyze the graph of equation (4). The general form of the graph is shown in the box just prior to this example. In our case, we have $a = 5$, $b = 3$, and

$$c = \sqrt{a^2 + b^2} = \sqrt{5^2 + 3^2} = \sqrt{34} \ (\approx 5.8)$$

Consequently, the vertices [for equation (4)] are $(0, \pm 5)$; the foci are $(0, \pm\sqrt{34})$; and the asymptotes are $y = \pm\frac{5}{3}x$. Figure 6 shows the graph of this hyperbola. Finally, by translating the graph in Figure 6 to the right four units and up three units, we obtain the graph of the original equation, as shown in Figure 7. You should verify for yourself that the information accompanying Figure 7 is correct.

FIGURE 6

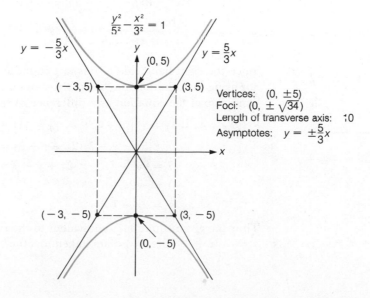

Vertices: $(0, \pm 5)$
Foci: $(0, \pm\sqrt{34})$
Length of transverse axis: 10
Asymptotes: $y = \pm\frac{5}{3}x$

FIGURE 7

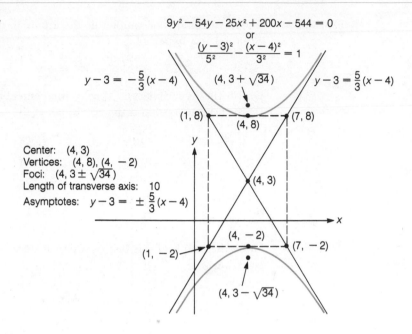

$$9y^2 - 54y - 25x^2 + 200x - 544 = 0$$
or
$$\frac{(y-3)^2}{5^2} - \frac{(x-4)^2}{3^2} = 1$$

$y - 3 = -\frac{5}{3}(x-4)$ $(4, 3 + \sqrt{34})$ $y - 3 = \frac{5}{3}(x-4)$

$(1, 8)$ $(4, 8)$ $(7, 8)$

Center: $(4, 3)$
Vertices: $(4, 8), (4, -2)$
Foci: $(4, 3 \pm \sqrt{34})$
Length of transverse axis: 10
Asymptotes: $y - 3 = \pm \frac{5}{3}(x-4)$

$(4, 3)$

$(4, -2)$

$(1, -2)$ $(7, -2)$

$(4, 3 - \sqrt{34})$

If you reread the example just completed, you'll see that it was not necessary to know in advance that the given equation represented a hyperbola. Rather, this fact emerged naturally after we completed the square. Indeed, completing the square is a useful technique for identifying the graph of any equation of the form

$$Ax^2 + Cy^2 + Dx + Ey + F = 0$$

EXAMPLE 3   Identify the graph of the equation

$$4x^2 - 32x - y^2 + 2y + 63 = 0$$

Solution   As before, we complete the squares:

$$4(x^2 - 8x) - (y^2 - 2y) = -63$$
$$4(x^2 - 8x + 16) - (y^2 - 2y + 1) = -63 + 4(16) - 1$$
$$4(x - 4)^2 - (y - 1)^2 = 0$$

Since the right-hand side of this last equation is 0, dividing both sides by 4 will not bring the equation into one of the standard forms. Indeed, if we factor the left-hand side of the equation as a difference of two squares, we obtain

$$[2(x - 4) - (y - 1)][2(x - 4) + (y - 1)] = 0$$
$$(2x - y - 7)(2x + y - 9) = 0$$

$2x - y - 7 = 0$ | $2x + y - 9 = 0$
$-y = -2x + 7$ | $y = -2x + 9$
$y = 2x - 7$

Thus the given equation is equivalent to the two linear equations $y = 2x - 7$ and $y = -2x + 9$. These two lines, taken together, constitute the graph. See Figure 8.

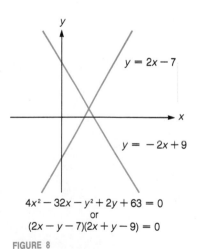

$y = 2x - 7$

$y = -2x + 9$

$4x^2 - 32x - y^2 + 2y + 63 = 0$
or
$(2x - y - 7)(2x + y - 9) = 0$

FIGURE 8

The two lines that we graphed in Figure 8 are actually the asymptotes for the hyperbola $4(x-4)^2 - (y-1)^2 = 1$. (Verify this for yourself.) For that reason, the graph in Figure 8 is referred to as a **degenerate hyperbola**. There are other cases similar to this that can arise in graphing equations of the form $Ax^2 + Cy^2 + Dx + Ey + F = 0$. For instance, as you can check for yourself by completing the squares, the graph of the equation

$$x^2 - 2x + 4y^2 - 16y + 17 = 0$$

consists of the single point $(1, 2)$. We refer to the graph in this case as a **degenerate ellipse**. Similarly, as you can check by completing the squares, the equation $x^2 - 2x + 4y^2 - 16y + 18 = 0$ has no graph; there are no points with coordinates that satisfy the equation.

We can obtain the equation of a tangent line to a hyperbola using the same ideas employed for the parabola and the ellipse in the previous sections. The result is this: The equation of the line tangent to the hyperbola $(x^2/a^2) - (y^2/b^2) = 1$ at the point (x_1, y_1) on the curve is

$$\frac{x_1 x}{a^2} - \frac{y_1 y}{b^2} = 1$$

(See Exercise 55 at the end of this section for an outline of the derivation.) As with the parabola and the ellipse, many of the interesting properties of the hyperbola are related to the tangent lines. For instance, the hyperbola has a reflection property that is similar to the reflection properties of the parabola and ellipse. To state this property, we first define a **focal radius** of a hyperbola as a line segment drawn from a focus to a point on the hyperbola. Then the **reflection property** of the hyperbola is the fact that the tangent line bisects the angle formed by the focal radii drawn to the point of tangency (as indicated in Figure 9). (See Exercise 58 for an outline of the proof.)

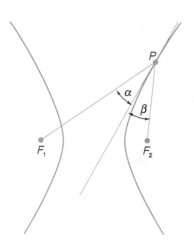

Reflection property of the hyperbola: $\alpha = \beta$. The tangent at P bisects the angle formed by the focal radii drawn to P.

FIGURE 9

▼ EXERCISE SET 8.4

A

For Exercises 1–24, graph the hyperbolas. In each case in which the hyperbola is nondegenerate, specify the following: vertices, foci, lengths of transverse and conjugate axes, eccentricity, and equations of the asymptotes. In Exercises 11–24, also specify the centers.

1. $x^2 - 4y^2 = 4$ **2.** $y^2 - x^2 = 1$

3. $y^2 - 4x^2 = 4$

4. $25x^2 - 9y^2 = 225$

5. $16x^2 - 25y^2 = 400$

6. $9x^2 - y^2 = 36$

7. $2y^2 - 3x^2 = 1$

8. $x^2 - y^2 = 9$

9. $4y^2 - 25x^2 = 100$

10. $x^2 - 3y^2 = 3$

11. $\dfrac{(x-5)^2}{5^2} - \dfrac{(y+1)^2}{3^2} = 1$

12. $\dfrac{(x-5)^2}{3^2} - \dfrac{(y+1)^2}{5^2} = 1$

13. $\dfrac{(y-2)^2}{2^2} - \dfrac{(x-1)^2}{1^2} = 1$

14. $\dfrac{(y-3)^2}{2^2} - \dfrac{x^2}{1^2} = 1$

15. $\dfrac{(x+3)^2}{4^2} - \dfrac{(y-4)^2}{4^2} = 1$

16. $\dfrac{(x+1)^2}{5^2} - \dfrac{(y+2)^2}{3^2} = 1$

17. $x^2 - y^2 + 2y - 5 = 0$

18. $16x^2 - 32x - 9y^2 + 90y - 353 = 0$

19. $x^2 - y^2 - 4x + 2y - 6 = 0$

20. $x^2 - 8x - y^2 + 8y - 25 = 0$

21. $y^2 - 25x^2 + 8y - 9 = 0$

22. $9y^2 - 18y - 4x^2 - 16x - 43 = 0$

23. $x^2 + 7x - y^2 - y + 12 = 0$

24. $9x^2 + 9x - 16y^2 + 4y + 2 = 0$

25. Let $P(x, y)$ be a point in the first quadrant on the hyperbola $(x^2/2^2) - (y^2/1^2) = 1$. Let Q be the point in the first quadrant with the same x-coordinate as P and lying on an asymptote to the hyperbola. Show that $PQ = \frac{1}{2}(x - \sqrt{x^2 - 4})$.

26. The distance PQ in Exercise 25 represents the vertical distance between the hyperbola and the asymptote. Complete the following table to see numerical evidence that this separation distance approaches zero as x gets larger and larger. (Round off each entry to one significant digit.)

x	10	50	100	500	1000	10000
PQ						

In Exercises 27–36, determine the equation of the hyperbola satisfying the given conditions. Write each answer in the form $Ax^2 - By^2 = C$ or in the form $Ay^2 - Bx^2 = C$.

27. Foci $(\pm 4, 0)$; vertices $(\pm 1, 0)$

28. Foci $(0, \pm 5)$; vertices $(0, \pm 3)$

29. Asymptotes $y = \pm \frac{1}{2}x$; vertices $(\pm 2, 0)$

30. Asymptotes $y = \pm x$; foci $(0, \pm 1)$

31. Asymptotes $y = \pm \sqrt{10}x/5$; foci $(\pm \sqrt{7}, 0)$

32. Length of transverse axis 6; eccentricity $\frac{4}{3}$; center $(0, 0)$; focal axis horizontal

33. Vertices $(0, \pm 7)$; passing through the point $(1, 9)$

34. Eccentricity 2; foci $(\pm 1, 0)$

35. Length of transverse axis 6; length of conjugate axis 2; foci on the y-axis; center at the origin

36. Asymptotes $y = \pm 2x$; passing through $(1, \sqrt{3})$

37. Show that the two asymptotes of the hyperbola $x^2 - y^2 = 16$ are perpendicular to each other.

38. (a) Verify that the point $P(6, 4\sqrt{3})$ lies on the hyperbola $16x^2 - 9y^2 = 144$.
    (b) In Example 1, we found that the foci of this hyperbola were $F_1(-5, 0)$ and $F_2(5, 0)$. Compute the lengths F_1P and F_2P, where P is the point $(6, 4\sqrt{3})$.
    (c) Verify that $|F_1P - F_2P| = 2a$.

39. (a) Verify that the point $P(5, 6)$ lies on the hyperbola $5y^2 - 4x^2 = 80$.
    (b) Find the foci.
    (c) Compute the lengths of the line segments $\overline{F_1P}$ and $\overline{F_2P}$, where P is the point $(5, 6)$.
    (d) Verify that $|F_1P - F_2P| = 2a$.

B

40. (a) Let e_1 denote the eccentricity of the hyperbola $(x^2/4^2) - (y^2/3^2) = 1$, and let e_2 denote the eccentricity of the hyperbola $(x^2/3^2) - (y^2/4^2) = 1$. Verify that
$$e_1^2 e_2^2 = e_1^2 + e_2^2$$
    (b) Let e_1 and e_2 denote the eccentricities of the hyperbolas $(x^2/A^2) - (y^2/B^2) = 1$ and $(y^2/B^2) - (x^2/A^2) = 1$, respectively. Verify that
$$e_1^2 e_2^2 = e_1^2 + e_2^2$$

41. (a) If the hyperbola $(x^2/a^2) - (y^2/b^2) = 1$ has perpendicular asymptotes, show that $a = b$. What is the eccentricity in this case?
    (b) Show that the asymptotes of the hyperbola $(x^2/a^2) - (y^2/a^2) = 1$ are perpendicular. What is the eccentricity of this hyperbola?

42. Derive equation (1) in this section from the equation that precedes it.

43. Let $P(x, y)$ be a point on the right-hand branch of the hyperbola $(x^2/a^2) - (y^2/b^2) = 1$. As usual, let F_2 denote the focus located at $(c, 0)$. The following steps outline a proof of the fact that the length of the line segment $\overline{F_2P}$ in this case is given by $F_2P = xe - a$
    (a) Explain why
$$\sqrt{(x + c)^2 + y^2} - \sqrt{(x - c)^2 + y^2} = 2a$$
    (b) In the preceding equation, add the quantity $\sqrt{(x - c)^2 + y^2}$ to both sides and then square both sides. Show that the result can be written
$$xc - a^2 = a\sqrt{(x - c)^2 + y^2}$$
or
$$xc - a^2 = a(F_2P)$$
    (c) Divide both sides of the preceding equation by a to show that $xe - a = F_2P$, as required.

44. Let $P(x, y)$ be a point on the right-hand branch of the hyperbola $(x^2/a^2) - (y^2/b^2) = 1$. As usual, let F_1 denote the focus located at $(-c, 0)$. Show that $F_1P = xe + a$. *Hint:* Use the result in Exercise 43 along with the fact that the right-hand branch is defined by the equation $F_1P - F_2P = 2a$.

45. Let P be a point on the right-hand branch of the hyperbola $x^2 - y^2 = k^2$. If d denotes the distance from P to the center of the hyperbola, show that

$$d^2 = (F_1P)(F_2P)$$

46. Let $P(x, y)$ be a point on the right-hand branch of the hyperbola $(x^2/a^2) - (y^2/b^2) = 1$. From P, a line segment is drawn perpendicular to the line $x = a/e$, meeting this line at D. If F_2 denotes (as usual) the focus located at $(c, 0)$, show that

$$\frac{F_2P}{PD} = e$$

The line $x = a/e$ is called the *directrix* of the hyperbola corresponding to the focus F_2.   *Hint:* Use the expression for F_2P developed in Exercise 43.

47. Consider the hyperbola $(x^2/a^2) - (y^2/b^2) = 1$, and let D be the point where the asymptote $y = (b/a)x$ meets the directrix $x = a/e$. If F denotes the focus corresponding to this directrix, show that the line segment $\overline{FD}$ is perpendicular to the asymptote.

48. Consider the hyperbola $(x^2/a^2) - (y^2/b^2) = 1$ and suppose that a perpendicular is drawn from the focus $F_2(c, 0)$ to the asymptote $y = (b/a)x$, meeting the asymptote at A.
(a) Show that $F_2A = b$.
(b) Show that the distance from the center of the hyperbola to A is equal to a.

49. By solving the system

$$\begin{cases} y = \frac{4}{3}x - 1 \\ 16x^2 - 9y^2 = 144 \end{cases}$$

show that the line and the hyperbola intersect in exactly one point. Draw a sketch of the situation. This demonstrates that a line that intersects a hyperbola in exactly one point does not have to be a tangent line.

50. Let $y = (b/a)x + d$, $d \neq 0$, be the equation of a line parallel to the asymptote $y = (b/a)x$ of the hyperbola $(x^2/a^2) - (y^2/b^2) = 1$. Show that the only point of intersection of this line with the hyperbola is given by

$$x = -\frac{a}{2}\left(\frac{b^2 + d^2}{bd}\right) \quad \text{and}$$

$$y = \frac{d^2 - b^2}{2d}$$

This shows that a line that intersects the hyperbola in exactly one point does not have to be a tangent line. (Contrast this to the situation with the ellipse.)

51. In this exercise, you will investigate the geometric significance of the eccentricity of a hyperbola.

(a) Use the definition of the eccentricity e to show that $e = \sqrt{1 + (b/a)^2}$. (This shows that the eccentricity is always greater than 1.)
(b) [C] Compute the eccentricity for each of the following hyperbolas.
   (i) $(0.0201)x^2 - y^2 = 0.0201$
   (ii) $3x^2 - y^2 = 3$
   (iii) $8x^2 - y^2 = 8$
   (iv) $15x^2 - y^2 = 15$
   (v) $99x^2 - y^2 = 99$
(c) [C] On the same set of axes, sketch the first-quadrant portion of each of the hyperbolas in part (b).
(d) Based on the pattern you observed in part (c), what would you say is the relationship between the eccentricity and the shape of a hyperbola?

In Exercises 52–54, find the equation of the line that is tangent to the hyperbola at the given point. Write your answer in the form $y = mx + b$.

52. $x^2 - 4y^2 = 16$; $(5, \frac{3}{2})$

53. $3x^2 - y^2 = 12$; $(4, 6)$

54. $16x^2 - 25y^2 = 400$; $(10, 4\sqrt{3})$

C
_____

55. We define a *tangent line to a hyperbola* as a line that is not parallel to an asymptote and that intersects the hyperbola in exactly one point. Show that the equation of the line tangent to the hyperbola $(x^2/a^2) - (y^2/b^2) = 1$ at the point (x_1, y_1) on the curve is

$$\frac{x_1 x}{a^2} - \frac{y_1 y}{b^2} = 1$$

Hint: Allow for signs, but follow exactly the same steps as were supplied in Exercise 64 of Exercise Set 8.3, where we treated the tangent to the ellipse. You should find that the slope in the present case is $m = (b^2 x_1/a^2 y_1)$. Explain why this slope cannot equal the slope of an asymptote as long as (x_1, y_1) is on the hyperbola.

56. (The results in this exercise will be used in Exercise 58 to prove the reflection property of the hyperbola.) Let $P(x_1, y_1)$ be a point on the right-hand branch of the hyperbola $(x^2/a^2) - (y^2/b^2) = 1$. Suppose that the tangent to the hyperbola at P meets the x-axis at A.
(a) Show that the coordinates of A are $(a^2/x_1, 0)$.
(b) Show that $F_1A = a(x_1 e + a)/x_1$.
(c) Show that $F_2A = a(x_1 e - a)/x_1$.

57. (The result in this exercise will be used in Exercise 58 to prove the reflection property of the hyperbola.) Refer to the following figure. If $F_1P/F_1A = F_2P/F_2A$, show that $\alpha = \beta$.

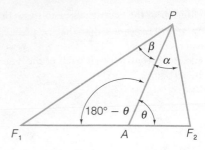

Hint: Use the law of sines in triangle AF_2P to show that

$$\frac{F_2P}{F_2A} = \frac{\sin \theta}{\sin \alpha}$$

Then carry out the same procedure in triangle AF_1P to show that

$$\frac{F_1P}{F_1A} = \frac{\sin \theta}{\sin \beta}$$

58. According to the reflection property of the hyperbola, the tangent to the hyperbola at P bisects the angle formed by the focal radii at P. This exercise outlines a proof of this property. The following figure shows a point $P(x_1, y_1)$ on the right-hand branch of the hyperbola $(x^2/a^2) - (y^2/b^2) = 1$. The tangent at P cuts the x-axis at A.

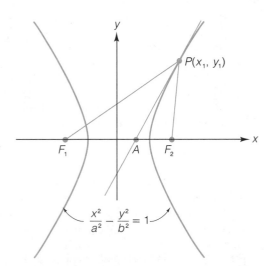

(a) Show that $F_1P/F_1A = x_1/a$.   *Hint:* This is easy if you use the expressions for F_1P and F_1A developed in Exercises 44 and 56, respectively.

(b) Show that $F_2P/F_2A = x_1/a$.   *Hint:* Use the expressions for F_2P and F_2A developed in Exercises 43 and 56, respectively.

(c) Use the results in parts (a) and (b) along with the result of Exercise 57 to conclude that angle F_1PA equals angle F_2PA, as required.

59. The *normal line* to a hyperbola at a point P on the hyperbola is the line through P that is perpendicular to the tangent at P. If the coordinates of P are (x_1, y_1), show that the equation of the normal line is

$$a^2 y_1 x + b^2 x_1 y = x_1 y_1 (a^2 + b^2)$$

60. Suppose the point $P(x_1, y_1)$ is on the hyperbola $(x^2/a^2) - (y^2/b^2) = 1$. A tangent is drawn at P, meeting the x-axis at A and the y-axis at B. Also, perpendiculars are drawn from P to the x- and y-axes, meeting these axes at C and D, respectively. If O denotes the origin, show that **(a)** $OA \cdot OC = a^2$;   **(b)** $OB \cdot OD = b^2$.

61. Suppose the point $P(x_1, y_1)$ is in the first quadrant on the hyperbola $(x^2/a^2) - (y^2/b^2) = 1$. The normal line through P meets the x- and y-axes at Q and R, respectively. If O denotes the origin, show that the area of triangle OQR is

$$\frac{x_1 y_1 (a^2 + b^2)^2}{2a^2 b^2}$$

62. Let $P(x_1, y_1)$ be a point in the first quadrant on the hyperbola $(x^2/a^2) - (y^2/b^2) = 1$. Let D be the point where the tangent at P meets the line $x = a/e$. Show that angle DF_2P is a right angle.

63. The tangent to the hyperbola $(x^2/a^2) - (y^2/b^2) = 1$ at the point P meets the asymptotes at A and B. Show that P is the midpoint of $\overline{AB}$.

64. At the point P on the hyperbola $(x^2/a^2) - (y^2/b^2) = 1$, a tangent line is drawn that meets the lines $x = a$ and $x = -a$ at S and T, respectively. Show that the circle with $\overline{ST}$ as a diameter passes through the two foci of the hyperbola.

65. Consider the following ellipse and hyperbola:

$$\left.\begin{array}{c} \dfrac{x^2}{a^2} + \dfrac{y^2}{b^2} = 1 \\[2ex] \dfrac{x^2}{a^2 - 2b^2} - \dfrac{y^2}{b^2} = 1 \end{array}\right\} \quad \text{where } a > \sqrt{2}\,b$$

(a) Show that the ellipse and the hyperbola are *confocal*, that is, that they have the same foci.

(b) In how many points do these curves intersect?

(c) Let e and E be the eccentricities of this ellipse and hyperbola, respectively. Also, let $C^2 = a^2 - b^2$. Show that the coordinates of the point in the first quadrant where the curves intersect are given by

$$x = \frac{C}{eE} \quad \text{and} \quad y = \frac{b^2}{C}$$

(d) Let P denote the intersection point determined in part (c). Show that the slopes of the tangents to the ellipse and hyperbola at P are $-e/E$ and E/e, respectively. What do you conclude from this?

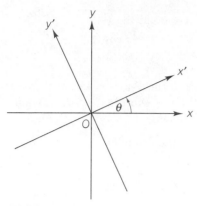

8.5 ▼ ROTATION OF AXES

In the previous section, we saw that the equation

$$Ax^2 + Cy^2 + Dx + Ey + F = 0$$

can represent, in general, one of three curves: a parabola, an ellipse, or a hyperbola. (We use the phrase "in general" here to allow for the so-called degenerate cases.) In the present section, we will find that the second-degree equation

$$Ax^2 + Bxy + Cy^2 + Dx + Ey + F = 0 \qquad (1)$$

also represents, in general, one of these three curves. The difference now is that due to the xy-term in equation (1), the axes of the curves will no longer be parallel or perpendicular to the x- and y-axes. To study curves defined by equation (1), it is useful to first introduce the technique known as **rotation of axes**.

Suppose that the x- and y-axes are rotated through a positive angle θ to yield a new x'-y' coordinate system, as shown in Figure 1. This procedure is referred to as a rotation of axes. We wish to obtain formulas relating the old and the new coordinates. Let P be a given point with coordinates (x, y) in the original coordinate system and (x', y') in the new coordinate system. In Figure 2, let r denote the distance OP and α the angle measured from the x-axis to $\overline{OP}$. From Figure 2, we have

$$\cos \alpha = \frac{\text{adjacent}}{\text{hypotenuse}} = \frac{x}{r} \qquad \text{and} \qquad \sin \alpha = \frac{\text{opposite}}{\text{hypotenuse}} = \frac{y}{r}$$

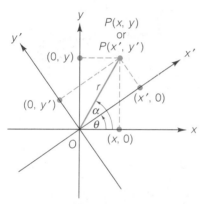

Thus,

$$x = r \cos \alpha \qquad \text{and} \qquad y = r \sin \alpha \qquad (2)$$

Again from Figure 2, we have

$$\cos(\alpha - \theta) = \frac{x'}{r} \qquad \text{and} \qquad \sin(\alpha - \theta) = \frac{y'}{r}$$

Thus,

$$x' = r \cos(\alpha - \theta) \qquad \text{and} \qquad y' = r \sin(\alpha - \theta)$$

or

$$x' = r \cos \alpha \cos \theta + r \sin \alpha \sin \theta$$

and

$$y' = r \sin \alpha \cos \theta - r \cos \alpha \sin \theta$$

With the aid of equations (2), this last pair of equations can be rewritten

$$x' = x \cos \theta + y \sin \theta \qquad \text{and} \qquad y' = -x \sin \theta + y \cos \theta$$

These two equations tell us how to express the new coordinates (x', y') in terms of the original coordinates (x, y) and the angle of rotation θ. On the other hand, it is also useful to express x and y in terms of x', y', and θ. This can be accomplished by treating the two equations we've just derived as a system of two equations in the unknowns x and y. As Exercise 41 at the end of this section asks you to verify, the results of solving this system for x and y are $x = x' \cos \theta - y' \sin \theta$ and $y = x' \sin \theta + y' \cos \theta$.

Formulas for Rotation of Axes

$$\begin{cases} x = x' \cos \theta - y' \sin \theta \\ y = x' \sin \theta + y' \cos \theta \end{cases} \qquad \begin{cases} x' = x \cos \theta + y \sin \theta \\ y' = -x \sin \theta + y \cos \theta \end{cases}$$

EXAMPLE 1 Suppose that the angle of rotation from the x-axis to the x'-axis is 45°. If the co-ordinates of a point P are $(2, 0)$ with respect to the x'-y' coordinate system, what are the coordinates of P with respect to the x-y system?

Solution Substitute the values $x' = 2$, $y' = 0$, and $\theta = 45°$ in the formulas

$$\begin{cases} x = x' \cos \theta - y' \sin \theta \\ y = x' \sin \theta + y' \cos \theta \end{cases}$$

This yields

$$x = 2 \cos 45° - 0 \sin 45° \qquad y = 2 \sin 45° + 0 \cos 45°$$

$$= 2\left(\frac{\sqrt{2}}{2}\right) = \sqrt{2} \qquad\qquad = 2\left(\frac{\sqrt{2}}{2}\right) = \sqrt{2}$$

Thus, the coordinates of P in the x-y system are $(\sqrt{2}, \sqrt{2})$. ▲

EXAMPLE 2 Suppose that the angle of rotation from the x-axis to the x'-axis is 45°. Write the equation $xy = 1$ in terms of the x'-y' coordinate system and then sketch the graph of this equation.

Solution With $\theta = 45°$, the rotation formulas for x and y become

$$\begin{cases} x = x' \cos 45° - y' \sin 45° \\ y = x' \sin 45° + y' \cos 45° \end{cases}$$

Thus we have

$$x = x'\left(\frac{\sqrt{2}}{2}\right) - y'\left(\frac{\sqrt{2}}{2}\right) = \frac{\sqrt{2}}{2}(x' - y')$$

and

$$y = x'\left(\frac{\sqrt{2}}{2}\right) + y'\left(\frac{\sqrt{2}}{2}\right) = \frac{\sqrt{2}}{2}(x' + y')$$

If we now substitute these expressions for x and y in the equation $xy = 1$, we obtain

$$\left[\frac{\sqrt{2}}{2}(x' - y')\right]\left[\frac{\sqrt{2}}{2}(x' + y')\right] = 1$$

$$\frac{1}{2}(x'^2 - y'^2) = 1$$

$$\frac{x'^2}{(\sqrt{2})^2} - \frac{y'^2}{(\sqrt{2})^2} = 1$$

This last equation represents a hyperbola in the x'-y' coordinate system. With respect to this x'-y' system, the hyperbola can be analyzed using the techniques developed in Section 8.4. The results (as you should verify for yourself) are as follows:

$$xy = 1$$

or

$$\frac{x'^2}{(\sqrt{2})^2} - \frac{y'^2}{(\sqrt{2})^2} = 1$$

$\left.\begin{array}{l}\text{Focal axis: } x'\text{-axis}\\ \text{Center: origin}\\ \text{Vertices: } (\pm\sqrt{2}, 0)\\ \text{Foci: } (\pm 2, 0)\\ \text{Asymptotes: } y' = \pm x'\end{array}\right\}$ These specifications are in terms of the x'-y' coordinate system.

As noted, the preceding specifications are in terms of the x'-y' coordinate system. However, since the original equation, $xy = 1$, is given in terms of the x-y system, we would like to express these specifications in terms of the same x-y coordinate system. This can be done by the method shown in Example 1. The results are as follows:

$$xy = 1$$

or

$$\frac{x'^2}{(\sqrt{2})^2} - \frac{y'^2}{(\sqrt{2})^2} = 1$$

$\left.\begin{array}{l}\text{Focal axis: } y = x\\ \text{Center: origin}\\ \text{Vertices: } (1, 1) \text{ and } (-1, -1)\\ \text{Foci: } (\sqrt{2}, \sqrt{2}) \text{ and }\\ \quad\quad (-\sqrt{2}, -\sqrt{2})\end{array}\right\}$ These specifications are in terms of the x-y coordinate system.

Figure 3 displays the graph of this hyperbola.

FIGURE 3
$xy = 1$

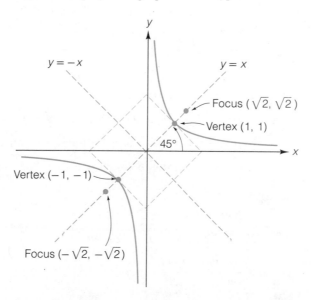

In Example 2, we saw that a rotation of 45° reduces the given equation to one of the standard forms with which we were already familiar. Now let us consider the situation in greater generality. We begin with the second-degree equation

$$Ax^2 + Bxy + Cy^2 + Dx + Ey + F = 0 \qquad (B \neq 0) \tag{3}$$

If we rotate the axes through an angle θ, equation (3) will, after some simplification, take on the form

$$A'x'^2 + B'x'y' + C'y'^2 + D'x' + E'y' + F' = 0 \tag{4}$$

for certain constants A', B', C', D', E', and F'. We wish to determine an angle of rotation θ for which $B' = 0$. The reason we want to do this is that if $B' = 0$, we will be able to analyze equation (4) using the techniques of the previous sections. We begin by recalling the rotation formulas:

$$x = x' \cos \theta - y' \sin \theta$$
$$y = x' \sin \theta + y' \cos \theta$$

Substituting these expressions for x and y in equation (3) yields

$$A(x' \cos \theta - y' \sin \theta)^2 + B(x' \cos \theta - y' \sin \theta)(x' \sin \theta + y' \cos \theta)$$
$$+ C(x' \sin \theta + y' \cos \theta)^2$$
$$+ D(x' \cos \theta - y' \sin \theta)$$
$$+ E(x' \sin \theta + y' \cos \theta) + F = 0$$

We can simplify this equation by performing the indicated operations and then collecting like terms. As Exercise 44 asks you to verify, the resulting equation is

$$A'x'^2 + B'x'y' + C'y'^2 + D'x' + E'y' + F' = 0 \tag{5}$$

where

$$A' = A \cos^2 \theta + B \sin \theta \cos \theta + C \sin^2 \theta$$
$$B' = 2(C - A) \sin \theta \cos \theta + B(\cos^2 \theta - \sin^2 \theta)$$
$$C' = A \sin^2 \theta - B \sin \theta \cos \theta + C \cos^2 \theta$$
$$D' = D \cos \theta + E \sin \theta$$
$$E' = E \cos \theta - D \sin \theta$$
$$F' = F$$

Thus, B' will be zero provided that

$$2(C - A) \sin \theta \cos \theta + B(\cos^2 \theta - \sin^2 \theta) = 0$$

By using the double-angle formulas, we can write this last equation as

$$(C - A) \sin 2\theta + B \cos 2\theta = 0$$

or

$$B \cos 2\theta = (A - C) \sin 2\theta$$

Now, dividing by $B \sin 2\theta$, we obtain

$$\frac{\cos 2\theta}{\sin 2\theta} = \frac{A - C}{B}$$

or

$$\cot 2\theta = \frac{A - C}{B}$$

We have now shown that if θ satisfies the condition $\cot 2\theta = (A - C)/B$, then equation (5) will contain no $x'y'$-term. Although we will not prove it here, it can be shown that there is always a value of θ in the range $0° < \theta < 90°$ for which $\cot 2\theta = (A - C)/B$. In subsequent examples, we will always choose θ in this range.

EXAMPLE 3   Graph the equation $2x^2 + \sqrt{3}\, xy + y^2 = 2$.

Solution   We first rotate the axes so that the new equation will contain no $x'y'$-term. To choose an appropriate value of θ, we require that

$$\cot 2\theta = \frac{A - C}{B} = \frac{2 - 1}{\sqrt{3}} = \frac{1}{\sqrt{3}}$$

Thus, $\cot 2\theta = 1/\sqrt{3}$, from which we conclude that $2\theta = 60°$, or $\theta = 30°$. With this value of θ, the rotation formulas become

$$x = x'\left(\frac{\sqrt{3}}{2}\right) - y'\left(\frac{1}{2}\right) \quad \text{and} \quad y = x'\left(\frac{1}{2}\right) + y'\left(\frac{\sqrt{3}}{2}\right)$$

Now we use these formulas to substitute for x and y in the given equation. This yields

$$2\left(\frac{x'\sqrt{3}}{2} - \frac{y'}{2}\right)^2 + \sqrt{3}\left(\frac{x'\sqrt{3}}{2} - \frac{y'}{2}\right)\left(\frac{x'}{2} + \frac{y'\sqrt{3}}{2}\right) + \left(\frac{x'}{2} + \frac{y'\sqrt{3}}{2}\right)^2 = 2$$

After simplification, this last equation becomes

$$y'^2 + 5x'^2 = 4$$

or

$$\frac{y'^2}{2^2} + \frac{x'^2}{(2/\sqrt{5})^2} = 1$$

We recognize this as the equation of an ellipse in the x'-y' coordinate system. The focal axis is the y'-axis. The values of a and b are 2 and $2/\sqrt{5}$, respectively. The ellipse can now be sketched as in Figure 4.   ▲

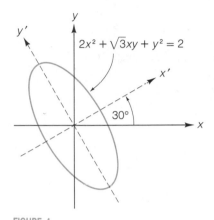

$2x^2 + \sqrt{3}xy + y^2 = 2$

$30°$

FIGURE 4

In Example 3, we were able to determine the angle θ directly, since we recognized the quantity $1/\sqrt{3}$ as the value of $\cot 60°$. However, this is the exception rather than the rule. In most problems, the value of $(A - C)/B$ is not so easily identified as the cotangent of one of the more familiar angles. The next examples demonstrate a technique that can be used in such cases. The technique relies on the following three trigonometric identities:

1. $\sec^2 \beta = 1 + \tan^2 \beta$

2. $\sin \theta = \sqrt{\dfrac{1 - \cos 2\theta}{2}}$ The positive square roots are appropriate, since $0° < \theta < 90°$.

3. $\cos \theta = \sqrt{\dfrac{1 + \cos 2\theta}{2}}$

EXAMPLE 4   Graph the equation $x^2 + 4xy - 2y^2 = 6$.

Solution   We have $A = 1$, $B = 4$, and $C = -2$. Therefore,

$$\cot 2\theta = \frac{A - C}{B} = \frac{1 - (-2)}{4} = \frac{3}{4}$$

Since $\cot 2\theta = \frac{3}{4}$, it follows that $\tan 2\theta = \frac{4}{3}$. Therefore,

$$\sec^2 2\theta = 1 + \left(\frac{4}{3}\right)^2 \quad \text{(Why?)}$$

$$= \frac{9}{9} + \frac{16}{9} = \frac{25}{9}$$

Thus,

$$\sec 2\theta = \pm \frac{5}{3}$$

At this point, we need to decide whether the positive or negative sign is appropriate. Since we are assuming that $0° < \theta < 90°$, the angle 2θ must lie in either the first or the second quadrant. To decide which, we note that the value determined for $\cot 2\theta$ was positive. That rules out the possibility that 2θ might lie in the second quadrant. We conclude that in this case, $0° < 2\theta < 90°$. Therefore, the sign of $\sec 2\theta$ must be positive, and we have

$$\sec 2\theta = \frac{5}{3} \qquad \text{or} \qquad \cos 2\theta = \frac{3}{5}$$

The values of $\sin \theta$ and $\cos \theta$ can now be obtained as follows:

$$\sin \theta = \sqrt{\frac{1 - \cos 2\theta}{2}} \qquad\qquad \cos \theta = \sqrt{\frac{1 + \cos 2\theta}{2}}$$

$$= \sqrt{\frac{1 - \frac{3}{5}}{2}} \qquad\qquad\qquad = \sqrt{\frac{1 + \frac{3}{5}}{2}}$$

$$= \frac{1}{\sqrt{5}} \quad \text{after simplifying} \qquad = \frac{2}{\sqrt{5}} \quad \text{after simplifying}$$

With these values for $\sin \theta$ and $\cos \theta$, the rotation formulas become

$$x = x'\left(\frac{2}{\sqrt{5}}\right) - y'\left(\frac{1}{\sqrt{5}}\right) = \frac{1}{\sqrt{5}}(2x' - y')$$

and

$$y = x'\left(\frac{1}{\sqrt{5}}\right) + y'\left(\frac{2}{\sqrt{5}}\right) = \frac{1}{\sqrt{5}}(x' + 2y')$$

We now substitute these expressions for x and y in the original equation $x^2 + 4xy - 2y^2 = 6$. This yields

$$\left[\frac{1}{\sqrt{5}}(2x' - y')\right]^2 + 4\left[\frac{1}{\sqrt{5}}(2x' - y')\right]\left[\frac{1}{\sqrt{5}}(x' + 2y')\right]$$

$$- 2\left[\frac{1}{\sqrt{5}}(x' + 2y')\right]^2 = 6$$

As Exercise 42 asks you to verify, this equation can be simplified to

$$2x'^2 - 3y'^2 = 6$$

or

$$\frac{x'^2}{(\sqrt{3})^2} - \frac{y'^2}{(\sqrt{2})^2} = 1$$

This last equation represents a hyperbola whose center is the origin of the x'-y' coordinate system and whose focal axis coincides with the x'-axis. We can sketch the hyperbola using the methods of the previous section, but it is first necessary

FIGURE 5
$x^2 + 4xy - 2y^2 = 6$

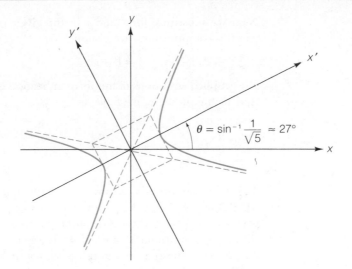

to know the angle θ between the x- and x'-axes. We have

$$\sin \theta = \frac{1}{\sqrt{5}}$$

$$\theta = \sin^{-1}\left(\frac{1}{\sqrt{5}}\right)$$

$$\theta \approx 27° \qquad \text{using a calculator}$$

Figure 5 shows the required graph. ▲

In Examples 3 and 4, we graphed equations of the form

$$Ax^2 + Bxy + Cy^2 + F = 0$$

The technique used in those examples is equally effective in graphing equations of the form $Ax^2 + Bxy + Cy^2 + Dx + Ey + F = 0$, in which the x- and y-terms are present. This is demonstrated in Example 5. Since the general technique employed in Example 5 is the same as in the previous examples, we will merely outline the procedure and the results in the solution, leaving the detailed calculations to Exercise 43 at the end of this section.

EXAMPLE 5 Graph the equation $16x^2 - 24xy + 9y^2 + 110x - 20y + 100 = 0$.

Outline of Solution

$$\cot 2\theta = \frac{A - C}{B} = -\frac{7}{24}$$

Now, proceeding as in the last example, we find that $\cos 2\theta = -\frac{7}{25}$, $\cos \theta = \frac{3}{5}$, and $\sin \theta = \frac{4}{5}$. Thus, the rotation formulas become

$$x = x'\left(\frac{3}{5}\right) - y'\left(\frac{4}{5}\right) = \frac{1}{5}(3x' - 4y')$$

and

$$y = x'\left(\frac{4}{5}\right) + y'\left(\frac{3}{5}\right) = \frac{1}{5}(4x' + 3y')$$

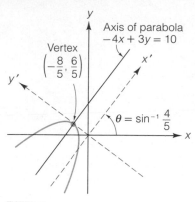

FIGURE 6

$16x^2 - 24xy + 9y^2 + 110x$
$\quad - 20y + 100 = 0$

Next we substitute for x and y in the given equation. After straightforward but lengthy computations, we obtain

$$y'^2 + 2x' - 4y' + 4 = 0$$

We graphed equations of this form in Section 8.2 by completing the square. Using that technique here, we have

$$y'^2 - 4y' = -2x' - 4$$
$$y'^2 - 4y' + 4 = -2x' - 4 + 4$$
$$(y' - 2)^2 = -2x'$$

This is the equation of a parabola. With respect to the x'-y' system, the vertex is $(0, 2)$ and the axis of the parabola is the line $y' = 2$. In terms of the x-y system, the vertex is $(-\frac{8}{5}, \frac{6}{5})$ and the axis of the parabola is the line $-4x + 3y = 10$. Finally, the angle of rotation is $\theta = \sin^{-1}(\frac{4}{5})$. Using a calculator, we find this to be approximately $53°$. The required graph is shown in Figure 6. ▲

▼ EXERCISE SET 8.5

A

In Exercises 1–3, an angle of rotation is specified, followed by the coordinates of a point in the x'-y' system. Find the coordinates of each point with respect to the x-y system.

1. $\theta = 30°$; $(x', y') = (\sqrt{3}, 2)$
2. $\theta = 60°$; $(x', y') = (-1, 1)$
3. $\theta = 45°$; $(x', y') = (\sqrt{2}, -\sqrt{2})$

In Exercises 4–6, an angle of rotation is specified, followed by the coordinates of a point in the x-y system. Find the coordinates of each point with respect to the x'-y' system.

4. $\theta = 45°$; $(x, y) = (0, -2)$
5. $\theta = \sin^{-1}(\frac{5}{13})$; $(x, y) = (-3, 1)$
6. $\theta = 15°$; $(x, y) = (1, 0)$

In Exercises 7–14, find $\sin \theta$ and $\cos \theta$, where θ is the (acute) angle of rotation that eliminates the $x'y'$ term. Note: *You are not asked to graph the equation.*

7. $25x^2 - 24xy + 18y^2 + 1 = 0$
8. $x^2 + 24xy + 8y^2 - 8 = 0$
9. $x^2 - 24xy + 8y^2 - 8 = 0$
10. $220x^2 + 119xy + 100y^2 = 0$
11. $x^2 - 2\sqrt{3}xy - y^2 = 3$
12. $5x^2 + 12xy - 4 = 0$
13. $161xy - 240y^2 - 1 = 0$
14. $4x^2 - 5xy + 4y^2 + 2 = 0$

15. Suppose that the angle of rotation is $45°$. Write the equation $2xy = 9$ in terms of the x'-y' coordinate system and then graph the equation.

16. Suppose that the angle of rotation is $45°$. Write the equation $5x^2 - 6xy + 5y^2 + 16 = 0$ in terms of the x'-y' system.

In Exercises 17–40, graph the equations.

17. $7x^2 + 8xy + y^2 - 1 = 0$
18. $2x^2 - \sqrt{3}xy + y^2 - 20 = 0$
19. $x^2 + 4xy + 4y^2 = 1$
20. $x^2 + 4xy + 4y^2 = 0$
21. $9x^2 - 24xy + 16y^2 - 400x - 300y = 0$
22. $8x^2 + 12xy + 13y^2 = 34$
23. $4xy + 3y^2 + 4x + 6y = 1$
24. $x^2 - 2xy + y^2 + x - y = 0$
25. $3x^2 - 2xy + 3y^2 - 6\sqrt{2}x + 2\sqrt{2}y + 4 = 0$
26. $x^2 + 3xy + y^2 = 1$
27. $(x - y)^2 = 8(y - 6)$
28. $4x^2 - 4xy + y^2 - 4x + 2y + 1 = 0$
29. $3x^2 + 4xy + 6y^2 = 7$
30. $x^2 + 2\sqrt{3}xy + 3y^2 + 12\sqrt{3}x - 12y - 24 = 0$
31. $17x^2 - 12xy + 8y^2 - 80 = 0$
32. $7x^2 - 2\sqrt{3}xy + 5y^2 = 32$
33. $3xy - 4y^2 + 18 = 0$
34. $x^2 + y^2 = 2xy + 4x + 4y - 8$
35. $(x + y)^2 + 4\sqrt{2}(x - y) = 0$
36. $41x^2 - 24xy + 9y^2 = 45$
37. $3x^2 - \sqrt{15}xy + 2y^2 = 3$
38. $3x^2 + 10xy + 3y^2 - 2\sqrt{2}x + 2\sqrt{2}y - 10 = 0$

39. $3x^2 - 2xy + 3y^2 + 2 = 0$

40. $(x + y)(x + y + 1) = 2$

B

41. Solve for x and y:

$$\begin{cases} (\cos\theta)x + (\sin\theta)y = x' \\ (-\sin\theta)x + (\cos\theta)y = y' \end{cases}$$

42. Simplify the equation:

$$\left[\frac{1}{\sqrt{5}}(2x' - y')\right]^2 + 4\left[\frac{1}{\sqrt{5}}(2x' - y')\right]\left[\frac{1}{\sqrt{5}}(x' + 2y')\right]$$
$$- 2\left[\frac{1}{\sqrt{5}}(x' + 2y')\right]^2 = 6$$

Answer: $2x'^2 - 3y'^2 = 6$

43. Refer to Example 5 in the text.
(a) Show that $\cos 2\theta = -\frac{7}{25}$.
(b) Show that $\cos\theta = \frac{3}{5}$ and $\sin\theta = \frac{4}{5}$.
(c) Make the substitutions $x = \frac{1}{5}(3x' - 4y')$ and $y = \frac{1}{5}(4x' + 3y')$ in the given equation $16x^2 - 24xy + 9y^2 + 110x - 20y + 100 = 0$ and show that the resulting equation simplifics to $y'^2 + 2x' - 4y' + 4 = 0$.

44. Make the substitutions $x = x'\cos\theta - y'\sin\theta$ and $y = x'\sin\theta + y'\cos\theta$ in the equation $Ax^2 + Bxy + Cy^2 + Dx + Ey + F = 0$ and show that the result is $A'x'^2 + B'x'y' + C'y'^2 + D'x' + E'y' + F' = 0$ where

$A' = A\cos^2\theta + B\sin\theta\cos\theta + C\sin^2\theta$

$B' = 2(C - A)\sin\theta\cos\theta + B(\cos^2\theta - \sin^2\theta)$

$C' = A\sin^2\theta - B\sin\theta\cos\theta + C\cos^2\theta$

$D' = D\cos\theta + E\sin\theta$

$E' = E\cos\theta - D\sin\theta$

$F' = F$

45. Show that $A + C = A' + C'$.

46. Complete the following steps to derive the equation $B'^2 - 4A'C' = B^2 - 4AC$.
(a) Show that $A' - C' = (A - C)\cos 2\theta + B\sin 2\theta$.
(b) Show that $B' = B\cos 2\theta - (A - C)\sin 2\theta$.
(c) Square the equations in parts (a) and (b), then add the two resulting equations to show that $(A' - C')^2 + B'^2 = (A - C)^2 + B^2$.
(d) Square the equation given in Exercise 45, then subtract the result from the equation in part (c). The result can be written $B'^2 - 4A'C' = B^2 - 4AC$, as required.

8.6 ▼ INTRODUCTION TO POLAR COORDINATES

FIGURE 1
FIGURE 1
This polar coordinate graph was discovered by Henri Berger, a student in one of the author's mathematics classes at U.C.L.A. in Spring of 1988.

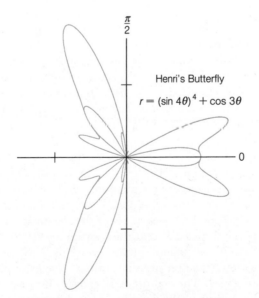

Henri's Butterfly

$r = (\sin 4\theta)^4 + \cos 3\theta$

Up until now, we have always specified the location of a point in the plane by means of a rectangular coordinate system. In this section, we introduce another coordinate system that can be used to locate points in the plane. This is the system of **polar coordinates**. We begin by drawing a half-line or ray emanating from

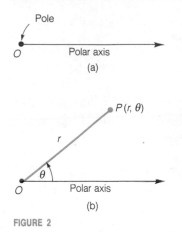

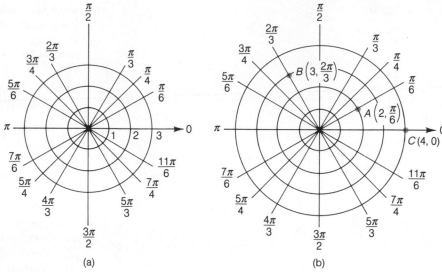

FIGURE 2

FIGURE 3

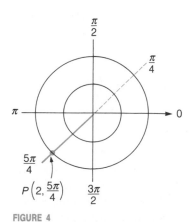

FIGURE 4

a fixed point O. The fixed point O is called the **pole** or **origin**, and the half-line is called the **polar axis**. As a matter of convention, the polar axis is usually depicted as being horizontal and extending to the right, as indicated in Figure 2(a). Now let P be any point in the plane. As indicated in Figure 2(b), we initially let r denote the distance from O to P and we let θ denote the angle measured from the polar axis counterclockwise to $\overline{OP}$. Then the ordered pair (r, θ) serves to locate the point P with respect to the pole and the polar axis. We refer to r and θ as the polar coordinates of P, and we write $P(r, \theta)$ to indicate that P is the point with polar coordinates (r, θ).

Plotting points in polar coordinates is facilitated by the use of polar coordinate graph paper, such as that shown in Figure 3(a). Figure 3(b) shows the points with polar coordinates $A(2, \pi/6)$, $B(3, 2\pi/3)$, and $C(4, 0)$.

There is a minor complication that arises in using polar rather than rectangular coordinates. Consider, for example, the point $C(4, 0)$ in Figure 3(b). This point could just as well have been labeled with the coordinates $(4, 2\pi)$ or $(4, 2k\pi)$ for any integral value of k. Similarly, the coordinates (r, θ) and $(r, \theta + 2k\pi)$ represent the same point for all integral values of k. This is in marked contrast to the situation with rectangular coordinates, where the coordinate representation of each point is unique. Also, what polar coordinates should we assign to the origin? For if $r = 0$ in Figure 2(b), we cannot really define an angle θ. The convention agreed on to cover this case is that the coordinates $(0, \theta)$ denote the origin for all values of θ. Finally, we point out that in working with polar coordinates, it is sometimes useful to let r take on negative values. For example, consider the point $P(2, 5\pi/4)$ in Figure 4. The coordinates $(2, 5\pi/4)$ indicate that to reach P from the origin, we go 2 units in the direction $5\pi/4$. Alternately, we can describe this as -2 units in the $\pi/4$ direction. (Refer again to Figure 4.) For reasons such as this, we will adhere to the convention that the polar coordinates (r, θ) and $(-r, \theta + \pi)$ represent the same point. (Since r can now, in fact, be positive, negative, or zero, r is referred to as a *directed distance*.)

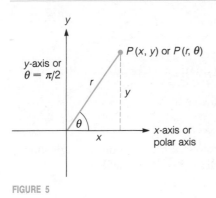

FIGURE 5

Relations between Polar and Rectangular Coordinates

It is often useful to consider simultaneously both rectangular and polar coordinates. To do this, we draw the two coordinate systems so that the origins coincide and the positive x-axis coincides with the polar axis (see Figure 5). Suppose now that a point P, other than the origin, has rectangular coordinates (x, y) and polar coordinates (r, θ), as indicated in Figure 5. We wish to find equations relating the two sets of coordinates. From Figure 5, we see that

$$x^2 + y^2 = r^2 \qquad \sin\theta = \frac{y}{r} \qquad \cos\theta = \frac{x}{r} \qquad \tan\theta = \frac{y}{x}$$

Although Figure 5 displays the point P in the first quadrant, it can be shown that these same equations hold when P is in any quadrant. For reference we summarize these equations as follows.

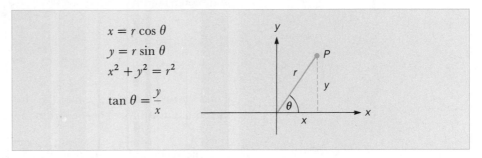

$$x = r\cos\theta$$
$$y = r\sin\theta$$
$$x^2 + y^2 = r^2$$
$$\tan\theta = \frac{y}{x}$$

EXAMPLE 1   The polar coordinates of a point are $(5, \pi/6)$. What are the rectangular coordinates?

Solution   We are given that $r = 5$ and $\theta = \pi/6$. Thus,

$$x = r\cos\theta = 5\cos\frac{\pi}{6} = 5\left(\frac{\sqrt{3}}{2}\right)$$

and

$$y = r\sin\theta = 5\sin\frac{\pi}{6} = 5\left(\frac{1}{2}\right)$$

The rectangular coordinates are therefore $\left(\dfrac{5\sqrt{3}}{2}, \dfrac{5}{2}\right)$.

The definition of the graph of an equation in polar coordinates is similar to the corresponding definition for rectangular coordinates. The **graph** of an equation in polar coordinates is the set of all points (r, θ) whose coordinates satisfy the given equation. It is often the case that the equation of a curve is simpler in one coordinate system than in another. The next two examples show instances of this.

EXAMPLE 2   Convert each polar equation to rectangular form.

(a) $r = \cos\theta + 2\sin\theta$ **(b)** $r^2 = \sin 2\theta$

Solution   **(a)** In view of the transformation equations $x = r\cos\theta$ and $y = r\sin\theta$, we multiply both sides of the given equation by r to obtain

$$r^2 = r\cos\theta + 2r\sin\theta$$
$$x^2 + y^2 = x + 2y$$

or

$$x^2 - x + y^2 - 2y = 0$$

This is the rectangular form of the given equation. *Question for review:* **What** is the graph of this last equation?

(b) Using the double-angle formula for sin 2θ, we have

$$r^2 = 2 \sin \theta \cos \theta$$

Now, in order to obtain the expressions $r \sin \theta$ and $r \cos \theta$ on the right-hand side of the equation, we multiply both sides by r^2. This yields

$$r^4 = 2(r \sin \theta)(r \cos \theta)$$

and, consequently,

$$(x^2 + y^2)^2 = 2yx$$

or

$$x^4 + 2x^2y^2 + y^4 - 2xy = 0$$

This is the rectangular form of the given equation. Notice how much simpler the equation is in its polar coordinate form. ▲

EXAMPLE 3   Convert the rectangular equation $x^2 + y^2 + ax = a\sqrt{x^2 + y^2}$ to polar form, expressing r as a function of θ. Assume that a is a constant.

Solution   Using the relations $x^2 + y^2 = r^2$ and $x = r \cos \theta$, we obtain

$$r^2 + ar \cos \theta = ar$$

Notice that this equation is satisfied by $r = 0$. In other words, the graph of this equation will pass through the origin. This is consistent with the fact that the original equation is satisfied when x and y are both zero. Now assume for the moment that $r \neq 0$. Then we can divide both sides of the last equation by r to obtain

$$r + a \cos \theta = a$$

or

$$r = a - a \cos \theta$$

This expresses r as a function of θ, as required. Notice that when $\theta = 0$, we obtain

$$r = a - a(1) = 0$$

That is, nothing has been lost in dividing through by r; the graph will still pass through the origin. ▲

Now let us turn to the problem of graphing curves defined by polar equations. As a first example, consider the equation

$$r = 2 \cos \theta$$

We need to keep in mind at this stage that r and θ are polar, not rectangular, coordinates. Thus, we should not expect the graph of $r = 2 \cos \theta$ to be the familiar cosine wave.

First we set up a table using convenient values of θ, as shown in Table 1.

TABLE 1

Values for $r = 2 \cos \theta$

θ	0	$\dfrac{\pi}{6}$	$\dfrac{\pi}{4}$	$\dfrac{\pi}{3}$	$\dfrac{\pi}{2}$	$\dfrac{2\pi}{3}$	$\dfrac{3\pi}{4}$	$\dfrac{5\pi}{4}$	π
$r = 2 \cos \theta$	2	$\sqrt{3} \approx 1.7$	$\sqrt{2} \approx 1.4$	1	0	-1	$-\sqrt{2} \approx -1.4$	$-\sqrt{3} \approx -1.7$	-2

As you can check, values of θ beyond π in this case merely lead to points already listed. For instance, with $\theta = 4\pi/3$, we have

$$r = 2 \cos \frac{4\pi}{3} = 2 \left(-\frac{1}{2} \right) = -1$$

However, the point $(-1, 4\pi/3)$ is the same point as $(1, \pi/3)$, according to our convention regarding negative values of r. In Figure 6, we have plotted the points obtained in Table 1 and connected them with a smooth curve. (The curve is in fact a circle, as we will show.)

As a check on our work in the example we just completed, we can convert the equation $r = 2 \cos \theta$ to rectangular form. Multiplying both sides by r, we have

$$r^2 = 2r \cos \theta$$

and, consequently,

$$x^2 + y^2 = 2x$$

or

$$x^2 - 2x + y^2 = 0$$

Completing the square now, we find

$$x^2 - 2x + 1 + y^2 = 1$$

or

$$(x - 1)^2 + y^2 = 1$$

This represents a circle with center at $(1, 0)$ and with radius 1, in agreement with Figure 6.

FIGURE 6

$r = 2 \cos \theta$

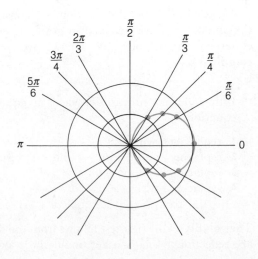

FIGURE 7
$r = 2 + 2 \cos \theta$

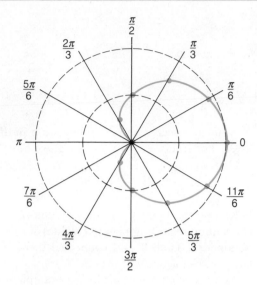

TABLE 2

Value for $r = 2 + 2 \cos \theta$

θ	0	$\dfrac{\pi}{6}$	$\dfrac{\pi}{3}$	$\dfrac{\pi}{2}$	$\dfrac{2\pi}{3}$	$\dfrac{5\pi}{6}$	π	$\dfrac{7\pi}{6}$	$\dfrac{4\pi}{3}$	$\dfrac{3\pi}{2}$	$\dfrac{5\pi}{3}$	$\dfrac{11\pi}{6}$	2π
$r = 2 + 2 \cos \theta$	4	3.7	3	2	1	0.3	0	0.3	1	2	3	3.7	4

EXAMPLE 4　　Graph the equation $r = 2 + 2 \cos \theta$.

Solution　　Again we need to keep in mind that we are using polar rather than rectangular coordinates. Thus, even though we have just seen that $r = 2 \cos \theta$ represents a circle, we cannot expect the graph of $r = 2 + 2 \cos \theta$ to be simply a translate of the circle. As before, we begin by setting up a table. This time, however, we find that values of θ beyond π yield new points on the graph. Thus, we continue the table to $\theta = 2\pi$. In Figure 7, we have plotted the points obtained from Table 2 and connected them with a smooth curve. The resulting heart-shaped curve is known as a **cardioid**.　　　　　　　　　　　　　　　　　　　▲

　　We saw in Chapter 2 that symmetry considerations can often be used to lessen the amount of work involved in graphing equations. This is also true for polar equations. In the box that follows, we list four tests for symmetry in polar coordinates.

Symmetry Tests in Polar Coordinates

If the following substitution yields an equivalent equation	Then the graph is symmetric about:
1. replacing θ with $-\theta$	the x-axis
2. replacing r and θ with $-r$ and $-\theta$, respectively	the y-axis
3. replacing θ with $\pi - \theta$	the y-axis
4. replacing r with $-r$	the origin

The validity of the first test follows from the fact that the points (r, θ) and $(r, -\theta)$ are reflections of each other about the x-axis [see Figure 8(a)]. (This first test

FIGURE 8

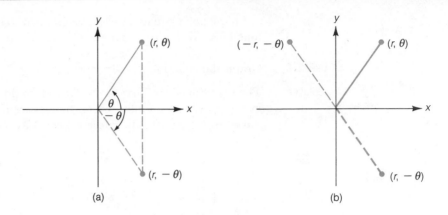

(a)   (b)

could have been used to reduce the labor involved in graphing the cardioid $r = 2 + 2 \cos \theta$ in Figure 7.) The validity of test 2 follows from the fact that the points (r, θ) and $(-r, -\theta)$ are reflections of each other about the y-axis, as indicated in Figure 8(b). The other tests can be justified in a similar manner.

EXAMPLE 5   Graph the equation $r^2 = 4 \cos 2\theta$.

Solution   As θ varies from 0 to π, the values of 2θ run from 0 to 2π, and hence $\cos 2\theta$ runs through one complete cycle of values. Thus, in setting up a table to graph this equation, we do not need to consider values of θ beyond π. Furthermore, we do not need to consider values of θ in the interval $\pi/4 < \theta < 3\pi/4$ because $\cos 2\theta$ is negative there, whereas r^2 is always nonnegative. We begin by setting up a table, using a calculator as necessary. See Table 3. If we plot the points in Table 3 and join them with a smooth curve, we obtain the graph in Figure 9(a).

TABLE 3

θ	0	$\dfrac{\pi}{12}$	$\dfrac{\pi}{8}$	$\dfrac{\pi}{6}$	$\dfrac{\pi}{4}$
$r = \pm 2\sqrt{\cos 2\theta}$	± 2	± 1.86	± 1.68	± 1.41	0

Rather than set up another table with values of θ running from $3\pi/4$ to π, we can rely on symmetry to complete the graph. According to the first two symmetry tests, the curve is symmetric about both the x-axis and the y-axis. Thus, we obtain the graph shown in Figure 9(b). This curve is called a **lemniscate**.

FIGURE 9

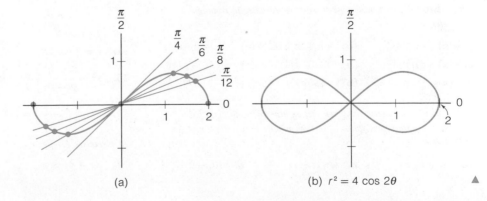

(a)   (b) $r^2 = 4 \cos 2\theta$

All the polar equations we have graphed so far have involved the trigonometric functions. As the next example indicates, this need not always be the case.

EXAMPLE 6   Graph the equation $r = \theta/\pi$ for $\theta \geq 0$.

Solution   The equation shows that as θ increases, so does r. We set up a table of values as in the previous examples. Using the entries in Table 4, we construct the graph shown in Figure 10. The curve is known as the **spiral of Archimedes**.

TABLE 4

θ	0	$\dfrac{\pi}{4}$	$\dfrac{\pi}{2}$	$\dfrac{3\pi}{4}$	π	$\dfrac{5\pi}{4}$	$\dfrac{3\pi}{2}$	$\dfrac{7\pi}{4}$	2π	$\dfrac{5\pi}{2}$	3π	4π
r	0	$\dfrac{1}{4}$	$\dfrac{1}{2}$	$\dfrac{3}{4}$	1	$\dfrac{5}{4}$	$\dfrac{3}{2}$	$\dfrac{7}{4}$	2	$\dfrac{5}{2}$	3	4

FIGURE 10
$r = \theta/\pi$ $(\theta \geq 0)$

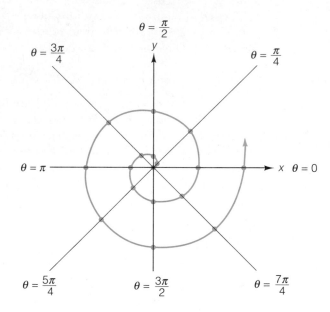

▼ EXERCISE SET 8.6

A

In Exercises 1–3, convert the given polar coordinates to rectangular coordinates.

1. **(a)** $(3, 2\pi/3)$ **(b)** $(4, 11\pi/6)$ **(c)** $(4, -\pi/6)$

2. **(a)** $(5, \pi/4)$ **(b)** $(-5, \pi/4)$ **(c)** $(-5, -\pi/4)$

3. **(a)** $(1, \pi/2)$ **(b)** $(1, 5\pi/2)$ **(c)** $(-1, \pi/8)$

In Exercises 4–6, convert the given rectangular coordinates to polar coordinates. Express your answers in such a way that r is nonnegative and $0 \leq \theta < 2\pi$.

4. $(3, \sqrt{3})$ **5.** $(-1, -1)$ **6.** $(0, -2)$

In Exercises 7–16, convert the polar equations to rectangular form.

7. $r = 2\cos\theta$ **8.** $2\sin\theta - 3\cos\theta = r$

9. $r = \tan\theta$ **10.** $r = 4$

11. $r = 3\cos 2\theta$ **12.** $r = 4\sin 2\theta$

13. $r^2 = \dfrac{8}{2 - \sin^2\theta}$ **14.** $r^2 = \dfrac{1}{3 + \cos^2\theta}$

15. $r\cos\left(\theta - \dfrac{\pi}{6}\right) = 2$ **16.** $r\sin\left(\theta + \dfrac{\pi}{4}\right) = 6$

In Exercises 17–24, convert the rectangular equations to polar form.

17. $3x - 4y = 2$ **18.** $x^2 + y^2 = 25$

19. $y^2 = x^3$

20. $y - x^2$

21. $2xy = 1$

22. $x^2 + 4x + y^2 + 4y = 0$

23. $9x^2 + y^2 = 9$

24. $x^2(x^2 + y^2) = y^2$

25. Notice that the rectangular form of the polar equation $r \cos \theta = a$ is $x = a$. Thus, the graph of $r \cos \theta = a$ is a vertical line with an x-intercept of a. Use this observation to graph the following polar equations.
(a) $r \cos \theta = 3$ **(b)** $r \cos \theta = -2$

26. Notice that the rectangular form of the polar equation $r \sin \theta = a$ is $y = a$. Thus, the graph of $r \sin \theta = a$ is a horizontal line with a y-intercept of a. Use this observation to graph the following polar equations.
(a) $r \sin \theta = 4$ **(b)** $r \sin \theta = -1$

In Exercises 27–51, graph the given equations.

27. $r \cos \theta = 5$ *Hint:* See Exercise 25

28. $r \sin \theta = \pi$

29. $r = 3 \cos \theta$

30. $r = 2 \sin \theta$

31. $r = 1 - \cos \theta$

32. $r = 1 + \cos \theta$

33. $r^2 = 2 \sin 2\theta$

34. $r^2 = 2 \cos 2\theta$

35. $r = 1$

36. $r = 2\theta/\pi, \quad \theta \geq 0$

37. $\theta = 1$

38. $\theta = \pi/6$

39. $r = 4 \sin \theta + 2 \cos \theta$ *Hint:* First convert to rectangular form.

40. $r(2 \sin \theta + \cos \theta) = 1$ (Use the hint in the previous exercise.)

41. $r = 2 \sin 2\theta$ (*four-leafed rose*)

42. $r = \sin 3\theta$ (*three-leafed rose*)

43. $r = 2 + \sin \theta$ (*limaçon*)

44. $r = 1 + 2 \sin \theta$ (*limaçon* with an inner loop)

45. $r = 1 - 2 \cos \theta$ (*limaçon* with an inner loop)

46. $r = 2 + 2 \sin \theta$ (*cardioid*)

17. $r^2 - 4 \sin 2\theta$ (*lemniscate*)

48. $r = 8 \tan \theta$ (*kappa curve*)

49. $r = \csc \theta + 2$ (*conchoid of Nicomedes*)

50. $r = e^\theta$ (*logarithmic spiral*)

51. $r = \sec^2(\theta/2)$ (*parabola*)

B

52. Show that the distance between the two points $P_1(r_1, \theta_1)$ and $P_2(r_2, \theta_2)$ is given by
$$d = \sqrt{r_1^2 + r_2^2 - 2r_1 r_2 \cos(\theta_1 - \theta_2)}$$

53. By converting the polar equation
$$r = a \cos \theta + b \sin \theta$$
to rectangular form, show that the graph is a circle, and find the center and the radius.

54. Show that the rectangular form of the equation $r = a \sin 3\theta$ is $(x^2 + y^2)^2 = ay(3x^2 - y^2)$.

55. Show that the rectangular form of the equation $r = ab/(1 - a \cos \theta) \quad (a < 1)$ is $(1 - a^2)x^2 + y^2 - 2a^2 bx - a^2 b^2 = 0$.

56. Show that the polar form of the equation $(x^2/a^2) - (y^2/b^2) = 1$ is $r^2 = a^2 b^2/(b^2 \cos^2 \theta - a^2 \sin^2 \theta)$.

57. Let k denote a positive constant, and let F_1 and F_2 denote the points with rectangular coordinates $(-k, 0)$ and $(k, 0)$, respectively. A curve known as the *lemniscate of Bernoulli* is defined as the set of points $P(x, y)$ such that $(F_1 P) \times (F_2 P) = k^2$.
(a) Show that the rectangular equation of the curve is $(x^2 + y^2)^2 = 2k^2(x^2 - y^2)$.
(b) Show that the polar equation is $r^2 = 2k^2 \cos 2\theta$.
(c) Graph the equation $r^2 = 2k^2 \cos 2\theta$. *Hint:* See Example 5 in the text.

58. A rigid wire of length $2k$ is constrained to move in such a way that its ends are always on the coordinate axes. Let P denote the point where the perpendicular from the origin to the wire meets the wire. Then as the wire moves, P moves and traces out a certain curve.
(a) If the polar coordinates of P are (r, θ), show that $r = k \sin 2\theta$.
(b) Graph the curve $r = k \sin 2\theta$, which represents the path of P.
(c) Show that in rectangular form, the equation becomes $(x^2 + y^2)^3 = 4k^2 x^2 y^2$.

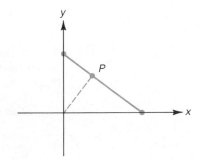

▼ **CHAPTER EIGHT   SUMMARY OF PRINCIPAL TERMS AND FORMULAS**

TERM OR FORMULA	PAGE REFERENCE	COMMENT
1. Angle of inclination	451	The angle of inclination of a line is the angle between the x-axis and the line, measured counterclockwise from the positive side or positive direction of the x-axis to the line.
2. $m = \tan \theta$	452	The slope of a line is equal to the tangent of the angle of inclination.
3. $d = \dfrac{|mx_0 + b - y_0|}{\sqrt{1 + m^2}}$	453	This is the formula for the (perpendicular) distance d from the point (x_0, y_0) to the line $y = mx + b$.
4. $d = \dfrac{|Ax_0 + By_0 + C|}{\sqrt{A^2 + B^2}}$	453, 456	This is the formula for the (perpendicular) distance d from the point (x_0, y_0) to the line $Ax + By + C = 0$.
5. Parabola	460	A parabola is the set of all points in the plane equally distant from a fixed line and a fixed point not on the line. The fixed line is called the *directrix*, and the fixed point is called the *focus*.
6. Axis (of a parabola)	462	This is the line drawn through the focus of the parabola, perpendicular to the directrix.
7. Vertex (of a parabola)	462	This is the point at which the parabola intersects its axis.
8. Focal chord (of a parabola)	463	A focal chord of a parabola is a line segment passing through the focus, with endpoints on the parabola.
9. Focal width (of a parabola)	463	The focal width of a parabola is the length of the focal chord that is perpendicular to the axis of the parabola. For a given value of p, the two parabolas $x^2 = 4py$ and $y^2 = 4px$ have the same focal width; it is $4p$.
10. Tangent line (to a parabola)	466	A line that is not parallel to the axis of a parabola is tangent to the parabola provided that it intersects the parabola in exactly one point. See the figure in the box on page 466.
11. Conic sections	467	These are the curves that are formed when a plane intersects the surface of a right circular cone. As indicated in Figure 14 in Section 8.2, these curves are the circle, the ellipse, the hyperbola, and the parabola.
12. Ellipse	471	An ellipse is the set of all points in the plane, the sum of whose distances from two fixed points is constant. Each fixed point is called a *focus* of the ellipse.
13. Eccentricity (of an ellipse)	473, 474	The eccentricity is a number that measures how much the ellipse deviates from being a circle. See Figure 7 in Section 8.3. The eccentricity e is defined by the formula $e = c/a$, where c and a are defined by the following conventions. The distance between the foci is denoted by $2c$. The sum of the distances from a point on the ellipse to the two foci is denoted by $2a$.
14. Focal axis (of an ellipse)	473	This is the line passing through the two foci.
15. Center (of an ellipse)	473	This is the midpoint of the line segment joining the foci.
16. Vertices (of an ellipse)	473	The two points at which an ellipse meets its focal axis are called the *vertices* of the ellipse.

TERM OR FORMULA	PAGE REFERENCE	COMMENT
17. Major axis (of an ellipse)	473	This is the line segment joining the two vertices of the ellipse.
18. Minor axis (of an ellipse)	473	This is the line segment through the center of the ellipse, perpendicular to the major axis, and with endpoints on the ellipse.
19. $\dfrac{x^2}{a^2} + \dfrac{y^2}{b^2} = 1$ $(a > b)$	474	This is the standard form for the equation of an ellipse with foci $(\pm c, 0)$. The specifications for this ellipse are summarized in the box on page 474.
$\dfrac{x^2}{b^2} + \dfrac{y^2}{a^2} = 1$ $(a > b)$	477	This is the standard form for the equation of an ellipse with foci $(0, \pm c)$. The specifications for this ellipse are summarized in the box on page 477.
20. Tangent line to an ellipse	478	A tangent to an ellipse is a line that intersects the ellipse in exactly one point. See Figure 12 in Section 8.3.
21. $\dfrac{x_1 x}{a^2} + \dfrac{y_1 y}{b^2} = 1$	478	This is the equation of the line that is tangent to the ellipse $(x^2/a^2) + (y^2/b^2) = 1$ at the point (x_1, y_1) on the ellipse.
22. Hyperbola	484	A hyperbola is the set of all points in the plane, the difference of whose distances from two fixed points is a positive constant. The two fixed points are the *foci*, and the line passing through the foci is the *focal axis*.
23. $\dfrac{x^2}{a^2} - \dfrac{y^2}{b^2} = 1$	487	This is the standard form for the equation of a hyperbola with foci $(\pm c, 0)$.
$\dfrac{y^2}{a^2} - \dfrac{x^2}{b^2} = 1$	488	This is the standard form for the equation of a hyperbola with foci $(0, \pm c)$.
24. Asymptote	486	A line is said to be an *asymptote* for a curve if the distance between the line and the curve approaches zero as we move out farther and farther along the line. The asymptotes for the hyperbola $(x^2/a^2) - (y^2/b^2) = 1$ are the two lines $y = \pm (b/a)x$. For the hyperbola $(y^2/a^2) - (x^2/b^2) = 1$, the asymptotes are $y = \pm (a/b)x$.
25. Focal axis (of a hyperbola)	487	This is the line passing through the foci.
26. Vertices (of a hyperbola)	487	The two points at which the hyperbola intersects its focal axis are called *vertices*. See Figure 4 on page 487 and the boxed figure on page 488.
27. Center (of a hyperbola)	487	This is the point on the focal axis midway between the foci. See Figure 4 on page 487 and the boxed figure on page 488.
28. Transverse axis	487	This is the line segment joining the two vertices of a hyperbola. See Figure 4 on page 487 and the boxed figure on page 488.
29. Conjugate axis	487	This is the line segment perpendicular to the transverse axis of the hyperbola, passing through the center and extending a distance $b(= \sqrt{c^2 - a^2})$ on either side of the center. See Figure 4 on page 487 and the boxed figure on page 488.
30. Eccentricity (of a hyperbola)	487	For both of the standard forms for the hyperbola, the eccentricity e is defined by $e = c/a$.
31. $\dfrac{x_1 x}{a^2} - \dfrac{y_1 y}{b^2} = 1$	491	This is the equation of the tangent to the hyperbola $(x^2/a^2) - (y^2/b^2) = 1$ at the point (x_1, y_1) on the curve. (See Exercise 55 in Exercise Set 8.4 for the definition of a tangent to a hyperbola.)

TERM OR FORMULA	PAGE REFERENCE	COMMENT
32. $\begin{cases} x = x' \cos \theta - y' \sin \theta \\ y = x' \sin \theta + y' \cos \theta \end{cases}$ $\begin{cases} x' = x \cos \theta + y \sin \theta \\ y' = -x \sin \theta + y \cos \theta \end{cases}$	496	These are the formulas that relate the coordinates of a point in the x-y system to the coordinates in the rotated x'-y' system. See Figures 1 and 2 in Section 8.5.
33. $\cot 2\theta = \dfrac{A - C}{B}$	498	This formula determines an angle of rotation θ. When the equation $$Ax^2 + Bxy + Cy^2 + Dx + Ey + F = 0 \quad (B \ne 0)$$ is written in the x'-y' system, the resulting equation does not contain an $x'y'$-term. The graph can then be analyzed by means of the technique of completing the square. See Examples 3, 4, and 5 in Section 8.5.
34. $x = r \cos \theta$ $y = r \sin \theta$ $x^2 + y^2 = r^2$ $\tan \theta = \dfrac{y}{x}$	505	These formulas relate the rectangular coordinates of a point (x, y) to its polar coordinates (r, θ). See the boxed figure on page 505.

▼ CHAPTER EIGHT   REVIEW EXERCISES

NOTE   Exercises 1–20 constitute a chapter test on the fundamentals, based on group A problems.

1. Find the focus and the directrix of the parabola $y^2 = -12x$, and sketch the graph.

2. Graph the hyperbola $x^2 - 4y^2 = 4$. Specify the foci and the asymptotes.

3. (a) Determine an angle of rotation θ so that there is no $x'y'$-term present when the equation
$$x^2 + 2\sqrt{3}\,xy + 3y^2 - 12\sqrt{3}\,x + 12y = 0$$
is transformed to the x'-y' coordinate system.
   (b) Graph the equation
$$x^2 + 2\sqrt{3}\,xy + 3y^2 - 12\sqrt{3}\,x + 12y = 0$$

4. Determine the angle of inclination for the line $y = (1/\sqrt{3})x - 4$.

5. Find the equation of the line that is tangent to the parabola $x^2 = 2y$ at the point $(4, 8)$. Write your answer in the form $y = mx + b$.

6. Find the equation of the line with a positive slope that is tangent to the circle $(x - 2)^2 + (y - 1)^2 = 9$ at one of its y-intercepts. Write your answer in the form $Ax + By + C = 0$.

7. The foci of an ellipse are $(0, \pm 2)$, and the eccentricity is $\frac{1}{2}$. Determine the equation of the ellipse. Write your answer in standard form.

8. From the point $(-4, 0)$, tangents are drawn to the circle $x^2 + y^2 = 1$. Find the slopes of the tangents.

9. The x-intercept of a line is 2, and its angle of inclination is $60°$. Find the equation of the line. Write your answer in the form $Ax + By + C = 0$.

10. Determine the equation of the hyperbola with foci $(\pm 2, 0)$ and with asymptotes $y = \pm (1/\sqrt{3})x$. Write your answer in standard form.

11. Graph the equation $r = 4 \cos 2\theta$.

12. Let F_1 and F_2 denote the foci of the hyperbola $5x^2 - 4y^2 = 80$.
   (a) Verify that the point P with coordinates $(6, 5)$ lies on the hyperbola.
   (b) Compute the quantity $(F_1 P - F_2 P)^2$.

13. Graph the ellipse $4x^2 + 25y^2 = 100$. Specify the foci and the lengths of the major and minor axes.

14. Convert the polar equation $r^2 = \cos 2\theta$ to rectangular form.

15. Compute the distance from the point $(-1, 0)$ to the line $2x - y - 1 = 0$.

16. Graph the equation $16x^2 + y^2 - 64x + 2y + 65 = 0$.

17. Graph the equation $r = 2(1 - \cos \theta)$.

18. Graph the equation $\dfrac{(x + 4)^2}{3^2} - \dfrac{(y - 4)^2}{1^2} = 1$.

19. Graph the parabola $(x - 1)^2 = 8(y - 2)$. Specify the focal width and the vertex.

20. Determine the equation of the line that is tangent to the ellipse $x^2 + 3y^2 = 52$ at the point $(-2, 4)$. Write your answer in the form $y = mx + b$.

In Exercises 21–35, refer to the following figure and show that the given statements are correct.

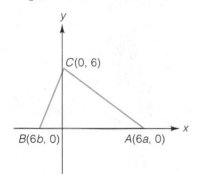

21. The equations of the lines forming the sides of triangle ABC are $x + by = 6b$, $x + ay = 6a$, and $y = 0$.

22. The equations of the lines forming the medians of triangle ABC are $2x + (a + b)y = 6(a + b)$, $x - (a - 2b)y = 6b$, and $x - (b - 2a)y = 6a$. (A *median* is a line segment drawn from a vertex of a triangle to the midpoint of the opposite side.)

23. Each pair of medians of triangle ABC intersect at the point $G(2a + 2b, 2)$. (The point G is called the *centroid* of triangle ABC.)

24. The equations of the lines forming the altitudes of triangle ABC are $y = ax - 6ab$, $y = bx - 6ab$, and $x = 0$. (An *altitude* is a line segment drawn from a vertex to the opposite side, perpendicular to that side.)

25. Each pair of altitudes intersect at the point $H(0, -6ab)$. (The point H is the *orthocenter* of triangle ABC.)

26. The equations of the perpendicular bisectors of the sides of triangle ABC are $x = 3a + 3b$, $bx - y = 3b^2 - 3$, and $ax - y = 3a^2 - 3$.

27. Each pair of perpendicular bisectors intersect at the point $O(3a + 3b, 3ab + 3)$. (The point O is called the *circumcenter* of triangle ABC.)

28. The distance from the circumcenter O to each vertex is $3\sqrt{(a^2 + 1)(b^2 + 1)}$. (This distance, denoted by R, is the *circumradius* of the triangle ABC. Note that the circle with center O and radius R passes through the points A, B, and C.)

29. In triangle ABC, let p, q, and r denote the lengths of $\overline{BC}$, $\overline{AC}$, and $\overline{AB}$, respectively. Then

$$\text{area of triangle } ABC = \frac{pqr}{4R}$$

30. $AH^2 + BC^2 = 4(OA)^2$

31. $OH^2 = 9R^2 - (p^2 + q^2 + r^2)$

32. $GH^2 = 4R^2 - \frac{4}{9}(p^2 + q^2 + r^2)$

33. $HA^2 + HB^2 + HC^2 = 12R^2 - (p^2 + q^2 + r^2)$

34. The points H, G, and O are collinear. (The line through these three points is the *Euler line* of triangle ABC.)

35. $GH = 2(GO)$

In Exercises 36–38, find the angle of inclination for each line. Use a calculator to express your answers in degrees. (Round off to one decimal place.)

36. $y = 4x - 3$ **37.** $2x + 3y = 6$ **38.** $y = 2x$

39. Find the distance from the point $(-1, -3)$ to the line $5x + 6y = 30$.

40. Find the distance from the point $(2, 1)$ to the line $y = \frac{1}{2}x + 4$.

41. The vertices of an equilateral triangle are $(\pm 6, 0)$ and $(0, 6\sqrt{3})$. Verify that the sum of the three distances from the point $(1, 2)$ to the sides of the triangle is equal to the height of the triangle. (It can be shown that, for any point inside an equilateral triangle, the sum of the distances to the sides is equal to the height of the triangle. This is *Viviani's theorem*.)

42. From the point $(-12, -1)$, a tangent line is drawn to the circle $x^2 + y^2 = 20$. Find the slope of this line, given that its y-intercept is positive.

In Exercises 43–46, find the equation of the parabola satisfying the given conditions. In each case, assume that the vertex is $(0, 0)$.

43. **(a)** The focus is $(4, 0)$.   **(b)** The focus is $(0, 4)$.

44. The focus lies on the x-axis, and the curve passes through the point $(3, 1)$.

45. The parabola is symmetric about the y-axis, the y-coordinate of the focus is positive, and the length of the focal chord perpendicular to the y-axis is 12.

46. The focus of the parabola is the center of the circle $x^2 - 8x + y^2 + 15 = 0$.

In Exercises 47–49, find the equation of the ellipse satisfying the given conditions. Write your answer in the form $Ax^2 + By^2 = C$.

47. Foci $(\pm 2, 0)$; endpoints of major axis $(\pm 8, 0)$

48. Foci $(0, \pm 1)$; endpoints of minor axis $(\pm 4, 0)$

49. Eccentricity $\frac{4}{5}$; one end of minor axis $(-6, 0)$; center at the origin

50. For any point P on an ellipse, the sum of the distances from $(1, 2)$ and $(-1, -2)$ is 12. Find the equation of the ellipse. *Hint:* Use the distance formula and the definition of an ellipse.

In Exercises 51–54, find the equation of the hyperbola satisfying the given conditions. Write each answer in the form $Ax^2 - By^2 = C$ or in the form $Ay^2 - Bx^2 = C$.

51. Foci $(\pm 6, 0)$; vertices $(\pm 2, 0)$

52. Asymptotes $y = \pm 2x$; foci $(0, \pm 3)$

53. Eccentricity 4; foci $(\pm 3, 0)$

54. Length of transverse axis 3; eccentricity $\frac{5}{4}$; center $(0, 0)$; focal axis horizontal

In Exercises 55–60, graph the parabolas, and in each case specify the vertex, the focus, the directrix, and the focal width.

55. $x^2 = 10y$ **56.** $x^2 = 5y$

57. $x^2 = -12(y - 3)$ **58.** $x^2 = -8(y + 1)$

59. $(y - 1)^2 = -4(x - 1)$ **60.** $(y + 3)^2 = 2(x - 1)$

In Exercises 61–66, graph the ellipses, and in each case specify the center, the foci, the lengths of the major and minor axes, and the eccentricity.

61. $x^2 + 2y^2 = 4$ **62.** $4x^2 + 9y^2 = 144$

63. $49x^2 + 9y^2 = 441$ **64.** $9x^2 + y^2 = 9$

65. $\dfrac{(x - 1)^2}{5^2} + \dfrac{(y + 2)^2}{3^2} = 1$

66. $\dfrac{(x + 3)^2}{3^2} + \dfrac{y^2}{3^2} = 1$

In Exercises 67–72, graph the hyperbolas. In each case specify the center, the vertices, the foci, the equations of the asymptotes, and the eccentricity.

67. $x^2 - 2y^2 = 4$ **68.** $4x^2 - 9y^2 = 144$

69. $49y^2 - 9x^2 = 441$ **70.** $9y^2 - x^2 = 9$

71. $\dfrac{(x - 1)^2}{5^2} - \dfrac{(y + 2)^2}{3^2} = 1$

72. $\dfrac{(y + 3)^2}{3^2} - \dfrac{x^2}{3^2} = 1$

In Exercises 73–86, use the technique of completing the square to graph the given equation. If the graph is a parabola, specify the vertex, axis, focus, and directrix. If the graph is an ellipse, specify the center, foci, and lengths of the major and minor axes. If the graph is a hyperbola, specify the center, vertices, foci, and equations of the asymptotes. Finally, if the equation has no graph, say so.

73. $3x^2 + 4y^2 - 6x + 16y + 7 = 0$

74. $y^2 - 16x - 8y + 80 = 0$

75. $y^2 + 4x + 2y - 15 = 0$

76. $16x^2 + 64x + 9y^2 - 54y + 1 = 0$

77. $16x^2 - 32x - 9y^2 + 90y - 353 = 0$

78. $x^2 + 6x - 12y + 33 = 0$

79. $5x^2 + 3y^2 - 40x - 36y + 188 = 0$

80. $x^2 - y^2 - 4x + 2y - 6 = 0$

81. $9x^2 - 90x - 16y^2 + 32y + 209 = 0$

82. $x^2 + 2y - 12 = 0$

83. $y^2 - 25x^2 + 8y - 9 = 0$

84. $x^2 + 16y^2 - 160y + 384 = 0$

85. $16x^2 + 25y^2 - 64x - 100y + 564 = 0$

86. $16x^2 - 25y^2 - 64x + 100y - 36 = 0$

87. Let F_1 and F_2 denote the foci of the hyperbola $5x^2 - 4y^2 = 80$.
   (a) Verify that the point P with coordinates $(6, 5)$ lies on the hyperbola.
   (b) Compute the quantity $(F_1P - F_2P)^2$.

88. Show that the coordinates of the vertex of the parabola $Ax^2 + Dx + Ey + F = 0$ are given by

$$x = -\frac{D}{2A} \quad \text{and} \quad y = \frac{D^2 - 4AF}{4AE}$$

89. If the equation $Ax^2 + Cy^2 + Dx + Ey + F = 0$ represents an ellipse or a hyperbola, show that the center is the point $(-D/2A, -E/2C)$.

90. The figure shows the parabola $x^2 = 4py$ and a circle with center at the origin and diameter $3p$. If V and F denote the vertex and focus of the parabola, respectively, show that the common chord of the circle and parabola bisects the line segment $\overline{VF}$.

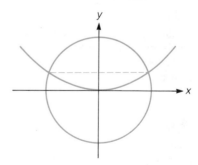

In Exercises 91–96, find the equations of the tangent lines to the given curves at the indicated points. (Write your answers in the form $y = mx + b$).

91. $x^2 = 4y$; $(-8, 16)$ **92.** $y^2 = 12x$; $(3, -6)$

93. $x^2 + 3y^2 = 52$; $(2, 4)$

94. $3x^2 + y^2 = 148$; $(4, -10)$

95. $12x^2 - 9y^2 = 108$; $(6, 6)$

96. $x^2 - 60y^2 = 4$; $(8, 1)$

97. The following figure shows an ellipse and a parabola. As indicated in the figure, the curves are symmetric about the x-axis and they both have an x-intercept of 5. Find the equation of the ellipse and the parabola, given that the point (3, 0) is a focus for both curves.

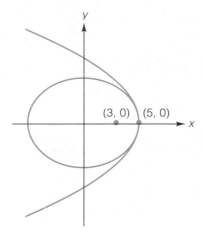

98. The area A of an ellipse $(x^2/a^2) + (y^2/b^2) = 1$ is given by the formula $A = \pi ab$. Use this formula to compute the area of the ellipse $5x^2 + 6y^2 = 60$.

99. As you know, there is a simple expression for the circumference of a circle of radius a, namely, $2\pi a$. However, there is no similar type of elementary expression for the circumference of an ellipse. (The circumference of an ellipse can be computed to as many decimal places as required using the methods of calculus.) Nevertheless, there are some interesting elementary formulas that allow us to approximate the circumference of an ellipse quite closely. Three such formulas follow, along with the names of their discoverers and the dates of discovery. Each formula yields an approximate value for the circumference of the ellipse $(x^2/a^2) + (y^2/b^2) = 1$.

$C_1 = \pi[a + b + \tfrac{1}{2}(\sqrt{a} - \sqrt{b})^2]$
   Giuseppe Peano, 1887

$C_2 = \pi[3(a + b) - \sqrt{(a + 3b)(3a + b)}]$
   Srinivasa Ramanujan, 1914

$C_3 = \dfrac{\pi}{2}(a + b + \sqrt{2(a^2 + b^2)})$
   R. A. Johnson, 1930

Use these formulas to complete the following table of approximations for the circumference of the ellipse $(x^2/5^2) + (y^2/3^2) = 1$. Round off the values of C_1, C_2, and C_3 to six decimal places. To complete the right-hand column of the table, you need two facts. First, the actual circumference of the ellipse, rounded off to six decimal places, is 25.526999. Second, percentage error in an approximation is given by

$$\frac{|\text{true value} - \text{approximation}|}{\text{true value}} \times 100$$

Round off the percentage errors to two significant digits. Which of the three approximations is the best?

	APPROXIMATION OBTAINED	PERCENTAGE ERROR
C_1		
C_2		
C_3		

In Exercises 100–103, find $\sin \theta$ and $\cos \theta$, where θ is the (acute) angle of the rotation that eliminates the $x'y'$-term. Then graph the equation.

100. $4x^2 + 2xy + 4y^2 - 15 = 0$

101. $4x^2 + 4xy + y^2 + 20x - 10y = 0$

102. $2x^2 + 4xy - y^2 - 12 = 0$

103. $13x^2 + 10xy + 13y^2 + 16x - 16y - 272 = 0$

104. In the following figure, V is the vertex of the parabola $y = ax^2 + bx + c$. If r_1 and r_2 are the roots of the equation $ax^2 + bx + c = 0$, show that
$$VO^2 - VA^2 = r_1 r_2$$

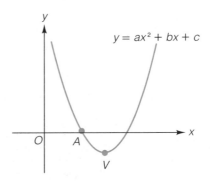

In Exercises 105–112, graph the polar equations.

105. (a) $r = 2 - 2\cos\theta$ (b) $r = 2 - 2\sin\theta$
106. (a) $r^2 = 4\cos 2\theta$ (b) $r^2 = 4\sin 2\theta$
107. (a) $r = 2\cos\theta - 1$ (b) $r = 2\sin\theta - 1$
108. (a) $r = 2\cos\theta$ (b) $r = 2\cos\theta + 1$
109. (a) $r = 4\sin 2\theta$ (b) $r = 4\cos 2\theta$
110. (a) $r = 4\sin 3\theta$ (b) $r = 4\cos 3\theta$
111. (a) $r = 1 + 2\sin(\theta/2)$ (b) $r = 1 - 2\cos(\theta/2)$
112. (a) $r = (1.5)^\theta$ $(\theta \geq 0)$ (b) $r = (1.5)^\theta$ $(\theta \leq 0)$

113. Show that the equation of a line tangent to the circle $(x - h)^2 + (y - k)^2 = r^2$ at the point (a, b) on the circle is
$$(a - h)(x - h) + (b - k)(y - k) = r^2$$

114. Find the equation of a line that passes through the point of intersection of the lines $bx + ay = ab$ and $ax + by = ab$ and also through the point (a, b). Write your answer in the form $Ax + By + C = 0$.

115. The equations of the sides of a triangle are $2px - 2y = qr$, $2qx - 2y = pr$, and $2rx - 2y = pq$.
   (a) Find the vertices of the triangle.
   (b) Show that the area of the triangle is

$$\frac{|(p - q)(q - r)(r - p)|}{8}$$

116. Suppose that the asymptote $y = (b/a)x$ of the hyperbola $(x^2/a^2) - (y^2/b^2) = 1$ makes an acute angle θ with the positive x-axis. Show that the eccentricity of the hyperbola is $\sec \theta$.

117. In the following figure, A and B are fixed points, and the point $P(x, y)$ moves in such a way that $\tan \theta \tan \beta = 2$. Show that the point P lies on the ellipse $2x^2 + y^2 = 2a^2$.

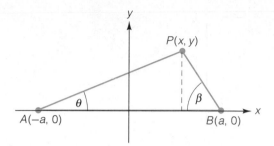

CHAPTER NINE

ROOTS OF POLYNOMIAL EQUATIONS

INTRODUCTION In this chapter, we continue the work begun in Chapter 1 solving polynomial equations. We begin in Section 9.1 with the long division process for polynomials. In this section, you will also see how synthetic division is used to abbreviate the long division process. Section 9.2 presents two theorems about polynomials: the remainder theorem and the factor theorem. It is in the application of the remainder theorem that you will begin to appreciate the utility of synthetic division. The two theorems are then applied in the next section (Section 9.3) to develop some fundamental results regarding polynomial equations and their solutions. You can view much of the material in this section as a kind of generalization of what you already know about quadratic equations. The last two sections of the chapter present additional results that are useful in actually solving polynomial equations or in determining the nature of their solutions.

9.1 ▼ DIVISION OF POLYNOMIALS

Although the process of long division for polynomials is often taught in elementary algebra courses, it usually does not receive sufficient emphasis there. As with ordinary long division for numbers, the process is best learned by first watching someone do examples and then practicing on your own. The terms *quotient*, *remainder*, *divisor*, and *dividend* will be used here in the same way they are used in ordinary division of numbers. For instance, when 7 is divided by 2, the quotient is 3 and the remainder is 1. We write this

$$\frac{7}{2} = 3 + \frac{1}{2}$$

or, equivalently,

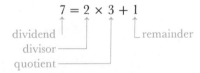

The process of long division for polynomials follows the same four-step cycle used in ordinary long division of numbers: divide, multiply, subtract, bring down. As a first example, we divide $2x^2 - 7x + 8$ by $x - 2$. Notice that in setting up the division, we write both the dividend and the divisor in decreasing powers of x.

$$\begin{array}{r} 2x - 3 \\ x - 2 \overline{\smash{)}\ 2x^2 - 7x + 8} \\ \underline{2x^2 - 4x} \\ -3x + 8 \\ \underline{-3x + 6} \\ 2 \end{array}$$

1. Divide the first term of the dividend by the first term of the divisor: $\frac{2x^2}{x} = 2x$. The result becomes the first term of the quotient, as shown.

2. Multiply the divisor $x - 2$ by the term $2x$ obtained in the previous step. This yields the quantity $2x^2 - 4x$, which is written below the dividend, as shown.

3. From the quantity $2x^2 - 7x$ in the dividend, subtract the quantity $2x^2 - 4x$. This yields $-3x$.

4. Bring down the $+8$ in the dividend, as shown. The resulting quantity, $-3x + 8$, is now treated as the dividend and the entire process is repeated.

We've now found that when $2x^2 - 7x + 8$ is divided by $x - 2$, the quotient is $2x - 3$ and the remainder is 2. This is summarized by writing either

$$\frac{2x^2 - 7x + 8}{x - 2} = 2x - 3 + \frac{2}{x - 2} \tag{1}$$

or, after multiplying through by $x - 2$,

$$\underbrace{2x^2 - 7x + 8}_{\text{dividend}} = \underbrace{(x - 2)}_{\text{divisor}} \underbrace{(2x - 3)}_{\text{quotient}} + \underbrace{2}_{\text{remainder}} \tag{2}$$

There are two observations to be made here. First, notice that equation (2) is valid for all real numbers x, whereas equation (1) carries the implicit restriction that x may not equal 2. For this reason, we often prefer to write our results in the form of equation (2). Second, notice that the degree of the remainder is less than the degree of the divisor. This is very similar to the situation with ordinary division of positive integers, where, as you know, the remainder is always less than the divisor.

As another example of the long division process, we divide $3x^4 - 2x^3 + 2$ by $x^2 - 1$. Notice in what follows that we've inserted the terms in the divisor and dividend that have coefficients of zero. These terms act as place holders.

$$
\require{enclose}
\begin{array}{r}
3x^2 - 2x + 3 \\
x^2 + 0x - 1 \enclose{longdiv}{3x^4 - 2x^3 + 0x^2 + 0x + 2} \\
\underline{3x^4 + 0x^3 - 3x^2} \\
-2x^3 + 3x^2 + 0x \\
\underline{-2x^3 + 0x^2 + 2x} \\
3x^2 - 2x + 2 \\
\underline{3x^2 + 0x - 3} \\
-2x + 5 \\
\end{array}
$$

We can write this result as

$$\frac{3x^4 - 2x^3 + 2}{x^2 - 1} = 3x^2 - 2x + 3 + \frac{-2x + 5}{x^2 - 1}$$

or, multiplying through by $x^2 - 1$,

$$\underbrace{3x^4 - 2x^3 + 2}_{\text{dividend}} = \underbrace{(x^2 - 1)}_{\text{divisor}} \underbrace{(3x^2 - 2x + 3)}_{\text{quotient}} + \underbrace{(-2x + 5)}_{\text{remainder}}$$

Notice that this last equation holds for all values of x, whereas the previous equation carries the restrictions that x may be neither 1 nor -1. Also, as in our previous example, observe that the degree of the remainder is less than that of the divisor.

There is a theorem, commonly referred to as the **division algorithm**, that summarizes rather nicely the key results of the long division process. We state the theorem here without proof.

The Division Algorithm

Let $p(x)$ and $d(x)$ be polynomials, and assume that $d(x)$ is not the zero polynomial. Then there are unique polynomials $q(x)$ and $R(x)$ such that

$$p(x) = d(x) \cdot q(x) + R(x)$$

where either $R(x)$ is the zero polynomial or the degree of $R(x)$ is less than the degree of $d(x)$.

The polynomials $p(x)$, $d(x)$, $q(x)$, and $R(x)$ are referred to, respectively, as the **dividend**, **divisor**, **quotient**, and **remainder**. When $R(x) = 0$, we have $p(x) = d(x) \cdot q(x)$ and we say that $d(x)$ and $q(x)$ are **factors** of $p(x)$. Also, since $d(x)$ is not the zero polynomial, notice that the equation $p(x) = d(x) \cdot q(x) + R(x)$ implies that the degree of $q(x)$ is less than or equal to the degree of $p(x)$. (Why?)

EXAMPLE 1 Let $p(x) = x^3 + 2x^2 - 4$ and $d(x) = x - 3$. Use the long division process to find the polynomials $q(x)$ and $R(x)$ such that

$$p(x) = d(x) \cdot q(x) + R(x)$$

where either $R(x) = 0$ or the degree of $R(x)$ is less than the degree of $d(x)$.

Solution After inserting the term $0x$ in the dividend $p(x)$, we use long division to divide $p(x)$ by $d(x)$:

$$
\require{enclose}
\begin{array}{r}
x^2 + 5x + 15 \\
x - 3 \enclose{longdiv}{x^3 + 2x^2 + 0x - 4} \\
\underline{x^3 - 3x^2} \\
5x^2 + 0x \\
\underline{5x^2 - 15x} \\
15x - 4 \\
\underline{15x - 45} \\
41
\end{array}
$$

We now have

$$\underbrace{x^3 + 2x^2 - 4}_{p(x)} = \underbrace{(x - 3)}_{d(x)}\underbrace{(x^2 + 5x + 15)}_{q(x)} + \underbrace{41}_{R(x)}$$

Thus, $q(x) = x^2 + 5x + 15$ and $R(x) = 41$. Notice that the degree of $R(x)$ is less than the degree of $d(x)$. ▲

The long division procedure for polynomials can be streamlined when the divisor is of the form $x - a$. This shortened version, known as **synthetic division**, will be useful in subsequent sections when we are solving polynomial equations.

We can explain the idea behind synthetic division by using the long division carried out in Example 1. The basic idea is that in the long division process, it is the *coefficients* of the various polynomials that carry all the necessary information. In our example, for instance, the quotient and remainder can be abbreviated by writing down a sequence of four numbers:

$$1 \qquad 5 \qquad 15 \qquad 41$$

By studying the long division process, you will find that these numbers are obtained through the following four steps:

STEP 1 Write down the first coefficient of the dividend. This will be the first coefficient of the quotient.

Result $\boxed{1}$.

STEP 2 Multiply the 1 obtained in the previous step by the -3 in the divisor. Then subtract this from the second coefficient of the dividend:

$$-3 \times 1 = -3 \qquad 2 - (-3) = 5 \qquad\qquad\text{\textit{Result} } \boxed{5}$$

STEP 3 Multiply the 5 obtained in the previous step by the -3 in the divisor. Then subtract this from the third coefficient of the dividend:

$$-3 \times 5 = -15 \qquad 0 - (-15) = 15 \qquad\qquad\text{\textit{Result} } \boxed{15}$$

STEP 4 Multiply the 15 obtained in the previous step by the -3 in the divisor. Then subtract this from the fourth coefficient of the dividend:

$$-3 \times 15 = -45 \qquad -4 - (-45) = 41 \qquad\qquad\text{\textit{Result} } \boxed{41}$$

A convenient format for setting up this process involves writing the constant term of the divisor and the coefficients of the dividend as follows:

$$\underline{-3}\ \Big|\quad 1 \qquad 2 \qquad 0 \qquad -4$$

Now, using this format, let us again go through the four steps we have just described:

STEP 1 Bring down the 1.

$$\begin{array}{r|rrrr} -3 & 1 & 2 & 0 & -4 \\ \hline & 1 & & & \end{array}$$

STEP 2 $-3 \times 1 = -3$
$2 - (-3) = 5$

$$\begin{array}{r|rrrr} -3 & 1 & 2 & 0 & -4 \\ & & -3 & & \\ \hline & 1 & 5 & & \end{array}$$

STEP 3 $-3 \times 5 = -15$
$0 - (-15) = 15$

$$\begin{array}{r|rrrr} -3 & 1 & 2 & 0 & -4 \\ & & -3 & -15 & \\ \hline & 1 & 5 & 15 & \end{array}$$

STEP 4 $-3 \times 15 = -45$
$-4 - (-45) = 41$

$$\begin{array}{r|rrrr} -3 & 1 & 2 & 0 & -4 \\ & & -3 & -15 & -45 \\ \hline & 1 & 5 & 15 & 41 \end{array}$$

Although we have now obtained the required sequence of numbers, 1   5   15   41, there is one further simplification that can be made. In steps 2 through 4, we can add instead of subtract if we use 3 instead of -3 in the initial format. (You will see the motivation for this in the next section when we discuss the remainder theorem.) With this change, let us now summarize the technique of synthetic division using the example with which we've been working. The method is applicable for any polynomial division in which the form of the divisor is $x - a$.

To Divide $x^3 + 2x^2 - 4$ by $x - 3$ Using Synthetic Division

			COMMENTS
Format	$3\rfloor$ 1 2 0 -4		Since the divisor is $x - 3$, the format begins with 3. The coefficients from the dividend are written in the order corresponding to decreasing powers of x. A zero coefficient is inserted as a place holder.
Procedure	$3\rfloor$ 1 2 0 -4 3 15 45 $\overline{1\ \ 5\ \ 15\ \ 41}$		**Step 1:** Bring down the 1 **Step 2:** $3 \times 1 = 3; 2 + 3 = 5$ **Step 3:** $3 \times 5 = 15; 0 + 15 = 15$ **Step 4:** $3 \times 15 = 45; -4 + 45 = 41$
Answer	Quotient: $x^2 + 5x + 15$ Remainder: 41		The degree of the first term in the quotient is one less than the degree of the first term of the dividend.

EXAMPLE 2 Use synthetic division to divide $x^3 - 6x + 4$ by $x - 2$.

Solution

$$2\rfloor \quad 1 \quad\ 0 \quad -6 \quad\ \ 4$$
$$ 2 \quad\ \ 4 \quad -4$$
$$\overline{ \quad 1 \quad\ 2 \quad -2 \quad\ \ 0}$$

Looking at the third row of numbers in the synthetic division we've carried out, we see that the quotient is $x^2 + 2x - 2$ and the remainder is 0. In other words, both $x - 2$ and $x^2 + 2x - 2$ are factors of $x^3 - 6x + 4$, and we have

$$x^3 - 6x + 4 = (x - 2)(x^2 + 2x - 2) \qquad\qquad \blacktriangle$$

As we have just seen, when the divisor is $x - 2$, we write $+2$ in the synthetic division format. In general, if the divisor is $x - a$, we write a. The next example shows what to do when the form of the divisor is $x + a$.

EXAMPLE 3 Use synthetic division to divide $x^4 - 2x^3 + 5x^2 - 4x + 3$ by $x + 1$.

Solution We first need to write the divisor $x + 1$ in the form $x - a$. We have

$$x + 1 = x - (-1)$$

In other words, a is -1, and this is the value we use to set up the synthetic division format. The format then is

$$-1\rfloor \quad 1 \quad -2 \quad\ \ 5 \quad -4 \quad\ \ 3$$

Now we carry out the synthetic division procedure:

$$-1\rfloor \quad 1 \quad -2 \quad\ \ 5 \quad -4 \quad\ \ 3$$
$$ -1 \quad\ \ 3 \quad -8 \quad\ 12$$
$$\overline{ \quad 1 \quad -3 \quad\ \ 8 \quad -12 \quad 15}$$

The quotient is therefore $x^3 - 3x^2 + 8x - 12$, and the remainder is 15. We can summarize this result by writing

$$x^4 - 2x^3 + 5x^2 - 4x + 3 = (x + 1)(x^3 - 3x^2 + 8x - 12) + 15$$

(Notice that the degree of the remainder is less than the degree of the divisor, in agreement with the division algorithm.) ▲

▼ EXERCISE SET 9.1

A

In Exercises 1–20, use long division to find the quotient and remainder. Also, write each answer in the form $p(x) = d(x)q(x) + R(x)$, as in equation (2) in the text.

1. $\dfrac{x^2 - 8x + 4}{x - 3}$

2. $\dfrac{x^3 - 4x^2 + x - 2}{x - 5}$

3. $\dfrac{x^2 - 6x - 2}{x + 5}$

4. $\dfrac{3x^2 + 4x - 1}{x - 1}$

5. $\dfrac{6x^3 - 2x + 3}{2x + 1}$

6. $\dfrac{x^4 - 4x^3 + 6x^2 - 4x + 1}{x - 1}$

7. $\dfrac{x^5 + 2}{x + 3}$

8. $\dfrac{4x^3 - x^2 + 8x - 1}{x^2 - x + 1}$

9. $\dfrac{x^6 - 64}{x - 2}$

10. $\dfrac{x^6 + 64}{x - 2}$

11. $\dfrac{5x^4 - 3x^2 + 2}{x^2 - 3x + 5}$

12. $\dfrac{8x^6 - 36x^4 + 54x^2 - 27}{2x^2 - 3}$

13. $\dfrac{3y^3 - 4y^2 - 3}{y^2 + 5y + 2}$

14. $\dfrac{4y^4 - y^3 + 2y - 1}{2y^2 - 3y - 4}$

15. $\dfrac{t^4 - 4t^3 + 4t^2 - 16}{t^2 - 2t + 4}$

16. $\dfrac{2t^5 - 6t^4 - t^2 + 2t + 3}{t^3 - 2}$

17. $\dfrac{z^5 - 1}{z - 1}$

18. $\dfrac{1 + z + z^2 + z^3}{1 + z + z^2}$

19. $\dfrac{ax^2 + bx + c}{x - r}$

20. $\dfrac{ax^3 + bx^2 + cx + d}{x - r}$

In Exercises 21–40, use synthetic division to find the quotients and remainders. Also, in each case, write the result of the division in the form $p(x) = d(x)q(x) + R(x)$, as in equation (2) in the text.

21. $\dfrac{x^2 - 6x - 2}{x - 5}$

22. $\dfrac{3x^2 + 4x - 1}{x - 1}$

23. $\dfrac{4x^2 - x - 5}{x + 1}$

24. $\dfrac{x^2 - 1}{x + 2}$

25. $\dfrac{6x^3 - 5x^2 + 2x + 1}{x - 4}$

26. $\dfrac{x^4 - 4x^3 + 6x^2 - 4x + 1}{x - 1}$

27. $\dfrac{x^3 - 1}{x - 2}$

28. $\dfrac{x^3 - 8}{x - 2}$

29. $\dfrac{x^5 - 1}{x + 2}$

30. $\dfrac{x^3 - 8x^2 - 1}{x + 3}$

31. $\dfrac{x^4 - 6x^3 + 2}{x + 4}$

32. $\dfrac{3x^3 - 2x^2 + x + 1}{x - \frac{1}{2}}$

33. $\dfrac{x^3 - 4x^2 - 3x + 6}{x - 10}$

34. $\dfrac{1 + 3x + 3x^2 + x^3}{x + 1}$

35. $\dfrac{x^3 - x^2}{x + 5}$

36. $\dfrac{x^4 + 3x^2 + 12}{x - 3}$

37. $\dfrac{14 - 27x - 27x^2 + 54x^3}{x - \frac{2}{3}}$

38. $\dfrac{14 - 27x - 27x^2 + 54x^3}{x + \frac{2}{3}}$

39. $\dfrac{5x^4 - 4x^3 + 3x^2 - 2x + 1}{x + \frac{1}{2}}$

40. (a) $\dfrac{x^2 - a^2}{x - a}$ (b) $\dfrac{x^3 - a^3}{x - a}$

(c) $\dfrac{x^4 - a^4}{x - a}$ (d) $\dfrac{x^5 - a^5}{x - a}$

B

In Exercises 41–44, use synthetic division to determine the quotient $q(x)$ and the remainder $R(x)$ in each case.

41. $\dfrac{6x^2 - 8x + 1}{3x - 4}$ *Hint:* Divide both numerator and denominator by 3. (Why?)

42. $\dfrac{4x^3 + 6x^2 - 6x - 5}{2x - 3}$

43. $\dfrac{6x^3 + 1}{2x + 1}$

44. $\dfrac{5x^3 - 3x^2 + 1}{3x + 1}$

45. When $x^3 + kx + 1$ is divided by $x + 1$, the remainder is -4. Find k.

46. (a) Show that when $x^3 + kx + 6$ is divided by $x + 3$, the remainder is $-21 - 3k$.

(b) Determine a value of k such that $x + 3$ will be a factor of $x^3 + kx + 6$.

47. When $x^2 + 2px - 3q^2$ is divided by $x - p$, the remainder is zero. Show that $p^2 = q^2$.

48. Given that $x - 3$ is a factor of $x^3 - 2x^2 - 4x + 3$, solve the equation $x^3 - 2x^2 - 4x + 3 = 0$.

The process of synthetic division applies equally well when some or all of the coefficients are nonreal complex numbers. In Exercises 49–52, use synthetic division to determine the quotient $q(x)$ and the remainder $R(x)$ in each case.

49. $\dfrac{x^2 - 4x + 1}{x - i}$

50. $\dfrac{x^3 - 2x^2 - 4}{x - 3i}$

51. $\dfrac{x^2 - 2x + 2}{x - (1 + i)}$

52. $\dfrac{x^3 - x^2 + 4x - 4}{x + 2i}$

In Exercises 53–56, determine k so that $d(x)$ is a factor of $p(x)$.

53. $d(x) = 2x + 1$; $p(x) = 2x^3 - 9x^2 - 21x + k$

54. $d(x) = 4x - 3$; $p(x) = 4x^3 + x^2 + x + k$

55. $d(x) = x^2 + x + 3$; $p(x) = x^3 + 2x^2 + kx + 3$

56. $d(x) = 6x^2 + x + 1$; $p(x) = 18x^3 + 9x^2 + 4x + k$

57. Given: the identity $f(x) = d(x)q(x) + R(x)$ holds for the following polynomials.

$f(x) = 2x^5 + 5x^4 - 8x^3 + 7x^2 - 9$ $d(x) = x^2 - 3$

$q(x) = 2x^3 + 5x^2 - 2x + 22$ $R(x) = -6x + 57$

Evaluate $f(\sqrt{3})$.   *Hint (of sorts):* There's an easy way and a tedious way.

58. Given: the identity $f(t) = d(t)q(t) + R(t)$ holds for the following polynomials.

$f(t) = t^5 - 3t^4 + 2t^3 - 5t^2 + 6t - 7$ $d(t) = t - 4$

$q(t) = t^4 + t^3 + 6t^2 + 19t + 82$ $R(t) = 321$

Evaluate $f(4)$.

59. Find the remainder when $t^5 - 5a^4t + 4a^5$ is divided by $t - a$.

60. When $f(x)$ is divided by $(x - a)(x - b)$, the remainder is $Ax + B$. Apply the division algorithm to show that

$$A = \frac{f(a) - f(b)}{a - b} \quad \text{and} \quad B = \frac{bf(a) - af(b)}{b - a}$$

9.2 ▼ THE REMAINDER THEOREM AND THE FACTOR THEOREM

The techniques for solving polynomial equations of degree 2 were discussed in Chapter 1. Now we want to extend those ideas. Our focus in this section and in the remainder of the chapter is on solving polynomial equations of any degree, that is, equations of the form

$$f(x) = a_n x^n + a_{n-1} x^{n-1} + \cdots + a_1 x + a_0 = 0 \tag{1}$$

Here, as in Chapter 1, a **root** or **solution** of equation (1) is a number r that when substituted for x leads to a true statement. Thus, r is a root of equation (1) provided that $f(r) = 0$. We also refer to the number r in this case as a **zero** of the function f.

EXAMPLE 1   **(a)** Is -3 a zero of the function f defined by $f(x) = x^4 + x^2 - 6$?
(b) Is $\sqrt{2}$ a root of the equation $x^4 + x^2 - 6 = 0$?

Solution   **(a)** By definition, -3 will be a zero of f if $f(-3) = 0$. We have

$$f(-3) = (-3)^4 + (-3)^2 - 6 = 81 + 9 - 6 = 84 \neq 0$$

Thus, -3 is not a zero of the function f.
(b) To check if $\sqrt{2}$ is a root of the given equation, we have

$$(\sqrt{2})^4 + (\sqrt{2})^2 - 6 \overset{?}{=} 0$$
$$4 + 2 - 6 \overset{?}{=} 0$$
$$0 \overset{?}{=} 0 \quad \text{True}$$

Thus, $\sqrt{2}$ is a root of the equation $x^4 + x^2 - 6 = 0$. ▲

There are cases in which a root of an equation is what we call a **repeated root**. Consider, for instance, the equation $x(x-1)(x-1) = 0$. We have

$$x(x-1)(x-1) = 0$$

$x = 0$	$x - 1 = 0$	$x - 1 = 0$
	$x = 1$	$x = 1$

The roots of the equation are therefore 0, 1, and 1. The repeated root here is $x = 1$. We say in this case that 1 is a **double root** or, equivalently, that 1 is a **root of multiplicity 2**. More generally, if a root is repeated k times, we call it a **root of multiplicity k**.

EXAMPLE 2   State the multiplicity of each root of the equation

$$(x-4)^2(x-5)^3 = 0$$

Solution   We have $(x-4)(x-4)(x-5)(x-5)(x-5) = 0$. By setting each factor equal to zero, we obtain the roots 4, 4, 5, 5, and 5. From this we see that the root 4 has multiplicity 2, while the root 5 has multiplicity 3. Notice that it is not really necessary to write out all the factors as we did here; the exponents of the factors in the original equation give us the required multiplicities.   ▲

There are two simple but important theorems that will form the basis for much of our subsequent work with polynomials. These are the *remainder theorem* and the *factor theorem*. We begin with a statement of the remainder theorem.

The Remainder Theorem

> When a polynomial $f(x)$ is divided by $x - r$, the remainder is $f(r)$.

Before turning to a proof of the remainder theorem, let us see what the theorem is saying in two particular cases. First, suppose that we divide the polynomial $f(x) = 2x^2 - 3x + 4$ by $x - 1$. Then according to the remainder theorem, the remainder in this case should be the number $f(1)$. Let's check:

$$
\begin{array}{r|rrr}
1 & 2 & -3 & 4 \\
  &   & 2 & -1 \\
\hline
  & 2 & -1 & ③
\end{array}
\qquad
\begin{aligned}
f(x) &= 2x^2 - 3x + 4 \\
f(1) &= 2(1)^2 - 3(1) + 4 \\
f(1) &= ③
\end{aligned}
$$

As the calculations show, the remainder is indeed equal to $f(1)$. As a second example, let us divide the polynomial $g(x) = ax^2 + bx + c$ by $x - r$. According to the remainder theorem, the remainder should be $g(r)$. Again, let us check:

$$
\begin{array}{r|rrr}
r & a & b & c \\
  &   & ar & ar^2 + br \\
\hline
  & a & ar + b & ar^2 + br + c
\end{array}
\qquad
\begin{aligned}
g(x) &= ax^2 + bx + c \\
\\
g(r) &= ar^2 + br + c
\end{aligned}
$$

The calculations show that the remainder is equal to $g(r)$, as we wished to check.

In the example just concluded, we verified that the remainder theorem holds for any quadratic polynomial $g(x) = ax^2 + bx + c$. A general proof of the remainder theorem can easily be given along these same lines. The only drawback is that it becomes slightly cumbersome to carry out the synthetic division process when the dividend is

$$a_n x^n + a_{n-1} x^{n-1} + \cdots + a_1 x + a_0$$

For this reason, mathematicians often prefer to base the proof of the remainder theorem on the division algorithm given in the previous section. This is the path we will follow here.

To prove the remainder theorem, we must show that when the polynomial $f(x)$ is divided by $x - r$, the remainder is $f(r)$. Now according to the division algorithm, we can write

$$f(x) = (x - r) \cdot q(x) + R(x) \tag{2}$$

In this identity, either $R(x)$ is the zero polynomial or the degree of $R(x)$ is less than that of $x - r$. But since the degree of $x - r$ is 1, we must have in this case that the degree of $R(x)$ is zero. Thus, in *either* case, the remainder $R(x)$ is a constant. Denoting this constant by c, we can rewrite equation (2) as

$$f(x) = (x - r) \cdot q(x) + c$$

Now if we set $x = r$ in this identity, we obtain

$$f(r) = (r - r) \cdot q(r) + c = c$$

We have now shown that $f(r) = c$. But by definition, c is the remainder $R(x)$. Thus, $f(r) = R(x)$. This proves the remainder theorem.

EXAMPLE 3 Let $f(x) = 2x^3 - 5x^2 + x - 6$. Use the remainder theorem to evaluate $f(3)$.

Solution According to the remainder theorem, $f(3)$ is the remainder when $f(x)$ is divided by $x - 3$. Using synthetic division, we have

$$
\begin{array}{r|rrrr}
3 & 2 & -5 & 1 & -6 \\
  &   & 6 & 3 & 12 \\
\hline
  & 2 & 1 & 4 & 6
\end{array}
$$

The remainder is 6, and therefore $f(3) = 6$. ▲

From our experience with quadratic equations, we know that there is a close connection between factoring a quadratic polynomial $f(x)$ and solving the polynomial equation $f(x) = 0$. The factor theorem states this relationship between roots and factors in a precise form. Furthermore, the factor theorem tells us that this relationship holds for polynomials of all degrees, not just quadratics.

The Factor Theorem

> Let $f(x)$ be a polynomial. If $f(r) = 0$, then $x - r$ is a factor of $f(x)$. Conversely, if $x - r$ is a factor of $f(x)$, then $f(r) = 0$.

In terms of roots, we can summarize the factor theorem by saying that r is a root of the equation $f(x) = 0$ if and only if $x - r$ is a factor of $f(x)$. To prove the factor theorem, let us begin by assuming that $f(r) = 0$. We want to show that $x - r$ is a factor of $f(x)$. Now, according to the remainder theorem, if $f(x)$ is divided by $x - r$, the remainder is $f(r)$. So we can write

$$f(x) = (x - r) \cdot q(x) + f(r) \qquad \text{for some polynomial } q(x)$$

But since $f(r)$ is zero, this equation becomes

$$f(x) = (x - r) \cdot q(x)$$

This last equation tells us that $x - r$ is a factor of $f(x)$, as we wished to prove. Now, conversely, let us assume that $x - r$ is a factor of $f(x)$. We want to show that $f(r) = 0$. Since $x - r$ is a factor of $f(x)$, we can write

$$f(x) = (x - r) \cdot q(x) \qquad \text{for some polynomial } q(x)$$

If we now let $x = r$ in this last equation, we obtain

$$f(r) = (r - r) \cdot q(r) = 0$$

as we wished to show.

The example that follows indicates how the factor theorem can be used to solve equations.

EXAMPLE 4   Solve the equation $x^3 - 2x + 1 = 0$, given that one root is $x = 1$.

Solution   Since $x = 1$ is a root, the factor theorem tells us that $x - 1$ is a factor of $x^3 - 2x + 1$. In other words,

$$x^3 - 2x + 1 = (x - 1) \cdot q(x) \qquad \text{for some polynomial } q(x)$$

To determine this other factor $q(x)$, we divide $x^3 - 2x + 1$ by $x - 1$:

$$
\begin{array}{r|rrrr}
1 & 1 & 0 & -2 & 1 \\
  &   & 1 & 1 & -1 \\
\hline
  & 1 & 1 & -1 & 0 \\
\end{array}
$$

Thus, $q(x) = x^2 + x - 1$ and we have the factorization

$$x^3 - 2x + 1 = (x - 1)(x^2 + x - 1)$$

Using this identity, the original equation becomes

$$(x - 1)(x^2 + x - 1) = 0$$

Now the problem is reduced to solving the linear equation $x - 1 = 0$ and the quadratic equation $x^2 + x - 1 = 0$. The linear equation yields $x = 1$, but we already knew that 1 was a root. So if we are to find any additional roots, they must come from the equation $x^2 + x - 1 = 0$. Using the quadratic formula, we obtain

$$x = \frac{-1 \pm \sqrt{(1)^2 - 4(1)(-1)}}{2(1)} = \frac{-1 \pm \sqrt{5}}{2}$$

We now have three roots of the cubic equation $x^3 - 2x + 1 = 0$. These are 1, $\frac{1}{2}(-1 + \sqrt{5})$, and $\frac{1}{2}(-1 - \sqrt{5})$. As you will see in the next section, a cubic equation can have at most three roots. So in this case, we have determined all the roots; that is, we have solved the equation.   ▲

Before going on to other examples, let's take a moment to summarize the technique we used in Example 4. We want to solve a polynomial equation $f(x) = 0$, given that one root is $x = r$. Since r is a root, the factor theorem tells us that $x - r$ is a factor of $f(x)$. Then, with the aid of synthetic division, we obtain a factorization

$$f(x) = (x - r) \cdot q(x) \qquad \text{for some polynomial } q(x)$$

This gives rise to the two equations, $x - r = 0$ and $q(x) = 0$. Since the first of these only reasserts that $x = r$ is a root, we try to solve the second equation, $q(x) = 0$. We refer to the equation $q(x) = 0$ as the **reduced equation**. Example 4 showed you the idea behind this terminology; the degree of $q(x)$ is one less than that of $f(x)$. If, as in Example 4, the reduced equation happens to be a quadratic equation, then we can always determine the remaining roots by factoring or by the quadratic formula. In subsequent sections, we will look at techniques that are helpful in cases where $q(x)$ is not quadratic.

EXAMPLE 5 Solve the equation $x^4 + 2x^3 - 7x^2 - 20x - 12 = 0$, given that $x = 3$ and $x = -2$ are roots.

Solution Since $x = 3$ is a root, the factor theorem tells us that $x - 3$ is a factor of the polynomial $x^4 + 2x^3 - 7x^2 - 20x - 12$. That is,

$$x^4 + 2x^3 - 7x^2 - 20x - 12 = (x - 3) \cdot q(x) \qquad \text{for some polynomial } q(x)$$

As in Example 4, we can find $q(x)$ by synthetic division. We have

$$
\begin{array}{r|rrrr}
3 & 1 & 2 & -7 & -20 & -12 \\
  &   & 3 & 15 & 24 & 12 \\
\hline
  & 1 & 5 & 8 & 4 & 0
\end{array}
$$

Thus, $q(x) = x^3 + 5x^2 + 8x + 4$, and our original equation is equivalent to

$$(x - 3)(x^3 + 5x^2 + 8x + 4) = 0$$

Now, $x = -2$ is also a root of this equation. But $x = -2$ surely is not a root of the equation $x - 3 = 0$. Therefore, $x = -2$ must be a root of the reduced equation $x^3 + 5x^2 + 8x + 4 = 0$. Again the factor theorem is applicable. Since $x = -2$ is a root of the reduced equation, $x + 2$ must be a factor of $x^3 + 5x^2 + 8x + 4$. That is,

$$x^3 + 5x^2 + 8x + 4 = (x + 2) \cdot p(x) \qquad \text{for some polynomial } p(x)$$

We can use synthetic division to determine $p(x)$:

$$
\begin{array}{r|rrrr}
-2 & 1 & 5 & 8 & 4 \\
   &   & -2 & -6 & -4 \\
\hline
   & 1 & 3 & 2 & 0
\end{array}
$$

Thus, $p(x) = x^2 + 3x + 2$, and our reduced equation can be written

$$(x + 2)(x^2 + 3x + 2) = 0$$

This gives rise to a second reduced equation,

$$x^2 + 3x + 2 = 0$$

In this case, the roots are readily obtained by factoring. We have

$$x^2 + 3x + 2 = 0$$
$$(x + 2)(x + 1) = 0$$

$$x + 2 = 0 \quad\Big|\quad x + 1 = 0$$
$$x = -2 \quad\Big|\quad x = -1$$

Now, of the two roots we've just found, the -2 happens to be one of the roots that we were initially given in the statement of the problem. The -1, on the other hand, is a distinct additional root. In summary, then, we have three distinct roots: 3, -2, and -1. The root -2 has multiplicity 2. As you will see in the next section, a fourth-degree equation can have at most four (not necessarily distinct) roots. So in the case at hand, we have found all the roots of the given equation. ▲

EXAMPLE 6 In each case, find a polynomial equation $f(x) = 0$ satisfying the given conditions. If there is no such equation, say so.

(a) The numbers -1, 4, and 5 are roots.
(b) A factor of $f(x)$ is $x - 3$, and -4 is a root of multiplicity 2.
(c) The degree of f is 4, the number -5 is a root of multiplicity 3, and 6 is a root of multiplicity 2.

Solution (a) If $f(x)$ is any polynomial containing the factors $(x + 1)$, $(x - 4)$, and $(x - 5)$, then the equation $f(x) = 0$ will certainly be satisfied when $x = -1$, $x = 4$, or $x = 5$. The simplest polynomial equation in this case is therefore

$$(x + 1)(x - 4)(x - 5) = 0$$

This is a polynomial equation with the required roots. If required, we can carry out the multiplication on the left-hand side of the equation. As you can check, this yields $x^3 - 8x^2 + 11x + 20 = 0$.

(b) According to the factor theorem, since -4 is a root, $x + 4$ must be a factor of $f(x)$. In fact, since -4 is a root of multiplicity 2, the quantity $(x + 4)^2$ must be a factor of $f(x)$. The following equation therefore satisfies the given conditions:

$$(x - 3)(x + 4)^2 = 0$$

(c) We are given two roots, one with multiplicity 3, the other with multiplicity 2. Thus the degree of $f(x)$ must be at least $3 + 2 = 5$. (Why?) But this then contradicts the given condition that the degree of f should be 4. Consequently, there is no polynomial equation that satisfies the given conditions. ▲

▼ **EXERCISE SET 9.2**

A

In Exercises 1–6, determine whether the given value for the variable is a root of the equation.

1. $12x - 8 = 112$; $x = 10$

2. $12x^2 - x - 20 = 0$; $x = \frac{5}{4}$

3. $x^2 - 2x - 4 = 0$; $x = 1 - \sqrt{5}$

4. $1 - x + x^2 - x^3 = 0$; $x = -1$

5. $2x^2 - 3x + 1 = 0$; $x = \frac{1}{2}$

6. $(x - 1)(x - 2)(x - 3) = 0$; $x = 4$

In Exercises 7–14, determine whether the given value is a zero of the function.

7. $f(x) = 3x - 2$; $x = \frac{2}{3}$

8. $g(x) = 1 + x^2$; $x = -1$

9. $h(x) = 5x^3 - x^2 + 2x + 8$; $x = -1$

10. $F(x) = -2x^5 + 3x^4 + 8x^3$; $x = 0$

11. $f(t) = 1 + 2t + t^3 - t^5$; $t = 2$

12. $f(t) = 1 + 2t + t^3 - t^5$; $t = \sqrt{2}$

13. $f(x) = 2x^3 - 3x + 1$; **(a)** $x = \frac{1}{2}(\sqrt{3} - 1)$;
    (b) $x = \frac{1}{2}(\sqrt{3} + 1)$

14. $g(x) = x^4 + 8x^3 + 9x^2 - 8x - 10$; **(a)** $x = 1$;
    (b) $x = \sqrt{6} - 4$; **(c)** $x = \sqrt{6} + 4$

15. List the distinct roots of each of the following equations. In the case of a repeated root, give its multiplicity.
    (a) $(x - 1)(x - 2)^3(x - 3) = 0$
    (b) $(x - 1)(x - 1)(x - 1) = 0$
    (c) $(x - 5)^6(x + 1)^4 = 0$ **(d)** $x^5(x - 1) = 0$

16. In this exercise, we verify that the remainder theorem is valid for the cubic polynomial

    $$g(x) = ax^3 + bx^2 + cx + d$$

    (a) Compute $g(r)$.
    (b) Using synthetic division, divide $g(x)$ by $x - r$. Check that the remainder you obtain is the same as the answer in part (a).

In Exercises 17–24, use the remainder theorem (as in Example 3) to evaluate $f(x)$ for the given value of x.

17. $f(x) = 4x^3 - 6x^2 + x - 5$; $x = -3$

18. $f(x) = 2x^3 - x - 4$; $x = 4$

19. $f(x) = 6x^4 + 5x^3 - 8x^2 - 10x - 3$; $x = \frac{1}{2}$

20. $f(x) = x^5 - x^4 - x^3 - x^2 - x - 1$; $x = -2$

21. $f(x) = x^2 + 3x - 4$; $x = -\sqrt{2}$

22. $f(x) = 3x^3 + 8x^2 - 12$; $x - 5$

23. $f(x) = \frac{1}{2}x^3 - 5x^2 - 13x - 10$; $x = 12$

24. $f(x) = x^7 - 7x^6 + 5x^4 + 1$; $x = -3$

In Exercises 25–38, you are given a polynomial equation and one or more roots. Solve each equation using the method shown in Examples 4 and 5. To help you decide if you have found all the roots in each case, you may rely on the following theorem, discussed in the next section: A polynomial equation of degree n has at most n (not necessarily distinct) roots.

25. $x^3 - 4x^2 - 9x + 36 = 0$; -3 is a root.

26. $x^3 + 7x^2 + 11x + 5 = 0$; -1 is a root.

27. $x^3 + x^2 - 7x + 5 = 0$; 1 is a root.

28. $x^3 + 8x^2 - 3x - 24 = 0$; -8 is a root.

29. $3x^3 - 5x^2 - 16x + 12 = 0$; -2 is a root.

30. $2x^3 - 5x^2 - 46x + 24 = 0$; 6 is a root.

31. $2x^3 + x^2 - 5x - 3 = 0$; $-\frac{3}{2}$ is a root.

32. $6x^4 - 19x^3 - 25x^2 + 18x + 8 = 0$; 4 and $-\frac{1}{3}$ are roots.

33. $x^4 - 15x^3 + 75x^2 - 125x = 0$; 5 is a root.

34. $2x^3 + 5x^2 - 8x - 20 = 0$; 2 is a root.

35. $x^4 + 2x^3 - 23x^2 - 24x + 144 = 0$; -4 and 3 are roots.

36. $6x^5 + 5x^4 - 29x^3 - 25x^2 - 5x = 0$; $\sqrt{5}$ and $-\frac{1}{3}$ are roots.

37. $x^3 + 7x^2 - 19x - 9 = 0$; -9 is a root.

38. $4x^5 - 15x^4 + 8x^3 + 19x^2 - 12x - 4 = 0$; 1, 2, and -1 are roots.

In Exercises 39 and 40, use the given information to solve each equation.

39. $f(x) = 4x^4 - 12x^3 + 5x^2 + 6x + 1 = 0$; $2x^2 - 3x - 1$ is a factor of $f(x)$.

40. $f(x) = 3x^5 - 3x^4 - x^3 - 24x^2 + 24x + 8 = 0$; $3x^2 - 3x - 1$ is a factor of $f(x)$.

For Exercises 41 and 42, refer to the following computer-generated tables for the functions f and g.

$f(t) = t^3 - 4t + 3$			$g(t) = t^5 + 2t^4 + t^3 + 2t^2 - t - 2$	
t	$f(t)$		t	$g(t)$
0.500	1.125000		-3.00	-89.0
0.750	0.421875		-2.50	-22.15625
1.000	0.000000		-2.25	-7.4228515625
1.250	-0.046875		-2.00	0.0
1.500	0.375000		-1.75	2.8603515625

41. **(a)** What is the remainder when $f(t)$ is divided by $t - \frac{1}{2}$?
    (b) What is the remainder when $f(t)$ is divided by $t - 1.25$?
    (c) Specify a linear factor of $f(t)$.
    (d) Solve the equation $f(t) = 0$.

12. **(a)** What is the remainder when $g(t)$ is divided by $t + 3$?
    (b) What is the remainder when $g(t)$ is divided by $t + \frac{5}{2}$?
    (c) Specify a linear factor of $g(t)$.

In Exercises 43–50, find a polynomial equation $f(x) = 0$ satisfying the given conditions. If no such equation is possible, state this.

43. Degree 3; the coefficient of x^3 is 1; three roots are 3, -4, and 5.

44. Degree 3; the coefficients are integers; $\frac{1}{2}$, $\frac{2}{3}$, and $-\frac{3}{4}$ are roots.

45. Degree 3; -1 is a root of multiplicity two; $x + 6$ is a factor of $f(x)$.

46. Degree 3; 4 is a root of multiplicity two; -1 is a root of multiplicity two.

47. Degree 4; $\frac{1}{2}$ is a root of multiplicity three; $x^2 - 3x - 4$ is a factor of $f(x)$.

48. Degree 3; -2 and -3 are roots.

49. Degree 4; $x^2 - 3x + 1$ and $x + 6$ are factors of $f(x)$.

50. Degree 4; the coefficients are integers; $\frac{1}{2}$ is a root of multiplicity two; $2x^2 - 4x - 1$ is a factor of $f(x)$.

C *In Exercises 51–54, use the remainder theorem (as in Example 3) to evaluate $f(x)$ for the given value of x. Use a calculator, and round off your answers to two decimal places.*

51. $f(x) = x^3 - 3x^2 + 12x + 9$; $x = 1.16$

52. $f(x) = x^3 - 2x - 5$;
   (a) $x = 2.09$
   (b) $x = 2.094$
   (c) $x = 2.0945$

53. $f(x) = x^3 - 5x - 2$;
   (a) $x = 2.41$
   (b) $x = 2.42$

54. $f(x) = x^4 - 2x^3 - 5x^2 + 10x - 3$;
   (a) $x = -2.3$
   (b) $x = -2.302$
   (c) $x = -2.30277$

B

In Exercises 55–58, determine whether the given value is a zero of the function.

55. $f(x) = \frac{1}{2}x^2 + bx + c$; $x = -b - \sqrt{b^2 - 2c}$

56. $Q(x) = ax^2 + bx + c$; $x = (-b + \sqrt{b^2 - 4ac})/2a$
   Hint: Look before you leap!

57. $F(x) = 2x^4 + 4x + 1$; $x = \frac{1}{2}(-\sqrt{2} + \sqrt{2\sqrt{2} - 2})$

58. $f(x) = x^3 - 3x^2 + 3x - 3$; **(a)** $x = \sqrt[3]{2} - 1$;
   (b) $x = \sqrt[3]{2} + 1$

59. Determine values for a and b such that $x - 1$ is a factor of both $x^3 + x^2 + ax + b$ and $x^3 - x^2 - ax + b$.

60. Determine a quadratic equation with the given roots.
   (a) a/b, $-b/a$ **(b)** $-a + 2\sqrt{2b}$, $-a - 2\sqrt{2b}$

61. One root of the equation $x^2 + bx + 1 = 0$ is twice the other; find b. (There are two answers.)

62. Determine a value for a such that one root of the equation $ax^2 + x - 1 = 0$ is five times the other.

C

63. Solve the equation $x^3 - 12x + 16 = 0$, given that one of the roots has multiplicity 2.

9.3 ▼ THE FUNDAMENTAL THEOREM OF ALGEBRA

Every equation of algebra has as many solutions as the exponent of the highest term indicates.

Albert Girard (1629) as quoted in *The History of Mathematics, An Introduction,* by David M. Burton (Boston: Allyn and Bacon, 1985)

Does every polynomial equation (of degree at least 1) have a root? Or, to put the question another way, is it possible to write a polynomial equation that has no solution? Certainly, if we consider only real roots (i.e., roots that are real numbers), then it is easy to specify an equation with no real root:

$$x^2 = -1$$

This equation has no real root because the square of a real number is never negative. On the other hand, if we expand our base of operations from the real number system to the complex number system, then $x^2 = -1$ does indeed have a root. In fact, both i and $-i$ are roots in this case.

As it turns out, the situation just described for the equation $x^2 = -1$ holds quite generally. That is, within the complex number system, every polynomial equation of degree at least 1 has at least one root. This is the substance of a remarkable theorem that was first proved by the great mathematician Carl Friedrich Gauss in 1799.* (Gauss was only 22 years old at the time.) Although there are many fundamental theorems in algebra, Gauss's result has come to be known as *the* **fundamental theorem of algebra**.

---

* In the opinion of George F. Simmons in his text, *Calculus with Analytic Geometry* (New York: McGraw-Hill, 1985), "Gauss was the greatest of all mathematicians and perhaps the most richly gifted genius of whom there is any record. This gigantic figure, towering at the beginning of the nineteenth century, separates the modern era in mathematics from all that went before."

The Fundamental Theorem of Algebra

Every polynomial equation of the form

$$a_n x^n + a_{n-1} x^{n-1} + \cdots + a_1 x + a_0 = 0 \qquad (n \geq 1, a_n \neq 0)$$

has at least one root among the complex numbers. (This root may be a real number.)

Although most of this section deals with polynomials with real coefficients, Gauss's theorem still applies in cases in which some or all of the coefficients are nonreal complex numbers. The proof of Gauss's theorem is usually given in the post-calculus course called complex variables.

The fundamental theorem of algebra asserts that every polynomial equation of degree at least 1 has a root. There are two initial observations to be made here. First, notice that the theorem says nothing about actually finding the root. Second, notice that the theorem deals only with polynomial equations. Indeed, it is easy to specify a nonpolynomial equation that does not have a root. Such an equation is $1/x = 0$. The expression on the left-hand side of this equation can never be zero because the numerator is 1.

EXAMPLE 1   Which of the following equations has at least one root?

(a) $x^3 - 17x^2 + 6x - 1 = 0$ **(b)** $\sqrt{2}\, x^{47} - \pi x^{25} + \sqrt{3} = 0$
(c) $x^2 - 2ix + (3 + i) = 0$

Solution   All three equations are polynomial equations. So, according to the fundamental theorem of algebra, each equation has at least one root.   ▲

In Chapter 1 we used factoring as a tool for solving quadratic equations. The next theorem, a consequence of the fundamental theorem of algebra, tells us that (in principle, at least) any polynomial of degree n can be factored into a product of n linear factors. In proving the **linear factors theorem**, we will need to use the factor theorem. If you reread the proof of that theorem in the previous section, you will see that it makes no difference whether the number r appearing in the factor $x - r$ is a real number or a nonreal complex number. Thus, the factor theorem is valid in either case.

The Linear Factors Theorem

Let $f(x) = a_n x^n + a_{n-1} x^{n-1} + \cdots + a_1 x + a_0$ $(n \geq 1, a_n \neq 0)$. Then $f(x)$ can be expressed as a product of n linear factors:

$$f(x) = a_n(x - r_1)(x - r_2) \cdots (x - r_n)$$

(The complex numbers r_k appearing in these factors are not necessarily all distinct, and some or all of the r_k may be real numbers.)

PROOF OF THE LINEAR FACTORS THEOREM   According to the fundamental theorem of algebra, the equation $f(x) = 0$ has a root. Let us call this root r_1. By the factor theorem, $x - r_1$ is a factor of $f(x)$, and we can write

$$f(x) = (x - r_1) \cdot Q_1(x)$$

for some polynomial $Q_1(x)$ that has degree $n - 1$ and leading coefficient a_n. If the degree of $Q_1(x)$ happens to be zero, we are done. On the other hand, if the degree of $Q_1(x)$ is at least 1, another application of the fundamental theorem of algebra followed by the factor theorem gives us

$$Q_1(x) = (x - r_2) \cdot Q_2(x)$$

where the degree of $Q_2(x)$ is $n - 2$ and the leading coefficient of $Q_2(x)$ is a_n. We now have

$$f(x) = (x - r_1)(x - r_2) \cdot Q_2(x)$$

We continue this process until the quotient is $Q_n(x) = a_n$. As a result, we obtain

$$f(x) = (x - r_1)(x - r_2) \cdots (x - r_n)a_n$$
$$= a_n(x - r_1)(x - r_2) \cdots (x - r_n)$$

as we wished to show.

The linear factors theorem tells us that any polynomial can be expressed as a product of linear factors. The theorem gives us no information, however, as to how those factors can actually be obtained. The next example demonstrates a case in which the factors are readily obtainable; this is always the case with quadratic polynomials.

EXAMPLE 2   Express each of the following second-degree polynomials in the form $a_n(x - r_1)(x - r_2)$.

(a) $3x^2 - 5x - 2$ **(b)** $x^2 - 4x + 5$

Solution   **(a)** A factorization for $3x^2 - 5x - 2$ can be found by simple trial and error. We have

$$3x^2 - 5x - 2 = (3x + 1)(x - 2)$$

We now write the factor $3x + 1$ as $3(x + \frac{1}{3})$. This, in turn, can be written $3[x - (-\frac{1}{3})]$. The final factorization is then

$$3x^2 - 5x - 2 = 3[x - (-\frac{1}{3})](x - 2)$$

(b) From the factor theorem, or from our more elementary work with quadratic equations, we know that if r_1 and r_2 are the roots of the equation $x^2 - 4x + 5 = 0$, then $x - r_1$ and $x - r_2$ are the factors of $x^2 - 4x + 5$. That is, $x^2 - 4x + 5 = (x - r_1)(x - r_2)$. The values for r_1 and r_2 in this case are readily obtained by using the quadratic formula. As you can check, the results are

$$r_1 = 2 + i \qquad \text{and} \qquad r_2 = 2 - i$$

The required factorization is therefore

$$x^2 - 4x + 5 = [x - (2 + i)][x - (2 - i)] \qquad \blacktriangle$$

Using the linear factors theorem, we can show that every polynomial equation of degree n $(n \geq 1)$ has exactly n roots. To help you follow the reasoning, we make two preliminary comments. First, we agree that a root of multiplicity k will be counted as k roots. For example, although the third-degree equation $(x - 1)(x - 4)^2 = 0$ has only two distinct roots, namely, 1 and 4, it has three roots *if* we agree to count the repeated root 4 two times. The second preliminary comment concerns the *zero-product property*, which states that $pq = 0$ if and only if $p = 0$ or $q = 0$. When we stated this in Section 1.6, we were working within the real number system. However, the property is also valid within the complex number system. Now let's state and prove our theorem.

Theorem

> Every polynomial equation of degree $n \geq 1$ has exactly n roots, where a root of multiplicity k is counted k times.

PROOF OF THEOREM   Using the linear factors theorem, we can write the nth-degree polynomial equation $f(x) = a_n x^n + a_{n-1} x^{n-1} + \cdots + a_0 = 0$ as

$$f(x) = a_n(x - r_1)(x - r_2) \cdots (x - r_n) = 0 \qquad (a_n \neq 0) \tag{1}$$

By the factor theorem, each of the numbers $r_1, r_2, \ldots, r_n$ is a root. Some of these numbers may in fact be equal; in other words, we may have repeated roots in this list. In any case, if we agree to count a root of multiplicity k as k roots, then we obtain exactly n roots from the list $r_1, r_2, \ldots, r_n$. Furthermore, the equation $f(x) = 0$ can have no other roots, as we now show. Suppose that r is any number distinct from all the numbers $r_1, r_2, \ldots, r_n$. Replacing x with r in equation (1) yields

$$f(r) = a_n(r - r_1)(r - r_2) \cdots (r - r_n)$$

But the expression on the right-hand side of this last equation cannot be zero, because none of the factors is zero. Thus, $f(r)$ is not zero, and so r is not a root. This completes the proof of the theorem.

EXAMPLE 3

Find a polynomial $f(x)$ with leading coefficient 1 such that the equation $f(x) = 0$ has only those roots specified in Table 1. What is the degree of this polynomial?

Solution

The expressions $(x - 3)^2$, $(x + 2)$, and $(x - 0)^2$ all must appear as factors of $f(x)$ and, furthermore, no other linear factor can appear. The form of $f(x)$ is therefore

$$f(x) = a_n(x - 3)^2(x + 2)(x - 0)^2$$

Since the leading coefficient a_n is to be 1, we can rewrite this last equation as

$$f(x) = x^2(x - 3)^2(x + 2)$$

This is the required polynomial. The degree here is 5. This can be seen either by multiplying out the factors or by simply adding the multiplicities of the roots in Table 1.   ▲

TABLE 1

ROOT	MULTIPLICITY
3	2
−2	1
0	2

EXAMPLE 4

Find a quadratic function f that has zeros of 3 and 5 and a graph that passes through the point $(2, -9)$.

Solution

The general form of a quadratic function with 3 and 5 as zeros is $f(x) = a_n(x - 3)(x - 5)$. Since the graph passes through $(2, -9)$, we have

$$-9 = a_n(2 - 3)(2 - 5) = a_n(3)$$

$$-3 = a_n$$

The required function is therefore

$$f(x) = -3(x - 3)(x - 5)$$

If we wish, we can carry out the multiplication and rewrite this as

$$f(x) = -3x^2 + 24x - 45$$

▲

The next example shows how we can use the factored form of a polynomial $f(x)$ to determine the relationships between the coefficients of the polynomial and the roots of the equation $f(x) = 0$.

EXAMPLE 5 Let r_1 and r_2 be the roots of the equation $x^2 + bx + c = 0$. Show that

$$r_1 r_2 = c \qquad \text{and} \qquad r_1 + r_2 = -b$$

Solution Since r_1 and r_2 are the roots of the equation $x^2 + bx + c = 0$, we have the identity

$$x^2 + bx + c = (x - r_1)(x - r_2)$$

After multiplying out the right-hand side, we can rewrite this identity as

$$x^2 + bx + c = x^2 - (r_1 + r_2)x + r_1 r_2$$

By equating coefficients, we readily obtain $r_1 r_2 = c$ and $r_1 + r_2 = -b$, as required.

▲

The technique used in Example 5 can be used to obtain similar relationships between the roots and the coefficients of polynomial equations of any given degree. In Table 2, for instance, we show the relationships obtained in Example 5, along with the corresponding relationships that can be derived for a cubic equation. (Exercise 31 at the end of this section asks you to verify the results for the cubic equation.)

We conclude this section with some remarks concerning the solving of polynomial equations by formulas. You know that the roots of the quadratic equation $ax^2 + bx + c = 0$ are given by the formula

$$x = \frac{-b \pm \sqrt{b^2 - 4ac}}{2a}$$

The question is, are there similar formulas for the solutions of higher-degree equations? By "similar" we mean a formula involving the coefficients and radicals. To answer this question, we look at a bit of history. As early as 1700 B.C., Babylonian

TABLE 2

EQUATION	ROOTS	RELATIONSHIPS BETWEEN ROOTS AND COEFFICIENTS
$x^2 + bx + c = 0$	r_1, r_2	$r_1 + r_2 = -b$ $r_1 r_2 = c$
$x^3 + bx^2 + cx + d = 0$	r_1, r_2, r_3	$r_1 + r_2 + r_3 = -b$ $r_1 r_2 + r_2 r_3 + r_3 r_1 = c$ $r_1 r_2 r_3 = -d$

mathematicians were able to solve quadratic equations. This is clear from the study of the clay tablets with cuneiform numerals that archeologists have found. The ancient Greeks were also able to solve quadratic equations. Like the Babylonians, the Greeks worked without the aid of algebra as we know it. The mathematicians of ancient Greece used geometric constructions to solve equations. Of course, since all quantities were interpreted geometrically, negative roots were never considered.

The general quadratic formula was known to the Moslem mathematicians sometime before A.D. 1000. For the next 500 years, mathematicians searched for, but did not discover, a formula to solve the general cubic equation. Indeed, in 1494, Luca Pacioli stated in his text *Summa di Arithmetica* that the general cubic equation could not be solved by the algebra techniques then available. All of this was to change, however, within the next several decades.

Around 1515, the Italian mathematician Scipione del Ferro solved the cubic equation $x^3 + px + q = 0$ using algebraic techniques. This essentially constituted a solution of the seemingly more general equation $x^3 + bx^2 + cx + d = 0$. The reason for this is that if we make the substitution $x = y - b/3$ in the latter equation, the result is a cubic equation with no y^2-term. By 1540, the Italian mathematician Ludovico Ferrari had solved the general fourth-degree equation. Actually, at that time in Renaissance Italy, there was considerable controversy as to exactly who discovered the various formulas first. Details of the dispute can be found in any text on the history of mathematics. But for our purposes here, the point is simply that by the middle of the sixteenth century, all polynomial equations of degree 4 or less could be solved by the formulas that had been discovered. The common feature of these formulas was that they involved the coefficients, the four basic operations of arithmetic, and various radicals. For example, a formula for the solution of the equation $x^3 + px + q = 0$ is as follows:

$$x = \sqrt[3]{\frac{-q}{2} + \sqrt{\frac{q^2}{4} + \frac{p^3}{27}}} + \sqrt[3]{\frac{-q}{2} - \sqrt{\frac{q^2}{4} + \frac{p^3}{27}}}$$

To get some idea of the practical difficulties inherent in computing with this formula, try using it to show that $x = -2$ is a root of the cubic equation $x^3 + 4x + 16 = 0$.

For more than 200 years after the cubic and quartic (fourth-degree) equations had been solved, mathematicians continued to search for a formula that would yield the solutions of the general fifth-degree equation. The first breakthrough, if it can be called that, occurred in 1770, when the French mathematician Joseph Louis Lagrange found a technique that served to unify and summarize all of the previous methods used for the equations of degrees 2, 3, and 4. However, Lagrange then showed that his technique could not work in the case of the general fifth-degree equation. While we will not describe the details of Lagrange's work here, it is worth pointing out that he relied on the types of relationships between roots and coefficients that we looked at in Table 2 of this section.

Finally, in 1828, the Norwegian mathematician Niels Henrik Abel proved that for the general polynomial equation of degree 5 or higher, there could be no formula yielding the solutions in terms of the coefficients and radicals. This is not to say that such equations do not possess solutions. In fact they must, as we saw earlier in this section. It is just that we cannot in every case express the solutions in terms of the coefficients and radicals. For example, it can be shown that the equation $x^5 - 6x + 3 = 0$ has a real root between 0 and 1, but this number cannot be expressed in terms of the coefficients and radicals. (We can, however, compute the root to as many decimal places as we wish, as you will see in the next section.) In 1830, the French mathematician Evariste Galois completed matters by giving conditions for determining exactly which polynomial equations can be solved in terms of coefficients and radicals.

▼ EXERCISE SET 9.3

A

According to the fundamental theorem of algebra, which of the equations in Exercises 1 and 2 has at least one root?

1. (a) $x^5 - 14x^4 + 8x + 53 = 0$
    (b) $(4.17)x^3 + (2.06)x^2 + (0.01)x + 1.23 = 0$
    (c) $ix^2 + (2 + 3i)x - 17 = 0$
    (d) $x^{2.1} + 3x^{0.3} + 1 = 0$
2. (a) $\sqrt{3}x^{17} + \sqrt{2}x^{13} + \sqrt{5} = 0$
    (b) $17x^{\sqrt{3}} + 13x^{\sqrt{2}} + \sqrt{5} = 0$
    (c) $1/(x^2 + 1) = 0$ (d) $2^{3x} - 2^x - 1 = 0$

In Exercises 3–10, express each polynomial in the form $a_n(x - r_1)(x - r_2) \cdots (x - r_n)$.

3. $x^2 - 2x - 3$
4. $x^3 - 2x^2 - 3x$
5. $4x^2 + 23x - 6$
6. $6x^2 + x - 12$
7. $x^2 - 5$
8. $x^2 + 5$
9. $x^2 - 10x + 26$
10. $x^3 + 2x^2 - 3x - 6$

In Exercises 11–16, find a polynomial $f(x)$ with leading coefficient 1 such that the equation $f(x) = 0$ has the given roots and no others. If the degree of $f(x)$ is 7 or more, express $f(x)$ in factored form; otherwise, express $f(x)$ in the form $a_nx^n + a_{n-1}x^{n-1} + \cdots + a_1x + a_0$.

11.

Root	1	-3
Multiplicity	2	1

12.

Root	0	4
Multiplicity	2	1

13.

Root	2	-2	$2i$	$-2i$
Multiplicity	1	1	1	1

14.

Root	$2 + i$	$2 - i$
Multiplicity	1	1

15.

Root	$\sqrt{3}$	$-\sqrt{3}$	$4i$	$-4i$
Multiplicity	2	2	1	1

16.

Root	5	1	$1 - i$	$1 + i$
Multiplicity	2	3	1	1

In Exercises 17–20, express the polynomial $f(x)$ in the form $a_nx^n + a_{n-1}x^{n-1} + \cdots + a_1x + a_0$.

17. Find a quadratic function that has zeros -4 and 9 and a graph that passes through the point $(3, 5)$.
18. Find a quadratic function that has a maximum value of 2 and that has -2 and 4 as zeros.
19. Find a third-degree polynomial function that has zeros -5, 2, and 3 and a graph that passes through the point $(0, 1)$.
20. Find a fourth-degree polynomial function that has zeros $\sqrt{2}$, $-\sqrt{2}$, 1, and -1 and a graph that passes through $(2, -20)$.

In Exercises 21–27, find a quadratic equation with the given roots and no others. Write your answers in the form $Ax^2 + Bx + C = 0$.
Suggestion: Make use of Table 2.

21. $r_1 = -i, r_2 = -\sqrt{3}$ 22. $r_1 = 1 + i\sqrt{3}, r_2 = 1 - i\sqrt{3}$
23. $r_1 = 9, r_2 = -6$ 24. $r_1 = 5, r_2 = \frac{3}{4}$
25. $r_1 = 1 + \sqrt{5}, r_2 = 1 - \sqrt{5}$
26. $r_1 = 6 - 5i, r_2 = 6 + 5i$
27. $r_1 = a + \sqrt{b}, r_2 = a - \sqrt{b}$ $(b > 0)$
28. True or false: mark **T** if the statement is true without exception; otherwise mark **F**.
    (a) Every equation has a root.
    (b) Every polynomial equation of degree at least 1 has a root.
    (c) Every polynomial equation of degree 4 has four distinct roots.
    (d) No cubic equation can have a root of multiplicity 4.
    (e) The degree of the polynomial
        $$x(x - 1)(x - 2)(x - 3)$$
        is 3.
    (f) The degree of the polynomial $6(x + 1)^2(x - 5)^4$ is 6.

(g) Every polynomial of degree n, where $n \geq 1$, can be written in the form $(x - r_1)(x - r_2) \cdots (x - r_n)$.

(h) The sum of the roots of the polynomial equation $x^2 - px + q = 0$ is p.

(i) The product of the roots of the polynomial equation $2x^3 - x^2 + 3x - 1 = 0$ is 1. *Hint:* Use Table 2.

(j) Although a polynomial equation of degree n may have n distinct roots, the fundamental theorem of algebra tells us how to find only one of the roots.

(k) Every polynomial equation of degree $n \geq 1$ has at least one real root.

(l) If all the coefficients in a polynomial equation are real, then at least one of the roots must be real.

(m) Every cubic equation whose roots are $\sqrt{5}$, $\sqrt{6}$, and $\sqrt{7}$ can be written in the form
$$a_n(x - \sqrt{5})(x - \sqrt{6})(x - \sqrt{7}) = 0$$

B

29. Express the polynomial $x^4 + 64$ as a product of four linear factors. *Hint:* First add and subtract the quantity $16x^2$ so that you can use the difference-of-squares factoring formula.

30. Suppose that p and q are positive integers with $p > q$. Find a quadratic equation with integer coefficients whose roots are $\sqrt{p}/(\sqrt{p} \pm \sqrt{p - q})$.

31. Let r_1, r_2, and r_3 be the roots of the equation $x^3 + bx^2 + cx + d = 0$. Use the method shown in Example 5 to verify the following relationships:

$r_1 + r_2 + r_3 = -b$

$r_1 r_2 + r_2 r_3 + r_3 r_1 = c$

$r_1 r_2 r_3 = -d$

32. Let r_1, r_2, r_3, and r_4 be the roots of the equation $x^4 + bx^3 + cx^2 + dx + e = 0$. Use the method shown in Example 5 to prove the following facts:

$r_1 + r_2 + r_3 + r_4 = -b$

$r_1 r_2 + r_2 r_3 + r_3 r_4 + r_4 r_1 + r_2 r_4 + r_3 r_1 = c$

$r_1 r_2 r_3 + r_2 r_3 r_4 + r_3 r_4 r_1 + r_4 r_1 r_2 = -d$

$r_1 r_2 r_3 r_4 = e$

33. Solve the equation $x^3 - 4x^2 - 9x + 36 = 0$, given that the sum of two of the roots is 0. *Suggestion:* Use Table 2.

34. Solve the equation $x^3 - 75x + 250 = 0$, given that two of the roots are equal. *Suggestion:* Use Table 2.

35. Let α and β be the roots of the polynomial equation $x^2 + bx + c = 0$.

(a) Show that $\alpha^2 + \beta^2 = b^2 - 2c$. *Hint:* Use Table 2 along with the identity $\alpha^2 + \beta^2 = (\alpha + \beta)^2 - 2\alpha\beta$.

(b) Show that $\dfrac{1}{\alpha^2} + \dfrac{1}{\beta^2} = \dfrac{b^2 - 2c}{c^2}$.

(c) Show that $\alpha^3 + \beta^3 = -b(b^2 - 3c)$.

C

36. (a) Let r_1, r_2, r_3, and r_4 be four real roots of the equation $x^4 + ax^2 + bx + c = 0$. Show that $r_1 + r_2 + r_3 + r_4 = 0$. *Hint:* Use the first formula in Exercise 32.

(b) Suppose a circle intersects the parabola $y = x^2$ in the points $(x_1, y_1), \ldots, (x_4, y_4)$. Show that $x_1 + x_2 + x_3 + x_4 = 0$. *Hint:* Use the result in part (a).

37. (a) Let r_1, r_2, and r_3 be three distinct numbers that are roots of the equation $f(x) = Ax^2 + Bx + C = 0$. Show that $f(x) = 0$ for all values of x. *Hint:* You need to show that $A = B = C = 0$. First show that both A and B are zero as follows. If either A or B were nonzero, then the equation $f(x) = 0$ would be a polynomial equation of degree at most 2 with three distinct roots. Why is that impossible?

(b) Use the result in part (a) to prove the following identity:

$$\frac{a^2 - x^2}{(a - b)(a - c)} + \frac{b^2 - x^2}{(b - c)(b - a)}$$
$$+ \frac{c^2 - x^2}{(c - a)(c - b)} - 1 = 0$$

Hint: Let $f(x)$ denote the quadratic expression on the left-hand side of the equation. Compute $f(a)$, $f(b)$, and $f(c)$.

38. Prove that the following equation is an identity:

$$\frac{(x - a)(x - b)c^2}{(c - a)(c - b)} + \frac{(x - b)(x - c)a^2}{(a - b)(a - c)}$$
$$+ \frac{(x - c)(x - a)b^2}{(b - c)(b - a)} - x^2 = 0$$

Hint: Use the result in Exercise 37(a).

9.4 ▼ RATIONAL AND IRRATIONAL ROOTS

As we saw in the previous section, not every polynomial equation has a real root. Furthermore, even if a polynomial equation does possess a real root, that root isn't necessarily a rational number. (The equation $x^2 = 2$ provides a simple example.) If a polynomial equation with integer coefficients does have a rational root, however, we can find that root by applying the **rational roots theorem**, which we now state.

The Rational Roots Theorem

Consider the polynomial equation

$$a_n x^n + a_{n-1} x^{n-1} + \cdots + a_1 x + a_0 = 0 \qquad (n \geq 1, \, a_n \neq 0)$$

and suppose that all the coefficients are integers. Let p/q be a rational number, where p and q have no common factors other than ± 1. If p/q is a root of the equation, then p is a factor of a_0 and q is a factor of a_n.

A proof of the rational roots theorem is outlined in Exercise 39 at the end of this section. For the moment, though, let's just see why the theorem is plausible. Suppose that the two rational numbers a/b and c/d are the roots of a certain quadratic equation. Then, from our experience with quadratics (or by the linear factors theorem), we know that the equation can be written in the form

$$k\left(x - \frac{a}{b}\right)\left(x - \frac{c}{d}\right) = 0 \tag{1}$$

where k is a constant. Now, as Exercise 28 will ask you to check, if we carry out the multiplication and clear of fractions, equation (1) becomes

$$(kbd)x^2 - (kad + kbc)x + kac = 0 \tag{2}$$

Observe that a and c (the numerators of the two roots) are factors of the constant term kac in equation (2), just as the rational roots theorem asserts. Furthermore, b and d (the denominators of the roots) are factors of the coefficient of the x^2-term in equation (2), again as the theorem asserts.

The following example shows how the rational roots theorem can be used to solve a polynomial equation.

EXAMPLE 1 Find the rational roots (if any) of the equation $2x^3 - x^2 - 9x - 4 = 0$. Then solve the equation.

Solution First we list the factors of a_0, the factors of a_n, and the possibilities for rational roots:

$$\text{factors of } a_0 = -4: \quad \pm 1, \, \pm 2, \, \pm 4$$
$$\text{factors of } a_3 = 2: \quad \pm 1, \, \pm 2$$
$$\text{possible rational roots:} \quad \pm\frac{1}{1}, \, \pm\frac{1}{2}, \, \pm\frac{2}{1}, \, \pm\frac{4}{1}$$

Now we can use synthetic division to test whether or not any of these possibilities is a root. (A zero remainder will tell us that we have a root.) As you can check, the first three possibilities $(1, -1, \text{and } \frac{1}{2})$ are not roots. However, using $-\frac{1}{2}$, we have

$$
\begin{array}{r|rrrr}
-1/2 & 2 & -1 & -9 & -4 \\
     &   & -1 & 1 & 4 \\
\hline
     & 2 & -2 & -8 & 0
\end{array}
$$

Thus, $x = -\frac{1}{2}$ is a root. We could now continue to check the remaining possibilities in this same manner. At this point, however, it is simpler to consider the reduced equation $2x^2 - 2x - 8 = 0$, or $x^2 - x - 4 = 0$. Since this is a quadratic equation, it can be solved directly. We have

$$ x = \frac{-(-1) \pm \sqrt{(-1)^2 - 4(1)(-4)}}{2(1)} = \frac{1 \pm \sqrt{17}}{2} $$

We have now determined three distinct roots. Since the degree of the original equation is 3, there can be no other roots. We conclude that $x = -\frac{1}{2}$ is the only rational root. The three roots of the equation are $-\frac{1}{2}, (1 \pm \sqrt{17})/2$. ▲

As Example 1 indicates, the number of possibilities for rational roots can be relatively large, even for rather simple equations. The next theorem we develop allows us to reduce the number of possibilities. We say that a real number B is an **upper bound** for the roots of an equation if every real root is less than or equal to B. Similarly, a real number b is a **lower bound** if every real root is greater than or equal to b. The following theorem tells us how synthetic division can be used in determining upper and lower bounds for roots.

The Upper and Lower Bound Theorem for Real Roots

> Consider the polynomial equation
>
> $$ f(x) = a_n x^n + a_{n-1} x^{n-1} + \cdots + a_1 x + a_0 = 0 $$
>
> where all of the coefficients are real numbers and a_n is positive.
>
> **1.** If we use synthetic division to divide $f(x)$ by $x - B$, where $B > 0$, and we obtain a third row containing no negative numbers, then B is an upper bound for the real roots of $f(x) = 0$.
>
> **2.** If we use synthetic division to divide $f(x)$ by $x - b$, where $b < 0$, and we obtain a third row in which the numbers are alternately positive and negative, then b is a lower bound for the real roots of $f(x) = 0$. (In determining whether the signs alternate in the third row, zeros are ignored.)

We will prove the first part of this theorem. A proof of the second part can be developed along similar lines. To prove the first part of the theorem, we use the division algorithm to write

$$ f(x) = (x - B) \cdot Q(x) + R \tag{3} $$

The remainder R here is a constant that may be zero. To show that B is an upper bound, we must show that any number greater than B is not a root. Toward this end, let p be a number that is greater than B. Note that p must be positive, since B is positive. Then with $x = p$, equation (3) becomes

$$ f(p) = (p - B) \cdot Q(p) + R \tag{4} $$

We are now going to show that the right-hand side of equation (4) is positive. This will tell us that p is not a root. First, look at the factor $(p - B)$. This is positive, since p is greater than B. Next consider $Q(p)$. By hypothesis, the coefficients of $Q(x)$ are all nonnegative. Furthermore, the leading coefficient of $Q(x)$ is a number a_n that is positive. Since p is also positive, it follows that $Q(p)$ must be positive. Finally, the number R is nonnegative because, in the synthetic division of $f(x)$ by $x - B$, all the numbers in the third row are nonnegative. It now follows that the right-hand side of equation (4) is positive. Consequently, $f(p)$ is not zero and p is not a root of the equation $f(x) = 0$. This is what we wished to show.

EXAMPLE 2   Determine the rational roots, or show that none exist, for the equation

$$\frac{1}{4}x^4 - \frac{3}{4}x^3 + \frac{17}{4}x^2 + 4x + 5 = 0$$

Solution   We will use the rational roots theorem along with the upper and lower bound theorem. First of all, if we are to apply the rational roots theorem, then our equation must have integer coefficients. In view of this, we multiply both sides of the given equation by 4 to obtain

$$x^4 - 3x^3 + 17x^2 + 16x + 20 = 0$$

As in the previous example, we list the factors of a_0, the factors of a_n, and the possibilities for rational roots:

factors of $a_0 = 20$: $\pm 1, \pm 2, \pm 4, \pm 5, \pm 10, \pm 20$

factors of $a_4 = 1$: ± 1

possible rational roots: $\pm 1, \pm 2, \pm 4, \pm 5, \pm 10, \pm 20$

Our strategy here will be to first check for positive rational roots, beginning with 1 and working upward. The checks for $x = 1$, $x = 2$, and $x = 4$ are as follows:

$1\rfloor$	1	-3	17	16	20		$2\rfloor$	1	-3	17	16	20		$4\rfloor$	1	-3	17	16	20
		1	-2	15	31				2	-2	30	92				4	4	84	400
	1	-2	15	31	51			1	-1	15	46	112			1	1	21	100	420

As you can see, none of the remainders here is zero. However, notice that in the division corresponding to $x = 4$, all the numbers appearing in the third row are nonnegative. It therefore follows that 4 is an upper bound for the roots of the given equation. In view of this, we needn't bother to check the remaining values $x = 5$, $x = 10$, and $x = 20$, since none of those can be roots. At this point, we can conclude that the given equation has no positive rational roots.

Next we check for negative rational roots, beginning with -1 and working downward (if necessary). Checking $x = -1$, we have

$-1\rfloor$	1	-3	17	16	20
		-1	4	-21	5
	1	-4	21	-5	25

Two conclusions can be drawn from this synthetic division. First, $x = -1$ is not a root of the equation. Second, -1 is a lower bound for the roots because the signs in the third row of the synthetic division alternate. This means that we needn't bother to check whether any of the numbers -2, -4, -5, -15, or -20 are roots; none of them can be roots, since they are all less than -1.

Let us summarize our results. We have shown that the given equation has no positive rational roots and no negative rational roots. Furthermore, by inspection we see that zero is not a root of the equation. Thus, the given equation possesses no rational roots.   ▲

We conclude this section by demonstrating a method for approximating irrational roots. The method depends on the **location theorem**.

The Location Theorem

Let $f(x)$ be a polynomial, all of whose coefficients are real numbers. If a and b are real numbers such that $f(a)$ and $f(b)$ have opposite signs, then the equation $f(x) = 0$ has at least one real root between a and b.

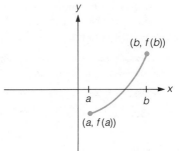

FIGURE 1

Figure 1 indicates why this theorem is plausible. If the point $(a, f(a))$ lies below the x-axis, and $(b, f(b))$ lies above the x-axis, then it certainly seems that the graph of f must cross the x-axis at some point x_0 between a and b. At this intercept, we have $f(x_0) = 0$; that is, x_0 is a root of the equation $f(x) = 0$. (The location theorem is a special case of the so-called intermediate value theorem; the proof is usually discussed in calculus courses.)

Our technique for approximating (or "locating") irrational roots uses the **method of successive approximations**. We will demonstrate this method in Example 3.

EXAMPLE 3   The following equation has exactly one positive root. Locate this root between successive hundredths.

$$f(x) = x^3 + 2x - 4 = 0$$

Solution   First we need to find two numbers a and b such that $f(a)$ and $f(b)$ have opposite signs. By inspection (or by trial and error), we find that $f(1) = -1$ and $f(2) = 8$. Thus, according to the location theorem, the equation $f(x) = 0$ has a real root in the interval $(1, 2)$.

Now that we have located the root between successive integers, we can locate it between successive tenths. We compute $f(1.0)$, $f(1.1)$, $f(1.2)$, and so on, up through $f(2.0)$ if necessary, until we find a sign change. As you can verify, using synthetic division and a calculator, the results are as follows:

$$f(1.0) = -1 \qquad \left. \begin{matrix} f(1.1) = -0.469 \\ f(1.2) = 0.128 \end{matrix} \right\} \longleftarrow \text{sign change}$$

This shows that the root lies between 1.1 and 1.2.

Having located the root between successive tenths, we follow a similar process to locate the root between successive hundredths. Using synthetic division and a calculator, we obtain these results:

$$\begin{matrix} f(1.10) = -0.469 & f(1.13) \approx -0.297 & f(1.16) \approx -0.119 \\ f(1.11) \approx -0.412 & f(1.14) \approx -0.238 & \left. f(1.17) \approx -0.058 \right. \\ f(1.12) \approx -0.355 & f(1.15) \approx -0.179 & \left. f(1.18) \approx 0.003 \right\} \end{matrix} \longleftarrow \text{sign change}$$

This shows that the root lies between 1.17 and 1.18, so we've located the root between successive hundredths, as required.   ▲

The procedure described in Example 3 could be continued to yield closer and closer approximations for the required root. We note in passing that such calculations are easily handled with the aid of a programmable calculator or a computer. [See, for example, page 47 in the paperback text *Calculus by Calculator* by Maurice D. Weir (Englewood Cliffs, N.J.: Prentice-Hall, 1982).]

▼ EXERCISE SET 9.4

A

In Exercises 1–6, list the possibilities for rational roots.

1. $4x^3 - 9x^2 - 15x + 3 = 0$
2. $x^4 - x^3 + 10x^2 - 24 = 0$
3. $8x^5 - x^2 + 9 = 0$
4. $18x^4 - 10x^3 + x^2 - 4 = 0$
5. $\frac{2}{3}x^3 - x^2 - 5x + 2 = 0$
6. $\frac{1}{2}x^4 - 5x^3 + \frac{4}{3}x^2 + 8x - \frac{1}{3} = 0$

In Exercises 7–12, show that each equation has no rational roots.

7. $x^3 - 3x + 1 = 0$
8. $x^3 + 8x^2 - 1 = 0$
9. $x^3 + x^2 - x + 1 = 0$
10. $x^4 + 4x^3 + 4x^2 - 16 = 0$
11. $12x^4 - x^2 - 6 = 0$
12. $4x^5 - x^4 - x^3 - x^2 + x - 8 = 0$

For Exercises 13–25, find the rational roots of each equation and then solve the equation. (Use the rational roots theorem and the upper and lower bound theorem, as in Example 2.)

13. $x^3 + 3x^2 - x - 3 = 0$
14. $2x^3 - 5x^2 - 3x + 9 = 0$
15. $4x^3 + x^2 - 20x - 5 = 0$
16. $3x^3 - 16x^2 + 17x - 4 = 0$
17. $9x^3 + 18x^2 + 11x + 2 = 0$
18. $4x^3 - 10x^2 - 25x + 4 = 0$
19. $x^4 + x^3 - 25x^2 - x + 24 = 0$
20. $10x^4 + 107x^3 + 301x^2 + 171x + 23 = 0$
21. $x^4 - 4x^3 + 6x^2 - 4x + 1 = 0$
22. $24x^3 - 46x^2 + 29x - 6 = 0$
23. $x^3 - \frac{5}{2}x^2 - 23x + 12 = 0$
24. $x^3 - \frac{17}{3}x^2 - \frac{10}{3}x + 8 = 0$
25. $2x^4 - \frac{9}{10}x^3 - \frac{29}{10}x^2 + \frac{27}{20}x - \frac{3}{20} = 0$

In Exercises 26 and 27, determine integral upper and lower bounds for the real roots of the equations. (Follow the method used within the solution of Example 2.)

26. (a) $x^3 + 2x^2 - 5x + 20 = 0$
    (b) $x^5 - 3x^2 + 100 = 0$

27. (a) $5x^4 - 10x - 12 = 0$
    (b) $3x^4 - 4x^3 + 5x^2 - 2x - 4 = 0$
    (c) $2x^4 - 7x^3 - 5x^2 + 28x - 12 = 0$

28. Referring to equation (1) in this section, multiply out the left-hand side and then clear the equation of fractions. Check that your result agrees with equation (2).

C *In Exercises 29–34, each equation has exactly one positive root. In each case, locate the root between successive hundredths. Use a calculator.*

29. $x^3 + x - 1 = 0$
30. $x^3 - 2x - 5 = 0$
31. $x^5 - 200 = 0$
32. $x^3 - 3x^2 + 3x - 26 = 0$
33. $x^3 - 8x^2 + 21x - 22 = 0$
34. $2x^4 - x^3 - 12x^2 - 16x - 8 = 0$

C *In Exercises 35–38, each equation has exactly one negative root. In each case, use a calculator to locate the root between successive hundredths.*

35. $x^3 + x^2 - 2x + 1 = 0$
36. $x^5 + 100 = 0$
37. $x^3 + 2x^2 + 2x + 101 = 0$
38. $x^4 + 4x^3 - 6x^2 - 8x - 3 = 0$

B

39. This exercise outlines a proof of the rational roots theorem. At one point in the proof, we will need to rely on the following fact, which is proved in courses on number theory.

    FACT FROM NUMBER THEORY Suppose that A, B, and C are integers and that A is a factor of the number BC. Then if A has no factor in common with C (other than ± 1), A must be a factor of B.

    (a) Let $A = 2$, $B = 8$, and $C = 5$. Verify that the fact from number theory is correct here.
    (b) Let $A = 20$, $B = 8$, and $C = 5$. Note that A is a factor of BC, but A is not a factor of B. Why doesn't this contradict the fact from number theory?
    (c) Now we're ready to prove the rational roots theorem. We begin with a polynomial equation with integer coefficients:
    $$a_n x^n + a_{n-1}x^{n-1} + \cdots + a_1 x + a_0 = 0$$
    $$(n \geq 1, a_n \neq 0)$$

    We assume that the rational number p/q is a root of the equation and that p and q have no common factors other than 1. Why is the following equation now true?

$$a_n\left(\frac{p}{q}\right)^n + a_{n-1}\left(\frac{p}{q}\right)^{n-1} + \cdots + a_1\left(\frac{p}{q}\right) + a_0 = 0$$

(d) Show that the last equation in part (c) can be written

$$p(a_n p^{n-1} + a_{n-1} q p^{n-2} + \cdots + a_1 q^{n-1}) = -a_0 q^n$$

Since p is a factor of the left-hand side of this last equation, p must also be a factor of the right-hand side. That is, p must be a factor of $a_0 q^n$. But since p and q have no common factors, neither do p and q^n. Our fact from number theory now tells us that p must be a factor of a_0, as we wished to show. (The proof that q is a factor of a_n is carried out in a similar manner.)

40. The location theorem asserts that the polynomial equation $f(x) = 0$ has a root in the open interval (a, b) whenever $f(a)$ and $f(b)$ have unlike signs. If $f(a)$ and $f(b)$ have the same sign, can the equation $f(x) = 0$ have a root between a and b? *Hint:* Look at the graph of $f(x) = x^2 - 2x + 1$ with $a = 0$ and $b = 2$.

C *Use a calculator for Exercises 41–43. Round off your answers to two decimal places.*

41. On the same set of axes, sketch the graphs of $y = x^3$ and $y = 1 - 3x$. Find the x-coordinate of the point where the graphs intersect.

42. On the same set of axes, sketch the graphs of $y = x^3$ and $y = x + 1$. Find the x-coordinate of the point where the graphs intersect.

43. Find the x-coordinate of the point where the curves $y = x^2 - 1$ and $y = \sqrt{x}$ meet.

44. In a note that appeared in *The Two-Year College Mathematics Journal* (Vol. 12, 1981, pp. 334–336), Professors Warren Page and Leo Chosid explain how the process of testing for rational roots can be shortened. In essence, their result is as follows. Suppose that we have a polynomial with integer coefficients and we are testing for a possible root p/q. Then, if a noninteger is generated at any point in the synthetic division process, p/q cannot be a root of the polynomial. For example, suppose we want to know if $\frac{4}{3}$ is a root of $6x^4 - 10x^3 + 2x^2 - 9x + 8 = 0$. The first few steps of the synthetic division are as follows.

$$
\begin{array}{r|rrrrr}
\frac{4}{3} & 6 & -10 & 2 & -9 & 8 \\
  &   & 8 & -\frac{8}{3} & & \\
\hline
  & 6 & -2 & & &
\end{array}
$$

Since the noninteger $-\frac{8}{3}$ has been generated in the synthetic division process, the process can be stopped; $\frac{4}{3}$ is not a root of the polynomial. Use this idea to shorten

your work in testing to see if the numbers $\frac{3}{4}$, $\frac{1}{8}$, and $-\frac{3}{2}$ are roots of the equation $8x^5 - 5x^4 + 3x^2 - 2x - 6 = 0$.

45. In a note that appeared in *The College Mathematics Journal* (Vol. 20, 1989, pp. 139–141), Professor Don Redmond proved the following interesting result.

Consider the polynomial equation $f(x) = a_n x^n + a_{n-1} x^{n-1} + \cdots + a_1 x + a_0 = 0$, and suppose that the degree of $f(x)$ is at least two and that all of the coefficents are integers. If the three numbers a_0, a_n, and $f(1)$ are all odd, then the given equation has no rational roots.

Use this result to show that the following equations have no rational roots.
(a) $9x^5 - 8x^4 + 3x^2 - 2x + 27 = 0$
(b) $5x^5 + 5x^4 - 11x^2 - 3x - 25 = 0$

C

In Exercises 46–50, you need to know that a prime number *is a positive integer greater than 1 with no factors other than itself and 1. Thus, the first seven prime numbers are 2, 3, 5, 7, 11, 13, and 17.*

46. Find all prime numbers p for which the equation $x^2 + x - p = 0$ has a rational root.

47. Find all prime numbers p for which the equation $x^3 + x^2 + x - p = 0$ has at least one rational root. For each value of p that you find, find the corresponding *real* roots of the equation.

48. Consider the equation $x^3 + px - q = 0$, where p and q are prime numbers. Observe that there are only four possible rational roots here: 1, -1, q, and $-q$.
(a) Show that if $x = 1$ is a root, then we must have $q = 3$ and $p = 2$. What are the remaining roots in this case?
(b) Show that none of the numbers -1, q, and $-q$ can be a root of the equation. *Hint:* For each case, assume the contrary and deduce a contradiction.

49. Consider the equation $x^2 + x - pq = 0$, where p and q are prime numbers. If this equation has rational roots, show that these roots must be -3 and 2. *Suggestion:* The possible rational roots are ± 1, $\pm p$, $\pm q$, and $\pm pq$. In each case, assume that the given number is a root and see where that leads.

50. If p and q are prime numbers, show that the equation $x^3 + px - pq = 0$ has no rational roots.

51. Find all integral values of b for which the equation $x^3 - b^2 x^2 + 3bx - 4 = 0$ has a rational root.

52. Let $P(a, b)$ be a point on the first-quadrant portion of the curve $y = x^2$ such that the distance of P from the origin is equal to ab. (Assume that $a \neq 0$.)
(a) C By using the method demonstrated in this section, find the value of a; round off your answer to two decimal places.
(b) Find the exact value of a.

9.5 ▼ CONJUGATE ROOTS AND DESCARTES'S RULE OF SIGNS

As you know from earlier work involving quadratic equations with real coefficients, when nonreal complex roots occur, they occur in conjugate pairs. For instance, as you can check by means of the quadratic formula, the roots of the equation $x^2 - 2x + 5 = 0$ are $1 + 2i$ and $1 - 2i$. The **conjugate roots theorem** tells us that the situation is the same for all polynomial equations with real coefficients.

The Conjugate Roots Theorem

> Let $f(x)$ be a polynomial, all of whose coefficients are real numbers. Suppose that $a + bi$ is a root of the equation $f(x) = 0$, where a and b are real and $b \neq 0$. Then $a - bi$ is also a root of the equation.

To prove the conjugate roots theorem, we use four of the properties of complex conjugates listed in Chapter 1:

property 1: $\bar{z}_1 \bar{z}_2 = \overline{z_1 z_2}$

property 2: $(\bar{z})^m = \overline{z^m}$

property 3: $\bar{r} = r$ for every real number r

property 4: $\bar{z}_1 + \bar{z}_2 = \overline{z_1 + z_2}$

Using these properties, we can prove the theorem as follows. We begin with a polynomial with real coefficients:

$$f(x) = a_n x^n + a_{n-1} x^{n-1} + \cdots + a_1 x + a_0$$

We must show that if $z = a + bi$ is a root of $f(x) = 0$, then $\bar{z} = a - bi$ is also a root. We have

$$
\begin{aligned}
f(\bar{z}) &= a_n \bar{z}^n + a_{n-1} \bar{z}^{n-1} + \cdots + a_1 \bar{z} + a_0 \\
&= \bar{a}_n \overline{z^n} + \overline{a_{n-1}} \overline{z^{n-1}} + \cdots + \bar{a}_1 \bar{z} + \bar{a}_0 \quad \text{properties 3 and 2} \\
&= \overline{a_n z^n} + \overline{a_{n-1} z^{n-1}} + \cdots + \overline{a_1 z} + \bar{a}_0 \quad \text{property 1} \\
&= \overline{a_n z^n + a_{n-1} z^{n-1} + \cdots + a_1 z + a_0} \quad \text{property 4} \\
&= \overline{f(z)} = \bar{0} \quad \text{\small $f(z) = 0$, since z is a root} \\
&= 0 \quad \text{property 3}
\end{aligned}
$$

We have now shown that $f(\bar{z}) = 0$, given that $f(z) = 0$. Thus, $\bar{z}$ is a root, as we wished to show.

Although the conjugate roots theorem concerns nonreal complex roots, it can nevertheless be used to obtain information about real roots, as the next two examples demonstrate.

EXAMPLE 1 Solve the equation $f(x) = 2x^4 - 3x^3 + 12x^2 + 22x - 60 = 0$, given that one root is $1 + 3i$.

Solution Since all the coefficients of $f(x)$ are real numbers, we know that the conjugate of $1 + 3i$ must also be a root. Thus, $1 + 3i$ and $1 - 3i$ are roots, from which it follows that $[x - (1 + 3i)]$ and $[x - (1 - 3i)]$ are factors of $f(x)$. As you can check, the product of these two factors is $x^2 - 2x + 10$. Thus, we must have

$$f(x) = (x^2 - 2x + 10) \cdot Q(x)$$

for some polynomial $Q(x)$. We compute $Q(x)$ using long division:

$$
\begin{array}{r}
2x^2 + x - 6 \\
x^2 - 2x + 10 \overline{\smash{\big)}\ 2x^4 - 3x^3 + 12x^2 + 22x - 60} \\
\underline{2x^4 - 4x^3 + 20x^2} \\
x^3 - 8x^2 + 22x \\
\underline{x^3 - 2x^2 + 10x} \\
-6x^2 + 12x - 60 \\
\underline{-6x^2 + 12x - 60} \\
0
\end{array}
$$

Thus, $Q(x) = 2x^2 + x - 6$, and the original equation becomes

$$f(x) = (x^2 - 2x + 10)(2x^2 + x - 6) = 0$$

We can now find any additional roots by solving the equation $2x^2 + x - 6 = 0$. We have

$$2x^2 + x - 6 = 0$$
$$(2x - 3)(x + 2) = 0$$

$$2x - 3 = 0 \qquad \qquad x + 2 = 0$$
$$x = \frac{3}{2} \qquad \qquad x = -2$$

We now have four distinct roots of the original equation: $1 + 3i$, $1 - 3i$, $\frac{3}{2}$, and -2. Since the degree of the equation is 4, there can be no other roots. ▲

EXAMPLE 2   Show that the equation $x^3 - 2x^2 + x - 1 = 0$ has at least one irrational root.

Solution   Allowing for multiple roots, the given equation has three roots. Therefore, since nonreal complex roots occur in pairs, the equation can have either two nonreal complex roots or no nonreal complex roots. In either case, then, the remaining root or roots must be real. Furthermore, the equation has no rational roots; this is easily checked using the rational roots theorem. It follows now that the equation must have at least one irrational root. (The reasoning here relies on the fact that each real number is either rational or irrational, but not both.) ▲

There is a theorem, similar to the conjugate roots theorem, that tells us about irrational roots of the form $a + b\sqrt{c}$. As background for this theorem, let us look at two preliminary examples. First, we consider the equation $x^2 - 2x - 5 = 0$. As you can check, the roots in this case are $1 + \sqrt{6}$ and $1 - \sqrt{6}$. However, it is not true in general that irrational roots such as these always occur in pairs. Consider as a second example the quadratic equation

$$(x + 2)(x - \sqrt{3}) = 0$$

or

$$x^2 + (2 - \sqrt{3})x - 2\sqrt{3} = 0$$

Here, one of the roots is $\sqrt{3}$, yet $-\sqrt{3}$ is not a root. This type of behavior can occur in polynomial equations where not all of the coefficients are rational. On the other hand, when the coefficients are all rational, we do have the following theorem. (See Exercise 41 at the end of this section for a proof.)

Theorem

> Let $f(x)$ be a polynomial in which all the coefficients are rational. Suppose that $a + b\sqrt{c}$ is a root of the equation $f(x) = 0$, where a, b, and c are rational and $\sqrt{c}$ is irrational. Then $a - b\sqrt{c}$ is also a root of the equation.

EXAMPLE 3 Find a quadratic equation with rational coefficients and a leading coefficient of 1 such that one of the roots is $r_1 = 4 + 5\sqrt{3}$.

Solution If one root is $r_1 = 4 + 5\sqrt{3}$, then the other is $r_2 = 4 - 5\sqrt{3}$. We denote the required equation by $x^2 + bx + c = 0$. The, according to Table 2 in Section 9.3, we have

$$b = -(r_1 + r_2) = -[(4 + 5\sqrt{3}) + (4 - 5\sqrt{3})] = -8$$

and

$$c = r_1 r_2 = (4 + 5\sqrt{3})(4 - 5\sqrt{3}) = 16 - 75 = -59$$

The required equation is therefore $x^2 - 8x - 59 = 0$. This answer can also be obtained without using the table. Since the roots are $4 \pm 5\sqrt{3}$, we can write the required equation as $[x - (4 + 5\sqrt{3})][x - (4 - 5\sqrt{3})] = 0$. As you can now check by multiplying out the two factors, this equation is equivalent to $x^2 - 8x - 59 = 0$, as obtained previously. ▲

We conclude this section with a discussion of **Descartes's rule of signs**. This rule, published by Descartes in 1637, provides us with information about the types of roots an equation can have, even before we attempt to solve the equation. In order to state Descartes's rule of signs, we first explain what is meant by a variation in sign in a polynomial with real coefficients. Suppose that $f(x)$ is a polynomial with real coefficients, written in descending (or ascending) powers of x. For example, let $f(x) = 2x^3 - 4x^2 - 3x + 1$. Then we say that there is a **variation in sign** if two successive coeffcients have opposite signs. In the case of $f(x) = 2x^3 - 4x^2 - 3x + 1$, there are two variations in sign, the first occurring as we go from 2 to -4 and the second occurring as we go from -3 to 1. In looking for variations in sign, we ignore terms with zero coefficients. Table 1 shows a few more examples of how we count variations in sign.

We now state Descartes's rule of signs and look at some examples. Altough the proof of this theorem is not difficult, it is rather lengthy, and we shall omit it here.

TABLE 1

POLYNOMIAL	NUMBER OF VARIATIONS IN SIGN
$x^2 + 4x$	0
$-3x^5 + x^2 + 1$	1
$x^3 + 3x^2 - x + 6$	2

Descartes's Rule of Signs

> Let $f(x)$ be a polynomial, all of whose coefficients are real numbers, and consider the equation $f(x) = 0$. Then:
>
> **a.** The number of positive roots either is equal to the number of variations in sign of $f(x)$ or is less than that by an even integer.
> **b.** The number of negative roots either is equal to the number of variations in sign of $f(-x)$ or is less than that by an even integer.

EXAMPLE 4   Use Descartes's rule of signs to obtain information regarding the roots of the equation $x^3 + 8x + 5 = 0$.

Solution   Let $f(x) = x^3 + 8x + 5$. Then, since there are no variations in sign for $f(x)$, we see from part (a) of Descartes's rule that the given equation has no positive roots. Next we compute $f(-x)$ to learn about the possibilities for negative roots: we have $f(-x) = -x^3 - 8x + 5$. So $f(-x)$ has one sign change, and consequently [by part (b) of Descartes's rule] the original equation has one negative root. Furthermore, notice that zero is not a root of the equation. Thus, the equation has only one real root, a negative root. Since the equation has a total of three roots, we can conclude that we have one negative root and two nonreal complex roots. The two nonreal roots will be complex conjugates.   ▲

EXAMPLE 5   Use Descartes's rule to obtain information regarding the roots of the equation $x^4 + 3x^2 - 7x - 5 = 0$.

Solution   Let $f(x) = x^4 + 3x^2 - 7x - 5$. Then $f(x)$ has one variation in sign. So according to part (a) of Descartes's rule, the equation has one positive root. That leaves us three roots still to account for, since the degree of the equation is 4. We have $f(-x) = x^4 + 3x^2 + 7x - 5$. Since $f(-x)$ has one sign change, we know from part (b) of Descartes's rule that the equation has one negative root. Noting now that zero is not a root, we conclude that the two remaining roots must be nonreal complex roots. In summary, then, the equation has one positive root, one negative root, and two nonreal complex (conjugate) roots.   ▲

EXAMPLE 6   Use Descartes's rule to obtain information regarding the roots of the equation $f(x) = x^3 - x^2 + 3x + 2 = 0$.

Solution   Since $f(x)$ has two variations in sign, the given equation has either two positive roots or no positive roots. To see how many negative roots are possible, we compute $f(-x)$:

$$f(-x) = (-x)^3 - (-x)^2 + 3(-x) + 2$$
$$= -x^3 - x^2 - 3x + 2$$

Since $f(-x)$ has one variation in sign, we conclude from part (b) of Descartes's rule that the equation has exactly one negative root. In summary then, there are two possibilities:

Either: one negative root and two positive roots;

or: one negative root and two nonreal complex roots.   ▲

By using Descartes's rule in Examples 4 and 5, we were able to determine the exact numbers of positive roots and negative roots for the given equations. As Example 6 indicates, however, there are cases in which a direct application of Descartes's rule provides several distinct possibilities for the types of roots, rather than a single definitive result. (Exercises 75 and 76 in the Review Exercises for this chapter illustrate a technique that is sometimes useful in gaining additional information from Descartes's rule. In particular, Exercise 76 will show you that the equation in Example 6 has no positive roots.)

▼ EXERCISE SET 9.5

A

In Exercises 1–16, an equation is given, followed by one or more roots of the equation. In each case, determine the remaining roots.

1. $x^2 - 14x + 53 = 0$; $x = 7 - 2i$
2. $x^2 - x - \frac{1535}{4} = 0$; $x = \frac{1}{2} + 8\sqrt{6}$
3. $x^3 - 13x^2 + 59x - 87 = 0$; $x = 5 + 2i$
4. $x^4 - 10x^3 + 30x^2 - 10x - 51 = 0$; $x = 4 + i$
5. $x^4 + 10x^3 + 38x^2 + 66x + 45 = 0$; $x = -2 + i$
6. $2x^3 + 11x^2 + 30x - 18 = 0$; $x = -3 - 3i$
7. $4x^3 - 47x^2 + 232x + 61 = 0$; $x = 6 - 5i$
8. $9x^4 + 18x^3 + 20x^2 - 32x - 64 = 0$; $x = -1 + \sqrt{3}\,i$
9. $4x^4 - 32x^3 + 81x^2 - 72x + 162 = 0$; $x = 4 + \sqrt{2}\,i$
10. $2x^4 - 17x^3 + 137x^2 - 57x - 65 = 0$; $x = 4 - 7i$
11. $x^4 - 22x^3 + 140x^2 - 128x - 416 = 0$; $x = 10 + 2i$
12. $4x^4 - 8x^3 + 24x^2 - 20x + 25 = 0$; $x = \frac{1}{2}(1 + 3i)$
13. $15x^3 - 16x^2 + 9x - 2 = 0$; $x = \frac{1}{3}(1 + \sqrt{2}\,i)$
14. $x^5 - 5x^4 + 30x^3 + 18x^2 + 92x - 136 = 0$;
    $x = -1 + i\sqrt{3}$, $x = 3 - 5i$
15. $x^7 - 3x^6 - 4x^5 + 30x^4 + 27x^3 - 13x^2 - 64x + 26 = 0$;
    $x = 3 - 2i$, $x = -1 + i$, $x = 1$
16. $x^6 - 2x^5 - 2x^4 + 2x^3 + 2x + 1 = 0$; $x = 1 + \sqrt{2}$

In Exercises 17–20, find a quadratic equation with rational coefficients, one of whose roots is the given number. Write your answer so that the coefficient of x^2 is 1. Use either of the methods shown in Example 3.

17. $r_1 = 1 + \sqrt{6}$
18. $r_1 = 2 - \sqrt{3}$
19. $r_1 = \frac{1}{3}(2 + \sqrt{10})$
20. $r_1 = \frac{1}{2} + \frac{1}{4}\sqrt{5}$

In Exercises 21–36, use Descartes's rule of signs to obtain information regarding the roots of the equations.

21. $x^3 + 5 = 0$
22. $x^4 + x^2 + 1 = 0$
23. $2x^5 + 3x + 4 = 0$
24. $x^3 + 8x - 3 = 0$
25. $5x^4 + 2x - 7 = 0$
26. $x^3 - 4x^2 + x - 1 = 0$
27. $x^3 - 4x^2 - x - 1 = 0$
28. $x^8 + 4x^6 + 3x^4 + 2x^2 + 5 = 0$
29. $3x^8 + x^6 - 2x^2 - 4 = 0$
30. $12x^4 - 5x^3 - 7x^2 - 4 = 0$
31. $x^9 - 2 = 0$
32. $x^9 + 2 = 0$
33. $x^8 - 2 = 0$
34. $x^8 + 2 = 0$
35. $x^6 + x^2 - x - 1 = 0$
36. $x^7 + x^2 - x - 1 = 0$

B

37. Consider the equation $x^4 + cx^2 + dx - e = 0$, where c, d, and e are positive. Show that the equation has one positive root, one negative root, and two nonreal complex roots.

38. Consider the equation $x^n - 1 = 0$.
    (a) Show that the equation has $n - 2$ nonreal complex roots when n is even.
    (b) How many nonreal complex roots are there when n is odd?

39. Find the polynomial $f(x)$ of lowest degree with rational coefficients and with leading coefficient 1, such that $\sqrt{3} + 2i$ is a root of the equation $f(x) = 0$.

40. Find a quadratic polynomial $f(x)$ with integer coefficients, such that $(2 + \sqrt{3})/(2 - \sqrt{3})$ is a root of the equation $f(x) = 0$. *Hint:* First rationalize the given root.

C

41. Let $f(x)$ be a polynomial, with rational coefficients. Suppose that $a + b\sqrt{c}$ is a root of $f(x) = 0$, where a, b, and c are rational and $\sqrt{c}$ is irrational. Complete the following steps to prove that $a - b\sqrt{c}$ is also a root of the equation $f(x) = 0$.
    (a) If $b = 0$, we're done. Why?
    (b) (From now on we'll assume that $b \neq 0$.) Let $d(x) = [x - (a + b\sqrt{c})][x - (a - b\sqrt{c})]$. Explain why $d(a + b\sqrt{c}) = 0$.
    (c) Verify that $d(x) = (x - a)^2 - b^2c$. Thus, $d(x)$ is a quadratic polynomial with rational coefficients.
    (d) Now suppose that we use the long division process to divide the polynomial $f(x)$ by the quadratic polynomial $d(x)$. We'll obtain a quotient $Q(x)$ and a remainder. Since the degree of $d(x)$ is 2, our remainder will be of degree 1 or less. In other words, the general form of this remainder will be $Cx + D$. Furthermore, C and D will have to be rational, because all of the coefficients in $f(x)$ and in $d(x)$ are rational. In summary, we have the identity

    $$f(x) = d(x) \cdot Q(x) + (Cx + D)$$

    Now make the substitution $x = a + b\sqrt{c}$ in this identity, and conclude that $C = D = 0$.
    (e) Using the result in part (d), we have

    $$f(x) = [x - (a + b\sqrt{c})][x - (a - b\sqrt{c})] \cdot Q(x)$$

    Let $x = a - b\sqrt{c}$ in this last identity and conclude that $a - b\sqrt{c}$ is a root of $f(x) = 0$, as required.

42. Find a polynomial $f(x)$ with integer coefficients, such that one root of $f(x) = 0$ is $x = \sqrt{2} + \sqrt[3]{2}$.

▼ CHAPTER NINE   SUMMARY OF PRINCIPAL TERMS AND THEOREMS

TERMS OR THEOREM	PAGE REFERENCE	COMMENT
1. Division algorithm	521	This theorem summarizes the results of the long division process for polynomials. Suppose that $p(x)$ and $d(x)$ are polynomials and $d(x)$ is not the zero polynomial. Then according to the *division algorithm*, there are unique polynomials $q(x)$ and $R(x)$ such that $$p(x) = d(x) \cdot q(x) + R(x)$$ where either $R(x)$ is the zero polynomial or the degree of $R(x)$ is less than the degree of $d(x)$.
2. Root, solution, zero	525	A *root*, or *solution*, of a polynomial equation $f(x) = 0$ is a number r such that $f(r) = 0$. The root r is also called a *zero* of the function f.
3. The remainder theorem	526	The *remainder theorem* asserts that when a polynomial $f(x)$ is divided by $x - r$, the remainder is $f(r)$. Example 3 in Section 9.2 shows how this theorem can be used with synthetic division to evaluate a polynomial.
4. The factor theorem	527	The *factor theorem* makes two statements about a polynomial $f(x)$. First, if $f(r) = 0$, then $x - r$ is a factor of $f(x)$. And second, if $x - r$ is a factor of $f(x)$, then $f(r) = 0$.
5. The fundamental theorem of algebra	532, 533	Let $f(x)$ be a polynomial of degree 1 or greater. The *fundamental theorem of algebra* asserts that the equation $f(x) = 0$ has at least one root among the complex numbers. (See Example 1 in Section 9.3.)
6. The linear factors theorem	533	Let $f(x) = a_n x^n + a_{n-1} x^{n-1} + \cdots + a_0$, where $n \geq 1$ and $a_n \neq 0$. Then this theorem asserts that $f(x)$ can be expressed as a product of n linear factors: $$f(x) = a_n(x - r_1)(x - r_2) \cdots (x - r_n)$$ (The complex numbers $r_1, r_2, \ldots, r_n$ are not necessarily all distinct, and some or all of them may be real numbers.) On page 535, the linear factors theorem is used to prove that every polynomial equation of degree $n \geq 1$ has exactly n roots, where a root of multiplicity k is counted k times.
7. The rational roots theorem	540	Given a polynomial equation, this theorem tells us which rational numbers are candidates for roots of the equation. For a statement of the theorem, see page 540. A proof of the theorem is outlined in Exercise 39, Exercise Set 9.4. For an example of how the theorem is applied, see Example 1 in Section 9.4.
8. Upper bound (for roots); lower bound (for roots)	541	A real number B is an upper bound for the roots of an equation if every real root is less than or equal to B. Similarly, a real number b is a lower bound for the roots if every real root is greater than or equal to b.
9. The upper and lower bound theorem for roots	541	This theorem tells how synthetic division can be used in determining upper and lower bounds for roots of equations. For the statement and proof of the theorem, see page 541 in Section 9.4. For a demonstration of how the theorem is applied, see Example 2 in Section 9.4.
10. The location theorem	543	Let $f(x)$ be a polynomial with real coefficients. If a and b are real numbers such that $f(a)$ and $f(b)$ have opposite signs, then the equation $f(x) = 0$ has at least one root between a and b. To see how this theorem is applied, see Example 3 in Section 9.4.

TERMS OR THEOREM	PAGE REFERENCE	COMMENT
11. The conjugate roots theorem	546	Let $f(x)$ be a polynomial with real coefficients, and suppose that $a + bi$ is a root of the equation $f(x) = 0$, where a and b are real numbers and $b \neq 0$. Then this theorem asserts that $a - bi$ is also a root of the equation. (In other words, for polynomial equations in which all of the coefficients are real numbers, when complex nonreal roots occur, they occur in conjugate pairs.) For illustrations of how this theorem is applied, see Examples 1 and 2 in Section 9.5.
12. Variation in sign	548	Suppose that $f(x)$ is a polynomial with real coefficients, written in descending or ascending powers of x. Then a variation in sign occurs whenever two successive coefficients have opposite signs. For examples, see Table 1 in Section 9.5.
13. Descartes's rule of signs	548	Let $f(x)$ be a polynomial, all of whose coefficients are real numbers, and consider the equation $f(x) = 0$. Then, according to Descartes's rule: **a.** The number of positive roots either is equal to the number of variations in sign of $f(x)$ or is less than that by an even integer. **b.** The number of negative roots either is equal to the number of variations in sign of $f(-x)$ or is less than that by an even integer. Examples 4–6 in Section 9.5 show how Descartes's rule is applied.

▼ CHAPTER NINE   REVIEW EXERCISES

NOTE   Exercises 1–15 constitute a chapter test on the fundamentals, based on group A problems.

1. Let $f(x) = 6x^4 - 5x^3 + 7x^2 - 2x - 2$. Make use of the remainder theorem and synthetic division to compute $f(\frac{1}{2})$.

2. Solve the equation $x^3 + x^2 - 11x - 15 = 0$, given that one of the roots is -3.

3. List the possibilities for the rational roots of the equation $2x^5 - 4x^3 + x - 6 = 0$.

4. Find a quadratic function with zeros 1 and -8 and whose graph has a y-intercept of -24.

5. Use synthetic division to divide $4x^3 + x^2 - 8x + 3$ by $x + 1$.

6. (a) State the factor theorem.
   (b) State the fundamental theorem of algebra.

7. (a) The equation $x^3 - 2x^2 - 1 = 0$ has just one positive root. Use the upper and lower bound theorem to determine the smallest integer that is an upper bound for that root.
   (b) Locate the root between successive tenths. (Use a calculator.)

8. Solve the equation $x^5 - 6x^4 + 11x^3 + 16x^2 - 50x + 52 = 0$, given that two of the roots are $1 + i$ and $3 - 2i$.

9. Let $p(x) = x^4 + 2x^3 - x + 6$ and $d(x) = x^2 + 1$. Use the long division process to find polynomials $q(x)$ and $R(x)$ such that $p(x) = d(x) \cdot q(x) + R(x)$.

10. Express the polynomial $2x^2 - 6x + 5$ in the factored form $a_n(x - r_1)(x - r_2)$.

11. Consider the equation $x^4 - x^3 + 24 = 0$.
    (a) List the possibilities for rational roots.
    (b) Use the upper and lower bound theorem to show that 2 is an upper bound for the roots.
    (c) In view of parts (a) and (b), what possibilities now remain for positive rational roots?
    (d) Which (if any) of the possibilities in part (c) are actually roots?

12. (a) Find the rational roots of the cubic equation $2x^3 - x^2 - x - 3 = 0$.
    (b) Find all solutions of the equation in part (a).

13. Use Descartes's rule of signs to obtain information regarding the roots of the following equation: $3x^4 + x^2 - 5x - 1 = 0$.

14. Find a cubic polynomial $f(x)$ with integer coefficients, such that $1 - 3i$ and -2 are roots of the equation $f(x) = 0$.

15. Find a polynomial $f(x)$ with leading coefficient 1, such that the equation $f(x) = 0$ has the following roots and no other:

ROOT	MULTIPLICITY
2	1
$3i$	3
$1 + \sqrt{2}$	2

Write your answer in the form

$$a_n(x - r_1)(x - r_2) \cdots (x - r_n)$$

In Exercises 16–20, you are given a binomial expression of the form $x + r$ or $x - r$, followed by a polynomial $f(x)$. In each case, determine whether the binomial expression is a factor of $f(x)$.

16. **(a)** $x + \frac{1}{3}$; $f(x) = 3x^3 - 4x^2 + 4x - 1$
   (b) $x - \frac{1}{3}$; $f(x)$ as in part (a)

17. **(a)** $x - 2$; $f(x) = x^4 + 2x^3 - x^2 + 4x + 12$
   (b) $x + 2$; $f(x)$ as in part (a)

18. **(a)** $x + 4$; $f(x) = x^4 - 4x^3 - 2x^2 + 7x + 4$
   (b) $x - 4$; $f(x)$ as in part (a)

19. **(a)** $x - 1$; $f(x) = x^4 + 3x^3 - x^2 + 3x - 18$
   (b) $x + 3$; $f(x)$ as in part (a)

20. **(a)** $x - \sqrt{3}$; $f(x) = x^6 - 3x^4 - 16x^2 + 48$
   (b) $x + \sqrt{3}$; $f(x)$ as in part (a)
   (c) $x - 2$; $f(x)$ as in part (a)
   (d) $x + 2$; $f(x)$ as in part (a)

In Exercises 21 and 22, you are given polynomials $p(x)$ and $d(x)$. In each case, use the long division process to determine polynomials $q(x)$ and $R(x)$ such that

$$p(x) = d(x) \cdot q(x) + R(x)$$

where either $R(x) = 0$ or the degree of $R(x)$ is less than the degree of $d(x)$.

21. $p(x) = x^4 + 3x^3 - x^2 - 5x + 1$; $d(x) = x + 2$

22. $p(x) = 4x^4 + 2x + 1$; $d(x) = 2x^2 + 1$

In Exercises 23–28, use synthetic division to find the quotients and the remainders.

23. $\dfrac{x^4 - 2x^2 + 8}{x - 3}$

24. $\dfrac{x^3 - 1}{x - 2}$

25. $\dfrac{2x^3 - 5x^2 - 6x - 3}{x + 4}$

26. $\dfrac{x^3 + x - 3\sqrt{2}}{x - \sqrt{2}}$

27. $\dfrac{5x^2 - 19x - 4}{x + 0.2}$

28. $\dfrac{x^3 - 3a^2x^2 - 4a^4x + 9a^6}{x - a^2}$

In Exercises 29–36, use synthetic division and the remainder theorem to find the indicated values of the functions.

29. $f(x) = x^5 - 10x + 4$; $f(10)$

30. $f(x) = x^4 + 2x^3 - x$; $f(-2)$

31. $f(x) = x^3 - 10x^2 + x - 1$; $f(\frac{1}{10})$

32. $f(x) = x^4 - 2a^2x^2 + 3a^3x - a^4$; $f(-a)$

33. $f(x) = x^3 + 3x^2 + 3x + 1$; $f(a - 1)$

34. $f(x) = x^3 - 1$; $f(1.1)$

C **35.** $f(x) = x^4 + 4x^3 - 6x^2 - 8x - 2$
   (a) $f(-0.3)$ (Round off the result to two decimal places.)
   (b) $f(-0.39)$ (Round off the result to three decimal places.)
   (c) $f(-0.394)$ (Round off the result to five decimal places.)

C **36.** $f(-4.907)$, where f is the function in Exercise 35. (Round off the result to three decimal places.)

37. Find a value for a such that 3 is a root of the equation $x^3 - 4x^2 - ax - 6 = 0$.

38. For which values of b will -1 be a root of the equation $x^3 + 2b^2x^2 + x - 48 = 0$?

39. For which values of a will $x - 1$ be a factor of the polynomial $a^2x^3 + 3ax^2 + 2$?

40. Use synthetic division to verify that $\sqrt{2} - 1$ is a root of the equation $x^6 + 14x^3 - 1 = 0$.

41. Let $f(x) = ax^3 + bx^2 + cx + d$ and suppose that r is a root of the equation $f(x) = 0$.
   (a) Show that $r - h$ is a root of the equation $f(x + h) = 0$.
   (b) Show that $-r$ is a root of the equation $f(-x) = 0$.
   (c) Show that kr is a root of the equation $f(x/k) = 0$.

42. Suppose that r is a root of the equation $a_2x^2 + a_1x + a_0 = 0$. Show that mr is a root of the quadratic equation $a_2x^2 + ma_1x + m^2a_0 = 0$.

In Exercises 43–48, list the possibilities for the rational roots of the equations.

43. $x^5 - 12x^3 + x - 18 = 0$

44. $x^5 - 12x^3 + x - 17 = 0$

45. $2x^4 - 125x^3 + 3x^2 - 8 = 0$

46. $\frac{3}{5}x^3 - 8x^2 - \frac{1}{2}x + \frac{3}{2} = 0$

47. $x^3 + x - p = 0$, where p is a prime number.

48. $x^3 + x - pq = 0$, where both p and q are prime numbers.

In Exercises 49–56, each equation has at least one rational root. Solve the equations. Suggestion: Use the upper and lower bound theorem to eliminate some of the possibilities for rational roots.

49. $2x^3 + x^2 - 7x - 6 = 0$

50. $x^3 + 6x^2 - 8x - 7 = 0$

51. $2x^3 - x^2 - 14x + 10 = 0$

52. $2x^3 + 12x^2 + 13x + 15 = 0$

53. $\frac{3}{2}x^3 + \frac{1}{2}x^2 + \frac{1}{2}x - 1 = 0$

54. $x^4 - 2x^3 - 13x^2 + 38x - 24 = 0$

55. $x^5 + x^4 - 14x^3 - 14x^2 + 49x + 49 = 0$

56. $8x^5 + 12x^4 + 14x^3 + 13x^2 + 6x + 1 = 0$

57. Solve the equation $x^3 - 9x^2 + 24x - 20 = 0$ using the fact that one of the roots has multiplicity 2.

58. One root of the equation $x^2 + kx + 2k = 0$ $(k \neq 0)$ is twice the other. Find k and find the roots of the equation.

59. State each of the following theorems.
  (a) The division algorithm
  (b) The remainder theorem
  (c) The factor theorem
  (d) The fundamental theorem of algebra

60. Find a quadratic equation with roots $a - \sqrt{a^2 - 1}$ and $a + \sqrt{a^2 - 1}$ $(a > 1)$.

In Exercises 61–64, write each polynomial in the form $a_n(x - r_1)(x - r_2) \cdots (x - r_n)$.

61. $6x^2 + 7x - 20$ **62.** $x^2 + x - 1$

63. $x^4 - 4x^3 + 5x - 20$ **64.** $x^4 - 4x^2 - 5$

Each of Exercises 65–68 gives an equation, followed by one or more roots. Solve the equation.

65. $x^3 - 7x^2 + 25x - 39 = 0$; $x = 2 - 3i$

66. $x^3 + 6x^2 - 24x + 160 = 0$; $x = 2 + 2i\sqrt{3}$

67. $x^4 - 2x^3 - 4x^2 + 14x - 21 = 0$; $x = 1 + i\sqrt{2}$

68. $x^5 + x^4 - x^3 + x^2 + x - 1 = 0$; $x = \frac{1}{2}(1 + i\sqrt{3})$, $x = \frac{1}{2}(-1 - \sqrt{5})$

In Exercises 69–74, use Descartes's rule of signs to obtain information regarding the roots of the equations.

69. $x^3 + 8x - 7 = 0$ **70.** $3x^4 + x^2 + 4x - 2 = 0$

71. $x^3 + 3x + 1 = 0$ **72.** $2x^6 + 3x^2 + 6 = 0$

73. $x^4 - 10 = 0$ **74.** $x^4 + 5x^2 - x + 2 = 0$

75. Consider the equation $x^3 + x^2 + x + 1 = 0$
  (a) Use Descartes's rule to show that the equation has either one or three negative roots.
  (b) Now show that the equation cannot have three negative roots. *Hint:* Multiply both sides of the equation by $x - 1$. Then simplify the left-hand side and reapply Descartes's rule to the new equation.
  (c) Actually, the original equation can be solved using only the basic algebraic techniques discussed in Chapter 1. Solve the equation in this manner.

76. Use Descartes's rule to show that the equation $x^3 - x^2 + 3x + 2 = 0$ has no positive roots. *Hint:* Multiply both sides of the equation by $x + 1$ and apply Descartes's rule to the resulting equation.

C **77.** Let P be the point in the first quadrant where the curve $y = x^3$ intersects the circle $x^2 + y^2 = 1$. Locate the x-coordinate of P within successive hundredths.

C **78.** Let P be the point in the first quadrant where the parabola $y = 4 - x^2$ intersects the curve $y = x^3$. Locate the x-coordinate of P within successive hundredths.

79. Consider the equation $x^3 - 36x - 84 = 0$.
  (a) Use Descartes's rule to check that this equation has exactly one positive root.
  (b) Use the upper and lower bound theorem to show that 7 is an upper bound for the positive root.
  (c) C Locate the positive root within successive hundredths.

80. Consider the equation $x^3 - 3x + 1 = 0$.
  (a) Use Descartes's rule to check that this equation has exactly one negative root.
  (b) Use the upper and lower bound theorem to show that -2 is a lower bound for the negative root.
  (c) C Locate the negative root within successive hundredths.

In Exercises 81–84, find polynomial equations that have integer coefficients and the given values as roots.

81. $4 - \sqrt{5}$

82. $a + b$ and $a - b$ (a and b are integers)

83. $6 - 2i$ and $\sqrt{5}$

84. $\dfrac{5 + \sqrt{6}}{5 - \sqrt{6}}$ *Hint:* First rationalize the expression.

85. Find a fourth-degree polynomial equation with integer coefficients, such that $x = 1 + \sqrt{2} + \sqrt{3}$ is a root. *Hint:* Begin by writing the given relationship as $x - 1 = \sqrt{2} + \sqrt{3}$; then square both sides.

86. Find a cubic equation with integer coefficients, such that $x = 1 + \sqrt[3]{2}$ is a root. Is $1 - \sqrt[3]{2}$ also a root of the equation?

In Exercises 87–91, first determine the zeros of each function; then sketch the graph.

87. $y = x^3 - 2x^2 - 3x$ **88.** $y = x^4 + 3x^3 + 3x^2 + x$

89. $y = x^4 - 4x^2$ **90.** $y = x^3 + 6x^2 + 5x - 12$

91. $y = x^5 - 8x^4 + 25x^3 - 38x^2 + 28x - 8$

In Exercises 92–99, mark **T** if the statement is true without exception; otherwise mark **F**.

92. Every polynomial equation of degree at least 1 has at least one real root.

93. Every polynomial of degree n, where $n \geq 1$, can be written in the form

$$a_n(x - r_1)(x - r_2) \cdots (x - r_n)$$

94. According to the rational roots theorem, there are four possibilities for the rational roots of the equation $x^5 + 6x^2 - 2 = 0$.

95. According to the rational roots theorem, there are only two possibilities for the rational roots of the equation $\frac{1}{3}x^3 - 5x + 1 = 0$.

96. The sum of the roots of the equation $x^2 - 12x + 16 = 0$ is -12.

97. According to the location theorem, if $f(x)$ is a polynomial and $f(a)$ and $f(b)$ have the same signs, then the equation $f(x) = 0$ has at least one root between a and b.

98. Let $f(x)$ be a polynomial with real coefficients, and let a and b be real numbers. If $f(a)$ and $f(b)$ have opposite signs, then the equation $f(x) = 0$ has a root between $f(a)$ and $f(b)$.

99. According to Descartes's rule, the equation $x^7 + 4x + 3 = 0$ has no positive real root.

CHAPTER TEN

ADDITIONAL TOPICS IN ALGEBRA AND TRIGONOMETRY

INTRODUCTION   In this final chapter, we develop several additional topics in algebra and trigonometry. We begin in Section 10.1 with the principle of mathematical induction. This gives us a framework for proving statements about the natural numbers. In Section 10.2 we discuss the binomial theorem, which is used to analyze and expand expressions of the form $(a + b)^n$. As you will see, the proof of the binomial theorem requires the use of mathematical induction. Section 10.3 introduces the related (but distinct) concepts of sequences and series. Then, in the next two sections (Sections 10.4 and 10.5), we study arithmetic and geometric sequences and series. Section 10.5 deals with the sum of an infinite geometric series, a topic that is closely related to the idea of limits, which is the starting point for calculus. In Sections 10.6 and 10.7, we introduce the important concept of a vector in the plane. Finally, in Section 10.8, we explain the trigonometric form for complex numbers. The discussion in this section includes DeMoivre's theorem and the nth roots of complex numbers.

10.1 ▼ MATHEMATICAL INDUCTION

Mathematical induction is not a method of discovery but a technique of proving rigorously what has already been discovered.

David M. Burton in his text *The History of Mathematics, An Introduction* (Boston: Allyn and Bacon, 1985)

In Pascal's treatise [*Traité du triangle arithmétique*, written in 1653] . . . appears one of the earliest acceptable statements of the method of mathematical induction.

Howard Eves in his text *An Introduction to the History of Mathematics*, 5th ed. (Philadelphia: Saunders, 1983)

TABLE 1

n	$1 + 3 + 5 + \cdots + (2n - 1)$	
1	1	$= 1$
2	$1 + 3$	$= 4$
3	$1 + 3 + 5$	$= 9$
4	$1 + 3 + 5 + 7$	$= 16$
5	$1 + 3 + 5 + 7 + 9$	$= 25$

Is mathematics an experimental science? The answer to this question is both yes and no, as the following example illustrates. Consider the problem of determining a formula for the sum of the first n odd natural numbers:

$$1 + 3 + 5 + \cdots + (2n - 1)$$

We begin by doing some calculations in the hope that this may shed some light on the problem. Table 1 shows the results of calculating the sum of the first n odd natural numbers for values of n ranging from 1 to 5. Upon inspecting the table, we observe that each sum in the right-hand column is the square of the corresponding entry in the left-hand column. For instance, for $n = 5$, we see that

$$\underbrace{1 + 3 + 5 + 7 + 9}_{\text{five terms}} = 5^2$$

Now let us try the next case, where $n = 6$, and see if the pattern persists. That is, we want to know if it is true that

$$\underbrace{1 + 3 + 5 + 7 + 9 + 11}_{\text{six terms}} = 6^2$$

As you can easily check, this last equation is true. Thus, based on the experimental, or empirical, evidence, we are led to the following conjecture.

CONJECTURE The sum of the first n odd natural numbers is n^2. That is, $1 + 3 + 5 + \cdots + (2n - 1) = n^2$, for each natural number n.

At this point, the "law" we've discovered is indeed really only a conjecture. After all, we've checked it only for values of n ranging from 1 to 6. It is conceivable at this point (although we may feel it is unlikely) that the conjecture is false for certain values of n. For the conjecture to be useful, we must be able to prove that it holds without exception for *all* natural numbers n. In fact, we will subsequently prove that this conjecture is valid. But before explaining the method of proof to be used, let us look at one more example.

Again, let n denote a natural number. Then consider the following question: Which quantity is the larger, 2^n or $(n + 1)^2$? As before, we begin by doing some calculations. This is the experimental stage of our work. According to Table 2, the quantity $(n + 1)^2$ is larger than 2^n for each value of n up through $n = 5$. Thus, we make the following conjecture.

CONJECTURE $(n + 1)^2 > 2^n$ for all natural numbers n.

TABLE 2

n	2^n	$(n + 1)^2$
1	2	4
2	4	9
3	8	16
4	16	25
5	32	36

Again, we note that this is only a conjecture at this point. Indeed, if we try the case where $n = 6$, we find that the pattern does not persist. That is, when $n = 6$, we find that 2^n is 64, while $(n + 1)^2$ is only 49. So in this example, the conjecture is not true in general; we have found a value of n for which it fails.

The preceding examples show that experimentation does have a place in mathematics, but we must be careful with the results. Where experimentation leads to a conjecture, proof is required before the conjecture can be viewed as a valid law. For the remainder of this section, we shall discuss one such method of proof, that of *mathematical induction*.

In order to state the principle of mathemtical induction, we first introduce some notation. Suppose that for each natural number n we have a statement P_n to be proved. Consider, for instance, the first conjecture we arrived at:

$$1 + 3 + 5 + \cdots + (2n - 1) = n^2$$

If we denote this statement by P_n, then we have, for example, that

P_1 is the statement that $1 = 1^2$

P_2 is the statement that $1 + 3 = 2^2$

P_3 is the statement that $1 + 3 + 5 = 3^2$

With this notation, we can now state the **principle of mathematical induction**.

Principle of Mathematical Induction

Suppose that for each natural number n, we have a statement P_n for which the following two conditions hold:

1. P_1 is true.

2. For each natural number k, if P_k is true, then P_{k+1} is true.

Then all the statements are true; that is, P_n is true for all natural numbers n.

The idea behind mathematical induction is a simple one. Think of each statement P_n as the rung of a ladder to be climbed. Then we can make the analogy shown in Table 3.

TABLE 3

MATHEMATICAL INDUCTION		LADDER ANALOGY	
Hypotheses	1. P_1 is true. 2. If P_k is true then P_{k+1} is true, for any k.	Hypotheses	1′. You can reach the first rung. 2′. If you are on the kth rung you can reach the $(k+1)$st rung, for any k.
Conclusion	3. P_n is true for all n.	Conclusion	3′. You can climb the entire ladder.

According to the principle of mathematical induction, we can prove that a statement or formula P_n is true for all n if we carry out the following two steps.

STEP 1 Show that P_1 is true.

STEP 2 Assume that P_k is true, and on the basis of this assumption, show that P_{k+1} is true.

In step 2, the assumption that P_k is true is referred to as the **induction hypothesis**. (In computer science, step 1 is sometimes referred to as the **initialization step**.) Let us now turn to some examples of proof by mathematical induction.

EXAMPLE 1   Use mathematical induction to prove that

$$1 + 3 + 5 + \cdots + (2n - 1) = n^2$$

for all natural numbers n.

Solution   Let P_n denote the statement that $1 + 3 + 5 + \cdots + (2n - 1) = n^2$. Then we want to show that P_n is true for all natural numbers n.

STEP 1 We must check that P_1 is true. But P_1 is just the statement that $1 = 1^2$, which is true.

STEP 2 Assuming that P_k is true, we must show that P_{k+1} is true. Thus, we assume that

$$1 + 3 + 5 + \cdots + (2k - 1) = k^2 \tag{1}$$

That is the induction hypothesis. We must now show that

$$1 + 3 + 5 + \cdots + (2k - 1) + [2(k + 1) - 1] = (k + 1)^2 \tag{2}$$

In order to derive equation (2) from equation (1), we add the quantity $[2(k + 1) - 1]$ to both sides of equation (1). (The motivation for this stems from the observation that the left-hand sides of equations (1) and (2) differ only by the quantity $[2(k + 1) - 1]$.) We obtain

$$\begin{aligned}
1 + 3 + 5 + \cdots + (2k - 1) + [2(k + 1) - 1] &= k^2 + [2(k + 1) - 1] \\
&= k^2 + 2k + 1 \\
&= (k + 1)^2
\end{aligned}$$

This last equation is what we wanted to show. And having now carried out steps 1 and 2, we conclude by the principle of mathematical induction that P_n is true for all natural numbers n.   ▲

EXAMPLE 2 Use mathematical induction to prove that

$$2^3 + 4^3 + 6^3 + \cdots + (2n)^3 = 2n^2(n+1)^2$$

for all natural numbers n.

Solution Let P_n denote the statement that

$$2^3 + 4^3 + 6^3 + \cdots + (2n)^3 = 2n^2(n+1)^2$$

Then we want to show that P_n is true for all natural numbers n.

STEP 1 We must check that P_1 is true. P_1 is the statement that

$$2^3 = 2(1^2)(1+1)^2 \qquad \text{or} \qquad 8 = 8$$

Thus, P_1 is true.

STEP 2 Assuming that P_k is true, we must show that P_{k+1} is true. Thus, we assume that

$$2^3 + 4^3 + 6^3 + \cdots + (2k)^3 = 2k^2(k+1)^2 \tag{3}$$

We must now show that

$$2^3 + 4^3 + 6^3 + \cdots + (2k)^3 + [2(k+1)]^3 = 2(k+1)^2(k+2)^2 \tag{4}$$

Adding $[2(k+1)]^3$ to both sides of equation (3) yields

$$
\begin{aligned}
2^3 + 4^3 + 6^3 + \cdots + (2k)^3 + [2(k+1)]^3 &= 2k^2(k+1)^2 + [2(k+1)]^3 \\
&= 2k^2(k+1)^2 + 8(k+1)^3 \\
&= 2(k+1)^2[k^2 + 4(k+1)] \\
&= 2(k+1)^2(k^2 + 4k + 4) \\
&= 2(k+1)^2(k+2)^2
\end{aligned}
$$

We have now derived equation (4) from equation (3), as we wished to do. Having carried out steps 1 and 2, we conclude by the principle of mathematical induction that P_n is true for all natural numbers n. ▲

EXAMPLE 3 As indicated in Table 4, the number 3 is a factor of $2^{2n} - 1$ when $n = 1, 2, 3$, and 4. Use mathematical induction to show that 3 is a factor of $2^{2n} - 1$ for all natural numbers n.

Solution Let P_n denote the statement that 3 is a factor of $2^{2n} - 1$. We want to show that P_n is true for all natural numbers n.

STEP 1 We must check that P_1 is true. But P_1 in this case is just the statement that 3 is a factor of $2^{2(1)} - 1$; that is, 3 is a factor of 3, which is surely true.

STEP 2 Assuming that P_k is true, we must show that P_{k+1} is true. Thus, we assume that

$$3 \text{ is a factor of } 2^{2k} - 1 \tag{5}$$

and we must show that

$$3 \text{ is a factor of } 2^{2(k+1)} - 1$$

TABLE 4

n	$2^{2n} - 1$
1	$3 \ (= 3 \cdot 1)$
2	$15 \ (= 3 \cdot 5)$
3	$63 \ (= 3 \cdot 21)$
4	$255 \ (= 3 \cdot 85)$

The strategy here will be to rewrite the expression $2^{2(k+1)} - 1$ in such a way that the induction hypothesis, statement (5), can be applied. We have

$$
\begin{aligned}
2^{2(k+1)} - 1 &= 2^{2k+2} - 1 \\
&= 2^2 \cdot 2^{2k} - 1 \\
&= 4 \cdot 2^{2k} - 4 + 3 \\
&= 4(2^{2k} - 1) + 3 \tag{6}
\end{aligned}
$$

Now, look at the right-hand side of equation (6). By the induction hypothesis, 3 is a factor of $2^{2k} - 1$. Thus, 3 is a factor of $4(2^{2k} - 1)$, from which it certainly follows that 3 is a factor of $4(2^{2k} - 1) + 3$. In summary, then, 3 is a factor of the right-hand side of equation (6). Consequently, 3 must be a factor of the left-hand side of equation (6), which is what we wished to show. Having now completed steps 1 and 2, we conclude by the principle of mathematical induction that P_n is true for all natural numbers n. In other words, 3 is a factor of $2^{2n} - 1$ for all natural numbers n. ▲

There are instances in which a given statement P_n is false for certain initial values of n, but true thereafter. An example of this is provided by the statement

$$2^n > (n+1)^2$$

As you can easily check, this statement is false for $n = 1, 2, 3, 4$, and 5. But, as Example 4 shows, the statement is true for $n \geq 6$. In Example 4, we adapt the principle of mathematical induction by beginning in step 1 with a consideration of P_6 rather than P_1.

EXAMPLE 4 Use mathematical induction to prove that

$$2^n > (n+1)^2 \qquad \text{for all natural numbers } n \geq 6$$

Solution **STEP 1** We must first check that P_6 is true. But P_6 is simply the assertion that

$$2^6 > (6+1)^2 \qquad \text{or} \qquad 64 > 49$$

Thus, P_6 is true.

STEP 2 Assuming that P_k is true, where $k \geq 6$, we must show that P_{k+1} is true. Thus, we assume that

$$2^k > (k+1)^2 \qquad \text{where } k \geq 6 \tag{7}$$

We must show that

$$2^{k+1} > (k+2)^2$$

Multiplying both sides of inequality (7) by 2 gives us

$$2(2^k) > 2(k+1)^2 = 2k^2 + 4k + 2$$

This can be rewritten

$$2^{k+1} > k^2 + 4k + (k^2 + 2)$$

However, since $k \geq 6$, it is certainly true that

$$k^2 + 2 > 4$$

We therefore have

$$2^{k+1} > k^2 + 4k + 4 \qquad \text{or} \qquad 2^{k+1} > (k+2)^2$$

as we wished to show. Having now completed steps 1 and 2, we conclude that P_n is true for all natural numbers $n \geq 6$. ▲

▼ EXERCISE SET 10.1

A

In Exercises 1–18, use the principle of mathematical induction to show that the statements are true for all natural numbers.

1. $1 + 2 + 3 + \cdots + n = \frac{1}{2}n(n+1)$

2. $2 + 4 + 6 + \cdots + 2n = n(n+1)$

3. $1 + 4 + 7 + \cdots + (3n-2) = \frac{1}{2}n(3n-1)$

4. $5 + 9 + 13 + \cdots + (4n+1) = n(2n+3)$

5. $1^2 + 2^2 + 3^2 + \cdots + n^2 = \frac{1}{6}n(n+1)(2n+1)$

6. $2^2 + 4^2 + 6^2 + \cdots + (2n)^2 = \frac{2}{3}n(n+1)(2n+1)$

7. $1^2 + 3^2 + 5^2 + \cdots + (2n-1)^2 = \frac{1}{3}n(2n-1)(2n+1)$

8. $2 + 2^2 + 2^3 + \cdots + 2^n = 2^{n+1} - 2$

9. $3 + 3^2 + 3^3 + \cdots + 3^n = \frac{1}{2}(3^{n+1} - 3)$

10. $e^x + e^{2x} + e^{3x} + \cdots + e^{nx} = \dfrac{e^{(n+1)x} - e^x}{e^x - 1} \quad (x \neq 0)$

11. $1^3 + 2^3 + 3^3 + \cdots + n^3 = [\frac{1}{2}n(n+1)]^2$

12. $2^3 + 4^3 + 6^3 + \cdots + (2n)^3 = 2n^2(n+1)^2$

13. $1^3 + 3^3 + 5^3 + \cdots + (2n-1)^3 = n^2(2n^2 - 1)$

14. $1 \cdot 2 + 3 \cdot 4 + 5 \cdot 6 + \cdots + (2n-1)(2n)$
    $$= \frac{1}{3}n(n+1)(4n-1)$$

15. $1 \cdot 3 + 3 \cdot 5 + 5 \cdot 7 + \cdots + (2n-1)(2n+1)$
    $$= \frac{1}{3}n(4n^2 + 6n - 1)$$

16. $\dfrac{1}{1 \times 3} + \dfrac{1}{2 \times 4} + \dfrac{1}{3 \times 5} + \cdots + \dfrac{1}{n(n+2)}$
    $$= \frac{n(3n+5)}{4(n+1)(n+2)}$$

17. $1 + \dfrac{3}{2} + \dfrac{5}{2^2} + \dfrac{7}{2^3} + \cdots + \dfrac{2n-1}{2^{n-1}} = 6 - \dfrac{2n+3}{2^{n-1}}$

18. $1 + 2 \cdot 2 + 3 \cdot 2^2 + 4 \cdot 2^3 + \cdots + n \cdot 2^{n-1}$
    $$= (n-1)2^n + 1$$

19. Show that $n \leq 2^{n-1}$ for all natural numbers n.

20. Show that 3 is a factor of $n^3 + 2n$ for all natural numbers n.

21. Show that $n^2 + 4 < (n+1)^2$ for all natural numbers $n \geq 2$.

22. Show that $n^3 > (n+1)^2$ for all natural numbers $n \geq 3$.

Ⓒ *In Exercises 23–26, prove that the statement is true for all natural numbers in the specified range. Use a calculator to carry out step 1.*

23. $(1.5)^n > 2n, \; n \geq 7$

24. $(1.25)^n > n, \; n \geq 11$

25. $(1.1)^n > n, \; n \geq 39$

26. $(1.1)^n > 5n, \; n \geq 60$

B

27. Let $f(n) = \dfrac{1}{1 \cdot 2} + \dfrac{1}{2 \cdot 3} + \dfrac{1}{3 \cdot 4} + \cdots + \dfrac{1}{n(n+1)}$.

    (a) Complete the following table.

n	1	2	3	4	5
$f(n)$					

    (b) On the basis of the results in the table, what would you guess to be the value of $f(6)$? Comput $f(6)$ to see if this is correct.

    (c) Make a conjecture about the value of $f(n)$, and prove it using mathematical induction.

28. Let $f(n) = \dfrac{1}{1 \times 3} + \dfrac{1}{3 \times 5} + \dfrac{1}{5 \times 7} + \cdots$
    $$+ \frac{1}{(2n-1)(2n+1)}.$$

    (a) Complete the following table.

n	1	2	3	4
$f(n)$				

(b) On the basis of the results in the table, what would you guess to be the value of $f(5)$? Compute $f(5)$ to see if your guess is correct.

(c) Make a conjecture about the value of $f(n)$, and prove it using mathematical induction.

29. Suppose that a function f satisfies the following conditions: $f(1) = 1$ and

$$f(n) = f(n-1) + 2\sqrt{f(n-1)} + 1 \qquad \text{for } n \geq 2$$

(a) Complete the table.

n	1	2	3	4	5
$f(n)$					

(b) On the basis of the results in the table, what would you guess to be the value of $f(6)$? Compute $f(6)$ to see if your guess is correct.

(c) Make a conjecture about the value of $f(n)$ when n is a natural number, and prove the conjecture using mathematical induction.

30. This exercise demonstrates the necessity of carrying out both step 1 and step 2 before considering an induction proof valid.
(a) Let P_n denote the statement that $n^2 + 1$ is even. Check that P_1 is true. Then give an example showing that P_n is not true for all n.
(b) Let Q_n denote the statement that $n^2 + n$ is odd. Show that step 2 of an induction proof can be completed in this case, but not step 1.

31. A *prime number* is a natural number that has no factors other than itself and 1. For technical reasons, 1 is not considered a prime. Thus, the list of the first seven primes looks like this: 2, 3, 5, 7, 11, 13, 17. Let P_n be the statement that $n^2 + n + 11$ is prime. Check that P_n is true for all values of n less than 10. Check that P_{10} is false.

32. Prove that if $x \neq 1$,

$$1 + 2x + 3x^2 + \cdots + nx^{n-1} = \frac{1-x^n}{(1-x)^2} - \frac{nx^n}{1-x}$$

for all natural numbers n.

33. If $r \neq 1$, show that $1 + r + r^2 + \cdots + r^{n-1} = \frac{(r^n - 1)}{(r-1)}$ for all natural numbers n.

34. Use mathematical induction to show that $x^n - 1 = (x-1)(1 + x + x^2 + \cdots + x^{n-1})$ for all natural numbers n.

35. Prove that 5 is a factor of $n^5 - n$ for all natural numbers $n \geq 2$.

36. Prove that 4 is a factor of $5^n + 3$ for all natural numbers n.

37. Prove that 5 is a factor of $2^{2n+1} + 3^{2n+1}$ for all nonnegative integers n.

38. Prove that 8 is a factor of $3^{2n} - 1$ for all natural numbers n.

39. Prove that 3 is a factor of $2^{n+1} + (-1)^n$ for all nonnegative integers n.

40. Prove that 6 is a factor of $n^3 + 3n^2 + 2n$ for all natural numbers n.

41. Use mathematical induction to show that $x - y$ is a factor of $x^n - y^n$ for all natural numbers n. *Suggestion for step 2:* Verify and then use the fact that $x^{k+1} - y^{k+1} = x^k(x-y) + (x^k - y^k)y$.

In Exercises 42–44, use mathematical induction to prove that the formulas hold for all natural numbers n.

42. $\log_{10}(a_1 a_2 \cdots a_n) = \log_{10} a_1 + \log_{10} a_2 + \cdots + \log_{10} a_n$

43. $(1 + p)^n \geq 1 + np$, where $p > -1$

44. $\sin(x + n\pi) = (-1)^n \sin x$

C

In Exercises 45–47, use mathematical induction to prove that the statements hold for all natural numbers n.

45. $1^5 + 2^5 + 3^5 + \cdots + n^5 = \dfrac{n^2(n+1)^2(2n^2 + 2n - 1)}{12}$

46. $(\cos x)(\cos 2x)(\cos 4x) \cdots (\cos 2^{n-1}x) = \dfrac{\sin 2^n x}{2^n \sin x}$

47. $\dfrac{0^3 + 1^3 + 2^3 + \cdots + n^3}{n^3 + n^3 + n^3 + \cdots + n^3} = \dfrac{1}{4} + \dfrac{1}{4n}$

48. **(a)** Without using induction, prove the following trigonometric identity.

$$\sin^2[(n+1)\theta] - \sin^2 n\theta = \sin[(2n+1)\theta]\sin\theta$$

Hint: On the left-hand side, use difference-of-squares factoring, followed by the sum-to-product formulas.

(b) Use mathematical induction to prove that the following equation holds for all natural numbers n:

$$\sin\theta + \sin 3\theta + \cdots + \sin[(2n-1)\theta] = \frac{\sin^2 n\theta}{\sin\theta}$$

Hint: In the induction step, use the identity in part (a).

(c) [C] Use the identity in part (b) to evaluate the sum

$$\sin 1 + \sin 3 + \sin 5 + \cdots + \sin 25$$

Round off your answer to four decimal places.

10.2 ▼ THE BINOMIAL THEOREM

If you look back at Section 1.5, you will see that two of the special products listed there are

$$(a + b)^2 = a^2 + 2ab + b^2$$

and

$$(a + b)^3 = a^3 + 3a^2b + 3ab^2 + b^3$$

Our present goal is to develop a general formula, known as the *binomial theorem*, for expanding any product of the form $(a + b)^n$, when n is a natural number.

We begin by looking for patterns in the expansion of $(a + b)^n$. To do this, let's list the expansions of $(a + b)^n$ for $n = 1, 2, 3, 4,$ and 5. (Exercises 1 and 2 ask you to verify these results simply by repeated multiplication.)

$$(a + b)^1 = a + b$$
$$(a + b)^2 = a^2 + 2ab + b^2$$
$$(a + b)^3 = a^3 + 3a^2b + 3ab^2 + b^3$$
$$(a + b)^4 = a^4 + 4a^3b + 6a^2b^2 + 4ab^3 + b^4$$
$$(a + b)^5 = a^5 + 5a^4b + 10a^3b^2 + 10a^2b^3 + 5ab^4 + b^5$$

After surveying these results, we note the following patterns.

PROPERTY SUMMARY **PATTERNS OBSERVED IN $(a + b)^n$ for $n = 1, 2, 3, 4, 5$**

GENERAL STATEMENT	EXAMPLE
There are $n + 1$ terms.	There are $4\ (= 3 + 1)$ terms in the expansion of $(a + b)^3$.
The expansion begins with a^n and ends with b^n.	$(a + b)^3$ begins with a^3 and ends with b^3.
The sum of the exponents in each term is n.	The sum of the exponents in each term of $(a + b)^3$ is 3.
The exponents of a decrease by 1 from term to term.	$(a + b)^3 = a^{③} + 3a^{②}b + 3a^{①}b^2 + a^{⓪}b^3$
The exponents of b increase by 1 from term to term.	$(a + b)^3 = a^3b^{⓪} + 3a^2b^{①} + 3ab^{②} + b^{③}$
When n is even, the coefficients are symmetric about the middle term.	The sequence of coefficients for $(a + b)^4$ is 1, 4, 6, 4, 1.
When n is odd, the coefficients are symmetric about the two middle terms.	The sequence of coefficients for $(a + b)^5$ is 1, 5, 10, 10, 5, 1.

The patterns we have just observed for $(a + b)^n$ persist for all natural numbers n. (This will follow from the binomial theorem, which is proved at the end of this section.) Thus, for example, the form of $(a + b)^6$ must be as follows:

$$(a + b)^6 = a^6 + \underline{?}a^5b + \underline{?}a^4b^2 + \underline{?}a^3b^3 + \underline{?}a^2b^4 + \underline{?}ab^5 + b^6$$

The problem now is to find the proper coefficient for each term. To do this, we need to discover additional patterns in the expansion of $(a + b)^n$.

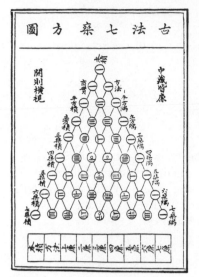

FIGURE 1
"Pascal's" triangle—by Chu-Shi-Kie, A.D. 1303

We have already written out the expansions of $(a + b)^n$ for values of n ranging from 1 to 5. Let us now write only the coefficients appearing in those expansions. The resulting triangular array of numbers is known as **Pascal's triangle**.* For reasons of symmetry, we begin with $(a + b)^0$ rather than $(a + b)^1$.

$$
\begin{array}{llccccccc}
(a + b)^0 & \dots\dots\dots\dots\dots & & & & 1 & & & \\
(a + b)^1 & \dots\dots\dots\dots & & & 1 & & 1 & & \\
(a + b)^2 & \dots\dots\dots & & 1 & & 2 & & 1 & \\
(a + b)^3 & \dots\dots\dots & 1 & & 3 & & 3 & & 1 \\
(a + b)^4 & \dots\dots & 1 & 4 & & 6 & & 4 & 1 \\
(a + b)^5 & \dots\dots & 1 & 5 & 10 & & 10 & 5 & 1 \\
\end{array}
$$

The key observation regarding Pascal's triangle is this: *Each entry in the array* (other than the 1's along the sides) *is the sum of the two numbers diagonally above it.* For instance, the 6 that appears in the fifth row is the sum of the two 3's diagonally above it. Using this observation, we can form as many additional rows as we please. The coefficients for $(a + b)^n$ will then appear in the $(n + 1)$st row of the array.† For instance, to obtain the row corresponding to $(a + b)^6$, we have

$$
\begin{array}{lcccccccc}
\text{sixth row,} & (a + b)^5: & 1 & 5 & 10 & 10 & 5 & 1 \\
\text{seventh row,} & (a + b)^6: & 1 & 6 & 15 & 20 & 15 & 6 & 1 \\
\end{array}
$$

Thus, the sequence of coefficients for $(a + b)^6$ is 1, 6, 15, 20, 15, 6, 1. This answers the question raised earlier about the expansions of $(a + b)^6$. We have

$$(a + b)^6 = a^6 + 6a^5b + 15a^4b^2 + 20a^3b^3 + 15a^2b^4 + 6ab^5 + b^6$$

For analytical work or for larger values of the exponent n, it is inefficient to rely on Pascal's triangle. For this reason, we point out another pattern in the expansions of $(a + b)^n$.

> In the expansion of $(a + b)^n$, the coefficient of any term after the first can be generated as follows. In the *previous* term, multiply the coefficient by the exponent of a and then divide by the number of that previous term.

To see how this observation is used, let's compute the second, third, and fourth coefficients in the expansion of $(a + b)^6$. To compute the coefficient of the second term, we go back to the first term, which is a^6. We have

$$
\text{coefficient of second term} = \frac{1 \cdot 6}{1} = 6
$$

where the numerator shows the coefficient of first term and the exponent of a in first term, and the denominator shows the number of first term.

---

* The array is named after Blaise Pascal, a seventeenth-century French mathematician and philosopher. However, as Figure 1 indicates, the Pascal triangle was known to Chinese mathematicians centuries earlier.

† That these numbers actually are the appropriate coefficients follows from the binomial theorem, which is proved at the end of this section.

Thus, the second term is $6a^5b$ and, consequently, we have

coefficient of second term

exponent of a in second term

$$\text{coefficient of third term} = \frac{6 \cdot 5}{2} = 15$$

number of second term

Continuing now with this method, you should check for yourself that the coefficient of the fourth term in the expansion of $(a + b)^6$ is 20.

Note We now know that the first four coefficients are 1, 6, 15, and 20. By symmetry, it follows that the complete sequence of coefficients for this expansion is 1, 6, 15, 20, 15, 6, 1. No additional calculation for the coefficients is necessary.

EXAMPLE 1 Expand $(2x - y^2)^7$.

Solution First we write the expansion of $(a + b)^7$ using the method explained just prior to this example, or using Pascal's triangle. As you should check for yourself, the expansion is

$$(a + b)^7 = a^7 + 7a^6b + 21a^5b^2 + 35a^4b^3 + 35a^3b^4 + 21a^2b^5 + 7ab^6 + b^7$$

Now we make the substitutions $a = 2x$ and $b = -y^2$. This yields

$$
\begin{aligned}
[2x + (-y^2)]^7 &= (2x)^7 + 7(2x)^6(-y^2) + 21(2x)^5(-y^2)^2 \\
&\quad + 35(2x)^4(-y^2)^3 + 35(2x)^3(-y^2)^4 \\
&\quad + 21(2x)^2(-y^2)^5 + 7(2x)(-y^2)^6 + (-y^2)^7 \\
&= 128x^7 - 448x^6y^2 + 672x^5y^4 - 560x^4y^6 + 280x^3y^3 \\
&\quad - 84x^2y^{10} + 14xy^{12} - y^{14}
\end{aligned}
$$

This is the required expansion. Notice how the signs alternate in the final answer; this is characteristic of all expansions of the form $(a - b)^n$. ▲

In preparation for the binomial theorem, we introduce two notations that are used not only in connection with the binomial theorem, but in many other areas of mathematics as well. The first of these notations is $n!$ (read "n factorial").

DEFINITION

The Factorial Symbol

EXAMPLES

$$n! = 1 \cdot 2 \cdot 3 \cdots n$$

where n is a natural number

$$0! = 1$$

$$3! = 1 \cdot 2 \cdot 3 = 6$$

$$\frac{6!}{4!} = \frac{6 \cdot 5 \cdot 4 \cdot 3 \cdot 2 \cdot 1}{4 \cdot 3 \cdot 2 \cdot 1}$$

$$= 6 \times 5 = 30$$

EXAMPLE 2 Simplify the expression $\dfrac{(n + 1)!}{(n - 1)!}$.

Solution $$\frac{(n + 1)!}{(n - 1)!} = \frac{(n + 1) \cdot n \cdot (n - 1)!}{(n - 1)!} = (n + 1) \cdot n = n^2 + n$$ ▲

The second notation that we introduce in preparation for the binomial theorem is $\binom{n}{k}$. This notation is read "n choose k." (The reason for this is that it can be shown $\binom{n}{k}$ is equal to the number of ways of choosing a subset of k elements from a set with n elements.)

DEFINITION

The Binomial Coefficient $\binom{n}{k}$

EXAMPLE

Let n and k be nonnegative integers with $k \leq n$. Then the binomial coefficient $\binom{n}{k}$ is defined by

$$\binom{n}{k} = \frac{n!}{k!(n-k)!}$$

$$\binom{5}{2} = \frac{5!}{2!(5-2)!}$$

$$= \frac{5!}{2!3!}$$

$$= \frac{5 \cdot 4 \cdot 3 \cdot 2 \cdot 1}{(2 \cdot 1)(3 \cdot 2 \cdot 1)}$$

$$= \frac{5 \cdot 4}{2 \cdot 1} = 10$$

The binomial coefficients are so named because they are indeed the coefficients in the expansion of $(a + b)^n$. More precisely, the relationship is this:

The coefficients in the expansion of $(a + b)^n$ are the $n + 1$ numbers

$$\binom{n}{0}, \binom{n}{1}, \binom{n}{2}, \ldots, \binom{n}{n}.$$

Subsequently, we will see why this statement is true. For now, however, let us look at an example. Consider the binomial coefficients $\binom{3}{0}, \binom{3}{1}, \binom{3}{2}$, and $\binom{3}{3}$. According to our statement, these four quantities should be the coefficients in the expansion of $(a + b)^3$. Let us check:

$$\binom{3}{0} = \frac{3!}{0!(3-0)!} = \frac{3!}{1(3!)} = 1$$

$$\binom{3}{1} = \frac{3!}{1!(3-1)!} = \frac{3 \cdot 2 \cdot 1}{1(2 \cdot 1)} = 3$$

$$\binom{3}{2} = \frac{3!}{2!(3-2)!} = \frac{3 \cdot 2 \cdot 1}{(2 \cdot 1)1} = 3$$

$$\binom{3}{3} = \frac{3!}{3!(3-3)!} = \frac{3!}{3!0!} = 1$$

The values of $\binom{3}{0}, \binom{3}{1}, \binom{3}{2}$, and $\binom{3}{3}$ are thus 1, 3, 3, and 1, respectively. But these last four numbers are indeed the coefficients in the expansion of $(a + b)^3$, as we wished to check.

We are now in a position to state the binomial theorem, after which we will look at several applications. Finally, at the end of this section, we will use mathematical induction to prove the theorem. In the statement of the theorem that follows, we are assuming that the exponent n is a natural number.

Binomial Theorem

$$(a + b)^n = \binom{n}{0}a^n + \binom{n}{1}a^{n-1}b + \binom{n}{2}a^{n-2}b^2 + \cdots$$
$$+ \binom{n}{n-1}ab^{n-1} + \binom{n}{n}b^n$$

One of the uses of the binomial theorem is in identifying specific terms in an expansion without computing the entire expansion. This is particularly helpful when the exponent n is relatively large. Looking back at the statement of the binomial theorem, there are three observations we can make. First, the coefficient of the rth term is $\binom{n}{r-1}$. For instance, the coefficient of the third term is $\binom{n}{3-1} = \binom{n}{2}$. The second observation is that the exponent for a in the rth term is $n - (r - 1)$. For instance, the exponent for a in the third term is $n - (3 - 1) = n - 2$. Finally, we observe that the exponent for b in the rth term is $r - 1$, the same quantity that appears in the lower position of the corresponding binomial coefficient. For instance, the exponent for b in the third term is $r - 1 = 3 - 1 = 2$. We summarize these three observations with the following statement.

The rth term in the expansion of $(a + b)^n$ is

$$\binom{n}{r-1}a^{n-r+1}b^{r-1}$$

EXAMPLE 3   Find the fifteenth term in the expansion of $\left(x^2 - \dfrac{1}{x}\right)^{18}$.

Solution   Using the values $r = 15$, $n = 18$, $a = x^2$, and $b = -1/x$, we have

$$\binom{n}{r-1}a^{n-r+1}b^{r-1} = \binom{18}{15-1}(x^2)^{18-15+1}\left(\frac{-1}{x}\right)^{15-1}$$
$$= \binom{18}{14}x^8 \cdot \frac{1}{x^{14}}$$
$$= \frac{18 \times 17 \times 16 \times 15 \times (14!)}{14!(4 \times 3 \times 2 \times 1)}x^{-6}$$
$$= \frac{18 \times 17 \times 16 \times 15}{4 \times 3 \times 2 \times 1}x^{-6}$$
$$= 3060x^{-6} \qquad \text{as required}$$

EXAMPLE 4   Find the coefficient of the term containing x^4 in the expression of $(x + y^2)^{30}$.

Solution   Again we use the fact that the rth term in the expansion of $(a + b)^n$ is $\binom{n}{r-1}a^{n-r+1}b^{r-1}$. In this case, n is 30 and x plays the role of a. The exponent for x is then $n - r + 1$ or $30 - r + 1$. To see when this exponent is 4, we write

$$30 - r + 1 = 4 \qquad \text{and therefore} \qquad r = 27$$

The required coefficient is therefore $\binom{30}{27-1}$. We then have

$$\binom{30}{26} = \frac{30!}{26!(30-26)!} = \frac{30 \times 29 \times 28 \times 27}{4 \times 3 \times 2 \times 1}$$

After carrying out the indicated arithmetic, we find that $\binom{30}{26} = 27{,}405$. This is the required coefficient.   ▲

EXAMPLE 5   Find the coefficient of the term containing a^9 in the expansion of $(a + 2\sqrt{a})^{10}$.

Solution   The rth term in this expansion will be

$$\binom{10}{r-1}a^{10-r+1}(2\sqrt{a})^{r-1}$$

We can rewrite this as

$$\binom{10}{r-1}a^{10-r+1}(2^{r-1})(a^{1/2})^{r-1}$$

or

$$2^{r-1}\binom{10}{r-1}a^{10-r+1+(r-1)/2}$$

This shows that the general form of the coefficient we wish to find is $2^{r-1}\binom{10}{r-1}$. We now need to determine r when the exponent of a is 9. Thus, we require that

$$10 - r + 1 + \frac{r-1}{2} = 9$$

$$-r + \frac{r-1}{2} = -2$$

$$-2r + r - 1 = -4 \quad \text{multiplying by 2}$$

$$r = 3$$

The required coefficient is now obtained by substituting $r = 3$ in the expression $2^{r-1}\binom{10}{r-1}$. Thus, the required coefficient is

$$2^2\binom{10}{2} = \frac{4 \cdot 10!}{2!(10-2)!} = 2 \cdot 10 \cdot 9 = 180$$

   ▲

There are three simple identities involving the binomial coefficients that will simplify our proof of the binomial theorem.

IDENTITY 1 $\dbinom{r}{0} = 1$ for all nonnegative integers r

IDENTITY 2 $\dbinom{r}{r} = 1$ for all nonnegative integers r

IDENTITY 3 $\dbinom{k}{r} + \dbinom{k}{r-1} = \dbinom{k+1}{r}$ for all natural numbers k and r with $r \le k$

All three of these identities can be proved directly from the definitions of the binomial coefficients, without the need for mathematical induction. The proofs of the first two are straightforward, and we omit them here. The proof of identity 3 runs as follows:

$$\binom{k}{r} + \binom{k}{r-1} = \frac{k!}{r!(k-r)!} + \frac{k!}{(r-1)!(k-r+1)!}$$

$$= \frac{k!}{r(r-1)!(k-r)!} + \frac{k!}{(r-1)!(k-r+1)(k-r)!}$$

Now, the common denominator on the right-hand side of the last equation is $r(r-1)!(k-r+1)(k-r)!$ Thus, we have

$$\binom{k}{r} + \binom{k}{r-1} = \frac{k!(k-r+1)}{r(r-1)!(k-r+1)(k-r)!} + \frac{k!r}{r(r-1)!(k-r+1)(k-r)!}$$

$$= \frac{k!(k-r+1) + k!r}{r(r-1)!(k-r+1)(k-r)!}$$

$$= \frac{k!(k-r+1+r)}{r(r-1)!(k-r+1)(k-r)!} = \frac{k!(k+1)}{r!(k-r+1)!}$$

$$= \frac{(k+1)!}{r![(k+1)-r]!}$$

$$= \binom{k+1}{r}   \text{as required}$$

Taken together, the three identities show why the $(n+1)$st row of Pascal's triangle consists of the numbers $\dbinom{n}{0}, \dbinom{n}{1}, \dbinom{n}{2}, \ldots, \dbinom{n}{n}$. Identities 1 and 2 tell us that this row of numbers begins and ends with 1. Identity 3 is then just a statement of the fact that each entry in the row, other than the initial and final 1, is generated by adding the two entries diagonally above it.

We conclude this section by using mathematical induction to prove the binomial theorem. The statement P_n that we wish to prove for all natural numbers n is this:

$$(a+b)^n = \binom{n}{0}a^n + \binom{n}{1}a^{n-1}b + \cdots + \binom{n}{n-1}ab^{n-1} + \binom{n}{n}b^n$$

First, we check that P_1 is true. The statement P_1 asserts that

$$(a + b)^1 = \binom{1}{0}a^1 + \binom{1}{1}a^0b$$

However, in view of identities 1 and 2, this last equation becomes

$$(a + b)^1 = 1 \cdot a + 1 \cdot b$$

which is surely true. Now let us assume that P_k is true and, on the basis of this assumption, show that P_{k+1} is true. The statement P_k is

$$(a + b)^k = \binom{k}{0}a^k + \binom{k}{1}a^{k-1}b + \cdots + \binom{k}{k-1}ab^{k-1} + \binom{k}{k}b^k$$

Multiplying both sides of this equation by the quantity $(a + b)$ yields

$$(a + b)^{k+1} = (a + b)\left[\binom{k}{0}a^k + \binom{k}{1}a^{k-1}b + \cdots + \binom{k}{k-1}ab^{k-1} + \binom{k}{k}b^k\right]$$

$$= a\left[\binom{k}{0}a^k + \binom{k}{1}a^{k-1}b + \cdots + \binom{k}{k-1}ab^{k-1} + \binom{k}{k}b^k\right]$$

$$+ b\left[\binom{k}{0}a^k + \binom{k}{1}a^{k-1}b + \cdots + \binom{k}{k-1}ab^{k-1} + \binom{k}{k}b^k\right]$$

$$= \binom{k}{0}a^{k+1} + \binom{k}{1}a^kb + \cdots + \binom{k}{k-1}a^2b^{k-1} + \binom{k}{k}ab^k$$

$$+ \binom{k}{0}a^kb + \binom{k}{1}a^{k-1}b^2 + \cdots + \binom{k}{k-1}ab^k + \binom{k}{k}b^{k+1}$$

$$= \binom{k}{0}a^{k+1} + \left[\binom{k}{1} + \binom{k}{0}\right]a^kb + \cdots + \left[\binom{k}{k} + \binom{k}{k-1}\right]ab^k + \binom{k}{k}b^{k+1}$$

We can now make some substitutions on the right-hand side of this last equation. The initial binomial coefficient $\binom{k}{0}$ can be replaced by $\binom{k+1}{0}$, for both are equal to 1 according to identity 1. Similarly, the binomial coefficient $\binom{k}{k}$ appearing at the end of the equation can be replaced by $\binom{k+1}{k+1}$, since both are equal to 1 according to identity 2. Finally, we can use identity 3 to simplify each of the sums in the brackets. We obtain

$$(a + b)^{k+1} = \binom{k+1}{0}a^{k+1} + \binom{k+1}{1}a^kb + \cdots + \binom{k+1}{k}ab^k + \binom{k+1}{k+1}b^{k+1}$$

But this last equation is just the statement P_{k+1}; that is, we have derived P_{k+1} from P_k, as we wished to do. The induction proof is now complete.

▼ EXERCISE SET 10.2

A

In Exercises 1 and 2, verify each statement directly, without using the techniques developed in this section.

1. **(a)** $(a + b)^2 = a^2 + 2ab + b^2$
   (b) $(a + b)^3 = a^3 + 3a^2b + 3ab^2 + b^3$
   Hint: $(a + b)^3 = (a + b)(a + b)^2$

2. **(a)** $(a + b)^4 = a^4 + 4a^3b + 6a^2b^2 + 4ab^3 + b^4$
   Hint: Use the result in Exercise 1(b) and the fact that $(a + b)^4 = (a + b)(a + b)^3$.
   (b) $(a + b)^5 = a^5 + 5a^4b + 10a^3b^2 + 10a^2b^3 + 5ab^4 + b^5$

In Exercises 3–28, carry out the indicated expansions.

3. $(a + b)^9$
4. $(a - b)^9$
5. $(2A + B)^3$
6. $(1 + 2x)^6$
7. $(1 - 2x)^6$
8. $(3x^2 - y)^5$
9. $(\sqrt{x} + \sqrt{y})^4$
10. $(\sqrt{x} - \sqrt{y})^4$
11. $(x^2 + y^2)^5$
12. $(5A - B^2)^3$
13. $[1 - (1/x)]^6$
14. $(3x + y^2)^4$
15. $[(x/2) - (y/3)]^3$
16. $(1 - z^2)^7$
17. $(ab^2 + c)^7$
18. $[x - (1/x)]^8$
19. $(x + \sqrt{2})^8$
20. $(4A - \frac{1}{2})^5$
21. $(\sqrt{2} - 1)^3$
22. $(1 + \sqrt{5})^4$
23. $(\sqrt{2} + \sqrt{3})^5$
24. $(\frac{1}{2} - 2a)^6$
25. $(2\sqrt[3]{2} - \sqrt[3]{4})^3$
26. $(x + y + 1)^4$ *Suggestion:* Rewrite the expression as $[(x + y) + 1]^4$.
27. $(x^2 - 2x - 1)^5$ *Suggestion:* Rewrite the expression as $[x^2 - (2x + 1)]^5$.
28. $[x^2 - 2x - (1/x)]^6$

In Exercises 29–38, evaluate or simplify each expression.

29. $5!$
30. **(a)** $3! + 2!$
    (b) $(3 + 2)!$
31. $\binom{7}{3}\binom{3}{2}$
32. $\dfrac{20!}{18!}$
33. **(a)** $\binom{5}{3}$ **(b)** $\binom{5}{4}$
34. **(a)** $\binom{7}{7}$ **(b)** $\binom{7}{0}$
35. $\dfrac{(n + 2)!}{n!}$
36. $\dfrac{n[(n - 2)!]}{(n + 1)!}$
37. $\binom{6}{4} + \binom{6}{3} - \binom{7}{4}$
38. $(3!)! + (3!)^2$
39. Find the fifteenth term in the expansion of $(a + b)^{16}$.
40. Find the third term in the expansion of $(a - b)^{30}$.

41. Find the one hundredth term in the expansion of $(1 + x)^{100}$.
42. Find the twenty-third term in the expansion of $[x - (1/x^2)]^{25}$.
43. Find the coefficient of the term containing a^4 in the expansion of $(\sqrt{a} - \sqrt{x})^{10}$.
44. Find the coefficient of the term containing a^4 in the expansion of $(3a - 5x)^{12}$.
45. Find the coefficient of the term containing y^8 in the expansion of $[(x/2) - 4y]^9$.
46. Find the coefficient of the term containing x^6 in the expansion of $[x^2 + (1/x)]^{12}$.
47. Find the coefficient of the term containing x^3 in the expansion of $(1 - \sqrt{x})^8$.
48. Find the coefficient of the term containing a^8 in the expansion of $[a - (2/\sqrt{a})]^{14}$.
49. Find the term that does not contain A in the expansion of $[(1/A) + 3A^2]^{12}$.
50. Find the coefficient of B^{-10} in the expansion of $[(B^2/2) - (3/B^3)]^{10}$.

B

51. Show that the coefficient of x^n in the expansion of $(1 + x)^{2n}$ is $(2n)!/(n!)^2$.
52. Find n so that the coefficients of the eleventh and thirteenth terms in $(1 + x)^n$ are the same.
53. **(a)** Complete the following table.

k	0	1	2	3	4	5	6	7	8
$\binom{8}{k}$									

   (b) Use the results in part (a) to verify that
   $$\binom{8}{0} + \binom{8}{1} + \binom{8}{2} + \cdots + \binom{8}{8} = 2^8.$$

   (c) By taking $a = b = 1$ in the expansion of $(a + b)^n$, show that $\binom{n}{0} + \binom{n}{1} + \binom{n}{2} + \cdots + \binom{n}{n} = 2^n$.

C

54. Two real numbers A and B are defined by $A = \sqrt[99]{99!}$ and $B = \sqrt[100]{100!}$. Which number is larger, A or B?
   Hint: Compare A^{9900} and B^{9900}.

55. This exercise outlines a proof of the identity

$$\binom{n}{0}^2 + \binom{n}{1}^2 + \binom{n}{2}^2 + \cdots + \binom{n}{n}^2 = \binom{2n}{n}$$

(a) Verify that

$$(1 + x)^n \left(1 + \frac{1}{x}\right)^n = \frac{(1 + x)^{2n}}{x^n} \tag{1}$$

(This requires only basic algebra, not the binomial theorem.)

(b) Show that the coefficient of the term independent of x on the right side of equation (1) is $\binom{2n}{n}$.

(c) Use the binomial theorem to expand $(1 + x)^n$. Then show that the coefficient of the term independent of x on the left side of equation (1) is

$$\binom{n}{0}^2 + \binom{n}{1}^2 + \binom{n}{2}^2 + \cdots + \binom{n}{n}^2$$

10.3 ▼ INTRODUCTION TO SEQUENCES AND SERIES

This section and the next two sections in this chapter deal with numerical sequences. We will begin with a somewhat informal definition of this concept. Then, after looking at some examples and terminology, we will present a more formal definition. A **numerical sequence** is an ordered list of numbers. Here are four examples:

example A: $1, \sqrt{2}, 10$
example B: $2, 4, 6, 8, \ldots$
example C: $1, \frac{1}{2}, \frac{1}{4}, \frac{1}{8}, \ldots$
example D: $1, 1, 1, 1, \ldots$

The individual entries in a numerical sequence are called the **terms** of the sequence. In this chapter, the terms in each sequence will always be real numbers, so for convenience, we will drop the adjective *numerical* and refer simply to *sequences*. (It is worth pointing out, however, that in more advanced courses, sequences are studied in which the individual terms are functions.) Any sequence possessing only a finite number of terms is called a **finite sequence**. Thus, the sequence in example A is a finite sequence. On the other hand, examples B, C, and D are examples of what we call **infinite sequences**; each contains infinitely many terms. As example D indicates, it is not necessary that all the terms in a sequence be distinct. In this chapter, all the sequences we discuss will be infinite sequences.

In examples B, C, and D, the three dots are read "and so on." In using this notation, we are assuming that it is clear what the subsequent terms of the sequence are. Toward this end, we often specify a formula for the nth term in a sequence. Example B in this case would appear this way:

$$2, 4, 6, 8, \ldots, 2n, \ldots$$

A letter with subscripts is often used to denote the various terms in a sequence. For instance, if we denote the sequence in example B by $a_1, a_2, a_3, \ldots$, then we have $a_1 = 2$, $a_2 = 4$, $a_3 = 6$, and, in general, $a_n = 2n$.

EXAMPLE 1   Consider the sequence $a_1, a_2, a_3, \ldots$, in which the nth term a_n is given by

$$a_n = \frac{n}{n+1}$$

Compute the first three terms of the sequence, as well as the one thousandth term.

Solution   To obtain the first term, we replace n by 1 in the given formula. This yields

$$a_1 = \frac{1}{1+1} = \frac{1}{2}$$

The other terms are similarly obtained. We have

$$a_2 = \frac{2}{2+1} = \frac{2}{3}$$

$$a_3 = \frac{3}{3+1} = \frac{3}{4}$$

$$a_{1000} = \frac{1000}{1000+1} = \frac{1000}{1001}$$

▲

In the example just concluded, we were given an explicit formula for the nth term of the sequence. The next example shows a different way of specifying a sequence, in which each term except the first is defined in terms of the previous term. This is an example of a **recursive definition**. (Recursive definitions are particularly useful in computer programming.)

EXAMPLE 2   Compute the first three terms of the sequence $b_1, b_2, b_3, \ldots$, which is defined recursively by

$$b_1 = 4$$
$$b_n = 2(b_{n-1} - 1) \qquad \text{for } n \geq 2$$

Solution   We are given the first term: $b_1 = 4$. To find b_2, we replace n by 2 in the formula $b_n = 2(b_{n-1} - 1)$ to obtain

$$b_2 = 2(b_1 - 1) = 2(4 - 1) = 6$$

Thus, $b_2 = 6$. Next we use this value of b_2 in the formula $b_n = 2(b_{n-1} - 1)$ to obtain b_3. Replacing n by 3 in this formula yields

$$b_3 = 2(b_2 - 1) = 2(6 - 1) = 10$$

We have now found the first three terms of the sequence: $b_1 = 4$, $b_2 = 6$, and $b_3 = 10$.

▲

If you think about the central idea behind the concept of a sequence, you can see that a function is involved: For each input n, we have an output a_n. For this reason, the formal definition of a sequence is phrased as follows.

DEFINITION

Sequence

A **sequence** is a function whose domain is the set of natural numbers.

If we denote the function by f for the moment, then $f(1)$ would denote what we have been calling a_1, the first term of the sequence. Similarly, $f(2) = a_2, f(3) = a_3$, and so on.

EXAMPLE 3   Consider the sequence with nth term

$$a_n = \frac{1}{2^n}$$

Graph this sequence for $n = 1, 2, 3,$ and 4.

Solution   First, we compute the required values of a_n, as shown in the following table.

n	1	2	3	4
a_n	$\frac{1}{2}$	$\frac{1}{4}$	$\frac{1}{8}$	$\frac{1}{16}$

Now, locating the inputs n on the horizontal axis and the outputs a_n on the vertical axis, we obtain the graph shown in Figure 1.   ▲

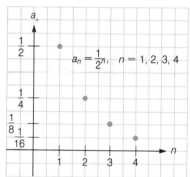

FIGURE 1

The graph of the sequence
$$a_n = \frac{1}{2^n}, \ n = 1, 2, 3, 4$$

We will often be interested in the sum of certain terms of a sequence. Consider, for example, the sequence

$$10, 20, 30, 40, \ldots$$

in which the nth term is $10n$. The sum of the first four terms in this sequence is

$$10 + 20 + 30 + 40 = 100$$

More generally, the sum of the first n terms in this sequence is indicated by

$$10 + 20 + 30 + \cdots + 10n$$

This expression is an example of a **finite series**, which simply means a sum of a finite number of terms.

We can indicate the sum of the first n terms of the sequence $a_1, a_2, a_3, \ldots$ by

$$a_1 + a_2 + a_3 + \cdots + a_n$$

Another way to indicate this sum uses what is called **sigma notation**, which we now introduce. The capital Greek letter sigma is written Σ. We define the notation $\sum_{k=1}^{n} a_k$ by the equation

$$\sum_{k=1}^{n} a_k = a_1 + a_2 + a_3 + \cdots + a_n$$

For example, $\sum_{k=1}^{3} a_k$ stands for the sum $a_1 + a_2 + a_3$, the idea in this case being to replace the subscript k successively by 1, 2, and 3 and then add the results.

For a more concrete example, let us evaluate the expression $\sum_{k=1}^{4} k^2$. We have

$$\sum_{k=1}^{4} k^2 = 1^2 + 2^2 + 3^3 + 4^2$$

$$= 1 + 4 + 9 + 16 = 30$$

There is nothing special about the choice of the letter k in the expression $\sum_{k=1}^{4} k^2$. For instance, we could equally well write

$$\sum_{j=1}^{4} j^2 = 1^2 + 2^2 + 3^2 + 4^2 = 30$$

The letter k in the expression $\sum_{k=1}^{4} k^2$ is called the **index of summation**. Similarly, the letter j appearing in $\sum_{j=1}^{4} j^2$ is the index of summation in that case. As we have seen, the choice of the letter used for thc indcx of summation has no effect on the value of the indicated sum. For this reason, the index of summation is referred to as a *dummy variable*. The next two examples provide further practice with the sigma notation.

EXAMPLE 4   Express each of the following sums without sigma notation.

(a) $\sum_{k=1}^{3} (3k - 2)^2$ **(b)** $\sum_{i=1}^{4} ix^{i-1}$ **(c)** $\sum_{j=1}^{5} (a_{j+1} - a_j)$

Solution   **(a)** The notation $\sum_{k=1}^{3} (3k - 2)^2$ directs us to replace k successively by 1, 2, and 3 in the expression $(3k - 2)^2$ and then add the results. We thus obtain

$$\sum_{k=1}^{3} (3k - 2)^2 = 1^2 + 4^2 + 7^2 = 66$$

(b) The notation $\sum_{i=1}^{4} ix^{i-1}$ directs us to replace i successively by 1, 2, 3, and 4 in the expression ix^{i-1} and then add the results. We have

$$\sum_{i=1}^{4} ix^{i-1} = 1x^0 + 2x^1 + 3x^2 + 4x^3$$

$$= 1 + 2x + 3x^2 + 4x^3$$

(c) To expand $\sum_{j=1}^{5} (a_{j+1} - a_j)$, we replace j successively by 1, 2, 3, 4, and 5 in the expression $(a_{j+1} - a_j)$ and then add. We obtain

$$\sum_{j=1}^{5} (a_{j+1} - a_j) = (a_2 - a_1) + (a_3 - a_2) + (a_4 - a_3) + (a_5 - a_4) + (a_6 - a_5)$$

$$= a_2 - a_1 + a_3 - a_2 + a_4 - a_3 + a_5 - a_4 + a_6 - a_5$$

Combining like terms, we have

$$\sum_{j=1}^{5} (a_{j+1} - a_j) = a_6 - a_1$$

Sums such as $\sum_{j=1}^{5} (a_{j+1} - a_j)$ are known as **collapsing** or **telescoping sums**.

▲

EXAMPLE 5   Use sigma notation to rewrite each sum.

(a) $\dfrac{x}{1!} + \dfrac{x^2}{2!} + \dfrac{x^3}{3!} + \cdots + \dfrac{x^{12}}{12!}$ (b) $\dfrac{x}{2!} + \dfrac{x^2}{3!} + \dfrac{x^3}{4!} + \cdots + \dfrac{x^n}{(n+1)!}$

Solution

(a) Since the exponents on x run from 1 to 12, we choose a dummy variable, say k, running from 1 to 12. Also, we notice that if the numerator of a given term in the sum is x^k, then the corresponding denominator is $k!$. Consequently, the sum can be written

$$\frac{x}{1!} + \frac{x^2}{2!} + \frac{x^3}{3!} + \cdots + \frac{x^{12}}{12!} = \sum_{k=1}^{12} \frac{x^k}{k!}$$

(b) Since the exponents on x run from 1 to n, we choose a dummy variable, say k, running from 1 to n. (Note that both of the letters n and x would be inappropriate here as dummy variables.) Also, we notice that if the numerator of a given term in the sum is x^k, then the corresponding denominator is $(k+1)!$. Thus, the given sum can be written

$$\frac{x}{2!} + \frac{x^2}{3!} + \frac{x^3}{4!} + \cdots + \frac{x^n}{(n+1)!} = \sum_{k=1}^{n} \frac{x^k}{(k+1)!}$$ ▲

▼ EXERCISE SET 10.3

A

In Exercises 1–20, compute the first five terms in each sequence. (In Exercises 1–12, where the nth term is given by an explicit formula, for instance, $a_n = n/(n+1)$, assume that the equation holds for all natural numbers n.)

1. $a_n = \dfrac{n}{n+1}$ **2.** $a_n = n^3$

3. $a_n = (n-1)^2$ **4.** $b_n = n!$

5. $b_n = [1 + (1/n)]^n$ **6.** $b_n = \dfrac{n^n}{n!}$

7. $u_n = (-1)^n$ **8.** $a_n = (-1)^n\sqrt{n^n}$

9. $a_n = (-1)^{n+1}/n!$ **10.** $S_n = 1/[(n+1)!]$

11. $a_n = 3n$ **12.** $b_n = 3$

13. $a_1 = 1;\ a_n = (1 + a_{n-1})^2,\ \ n \geq 2$

14. $a_1 = 2;\ a_n = \sqrt{a_{n-1}^2 + 1},\ \ n \geq 2$

15. $a_1 = 2;\ a_2 = 2;\ a_n = a_{n-1}a_{n-2},\ \ n \geq 3$

16. $F_1 = 1;\ F_2 = 1;\ F_n = F_{n-1} + F_{n-2},\ \ n \geq 3$

17. $a_1 = 1;\ a_{n+1} = na_n,\ \ n \geq 1$

18. $a_1 = 1;\ a_2 = 2;\ a_n = a_{n-1}/a_{n-2},\ \ n \geq 3$

19. $a_1 = 0;\ a_n = 2^{a_{n-1}},\ \ n \geq 2$

20. $a_1 = 0;\ a_2 = 1;\ a_n = \tfrac{1}{2}(a_{n-1} + a_{n-2}),\ \ n \geq 3$

In Exercises 21–26, graph the sequences for the indicated values of n.

21. $a_n = (-1)^n,\ n = 1, 2, 3, 4$

22. $a_n = 2n - 1,\ n = 1, 2, 3$

23. $a_n = 1/n,\ n = 1, 2, 3, 4$

24. $a_n = (-1)^n/n,\ n = 1, 2, 3, 4$

25. $a_n = (n-1)/(n+1),\ n = 1, 2, 3$

26. $a_n = [1 + (1/n)]^n,\ n = 1, 2, 3$

In Exercises 27–32, find the sum of the first five terms in the sequence with an nth term as given.

27. $a_n = 2^n$ **28.** $b_n = 2^{-n}$ **29.** $a_n = n^2 - n$

30. $b_n = (n-1)!$ **31.** $a_n = (-1)^n/n!$

32. $a_1 = 1;\ a_n = \dfrac{1}{n-1} - \dfrac{1}{n+1}\ \ (n \geq 2)$

33. Find the sum of the first five terms of the sequence that is defined recursively by $a_1 = 1;\ a_2 = 2;\ a_n = a_{n-1}^2 + a_{n-2}^2\ \ (n \geq 3)$.

34. Find the sum of the first four terms of the sequence defined recursively by $a_1 = 2;\ a_n = n/a_{n-1}\ \ (n \geq 2)$.

35. Find the sum of the first four terms of the sequence defined recursively by $a_1 = 2;\ a_n = (a_{n-1})^2\ \ (n \geq 2)$.

36. Find the sum of the first six terms of the sequence defined recursively by $a_1 = 1$; $a_2 = 2$; $a_{n+1} = a_n a_{n-1}$ $(n \geq 2)$.

In Exercises 37–48, express each of the sums without using sigma notation. Simplify your answers where possible.

37. $\displaystyle\sum_{k=1}^{3} (k-1)$

38. $\displaystyle\sum_{k=1}^{5} k$

39. $\displaystyle\sum_{k=4}^{5} k^2$

40. $\displaystyle\sum_{k=2}^{6} (1-2k)$

41. $\displaystyle\sum_{n=1}^{3} x^n$

42. $\displaystyle\sum_{n=1}^{3} (n-1)x^{n-2}$

43. $\displaystyle\sum_{n=1}^{4} \frac{1}{n}$

44. $\displaystyle\sum_{n=0}^{4} 3^n$

45. $\displaystyle\sum_{j=1}^{9} \log_{10} \frac{j}{j+1}$

46. $\displaystyle\sum_{j=2}^{5} \log_{10} j$

47. $\displaystyle\sum_{j=1}^{6} \left(\frac{1}{j} - \frac{1}{j+1} \right)$

48. $\displaystyle\sum_{j=1}^{5} (x^{j+1} - x^j)$

In Exercises 49–58, rewrite the sums using sigma notation.

49. $5 + 5^2 + 5^3 + 5^4$

50. $5 + 5^2 + 5^3 + \cdots + 5^n$

51. $x + x^2 + x^3 + x^4 + x^5 + x^6$

52. $x + 2x^2 + 3x^3 + 4x^4 + 5x^5 + 6x^6$

53. $\frac{1}{1} + \frac{1}{2} + \frac{1}{3} + \cdots + \frac{1}{12}$

54. $\frac{1}{1} + \frac{1}{2} + \frac{1}{3} + \cdots + \frac{1}{n}$

55. $2 - 2^2 + 2^3 - 2^4 + 2^5$

56. $\binom{10}{3} + \binom{10}{4} + \binom{10}{5} + \cdots + \binom{10}{10}$

57. $1 - 2 + 3 - 4 + 5$

58. $\frac{1}{2} - \frac{1}{4} + \frac{1}{8} - \frac{1}{16} + \frac{1}{32} - \frac{1}{64} + \frac{1}{128}$

B

59. The *Fibonacci numbers* $F_1, F_2, F_3, \ldots$ are defined as follows:

$$F_1 = 1; \quad F_2 = 1; \quad F_{n+2} = F_n + F_{n+1} \quad (n \geq 1)$$

(a) Verify that the first ten Fibonacci numbers are as follows.

F_1	F_2	F_3	F_4	F_5	F_6	F_7	F_8	F_9	F_{10}
1	1	2	3	5	8	13	21	34	55

(b) Complete the following table.

n	$F_1 + F_2 + F_3 + \cdots + F_n$	$F_{n+2} - 1$
1		
2		
3		
4		
5		

(c) Use mathematical induction to prove that

$$F_1 + F_2 + F_3 + \cdots + F_n = F_{n+2} - 1$$

for all natural numbers n.

(d) Use mathematical induction to prove that

$$F_n \geq n \qquad \text{for all natural numbers } n \geq 5$$

(e) Use mathematical induction to show that

$$F_1^2 + F_2^2 + F_3^2 + \cdots + F_n^2 = F_n F_{n+1}$$

for all natural numbers n.

(f) Use mathematical induction to show that

$$F_{n+1}^2 = F_n F_{n+2} + (-1)^n$$

for all natural numbers n. *Hint for Step 2:* Add $F_{k+1}F_{k+2}$ to both sides of the equation in the induction hypothesis. Then factor F_{k+1} from the left-hand side and factor F_{k+2} from the first two terms on the right-hand side.

(g) This part of the exercise requires a knowledge of matrix multiplication. Show that

$$\begin{pmatrix} 1 & 1 \\ 1 & 0 \end{pmatrix}^n = \begin{pmatrix} F_{n+1} & F_n \\ F_n & F_{n-1} \end{pmatrix} \qquad \text{for } n \geq 2$$

10.4 ▼ ARITHMETIC SEQUENCES AND SERIES

One of the most natural ways to generate a sequence is to begin with a fixed number a and then repeatedly add a fixed constant d. This yields the sequence

$$a, \; a + d, \; a + 2d, \; a + 3d, \ldots$$

Such a sequence is called an **arithmetic sequence** or **arithmetic progression**. Notice that the difference between any two consecutive terms is the constant d. We call d the **common difference**. Here are several examples of arithmetic sequences:

example A: $1, 2, 3, \ldots$
example B: $3, 7, 11, 15, \ldots$
example C: $10, 5, 0, -5, \ldots$

In example A, the first term is $a = 1$ and the common difference is $d = 1$. For example B, we have $a = 3$. The value of d in this example is found by subtracting any two consecutive terms; thus, $d = 4$. Finally, in example C, we have $a = 10$ and $d = -5$. Notice that when the common difference is negative, the terms of the sequence decrease.

There is a simple formula for the nth term in an arithmetic sequence

$$a, \ a + d, \ a + 2d, \ a + 3d, \ldots$$

In this sequence, notice that

$$a_1 = a + 0d$$
$$a_2 = a + 1d$$
$$a_3 = a + 2d$$
$$a_4 = a + 3d$$

Following this pattern, it appears that the formula for a_n should be

$$a_n = a + (n - 1)d$$

Indeed, this is the correct formula, and Exercise 32 asks you to verify it using mathematical induction.

---

nth Term of an Arithmetic Sequence

The nth term of an arithmetic sequence $a, \ a + d, \ a + 2d, \ldots$ is given by

$$a_n = a + (n - 1)d$$

EXAMPLE 1   Determine the one hundredth term of the arithmetic sequence

$$7, \ 10, \ 13, \ 16, \ldots$$

Solution   The first term is $a = 7$, and the common difference is $d = 3$. Substituting these values in the formula $a_n = a + (n - 1)d$ yields

$$a_n = 7 + (n - 1)3 = 3n + 4$$

To find the one hundredth term, we replace n by 100 in this last equation to obtain

$$a_{100} = 3(100) + 4 = 304 \qquad \text{as required} \qquad \blacktriangle$$

EXAMPLE 2   Determine the arithmetic sequence in which the second term is -2 and the eighth term is 40.

Solution   We are given that the second term is -2. Using this information in the formula $a_n = a + (n - 1)d$, we have

$$-2 = a + (2 - 1)d = a + d$$

This gives us one equation in two unknowns. We are also given that the eighth term is 40. Therefore,

$$40 = a + (8 - 1)d = a + 7d$$

We now have a system of two equations in two unknowns:

$$\begin{cases} -2 = a + d \\ 40 = a + 7d \end{cases}$$

Subtracting the first equation from the second gives us

$$42 = 6d \qquad \text{or} \qquad 7 = d$$

To find a, we replace d by 7 in the first equation of the system. This yields

$$-2 = a + 7 \qquad \text{or} \qquad -9 = a$$

We have now determined the sequence, since we know that the first term is -9 and the common difference is 7. The first four terms of the sequence are -9, -2, 5, and 12.   ▲

Next we would like to derive a formula for the sum of the first n terms of an arithmetic sequence. Such a sum is referred to as an **arithmetic series**. If we use S_n to denote the required sum, we have

$$S_n = a + (a + d) + \cdots + [a + (n - 2)d] + [a + (n - 1)d] \qquad (1)$$

Of course, we must obtain the same sum if we add the terms from right to left rather than left to right. That is, we must have

$$S_n = [a + (n - 1)d] + [a + (n - 2)d] + \cdots + (a + d) + a \qquad (2)$$

Let us now add equations (1) and (2). Adding the left-hand sides is easy; we obtain $2S_n$. Now we add the corresponding terms on the right-hand sides. For the first terms, we have

$$\underset{\substack{\text{first term} \\ \text{in equation (1)}}}{a} \quad + \quad \underset{\substack{\text{first term} \\ \text{in equation (2)}}}{\underbrace{[a + (n - 1)d]}} = 2a + (n - 1)d$$

Next we add the second terms:

$$\underset{\substack{\text{second term} \\ \text{in equation (1)}}}{(a + d)} \quad + \quad \underset{\substack{\text{second term} \\ \text{in equation (2)}}}{\underbrace{[a + (n - 2)d]}} = 2a + d + (n - 2)d = 2a + (n - 1)d$$

Notice that the sum of the second terms is again $2a + (n - 1)d$, the same quantity we arrived at with the first terms. As you can check, this pattern continues all the way through to the last terms. For instance,

$$\underbrace{[a + (n-1)d]}_{\substack{\text{last term} \\ \text{in equation (1)}}} + \underset{\substack{\uparrow \\ \text{last term} \\ \text{in equation (2)}}}{a} = 2a + (n-1)d$$

We conclude from these observations that by adding the right-hand sides of equations (1) and (2), the quantity $2a + (n-1)d$ is added a total of n times. Therefore,

$$2S_n = n[2a + (n-1)d]$$
$$S_n = \frac{n}{2}[2a + (n-1)d]$$

This gives us the desired formula for the sum of the first n terms in an arithmetic sequence. There is an alternate form of this formula, which now follows rather quickly:

$$S_n = \frac{n}{2}[2a + (n-1)d]$$

$$= \frac{n}{2}\{a + [a + (n-1)d]\}$$

$$= \frac{n}{2}(a + a_n) = n\left(\frac{a + a_n}{2}\right)$$

This last equation is easy to remember. It says that the sum of an arithmetic series is obtained by averaging the first and last terms and then multiplying this average by n, the number of terms. For reference, we summarize both formulas as follows.

Formulas for the Sum of an Arithmetic Series

$$S_n = \frac{n}{2}[2a + (n-1)d]$$

$$S_n = n\left(\frac{a + a_n}{2}\right)$$

EXAMPLE 3　Find the sum of the first 30 terms of the arithmetic sequence 2, 6, 10, 14, ...

Solution　We have $a = 2$, $d = 4$, and $n = 30$. Substituting these values in the formula $S_n = (n/2)[2a + (n-1)d]$ then yields

$$S_{30} = \frac{30}{2}[2(2) + (30-1)4]$$

$$= 15[4 + 29(4)]$$
$$= 1800 \quad \text{(Check the arithmetic)} \qquad \blacktriangle$$

EXAMPLE 4　In a certain arithmetic sequence, the first term is 6 and the fortieth term is 71. Find the sum of the first 40 terms and also the common difference for the sequence.

Solution We have $a = 6$ and $a_{40} = 71$. Using these values in the formula $S_n = n(a + a_n)/2$ yields

$$S_{40} = \frac{40}{2}(6 + 71) = 20(77) = 1540$$

The sum of the first 40 terms is thus 1540. The value of d can now be found by using the formula $S_n = (n/2)[2a + (n - 1)d]$. We obtain

$$1540 = \frac{40}{2}[2(6) + (40 - 1)d]$$

$$= 20(12 + 39d)$$

$$= 240 + 780d$$

$$1300 = 780d$$

$$d = \frac{1300}{780} = \frac{5}{3}$$

The required value of d is therefore $\frac{5}{3}$. ▲

EXAMPLE 5 Show that $\sum_{k=1}^{50} (3k - 2)$ represents an arithmetic series, and compute the sum.

Solution There are two different ways to see that $\sum_{k=1}^{50} (3k - 2)$ is an arithmetic series. One way is simply to write out the first few terms and look at the pattern. We then have

$$\sum_{k=1}^{50} (3k - 2) = 1 + 4 + 7 + 10 + \cdots + 148$$

From this it is clear that we are indeed summing the terms in an arithmetic sequence in which $d = 3$ and $a = 1$.

Another, more formal way to show that $\sum_{k=1}^{50} (3k - 2)$ represents an arithmetic series is to prove that the difference between successive terms in the indicated sum is a constant. Now, the form of a typical term in this sum is $(3k - 2)$. Thus, the form of the next term must be $[3(k + 1) - 2]$. The difference between these terms is then

$$[3(k + 1) - 2] - (3k - 2) = 3k + 3 - 2 - 3k + 2 = 3$$

The difference therefore is constant, as we wished to show.

To evaluate $\sum_{k=1}^{50} (3k - 2)$, we can use either of the two formulas for the sum of an arithmetic series. Using the formulas $S_n = (n/2)[2a + (n - 1)d]$, we obtain

$$S_{50} = \frac{50}{2}[2(1) + 49(3)] = 25(149) = 3725$$

Thus, the required sum is 3725. You should check for yourself that the same value is obtained using the formula $S_n = n(a + a_n)/2$. ▲

▼ EXERCISE SET 10.4

A

1. Find the common difference d for each of the following arithmetic sequences.
   (a) $1, 3, 5, 7, \ldots$
   (b) $10, 6, 2, -2, \ldots$
   (c) $\frac{2}{3}, 1, \frac{4}{3}, \frac{5}{3}, \ldots$
   (d) $1, 1 + \sqrt{2}, 1 + 2\sqrt{2}, 1 + 3\sqrt{2}, \ldots$

2. Which of the following are arithmetic sequences?
   (a) $2, 4, 8, 16, \ldots$
   (b) $5, 9, 13, 17, \ldots$
   (c) $3, \frac{11}{5}, \frac{7}{5}, \frac{3}{5}, \ldots$
   (d) $-1, -1, -1, -1, \ldots$
   (e) $-1, 1, -1, 1, \ldots$

In Exercises 3–8, find the indicated term in each sequence.

3. $10, 21, 32, 43, \ldots$; a_{12}
4. $7, 2, -3, -8, \ldots$; a_{20}
5. $6, 11, 16, 21, \ldots$; a_{100}
6. $\frac{2}{5}, \frac{4}{5}, \frac{6}{5}, \frac{8}{5}, \ldots$; a_{30}
7. $-1, 0, 1, 2, \ldots$; a_{1000}
8. $42, 1, -40, -81, \ldots$; a_{15}

9. The fourth term in an arithmetic sequence is -6, and the tenth term is 5. Find the common difference and the first term.

10. The fifth term in an arithmetic sequence is $\frac{1}{2}$, and the twentieth term is $\frac{7}{8}$. Find the first three terms of the sequence.

11. The sixtieth term in an arithmetic sequence is 105, and the common difference is 5. Find the first term.

12. Find the common difference in an arithmetic sequence in which $a_{10} - a_{20} = 70$.

13. Find the common difference in an arithmetic sequence in which $a_{15} - a_7 = -1$.

14. Find the sum of the first 16 terms in the sequence 2, 11, 20, 29, ...

15. Find the sum of the first 1000 terms in the sequence 1, 2, 3, 4, ...

16. Find the sum of the first 50 terms in an arithmetic series whose first term is -8 and whose fiftieth term is 139.

17. Find the sum: $\dfrac{\pi}{3} + \dfrac{2\pi}{3} + \pi + \dfrac{4\pi}{3} + \cdots + \dfrac{13\pi}{3}$.

18. Find the sum: $\dfrac{1}{e} + \dfrac{3}{e} + \dfrac{5}{e} + \cdots + \dfrac{21}{e}$.

19. Determine the first term of an arithmetic sequence in which the common difference is 5 and the sum of the first 38 terms is 3534.

20. The sum of the first 12 terms in an arithmetic sequence is 156. What is the sum of the first and twelfth terms?

21. In a certain arithmetic sequence, the first term is 4 and the sixteenth term is -100. Find the sum of the first 16 terms and also the common difference for the sequence.

22. The fifth and fiftieth terms of an arithmetic sequence are 3 and 30, respectively. Find the sum of the first ten terms.

23. The eighth term in an arithmetic sequence is 5, and the sum of the first ten terms is 20. Find the common difference and the first term of the sequence.

In Exercises 24–26, find each sum.

24. $\displaystyle\sum_{i=1}^{10} (2i - 1) = 1 + 3 + 5 + \cdots + 19$

25. $\displaystyle\sum_{k=1}^{20} (4k + 3)$

26. $\displaystyle\sum_{n=5}^{100} (2n - 1)$

27. The sum of three consecutive terms in an arithmetic sequence is 30, and their product is 360. Find the three terms.   *Suggestion:* Let x denote the *middle* term and d the common difference.

28. The sum of three consecutive terms in an arithmetic sequence is 21, and the sum of their squares is 197. Find the three terms.

29. The sum of three consecutive terms in an arithmetic sequence is 6, and the sum of their cubes is 132. Find the three terms.

30. In a certain arithmetic sequence, $a = -4$ and $d = 6$. If $S_n = 570$, find n.

31. Let $a_1 = \dfrac{1}{1 + \sqrt{2}}$, $a_2 = -1$, and $a_3 = \dfrac{1}{1 - \sqrt{2}}$.
    (a) Show that $a_2 - a_1 = a_3 - a_2$.
    (b) Find the sum of the first six terms in the arithmetic sequence
    $$\frac{1}{1 + \sqrt{2}}, \quad -1, \quad \frac{1}{1 - \sqrt{2}}, \ldots$$

32. Using mathematical induction, prove that the nth term of the sequence $a, a + d, a + 2d, \ldots$ is given by
    $$a_n = a + (n - 1)d$$

B

33. Let b denote a positive constant. Find the sum of the first n terms in the sequence
    $$\frac{1}{1 + \sqrt{b}}, \quad \frac{1}{1 - b}, \quad \frac{1}{1 - \sqrt{b}}, \ldots$$

34. The sum of the first n terms in a certain arithmetic sequence is given by $S_n = 3n^2 - n$. Show that the rth term is given by $a_r = 6r - 4$.

35. Let $a_1, a_2, a_3, \ldots$ be an arithmetic sequence, and let S_k denote the sum of the first k terms. If $S_n/S_m = n^2/m^2$,

show that

$$\frac{a_n}{a_m} = \frac{2n-1}{2m-1}$$

36. If the common difference in an arithmetic sequence is twice the first term, show that

$$\frac{S_n}{S_m} = \frac{n^2}{m^2}$$

37. The lengths of the sides of a right triangle form three consecutive terms in an arithmetic sequence. Show that the triangle is similar to the 3-4-5 right triangle.

38. Suppose that $1/a$, $1/b$, and $1/c$ are three consecutive terms in an arithmetic sequence. Show that:

(a) $\dfrac{a}{c} = \dfrac{a-b}{b-c}$ **(b)** $b = \dfrac{2ac}{a+c}$

39. Suppose that a, b, and c are three positive numbers with $a > c > 2b$. If $1/a$, $1/b$, and $1/c$ are consecutive terms in an arithmetic sequence, show that

$$\ln(a+c) + \ln(a-2b+c) = 2\ln(a-c)$$

10.5 ▼ GEOMETRIC SEQUENCES AND SERIES

A **geometric sequence** or **geometric progression**, is a sequence of the form

$$a, ar, ar^2, ar^3, \ldots \qquad \text{where } a \text{ and } r \text{ are constants}$$

As you can see, each term after the first in a geometric sequence is obtained by multiplying the previous term by r. The number r is called the **common ratio**, since the ratio of any term to the previous one is always r. For instance, the ratio of the fourth term to the third is $ar^3/ar^2 = r$. Here are two examples of geometric sequences:

$$1, \frac{1}{2}, \frac{1}{4}, \frac{1}{8}, \ldots$$

$$10, -100, 1000, -10000, \ldots$$

In the first example, we have $a = 1$ and $r = \frac{1}{2}$; in the second example, we have $a = 10$ and $r = -10$.

EXAMPLE 1 In a certain geometric sequence, the first term is 2, the third term is 3, and the common ratio is negative. Find the second term.

Solution Let x denote the second term, so that the sequence begins

$$2, x, 3, \ldots$$

By definition, the ratios $3/x$ and $x/2$ must be equal. Thus, we have

$$\frac{3}{x} = \frac{x}{2}$$
$$x^2 = 6$$
$$x = \pm\sqrt{6}$$

Now, the second term must be negative, since the first term is positive and the common ratio is negative. Thus, the second term is $x = -\sqrt{6}$. ▲

The formula for the nth term of a geometric sequence is easily deduced by considering Table 1. The table indicates that the exponent on r is one less than the value of n in each case. On the basis of this observation, it appears that the nth term must be given by $a_n = ar^{n-1}$. Indeed, it can be shown by mathematical induction that this formula does hold for all natural numbers n. (Exercise 30 asks you to carry out the proof.) We summarize this result as follows.

TABLE 1

n	a_n
1	ar^0
2	ar^1
3	ar^2
4	ar^3
⋮	⋮

nth Term of a Geometric Sequence

> The nth term of the geometric sequence a, ar, ar^2, ... is given by
>
> $$a_n = ar^{n-1}$$

EXAMPLE 2 Find the seventh term in the geometric sequence 2, 6, 18, ...

Solution We can find the common ratio r by dividing the second term by the first. Thus, $r = 3$. Now, using $a = 2$, $r = 3$, and $n = 7$ in the formula $a_n = ar^{n-1}$, we have

$$a_7 = 2(3)^6 = 2(729) = 1458$$

The seventh term of the sequence is therefore 1458. ▲

Suppose that we begin with a geometric sequence a, ar, ar^2, ..., in which $r \neq 1$. If we add the first n terms and denote the sum by S_n, we have

$$S_n = a + ar + ar^2 + \cdots + ar^{n-2} + ar^{n-1} \tag{1}$$

This sum is called a **finite geometric series**. We would like to find a formula for S_n. To do this, we multiply equation (1) by r to obtain

$$rS_n = ar + ar^2 + ar^3 + \cdots + ar^{n-1} + ar^n \tag{2}$$

We now subtract equation (2) from equation (1). This yields (after combining like terms)

$$S_n - rS_n = a - ar^n$$
$$S_n(1 - r) = a(1 - r^n)$$
$$S_n = \frac{a(1 - r^n)}{1 - r} \qquad (r \neq 1)$$

This is the formula for the sum of a finite geometric series. We summarize this result in the box that follows.

Formula for the Sum of a Geometric Series

> Let S_n denote the sum $a + ar + ar^2 + \cdots + ar^{n-1}$, and assume that $r \neq 1$. Then
>
> $$S_n = \frac{a(1 - r^n)}{1 - r}$$

EXAMPLE 3 Evaluate the sum $\dfrac{1}{2^1} + \dfrac{1}{2^2} + \dfrac{1}{2^3} + \cdots + \dfrac{1}{2^{10}}$.

Solution This is a finite geometric series with $a = \frac{1}{2}$, $r = \frac{1}{2}$, and $n = 10$. Using these values in the formula for S_n yields

$$S_{10} = \frac{\frac{1}{2}[1 - (\frac{1}{2})^{10}]}{1 - \frac{1}{2}}$$

$$= \frac{\frac{1}{2}[1 - \frac{1}{1024}]}{\frac{1}{2}} = 1 - \frac{1}{1024}$$

$$= \frac{1023}{1024} \qquad \text{as required}$$

▲

We would now like to attach a meaning to certain expressions of the form

$$a + ar + ar^2 + \cdots$$

Such an expression is called an **infinite geometric series**. The three dots indicate (intuitively at least) that the additions are to be carried out indefinitely, without end. To see how to proceed here, let us look at some examples involving finite geometric series. In particular, we will consider the series

$$\frac{1}{2^1} + \frac{1}{2^2} + \frac{1}{2^3} + \cdots + \frac{1}{2^n}$$

for increasing values of n. The idea is to look for a pattern as n grows ever larger. Let $S_1 = \frac{1}{2}$, $S_2 = \frac{1}{2^1} + \frac{1}{2^2}$, $S_3 = \frac{1}{2^1} + \frac{1}{2^2} + \frac{1}{2^3}$, and, in general,

$$S_n = \frac{1}{2^1} + \frac{1}{2^2} + \cdots + \frac{1}{2^n}$$

Then we can compute S_n for any given value of n by means of the formula for the sum of a finite geometric series. From Table 2, which displays the results of such calculations, it seems clear that as n grows larger and larger, the value of S_n grows ever closer to 1. More precisely (but leaving the details for calculus), it can be shown that the value of S_n can be made as close to 1 as we please, provided only that n is sufficiently large. For this reason, we say that the *sum of the infinite geometric series* $\frac{1}{2^1} + \frac{1}{2^2} + \frac{1}{2^3} + \cdots$ is 1. That is,

$$\frac{1}{2^1} + \frac{1}{2^2} + \frac{1}{2^3} + \cdots = 1$$

We can arrive at this last result another way. First we compute the sum of the finite geometric series $\frac{1}{2^1} + \frac{1}{2^2} + \cdots + \frac{1}{2^n}$. As you can check, the result is

$$S_n = 1 - \left(\frac{1}{2}\right)^n$$

Now, as n grows larger and larger, the value of $\left(\frac{1}{2}\right)^n$ gets closer and closer to zero. Thus, as n grows ever larger, the value of S_n will more and more resemble $1 - 0$, or 1.

Now let us repeat our reasoning to obtain a formula for the sum of the infinite geometric series

$$a + ar + ar^2 + \cdots \qquad \text{where } |r| < 1$$

First we consider the finite geometric series

$$a + ar + ar^2 + \cdots + ar^{n-1}$$

The sum S_n in this case is

$$S_n = \frac{a(1 - r^n)}{1 - r}$$

We want to know how S_n behaves as n grows ever larger. This is where the assumption $|r| < 1$ is crucial. Just as $\left(\frac{1}{2}\right)^n$ approaches zero as n grows larger and

TABLE 2

n	$S_n = \frac{1}{2} + \frac{1}{2^2} + \cdots + \frac{1}{2^n}$
1	0.5
2	0.75
5	0.96875
10	0.999023437 ...
15	0.999969482 ...
20	0.999999046 ...
25	0.999999970 ...

larger, so will r^n approach zero as n grows larger and larger. Thus, as n grows ever larger, the sum S_n will more and more resemble

$$\frac{a(1-0)}{1-r} = \frac{a}{1-r}$$

For this reason, we say that the sum of the infinite geometric series is $\dfrac{a}{1-r}$.

We will make free use of this result in the subsequent examples. However, a more rigorous development of infinite series properly belongs to calculus.

Formula for the Sum of an Infinite Geometric Series

> Suppose that $|r| < 1$. Then the sum S of the infinite geometric series $a + ar + ar^2 + \cdots$ is given by
>
> $$S = \frac{a}{1-r}$$

EXAMPLE 4   Find the sum of the infinite geometric series $1 + \frac{2}{3} + \frac{4}{9} + \cdots$.

Solution   In this case, we have $a = 1$ and $r = \frac{2}{3}$. Thus,

$$S = \frac{a}{1-r} = \frac{1}{1-\frac{2}{3}} = \frac{1}{\frac{1}{3}} = 3$$

The sum of the series is 3.   ▲

EXAMPLE 5   Find a fraction equivalent to the repeating decimal $0.2\overline{35}$.

Solution   Let $S = 0.2\overline{35}$. Then we have

$$S = 0.2353535\ldots$$

$$= \frac{2}{10} + \frac{35}{1000} + \frac{35}{100,000} + \frac{35}{10,000,000} + \cdots$$

Now, the expression following $\frac{2}{10}$ on the right-hand side of this last equation is an infinite geometric series in which $a = \frac{35}{1000}$ and $r = \frac{1}{100}$. Thus,

$$S = \frac{2}{10} + \frac{a}{1-r}$$

$$= \frac{2}{10} + \frac{\frac{35}{1000}}{1-\frac{1}{100}} = \frac{233}{990} \qquad \text{(Check the arithmetic!)}$$

The given decimal is therefore equivalent to 233/990.   ▲

▼ **EXERCISE SET 10.5**

A

1. Find the second term in a geometric sequence in which the first term is 9, the third term is 4, and the common ratio is positive.

2. Find the fifth term in a geometric sequence in which the fourth term is 4, the sixth term is 6, and the common ratio is negative.

3. The product of the first three terms in a geometric sequence is 8000. If the first term is 4, find the second and third terms.

In Exercises 4–8, find the indicated term for the geometric sequence given.

4. $9, 81, 729, \ldots; a_7$

5. $-1, 1, -1, 1, \ldots; a_{100}$

6. $\frac{1}{2}, \frac{1}{4}, \frac{1}{8}, \ldots; a_9$ **7.** $\frac{2}{3}, \frac{4}{9}, \frac{8}{27}, \ldots; a_8$

8. $1, -\sqrt{2}, 2, \ldots; a_6$

9. Find the common ratio in a geometric sequence in which the first term is 1 and the seventh term is 4096.

10. Find the first term in a geometric sequence in which the common ratio is $\frac{4}{3}$ and the tenth term is $\frac{16}{9}$.

11. Find the sum of the first ten terms of the sequence 7, 14, 28, ...

12. Find the sum of the first five terms of the sequence $-\frac{1}{2}, \frac{3}{10}, -\frac{9}{50}, \ldots$.

13. Find the sum: $1 + \sqrt{2} + 2 + \cdots + 32$.

14. Find the sum of the first 12 terms in the sequence $-4, -2, -1, \ldots$.

In Exercises 15–17, evaluate each sum.

15. $\displaystyle\sum_{k=1}^{6} \left(\frac{3}{2}\right)^k$ **16.** $\displaystyle\sum_{k=1}^{6} \left(\frac{2}{3}\right)^{k+1}$ **17.** $\displaystyle\sum_{k=2}^{6} \left(\frac{1}{10}\right)^k$

In Exercises 18–22, determine the sum of each infinite geometric series.

18. $\dfrac{1}{4} + \dfrac{1}{4^2} + \dfrac{1}{4^3} + \cdots$ **19.** $\dfrac{2}{3} - \dfrac{4}{9} + \dfrac{8}{27} - \cdots$

20. $\dfrac{9}{10} + \dfrac{9}{100} + \dfrac{9}{1000} + \cdots$

21. $1 + \dfrac{1}{1.01} + \dfrac{1}{(1.01)^2} + \cdots$ **22.** $-1 - \dfrac{1}{\sqrt{2}} - \dfrac{1}{2} - \cdots$

In Exercises 23–27, express each repeating decimal as a fraction.

23. $0.555\ldots$ **24.** $0.\overline{47}$ **25.** $0.1\overline{23}$

26. $0.050505\ldots$ **27.** $0.\overline{432}$

B

28. The lengths of the sides in a right triangle form three consecutive terms of a geometric sequence. Find the

common ratio of the sequence. (There are two distinct answers.)

29. The product of three consecutive terms in a geometric sequence is -1000, and their sum is 15. Find the common ratio. (There are two answers.)
    Suggestion: Denote the terms by a/r, a, and ar.

30. Use mathematical induction to prove that the nth term of the geometric sequence $a, ar, ar^2, \ldots$ is ar^{n-1}.

31. Show that the sum of the following infinite geometric series is $\frac{3}{2}$:
$$\frac{\sqrt{3}}{\sqrt{3} + 1} + \frac{\sqrt{3}}{\sqrt{3} + 3} + \cdots$$

32. Let A_1 denote the area of an equilateral triangle, each side of which is 1 unit long. A second equilateral triangle is formed by joining the midpoints of the sides of the first triangle. Let A_2 denote the area of this second triangle. This process is then repeated to form a third triangle with area A_3; and so on. Find the sum of the areas: $A_1 + A_2 + A_3 + \cdots$.

33. Let $a_1, a_2, a_3, \ldots$ be a geometric sequence such that $r \neq 1$. Let $S = a_1 + a_2 + a_3 + \cdots + a_n$, and let $T = \dfrac{1}{a_1} + \dfrac{1}{a_2} + \cdots + \dfrac{1}{a_n}$. Show that $\dfrac{S}{T} = a_1 a_n$.

34. Suppose that a, b, and c are three consecutive terms in a geometric sequence. Show that $\dfrac{1}{a+b}, \dfrac{1}{2b}$, and $\dfrac{1}{c+b}$ are three consecutive terms in an arithmetic sequence.

35. A ball is dropped from a height of 6 ft. Assuming that on each bounce the ball rebounds to one-third of its previous height, find the total distance traveled by the ball.

10.6 ▼ VECTORS IN THE PLANE, A GEOMETRIC APPROACH

The idea of a parallelogram of velocities may be found in various ancient Greek authors, and the concept of a parallelogram of forces was not uncommon in the sixteenth and seventeenth centuries. By the early nineteenth century parallelograms of physical entities frequently appeared in treatises, and this usage indirectly led to vector analysis, for this area provided a striking example of how vectorial entities could be used for physical applications.

Michael J. Crowe in *A History of Vector Analysis* (Notre Dame, Ind.: University of Notre Dame Press, 1967)

Certain quantities, such as temperature, length, and mass, can be specified by means of a single number (assuming that a system of units has been agreed on). We call these quantities **scalars**. On the other hand, quantities such as force and velocity are characterized by both a *magnitude* (a number) and a *direction*. We call these quantities **vectors**.

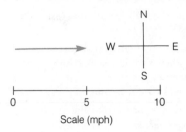

FIGURE 1

A vector representing a wind velocity of 5 mph from the west.

Geometrically, a vector is a directed line segment or arrow. The vector in Figure 1, for instance, represents a wind velocity of 5 mph from the west. The length of this vector represents the magnitude of the wind velocity, while the direction of the vector indicates the direction of the wind velocity. As another example, the vector in Figure 2 represents a force acting on an object: the magnitude of the force is 3 pounds, and the force acts at an angle of 135° with the horizontal.

In a moment, we are going to discuss the important concept of vector addition, but first let us agree on some matters of notation. Suppose that we have a vector drawn from a point P to point Q, as shown in Figure 3.

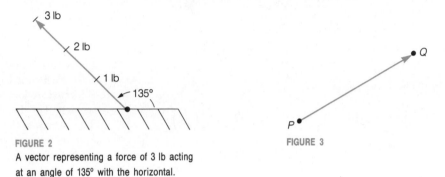

FIGURE 2

A vector representing a force of 3 lb acting at an angle of 135° with the horizontal.

FIGURE 3

The point P in Figure 3 is called the **initial point** of the vector, and Q is the **terminal point**. We can denote this vector by

$$\overrightarrow{PQ}$$

The length of the vector $\overrightarrow{PQ}$ is denoted by $|\overrightarrow{PQ}|$. On the printed page, vectors are often indicated by boldface letters, such as **a**, **A**, and **v**.

A word about notation. In Chapter 1 you saw that the notation (a, b) can denote either a point in the *x-y* plane or an open interval on the number line. In each instance, however, there was no danger of confusion; the context made it clear which meaning was intended. Now we have a similar situation with the notation for the length of a vector. As has perhaps already occurred to you, the same vertical bars that we are using to denote the length of a vector are also used, in another context, to denote the absolute value of a real number. Again, it will be clear from the context which meaning is intended. Some books avoid this situation by using double bars to indicate the length of a vector: $\|\mathbf{v}\|$. In this text, we use the notation $|\mathbf{v}|$ simply because that is the one found in most calculus books.

If two vectors **a** and **b** have the same length and the same direction, we say that they are *equal* and we write **a** = **b** (see Figure 4). Notice that our definition

FIGURE 4

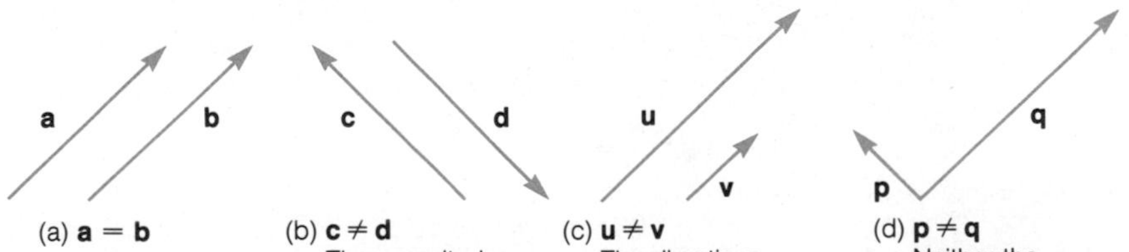

(a) **a** = **b**

(b) **c** ≠ **d**
The magnitudes are the same, but the directions are not.

(c) **u** ≠ **v**
The directions are the same, but the magnitudes are not.

(d) **p** ≠ **q**
Neither the magnitudes nor the directions are the same.

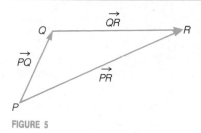

FIGURE 5

for vector equality involves magnitude and direction, but not location. Thus, when it is convenient to do so, we are free to move a given vector to another location, provided we do not alter the magnitude or direction.

As motivation for the definition of vector addition, let us suppose that an object moves from a point P to a point Q. Then we can represent this *displacement* by the vector $\overrightarrow{PQ}$. (Indeed, the word *vector* is derived from the Latin *vectus*, meaning "carried.") Now suppose that after moving from P to Q, the object moves from Q to R. Then, as you can see in Figure 5, the net effect is a displacement from P to R. We say in this case that the vector $\overrightarrow{PR}$ is the **sum** or **resultant** of the vectors $\overrightarrow{PQ}$ and $\overrightarrow{QR}$, and we write

$$\overrightarrow{PQ} + \overrightarrow{QR} = \overrightarrow{PR}$$

These ideas are formalized in the definition that follows.

DEFINITION

Vector Addition

Let **u** and **v** be two vectors. Position **v** (without changing its magnitude or direction) so that its initial point coincides with the terminal point of **u**, as in Figure 6(a). Then, as indicated in Figure 6(b), the vector **u** + **v** is the directed line segment from the initial point of **u** to the terminal point of **v**. The vector **u** + **v** is called the **sum** or **resultant** of **u** and **v**.

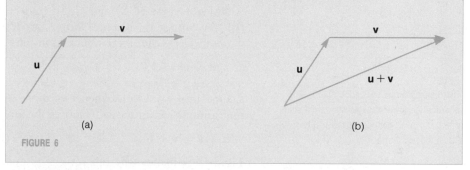

(a)   (b)

FIGURE 6

EXAMPLE 1   Refer to Figure 7.

(a) Determine the initial and terminal points of **u** + **v**.
(b) Compute $|\mathbf{u} + \mathbf{v}|$.

Solution   (a) According to the definition, we first need to move **v** (without changing its length or direction) so that its initial point coincides with the terminal point

FIGURE 7

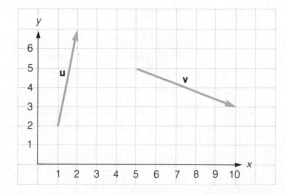

FIGURE 8

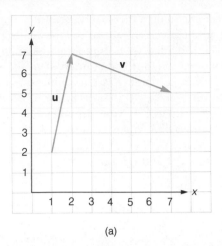
(a)

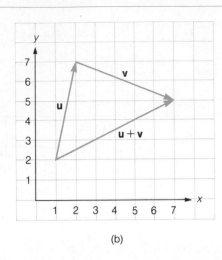
(b)

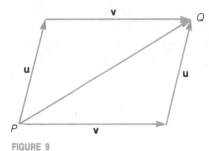

FIGURE 9

The upper triangle shows the sum $\mathbf{u} + \mathbf{v}$, whereas the lower triangle shows the sum $\mathbf{v} + \mathbf{u}$. Since in both cases the resultant is $\overrightarrow{PQ}$, it follows that $\mathbf{u} + \mathbf{v} = \mathbf{v} + \mathbf{u}$.

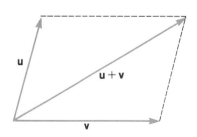

FIGURE 10

The parallelogram law for vector addition.

of $\mathbf{u}$. From Figure 7, we see that this can be accomplished by moving each point of $\mathbf{v}$ three units in the negative x-direction and two units in the positive y-direction. Figure 8(a) shows the new location of $\mathbf{v}$, and Figure 8(b) indicates the sum $\mathbf{u} + \mathbf{v}$. From Figure 8(b), we see that the initial and terminal points of $\mathbf{u} + \mathbf{v}$ are $(1, 2)$ and $(7, 5)$, respectively.

(b) We can use the distance formula to determine $|\mathbf{u} + \mathbf{v}|$. Using the points $(1, 2)$ and $(7, 5)$ that were obtained in part (a), we have

$$|\mathbf{u} + \mathbf{v}| = \sqrt{(7-1)^2 + (5-2)^2} = \sqrt{45} = \sqrt{9 \cdot 5} = 3\sqrt{5} \qquad \blacktriangle$$

One important consequence of our definition for vector addition is that this operation is commutative. That is, for any two vectors $\mathbf{u}$ and $\mathbf{v}$, we have

$$\mathbf{u} + \mathbf{v} = \mathbf{v} + \mathbf{u}$$

Figure 9 indicates why this is so.

In view of Figure 9, vector addition can also be carried out by using the **parallelogram law**: To determine $\mathbf{u} + \mathbf{v}$, position $\mathbf{u}$ and $\mathbf{v}$ so that their initial points coincide. Then, as indicated in Figure 10, the vector $\mathbf{u} + \mathbf{v}$ is the directed diagonal of the parallelogram determined by $\mathbf{u}$ and $\mathbf{v}$.

It is a fact—and it has been verified experimentally—that if two forces $\mathbf{F}$ and $\mathbf{G}$ act on an object, the net effect is the same as if just the resultant $\mathbf{F} + \mathbf{G}$ acted on the object. In Example 2, we use the parallelogram law to compute the resultant of two forces. Note that the units of force used in this example are **newtons** (N) where $1 \, \text{N} \approx 0.2248 \, \text{lb}$.

EXAMPLE 2 Two forces $\mathbf{F}$ and $\mathbf{G}$ act on an object. As indicated in Figure 11, the force $\mathbf{G}$ acts horizontally to the right with a magnitude of 12 N, while $\mathbf{F}$ acts vertically upward with a magnitude of 16 N. Determine the magnitude and direction of the resultant force.

Solution We complete the parallelogram, as shown in Figure 12. Now we need to calculate the length of $\mathbf{F} + \mathbf{G}$ and the angle θ. Applying the Pythagorean theorem in Figure 12, we have

$$|\mathbf{F} + \mathbf{G}| = \sqrt{12^2 + 16^2} = \sqrt{144 + 256} = \sqrt{400} = 20$$

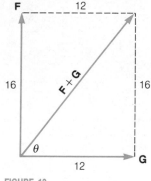

FIGURE 11

FIGURE 12

Also from Figure 12, we have

$$\tan \theta = \frac{16}{12} = \frac{4}{3}$$

Consequently,

$$\theta = \tan^{-1} \frac{4}{3} \approx 53.1°$$

Now we can summarize our results. The magnitude of $\mathbf{F} + \mathbf{G}$ is 20 N, and the angle θ between $\mathbf{F} + \mathbf{G}$ and the horizontal is (approximately) 53.1°. ▲

In Example 2, we determined the resultant for two perpendicular forces. The next example shows how to compute the resultant when the forces are not perpendicular. Our calculations will make use of both the law of sines and the law of cosines.

EXAMPLE 3 Determine the resultant of the two forces in Figure 13. (Round off the answers to one decimal place.)

Solution As in the previous example, we complete the parallelogram. In Figure 14, the angle in the lower right-hand corner of the parallelogram is 140°. This is because the sum of two adjacent angles in any parallelogram is always 180°. Letting d denote the length of the diagonal in Figure 14, we can use the law of cosines to write

$$d^2 = 15^2 + 5^2 - 2(15)(5) \cos 140°$$

$$d = \sqrt{250 - 150 \cos 140°} = \sqrt{250 + 150 \cos 40°} \qquad \text{(Why?)}$$

$$= \sqrt{25(10 + 6 \cos 40°)} = 5\sqrt{10 + 6 \cos 40°}$$

$$\approx 19.1 \quad \text{using a calculator}$$

So the magnitude of the resultant is 19.1 N (to one decimal place). To specify the direction of the resultant, we need to determine the angle θ in Figure 14.

FIGURE 13

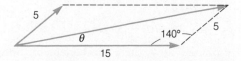

FIGURE 14

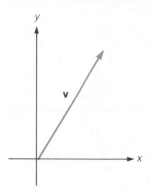

FIGURE 15

Using the law of sines, we have

$$\frac{\sin \theta}{5} = \frac{\sin 140°}{d}$$

and, consequently,

$$\sin \theta = \frac{5 \sin 40°}{d} \qquad \text{(Why?)}$$

$$= \frac{5 \sin 40°}{5\sqrt{10 + 6 \cos 40°}} = \frac{\sin 40°}{\sqrt{10 + 6 \cos 40°}}$$

Using a calculator now, we obtain

$$\theta = \sin^{-1}\left(\frac{\sin 40°}{\sqrt{10 + 6 \cos 40°}}\right) \approx 9.7°$$

In summary, the magnitude of the resultant force is about 19.1 N, and the angle θ between the resultant and the 15-N force is approximately 9.7°. ▲

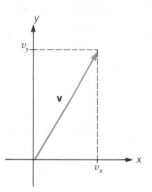

FIGURE 16

As background for the next example, we introduce the notion of *components* of a vector. (You will see this concept again in the next section, but in a more algebraic context.) Suppose that the initial point of a vector **v** is located at the origin of a rectangular coordinate system, as shown in Figure 15. Now suppose we draw perpendiculars from the terminal point of **v** to the axes, as indicated in Figure 16. Then the coordinates v_x and v_y in Figure 16 are called the **components** of the vector **v** in the *x*- and *y*-directions, respectively. For an example involving components, refer back to Figure 12. The horizontal component of the vector **F** + **G** is 12 N, and the vertical component is 16 N.

EXAMPLE 4 Determine the horizontal and vertical components of the velocity vector **v** in Figure 17.

Solution   From Figure 17, we can write

$$\cos 30° = \frac{\text{adjacent}}{\text{hypotenuse}} = \frac{v_x}{70}$$

FIGURE 17

and, consequently,

$$v_x = (\cos 30°)(70) = \frac{\sqrt{3}}{2}(70)$$

$$= 35\sqrt{3} \approx 61 \text{ cm/sec} \qquad \text{(to two significant digits)}$$

Similarly, we have

$$\sin 30° = \frac{v_y}{70}$$

and therefore,

$$v_y = (\sin 30°)(70) = 35 \text{ cm/sec}$$

In summary, now, the *x*-component of the velocity is about 61 cm/sec, and the *y*-component is 35 cm/sec. ▲

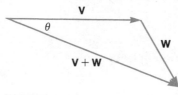

FIGURE 18

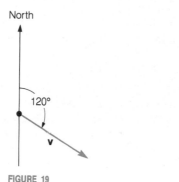

FIGURE 19

Our last example in this section will indicate how vectors are used in navigation. First, however, let us introduce some terminology. Suppose that an airplane has a **heading** of due east. This means that the airplane is pointed due east, and if there were no wind effects, the plane would indeed travel due east with respect to the ground. The **air speed** is the speed of the airplane relative to the air, whereas the **ground speed** is the plane's speed relative to the ground. Again, if there were no wind effects, then the air speed and the ground speed would be equal. Now suppose that the heading and air speed of an airplane are represented by the velocity vector **V** in Figure 18. (The direction of **V** is the heading; the magnitude of **V** is the air speed.) Also suppose that the wind velocity is represented by the vector **W** in Figure 18. Then the vector sum **V** + **W** represents the actual velocity of the plane with respect to the ground. The direction of **V** + **W** is called the **course**; it is the direction in which the airplane is moving with respect to the ground. The magnitude of **V** + **W** is the ground speed (which was defined previously). The angle θ in Figure 18 is called, naturally enough, the **drift angle**.

In navigation, directions are given in terms of the angle measured clockwise from true north. For example, the direction of the velocity vector **v** in Figure 19 is 120°. We say in this case that the **bearing** of **v** is 120°.

EXAMPLE 5 The heading and air speed of an airplane are 60° and 250 mph, respectively. If the wind is 40 mph from 150°, find the ground speed, the drift angle, and the course.

Solution In Figure 20(a), the vector $\overrightarrow{OA}$ represents the air speed of 250 mph and the heading of 60°. The vector $\overrightarrow{OW}$ represents a wind of 40 mph from 150°. Also, angle $WOA = 90°$. (Why?) In Figure 20(b), we have completed the parallelogram to obtain the vector sum $\overrightarrow{OA} + \overrightarrow{OW} = \overrightarrow{OB}$. The length of $\overrightarrow{OB}$ represents the ground speed, θ is the drift angle, and α is the bearing of the course. Because triangle BOA is a right triangle, we have

$$\tan \theta = \frac{40}{250} = \frac{4}{25} \qquad |\overrightarrow{OB}| = \sqrt{250^2 + 40^2}$$
$$\theta \approx 9.1° \qquad\qquad\quad \approx 253.2$$

From these calculations, we conclude that the ground speed is approximately 253.2 mph and the drift angle is about 9.1°. We still need to compute the bearing

FIGURE 20

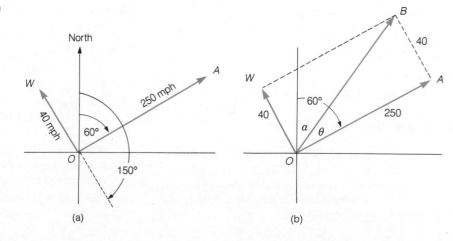

(a)

(b)

of the course. From Figure 20(b), we have

$$\alpha = 60° - \theta \approx 60° - 9.1° = 50.9°$$

Thus, the bearing of the course is 50.9° (to one decimal place).   ▲

▼ EXERCISE SET 10.6

A

In Exercises 1–26, assume that the coordinates of the points P, Q, R, S, and O are as follows:

$$P(-1, 3) \qquad Q(4, 6) \qquad R(4, 3) \qquad S(5, 9) \qquad O(0, 0)$$

For each exercise, draw the indicated vector (using graph paper) and compute its magnitude. In Exercises 7–20, compute the sums using the definition given on page 589. In Exercises 21–26, use the parallelogram law to compute the sums.

1. $\overrightarrow{PQ}$
2. $\overrightarrow{QP}$
3. $\overrightarrow{SQ}$
4. $\overrightarrow{QS}$
5. $\overrightarrow{OP}$
6. $\overrightarrow{PO}$
7. $\overrightarrow{PQ} + \overrightarrow{QS}$
8. $\overrightarrow{SQ} + \overrightarrow{QP}$
9. $\overrightarrow{OP} + \overrightarrow{PQ}$
10. $\overrightarrow{OS} + \overrightarrow{SQ}$
11. $(\overrightarrow{OS} + \overrightarrow{SQ}) + \overrightarrow{QP}$
12. $(\overrightarrow{OS} + \overrightarrow{SP}) + \overrightarrow{PR}$
13. $\overrightarrow{OP} + \overrightarrow{QS}$
14. $\overrightarrow{QS} + \overrightarrow{PO}$
15. $\overrightarrow{SR} + \overrightarrow{PO}$
16. $\overrightarrow{OS} + \overrightarrow{QO}$
17. $\overrightarrow{OP} + \overrightarrow{RQ}$
18. $\overrightarrow{OP} + \overrightarrow{QR}$
19. $\overrightarrow{SQ} + \overrightarrow{RO}$
20. $\overrightarrow{SQ} + \overrightarrow{OR}$
21. $\overrightarrow{OP} + \overrightarrow{OR}$
22. $\overrightarrow{OP} + \overrightarrow{OQ}$
23. $\overrightarrow{RP} + \overrightarrow{RS}$
24. $\overrightarrow{QP} + \overrightarrow{QR}$
25. $\overrightarrow{SO} + \overrightarrow{SQ}$
26. $\overrightarrow{SQ} + \overrightarrow{SR}$

In Exercises 27–32, the vectors **F** and **G** denote two forces that act on an object: **G** acts horizontally to the right, and **F** acts vertically upward. In each case, use the information that is given to compute $|\mathbf{F} + \mathbf{G}|$ and θ, where θ is the angle between **G** and the resultant.

27. $|\mathbf{F}| = 4$ N, $|\mathbf{G}| = 5$ N
28. $|\mathbf{F}| = 15$ N, $|\mathbf{G}| = 6$ N
29. $|\mathbf{F}| = |\mathbf{G}| = 9$ N
30. $|\mathbf{F}| = 28$ N, $|\mathbf{G}| = 1$ N
31. $|\mathbf{F}| = 3.22$ N, $|\mathbf{G}| = 7.21$ N
32. $|\mathbf{F}| = 4.06$ N, $|\mathbf{G}| = 26.83$ N

In Exercises 33–38, the vectors **F** and **G** represent two forces acting on an object, as indicated in the following figure. In each case, use the given information to compute (to two decimal places) the magnitude and direction of the resultant. (Give the direction of the resultant by specifying the angle θ between **F** and the resultant.)

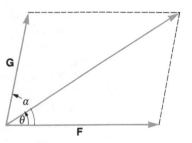

33. $|\mathbf{F}| = 5$ N, $|\mathbf{G}| = 4$ N, $\alpha = 80°$
34. $|\mathbf{F}| = 8$ N, $|\mathbf{G}| = 10$ N, $\alpha = 60°$
35. $|\mathbf{F}| = 16$ N, $|\mathbf{G}| = 25$ N, $\alpha = 35°$
36. $|\mathbf{F}| = 4.24$ N, $|\mathbf{G}| = 9.01$ N, $\alpha = 45°$
37. $|\mathbf{F}| = 50$ N, $|\mathbf{G}| = 25$ N, $\alpha = 130°$
38. $|\mathbf{F}| = 1.26$ N, $|\mathbf{G}| = 2.31$ N, $\alpha = 160°$

In Exercises 39–46, the initial point for each vector is the origin, and θ denotes the angle (measured counterclockwise) from the x-axis to the vector. In each case, compute the horizontal and vertical components of the given vector. (Round off your answers to two decimal places.)

39. The magnitude of **V** is 16 cm/sec, and $\theta = 30°$.
40. The magnitude of **V** is 40 cm/sec, and $\theta = 60°$.
41. The magnitude of **F** is 14 N, and $\theta = 75°$.
42. The magnitude of **F** is 23.12 N, and $\theta = 52°$.
43. The magnitude of **V** is 1 cm/sec, and $\theta = 135°$.
44. The magnitude of **V** is 12 cm/sec, and $\theta = 120°$.
45. The magnitude of **F** is 1.25 N, and $\theta = 145°$.
46. The magnitude of **F** is 6.34 N, and $\theta = 175°$.

In Exercises 47–50, use the given flight data to compute the ground speed, the drift angle, and the course. (Round off your answers to two decimal places.)

47. The heading and air speed are 30° and 300 mph, respectively; the wind is 25 mph from 120°.
48. The heading and air speed are 45° and 275 mph, respectively; the wind is 50 mph from 135°.
49. The heading and air speed are 100° and 290 mph, respectively; the wind is 45 mph from 190°.
50. The heading and air speed are 90° and 220 mph, respectively; the wind is 80 mph from **(a)** 180°; **(b)** 90°.

B

51. A block weighing 12 lb rests on an inclined plane, as indicated in the following figure. Determine the components of the weight perpendicular to and parallel to the plane. Round off your answers to two decimal places.   *Hint:* In Figure (b), the component of the weight perpendicular to the plane is $|\overrightarrow{OR}|$; the component parallel to the plane is $|\overrightarrow{OP}|$. Why does angle QOR equal 35°?

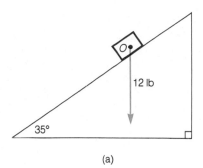

(a)

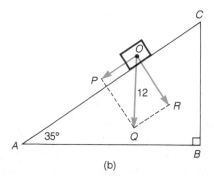

(b)

In Exercises 52 and 53, you are given the weight of a block on an inclined plane, along with the angle θ that the inclined plane makes with the horizontal. In each case, determine the components of the weight perpendicular to and parallel to the plane. (Round off your answers to two decimal places where necessary.)

52. 15 lb; $\theta = 30°$ **53.** 12 lb; $\theta = 10°$

54. A block rests on an inclined plane that makes an angle of 20° with the horizontal. The component of the weight parallel to the plane is 34.2 lb.
   (a) Determine the weight of the block. (Round off your answer to one decimal place.)
   (b) Determine the component of the weight perpendicular to the plane. (Round off your answer to one decimal place.)

55. In Section 10.7, we will see that vector addition is associative. That is, for any three vectors **A**, **B**, and **C**, we have $(\mathbf{A} + \mathbf{B}) + \mathbf{C} = \mathbf{A} + (\mathbf{B} + \mathbf{C})$. In this exercise, you are going to check that this property holds in a particular case. Let **A**, **B**, and **C** be the vectors whose initial and terminal points are as follows:

VECTOR	INITIAL POINT	TERMINAL POINT
A	$(-1, 2)$	$(2, 4)$
B	$(1, 2)$	$(3, 0)$
C	$(6, 2)$	$(4, -3)$

   (a) Use the definition of vector addition on page 589 to determine the initial and terminal points of $(\mathbf{A} + \mathbf{B}) + \mathbf{C}$.   *Suggestion:* Use graph paper.
   (b) Use the definition of vector addition to determine the initial and terminal points of $\mathbf{A} + (\mathbf{B} + \mathbf{C})$. [Your answers should agree with those in part (a).]

10.7 ▼ VECTORS IN THE PLANE, AN ALGEBRAIC APPROACH

A great many of the mathematical ideas that apply to physics and engineering are collected in the concept of vector spaces. This branch of mathematics has applications in such practical problems as calculating the vibrations of bridges and airplane wings. Logical extensions to spaces of infinitely many dimensions are widely used in modern theoretical physics as well as in many branches of mathematics itself.

From *The Mathematical Sciences*, edited by the Committee on Support of Research in the Mathematical Sciences with the collaboration of George Boehm (Cambridge, Mass.: The MIT Press, 1969)

The geometric concept of a vector in the plane can be recast in an algebraic setting. This is useful both for computational purposes and (as our opening quotation implies) for more advanced work.

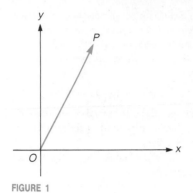

FIGURE 1

Consider an x-y coordinate system and a vector $\overrightarrow{OP}$ with initial point at the origin, as shown in Figure 1. We call $\overrightarrow{OP}$ the **position vector** (or **radius vector**) of the point P. Most of our work in this section will involve such position vectors. There is no loss of generality in focusing on these types of vectors, for, as indicated in Figure 2, each vector $\mathbf{v}$ in the plane is equal to a unique position vector $\overrightarrow{OP}$.

If the coordinates of the point P are (a, b), we call a and b the **components** of the vector $\overrightarrow{OP}$, and we use the notation

$$\langle a, b \rangle$$

to denote this vector (see Figure 3).

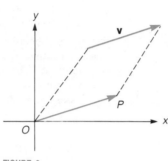

FIGURE 2

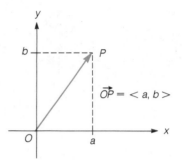

FIGURE 3
The vector $\overrightarrow{OP}$ is denoted by $\langle a, b \rangle$. The coordinates a and b are the components of the vector.

In the previous section, we said that two vectors are equal provided they have the same length and the same direction. For vectors $\langle a, b \rangle$ and $\langle c, d \rangle$, this implies that

$$\langle a, b \rangle = \langle c, d \rangle \qquad \text{if and only if} \qquad a = c \quad \text{and} \quad b = d$$

(Notice the similarity here to equality of complex numbers. Two complex numbers are equal provided their corresponding real and imaginary parts are equal; two vectors are equal provided their corresponding components are equal.)

It's easy to calculate the length of a vector $\mathbf{v}$ when its components are given. Suppose that $\mathbf{v} = \langle v_1, v_2 \rangle$. Applying the Pythagorean theorem in Figure 4, we have

$$|\mathbf{v}|^2 = v_1^2 + v_2^2$$

and, consequently,

$$|\mathbf{v}| = \sqrt{v_1^2 + v_2^2}$$

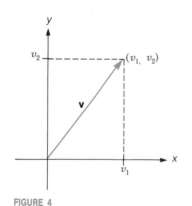

FIGURE 4

Although Figure 4 shows the point (v_1, v_2) in the first quadrant, you can check for yourself that the same formula results when (v_1, v_2) is located in any of the other three quadrants. We therefore have the following general formula.

The Length of a Vector

If $\mathbf{v} = \langle v_1, v_2 \rangle$, then

$$|\mathbf{v}| = \sqrt{v_1^2 + v_2^2}$$

FIGURE 5

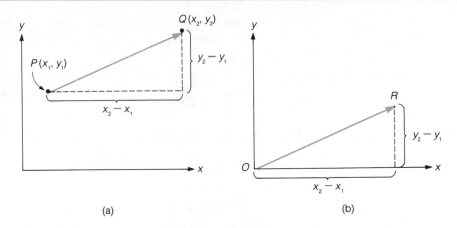

(a)   (b)

EXAMPLE 1   Compute the length of a vector $\mathbf{v} = \langle 2, -4 \rangle$.

Solution $|\mathbf{v}| = \sqrt{v_1^2 + v_2^2}$

$\qquad = \sqrt{2^2 + (-4)^2} = \sqrt{20} = 2\sqrt{5}$ ▲

Even if the initial point of a vector $\mathbf{v}$ is not the origin, we can still find the *components* of $\mathbf{v}$ by determining the components of the equivalent position vector. The formula that we are going to derive for this is as follows.

If the coordinates of the points P and Q are $P(x_1, y_1)$ and $Q(x_2, y_2)$, then

$\overrightarrow{PQ} = \langle x_2 - x_1, y_2 - y_1 \rangle$

To derive this formula, we first construct the right triangle shown in Figure 5(a). Now we let R denote the point $(x_2 - x_1, y_2 - y_1)$ and draw the position vector $\overrightarrow{OR}$ shown in Figure 5(b). Since the right triangles in Figures 5(a) and 5(b) are congruent and have corresponding legs that are parallel, we have

$\overrightarrow{PQ} = \overrightarrow{OR} = \langle x_2 - x_1, y_2 - y_1 \rangle$

as required. We can summarize this result as follows. For any vector $\mathbf{v}$, we have

x-component of $\mathbf{v}$:

$\qquad\qquad$ x-coordinate of terminal point $-$ x-coordinate of initial point

y-component of $\mathbf{v}$:

$\qquad\qquad$ y-coordinate of terminal point $-$ y-coordinate of initial point

EXAMPLE 2   Let P and Q be the points $P(3, 1)$ and $Q(7, 3)$. Find the components of $\overrightarrow{PQ}$.

Solution x-component of $\overrightarrow{PQ} = x$-coordinate of $Q - x$-coordinate of P

$\qquad = 7 - 3 = 4$

y-component of $\overrightarrow{PQ} = y$-coordinate of $Q - y$-coordinate of P

$\qquad = 3 - 1 = 2$

Consequently,

$\overrightarrow{PQ} = \langle 4, 2 \rangle$ ▲

Vector addition is particularly simple to carry out when the vectors are in component form. Indeed, we can use the parallelogram law to verify the following result.

Theorem

If $\mathbf{u} = \langle u_1, u_2 \rangle$ and $\mathbf{v} = \langle v_1, v_2 \rangle$, then $\mathbf{u} + \mathbf{v} = \langle u_1 + v_1, u_2 + v_2 \rangle$.

This theorem tells us that vector addition can be carried out *componentwise*; in other words, to add two vectors, just add the corresponding components. For example,

$$\langle 1, 2 \rangle + \langle 3, 7 \rangle = \langle 4, 9 \rangle$$

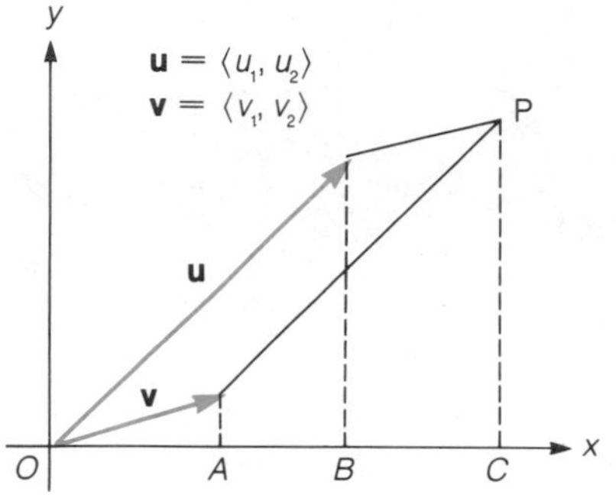

To see why this theorem is valid, consider Figure 6. (*Caution:* In reading the derivation that follows, don't confuse notation such as OA with $\overrightarrow{OA}$; recall that OA denotes the length of the line segment $\overline{OA}$.) Since the x-component of $\mathbf{u}$ is u_1, we have

$$OB = u_1$$

Also,

$$BC = v_1 \qquad \text{(Why?)}$$

Therefore,

$$OC = OB + BC$$
$$= u_1 + v_1$$

FIGURE 6

But OC is the x-component of the vector $\overrightarrow{OP}$ and, by the parallelogram law, $\overrightarrow{OP} = \mathbf{u} + \mathbf{v}$. In other words, the x-component of $\mathbf{u} + \mathbf{v}$ is $u_1 + v_1$, as we wished to show. The fact that the y-component of $\mathbf{u} + \mathbf{v}$ is $u_2 + v_2$ is proved in a similar fashion.

The vector $\langle 0, 0 \rangle$ is called the **zero vector**, and it is denoted by $\mathbf{0}$. Notice that for any vector $\mathbf{v} = \langle v_1, v_2 \rangle$, we have

$$\mathbf{v} + \mathbf{0} = \mathbf{v} \qquad \text{because} \qquad \langle v_1, v_2 \rangle + \langle 0, 0 \rangle = \langle v_1, v_2 \rangle$$

and

$$\mathbf{0} + \mathbf{v} = \mathbf{v} \qquad \text{because} \qquad \langle 0, 0 \rangle + \langle v_1, v_2 \rangle = \langle v_1, v_2 \rangle$$

So for the operation of vector addition, the zero vector plays the same role as does the real number zero in addition of real numbers. There are, in fact, several other ways in which vector addition resembles ordinary addition of real numbers. We'll return to this point again near the end of this section.

In the box that follows, we define an operation called **scalar multiplication**, in which a vector is "multiplied" by a real number (a **scalar**) to obtain another vector.

DEFINITION

Scalar Multiplication

	EXAMPLES
For each real number k and each vector $\mathbf{v} = \langle x, y \rangle$, we define a vector $k\mathbf{v}$ by the equation $$k\mathbf{v} = k\langle x, y \rangle = \langle kx, ky \rangle$$	If $\mathbf{v} = \langle 2, 1 \rangle$, then $$2\mathbf{v} = \langle 4, 2 \rangle$$ $$3\mathbf{v} = \langle 6, 3 \rangle$$ $$-1\mathbf{v} = \langle -2, -1 \rangle$$

FIGURE 7

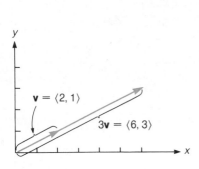

(a) The vectors **v** and 3**v** have the
same direction. The length of
3**v** is three times that of **v**.

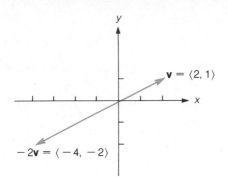

(b) The vectors **v** and $-2\mathbf{v}$ have
opposite directions. The length
of $-2\mathbf{v}$ is twice that of **v**.

In geometric terms, the length of $k\mathbf{v}$ is $|k|$ times the length of **v**. The vectors **v** and $k\mathbf{v}$ have the same direction if $k > 0$ and opposite directions if $k < 0$. For example, let $\mathbf{v} = \langle 2, 1 \rangle$. In Figure 7(a) we show the vectors **v** and 3**v**, while in Figure 7(b) we show **v** and $-2\mathbf{v}$.

EXAMPLE 3   Let $\mathbf{v} = \langle 3, 4 \rangle$ and $\mathbf{w} = \langle -1, 2 \rangle$. Compute each of the following.

(a) $\mathbf{v} + \mathbf{w}$ **(b)** $-2\mathbf{v} + 3\mathbf{w}$ **(c)** $\left| -2\mathbf{v} + 3\mathbf{w} \right|$

Solution   **(a)** $\mathbf{v} + \mathbf{w} = \langle 3, 4 \rangle + \langle -1, 2 \rangle = \langle 3 - 1, 4 + 2 \rangle = \langle 2, 6 \rangle$

(b) $-2\mathbf{v} + 3\mathbf{w} = -2\langle 3, 4 \rangle + 3\langle -1, 2 \rangle = \langle -6, -8 \rangle + \langle -3, 6 \rangle$
$= \langle -9, -2 \rangle$

(c) $\left| -2\mathbf{v} + 3\mathbf{w} \right| = \left| \langle -9, -2 \rangle \right| = \sqrt{(-9)^2 + (-2)^2} = \sqrt{85}$ ▲

For each vector **v**, we define a vector $-\mathbf{v}$, called the **negative** of **v**, by the equation

$$-\mathbf{v} = -1\mathbf{v}$$

Thus, if $\mathbf{v} = \langle a, b \rangle$, then $-\mathbf{v} = \langle -a, -b \rangle$. As indicated in Figure 8, the vectors **v** and $-\mathbf{v}$ have the same length but opposite directions.

We can use the ideas in the preceding paragraph to define **vector subtraction**. Given two vectors **u** and **v**, we define a vector $\mathbf{u} - \mathbf{v}$ by the equation

$$\mathbf{u} - \mathbf{v} = \mathbf{u} + (-\mathbf{v}) \tag{1}$$

First let's see what equation (1) is saying in terms of components. Then we will indicate a nice geometric interpretation of vector subtraction.

If $\mathbf{u} = \langle u_1, u_2 \rangle$ and $\mathbf{v} = \langle v_1, v_2 \rangle$, then equation (1) tells us that

$$\mathbf{u} - \mathbf{v} = \langle u_1, u_2 \rangle + \langle -v_1, -v_2 \rangle$$
$$= \langle u_1 + (-v_1), u_2 + (-v_2) \rangle$$

That is,

$$\mathbf{u} - \mathbf{v} = \langle u_1 - v_1, u_2 - v_2 \rangle \tag{2}$$

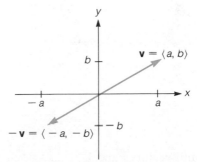

FIGURE 8

In other words, to subtract two vectors, just subtract the corresponding components.

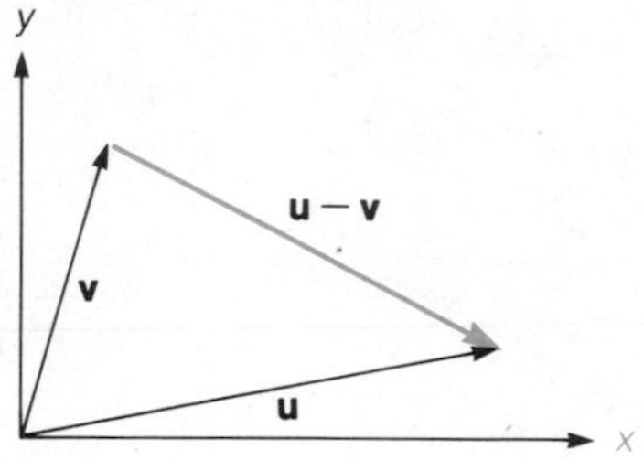

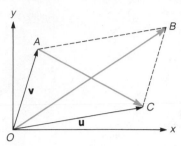

FIGURE 9

FIGURE 10

$\mathbf{u} + \mathbf{v}$ and $\mathbf{u} - \mathbf{v}$ are the directed diagonals of parallelogram $OABC$: $\mathbf{u} + \mathbf{v} = \overrightarrow{OB}$ and $\mathbf{u} - \mathbf{v} = \overrightarrow{AC}$.

Vector subtraction can be interpreted geometrically. According to the formula in the box on page 597, the right-hand side of equation (2) represents a vector drawn from the terminal point of **v** to the terminal point of **u**. Figure 9 summarizes this fact, and Figure 10 provides a geometric comparison of vector addition and vector subtraction.

EXAMPLE 4   Let $\mathbf{u} = \langle 5, 3 \rangle$ and $\mathbf{v} = \langle -1, 2 \rangle$. Compute $3\mathbf{u} - \mathbf{v}$.

Solution $3\mathbf{u} - \mathbf{v} = 3\langle 5, 3 \rangle - \langle -1, 2 \rangle$
$= \langle 15, 9 \rangle - \langle -1, 2 \rangle = \langle 16, 7 \rangle$ ▲

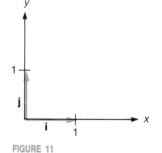

FIGURE 11

The next three examples deal with unit vectors. By definition, any vector with a length of 1 is called a **unit vector**. Two particularly useful unit vectors are

$$\mathbf{i} = \langle 1, 0 \rangle \qquad \text{and} \qquad \mathbf{j} = \langle 0, 1 \rangle$$

These are shown in Figure 11.

Any vector $\mathbf{v} = \langle x, y \rangle$ can be uniquely expressed in terms of the unit vectors **i** and **j** as follows:

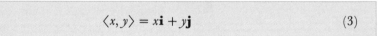

$$\langle x, y \rangle = x\mathbf{i} + y\mathbf{j} \tag{3}$$

To verify equation (3), we have

$$x\mathbf{i} + y\mathbf{j} = x\langle 1, 0 \rangle + y\langle 0, 1 \rangle$$
$$= \langle x, 0 \rangle + \langle 0, y \rangle$$
$$= \langle x, y \rangle$$

EXAMPLE 5   (a) Express the vector $\langle 3, -7 \rangle$ in terms of the unit vectors **i** and **j**.
(b) Express the vector $\mathbf{v} = -4\mathbf{i} + 5\mathbf{j}$ in component form.

Solution   (a) Using equation (3), we can write

$$\langle 3, -7 \rangle = 3\mathbf{i} + (-7)\mathbf{j} = 3\mathbf{i} - 7\mathbf{j}$$

(b) $\mathbf{v} = -4\mathbf{i} + 5\mathbf{j} = -4\langle 1, 0 \rangle + 5\langle 0, 1 \rangle$
$= \langle -4, 0 \rangle + \langle 0, 5 \rangle = \langle -4, 5 \rangle$

Thus, the component form of **v** is $\langle -4, 5 \rangle$. ▲

EXAMPLE 6   Find a unit vector $\mathbf{u}$ that has the same direction as the vector $\mathbf{v} = \langle 3, 4 \rangle$.

Solution   First, let us determine the length of $\mathbf{v}$:

$$|\mathbf{v}| = \sqrt{3^2 + 4^2} = \sqrt{25} = 5$$

So we want a vector whose length is one-fifth that of $\mathbf{v}$ and whose direction is the same as that of $\mathbf{v}$. Such a vector is

$$\mathbf{u} = \frac{1}{5}\mathbf{v} = \frac{1}{5}\langle 3, 4 \rangle = \left\langle \frac{3}{5}, \frac{4}{5} \right\rangle$$

(You should check for yourself now that the length of this vector is 1.)   ▲

EXAMPLE 7   The angle from the positive x-axis to the unit vector $\mathbf{u}$ is $\pi/3$, as indicated in Figure 12. Determine the components of $\mathbf{u}$.

Solution   Let P denote the terminal point of $\mathbf{u}$. Since P lies on the unit circle, the coordinates of P are by definition $(\cos(\pi/3), \sin(\pi/3))$. We therefore have

$$\mathbf{u} = \left\langle \frac{1}{2}, \frac{\sqrt{3}}{2} \right\rangle$$ ▲

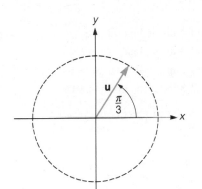

FIGURE 12

It was mentioned earlier that there are several ways in which vector addition resembles ordinary addition of real numbers. In the previous section, for example, we saw that vector addition is commutative. That is, for any two vectors $\mathbf{u}$ and $\mathbf{v}$,

$$\mathbf{u} + \mathbf{v} = \mathbf{v} + \mathbf{u}$$

By using components, we can easily verify that this property holds. (In the previous section, we used a geometric argument to establish this property.) We begin by letting $\mathbf{u} = \langle u_1, u_2 \rangle$ and $\mathbf{v} = \langle v_1, v_2 \rangle$. Then we have

$$\begin{aligned}
\mathbf{u} + \mathbf{v} &= \langle u_1, u_2 \rangle + \langle v_1, v_2 \rangle \\
&= \langle u_1 + v_1, u_2 + v_2 \rangle \\
&= \langle v_1 + u_1, v_2 + u_2 \rangle \quad \text{Addition of real numbers is commutative.} \\
&= \langle v_1, v_2 \rangle + \langle u_1, u_2 \rangle \\
&= \mathbf{v} + \mathbf{u}
\end{aligned}$$

which is what we wanted to show.

There are a number of other properties of vector addition and scalar multiplication that can be proved in a similar fashion. In the following box, we list a particular collection of these properties, known as the **vector space properties**. (Exercises 55–58 ask that you verify these properties by using components, just as we did for the commutative property.)

Properties of Vector Addition and Scalar Multiplication

For all vectors $\mathbf{u}$, $\mathbf{v}$, and $\mathbf{w}$, and for all scalars (real numbers) a and b, the following properties hold.

1. $\mathbf{u} + (\mathbf{v} + \mathbf{w}) = (\mathbf{u} + \mathbf{v}) + \mathbf{w}$
2. $\mathbf{0} + \mathbf{v} = \mathbf{v} + \mathbf{0} = \mathbf{v}$
3. $\mathbf{v} + (-\mathbf{v}) = \mathbf{0}$
4. $\mathbf{u} + \mathbf{v} = \mathbf{v} + \mathbf{u}$
5. $a(\mathbf{u} + \mathbf{v}) = a\mathbf{u} + a\mathbf{v}$
6. $(a + b)\mathbf{v} = a\mathbf{v} + b\mathbf{v}$
7. $(ab)\mathbf{v} = a(b\mathbf{v})$
8. $1\mathbf{v} = \mathbf{v}$

▼ EXERCISE SET 10.7

A

In Exercises 1–6, sketch each vector in an x-y coordinate system, and compute the length of the vector.

1. $\langle 4, 3 \rangle$ **2.** $\langle 5, 12 \rangle$ **3.** $\langle -4, 2 \rangle$
4. $\langle -6, -6 \rangle$ **5.** $\langle \frac{3}{4}, -\frac{1}{2} \rangle$ **6.** $\langle -3, 0 \rangle$

In Exercises 7–12, the coordinates of two points P and Q are given. In each case, determine the components of the vector $\overrightarrow{PQ}$. Write your answers in the form $\langle a, b \rangle$.

7. $P(2, 3)$ and $Q(3, 7)$
8. $P(5, 1)$ and $Q(4, 9)$
9. $P(-2, -3)$ and $Q(-3, -2)$
10. $P(0, -4)$ and $Q(0, -8)$
11. $P(-5, 1)$ and $Q(3, -4)$
12. $P(1, 0)$ and $Q(0, 1)$

In Exercises 13–32, assume that the vectors $\mathbf{a}$, $\mathbf{b}$, $\mathbf{c}$, and $\mathbf{d}$ are defined as follows:

$\mathbf{a} = \langle 2, 3 \rangle$ $\mathbf{b} = \langle 5, 4 \rangle$ $\mathbf{c} = \langle 6, -1 \rangle$ $\mathbf{d} = \langle -2, 0 \rangle$

Compute each of the indicated quantities.

13. $\mathbf{a} + \mathbf{b}$ **14.** $\mathbf{c} + \mathbf{d}$
15. $2\mathbf{a} + 4\mathbf{b}$ **16.** $-2\mathbf{c} + 2\mathbf{d}$
17. $|\mathbf{b} + \mathbf{c}|$ **18.** $|5\mathbf{b} + 5\mathbf{c}|$
19. $|\mathbf{a} + \mathbf{c}| - |\mathbf{a}| - |\mathbf{c}|$ **20.** $1/|\mathbf{d}|$
21. $\mathbf{a} + (\mathbf{b} + \mathbf{c})$ **22.** $(\mathbf{a} + \mathbf{b}) + \mathbf{c}$
23. $3\mathbf{a} + 4\mathbf{a}$ **24.** $|4\mathbf{b} + 5\mathbf{b}|$
25. $\mathbf{a} - \mathbf{b}$ **26.** $\mathbf{b} - \mathbf{c}$
27. $3\mathbf{b} - 4\mathbf{d}$ **28.** $\dfrac{1}{|3\mathbf{b} - 4\mathbf{d}|}(3\mathbf{b} - 4\mathbf{a})$
29. $\mathbf{a} - (\mathbf{b} + \mathbf{c})$ **30.** $(\mathbf{a} - \mathbf{b}) - \mathbf{c}$
31. $|\mathbf{c} + \mathbf{d}|^2 - |\mathbf{c} - \mathbf{d}|^2$
32. $|\mathbf{a} + \mathbf{b}|^2 + |\mathbf{a} - \mathbf{b}|^2 - 2|\mathbf{a}|^2 - 2|\mathbf{b}|^2$

In Exercises 33–38, express each vector in terms of the unit vectors $\mathbf{i}$ and $\mathbf{j}$.

33. $\langle 3, 8 \rangle$ **34.** $\langle 4, -2 \rangle$
35. $\langle -8, -6 \rangle$ **36.** $\langle -9, 0 \rangle$
37. $3\langle 5, 3 \rangle + 2\langle 2, 7 \rangle$
38. $|\langle 12, 5 \rangle|\langle 3, 4 \rangle + |\langle 3, 4 \rangle|\langle 12, 5 \rangle$

In Exercises 39–42, express each vector in the form $\langle a, b \rangle$.

39. $\mathbf{i} + \mathbf{j}$ **40.** $\mathbf{i} - 2\mathbf{j}$
41. $5\mathbf{i} - 4\mathbf{j}$ **42.** $\dfrac{1}{|\mathbf{i} + \mathbf{j}|}(\mathbf{i} + \mathbf{j})$

In Exercises 43–48, find a unit vector having the same direction as the given vector.

43. $\langle 4, 8 \rangle$ **44.** $\langle -3, 3 \rangle$ **45.** $\langle 6, -3 \rangle$
46. $\langle -12, 5 \rangle$ **47.** $8\mathbf{i} - 9\mathbf{j}$ **48.** $\langle 7, 3 \rangle - \mathbf{i} + \mathbf{j}$

In Exercises 49–54, you are given an angle θ measured counterclockwise from the positive x-axis to a unit vector $\mathbf{u} = \langle u_1, u_2 \rangle$. In each case, determine the components u_1 and u_2.

49. $\theta = \pi/6$ **50.** $\theta = \pi/4$ **51.** $\theta = 2\pi/3$
52. $\theta = 3\pi/4$ **53.** $\theta = 5\pi/6$ **54.** $\theta = 3\pi/2$

B

In Exercises 55–58, let $\mathbf{u} = \langle u_1, u_2 \rangle$, $\mathbf{v} = \langle v_1, v_2 \rangle$, and $\mathbf{w} = \langle w_1, w_2 \rangle$.

55. Verify properties 1 and 2 in the box on page 601.
56. Verify properties 3 and 4 in the box on page 601.
57. Verify properties 5 and 6 in the box on page 601.
58. Verify properties 7 and 8 in the box on page 601.

*In Exercises 59–75, we study the dot product of two vectors. Given two vectors $\mathbf{A} = \langle x_1, y_1 \rangle$ and $\mathbf{B} = \langle x_2, y_2 \rangle$, we define the **dot product** $\mathbf{A} \cdot \mathbf{B}$ as follows:*

$\mathbf{A} \cdot \mathbf{B} = x_1 x_2 + y_1 y_2$

*For example, if $\mathbf{A} = \langle 3, 4 \rangle$ and $\mathbf{B} = \langle -2, 5 \rangle$, then $\mathbf{A} \cdot \mathbf{B} = (3)(-2) + (4)(5) = 14$. Notice that the dot product of two vectors is a real number. For this reason, the dot product is also known as the **scalar product**. For Exercises 59–61, the vectors $\mathbf{u}$, $\mathbf{v}$, and $\mathbf{w}$ are defined as follows:*

$\mathbf{u} = \langle -4, 5 \rangle$ $\mathbf{v} = \langle 3, 4 \rangle$ $\mathbf{w} = \langle 2, -5 \rangle$

59. (a) Computer $\mathbf{u} \cdot \mathbf{v}$ and $\mathbf{v} \cdot \mathbf{u}$.
(b) Compute $\mathbf{v} \cdot \mathbf{w}$ and $\mathbf{w} \cdot \mathbf{v}$.
(c) Show that for any two vectors $\mathbf{A}$ and $\mathbf{B}$, we have $\mathbf{A} \cdot \mathbf{B} = \mathbf{B} \cdot \mathbf{A}$. That is, show that the dot product is commutative. *Hint:* Let $\mathbf{A} = \langle x_1, y_1 \rangle$ and $\mathbf{B} = \langle x_2, y_2 \rangle$.

60. (a) Compute $\mathbf{v} + \mathbf{w}$.
(b) Compute $\mathbf{u} \cdot (\mathbf{v} + \mathbf{w})$.
(c) Compute $\mathbf{u} \cdot \mathbf{v} + \mathbf{u} \cdot \mathbf{w}$.
(d) Show that for any three vectors $\mathbf{A}$, $\mathbf{B}$, and $\mathbf{C}$, we have $\mathbf{A} \cdot (\mathbf{B} + \mathbf{C}) = \mathbf{A} \cdot \mathbf{B} + \mathbf{A} \cdot \mathbf{C}$.

61. (a) Compute $\mathbf{v} \cdot \mathbf{v}$ and $|\mathbf{v}|^2$.
(b) Compute $\mathbf{w} \cdot \mathbf{w}$ and $|\mathbf{w}|^2$.

62. Show that for any vector $\mathbf{A}$, we always have $|\mathbf{A}|^2 = \mathbf{A} \cdot \mathbf{A}$. That is, the square of the length of a vector is equal to the dot product of the vector with itself.
Hint: Let $A = \langle x, y \rangle$.

Let θ $(0 \leq \theta \leq \pi)$ denote the angle between the two nonzero vectors **A** *and* **B**. *Then, it can be shown that the cosine of θ is given by the formula*

$$\cos \theta = \frac{\mathbf{A} \cdot \mathbf{B}}{|\mathbf{A}| |\mathbf{B}|}$$

(See Exercise 75 for the derivation of this result.) In Exercises 63–68, use this formula to find the cosine of the angle between the given pair of vectors. Also, in each case use a calculator to compute the angle. Express the angle using degrees and using radians. Round off the values to two decimal places.

63. $\mathbf{A} = \langle 4, 1 \rangle$ and $\mathbf{B} = \langle 2, 6 \rangle$

64. $\mathbf{A} = \langle 3, -1 \rangle$ and $\mathbf{B} = \langle -2, 5 \rangle$

65. $\mathbf{A} = \langle 5, 6 \rangle$ and $\mathbf{B} = \langle -3, -7 \rangle$

66. $\mathbf{A} = \langle 3, 0 \rangle$ and $\mathbf{B} = \langle 1, 4 \rangle$

67. **(a)** $\mathbf{A} = \langle -8, 2 \rangle$ and $\mathbf{B} = \langle 1, -3 \rangle$

　　(b) $\mathbf{A} = \langle -8, 2 \rangle$ and $\mathbf{B} = \langle -1, 3 \rangle$

68. **(a)** $\mathbf{A} = \langle 7, 12 \rangle$ and $\mathbf{B} = \langle 1, 2 \rangle$

　　(b) $\mathbf{A} = \langle 7, 12 \rangle$ and $\mathbf{B} = \langle -1, -2 \rangle$

69. **(a)** Compute the cosine of the angle between the vectors $\langle 2, 5 \rangle$ and $\langle -5, 2 \rangle$.

　　(b) What can you conclude from your answer in part (a)?

　　(c) Draw a sketch to check your conclusion in part (b).

70. Follow Exercise 69, but use the vectors $\langle 6, -8 \rangle$ and $\langle -4, -3 \rangle$.

71. Suppose that **A** and **B** are nonzero vectors and $\mathbf{A} \cdot \mathbf{B} = 0$. Explain why **A** and **B** are perpendicular.

72. Find a value for t such that the vectors $\langle 15, -3 \rangle$ and $\langle -4, t \rangle$ are perpendicular.

73. Find a unit vector that is perpendicular to the vector $\langle -12, 5 \rangle$. (There are two answers.)

C

74. Suppose that **A** and **B** are nonzero vectors such that $|\mathbf{A} + \mathbf{B}|^2 = |\mathbf{A}|^2 + |\mathbf{B}|^2$. Show that **A** and **B** are perpendicular. *Hint:* Let $\mathbf{A} = \langle x_1, y_1 \rangle$ and $\mathbf{B} = \langle x_2, y_2 \rangle$. After making some calculations, you should be able to apply the result in Exercise 71.

75. Refer to the accompanying figure. In this exercise we are going to derive the following formula for the cosine of the angle θ between two nonzero vectors **A** and **B**:

$$\cos \theta = \frac{\mathbf{A} \cdot \mathbf{B}}{|\mathbf{A}| |\mathbf{B}|}$$

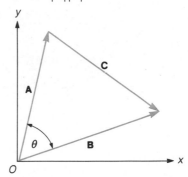

　　(a) Let $\mathbf{A} = \langle x_1, y_1 \rangle$ and $\mathbf{B} = \langle x_2, y_2 \rangle$. Compute the length of the vector **C** in the figure using the fact that $\mathbf{C} = \langle x_2 - x_1, y_2 - y_1 \rangle$.

　　(b) Using your result in part (a), show that
$$|\mathbf{C}|^2 = |\mathbf{A}|^2 + |\mathbf{B}|^2 - 2(\mathbf{A} \cdot \mathbf{B})$$

　　(c) According to the law of cosines, we have
$$|\mathbf{C}|^2 = |\mathbf{A}|^2 + |\mathbf{B}|^2 - 2|\mathbf{A}| |\mathbf{B}| \cos \theta$$

　　　　Set this expression for $|\mathbf{C}|^2$ equal to the expression obtained in part (b), and then solve for $\cos \theta$ to obtain the required formula.

10.8 ▼ TRIGONOMETRIC FORM FOR COMPLEX NUMBERS

Go to Mr. DeMoivre; he knows these things better than I do.

Isaac Newton [according to Boyer's A History of Mathematics *(New York: Wiley, 1968)]*

In this section, we explore one of the many important connections between trigonometry and the complex number system. We begin with an observation that was first made in the year 1797 by the Norwegian surveyor and mathematician Caspar Wessel. Wessel realized, essentially, that the complex numbers could be visualized as points in the *x-y* plane, the complex number $a + bi$ being identified with the point (a, b). In this context, we often refer to the *x-y* plane as the **complex plane**, and we refer to the complex numbers as points in this plane.

EXAMPLE 1 Plot the point $2 + 3i$ in the complex plane.

Solution The complex number $2 + 3i$ is identified with the point $(2, 3)$. See Figure 1. ▲

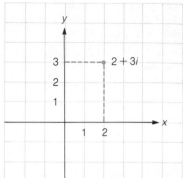

As indicated in Figure 2, the distance from the origin to the point $a + bi$ is denoted by r. We call the distance r the **modulus** of the complex number $a + bi$. The angle θ in Figure 2 (measured counterclockwise from the positive x-axis) is referred to as the **argument** of the complex number $a + bi$. (Using the terminology of Section 8.6, r and θ are the **polar coordinates** of the point $a + bi$.)

From Figure 2, we have the following three equations relating the quantities a, b, r, and θ. (Although Figure 2 shows $a + bi$ in the first quadrant, the equations remain valid for the other quadrants as well.)

$$r = \sqrt{a^2 + b^2} \tag{1}$$

$$a = r \cos \theta \tag{2}$$

$$b = r \sin \theta \tag{3}$$

FIGURE 1

If we have a complex number $z = a + bi$, we can use equations (2) and (3) to write

$$z = (r \cos \theta) + (r \sin \theta)i = r(\cos \theta + i \sin \theta)$$

That is,

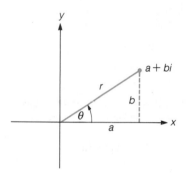

$$z = r(\cos \theta + i \sin \theta) \tag{4}$$

The expression that appears on the right-hand side of equation (4) is called the **trigonometric** (or **polar**) **form** of the complex number z. In contrast to this, the expression $a + bi$ is referred to as the **rectangular form** of the complex number z.

FIGURE 2

EXAMPLE 2 Express the complex number $z = 3\left(\cos \dfrac{\pi}{3} + i \sin \dfrac{\pi}{3} \right)$ in rectangular form.

Solution
$$z = 3\left(\cos \frac{\pi}{3} + i \sin \frac{\pi}{3} \right)$$
$$= 3\left(\frac{1}{2} + i \frac{\sqrt{3}}{2} \right) = \frac{3}{2} + \frac{3\sqrt{3}}{2}i$$

The rectangular form is therefore $\frac{3}{2} + \frac{3}{2}\sqrt{3}\,i$. ▲

EXAMPLE 3 Find the trigonometric form for the complex number $-\sqrt{2} + i\sqrt{2}$.

Solution We are asked to write the given number in the form $r(\cos \theta + i \sin \theta)$, so we need to find r and θ. Using equation (1) and the values $a = -\sqrt{2}$, $b = \sqrt{2}$, we have

$$r = \sqrt{(-\sqrt{2})^2 + (\sqrt{2})^2} = \sqrt{2 + 2} = \sqrt{4} = 2$$

Now that we know r, we can use equations (2) and (3) to determine θ. From equation (2), we obtain

$$\cos \theta = \frac{a}{r} = \frac{-\sqrt{2}}{2} \tag{5}$$

Similarly, equation (3) gives us

$$\sin \theta = \frac{b}{r} = \frac{\sqrt{2}}{2} \tag{6}$$

One angle satisfying both of equations (5) and (6) is $\theta = 3\pi/4$. (There are other angles, and we'll return to this point in a moment.) In summary, then, we have $r = 2$ and $\theta = 3\pi/4$, so the required trigonometric form is

$$2\left(\cos \frac{3\pi}{4} + i \sin \frac{3\pi}{4} \right) \qquad\qquad \blacktriangle$$

In the example we just completed, we noted that $\theta = 3\pi/4$ was only one angle satisfying the conditions $\cos \theta = -\sqrt{2}/2$ and $\sin \theta = \sqrt{2}/2$. Another such angle is $(3\pi/4) + 2\pi$. Indeed, any angle of the form $(3\pi/4) + 2\pi k$, where k is an integer, would do just as well. The upshot of this is that θ, the argument of a complex number, is not uniquely determined. In Example 3, we followed a common convention in converting to trigonometric form: we picked θ in the interval $0 \leq \theta < 2\pi$. Furthermore, although it won't cause us any difficulties in this section, you might also note that the argument θ is undefined for the complex number $0 + 0i$. (Why?)

We are now ready to derive a formula that will make it easy to multiply two complex numbers in trigonometric form. Suppose that the two complex numbers are

$$r(\cos \alpha + i \sin \alpha) \qquad \text{and} \qquad R(\cos \beta + i \sin \beta)$$

Then their product is

$$rR[(\cos \alpha + i \sin \alpha)(\cos \beta + i \sin \beta)]$$
$$= rR[(\cos \alpha \cos \beta - \sin \alpha \sin \beta) + i(\sin \alpha \cos \beta + \cos \alpha \sin \beta)]$$
$$= rR[\cos(\alpha + \beta) + i \sin(\alpha + \beta)] \quad \text{using the addition formulas from Section 6.5}$$

Notice that the modulus of the product is rR, which is the product of the two original moduli. Also, the argument is $\alpha + \beta$, which is the *sum* of the two original arguments. So, to multiply two complex numbers, just multiply the moduli and add the arguments. There is a similar rule for obtaining the quotient of two complex numbers: Divide their moduli and subtract the arguments. These two rules are stated more precisely in the box that follows. (For a proof of the division rule, see Exercise 75.)

Let $z = r(\cos \alpha + i \sin \alpha)$ and $w = R(\cos \beta + i \sin \beta)$. Then

$$zw = rR[\cos(\alpha + \beta) + i \sin(\alpha + \beta)]$$

Also, if $R \neq 0$, then

$$\frac{z}{w} = \frac{r}{R}[\cos(\alpha - \beta) + i \sin(\alpha - \beta)]$$

EXAMPLE 4   Let $z = 8\left(\cos\dfrac{5\pi}{3} + i\sin\dfrac{5\pi}{3}\right)$ and $w = 4\left(\cos\dfrac{2\pi}{3} + i\sin\dfrac{2\pi}{3}\right)$. Compute **(a)** zw; **(b)** z/w. Express each answer in both trigonometric and rectangular form.

Solution   **(a)** $zw = (8)(4)\left[\cos\left(\dfrac{5\pi}{3} + \dfrac{2\pi}{3}\right) + i\sin\left(\dfrac{5\pi}{3} + \dfrac{2\pi}{3}\right)\right]$

$$= 32\left(\cos\dfrac{7\pi}{3} + i\sin\dfrac{7\pi}{3}\right) = 32\left(\cos\dfrac{\pi}{3} + i\sin\dfrac{\pi}{3}\right) \quad \text{trigonometric form}$$

$$= 32\left(\dfrac{1}{2} + i\dfrac{\sqrt{3}}{2}\right) = 16 + 16\sqrt{3}\,i \quad \text{rectangular form}$$

(b) $\dfrac{z}{w} = \dfrac{8}{4}\left[\cos\left(\dfrac{5\pi}{3} - \dfrac{2\pi}{3}\right) + i\sin\left(\dfrac{5\pi}{3} - \dfrac{2\pi}{3}\right)\right]$

$$= 2(\cos\pi + i\sin\pi) \quad \text{trigonometric form}$$
$$= 2(-1 + i\cdot 0) = -2 \quad \text{rectangular form} \qquad \blacktriangle$$

EXAMPLE 5   Compute z^2, where $z = r(\cos\theta + i\sin\theta)$.

Solution $z^2 = [r(\cos\theta + i\sin\theta)][r(\cos\theta + i\sin\theta)]$
$\quad\;\; = r^2[\cos(\theta + \theta) + i\sin(\theta + \theta)]$
$\quad\;\; = r^2(\cos 2\theta + i\sin 2\theta) \qquad\qquad\qquad\qquad\qquad \blacktriangle$

The result in Example 5 is a particular case of an important theorem attributed to Abraham DeMoivre (1667–1754). In the box that follows, we state *DeMoivre's theorem*. (The theorem can be proved using mathematical induction; Exercise 79 asks you to carry out the proof.)

DeMoivre's Theorem

Let n be a natural number. Then
$$[r(\cos\theta + i\sin\theta)]^n = r^n(\cos n\theta + i\sin n\theta)$$

EXAMPLE 6   Use DeMoivre's theorem to compute $(-\sqrt{2} + i\sqrt{2})^5$. Express your answer in rectangular form.

Solution   In Example 3, we saw that the trigonometric form of $-\sqrt{2} + i\sqrt{2}$ is given by

$$-\sqrt{2} + i\sqrt{2} = 2\left(\cos\dfrac{3\pi}{4} + i\sin\dfrac{3\pi}{4}\right)$$

Therefore,

$$(-\sqrt{2} + i\sqrt{2})^5 = 2^5\left(\cos\dfrac{15\pi}{4} + i\sin\dfrac{15\pi}{4}\right)$$

$$= 32\left[\dfrac{\sqrt{2}}{2} + i\left(-\dfrac{\sqrt{2}}{2}\right)\right]$$

$$= 16\sqrt{2} - 16\sqrt{2}\,i \qquad\qquad\qquad \blacktriangle$$

The next two examples show how DeMoivre's theorem is used in computing roots. If n is a natural number and $z^n = w$, then we say that z is an **nth root** of w. The work in the examples also relies on the following observation about equality between nonzero complex numbers in trigonometric form: If $r(\cos\theta + i\sin\theta) = R(\cos A + i\sin A)$, then $r = R$ and $\theta = A + 2\pi k$, where k is an integer.

EXAMPLE 7   Find the cube roots of $8i$.

Solution   First we express $8i$ in trigonometric form. As can be seen from Figure 3,

$$8i = 8\left(\cos\frac{\pi}{2} + i\sin\frac{\pi}{2}\right)$$

Now we let $z = r(\cos\theta + i\sin\theta)$ denote a cube root of $8i$. Then the equation $z^3 = 8i$ becomes

$$r^3(\cos 3\theta + i\sin 3\theta) = 8\left(\cos\frac{\pi}{2} + i\sin\frac{\pi}{2}\right) \qquad (7)$$

From equation (7), we conclude that $r^3 = 8$ and, consequently, that $r = 2$. Also from equation (7), we have

$$3\theta = \frac{\pi}{2} + 2\pi k$$

or

$$\theta = \frac{\pi}{6} + \frac{2\pi k}{3} \quad \text{dividing by 3} \qquad (8)$$

If we let $k = 0$, equation (8) yields $\theta = \pi/6$. Thus, one of the cube roots of $8i$ is

$$2\left(\cos\frac{\pi}{6} + i\sin\frac{\pi}{6}\right) = 2\left(\frac{\sqrt{3}}{2} + i\cdot\frac{1}{2}\right) = \sqrt{3} + i$$

Next we let $k = 1$ in equation (8). As you can check, this yields $\theta = 5\pi/6$. So another cube root of $8i$ is

$$2\left(\cos\frac{5\pi}{6} + i\sin\frac{5\pi}{6}\right) = 2\left(\frac{-\sqrt{3}}{2} + i\cdot\frac{1}{2}\right) = -\sqrt{3} + i$$

Similarly, using $k = 2$ in equation (8) yields $\theta = 3\pi/2$. (Verify this.) Consequently, a third cube root is

$$2\left(\cos\frac{3\pi}{2} + i\sin\frac{3\pi}{2}\right) = 2[0 + i(-1)] = -2i$$

We have now found three distinct cube roots of $8i$. If we were to continue the process, using $k = 3$, for example, we would find that no additional roots are obtained in this manner. We therefore conclude that there are exactly three cube roots of $8i$. We have plotted these cube roots in Figure 4. Notice that the points lie equally spaced on a circle of radius 2 about the origin. ▲

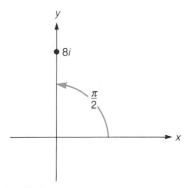

FIGURE 3

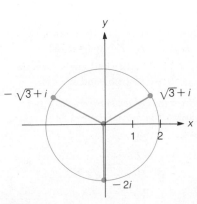

FIGURE 4

When we used DeMoivre's theorem in Example 7, we found that the number $8i$ had three distinct cube roots. Along these same lines, it is true in general that any nonzero number $a + bi$ possesses exactly n distinct nth roots. This is true even if $b = 0$. In the next example, for instance, we compute the five fifth roots of 2.

EXAMPLE 8   Compute the five fifth roots of 2.

Solution   We will follow the procedure we used in Example 7. In trigonometric form, the number 2 becomes $2(\cos 0 + i \sin 0)$. Now we let $z = r(\cos \theta + i \sin \theta)$ denote a fifth root of 2. Then the equation $z^5 = 2$ becomes

$$r^5(\cos 5\theta + i \sin 5\theta) = 2(\cos 0 + i \sin 0) \tag{9}$$

From equation (9), we see that $r^5 = 2$ and, consequently, $r = 2^{1/5}$. Also from equation (9), we have

$$5\theta = 0 + 2\pi k \qquad \text{or} \qquad \theta = \frac{2\pi k}{5}$$

Using the values $k = 0, 1, 2, 3,$ and 4 in succession, we obtain the following results:

k	0	1	2	3	4
θ	0	$2\pi/5$	$4\pi/5$	$6\pi/5$	$8\pi/5$

The five fifth roots of 2 are therefore

$$z_1 = 2^{1/5}(\cos 0 + i \sin 0) = 2^{1/5} \qquad z_4 = 2^{1/5}\left(\cos \frac{6\pi}{5} + i \sin \frac{6\pi}{5}\right)$$

$$z_2 = 2^{1/5}\left(\cos \frac{2\pi}{5} + i \sin \frac{2\pi}{5}\right)$$

$$z_5 = 2^{1/5}\left(\cos \frac{8\pi}{5} + i \sin \frac{8\pi}{5}\right)$$

$$z_3 = 2^{1/5}\left(\cos \frac{4\pi}{5} + i \sin \frac{4\pi}{5}\right)$$

In Figure 5, we have plotted these five fifth roots. Notice that the points are equally spaced on a circle of radius $2^{1/5}$ about the origin. ▲

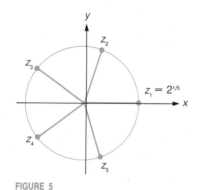

FIGURE 5

▼ EXERCISE SET 10.8

A

In Exercises 1–8, plot each point in the complex plane.

1. $4 + 2i$ **2.** $4 - 2i$ **3.** $-5 + i$

4. $-3 - 5i$ **5.** $1 - 4i$ **6.** i

7. $-i$ **8.** $1 (= 1 + 0i)$

In Exercises 9–18, convert each complex number to rectangular form.

9. $2(\cos \frac{1}{4}\pi + i \sin \frac{1}{4}\pi)$ **10.** $6(\cos \frac{5}{3}\pi + i \sin \frac{5}{3}\pi)$

11. $4(\cos \frac{5}{6}\pi + i \sin \frac{5}{6}\pi)$ **12.** $3(\cos \frac{3}{2}\pi + i \sin \frac{3}{2}\pi)$

13. $\sqrt{2}(\cos 225° + i \sin 225°)$ **14.** $\frac{1}{2}(\cos 240° + i \sin 240°)$

15. $\sqrt{3}(\cos \frac{1}{2}\pi + i \sin \frac{1}{2}\pi)$ **16.** $5(\cos \pi + i \sin \pi)$

17. $4(\cos 75° + i \sin 75°)$ *Hint:* Use the addition formulas from Section 6.5 to evaluate $\cos 75°$ and $\sin 75°$.

18. $2(\cos \frac{1}{8}\pi + i \sin \frac{1}{8}\pi)$ *Hint:* Use the half-angle formulas from Section 6.6 to evaluate $\cos(\pi/8)$ and $\sin(\pi/8)$.

In Exercises 19–28, convert from rectangular to trigonometric form. (In each case, choose an argument θ such that $0 \le \theta < 2\pi$.)

19. $\frac{1}{2}\sqrt{3} + \frac{1}{2}i$ **20.** $\sqrt{2} + \sqrt{2}\,i$

21. $-1 + \sqrt{3}\,i$ **22.** -4

23. $-2\sqrt{3} - 2i$ **24.** $-3\sqrt{2} - 3\sqrt{2}\,i$

25. $-6i$ **26.** $-4 - 4\sqrt{3}\,i$

27. $\frac{1}{4}\sqrt{3} - \frac{1}{4}i$ **28.** 16

In Exercises 29–54, carry out the indicated operations. Express your results in rectangular form for those cases in which the trigonometric functions are readily evaluated without tables or a calculator.

29. $2(\cos 22° + i \sin 22°) \times 3(\cos 38° + i \sin 38°)$

30. $4(\cos 5° + i \sin 5°) \times 6(\cos 130° + i \sin 130°)$

31. $\sqrt{2}(\cos \frac{1}{3}\pi + i \sin \frac{1}{3}\pi) \times \sqrt{2}(\cos \frac{4}{3}\pi + i \sin \frac{4}{3}\pi)$

32. $(\cos \frac{1}{5}\pi + i \sin \frac{1}{5}\pi) \times (\cos \frac{1}{20}\pi + i \sin \frac{1}{20}\pi)$

33. $3(\cos \frac{1}{7}\pi + i \sin \frac{1}{7}\pi) \times \sqrt{2}(\cos \frac{1}{7}\pi + i \sin \frac{1}{7}\pi)$

34. $\sqrt{3}(\cos 3° + i \sin 3°) \times \sqrt{3}(\cos 38° + i \sin 38°)$

35. $6(\cos 50° + i \sin 50°) \div 2(\cos 5° + i \sin 5°)$

36. $\sqrt{3}(\cos 140° + i \sin 140°) \div 3(\cos 5° + i \sin 5°)$

37. $2^{4/3}(\cos \frac{5}{12}\pi + i \sin \frac{5}{12}\pi) \div 2^{1/3}(\cos \frac{1}{4}\pi + i \sin \frac{1}{4}\pi)$

38. $\sqrt{6}(\cos \frac{16}{9}\pi + i \sin \frac{16}{9}\pi) \div \sqrt{2}(\cos \frac{1}{9}\pi + i \sin \frac{1}{9}\pi)$

39. $(\cos \frac{2}{5}\pi + i \sin \frac{2}{5}\pi) \div (\cos \frac{2}{5}\pi + i \sin \frac{2}{5}\pi)$

40. $(\cos \frac{2}{5}\pi + i \sin \frac{2}{5}\pi) \div (\cos \frac{1}{10}\pi + i \sin \frac{1}{10}\pi)$

41. $[3(\cos \frac{1}{3}\pi + i \sin \frac{1}{3}\pi)]^5$

42. $[\sqrt{2}(\cos \frac{5}{6}\pi + i \sin \frac{5}{6}\pi)]^4$

43. $[\frac{1}{2}(\cos \frac{1}{24}\pi + i \sin \frac{1}{24}\pi)]^6$

44. $[\sqrt{3}(\cos 70° + i \sin 70°)]^3$

45. $[2^{1/5}(\cos 63° + i \sin 63°)]^{10}$

46. $[2(\cos \frac{1}{5}\pi + i \sin \frac{1}{5}\pi)]^3$

47. $2(\cos 200° + i \sin 200°) \times \sqrt{2}(\cos 20° + i \sin 20°)$
$\times \frac{1}{2}(\cos 5° + i \sin 5°)$

48. $\left[\dfrac{\cos(\pi/8) + i \sin(\pi/8)}{\cos(-\pi/8) + i \sin(-\pi/8)} \right]^5$

49. $[\frac{1}{2}(1 - \sqrt{3}\,i)]^5$ *Hint:* Convert to trigonometric form.

50. $(1 - i)^3$ **51.** $(-2 - 2i)^5$

52. $(-\frac{1}{2} + \frac{1}{2}\sqrt{3}\,i)^6$ **53.** $(-2\sqrt{3} - 2i)^4$

54. $(1 + i)^{16}$

In Exercises 55–61, use DeMoivre's theorem to find the indicated roots. Express the results in rectangular form.

55. Cube roots of $-27i$ **56.** Cube roots of 2

57. Eighth roots of 1 **58.** Square roots of i

59. Cube roots of 64 **60.** Square roots of $-\frac{1}{2} - \frac{1}{2}\sqrt{3}\,i$

61. Sixth roots of 729

C *Use a calculator to complete Exercises 62–65.*

62. Compute $(9 + 9i)^6$. **63.** Compute $(7 - 7i)^8$.

64. Compute the cube roots of $1 + 2i$. Express your answers in rectangular form, with the real and imaginary parts rounded off to two decimal places.

65. Compute the fifth roots of i. Express your answers in rectangular form, with the real and imaginary parts rounded off to two decimal places.

B

In Exercises 66–68, find the indicated roots. Express the results in rectangular form.

66. Find the fourth roots of i. *Hint:* Use the half-angle formulas from Section 6.6.

67. Find the fourth roots of $8 - 8\sqrt{3}\,i$. *Hint:* Use the addition formulas or the half-angle formulas.

68. Find the square roots of $7 + 24i$. *Hint:* You'll need to use the half-angle formulas from Section 6.6.

69. **(a)** Compute the three cube roots of 1.
(b) Let z_1, z_2, and z_3 denote the three cube roots of 1. Verify that $z_1 + z_2 + z_3 = 0$ and also that $z_1 z_2 + z_2 z_3 + z_3 z_1 = 0$.

70. **(a)** Compute the four fourth roots of 1.
(b) Verify that the sum of these four fourth roots is 0.

71. Evaluate $(-\frac{1}{2} + \frac{1}{2}\sqrt{3}\,i)^5 + (-\frac{1}{2} - \frac{1}{2}\sqrt{3}\,i)^5$.
Hint: Use DeMoivre's theorem.

72. Show that $(-\frac{1}{2} + \frac{1}{2}\sqrt{3}\,i)^6 + (-\frac{1}{2} - \frac{1}{2}\sqrt{3}\,i)^6 = 2$.

73. Compute $(\cos \theta + i \sin \theta)(\cos \theta - i \sin \theta)$.

74. In the identity $(\cos \theta + i \sin \theta)^2 = \cos 2\theta + i \sin 2\theta$, carry out the actual multiplication on the left-hand side of the equation. Then equate the corresponding real parts and the corresponding imaginary parts from each side of the equation that results. What do you obtain?

75. Show that

$$\frac{r(\cos \alpha + i \sin \alpha)}{R(\cos \beta + i \sin \beta)} = \frac{r}{R}[\cos(\alpha - \beta) + i \sin(\alpha - \beta)]$$

Suggestion: Begin with the quantity on the left side and multiply it by $(\cos \beta - i \sin \beta)/(\cos \beta - i \sin \beta)$.

76. Show that

$$1 + \cos \theta + i \sin \theta = 2 \cos\left(\frac{\theta}{2}\right)\left[\cos\left(\frac{\theta}{2}\right) + i \sin\left(\frac{\theta}{2}\right)\right]$$

77. If $z = r(\cos \theta + i \sin \theta)$, show that

$$\frac{1}{z} = \frac{1}{r}(\cos \theta - i \sin \theta)$$

Hint: $1/z = [1(\cos 0 + i \sin 0)]/[r(\cos \theta + i \sin \theta)]$.

78. Show that $\dfrac{1 + \sin \theta + i \cos \theta}{1 + \sin \theta - i \cos \theta} = \sin \theta + i \cos \theta$.
Assume that $\theta \neq \frac{3}{2}\pi + 2\pi k$, where k is an integer.
Hint: Work with the left-hand side; first "rationalize" the denominator by multiplying by

$$\frac{(1 + \sin \theta) + i \cos \theta}{(1 + \sin \theta) + i \cos \theta}$$

79. Prove DeMoivre's theorem, using mathematical induction.

▼ **CHAPTER TEN   SUMMARY OF PRINCIPAL TERMS AND FORMULAS**

TERM OR FORMULA	PAGE REFERENCE	COMMENT
1. Principle of mathematical induction	557	This can be stated as follows. Suppose that for each natural number n we have a statement P_n. Suppose P_1 is true. Also suppose that P_{k+1} is true whenever P_k is true. Then according to the *principle of mathematical induction*, all the statements are true; that is, P_n is true for all natural numbers n. The principle of mathematical induction has the status of an axiom; that is, we accept its validity without proof.
2. Pascal's triangle	564	Pascal's triangle refers to the triangular array of numbers displayed on page 564 Additional rows can be added to the triangle according to the following rule: Each entry in the array (other than the 1's along the sides) is the sum of the two numbers diagonally above it. The numbers in the nth row of Pascal's triangle are the coefficients of the terms in the expansion of $(a + b)^{n-1}$.
3. $n!$ (read: n factorial)	565	This denotes the product of the first n natural numbers. For example, $4! = (4)(3)(2)(1) = 24$.
4. $\binom{n}{k}$ (read: n choose k)	566	Let n and k be nonnegative integers with $k \leq n$. Then the *binomial coefficient* $\binom{n}{k}$ is defined by $$\binom{n}{k} = \frac{n!}{k!(n-k)!}$$
5. Binomial theorem	567	The *binomial theorem* is a formula that allows us to analyze and expand expressions of the form $(a + b)^n$. If we use the sigma notation, then the statement of the binomial theorem that appears on page 567 can be abbreviated to read as follows: $$(a + b)^n = \sum_{k=0}^{n} \binom{n}{k} a^{n-k} b^k$$
6. Sequence	572, 574	In the context of the present chapter, a *sequence* is an ordered list of real numbers: $a_1, a_2, a_3, \ldots$. More generally, a sequence is a function whose domain is the set of natural numbers.
7. $\displaystyle\sum_{k=1}^{n} a_k$	574	This expression stands for the sum $a_1 + a_2 + a_3 + \cdots + a_n$. The letter k in the expression is referred to as the *index of summation*.
8. Arithmetic sequence	578	A sequence in which the successive terms differ by a constant is called an *arithmetic sequence*. The general form of an arithmetic sequence is $$a, a + d, a + 2d, a + 3d, \ldots$$ In this sequence, d is referred to as the *common difference*.
9. $a_n = a + (n-1)d$	578	This is the formula for the nth term of an arithmetic sequence.
10. $S_n = \dfrac{n}{2}[2a + (n-1)d]$ $S_n = n\left(\dfrac{a + a_n}{2}\right)$	580	These are the formulas for the sum of the first n terms of an arithmetic sequence.
11. Geometric sequence	583	A sequence in which the ratio of successive terms is constant is called a *geometric sequence*. The general form of a geometric sequence is $$a, ar, ar^2, \ldots$$ The number r is referred to as the *common ratio*.

TERM OR FORMULA	PAGE REFERENCE	COMMENT
12. $a_n = ar^{n-1}$	584	This is the formula for the nth term of a geometric sequence.
13. $S_n = \dfrac{a(1-r^n)}{1-r}$	584	This is the formula for the sum of the first n terms of a geometric sequence.
14. $S = \dfrac{a}{1-r}$	586	This is the formula for the sum of the infinite geometric series $a + ar + ar^2 + \cdots$, where $|r| < 1$.
15. Vector	588	Geometrically, a *vector* in the plane is a directed line segment. Vectors can be used to represent quantities such as force and velocity that have both magnitude and direction. For examples, see Figures 1 and 2 in Section 10.6. Algebraically, a vector in the plane is an ordered pair of real numbers, denoted by $\langle x, y \rangle$. (See Figure 3 in Section 10.7.) The numbers x and y are called the *components* of the vector $\langle x, y \rangle$.
16. Vector equality	588 596	Geometrically, two vectors are said to be equal provided they have the same length and the same direction. In terms of components, this means that two vectors are equal if and only if their corresponding components are equal.
17. Vector addition	589 590 596	Geometrically, two vectors can be added by using the parallelogram law, as in Figure 10 in Section 10.6. Algebraically, this is equivalent to the following componentwise formula for vector addition: $$\langle a, b \rangle + \langle c, d \rangle = \langle a + c, b + d \rangle$$
18. $\|\mathbf{v}\| = \sqrt{v_1^2 + v_2^2}$	588 596	The expression on the left-hand side of the equation denotes the *length* or *magnitude* of the vector $\mathbf{v}$. The expression on the right-hand side of the equation tells us how to compute the length of $\mathbf{v}$ in terms of its respective x- and y-components, v_1 and v_2.
19. $\overrightarrow{PQ} = \langle x_2 - x_1, y_2 - y_1 \rangle$	597	This formula gives the components of the vector $\overrightarrow{PQ}$, where P and Q are the points $P(x_1, y_1)$ and $Q(x_2, y_2)$. See Figure 5 in Section 10.7 for the derivation of this formula.
20. The zero vector	598	The *zero vector*, denoted by $\mathbf{0}$, is the vector $\langle 0, 0 \rangle$. The zero vector plays the same role in vector addition as does the real number zero in addition of real numbers. For any vector $\mathbf{v}$, we have $$\mathbf{0} + \mathbf{v} = \mathbf{v} \mid \mathbf{0} - \mathbf{v}$$
21. Scalar multiplication	598	For each real number k and each vector $\mathbf{v} = \langle x, y \rangle$, the vector $k\mathbf{v}$ is defined by the equation $$k\mathbf{v} = \langle kx, ky \rangle$$ Geometrically, the length of $k\mathbf{v}$ is $|k|$ times the length of $\mathbf{v}$. The operation whereby the vector $k\mathbf{v}$ is formed from the scalar k and the vector $\mathbf{v}$ is called *scalar multiplication*.
22. $\mathbf{i} = \langle 1, 0 \rangle$ $\mathbf{j} = \langle 0, 1 \rangle$	600	These two equations define the *unit vectors* $\mathbf{i}$ and $\mathbf{j}$. See Figure 11 in Section 10.7. Any vector $\langle x, y \rangle$ can be expressed in terms of the unit vectors $\mathbf{i}$ and $\mathbf{j}$ as follows: $$\langle x, y \rangle = x\mathbf{i} + y\mathbf{j}$$

TERM OR FORMULA	PAGE REFERENCE	COMMENT
23. The complex plane	603	Refer to the following figure. A complex number $a + bi$ can be identified with the point (a, b) in the x-y plane. In this context, the x-y plane is referred to as the *complex plane*. The distance r in the figure is the *modulus* of the complex number $a + bi$; the angle θ is an *argument* of $a + bi$.

24. $z = r(\cos \theta + i \sin \theta)$	604	The expression on the right-hand side is the *trigonometric form* of the complex number $z = a + bi$. (The expression $a + bi$ is the *rectangular form* of the complex number z.) The trigonometric form and the rectangular form are related by the following equations: $$r = \sqrt{a^2 + b^2} \qquad x = r \cos \theta \qquad y = r \sin \theta$$
25. DeMoivre's theorem	606	Let n be a natural number and let $z = r(\cos \theta + i \sin \theta)$. Then DeMoivre's theorem states that $$z^n = r^n(\cos n\theta + i \sin n\theta)$$ For applications of this theorem, see Examples 6, 7, and 8 in Section 10.8.

▼ CHAPTER TEN   REVIEW EXERCISES

NOTE   Exercises 1–20 constitute a chapter test on the fundamentals, based on group A problems.

1.   Use the principle of mathematical induction to show that the following formula is valid for all natural numbers n.

$$1^2 + 2^2 + 3^2 + \cdots + n^2 = \frac{n(n + 1)(2n + 1)}{6}$$

2.   Express each of the following sums without using sigma notation, and then evaluate each sum.

(a) $\sum_{k=0}^{2} (10k - 1)$ (b) $\sum_{k=1}^{3} (-1)^k k^2$

3.   (a) Write the formula for the sum S_n of a finite geometric series.

(b) Evaluate the sum $\dfrac{3}{2} + \dfrac{3^2}{2^2} + \dfrac{3^3}{2^3} + \cdots + \dfrac{3^{10}}{2^{10}}$.

4.   (a) Determine the coefficient of the term containing a^3 in the expansion of $(a - 2b^3)^{11}$.

(b) Find the fifth term of the expansion in part (a).

5.   Expand the expression $(3x^2 + y^3)^5$.

6.   Determine the sum of the first 12 terms of an arithmetic sequence in which the first term is 8 and the twelfth term is $\frac{43}{2}$.

7.   Find the sum of the following infinite geometric series: $\frac{7}{10} + \frac{7}{100} + \frac{7}{1000} + \cdots$.

8.   A sequence is defined recursively as follows: $a_1 = 1$, $a_2 = 1$, and $a_n = (a_{n-1})^2 + a_{n-2}$ for $n \geq 3$. Determine the fourth and the fifth terms in this sequence.

9.   In a certain geometric sequence, the third term is 4 and the fifth term is 10. Find the sixth term, given that the common ratio is negative.

10. What is the twentieth term in the arithmetic sequence $-61, -46, -31, \ldots$?

11. Two forces **F** and **G** act on an object. The force **G** acts horizontally with a magnitude of 2 N, and **F** acts vertically upward with a magnitude of 4 N.
   (a) Find the magnitude of the resultant.
   (b) Find $\tan \theta$, where θ is the angle between **G** and the resultant.

12. Two forces act on an object, as shown in the accompanying figure.
   (a) Find the magnitude of the resultant. (Leave your answer in terms of radicals and the trigonometric functions.)
   (b) Find $\sin \theta$, where θ is the angle between the 12-N force and the resultant.

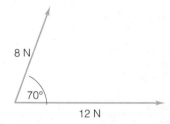

13. The heading and air speed of an airplane are 40° and 300 mph, respectively. If the wind is 50 mph from 130°, find the ground speed and the tangent of the drift angle. (Leave your answer in terms of radicals and the trigonometric functions.)

14. Let $\mathbf{A} = \langle 2, 4 \rangle$, $\mathbf{B} = \langle 3, -1 \rangle$, and $\mathbf{C} = \langle 4, -4 \rangle$.
   (a) Find $2\mathbf{A} + 3\mathbf{B}$.   **(b)** Find $|2\mathbf{A} + 3\mathbf{B}|$.
   (c) Express $\mathbf{C} - \mathbf{B}$ in terms of **i** and **j**.

15. Let P and Q be the points $(4, 5)$ and $(-7, 2)$, respectively. Find a unit vector having the same direction as $\overrightarrow{PQ}$.

16. Find the rectangular form for the complex number $z = 2(\cos \frac{2}{3}\pi + i \sin \frac{2}{3}\pi)$.

17. Find the trigonometric form of the complex number $\sqrt{2} - \sqrt{2}\,i$.

18. Complete the statement: According to DeMoivre's theorem, if n is a natural number, then

$$[r(\cos \theta + i \sin \theta)]^n = \underline{\hspace{2cm}}.$$

19. Let $z = 3[\cos(2\pi/9) + i \sin(2\pi/9)]$ and $w = 5[\cos(\pi/9) + i \sin(\pi/9)]$. Compute the product zw and express your answer in rectangular form.

20. Compute the cube roots of $64i$.

In Exercises 21–30, use the principle of mathematical induction to show that the statements are true for all natural numbers.

21. $5 + 10 + 15 + \cdots + 5n = \frac{5}{2}n(n + 1)$

22. $10 + 10^2 + 10^3 + \cdots + 10^n = \frac{10}{9}(10^n - 1)$

23. $1 \cdot 2 + 2 \cdot 3 + 3 \cdot 4 + \cdots + n(n + 1) = \frac{1}{3}n(n + 1)(n + 2)$

24. $\dfrac{1}{2} + \dfrac{2}{2^2} + \dfrac{3}{2^3} + \cdots + \dfrac{n}{2^n} = 2 - \dfrac{2 + n}{2^n}$

25. $1 + 3 \cdot 2 + 5 \cdot 2^2 + 7 \cdot 2^3 + \cdots + (2n - 1) \cdot 2^{n-1}$
$$= 3 + (2n - 3) \cdot 2^n$$

26. $\dfrac{1}{1 \cdot 4} + \dfrac{1}{4 \cdot 7} + \dfrac{1}{7 \cdot 10} + \cdots + \dfrac{1}{(3n - 2)(3n + 1)} = \dfrac{n}{3n + 1}$

27. $1 + 2^2 \cdot 2 + 3^2 \cdot 2^2 + 4^2 \cdot 2^3 + \cdots + n^2 \cdot 2^{n-1}$
$$= (n^2 - 2n + 3)2^n - 3$$

28. 9 is a factor of $n^3 + (n + 1)^3 + (n + 2)^3$

29. 3 is a factor of $7^n - 1$

30. 8 is a factor of $9^n - 1$

In Exercises 31–40, expand the given expressions.

31. $(3a + b^2)^4$ **32.** $(5a - 2b)^3$

33. $(x + \sqrt{x})^4$ **34.** $(1 - \sqrt{3})^6$

35. $(x^2 - 2y^2)^5$ **36.** $\left(\dfrac{1}{a} + \dfrac{2}{b}\right)^3$

37. $\left(1 + \dfrac{1}{x}\right)^5$ **38.** $\left(x^3 + \dfrac{1}{x^2}\right)^6$

39. $(a\sqrt{b} - b\sqrt{a})^4$ **40.** $(x^{-2} + y^{5/2})^8$

41. Find the fifth term in the expansion of $(3x + y^2)^5$.

42. Find the eighth term in the expansion of $(2x - y)^9$.

43. Find the coefficient of the term containing a^5 in the expansion of $(a - 2b)^7$.

44. Find the coefficient of the term containing b^8 in the expansion of $[2a - (b/3)]^{10}$.

45. Find the coefficient of the term containing x^3 in the expansion of $(1 + \sqrt{x})^8$.

46. Expand $(1 + \sqrt{x} + x)^6$. *Suggestion:* Rewrite the expression as $[(1 + \sqrt{x}) + x]^6$.

In Exercises 47–52, verify each assertion by computing the indicated binomial coefficients.

47. $\dbinom{2}{0}^2 + \dbinom{2}{1}^2 + \dbinom{2}{2}^2 = \dbinom{4}{2}$

48. $\dbinom{3}{0}^2 + \dbinom{3}{1}^2 + \dbinom{3}{2}^2 + \dbinom{3}{3}^2 = \dbinom{6}{3}$

49. $\dbinom{4}{0}^2 + \dbinom{4}{1}^2 + \dbinom{4}{2}^2 + \dbinom{4}{3}^2 + \dbinom{4}{4}^2 = \dbinom{8}{4}$

50. $\dbinom{2}{0} + \dbinom{2}{1} + \dbinom{2}{2} = 2^2$

51. $\dbinom{3}{0} + \dbinom{3}{1} + \dbinom{3}{2} + \dbinom{3}{3} = 2^3$

52. $\dbinom{4}{0} + \dbinom{4}{1} + \dbinom{4}{2} + \dbinom{4}{3} + \dbinom{4}{4} = 2^4$

In Exercises 53–58, compute the first five terms in each sequence. (In Exercises 53–55, in which the nth term is defined by a formula, assume that the formula holds for all natural numbers n.)

53. $a_n = 2n/(n + 1)$

54. $a_n = (3n - 2)/(3n + 2)$

55. $a_n = (-1)^n \left(1 - \dfrac{1}{n + 1}\right)$

56. $a_0 = 4;\ a_n = 2a_{n-1}$ for $n \geq 1$

57. $a_0 = -3;\ a_n = 4a_{n-1}$ for $n \geq 1$

58. $a_0 = 1;\ a_1 = 2;\ a_n = 3a_{n-1} + 2a_{n-2}$ for $n \geq 2$

In Exercises 59 and 60, express each sum without using sigma notation. Simplify each answer.

59. $\displaystyle\sum_{k=1}^{5} (2k + 3)$

60. $\displaystyle\sum_{k=0}^{8} \left(\dfrac{1}{k + 1} - \dfrac{1}{k + 2}\right)$

In Exercises 61 and 62, rewrite each sum using sigma notation.

61. $\dfrac{5}{3} + \dfrac{5}{3^2} + \dfrac{5}{3^3} + \dfrac{5}{3^4} + \dfrac{5}{3^5}$

62. $\dfrac{1}{2} - \dfrac{2}{2^2} + \dfrac{3}{2^3} - \dfrac{4}{2^4} + \dfrac{5}{2^5} - \dfrac{6}{2^6}$

In Exercises 63–66, find the indicated term in each sequence.

63. $a_{14};\ 5, 9, 13, 17, \ldots$

64. $a_{20};\ 5, \frac{9}{2}, 4, \frac{7}{2}, \ldots$

65. $a_{12};\ 10, 5, \frac{5}{2}, \frac{5}{4}, \ldots$

66. $a_{10};\ \sqrt{2} + 1, 1, \sqrt{2} - 1, 3 - 2\sqrt{2}, \ldots$

67. Find the sum of the first 100 terms in the sequence $-5, -2, 1, 4, \ldots$.

68. Find the sum of the first 45 terms in the sequence $10, \frac{29}{3}, \frac{28}{3}, 9, \ldots$.

69. Find the sum of the first 10 terms in the sequence $7, 70, 700, \ldots$.

70. Find the sum of the first 12 terms in the sequence $\frac{1}{3}, -\frac{2}{9}, \frac{4}{27}, -\frac{8}{81}, \ldots$.

In Exercises 71–74, find the sum of each infinite geometric series.

71. $\frac{3}{5} + \frac{3}{25} + \frac{3}{125} + \cdots$

72. $\frac{7}{10} + \frac{7}{100} + \frac{7}{1000} + \cdots$

73. $\frac{1}{9} - \frac{1}{81} + \frac{1}{729} - \cdots$

74. $1 + \dfrac{1}{1 + \sqrt{2}} + \dfrac{1}{(1 + \sqrt{2})^2} + \cdots$

75. Find a fraction equivalent to $0.\overline{45}$.

76. Find a fraction equivalent to $0.2\overline{13}$.

In Exercises 77–80, verify each equation using the formula for the sum of an arithmetic series:

$$S_n = \frac{n}{2}[2a + (n - 1)d]$$

(The formulas given in Exercises 77–80, appear in Elements of Algebra *by Leonhard Euler, first published in 1770.)*

77. $1 + 2 + 3 + \cdots + n = n + \dfrac{n(n - 1)}{2}$

78. $1 + 3 + 5 + \cdots + $ (to n terms) $= n + \dfrac{2n(n - 1)}{2}$

79. $1 + 4 + 7 + \cdots + $ (to n terms) $= n + \dfrac{3n(n - 1)}{2}$

80. $1 + 5 + 9 + \cdots + $ (to n terms) $= n + \dfrac{4n(n - 1)}{2}$

C **81.** In this exercise, you will use the following (remarkably simple) formula for approximating sums of powers of integers:

$$1^k + 2^k + 3^k + \cdots + n^k \approx \frac{(n + \frac{1}{2})^{k+1}}{k + 1} \tag{1}$$

[This formula appears in the article "Sums of Powers of Integers," by B. L. Burrows and R. F. Talbot, published in the *American Mathematical Monthly* **91** (1984) 394.]

(a) Use formula (1) to estimate the sum $1^2 + 2^2 + 3^2 + \cdots + 50^2$. Round off your answer to the nearest integer.

(b) Compute the exact value of the sum in part (a) using the formula $\displaystyle\sum_{k=1}^{n} k^2 = \dfrac{n(n + 1)(2n + 1)}{6}$. (This formula can be proved using mathematical induction.) Then compute the percent error for the approximation obtained in part (a). The percent error is given by

$$\frac{|\text{actual value} - \text{approximate value}|}{\text{actual value}} \times 100$$

(c) Use formula (1) to estimate the sum $1^4 + 2^4 + 3^4 + \cdots + 200^4$. Round off your answer to six significant digits.

(d) The following formula for the sum $1^4 + 2^4 + \cdots + n^4$ can be proved using mathematical induction:

$$\sum_{k=1}^{n} k^4 = \frac{n(n + 1)(2n + 1)(3n^2 + 3n - 1)}{30}$$

Use this formula to compute the sum in part (c). Round off your answer to six significant digits. Then use this result to compute the percent error for the approximation in part (c).

C **82.** According to *Stirling's formula* [named after James Stirling (1692–1770)], the quantity $n!$ can be approximated as follows:

$$n! \approx \sqrt{2\pi n} \left(\frac{n}{e}\right)^n$$

In this formula, e is the constant 2.718 . . . (discussed in Section 4.2). Use a calculator to complete the following table. Round off your answers to five significant digits. As you will see, the numbers in the right-hand column approach 1 as n increases. This shows that in a certain sense, the approximation improves as n increases.

n	$n!$	$\sqrt{2\pi n}\left(\dfrac{n}{e}\right)^n$	$\dfrac{n!}{\sqrt{2\pi n}(n/e)^n}$
10			
20			
30			
40			
50			
60			
65			

83. Two forces **F** and **G** act on an object. The force **G** acts horizontally to the right with a magnitude of 15 N, while **F** acts vertically upward with a magnitude of 20 N. Determine the magnitude and direction of the resultant force. (Use a calculator to determine the angle between the horizontal and the resultant; round off the result to one decimal place.)

84. Determine the resultant of the two forces in the accompanying figure. (Use a calculator, and round off the values you obtain for the magnitude and direction to one decimal place.)

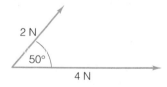

85. Determine the horizontal and vertical components of the velocity vector **v** in the following figure. (Use a calculator, and round off your answers to one decimal place.)

86. The heading and air speed of an airplane are 50° and 220 mph, respectively. If the wind is 60 mph from 140°, find the ground speed, the drift angle, and the course. (Use a calculator, and round off your answers to one decimal place.)

87. A block rests on an inclined plane that makes an angle of 24° with the horizontal. The component of the weight parallel to the plane is 14.8 pounds. Determine the weight of the block and the component of the weight perpendicular to the plane. (Use a calculator, and round off your answers to one decimal place.)

88. Find the length of the vector $\langle 20, 99 \rangle$.

89. For which values of b will the vectors $\langle 2, 6 \rangle$ and $\langle -5, b \rangle$ have the same length?

90. The coordinates of the points A and B are $(2, 6)$ and $(-7, 4)$, respectively. Find the components of the following vectors. Write your answers in the form $\langle x, y \rangle$.

(a) $\overrightarrow{AB}$ (b) $\overrightarrow{BA}$ (c) $3\overrightarrow{AB}$ (d) $\dfrac{1}{|\overrightarrow{AB}|}\overrightarrow{AB}$

In Exercises 91–102, compute each of the indicated quantities, given that the vectors **a**, **b**, **c**, *and* **d**, *are defined as follows:*

$$\mathbf{a} = \langle 3, 5 \rangle \qquad \mathbf{b} = \langle 7, 4 \rangle \qquad \mathbf{c} = \langle 2, -1 \rangle \qquad \mathbf{d} = \langle 0, 3 \rangle$$

91. $\mathbf{a} + \mathbf{b}$ **92.** $\mathbf{b} - \mathbf{d}$

93. $3\mathbf{c} + 2\mathbf{a}$ **94.** $|\mathbf{a}|$

95. $|\mathbf{b} + \mathbf{d}|^2 - |\mathbf{b} - \mathbf{d}|^2$ **96.** $\mathbf{a} + (\mathbf{b} + \mathbf{c})$

97. $(\mathbf{a} + \mathbf{b}) + \mathbf{c}$ **98.** $\mathbf{a} - (\mathbf{b} - \mathbf{c})$

99. $(\mathbf{a} - \mathbf{b}) - \mathbf{c}$

100. $|\mathbf{a} + \mathbf{b}|^2 + |\mathbf{a} - \mathbf{b}|^2 - 2|\mathbf{a}|^2 - 2|\mathbf{b}|^2$

101. $4\mathbf{c} + 2\mathbf{a} - 3\mathbf{b}$ **102.** $|4\mathbf{c} + 2\mathbf{a} - 3\mathbf{b}|$

103. Express the vector $\langle 7, -6 \rangle$ in terms of **i** and **j**.

104. Express the vector $4\mathbf{i} - 6\mathbf{j}$ in the form $\langle a, b \rangle$.

105. Find a unit vector having the same direction as $\langle 6, 4 \rangle$.

In Exercises 106 and 107, you are given an angle θ measured counterclockwise from the positive x-axis to a unit vector $\langle u_1, u_2 \rangle$. In each case, determine the components u_1 and u_2.

106. $\theta = 7\pi/4$

107. $\theta = \pi/12$ *Hint:* Use the addition formulas from Section 6.5.

108. Prove properties 5 and 6 in the box on page 601. *Hint:* For property 5, let $\mathbf{u} = \langle u_1, u_2 \rangle$ and $\mathbf{v} = \langle v_1, v_2 \rangle$. Compute the quantities on each side of the equation in property 5 and compare the results.

In Exercises 109–112, convert each complex number to rectangular form.

109. $3(\cos \frac{1}{3}\pi + i \sin \frac{1}{3}\pi)$ **110.** $\cos \frac{1}{6}\pi + i \sin \frac{1}{6}\pi$

111. $2^{1/4}(\cos \frac{7}{4}\pi + i \sin \frac{7}{4}\pi)$

112. $5[\cos(-\frac{1}{4}\pi) + i \sin(-\frac{1}{4}\pi)]$

In Exercises 113–116, express the complex numbers in trigonometric form.

113. $\frac{1}{2}(1 + \sqrt{3}\,i)$

114. $3i$

115. $-3\sqrt{2} - 3\sqrt{2}\,i$

116. $2\sqrt{3} - 2i$

In Exercises 117–126, carry out the indicated operations. Express your results in rectangular form for those cases in which the trigonometric functions are readily evaluated without tables or a calculator.

117. $5(\cos \frac{1}{7}\pi + i \sin \frac{1}{7}\pi) \times 2(\cos \frac{3}{28}\pi + i \sin \frac{3}{28}\pi)$

118. $4(\cos \frac{1}{12}\pi + i \sin \frac{1}{12}\pi) \times 3(\cos \frac{1}{12}\pi + i \sin \frac{1}{12}\pi)$

119. $8(\cos \frac{1}{12}\pi + i \sin \frac{1}{12}\pi) \div 4(\cos \frac{1}{3}\pi + i \sin \frac{1}{3}\pi)$

120. $4(\cos 32° + i \sin 32°) \div 2^{1/2}(\cos 2° + i \sin 2°)$

121. $(\cos \frac{1}{9}\pi + i \sin \frac{1}{9}\pi) \times 3(\cos \frac{4}{9}\pi + i \sin \frac{4}{9}\pi)$

122. $5(\cos 3° + i \sin 3°) \div 4(\cos 5° + i \sin 5°)$

123. $[3^{1/4}(\cos \frac{1}{36}\pi + i \sin \frac{1}{36}\pi)]^{12}$

124. $[2(\cos \frac{2}{15}\pi + i \sin \frac{2}{15}\pi)]^5$

125. $(\sqrt{3} + i)^{10}$

126. $(\sqrt{2} - \sqrt{2}\,i)^{15}$

In Exercises 127–130, use DeMoivre's theorem to find the indicated roots. Express your results in rectangular form.

127. Sixth roots of 1

128. Cube roots of $-64i$

129. Square roots of $\sqrt{2} - \sqrt{2}\,i$

130. Fourth roots of $1 + \sqrt{3}\,i$

131. Find the five fifth roots of $1 + i$. Express the roots in rectangular form. (Round off each decimal to two places in the final answer.)

132. If $z = r(\cos \theta + i \sin \theta)$, show that
$$r[\cos(\theta + \pi) + i \sin(\theta + \pi)] = -z$$

133. Trigonometric identities for $\cos n\theta$ and $\sin n\theta$ can be derived using DeMoivre's theorem. For example, according to DeMoivre's theorem, $(\cos \theta + i \sin \theta)^3 = \cos 3\theta + i \sin 3\theta$. Now expand the expression on the left-hand side of this last equation, and then equate the corresponding real and imaginary parts in the equation that results. What identities do you obtain?

134. The sum of three consecutive terms in a geometric sequence is 13, and the sum of the reciprocals is $\frac{13}{9}$. What are the possible values for the common ratio? (There are four answers.)

135. The nonzero numbers a, b, and c are consecutive terms in a geometric sequence, and $a + b + c = 70$. Furthermore, $4a$, $5b$, and $4c$ are consecutive terms in an arithmetic sequence. Find a, b, and c.

136. The positive numbers a, b, and c are consecutive terms in a geometric sequence, and a, $2b$, and c are consecutive terms in an arithmetic sequence. Show that the common ratio in the geometric sequence must be either $2 + \sqrt{3}$ or $2 - \sqrt{3}$.

137. If the numbers $\dfrac{1}{b + c}, \dfrac{1}{c + a}$, and $\dfrac{1}{a + b}$ are consecutive terms in an arithmetic sequence, show that a^2, b^2, and c^2 are also consecutive terms in an arithmetic sequence.

138. If a, b, and c are consecutive terms in a geometric sequence, prove that $\dfrac{1}{a + b}, \dfrac{1}{2b}$, and $\dfrac{1}{c + b}$ are consecutive terms in an arithmetic sequence.

139. **(a)** Find a value for x such that $3 + x$, $4 + x$, and $5 + x$ are consecutive terms in a geometric sequence.

(b) Given three numbers a, b, and c, find a value for x (in terms of a, b, and c) such that $a + x$, $b + x$, and $c + x$ are consecutive terms in a geometric sequence.

140. If $\ln(A + C) + \ln(A + C - 2B) = 2\ln(A - C)$, show that $1/A$, $1/B$, and $1/C$ are consecutive terms in an arithmetic sequence.

141. Let $a_1, a_2, a_3, \ldots$ be an arithmetic sequence with common difference d, and let r ($\neq 1$) be a real number. In this exercise we develop a formula for the sum of the series
$$a_1 + ra_2 + r^2 a_3 + \cdots + r^{n-1} a_n$$

The method we use here is essentially the same as the method used in the text to derive the formula for the sum of a finite geometric series. So, as background for this exercise, you should review the derivation on page 584 for the sum of a geometric series.

(a) Let S denote the required sum. Show that
$$S - rS = a_1 + rd + r^2 d + \cdots + r^{n-1} d - r^n a_n$$
$$= a_1 + \frac{d(r - r^n)}{1 - r} - r^n a_n$$

(b) Show that
$$S = \frac{a_1 - r^n a_n}{1 - r} + \frac{d(r - r^n)}{(1 - r)^2}$$

142. Use the formula in Exercise 141(b) to find the sum of each of the following series.

(a) $1 + 2 \times 2 + 2^2 \times 3 + 2^3 \times 4 + \cdots + 2^{13} \times 14$

(b) $2 + 4 \times 5 + 4^2 \times 8 + 4^3 \times 11 + \cdots + 4^6 \times 20$

(c) $3 - \dfrac{1}{2} \cdot 5 + \dfrac{1}{2^2} \cdot 7 - \dfrac{1}{2^3} \cdot 9 + \cdots +$ (to 10 terms)

APPENDIX

A.1 ▼ SIGNIFICANT DIGITS AND CALCULATORS

Many of the numbers that we use in scientific work and in daily life are approximations. In some cases the approximations arise because the numbers were obtained through measurements or experiments. Consider, for example, the following statement from an astronomy textbook:

> The diameter of the Moon is 3476 km

We interpret this statement as meaning that the actual diameter D is closer to 3476 km than it is to either 3475 km or 3477 km. In other words,

$$3475.5 \text{ km} \leq D \leq 3476.5 \text{ km}$$

The interval [3475.5, 3476.5] in this example provides information about the accuracy of the measurement. Another way to indicate accuracy in an approximation is by specifying the number of *significant digits* it contains. The measurement 3476 km has four significant digits. In general, the number of significant digits in a given number is found as follows.

Significant Digits

	EXAMPLES	
The number of significant digits in a given number is determined by counting the digits from left to right, beginning with the left-most nonzero digit.	Number	Number of Significant Digits
	1.43	3
	0.52	2
	0.05	1
	4837	4
	4837.0	5

Numbers obtained through measurements are not the only source of approximations in scientific work. For example, to five significant digits, we have the following approximation for the irrational number π:

$$\pi \approx 3.1416$$

This statement tells us that π is closer to 3.1416 than it is to either 3.1415 or 3.1417. In other words,

$$3.14155 \leq \pi \leq 3.14165$$

Table 1 provides some additional examples of the ideas we've introduced.

TABLE 1

NUMBER	NUMBER OF SIGNIFICANT DIGITS	RANGE OF MEASUREMENT
37	2	[36.5, 37.5]
37.0	3	[36.95, 37.05]
268.1	4	[268.05, 268.15]
1.036	4	[1.0355, 1.0365]
0.036	2	[0.0355, 0.0365]

There is an ambiguity involving zero that can arise in counting significant digits. Suppose that someone measures the width w of a rectangle and reports the result as 30 cm. How many significant digits are there? If the value 30 cm was obtained by measuring to the nearest 10 cm, then only the digit 3 is significant, and we can conclude only that the width w lies in the range 25 cm $\leq w \leq$ 35 cm. On the other hand, if the 30 cm was obtained by measuring to the nearest 1 cm, then both the digits 3 and 0 are significant and we have 29.5 cm $\leq w \leq$ 30.5 cm.

By using **scientific notation** we can avoid the type of ambiguity discussed in the previous paragraph. A number written in the form

$$b \times 10^n \qquad \text{where } 1 \leq b < 10 \text{ and } n \text{ is an integer}$$

is said to be expressed in scientific notation. For the example in the previous paragraph, then, we would write

$$w = 3 \times 10^1 \text{ cm} \qquad \text{if the measurement were to the nearest 10 cm}$$

and

$$w = 3.0 \times 10^1 \text{ cm} \qquad \text{if the measurement were to the nearest 1 cm}$$

As the figures in Table 2 indicate, for a number $b \times 10^n$ in scientific notation, the number of significant digits is just the number of digits in b. (This is one of the advantages in using scientific notation; the number of significant digits, and hence the accuracy of the measurement, is readily apparent.)

TABLE 2

MEASUREMENT	NUMBER OF SIGNIFICANT DIGITS	RANGE OF MEASUREMENT
Mass of the earth:		
6×10^{27} g	1	$[5.5 \times 10^{27} \text{ g}, 6.5 \times 10^{27} \text{ g}]$
6.0×10^{27} g	2	$[5.95 \times 10^{27} \text{ g}, 6.05 \times 10^{27} \text{ g}]$
5.974×10^{27} g	4	$[5.9735 \times 10^{27} \text{ g}, 5.9745 \times 10^{27} \text{ g}]$
Mass of a proton:		
1.67×10^{-24} g	3	$[1.665 \times 10^{-24} \text{ g}, 1.675 \times 10^{-24} \text{ g}]$

There are a number of exercises in this text in which a calculator either is required or is extremely useful. Although you could work many of these exercises using tables instead of a calculator, this author recommends that you purchase a scientific calculator. *Scientific calculator* is a generic term describing a calculator with (at least) the following features or functions beyond the usual arithmetic functions:

1. Memory
2. Scientific notation
3. Powers and roots
4. Logarithms (base ten and base e)
5. Trigonometric functions
6. Inverse functions

Because of the variety and differences in the calculators that are available, no specific instructions are provided in this appendix for operating a calculator. When you buy a calculator, read the owner's manual carefully and work through some of the examples in it. In general, learn to use the memory capabilities of the calculator so that, as far as possible, you don't need to write down the results of the intermediate steps in a given calculation.

Many of the numerical exercises in the text ask that you round off the answers to a specified number of decimal places. Our rules for rounding off are as follows.

Rules for Rounding Off a Number (With More Than n Decimal Places) to n Decimal Places

1. If the digit in the $(n + 1)$st decimal place is greater than 5, increase the digit in the nth place by 1. If the digit in the $(n + 1)$st place is less than 5, leave the nth digit unchanged.

2. If the digit in the $(n + 1)$st decimal place is 5 and there is at least one nonzero digit to the right of this 5, increase the digit in the nth decimal place by 1.

3. If the digit in the $(n + 1)$st decimal place is 5 and there are no nonzero digits to the right of this 5, then increase the digit in the nth decimal place by 1 only if this results in an even digit.

The examples in Table 3 illustrate the use of these rules.

TABLE 3

NUMBER	ROUNDED TO ONE DECIMAL PLACE	ROUNDED TO THREE DECIMAL PLACES
4.3742	4.4	4.374
2.0515	2.1	2.052
2.9925	3.0	2.992

These same rules can be adapted for rounding off a result to a specified number of significant digits. As examples of this, we have

2347	rounded off to two significant digits is $2300 = 2.3 \times 10^3$
2347	rounded off to three significant digits is $2350 = 2.35 \times 10^3$
975	rounded off to two significant digits is $980 = 9.8 \times 10^2$
0.985	rounded off to two significant digits is $0.98 = 9.8 \times 10^{-1}$

In calculator exercises that ask you to round off your answers, it's important that you postpone rounding until the final calculation is carried out. For example, suppose that you are required to determine the hypotenuse x of the right triangle in Figure 1 to two significant digits. Using the Pythagorean theorem, we have

$$x = \sqrt{(1.36)^2 + (2.46)^2}$$
$$= 2.8 \quad \text{(using a calculator and rounding off the final result to two significant digits)}$$

On the other hand, if we first round off each of the given lengths to two significant digits, we obtain

$$x = \sqrt{(1.4)^2 + (2.5)^2}$$
$$= 2.9 \quad \text{(to two significant digits)}$$

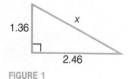

1.36

x

2.46

FIGURE 1

TABLE 4

NUMBER	RANGE OF MEASUREMENT
1.36	[1.355, 1.365]
2.46	[2.455, 2.465]

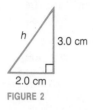

FIGURE 2

Guidelines for Computing with Approximations

This last result is inappropriate, and we can see why as follows. As Table 4 shows, the maximum possible values for the sides are 1.365 and 2.465, respectively.

Thus, the maximum possible value for the hypotenuse must be

$$\sqrt{(1.365)^2 + (2.465)^2} = 2.817\ldots \qquad \text{(calculator display)}$$
$$= 2.8 \qquad \text{(to two significant digits)}$$

This shows that the value 2.9 is indeed inappropriate, as we stated previously.

An error often made by people working with calculators and approximations is to report a final answer with a greater degree of accuracy than the data warrant. Consider, for example, the right triangle in Figure 2. Using the Pythagorean theorem and a calculator with an eight-digit display, we obtain

$$h = 3.6055513 \text{ cm}$$

This value for h is inappropriate, since common sense tells us that the answer should be no more accurate than the data used to obtain that answer. In particular, since the given sides of the triangle apparently were measured only to the nearest tenth of a centimeter, we certainly should not expect any improvement in accuracy for the resulting value of the hypotenuse. An appropriate form for the value of h here would be $h = 3.6$ cm. In general, for calculator exercises in this text that do not specify a required number of decimal places or significant digits in the final results, you should use the following guidelines.

1. *For adding and subtracting:* Round off the final result so that it contains only as many decimal places as there are in the data with the fewest decimal places.

2. *For multiplying and dividing:* Round off the final result so that it contains only as many significant digits as there are in the data with the fewest significant digits.

3. *For powers and roots:* In computing a power or a root of a real number b, round off the result so that it contains as many significant digits as there are in b.

A.2 ▼ SOME PROOFS FOR SECTION 1.1

In this section, we list the most basic properties of the real number system. Then we use those properties to prove the familiar rules of algebra for working with signed numbers.

The set of real numbers is **closed** with respect to the operations of addition and multiplication. This just means that when we add or multiply two real numbers, the result (i.e., the **sum** or **product**, respectively) is again a real number. Some of the other basic properties and definitions for the real number system are listed in the following box. Note that the lowercase letters a, b, and c denote arbitrary real numbers.

PROPERTY SUMMARY	SOME FUNDAMENTAL PROPERTIES OF THE REAL NUMBERS
Commutative properties	$a + b = b + a \qquad ab = ba$
Associative properties	$a + (b + c) = (a + b) + c$ $a(bc) = (ab)c$
Identity properties	**1.** There is a unique real number 0 (called **zero** or the **additive identity**) such that $a + 0 = a \qquad$ and $\qquad 0 + a = a$ **2.** There is a unique real number 1 (called **one** or the **multiplicative identity**) such that $a \cdot 1 = a \qquad$ and $\qquad 1 \cdot a = a$
Inverse properties	**1.** For each real number a, there is a real number $-a$ (called the **additive inverse** of a or the **negative** of a) such that $a + (-a) = 0 \qquad$ and $\qquad (-a) + a = 0$ **2.** For each real number $a \neq 0$, there is a real number denoted by $\dfrac{1}{a}$ (or $1/a$ or a^{-1}), and called the **multiplicative inverse** or **reciprocal** of a, such that $a \cdot \dfrac{1}{a} = 1 \qquad$ and $\qquad \dfrac{1}{a} \cdot a = 1$
Distributive properties	$a(b + c) = ab + ac \qquad (b + c)a = ba + ca$

After reading this list of properties for the first time, many students ask the natural question, "Why do we even bother to list such obvious properties?" One reason is that all the other laws of arithmetic and algebra (including the "not-so-obvious" ones) can be derived from our rather short list. For example, the rule $0 \cdot a = 0$ can be proved using the distributive property, as can the rule that the product of two negative numbers is a positive number. We will now state and prove these properties as well as several others.

Theorem

Let a and b be real numbers. Then

(a) $a \cdot 0 = 0$ **(b)** $-a = (-1)a$ **(c)** $-(-a) = a$
(d) $a(-b) = -ab$ **(e)** $(-a)(-b) = ab$

PROOF OF PART (a)

$$a \cdot 0 = a \cdot (0 + 0) \qquad \text{additive identity property}$$
$$= a \cdot 0 + a \cdot 0 \qquad \text{distributive property}$$

Now, since $a \cdot 0$ is a real number, it has an additive inverse, $-(a \cdot 0)$. Adding this to both sides of the last equation, we obtain

$$a \cdot 0 + [-(a \cdot 0)] = (a \cdot 0 + a \cdot 0) + [-(a \cdot 0)]$$

$a \cdot 0 + [-(a \cdot 0)] = a \cdot 0 + \{a \cdot 0 + [-(a \cdot 0)]\}$ associative property of addition

$\qquad\qquad\quad 0 = a \cdot 0 + 0$ additive inverse property

$\qquad\qquad\quad 0 = a \cdot 0$ additive identity property

Thus $a \cdot 0 = 0$, as we wished to show.

PROOF OF PART (b)

$0 = 0 \cdot a$ using part (a) and the commutative property of multiplication

$ = [1 + (-1)]a$ additive inverse property

$ = 1 \cdot a + (-1)a$ distributive property

$ = a + (-1)a$ multiplicative identity property

Now, by adding $-a$ to both sides of this last equation, we obtain

$-a + 0 = -a + [a + (-1)a]$

$\quad -a = (-a + a) + (-1)a$ additive identity property and associative property of addition

$ = 0 + (-1)a$ additive inverse property

$ = (-1)a$ additive identity property

This last equation asserts that $-a = (-1)a$, as we wished to show.

PROOF OF PART (c)

$-(-a) + (-a) = 0$ additive inverse property

By adding a to both sides of this last equation, we obtain

$[-(-a) + (-a)] + a = 0 + a$

$\quad -(-a) + (-a + a) = a$ associative property of addition and additive identity property

$\qquad\quad -(-a) + 0 = a$ additive inverse property

$\qquad\qquad\quad -(-a) = a$ additive identity property

This last equation states that $-(-a) = a$, as we wished to show.

PROOF OF PART (d)

$a(-b) = a[(-1)b]$ using part (b)

$ = [a(-1)]b$ associative property of multiplication

$ = [(-1)a]b$ commutative property of multiplication

$ = (-1)(ab)$ associative property of multiplication

$ = -(ab)$ using part (b)

Thus $a(-b) = -ab$, as we wished to show.

PROOF OF PART (e)

$$\begin{aligned}
(-a)(-b) &= -[(-a)b] & \text{using part (d)}\\
&= -[b(-a)] & \text{commutative property of multiplication}\\
&= -[-(ba)] & \text{using part (d)}\\
&= ba & \text{using part (c)}\\
&= ab & \text{commutative property of multiplication}
\end{aligned}$$

We've now shown that $(-a)(-b) = ab$, as required.

A.3 ▼ $\sqrt{2}$ IS IRRATIONAL

The proof is by *reductio ad absurdum*, and *reductio ad absurdum*, which Euclid loved so much, is one of a mathematician's finest weapons.

G. H. Hardy (1877–1947)

We will use an *indirect proof* to show that the square root of two is an irrational number. The strategy is as follows.

1. We suppose that $\sqrt{2}$ is a rational number.
2. Using (1) and the usual rules of logic and algebra, we derive a contradiction.
3. On the basis of the contradiction in (2), we conclude that the supposition in (1) is untenable; that is, we conclude that $\sqrt{2}$ is irrational.

In carrying out the proof, we'll assume as known the following three statements.

If x is an even natural number, then $x = 2k$, for some natural number k.

Any rational number can be written in the form a/b, where the integers a and b have no common integral factors other than ± 1. (In other words, any fraction can be reduced to lowest terms.)

If x is a natural number and x^2 is even, then x is even.

Our indirect proof now proceeds as follows. Suppose that $\sqrt{2}$ were a rational number. Then we would be able to write

$$\sqrt{2} = \frac{a}{b} \qquad \begin{array}{l}\text{where } a \text{ and } b \text{ are natural numbers with no}\\ \text{common factor other than 1}\end{array} \qquad (1)$$

Since both sides of equation (1) are positive, we can square both sides to obtain the equivalent equation

$$2 = \frac{a^2}{b^2}$$

or

$$2b^2 = a^2 \qquad (2)$$

Since the left-hand side of equation (2) is an even number, the right-hand side must be even. But if a^2 is even, then a is even, and so

$$a = 2k, \qquad \text{for some natural number } k$$

Using this last equation to substitute for a in equation (2), we have

$$2b^2 = (2k)^2 = 4k^2$$

or

$$b^2 = 2k^2$$

Hence (reasoning as before) b^2 is even, and therefore b is even. But then we have that both b and a are even, contrary to our hypothesis that b and a have no common factor other than 1. We conclude from this that equation (1) cannot hold; that is, there is no rational number a/b such that $\sqrt{2} = a/b$. Thus $\sqrt{2}$ is irrational, as we wished to prove.

A.4 ▼ WORKING WITH FRACTIONAL EXPRESSIONS

But if you do use a rule involving mechanical calculation, be patient, accurate, and systematically neat in the working. It is well known to mathematical teachers that quite half the failures in algebraic exercises arise from arithmetical inaccuracy and slovenly arrangement.

George Chrystal (1893)

In algebra, as in arithmetic, we say that a fraction is **reduced to lowest terms** or **simplified** when the numerator and denominator contain no common factors (other than 1 and -1). The factoring techniques developed in Section 1.5 are used to reduce fractions. For example, to reduce the fraction $\dfrac{x^2 - 9}{x^2 + 3x}$, we write

$$\frac{x^2 - 9}{x^2 + 3x} = \frac{(x - 3)(x + 3)}{x(x + 3)} = \frac{x - 3}{x}$$

In Example 1 we display two more instances in which factoring is used to reduce a fraction. After that, Example 2 indicates how these skills are used to multiply and divide fractional expressions.

EXAMPLE 1 Simplify: **(a)** $\dfrac{x^3 - 8}{x^2 - 2x}$ **(b)** $\dfrac{x^2 - 6x + 8}{a(x - 2) + b(x - 2)}$

Solution **(a)** $\dfrac{x^3 - 8}{x^2 - 2x} = \dfrac{(x - 2)(x^2 + 2x + 4)}{x(x - 2)} = \dfrac{x^2 + 2x + 4}{x}$

(b) $\dfrac{x^2 - 6x + 8}{a(x - 2) + b(x - 2)} = \dfrac{(x - 2)(x - 4)}{(x - 2)(a + b)} = \dfrac{x - 4}{a + b}$ ▲

EXAMPLE 2 Carry out the indicated operations and simplify:

(a) $\dfrac{x^3}{2x^2 + 3x} \cdot \dfrac{12x + 18}{4x^2 - 6x}$ **(b)** $\dfrac{2x^2 - x - 6}{x^2 + x + 1} \cdot \dfrac{x^3 - 1}{4 - x^2}$

(c) $\dfrac{x^2 - 144}{x^2 - 4} \div \dfrac{x + 12}{x + 2}$

Solution **(a)** $\dfrac{x^3}{2x^2 + 3x} \cdot \dfrac{12x + 18}{4x^2 - 6x} = \dfrac{x^3}{x(2x + 3)} \cdot \dfrac{6(2x + 3)}{2x(2x - 3)}$

$$= \frac{6x^3}{2x^2(2x - 3)} = \frac{3x}{2x - 3}$$

(b) $\dfrac{2x^2 - x - 6}{x^2 + x + 1} \cdot \dfrac{x^3 - 1}{4 - x^2} = \dfrac{(2x + 3)(x - 2)}{(x^2 + x + 1)} \cdot \dfrac{(x - 1)(x^2 + x + 1)}{(2 - x)(2 + x)}$

$$= \frac{-(2x + 3)(x - 1)}{2 + x} \qquad \text{Using the fact that } \frac{x - 2}{2 - x} = -1$$

$$= \frac{-2x^2 - x + 3}{2 + x}$$

(c) $\dfrac{x^2 - 144}{x^2 - 4} \div \dfrac{x + 12}{x + 2} = \dfrac{x^2 - 144}{x^2 - 4} \cdot \dfrac{x + 2}{x + 12}$

$$= \dfrac{(x - 12)(x + 12)}{(x - 2)(x + 2)} \cdot \dfrac{x + 2}{x + 12} = \dfrac{x - 12}{x - 2}$$

▲

As in arithmetic, to add or subtract two fractions, the denominators must be the same. The rules in this case are

$$\dfrac{a}{b} + \dfrac{c}{b} = \dfrac{a + c}{b} \qquad \dfrac{a}{b} - \dfrac{c}{b} = \dfrac{a - c}{b}$$

For example, we have

$$\dfrac{4x - 1}{x + 1} - \dfrac{2x - 1}{x + 1} = \dfrac{4x - 1 - (2x - 1)}{x + 1} = \dfrac{4x - 1 - 2x + 1}{x + 1} = \dfrac{2x}{x + 1}$$

Fractions with unlike denominators are added or subtracted by first converting to a common denominator. For instance, to add $9/a$ and $10/a^2$ we write

$$\dfrac{9}{a} + \dfrac{10}{a^2} = \dfrac{9}{a} \cdot \dfrac{a}{a} + \dfrac{10}{a^2} = \dfrac{9a}{a^2} + \dfrac{10}{a^2} = \dfrac{9a + 10}{a^2}$$

Notice that the common denominator used was a^2. This is the **least common denominator**. In fact, other common denominators (such as a^3 or a^4) could be used here, but that would be less efficient. In general, the least common denominator for a given group of fractions is chosen as follows. Write down a product involving the irreducible factors from each denominator. The power of each factor should be equal to (but not greater than) the highest power of that factor appearing in any of the individual denominators. For example, the least common denominator for the two fractions $\dfrac{1}{(x + 1)^2}$ and $\dfrac{1}{(x + 1)(x + 2)}$ is $(x + 1)^2(x + 2)$. In the example that follows, notice how factoring the denominators facilitates the choice of the least common denominator.

EXAMPLE 3 Combine into a single fraction and simplify:

(a) $\dfrac{3}{4x} + \dfrac{7x}{10y^2} - 2$ **(b)** $\dfrac{x}{x^2 - 9} - \dfrac{1}{x + 3}$ **(c)** $\dfrac{15}{x^2 + x - 6} + \dfrac{x + 1}{2 - x}$

Solution **(a)** Denominators: $2^2 \cdot x$; $2 \cdot 5 \cdot y^2$; 1
Least common denominator: $2^2 \cdot 5xy^2 = 20xy^2$

$$\dfrac{3}{4x} + \dfrac{7x}{10y^2} - \dfrac{2}{1} = \dfrac{3}{4x} \cdot \dfrac{5y^2}{5y^2} + \dfrac{7x}{10y^2} \cdot \dfrac{2x}{2x} - \dfrac{2}{1} \cdot \dfrac{20xy^2}{20xy^2}$$

$$= \dfrac{15y^2 + 14x^2 - 40xy^2}{20xy^2}$$

(b) Denominators: $x^2 - 9 = (x - 3)(x + 3)$; $x + 3$
Least common denominator: $(x - 3)(x + 3)$

$$\frac{x}{x^2 - 9} - \frac{1}{x + 3} = \frac{x}{(x - 3)(x + 3)} - \frac{1}{x + 3}$$

$$= \frac{x}{(x - 3)(x + 3)} - \frac{1}{x + 3} \cdot \frac{x - 3}{x - 3}$$

$$= \frac{x - (x - 3)}{(x - 3)(x + 3)} = \frac{3}{(x - 3)(x + 3)}$$

(c) $\dfrac{15}{x^2 + x - 6} + \dfrac{x + 1}{2 - x} = \dfrac{15}{(x - 2)(x + 3)} - \dfrac{x + 1}{x - 2}$ Using the fact that $2 - x = -(x - 2)$

$$= \frac{15}{(x - 2)(x + 3)} - \frac{x + 1}{x - 2} \cdot \frac{x + 3}{x + 3}$$

$$= \frac{15 - (x^2 + 4x + 3)}{(x - 2)(x + 3)} = \frac{-x^2 - 4x + 12}{(x - 2)(x + 3)}$$

$$= \frac{(x - 2)(-x - 6)}{(x - 2)(x + 3)} = \frac{-x - 6}{x + 3}$$

Notice in the last line that we were able to simplify the answer by factoring the numerator and reducing the fraction. (This is why we prefer to leave the least common denominator in factored form, rather than multiply it out, in this type of problem.) ▲

EXAMPLE 4   Simplify: $\dfrac{\dfrac{1}{3a} - \dfrac{1}{4b}}{\dfrac{5}{6a^2} + \dfrac{1}{b}}$.

Solution   The least common denominator for the four individual fractions is $12a^2b$. Multiplying the given expression by $\dfrac{12a^2b}{12a^2b}$, which equals 1, yields

$$\frac{12a^2b}{12a^2b} \cdot \frac{\dfrac{1}{3a} - \dfrac{1}{4b}}{\dfrac{5}{6a^2} + \dfrac{1}{b}} = \frac{4ab - 3a^2}{10b + 12a^2}$$

▲

EXAMPLE 5   Simplify $(x^{-1} + y^{-1})^{-1}$. (The answer is *not* $x + y$.)

Solution   After applying the definition of negative exponents to rewrite the given expression, we'll use the method shown in Example 4.

$$(x^{-1} + y^{-1})^{-1} = \left(\frac{1}{x} + \frac{1}{y} \right)^{-1}$$

$$= \frac{1}{\dfrac{1}{x} + \dfrac{1}{y}}$$

$$= \frac{xy}{xy} \cdot \frac{1}{\dfrac{1}{x} + \dfrac{1}{y}} = \frac{xy}{y + x}$$

▲

EXAMPLE 6 Simplify $\dfrac{\dfrac{1}{x+h}-\dfrac{1}{x}}{h}$. (This type of expression occurs in calculus.)

Solution

$$\frac{\dfrac{1}{x+h}-\dfrac{1}{x}}{h}=\frac{(x+h)x}{(x+h)x}\cdot\frac{\dfrac{1}{x+h}-\dfrac{1}{x}}{h}$$

$$=\frac{x-(x+h)}{(x+h)xh}=\frac{-h}{(x+h)xh}=-\frac{1}{(x+h)x}$$

EXAMPLE 7 Simplify: $\dfrac{x-\dfrac{1}{x^2}}{\dfrac{1}{x^2}-1}$.

Solution

$$\frac{x-\dfrac{1}{x^2}}{\dfrac{1}{x^2}-1}=\frac{x^2}{x^2}\cdot\frac{x-\dfrac{1}{x^2}}{\dfrac{1}{x^2}-1}$$

$$=\frac{x^3-1}{1-x^2}=-\frac{x^3-1}{x^2-1}$$

$$=-\frac{(x-1)(x^2+x+1)}{(x-1)(x+1)}=-\frac{x^2+x+1}{x+1}$$

▼ EXERCISE SET A.4

A

In Exercises 1–12, reduce the fractions to lowest terms.

1. $\dfrac{x^2-9}{x+3}$

2. $\dfrac{25-x^2}{x-5}$

3. $\dfrac{x+2}{x^4-16}$

4. $\dfrac{x^2-x-20}{2x^2+7x-4}$

5. $\dfrac{x^2+2x+4}{x^3-8}$

6. $\dfrac{a+b}{ax^2+bx^2}$

7. $\dfrac{9ab-12b^2}{6a^2-8ab}$

8. $\dfrac{a^3b^2-27b^5}{(ab-3b^2)^2}$

9. $\dfrac{a^3+a^2+a+1}{a^2-1}$

10. $\dfrac{(x-y)^2(a+b)}{(x^2-y^2)(a^2+2ab+b^2)}$

11. $\dfrac{x^3-y^3}{(x-y)^3}$

12. $\dfrac{x^4-y^4}{(x^4y+x^2y^3+x^3y^2+xy^4)(x-y)^2}$

In Exercises 13–63, carry out the indicated operations and simplify where possible.

13. $\dfrac{2}{x-2}\cdot\dfrac{x^2-4}{x+2}$

14. $\dfrac{ax+3}{2a+1}\div\dfrac{a^2x^2+3ax}{4a^2-1}$

15. $\dfrac{x^2-x-2}{x^2+x-12}\cdot\dfrac{x^2-3x}{x^2-4x+4}$

16. $(3t^2+4tx+x^2)\div\dfrac{3t^2-2tx-x^2}{t^2-x^2}$

17. $\dfrac{x^3+y^3}{x^2-4xy+3y^2}\div\dfrac{(x+y)^3}{x^2-2xy-3y^2}$

18. $\dfrac{a^2-a-42}{a^4+216a}\div\dfrac{a^2-49}{a^3-6a^2+36a}$

19. $\dfrac{x^2+xy-2y^2}{x^2-5xy+4y^2}\cdot\dfrac{x^2-7xy+12y^2}{x^2+5xy+6y^2}$

20. $\dfrac{x^4y^2-xy^5}{x^4-2x^2y^2+y^4}\cdot\dfrac{x^2+2xy+y^2}{x^3y^2+x^2y^3+xy^4}$

21. $\dfrac{4}{x}-\dfrac{2}{x^2}$

22. $\dfrac{1}{3x}+\dfrac{1}{5x^2}-\dfrac{1}{30x^3}$

23. $\dfrac{6}{a} - \dfrac{a}{6}$

24. $\dfrac{1}{a} + \dfrac{1}{b} + \dfrac{1}{c}$

25. $\dfrac{1}{x+3} + \dfrac{3}{x+2}$

26. $\dfrac{4}{x-4} - \dfrac{4}{x+1}$

27. $\dfrac{3x}{x-2} - \dfrac{6}{x^2-4}$

28. $1 + \dfrac{1}{x} - \dfrac{1}{x^2}$

29. $\dfrac{a}{x-1} + \dfrac{2ax}{(x-1)^2} + \dfrac{3ax^2}{(x-1)^3}$

30. $\dfrac{a^2+5a-4}{a^2-16} - \dfrac{2a}{2a^2+8a}$

31. $\dfrac{x}{x^2-9} + \dfrac{x-1}{x^2-5x+6}$

32. $\dfrac{1}{x-1} + \dfrac{1}{1-x}$

33. $\dfrac{4}{x-5} - \dfrac{4}{5-x}$

34. $\dfrac{x}{x+a} + \dfrac{a}{a-x}$

35. $\dfrac{a^2+b^2}{a^2-b^2} + \dfrac{a}{a+b} + \dfrac{b}{b-a}$

36. $\dfrac{3}{2x+2} - \dfrac{5}{x^2-1} + \dfrac{1}{x+1}$

37. $\dfrac{1}{x^2+x-20} - \dfrac{1}{x^2-8x+16}$

38. $\dfrac{4}{6x^2+5x-4} + \dfrac{1}{3x^2+4x} - \dfrac{1}{2x-1}$

39. $\dfrac{2q+p}{2p^2-9pq-5q^2} - \dfrac{p+q}{p^2-5pq}$

40. $\dfrac{1}{x-1} + \dfrac{1}{x^2-1} + \dfrac{1}{x^3-1}$

41. $\dfrac{x}{(x-y)(x-z)} + \dfrac{y}{(y-z)(y-x)} + \dfrac{z}{(z-x)(z-y)}$

42. $\dfrac{3x}{x-2} + \dfrac{4x^2}{x^2+2x+1} - \dfrac{x^2}{x^2-x-2}$

43. $\dfrac{y+z}{x^2-xy-xz+yz} - \dfrac{x+z}{xy-xz-y^2+yz} + \dfrac{x+y}{xy-yz-xz+z^2}$

44. $\left(x+\dfrac{1}{x}\right) \cdot \left(1-\dfrac{1}{x}\right)^2 \div \left(x-\dfrac{1}{x}\right)$

45. $\dfrac{x^2+x-a-1}{a^2-x^2} + \dfrac{x+1}{x+a} \div \dfrac{x-a}{x-1}$

46. $\dfrac{1}{2ax-2a^2} - \dfrac{1}{2ax+2a^2}$

47. $\dfrac{x^2-2ax+a^2}{px+q}\left(\dfrac{p}{x-a} + \dfrac{ap+q}{(x-a)^2}\right)$

48. $\dfrac{\dfrac{1}{a}+\dfrac{1}{b}}{\dfrac{1}{a}-\dfrac{1}{b}}$

49. $\dfrac{1+\dfrac{4}{x}}{\dfrac{3}{x}-2}$

50. $\dfrac{\dfrac{1}{a}}{1+\dfrac{b}{c}}$

51. $\dfrac{a-\dfrac{1}{a}}{1+\dfrac{1}{a}}$

52. $\dfrac{\dfrac{1}{x^2}-\dfrac{1}{y^2}}{\dfrac{1}{x}+\dfrac{1}{y}}$

53. $\dfrac{\dfrac{1}{2+h}-\dfrac{1}{2}}{h}$

54. $\dfrac{\dfrac{3}{x^2+h}-\dfrac{3}{x^2}}{h}$

55. $\dfrac{\dfrac{a}{x^2}+\dfrac{x}{a^2}}{a^2-ax+x^2}$

56. $\dfrac{x+\dfrac{xy}{y-x}}{\dfrac{y^2}{x^2-y^2}+1}$

B

57. $\dfrac{x+y}{2(x^2+y^2)} - \dfrac{1}{2(x+y)} + \dfrac{x-y}{x^2-y^2} - \dfrac{x^3-y^3}{x^4-y^4}$

58. $\dfrac{4y}{x^2-y^2} + \dfrac{1}{x+y} + \dfrac{1}{y-x} - \dfrac{2y}{x^2+y^2}$

59. $\dfrac{\dfrac{a+b}{a-b}+\dfrac{a-b}{a+b}}{\dfrac{a-b}{a+b}-\dfrac{a+b}{a-b}} \cdot \dfrac{ab^3-a^3b}{a^2+b^2}$

60. $\dfrac{\sqrt{x+a}}{\sqrt{x-a}} - \dfrac{\sqrt{x-a}}{\sqrt{x+a}}$

C

61. $\dfrac{ap+q}{ax-bx-a^2+ab} + \dfrac{bp+q}{bx-ax-b^2+ab}$

62. $\dfrac{b-c}{a^2-ab-ac+bc} + \dfrac{a-c}{b^2-bc-ab+ac} + \dfrac{a-b}{c^2-ac-bc+ab}$

63. $\dfrac{\dfrac{1}{a}-\dfrac{a-x}{a^2+x^2}}{\dfrac{1}{x}-\dfrac{x-a}{x^2+a^2}} + \dfrac{\dfrac{1}{a}-\dfrac{a+x}{a^2+x^2}}{\dfrac{1}{x}-\dfrac{x+a}{x^2+a^2}}$

64. Simplify: $\dfrac{\left(a+\dfrac{1}{b}\right)^a\left(a-\dfrac{1}{b}\right)^b}{\left(b+\dfrac{1}{a}\right)^a\left(b-\dfrac{1}{a}\right)^b}$

65. Simplify:

$$\dfrac{x^2-qr}{(p-q)(p-r)} + \dfrac{x^2-rp}{(q-r)(q-p)} + \dfrac{x^2-pq}{(r-p)(r-q)}$$

66. Simplify: $\left(\dfrac{x^{-2}+y^{-2}}{x^{-2}-y^{-2}} - \dfrac{x^{-2}-y^{-2}}{x^{-2}+y^{-2}}\right)$

$$\div\left[\left(\dfrac{x+y}{x-y}+\dfrac{x-y}{x+y}\right)\left(\dfrac{x^2}{y^2}+\dfrac{y^2}{x^2}-2\right)\right]^{-1}$$

67. Consider the three fractions

$$\dfrac{b-c}{1+bc}, \qquad \dfrac{c-a}{1+ca}, \qquad \text{and} \qquad \dfrac{a-b}{1+ab}$$

(a) If $a=1$, $b=2$, and $c=3$, find the sum of the three fractions. Also compute their product. What do you observe?

(b) Show that the sum and the product of the three given fractions are, in fact, always equal.

68. Evaluate the expression

$$\frac{y-a}{x-a} - \frac{y+a}{x+a} + \frac{y-b}{x-b} - \frac{y+b}{x+b}$$

when $x = \sqrt{ab}$. *Hint:* Keep the first and second terms separate from the third and fourth as long as possible.

69. Show that $(1 + a^{x-y})^{-1} + (1 + a^{y-x})^{-1} = 1$.

70. Simplify:

$$\frac{x^3 y^{-3} - y^3 x^{-3}}{(xy^{-1} - yx^{-1})(xy^{-1} + yx^{-1} - 1)} \cdot \frac{y^{-1} - x^{-1}}{x^{-2} + y^{-2} + x^{-1}y^{-1}}$$

A.5 ▼ ANGLES IN ELEMENTARY GEOMETRY

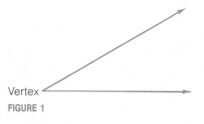

Vertex

FIGURE 1

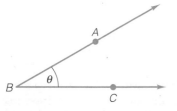

Notations for the angle at B:
$\angle B$, θ, $\angle ABC$, $\angle CBA$

FIGURE 2

In elementary geometry, an *angle* is a figure formed by two rays with a common endpoint. As indicated in Figure 1, the common endpoint is called the **vertex** of the angle. There are several common conventions used in naming angles. Figure 2 indicates some of these. In Figure 2, the symbol θ is the lowercase Greek letter *theta*. Greek letters are often used to name angles. (For reference, the Greek alphabet is given in the endpapers at the back of this book.) The $\angle$ symbol that you see in Figure 2 stands for the word "angle." When three letters are used to name an angle, as with $\angle ABC$ in Figure 2, the middle letter always indicates the vertex of the angle.

There are several ways to specify the size of (i.e., the amount of rotation in) an angle. In this appendix, we use the familiar units of *degrees*. Recall that 360 degrees is the measure of an angle obtained by rotating a ray through one complete circle. The symbol for degrees is °. Figure 3 displays angles of various degrees, and Table 1 summarizes some of the terminology for angles.

Let us agree on the following convention. Suppose, for example, that the measure of an angle, say, angle B, is 70°. We can write this

measure $\angle B = 70°$ or, more concisely,   m $\angle B = 70°$

For ease of notation and speech, however, we will often simply write

$\angle B = 70°$

The next example concerns *complementary angles* and *supplementary angles*. Recall the definitions. Two angles are said to be **complementary** provided that their sum is 90°. Two angles are **supplementary** provided that their sum is 180°.

FIGURE 3

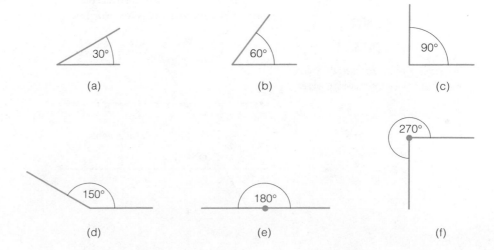

| (a) 30° | (b) 60° | (c) 90° |
| (d) 150° | (e) 180° | (f) 270° |

	TERM	DEFINING CONDITION	EXAMPLES
TABLE 1 **Terminology for Angles**	Acute angle	An angle whose measure is (strictly) between 0° and 90°	Figures 3(a) and (b)
	Right angle	An angle whose measure is 90°	Figure 3(c)
	Obtuse angle	An angle whose measure is (strictly) between 90° and 180°	Figure 3(d)
	Straight angle	An angle whose measure is 180°	Figure 3(e)

EXAMPLE 1 (a) Find a pair of complementary angles such that one angle is five times the other.
(b) Find a pair of supplementary angles that differ by 20°.

Solution (a) Denote the two angles by x and $90° - x$ (so that their sum is 90°). Then we have

$$5x = 90° - x$$
$$6x = 90°$$
$$x = 15° \quad \text{and therefore} \quad 90° - x = 75°$$

The two angles are therefore 15° and 75°.

(b) Denote the two angles by x and $180° - x$ (so that their sum is 180°). Then we have

$$x - (180° - x) = 20°$$
$$2x = 200°$$
$$x = 100° \quad \text{and therefore} \quad 180° - x = 80°$$

The two angles are therefore 100° and 80°. ▲

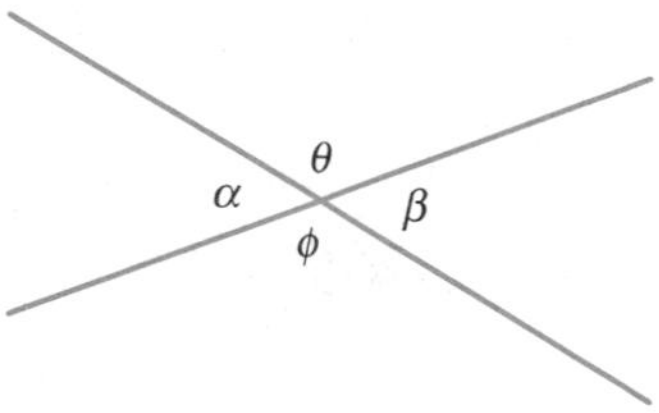

In the front endpapers of this text, we list some of the basic results about angles and triangles that you need to know for precalculus and calculus. In addition to these, you'll also need to be aware of several facts regarding the angles formed by intersecting lines. Before stating these theorems, we explain the terminology used in this context. Figure 4 shows two intersecting lines and the four angles that are determined by the intersection. The angles α and β in Figure 4 are a pair of **vertical angles**; they're opposite, rather than adjacent to, one another. Likewise, the angles θ and ϕ are also vertical angles. In Figure 5, the angles α and β are a pair of **alternate interior angles**. [They are on alternate sides of the line cut-

FIGURE 4
The angles α and β are vertical angles; so are θ and ϕ.

FIGURE 5
The angles α and β are alternate angles, while β and θ are corresponding angles.

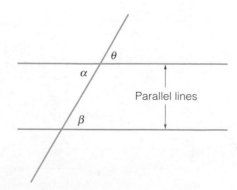

ting the parallel lines, and they are between (interior to) the parallel lines. The angles β and θ in Figure 5 are **corresponding angles**. (They are on the same side of the line cutting the parallel lines and they are on corresponding sides of the parallel lines.) In the box that follows, we summarize the relationships between the angles formed by intersecting lines.

Angles Formed by Intersecting Lines

Theorem 1	When two lines intersect, the vertical angles are equal. Thus, in Figure 4, $\alpha = \beta$ and $\theta = \phi$.
Theorem 2	Suppose that two parallel lines are cut by a third line, as in Figure 5. Then the alternate interior angles are equal, as are the corresponding angles. Thus, in Figure 5, $\alpha = \beta$ (alternate interior angles) and $\beta = \theta$ (corresponding angles).

EXAMPLE 2   In Figure 6, the dashed line is parallel to the base of the triangle. Determine angles α, β, and θ.

Solution   Since the 70° angle and α are alternate interior angles, we have $\alpha = 70°$. Also, $\beta = 70°$ because β and the given 70° angle are corresponding angles. (Alternately, α and β are vertical angles, and since $\alpha = 70°$, β must also be 70°. Finally, to find θ, note that the given 80° angle along with θ and β are the three angles of a triangle, and therefore

$$\theta + \beta + 80° = 180°$$

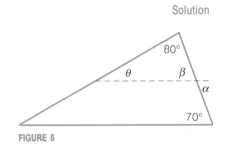

FIGURE 6

But β is 70° and therefore $\theta = 180° - 80° - 70° = 30°$.   ▲

TABLES

TABLE 1

Squares and Square Roots

NO.	SQ.	SQ. RT.	NO.	SQ.	SQ. RT.
1	1	1.000	51	2,601	7.141
2	4	1.414	52	2,704	7.211
3	9	1.732	53	2,809	7.280
4	16	2.000	54	2,916	7.348
5	25	2.236	55	3,025	7.416
6	36	2.449	56	3,136	7.483
7	49	2.646	57	3,249	7.550
8	64	2.828	58	3,364	7.616
9	81	3.000	59	3,481	7.681
10	100	3.162	60	3,600	7.746
11	121	3.317	61	3,721	7.810
12	144	3.464	62	3,844	7.874
13	169	3.606	63	3,969	7.937
14	196	3.742	64	4,096	8.000
15	225	3.873	65	4,225	8.062
16	256	4.000	66	4,356	8.124
17	289	4.123	67	4,489	8.185
18	324	4.243	68	4,624	8.246
19	361	4.359	69	4,761	8.307
20	400	4.472	70	4,900	8.367
21	441	4.583	71	5,041	8.426
22	484	4.690	72	5,184	8.485
23	529	4.796	73	5,329	8.544
24	576	4.899	74	5,476	8.602
25	625	5.000	75	5,625	8.660
26	676	5.099	76	5,776	8.718
27	729	5.196	77	5,929	8.775
28	784	5.292	78	6,084	8.832
29	841	5.385	79	6,241	8.888
30	900	5.477	80	6,400	8.944
31	961	5.568	81	6,561	9.000
32	1,024	5.657	82	6,724	9.055
33	1,089	5.745	83	6,889	9.110
34	1,156	5.831	84	7,056	9.165
35	1,225	5.916	85	7,225	9.220
36	1,296	6.000	86	7,396	9.274
37	1,369	6.083	87	7,569	9.327
38	1,444	6.164	88	7,744	9.381
39	1,521	6.245	89	7,921	9.434
40	1,600	6.325	90	8,100	9.487
41	1,681	6.403	91	8,281	9.539
42	1,764	6.481	92	8,464	9.592
43	1,849	6.557	93	8,649	9.644
44	1,936	6.633	94	8,836	9.695
45	2,025	6.708	95	9,025	9.747
46	2,116	6.782	96	9,216	9.798
47	2,209	6.856	97	9,409	9.849
48	2,304	6.928	98	9,604	9.899
49	2,401	7.000	99	9,801	9.950
50	2,500	7.071	100	10,000	10.000

TABLE 2

Exponential Functions

x	e^x	e^{-x}	x	e^x	e^{-x}
0.00	1.0000	1.0000	1.5	4.4817	0.2231
0.01	1.0101	0.9901	1.6	4.9530	0.2019
0.02	1.0202	0.9802	1.7	5.4739	0.1827
0.03	1.0305	0.9704	1.8	6.0496	0.1653
0.04	1.0408	0.9608	1.9	6.6859	0.1496
0.05	1.0513	0.9512	2.0	7.3891	0.1353
0.06	1.0618	0.9418	2.1	8.1662	0.1225
0.07	1.0725	0.9324	2.2	9.0250	0.1108
0.08	1.0833	0.9231	2.3	9.9742	0.1003
0.09	1.0942	0.9139	2.4	11.023	0.0907
0.10	1.1052	0.9048	2.5	12.182	0.0821
0.11	1.1163	0.8958	2.6	13.464	0.0743
0.12	1.1275	0.8869	2.7	14.880	0.0672
0.13	1.1388	0.8781	2.8	16.445	0.0608
0.14	1.1503	0.8694	2.9	18.174	0.0550
0.15	1.1618	0.8607	3.0	20.086	0.0498
0.16	1.1735	0.8521	3.1	22.198	0.0450
0.17	1.1853	0.8437	3.2	24.533	0.0408
0.18	1.1972	0.8353	3.3	27.113	0.0369
0.19	1.2092	0.8270	3.4	29.964	0.0334
0.20	1.2214	0.8187	3.5	33.115	0.0302
0.21	1.2337	0.8106	3.6	36.598	0.0273
0.22	1.2461	0.8025	3.7	40.447	0.0247
0.23	1.2586	0.7945	3.8	44.701	0.0224
0.24	1.2712	0.7866	3.9	49.402	0.0202
0.25	1.2840	0.7788	4.0	54.598	0.0183
0.30	1.3499	0.7408	4.1	60.340	0.0166
0.35	1.4191	0.7047	4.2	66.686	0.0150
0.40	1.4918	0.6703	4.3	73.700	0.0136
0.45	1.5683	0.6376	4.4	81.451	0.0123
0.50	1.6487	0.6065	4.5	90.017	0.0111
0.55	1.7333	0.5769	4.6	99.484	0.0101
0.60	1.8221	0.5488	4.7	109.95	0.0091
0.65	1.9155	0.5220	4.8	121.51	0.0082
0.70	2.0138	0.4966	4.9	134.29	0.0074
0.75	2.1170	0.4724	5.0	148.41	0.0067
0.80	2.2255	0.4493	5.5	244.69	0.0041
0.85	2.3396	0.4274	6.0	403.43	0.0025
0.90	2.4596	0.4066	6.5	665.14	0.0015
0.95	2.5857	0.3867	7.0	1096.6	0.0009
1.0	2.7183	0.3679	7.5	1808.0	0.0006
1.1	3.0042	0.3329	8.0	2981.0	0.0003
1.2	3.3201	0.3012	8.5	4914.8	0.0002
1.3	3.6693	0.2725	9.0	8103.1	0.0001
1.4	4.0552	0.2466	10.0	22026	0.00005

TABLE 3
Logarithms to the Base Ten

x	0	1	2	3	4	5	6	7	8	9
1.0	.0000	.0043	.0086	.0128	.0170	.0212	.0253	.0294	.0334	.0374
1.1	.0414	.0453	.0492	.0531	.0569	.0607	.0645	.0682	.0719	.0755
1.2	.0792	.0828	.0864	.0899	.0934	.0969	.1004	.1038	.1072	.1106
1.3	.1139	.1173	.1206	.1239	.1271	.1303	.1335	.1367	.1399	.1430
1.4	.1461	.1492	.1523	.1553	.1584	.1614	.1644	.1673	.1703	.1732
1.5	.1761	.1790	.1818	.1847	.1875	.1903	.1931	.1959	.1987	.2014
1.6	.2041	.2068	.2095	.2122	.2148	.2175	.2201	.2227	.2253	.2279
1.7	.2304	.2330	.2355	.2380	.2405	.2430	.2455	.2480	.2504	.2529
1.8	.2553	.2577	.2601	.2625	.2648	.2672	.2695	.2718	.2742	.2765
1.9	.2788	.2810	.2833	.2856	.2878	.2900	.2923	.2945	.2967	.2989
2.0	.3010	.3032	.3054	.3075	.3096	.3118	.3139	.3160	.3181	.3201
2.1	.3222	.3243	.3263	.3284	.3304	.3324	.3345	.3365	.3385	.3404
2.2	.3424	.3444	.3464	.3483	.3502	.3522	.3541	.3560	.3579	.3598
2.3	.3617	.3636	.3655	.3674	.3692	.3711	.3729	.3747	.3766	.3784
2.4	.3802	.3820	.3838	.3856	.3874	.3892	.3909	.3927	.3945	.3962
2.5	.3979	.3997	.4014	.4031	.4048	.4065	.4082	.4099	.4116	.4133
2.6	.4150	.4166	.4183	.4200	.4216	.4232	.4249	.4265	.4281	.4298
2.7	.4314	.4330	.4346	.4362	.4378	.4393	.4409	.4425	.4440	.4456
2.8	.4472	.4487	.4502	.4518	.4533	.4548	.4564	.4579	.4594	.4609
2.9	.4624	.4639	.4654	.4669	.4683	.4698	.4713	.4728	.4742	.4757
3.0	.4771	.4786	.4800	.4814	.4829	.4843	.4857	.4871	.4886	.4900
3.1	.4914	.4928	.4942	.4955	.4969	.4983	.4997	.5011	.5024	.5038
3.2	.5051	.5065	.5079	.5092	.5105	.5119	.5132	.5145	.5159	.5172
3.3	.5185	.5198	.5211	.5224	.5237	.5250	.5263	.5276	.5289	.5302
3.4	.5315	.5328	.5340	.5353	.5366	.5378	.5391	.5403	.5416	.5428
3.5	.5441	.5453	.5465	.5478	.5490	.5502	.5514	.5527	.5539	.5551
3.6	.5563	.5575	.5587	.5599	.5611	.5623	.5635	.5647	.5658	.5670
3.7	.5682	.5694	.5705	.5717	.5729	.5740	.5752	.5763	.5775	.5786
3.8	.5798	.5809	.5821	.5832	.5843	.5855	.5866	.5877	.5888	.5899
3.9	.5911	.5922	.5933	.5944	.5955	.5966	.5977	.5988	.5999	.6010
4.0	.6021	.6031	.6042	.6053	.6064	.6075	.6085	.6096	.6107	.6117
4.1	.6128	.6138	.6149	.6160	.6170	.6180	.6191	.6201	.6212	.6222
4.2	.6232	.6243	.6253	.6263	.6274	.6284	.6294	.6304	.6314	.6325
4.3	.6335	.6345	.6355	.6365	.6375	.6385	.6395	.6405	.6415	.6425
4.4	.6435	.6444	.6454	.6464	.6474	.6484	.6493	.6503	.6513	.6522
4.5	.6532	.6542	.6551	.6561	.6571	.6580	.6590	.6599	.6609	.6618
4.6	.6628	.6637	.6646	.6656	.6665	.6675	.6684	.6693	.6702	.6712
4.7	.6721	.6730	.6739	.6749	.6758	.6767	.6776	.6785	.6794	.6803
4.8	.6812	.6821	.6830	.6839	.6848	.6857	.6866	.6875	.6884	.6893
4.9	.6902	.6911	.6920	.6928	.6937	.6946	.6955	.6964	.6972	.6981
5.0	.6990	.6998	.7007	.7016	.7024	.7033	.7042	.7050	.7059	.7067
5.1	.7076	.7084	.7093	.7101	.7110	.7118	.7126	.7135	.7143	.7152
5.2	.7160	.7168	.7177	.7185	.7193	.7202	.7210	.7218	.7226	.7235
5.3	.7243	.7251	.7259	.7267	.7275	.7284	.7292	.7300	.7308	.7316
5.4	.7324	.7332	.7340	.7348	.7356	.7364	.7372	.7380	.7388	.7396
x	0	1	2	3	4	5	6	7	8	9

TABLE 3

Logarithms to the Base

Ten (*continued*)

x	0	1	2	3	4	5	6	7	8	9
5.5	.7404	.7412	.7419	.7427	.7435	.7443	.7451	.7459	.7466	.7474
5.6	.7482	.7490	.7497	.7505	.7513	.7520	.7528	.7536	.7543	.7551
5.7	.7559	.7566	.7574	.7582	.7589	.7597	.7604	.7612	.7619	.7627
5.8	.7634	.7642	.7649	.7657	.7764	.7672	.7679	.7686	.7694	.7701
5.9	.7709	.7716	.7723	.7731	.7738	.7745	.7752	.7760	.7767	.7774
6.0	.7782	.7789	.7796	.7803	.7810	.7818	.7825	.7832	.7839	.7846
6.1	.7853	.7860	.7868	.7875	.7882	.7889	.7896	.7903	.7910	.7917
6.2	.7924	.7931	.7938	.7945	.7952	.7959	.7966	.7973	.7980	.7987
6.3	.7993	.8000	.8007	.8014	.8021	.8028	.8035	.8041	.8048	.8055
6.4	.8062	.8069	.8075	.8082	.8089	.8096	.8102	.8109	.8116	.8122
6.5	.8129	.8136	.8142	.8149	.8156	.8162	.8169	.8176	.8182	.8189
6.6	.8195	.8202	.8209	.8215	.8222	.8228	.8235	.8241	.8248	.8254
6.7	.8261	.8267	.8274	.8280	.8287	.8293	.8299	.8306	.8312	.8319
6.8	.8325	.8331	.8338	.8344	.8351	.8357	.8363	.8370	.8376	.8382
6.9	.8388	.8395	.8401	.8407	.8414	.8420	.8426	.8432	.8439	.8445
7.0	.8451	.8457	.8463	.8470	.8476	.8482	.8488	.8494	.8500	.8506
7.1	.8513	.8519	.8525	.8531	.8537	.8543	.8549	.8555	.8561	.8567
7.2	.8573	.8579	.8585	.8591	.8597	.8603	.8609	.8615	.8621	.8627
7.3	.8633	.8639	.8645	.8651	.8657	.8663	.8669	.8675	.8681	.8686
7.4	.8692	.8698	.8704	.8710	.8716	.8722	.8727	.8733	.8739	.8745
7.5	.8751	.8756	.8762	.8768	.8774	.8779	.8785	.8791	.8797	.8802
7.6	.8808	.8814	.8820	.8825	.8831	.8837	.8842	.8848	.8854	.8859
7.7	.8865	.8871	.8876	.8882	.8887	.8893	.8899	.8904	.8910	.8915
7.8	.8921	.8927	.8932	.8938	.8943	.8949	.8954	.8960	.8965	.8971
7.9	.8976	.8982	.8987	.8993	.8998	.9004	.9009	.9015	.9020	.9025
8.0	.9031	.9036	.9042	.9047	.9053	.9058	.9063	.9069	.9074	.9079
8.1	.9085	.9090	.9096	.9101	.9106	.9112	.9117	.9122	.9128	.9133
8.2	.9138	.9143	.9149	.9154	.9159	.9165	.9170	.9175	.9180	.9186
8.3	.9191	.9196	.9201	.9206	.9212	.9217	.9222	.9227	.9232	.9238
8.4	.9243	.9248	.9253	.9258	.9263	.9269	.9274	.9279	.9284	.9289
8.5	.9294	.9299	.9304	.9309	.9315	.9320	.9325	.9330	.9335	.9340
8.6	.9345	.9350	.9355	.9360	.9365	.9370	.9375	.9380	.9385	.9390
8.7	.9395	.9400	.9405	.9410	.9415	.9420	.9425	.9430	.9435	.9440
8.8	.9445	.9450	.9455	.9460	.9465	.9469	.9474	.9479	.9484	.9489
8.9	.9494	.9499	.9504	.9509	.9513	.9518	.9523	.9528	.9533	.9538
9.0	.9542	.9547	.9552	.9557	.9562	.9566	.9571	.9576	.9581	.9586
9.1	.9590	.9595	.9600	.9605	.9609	.9614	.9619	.9624	.9628	.9633
9.2	.9638	.9643	.9647	.9652	.9657	.9661	.9666	.9671	.9675	.9680
9.3	.9685	.9689	.9694	.9699	.9703	.9708	.9713	.9717	.9722	.9727
9.4	.9731	.9736	.9741	.9745	.9750	.9754	.9759	.9763	.9768	.9773
9.5	.9777	.9782	.9786	.9791	.9795	.9800	.9805	.9809	.9814	.9818
9.6	.9823	.9827	.9832	.9836	.9841	.9845	.9850	.9854	.9859	.9863
9.7	.9868	.9872	.9877	.9881	.9886	.9890	.9894	.9899	.9903	.9908
9.8	.9912	.9917	.9921	.9926	.9930	.9934	.9939	.9943	.9948	.9952
9.9	.9956	.9961	.9965	.9969	.9974	.9978	.9983	.9987	.9991	.9996
x	0	1	2	3	4	5	6	7	8	9

TABLE 4
Natural Logarithms

n	ln n	n	ln n	n	ln n
		4.5	1.5041	9.0	2.1972
0.1	−2.3026	4.6	1.5261	9.1	2.2083
0.2	−1.6094	4.7	1.5476	9.2	2.2192
0.3	−1.2040	4.8	1.5686	9.3	2.2300
0.4	−0.9163	4.9	1.5892	9.4	2.2407
0.5	−0.6931	5.0	1.6094	9.5	2.2513
0.6	−0.5108	5.1	1.6292	9.6	2.2618
0.7	−0.3567	5.2	1.6487	9.7	2.2721
0.8	−0.2231	5.3	1.6677	9.8	2.2824
0.9	−0.1054	5.4	1.6864	9.9	2.2925
1.0	0.0000	5.5	1.7047	10	2.3026
1.1	0.0953	5.6	1.7228	11	2.3979
1.2	0.1823	5.7	1.7405	12	2.4849
1.3	0.2624	5.8	1.7579	13	2.5649
1.4	0.3365	5.9	1.7750	14	2.6391
1.5	0.4055	6.0	1.7918	15	2.7081
1.6	0.4700	6.1	1.8083	16	2.7726
1.7	0.5306	6.2	1.8245	17	2.8332
1.8	0.5878	6.3	1.8405	18	2.8904
1.9	0.6419	6.4	1.8563	19	2.9444
2.0	0.6931	6.5	1.8718	20	2.9957
2.1	0.7419	6.6	1.8871	25	3.2189
2.2	0.7885	6.7	1.9021	30	3.4012
2.3	0.8329	6.8	1.9169	35	3.5553
2.4	0.8755	6.9	1.9315	40	3.6889
2.5	0.9163	7.0	1.9459	45	3.8067
2.6	0.9555	7.1	1.9601	50	3.9120
2.7	0.9933	7.2	1.9741	55	4.0073
2.8	1.0296	7.3	1.9879	60	4.0943
2.9	1.0647	7.4	2.0015	65	4.1744
3.0	1.0986	7.5	2.0149	70	4.2485
3.1	1.1314	7.6	2.0281	75	4.3175
3.2	1.1632	7.7	2.0412	80	4.3820
3.3	1.1939	7.8	2.0541	85	4.4427
3.4	1.2238	7.9	2.0669	90	4.4998
3.5	1.2528	8.0	2.0794	95	4.5539
3.6	1.2809	8.1	2.0919	100	4.6052
3.7	1.3083	8.2	2.1041		
3.8	1.3350	8.3	2.1163		
3.9	1.3610	8.4	2.1282		
4.0	1.3863	8.5	2.1401		
4.1	1.4110	8.6	2.1518		
4.2	1.4351	8.7	2.1633		
4.3	1.4586	8.8	2.1748		
4.4	1.4816	8.9	2.1861		

TABLE 5
Values of Trigonometric Functions

ANGLE θ

Degrees	Radians	sin θ	csc θ	tan θ	cot θ	sec θ	cos θ		
0° 00′	.0000	.0000	No value	.0000	No value	1.000	1.0000	1.5708	90° 00′
10	029	029	343.8	029	343.8	000	000	679	50
20	058	058	171.9	058	171.9	000	000	650	40
30	087	087	114.6	087	114.6	000	1.0000	621	30
40	116	116	85.95	116	85.94	000	.9999	592	20
50	145	145	68.76	145	68.75	000	999	563	10
1° 00′	.0175	.0175	57.30	.0175	57.29	1.000	.9998	1.5533	89° 00′
10	204	204	49.11	204	49.10	000	998	504	50
20	233	233	42.98	233	42.96	000	997	475	40
30	262	262	38.20	262	38.19	000	997	446	30
40	291	291	34.38	291	34.37	000	996	417	20
50	320	320	31.26	320	31.24	001	995	388	10
2° 00′	.0349	.0349	28.65	.0349	28.64	1.001	.9994	1.5359	88° 00′
10	378	378	26.45	378	26.43	001	993	330	50
20	407	407	24.56	407	24.54	001	992	301	40
30	436	436	22.93	437	22.90	001	990	272	30
40	465	465	21.49	466	21.47	001	989	243	20
50	495	494	20.23	495	20.21	001	988	213	10
3° 00′	.0524	.0523	19.11	.0524	19.08	1.001	.9986	1.5184	87° 00′
10	553	552	18.10	553	18.07	002	985	155	50
20	582	581	17.20	582	17.17	002	983	126	40
30	611	610	16.38	612	16.35	002	981	097	30
40	640	640	15.64	641	15.60	002	980	068	20
50	669	669	14.96	670	14.92	002	978	039	10
4° 00′	.0698	.0698	14.34	.0699	14.30	1.002	.9976	1.5010	86° 00′
10	727	727	13.76	729	13.73	003	974	981	50
20	756	756	13.23	758	13.20	003	971	952	40
30	785	785	12.75	787	12.71	003	969	923	30
40	814	814	12.29	816	12.25	003	967	893	20
50	844	843	11.87	846	11.83	004	964	864	10
5° 00′	.0873	.0872	11.47	.0875	11.43	1.004	.9962	1.4835	85° 00′
10	902	901	11.10	904	11.06	004	959	806	50
20	931	929	10.76	934	10.71	004	957	777	40
30	960	958	10.43	963	10.39	005	954	748	30
40	.0989	.0987	10.13	.0992	10.08	005	951	719	20
50	.1018	.1016	9.839	.1022	9.788	005	948	690	10
6° 00′	.1047	.1045	9.567	.1051	9.514	1.006	.9945	1.4661	84° 00′
10	076	074	9.309	080	9.255	006	942	632	50
20	105	103	9.065	110	9.010	006	939	603	40
30	134	132	8.834	139	8.777	006	936	573	30
40	164	161	8.614	169	8.556	007	932	544	20
50	193	190	8.405	198	8.345	007	929	515	10
7° 00′	.1222	.1219	8.206	.1228	8.144	1.008	.9925	1.4486	83° 00′
		cos θ	sec θ	cot θ	tan θ	csc θ	sin θ	Radians	Degrees

ANGLE θ

TABLE 5
Values of Trigonometric Functions (*continued*)

ANGLE θ									
Degrees	Radians	sin θ	csc θ	tan θ	cot θ	sec θ	cos θ		
7° 00′	.1222	.1219	8.206	.1228	8.144	1.008	.9925	1.4486	83° 00′
10	251	248	8.016	257	7.953	008	922	457	50
20	280	276	7.834	287	7.770	008	918	428	40
30	309	305	7.661	317	7.596	009	914	399	30
40	338	334	7.496	346	7.429	009	911	370	20
50	367	363	7.337	376	7.269	009	907	341	10
8° 00′	.1396	.1392	7.185	.1405	7.115	1.010	.9903	1.4312	82° 00′
10	425	421	7.040	435	6.968	010	899	283	50
20	454	449	6.900	465	827	011	894	254	40
30	484	478	765	495	691	011	890	224	30
40	513	507	636	524	561	012	886	195	20
50	542	536	512	554	435	012	881	166	10
9° 00′	.1571	.1564	6.392	.1584	6.314	1.012	.9877	1.4137	81° 00′
10	600	593	277	614	197	013	872	108	50
20	629	622	166	644	6.084	013	868	079	40
30	658	650	6.059	673	5.976	014	863	050	30
40	687	679	5.955	703	871	014	858	1.4021	20
50	716	708	855	733	769	015	853	1.3992	10
10° 00′	.1745	.1736	5.759	.1763	5.671	1.015	.9848	1.3963	80° 00′
10	774	765	665	793	576	016	843	934	50
20	804	794	575	823	485	016	838	904	40
30	833	822	487	853	396	017	833	875	30
40	862	851	403	883	309	018	827	846	20
50	891	880	320	914	226	018	822	817	10
11° 00′	.1920	.1908	5.241	.1944	5.145	1.019	.9816	1.3788	79° 00′
10	949	937	164	.1974	5.066	019	811	759	50
20	.1978	965	089	.2004	4.989	020	805	730	40
30	.2007	.1994	5.016	035	915	020	799	701	30
40	036	.2022	4.945	065	843	021	793	672	20
50	065	051	876	095	773	022	787	643	10
12° 00′	.2094	.2079	4.810	.2126	4.705	1.022	.9781	1.3614	78° 00′
10	123	108	745	156	638	023	775	584	50
20	153	136	682	186	574	024	769	555	40
30	182	164	620	217	511	024	763	526	30
40	211	193	560	247	449	025	757	497	20
50	240	221	502	278	390	026	750	468	10
13° 00′	.2269	.2250	4.445	.2309	4.331	1.026	.9744	1.3439	77° 00′
10	298	278	390	339	275	027	737	410	50
20	327	306	336	370	219	028	730	381	40
30	356	334	284	401	165	028	724	352	30
40	385	363	232	432	113	029	717	323	20
50	414	391	182	462	061	030	710	294	10
14° 00′	.2443	.2419	4.134	.2493	4.011	1.031	.9703	1.3265	76° 00′
		cos θ	sec θ	cot θ	tan θ	csc θ	sin θ	Radians	Degrees
								ANGLE θ	

TABLE 5
Values of Trigonometric Functions (*continued*)

ANGLE θ

Degrees	Radians	sin θ	csc θ	tan θ	cot θ	sec θ	cos θ		
14° 00′	.2443	.2419	4.134	.2493	4.011	1.031	.9703	1.3265	76° 00′
10	473	447	086	524	3.962	031	696	235	50
20	502	476	4.039	555	914	032	689	206	40
30	531	504	3.994	586	867	033	681	177	30
40	560	532	950	617	821	034	674	148	20
50	589	560	906	648	776	034	667	119	10
15° 00′	.2618	.2588	3.864	.2679	3.732	1.035	.9659	1.3090	75° 00′
10	647	616	822	711	689	036	652	061	50
20	676	644	782	742	647	037	644	032	40
30	705	672	742	773	606	038	636	1.3003	30
40	734	700	703	805	566	039	628	1.2974	20
50	763	728	665	836	526	039	621	945	10
16° 00′	.2793	.2756	3.628	.2867	3.487	1.040	.9613	1.2915	74° 00′
10	822	784	592	899	450	041	605	886	50
20	851	812	556	931	412	042	596	857	40
30	880	840	521	962	376	043	588	828	30
40	909	868	487	.2944	340	044	580	799	20
50	938	896	453	.3026	305	045	572	770	10
17° 00′	.2967	.2924	3.420	.3057	3.271	1.046	.9563	1.2741	73° 00′
10	.2996	952	388	089	237	047	555	712	50
20	.3025	.2979	357	121	204	048	546	683	40
30	054	.3007	326	153	172	048	537	654	30
40	083	035	295	185	140	049	528	625	20
50	113	062	265	217	108	050	520	595	10
18° 00′	.3142	.3090	3.236	.3249	3.078	1.051	.9511	1.2566	72 °00′
10	171	118	207	281	047	052	502	537	50
20	200	145	179	314	3.018	053	492	508	40
30	229	173	152	346	2.989	054	483	479	30
40	258	201	124	378	960	056	474	450	20
50	287	228	098	411	932	057	465	421	10
19° 00′	.3316	.3256	3.072	.3443	2.904	1.058	.9455	1.2392	71° 00′
10	345	283	046	476	877	059	446	363	50
20	374	311	3.021	508	850	060	436	334	40
30	403	338	2.996	541	824	061	426	305	30
40	432	365	971	574	798	062	417	275	20
50	462	393	947	607	773	063	407	246	10
20 °00′	.3491	.3420	2.924	.3640	2.747	1.064	.9397	1.2217	70 °00′
10	520	448	901	673	723	065	387	188	50
20	549	475	878	706	699	066	377	159	40
30	578	502	855	739	675	068	367	130	30
40	607	529	833	772	651	069	356	101	20
50	636	557	812	805	628	070	346	072	10
21° 00′	.3665	.3584	2.790	.3839	2.605	1.071	.9336	1.2043	69° 00′
		cos θ	sec θ	cot θ	tan θ	csc θ	sin θ	Radians	Degrees

ANGLE θ

TABLE 5

Values of Trigonometric Functions (*continued*)

ANGLE θ

Degrees	Radians	sin θ	csc θ	tan θ	cot θ	sec θ	cos θ		
21° 00′	.3665	.3584	2.790	.3839	2.605	1.071	.9336	1.2043	69° 00′
10	694	611	769	872	583	072	325	1.2014	50
20	723	638	749	906	560	074	315	1.1985	40
30	752	665	729	939	539	075	304	956	30
40	782	692	709	.3973	517	076	293	926	20
50	811	719	689	.4006	496	077	283	897	10
22° 00′	.3840	.3746	2.669	.4040	2.475	1.079	.9272	1.1868	68° 00′
10	869	773	650	074	455	080	261	839	50
20	898	800	632	108	434	081	250	810	40
30	927	827	613	142	414	082	239	781	30
40	956	854	595	176	394	084	228	752	20
50	985	881	577	210	375	085	216	723	10
23° 00′	.4014	.3907	2.559	.4245	2.356	1.086	.9205	1.1694	67° 00′
10	043	934	542	279	337	088	194	665	50
20	072	961	525	314	318	089	182	636	40
30	102	.3987	508	348	300	090	171	606	30
40	131	.4014	491	383	282	092	159	577	20
50	160	041	475	417	264	093	147	548	10
24° 00′	.4189	.4067	2.459	.4452	2.246	1.095	.9135	1.1519	66° 00′
10	218	094	443	487	229	096	124	490	50
20	247	120	427	522	211	097	112	461	40
30	276	147	411	557	194	099	100	432	30
40	305	173	396	592	177	100	088	403	20
50	334	200	381	628	161	102	075	374	10
25° 00′	.4363	.4226	2.366	.4663	2.145	1.103	.9063	1.1345	65° 00′
10	392	253	352	699	128	105	051	316	50
20	422	279	337	734	112	106	038	286	40
30	451	305	323	770	097	108	026	257	30
40	480	331	309	806	081	109	013	228	20
50	509	358	295	841	066	111	.9001	199	10
26° 00′	.4538	.4384	2.281	.4877	2.050	1.113	.8988	1.1170	64° 00′
10	567	410	268	913	035	114	975	141	50
20	596	436	254	950	020	116	962	112	40
30	625	462	241	.4986	2.006	117	949	083	30
40	654	488	228	.5022	1.991	119	936	054	20
50	683	514	215	059	977	121	923	1.1025	10
27° 00′	.4712	.4540	2.203	.5095	1.963	1.122	.8910	1.0996	63° 00′
		cos θ	sec θ	cot θ	tan θ	csc θ	sin θ	Radians	Degrees

ANGLE θ

TABLE 5

Values of Trigonometric Functions (*continued*)

Degrees	Radians	sin θ	csc θ	tan θ	cot θ	sec θ	cos θ		
27° 00′	.4712	.4540	2.203	.5095	1.963	1.122	.8910	1.0996	63° 00′
10	741	566	190	132	949	124	897	966	50
20	771	592	178	169	935	126	884	937	40
30	800	617	166	206	921	127	870	908	30
40	829	643	154	243	907	129	857	879	20
50	858	669	142	280	894	131	843	850	10
28° 00′	.4887	.4695	2.130	.5317	1.881	1.133	.8829	1.0821	62° 00′
10	916	720	118	354	868	134	816	792	50
20	945	746	107	392	855	136	802	763	40
30	.4974	772	096	430	842	138	788	734	30
40	.5003	797	085	467	829	140	774	705	20
50	032	823	074	505	816	142	760	676	10
29° 00′	.5061	.4848	2.063	.5543	1.804	1.143	.8746	1.0647	61° 00′
10	091	874	052	581	792	145	732	617	50
20	120	899	041	619	780	147	718	588	40
30	149	924	031	658	767	149	704	559	30
40	178	950	020	696	756	151	689	530	20
50	207	.4975	010	735	744	153	675	501	10
30° 00′	.5236	.5000	2.000	.5774	1.732	1.155	.8660	1.0472	60° 00′
10	265	025	1.990	812	720	157	646	443	50
20	294	050	980	851	709	159	631	414	40
30	323	075	970	890	698	161	616	385	30
40	352	100	961	930	686	163	601	356	20
50	381	125	951	.5969	675	165	587	327	10
31° 00′	.5411	.5150	1.942	.6009	1.664	1.167	.8572	1.0297	59° 00′
10	440	175	932	048	653	169	557	268	50
20	469	200	923	088	643	171	542	239	40
30	498	225	914	128	632	173	526	210	30
40	527	250	905	168	621	175	511	181	20
50	556	275	896	208	611	177	496	152	10
32° 00′	.5585	.5299	1.887	.6249	1.600	1.179	.8480	1.0123	58° 00′
10	614	324	878	289	590	181	465	094	50
20	643	348	870	330	580	184	450	065	40
30	672	373	861	371	570	186	434	036	30
40	701	398	853	412	560	188	418	1.0007	20
50	730	422	844	453	550	190	403	.9977	10
33° 00′	.5760	.5446	1.836	.6494	1.540	1.192	.8387	.9948	57° 00′
		cos θ	sec θ	cot θ	tan θ	csc θ	sin θ	Radians	Degrees

ANGLE θ

TABLE 5

Values of Trigonometric Functions (continued)

ANGLE θ

Degrees	Radians	sin θ	csc θ	tan θ	cot θ	sec θ	cos θ		
33° 00'	.5760	.5446	1.836	.6494	1.540	1.192	.8387	.9948	57° 00'
10	789	471	328	536	530	195	371	919	50
20	818	495	820	577	520	197	355	890	40
30	847	519	812	619	511	199	339	861	30
40	876	544	804	661	501	202	323	832	20
50	905	568	796	703	492	204	307	803	10
34° 00'	.5934	.5592	1.788	.6745	1.483	1.206	.8290	.9774	56° 00'
10	963	616	781	787	473	209	274	745	50
20	.5992	640	773	830	464	211	258	716	40
30	.6021	644	766	873	455	213	241	687	30
40	050	688	758	916	446	216	225	657	20
50	080	712	751	.6959	437	218	208	628	10
35° 00'	.6109	.5736	1.743	.7002	1.428	1.221	.8192	.9599	55° 00'
10	138	760	736	046	419	223	175	570	50
20	167	783	729	089	411	226	158	541	40
30	196	807	722	133	402	228	141	512	30
40	225	831	715	177	393	231	124	483	20
50	254	854	708	221	385	233	107	454	10
36° 00'	.6283	.5878	1.701	.7265	1.376	1.236	.8090	.9425	54° 00'
10	312	901	695	310	368	239	073	396	50
20	341	925	688	355	360	241	056	367	40
30	370	948	681	400	351	244	039	338	30
40	400	972	675	445	343	247	021	308	20
50	429	.5995	668	490	335	249	.8004	279	10
37° 00'	.6458	.6018	1.662	.7536	1.327	1.252	.7986	.9250	53° 00'
10	487	041	655	581	319	255	969	221	50
20	516	065	649	627	311	258	951	192	40
30	545	088	643	673	303	260	934	163	30
40	574	111	636	720	295	263	916	134	20
50	603	134	630	766	288	266	898	105	10
38° 00'	.6632	.6157	1.624	.7813	1.280	1.269	.7880	.9076	52° 00'
10	661	180	618	860	272	272	862	047	50
20	690	202	612	907	265	275	844	.9018	40
30	720	225	606	.7954	257	278	826	.8988	30
40	749	248	601	.8002	250	281	808	959	20
50	778	271	595	050	242	284	790	930	10
39° 00'	.6807	.6293	1.589	.8098	1.235	1.287	.7771	.8901	51° 00'
		cos θ	sec θ	cot θ	tan θ	csc θ	sin θ	Radians	Degrees

ANGLE θ

TABLE 5

Values of Trigonometric Functions (*continued*)

ANGLE θ									
Degrees	Radians	sin θ	csc θ	tan θ	cot θ	sec θ	cos θ		
39° 00′	.6807	.6293	1.589	.8098	1.235	1.287	.7771	.8901	51° 00′
10	836	316	583	146	228	290	753	872	50
20	865	338	578	195	220	293	735	843	40
30	894	361	572	243	213	296	716	814	30
40	923	383	567	292	206	299	698	785	20
50	952	406	561	342	199	302	679	756	10
40° 00′	.6981	.6428	1.556	.8391	1.192	1.305	.7660	.8727	50° 00′
10	.7010	450	550	441	185	309	642	698	50
20	039	472	545	491	178	312	623	668	40
30	069	494	540	541	171	315	604	639	30
40	098	517	535	591	164	318	585	610	20
50	127	539	529	642	157	322	566	581	10
41° 00′	.7156	.6561	1.524	.8693	1.150	1.325	.7547	.8552	49° 00′
10	185	583	519	744	144	328	528	523	50
20	214	604	514	796	137	332	509	494	40
30	243	626	509	847	130	335	490	465	30
40	272	648	504	899	124	339	470	436	20
50	301	670	499	.8952	117	342	451	407	10
42° 00′	.7330	.6691	1.494	.9004	1.111	1.346	.7431	.8378	48° 00′
10	359	713	490	057	104	349	412	348	50
20	389	734	485	110	098	353	392	319	40
30	418	756	480	163	091	356	373	290	30
40	447	777	476	217	085	360	353	261	20
50	476	799	471	271	079	364	333	232	10
43° 00′	.7505	.6820	1.466	.9325	1.072	1.367	.7314	.8203	47° 00′
10	534	841	462	380	066	371	294	174	50
20	563	862	457	435	060	375	274	145	40
30	592	884	453	490	054	379	254	116	30
40	621	905	448	545	048	382	234	087	20
50	650	926	444	601	042	386	214	058	10
44° 00′	.7679	.6947	1.440	.9657	1.036	1.390	.7193	.8029	46° 00′
10	709	967	435	713	030	394	173	.7999	50
20	738	.6988	431	770	024	398	153	970	40
30	767	.7009	427	827	018	402	133	941	30
40	796	030	423	884	012	406	112	912	20
50	825	050	418	.9942	006	410	092	883	10
45° 00′	.7854	.7071	1.414	1.000	1.000	1.414	.7071	.7854	45° 00′
		cos θ	sec θ	cot θ	tan θ	csc θ	sin θ	Radians	Degrees

ANGLE θ

ANSWERS TO SELECTED EXERCISES

CHAPTER ONE
ALGEBRA AND COORDINATE
GEOMETRY FOR PRECALCULUS

1.1 The Real Numbers

1. **(a)** natural number, integer, rational number **(b)** integer, rational number **3.** **(a)** rational number **(b)** irrational number **5.** **(a)** natural number, integer, rational number **(b)** rational number **7.** **(a)** rational number **(b)** rational number **9.** irrational number **11.** natural number, integer, rational number **13.** **(a)** $\frac{27}{5}$ **(b)** $\frac{49}{9}$ **15.** **(a)** $\frac{99}{100}$ **(b)** 1

17.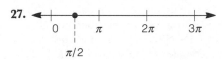

19.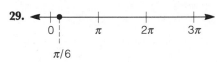

21.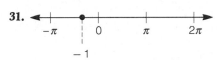

23.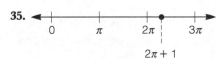

25.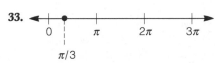

27.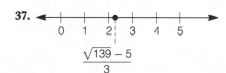

29.

31.

33.

35.

37.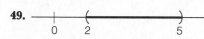

39. false **41.** true **43.** false
45. false **47.** true

49.

51.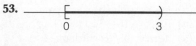

53.

55.

57.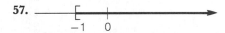

59.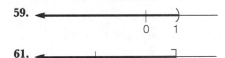

61.

63. $\frac{19}{99}$ **65.** $\frac{103}{330}$ **67.** **(a)** 3.863703
(b) 3.162277 **(c)** 1.847759
(d) 2.000000
69. **(a)** $a = \sqrt{2}, b = -\sqrt{2}$
(b) $a = \sqrt{2}, b = \sqrt{3}$
71. **(a)** $a = \sqrt{12}, b = \sqrt{3}$
   (b) $a = \sqrt{2}, b = \sqrt{3}$
73. **(a)** If A is rational, then it already represents an irrational number raised to an irrational power.
(b) Using the hint and rules of exponents, we obtain $(\sqrt{2})^2 = 2$, which is rational.

1.2 Absolute Value

1. 3 **3.** 6 **5.** 2 **7.** 0 **9.** 0 **11.** 17
13. 1 **15.** 0 **17.** $\sqrt{2} - 1$ **19.** $x - 3$
21. $-x + 3$ or $3 - x$ **23.** $t^2 + 1$
25. $\sqrt{3} + 4$ **27.** **(a)** $-2x + 7$
(b) $2x - 7$ **(c)** 1 **(d)** 1 **(e)** 1
29. $3x + 11$ **31.** $|x - 4| = 8$
33. $|x - (-3)| = 1$ or $|x + 3| = 1$
35. **(a)** $|x - 2| = \frac{1}{2}$ **(b)** $|x - 2| < \frac{1}{2}$
(c) $|x - 2| \geq \frac{1}{2}$ **(d)** $|x - 2| > \frac{1}{2}$
37. $|y| < 3$ **39.** $|x^2 - a^2| < M$
41. $|x - a| < \delta$

43.

45.

47.

49.

51.

53.

55. (a)

(b)

(c) The interval in part (b) does not include the number 2.
57. (a) 1 **(b)** -6 **(c)** -6
63. (a) Property 1(b)
(b) $a + b \leq |a| + |b|$
(c) $(-a) + (-b) \leq |a| + |b|$, so $-(a + b) \leq |a| + |b|$
(d) Since $a + b \leq |a| + |b|$ and $-(a + b) \leq |a| + |b|$, then $|a + b| \leq |a| + |b|$ since $|a + b|$ is either $a + b$ or $-(a + b)$

1.3 Integer Exponents and nth Roots

1. (a) x^{15} **(b)** x^{36} **(c)** $(x + 1)^{15}$
(d) $(x + 1)^{36}$ **3. (a)** a^6 **(b)** $(a + 1)^6$
(c) $(a + 2)(a + 1)^{10}$ **5. (a)** 1 **(b)** 1
(c) 1 **7. (a)** $\frac{11}{100}$ **(b)** $\frac{100}{11}$ **(c)** 100
9. (a) $1/a^6 b^3$ **(b)** a^7/b **(c)** $a^6 b^2/c^6$
11. $\frac{16}{49}$ **13.** $y^{12}/x^6 z^{12}$ **15.** $y^{20}/x^6 z^{16}$
17. $1/a^8 b^{12} c^{16}$ **19.** y^6 **21. (a)** false
(b) true **(c)** true **23. (a)** false
(b) true **25. (a)** false **(b)** false
27. (a) -4 **(b)** undefined **29. (a)** $\frac{2}{5}$
(b) $-\frac{2}{5}$ **31. (a)** undefined
(b) undefined **33. (a)** $\frac{4}{3}$ **(b)** $-\frac{3}{5}$
35. (a) -2 **(b)** 2 **37. (a)** $3\sqrt{2}$
(b) $3\sqrt[3]{2}$ **39. (a)** $7\sqrt{2}$
(b) $-2\sqrt[5]{2}$ **41. (a)** $\frac{5}{2}$ **(b)** $\frac{2}{5}$
43. (a) $3\sqrt{2}$ **(b)** $3\sqrt[3]{2}$
45. (a) $-4\sqrt{2}$ **(b)** $5\sqrt[4]{2}$
47. (a) 0.3 **(b)** 0.2 **49.** $-14\sqrt{6}$
51. $2\sqrt{2}$ **53. (a)** $6x$ **(b)** $-6y$
55. (a) $ab\sqrt{ab}$ **(b)** $a^2 b^2$
57. $6ab^2 c^2 \sqrt{2ac}$ **59.** $-2ab\sqrt[4]{b}$

61. $3ab\sqrt{2a}$ **63.** $2a^4 \sqrt[3]{2b^2}/c^3$
65. $-a^2 \sqrt[6]{5}/b$ **67. (a)** $4\sqrt{7}/7$ **(b)** $\sqrt{3}$
(c) $\sqrt{10}/5$ **69. (a)** $-\frac{1}{4}(1 - \sqrt{5})$
(b) $-\frac{1}{4}(1 + \sqrt{5})$ **(c)** $-\frac{1}{2}(3 + \sqrt{5})$
71. $61\sqrt{5}/5$ **73.** $\sqrt[3]{5}/5$ **75.** $\sqrt[4]{27}$
77. $\sqrt[4]{8a^3 b^3}/2ab^2$ **79.** $3\sqrt[5]{2ab}/2ab^2$
81. $x(\sqrt{x} + 2)/(x - 4)$
83. $(x - 2\sqrt{ax} + a)/(x - a)$
85. $1/(\sqrt{x} + \sqrt{5})$ **87.** $1/(\sqrt{2 + h} + \sqrt{2})$
89. 9^{10} **91. (a)** To six decimal places, both values are 1.645751.
93. For the first of the two given equations, the common value of both sides to six places (not rounding off) is 1.414213. For the second equation, the value of both sides to six decimal places is 3.141592.
95. $(a + b)/(a - b)$ **97.** x^{2c}

1.4 Rational Exponents

1. 4 **3.** $\frac{1}{6}$ **5.** undefined **7.** 5 **9.** 2
11. 4 **13.** -2 **15.** -10 **17.** $\frac{1}{7}$
19. undefined **21.** $\frac{1}{216}$ **23.** 25
25. -1 **27.** $\frac{255}{16}$ **29.** $-\frac{28976}{243}$
31. $6a^{7/12}$ **33.** $2(4^{1/4})a^{1/12}$ or $2^{3/2}a^{1/12}$
35. $x^2 + 1$ **37. (a)** $2^{1/3}3^{5/6}$
(b) $\sqrt[6]{972}$ **39. (a)** $2^{7/12}3^{1/3}$
(b) $\sqrt[12]{10368}$ **41. (a)** $x^{2/3}y^{4/5}$
(b) $\sqrt[15]{x^{10}y^{12}}$ **43. (a)** $x^{(a+b)/3}$
(b) $\sqrt[3]{x^{a+b}}$ **45.** $(x + 1)^{2/3}$
47. $(x + y)^{2/5}$ **49.** $2x^{1/6}$
51. $x^{1/6}y^{1/8}$ **53.** $9^{10/9}$
55.

n	2	5	10	100	10^3	10^4	10^5	10^6
$n^{1/n}$	1.4142	1.3797	1.2589	1.0471	1.0069	1.0009	1.0001	1.0000

57. (a) $2^{3/2}$ **(b)** $5^{1/2}$ **(c)** $2^{1/2}$
(d) $(1/2)^{1/3}$ **(e)** $10^{1/10}$ **59. (a)** $2^{1/2}$
(b) $(\sqrt{2})^2$ **61.** $(-0.5)^{1/3}$

1.5 Polynomials and Factoring

1. (a) $(-\infty, \infty)$ **(b)** $[0, \infty)$
3. (a) $(-\infty, \infty)$ **(b)** $(-\infty, \infty)$
5. (a) $[0, \infty)$ **(b)** $(0, 1) \cup (1, \infty)$
7. (a) degree: 0; coefficient: 4
(b) degree: 3; coefficient: 4
(c) degree: 6; coefficients: 1, 4, -1, 2

11. (a) $x^2 - y^2$ **(b)** $x^4 - 25$
13. $A^2 - 16$ **15.** $ab - c$
17. $x^2 - 16x + 64$ **19.** $2^{2m} + 2^{m+1} + 1$
21. $x + 2\sqrt{xy} + y$
23. $8x^3 + 12x^2 y + 6xy^2 + y^3$
25. $a^3 + 3a^2 + 3a + 1$ **27.** $x^3 - y^3$
29. $x^3 + 1$ **31.** $x - y$
33. (a) $(x + 8)(x - 8)$ **(b)** $7x^2(x^2 + 2)$
(c) $z(11 + z)(11 - z)$
(d) $(ab + c)(ab - c)$
35. (a) $(x + 3)(x - 1)$
(b) $(x - 3)(x + 1)$ **(c)** irreducible
(d) $(-x + 3)(x + 1)$ or $-(x - 3)(x + 1)$
37. (a) $(x + 1)(x^2 - x + 1)$
(b) $(x + 6)(x^2 - 6x + 36)$
(c) $8(5 - x^2)(25 + 5x^2 + x^4)$
(d) $(4ax - 5)(16a^2 x^2 + 20ax + 25)$
39. $2x(1 - x)(1 + x)$
41. $x^3(10 + x)(10 - x)$
43. $x^2(2x - 3)(x + 3)$ **45.** $x(2x - 5)^2$
47. $(xz + t)(xz + y)$
49. $(a^2 + b^2)(t^2 - c)$
51. $x(x - 18)(x + 5)$
53. $(4x + 3y)(x - 8y)$ **55.** irreducible
57. $(1 + x + y)(1 - x - y)$
59. $(x - 1)(x + 1)(x^2 + 1)(x^4 + 1)$
61. $(x + 1)^3$ **63.** $(3x + 4)^3$
65. $(x - 3)(x + 3)(x - 4)(x + 4)$
67. irreducible **69.** $x(x + 17)(x - 15)$
71. $(x + a)(x^2 - xa + a^2 + 1)$
73. $(a - b - c)(a + b + c)(a^2 + b^2 + 2bc + c^2)$
75. $(x - 2y)(x^2 + 2xy + 4y^2)(a - 2b)(a + 2b)$
77. $(ax + y)(x + by)$
79. $(a + 2)^2(9a - 8)(a - 2)$
81. $-x(x + 1)^{1/2}$

83. $x(x + 1)^{-3/2}$ or $x/(x + 1)^{3/2}$
85. $a^2(a^2 - x^2)^{-1/2}$ or $a^2/(a^2 - x^2)^{1/2}$

87. (a)

x	$\dfrac{x^2 - 16}{x - 4}$
3.9	7.9
3.99	7.99
3.999	7.999
3.9999	7.9999
3.99999	7.99999

x	$\dfrac{x^2 - 16}{x - 4}$
4.1	8.1
4.01	8.01
4.001	8.001
4.0001	8.0001
4.00001	8.00001

(b) 8

(c) $\dfrac{x^2 - 16}{x - 4} = \dfrac{(x + 4)(x - 4)}{x - 4} = x + 4$,

which approaches 8 as x approaches 4.
89. $(x^2 - 4x + 8)(x^2 + 4x + 8)$
91. $2(x - t)(x + y + z + t)$
93. $3(b - a)(c - a)(c - b)$

1.6 Quadratic Equations

1. yes **3.** no **5.** no **7.** $6, -1$
9. ± 10 **11.** $\frac{6}{5}$ **13.** $-\frac{1}{5}, \frac{3}{2}$ **15.** $1, -3$
17. $-\frac{1}{2}, 8$ **19.** $\frac{1}{2}, 6$ **21.** $-13, 12$
23. $-1, -9$ **25.** $\frac{1}{2}(1 \pm \sqrt{21})$
27. $\frac{1}{4}(-3 \pm \sqrt{41})$ **29.** $-\frac{1}{6}, -\frac{5}{2}$
31. $\frac{1}{4}(1 \pm \sqrt{41})$ **33.** $\frac{1}{6}(6 \pm \sqrt{42})$
35. $\approx -1.47, -1.53$ **37.** $-67, -89$
39. $\frac{2}{3}, 1$ **41.** $\frac{1}{2}, -1$
43. $\frac{1}{3}(-2 \pm \sqrt{13})$ **45.** $\pm 2\sqrt{6}$
47. $\frac{1}{2}(1 \pm \sqrt{5})$ **49.** $\frac{1}{3}(3 \pm \sqrt{15})$
51. $\sqrt{5}/2, -2\sqrt{5}/5$
53. $x = 1/2y$ or $x = 1/y$
55. $x = 0$ or $x = p + q$
57. $x = -5a/4$ or $x = 4a/3$
59. $r = \frac{1}{2}(\ h \pm \sqrt{h^2 + 40}\ )$
61. $t = 0$ or $t = v_0/16$
63. $x = 0, -2b, b$ **65.** two **67.** two
69. one **71.** two **73.** $k = 36$
75. $k = \pm 2\sqrt{5}$ **77. (a)** 24
(b) no real solutions **79.** -1
81. $\pm 2, \pm 3$ **83.** $-4, -\frac{20}{9}$
85. no real solution **87.** 2
89. no real solution **91.** $\frac{5}{2}, -4$
93. $\frac{1}{3}, \frac{1}{2}$ **95.** 1 **97.** $-3 \pm 2\sqrt{2}$
99. any real number except -1 or -3
101. $a + b$ **103.** $a^2b^2/(2a + b)^2$
105. $\frac{1}{2}(1 \pm \sqrt{21})$ **107.** $-a/15$
109. 1 m **111. (a)** 266 people
(b) 8 days **113.** $L/4$ in. and $3L/4$ in.
117. $-1, 0, \frac{1}{2}(1 - \sqrt{5})$
119. $8 - 2\sqrt{6\pi - 16} \approx 4.62$ cm

1.7 Inequalities

1. $(-\infty, -1)$ **3.** $[\frac{1}{3}, \infty)$
5. $(-\infty, -\frac{9}{2})$ **7.** $[-6, \infty)$
9. $(-\infty, \frac{5}{2})$ **11.** $[-\frac{17}{23}, \infty)$ **13.** $[4, 6]$
15. $[-\frac{1}{2}, 1]$ **17.** $(3.98, 3.998)$
19. (a) $[-\frac{1}{2}, \frac{1}{2}]$
(b) $(-\infty, -\frac{1}{2}] \cup [\frac{1}{2}, \infty)$
21. (a) $(-\infty, 0) \cup (0, \infty)$ **(b)** none
23. (a) $(1, 3)$ **(b)** $(-\infty, 1) \cup (3, \infty)$
25. (a) $[-4, 6]$ **(b)** $[-1, \frac{3}{2}]$
(c) $(-\infty, -1) \cup (\frac{3}{2}, \infty)$
27. $(a - c, a + c)$ **29.** $(-10, 14)$
31. $(-11, 1)$ **33. (a)** $(-2h, h)$
(b) $(h, -2h)$ **35.** $(-3, 2)$
37. $(-\infty, 2) \cup (9, \infty)$
39. $(-\infty, 4] \cup [5, \infty)$
41. $(-\infty, -4] \cup [4, \infty)$
43. no solution
45. $(-7, -6) \cup (0, \infty)$ **47.** $(-\infty, \infty)$
49. $[-4, -3] \cup [1, \infty]$
51. $(-\infty, -\frac{1}{3}) \cup (\frac{1}{3}, 2) \cup (2, \infty)$
53. $[-4, -3] \cup [3, 4]$
55. $(-2, -1) \cup (1, \infty)$ **57.** $(-1, 1]$
59. $(-\infty, \frac{3}{2}) \cup [2, \infty)$
61. $(-\frac{7}{2}, -\frac{4}{3}) \cup (-1, 0) \cup (1, \infty)$
63. $[-1, 1) \cup (2, 4]$
65. $(-1, 2 - \sqrt{5}) \cup (1, 2 + \sqrt{5})$
67. $(-\frac{3}{2}, 0)$ **69.** $-297° \le F \le 234°$,
or $[-297°, 234°]$
71. $\frac{5}{16} \le t \le \frac{5}{8}$, or $[\frac{5}{16}, \frac{5}{8}]$
73. (a) $(-\infty, -1] \cup [5, \infty)$
(b) $(-\infty, -1] \cup (5, \infty)$
75. $(-\infty, -2] \cup [2, \infty)$
77. $(-\infty, 0) \cup (\frac{1}{2}, \infty)$
79. $(\frac{1}{5}, \frac{4}{5})$ **81.** $0 < r < 2$

83.

a	b	$\sqrt{ab}$	$\dfrac{a + b}{2}$	$\sqrt{\dfrac{a^2 + b^2}{2}}$	largest	smallest
1	2	1.4142	1.5	1.5811	R.M.	G.M.
1	3	1.7320	2.0	2.2361	R.M.	G.M.
1	4	2.0000	2.5	2.9155	R.M.	G.M.
2	3	2.4495	2.5	2.5495	R.M.	G.M.
3	4	3.4641	3.5	3.5355	R.M.	G.M.
9	10	9.4868	9.5	9.5131	R.M.	G.M.
99	100	99.4987	99.5	99.5012	R.M.	G.M.
999	1000	999.4999	999.5	999.5001	R.M.	G.M.

85. (a) Because $a + b$ is the diameter of the circle, $CE = EF = (a + b)/2$, since they are radii of the circle.

(c) That $DG \le EH$ is clear, as is $EH \le DH$, since $\overline{DH}$ is the hypotenuse of $\triangle DEH$. So
$\sqrt{ab} \le (a + b)/2 \le \sqrt{(a^2 + b^2)/2}$.
87. (c) $x = \frac{15}{2}$, in which case the rectangle is $\frac{15}{2}$ ft by $\frac{15}{2}$ ft **89.** $c = 2$

1.8 Rectangular Coordinates

1.

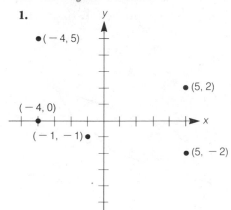

3. (a)

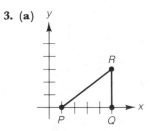

(b) 6
5. (a) 5 **(b)** 13 **7. (a)** 10 **(b)** 9
9. 4 **11. (a)** $(4, \frac{1}{2})$ **(b)** $(-6, 7)$
13. (a) yes **(b)** yes **(c)** no
15. 0; the three points are collinear.

17. center is $(3, 1)$; radius $= 5$; $y = 5$ or -3 **19.** center is $(0, 0)$; radius $= \sqrt[4]{2}$; $y = \pm\sqrt[4]{2}$
21. center is $(-4, 3)$; radius $= 1$; no y-intercepts **23.** center is $(-3, \frac{1}{3})$; radius $= \sqrt{2}$; no y-intercepts
25. center is $(\frac{1}{2}, \frac{3}{2})$; radius $= 1$; $y = \frac{1}{2}(3 \pm \sqrt{3})$
27. $(x - 1)^2 + (y - 1)^2 = 34$
29. (a) $(6, 5)$ (b) $(\frac{1}{2}, -\frac{3}{2})$ (c) $(1, -4)$ **31.** $(11, -8)$

33. (a)

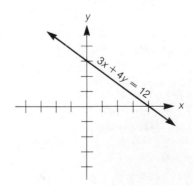

(b) $(0, \frac{7}{2})$ (c) The diagonals bisect each other.
35. (a) 128 (b) 96 (c) $\frac{4}{3}$
37. $(x - 3)^3 + (y - 2)^2 = 169$
39. $(x - 3)^2 + (y - 5)^2 = 9$
41. $a = \sqrt{2}, b = \sqrt{3}, c = 2, d = \sqrt{5}$, $e = \sqrt{6}, f = \sqrt{7}, g = 2\sqrt{2}$
43. $(x - 3)^2 + (y - 2)^2 = 13$
45. $t = 7$ or $t = -3$
47. (b) $M_{OB} = \left(\dfrac{a + b}{2}, \dfrac{c}{2}\right)$; $M_{AC} = \left(\dfrac{a + b}{2}, \dfrac{c}{2}\right)$
55. (a) 8 sq. units (b) $\frac{21}{2}$ or $10\frac{1}{2}$ sq. units
57. (c) $u = 5, v = 12, w = 13$ (many answers are possible)

1.9 The Complex Number System

1.

i^2	i^3	i^4	i^5	i^6	i^7	i^8
-1	$-i$	1	i	-1	$-i$	1

3. (a) real: 4; imaginary: 5
(b) real: 4; imaginary: -5
(c) real: $\frac{1}{2}$; imaginary: -1
(d) real: 0; imaginary: 16
5. $c = 4, d = -3$

7. (a) $14 - 4i$ (b) $-4 - 8i$
9. (a) $19 - 17i$ (b) $19 - 17i$
(c) $\frac{11}{26} - \frac{23}{26}i$ (d) $\frac{11}{25} + \frac{23}{25}i$
11. (a) $11 - i$ (b) $11 - 7i$ (c) 4
13. $4 - 2i$ **15.** $30 - 19i$ **17.** 13
19. $-191 - 163i$ **21.** $19 - 4i$
23. $-70 + 84i$ **25.** $539 + 1140i$
27. $-46 + 9i$ **29.** $\frac{6}{97} + \frac{35}{97}i$
31. $\frac{6}{97} - \frac{35}{97}i$ **33.** $-\frac{5}{13} + \frac{12}{13}i$ **35.** -4
37. $\frac{1}{26} + \frac{5}{26}i$ **39.** $-i$ **41.** $12i$
43. $-3\sqrt{5}i$ **45.** -35 **47.** $12\sqrt{2}i$
49. (a) $(a + c) + (b + d)i$
(b) $(a - c) + (b - d)i$
(c) $(ac - bd) + (bc + ad)i$
(d) $\dfrac{ac + bd}{c^2 + d^2} + \dfrac{bc - ad}{c^2 + d^2}i$
57. real: $\dfrac{2a^2 - 2b^2}{a^2 + b^2}$; imaginary: 0
59. 0

Chapter 1 Review Exercises

1. (a) $x(1 + 9x)(1 - 9x)$
(b) $(x^2 + a^2)(x + b)(x^2 - bx + b^2)$
(c) $2(x^2 + 4)^{-3/2}(x^4 + 6x^2 + 16)$
2. $|x - 3| < 5$ **3.** (a) $6\sqrt[3]{2}$
(b) a^3/b^6c^6 (c) $\frac{67}{8}$ **4.** (a) $\sqrt{10}/5$
(b) $-1 + \sqrt{2}$ (c) $2\sqrt[3]{2}$ **5.** $2\sqrt{65}$
6. (a) $-\frac{1}{4}, \frac{3}{2}$ (b) $\frac{1}{6}(-1 \pm \sqrt{13})$
(c) -1 **7.** $2x + 5$ **8.** $-1 - 2i$
9. (a) $-20 + 4i$ (b) $-\frac{10}{13} - \frac{11}{13}i$
(c) $5 - 15i$ **10.** $k = 2\sqrt{15}$
11. $(x + 1)^2 + (y + 2)^2 = 34$
12. (a) $[-\frac{8}{3}, -\frac{4}{3}]$ (b) $(5, 7)$
(c) $(-\infty, 1] \cup [3, \infty)$
13. (a) $(-\infty, -\frac{7}{3}) \cup (0, \frac{1}{2})$
(b) $(-\infty, -3] \cup (4, \infty)$
(c) $(-\infty, 0) \cup (1, \infty)$ **14.** $\frac{52}{9}$
15. (a) $a^{1/2}b^{3/4}$ (b) $\sqrt[4]{a^2b^3}$
17. $(x + a)(x + y)$ **19.** $(x - 9)^2$
21. $8a^2(x^2 + 2a)$
23. $2(3x - 2)(2x + 1)$
25. $(x - b)(2x + a)$
27. $(x + 3)(x - 3)(x^2 + 4x - 7)$
29. $4x(y - 1)$ **31.** $4x(3x - 1)(x + 4)$
33. $(a - b)(a + b + c + ab)$ **35.** 8
37. 1 **39.** $\frac{1}{16}$ **41.** $\frac{1}{9}$ **43.** ab^3c^4
45. $2a^2b^4$ **47.** $5\sqrt[3]{2}$ **49.** $5ab\sqrt{6b}$
51. $|t|$ **53.** $2|x|$ **55.** (a) $x^{13/12}$
(b) $x^{12}\sqrt{x}$ **57.** (a) $t^{17/30}$ (b) $\sqrt[30]{t^{17}}$
59. $2\sqrt{3}$ **61.** $\frac{1}{3}(\sqrt{6} + \sqrt{3})$
63. $(a^2 + \sqrt{a^4 - x^4})/x^2$ **65.** $1/(\sqrt{x} + 5)$

67. $x^2/(a^2 + \sqrt{a^4 - x^4})$ **69.** $81x^2 - y^2$
71. $9x^4 + 6x^2y^2 + y^4$ **73.** $1 - 27a^3$
75. x **77.** $x + y$ **79.** $5 + 6i$ **81.** 6
83. $\frac{1}{2}(1 - \sqrt{3}i)$ **85.** $3\sqrt{2} - 4\sqrt{2}i$
89. $|x - 6| = 2$ **91.** $|a - b| = 3$
93. $|x| > 10$ **95.** $\sqrt{6} - 2$
97. $x^4 + x^2 + 1$ **99.** (a) $-2x + 5$
(b) 1 (c) $2x - 5$ **101.** (a) T (b) F
(c) F (d) T (e) F (f) F
103. $-1 + \sqrt{2}$ cm **105.** $x > 16$ cm
107. $11, -9$ **109.** $-\frac{3}{4}, 6$
111. $\frac{14}{11}, -1$ **113.** $0, 4$
115. $\pm\sqrt{4 \pm \sqrt{5}}$ **117.** $-6, -8$
119. $\frac{1}{4}$ **121.** 2 **123.** 16 **125.** $\frac{1}{2}$
127. 10 **129.** $-8b, 4b$ **131.** $1/2y$
133. $(a + b)/ab, -2/a$ **135.** $-a, -b$
137. $[9, 12]$ **139.** $(-\infty, 0] \cup [15, \infty)$
141. $[-8, \infty)$
143. $(-\infty, -12) \cup (5, \infty)$
145. $(-\infty, -1) \cup [1, 9]$
147. $(-\infty, -\frac{1}{2}) \cup [\frac{1}{6}, \infty)$
149. $\left(-\infty, -\sqrt{\frac{1}{2}(3 + \sqrt{5})}\right)$
$\cup \left(-\sqrt{\frac{1}{2}(3 - \sqrt{5})}, 0\right)$
$\cup \left(0, \sqrt{\frac{1}{2}(3 - \sqrt{5})}\right)$
$\cup \left(\sqrt{\frac{1}{2}(3 + \sqrt{5})}, \infty\right)$
151. $k \leq \frac{9}{5}$ **153.** $k < 0$ or $k \geq 1$
157. (a) $\frac{1}{4}$ sec (on the way up) and $\frac{15}{4}$ sec (on the way down)
(b) between 1.75 sec and 2.25 sec
(c) the second ball (the one thrown from 100 ft)
159. $r/(1 + \sqrt{2})$

CHAPTER TWO
FUNCTIONS AND GRAPHS

2.1 Graphs and Symmetry

1. x-intercept: 4; y-intercept: 3

3. x-intercept: 2; y-intercept: -4

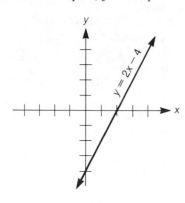

5. x-intercept: 1; y-intercept: 1

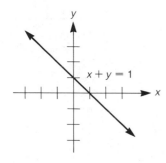

7. (a) y-intercept: 2;
x-intercepts: $-1, -2$
(b) y-intercept: 3; x-intercept: none
9. (a) y-intercept: -1;
x-intercepts: $(-1 \pm \sqrt{5})/2$
(b) y-intercept: 1; x-intercept: none
11. y-intercept: 3; x-intercept: 2
13. y-intercept: -2; x-intercept: $\frac{10}{3}$
15. y-intercept: -8; x-intercept: 2
17. (a) y-intercept: none;
x-intercept: none **(b)** y-intercept: 3;
x-intercept: none
(c) y-intercepts: $3 \pm \sqrt{5}$; x-intercept: 2
(d) y-intercepts: $3 \pm 2\sqrt{3}$;
x-intercepts: $2 \pm \sqrt{7}$
19. y-intercept: none; x-intercept: 5
21. y-intercept: 0; x-intercepts:
$-1 + 2\sqrt{3} \approx 2.46, -1 - 2\sqrt{3} \approx -4.46$
23. y-intercept: -3; x-intercepts:
$\sqrt{3} \approx 1.73, -\sqrt{3} \approx -1.73$

25. (a) reflection about the x-axis

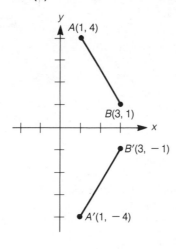

(b) reflection about the y-axis

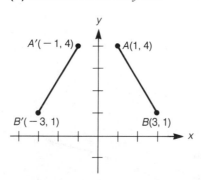

(c) reflection about the origin

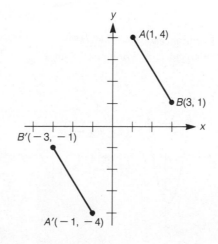

27. (a) reflection about the x-axis

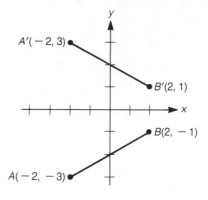

(b) reflection about the y-axis

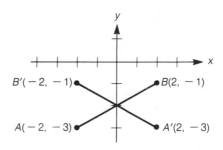

(c) reflection about the origin

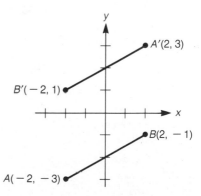

29. (a) x-axis: no; y-axis: yes; origin: no
(b) x-axis: yes; y-axis: no; origin: no
31. (a) x-axis: no; y-axis: yes; origin: no
(b) x-axis: no; y-axis: no; origin: no
(c) x-axis: yes; y-axis: no; origin: no
(d) x-axis: no; y-axis: no; origin: yes
33. x-axis: yes; y-axis: yes; origin: yes
35. x-axis: no; y-axis: no; origin: no
37. x-axis: no; y-axis: no; origin: no

39. x- and y-intercepts: 0; symmetry: y-axis

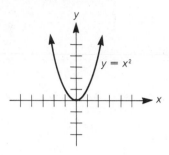

41. x- and y-intercepts: none; symmetry: origin

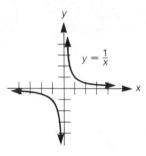

43. x- and y-intercepts: 0; symmetry: y-axis

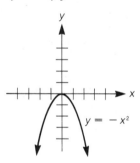

45. x- and y-intercepts: none; symmetry: origin

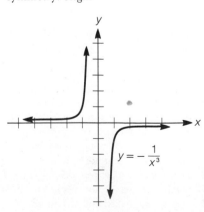

47. x- and y-intercepts: 0; symmetry: y-axis

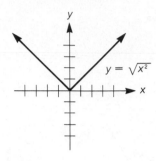

49. x-intercept: 1; y-intercept: 1; symmetry: none

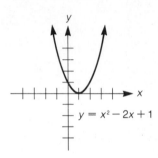

51. x-intercept: 2; y-intercept: none; symmetry: none

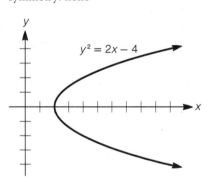

53. x-intercepts: $\frac{1}{4}(-1 \pm \sqrt{33})$; y-intercept: -4; symmetry: none

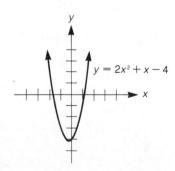

55. **(a)** x-intercept: 2; y-intercept: -6; symmetry: none

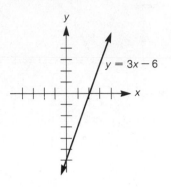

(b) x-intercept: 2; y-intercept: 6; symmetry: none

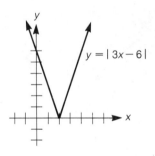

57. domain: $(-\infty, 4) \cup (4, \infty)$

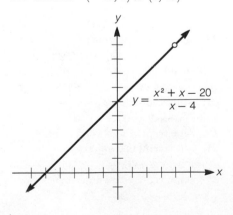

59. domain: $(-\infty, 3) \cup (3, \infty)$

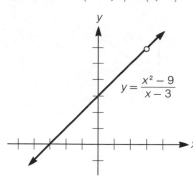

$$y = \frac{x^2 - 9}{x - 3}$$

61. (a)

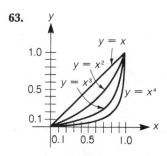

$s = 16t^2$

(b) $0 \le s \le 16$ **(c)** $16 \le s \le 64$

63.

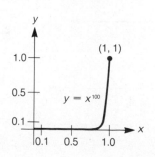

$y = x$
$y = x^2$
$y = x^3$
$y = x^4$

The pattern seems to be that as n gets larger, the graph of $y = x^n$ flattens out more and more, like in the figure. We could guess that $y = x^{100}$ would look something like this:

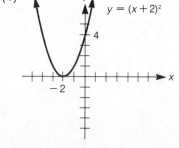

$(1, 1)$

$y = x^{100}$

65. (a)

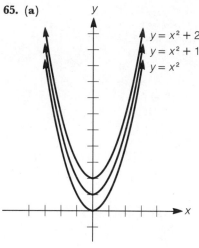

$y = x^2 + 2$
$y = x^2 + 1$
$y = x^2$

(b) The graph is shifted $|K|$ units (up if $K > 0$, down if $K < 0$).

67. (a)

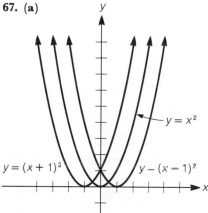

$y = x^2$

$y = (x + 1)^2$ $y = (x - 1)^2$

(b)

$y = (x + 2)^2$

4

-2

69. (a) 500 bacteria   **(b)** 1.5 hours
(c) 3.5 hours   **(d)** $3 \le t \le 4$
71. x-intercepts: $0, \pm 3\sqrt{3}$;
y-intercept: 0; symmetry: origin

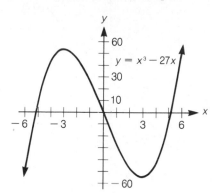

60

$y = x^3 - 27x$

30

10

-6 -3 3 6

-60

73.

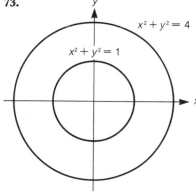

$x^2 + y^2 = 4$

$x^2 + y^2 = 1$

75.

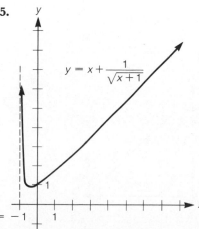

$$y = x + \frac{1}{\sqrt{x + 1}}$$

1

$x = -1$ 1

2.2 Equations of Lines

1. (a) -2 **(b)** 3 **(c)** $-\frac{7}{3}$ **(d)** 3
3. (a) $m = 1$

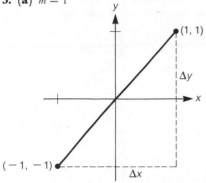

(b) $m = 0$

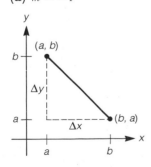

(c) $m = -1$

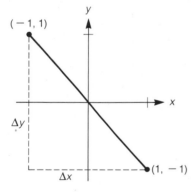

(d) $m = -1$

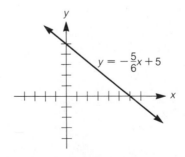

5. $m_3 < m_2 < m_4 < m_1$
7. no (slopes are $\frac{1}{4}$, $-\frac{1}{6}$)
9. yes (slopes are 3)
11. (a) $y = -5x - 9$ **(b)** $y = 4x - 20$
(c) $y = \frac{1}{3}x + \frac{4}{3}$ **(d)** $y = -x + 1$
13. (a) $y = 2x$ **(b)** $y = -2x - 4$
(c) $y = \frac{1}{7}x - \frac{11}{7}$ **15.** $x = -3$
17. $y = 4$ **19.** y-axis
21. (a) $y = -4x + 7$ **(b)** $y = 2x + \frac{3}{2}$
(c) $y = -\frac{4}{3}x + 14$
23. $y = 4x + 11$

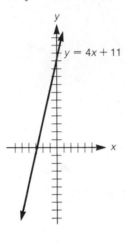

(b) $y = \frac{1}{2}x - \frac{5}{4}$

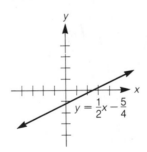

(c) $y = -\frac{5}{6}x + 5$

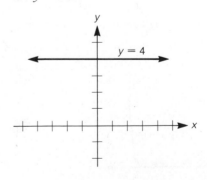

(d) $y = \frac{3}{4}x + \frac{3}{2}$

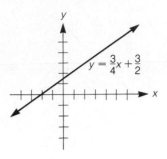

(e) $y = 4x - 2$

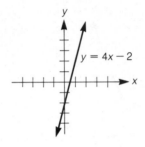

25. $y = x + 2$

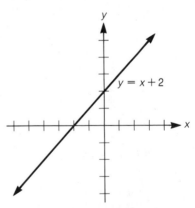

27. $y - 4 = 0$

29. (a) $(x/5) + (y/3) = 1$; area $= \frac{15}{2}$;
perimeter $= 8 + \sqrt{34}$
(b) $(x/5) - (y/3) = 1$; area $= \frac{15}{2}$;
perimeter $= 8 + \sqrt{34}$
31. (a) neither **(b)** parallel
(c) perpendicular **(d)** perpendicular
(e) neither **(f)** parallel
33. $y = \frac{2}{5}x + \frac{12}{5}$; $2x - 5y + 12 = 0$
35. $y = -\frac{4}{3}x + \frac{16}{3}$; $4x + 3y - 16 = 0$
37. $y = \frac{3}{5}x + 11$; $3x - 5y + 55 = 0$

39. (a)

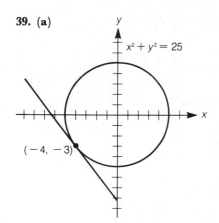

(b) $y = -\frac{4}{3}x - \frac{25}{3}$

47.

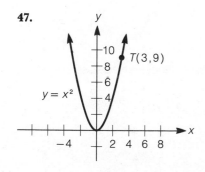

x	2.5	2.9	2.99	2.999	2.9999
y	6.25	8.41	8.9401	8.994001	8.99940001
Δx	0.5	0.1	0.01	0.001	0.0001
Δy	2.75	0.59	0.0599	0.005999	0.00059999
m	5.5	5.9	5.99	5.999	5.9999

Both slopes are 6.

49. $\left(-\frac{1}{2}, -\frac{1}{8}\right)$ **51.** $\left(8, \frac{1}{8}\right)$
53. 44.1 sq. units

55. (a)

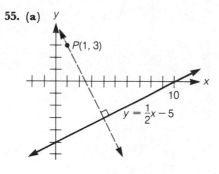

(b) $y = -2x + 5$ **(c)** $(4, -3)$
(d) $3\sqrt{5}$ **59. (a)** $x = 410 - \frac{2}{3}P$
(b) 230 units **(c)** \$307.50
63. (a) $y = -mx - b$
(b) $y = -mx + b$ **(c)** $y = mx - b$
65. (a) $-b/a$ **67. (a)** $y = 2x$
(b) $y = \sqrt{m/n}\,x$

2.3 The Definition of a Function

1. (a) $(-\infty, \infty)$ **(b)** $\left(-\infty, \frac{1}{5}\right]$
(c) $\left(-\infty, \frac{1}{5}\right) \cup \left(\frac{1}{5}, \infty\right)$
3. (a) $(-\infty, 4) \cup (4, \infty)$
(b) $(-\infty, -2) \cup (-2, 2) \cup (2, \infty)$
(c) $(-\infty, -2) \cup (2, \infty)$
5. (a) $(-\infty, -1] \cup [5, \infty)$
(b) $(-\infty, -1) \cup (5, \infty)$
(c) $(-\infty, \infty)$

7. (a) $(-\infty, 1) \cup (1, \infty)$
(b) $(-\infty, 1) \cup (1, \infty)$ **9. (a)** $[4, \infty)$
(b) $(-\infty, \infty)$ **11.** f, g, F, H
13. (a) range of $f = \{1, 2, 3\}$;
range of $g = \{2, 3\}$; range of $F = \{1\}$;
range of $H = \{1, 2\}$
(b) range of $g = \{i, j\}$;
range of $F = \{i, j\}$; range of $G = \{k\}$
15. (a) $y = (x - 3)^2$ **(b)** $y = x^2 - 3$
(c) $y = (3x)^2$ **(d)** $y = 3x^2$
17. (a) -1 **(b)** 1 **(c)** 5 **(d)** $-\frac{5}{4}$
(e) $z^2 - 3z + 1$ **(f)** $x^2 - x - 1$
(g) $a^2 - a - 1$ **(h)** $x^2 + 3x + 1$
(i) 1 **(j)** $4 - 3\sqrt{3}$ **(k)** $1 - \sqrt{2}$ **(l)** 2
19. (a) $12x^2$ **(b)** $6x^2$ **(c)** $3x^4$
(d) $9x^4$ **(e)** $3x^2/4$ **(f)** $3x^2/2$
21. (a) 1 **(b)** -7 **(c)** -3 **(d)** $-\frac{7}{18}$
(e) $-2x^2 - 4x - 1$
(f) $1 - 2x^2 - 4xh - 2h^2$
(g) $-4xh - 2h^2$ **(h)** $-4x - 2h$
23. (a) domain: $(-\infty, 2) \cup (2, \infty)$;
range: $(-\infty, 2) \cup (2, \infty)$ **(b)** $\frac{1}{2}$
(c) 0 **(d)** 1 **(e)** $(2x^2 - 1)/(x^2 - 2)$
(f) $(2 - x)/(1 - 2x)$
(g) $(2a - 1)/(a - 2)$
(h) $(2x - 3)/(x - 3)$
25. (a) $d(1) = 80$; $d(\frac{3}{2}) = 108$;
$d(2) = 128$; $d(t_0) = -16t_0^2 + 96t_0$
(b) $0, 6$ **(c)** $\frac{1}{4}(12 \pm \sqrt{143})$
27. (a) 1 **(b)** $|x|$ **29. (a)** $2x + h$
(b) $4x + 2h - 3$ **(c)** $3x^2 + 3xh + h^2$
31. (a) $-1/[(x - 1)(a - 1)]$
(b) $-1/[2(x - 1)]$
(c) $-1/[(x - 1)(x + h - 1)]$
(d) $-1/[2(2 + h)]$ **33. (a)** $\frac{11}{5}$
(b) ± 2 **(c)** $0, 1$ **(d)** $\frac{3}{2}, -1$
35. (a) \$125.51 **(b)** \$363.76
37. (a)

n	2	3	4	5	6	7	8
$g(n)$	1.4142	1.4422	1.4142	1.3797	1.3480	1.3205	1.2968

(b) 15 **39. (a)** $0, \frac{1}{3}, \frac{1}{2}$; no
43. $a = \frac{1}{2}$, $b = -\frac{3}{2}$ **49.** $-\frac{1}{3}$
51. (a) 0 **(b)** 1 **(c)** 2 **(d)** 6
(e) -1 **(f)** -2 **(g)** -6 **(h)** $\frac{1}{2}$
53. 0 **55. (a)** x **(b)** $\frac{22}{7}$

57. The problem here is the word "nearest." By the rule, $G(4) = 3$, but also $G(4) = 5$, since both 3 and 5 are equally "near" 4. So G is not a function, since it assigns more than one value to $x = 4$. To alter the definition of G, one could define $G(x)$ to be the closest prime number less than or *equal* to x (or, for that matter, greater than or equal to x). This would provide G with a way of "deciding" between 3 and 5, in the previous example.
59. 41, 43, 47, 53; yes ($x = 41$)

2.4 The Graph of a Function

1. (a) positive **(b)** 4, 1, 2, 0
(c) $f(2)$ **(d)** -3 **(e)** 3
(f) domain $= [-2, 4]$, range $= [-2, 4]$
3. (a) $g(-2)$ **(b)** 5 **(c)** $f(2) - g(2)$
(d) $-2, 3$ **(e)** f
5. range of $f = [0, 4]$;
range of $g = [-3, 3]$

7.

FUNCTION	$|x|$	x^2	x^3
Turning Point	$(0, 0)$	$(0, 0)$	none
Maximum Value	none	none	none
Minimum Value	0	0	none
Interval(s) Increasing	$(0, \infty)$	$(0, \infty)$	$(-\infty, \infty)$
Interval(s) Decreasing	$(-\infty, 0)$	$(-\infty, 0)$	none

9. (a) yes **(b)** no **(c)** no **(d)** yes
11. $m = 7$

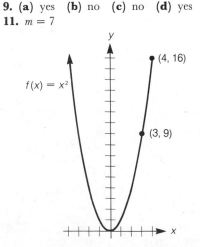

13. $(1, T(1))$ and $(4, T(4))$

15. (a)

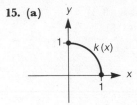

(b)

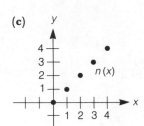

(c)

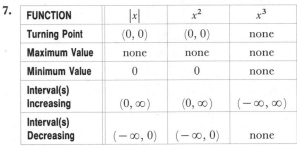

(d)

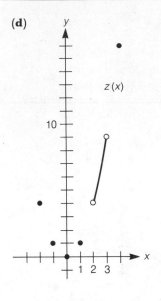

17.

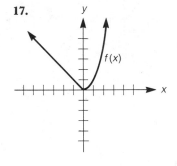

19.

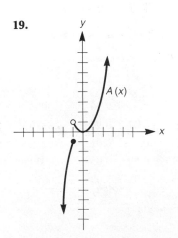

21.

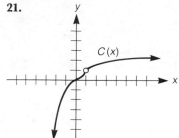

29.

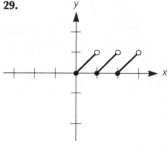

37. (a)

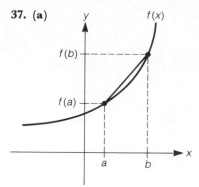

31. (a) domain: $(-\infty, -5) \cup (-5, \infty)$

23. (a)

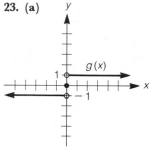

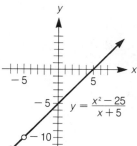

(b) between 10 and 11
(c) $g(x) = x^3$ **(d)** **(i)** 2.1;
(ii) 2.01; **(iii)** 2.001; target value: 2
39. (a) 48

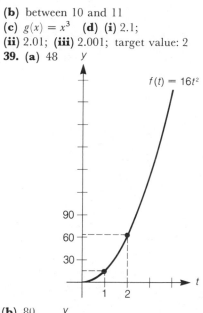

(b)

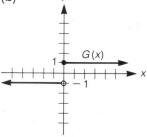

(b)

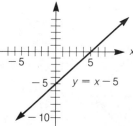

(b) 80

25.

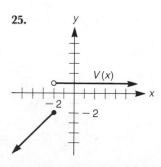

33. $t = 5$

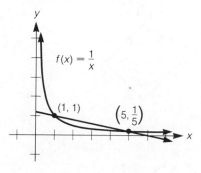

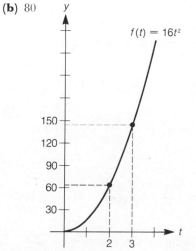

35. $(x + h, f(x + h))$

27.

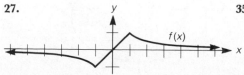

(c) 64 ft/sec   **(d)** **(i)** 33.6;
(ii) 32.016; **(iii)** 32.00016;
target value: 32

41. (a)

t	0	0.25	0.5	0.75	1	1.25	1.5	1.75	2	2.25	2.5	2.75	3
$S(t)$	0	0.25	0.5	0.75	1	1	1	1	1	1	1	1	1

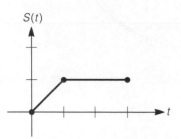

(b)

t	3	3.25	3.5	3.75	4	4.25	4.5	4.75	5
$S(t)$	1	0.75	0.5	0.25	0	−0.25	−0.5	−0.75	−1

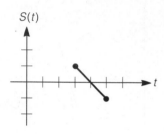

(c)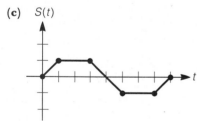

(d)

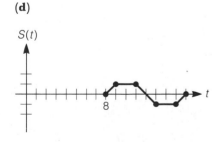

The shape of the graph is identical to that in part (c).

2.5 Techniques in Graphing

1. (a) C **(b)** F **(c)** I **(d)** A
(e) J **(f)** K **(g)** D **(h)** B **(i)** E
(j) H **(k)** G

3. (a)
$|x-2|+|y|=2$

(b) $|x+2|+|y|=2$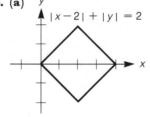

(c) $|x|+|y-2|=2$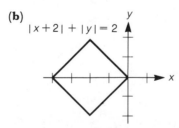

(d)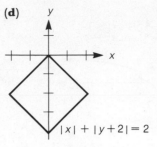
$|x|+|y+2|=2$

(e) $|x-2|+|y+2|=2$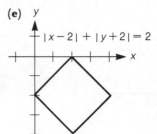

(f) $|x+2|+|y+2|=2$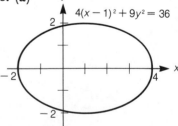

5. (a)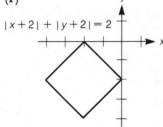
$4(x-1)^2+9y^2=36$

(b)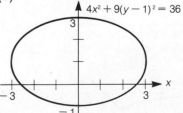
$4x^2+9(y-1)^2=36$

(c)

$4(x + 1)^2 + 9(y + 1)^2 = 36$

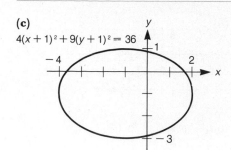

(d)

$4(x - 1)^2 + 9(y - 1)^2 = 36$

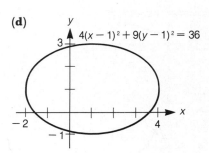

7.

$y = x^2 - 3$

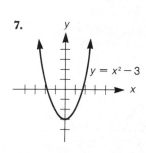

9.

$y = (x + 4)^2$

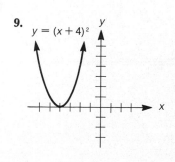

11.

$y = (x - 4)^2$

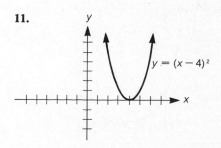

13.

$y = -x^2$

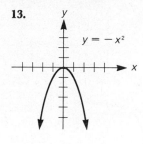

15.

$y = -(x - 3)^2$

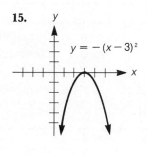

17.

$y = \sqrt{x - 3}$

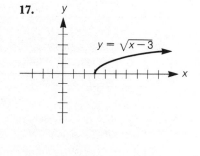

19.

$y = -\sqrt{x + 1}$

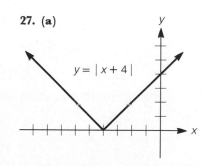

21.

$y = \dfrac{1}{x + 2} + 2$

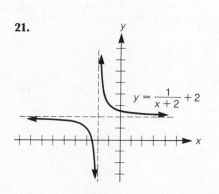

23.

$y = (x - 2)^3$

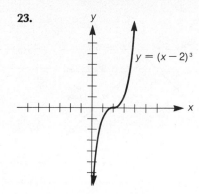

25.

$y = -x^3 + 4$

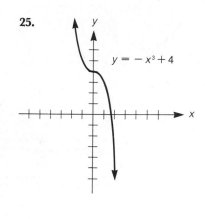

27. (a)

$y = |x + 4|$

(b)

$y = |4 - x|$

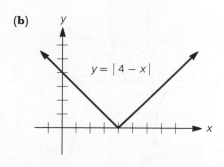

(c)

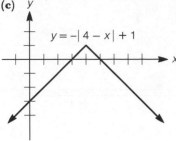

$y = -|4 - x| + 1$

37.

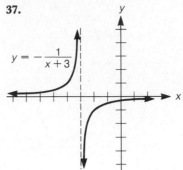

$y = -\dfrac{1}{x+3}$

(b)

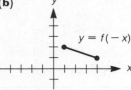

$y = f(-x)$

(c)

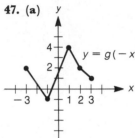

$y = -f(-x)$

29.

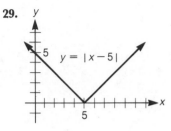

$y = |x - 5|$

5

5

39.

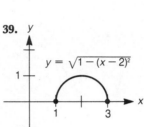

$y = \sqrt{1 - (x-2)^2}$

1

1   3

47. (a)

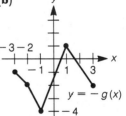

$y = g(-x)$

4

2

-3 1 2 3

31.

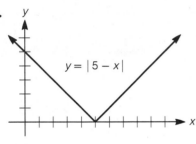

$y = |5 - x|$

41.

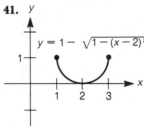

$y = 1 - \sqrt{1 - (x-2)^2}$

1

1   2   3

(b)

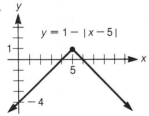

$-3 -2$

-1 1   3

$y = -g(x)$

-4

33.

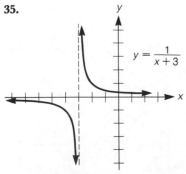

$y = 1 - |x - 5|$

1

5

-4

43.

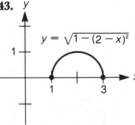

$y = \sqrt{1 - (2 - x)^2}$

1

1   3

(c)

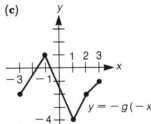

1 2 3

-3 -1

$y = -g(-x)$

-4

35.

$y = \dfrac{1}{x + 3}$

45. (a)

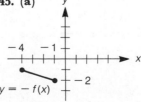

-4 -1

-2

$y = -f(x)$

49. (a)

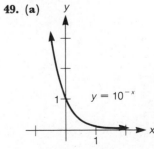

1

$y = 10^{-x}$

1

(b)

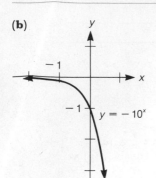

$y = -10^x$

(c)

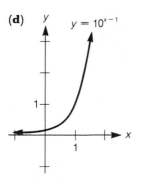

$y = -10^{-x}$

(d)

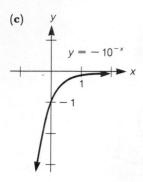

$y = 10^{x-1}$

(e)

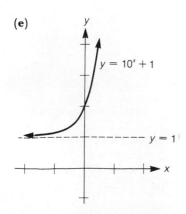

$y = 10^x + 1$

$y = 1$

(f)

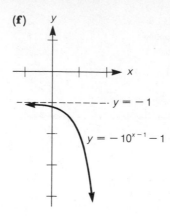

$y = -1$

$y = -10^{x-1} - 1$

51. (a) $(a + 3, b)$ **(b)** $(a, b - 3)$
(c) $(a + 3, b - 3)$ **(d)** $(a, -b)$
(e) $(-a, b)$ **(f)** $(-a, -b)$
(g) $(-a + 3, b)$
(h) $(-a + 3, -b + 1)$

53.

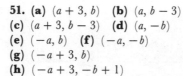

$g(x) = \dfrac{x}{x - 1}$

$y = 1$

$x = 1$

55.

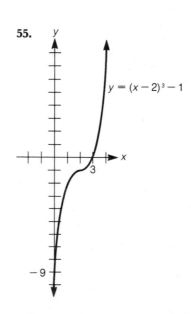

$y = (x - 2)^3 - 1$

57. $[1, 2)$
61. (a) increasing:
$(-\infty, -4) \cup (-2, \infty)$;
decreasing: $(-4, -2)$
(b) increasing: $(-\infty, 2) \cup (4, \infty)$;
decreasing: $(2, 4)$

2.6 Methods of Combining Functions

1. (a) $x^2 - x - 7$ **(b)** $-x^2 + 5x + 5$
(c) 5 **3. (a)** $x^2 - 2x - 8$
(b) $-x^2 + 2x + 8$ **5. (a)** $4x - 2$
(b) $4x - 2$ **(c)** -4
7. (a) $\dfrac{-x^4 + 22x^2 - 4x - 80}{2x^3 - x^2 - 18x + 9}$
(b) $-80/9$
9. (a) $-x^5 + 9x^3 + 2x^2 - 18$
(b) $-x^5 + 9x^3 + 2x^2 - 18$ **(c)** -24
11. (a) $-6x - 14$ **(b)** -74
(c) $-6x - 7$ **(d)** -67
13. (a) $(f \circ g)(x) = -x$;
$(f \circ g)(-2) = 2$; $(g \circ f)(x) = 2 - x$;
$(g \circ f)(-2) = 4$ **(b)** $(f \circ g)(x) =$
$9x^2 - 3x - 6$; $(f \circ g)(-2) = 36$;
$(g \circ f)(x) = -3x^2 + 9x + 14$;
$(g \circ f)(-2) = -16$ **(c)** $(f \circ g)(x) =$
$(1 - x^4)/3$; $(f \circ g)(-2) = -5$;
$(g \circ f)(x) = 1 - (x^4/81)$;
$(g \circ f)(-2) = 65/81$ **(d)** $(f \circ g)(x) =$
$2^{x^2 + 1}$; $(f \circ g)(-2) = 32$;
$(g \circ f)(x) = 2^{2x} + 1$; $(g \circ f)(-2) =$
$17/16$ **(e)** $(f \circ g)(x) = 3x^5 - 4x^2$;
$(f \circ g)(-2) = -112$; $(g \circ f)(x) =$
$3x^5 - 4x^2$; $(g \circ f)(-2) = -112$
(f) $(f \circ g)(x) = x$; $(f \circ g)(-2) = -2$;
$(g \circ f)(x) = x$; $(g \circ f)(-2) = -2$
15. (a) $(-x + 7)/6x$ **(b)** $(-t + 7)/6t$
(c) $5/12$ **(d)** $(1 - 6x)/7$
(e) $(1 - 6y)/7$ **(f)** $-11/7$
17. (a) $M(7) = \frac{13}{5}$, $M[M(7)] = 7$
(b) x **(c)** 7 **19. (a)** 1 **(b)** -3
(c) -1 **(d)** 2 **(e)** 2 **(f)** -3
21. (a) Both are $4x^3 - 3x^2 + 6x - 1$.
(b) Both are $ax^2 + bx + c$.
(c) $(f \circ I)(x) = (I \circ f)(x) = f(x)$ for *any*
function $f(x)$.

23.

x	0	1	2	3	4
$(f \circ g)(x)$	1	3	2	und.	2

x	-1	0	1	2	3	4
$(g \circ f)(x)$	0	0	3	4	2	und.

25. $(f \circ g)(x) = 6x - 7$

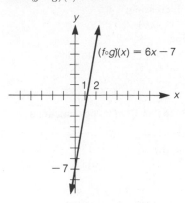

(b) $(g \circ f)(x) = 6x - 1$

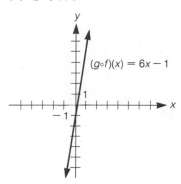

27. (a) domain: $[0, \infty)$;
range: $[-3, \infty)$

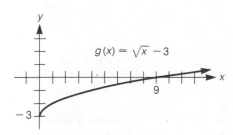

(b) domain: $(-\infty, \infty)$;
range: $(-\infty, \infty)$

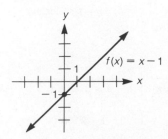

(c) $(f \circ g)(x) = \sqrt{x} - 4$
domain: $[0, \infty)$;
range: $[-4, \infty)$

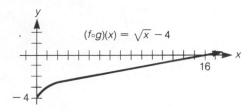

(d) $g[f(x)] = \sqrt{x - 1} - 3$
domain: $[1, \infty)$

(e)

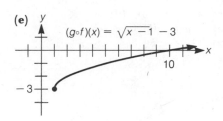

29. $(f \circ g)(x)$, where $f(x) = x^4$ and $g(x) = 3x - 1$

31. (a) $(f \circ g)(x)$, where $f(x) = \sqrt[3]{x}$ and $g(x) = 3x + 4$
(b) $(f \circ g)(x)$, where $f(x) = |x|$ and $g(x) = 2x - 3$
(c) $(f \circ g)(x)$, where $f(x) = x^5$ and $g(x) = ax + b$
(d) $(f \circ g)(x)$, where $f(x) = 1/x$ and $g(x) = \sqrt{x}$
33. (a) $(b \circ c)(x)$ **(b)** $(a \circ d)(x)$
(c) $(c \circ d)(x)$ **(d)** $(c \circ b)(x)$
(e) $(c \circ a)(x)$ **(f)** $(a \circ c)(x)$
(g) $(b \circ d)(x)$ or $(d \circ b)(x)$
35. $(C \circ f)(t) = 2\pi/(t^2 + 1)$; $\pi/5$ ft
37. (a) $100 + 450t - 25t^2$ **(b)** \$1225
(c) No, it is \$1900. **39.** $(x + 6)/4$
41. $a = -\frac{1}{2}, b = \frac{1}{2}$ **43. (a)** $2x + 2a - 2$
(b) $4x + 4a - 4$ **45. (a)** $(x^2/2) + 1$
(b) $(x + 1)^2/2$ **(c)** $(x^2/4) + 1$
(d) $(x + 1)^2/4$ **(e)** $(x^2 + 1)/2$
47. (a) $p(x) = (g \circ f \circ h)(x)$
(b) $q(x) = (h \circ g \circ f)(x)$
(c) $r(x) = (f \circ g \circ h)(x)$
(d) $s(x) = (h \circ f \circ g)(x)$
49. (a) $(x + 1)/\sqrt{x^2 + 2x + 2}$
(b) $10\sqrt{101}/101$

51. (a)

$\circ$	i	a	b	c
i	i	a	b	c
a	a	i	c	b
b	b	c	i	a
c	c	b	a	i

(b) yes **(c)** i, i, i, c **(d)** yes **(e)** $\{i\}$
(f) They both equal i. **(g)** yes

2.7 Inverse Functions

3. (a) 4 **(b)** -1 **(c)** $\sqrt{2}$ **(d)** $t + 1$
(e) $f(0) = 1, f^{-1}(1) = 0$
(f) $f(-1) = -2, f^{-1}(-2) = -1$
5. (a) $(x + 1)/3$

(c)

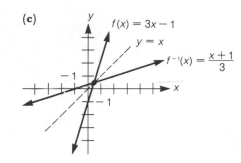

7. (a) $x^2 + 1$

(c)

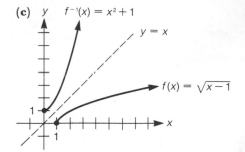

9. (a) domain: $(-\infty, 3) \cup (3, \infty)$;
range: $(-\infty, 1) \cup (1, \infty)$
(b) $(3x + 2)/(x - 1)$
(c) domain: $(-\infty, 1) \cup (1, \infty)$;
range: $(-\infty, 3) \cup (3, \infty)$;
domain of f^{-1} = range of f;
range of f^{-1} = domain of f
11. $\sqrt[3]{(x - 1)/2}$

13.

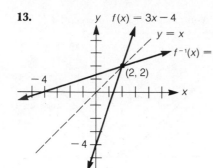

$f(x) = 3x - 4$
$y = x$
$f^{-1}(x) = \dfrac{x+4}{3}$
$(2, 2)$

15. (a) $\sqrt[3]{x+1} + 3$

(b)

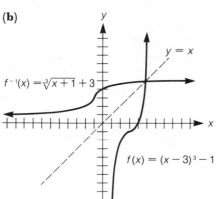

$y = x$
$f^{-1}(x) = \sqrt[3]{x+1} + 3$
$f(x) = (x-3)^3 - 1$

17. (a)

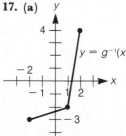

$y = g^{-1}(x)$

(b)

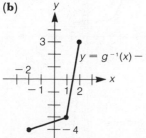

$y = g^{-1}(x) - 1$

(c)

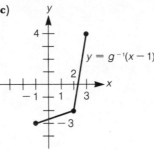

$y = g^{-1}(x-1)$

(d)

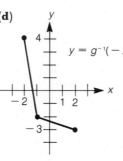

$y = g^{-1}(-x)$

(e)

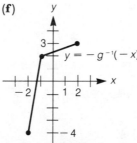

$y = -g^{-1}(x)$

(f)

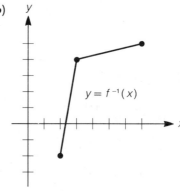

$y = -g^{-1}(-x)$

19. no **21.** yes **23.** yes **25.** no
27. no **29.** yes **33. (a)** $x = 7$
(b) $x = -3$ **35.** $t = \frac{9}{4}$
37. (a) $f^{-1}(x) = x^2$; domain: $[0, \infty)$
(b) (i) f; **(ii)** f^{-1}; **(iii)** f;
(iv) f^{-1}; **(v)** f; **(vi)** f^{-1};
(vii) f^{-1}; **(viii)** f **43.** $y = 2x - 6$
45. $F[f^{-1}(x) + 1] = x$ **47.** $(-1, 2)$
49. (a) $(-b, -a)$

Chapter 2 Review Exercises

1. (a) $(-\infty, 3]$
(b) $(-\infty, \frac{1}{2}) \cup (\frac{1}{2}, \infty)$
2. (a) $3x^2 - 6x - 1$ **(b)** $12x^2 + 4x - 1$
(c) 7 **3.** $y = -\frac{3}{2}x + \frac{9}{2}$
4. x-axis: no; y-axis: no; origin: yes
5. (a) $-1/ax$ **(b)** $1 - 4x - 2h$
6. slope $= \frac{3}{5}$; y-intercept $= -3$

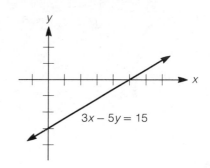

$3x - 5y = 15$

7. (a) $g^{-1}(x) = 1/(3x + 5)$

(b)

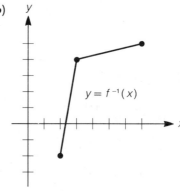

$y = f^{-1}(x)$

8. $y = \frac{5}{6}x$
9. (a) x-intercepts: $-5, 1$;
y-intercept: -1

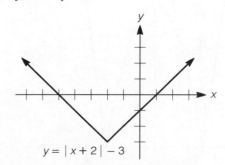

$y = |x + 2| - 3$

(b) x-intercept: -1; y-intercept: $-\frac{1}{2}$

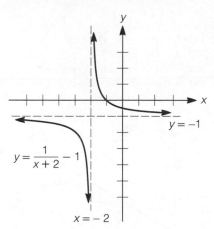

$$y = \frac{1}{x+2} - 1$$

$$x = -2$$

$$y = -1$$

10. (a) $(-3, 3)$, $(-2, -1)$ **(b)** 3
(c) $x = -1$ **(d)** $(-5, -3) \cup (-2, 2)$
11. (a) 5 **(b)** $7 - 4\sqrt{2}$

12.

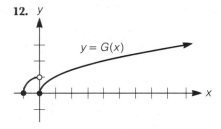

$$y = G(x)$$

13. $10 + h$
14. x-intercepts: $-3, 3$; y-intercept: 9

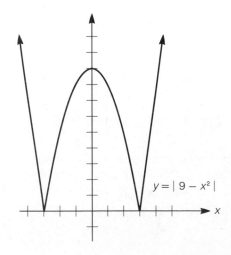

$$y = |9 - x^2|$$

15.

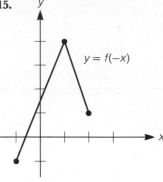

$$y = f(-x)$$

16. $t = -3$ **17.** $y = -2x - 6$
19. $y = \frac{1}{4}x - \frac{5}{2}$ **21.** $y = 2x + 8$
23. $y = -2$ **25.** $y = x + 1$
27. $y = \frac{9}{4}x + \frac{7}{2}$ **29.** $y = -2x$
31. $y = \frac{3}{4}x$ **33.** $y = -x + 1$ or
$y = \frac{1}{2}x - 2$ (both satisfy the conditions)
35. (a) E **(b)** C **(c)** L **(d)** A
(e) J **(f)** G **(g)** B **(h)** M **(i)** K
(j) D **(k)** I **(l)** H **(m)** N **(n)** F
37. x-intercepts: $1 \pm \sqrt{2}$; y-intercept: 1

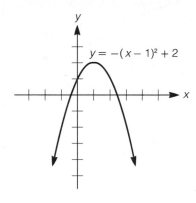

$$y = -(x - 1)^2 + 2$$

39. x-intercept: none; y-intercept: 1

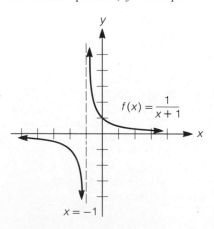

$$f(x) = \frac{1}{x + 1}$$

$$x = -1$$

41. x-intercept: -3; y-intercept: 3

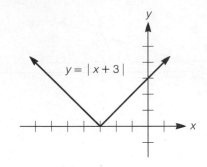

$$y = |x + 3|$$

43. x-intercepts: ± 1; y-intercept: 1

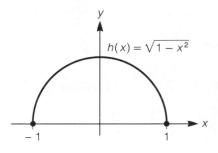

$$h(x) = \sqrt{1 - x^2}$$

$$-1 \qquad 1$$

45. x-intercept: 0; y-intercept: 0

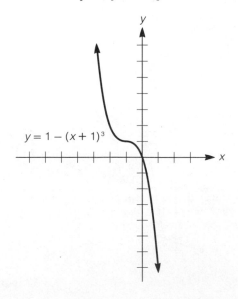

$$y = 1 - (x + 1)^3$$

47. x-intercept: 0; y-intercept: 0

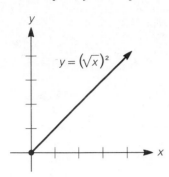

49. x-intercept: 1; y-intercept: none

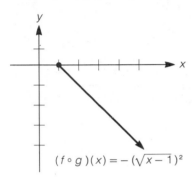

51. x-intercepts: -1, 0; y-intercept: 0

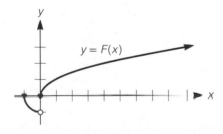

53. x-intercept: none; y-intercept: none

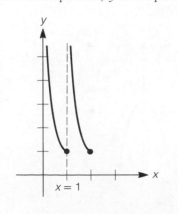

55. x-intercept: $\frac{1}{2}$; y-intercept: -1

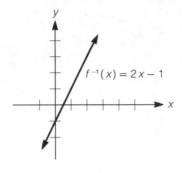

57. x-intercept: 0; y-intercept: 0

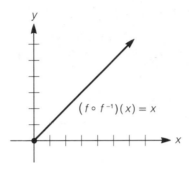

59. x-axis: no; y-axis: yes; origin: no
61. x-axis: no; y-axis: no; origin: no
63. x-axis: no; y-axis: no; origin: yes
65. $(-\infty, -3) \cup (-3, 3) \cup (3, \infty)$
67. $(-\infty, 4]$ **69.** $(-\infty, \infty)$
71. $(-\infty, -1] \cup [3, \infty)$
73. $(-\infty, 0) \cup (0, \infty)$
75. $(-\infty, \frac{1}{3}) \cup (\frac{1}{3}, \infty)$
77. $(-\infty, 0) \cup (0, \infty)$
79. $(-\infty, 2) \cup (2, \infty)$
81. $a(x) = (f \circ g)(x)$
83. $c(x) = (G \circ g)(x)$
85. $A(x) = (g \circ f \circ G)(x)$
87. $C(x) = (g \circ G \circ G)(x)$

89. 12   **91.** $-\frac{9}{19}$ **93.** $t^2 + t$
95. $x^2 - 5x + 6$ **97.** -4 **99.** 6
101. $x^4 - x^2$ **103.** $-2x^3 + 3x^2 - x$
105. $4x^2 - 2x$ **107.** $-2x^2 + 2x + 1$
109. $(2x + 2)/(2x - 5)$
111. $2x + h - 1$
113. $F^{-1}(x) = (4x + 3)/(1 - x)$
115. x **117.** x **119.** $(1 + x)/2$
121. $\frac{22}{7}$ **123.** negative   **125.** -1
127. -1 **129.** $(0, -2)$ and $(5, 1)$
131. $(-6, 0) \cup (5, 8)$
133. 0 (at $x = 2$)   **135.** no   **137.** $x = 4$
139. (a) $x = 10$ **(b)** $x = 0$
141. (a) 5   **(b)** -3 **(c)** 4   **(d)** $\frac{1}{4}$
143. $(f \circ f)(10)$ **145.** $[0, 4]$ **147.** 5
149. $(1, 3) \cup (6, 10)$ **151.** $(4, 7)$
157. (a) $(x - 1)^2 + (y - 1)^2 = 1$
(b) $\overline{AT}$: $y = \frac{9}{8}x$; $\overline{BU}$: $y = -\frac{1}{4}x + 1$;
$\overline{SC}$: $y = -3x + 3$
(c) $\overline{AT}$ and $\overline{CS}$ intersection: $(\frac{8}{11}, \frac{9}{11})$;
$\overline{AT}$ and $\overline{BU}$ intersection: $(\frac{8}{11}, \frac{9}{11})$;
notice that the point of intersection is
the same in both cases.

CHAPTER THREE
POLYNOMIAL AND RATIONAL
FUNCTIONS. APPLICATIONS TO
OPTIMIZATION

3.1 Linear Functions

1. $f(x) = \frac{2}{3}x + \frac{2}{3}$ **3.** $g(x) = \sqrt{2}\,x$
5. $f(x) = x - \frac{7}{2}$ **7.** $f(x) = \sqrt{3}$
9. $f(x) = \frac{1}{2}x - 2$ **11.** yes
13. $V(t) = -2{,}375t + 20{,}000$
15. (a) $V(t) = -12{,}000t + 60{,}000$

(b)

END OF YEAR	YEARLY DEPRECIATION	ACCUMULATED DEPRECIATION	VALUE V
0	0	0	60,000
1	12,000	12,000	48,000
2	12,000	24,000	36,000
3	12,000	36,000	24,000
4	12,000	48,000	12,000
5	12,000	60,000	0

17. (a) $530 **(b)** $538 **(c)** $8/fan
19. (a) $50 **(b)** $50/player
(c) The answers are the same.
21. (a) $\frac{4}{5}$ ft/sec **(b)** 0 cm/sec
(c) 8 mph **23. (a)** B is traveling faster.
(b) A is farther to the right.
(c) $t = 8$ sec

25. (a)

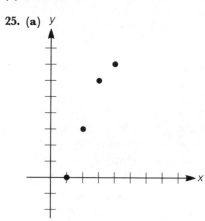

(b) $m \approx 2.5, y \approx -2$

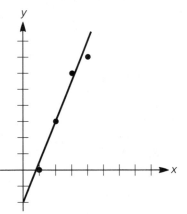

(c)

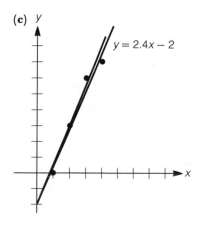

27. (a) no **(b)** 4195000
29. $f(x) = 3x + 1$
33. (a) $\sum x = 10$; $\sum y = 16$
(b) $\sum x^2 = 30$; $\sum xy = 52$
(c) $b = -2$; $m = 2.4$
35. $y = 2.4x - 0.4$
37. $y = 0.084x + 37.241$

3.2 Quadratic Functions

1. vertex: $(-2, 0)$; axis: $x = -2$;
minimum: 0; x-intercept: -2;
y-intercept: 4

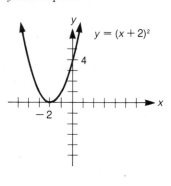

3. vertex: $(-2, 0)$; axis: $x = -2$;
minimum: 0; x-intercept: -2;
y-intercept: 8

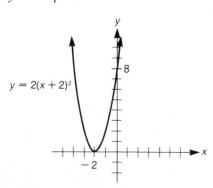

5. vertex: $(-2, 4)$; axis: $x = -2$;
maximum: 4; x-intercepts: $-2 \pm \sqrt{2}$;
y-intercept: -4

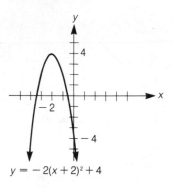

$y = -2(x + 2)^2 + 4$

7. vertex: $(2, -4)$; axis: $x = 2$;
minimum: -4; x-intercepts: 0, 4;
y-intercept: 0

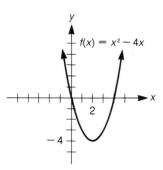

9. vertex: $(0, 1)$; axis: $x = 0$;
maximum: 1; x-intercepts: ± 1;
y-intercept: 1

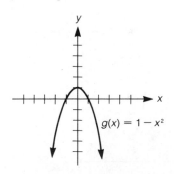

11. vertex: $(1, -4)$; axis: $x = 1$; minimum: -4; x-intercepts: $3, -1$; y-intercept: -3

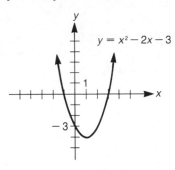

13. vertex: $(3, 11)$; axis: $x = 3$; maximum: 11; x-intercepts: $3 \pm \sqrt{11}$; y-intercept: 2

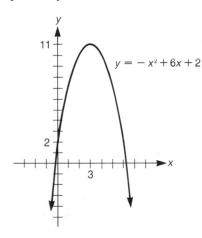

15. vertex: $(0, 0)$; axis: $t = 0$; minimum: 0; t-intercept: 0; s-intercept: 0

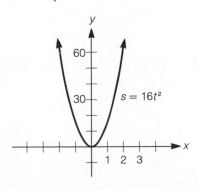

17. vertex: $\left(\frac{1}{6}, \frac{9}{4}\right)$; axis: $t = \frac{1}{6}$; maximum: $\frac{9}{4}$; t-intercepts: $-\frac{1}{3}, \frac{2}{3}$; s-intercept: 2

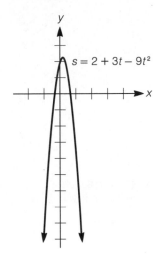

19. $x = 1$ will yield a minimum output value.
21. $x = \frac{3}{2}$ will yield a maximum output value.
23. $x = 0$ will yield a minimum output value.
25. minimum value of -13

27. maximum value of $\frac{25}{8}$
29. maximum value of 1000
31. 5 units   **33.** quadratic
35. neither linear nor quadratic
37. linear   **39. (a)** minimum at $(3, 8)$
(b) minimum at $(3, 4)$
(c) minimum at $(\pm\sqrt{3}, 64)$
41. (a) 4 (at $x = 2$)   **(b)** $2\sqrt[3]{2}$ (at $x = 2$)
(c) 16 (at $x = \pm\sqrt{2}$)
43. $y = -\frac{1}{2}(x - 2)^2 + 2$
45. $y = \frac{1}{4}(x - 3)^2 - 1$ **49.** $c = 1 + \sqrt{2}$
51. $x = (a + b)/2$

3.3 Applied Functions: Setting up Equations

1. (a) $p(x) = 2x + 2\sqrt{144 - x^2}$
(b) $A(x) = x\sqrt{144 - x^2}$
3. (a) $D(x) = \sqrt{x^4 + 3x^2 + 1}$
(b) $M(x) = (x^2 + 1)/x$
5. (a) $A(y) = \pi y^2/4$
(b) $A(y) = \pi^2 y^2/16$
7. (a) $P(x) = 16x - x^2$
(b) $S(x) = 2x^2 - 32x + 256$
(c) $D(x) = (16 - x)^3 - x^3$
(d) $A(x) = 8$ (it's constant)
9. $R(x) = -\frac{1}{4}x^2 + 8x$
11. (a)

x	1	2	3	4	5	6	7
$P(x)$	17.88	19.49	20.83	21.86	22.49	22.58	21.75

(b) 22.58, $x = 6$ **(c)** 22.63
13. (a) $h(s) = \sqrt{3}\,s$ **(b)** $A(s) = \sqrt{3}\,s^2$
(c) $4\sqrt{3}$ cm   **(d)** $25\sqrt{3}/4$ in.^{2}
15. $V(r) = 2\pi r^3$ **17. (a)** $h(r) = 12/r^2$
(b) $S(r) = 2\pi r^2 + (24\pi/r)$
19. $V(S) = S\sqrt{\pi S}/6\pi$
21. $A(x) = \frac{1}{2}x\sqrt{400 - x^2}$
23. $d(x) = (x + 4)\sqrt{x^2 + 25}/x$
25. $x = 25$, $L = 25$

x	5	10	20	24	24.8	24.9	25	25.1	25.2	45
$A(x)$	225	400	600	624	624.96	624.99	625	624.99	624.96	225

27. (a) Table 1

x	1	2	3	4
A	7.5	12	10.5	0

$x = 2$ yields the largest area.

Table 2

x	1.75	2.00	2.25	2.50	2.75
A	11.3203	12.0000	12.3047	12.1875	11.6016

$x = 2.25$ yields the largest area.

Table 3

x	2.15	2.20	2.25	2.30	2.35
A	12.2308	12.2760	12.3047	12.3165	12.3111

$x = 2.30$ yields the largest area.

27. (b) $x = 2.30$, $A = 12.316806$
29. (a) $V(r) = \frac{1}{3}\sqrt{3}\pi r^3$ **(b)** $S(r) = 2\pi r^2$
31. (a) $r(h) = 3h/\sqrt{h^2 - 9}$
(b) $h(r) = 3r/\sqrt{r^2 - 9}$
33. $A(x) = [4x^2 + \pi(14 - x)^2]/16\pi$
35. $A(r) = r(1 - 4\pi r)/4$
37. $A(x) = \pi x^2/3$
39. $A(h) = \pi R^2 - h\sqrt{2Rh - h^2}$
41. $V(x) = 4x^3 - 28x^2 + 48x$
43. (a) $A(r) = 32r - 2r^2 - \frac{1}{2}\pi r^2$
(b) downward, passes through the origin
45. (a) $y(s) = 3s/\sqrt{1 - s^2}$
(b) $s(y) = y/\sqrt{y^2 + 9}$
(c) $z(s) = 3/\sqrt{1 - s^2}$
(d) $s(z) = \sqrt{z^2 - 9}/z$
47. (a) $m(a) = (a^2 + 1)/a$
49. $A(x) = (8x - x^2)/4$
51. $A(m) = (2m^2 - 8m + 8)/(m^2 - 4m)$
53. $A(m) = -(ma - b)^2/2m$

3.4 Maximum and Minimum Problems

1. $\frac{25}{4}$ **3.** $\frac{1}{2}$ **5.** $\frac{25}{4}$ m by $\frac{25}{4}$ m
7. 1250 in.^{2} **9. (a)** 18 **(b)** $\frac{23}{4}$ **(c)** $\frac{47}{8}$
(d) $\frac{95}{16}$ **11. (a)** 16 ft, 12 ft
(b) 16 ft, 1 sec **(c)** $\frac{1}{4}$ sec, $\frac{7}{4}$ sec
13. $\left(\frac{7}{2}, \frac{1}{2}(2 + \sqrt{6})\right)$; $d = \sqrt{7}/2$
15. (a) $\frac{1}{2}$ **(b)** $\frac{1}{4}$ **17.** 125 ft by 250 ft
19. 40 **21.** $x = 60$, $R = \$900$, $p = \$15$
23. (a) $\frac{36}{13}$ **(b)** $\sqrt{\frac{36}{13}} = \frac{6}{13}\sqrt{13}$
25. (a) $\frac{225}{2}$ **27.** $x = \frac{1}{2}\sqrt{2}$, area $= \frac{1}{2}$
29. $\frac{49}{12}$ **31.** $2R^2$ **33.** 100 yd by 150 yd
37. (a) $p(x) = -2x + 500$
(b) \$31,250 when $p = \$250$
39. $\pm\sqrt{2}/2$ **41.** $\pm\sqrt{3}$
43. (a) $A(x) = \left(\frac{1}{4\pi} + \frac{1}{16}\right)x^2 - 2x + 16$

(b) $x = \dfrac{16\pi}{4 + \pi}$ **(c)** $\dfrac{\pi}{4}$
45. (a) $A(x) = [(4 + \pi)/16\pi]x^2$
$\qquad\qquad - (L/8)x + (L^2/16)$
(b) $x = \pi L/(4 + \pi)$ **(c)** $\pi/4$
47. 2 **51.** 2

3.5 Polynomial Functions

1. This graph has four turning points, but a polynomial function of degree 3 can have at most two turning points.
3. As $|x|$ gets very large, our function should be similar to $f(x) = a_3 x^3$. But $f(x)$ does not have a parabolic shape, like the given graph.
5. As $|x|$ gets very large, with x negative, the graph should resemble $2x^5$. But the y-values of $2x^5$ are always negative when x is negative, contrary to the given graph.
7. This graph has a cusp, which cannot occur in the graph of a polynomial function.
9. x-intercept: none; y-intercept: 5

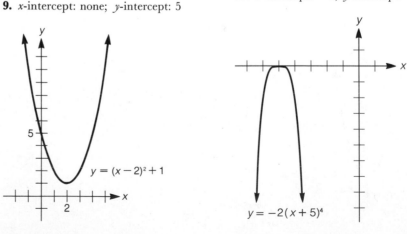

$y = (x - 2)^2 + 1$

11. x-intercept: 1; y-intercept: -1

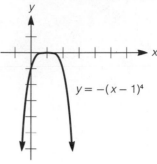

$y = -(x - 1)^4$

13. x-intercept: $4 + \sqrt[3]{2}$; y-intercept: -66

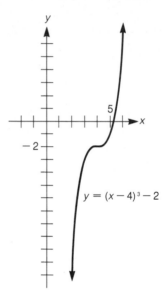

$y = (x - 4)^3 - 2$

15. x-intercept: -5; y-intercept: -1250

$y = -2(x + 5)^4$

17. x-intercept: -1; y-intercept: $\frac{1}{2}$

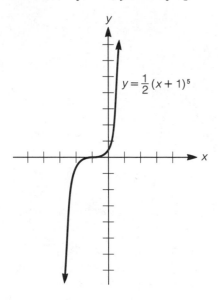

$y = \frac{1}{2}(x+1)^5$

19. x-intercept: 0; y-intercept: 0

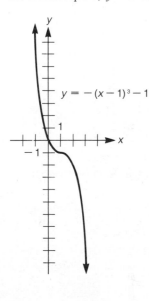

$y = -(x-1)^3 - 1$

21.

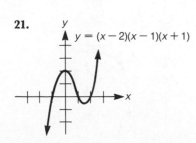

$y = (x-2)(x-1)(x+1)$

23.

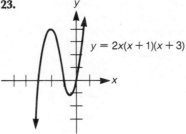

$y = 2x(x+1)(x+3)$

25.

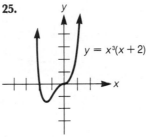

$y = x^3(x+2)$

27.

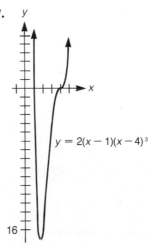

$y = 2(x-1)(x-4)^3$

16

29.

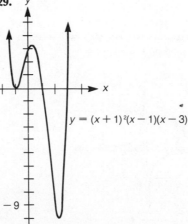

$y = (x+1)^2(x-1)(x-3)$

-9

31.

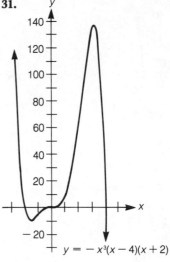

$y = -x^3(x-4)(x+2)$

33.

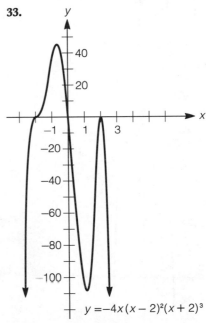

$y = -4x(x-2)^2(x+2)^3$

35. from left to right: $f(x)$, $g(x)$, $h(x)$, $F(x)$, $G(x)$, $H(x)$ **37.** $[0, 0.68]$ **39.** no
41. yes, at $(100, 100)$

43.

$y = x^3(2-x)$

45.

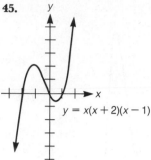

$y = x(x + 2)(x - 1)$

47.

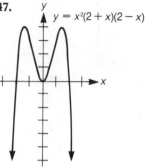

$y = x^2(2 + x)(2 - x)$

49. (a)

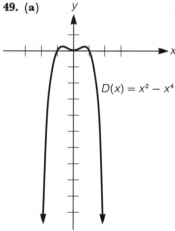

$D(x) = x^2 - x^4$

(b) $(\pm\sqrt{2}/2, \frac{1}{4})$, $(0, 0)$
(c) Maximum vertical distance is $\frac{1}{4}$.

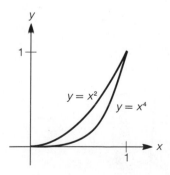

$y = x^2$
$y = x^4$

51. (c) $(0, 6)$
(d) maximum $f(r) \approx 272875$,
maximum $V \approx 522$ cm^3

r	f(r)
0.0	0
0.5	88
1.0	1382
1.5	6745
2.0	20213
2.5	45878
3.0	86339
3.5	140700
4.0	202129
4.5	254971
5.0	271414
5.5	207720
6.0	0

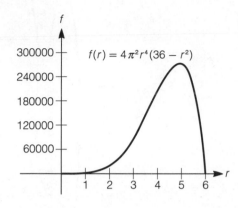

$f(r) = 4\pi^2 r^4(36 - r^2)$

3.6 Graphs of Rational Functions

1. domain: $(-\infty, 3) \cup (3, \infty)$;
x-intercept: -5; y-intercept: $-\frac{5}{4}$
3. domain: $(-\infty, -2) \cup (-2, 3) \cup (3, \infty)$; x-intercepts: 9, -1;
y-intercept: $\frac{3}{2}$
5. domain: $(-\infty, 0) \cup (0, \infty)$;
x-intercepts: ± 2, 1; y-intercept: none
7. x-intercept: none; y-intercept: $\frac{1}{4}$;
horizontal asymptote: $y = 0$ (x-axis);
vertical asymptote: $x = -4$

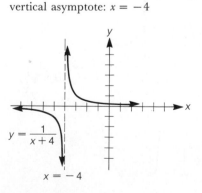

$y = \frac{1}{x + 4}$

$x = -4$

9. x-intercept: none; y-intercept: $\frac{3}{2}$;
horizontal asymptote: $y = 0$ (x-axis);
vertical asymptote: $x = -2$

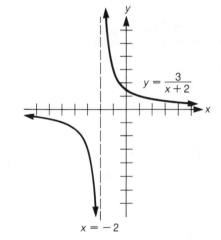

$y = \frac{3}{x + 2}$

$x = -2$

11. x-intercept: none; y-intercept: $\frac{2}{3}$;
horizontal asymptote: $y = 0$ (x-axis);
vertical asymptote: $x = 3$

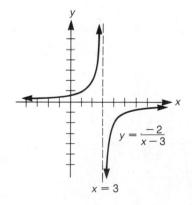

$y = \frac{-2}{x - 3}$

$x = 3$

13. x-intercept: 3; y-intercept: 3; horizontal asymptote: $y = 1$; vertical asymptote: $x = 1$

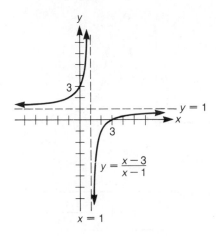

15. x-intercept: $\frac{1}{2}$; y-intercept: -2; horizontal asymptote: $y = 2$; vertical asymptote: $x = -\frac{1}{2}$

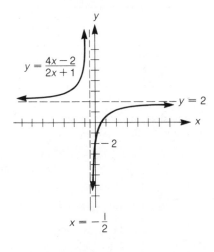

17. x-intercept: none; y-intercept: $\frac{1}{4}$; horizontal asymptote: $y = 0$ (x-axis); vertical asymptote: $x = 2$

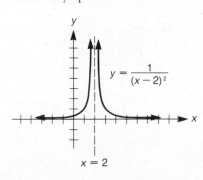

19. x-intercept: none; y-intercept: 3; horizontal asymptote: $y = 0$ (x-axis); vertical asymptote: $x = -1$

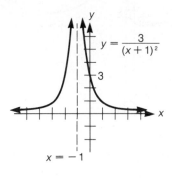

21. x-intercept: none; y-intercept: $\frac{1}{8}$; horizontal asymptote: $y = 0$ (x-axis); vertical asymptote: $x = -2$

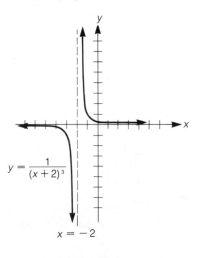

23. x-intercept: none; y-intercept: $-\frac{4}{125}$; horizontal asymptote: $y = 0$ (x-axis); vertical asymptote: $x = -5$

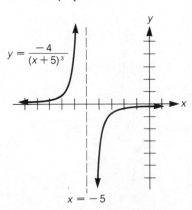

25. x-intercept: 0; y-intercept: 0; horizontal asymptote: $y = 0$ (x-axis); vertical asymptotes: $x = -2$, $x = 2$

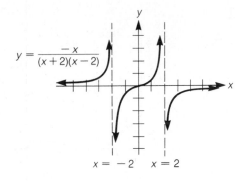

27. x-intercept: 0; y-intercept: 0; horizontal asymptote: $y = 0$ (x-axis); vertical asymptotes: $x = -3$, $x = 1$

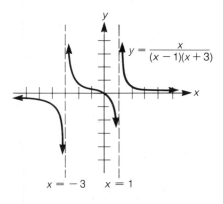

29. (a) x-intercepts: 2, 4; y-intercept: none; horizontal asymptote: $y = 1$; vertical asymptotes: $x = 0$, $x = 1$

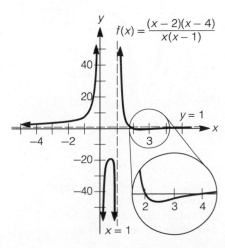

(b) x-intercepts: 2, 4; y-intercept: none;
horizontal asymptote: $y = 1$;
vertical asymptotes: $x = 0$, $x = 3$

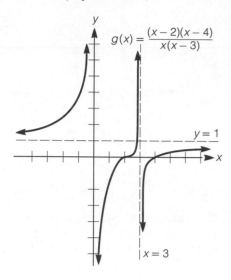

$$g(x) = \frac{(x-2)(x-4)}{x(x-3)}$$

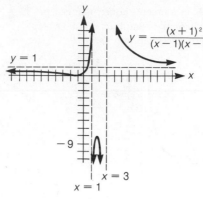

$$y = \frac{(x+1)^2}{(x-1)(x-3)}$$

$y = 1$

-9

$x = 3$
$x = 1$

37. (a)

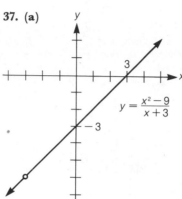

$$y = \frac{x^2 - 9}{x + 3}$$

(b)

$$y = \frac{x^2 - 5x + 6}{x^2 - 2x - 3}$$

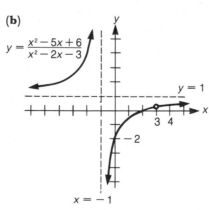

$y = 1$

$3 \quad 4$

-2

$x = -1$

31. x-intercepts: -2, 4; y-intercept: $-\frac{8}{3}$;
horizontal asymptote: $y = 1$;
vertical asymptotes: $x = 1$, $x = 3$;
crosses the horizontal asymptote: $x = \frac{11}{2}$

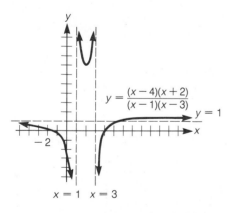

$$y = \frac{(x-4)(x+2)}{(x-1)(x-3)}$$

$y = 1$

-2

$x = 1 \quad x = 3$

(c)

$$y = \frac{(x-1)(x-2)(x-3)}{(x-1)(x-2)(x-3)(x-4)}$$

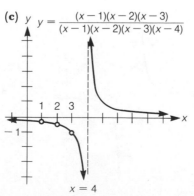

$1 \quad 2 \quad 3$

-1

$x = 4$

33. x-intercept: -1; y-intercept: $\frac{1}{3}$;
horizontal asymptote: $y = 1$;
vertical asymptotes: $x = 1$, $x = 3$;
crosses the horizontal asymptote: $x = \frac{1}{3}$

39. horizontal asymptote: $y = 0$;
vertical asymptote: $x = 3$;
low point: $\left(-3, -\frac{1}{12}\right)$

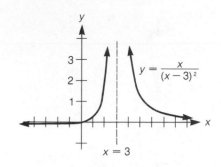

3

2

1

$$y = \frac{x}{(x-3)^2}$$

$x = 3$

41. (b)

x	$x + 4$	$\dfrac{x^2 + x - 6}{x - 3}$
10	14	14.8571
100	104	104.0619
1000	1004	1004.0060

x	$x + 4$	$\dfrac{x^2 + x - 6}{x - 3}$
-10	-6	-6.4615
-100	-96	-96.0583
-1000	-996	-996.060

(c) vertical asymptote: $x = 3$;
x-intercepts: -3, 2; y-intercept: 2

(d)

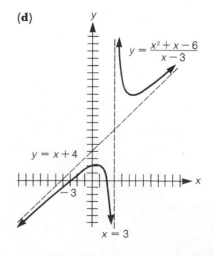

$$y = \frac{x^2 + x - 6}{x - 3}$$

$y = x + 4$

-3

$x = 3$

(e) $(3 + \sqrt{6}, 7 + 2\sqrt{6})$,
$(3 - \sqrt{6}, 7 - 2\sqrt{6})$

43.

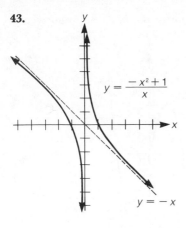

$$y = \frac{-x^2 + 1}{x}$$

$$y = -x$$

Chapter 3 Review Exercises

1. -5

2. (a) Minimum value is -13 when $x = -1$.

(b) Maximum value is 9 when $t = \pm\sqrt{3}$.

3. \$20,000

4.

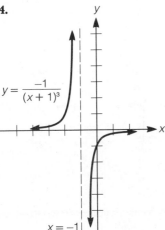

$$y = \frac{-1}{(x+1)^3}$$

$$x = -1$$

5.

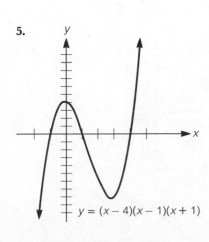

$$y = (x-4)(x-1)(x+1)$$

6. vertex: $(-2, -9)$;

x-intercepts: $-5, 1$; y-intercept: -5;

axis of symmetry: $x = -2$

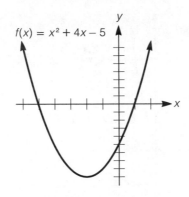

$$f(x) = x^2 + 4x - 5$$

7. $V(t) = -180t + 1000$

8. crosses the y-axis at $(0, 162)$

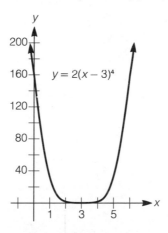

$$y = 2(x-3)^4$$

9. x-intercept: $-\frac{5}{3}$; y-intercept: $\frac{5}{2}$;

horizontal asymptote: $y - 3$;

vertical asymptote: $x = -2$

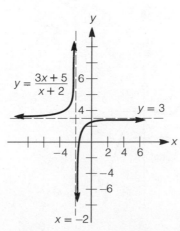

$$y = \frac{3x+5}{x+2}$$

$$y = 3$$

$$x = -2$$

10. 18 cm^2

11.

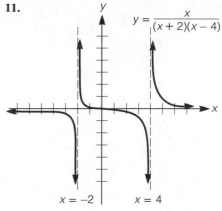

$$y = \frac{x}{(x+2)(x-4)}$$

$$x = -2 \qquad x = 4$$

12. $m(x) = x^2 - x + 1$

13. $P(w) = 2w + (2\sqrt{144 - \pi^2 w^2})/\pi$

14. (a) This graph *looks* like x^3, but $-\frac{1}{3}x^3$ should decrease as x gets large, not increase.

(b) This graph has four turning points, but $-\frac{1}{3}x^3$ can have at most two turning points.

15. $f(x) = x + 2$ **17.** $f(x) = \frac{3}{8}x - \frac{5}{2}$

19. $f(x) = -\frac{3}{4}x + \frac{11}{4}$

21. vertex: $(-1, -4)$;

x-intercepts: $1, -3$; y-intercept: -3

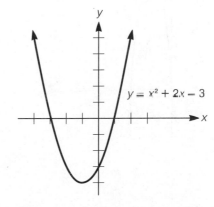

$$y = x^2 + 2x - 3$$

23. vertex: $(\sqrt{3}, 6)$;
x-intercepts: $\sqrt{3} + \sqrt{6}$, $\sqrt{3} - \sqrt{6}$;
y-intercept: 3

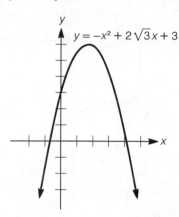

25. vertex: $(2, 12)$; x-intercepts: 0, 4;
y-intercept: 0

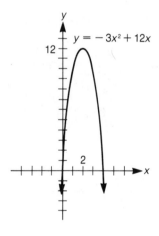

27. 5   **29.** $\frac{3}{4}$ **31.** (a) $v_0^2/64$ ft when
$t = v_0/32$ sec   (b) $v_0/16$ sec
33. (a) $d(x) = \sqrt{x^4 - 3x^2 + 4}$
(b) $\left(-\frac{1}{2}\sqrt{6}, \frac{3}{2}\right)$ **35.** $b = \frac{17}{3}, -11$
37. 1   **39.** $\frac{225}{4}$ cm²   **41.** $a = 1$
43. 400 units at \$80

45. x-intercepts: -4, 2; y-intercept: -8;
vertex: $(-1, -9)$

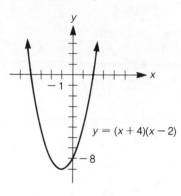

47. x-intercept: -5; y-intercept: -125

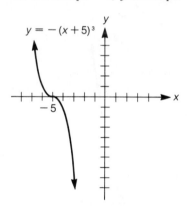

49. x-intercepts: -1, 0; y-intercept: 0

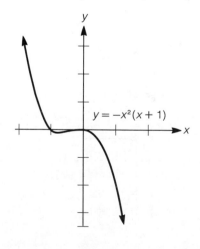

51. x-intercepts: 0, 2, -2; y-intercept: 0

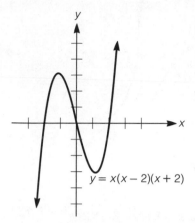

53. x-intercept: $-\frac{1}{3}$; y-intercept: none;
horizontal asymptote: $y = 3$;
vertical asymptote: $x = 0$

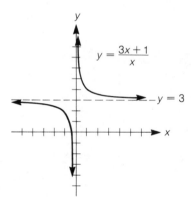

55. x-intercept: none; y-intercept: -1;
horizontal asymptote: $y = 0$;
vertical asymptote: $x = 1$

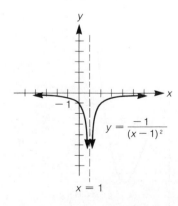

57. x-intercept: 2; y-intercept: $\frac{2}{3}$;
horizontal asymptote: $y = 1$;
vertical asymptote: $x = 3$

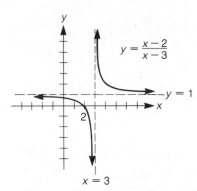

$y = \dfrac{x-2}{x-3}$

$y = 1$

$x = 3$

59. x-intercept: 1; y-intercept: $\frac{1}{4}$;
horizontal asymptote: $y = 1$;
vertical asymptote: $x = 2$

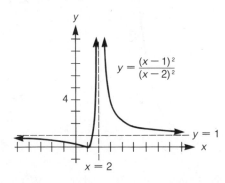

$y = \dfrac{(x-1)^2}{(x-2)^2}$

$y = 1$

$x = 2$

61. (a) 1 **(b)** $\sqrt{13}$ **(c)** $\pm\sqrt{2}/2$
63. $k = 6$
65. $(-\infty, 4 - 2\sqrt{3}] \cup [4 + 2\sqrt{3}, \infty)$
67. $A(m) = m/2$
69. $A(x) = (1-x)\sqrt{1-x^2}$
71. (a)

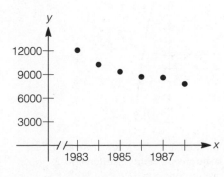

(b) $m \approx -770$;
y-intercept ≈ 1.5 million

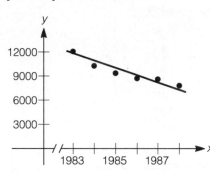

(c) 2100 refugees
(d) $f^{-1}(x) = (1541090 - x)/771.4$
(e) 1996
73. (a) The graphs are reflections of
each other about the y-axis.

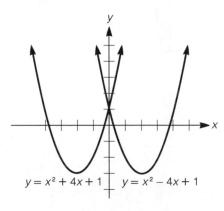

$y = x^2 + 4x + 1$ | $y = x^2 - 4x + 1$

(b) $f(-x) = ax^2 - bx + c$
(c) Reversing the sign of b will reflect
the graph of $f(x) = ax^2 + bx + c$ across
the y-axis.

CHAPTER FOUR
EXPONENTIAL AND
LOGARITHMIC FUNCTIONS

4.1 Exponential Functions

1. (a) 10^9 **(b)** 10^{15} **3. (a)** $x = 3$
(b) $t = \frac{3}{2}$ **(c)** $y = \frac{1}{4}$ **(d)** $z = \frac{5}{2}$
5. $(-\infty, \infty)$ **7.** $(-\infty, \infty)$

9.

$y = 2^{-x}$ $y = 2^x$

$(0, 1)$

11.

$y = 3^x$

1

-1

$y = -3^x$

13.

$y = 3^x$

$y = 2^x$

$(0, 1)$

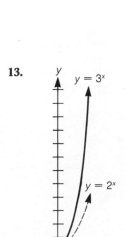

15.

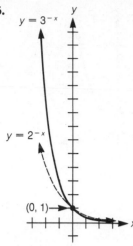

17. domain: $(-\infty, \infty)$; range: $(-\infty, 1)$; x-intercept: 0; y-intercept: 0; asymptote: $y = 1$

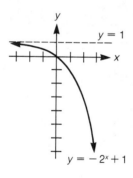

19. domain: $(-\infty, \infty)$; range: $(1, \infty)$; x-intercept: none; y-intercept: 2; asymptote: $y = 1$

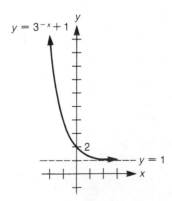

21. domain: $(-\infty, \infty)$; range: $(0, \infty)$; x-intercept: none; y-intercept: $\frac{1}{2}$; asymptote: $y = 0$

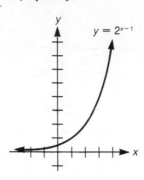

23. domain: $(-\infty, \infty)$; range: $(1, \infty)$; x-intercept: none; y-intercept: 4; asymptote: $y = 1$

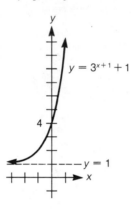

25. $-\frac{1}{3}$ **27.** $\frac{3}{2}$, 1

31. (a)

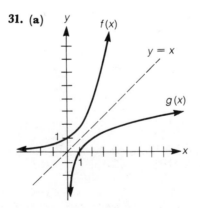

(b) domain: $(0, \infty)$; range: $(-\infty, \infty)$; intercept: $x = 1$ (no y-intercept); asymptote: $x = 0$

33. (a) 1.4   **(b)** 1.15   **(c)** 1.5

35.

x	$\log_{10} x$
1	0
2	0.3
3	0.48
4	0.6
5	0.7
6	0.78
7	0.85
8	0.9
9	0.95
10	1

4.2 The Exponential Function $y = e^x$

1. domain: $(-\infty, \infty)$; range: $(0, \infty)$; x-intercept: none; y-intercept: 1; asymptote: $y = 0$

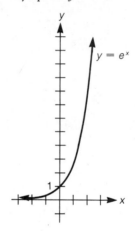

3. domain: $(-\infty, \infty)$; range: $(-\infty, 0)$; x-intercept: none; y-intercept: -1; asymptote: $y = 0$

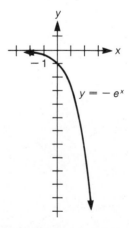

5. domain: $(-\infty, \infty)$; range: $(1, \infty)$; x-intercept: none; y-intercept: 2; asymptote: $y = 1$

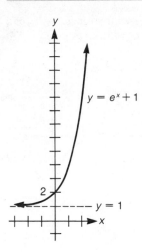

7. domain: $(-\infty, \infty)$; range: $(1, \infty)$; x-intercept: none; y-intercept: $e + 1$; asymptote: $y = 1$

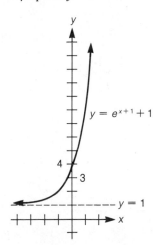

9. domain: $(-\infty, \infty)$; range: $(-\infty, 0)$; x-intercept: none; y-intercept: $-1/e^2$; asymptote: $y = 0$

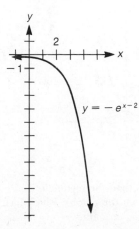

11. domain: $(-\infty, \infty)$; range: $(-\infty, e)$; x-intercept: 1; y-intercept: $e - 1$; asymptote: $y = e$

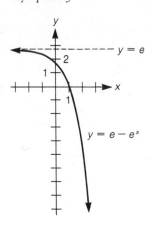

13.

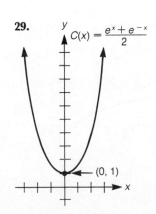

15. 2 **17.** $\sqrt[3]{4}$ **19.** **(a)** $\sqrt{3}$
(b) 486,000 **21.** 6400 bacteria
23. $\left(\frac{1}{2}\right)^{7/8} \approx 0.55$ g **25.** 97.15%

29.

$$C(x) = \frac{e^x + e^{-x}}{2}$$

$(0, 1)$

31.

n	$1 + \dfrac{1}{n}$	$\left(1 + \dfrac{1}{n}\right)^n$
1	2	2.0000
10	1.1	2.5937
100	1.01	2.7048
1000	1.001	2.7169
10000	1.0001	2.7181
100000	1.00001	2.7183

33. **(a)** $2^{27/16} \approx 3.22$ billion
(b) too low

35. **(a)**

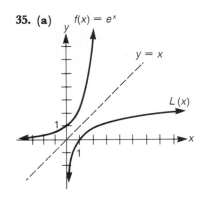

(b) domain: $(0, \infty)$; range: $(-\infty, \infty)$; x-intercept: 1; asymptote: $x = 0$
(c) **(i)** x-intercept: 1; asymptote: $x = 0$

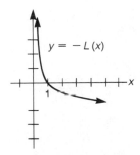

(ii) x-intercept: -1; asymptote: $x = 0$

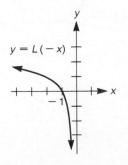

(iii) x-intercept: 2; asymptote: $x = 1$

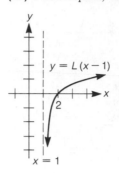

(b) domain: $(-1, \infty)$; range: $(-\infty, \infty)$; x-intercept: 0; y-intercept: 0; asymptote: $x = -1$

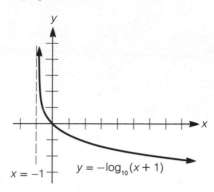

(c) domain: $(-\infty, 0)$; range: $(-\infty, \infty)$; x-intercept: $-e$; y-intercept: none; asymptote: $x = 0$

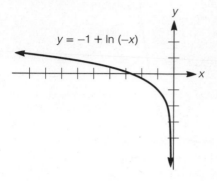

25. (a) 4 **(b)** -1 **(c)** $\frac{1}{2}$
27. $\log_{10} 25 \approx 1.40$
29. $\pm\sqrt{1 + \log_{10} 4} \approx \pm 1.27$
31. $\frac{1}{2}(-3 + \ln 10) \approx -0.35$
33. $\frac{1}{4}(1 - \ln 12.405) \approx -0.38$
35. (a) $k \approx -1.54 \times 10^{-10}$
(b) 99.9999846% **37. (a)** $k = \ln \frac{1}{2}$
(b) 3.32 years **39. (a)** 0.51 hr
(b) 2.89 hr
41. $f^{-1}(x) = -1 + \ln x$; x-intercept: e; asymptote: $x = 0$

4.3 Logarithmic Functions

1. (a) no **(b)** yes **(c)** yes
3. (a) $f^{-1}(x) = (4x + 1)/(2 - 3x)$
(b) $(3x + 4)/(2x - 1)$ **(c)** $\frac{1}{2}$ **(d)** -4
5. $(-1, 3)$ and $(6, -1)$
7. (a) $\log_3 9 = 2$ **(b)** $\log_{10} 1000 = 3$
(c) $\log_7 343 = 3$ **(d)** $\log_2 \sqrt{2} = \frac{1}{2}$
9. (a) $2^5 = 32$ **(b)** $10^0 = 1$
(c) $e^{1/2} = \sqrt{e}$ **(d)** $3^{-4} = \frac{1}{81}$
(e) $t^v = u$ **11.** $\log_5 30$ **13. (a)** $\frac{3}{2}$
(b) $-\frac{5}{2}$ **(c)** $\frac{3}{2}$ **15. (a)** $x = 2$
(b) $x = \frac{1}{5}$ **17. (a)** $(0, \infty)$
(b) $(-\infty, \frac{3}{4})$ **(c)** $(-\infty, 0) \cup (0, \infty)$
(d) $(0, \infty)$ **(e)** $(-\infty, -5) \cup (5, \infty)$
19. domain: $(0, e) \cup (e, \infty)$;
range: $(-\infty, 0) \cup (0, \infty)$
21. (a) domain: $(-1, \infty)$;
range: $(-\infty, \infty)$; x-intercept: 0;
y-intercept: 0; asymptote: $x = -1$

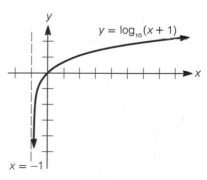

23. (a) domain: $(0, \infty)$;
range: $(-\infty, \infty)$; x-intercept: 1;
y-intercept: none; asymptote: $x = 0$

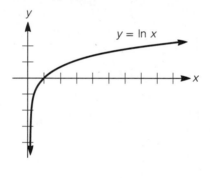

(b) domain: $(-\infty, 0)$; range: $(-\infty, \infty)$;
x-intercept: -1; y-intercept: none;
asymptote: $x = 0$

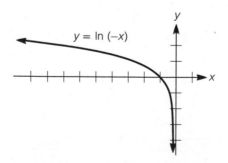

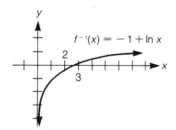

43.

The area is less than the rectangle (shown dashed on the graph), which has an area of two square units.

45. $\ln 6$ **47.** 10^{30}
49. $t = [\ln(A_0/B_0)]/(k_2 - k_1)$; this represents the time it takes for two radioactive substances to decay to the same mass.
51. (a) $P(10) = 4$; $P(18) = 7$; $P(19) = 8$

(b)

x	$P(x)$	$\dfrac{x}{\ln x}$	$\dfrac{P(x)}{x/\ln x}$
10^2	25	21.715	1.151
10^4	1229	1085.736	1.132
10^6	78498	72382.414	1.084
10^8	5761455	5428681.024	1.061
10^9	50847534	48254942.43	1.054
10^{10}	455052512	434294481.9	1.048

(c)

x	$P(x)$	$\dfrac{x}{\ln x - 1.08366}$	$\dfrac{P(x)}{x/(\ln x - 1.08366)}$
10^2	25	28	0.8804
10^4	1229	1231	0.9988
10^6	78498	78543	0.9994
10^8	5761455	5768004	0.9989
10^9	50847534	50917519	0.9986
10^{10}	455052512	455743004	0.9985

53.

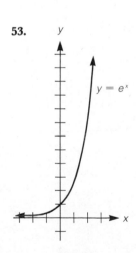

$y = e^x$

4.4 Properties of Logarithms

1. 1   **3.** $\frac{1}{2}$ **5.** 4   **7.** 0   **9.** 4
11. $\log_{10} 60$ **13.** $\log_5 20$
15. (a) $\ln 6$ **(b)** $\ln(\frac{3}{16384})$
17. $\log_b[4(1 + x)^3/(1 - x)^{3/2}]$
19. $\log_{10}[27\sqrt{x + 1}/(x^2 + 1)^6]$

21. (a) $2\log_{10} x - \log_{10}(1 + x^2)$
(b) $2\ln x - \frac{1}{2}\ln(1 + x^2)$
23. (a) $\frac{1}{2}\log_{10}(3 + x) + \frac{1}{2}\log_{10}(3 - x)$
(b) $\frac{1}{2}\ln(2 + x) + \frac{1}{2}\ln(2 - x)$
  $- \ln(x - 1) - \frac{3}{2}\ln(x + 1)$

25. (a) $\frac{1}{2}\log_b x - \frac{1}{2}$
(b) $\ln(1 + x^2) + \ln(1 + x^4) + \ln(1 + x^6)$
27. (a) $a + 2b + 3c$
(b) $\frac{1}{2}(1 + a + b + c)$
(c) $1 + a - \frac{1}{2}b - \frac{1}{2}c$
(d) $2 + 2a - 4b - \frac{1}{3}c$ **(e)** $5a + 5b - c$
29. $\frac{1}{2}(\ln 5 - \ln 2 + 1)$
31. $\ln 2 - \ln 3 - 1$ **33.** $(\ln 9)/(\ln 2)$
35. $(\ln 5 + \ln 2)/(\ln 5 - \ln 2)$ **37.** $\frac{1}{2}$
39. 3   **41.** $\frac{203}{99}$ **43.** 3   **45.** 7
47. (a) $x = 10^y/(3 \cdot 10^y - 1)$
(b) $x = (1 - y)/2$

49. $(\log_{10} 5)/(\log_{10} 2)$
51. $(\log_{10} 3)/(\log_{10} e)$
53. $(\log_{10} 2)/(\log_{10} b)$
55. $(\ln 6)/(\ln 10)$ **57.** $1/(\ln 10)$
59. $[\ln(\ln x) - \ln(\ln 10)]/(\ln 10)$
61. (a) true   **(b)** true   **(c)** true
(d) false   **(e)** true   **(f)** false   **(g)** true
(h) false   **(i)** true   **(j)** false   **(k)** false
(l) true   **(m)** true   **75.** 0.123456
81. x^3 **83.** $\frac{1}{2}$ **85.** $x = \beta e^{\alpha/3}$
93. $f^{-1}(x) = (e^{2x} - 1)/2e^x$
95.

x	10^2	10^3	10^6	10^{20}	10^{50}	10^{99}
y	0.4234	0.6589	0.9654	1.3428	1.5573	1.6918

4.5 Applications

1. $1009.98   **3.** 8.45%   **5.** $767.27
7. (a) $3869.68   **(b)** $4006.39
9. 13 quarters   **11.** $3,487.50
13. $2610.23   **15.** 5.83%
17. the first investment ($13,382.26)
19. (a) 14 years   **(b)** 13.86 years
(c) 1.01%   **21.** $26.5 trillion
23. (a) 14 years

(b)

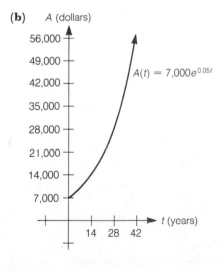

$A(t) = 7{,}000e^{0.05t}$

25.

	1975 POPULATION (billions)	% POPULATION IN 1975	GROWTH RATE (% per year)	YEAR 2000 POPULATION (billions)	% OF WORLD POPULATION IN 2000
World	4.090	100	1.8	6.414	100
More Dev.	1.131	27.7	0.6	1.314	20.5
Less Dev.	2.959	72.3	2.1	5.002	78.0

27.

	1975 POPULATION (billions)	GROWTH RATE (% per year)	YEAR 2000 POPULATION (billions)	% INCREASE IN POPULATION
Low	4.043	1.5	5.883	45.5
Medium	4.090	1.8	6.414	56.8
High	4.134	2.0	6.816	64.9

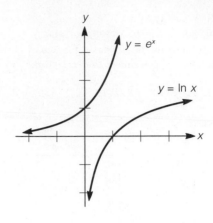

29. (a) $k \approx 0.0200$
(b) $N \approx 170,853,155$ **(c)** slower
31. (a) 0.68% **(b)** 0.96%
(c) $19,751,512$ **(d)** higher
33. (a)

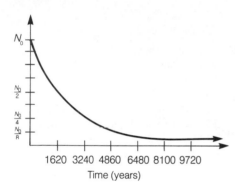

(b)

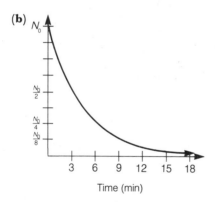

35. (a) -0.0533 **(b)** 58.7%; 0.5%
37. (a) $k \approx -0.0248$ **(b)** 279 years
(c) 280 years   **39. (a)** 2020 **(b)** 2035
(c) 2019 **41. (a)** 201 years
(b) 132 years
43. (a) $E = 10^{11.4} \cdot 10^{1.5M}$ **(b)** 31.6
45. -1.4748×10^{-11}
47. 4.181 billion years
49. $15,505$ years

Chapter 4 Review Exercises

1. $\log_5 126$ is larger.
2. domain: $(-\infty, \infty)$; range: $(-3, \infty)$;
x-intercept: -1; y-intercept: -2;
asymptote: $y = -3$

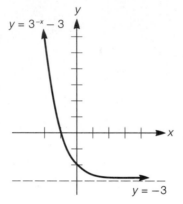

3. $(4 \ln 1.5)/(\ln 1.25)$ hr
4. $(\ln 2)/(\ln 10)$
5. The function f is not one-to-one.

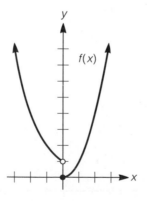

6. 10^{18} **7.** $(e + 1)/(e - 1)$
8. $(\ln 2)/(\ln 1.08) \approx 9$ years
9. $y = e^x$ has domain $(-\infty, \infty)$, range
$(0, \infty)$; $y = \ln x$ has domain $(0, \infty)$,
range $(-\infty, \infty)$.

10. 2 **11.** $-\frac{3}{2}$ **12.** $2a + \frac{1}{2}b - 3c$
13. $\frac{1}{13} \ln \frac{1}{2} \approx -0.05$
14. $\log_{10}[x^3/(1-x)]$ **15.** $2 - \ln(\frac{12}{5})$
16. $f^{-1}(x) = -1 + \ln x$; domain: $(0, \infty)$
17. the year 2000   **18.** $\frac{3}{2} + \ln 10$
19. 7 years

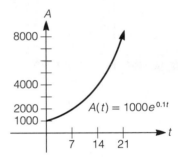

20. 0
21. horizontal asymptote: $y = 0$;
vertical asymptote: none; x-intercept:
none; y-intercept: 1

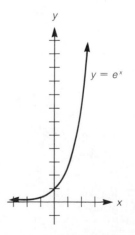

23. horizontal asymptote: none; vertical asymptote: $x = 0$; x-intercept: 1; y-intercept: none

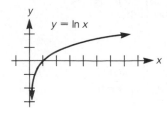

25. horizontal asymptote: $y = 1$; vertical asymptote: none; x-intercept: none; y-intercept: 3

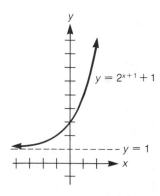

27. horizontal asymptote: $y = 0$; vertical asymptote: none; x-intercept: none; y-intercept: 1

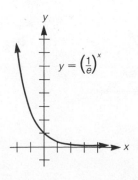

29. horizontal asymptote: $y = 1$; vertical asymptote: none; x-intercept: none; y-intercept: $e + 1$

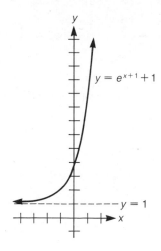

31. horizontal asymptote: none; vertical asymptote: none; x-intercept: 0; y-intercept: 0

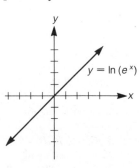

33. 4 **35.** 3 **37.** $\frac{2}{3}$ **39.** $\sqrt[3]{3}$
41. $\frac{1}{10}(1 - 2 \ln 3)$ **43.** $\frac{200}{99}$ **45.** 2
47. $(0, \infty)$ **49.** 1 **51.** $\frac{1}{2}$ **53.** $\frac{1}{5}$
55. -1 **57.** 16 **59.** 4 **61.** 2
63. 2 **65.** $\frac{9}{14}$ **67.** $2a + 3b + \frac{1}{2}c$
69. $8a + 4b$ **71.** 2 and 3 **73.** 2 and 3
75. -3 and -2
77. **(a)** third quadrant

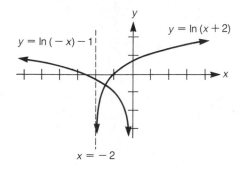

(b) $-2e/(e + 1) \approx -1.46$
79. $k = (\ln \frac{1}{2})/T$ **81.** 6.25%
83. $[d \ln(c/b)]/\ln(\frac{1}{2})$ days **85.** $\log_{10} 2$
87. $\ln 10$ **89.** $\ln(x^a y^b)$
91. $\frac{1}{2} \ln(x - 3) + \frac{1}{2} \ln(x + 4)$
93. $3 \log_{10} x - \frac{1}{2} \log_{10}(1 + x)$
95. $\frac{1}{3} \log_{10} x - \frac{2}{3}$
97. $3 \ln(1 + 2e) - 3 \ln(1 - 2e)$
99. $(\ln 2)/\ln[1 + (R/100)]$ years
101. 9.92% **103.** $(100 \ln 2)/R$ years
105. **(a)** $7\frac{3}{4}$ years
(b) $(\ln n)/\ln(1 + (R/400))$ years
107. **(a)** $(0, \infty)$ **(b)** $[1, \infty)$
109. **(a)** $(-\infty, -3) \cup (-3, 5) \cup (5, \infty)$
(b) $(-\infty, -3) \cup (5, \infty)$
111. $(-\infty, -1) \cup (1, \infty)$ **113.** -0.7
115. -2.2 **117.** 1.3 **119.** 2.7
121. 1.6
123. **(a)**

REGION	1990 POPULATION (millions)	GROWTH RATE (%)	2025 POPULATION
North America	275.2	0.7	351.6
Soviet Union	291.3	0.7	372.2
Europe	499.5	0.2	535.7
Nigeria	113.3	3.1	335.3

(b) 222.0 million **(c)** 193.5 million
(d) support **125.** **(a)** 1.0986
(b) 0.00112% **(c)** $\ln \sqrt{3} \approx 0.5493$;
$\ln 9 \approx 2.1972$; $\ln \frac{1}{3} \approx -1.0986$

CHAPTER FIVE
TRIGONOMETRIC FUNCTIONS
OF ANGLES

5.1 Trigonometric Functions of Acute Angles

1. (b) $\sin \theta = \frac{15}{17}$; $\cos \beta = \frac{15}{17}$
(c) $\cos \theta = \frac{8}{17}$; $\sin \beta = \frac{8}{17}$
(d) $\tan \theta = \frac{15}{8}$; $\csc \theta = \frac{17}{15}$; $\sec \theta = \frac{17}{8}$;
$\cot \theta = \frac{8}{15}$ **(e)** $\tan \beta = \frac{8}{15}$; $\csc \beta = \frac{17}{8}$;
$\sec \beta = \frac{17}{15}$; $\cot \beta = \frac{15}{8}$
3. (a) $\sin A = 2\sqrt{13}/13$;
$\cos A = 3\sqrt{13}/13$; $\tan A = \frac{2}{3}$
(c) $\cos B = 2/\sqrt{13}$ **5.** $\sin A = \frac{5}{13}$;
$\cos A = \frac{12}{13}$; $\tan A = \frac{5}{12}$; $\sec A = \frac{13}{12}$;
$\csc A = \frac{13}{5}$; $\cot A = \frac{12}{5}$
7. (a) $\sin B = \frac{4}{5}$; $\cos A = \frac{4}{5}$
(b) $\cos B = \frac{3}{5}$; $\sin A = \frac{3}{5}$ **(c)** 1
9. (a) $\sin A = \sqrt{5}/4$; $\cos A = \sqrt{11}/4$
(b) $\tan B = \sqrt{11}/\sqrt{5} = \sqrt{55}/5$;
$\sin B = \sqrt{11}/4$; $\cos B = \sqrt{5}/4$
23. (a) $\sin \theta = \dfrac{2x\sqrt{4x^2 + 9}}{4x^2 + 9}$;
$\cos \theta = \dfrac{3\sqrt{4x^2 + 9}}{4x^2 + 9}$; $\tan \theta = \dfrac{2x}{3}$
(b) $\sin^2 \theta = \dfrac{4x^2}{4x^2 + 9}$; $\cos^2 \theta = \dfrac{9}{4x^2 + 9}$;
$\tan^2 \theta = \dfrac{4x^2}{9}$
(c) $\sin(90° - \theta) = \dfrac{3\sqrt{4x^2 + 9}}{4x^2 + 9}$;
$\cos(90° - \theta) = \dfrac{2x\sqrt{4x^2 + 9}}{4x^2 + 9}$;
$\tan(90° - \theta) = \dfrac{3}{2x}$
25. (a) $\sin \beta = \sqrt{16x^2 - 1}/4x$;
$\cos \beta = 1/4x$; $\tan \beta = \sqrt{16x^2 - 1}$
(b) $\csc \beta = \dfrac{4x\sqrt{16x^2 - 1}}{16x^2 - 1}$; $\sec \beta = 4x$;
$\cot \beta = \dfrac{\sqrt{16x^2 - 1}}{16x^2 - 1}$
(c) $\sin(90° - \beta) = 1/4x$;
$\cos(90° - \beta) = \sqrt{16x^2 - 1}/4x$;
$\tan(90° - \beta) = \sqrt{16x^2 - 1}/(16x^2 - 1)$
27. $\sin B = \sqrt{33}/7$; $\tan B = \sqrt{33}/4$;
$\sec B = \frac{7}{4}$; $\csc B = 7\sqrt{33}/33$;
$\cot B = 4\sqrt{33}/33$
29. $\cos \theta = \sqrt{13}/5$; $\tan \theta = 2\sqrt{39}/13$;

$\sec \theta = 5\sqrt{13}/13$; $\csc \theta = 5\sqrt{3}/6$;
$\cot \theta = \sqrt{39}/6$
31. $\sin A = \frac{1}{6}(2\sqrt{3} - \sqrt{6})$;
$\cos A = \frac{1}{6}(2\sqrt{3} + \sqrt{6})$;
$\sec A = 2\sqrt{3} - \sqrt{6}$; $\csc A = 2\sqrt{3} + \sqrt{6}$;
$\cot A = 3 + 2\sqrt{2}$

33. $\cos \theta = \sqrt{4 - x^2}/2$;
$\tan \theta = x\sqrt{4 - x^2}/(4 - x^2)$;
$\sec \theta = 2\sqrt{4 - x^2}/(4 - x^2)$; $\csc \theta = 2/x$;
$\cot \theta = \sqrt{4 - x^2}/x$
35. $\sin \theta = \sqrt{1 - x^4}$;
$\tan \theta = \sqrt{1 - x^4}/x^2$; $\sec \theta = 1/x^2$;
$\csc \theta = \sqrt{1 - x^4}/(1 - x^4)$;
$\cot \theta = x^2\sqrt{1 - x^4}/(1 - x^4)$
37. (a) $\cos 30° \approx .86603$;
$\cos 45° \approx .70711$; $\sin 60° \approx .86603$
(b) $\cos 30° = \sin 60° = \sqrt{3}/2 \approx .86603$;
$\cos 45° = \sqrt{2}/2 \approx .70711$
39. $\csc 25° \approx 2.36620$;
$\sec 25° \approx 1.10338$; $\cot 25° \approx 2.14451$
41. (a) $\sin \beta$ is larger.
(b) $\csc \theta$ is larger. **(c)** $\cos \theta$ is larger.
47. (a) $\angle ADC = 108°$; $\angle BDC = 72°$;
$\angle B = 36°$
(b) $\angle A = \angle B$ because both are 36°;
therefore $\triangle ACB$ is isosceles with
$AC = BC$. Also, $\angle BDC = \angle BCD = 72°$,
and therefore $\triangle BDC$ is isosceles with
$BC = BD$. Thus we have $AC = BC = BD$.
(c) $\angle DAC = \angle DCA = 36°$, so $\triangle DAC$
is isosceles with $AD = CD$. Therefore
$CD = 1$ because $AD = 1$.

5.2 Right-Triangle Applications

1. $AC = 30\sqrt{3}$ cm; $BC = 30$ cm
3. $AB = \frac{32}{3}\sqrt{3}$ cm; $BC = \frac{16}{3}\sqrt{3}$ cm
5. $BC \approx 9.6$ cm; $AC \approx 11.5$ cm
7. (a) $9\sqrt{3} \approx 15.59$ ft **(b)** 9 ft
9. 34 million miles **11. (a)** 1.50 in^2
(b) 11.28 cm^2 **13.** 1175.6 cm^2
15. 10,660 ft **17.** 136 m
19. $18(\sqrt{3} - 1)$ cm
21. (a) $\angle BOA = 90° - \theta$;
$\angle OAB = \theta$; $\angle BAP = 90° - \theta$;
$\angle BPA = \theta$ **(b)** $AO = \sin \theta$;
$AP = \cos \theta$; $OB = \sin^2 \theta$; $BP = \cos^2 \theta$

23. (a) 53.1° **(b)** 67.4° **(c)** 48.6°
25. $AC = 4 \sec \theta + 5 \csc \theta$
27. (a) $\sin \theta$ **(b)** $\cos \theta$ **(c)** $\tan \theta$
(d) $\sec \theta$ **(e)** $\cot \theta$ **(f)** $\csc \theta$
29. (b) 1080 miles

31. (b)

n	5	10	50	100	1,000	5,000	10,000
A_n	2.38	2.94	3.1333	3.1395	3.141572	3.1415918	3.1415924

(c) As n gets larger, A_n becomes closer
to the area of the circle, which is π.
35. $a \tan \alpha \sin \beta$ ft

5.3 Trigonometric Functions of General Angles

1. (a) reference angle: 70°

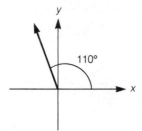

(b) reference angle: 60°

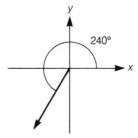

(c) reference angle: 60°

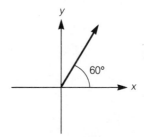

(d) reference angle: $60°$

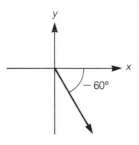

3. (a) 1 **(b)** 1 **(c)** -1 **(d)** 1
5. (a) -1 **(b)** undefined **(c)** -1
(d) 1 **7. (a)** $\frac{3}{4}$ **(b)** $\sin\theta = \sqrt{7}/4$;
$\tan\theta = \sqrt{7}/3$ **(c)** $\cos(\beta - 90°)$
9. (a) $-\sqrt{2}/2$ **(b)** $\frac{1}{2}$ **(c)** $-\sqrt{3}/3$
(d) $\sqrt{2}/2$ **11. (a)** -1 **(b)** $-\sqrt{3}/3$
(c) $-\sqrt{3}/3$ **(d)** 1
13. $\sin(-30°) = -\frac{1}{2}$; $\cos(-30°) = \sqrt{3}/2$;
$\tan(-30°) = -\sqrt{3}/3$;
$\sec(-30°) = 2\sqrt{3}/3$; $\csc(-30°) = -2$;
$\cot(-30°) = -\sqrt{3}$

15. (a)

θ	0°	90°	180°	270°	360°	450°	540°	630°	720°
$\sin\theta$	0	1	0	-1	0	1	0	-1	0
$\cos\theta$	1	0	-1	0	1	0	-1	0	1

(b)

θ	30°	60°	90°	120°	150°	180°	210°	240°	270°	300°	330°	360°
$\sin\theta$	$\frac{1}{2}$	$\frac{\sqrt{3}}{2}$	1	$\frac{\sqrt{3}}{2}$	$\frac{1}{2}$	0	$-\frac{1}{2}$	$-\frac{\sqrt{3}}{2}$	-1	$-\frac{\sqrt{3}}{2}$	$-\frac{1}{2}$	0

17. $\cos\theta = -2\sqrt{6}/5$; $\tan\theta = -\sqrt{6}/12$;
$\sec\theta = -5\sqrt{6}/12$; $\csc\theta = 5$;
$\cot\theta = -2\sqrt{6}$ **19.** $\sin\theta = -\frac{4}{5}$;
$\tan\theta = \frac{4}{3}$; $\sec\theta = -\frac{5}{3}$; $\csc\theta = -\frac{5}{4}$;
$\cot\theta = \frac{3}{4}$ **21.** $35\sqrt{3}/4$ cm^2
23. 302.7 cm^2 **25.** $90°, 450°, -270°$
(other answers possible)
27. (a) $\sin 10° \approx 0.17$; $\cos 10° \approx 0.99$
(b) $\sin 20° \approx 0.34$; $\cos 20° \approx 0.94$
(c) $\sin 30° \approx 0.50$; $\cos 30° \approx 0.87$
(d) $\sin 40° \approx 0.64$; $\cos 40° \approx 0.77$
(e) $\sin 50° \approx 0.77$; $\cos 50° \approx 0.64$
(f) $\sin 70° \approx 0.94$; $\cos 70° \approx 0.34$
(g) $\sin 80° \approx 0.99$; $\cos 80° \approx 0.17$
(h) $\sin 100° \approx 0.99$; $\cos 100° \approx -0.17$
(i) $\sin 130° \approx 0.77$; $\cos 130° \approx -0.64$

29. (a) $\dfrac{\sqrt{2}-\sqrt{6}}{4}$ **(b)** $\dfrac{-\sqrt{10+2\sqrt{5}}}{4}$
(c) $\dfrac{\sqrt{2}-\sqrt{6}}{2\sqrt{2+\sqrt{3}}}$ **(d)** $\dfrac{\sqrt{2}-\sqrt{6}}{4}$
(e) $\dfrac{\sqrt{10+2\sqrt{5}}}{4}$ **(f)** $\dfrac{-\sqrt{10+2\sqrt{5}}}{4}$
31. (a) $P(\cos\theta, \sin\theta)$; $Q(\cos\phi, \sin\phi)$
33. The restriction $0° < \theta < 180°$
guarantees that $\sin\theta$ will be positive,
and thus $\log_{10}\sin\theta$ is defined.

5.4 Algebra and the Trigonometric Functions

1. (a) $1 - 2\cos\theta + \cos^2\theta$
(b) $1 + 2\sin\theta\cos\theta$
(c) $(\cos\theta\sin\theta + 1)/\sin\theta$
3. (a) $(\tan\theta - 6)(\tan\theta + 1)$
(b) $(\sin B + \cos B)(\sin B - \cos B)$
(c) $(\cos A + 1)^2$ **5.** $-2\cos\theta - 1$
7. -1 **9.** $\sin A + \cos A$ **11.** $\csc\theta$

13. $\cos^2 B$ **15.** $\cos A + 4$ **17.** $\cot A$
19. $\tan\theta$ **21.** 0 **23.** 1
55. $m(\theta) = \tan\theta$

5.5 The Law of Sines and The Law of Cosines

1. $4\sqrt{6}$ cm **3.** $20\sin 50°$ cm
5. $a = 9.7$ cm, $c = 16.4$ cm
7. $A = 63.3°$, $C = 50.7°$, $c = 25.5$ cm
9. (a) $45°$ or $135°$ **(b)** $150°$
(c) $14.5°$ or $165.5°$
11. (c) $C = 105°$, $c \approx 1.93$
(d) $C = 15°$, $c \approx 0.52$

13. $a = \dfrac{2\sin 70°}{\sin 20°}$ cm; $b = \dfrac{2\sin 50°}{\sin 20°}$ cm;
$c = \dfrac{2\sin 50°\sin 70°}{\sin 20°\sin 85°}$ cm;
$d = \dfrac{2\sin 50°\sin 15°}{\sin 20°\sin 85°}$ cm **15. (a)** 21 cm
(b) 5.4 in. **17.** $\cos A = \frac{113}{140}$;
$\cos B = \frac{29}{40}$; $\cos C = -\frac{5}{28}$ **19.** 5.9 units
21. (a) $a \approx 4.2$ cm **(b)** $\angle C \approx 29.3°$
(c) $\angle B \approx 110.7°$ **23.** $PD = \sqrt{2-\sqrt{2}}$
25. (b) $a = 5, b = 3, c = 7$ (using
$m = 2, n = 1$) **27.** $x \approx 881.5$ ft
29. (b) $\sin B = 1/(\sqrt{6}-\sqrt{2})$;
$\sin C = 1/(\sqrt{6}+\sqrt{2})$ **31.** 12,495 ft
33. 2:40 P.M. **35.** A: 1.26 miles;
B: 1.60 miles **43.** $a = 5, b = 8$

Chapter 5 Review Problems

1. $\cos 30° = \sqrt{3}/2$; $\tan 60° = \sqrt{3}$;
$\sin^2 7° + \cos^2 7° = 1$
2. (a) $\sin(-270°) = 1$
(b) $\cos 540° = -1$ **(c)** $\cot 450° = 0$
3. $(2\cos\theta + 3)(\cos\theta + 4)$
4. $15\sqrt{2}/2$ cm^2 **5.** $a = 7$ cm
6. $\cos\theta = -1/4$; based on this, the
angle opposite the 4-cm side must be
obtuse (not acute), since its cosine is
negative. **7. (a)** $\sqrt{3}/2$ **(b)** 1
8. $\cos\theta = 2\sqrt{2}/3$; $\cot\theta = -2\sqrt{2}$
9. $\tan\theta = \frac{3}{4}$ **10.** $-2\tan\theta - 1$
11. 20 cm **12.** $5\sqrt{3}$ ft
14. $BP = \cos^2\theta$; $AB = \cos\theta\sin\theta$
16. 1 **17.** $\sqrt{2-\sqrt{3}/2}$
18. $PV = 2\sqrt{17} - 4\sqrt{2}$ cm
19. $CD = 25/(\cos 25°\sin 35°)$ cm
20. $36\sin 20°\cos 20°$ m^2 or
$18\sin 40°$ m^2 **21.** $b = \frac{35}{2}$
23. $\sin A = \sqrt{55}/8$, $\cot A = 3\sqrt{55}/55$
25. $a = \sqrt{41 + 20\sqrt{3}}$; area $= 5$
27. $\cos A = -\frac{1}{15}$, $\sin A = 4\sqrt{14}/15$
29. $a = \frac{1}{5}$ **31.** $c = 3$
33. $\sin A = \sqrt{2}/8$, $\cos A = \sqrt{62}/8$,
$B \approx 34.82°$ **35.** $c = 6$ **37.** $\sqrt{2}/2$
39. $-\sqrt{3}$ **41.** -2 **43.** -1
45. $\sqrt{2}/2$ **47.** 1 **49.** $-2\sqrt{3}/3$
51. 2 **57.** $\cos\theta = \dfrac{p^4 - q^4}{p^4 + q^4}$,
$\tan\theta = \dfrac{2p^2 q^2}{p^4 - q^4}$ **59.** $\alpha = 45°$, $\beta = 30°$

61. $\sin \theta = \frac{4}{5}$, $\tan \theta = \frac{4}{3}$
63. $\tan \theta = -\frac{24}{7}$ **65.** $\cot \theta = \frac{5}{12}$
67. $\tan(90° - \theta) = -3\sqrt{7}/7$
69. $\cos(180° - \theta) = -\frac{7}{9}$
73. $3600 \tan 35°$ cm^2
81. (a) $PN = \sin \theta$ (b) $ON = \cos \theta$
(c) $PT = \tan \theta$ (d) $OT = \sec \theta$
(e) $NA = 1 - \cos \theta$
(f) $NT = \sin \theta \tan \theta$
111. $\sin A \cos A$ **113.** $1/\cos A$
115. $(1 + \cos A)/(1 - \cos A)$
117. $(\sin A - \cos A)/\sin A$
119. $1 + \sin A \cos A$
125. (a) The area is $\frac{1}{2}cr$, since the base is c and the height is r. (b) area of $\triangle AOC = \frac{1}{2}br$; area of $\triangle BOC = \frac{1}{2}ar$

CHAPTER SIX
TRIGONOMETRIC FUNCTIONS OF REAL NUMBERS

6.1 Radian Measure

1. (a) $\pi/3$ radians (b) $5\pi/4$ radians
(c) $\pi/5$ radians (d) $5\pi/2$ radians
(e) 0 radians **3.** (a) $15°$ (b) $270°$
(c) $1080°$ (d) $18°$ (e) $90°$
(f) $540°/\pi$ **5.** smaller

7.

θ	0	$\pi/2$	π	$3\pi/2$	2π
$\cos \theta$	1	0	-1	0	1

9. $30° = \pi/6$ radians, $45° = \pi/4$ radians, $60° = \pi/3$ radians, $120° = 2\pi/3$ radians, $135° = 3\pi/4$ radians, $150° = 5\pi/6$ radians
11. (a) positive (b) positive
13. (a) negative (b) positive
15. 4π ft **17.** $\pi/2$ cm **19.** $\frac{1}{6}$
21. (a) 12π radians/sec
(b) 144π cm/sec (c) 72π cm/sec
23. (a) 6π radians/sec
(b) 150π cm/sec (c) 75π cm/sec
25. (a) $50\pi/3$ radians/sec
(b) 750π cm/sec (c) 375π cm/sec
27. (a) $175\pi/12$ cm^2 (b) $\pi/6$ in.^{2}
29. $\frac{50}{3}\pi - 25\sqrt{3}$ cm^2
31. 1 radian $\approx 57.3°$ **33.** 1200 rev
35. (a) 0.000073 radians/sec
(b) 1040 mph **37.** (b) $\theta = \pi/2$

6.2 Trigonometric Functions of Real Numbers

1.

θ	$\sin \theta$	$\cos \theta$	$\tan \theta$	$\csc \theta$	$\sec \theta$	$\cot \theta$
0	0	1	0	undef.	1	undef.
$\pi/6$	$\frac{1}{2}$	$\frac{1}{2}\sqrt{3}$	$\frac{1}{3}\sqrt{3}$	2	$\frac{2}{3}\sqrt{3}$	$\sqrt{3}$
$\pi/4$	$\frac{1}{2}\sqrt{2}$	$\frac{1}{2}\sqrt{2}$	1	$\sqrt{2}$	$\sqrt{2}$	1
$\pi/3$	$\frac{1}{2}\sqrt{3}$	$\frac{1}{2}$	$\sqrt{3}$	$\frac{2}{3}\sqrt{3}$	2	$\frac{1}{3}\sqrt{3}$
$\pi/2$	1	0	undef.	1	undef.	0
$2\pi/3$	$\frac{1}{2}\sqrt{3}$	$-\frac{1}{2}$	$-\sqrt{3}$	$\frac{2}{3}\sqrt{3}$	-2	$-\frac{1}{3}\sqrt{3}$
$3\pi/4$	$\frac{1}{2}\sqrt{2}$	$-\frac{1}{2}\sqrt{2}$	-1	$\sqrt{2}$	$-\sqrt{2}$	-1
$5\pi/6$	$\frac{1}{2}$	$-\frac{1}{2}\sqrt{3}$	$-\frac{1}{3}\sqrt{3}$	2	$-\frac{2}{3}\sqrt{3}$	$-\sqrt{3}$
π	0	-1	0	undef.	1	undef.

3. $\cos \theta = -\frac{4}{5}$; $\tan \theta = \frac{3}{4}$
5. $\tan t = -\sqrt{39}/13$ **7.** $\sec \alpha = 13/5$; $\cos \alpha = 5/13$; $\sin \alpha = 12/13$
13. $(\sqrt{7} \cos \theta)/7$ **15.** (a) $-\frac{2}{3}$ (b) $\frac{1}{4}$
(c) $\frac{1}{5}$ (d) $-\frac{1}{5}$ **17.** (a) $-(1 + 2\sqrt{2})/3$
(b) 1 **19.** (a) $\sqrt{2}/2$ (b) $\sqrt{3}/2$ (c) 1
21. $\cos^2 \theta$ **23.** $\sin^2 \theta$ **33.** $\frac{13}{31}$

39.

θ	$1 - \dfrac{\theta^2}{2}$	$\cos \theta$
0.1	0.995	0.995004...
0.2	0.980	0.980066...
0.3	0.955	0.955336...

43. (a) $0, \pi, 2\pi, 3\pi$ (b) all reals except $k\pi$, $k =$ any integer
(c) all reals except integers

45. (a)

t	0.2	0.4	0.6	0.8	1.0	1.2	1.4
$f(t)$	219.07	50.53	19.70	9.55	6.14	7.98	33.88

(b) 6.14, which occurs at $t = 1.0$
(d) $\tan^2 t + 9 \cot^2 t$
$= (\tan t - 3 \cot t)^2 + 6 \geq 0 + 6 \geq 6$
(e) Since $\pi/3 \approx 1.05$, our answer from part (b) is consistent with this result.

6.3 Graphs of the Sine and Cosine Functions

1. period: 2; amplitude: 1
3. period: 4; amplitude: 6
5. period: 4; amplitude: 2
7. period: 6; amplitude: $\frac{3}{2}$

9. (a) amplitude: 2; period: 2π; x-intercepts: $0, \pi, 2\pi$; increasing on $(0, \pi/2) \cup (3\pi/2, 2\pi)$

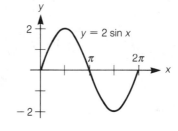

(b) amplitude: 1; period: π; x-intercepts: $0, \pi/2, \pi$; increasing on $(\pi/4, 3\pi/4)$

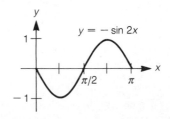

11. **(a)** amplitude: 1; period: π; x-intercepts: $\pi/4$, $3\pi/4$; increasing on $(\pi/2, \pi)$

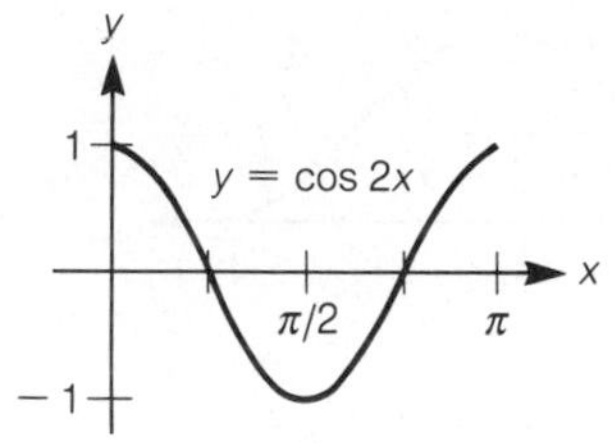

(b) amplitude: 2; period: π; x-intercepts: $\pi/4$, $3\pi/4$; increasing on $(\pi/2, \pi)$

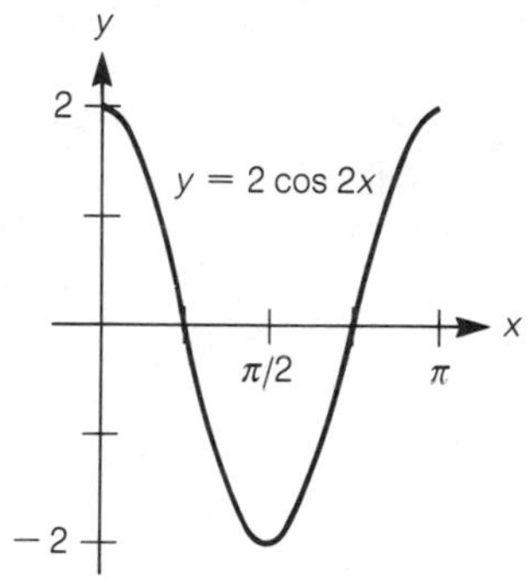

13. **(a)** amplitude: 3; period: 4; x-intercepts: 0, 2, 4; increasing on $(0, 1) \cup (3, 4)$

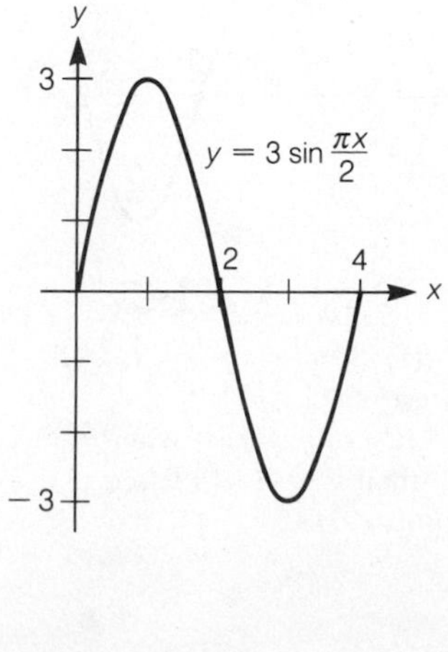

(b) amplitude: 3; period: 4; x-intercepts: 0, 2, 4; increasing on $(1, 3)$

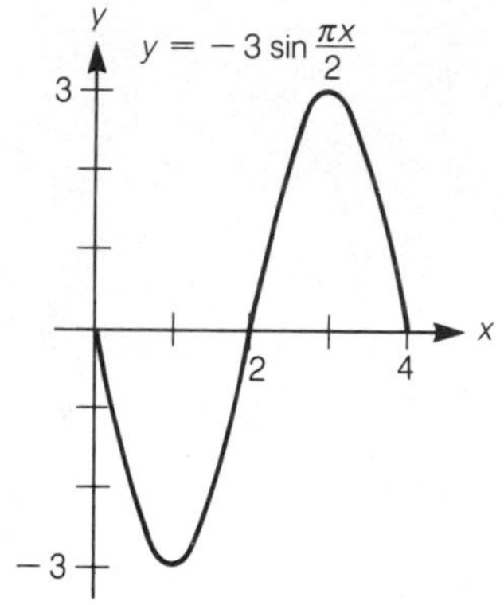

15. **(a)** amplitude: 1; period: 1; x-intercepts: $\frac{1}{4}$, $\frac{3}{4}$; increasing on $(\frac{1}{2}, 1)$

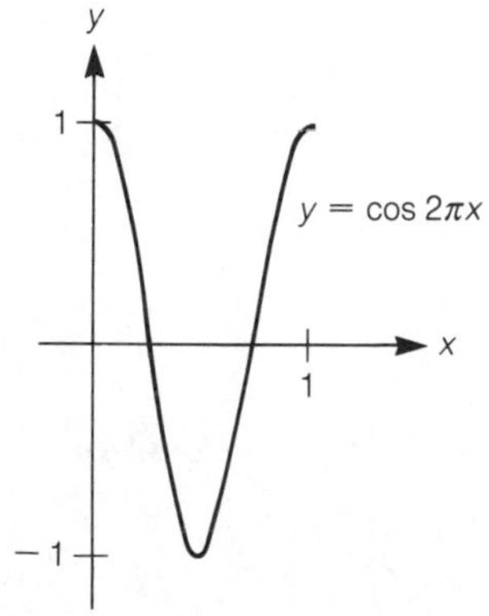

(b) amplitude: 4; period: 1; x-intercepts: $\frac{1}{4}$, $\frac{3}{4}$; increasing on $(0, \frac{1}{2})$

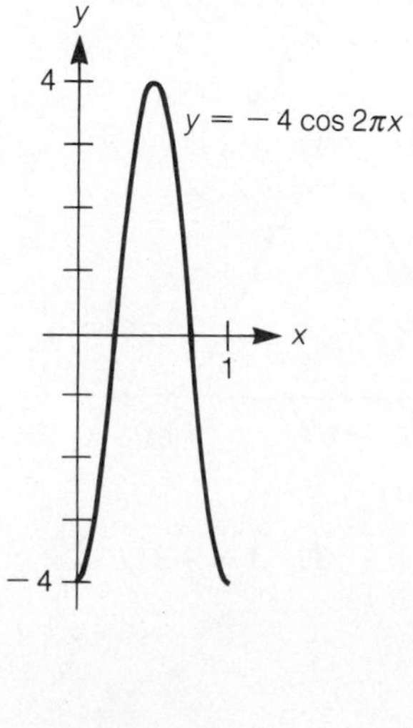

17. amplitude: 1; period: π; x-intercept: $3\pi/4$; increasing on $(0, \pi/4) \cup (3\pi/4, \pi)$

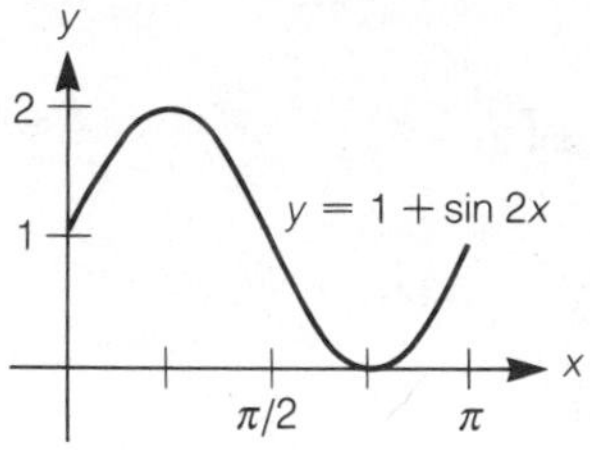

19. amplitude: 1; period: 6; x-intercepts: 0, 6; increasing on $(0, 3)$

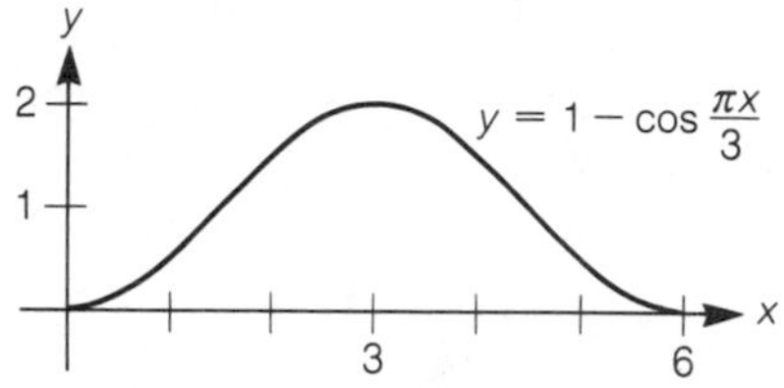

21. amplitude: 1; period: 2π; phase shift: $\pi/6$; x-intercepts: $\pi/6$, $7\pi/6$, $13\pi/6$; high point at $(2\pi/3, 1)$; low point at $(5\pi/3, -1)$

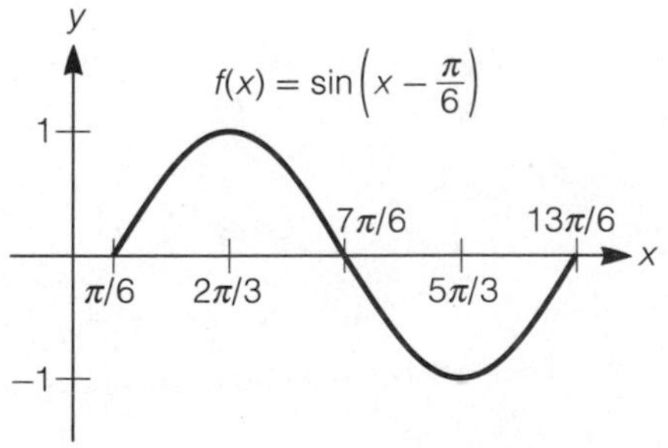

23. amplitude: 1; period: 2π; phase shift: $-\pi/4$; x-intercepts: $\pi/4$, $5\pi/4$; high point at $(3\pi/4, 1)$; low points at $(-\pi/4, -1)$, $(7\pi/4, -1)$

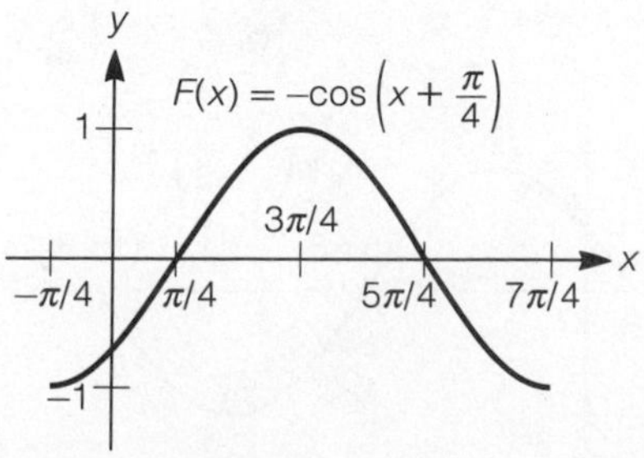

25. amplitude: 1; period: π; phase shift: $\pi/4$; x-intercepts: $\pi/4$, $3\pi/4$, $5\pi/4$; high point at $(\pi/2, 1)$; low point at $(\pi, -1)$

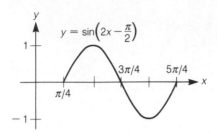

27. amplitude: 1; period: π; phase shift: $\pi/2$; x-intercepts: $3\pi/4$, $5\pi/4$; high points at $(\pi/2, 1)$, $(3\pi/2, 1)$; low point at $(\pi, -1)$

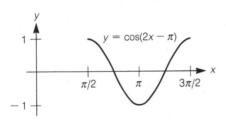

29. amplitude: 3; period: 4π; phase shift: $-\pi/3$; x-intercepts: $-\pi/3$, $5\pi/3$, $11\pi/3$; high point at $(2\pi/3, 3)$; low point at $(8\pi/3, -3)$

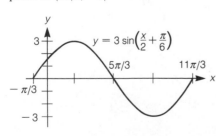

31. amplitude: 4; period: $2\pi/3$; phase shift: $\pi/12$; x-intercepts: $\pi/4$, $7\pi/12$; high points at $(\pi/12, 4)$, $(3\pi/4, 4)$; low point at $(5\pi/12, -4)$

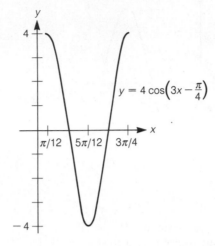

33. amplitude: $\frac{1}{2}$; period: 4; phase shift: 2π; x-intercepts: 2π, $2\pi + 2$, $2\pi + 4$; high point at $(2\pi + 1, \frac{1}{2})$; low point at $(2\pi + 3, -\frac{1}{2})$

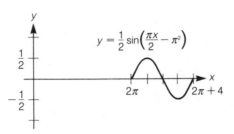

35. amplitude: 1; period: π; phase shift: $\pi/6$; x-intercepts: $\pi/6$, $7\pi/6$; high point at $(2\pi/3, 2)$; low points at $(\pi/6, 0)$, $(7\pi/6, 0)$

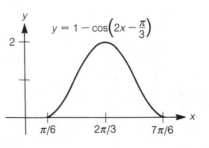

37. $A = 2$, $B = \frac{1}{2}$ **39.** $A = -3$, $B = \pi$
41. $A = -4$, $B = \frac{1}{5}$

43. amplitude: $\frac{1}{2}$; period: π

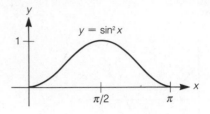

45. amplitude: $\frac{1}{2}$; period: π

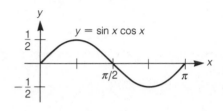

47. (a) $C(\cos \theta, \sin \theta)$
49. $P = (\pi/2, e) \approx (1.57, 2.71)$; $Q = (3\pi/2, 1/e) \approx (4.71, 0.37)$
51. (a) amplitude: 3; period: 6

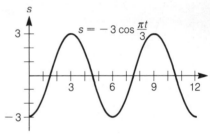

(b) At $t = 0$, 3, 6, 9, 12 sec
(c) At $t = \frac{3}{2}, \frac{9}{2}, \frac{15}{2}, \frac{21}{2}$ sec
(d) amplitude: π; period: 6

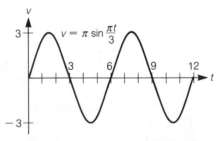

(e) $t = 0$, 3, 6, 9, 12 sec; $s = -3$, 3, -3, 3, -3 (f) $3 < t < 6$; $9 < t < 12$
(g) v is greatest when $t = \frac{3}{2}, \frac{15}{2}$ sec; then $v \approx 3.14$ ft/sec; v is least when $t = \frac{9}{2}, \frac{21}{2}$ sec; then $v \approx -3.14$ ft/sec. At all four times, $s = 0$.

(h) We draw the graph.

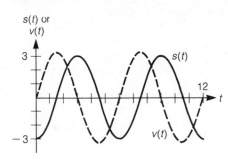

6.4 Graphs of the Tangent and the Reciprocal Functions

1. (a) x-intercept: $-\pi/4$; y-intercept: 1; asymptotes: $x = -3\pi/4$, $x = \pi/4$

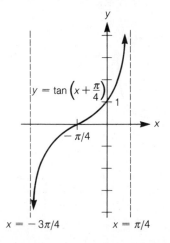

(b) x-intercept: $-\pi/4$; y-intercept: -1; asymptotes: $x = -3\pi/4$, $x = \pi/4$

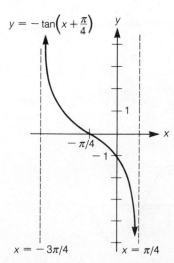

3. (a) x-intercept: 0; y-intercept: 0; asymptotes: $x = -3\pi/2$, $x = 3\pi/2$

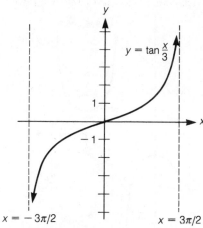

(b) x-intercept: 0; y-intercept: 0; asymptotes: $x = -3\pi/2$, $x = 3\pi/2$

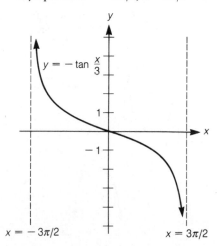

5. x-intercept: 0; y-intercept: 0; asymptotes: $x = -1$, $x = 1$

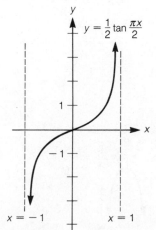

7. x-intercept: 1; y-intercept: none; asymptotes: $x = 0$, $x = 2$

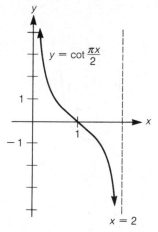

9. x-intercept: $3\pi/4$; y-intercept: 1; asymptotes: $x = \pi/4$, $x = 5\pi/4$

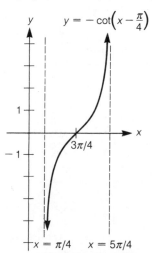

11. x-intercept: $\pi/4$; y-intercept: none; asymptotes: $x = 0$, $x = \pi/2$

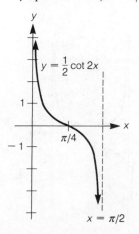

13. x-intercept: none; y-intercept: $-\sqrt{2}$;
asymptotes: $x = -3\pi/4$, $x = \pi/4$,
$x = 5\pi/4$

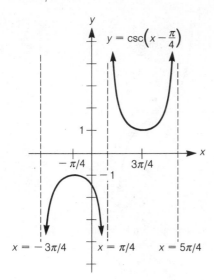

15. x-intercept: none; y-intercept: none;
asymptotes: $x = -2\pi$, $x = 0$, $x = 2\pi$

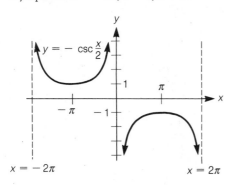

17. x-intercept: none; y-intercept: none;
asymptotes: $x = -1$, $x = 0$, $x = 1$

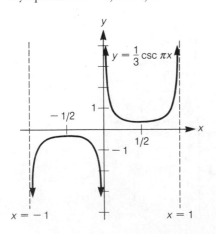

19. x-intercept: none; y-intercept: -1;
asymptotes: $x = -\pi/2$, $x = \pi/2$, $x = 3\pi/2$

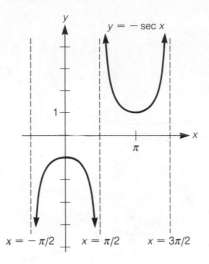

21. x-intercept: none; y-intercept: -1;
asymptotes: $x = \pi/2$, $x = 3\pi/2$, $x = 5\pi/2$

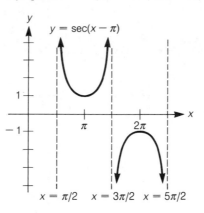

23. x-intercept: none; y-intercept: 3;
asymptotes: $x = -1$, $x = 1$, $x = 3$

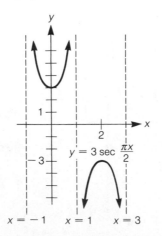

25. period: 2; amplitude: 1;
phase shift: $\frac{1}{6}$

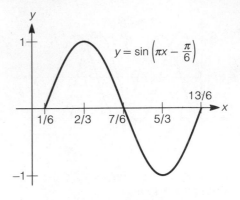

(b) period: 2; phase shift: $\frac{1}{6}$;
asymptotes: $x = \frac{1}{6}$, $x = \frac{7}{6}$, $x = \frac{11}{6}$

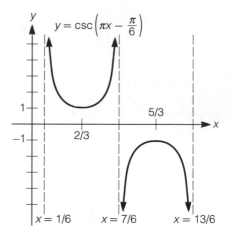

27. period: 2; amplitude: 1;
phase shift: $-\frac{1}{4}$

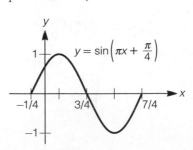

(b) period: 2; phase shift: $-\frac{1}{4}$;
asymptotes: $x = -\frac{1}{4}$, $x = \frac{3}{4}$, $x = \frac{7}{4}$

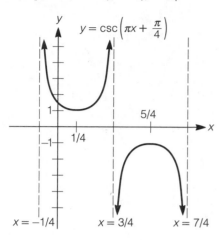

$y = \csc\left(\pi x + \frac{\pi}{4}\right)$

$x = -1/4$ $x = 3/4$ $x = 7/4$

29.

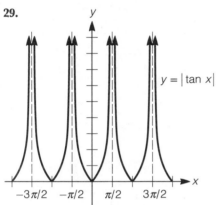

$y = |\tan x|$

$-3\pi/2$ $-\pi/2$ $\pi/2$ $3\pi/2$

31.

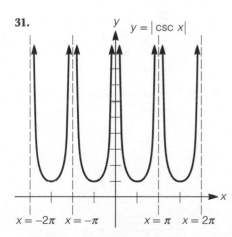

$y = |\csc x|$

$x = -2\pi$ $x = -\pi$ $x = \pi$ $x = 2\pi$

6.5 The Addition Formulas

1. (a) $\sin 3\theta$ **(b)** 1 **3. (a)** $\cos 5u$
(b) $\cos u$ **5.** $\sin B$ **7.** $\cos \theta$
9. $-\cos \theta$ **11.** $\sin \theta$ **13.** $(\sqrt{6} - \sqrt{2})/4$

15. $(\sqrt{6} + \sqrt{2})/4$ **17.** $\sqrt{2}\sin s$
19. $\sqrt{3}\sin\theta$ **21. (a)** $\frac{56}{65}$ **(b)** $\frac{16}{65}$
(c) $\frac{33}{65}$ **(d)** $-\frac{63}{65}$ **23. (a)** $-\frac{5}{13}$
(b) $\frac{119}{169}$ **25.** $\sin(s - t) = -\sqrt{2}/10$;
$\cos(s + t) = -\sqrt{2}/10$

35. (a)

t	1	2	3	4
$f(t)$	1.5	1.5	1.5	1.5

(b) $f(t) = 1.5$
45. (a) They both equal -1.1918.
(b) They both equal -1.3764.

6.6 Further Identities

1. (a) $\frac{4}{3}$ **(b)** $\frac{44}{117}$ **3. (a)** 7 **(b)** $-\frac{1}{7}$
5. (a) $\frac{24}{25}$ **(b)** $\frac{7}{25}$ **(c)** $\frac{24}{7}$ **7. (a)** $\frac{24}{25}$
(b) $-\frac{7}{25}$ **(c)** $-\frac{24}{7}$ **9. (a)** $\sqrt{10}/10$
(b) $3\sqrt{10}/10$ **(c)** $\frac{1}{3}$ **11. (a)** $\sqrt{5}/5$
(b) $2\sqrt{5}/5$ **(c)** $\frac{1}{2}$ **13. (a)** $-3\sqrt{7}/8$
(b) $-\frac{1}{8}$ **(c)** $\sqrt{8 + 2\sqrt{7}}/4$
(d) $\sqrt{8 - 2\sqrt{7}}/4$ **15. (a)** $4\sqrt{2}/9$
(b) $-\frac{7}{9}$ **(c)** $\sqrt{6}/3$ **(d)** $-\sqrt{3}/3$
17. (a) $\sqrt{2 - \sqrt{3}}/2$ **(b)** $\sqrt{2 + \sqrt{3}}/2$
(c) $2 - \sqrt{3}$ **19. (a)** $\sqrt{2 + \sqrt{3}}/2$
(b) $-\sqrt{2 - \sqrt{3}}/2$ **(c)** $-2 - \sqrt{3}$
21. $\sin 2\theta = \frac{2}{25}x\sqrt{25 - x^2}$;
$\cos 2\theta = 1 - \frac{2}{25}x^2$
23. $\sin 2\theta = \frac{1}{2}(x - 1)\sqrt{3 + 2x - x^2}$;
$\cos 2\theta = \frac{1}{2}(1 + 2x - x^2)$
25. $\frac{1}{8}(3 - 4\cos 2\theta + \cos 4\theta)$
27. $\frac{1}{8}(3 - 4\cos\theta + \cos 2\theta)$
55. (b) $(\sqrt{3} + \sqrt{2})/2$

6.7 Trigonometric Equations

1. yes **3.** no **5.** $\theta = (\pi/3) + 2\pi k$,
$\theta = (2\pi/3) + 2\pi k$, where $k =$ any integer
7. $\theta = (7\pi/6) + 2\pi k$, $\theta = (11\pi/6) + 2\pi k$,
where $k =$ any integer
9. $\theta = \pi + 2\pi k$, where $k =$ any integer
11. $\theta = (\pi/3) + \pi k$, where $k =$ any
integer **13.** $x = \pi k$, where $k =$ any
integer **15.** $\theta = (\pi/2) + \pi k$,
$\theta = (2\pi/3) + 2\pi k$, $\theta = (4\pi/3) + 2\pi k$,
where $k =$ any integer
17. $t = \pi k$, where $k =$ any integer
19. $x = (\pi/6) + 2\pi k$, $x = (5\pi/6) + 2\pi k$,
$x = (3\pi/2) + 2\pi k$, where $k =$ any integer
21. $t = (\pi/4) + 2\pi k$, $t = (3\pi/4) + 2\pi k$,
where $k =$ any integer

23. $0°, 120°, 240°$
25. $75°, 105°, 195°, 225°, 315°, 345°$
27. $60°, 180°, 300°$ **29.** $30°, 120°$,
$210°, 300°$ **31.** $14.5°, 165.5°$
33. $116.6°, 296.6°$ **35.** $128.2°, 231.8°$
37. $1.39, 4.90$ **39.** $0.46, 2.68$
41. $1.37, 4.51$ **43.** $0, \pi$ **45.** $2\pi/3, \pi$
47. $(3\pi/16) + (\pi/4)k$, where $k =$ any
integer **49.** $\pi/8, 3\pi/8, 5\pi/8, 7\pi/8, 9\pi/8$,
$11\pi/8, 13\pi/8, 15\pi/8, 0, \pi, 2\pi$
51. $\pi/10, \pi/2, 9\pi/10, 13\pi/10, 17\pi/10$
53. $\pi/3, 4\pi/3$ **55.** $60.45°$
59. (a) 1000.173 **(b)** 1001.022

6.8 The Inverse Trigonometric Functions

1. $\pi/3$ **3.** $\pi/3$ **5.** $-\pi/6$ **7.** $\pi/4$
9. undef. **11.** $\frac{1}{4}$ **13.** $\frac{3}{4}$ **15.** $-\pi/7$
17. $\pi/2$ **19.** 0 **21.** $\frac{4}{3}$ **23.** $\sqrt{2}/2$
25. $\frac{12}{5}$ **27.** $\frac{1}{2}$ **29.** $2\sqrt{2}/3$
31. (a) 0.84 **(b)** 0.84 **(c)** 1.26
(d) 0.90 **33.** $\sqrt{2}$
37. $(\theta/4) - \sin 2\theta$
$\quad = \frac{1}{4}\sin^{-1}(3x/2) - 3x\sqrt{4 - 9x^2}/2$
39. $\theta - \cos\theta$
$\quad = \tan^{-1}[(x - 1)/2] - 2/\sqrt{5 - 2x + x^2}$
41. $\frac{8}{17}$ **43.** $31\sqrt{218}/1090$ **45.** $-\sqrt{2}/2$
55. (a) $[-1, 1]$
(b) $f(0.1) = f(0.2) = f(0.3) = \pi/2 \approx 1.57$
(c) $f(\frac{1}{2}) = f(\sqrt{2}/2)$
$\qquad = f(\sqrt{3}/2) = \pi/2 \approx 1.57$
(d) $f(x) = \sin^{-1}x + \cos^{-1}x = \pi/2$
57. (a) $\frac{3}{4}$ **(b)** $\sqrt{15}/4$ **(c)** $2\sqrt{2}$
(d) $\frac{12}{5}$ **59.** $t = \frac{1}{2}, 1$

Chapter 6 Review Exercises

1. (a) $-\sqrt{3}/2$ **(b)** $-\sqrt{3}$
2. x-intercepts: $-\frac{1}{6}, \frac{1}{6}$;
high point at $(0, 3)$

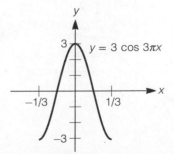

$y = 3\cos 3\pi x$

$-1/3$ $1/3$

3. $-\cos\theta$ **4. (a)** $-\frac{3}{5}$ **(b)** $\sqrt{5}-2$
5. (a) $5\sqrt{5}\,\pi/12$ cm **(b)** $25\pi/24$ cm^2
6. $\frac{1}{2}\csc x$
7. $\pi^2/180$ radian is larger than $3°$
8. $7\pi/6,\ 11\pi/6$ **9.** $\sqrt{5}/5$ **11.** $\frac{3}{4}$
12. $30°$ **13.** amplitude: 1; period: π;
phase shift: $\pi/2$

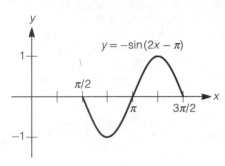

14. $\sqrt{18+12\sqrt{2}}/6$

15.

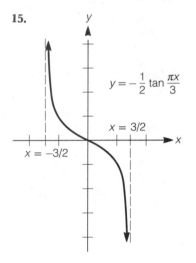

16. $A(\theta)=\pi-\theta-\sin\theta$
17. $y=\sin x$: domain: $[-\pi/2,\ \pi/2]$;
range: $[-1,1]$; $y=\sin^{-1}x$: domain:
$[-1,1]$; range: $[-\pi/2,\ \pi/2]$

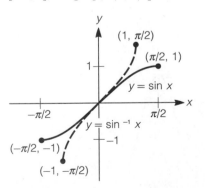

18. (a) $\pi/10$ **(b)** 0 **19.** $\sqrt{7}/4$
21. $135°$ **23.** $900°$ **25.** $150°$
27. $360°/\pi$ **29.** 2π radians
31. $\pi/180$ radians **33.** $\pi/24$ radians
35. $1/180$ radians **37.** -1 **39.** $2\sqrt{3}/3$
41. $-\sqrt{3}$ **43.** $\frac{1}{2}$ **45.** 1 **47.** -2
49. $s=2\pi$ cm; $A=16\pi$ cm^2
51. $A=\frac{1}{2}$ cm^2 **53.** $r=20/\pi$ cm,
$A=40/\pi$ cm^2 **55.** $r=3$ cm,
$\theta=4$ radians **57.** $7\pi/18$ radians
59. 0.841 **61.** -1.000 **63.** 0.012
69. $5\cos\theta$ **71.** $10\tan\theta$
73. $\frac{1}{5}\sqrt{5}\cos\theta$ **75.** $\sqrt{10}/10$ **77.** $\frac{527}{625}$
79. $2\sqrt{29}/29$ **81.** $\sqrt{2}/10$
83. $(\pi-2)/2$ cm^2 **85. (a)** $1.62=1.62$
(b) $1.61=1.61$ **(c)** $1.61=1.61$
95. $-\frac{16}{65}$ **97.** $\sin x$ **99.** 1 **101.** 0
103. $\cos x$ **105.** -3 **107.** $\frac{1104}{1105}$
109. $y=4\sin x$ **111.** $y=-2\cos 4x$
113. $\sqrt{2}$ **115.** x-intercepts: $\pi/8,\ 3\pi/8$;
high point at $(\pi/4,3)$; low points at
$(0,-3),\ (\pi/2,-3)$

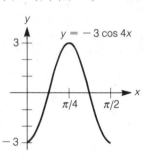

117. x-intercepts: $\frac{1}{2},\frac{5}{2},\frac{9}{2}$; high point at
$(\frac{3}{2},2)$; low point at $(\frac{7}{2},-2)$

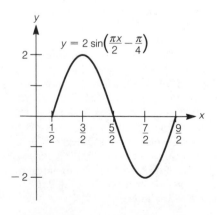

119. x-intercepts: $\frac{5}{2},\frac{11}{2}$; high points at
$(1,3),(7,3)$; low point at $(4,-3)$

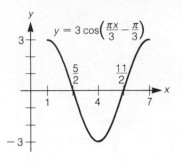

121. x-intercepts: $0,\ \pi/4,\ \pi/2$; high point
at $(3\pi/8,\frac{1}{2})$; low point at $(\pi/8,-\frac{1}{2})$

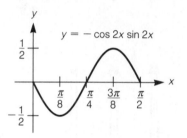

123. (a)

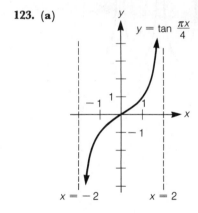

(b)

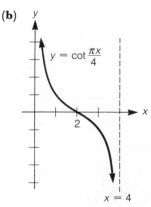

125. (a)

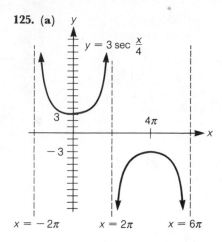

$y = 3 \sec \dfrac{x}{4}$

$x = -2\pi$ $x = 2\pi$ $x = 6\pi$

(b)

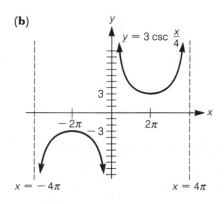

$y = 3 \csc \dfrac{x}{4}$

$x = -4\pi$ $x = 4\pi$

127.

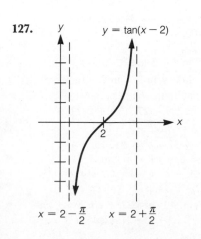

$y = \tan(x - 2)$

$x = 2 - \dfrac{\pi}{2}$ $x = 2 + \dfrac{\pi}{2}$

---

155. no

167. (a) The triangle where $a = 1$, $b = 1$, and $c = \sqrt{2}$ yields the largest value.

a	b	c	cos A + cos B
3	4	5	1.400000
5	12	13	1.307692
20	21	29	1.413793
8	15	17	1.352941
1	$\sqrt{3}$	2	1.366025
1	1	$\sqrt{2}$	1.414214
$\sqrt{2}$	$\sqrt{3}$	$\sqrt{5}$	1.407052
696	697	985	1.414213

169. $\pi/4,\ \pi/2,\ 5\pi/4,\ 3\pi/2$

171. $\pi/9,\ 2\pi/9,\ 7\pi/9,\ 8\pi/9,\ 13\pi/9,\ 14\pi/9$

173. $0,\ 2\pi/3,\ \pi,\ 4\pi/3$

175. $\pi/6,\ 5\pi/6,\ 3\pi/2$ **177.** $\pi/3,\ \pi,\ 5\pi/3$

179. $\pi/6,\ \pi/3,\ 2\pi/3,\ 5\pi/6,\ 7\pi/6,\ 4\pi/3,$ $5\pi/3,\ 11\pi/6$ **183.** $\sqrt{3}/2$ **185.** $\pi/6$

187. $\pi/6$ **189.** $\pi/3$ **191.** $2\pi/3$

193. $\frac{2}{7}$ **195.** $-\sqrt{2}/2$ **197.** $3\sqrt{2}/2$

199. $\frac{17}{7}$ **201.** $-\frac{4}{3}$ **203.** $3\sqrt{10}/10$

205. $(1 + 2\sqrt{6})/(6 + 2\sqrt{2} - \sqrt{3})$

CHAPTER SEVEN
SYSTEMS OF EQUATIONS

7.1 Systems of Two Linear Equations in Two Unknowns

1. (a) yes **(b)** no **(c)** yes **(d)** yes

3. yes **5.** no **7.** yes **9.** $(-6, \frac{1}{2})$

11. $(-\frac{5}{2}, \frac{13}{2})$ **13.** $(\frac{1}{5}, \frac{7}{10})$

15. $(-\frac{60}{13}, \frac{60}{13})$ **17.** $(\sqrt{6}, 1)$

19. $(-\frac{2}{9}, \frac{23}{27})$ **21.** $(-\frac{283}{242}, -\frac{3}{121})$

23. $(-\frac{226}{25}, -\frac{939}{50})$ **25.** $(\frac{5}{8}, 0)$

27. $(1, 2)$ **29.** $b = 3,\ c = 4$

31. $A = -28,\ B = -22$

33. $\frac{49}{20}$ sq. units

35. 80 cc of 10%; 120 cc of 35%

37. 8 lb of \$5.20; 8 lb of \$5.80

39. $\left(\dfrac{ab}{a + b}, \dfrac{ab}{a + b}\right)$; $a \ne \pm b$

41. $\left(\dfrac{a + b}{ab}, -\dfrac{1}{ab}\right)$; $a \ne 0,\ b \ne 0$

43. $(-\frac{1}{11}, \frac{1}{9})$ **45.** $(5, 2)$

47. $(1.32, -1.62)$ **49.** $(4.04, 0.44)$

51. 59 **53.** 12 in., 5 in.

55. (a) $(60, 40)$ **(b)** $(60, 40)$

---

57. $(0, a)$ **59.** $(e^3, e^{-2}) \approx (20.09, 0.14)$

61. $(1/b, 1/a)$ **63.** $k = -29/7,\ 2$

7.2 Gaussian Elimination

1. $(-3, -2, -1)$ **3.** $(-\frac{1}{60}, -\frac{2}{15}, \frac{3}{5})$

5. $(\frac{25}{36}, \frac{5}{9}, -\frac{1}{3})$ **7.** $(x, x/8, 0)$, for any real number x **9.** $(-1, 0, 1, -5)$

11. $(\frac{11}{3}, \frac{8}{3}, \frac{17}{3})$ **13.** $(1, 0, 1)$

15. $(2, 3, 1)$ **17.** inconsistent

19. $(-\frac{1}{7}(z + 1), \frac{5}{7}(z + 1), z)$, for any real number z

21. $\left(\dfrac{11 - 5z}{7}, \dfrac{-3z - 6}{7}, z\right)$, for any real number z **23.** $(4, 1, -3, 2)$

25. $\left(\dfrac{17 - 17z}{5}, \dfrac{8z - 3}{5}, z\right)$, for any real number z

27. $\left(\dfrac{12 + 10w}{11}, \dfrac{146 + 19w}{55}, \dfrac{159 + 61w}{55}, w\right)$, for any real number w

29. $\left(-\dfrac{5z}{12}, \dfrac{2z + 3}{3}, z\right)$, for any real number z

31. (a) $A = \frac{1}{4},\ B = -\frac{1}{4}$

(b) $A = \frac{1}{2},\ B = \frac{1}{2}$ **33.** $A = \frac{4}{3},\ B = \frac{5}{3}$

35. (a) $A = \frac{3}{8},\ B = -\frac{3}{8}$

(b) $A = \frac{3}{64},\ B = -\frac{3}{64},\ C = -\frac{3}{8}$

37. $A = \frac{1}{3},\ B = -\frac{1}{3},\ C = \frac{2}{3}$

39. $A = 4,\ B = 4$

41. $A = \frac{1}{2},\ B = \frac{1}{2},\ C = -\frac{1}{2},\ D = \frac{1}{2}$

43. $17x^2 + 17y^2 - 49x + 65y - 166 = 0$

45. $x = \ln a,\ y = \ln 2a,\ z = \ln(a/2)$

49. (a) $A = \dfrac{1}{a - b},\ B = \dfrac{1}{b - a}$

(b) $A = \dfrac{ap + q}{a - b},\ B = \dfrac{bp + q}{b - a}$

51. $A = \dfrac{a^2 + pa + q}{(a - b)(a - c)}$,

$B = \dfrac{b^2 + pb + q}{(b - a)(b - c)},\ C = \dfrac{c^2 + pc + q}{(c - a)(c - b)}$

53. 60 miles

7.3 Matrices

1. (a) 2×3 **(b)** 3×2 **3.** 5×4

5. coefficient matrix: $\begin{pmatrix} 2 & 3 & 4 \\ 5 & 6 & 7 \\ 8 & 9 & 10 \end{pmatrix}$

augmented matrix: $\begin{pmatrix} 2 & 3 & 4 & 10 \\ 5 & 6 & 7 & 9 \\ 8 & 9 & 10 & 8 \end{pmatrix}$

7. coefficient matrix:
$$\begin{pmatrix} 1 & 0 & 1 & 1 \\ 1 & 1 & 0 & 2 \\ 0 & 1 & 1 & 1 \\ 2 & -1 & -1 & 0 \end{pmatrix}$$

augmented matrix:
$$\begin{pmatrix} 1 & 0 & 1 & 1 & -1 \\ 1 & 1 & 0 & 2 & 0 \\ 0 & 1 & 1 & 1 & 1 \\ 2 & -1 & -1 & 0 & 2 \end{pmatrix}$$

9. $(-1, -2, 3)$ **11.** $(-5, 1, 3)$
13. $(3, 0, -7)$ **15.** $(8, 9, -1)$
17. $\left(\dfrac{9z + 5}{19}, \dfrac{31z - 6}{19}, z\right)$, for any
real number z
19. $(2, -1, 0, 3)$

21. no solution **23.** $\begin{pmatrix} 3 & 2 \\ 2 & 4 \end{pmatrix}$

25. $\begin{pmatrix} 6 & 4 \\ 4 & 8 \end{pmatrix}$ **27.** $\begin{pmatrix} 11 & -2 \\ 11 & 1 \end{pmatrix}$

29. $\begin{pmatrix} 2 & 3 \\ -1 & 4 \end{pmatrix}$ **31.** undefined

33. $\begin{pmatrix} 10 & -2 \\ -8 & 0 \\ 4 & 6 \end{pmatrix}$ **35.** $\begin{pmatrix} 2 & 4 & 11 \\ -12 & 16 & 19 \\ 14 & 12 & 43 \end{pmatrix}$

37. $\begin{pmatrix} -9 & 10 & 10 \\ 4 & -8 & -12 \\ 10 & 4 & 21 \end{pmatrix}$ **39.** undefined

41. $\begin{pmatrix} 0 & 0 & 0 \\ 0 & 0 & 0 \\ 0 & 0 & 0 \end{pmatrix}$ **43.** $\begin{pmatrix} 4 & 2 \\ 2 & 5 \end{pmatrix}$

45. undefined **47.** $\begin{pmatrix} 1 & 18 \\ -6 & 13 \end{pmatrix}$

49. $\begin{pmatrix} -16 & 75 \\ -25 & 34 \end{pmatrix}$

51. **(a)** $\begin{pmatrix} -13 & 1 & 40 \\ 43 & 17 & 0 \\ 89 & 61 & 60 \end{pmatrix}$

(b) $\begin{pmatrix} -13 & 1 & 40 \\ 43 & 17 & 0 \\ 89 & 61 & 60 \end{pmatrix}$

(c) $\begin{pmatrix} -52 & -82 & 61 \\ 87 & 141 & 0 \\ 216 & 318 & 165 \end{pmatrix}$

(d) $\begin{pmatrix} -52 & -82 & 61 \\ 87 & 141 & 0 \\ 216 & 318 & 165 \end{pmatrix}$

53. **(a)** $\begin{pmatrix} 16 & 20 \\ 24 & 28 \end{pmatrix}$ **(b)** $\begin{pmatrix} 18 & 26 \\ 18 & 26 \end{pmatrix}$

(c) $\begin{pmatrix} 14 & 14 \\ 30 & 30 \end{pmatrix}$ **(d)** $\begin{pmatrix} 18 & 26 \\ 18 & 26 \end{pmatrix}$

55. **(a)** $\begin{pmatrix} x \\ -y \end{pmatrix}$ **(b)** $\begin{pmatrix} -x \\ y \end{pmatrix}$

(c) $\begin{pmatrix} -x \\ -y \end{pmatrix}$, reflection about the origin.

57. **(a)** $f(A) = -2, f(B) = 29,$
$f(AB) = -58$; yes

7.4 The Inverse of a Square Matrix

1. $\begin{pmatrix} 4 & -1 \\ -5 & 2 \end{pmatrix} = A$

3. $\begin{pmatrix} 3 & 0 & -2 \\ 0 & 5 & 6 \\ 1 & 4 & -7 \end{pmatrix} = C$ **5.** $\begin{pmatrix} -5 & 9 \\ 4 & -7 \end{pmatrix}$

7. $\begin{pmatrix} -\frac{6}{23} & \frac{1}{23} \\ \frac{5}{23} & \frac{3}{23} \end{pmatrix}$

9. The inverse does not exist.

11. $\begin{pmatrix} 2 & -3 \\ 1 & 3 \end{pmatrix}$ **13.** $\begin{pmatrix} 2 & -1 \\ -3 & 2 \end{pmatrix}$

15. $\begin{pmatrix} \frac{6}{11} & 1 \\ -\frac{1}{11} & 0 \end{pmatrix}$

17. The inverse does not exist.

19. $\begin{pmatrix} -1 & 2 & -3 \\ 2 & 1 & 0 \\ 4 & -2 & 5 \end{pmatrix}$

21. $\begin{pmatrix} 5 & -\frac{10}{3} & 1 \\ 0 & \frac{1}{3} & 0 \\ 4 & -\frac{8}{3} & 1 \end{pmatrix}$

23. $\begin{pmatrix} 2 & 1 & 4 \\ 3 & 2 & 5 \\ 0 & -1 & 1 \end{pmatrix}$

25. The inverse does not exist.
27. **(a)** $(-1, 1)$ **(b)** $(-132, 48)$
29. **(a)** $(2, -1, 4)$ **(b)** $(1, 1, -2)$

31. **(a)** $I_3 = \begin{pmatrix} 1 & 0 & 0 \\ 0 & 1 & 0 \\ 0 & 0 & 1 \end{pmatrix}$, so $A^{-1} = A$.

(b) $(\frac{1}{2}, -1, 1)$

35. **(a)** $A^{-1} = \begin{pmatrix} -\frac{5}{2} & \frac{3}{2} \\ 2 & -1 \end{pmatrix}$,

$B^{-1} = \begin{pmatrix} 7 & -8 \\ -6 & 7 \end{pmatrix}$,

$B^{-1}A^{-1} = \begin{pmatrix} -\frac{67}{2} & \frac{37}{2} \\ 29 & -16 \end{pmatrix}$

(b) $\begin{pmatrix} -\frac{67}{2} & \frac{37}{2} \\ 29 & -16 \end{pmatrix}$, $(AB)^{-1} = B^{-1}A^{-1}$

7.5 Determinants and Cramer's Rule

1. **(a)** 29 **(b)** -29 **3.** **(a)** 0 **(b)** 0
5. -1 **7.** -60 **9.** 9
11. **(a)** 314 **(b)** 674 **(c)** part (b)
13. **(a)** 0 **(b)** 0 **(c)** 0 **(d)** 0 **15.** 0
17. -3 **19.** 6848 **21.** 17120
23. $(y - x)(z - x)(z - y)$ **25.** xy
27. 20 **29.** 120 **33.** $(1, 1, 2)$
35. $(2, -3, 6)$ **37.** $(0, 0, 0)$
39. $(13 - \frac{11}{3}y, y, 13 - 4y)$, for any real
number y **41.** $(1, 0, -10, 2)$
43. $x = 4, -4, -1$ **55.** **(a)** 1 **(b)** 1
59. $\left(\dfrac{k(k - b)(k - c)}{a(a - b)(a - c)}, \dfrac{k(k - a)(k - c)}{b(b - a)(b - c)},\right.$
$\left.\dfrac{k(k - a)(k - b)}{c(c - a)(c - b)}\right)$ **63.** **(b)** $\begin{pmatrix} -\frac{9}{61} & \frac{7}{61} \\ \frac{1}{61} & \frac{6}{61} \end{pmatrix}$

7.6 Nonlinear Systems of Equations

1. $(0, 0), (3, 9)$ **3.** $(\pm 2\sqrt{6}, 1)$
5. $(-1, -1)$ **7.** $(2, \pm 3), (-2, \pm 3)$
9. $(\pm 1, 0)$ **11.** $\left(\dfrac{-1 + \sqrt{65}}{8}, \dfrac{1 + \sqrt{65}}{2}\right),$
$\left(\dfrac{-1 - \sqrt{65}}{8}, \dfrac{1 - \sqrt{65}}{2}\right)$
13. $(\sqrt{17}/17, \pm 1), (-\sqrt{17}/17, \pm 1)$
15. $(1, 0), (4, -\sqrt{3})$ **17.** $(2, 4)$
19. $(100, 1000), (100, \frac{1}{1000}), (\frac{1}{100}, 1000),$
$(\frac{1}{100}, \frac{1}{1000})$
21. $\left(\dfrac{2 \ln 2 - 5 \ln 3}{\ln 2 - \ln 3}, \dfrac{3 \ln 2}{\ln 2 - \ln 3}\right)$
23. They both yield $\frac{1}{2}(1 + \sqrt{13})$.
25. $(1/a, 1/b)$ **27.** $(9, 14), (14, 9)$
29. $\frac{1}{2}(p - \sqrt{2d^2 - p^2})$ by
$\frac{1}{2}(p + \sqrt{2d^2 - p^2})$ **31.** $(5, \pm 4)$
33. $(p^2/A, q^2/A, r^2/A)$ and
$(-p^2/A, -q^2/A, -r^2/A)$, where
$A = \sqrt{p^2 + q^2 + r^2}$
35. 9 cm, 40 cm **37.** 3 cm by 20 cm
39. $(1, 2), (-1, -2), (2, 1), (-2, -1)$

41. $\left(\dfrac{1+\sqrt{13}}{2}, \dfrac{-1+\sqrt{13}}{2}\right)$,

$\left(\dfrac{-1+\sqrt{13}}{2}, \dfrac{1+\sqrt{13}}{2}\right)$,

$\left(\dfrac{1-\sqrt{13}}{2}, \dfrac{-1-\sqrt{13}}{2}\right)$,

$\left(\dfrac{-1-\sqrt{13}}{2}, \dfrac{1-\sqrt{13}}{2}\right)$

43. $(2,3), (-2,-3), (1,2), (-1,-2)$

45. $(q,p), \left(\dfrac{-1+\sqrt{3}}{2}q, (1-\sqrt{3})p\right)$,

$\left(\dfrac{-1-\sqrt{3}}{2}q, (1+\sqrt{3})p\right)$

47. $(e^{9/2}, e^3)$

7.7 Systems of Inequalities

1. (a) no **(b)** yes

3.

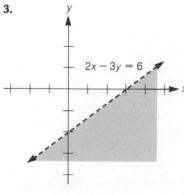

5.

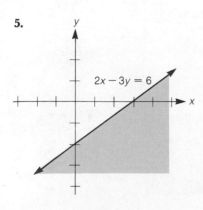

7.

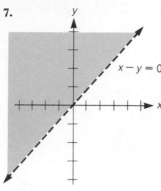

9.

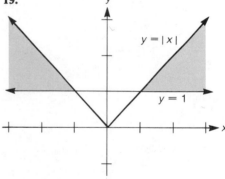

11.

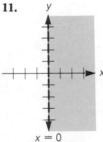

13.

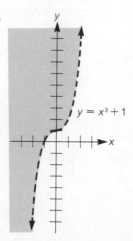

15.

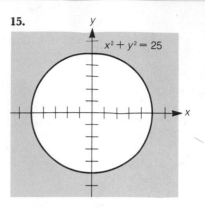

17.

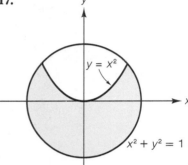

19.

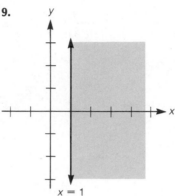

21.

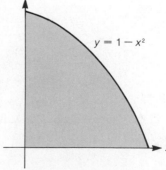

23. convex: yes; bounded: yes; vertices: $(0, 0)$, $(7, 0)$, $(3, 8)$, $(0, 5)$

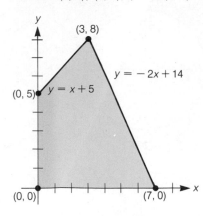

25. convex: yes; bounded: yes; vertices: $(0, 0)$, $(0, 4)$, $(3, 5)$, $(8, 0)$

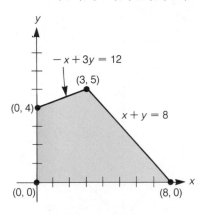

27. convex: yes; bounded: no; vertices: $(2, 7)$, $(8, 5)$

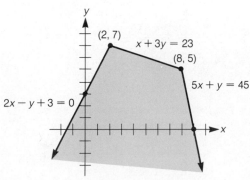

29. convex: yes; bounded: no; vertex: $(6, 0)$

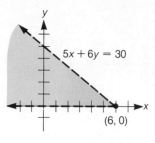

31. convex: yes; bounded: yes; vertices: $(0, 0)$, $(0, 5)$, $(6, 0)$

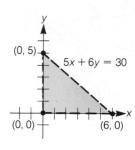

33. convex: yes; bounded: yes; vertices: $(5, 30)$, $(10, 30)$, $(20, 15)$, $(20, 20)$

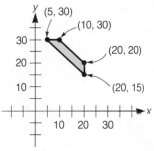

35. vertices: $(-e, 0)$, $(-e, e^{-e})$, $(0, 1)$, (e, e^{-e}), $(e, 0)$

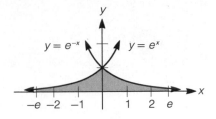

37.

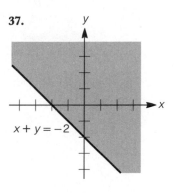

39.

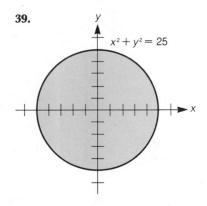

41.

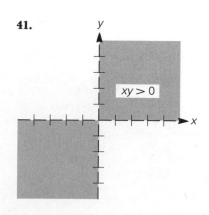

Chapter 7 Review Exercises

1. $(0, 3)$, $(-\frac{11}{4}, \frac{81}{16})$ **2.** $(3, -5)$

3. (a) $(1, -1, -3)$
(b) $D = -19$, $D_x = -19$, $D_y = 19$,
$D_z = 57$; the solution is $(1, -1, -3)$.

4. (a) $\begin{pmatrix} 2 & -10 \\ 3 & -5 \end{pmatrix}$

(b) $\begin{pmatrix} 8 & -4 \\ 7 & -6 \end{pmatrix}$

5. 12 sq. units   **6.** $(\frac{1}{116}, -\frac{1}{144})$

7. coefficient matrix: $\begin{pmatrix} 1 & 1 & -1 \\ 2 & -1 & 2 \\ 1 & -2 & 1 \end{pmatrix}$

augmented matrix:
$\begin{pmatrix} 1 & 1 & -1 & -1 \\ 2 & -1 & 2 & 11 \\ 1 & -2 & 1 & 10 \end{pmatrix}$

8. $(3, -3, 1)$ **9.** $y = -2x - 4$
10. $A = -\frac{3}{4}$, $B = \frac{3}{4}$, $C = -\frac{1}{2}$
11. (a) 8   **(b)** -8 **12.** -1120

13. $\left(\dfrac{5 + \sqrt{5}}{2}, \dfrac{5 - \sqrt{5}}{2}\right)$,

$\left(\dfrac{5 - \sqrt{5}}{2}, \dfrac{5 + \sqrt{5}}{2}\right)$,

$\left(\dfrac{-5 + \sqrt{5}}{2}, \dfrac{-5 - \sqrt{5}}{2}\right)$ and

$\left(\dfrac{-5 - \sqrt{5}}{2}, \dfrac{-5 + \sqrt{5}}{2}\right)$

14. $(1 - \frac{1}{5}C, -\frac{7}{5}C, C)$, where $C = $ any real number

15. (a) $\begin{pmatrix} 1 & -2 & 3 \\ 2 & -5 & 10 \\ -1 & 2 & -2 \end{pmatrix}$

(b) $(12, 38, -9)$

16.

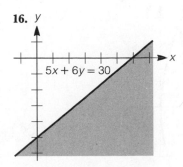
$5x + 6y = 30$

17. $P = -1$, $Q = -4$

18. neither bounded nor convex

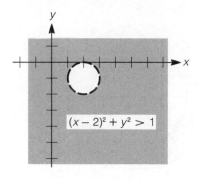
$(x - 2)^2 + y^2 > 1$

19. vertices: $(0, 0)$, $(0, 7)$, $(6, 10)$, $(\frac{261}{26}, \frac{225}{26})$, $(11, 0)$

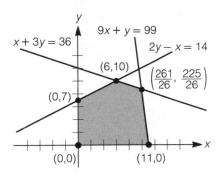
$x + 3y = 36$ $9x + y = 99$ $2y - x = 14$
$(6, 10)$ $\left(\frac{261}{26}, \frac{225}{26}\right)$
$(0, 7)$ $(0, 0)$ $(11, 0)$

20. $k = -\frac{13}{3}$ **21.** $(3, -5)$
23. $(-1, 4)$ **25.** $(-3, 15)$
27. $(-\frac{18}{5}, \frac{8}{5})$ **29.** $(\frac{2}{3}, -\frac{1}{5})$
31. $(-12, -8)$ **33.** $(\frac{1}{3}, -\frac{1}{4})$

35. $\left(\dfrac{-1}{a^2 - 3a + 1}, \dfrac{1 - a}{a^2 - 3a + 1}\right)$,
$a \ne \dfrac{3 \pm \sqrt{5}}{2}$

37. $(u^2, 1 - a^2)$ **39.** $(a^2 - b^2, a^2 + b^2)$

41. $\left(\dfrac{pq(p + q)}{p^2 + q^2}, \dfrac{p^3 - q^3}{p^2 + q^2}\right)$,
where $(p, q) \ne (0, 0)$

43. $\left(\dfrac{a}{a - b}, \dfrac{b}{a + b}\right)$, $ab \ne 0$, $9a - 8b \ne 0$,
$a \ne \pm b$ **45.** $(2, 3, 4)$
47. $(-1, -2, 0)$ **49.** no solution
51. $(x, 6 - 2x, -1)$, for any real number x **53.** no solution
55. $(2b - z, a - b, z)$, for any real number z **57.** $(4, 3, -1, 2)$
59. $A = \frac{1}{20}$, $B = -\frac{1}{20}$
61. $A = 2$, $B = -2$
63. $A = -\frac{5}{4}$, $B = \frac{5}{4}$
65. $A = \frac{1}{16}$, $B = -\frac{1}{16}$, $C = -\frac{1}{4}$

67. $A = 3$, $B = 1$, $C = 0$
69. $A = \frac{1}{48}$, $B = -\frac{1}{48}$, $C = \frac{1}{6}$
71. $A = 0$, $B = 1$, $C = a$
73. $A = b$, $B = -a$ **75.** 34   **77.** -56
79. 0   **81.** 24   **87.** $a = 2$, $b = 1$
89. (a) $(\frac{2}{3}, 2)$ **(b)** $(\frac{2}{3}, 2)$ **(c)** $(\frac{2}{3}, 2)$
(d) They all equal 2.
93. $(x + \frac{17}{6})^2 + (y + \frac{8}{3})^2 = \frac{245}{36}$

95. $\begin{pmatrix} 10 & -2 \\ 4 & 26 \end{pmatrix}$ **97.** $\begin{pmatrix} 8 & 4 \\ 4 & 32 \end{pmatrix}$

99. $\begin{pmatrix} 4 & -13 \\ 7 & 41 \end{pmatrix}$ **101.** $\begin{pmatrix} -3 & -14 \\ -4 & 3 \end{pmatrix}$

103. $\begin{pmatrix} 1 & 1 \\ 1 & 7 \end{pmatrix}$ **105.** $\begin{pmatrix} 1 & -11 \\ 6 & 36 \end{pmatrix}$

107. $\begin{pmatrix} 4 & 3 \\ 10 & 33 \end{pmatrix}$ **109.** $\begin{pmatrix} -42 & 58 \\ 5 & 20 \end{pmatrix}$

111. undefined   **113.** undefined

115. $\begin{pmatrix} 4 & -1 \\ 2 & 12 \end{pmatrix}$ **117.** $\begin{pmatrix} -4 & 13 \\ -7 & -41 \end{pmatrix}$

121. (a) $\begin{pmatrix} -9 & 5 \\ 2 & -1 \end{pmatrix}$

(b) $x = -47$, $y = 10$

123. (a) $\begin{pmatrix} 10 & -2 & 5 \\ 6 & -1 & 4 \\ 1 & 0 & 1 \end{pmatrix}$

(b) $x = 16$, $y = 15$, $z = 4$

125. $\begin{pmatrix} -65 & 20 & 9 & 133 \\ 3 & -1 & 0 & -6 \\ 26 & -8 & -4 & -53 \\ -23 & 7 & 3 & 47 \end{pmatrix}$

127. $D = -20$, $D_x = 12$, $D_y = 13$,
$D_z = -31$; $(-\frac{3}{5}, -\frac{13}{20}, \frac{31}{20})$
129. $D = 0$; no solution
131. $D = 0$; $(-5 - 4y, y, -8 - 5y)$,
for any real number y
133. $D = 149$, $D_x = 596$, $D_y = 149$,
$D_z = -149$, $D_w = 447$; $(4, 1, -1, 3)$
135. $(0, 0)$, $(6, 36)$ **137.** $(3, 0)$, $(-3, 0)$
139. $(5\sqrt{2}/2, \pm\sqrt{14}/2)$,
$(-5\sqrt{2}/2, \pm\sqrt{14}/2)$

141. $\left(\dfrac{-1 + \sqrt{5}}{2}, \dfrac{\sqrt{-2 + 2\sqrt{5}}}{2}\right)$

143. $\left(\pm\dfrac{\sqrt{-2 + 2\sqrt{17}}}{4}, \dfrac{-1 + \sqrt{17}}{4}\right)$

145. $(\sqrt{2}, 3\sqrt{2})$, $(-\sqrt{2}, -3\sqrt{2})$,
$(7\sqrt{22}/33, 31\sqrt{22}/33)$,
$(-7\sqrt{22}/33, -31\sqrt{22}/33)$

147. $(5, 2)$, $(5, -4)$, $(1, 2)$, $(1, -4)$

149. $as/(a + b)$ and $bs/(a + b)$

151. 24, 60, 120

153. $\frac{1}{2}(\sqrt{n + 2m} + \sqrt{n - 2m})$ and $\frac{1}{2}(\sqrt{n + 2m} - \sqrt{n - 2m})$, $\frac{1}{2}(\sqrt{n - 2m} - \sqrt{n + 2m})$ and $-\frac{1}{2}(\sqrt{n - 2m} + \sqrt{n + 2m})$

155. convex: no; bounded: no

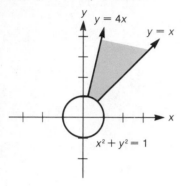

157. convex: yes; bounded: yes

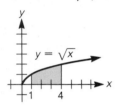

159. convex: no; bounded: no

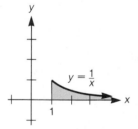

161.

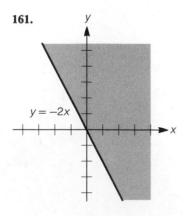

163.

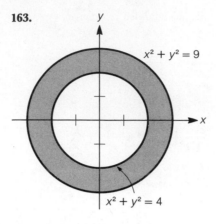

165. $(\tan^{-1}(b/a), \sqrt{a^2 + b^2})$

CHAPTER EIGHT
ANALYTIC GEOMETRY

8.1 The Basic Equations

1. $\sqrt{89}$ **3.** $5x + 4y - 4 = 0$

5. $2x + 3y - 20 = 0$

7. x-intercepts: 6, -4; y-intercepts: $\pm 2\sqrt{6}$

9. $13x - 7y - 35 = 0$ **11.** $12 + \sqrt{74}$

13. $\theta = \pi/3$ or $60°$

15. (a) $\theta = 1.37$ or $78.69°$ (b) $\theta = 1.77$ or $101.31°$

17. (a) $5\sqrt{2}/2$ (b) $5\sqrt{2}/2$

19. (a) $19\sqrt{41}/41$ (b) $19\sqrt{41}/41$

21. (a) $(x + 2)^2 + (y + 3)^2 = \frac{361}{13}$ (b) $\sqrt{5}$ **23.** $\frac{65}{2}$ **25.** $\frac{1}{5}(15 \pm 2\sqrt{30})$

27. $\frac{12}{5}$ **29.** $y = -3x + 12$

31. $y = \frac{1}{3}x + \frac{9}{2}$ **33.** (a) $m = -A/B$ and $b = -(Ax_0 + By_0 + C)/B$

35. (a) center: $(-5, 2)$; radius: $5\sqrt{2}$

37. $5\sqrt{2}$ **41.** (a) A: $y = \frac{1}{3}x$; B: $y = -x + 8$; C: $y = 2x - 10$ (b) The intersection point is $(6, 2)$ for all three bisectors.

47. (a) slope: $-A/B$; y-intercept: $-C/B$

51. $(x - \frac{26}{11})^2 + (y + \frac{25}{11})^2 = \frac{4964}{121}$

53. $(x - 4 - \frac{10}{13}\sqrt{13})^2 + (y - 6 - \frac{15}{13}\sqrt{13})^2 = 25$ or $(x - 4 + \frac{10}{13}\sqrt{13})^2 + (y - 6 + \frac{15}{13}\sqrt{13})^2 = 25$

8.2 The Parabola

1. focus: $(0, 1)$; directrix: $y = -1$; focal width: 4

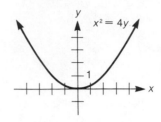

3. focus: $(-2, 0)$; directrix: $x = 2$; focal width: 8

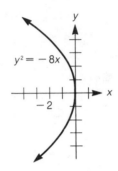

5. focus: $(0, -5)$; directrix: $y = 5$; focal width: 20

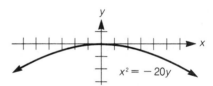

7. focus: $(-7, 0)$; directrix: $x = 7$; focal width: 28

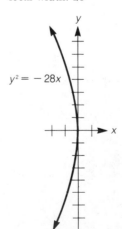

9. focus: $(0, \frac{3}{2})$; directrix: $y = -\frac{3}{2}$; focal width: 6

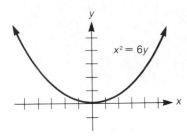

11. focus: $(0, \frac{7}{16})$; directrix: $y = -\frac{7}{16}$; focal width: $\frac{7}{4}$

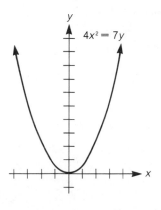

13. vertex: $(2, 3)$; focus: $(3, 3)$; directrix: $x = 1$; focal width: 4

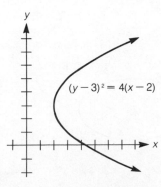

15. vertex: $(4, 2)$; focus: $(4, \frac{9}{4})$; directrix: $y = \frac{7}{4}$; focal width: 1

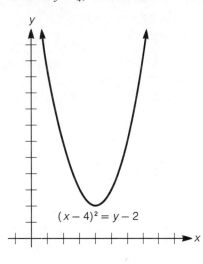

17. vertex: $(0, -1)$; focus: $(\frac{1}{4}, -1)$; directrix: $x = -\frac{1}{4}$; focal width: 1

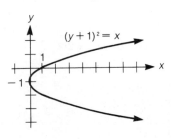

19. vertex: $(3, 0)$; focus: $(3, \frac{1}{8})$; directrix: $y = -\frac{1}{8}$; focal width: $\frac{1}{2}$

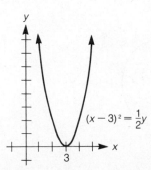

21. vertex: $(4, 1)$; focus: $(4, \frac{9}{8})$; directrix: $y = \frac{7}{8}$; focal width: $\frac{1}{2}$

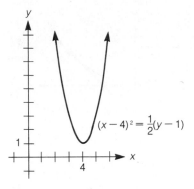

23. line of symmetry: $y = 1$

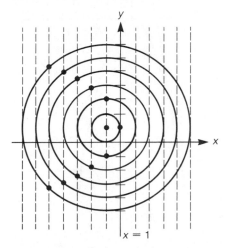

25. $x^2 = 12y$ **27.** $y^2 = 128x$
29. $y^2 = \frac{36}{5}x$ **31.** $y^2 = -9x$
33. $Q(-2, 1)$ **35.** $\frac{1369}{144}$

37.

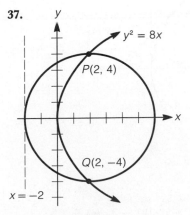

39. 11.25 ft   **41.** **(a)** $y = \frac{9}{4}x - 1$
(b) $y = -\frac{9}{8}x - \frac{1}{4}$ **43.** **(a)** $Q(-\frac{1}{8}, \frac{1}{64})$
(b) $M(\frac{15}{16}, \frac{257}{128})$ **(c)** $ST = \frac{1}{2}$

45. $y = x - 2$

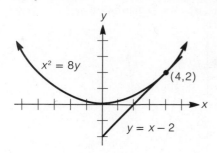

47. $y = 6x + 9$

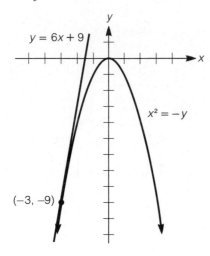

49. $y = -x + 2$

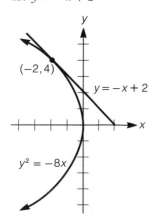

51. side: $8\sqrt{3}\,p$ units;
area: $48\sqrt{3}\,p^2$ sq. units
53. (a) -1 **(b)** -2 **(c)** -3
(d) The y-intercept of the tangent line
at (x_0, y_0) is $-y_0$. **55.** $\frac{1}{4}$ **57.** 12

8.3 The Ellipse

1. major axis: 6; minor axis: 4;
foci: $(\pm\sqrt{5}, 0)$; eccentricity: $\sqrt{5}/3$

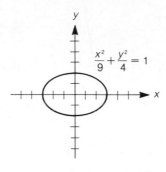

3. major axis: 8; minor axis: 2;
foci: $(\pm\sqrt{15}, 0)$; eccentricity: $\sqrt{15}/4$

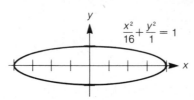

5. major axis: $2\sqrt{2}$; minor axis: 2;
foci: $(\pm 1, 0)$; eccentricity: $\sqrt{2}/2$

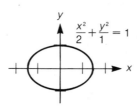

7. major axis: 8; minor axis: 6;
foci: $(0, \pm\sqrt{7})$; eccentricity: $\sqrt{7}/4$

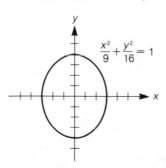

9. major axis: $2\sqrt{15}/3$;
minor axis: $2\sqrt{3}/3$; foci: $(0, \pm 2\sqrt{3}/3)$;
eccentricity: $2\sqrt{5}/5$

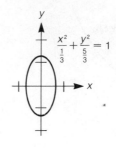

11. major axis: 4; minor axis: $2\sqrt{2}$;
foci: $(0, \pm\sqrt{2})$; eccentricity: $\sqrt{2}/2$

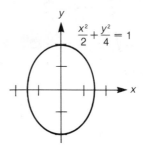

13. center: $(5, -1)$; major axis: 10;
minor axis: 6; foci: $(9, -1)$, $(1, -1)$;
eccentricity: $\frac{4}{5}$

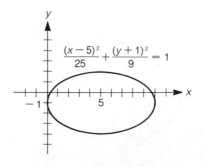

15. center: $(1, 2)$; major axis: 4; minor axis: 2; foci: $(1, 2 \pm \sqrt{3})$; eccentricity: $\sqrt{3}/2$

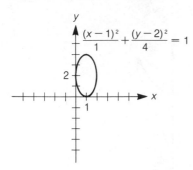

$$\frac{(x-1)^2}{1} + \frac{(y-2)^2}{4} = 1$$

17. center: $(-3, 0)$; major axis: 6; minor axis: 2; foci: $(-3 \pm 2\sqrt{2}, 0)$; eccentricity: $2\sqrt{2}/3$

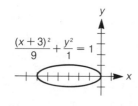

$$\frac{(x+3)^2}{9} + \frac{y^2}{1} = 1$$

19. center: $(1, -2)$; major axis: 4; minor axis: $2\sqrt{3}$; foci: $(2, -2)$, $(0, -2)$; eccentricity: $\frac{1}{2}$

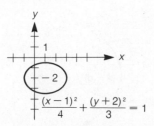

$$\frac{(x-1)^2}{4} + \frac{(y+2)^2}{3} = 1$$

21. center: $(4, 6)$; major axis: 0; minor axis: 0

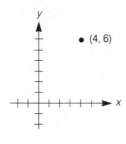

$\bullet \ (4, 6)$

23. no graph
25. $(x^2/5^2) + (y^2/4^2) = 1$, or $6x^2 + 25y^2 = 400$
27. $(x^2/16) + (y^2/15) = 1$, or $15x^2 + 16y^2 = 240$
29. $(x^2/21) + (y^2/25) = 1$, or $25x^2 + 21y^2 = 525$
31. $(x^2/3^2) + [y^2/(\frac{3}{2})^2] = 1$, or $x^2 + 4y^2 = 9$
33. (a) $y = -\frac{4}{3}x + \frac{38}{3}$ **(b)** $y = \frac{7}{9}x + \frac{76}{9}$
(c) $y = \frac{1}{15}x - \frac{76}{15}$
35. (a) $y = -6x + 26$ **(b)** $\frac{169}{3}$
37. (b) $3x + 10\sqrt{6}\,y - 75 = 0$
(c) $d_1 = 87/\sqrt{609}$; $d_2 = 63/\sqrt{609}$
(d) $d_1 d_2 = \dfrac{87}{\sqrt{609}} \cdot \dfrac{63}{\sqrt{609}} = \dfrac{5481}{609} = 9$,
which is b^2.

39.

x	0	0.1	0.2	0.3	0.4	0.5	0.6	0.7	0.8	0.9	1.0
y	± 4	± 3.98	± 3.92	± 3.82	± 3.67	± 3.46	± 3.2	± 2.86	± 2.4	± 1.74	0

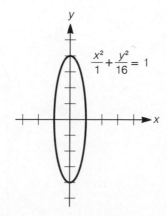

$$\frac{x^2}{1} + \frac{y^2}{16} = 1$$

45. $(ab/A, ab/A)$, $(ab/A, -ab/A)$, $(-ab/A, ab/A)$, $(-ab/A, -ab/A)$, where $A = \sqrt{a^2 + b^2}$

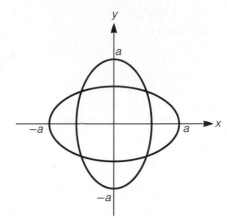

49. (b) $D\left(-\frac{20}{7}, -\frac{18}{7}\right)$
53. (a) We know that

$$\frac{x_1 x}{a^2} + \frac{y_1 y}{b^2} = 1$$

Multiply by $a^2 b^2$:

$$b^2 x_1 x + a^2 y_1 y = a^2 b^2$$

[Parts (b)–(d) are on the next page.]

(b) Since (h, k) lies on this line, replacing (x, y) with (h, k) results in the equation

$$b^2 x_1 h + a^2 y_1 k = a^2 b^2$$

(c) Repeating part (a), we have $b^2 x_2 x + a^2 y_2 y = a^2 b^2$. Now substitute (h, k) to get $b^2 x_2 h + a^2 y_2 k = a^2 b^2$.
(d) Replacing (x, y) with (x_1, y_1) and (x_2, y_2), respectively, results in the equations we obtained in (b) and (c). Thus this line must pass through the points (x_1, y_1) and (x_2, y_2).
59. (b) $(a/e, 0)$

5. vertices: $(5, 0)$, $(-5, 0)$;
transverse axis: 10; conjugate axis: 8;
asymptotes: $y = \pm\frac{4}{5}x$; foci: $(\pm\sqrt{41}, 0)$;
eccentricity: $\sqrt{41}/5$

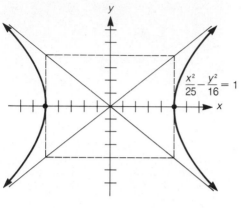

9. vertices: $(0, 5)$, $(0, -5)$;
transverse axis: 10; conjugate axis: 4;
asymptotes: $y = \pm\frac{5}{2}x$; foci: $(0, \pm\sqrt{29})$;
eccentricity: $\sqrt{29}/5$

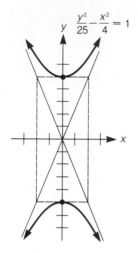

8.4 The Hyperbola

1. vertices: $(2, 0)$, $(-2, 0)$;
transverse axis: 4; conjugate axis: 2;
asymptotes: $y = \pm\frac{1}{2}x$; foci: $(\pm\sqrt{5}, 0)$;
eccentricity: $\sqrt{5}/2$

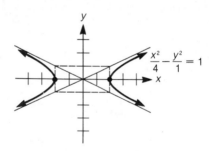

7. vertices: $(0, \pm\sqrt{2}/2)$;
transverse axis: $\sqrt{2}$;
conjugate axis: $2\sqrt{3}/3$;
asymptotes: $y = \pm(\sqrt{6}/2)x$;
foci: $(0, \pm\sqrt{30}/6)$; eccentricity: $\sqrt{15}/3$

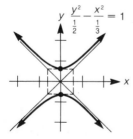

3. vertices: $(0, 2)$, $(0, -2)$;
transverse axis: 4; conjugate axis: 2;
asymptotes: $y = \pm 2x$; foci: $(0, \pm\sqrt{5})$;
eccentricity: $\sqrt{5}/2$

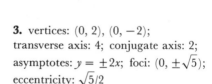

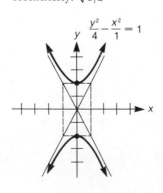

11. center: $(5, -1)$;
vertices: $(10, -1)$, $(0, -1)$;
transverse axis: 10; conjugate axis: 6;
asymptotes: $y = \frac{3}{5}x - 4$ and
$y = -\frac{3}{5}x + 2$; foci: $(5 \pm \sqrt{34}, -1)$;
eccentricity: $\sqrt{34}/5$

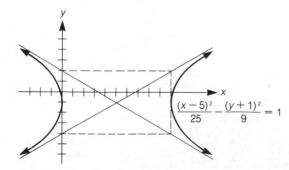

13. center: $(1, 2)$; vertices: $(1, 4)$, $(1, 0)$; transverse axis: 4; conjugate axis: 2; asymptotes: $y = 2x$ and $y = -2x + 4$; foci: $(1, 2 \pm \sqrt{5})$; eccentricity: $\sqrt{5}/2$

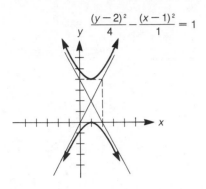

$$\frac{(y-2)^2}{4} - \frac{(x-1)^2}{1} = 1$$

15. center: $(-3, 4)$; vertices: $(1, 4)$, $(-7, 4)$; transverse axis: 8; conjugate axis: 8; asymptotes: $y = x + 7$ and $y = -x + 1$; foci: $(-3 \pm 4\sqrt{2}, 4)$; eccentricity: $\sqrt{2}$

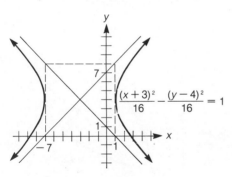

$$\frac{(x+3)^2}{16} - \frac{(y-4)^2}{16} = 1$$

17. center: $(0, 1)$; vertices: $(\pm 2, 1)$; transverse axis: 4; conjugate axis: 4; asymptotes: $y = x + 1$ and $y = -x + 1$; foci: $(\pm 2\sqrt{2}, 1)$; eccentricity: $\sqrt{2}$

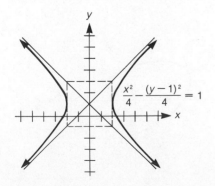

$$\frac{x^2}{4} - \frac{(y-1)^2}{4} = 1$$

19. center: $(2, 1)$; vertices: $(-1, 1)$, $(5, 1)$; transverse axis: 6; conjugate axis: 6; asymptotes: $y = x - 1$ and $y = -x + 3$; foci: $(2 \pm 3\sqrt{2}, 1)$; eccentricity: $\sqrt{2}$

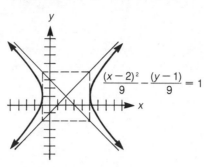

$$\frac{(x-2)^2}{9} - \frac{(y-1)}{9} = 1$$

21. center: $(0, -4)$; vertices: $(0, -9)$, $(0, 1)$; transverse axis: 10; conjugate axis: 2; asymptotes: $y = 5x - 4$ and $y = -5x - 4$; foci: $(0, -4 \pm \sqrt{26})$; eccentricity: $\sqrt{26}/5$

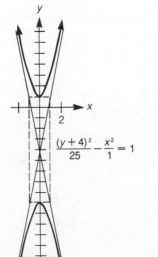

$$\frac{(y+4)^2}{25} - \frac{x^2}{1} = 1$$

23. degenerate: two lines

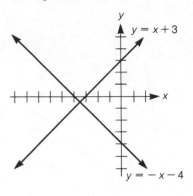

27. $15x^2 - y^2 = 15$ **29.** $x^2 - 4y^2 = 4$
31. $2x^2 - 5y^2 = 10$
33. $y^2 - 32x^2 = 49$ **35.** $y^2 - 9x^2 = 9$
37. The slopes of the asymptotes are 1 and -1; therefore the asymptotes are perpendicular. **39. (b)** $(0, \pm 6)$
(c) $F_1 P = \sqrt{(5 - 0)^2 + (6 - 6)^2} = 5$; $F_2 P = \sqrt{(5 - 0)^2 + (6 - (-6))^2} = \sqrt{25 + 144} = 13$
41. (a) $e = \sqrt{2}$ **(b)** $e = \sqrt{2}$
43. (a) By the definition of a hyperbola, the difference in the distances between a fixed point and each focus is $2a$.
49. $\left(\frac{51}{8}, \frac{15}{2}\right)$ is the intersection point.

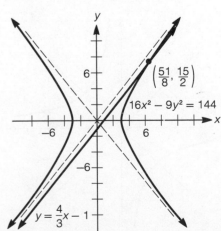

51. (b) (i) 1.01; **(ii)** 2; **(iii)** 3;
(iv) 4; **(v)** 10
(c)

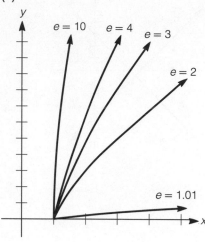

(d) It appears the eccentricity affects the width of the hyperbola. That is, larger eccentricities result in a much wider curve than eccentricities closer to 1.   **53.** $y = 2x - 2$ **65. (b)** four   **(d)** The tangents are perpendicular.

8.5 Rotation of Axes

1. $(\frac{1}{2}, 3\sqrt{3}/2)$ **3.** $(2, 0)$ **5.** $(-\frac{31}{13}, \frac{27}{13})$
7. $\sin \theta = \frac{4}{5}$; $\cos \theta = \frac{3}{5}$
9. $\sin \theta = \frac{3}{5}$; $\cos \theta = \frac{4}{5}$
11. $\sin \theta = \sqrt{3}/2$; $\cos \theta = \frac{1}{2}$
13. $\sin \theta = 7\sqrt{2}/34$; $\cos \theta = 23\sqrt{2}/34$
15. $x'^2 - y'^2 = 9$

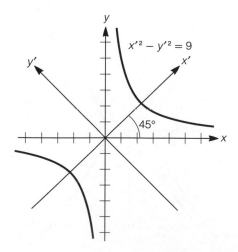

17.

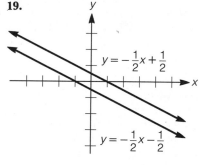

19.

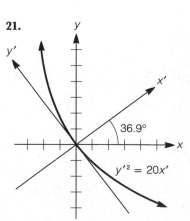

21.

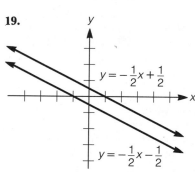

23. $\dfrac{\left(x' + \dfrac{2\sqrt{5}}{5}\right)^2}{1} - \dfrac{\left(y' + \dfrac{\sqrt{5}}{5}\right)^2}{4} = 1$

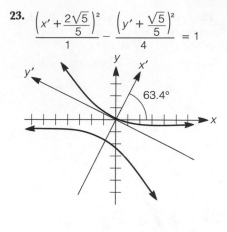

25.

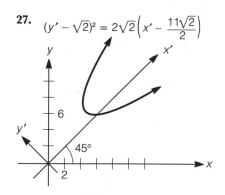

27. $(y' - \sqrt{2})^2 = 2\sqrt{2}\left(x' - \dfrac{11\sqrt{2}}{2}\right)$

29.

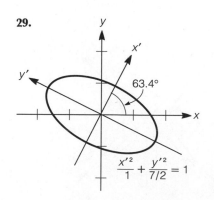

31.

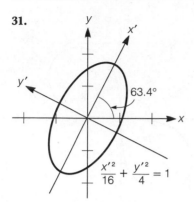

$$\frac{x'^2}{16} + \frac{y'^2}{4} = 1$$

63.4°

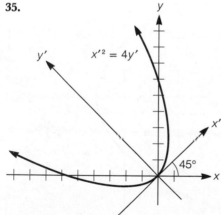

$$\frac{y'^2}{4} - \frac{x'^2}{36} = 1$$

18.4°

35.

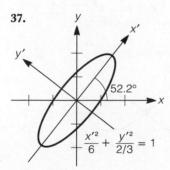

$x'^2 = 4y'$

45°

37.

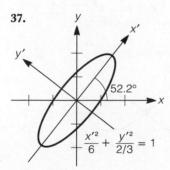

$$\frac{x'^2}{6} + \frac{y'^2}{2/3} = 1$$

52.2°

39. no graph

41. $x = x' \cos \theta - y' \sin \theta,$
$y = x' \sin \theta + y' \cos \theta$

8.6 Introduction to Polar Coordinates

1. (a) $(-\frac{3}{2}, 3\sqrt{3}/2)$ **(b)** $(2\sqrt{3}, -2)$
(c) $(2\sqrt{3}, -2)$ **3. (a)** $(0, 1)$
(b) $(0, 1)$ **(c)** $\left(\frac{1}{2}\sqrt{2 + \sqrt{2}}, \frac{1}{2}\sqrt{2 - \sqrt{2}}\right)$
5. $(\sqrt{2}, 5\pi/4)$ **7.** $(x - 1)^2 + y^2 = 1$
9. $x^4 + x^2y^2 - y^2 = 0$
11. $x^6 - 9x^4 + 3x^4y^2 + 18x^2y^2$
$\qquad\qquad + 3x^2y^4 - 9y^4 + y^6 = 0$
13. $(x^2/4) + (y^2/8) = 1$
15. $y = -\sqrt{3}\,x + 4$
17. $r = 2/(3 \cos \theta - 4 \sin \theta)$
19. $r = \tan^2 \theta \sec \theta$ **21.** $r^2 = 1/\sin 2\theta$
23. $r^2 = 9/(9 \cos^2 \theta + \sin^2 \theta)$
25. (a)

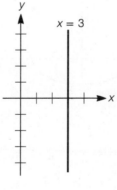

$x = 3$

(b)

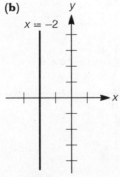

$x = -2$

27.

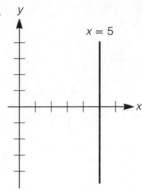

$x = 5$

29. $\theta = \pi/2$

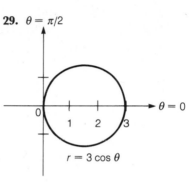

$r = 3 \cos \theta$

31. $\theta = \pi/2$

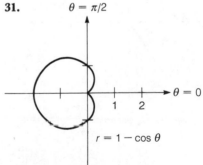

$r = 1 - \cos \theta$

33. $\theta = \pi/2$

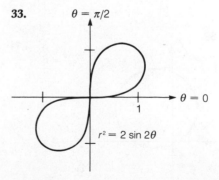

$r^2 = 2 \sin 2\theta$

35.

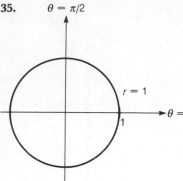

37.

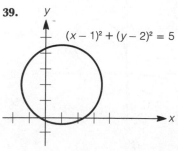

39.

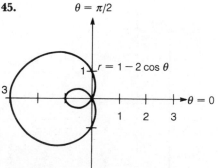

41.

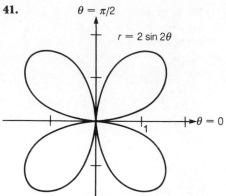

43.

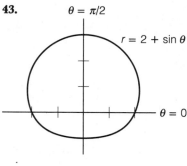

45.

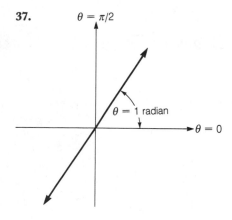

47.

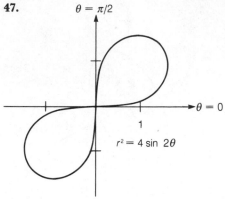

49.

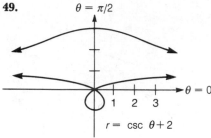

51.

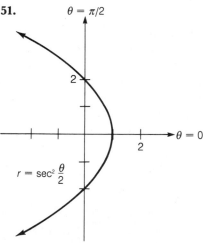

53. $(x - \frac{1}{2}a)^2 + (y - \frac{1}{2}b)^2 = \frac{1}{4}(a^2 + b^2)$;
center: $(a/2, b/2)$; radius: $\frac{1}{2}\sqrt{a^2 + b^2}$

57. (c)

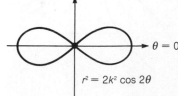

Chapter 8 Review Exercises

1. focus: $(-3, 0)$; directrix: $x = 3$

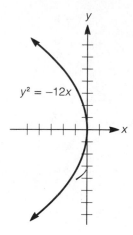

$y^2 = -12x$

2. foci: $(\pm\sqrt{5}, 0)$; asymptotes: $y = \pm\frac{1}{2}x$

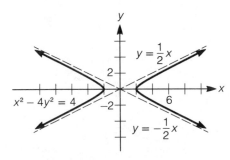

$y = \frac{1}{2}x$

$x^2 - 4y^2 = 4$

$y = -\frac{1}{2}x$

3. (a) $\theta = 60°$

(b)

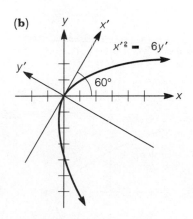

$x'^2 = 6y'$

$60°$

4. $\theta = 30°$ **5.** $y = 4x - 8$
6. $2x - \sqrt{5}y + (5 + \sqrt{5}) = 0$
7. $(x^2/12) + (y^2/16) = 1$ **8.** $\pm\sqrt{15}/15$
9. $\sqrt{3}x - y - 2\sqrt{3} = 0$
10. $(x^2/3) - (y^2/1) = 1$

11.

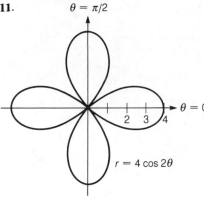

$\theta = \pi/2$

$\theta = 0$

$r = 4\cos 2\theta$

12. (b) 64
13. length of major axis: 10;
length of minor axis: 4;
foci: $(\pm\sqrt{21}, 0)$

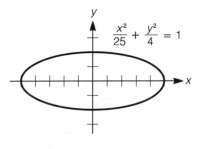

$\dfrac{x^2}{25} + \dfrac{y^2}{4} = 1$

14. $(x^2 + y^2)^2 = x^2 - y^2$ **15.** $3\sqrt{5}/5$

16.

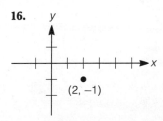

$(2, -1)$

17.

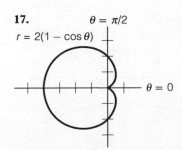

$\theta = \pi/2$

$r = 2(1 - \cos\theta)$

$\theta = 0$

18.

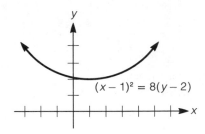

$\dfrac{(x + 4)^2}{9} - \dfrac{(y - 4)^2}{1} = 1$

-4

19. focal width: 8; vertex: $(1, 2)$

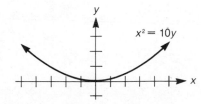

$(x - 1)^2 = 8(y - 2)$

20. $y = \frac{1}{6}x + \frac{13}{3}$ **37.** $\theta \approx 146.3°$
39. $53\sqrt{61}/61$ **43. (a)** $y^2 = 16x$
(b) $x^2 = 16y$ **45.** $x^2 = 12y$
47. $15x^2 + 16y^2 = 960$
49. $25x^2 + 9y^2 = 900$
51. $8x^2 - y^2 = 32$
53. $240x^2 - 16y^2 = 135$
55. vertex: $(0, 0)$; focus: $(0, \frac{5}{2})$;
directrix: $y = -\frac{5}{2}$; focal width: 10

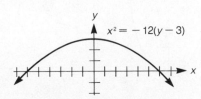

$x^2 = 10y$

57. vertex: $(0, 3)$; focus: $(0, 0)$;
directrix: $y = 6$; focal width: 12

$x^2 = -12(y - 3)$

59. vertex: $(1, 1)$; focus: $(0, 1)$;
directrix: $x = 2$; focal width: 4

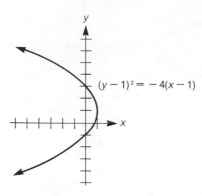

$(y - 1)^2 = -4(x - 1)$

61. center: $(0, 0)$; foci: $(\pm\sqrt{2}, 0)$;
major axis: 4; minor axis: $2\sqrt{2}$;
eccentricity: $\sqrt{2}/2$

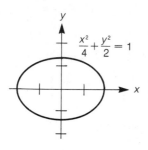

$\dfrac{x^2}{4} + \dfrac{y^2}{2} = 1$

63. center: $(0, 0)$; foci: $(0, \pm 2\sqrt{10})$;
major axis: 14; minor axis: 6;
eccentricity: $2\sqrt{10}/7$

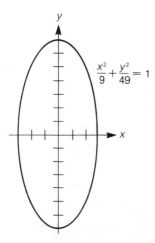

$\dfrac{x^2}{9} + \dfrac{y^2}{49} = 1$

65. center: $(1, -2)$;
foci: $(5, -2)$, $(-3, -2)$; major axis: 10;
minor axis: 6; eccentricity: $\frac{4}{5}$

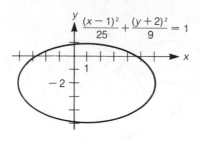

$\dfrac{(x-1)^2}{25} + \dfrac{(y+2)^2}{9} = 1$

67. center: $(0, 0)$; vertices: $(\pm 2, 0)$;
foci: $(\pm\sqrt{6}, 0)$;
asymptotes: $y = \pm(\sqrt{2}/2)x$;
eccentricity: $\sqrt{6}/2$

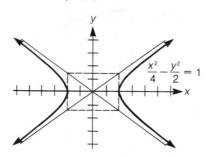

$\dfrac{x^2}{4} - \dfrac{y^2}{2} = 1$

69. center: $(0, 0)$; vertices: $(0, \pm 3)$;
foci: $(0, \pm\sqrt{58})$; asymptotes: $y = \pm\frac{3}{7}x$;
eccentricity: $\sqrt{58}/3$

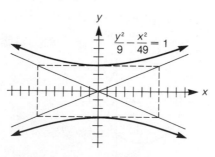

$\dfrac{y^2}{9} - \dfrac{x^2}{49} = 1$

71. center: $(1, -2)$;
vertices: $(6, -2)$, $(-4, -2)$;
foci: $(1 \pm \sqrt{34}, -2)$;
asymptotes: $y + 2 = \pm\frac{3}{5}(x - 1)$;
eccentricity: $\sqrt{34}/5$

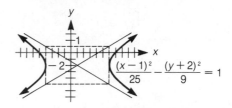

$\dfrac{(x-1)^2}{25} - \dfrac{(y+2)^2}{9} = 1$

73. *ellipse:* center: $(1, -2)$;
foci: $(0, -2)$, $(2, -2)$; major axis: 4;
minor axis: $2\sqrt{3}$

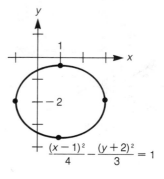

$\dfrac{(x-1)^2}{4} - \dfrac{(y+2)^2}{3} = 1$

75. *parabola:* vertex: $(4, -1)$;
axis: $y = -1$; focus: $(3, -1)$;
directrix: $x = 5$

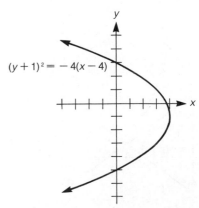

$(y + 1)^2 = -4(x - 4)$

77. *hyperbola:* center: $(1, 5)$;
vertices: $(-2, 5)$, $(4, 5)$;
foci: $(-4, 5)$, $(6, 5)$;
asymptotes: $y = \frac{4}{3}x + \frac{11}{3}$ and
$y = -\frac{4}{3}x + \frac{19}{3}$

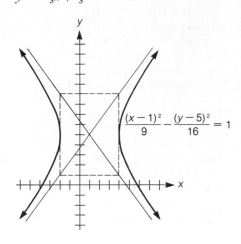

$$\frac{(x-1)^2}{9} - \frac{(y-5)^2}{16} = 1$$

79. *degenerate ellipse*

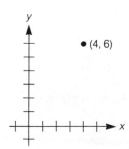

81. *degenerate hyperbola*

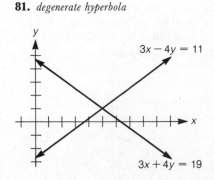

$3x - 4y = 11$

$3x + 4y = 19$

83. *hyperbola:* center: $(0, -4)$;
vertices: $(0, 1)$, $(0, -9)$;
foci: $(0, -4 \pm \sqrt{26})$;
asymptotes: $y = 5x - 4$ and
$y = -5x - 4$

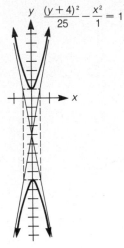

$$\frac{(y+4)^2}{25} - \frac{x^2}{1} = 1$$

85. no graph   **87. (b)** 64
91. $y = -4x - 16$ **93.** $y = -\frac{1}{6}x + \frac{13}{3}$
95. $y = \frac{4}{3}x - 2$
97. parabola: $y^2 = -8(x - 5)$;
ellipse: $(x^2/25) + (y^2/16) = 1$
99. C_2 would be the best approximation.

	APPROXIMATION OBTAINED	PERCENTAGE ERROR
C_1	25.531776	0.019
C_2	25.526986	0.000049
C_3	25.519489	0.029

101. $\sin \theta = \sqrt{5}/5$; $\cos \theta = 2\sqrt{5}/5$

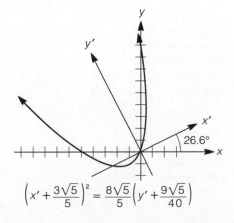

$$\left(x' + \frac{3\sqrt{5}}{5}\right)^2 = \frac{8\sqrt{5}}{5}\left(y' + \frac{9\sqrt{5}}{40}\right)$$

103. $\sin \theta = \sqrt{2}/2$; $\cos \theta = \sqrt{2}/2$

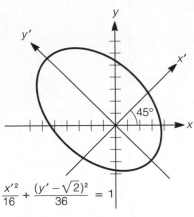

$$\frac{x'^2}{16} + \frac{(y' - \sqrt{2})^2}{36} = 1$$

105. (a) $\theta = \pi/2$

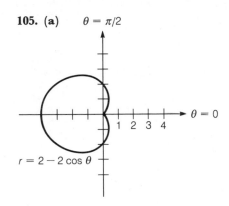

$r = 2 - 2\cos\theta$

(b) $\theta = \pi/2$

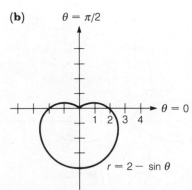

$r = 2 - \sin\theta$

107. (a)

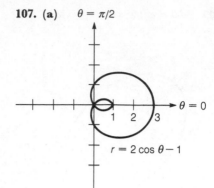

$\theta = \pi/2$

$r = 2\cos\theta - 1$

$\theta = 0$

(b)

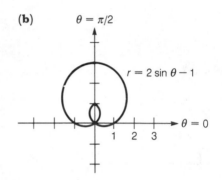

$\theta = \pi/2$

$r = 2\sin\theta - 1$

$\theta = 0$

109. (a)

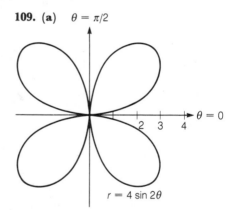

$\theta = \pi/2$

$\theta = 0$

$r = 4\sin 2\theta$

(b)

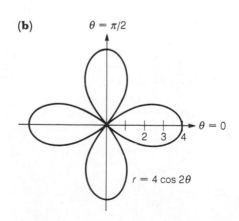

$\theta = \pi/2$

$\theta = 0$

$r = 4\cos 2\theta$

111. (a)

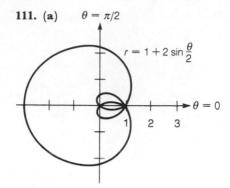

$\theta = \pi/2$

$r = 1 + 2\sin\dfrac{\theta}{2}$

$\theta = 0$

(b)

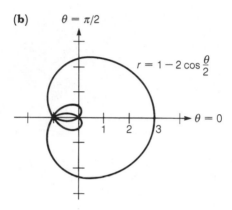

$\theta = \pi/2$

$r = 1 - 2\cos\dfrac{\theta}{2}$

$\theta = 0$

115. (a) $A(-\frac{1}{2}r, -\frac{1}{2}r(p+q))$,
$B(-\frac{1}{2}q, -\frac{1}{2}q(p+r))$,
$C(-\frac{1}{2}p, -\frac{1}{2}p(q+r))$

CHAPTER NINE
ROOTS OF POLYNOMIAL EQUATIONS

9.1 Division of Polynomials

1. quotient: $x - 5$; remainder: -11;
$x^2 - 8x + 4 = (x-3)(x-5) - 11$
3. quotient: $x - 11$; remainder: 53;
$x^2 - 6x - 2 = (x+5)(x-11) + 53$
5. quotient: $3x^2 - \frac{3}{2}x - \frac{1}{4}$;
remainder: $\frac{13}{4}$; $6x^3 - 2x + 3 =$
$(2x+1)(3x^2 - \frac{3}{2}x - \frac{1}{4}) + \frac{13}{4}$
7. quotient: $x^4 - 3x^3 + 9x^2 - 27x + 81$;
remainder: -241; $x^5 + 2 =$
$(x+3)(x^4 - 3x^3 + 9x^2 - 27x + 81) - 241$
9. quotient: $x^5 + 2x^4 + 4x^3 + 8x^2 +$
$16x + 32$; remainder: 0; $x^6 - 64 =$
$(x-2)(x^5 + 2x^4 + 4x^3 + 8x^2 +$
$16x + 32)$
11. quotient: $5x^2 + 15x + 17$;
remainder: $-24x - 83$; $5x^4 - 3x^2 + 2 =$
$(x^2 - 3x + 5)(5x^2 + 15x + 17) +$
$(-24x - 83)$

13. quotient: $3y - 19$;
remainder: $89y + 35$; $3y^3 - 4y^2 - 3 =$
$(y^2 + 5y + 2)(3y - 19) + (89y + 35)$
15. quotient: $t^2 - 2t - 4$; remainder: 0;
$t^4 - 4t^3 + 4t^2 - 16 =$
$(t^2 - 2t + 4)(t^2 - 2t - 4)$
17. quotient: $z^4 + z^3 + z^2 + z + 1$;
remainder: 0; $z^5 - 1 =$
$(z-1)(z^4 + z^3 + z^2 + z + 1)$
19. quotient: $ax + (ar + b)$;
remainder: $ar^2 + br + c$; $ax^2 + bx + c =$
$(x-r)[ax + (ar+b)] + (ar^2 + br + c)$
21. quotient: $x - 1$; remainder: -7;
$x^2 - 6x - 2 = (x-5)(x-1) - 7$
23. quotient: $4x - 5$; remainder: 0;
$4x^2 - x - 5 = (x+1)(4x-5)$
25. quotient: $6x^2 + 19x + 78$;
remainder: 313; $6x^3 - 5x^2 + 2x + 1 =$
$(x-4)(6x^2 + 19x + 78) + 313$
27. quotient: $x^2 + 2x + 4$; remainder: 7;
$x^3 - 1 = (x-2)(x^2 + 2x + 4) + 7$
29. quotient: $x^4 - 2x^3 + 4x^2 - 8x + 16$;
remainder: -33; $x^5 - 1 =$
$(x+2)(x^4 - 2x^3 + 4x^2 - 8x + 16) - 33$
31. quotient: $x^3 - 10x^2 + 40x - 160$;
remainder: 642; $x^4 - 6x^3 + 2 =$
$(x+4)(x^3 - 10x^2 + 40x - 160) + 642$
33. quotient: $x^2 + 6x + 57$;
remainder: 576; $x^3 - 4x^2 - 3x + 6 =$
$(x-10)(x^2 + 6x + 57) + 576$
35. quotient: $x^2 - 6x + 30$;
remainder: -150; $x^3 - x^2 =$
$(x+5)(x^2 - 6x + 30) - 150$
37. quotient: $54x^2 + 9x - 21$;
remainder: 0; $54x^3 - 27x^2 - 27x + 14 =$
$(x - \frac{2}{3})(54x^2 + 9x - 21)$
39. quotient: $5x^3 - \frac{13}{2}x^2 + \frac{25}{4}x - \frac{41}{8}$;
remainder: $\frac{57}{16}$;
$5x^4 - 4x^3 + 3x^2 - 2x + 1 =$
$(x + \frac{1}{2})(5x^3 - \frac{13}{2}x^2 + \frac{25}{4}x - \frac{41}{8}) + \frac{57}{16}$
41. $q(x) = 2x$; $R(x) = \frac{1}{3}$
43. $q(x) = 3x^2 - \frac{3}{2}x + \frac{3}{4}$; $R(x) = \frac{1}{8}$
45. $k = 4$ **49.** $q(x) = x + (-4 + i)$;
$R(x) = -4i$ **51.** $q(x) = x + (-1 + i)$;
$R(x) = 0$ **53.** $k = -8$ **55.** $k = 4$
57. $-6\sqrt{3} + 57$ **59.** 0

9.2 The Remainder Theorem and the Factor Theorem

1. yes **3.** yes **5.** yes **7.** yes **9.** yes
11. no **13. (a)** yes **(b)** no
15. (a) 1, 2 (multiplicity 3), 3
(b) 1 (multiplicity 3)

(c) 5 (multiplicity 6), -1 (multiplicity 4)
(d) 0 (multiplicity 5), 1   **17.** -170
19. -9 **21.** $-3\sqrt{2}-2$
23. -22 **25.** $-3, 4, 3$
27. $1, -1+\sqrt{6}, -1-\sqrt{6}$
29. $-2, \frac{2}{3}, 3$
31. $-\frac{3}{2}, \frac{1}{2}(1+\sqrt{5}), \frac{1}{2}(1-\sqrt{5})$
33. $0, 5$ **35.** $-4, 3$
37. $-9, 1+\sqrt{2}, 1-\sqrt{2}$
39. $\frac{1}{4}(3+\sqrt{17})$ (multiplicity 2),
$\frac{1}{4}(3-\sqrt{17})$ (multiplicity 2)
41. (a) 1.125 (b) -0.046875
(c) $t-1$ (d) $1, \frac{1}{2}(-1 \pm \sqrt{13})$
43. $x^3 - 4x^2 - 17x + 60 = 0$
45. $x^3 + 8x^2 + 13x + 6 = 0$
47. No such polynomial equation exists.
49. $ax^4 + (3a+b)x^3 + (-17a+3b)x^2 + (6a-17b)x + 6b = 0$, for any real numbers a and b, where $a \neq 0$
51. 20.44 **53.** (a) -0.05 (b) 0.07
55. yes   **57.** yes   **59.** $a = -1, b = -1$
61. $b = \pm 3\sqrt{2}/2$
63. 2 (multiplicity 2), -4

9.3 The Fundamental Theorem of Algebra

1. a, b, c **3.** $[x-(-1)](x-3)$
5. $4(x-\frac{1}{4})[x-(-6)]$
7. $[x-(-\sqrt{5})](x-\sqrt{5})$
9. $[x-(5+i)][x-(5-i)]$
11. $x^3 + x^2 - 5x + 3$ **13.** $x^4 - 16$
15. $x^6 + 10x^4 - 87x^2 + 144$
17. $f(x) = -\frac{5}{42}x^2 + \frac{25}{42}x + \frac{30}{7}$
19. $f(x) = \frac{1}{30}x^3 - \frac{19}{30}x + 1$
21. $x^2 + (i+\sqrt{3})x + i\sqrt{3} = 0$
23. $x^2 - 3x - 54 = 0$
25. $x^2 - 2x - 4 = 0$
27. $x^2 - 2ax + a^2 - b = 0$
29. $(x+2-2i)(x+2+2i)$
          $\times (x-2-2i)(x-2+2i)$
33. $4, 3, -3$

9.4 Rational and Irrational Roots

1. $\pm 1, \pm\frac{1}{2}, \pm\frac{1}{4}, \pm 3, \pm\frac{3}{2}, \pm\frac{3}{4}$
3. $\pm 1, \pm\frac{1}{2}, \pm\frac{1}{4}, \pm\frac{1}{8}, \pm 3, \pm\frac{3}{2}, \pm\frac{3}{4}, \pm\frac{3}{8}, \pm 9, \pm\frac{9}{2}, \pm\frac{9}{4}, \pm\frac{9}{8}$
5. $\pm 1, \pm\frac{1}{2}, \pm 2, \pm 3, \pm\frac{3}{2}, \pm 6$
13. $1, -1, -3$ **15.** $-\frac{1}{4}, -\sqrt{5}, \sqrt{5}$
17. $-1, -\frac{2}{3}, -\frac{1}{3}$

19. $1, -1, \frac{1}{2}(-1+\sqrt{97}), \frac{1}{2}(-1-\sqrt{97})$ **21.** 1 (multiplicity 4)
23. $\frac{1}{2}, 6, -4$ **25.** $\frac{1}{4}, \frac{1}{5}, -\sqrt{6}/2, \sqrt{6}/2$
27. (a) upper bound: 2, lower bound: -1
(b) upper bound: 2, lower bound: -1
(c) upper bound: 6, lower bound: -2
29. between 0.68 and 0.69
31. between 2.88 and 2.89
33. between 4.31 and 4.32
35. between -2.15 and -2.14
37. between -5.27 and -5.26
39. (b) The result guarantees only that A is a factor of B in the case where A and C have no factor in common. Here A and C have a common factor of 5, so the result does not apply.
(c) Since $x = p/q$ is a root of the equation, this statement must be true.
41. $x \approx 0.32$

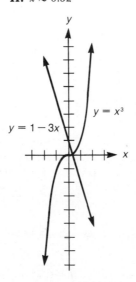

43. $x \approx 1.49$ **47.** $p = 3, x = 1$
51. $b = 2$

9.5 Conjugate Roots and Descartes's Rule of Signs

1. $7 + 2i$ **3.** $5 - 2i, 3$
5. $-2 - i, -3$ (multiplicity 2)
7. $6 + 5i, -\frac{1}{4}$ **9.** $4 - \sqrt{2}i, -\frac{3}{2}i, \frac{3}{2}i$
11. $10 - 2i, 1 + \sqrt{5}, 1 - \sqrt{5}$
13. $\frac{1}{3}(1 - i\sqrt{2}), \frac{2}{5}$
15. $3 + 2i, -1 - i, -1 + \sqrt{2}, -1 - \sqrt{2}$ **17.** $x^2 - 2x - 5 = 0$
19. $x^2 - \frac{4}{3}x - \frac{2}{3} = 0$

21. 2 complex roots, 1 negative real root
23. 4 complex roots, 1 negative real root
25. 2 complex roots, 1 positive real root, 1 negative real root
27. either 1 positive real root, 2 negative real roots or 1 positive real root, 2 complex roots
29. 1 positive real root, 1 negative real root, 6 complex roots
31. 1 positive real root, 8 complex roots
33. 1 positive real root, 1 negative real root, 6 complex roots
35. 1 positive real root, 1 negative real root, 4 complex roots
39. $f(x) = x^4 + 2x^2 + 49$
41. (a) If $b = 0$, then $a + b\sqrt{c} = a = a - b\sqrt{c}$, so $a - b\sqrt{c}$ is also a root.
(b) $d(a + b\sqrt{c}) = 0$ since the first factor of $d(x)$ is 0.

Chapter 9 Review Exercises

1. $f(\frac{1}{2}) = -\frac{3}{2}$ **2.** $-3, 1 + \sqrt{6}, 1 - \sqrt{6}$
3. $\pm 1, \pm\frac{1}{2}, \pm 2, \pm 3, \pm\frac{3}{2}, \pm 6$
4. $f(x) = 3x^2 + 21x - 24$
5. quotient: $4x^2 - 3x - 5$; remainder: 8
6. (a) Let $f(x)$ be a polynomial. If $f(r) = 0$, then $x - r$ is a factor of $f(x)$. Conversely, if $x - r$ is a factor of $f(x)$, then $f(r) = 0$.
(b) Every polynomial equation of the form
$$a_n x^n + a_{n-1}x^{n-1} + \cdots + a_1 x + a_0 = 0$$
$$(n \geq 1, a_n \neq 0)$$
has at least one root among the complex numbers. (This root may be a real number.)
7. (a) 3   (b) The root lies between 2.2 and 2.3.   **8.** $1 \pm i, 3 \pm 2i, -2$
9. $q(x) = x^2 + 2x - 1$; $R(x) = -3x + 7$
10. $2[x - (\frac{3}{2} + \frac{1}{2}i)][x - (\frac{3}{2} - \frac{1}{2}i)]$
11. (a) $\pm 1, \pm 2, \pm 3, \pm 4, \pm 6, \pm 8, \pm 12, \pm 24$ (c) $x = 1$
(d) There are no positive rational roots.
12. (a) $\frac{3}{2}$ (b) $\frac{3}{2}, \frac{1}{2}(-1 \pm i\sqrt{3})$
13. 1 positive real root, 1 negative real root, and 2 complex roots
14. $x^3 + 6x + 20 = 0$
15. $f(x) = (x-2)(x-3i)^3[x - (1+\sqrt{2})]^2$
17. (a) no   (b) yes
19. (a) no   (b) no
21. $q(x) = x^3 + x^2 - 3x + 1$; $R(x) = -1$
23. $q(x) = x^3 + 3x^2 + 7x + 21$; $R(x) = 71$
25. $q(x) = 2x^2 - 13x + 46$; $R(x) = -187$

27. $q(x) = 5x - 20$; $R(x) = 0$
29. $f(10) = 99,904$ **31.** $f(\frac{1}{10}) = -\frac{999}{1000}$
33. $f(a - 1) = a^3$
35. (a) $f(-0.3) \approx -0.24$
(b) $f(-0.39) \approx -0.007$
(c) $f(-0.394) \approx 0.00003$ **37.** $a = -5$
39. $a = -1$ or $a = -2$
43. $\pm 1, \pm 2, \pm 3, \pm 6, \pm 9, \pm 18$
45. $\pm 1, \pm \frac{1}{2}, \pm 2, \pm 4, \pm 8$
47. $\pm p, \pm 1$ **49.** $2, -\frac{3}{2}, -1$
51. $\frac{5}{2}, -1 + \sqrt{3}, -1 - \sqrt{3}$
53. $\frac{2}{3}, \frac{1}{2}(-1 + i\sqrt{3}), \frac{1}{2}(-1 - i\sqrt{3})$
55. $-1, -\sqrt{7}$ (multiplicity 2),
$\sqrt{7}$ (multiplicity 2)
57. 2 (multiplicity 2), 5
59. (a) Let $p(x)$ and $d(x)$ be
polynomials, where $d(x) \neq 0$. Then there
are unique polynomials $q(x)$ and $R(x)$
such that

$$p(x) = d(x) \cdot q(x) + R(x)$$

where either $R(x) = 0$ or the degree of
$R(x)$ is less than the degree of $d(x)$.
(b) When a polynomial $f(x)$ is divided
by $x - r$, the remainder is $f(r)$.
(c) Let $f(x)$ be a polynomial. If
$f(r) = 0$, then $x - r$ is a factor of $f(x)$.
Conversely, if $x - r$ is a factor of $f(x)$,
then $f(r) = 0$.
(d) Every polynomial equation of the
form

$$a_n x^n + a_{n-1} x^{n-1} + \cdots + a_1 x + a_0 = 0$$
$$(n \geq 1, a_n \neq 0)$$

has at least one root among the complex
numbers. (This root may be a real
number.)
61. $6(x - \frac{4}{3})(x - (-\frac{5}{2}))$
63. $(x - 4)[x - (-\sqrt[3]{-5})]$
$\times [x - (\frac{1}{2}\sqrt[3]{5} + \frac{1}{2}i\sqrt{3}\sqrt[3]{25})]$
$\times [x - (\frac{1}{2}\sqrt[3]{5} - \frac{1}{2}i\sqrt{3}\sqrt[3]{25})]$
65. $2 - 3i, 2 + 3i, 3$
67. $1 + i\sqrt{2}, 1 - i\sqrt{2}, \sqrt{7}, -\sqrt{7}$
69. 1 positive real root, 2 complex roots
71. 1 negative real root, 2 complex roots
73. 1 positive real root, 1 negative real
root, 2 complex roots
75. (c) $-1, i, -i$
77. between 0.82 and 0.83
79. (c) between 6.93 and 6.94
81. $x^2 - 8x + 11 = 0$
83. $x^4 - 12x^3 + 35x^2 + 60x - 200 = 0$
85. $x^4 - 4x^3 - 4x^2 + 16x - 8 = 0$

87. zeros: 0, 3, −1

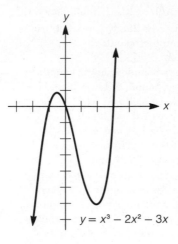

$y = x^3 - 2x^2 - 3x$

89. zeros: 0, −2, 2

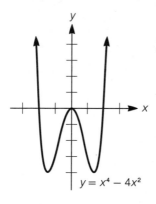

$y = x^4 - 4x^2$

91. zeros: 1, 2

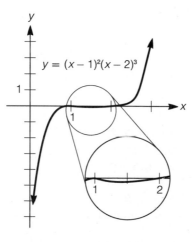

$y = (x - 1)^2(x - 2)^3$

93. T **95.** F **97.** F **99.** T

10.1 Mathematical Induction

27. (a)

n	1	2	3	4	5
$f(n)$	$\frac{1}{2}$	$\frac{2}{3}$	$\frac{3}{4}$	$\frac{4}{5}$	$\frac{5}{6}$

(b) $\frac{6}{7}$

(c) $f(n) = \dfrac{1}{1 \times 2} + \dfrac{1}{2 \times 3} + \cdots$

$$+ \frac{1}{n(n + 1)} = \frac{n}{n + 1}$$

29. (a)

n	1	2	3	4	5
$f(n)$	1	4	9	16	25

(b) 36 **(c)** $f(n) = n^2$

10.2 The Binomial Theorem

3. $a^9 + 9a^8 b + 36a^7 b^2 + 84a^6 b^3$
$+ 126a^5 b^4 + 126a^4 b^5 + 84a^3 b^6$
$+ 36a^2 b^7 + 9ab^8 + b^9$
5. $8A^3 + 12A^2 B + 6AB^2 + B^3$
7. $1 - 12x + 60x^2 - 160x^3 + 240x^4$
$\qquad - 192x^5 + 64x^6$
9. $x^2 + 4x\sqrt{xy} + 6xy + 4y\sqrt{xy} + y^2$
11. $x^{10} + 5x^8 y^2 + 10x^6 y^4 + 10x^4 y^6$
$\qquad + 5x^2 y^8 + y^{10}$
13. $1 - \dfrac{6}{x} + \dfrac{15}{x^2} - \dfrac{20}{x^3} + \dfrac{15}{x^4} - \dfrac{6}{x^5} + \dfrac{1}{x^6}$
15. $\dfrac{x^3}{8} - \dfrac{x^2 y}{4} + \dfrac{xy^2}{6} - \dfrac{y^3}{27}$
17. $a^7 b^{14} + 7a^6 b^{12} c + 21a^5 b^{10} c^2$
$\qquad + 35a^4 b^8 c^3 + 35a^3 b^6 c^4$
$\qquad + 21a^2 b^4 c^5 + 7ab^2 c^6 + c^7$
19. $x^8 + 8\sqrt{2}x^7 + 56x^6 + 112\sqrt{2}x^5$
$\qquad + 280x^4 + 224\sqrt{2}x^3 + 224x^2$
$\qquad + 64\sqrt{2}x + 16$
21. $5\sqrt{2} - 7$ **23.** $89\sqrt{3} + 109\sqrt{2}$
25. $12 - 24\sqrt[3]{2} + 12\sqrt[3]{4}$
27. $x^{10} - 10x^9 + 35x^8 - 40x^7 - 30x^6$
$+ 68x^5 + 30x^4 - 40x^3 - 35x^2 - 10x - 1$
29. 120 **31.** 105 **33. (a)** 10 **(b)** 5
35. $n^2 + 3n + 2$ **37.** 0 **39.** $120a^2 b^{14}$
41. $100x^{99}$ **43.** 45 **45.** 294,912
47. 28 **49.** 40,095

53. (a)

k	0	1	2	3	4	5	6	7	8
$\binom{8}{k}$	1	8	28	56	70	56	28	8	1

10.3 Introduction to Sequences and Series

1. $\frac{1}{2}, \frac{2}{3}, \frac{3}{4}, \frac{4}{5}, \frac{5}{6}$ **3.** $0, 1, 4, 9, 16$

5. $2, \frac{9}{4}, \frac{64}{27}, \frac{625}{256}, \frac{7776}{3125}$

7. $-1, 1, -1, 1, -1$

9. $1, -\frac{1}{2}, \frac{1}{6}, -\frac{1}{24}, \frac{1}{120}$

11. $3, 6, 9, 12, 15$

13. $1, 4, 25, 676, 458329$

15. $2, 2, 4, 8, 32$ **17.** $1, 1, 2, 6, 24$

19. $0, 1, 2, 4, 16$

21.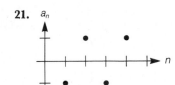

23.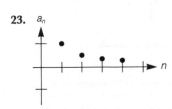

25.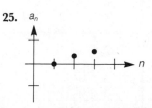

27. 62 **29.** 40 **31.** $-\frac{19}{30}$ **33.** 903

35. 278 **37.** 3 **39.** 41

41. $x + x^2 + x^3$ **43.** $\frac{25}{12}$ **45.** -1

47. $\frac{6}{7}$ **49.** $\sum_{j=1}^{4} 5^j$ **51.** $\sum_{j=1}^{6} x^j$

53. $\sum_{k=1}^{12} (1/k)$ **55.** $\sum_{j=1}^{5} (-1)^{j+1} 2^j$

57. $\sum_{j=1}^{5} (-1)^{j+1} j$

59. (b)

n	$F_1 + F_2 + \cdots + F_n$	$F_{n+2} - 1$
1	1	1
2	2	2
3	4	4
4	7	7
5	12	12

10.4 Arithmetic Sequences and Series

1. (a) 2 **(b)** -4 **(c)** $\frac{1}{3}$ **(d)** $\sqrt{2}$

3. 131 **5.** 501 **7.** 998

9. $d = \frac{11}{6}; a = -\frac{23}{2}$ **11.** -190

13. $-\frac{1}{8}$ **15.** $500,500$ **17.** $\frac{91}{3}\pi$ **19.** $\frac{1}{2}$

21. $S_{16} = -768, d = -\frac{104}{15}$

23. $d = \frac{6}{5}, a = -\frac{17}{5}$ **25.** 900

27. $2, 10, 18$ or $18, 10, 2$

29. $-1, 2, 5$ or $5, 2, -1$

31. (b) $-6 - 9\sqrt{2}$

33. $\dfrac{n}{2(1-b)}[2 + (n-3)\sqrt{b}]$

10.5 Geometric Sequences and Series

1. 6 **3.** $20, 100$ **5.** 1 **7.** $\frac{256}{6561}$

9. ± 4 **11.** 7161 **13.** $63 + 31\sqrt{2}$

15. $\frac{1995}{64}$ **17.** 0.011111 **19.** $\frac{2}{5}$

21. 101 **23.** $\frac{5}{9}$ **25.** $\frac{61}{495}$ **27.** $\frac{16}{37}$

29. $r = -\frac{1}{2}, -2$ **35.** 12 ft

10.6 Vectors in the Plane, a Geometric Approach

1. $|\overrightarrow{PQ}| = \sqrt{34}$

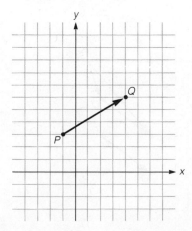

3. $|\overrightarrow{SQ}| = \sqrt{10}$

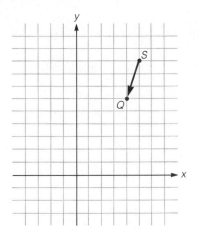

5. $|\overrightarrow{OP}| = \sqrt{10}$

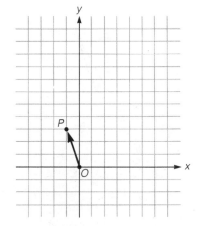

7. $|\overrightarrow{PQ} + \overrightarrow{QS}| = 6\sqrt{2}$

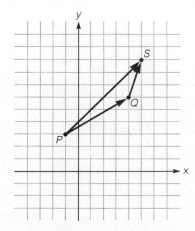

9. $\left|\overrightarrow{OP} + \overrightarrow{PQ}\right| = 2\sqrt{13}$

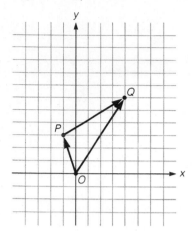

15. $\left|\overrightarrow{SR} + \overrightarrow{PO}\right| = 9$

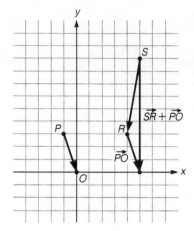

21. $\left|\overrightarrow{OP} + \overrightarrow{OR}\right| = 3\sqrt{5}$

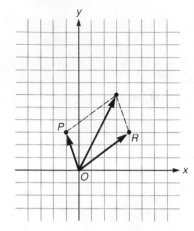

11. $\left|(\overrightarrow{OS} + \overrightarrow{SQ}) + \overrightarrow{QP}\right| = \sqrt{10}$

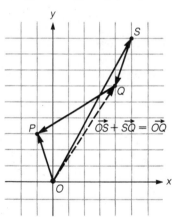

17. $\left|\overrightarrow{OP} + \overrightarrow{RQ}\right| = \sqrt{37}$

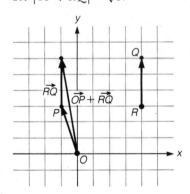

23. $\left|\overrightarrow{RP} + \overrightarrow{RS}\right| = 2\sqrt{13}$

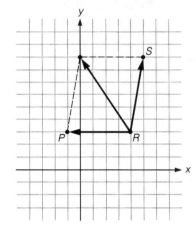

13. $\left|\overrightarrow{OP} + \overrightarrow{QS}\right| = 6$

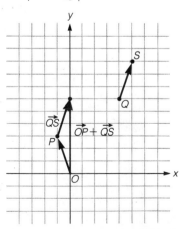

19. $\left|\overrightarrow{SQ} + \overrightarrow{RO}\right| = \sqrt{61}$

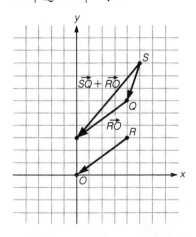

25. $\left|\overrightarrow{SO} + \overrightarrow{SQ}\right| = 6\sqrt{5}$

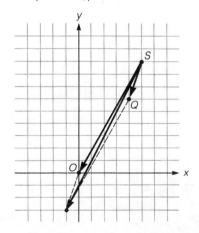

27. $|\mathbf{F} + \mathbf{G}| = \sqrt{41}$ N, $\theta \approx 38.7°$

29. $|\mathbf{F} + \mathbf{G}| = 9\sqrt{2}$ N, $\theta = 45°$

31. $|\mathbf{F} + \mathbf{G}| \approx 7.90$ N, $\theta \approx 24.1°$

33. $|\mathbf{F} + \mathbf{G}| \approx 6.92$ N, $\theta \approx 34.67°$

35. $|\mathbf{F} + \mathbf{G}| \approx 39.20$ N, $\theta \approx 21.46°$

37. $|\mathbf{F} + \mathbf{G}| \approx 38.96$ N, $\theta \approx 29.44°$

39. $V_x \approx 13.86$ cm/sec, $V_y = 8$ cm/sec

41. $F_x \approx 3.62$ N, $F_y \approx 13.52$ N

43. $V_x \approx -0.71$ cm/sec,
$V_y \approx 0.71$ cm/sec

45. $F_x \approx -1.02$ N, $F_y \approx 0.72$ N

47. ground speed: 301.04 mph;
drift angle: 4.76°; bearing: 25.24°

49. ground speed: 293.47 mph;
drift angle: 8.82°; bearing: 91.18°

51. perpendicular: 9.83 lb;
parallel: 6.88 lb

53. perpendicular: 11.82 lb;
parallel: 2.08 lb

55. (a) The initial point of $(\mathbf{A} + \mathbf{B}) + \mathbf{C}$ is $(-1, 2)$ and the terminal point is $(2, -3)$.
(b) The initial point of $\mathbf{A} + (\mathbf{B} + \mathbf{C})$ is $(-1, 2)$ and the terminal point is $(2, -3)$.

10.7 Vectors in the Plane, an Algebraic Approach

1. $|\langle 4, 3 \rangle| = 5$

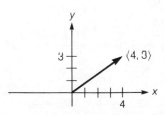

3. $|\langle -4, 2 \rangle| = 2\sqrt{5}$

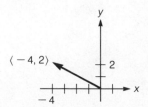

5. $|\langle \frac{3}{4}, -\frac{1}{2} \rangle| = \sqrt{13}/4$

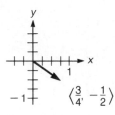

$\langle \frac{3}{4}, -\frac{1}{2} \rangle$

7. $\langle 1, 4 \rangle$ **9.** $\langle -1, 1 \rangle$ **11.** $\langle 8, -5 \rangle$

13. $\langle 7, 7 \rangle$ **15.** $\langle 24, 22 \rangle$ **17.** $\sqrt{130}$

19. $2\sqrt{17} - \sqrt{13} - \sqrt{37}$ **21.** $\langle 13, 6 \rangle$

23. $\langle 14, 21 \rangle$ **25.** $\langle -3, -1 \rangle$

27. $\langle 23, 12 \rangle$ **29.** $\langle -9, 0 \rangle$ **31.** -48

33. $3\mathbf{i} + 8\mathbf{j}$ **35.** $-8\mathbf{i} - 6\mathbf{j}$

37. $19\mathbf{i} + 23\mathbf{j}$ **39.** $\langle 1, 1 \rangle$

41. $\langle 5, -4 \rangle$ **43.** $\langle \frac{1}{5}\sqrt{5}, \frac{2}{5}\sqrt{5} \rangle$

45. $\langle \frac{2}{5}\sqrt{5}, -\frac{1}{5}\sqrt{5} \rangle$

47. $\frac{8}{145}\sqrt{145}\mathbf{i} - \frac{9}{145}\sqrt{145}\mathbf{j}$

49. $u_1 = \sqrt{3}/2, u_2 = \frac{1}{2}$

51. $u_1 = -\frac{1}{2}, u_2 = \sqrt{3}/2$

53. $u_1 = -\sqrt{3}/2, u_2 = \frac{1}{2}$

59. (a) $\mathbf{u} \cdot \mathbf{v} = 8$; $\mathbf{v} \cdot \mathbf{u} = 8$
(b) $\mathbf{v} \cdot \mathbf{w} = -14$; $\mathbf{w} \cdot \mathbf{v} = -14$

61. (a) $\mathbf{v} \cdot \mathbf{v} = 25$; $|\mathbf{v}|^2 = 25$
(b) $\mathbf{w} \cdot \mathbf{w} = 29$; $|\mathbf{w}|^2 = 29$

63. $\cos \theta = 7/\sqrt{170}$;
$\theta \approx 57.53°$ or $\theta \approx 1.00$

65. $\cos \theta = -57/\sqrt{3538}$;
$\theta \approx 163.39°$ or $\theta \approx 2.85$

67. (a) $\cos \theta = -7/\sqrt{170}$;
$\theta \approx 122.47°$ or $\theta \approx 2.14$
(b) $\cos \theta = 7/\sqrt{170}$;
$\theta \approx 57.53°$ or $\theta \approx 1.00$

69. (a) $\cos \theta = 0$
(b) Since the angle between the vectors is 90°, the vectors are perpendicular.

(c)

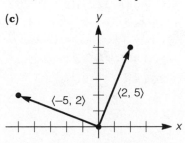

$\langle -5, 2 \rangle$ $\langle 2, 5 \rangle$

71. Since $\cos \theta = \mathbf{A} \cdot \mathbf{B}/|\mathbf{A}||\mathbf{B}|$, the fact that $\mathbf{A} \cdot \mathbf{B} = 0$ implies that $\cos \theta = 0$, and thus $\theta = 90°$. Hence the vectors are perpendicular.

73. $\langle \frac{5}{13}, \frac{12}{13} \rangle$ and $\langle -\frac{5}{13}, -\frac{12}{13} \rangle$

75. (a)
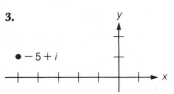
$$|\mathbf{C}| = \sqrt{x_1^2 + y_1^2 + x_2^2 + y_2^2 - 2x_1x_2 - 2y_1y_2}$$

10.8 Trigonometric Form for Complex Numbers

1.

$\bullet\, 4 + 2i$

3.

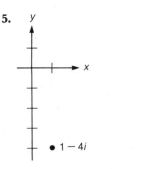

$\bullet\, -5 + i$

5.

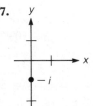

$\bullet\, 1 - 4i$

7.

$\bullet\, -i$

9. $\sqrt{2} + \sqrt{2}i$ **11.** $-2\sqrt{3} + 2i$

13. $-1 - i$ **15.** $\sqrt{3}i$

17. $(\sqrt{6} - \sqrt{2}) + (\sqrt{6} + \sqrt{2})i$

19. $\cos(\pi/6) + i\sin(\pi/6)$

21. $2[\cos(2\pi/3) + i\sin(2\pi/3)]$

23. $4[\cos(7\pi/6) + i\sin(7\pi/6)]$

25. $6[\cos(3\pi/2) + i\sin(3\pi/2)]$

27. $\frac{1}{2}[\cos(11\pi/6) + i\sin(11\pi/6)]$

29. $3 + 3\sqrt{3}i$ **31.** $1 - \sqrt{3}i$

33. $3\sqrt{2}[\cos(2\pi/7) + i\sin(2\pi/7)]$

35. $\frac{3}{2}\sqrt{2} + \frac{3}{2}\sqrt{2}i$ **37.** $\sqrt{3} + i$ **39.** 1

41. $\frac{243}{2} - \frac{243}{2}\sqrt{3}i$ **43.** $\frac{1}{128}\sqrt{2} + \frac{1}{128}\sqrt{2}i$

45. $-4i$ **47.** $-1 - i$ **49.** $\frac{1}{2} + \frac{1}{2}\sqrt{3}i$

51. $128 + 128i$ **53.** $-128 + 128\sqrt{3}\,i$

55. $3i, -\frac{3}{2}\sqrt{3} - \frac{3}{2}i, \frac{3}{2}\sqrt{3} - \frac{3}{2}i$

57. $1, \frac{1}{2}\sqrt{2} + \frac{1}{2}\sqrt{2}\,i, i, -\frac{1}{2}\sqrt{2} + \frac{1}{2}\sqrt{2}\,i,$
$-1, -\frac{1}{2}\sqrt{2} - \frac{1}{2}\sqrt{2}\,i, -i, \frac{1}{2}\sqrt{2} - \frac{1}{2}\sqrt{2}\,i$

59. $4, -2 + 2\sqrt{3}\,i, -2 - 2\sqrt{3}\,i$

61. $3, \frac{3}{2} + \frac{3}{2}\sqrt{3}\,i, -\frac{3}{2} + \frac{3}{2}\sqrt{3}\,i, -3,$
$-\frac{3}{2} - \frac{3}{2}\sqrt{3}\,i, \frac{3}{2} - \frac{3}{2}\sqrt{3}\,i$ **63.** $92,236,816$

65. $0.95 + 0.31i, i, -0.95 + 0.31i,$
$-0.59 - 0.81i, 0.59 - 0.81i$

67. $\frac{1}{2}(\sqrt{6} - \sqrt{2}) + \frac{1}{2}(\sqrt{6} + \sqrt{2})i,$
$-\frac{1}{2}(\sqrt{6} + \sqrt{2}) + \frac{1}{2}(\sqrt{6} - \sqrt{2})i,$
$\frac{1}{2}(\sqrt{2} - \sqrt{6}) - \frac{1}{2}(\sqrt{2} + \sqrt{6})i,$ and
$\frac{1}{2}(\sqrt{2} + \sqrt{6}) + \frac{1}{2}(\sqrt{2} - \sqrt{6})i$

69. (a) $1, -\frac{1}{2} + \frac{1}{2}\sqrt{3}\,i, -\frac{1}{2} - \frac{1}{2}\sqrt{3}\,i$

71. -1 **73.** 1

Chapter 10 Review Exercises

2. (a) $-1 + 9 + 19 = 27$
(b) $-1 + 4 - 9 = -6$

3. (a) $S_n = a(1 - r^n)/(1 - r)$ **(b)** $\frac{174075}{1024}$

4. (a) 42240 **(b)** $5280a^7b^{12}$

5. $243x^{10} + 405x^8y^3 + 270x^6y^6$
$\qquad + 90x^4y^9 + 15x^2y^{12} + y^{15}$

6. 177 **7.** $\frac{7}{9}$ **8.** $a_4 = 5, a_5 = 27$

9. $-5\sqrt{10}$ **10.** 224

11. (a) $2\sqrt{5}\,\text{N}$ **(b)** $\tan\theta = 2$

12. (a) $4\sqrt{13 - 12\cos 110°}\,\text{N}$

(b) $\sin\theta = 2\sin 110°/\sqrt{13 - 12\cos 110°}$

13. ground speed $= 50\sqrt{37}$ mph;
$\tan\theta = \frac{1}{6}$ **14. (a)** $\langle 13, 5\rangle$ **(b)** $\sqrt{194}$

(c) $\mathbf{i} - 3\mathbf{j}$ **15.** $\langle \frac{-11}{130}\sqrt{130}, \frac{-3}{130}\sqrt{130}\rangle$

16. $-1 + \sqrt{3}\,i$ **17.** $2(\cos\frac{7}{4}\pi + i\sin\frac{7}{4}\pi)$

18. $r^n(\cos n\theta + i\sin n\theta)$

19. $\frac{15}{2} + \frac{15}{2}\sqrt{3}\,i$

20. $2\sqrt{3} + 2i, -2\sqrt{3} + 2i,$ and $-4i$

31. $81a^4 + 108a^3b^2 + 54a^2b^4 + 12ab^6 + b^8$

33. $x^4 + 4x^3\sqrt{x} + 6x^3 + 4x^2\sqrt{x} + x^2$

35. $x^{10} - 10x^8y^2 + 40x^6y^4 - 80x^4y^6$
$\qquad + 80x^2y^8 - 32y^{10}$

37. $1 + \dfrac{5}{x} + \dfrac{10}{x^2} + \dfrac{10}{x^3} + \dfrac{5}{x^4} + \dfrac{1}{x^5}$

39. $a^4b^2 - 4a^3b^2\sqrt{ab} + 6a^3b^3$
$\qquad - 4a^2b^3\sqrt{ab} + a^2b^4$

41. $15xy^8$ **43.** 84 **45.** 28

47. $\dbinom{2}{0}^2 + \dbinom{2}{1}^2 + \dbinom{2}{2}^2 = 6; \dbinom{4}{2} = 6$

49. $\dbinom{4}{0}^2 + \dbinom{4}{1}^2 + \dbinom{4}{2}^2 + \dbinom{4}{3}^2$
$\qquad\qquad\qquad\qquad + \dbinom{4}{4}^2 = 70;$
$\dbinom{8}{4} = 70$

51. $\dbinom{3}{0} + \dbinom{3}{1} + \dbinom{3}{2} + \dbinom{3}{3} = 8 = 2^3$

53. $1, \frac{4}{3}, \frac{3}{2}, \frac{8}{5}, \frac{5}{3}$ **55.** $-\frac{1}{2}, \frac{2}{3}, -\frac{3}{4}, \frac{4}{5}, -\frac{5}{6}$

57. $-3, -12, -48, -192, -768$

59. $5 + 7 + 9 + 11 + 13 = 45$

61. $\displaystyle\sum_{k=1}^{5} (5/3^k)$ **63.** 57 **65.** $\frac{5}{1024}$

67. 14350 **69.** $7,777,777,777$ **71.** $\frac{3}{4}$

73. $\frac{1}{10}$ **75.** $\frac{5}{11}$ **81. (a)** $42,929$

(b) $42,925; 0.00932\%$

(c) 6.48040×10^{10}

(d) $6.48027 \times 10^{10}; 2 \times 10^{-3}\%$

83. magnitude: 25 N; direction: 53.1°

85. $v_x \approx 41.0$ cm/sec; $v_y \approx 28.7$ cm/sec

87. perpendicular: 33.2 lb; weight: 36.4 lb

89. $b = \pm\sqrt{15}$ **91.** $\langle 10, 9\rangle$

93. $\langle 12, 7\rangle$ **95.** 48 **97.** $\langle 12, 8\rangle$

99. $\langle -6, 2\rangle$ **101.** $\langle -7, -6\rangle$

103. $7\mathbf{i} - 6\mathbf{j}$ **105.** $\langle \frac{3}{13}\sqrt{13}, \frac{2}{13}\sqrt{13}\rangle$

107. $u_1 = \frac{1}{4}(\sqrt{6} + \sqrt{2}); u_2 = \frac{1}{4}(\sqrt{6} - \sqrt{2})$

109. $\frac{3}{2} + \frac{3}{2}\sqrt{3}\,i$ **111.** $2^{-1/4} - 2^{-1/4}i$

113. $1[\cos(\pi/3) + i\sin(\pi/3)]$

115. $6[\cos(5\pi/4) + i\sin(5\pi/4)]$

117. $5\sqrt{2} + 5\sqrt{2}\,i$ **119.** $\sqrt{2} - \sqrt{2}\,i$

121. $3[\cos(5\pi/9) + i\sin(5\pi/9)]$

123. $\frac{27}{2} + \frac{27}{2}\sqrt{3}\,i$ **125.** $512 - 512\sqrt{3}\,i$

127. $1, \frac{1}{2} + \frac{1}{2}\sqrt{3}\,i, -\frac{1}{2} + \frac{1}{2}\sqrt{3}\,i, -1,$
$-\frac{1}{2} - \frac{1}{2}\sqrt{3}\,i, \frac{1}{2} - \frac{1}{2}\sqrt{3}\,i$

129. $-\frac{1}{2}\sqrt{4 + 2\sqrt{2}} + \frac{1}{2}\sqrt{4 - 2\sqrt{2}}\,i,$
$\frac{1}{2}\sqrt{4 + 2\sqrt{2}} - \frac{1}{2}\sqrt{4 - 2\sqrt{2}}\,i$

131. $1.06 + 0.17i, 0.17 + 1.06i,$
$-0.95 + 0.49i, -0.76 - 0.76i,$
$0.49 - 0.95i$

133. $\cos 3\theta = \cos^3\theta - 3\cos\theta\sin^2\theta;$
$\sin 3\theta = 3\cos^2\theta\sin\theta - \sin^3\theta$

135. 40, 20, 10 or 10, 20, 40

139. (a) No such x exists.
(b) $x = (b^2 - ac)/(a + c - 2b)$

APPENDIX

A.4 Working With Fractional Expressions

1. $x - 3$ **3.** $\dfrac{1}{(x-2)(x^2+4)}$ **5.** $\dfrac{1}{x-2}$

7. $\dfrac{3b}{2a}$ **9.** $\dfrac{a^2+1}{a-1}$ **11.** $\dfrac{x^2+xy+y^2}{(x-y)^2}$

13. 2 **15.** $\dfrac{x^2+x}{(x+4)(x-2)}$

17. $\dfrac{x^2-xy+y^2}{(x-y)(x+y)}$ **19.** $\dfrac{x-3y}{x+3y}$

21. $\dfrac{4x-2}{x^2}$ **23.** $\dfrac{36-a^2}{6a}$

25. $\dfrac{4x+11}{(x+3)(x+2)}$ **27.** $\dfrac{3x^2+6x-6}{(x-2)(x+2)}$

29. $\dfrac{6ax^2-4ax+a}{(x-1)^3}$

31. $\dfrac{2x^2-3}{(x-3)(x+3)(x-2)}$ **33.** $\dfrac{8}{x-5}$

35. $\dfrac{2a}{a+b}$ **37.** $\dfrac{-9}{(x+5)(x-4)^2}$

39. $\dfrac{-p^2-pq-q^2}{p(2p+q)(p-5q)}$ **41.** 0 **43.** 0

45. $\dfrac{-1}{x+a}$ **47.** 1 **49.** $\dfrac{x+4}{3-2x}$

51. $a - 1$ **53.** $\dfrac{-1}{2(2+h)}$ **55.** $\dfrac{a+x}{x^2a^2}$

57. 0 **59.** $\dfrac{a^2-b^2}{2}$ **61.** $\dfrac{px+q}{(x-a)(x-b)}$

63. 0 **65.** -1 **67. (a)** Both equal $\frac{1}{42}$.

INDEX